PERSONAL NUTRITION

Eighth Edition

PERSONAL NUTRITION

MARIE A. BOYLE
COLLEGE OF SAINT ELIZABETH

SARA LONG ROTH
SOUTHERN ILLINOIS UNIVERSITY

∴ WADSWORTH
CENGAGE Learning·

Australia · Brazil · Japan · Korea · Mexico · Singapore · Spain · United Kingdom · United States

Personal Nutrition, **Eighth Edition**
Marie A. Boyle, Sara Long Roth

Publisher: Yolanda Cossio

Senior Acquisitions Editor: Peggy Williams

Senior Development Editor: Nedah Rose

Associate Development Editor: Elesha Feldman

Editorial Assistant: Sean Cronin

Technology Project Manager: Miriam Myers

Senior Marketing Manager: Laura McGinn

Marketing Communications Manager: Linda Yip

Senior Content Project Manager: Carol Samet

Creative Director: Rob Hugel

Art Director: John Walker

Print Buyer: Judy Inouye

Rights Acquisitions Specialist: Dean Dauphinais

Production Service: Lachina Publishing Services

Text Designer: Diane Beasley

Photo Researcher: Chris Althof, Bill Smith Group

Text Permission Researcher: Pablo D'Stair

Copy Editor: Lachina Publishing Services

Cover Designer: John Walker

Cover Image: © Masterfile 2011

Compositor: Lachina Publishing Services

For product information and technology assistance, contact us at
Cengage Learning Customer & Sales Support, 1-800-354-9706
For permission to use material from this text or product,
submit all requests online at **www.cengage.com/permissions**.
Further permissions questions can be emailed to
permissionrequest@cengage.com.

Library of Congress Control Number: 2011927040

ISBN-13: 978-1-111-57113-9

ISBN-10: 1-111-57113-9

Wadsworth
20 Davis Drive
Belmont, CA 94002-3098
USA

Cengage Learning is a leading provider of customized learning solutions with office locations around the globe, including Singapore, the United Kingdom, Australia, Mexico, Brazil, and Japan. Locate your local office at **www.cengage.com/global**.

Cengage Learning products are represented in Canada by Nelson Education, Ltd.

To learn more about Wadsworth, visit **www.cengage.com/Wadsworth**.

Purchase any of our products at your local college store or at our preferred online store **www.CengageBrain.com**.

Printed in Canada
1 2 3 4 5 6 7 15 14 13 12 11

In memory of Jesse and Dylan—
my twinkling stars in the night sky—
and to all those who surround
themselves with lives even more
fleeting than their own.

—Marie A. Boyle

This is to JKR . . . your love and friendship
make life a joyous experience.

—Sara Long Roth

About the Authors

Marie Boyle, PhD, RD, received her BA in psychology from the University of Southern Maine and her MS and PhD in nutrition from Florida State University in Tallahassee, Florida. She is author of the community nutrition textbook *Community Nutrition in Action: An Entrepreneurial Approach*. She also teaches online distance courses in Introductory Nutrition, Community Nutrition, Public Health Nutrition, and other areas for the Graduate Nutrition Programs at the College of Saint Elizabeth in Morristown, New Jersey and the University of Massachusetts in Amherst, Massachusetts. Her other professional activities include serving as an author and reviewer for the American Dietetic Association and Society for Nutrition Education. She coauthored the current Position Paper of the American Dietetic Association on *Addressing World Hunger, Malnutrition, and Food Insecurity* and serves as Editor-in-Chief of the *Journal of Hunger and Environmental Nutrition* by Taylor & Francis Publishers. She is a member of the American Dietetic Association, the American Public Health Association, and the Society for Nutrition Education.

Sara Long Roth, PhD, RD, is Professor in the Department of Animal Science, Food and Nutrition and Director, Didactic Program in Dietetics at Southern Illinois University Carbondale. Prior to obtaining her PhD in health education, she practiced as a clinical dietitian for 11 years. Her specialty areas are medical nutrition therapy, nutrition education, and food and nutrition assessment. She is an active leader in national, state, and district dietetic associations where she has served in numerous elected and appointed positions, including the Commission on Accreditation of Dietetics Education, and Commission on Dietetic Registration. Dr. Long is coauthor of *Understanding Nutrition Therapy and Pathophysiology, Medical Nutrition Therapy: A Case Study Approach, Nutritional Foundations and Clinical Applications: A Nursing Approach*, and *Essentials of Nutrition and Diet Therapy*. Dr. Long has received various awards and honors for teaching, including Outstanding Dietetic Educator (ADA) and Outstanding Educator for the College of Agricultural Sciences.

Contents in Brief

Contents

3 Anatomy for Nutrition's Sake 69

4 The Carbohydrates: Sugar, Starch, and Fiber 97

5 The Lipids: Fats and Oils   127

6 The Proteins and Amino Acids   161

7 The Vitamins: A Functional Approach 189

8 The Minerals and Water: A Functional Approach   229

11 Nutrition, Physical Activity, and Fitness 333

12 The Life Cycle: Conception through the Later Years   361

13 Food Safety and the Global Food Supply 399

Preface

This Eighth Edition of *Personal Nutrition* reflects the same vision we had in writing the first edition of this book more than 20 years ago—that is, to apply basic nutrition concepts to personal, everyday life. The text is designed to support the many one- to four-credit introductory nutrition courses available to students today from a variety of majors, and offers all readers the opportunity to develop practical skills in making decisions regarding their personal nutrition and health. Our challenge has been to teach facts about nutrition, to nurture critical thinking skills, and to motivate readers to apply what they learn in daily life.

Chapter Content

Chapter 1 introduces the basic nutrients the body needs and provides a personal invitation to eat well for optimum health. It assists the reader in becoming a sophisticated consumer of new information about nutrition, and explores the factors that affect food choices, including the media, advertising, and cultural factors. Chapter 2 describes in detail the new MyPlate graphic, food guide and related nutrition tools, and the new 2010 *Dietary Guidelines for Americans* needed to help make sound food choices. It provides sample food labels for understanding the nutrition information, terminology, and health claims found on labels. Chapter 2 also includes a section on various international and ethnic cuisines that highlights the multicultural heritage of our country. Chapter 3 provides a colorfully illustrated introduction to the workings of the human body, with an emphasis on the body's digestion and absorption of nutrients from foods. Chapters 4 through 8 present the nutrients and show how they all work together to nourish the body. Chapters 7 and 8 take a functional approach in presenting the roles of vitamins and minerals and spotlights the emerging importance of the antioxidant nutrients and phytochemicals; both chapters include colorful food photos depicting excellent food sources for individual vitamins and minerals. Chapter 9 discusses the impact of the beverages we drink on our nutritional health. Alcohol is covered in depth in Chapter 9 and provides students with important information on alcohol's relationship to nutrition and health, helping them make informed and responsible decisions. Chapter 10 discusses weight management issues and compares major weight-loss programs. Chapter 11 addresses the relationships between nutrition and personal fitness. Chapter 12 describes the special nutrition needs and concerns that arise during the various stages of the life cycle from conception through the older adult years. Chapter 13 addresses consumer concerns about the safety of our food supply, provides a glimpse at some of the problems and advantages of the newer food technologies, illustrates the global benefits derived when consumers choose locally grown, seasonal foods as much as possible, and presents a brief overview of hunger and food insecurity—both at home and abroad.

Features

The *Eat Well Be Well* features throughout the text motivate readers to make good health a priority and provide suggestions for making the best food and lifestyle choices for healthy living and disease prevention. This feature includes practical tips for today's student that offer health benefits for a lifetime. Topics include "Whole Grains for Health," "Nourish the Heart," "Never Say 'Diet,'" "An Eating Pattern for Longevity," and "Color Your Plate for Health."

The *Savvy Diner* feature provides practical suggestions for healthy eating and reinforces the recommendations made in the 2010 *Dietary Guidelines for Americans*. The *Savvy Diner* features include tips for supermarketing, choosing healthy portion sizes, consuming heart-healthy diets, eating more beans, preserving vitamins in foods, seasoning foods without excess salt, dining out defensively, eating for peak performance, creating tasteful meals for one, and practicing home food safety.

The *Nutrition Action* features are magazine-style essays that keep you abreast of current topics important to the nutrition-conscious consumer. The *Nutrition Action* features address topics such as fast food, smart snacking, the Mediterranean Diet, food allergies, medicinal herbs, diet and blood pressure, and aging well with physical activity. Several *Nutrition Action* sections have been updated to reflect the latest issues in the field. For example, "Carbohydrates—Friend or Foe?" helps consumers choose healthful carbohydrates while making sense of the carbohydrate debate; "Diet Confusion: Weighing the Evidence" helps readers make sense of the current weight-loss scene; and "Eat Fresh Eat Local" includes the earth-friendly benefits of eating fresh and locally grown organic foods.

The *Ask Yourself* sections at the beginning of each chapter contain a set of true/false questions designed to provide readers with a preview of the chapter's contents. Answers to the questions appear on the following page. *Scorecards* are hands-on features that allow readers to evaluate their own nutrition behaviors and knowledge in many areas. Some of the *Scorecards* assist readers in assessing their longevity, overall diet, fruit and vegetable consumption, weight status, physical activity habits, and food safety know-how.

The final special feature of each chapter is the *Spotlight*—all of which have been thoroughly updated. Each addresses a common concern people have about nutrition. *Spotlight* topics include nutrition and the media, ethnic cuisines, alternative sweeteners, diet and heart disease, the benefits derived from soy foods, nutrition and cancer prevention, fetal alcohol syndrome, eating disorders, popular fitness aids and supplements, osteoporosis, and child obesity. The Chapter 13 *Spotlight* covers the many factors that influence nutrition and food insecurity among the people of the world, and underscores that the practical suggestions offered throughout this book for attaining the ideals of personal nutrition are the very suggestions that best support the health of the whole earth as well. The *Spotlights* continue in their question and answer format to encourage the reader to ask further questions about nutrition issues. We encourage you to ask us questions, too, in care of the publisher.

The chapter *Study Cards*—new to this edition and located at the back of the text—provide a consistent resource for students to review core concepts and key points from each chapter of the text. The reverse side of the study card contains essay and multiple choice practice questions to help students prepare for exams. The *Nutrition on the Web* feature is located at the end of each chapter and contains an annotated list of World Wide Web addresses that provides links to reliable sources of nutrition and health information from the Internet related to the chapter's topics. Moreover, you can link with the Internet addresses presented in this book through the publisher's Nutrition Resource Center online at **www.cengagebrain.com/shop/ISBN/1111571139**. At the end of every chapter you will also find the *Menu of Online Study Tools*, which contains a list of helpful study tools (practice tests, flash cards, glossary, web links, animations, chapter summaries, and crossword puzzles) and video clips available on the student website. *Diet Analysis Plus Activity* contains an activity that links the chapter content to *Diet Analysis Plus*, the dietary analysis software that can be used with this text. These activities can be used for enrichment or assigned as homework or in-class discussion.

To get access, visit CengageBrain.com

The Appendixes have also been updated. Appendix A presents aids to calculations, including how to calculate the percentage of calories from fat in one's diet. Appendix A also provides a series of photos depicting the *Choose Your Foods: Exchange Lists for Diabetes*. Appendix B includes *Eating Well with Canada's Food Guide*, which interprets Canada's *Guidelines for Healthy Living*, including recommendations for physical activity; Appendix C includes the chapter reference notes; and Appendix D contains answers to the review questions found on the Study Cards. Appendix E includes our Table of Food Composition. The Glossary of terms that follows the Appendixes provides a quick reference to the nutrition terminology defined in the margins of the text and can be used as a review tool.

Notable Changes in the Eighth Edition

This edition incorporates the many changes that have taken place in the field of nutrition in recent years. Nutrition is a subject that is forever changing. Since the last edition was published, we have witnessed the emergence of new research findings regarding the essential nutrients, phytochemicals, and the prevention of chronic diseases such as heart disease and certain types of cancer, as well as the pervasive marketing of functional foods for disease prevention. Additionally, we have been challenged by the increasing cultural, ethnic, and generational diversity of our society, recent advances in biotechnology, and the parallel trends toward supersized food portions and obesity. Nutrition claims bombard us frequently in advertising and articles about diet and nutrition on television, radio, and the Internet, and in newspapers and magazines. It is important that consumers have the knowledge to evaluate the nutrition issues and controversies. This Eighth Edition of *Personal Nutrition* provides a sieve through which to separate the valid nutrition information from the rest.

For all chapters and special chapter features, we have:

- Peer-reviewed and updated chapter content.

- Created detachable chapter summary *Study Cards* for each chapter.

- Added learning objectives at the start of each chapter.

- Listed key summary points in the margins at the ends of certain major sections.

- Added additional math calculation practice examples to chapter margins.

- Created new figures and tables and updated others to enhance learning.

- Enhanced cross-referencing throughout the text.

- **Chapter 1**
  - Introduced the new *Healthy People 2020* health initiative for the United States

- **Chapter 2**
  - Introduced the key recommendations of the new 2010 *Dietary Guidelines for Americans* and the new MyPlate Food Guide
  - Included new examples of food group/MyPlate serving equivalents for things like pizza and casseroles

- **Chapter 3**
  - Added new material on digestive problems such as choking and heartburn
  - Strengthened "Probiotics" section in the *Eat Well Be Well* feature

- **Chapter 4**
  - Deleted the *Savvy Diner* feature and reorganized the material into the *Eat Well Be Well* feature "Whole Grains for Health"
  - Strengthened the chapter discussion regarding diabetes

- Updated material in chapter *Spotlight* to include new sweeteners approved for consumer use

- **Chapter 5**
  - Reorganized major "Fat in the Diet" section with incorporation of material from the *Nutrition Action* feature
  - Strengthened discussion of omega-3 fats in the diet
  - Updated recommendations for a heart-healthy lifestyle from major professional organizations and included a new Mediterranean Food Guide

- **Chapter 6**
  - Updated "The Vegetarian Diet" section
  - Added new Vegetarian Food Pyramid
  - Updated *Spotlight* discussion regarding health benefits of soy protein

- Reorganized old chapter on vitamins and minerals into two new chapters retaining a functional approach to the discussion of the vitamin and mineral roles in the body:

- **New Chapter 7**
  - Strengthened the Vitamin D discussion and included new Dietary Reference Intake values
  - Strengthened the antioxidant explanation; added diagram of free radical action
  - Reworked the vitamin function tables to decrease duplication of information
  - Differentiated between organic and inorganic micronutrients and the significance of that difference (in food preparation, storage, bioavailability)
  - Added *Eat Well Be Well* feature "Choosing a Vitamin-Mineral Supplement"
  - Added *Nutrition Action* feature "Medicinal Herbs"
  - Updated recommendations for reducing risk for cancer from major professional organizations

- **New Chapter 8**
  - Moved "water in the body" discussions from other chapters to Chapter 8
  - Reworked the mineral function tables to decrease duplication of information and included new Dietary Reference Intake values for calcium
  - Changed title and updated Nutrition Action feature "Diet and Blood Pressure—Beyond Just the Salt Shaker"
  - Added *Spotlight* feature "Osteoporosis"

- **Chapter 9**
  - Enhanced discussion of fructose- versus glucose-based sports drinks and sports performance
  - Added a cross-reference to discussion of soy milk in Chapter 6
  - Added discussion of milk products and trends in milk consumption

- **Chapter 10**
  - Added new figure on current prevalence of obesity
  - Revised opening discussion and section on "What Is a Healthy Weight?"
  - Added the Mifflin–St. Jeor equation for resting energy expenditure
  - Updated discussion and table comparing weight-loss diets
  - Revised discussion of eating disorders to include males and females

- **Chapter 11**
  - Updated photos of people exercising
  - Added new White House campaign on physical activity for adults
  - Added new First Lady Michelle Obama's "Let's Move" campaign directed at healthy children and families
  - Included new physical activity pyramid
  - Added explanation regarding body's use of ATP
  - Added discussion of hyponatremia

- **Chapter 12**
  - Added new Food Guide for Pregnant and Breastfeeding Women and updated weight gain recommendations during pregnancy
  - Added new *Spotlight* feature—"Addressing Weight Problems in Children and Adolescents"
  - Added a cross-reference to the Chapter 1 *Nutrition Action* feature on fast foods to the Teen section of Chapter 12
  - Added new Food Guide for Young Children
  - Added new figure depicting recommended eating patterns for older adults

- **Chapter 13**
  - Enhanced discussions of problem foods needing special handling for food safety
  - Added new figure on food safety and safe cooking temperatures
  - Expanded the *Eat Well Be Well* feature on local foods and sustainable food system approaches
  - Updated statistics and diagrams in chapter *Spotlight* on global hunger issues

We welcome you to the fascinating subject of nutrition. We hope that the book speaks to you personally and that you find it practical for your everyday use. We hope, too, that by reading it you may enhance your own personal nutrition and health.

Student and Instructor Resources

In addition to the text, an array of teaching and learning resources are available for both instructors and students.

Save up to 70% on course materials by purchasing or renting directly from **CengageBrain.com**. Purchase print textbooks, eTextbooks, individual eChapters, CourseMate, and other supplemental materials all for substantial savings over average retail prices.

CengageBrain.com also includes single sign-on access to Cengage Learning's broad range of online homework and study tools, including a selection of free or sample content to preview before purchasing.

Additional formats available on CengageBrain.com include:

- eTextbooks
- eChapters
- Loose-leaf
- Textbook rental
- CourseMate

Instructor's Manual/Test Bank The Instructor's Manual includes student chapter review and Internet exercise worksheets. Other features include learning objectives, chapter summaries, detailed lecture presentation outlines annotated with all available

To get access, visit CengageBrain.com

media and print resources, sample healthy recipes, Points to Consider to enrich lectures, and suggested student activities that help readers practice what they have learned. The updated Test Bank includes real-life vignettes (case studies) as well as multiple-choice, true/false, discussion, and matching items.

PowerLecture™ with ExamView® Computerized Testing and JoinIn™ Student Response System This one-stop lecture tool makes it easy for instructors to assemble, edit, and present custom lectures for your course, using Microsoft® PowerPoint®. The CD-ROM contains figures from Boyle and Long's text as PowerPoint-ready image files, ready-to-use PowerPoint lecture presentations with visuals, BBC® video clips, animations, and the Instructor's Manual/Test Bank in Microsoft® Word format. It also includes ExamView®, featuring Test Bank items in electronic format that allow you to create, customize, and deliver tests in minutes; and JoinIn™ Student Response System content, which allows you to pose questions and display the answers within the PowerPoint slides of your lecture in conjunction with the "clicker" hardware and student response system software of your choice.

CourseMate This feature brings course concepts to life with interactive learning, study, and exam preparation tools that support the textbook. Watch student comprehension soar as your class works with the textbook and the textbook-specific website. CourseMate goes beyond the book to deliver what you need! Learn more at cengage.com/coursemate.

To get access, visit CengageBrain.com

Diet Analysis Plus *Diet Analysis Plus* 10.0 is the market-leading diet assessment program used by colleges and universities. With a database of 22,000 foods, *Diet Analysis Plus* allows your students to easily create personalized profiles based on height, weight, age, sex, and activity level. Thoroughly revised and updated with the latest USDA standards, the software is available online. Visit http://www.cengagebrain.com/shop/ISBN/0538495081 for more information.

WebTutor™ Provides customizable, text-specific content that allows instructors to edit, reorganize, or delete content to meet their course needs. It offers quizzing, videos, pop-up tutors, and test bank materials along with direct access to Diet Analysis Plus, Global Nutrition Watch, and an interactive eBook. Visit www.cengage.com/webtutor to learn more.

Walk4Life® Pedometer The Walk4Life Elite Model pedometer is a practical tool that tracks steps, elapsed time, and distance, and includes a calorie counter and a clock—all for an affordable price when packaged with the text. It's a "must have" item, whether to be used as an activity in class or as a tool to encourage students to walk towards better fitness awareness.

Nutrition Applications Workbook A great resource companion, this workbook includes an extensive family case study project suitable for individual or group assignments, 20 brief case studies, 6 self-contained nutrient-focused calculation activities, and 8 personal diet projects to choose from—all providing critical thinking and exploration opportunities for students.

Personal Dietary Assessment This workbook includes 48 perforated pages of assignments that allow students to evaluate their current eating and lifestyle habits, identify ways to adjust their diet to better meet nutritional needs, and compare their diets to their identified needs generated by the *Diet Analysis Plus* software.

Transparency Acetates More than 100 colorful tables and figures from the text illustrate key facts and concepts.

Acknowledgments

We are grateful to the many individuals who have made contributions to the development of this eighth edition of *Personal Nutrition*. We thank our family and friends for their continued support and encouragement throughout this endeavor and countless others. We appreciate the insights provided by our colleagues—especially to all those individuals who have contributed their expertise to previous editions of this text—including Eleanor Whitney, Diane Morris, Gail Zyla, Kathleen Morgan, Kathy Roberts, and Brian K. Jones. Their insights are reflected in this new edition, still. We appreciate the hard work and expertise of the team of authors who prepared ancillary material for this edition. Thanks to Art Gilbert for his work on the instructor's manual; Charalee Allen for her excellent electronic slide presentations, student worksheets, and website content; Judy Kaufman for creating test bank questions; Karen Schuster for updating online quizzes and writing JoinIn questions; and Mary Puccini for preparing critical thinking questions for the website. Thanks to Shazia Nathoo at Axxya Systems for creating the food composition table found in Appendix E and the computerized diet analysis program that accompanies this book. Special thanks go to the editorial team and their staff: Peggy Williams, Senior Acquisitions Editor; Nedah Rose, Senior Development Editor; Bonnie Briggle, Lachina Publishing Services; and Carol Samet, Senior Content Project Manager. Their guidance ensured the highest quality of work throughout all facets of this production. We appreciate the work of Miriam Myers in creating and managing the technology products that accompany this text, and Elesha Feldman for her work on the student and instructor resources. Our gratitude goes to Laura McGinn, Senior Marketing Manager, and her team for their fine efforts in marketing this book. Our thanks to the many sales representatives who will introduce this new book to its readers. Our appreciation goes to other members of the production team: Jim Atherton, artist, Chris Althof of the Bill Smith Group for photo research, and everyone at Lachina Publishing Services for their hard work and diligence in producing a text to be proud of. Last, but not least, we owe much to our colleagues who provided expert reviews of the manuscript for this and earlier editions—not only for their ideas and suggestions, many of which made their way into the text, but also for their continued enthusiasm, support, and interest in *Personal Nutrition*. Thanks to all of you:

Debra Boardley, *University of Toledo*

Jean Freeland-Graves, *University of Texas*

Art Gilbert, *University of California, Santa Barbara*

Beth Goehring, MS MA, *Contra Costa College*

Joanne Gould, *Kean University*

Gustavo Ibarra, *Salt Lake Community College*

Judy Kaufman, *Monroe Community College*

Zaheer Ali Kirmani, *Sam Houston State University*

Elizabeth Koester Morton, *University of South Carolina*

Pam Massey, *University of Wisconsin—Fox Valley*

Lori Miller Kohler, *Minneapolis Community and Technical College*

Rosa A. Mo, *University of New Haven*

Kristen S. Montgomery, PHD RN, *University of North Carolina—Charlotte*

Esther Okeiyi, *North Carolina Central University*

Caren L. Rojas, MA RD, *Western Michigan University*

Robin S. Schenk, *Buffalo State, SUNY*

Karen Schuster, *Florida Community College, Jacksonville*

Denise L. Signorelli, *College of Southern Nevada*

Tricia Steffen, *Nova Southeastern University*

Melissa Wdowik, *University of North Carolina, Charlotte*

Sharman Willmore, *Cincinnati State Technical and Community College*

Lisa Wright, MT (ASCP) CMA, *Bristol Community College*

Linda O. Young, *University of Nebraska, Lincoln*

Maureen Zimmerman, *Mesa Community College*

—Marie Boyle
Sara Long Roth

The Basics of Understanding Nutrition

Ask Yourself . . .

Which of the following statements about nutrition are true, and which are false? For each false statement, what is true?

1. It is possible to have an appetite without being hungry.
2. Most people obtain information about nutrition from health professionals.
3. The way people choose to live and eat can affect their health and quality of life as they age.
4. Vitamins and minerals supply calories.
5. You can order a low-fat, balanced meal at a fast-food outlet.
6. Healthful diets cost more than relatively unhealthful diets.
7. When a person suffers from malnutrition, it means that he or she is taking in too few nutrients.
8. A nutritionist is a professional who is certified to advise people on nutrition.
9. The notion of eating insects repels people around the world.
10. The more current a dietary claim, the more you can trust its accuracy and reliability.

Answers found on the following page.

Stroll down the aisle of any supermarket, and you'll see all manner of foods touted with such claims as "low-fat," "low-carb," "low-calorie," and "fat-free." Flip through the pages of just about any magazine, and you're likely to find advice on how to lose weight. Walk into any gym, and you'll probably hear members discussing the merits of one performance-enhancing food or another. All this boils down to the fact that nutrition has become part and parcel of the American lifestyle.

It wasn't always that way, however. The field of **nutrition** is a relative newcomer on the scientific block. Although Hippocrates recognized diet as a component of health back in 400 B.C., only in the past hundred years or so have researchers begun to understand that carbohydrates, fats, and proteins are needed for normal growth. The next nutrition breakthrough—the discovery of the first vitamin—occurred in the early 1900s. It wasn't until 1928, when an organization called the American Institute of Nutrition was formed, that nutrition was officially recognized as a distinct field of study.[1]* It took several more decades before nutrition achieved its current status as one of the most talked-about scientific disciplines.

---

*Reference notes for each chapter are in Appendix C.

© Elena Elisseeva/istockphoto.com

nutrition the study of foods, their nutrients and other chemical components, their actions and interactions in the body, and their influence on health and disease.

Tell me what you eat, and I will tell you what you are.

ANTHELME BRILLAT-SAVARIN (1755–1826, FRENCH POLITICIAN AND GOURMET; AUTHOR OF *PHYSIOLOGY OF TASTE*)

Today we spend billions of dollars each year to investigate the many aspects of nutrition, a science that encompasses the study not only of vitamins, minerals, and other nutrients but also of such diverse subjects as the effects of alcohol, caffeine, and pesticides. In addition, nutrition scientists continually expand our understanding of the impact food has on our bodies by examining research in chemistry, physics, biology, biochemistry, genetics, immunology, and other nutrition-related fields. A number of other disciplines also make valuable contributions to the study of nutrition. These related fields include psychology, anthropology, epidemiology, geography, agriculture, ethics, economics, sociology, and philosophy.

Even though science has shown us that, to some extent, we really are what we eat, many consumers have become more confused than ever about how to translate the steady stream of new findings about nutrition into a lifestyle of healthful eating. Reportedly, Americans spend more than $40 billion a year on diet and health books. As Table 1-1 illustrates, people of all ages make food purchasing decisions based on

TABLE 1-1	Proportion of Food Shoppers Who Have Purchased a Product Based on a Specified Health-Related Claim on the Package					
			% by generation*			
Claim	**Total %**	**Gen Y**	**Gen X**	**Boomers**	**Matures**
Low-fat	63	52	61	64	66
Whole-grain	62	49	55	68	61
Low in saturated fats	55	44	49	55	60
Low calorie	52	42	52	54	52
High in calcium	51	58	55	47	48
High in vitamin C	51	57	59	49	46
Low sodium	48	31	46	47	56
Vitamin-rich or vitamin-fortified	47	51	52	47	41
Sugar-free	46	33	46	46	49
Reduces risk of heart disease	42	31	36	43	49
Low-carb	40	41	39	43	36
Reduces risk of cancer	26	19	19	24	35
No. of Interviews	**1,001**	**75**	**221**	**382**	**292**

*Generation birth years: Generation Y = 1981–1994; Generation X = 1965–1980; Baby Boomers = 1946–1964; Matures = Pre-1946.

SOURCE: A. E. Sloan, "Top Ten Global Food Trends," Food Technology 59 (2005): 31. Data from the Food Marketing Institute, *Trends in the United States: Consumer Attitudes and the Supermarket, 2005 Edition* (Washington, DC: The Research Department, Food Marketing Institute).

ASK YOURSELF ANSWERS: 1. True. **2.** False. Most Americans look first to television for nutrition information, then to magazines, and then to newspapers. **3.** True. **4.** False. Only protein, carbohydrate, and fat supply calories. **5.** True. **6.** False. People can save money when they switch from a typical high-fat diet to the grain-based, produce-rich diet recommended by health experts. **7.** False. Malnutrition can be caused either by taking in too few nutrients or by consuming excess nutrients. **8.** False. A nutritionist is a person who claims to specialize in the study of nutrition, but some are self-described experts whose training is questionable. A registered dietitian (RD), however, is recognized as a nutrition expert—with training in nutrition, food science, and diet planning. **9.** False. It's true that most people in the United States and Canada find the idea of eating insects repulsive, because the practice is not part of the North American culture. However, people in many other countries, because they have been brought up in cultures in which insects have long been a traditional food, consider dishes prepared with insects a delicacy. **10.** False. If a nutrition claim is too new, it may not have been adequately tested. Findings must be confirmed many times over by experiments and evaluated in light of other knowledge before they can be translated into recommendations for the public.

claims regarding nutrition and health.[2] Each additional nugget of nutrition news that comes along raises new concerns: Is caffeine bad for me? Should I take vitamin supplements? Do diet pills work? Can a sports drink improve my performance? Do pesticides really pose a hazard?

Some manufacturers and media outlets feed the confusion by offering health-conscious consumers unreliable products and misleading dietary advice, often making unsubstantiated claims for a number of nutritional products, including supplements touted as fat melters, muscle builders, and energy boosters. Unfortunately, misinformation runs rampant in the marketplace. Americans spend more than $30 billion annually on medical and nutritional **health fraud** and **quackery**, up from only $1 billion to $2 billion in the early 1960s.[3] Consider, for example, that college athletes alone may spend hundreds of dollars a month on nutritional supplements, even though most of the products pitched to serious exercisers are useless and, in some cases, potentially harmful. At the same time, the sale of weight-loss foods, products, and services—not all of them sound—has become a $40 billion industry.[4]

To be sure, the widespread interest in nutrition has generated some positive changes in the marketplace. Whereas the sale of fresh fruit, salads, and other low-fat items such as frozen yogurt in fast-food chains was virtually unheard of at one time, those eateries couldn't survive in the current nutrition-conscious environment without offering such healthful fare.[5] (See the Nutrition Action feature later in this chapter for tips on eating healthfully at fast-food outlets.) By the same token, food manufacturers have responded to consumer concerns about diet by developing new technologies, such as the creation of fat substitutes, to provide shoppers with an unprecedented number of choices at the supermarket.

With the amount of nutrition information and the number of food alternatives always on the rise, choosing a healthful diet can seem like a daunting task. Fortunately, you don't need a degree in nutrition to put the principles of the science to use in your own life. A basic understanding of nutrition can go a long way in helping you protect your health (and your wallet). This book lays the foundation you need to take nutrition science out of the laboratory and move it into your kitchen, both today and tomorrow. The first steps are understanding the nature of the nutrients themselves and exploring the current thrust of the field of nutrition.

The Nutrients in Foods

Almost any food you eat is mostly water, and some foods are as high as 99 percent water. The bulk of the solid materials consists of carbohydrate, fat, and protein. If you could remove these materials, you would be left with a tiny residue of minerals, vitamins, and other materials. Water, carbohydrate, fat, protein, vitamins, and some of the minerals are **nutrients**. Some of the other materials are not nutrients. The six classes of nutrients are carbohydrate, fat, protein, vitamins, minerals, and water.

A complete chemical analysis of your body would show that it is made of similar materials in roughly the same proportions as most foods. For example, if you weigh 150 pounds (and if that is a desirable weight for you), your body contains about 90 pounds of water and about 30 pounds of fat. The other 30 pounds consist mostly of protein, carbohydrate, and the major minerals of your bones—calcium and phosphorus. Vitamins, other minerals, and incidental extras constitute only a fraction of a pound.

Scientists use the term **essential nutrient** to describe the nutrients that the body must obtain from food. About 40 nutrients are known to be *essential*; that is, they are compounds that the body cannot make for itself but are indispensable to life processes. How can you be sure you're getting all the nutrients you need? The rest of this chapter, along with the diet-planning tools presented in Chapter 2, will help you design a diet that supports good health.

Some stores sell pills and potions touted as fat melters, energy boosters, and muscle builders.

© Najah Feanny/CORBIS SABA

The six classes of nutrients:

- carbohydrate
- fat
- protein
- vitamins
- minerals
- water

health fraud conscious deceit practiced for profit, such as the promotion of a false or an unproven product or therapy.

quackery A quack is a person who practices health fraud (quack = to boast loudly).

nutrients substances obtained from food and used in the body to promote growth, maintenance, and repair. The nutrients include carbohydrate, fat, protein, vitamins, minerals, and water.

essential nutrient a nutrient that must be obtained from food because the body cannot make it for itself.

The energy-yielding nutrients:

- carbohydrate
- fat
- protein

FIGURE 1-1

Caloric Values of Carbohydrate, Protein, Fat, and Alcohol

Calories per gram

- Carbohydrate: 4
- Protein: 4
- Fat: 9
- Alcohol: 7

energy the capacity to do work, such as moving or heating something.

calorie the unit used to measure energy.

vitamins organic, or carbon-containing, essential nutrients that are vital to life but needed only in relatively minute amounts (*vita* = life; *amine* = containing nitrogen).

minerals inorganic compounds, some of which are essential nutrients.

metabolism collective term for all the chemical and physical reactions occurring in living cells, including the reactions by which the body obtains and uses energy from foods.

water provides the medium for life processes.

The Energy-Yielding Nutrients

Upon being broken down in the body, or digested, three of the nutrients—carbohydrate, protein, and fat—yield the **energy** that the body uses to fuel its various activities. In contrast, vitamins, minerals, and water, once broken down in the body, do not yield energy but perform other tasks, such as maintenance and repair. Each gram of carbohydrate and protein consumed supplies your body with 4 calories, and each gram of fat provides 9 calories (see Figure 1-1). Only one other substance that people consume supplies calories, and that is alcohol, which provides 7 calories per gram. Alcohol is not considered a nutrient, however, because it does not help maintain or repair body tissues the way nutrients do.

The body uses energy from carbohydrate, fat, and protein to do work or generate heat. This energy is measured in **calories**—familiar to almost everyone as markers of how "fattening" foods are. If your body doesn't use (release) the energy you obtained from a food soon after you've eaten it, it stores that energy, usually as body fat, for later use. If a person eats excess amounts of protein, fat, or carbohydrate fairly regularly, the stored fat builds up over time and leads to obesity. Too much of any food, whether lean meat (a protein-rich food), potatoes (a high-carbohydrate food), or butter (a fatty food), can contribute excess calories that result in overweight.

Vitamins, Minerals, and Water

Unlike carbohydrate, fat, and protein, **vitamins** and **minerals** do not supply energy, or calories. Instead, they regulate the release of energy and other aspects of **metabolism**. As Table 1-2 shows, there are 13 vitamins, each with its special role to play. Vitamins are divided into two classes: water-soluble (the B vitamins and vitamin C) and fat-soluble (vitamins A, D, E, and K). This distinction has many implications for the kinds of foods that provide the different vitamins and how the body uses them, as you will see in Chapter 7.

The minerals also perform important functions. Some, such as calcium, make up the structure of bones and teeth. Others, including sodium, float about in the body's fluids, where they help regulate crucial bodily functions, such as heartbeat and muscle contractions.

Often neglected but equally vital, **water** is the medium in which all the body's processes take place. Some 60 percent of your body's weight consists of water, which carries materials to and from cells and provides the warm, nutrient-rich bath in which cells thrive. Water also transports hormonal messages from place to place. When energy-yielding nutrients release energy, they break down into water and other simple compounds. Without water, you could live only a few days.

Each day your body loses water in the form of sweat and urine. Therefore, you must replace large amounts of it—on the order of two to three quarts a day. To be sure, you don't need to *drink* that much water daily, because the foods and other beverages you consume do supply some of the water you need.

TABLE 1-2	The Vitamins and Minerals					
The Vitamins				**The Minerals**		
The water-soluble vitamins:		**The fat-soluble vitamins:**	**The major minerals:**		**The trace minerals:**	
B vitamins		Vitamin A	Calcium	Potassium	Chromium	Manganese
Thiamin	Vitamin B$_{12}$	Vitamin D	Chloride	Sodium	Copper	Molybdenum
Riboflavin	Folate	Vitamin E	Magnesium	Sulfur	Fluoride	Selenium
Niacin	Biotin	Vitamin K	Phosphorus		Iodine	Zinc
Vitamin B$_6$	Pantothenic acid				Iron	
Vitamin C						

Note: A number of trace minerals are currently under study to determine possible dietary requirements for humans. These include arsenic, boron, cadmium, cobalt, lead, lithium, nickel, silicon, tin, and vanadium.

▲ Remember that 1 gram is a very small amount. For instance, 1 teaspoon of sugar weighs roughly 4 grams.

Calorie Value of Carbohydrate, Fat, and Protein

If you know the number of grams of carbohydrate, fat, and protein in a food, you can calculate the number of calories in it. Simply multiply the carbohydrate grams by 4, the fat grams by 9, and the protein grams by 4. Add the totals together to obtain the number of calories. For example, a deluxe fast-food hamburger contains about 45 grams of carbohydrate, 27 grams of protein, and 39 grams of fat:

45 grams of carbohydrate × 4 calories	=	180 calories
39 grams of fat × 9 calories	=	351 calories
27 grams of protein × 4 calories	=	108 calories
Total:		639 calories

The percentage of your total energy intake from carbohydrate, fat, and protein can then be determined by dividing the number of calories from each energy nutrient by the total calories and then multiplying your answer by 100 to get the percentage:

$$\text{calories from carbohydrate} = \frac{45 \times 4 \text{ cal/g}}{639} = 0.281 \times 100 = 28\%$$

$$\text{calories from fat} = \frac{39 \times 9 \text{ cal/g}}{639} = 0.548 \times 100 = 55\%$$

$$\text{calories from protein} = \frac{27 \times 4 \text{ cal/g}}{639} = 0.168 \times 100 = 17\%$$

See Appendix A for help with figuring percentages and other calculations.

Nutrition and Health Promotion

In the past, scientists investigating the role that diet plays in health focused on the consequences of getting too little of one nutrient or another. Until the end of World War II, in fact, nutrition researchers concentrated on eliminating diseases caused by deficiency (lack or absence) of a particular vitamin or mineral.

These days, the focus is just the opposite. Deficiency diseases have been virtually eliminated in America because of our abundant food supply and the practice of fortifying food with essential nutrients (adding iodine to salt, for example). Nevertheless, diseases related to **malnutrition**, in the form of dietary excess and imbalance, run rampant. Five of the leading causes of death—heart disease, cancer, stroke, diabetes, and hypertension—have been linked to diet (see Table 1-3). Another four are associated with excessive alcohol consumption: accidents, suicide, homicide, and liver disease.[6] **Overnutrition** contributes to other ills as well, including obesity and dental disease. Because obesity and a sedentary lifestyle are linked with chronic diseases, such as diabetes, heart disease, and certain cancers, it can be projected that increased rates of obesity will lead to increased deaths each year, not to mention hospitalizations, disability, time lost from jobs, and poor quality of life for many Americans.[7]

This is not to say that diet is the sole culprit causing these diseases. Figure 1-2 puts nutrition (a factor you can control) in perspective with respect to heredity (a factor you can't control). It illustrates the point that some diseases are much more responsive to nutrition than others and that some are not responsive at all. At one extreme are diseases that can be completely cured by supplying missing nutrients, and at the other extreme are certain genetic (inherited) diseases that are unaltered by nutrition. Most diseases fall in between, being influenced by inherited susceptibility but responsive to dietary manipulations that help counteract the disease process. Thus, diabetes may be managed by means of a special diet; cardiovascular disease may respond favorably to a diet low in unhealthful fats and cholesterol.

malnutrition any condition caused by an excess, deficiency, or imbalance of calories or nutrients.

overnutrition calorie or nutrient overconsumption severe enough to cause disease or increased risk of disease; a form of malnutrition.

Many of the major killers, such as heart disease, some types of cancer, stroke, and diabetes, are influenced by a number of factors, including a person's genetic makeup, eating and physical activity habits, exposure to tobacco, and other lifestyle practices.[a] Blue shading indicates a cause of death in which diet can play a part; yellow shading indicates a cause of death in which alcohol use can play a part.

Rank	Cause of Death	Percentage of Total Deaths
1	Heart disease	25.4%
2	Cancer	23.2%
3	Stroke	5.6%
4	Chronic lower-respiratory diseases	5.3%
5[b]	Motor vehicle and other accidents	5.1%
6	Alzheimer's disease	3.1%
7	Diabetes mellitus	2.9%
8	Influenza and pneumonia	2.2%
9	Kidney disease	1.9%
10	Infections of the blood (septicemia)	1.4%
11	Suicide	1.4%
12	Chronic liver disease and cirrhosis	1.2%
13	Hypertension (high blood pressure)	1.0%
14	Parkinson's disease	0.8%
15	Homicide	0.8%

[a] Lifestyle related (67%); diet related (58%); alcohol related (9%).
[b] The leading cause of death for persons aged 15 to 24 is motor vehicle and other accidents, followed by homicide, suicide, cancer, and heart disease. About half of all accident fatalities are alcohol related.

SOURCE: Centers for Disease Control and Prevention, National Vital Statistics Report, 2010; available at www.cdc.gov/nchs.

A number of environmental, behavioral, social, and genetic factors work together to determine a person's likelihood of suffering from a **degenerative disease**. For example, diet notwithstanding, someone who smokes, doesn't exercise regularly, and has a parent who suffered a heart attack is more likely to end up with heart disease than a nonsmoker who works out regularly and does not have a close relative with heart disease. The way to alter disease risk is to concentrate on changing the daily habits that can be controlled. The results can be significant.

Researchers who monitored the habits and health of a group of some 7,000 Californians for nearly two decades were able to pinpoint seven common lifestyle elements

degenerative disease chronic disease characterized by deterioration of body organs as a result of misuse and neglect. Poor eating habits, smoking, lack of exercise, and other lifestyle habits often contribute to degenerative diseases including heart disease, cancer, osteoporosis, and diabetes.

FIGURE 1-2

Nutrition and Disease

Not all diseases are equally influenced by diet. Some are purely hereditary, such as sickle-cell anemia. Some may be inherited (or the tendency to develop them may be inherited) but may be influenced by diet, such as some forms of diabetes. Some are purely dietary, such as the vitamin and mineral deficiency diseases. Nutrition alone is certainly not enough to prevent many diseases, but it helps.

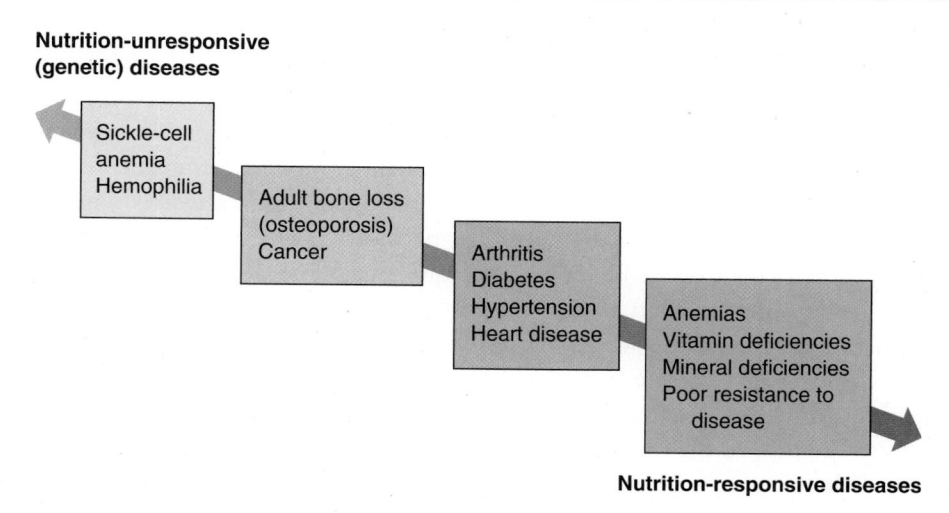

Nutrition-unresponsive (genetic) diseases

Sickle-cell anemia
Hemophilia

Adult bone loss (osteoporosis)
Cancer

Arthritis
Diabetes
Hypertension
Heart disease

Anemias
Vitamin deficiencies
Mineral deficiencies
Poor resistance to disease

Nutrition-responsive diseases

associated with optimal quality of life and longevity: avoiding excess alcohol; not smoking; maintaining a healthful weight; exercising regularly; sleeping seven to eight hours a night; eating breakfast; and eating nutritious, regular meals. In fact, after 20 years, those who had adhered to these healthful habits were only half as likely to have died as those who had unhealthful habits. The group with healthful habits was also half as likely to have suffered the types of disabilities that interfere with day-to-day living. Granted, the researchers speculated that some of the habits—for example, sleeping seven to eight hours a night—are not necessarily as beneficial as, say, the habit of exercising regularly. Rather, regular eating and sleeping habits are most likely to be signs that people make the time and have enough control of their lives to take care of their health.[8]

Nutrition shares responsibility with other lifestyle factors for maintaining good health. By the time you are 65 years old, you will have eaten about 100,000 pounds of food. Each bite may or may not have brought with it the nutrients you needed. The impact of the food you have eaten, together with your lifestyle habits, accumulates over a lifetime, and people who have lived and eaten differently all their lives are in widely different states of health by the time they reach 65. Researchers repeatedly report that people who regularly consume a variety of plant foods, such as fruits, vegetables, legumes, nuts, and whole grains, have reduced risks of heart disease, stroke, diabetes, certain cancers, and other chronic diseases.[9] The key to disease prevention and optimal health is not in eating or avoiding a certain food, but rather in creating a lifestyle that includes time for preparing nutritious meals and enjoying regular physical activity.

A person who practices good health habits can expect to delay the onset of even minimal disability by several years, compared with a person who practices few or none of them.[10] If you believe in accepting the things you can't control and controlling the things you can, you are in luck. Your nutritional health can be controlled, and it deserves your conscientious attention.

The findings in Figure 1-3 illustrate that you can change the probable length and quality of your life. Because nutrition is involved in at least half of the preceding lifestyle recommendations, it clearly plays a key role in maintaining good health. This chapter's Scorecard feature, The Longevity Game, further demonstrates this point.

Table 1-4 summarizes ways to improve health with sound nutrition practices. As you read Table 1-4, keep in mind that although everyone can benefit from eating a healthful diet that complies with the guidelines, some people stand to gain more

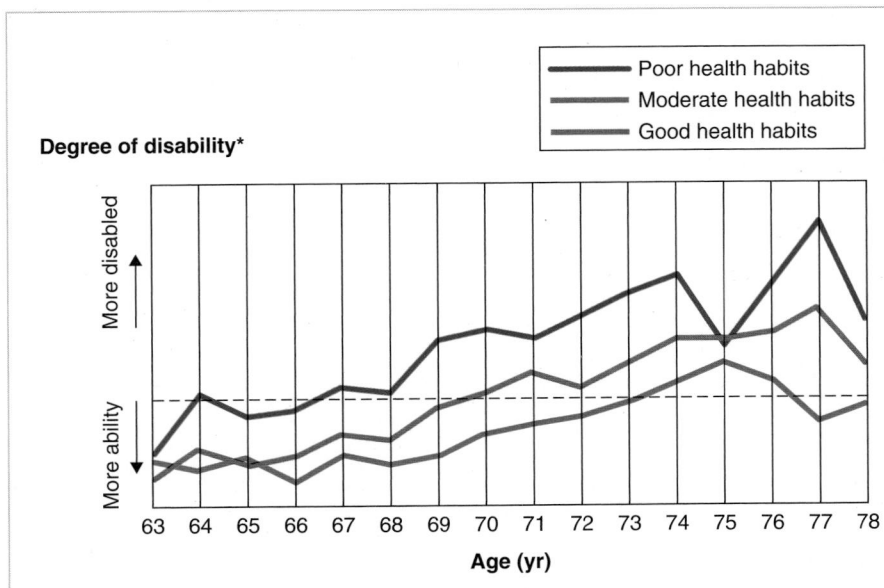

Degree of disability*

- Poor health habits
- Moderate health habits
- Good health habits

More disabled → More ability

Age (yr): 63 64 65 66 67 68 69 70 71 72 73 74 75 76 77 78

TABLE 1-4	Eating to Beat the Odds
Dietary Recommendation	**To Help Reduce the Risk of . . .**
Fat: Reduce total fat intake to 20%–35% of total calories, but derived mostly from oils. Reduce saturated fat intake to less than 10% of calories and the intake of cholesterol to less than 300 mg daily.[a,b]	Some types of cancer, obesity, heart disease, and possibly gallbladder disease.
Weight: Achieve and maintain a healthy weight.[a] Balance caloric intake with physical activity to manage weight.	Diabetes, high blood pressure, stroke, cancers (especially breast and uterine), osteoarthritis, and gallbladder disease.
Carbohydrates and Fiber: Increase consumption of fruits, vegetables, legumes, and whole grains.[a,c,e] Increase whole-grain intake by replacing refined grains with whole grains.[d]	Diabetes, heart disease, and some types of cancer.
Sodium and Potassium: Limit daily intake of salt.[a,f] Increase food sources of potassium such as fresh fruits and vegetables.[a,e]	High blood pressure and stroke.
Alcohol: Avoid completely or drink only in moderation.[g]	Heart disease, high blood pressure, liver disease, stroke, some forms of cancer, and malformations in babies born to mothers who drink alcohol during pregnancy.
Sugar: Choose and prepare foods and beverages with little added sugars or caloric sweeteners. Aim for less than 10% of total calories from sugars.[d]	Tooth decay and gum disease.
Calcium and Vitamin D: Maintain adequate calcium and vitamin D intakes.[e]	Osteoporosis (adult bone loss) and bone fractures.

[a] Pay particular attention to this guideline if you have glucose intolerance, high blood cholesterol, high triglyceride levels, or high blood pressure.

[b] The intake of saturated fat and cholesterol can be reduced by substituting fish, poultry without skin, lean meats, and low-fat or fat-free dairy products for fatty meats and whole-milk products; by choosing more vegetables, fruits, whole-grain cereals, and legumes; by using oils to replace solid fats where possible, and by limiting egg yolks and fried or other fatty foods.

[c] Increase vegetable and fruit intake; eat a variety of vegetables, especially dark-green and red and orange vegetables and beans and peas. Consume at least half of all grains as whole grains; increase whole-grain intake by replacing refined grains with whole grains. Limit the consumption of foods that contain refined grains, especially refined grain foods that contain solid fats, added sugars, and sodium. Specific recommendations are provided in Figure 2-5 on page 41.

[d] Refined grains (discussed in Chapter 4) are processed grains and grain products that are not whole grain. Solid fats are fats with a high content of saturated and/or *trans* fatty acids (discussed in Chapter 5), which are usually solid at room temperature. Added sugars are those incorporated into foods and beverages during processing or preparation, or consumed separately.

[e] Choose foods that provide more potassium, dietary fiber, calcium, and vitamin D, which are nutrients of concern in American diets. These foods include vegetables, fruits, whole grains, and milk and milk products. Specific nutrient recommendations are provided on Study Cards pages 27–29 at the back of this book.

[f] Reduce daily sodium intake to less than 2,300 milligrams and further reduce intake to 1,500 mg among persons who are 51 and older and those of any age who are African American or have hypertension, diabetes, or chronic kidney disease. Limit the use of salt in cooking, and avoid adding it to foods at the table. Highly processed, salty, salt-preserved, and salt-pickled foods should be consumed sparingly.

[g] *Moderate drinking* is defined as no more than one drink a day for the average-sized woman and no more than two drinks a day for the average-sized man. A *drink* is any alcoholic beverage that delivers ½ ounce of pure ethanol: 5 ounces of wine, 12 ounces of beer, or 1-1/2 ounces of hard liquor (whiskey, scotch, etc.). There are many circumstances in which people should avoid alcohol.

SOURCES: Adapted from U.S. Department of Health and Human Services and U.S. Department of Agriculture, *Dietary Guidelines for Americans*, 2010, 7th ed. (Washington, DC: U.S. Government Printing Office, 2010); and Institute of Medicine, *Dietary Reference Intakes for Energy, Carbohydrate, Fiber, Fat, Fatty Acids, Cholesterol, Protein, and Amino Acids* (Washington, DC: National Academy Press, 2002).

than others. Those who have high blood cholesterol levels, for instance, are already at risk for heart disease, thereby making it especially important for them to eat a healthful diet and maintain a healthy weight. By the same token, those who have close relatives with, say, diabetes, would do well to keep their weight down and pay particular attention to the other nutrition guidelines that help stave off that condition. (The chapters that follow explain the link between diet and chronic diseases in more detail and offer advice on how to follow each dietary recommendation.)

A National Agenda for Improving Nutrition and Health

Some people do things that are not good for their health. They overeat, smoke, refuse to wear a helmet when riding a bicycle, never wear seat belts when driving, fail to take their blood pressure medication—the list is endless. These behaviors reflect personal choices, habits, and customs that are influenced and modified by social forces. We call these *lifestyle behaviors*, and they can be changed if the individual is so

Need proof that good dietary choices can have a profound effect on your health and longevity? Look no further than the Okinawans! These people dwell on a group of islands—known collectively as Okinawa—that lie southwest of mainland Japan. Okinawans enjoy one of the longest life spans of anyone on earth, and they do so while maintaining a very high quality of life. This fact has sparked much interest in the Okinawan culture and resulted in the 25-year Okinawa Centenarian Study that began in 1976 to investigate what makes these people so healthy.[11] Researchers studied more than 600 Okinawan centenarians and numerous others in their 70s, 80s, and 90s.

When the study began, researchers were surprised to find that many centenarians were still very active, devoid of health problems, and looked years younger than their chronological age. Upon further investigation, they also discovered that these people have low levels of cancer-causing free radicals, low cardiovascular risks, extremely healthy bone densities, and a lower prevalence of dementia than those of the same age in other countries. Indeed, Okinawan elders have among the lowest mortality rates in the world from many chronic diseases, including cancer, stroke, osteoporosis, and heart disease. Although there is no magic bullet that results in the Okinawans' longevity and health, Okinawan centenarians have a number of variables in common.[12]

1 Enough Is Enough.

Okinawan elders have an average body mass index (BMI) that ranges from 18 to 22. The Okinawans stay lean by eating a low-calorie diet and practicing calorie control in a cultural habit known as *hara hachi bu* (eating until they are only 80 percent full) and keeping physically active every day. In contrast, middle-aged

Okinawan twin sisters—Gin Kanie and Kin Narita at age 106—celebrate with family and friends. Okinawan centenarians maintain optimistic attitudes and strong social bonds throughout their lives.

© Yoshida–Fujifotos/The Image Works

Okinawans, who have a less traditional lifestyle, have an average BMI of 26, the highest in Japan and similar to that of Americans. As discussed in Chapter 10, a BMI of 25 or greater is considered overweight and places you at greater risk for chronic diseases, especially heart disease and stroke.

2 Moderation and a Healthful Lifestyle Are Key Cultural Values.

Okinawan elders never smoke. They consume a diet that is 80 percent plant-based and naturally high in unrefined whole grains, soy, vegetables, and fruits—all of which are rich in antioxidants and phytochemicals. They consume higher intakes of good fats from fish, nuts, and vegetable oils, and they have rather low amounts of less healthy fats in their diet. They keep active every day throughout their lives with a variety of activities, such as gardening, traditional dance, and martial arts. In fact, many of those studied still participated in competitive games and karate past the age of 100!

Even though these concepts may seem rather simple, the average American is not following suit. With high stress, sedentary lifestyles, and diets high in saturated fats and low in protective nutrients such as antioxidants, it seems that the American way of life is the opposite of the Okinawan path. It might be hard to drastically change your entire lifestyle to follow that of the Okinawans; however, making some dietary changes would be a good place to start. Who knows, maybe you too can live to be a centenarian! Here is a general overview of the Okinawan elders' way of eating:

- They eat about 500 calories less per day than Americans.
- They eat plenty of whole grains, seven servings of vegetables, and about four servings of fruits a day.
- They eat seaweed, coldwater fish, and other seafood choices at least three times a week.
- They include soy foods in their daily diet, and consume poultry, eggs, pork, beef, and other meats in moderation.
- They drink plenty of water and tea, and drink alcohol in moderation.
- They consume very few sweets.

3 Psychological and Spiritual Health Matters.

Okinawans put family first. They keep socially engaged, maintaining strong bonds with friends and family. They have an easygoing approach to life. Centenarians score high on optimistic attitudes and adaptability. They possess a strong sense of purpose, which translates roughly to "that which makes one's life worth living."

motivated. **Health promotion** focuses on changing human behavior: getting people to eat healthful diets, be physically active, get regular rest, develop leisure-time hobbies for relaxation, strengthen social networks with family and friends, and achieve a balance among family, work, and play.[13]

The relative importance of certain dietary recommendations is underscored by their appearance in the U.S. Department of Health and Human Services' official health promotion strategy for improving the nation's health, laid out in the year 2010. Called *Healthy People 2020*, the plan of action presents a national health promotion and disease prevention agenda for the next 10 years.[14] The report sets health objectives for the year 2020 with four broad goals designed to help all Americans achieve their full potential by attaining high-quality, longer lives free of preventable disease, by eliminating health disparities, by creating social and physical environments that promote good health, and by promoting healthy behaviors across the lifespan. Table 1-5 lists the nutrition-related objectives considered to be top health priorities for the present decade.

The goals of the *Healthy People 2020* nutrition topic area promote health and reduce chronic disease risk through the consumption of healthful diets and achievement and maintenance of healthy body weights. Several *Healthy People 2020* objectives focus on risk reduction and specify targets for the intake of nutrients such as sodium and calcium and of foods such as fruits, vegetables, and whole-grain products. Other risk-reduction objectives set targets to reduce the prevalence of obesity, to reduce consumption of calories from solid fats and added sugars, to increase the proportion of schools that do not

TABLE 1-5	*Healthy People 2020* Nutrition-Related Objectives for the Nation[a]

Disease-Related Objectives
- Reduce rates of heart disease and stroke, hypertension, cancer, diabetes, osteoporosis, and tooth decay.

Nutrition and Weight Status Objectives
- Increase the proportion of adults who are at a healthy weight and reduce the proportion of children, adolescents, and adults who are considered obese.
- Increase the proportion of primary care physicians who regularly measure the body mass index of their patients.
- Increase proportion of worksites that offer nutrition or weight-management classes or counseling.
- Increase proportion of people who meet the recommended dietary intakes for saturated fat, sodium, and calcium in the diet.[b]
- Increase dietary intakes of fruits, vegetables, and whole grains.[c]
- Reduce consumption of calories from solid fats and added sugars in the diet.[c]
- Increase proportion of mothers who breastfeed, children whose intake of meals and snacks at school contributes to overall dietary quality, and schools that require daily physical education for all students.
- Reduce rates of iron deficiency in young children, women of childbearing age, and pregnant women.
- Reduce household food insecurity and in so doing reduce hunger.

Food Safety Objectives[d]
- Reduce infections caused by key pathogens transmitted commonly through food.
- Reduce severe allergic reactions to food among adults with a food allergy diagnosis.
- Increase the proportion of consumers who practice four essential food safety behaviors when handling food: washing hands, avoiding cross-contamination, cooking foods thoroughly, and chilling foods promptly.
- Improve food safety practices associated with foodborne illness in food service and retail establishments.

[a] The complete list of *Healthy People 2020* objectives may be viewed online at www.healthypeople.gov.
[b] Recommendations: < 10% of calories from saturated fat; 2,300 mg or less of sodium and 1,300 mg of calcium for children (ages 9 to 18); 1,000 mg of calcium for adults (ages 19 to 50); and 1,200 mg of calcium for adults over 50.
[c] See Table 2-5 and Figure 2-5 for specific recommendations.
[d] See Chapter 13 for more on the topic of food safety.

SOURCE: U.S. Department of Health and Human Services, Office of Disease Prevention and Health Promotion, *Healthy People 2020* (Washington, DC: U.S. Department of Health and Human Services, 2010), available at www.healthypeople.gov.

health promotion helping people achieve their maximum potential for good health.

sell or offer calorically sweetened beverages to students, and to reduce the proportion of people who engage in no leisure-time physical activity. Each *Healthy People* objective has a target for specific improvements to be achieved by the year 2020.

Healthy People 2020 is based on the accomplishments of 4 previous Healthy People initiatives, including most recently *Healthy People 2010: Objectives for Improving Health*. How are Americans doing in terms of meeting the Healthy People objectives? When *Healthy People 2010* was released in 2000, life expectancy was 76 years. Today, the average life expectancy at birth is 78 years, and death rates for heart disease, stroke, and certain types of cancer have declined.[15] Since the 1980s, however, the prevalence of overweight has soared, and the trend toward obesity continues (see Chapter 10). In fact, overweight has increased among all ethnic and age subgroups of the population. One contributing factor is that people seem to be taking fewer steps to control their weight by adopting sound dietary patterns and being physically active. Some 36 percent of adults engage in no leisure-time physical activity. Little or no improvements are noted for consumption of fruits, vegetables, and whole grains.

Table 1-6 shows the current status of the U.S. population for various nutrition and health-related lifestyle habits and compares it with the nation's goals for 2020.

TABLE 1-6	Status Report on Healthy Lifestyle Habits		
This table shows the percentage of adults practicing certain healthy lifestyle habits compared to the *Healthy People 2020* goals for the nation.			
Healthy Lifestyle Objective		**Current Estimate**	**2020 Target**
Increase the proportion of adults who are at a healthy weight (BMI[b] 18.5–24.9).		31%	34%[a]
Reduce the proportion of adults who are obese (BMI ≥ 30).		34%	31%[a]
Reduce the proportion of children and adolescents (ages 2–19) who are considered obese.		16%	14.6%[a]
Increase the proportion of schools that do not sell or offer calorically sweetened beverages to students.		9%	21%
In the diets of persons aged 2 years and older:			
• Increase the intake of fruits.		0.5 cup per 1,000 calories	0.9 cup per 1,000 calories
• Increase the intake of total vegetables.		0.8 cup per 1,000 calories	1.1 cup per 1,000 calories
• Increase the intake of dark-green vegetables, orange vegetables, and legumes.		0.1 cup per 1,000 calories	0.3 cup per 1,000 calories
• Increase the intake of whole grains (e.g., whole-wheat bread and oatmeal).		0.3 ounce per 1,000 calories	0.6 ounce per 1,000 calories
• Reduce consumption of calories from solid fats.		18.9%	16.7%
• Reduce consumption of calories from added sugars.		16%	10.8%
• Reduce consumption of saturated fat.		11.3%	9.5%
• Reduce consumption of sodium.		3,641 milligrams	2,300 milligrams
• Increase consumption of calcium.		1,118 milligrams	1,300 milligrams
Reduce iron deficiency among females of childbearing age.		10.4%	9.4 %[a]
Reduce the proportion of adults who engage in no leisure-time physical activity.		36.2%	32.6%[a]
Increase the proportion of adults who engage in aerobic physical activity of at least moderate intensity for at least 150 minutes/week, or 75 minutes/week of vigorous intensity, or an equivalent combination.		43.5%	47.9%[a]
Increase the proportion of adults who perform muscle-strengthening activities on 2 or more days of the week.		22%	24%[a]

[a] Target setting method: 10 percent improvement.
[b] Body mass index (BMI) is calculated as weight in kilograms (kg) divided by the square of height in meters (m)^2. [BMI = weight (kg)/height (m^2)]. To estimate BMI using pounds (lb) and inches (in.), divide weight in pounds by the square of height in inches. Then multiply the resulting number by 703 [BMI = weight (lb)/height (in.)^2 × 703].

SOURCE: U.S. Department of Health and Human Services, Office of Disease Prevention and Health Promotion, *Healthy People 2020* (Washington, DC: U.S. Department of Health and Human Services, 2010), available at www.healthypeople.gov.

You can't look into a crystal ball to find out how long you will live—but you can get a rough idea of the number of years you're likely to live based largely on your lifestyle today as well as certain givens, such as your family history. To do so, play the Longevity Game.

Start at the top line: age 78, the average life expectancy for adults in the United States today. For each of the 11 lifestyle areas, add or subtract years as instructed. If an area doesn't apply to you, go on to the next one. If you are not sure of the exact number to add or subtract, make a guess. Don't take the score too seriously, but do pay attention to those areas where you lose years; they could point to habits you might want to change.

START WITH	78
1. Exercise	_____
2. Relaxation	_____
3. Driving	_____
4. Blood pressure	_____
5. 65 and working	_____
6. Family history	_____
7. Smoking	_____
8. Drinking	_____
9. Gender	_____
10. Weight	_____
11. Age	_____
12. Seatbelts	_____
YOUR FINAL SCORE:	_____

1. **Exercise.** If you work out on a daily basis, add 3 years. If you don't get much exercise at home, on the job, or at play, subtract 3 years.

2. **Relaxation.** If you have a laid-back approach to life (you roll with the punches), add 1 year. If you're aggressive, hard-driving, or anxious (suffer from sleepless nights, bite your nails, etc.), subtract 1 year. If you consider yourself unhappy, subtract another year.

3. **Driving.** Drivers involved in an accident or violation in the past 3 years should subtract 2 years. If you have been convicted of driving while intoxicated (DWI), subtract 5 years. If your record is free of accidents and violations over the last 2 years, add 1 year.

4. **Blood pressure.** Although high blood pressure is a major risk factor to common conditions (heart attacks and strokes), it can be lowered effectively through changes in lifestyle and/or medication. The problem is that rises in blood pressure can't be felt, so many people don't know they have it and therefore never receive treatment. If your blood pressure is checked regularly with a normal reading, add 3 years. If you have high blood pressure, subtract 3.

5. **65 and working.** If you are at the traditional retirement age or older and still working, add 2 years.

6. **Family history.** If any grandparent has reached age 85, add 2 years; if all grandparents have reached age 80, add 4. If a parent or sibling suffered from or died of a stroke or heart attack before age 50, subtract 3.

7. **Smoking.** Cigarette smokers who finish more than two packs a day, subtract 8; one or two packs a day, subtract 6; one-half to one pack, subtract 3.

8. **Drinking.** If you drink three cocktails (or beers or glasses of wine) a day, subtract 1 year.

9. **Gender.** Women live longer than men. Females add 3 years; males subtract 3 years.

10. **Weight.** If you generally follow a healthy diet, you will probably remain healthy longer, entitling you to add 2 years.

11. **Age.** How long you have already lived can help predict how much longer you'll survive. If you're under 30, the jury is still out. But if your age is 30 to 39, add 2 years; 40 to 49, add 3 years; 50 to 69, add 4 years; 70 or over, add 5 years.

12. **Seatbelts.** Do you always buckle up, even if it's a short trip? If you use a seatbelt, add 1 year. If you don't, subtract 1 year.

SOURCE: From "The Longevity Game," by Northwestern Mutual Life Insurance Company. Reproduced with permission.

NOTE: The Longevity Game uses statistics from the U.S. Decennial Life Tables 1999 to 2001, U.S. Centers for Disease Control and Prevention (CDC). The outcome may not provide accurate results for individuals with chronic illnesses and conditions.

Although these are just a few of the goals for nutrition spelled out in *Healthy People 2020*, they represent some of the priorities for maintaining good health. Much of the practical information presented later in this chapter and in those that follow is aimed at guiding you toward developing eating and lifestyle habits that will help you achieve the goals.

Understanding Our Food Choices

The choices you make about what to eat can have a profound impact on your health, both now and in your later years. The healthy eater resists disease and other stresses better than a person with poor dietary habits and is more likely to enjoy an active, vigorous lifestyle for a greater number of years. Even so, the nutritional profile of various foods ranks as only one of many factors that influence your eating habits. Whether you realize it or not, each time you sit down to a meal you bring to the table such factors as your own personal preferences, cultural traditions, and economic considerations. These influences exert as great an impact on your eating habits as does **hunger**—the physiological need for food—and **appetite**—the psychological desire for food, which may be stimulated in response to the sight, smell, or thought of food even when you're not hungry. The following sections examine some of the most influential factors in making food choices.

Availability   Our diets are limited by the types and amounts of food available through the food supply, which in turn is influenced by many forces. Because we have the geographic area, climate, soil conditions, labor, and capital necessary to maintain a large agricultural industry, Americans enjoy what is arguably the most abundant food supply in the world. In addition, unlike many other less wealthy countries, the United States and Canada have the resources needed to import and distribute a wide variety of foods from other countries—everything from kiwi from New Zealand to mangoes from the tropics.

History has shown, however, that when it comes to health, an abundant food supply can be a double-edged sword. Access to many types of foods allows people to choose high-fat diets that are rich in meats and other fatty foods, which can contribute to increased rates of heart disease and other problems. That's one of the reasons why degenerative diseases are sometimes referred to as *diseases of affluence*.

Income, Food Prices, and Convenience   As most college students know firsthand, the amount of money available to spend on food can mean the difference between ordering pizza every night and resigning yourself to a steady diet of peanut butter and jelly sandwiches. Extremely low incomes can make it difficult for people to buy enough food to meet their minimum nutritional needs, thereby putting them at risk of **undernutrition**.

We know that a balanced diet of healthful foods is associated with reduced risk of chronic diseases and death, particularly from heart disease and cancer. One reason people may not choose healthful foods over foods that are low in nutrients and high in fat is cost. If you are like most Americans, you think it's "more expensive" to eat healthful foods like fruits and vegetables. But is the cost of a healthful diet really significantly higher than the standard American diet of "convenience" foods, snacks, bakery items, soft drinks, and other less nutritious foods? Compare the cost of convenience foods with the same foods made from scratch (most convenience foods cost more).

A consumer's *perception* of the cost of various foods can also play a role in his or her choices. For example, two barriers that prevent people from adopting healthful

© PhotoDisc/Getty Images

Numerous factors influence your food choices, including these:

- Hunger, appetite, and food habits
- Nutrition knowledge, health beliefs/ concerns, and practices
- Availability, convenience, and economy
- Advertising and the media
- Early experiences, social interactions, and cultural traditions
- Personal preference, taste, and psychological needs
- Values, such as political views, environmental concerns, and religious beliefs

hunger the physiological need for food.

appetite the psychological desire to eat, which is often but not always accompanied by hunger.

undernutrition severe underconsumption of calories or nutrients, leading to disease or increased susceptibility to disease; a form of malnutrition.

You *Can* Afford to Eat Nutritious Foods— Tips for Supermarketing

Want to save money and consume less fat and fewer calories? Healthful eating starts with healthful food shopping. With preparation and some label-reading skills, you can leave the grocery store with a supply of lower-fat, lower-calorie basics like these:[16]

- Fat-free or low-fat milk, yogurt, cheeses, and cottage cheese
- Light or reduced-fat margarines
- Eggs/egg substitutes
- Whole-grain breads, bagels, and English muffins
- Low-fat flour tortillas, soft corn tortillas
- Plain cereals (dry or cooked)
- Rice, pasta (remember, it's what you put on these grain products that increases the fat/calorie content)
- Fresh, frozen, or canned fruits packed in juice
- Fresh, frozen, or canned vegetables
- Dry beans and peas
- Skinless, white-meat chicken or turkey
- Fish and shellfish (not battered and fried)
- Beef: round, sirloin, chuck, loin, and extra lean ground beef
- Pork: leg, shoulder, tenderloin
- Low-fat or nonfat salad dressings, salsa, herbs, and spices

The next time you shop for groceries, keep in mind these smart shopping tips:[17]

1. Buy local foods and fresh foods in season. Use the local newspaper to find the best seasonal buys and special sale items.
2. Shop from a list to help avoid buying unnecessary items. Keep a running list in your kitchen, and note items that you need to replace.
3. Read the ingredients list and Nutrition Facts label on packaged foods; compare amounts of fat, sodium, calories, and nutrients in similar products. Ingredients are listed in order of quantity.
4. Use "sell by" and "best if used by" dates to ensure quality and freshness. Buy only the amount you or your family will eat before the food spoils.
5. Shop the perimeter of the grocery store to find many fresh whole foods: fresh produce; low-fat dairy products; lean meats, poultry, and fish; and whole-grain breads. Maneuver down the aisles only for specific items on your list, such as canned tomato products, spices, and canned or dry beans.

Perceived Barriers to Healthful Eating

- Healthful foods are not always available from fast-food and take-out restaurants.
- It costs more to eat healthful foods.
- I'm too busy to take the time to eat healthfully.
- I hear too much conflicting information about which foods are good for me and which foods are not.
- Healthful foods don't taste as good.
- The people I usually eat with do not eat healthful foods.

eating habits are the beliefs that it would be too expensive and inconvenient. See the margin list for the most frequently cited roadblocks to healthful eating.[18] In fact, 40 percent of consumers who answered one survey said that fruits, vegetables, seafood, and other elements of a low-fat, nutrient-rich diet would strain their budgets. In reality, some research has shown that switching from a high-fat diet to one that is lower in fat can *reduce* food costs.[19] Just cutting back on the amount of meat and poultry— the source of much of the fat in the American diet and where many of our food dollars are spent—goes a long way in trimming food budgets.

Research reported in the *Journal of the American Dietetic Association* found that not only is a diet of healthful foods less costly, but it also promotes weight loss.[20] In one study, overweight children and their parents were encouraged to consume more "nutrient-dense" foods like fruits and vegetables (high in nutrients and low in fat and calories) and less "empty-calorie" foods (low in nutrients and high in fat and/or calories). After one year, food costs decreased, and the overweight children and parents were 5 to 8 percent less overweight than they were at the beginning of the study![21]

Advertising and the Media
Television and radio commercials, as well as magazine and newspaper ads, play an extremely powerful role in influencing our food choices and our knowledge of nutrition. Given today's health-conscious environment, food manufacturers promote the nutritional merits of their products more than ever before. In fact, in some of the most popular women's magazines, the number of food and beverage ads containing nutrition claims has increased by nearly 100 percent in the last two decades.[22]

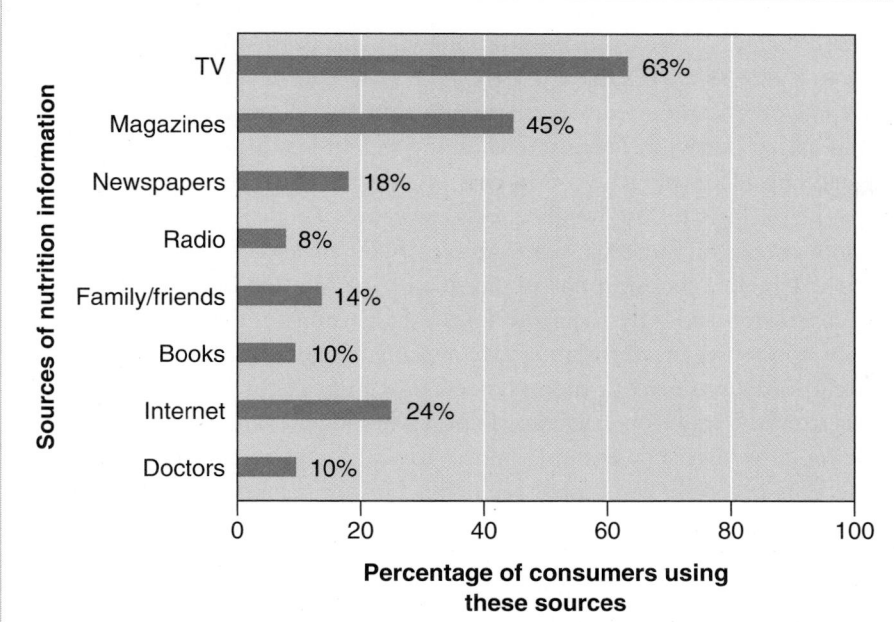

FIGURE 1-4

Consumer Sources of Nutrition Information*

*NOTE: Consumers could cite more than one source.

SOURCE: Reprinted with permission from the American Dietetic Association, *Nutrition and You: Trends 2008,* American Dietetic Association © Copyright 2008.

Unfortunately, advertising is not always created with the consumer's best interest in mind. Much of the food advertising that we're exposed to, from the earliest ages, is aimed at selling products that aren't the optimal choices for regular inclusion in a healthful diet. For example, the great majority of television commercials geared to children and aired on Saturday mornings promote high-fat, sugary foods such as candy and sugar-coated cereals.[23] In fact, the food industry spends more than $11 billion overall for food, beverage, and fast-food advertising. This amount dwarfs the almost $5 million spent by the government in promoting its campaign to encourage people of all ages to increase their consumption of fruits and vegetables.[24] As a result, commercials promoting good nutrition are relatively few and far between.

Along with advertising, the media rank among the most influential sources of diet and nutrition information, which in turn affects our food choices. Consider that most Americans look first to television as a source of nutrition information; then to magazines, newspapers, radio, family or friends, books, and the Internet; and finally to doctors (see Figure 1-4).[25] Unfortunately, the reliability of the information delivered by the media varies considerably. The Spotlight at the end of this chapter will help you learn how to evaluate the nutrition information you receive via the media.

social group a group of people, such as a family, who depend on one another and share a set of norms, beliefs, values, and behaviors.

Social and Cultural Factors

A person's social and cultural groups have a significant impact on his or her food choices. **Social groups** such as families, friends, and coworkers tend to exert the most influence. The family, particularly the wife/mother, plays one of the most powerful roles in determining our food choices. That makes sense, because the family is the first social group a person encounters, as well as the one to which he or she typically belongs for the longest period of time. The values, attitudes, and traditions of our family can have a lasting

The act of eating is complex. We derive many benefits—both physical and emotional—when we eat foods.

effect on our food choices. Think of the holiday food traditions in your own family. Treasured recipes or rituals surrounding holiday meals are often passed from one generation to the next.

Friends, coworkers, and members of other social networks also influence our food choices and eating behavior. For instance, many weight-loss programs feature group sessions made up of people who are in the same boat. Thus, they can support one another in their efforts to lose weight. Social pressure can also push us to eat meals we might not choose on our own. For example, if you are a guest in another country or in a friend's home, choosing not to partake of the food and drink that's offered might be considered rude. By the same token, it's natural to join your friends on a spontaneous trip for ice cream or pizza even when you're not hungry.

Culture also determines our food choices, to a large extent. Many of our eating habits arise from the traditions, belief systems, technologies, values, and norms of the culture in which we live. For example, in the United States and Canada, the idea of eating insects is generally considered repulsive. However, many people throughout the rest of the world relish dishes prepared with various bugs, including locust dumplings (northern Africa), red-ant chutney (India), water beetles in shrimp sauce (Laos), and fried caterpillars (South Africa).[26]

One of the ways people of different cultures come together to share their respective heritages is by sampling each other's traditional foods. American consumers are particularly fortunate because they typically don't have to travel far to taste the food of different cultures. **Ethnic cuisine**, ranging from Chinese to Mexican to Italian to Indian food, has become embedded in American culture. (The Spotlight in Chapter 2 discusses the eating patterns of different cultures in detail.)

Religion is one aspect of culture that affects the food choices of millions of people worldwide. The practice of giving and abstaining from food has long been used by many cultures as a way to show devotion, respect, and love to a supreme being or power. Also, dietary customs play an important role in the practice of many of the world's major religions. As Table 1-7 shows, many religions specify which foods their followers may eat and how those foods must be prepared.

Personal Values or Beliefs

Some people adopt a certain way of eating or making food choices based on a larger worldview. For instance, many environmentally conscious people, believing that raising animals for human consumption strains the world's supply of land and water, choose to abstain from meat as much as possible in an effort to preserve Earth's resources. Others may choose to boycott certain manufacturers' items for political reasons, perhaps because they disagree with a company's advertising practices.

See the Nutrition Action feature in Chapter 13 for more about the benefits of purchasing locally produced, seasonally available, and organically grown foods.

Over the past decade, many people have begun to make food choices based on sustainability issues and the benefits of purchasing locally produced, seasonally available, and organically grown food.[27] *Sustainability* in the food system is defined as a society's ability to shape its economic and social systems to maintain both natural resources and human life, and it involves building locally based, self-reliant food systems.[28] When consumers purchase foods that have been produced locally, a greater proportion of the profits remains with local farmers, providing them with a livable income while supporting local economies.[29] Additionally, the purchase of locally produced foods protects the environment by reducing the use of fossil fuel (for transporting food) and packaging materials.[30]

Other Factors That Affect Our Food Choices

One of the main reasons you choose to eat certain foods is your preference for certain tastes. Just about everyone enjoys sweet foods, for example, because humans are born with an affinity for sugar.[31] In addition, we usually prefer foods that have happy associations for us: foods prepared for special occasions, those given to us by a loved one when we

culture knowledge, beliefs, customs, laws, morals, art, and literature acquired by members of a society and passed along to succeeding generations.

ethnic cuisine the traditional foods eaten by the people of a particular culture.

TABLE 1-7	Dietary Practices of Selected Religious Groups
Religious Group	**Dietary Practices***
Buddhist	Dietary customs vary depending on sect. Many are lacto-ovovegetarians, because of restrictions on taking a life. Some eat fish, and most eat no beef or poultry. Monks fast at certain times of the month and avoid eating solid food after the noon hour.
Hindu	All foods thought to interfere with physical and spiritual development are avoided. Many are lacto-vegetarians and/or avoid alcohol. The cow is considered sacred—an animal dear to the Lord Krishna. Beef is never consumed, and often pork is avoided.
Jewish	*Kashrut* is the body of Jewish law dealing with foods. The purpose of following the complex dietary laws is to conform to the Divine Will as expressed in the Torah. The term *kosher* denotes all foods that are permitted for consumption. To "keep kosher" means to follow dietary laws in the home. A lengthy list of prohibited foods, called *treyf*, includes pork and shellfish. The laws define how birds and mammals must be slaughtered, how foods must be prepared, and when they may be consumed. For example, dairy foods and meat products cannot be eaten at the same meal. During Passover, special laws are observed, such as the elimination of any leavened foods.
Mormon	Alcoholic beverages and coffee, tea, and other beverages containing caffeine are avoided. Mormons are encouraged to limit meat intake and emphasize grains in the diet.
Muslim	Overeating is discouraged, and consuming only two-thirds of capacity is suggested. Dietary laws are called *halal*. Prohibited foods are called *haram*, and they include pork and birds of prey. Laws define how animals must be slaughtered. Alcoholic drinks are not allowed. Fasting is required from sunup to sundown during the month of Ramadan.
Roman Catholic	Meat is not consumed on Fridays during Lent (40 days before Easter). No food or beverages (except water) are to be consumed 1 hour before taking communion.
Seventh-Day Adventist	Most are lacto-ovovegetarians. If meat is consumed, pork is avoided. Tea, coffee, and alcoholic beverages are not allowed. Water is not consumed with meals, but is drunk before and after meals. Followers refrain from using seasonings and condiments. Overeating and snacking are discouraged.

*Many of the religious guidelines regarding food have practical applications for the society. For example, the Hindu prohibition against killing cattle respects the needs for Indian farmers to use cattle for power and cattle dung for fuel. Cows also supply milk to make dairy products.

SOURCE: Adapted from Religious Food Practices, accessed at www.eatethnic.com; and M. Boyle and D. Holben, *Community Nutrition in Action: An Entrepreneurial Approach,* 5th ed. (Belmont, CA: Wadsworth, 2010), 541.

were children, or those eaten by an admired role model. By the same token, intense aversions to certain foods—perhaps foods you were given when you were sick or foods you were forced to eat as a child—can be strong enough to last a lifetime. Your parents may have taught you to prefer certain foods and pass up others for reasons of their own, without even being aware they were doing so.

Food habits are also intimately tied to deep psychological needs, such as an infant's association of food with a parent's love. Yearnings, cravings, and addictions with profound meaning and significance sometimes surface as food behavior. Some people respond to stress—positive or negative—by eating; others use food to fill a void, such as lack of satisfying personal relationships or fulfilling work.

The influences on people's eating habits are as many and varied as the individuals themselves. Our food choices reflect our own unique cultural legacies, philosophies, and beliefs. To think of food as nothing more than a source of nutrients would deny food's rich symbolism and meaning and take away much of the pleasure of breaking bread with friends and family. As you read this book and consider ways to improve your own eating habits, take time to reflect on your unique background and think about how you can integrate your knowledge of nutrition into your cultural heritage and philosophies.

Even at many American-style restaurants, you can experience other cultures by sampling from the various ethnic cuisines found on the menu.

© Bob Daemmrich/Stock Boston, Inc.

Nutrition Action

Good and Fast—A Guide to Eating on the Run, Or Has Your Waistline Been Supersized?

Contrary to popular belief, your professors do not conspire to plan due dates for assignments, quizzes, and exams on the same day. As a student, your schedule can get hectic, especially at midterm and the end of a semester. In the working world, much like college, you will also have deadlines and scheduling conflicts that may leave little time for everyday activities like sitting down to a meal with friends or family. Even if your life is not hectic, chances are you've stood in line for a burger and fries, a slice of pizza, a taco, or a muffin and coffee at least once this week. Almost 60 percent of Americans eat out for lunch at least once a week,[32] and nearly 25 percent eat lunch out five or more times a week.[33]

Even though a meal eaten on the run fits easily into a busy schedule, it's not necessarily as simple to work it into the dietary guidelines recommended by major health organizations. Eating away from home is not an invitation to forget about good nutrition, eat more ("supersize"), or eat differently than you would at home.[34] Can you eat away from home and still eat healthfully? It is possible, but it takes some planning.

Fast food does not have to be an unhealthy option, as long as you don't supersize. Fast foods commonly contain more fat (including more saturated fat), less fiber, more cholesterol, and more calories than meals made at home.

Portion sizes of foods have been on the increase since the early 1970s.[35] In the 1950s, a single-serve bottle of a soft drink was 6 ounces. Today, a single-serve bottle is 20 ounces. A typical bagel purchased today is, at 4 to 7 ounces, almost twice the size bagels used to be, at 2 to 3 ounces.[36]

Along with the larger portions come more calories. Could these larger portions be contributing to the obesity epidemic in the United States? Yes! When people are served more food, they eat more.[37] Is it any coincidence, then, that as portion sizes have increased over the past two decades, the incidence of overweight and obesity among adults and children also has increased?[38] We don't think so.

The enlarging size of American food portions is linked to a practice used by the food industry called "value" marketing. Consumers are prompted by point-of-purchase displays and verbally by employ-

Cheeseburger
Small fries
Small soft drink
690 calories
24 g fat
8 g saturated fat

Cheeseburger
Supersize fries
Supersize soft drink
1,350 calories
43 g fat
13 g saturated fat

ees to spend a little extra money to "upgrade" to larger portions, leaving the customer with the feeling of "What a deal I got!"[39] For the food sellers, costs to increase portion sizes are small, but the profit margin is huge.

So, how do you, the typical harried college student, get fast meals without increasing the calories and fat in your daily diet? Remember these strategies next time you find yourself in a time crunch and in line at the closest fast-food restaurant.

Strategy 1: Don't supersize. It's a great marketing ploy to get you to buy more, but you will also be eating more fat, cholesterol, and salt—and weighing more as a result. Think small. Stick with smaller burgers, sliced meat, or grilled chicken sandwiches with mustard, catsup, lettuce, onion, pickles, and tomatoes.

Strategy 2: Think grilled, not fried. Frying foods adds about 50 percent more fat and/or calories compared with items that have not been fried. (At Wendy's, a grilled chicken sandwich has 8 grams of fat versus 18 for a deep-fried, breaded chicken sandwich.[40]) Order a baked potato that you can dress with low-fat margarine and/or low-fat or fat-free sour cream. If you can't live without French fries, get them on occasion; but most of the time, try to substitute a salad (with low-fat, fat-free, or regular dressing on the side) or a baked potato with low-fat toppings.

Strategy 3: Hold the mayo. Each spoonful of mayonnaise adds about 100 calories, nearly all of which are from fat. Most fast-food sandwiches contain more than just one spoonful of mayo in their toppings and special sauces. The same goes for tartar sauce. Order a fish sandwich without it, and you'll trim at least 70 fat-laden calories (the amount in just 1 tablespoon) from your meal. Ask for lots of lettuce and tomatoes and less sour cream and guacamole on your nachos and tacos. A tablespoon of either sour cream or guacamole adds about 25 calories to Mexican fare. A few extra chunks of tomato, in contrast, supply a negligible number of calories, no fat, and a good deal of vitamin C.

Strategy 4: Avoid all-you-can-eat restaurants. No explanation necessary. Remember moderation and variety.

Strategy 5: "Just say no." Did you know that a 16-ounce soft drink adds 200 calories to a meal and nothing else? A medium chocolate shake can add 350 calories to a meal, and a large shake can add 770 calories. Wash your meal down with low-fat milk instead of a milk shake, and you'll cut the fat and calorie count at least in half. Better yet, in the continuing effort to get enough fluids, drink water instead.

Strategy 6: Balance fast-food meals with other food choices during the day.[41] If you can't avoid fast food, then adjust your portion sizes and food choices at other meals. Increasing physical activity will help counterbalance extra calories.

Strategy 7: Split your order—share with a friend. Or split dessert. Frequently a few bites can satisfy a sweet tooth.

Strategy 8: Bring your lunch. You'll save money and time in addition to planning a healthful lunch in less time than you would probably spend in line or in the drive-through line. Try taking leftovers in a microwavable container. Make double batches when you cook, and put some in the freezer in single-serving containers for lunch.

Strategy 9: Choose grab-and-go foods. Keep these foods handy for a fast and healthful snack: baby carrots, fresh fruit, dried fruit, fruit juices (not fruit drinks), low-fat yogurt, string cheese, and low-fat popcorn.

Strategy 10: If all else fails, go for the obvious low-calorie choices. For example, Subway sells subs with 6 grams of fat or less. Don't let the word *chicken* or *fish* fool you. Many health-conscious consumers have heard the advice to choose skinless poultry and fish instead of relatively high-fat red meat—but when it comes to chicken nuggets and fish patties coated with batter and deep fried, it is a different story. Six chicken nuggets, for example, typically contain as many calories (about 300) as an entire burger. What's more, many chicken and fish sandwiches chalk up as much fat as a pint and a half of ice cream. Even rotisserie-style chicken contains a large amount of fat and calories if you don't remove the skin before eating it. When ordering a pizza, hold the sausage and pepperoni, and ask for mushrooms, green peppers, and onions instead. Pizza is an excellent source of calcium—the bone-building mineral that many Americans don't get enough of—as well as protein, carbohydrate, and a number of vitamins and minerals. Still, two slices of pepperoni pizza can easily contain 100 more calories and twice as much fat as the same amount topped with onions, green peppers, and mushrooms.

How Do You Tell If It's Nutrition Fact or Nutrition Fiction?

Most people want to eat a healthful diet and lower their chances of cancer and other chronic diseases. Frequently, however, information they read or hear about healthful eating from one source is contradicted by another. It's not always easy to distinguish accurate information from misinformation. Do you know which of the following statements are true or false? Check your answers with the correct answers at the end of this section.*

1. It is difficult for busy people to eat a balanced diet.
2. People who graduate from college are smart enough not to be victimized by nutrition misinformation.
3. Sugar is a major cause of hyperactivity in children.
4. No special training is legally required for a person to offer nutrition information to the public.
5. Most health food retailers have been educated about the research done regarding the products they sell.
6. Protein and/or amino acid supplements help bodybuilders, recreational weight lifters, and other athletes improve their performance by increasing muscle size.
7. Most nutrition-related books and magazine articles undergo prepublication review by experts.
8. Cigarette smoking is the leading cause of preventable death in the United States.
9. Herbal products are as safe and effective as many drugs prescribed by physicians.
10. It is illegal for manufacturers of dietary supplements to print false information on their product labels.

You've just watched a television commercial for a vitamin supplement that is guaranteed to produce a laundry list of benefits, including fewer colds, a better complexion, and a decreased risk of cancer. Should you buy it? You've just read a magazine article with a plan for quick weight loss. Should you believe it? Someone who plays the same sport as you do says that improving your diet will help your game. Where do you go for help? A friend just sent you an e-mail saying that bananas imported from Costa Rica are infected with necrotizing fasciitis. Should you pass this on to all of your friends?

We all find ourselves faced with such decisions at one time or another. It's crucial to know how to protect ourselves from nutrition misinformation. Health fraud costs consumers almost $30 billion each year, and money down the drain is just one problem stemming from misleading nutritional information. Although some fraudulent claims about nutrition are harmless and may make for a good laugh, others can be harmful. False claims about nutritional

*Adapted from Quackwatch. Consumer Health I.Q. Test. Available at http://www .quackwatch.org/04ConsumerEducation/iq.html..

products have been known to bring about malnutrition, birth defects, mental retardation, and, in extreme cases, even death.

Negative effects due to following false nutritional claims can happen in two ways. First, the product in question may cause direct harm. Even a seemingly innocuous substance such as vitamin A can cause severe liver damage over time if taken in large enough amounts. Second, using bogus nutritional remedies can cause problems because such remedies can build false hope and might keep a consumer from obtaining sound, scientifically tested medical treatment. A person who relies on a so-called anticancer diet as a cure for the disease, for example, might forgo possible lifesaving interventions such as surgery or chemotherapy.

The following questions and answers will help you learn to evaluate the nutrition information that you see in the media or on the Internet. They will help you develop the skills necessary to view nutrition claims with a skeptic's eye or, at the very least, to help you decide when to find a qualified professional to help you evaluate the information.

Judging by what I've read on television, it seems as though nutritionists are always changing their minds. One week the headlines say to take vitamin E to help prevent heart disease, and the next week they say that vitamin E may not, in fact, prevent the disease. Why is there so much controversy?

Part of the confusion stems from the way the media interpret findings of scientific research. A good case in point is the controversy over whether a high-fiber diet protects against colon and rectal cancers—diseases that affect some 148,000 Americans each year. (A detailed discussion of nutrition and cancer appears in the Chapter 7 Spotlight.)

The fiber and colon cancer connection dates back to the early 1970s, when scientists observed that colon cancer was extremely uncommon in areas of the world where the diet consisted largely of unrefined foods and little meat. Researchers theorized that dietary fiber may protect against colon cancer by binding bile (a chemical substance needed in fat digestion) and speeding the passage of wastes and potentially harmful compounds through the colon. Since then, other studies have also suggested that those who eat a high-fiber diet have a lower risk of colon and rectal cancers.[42]

However, a flurry of headlines later threatened to pull the pedestal out from under the popular fiber theory, asking, "Fiber: Is It Still the Right Choice?" A Harvard-based study published in the *New England Journal of Medicine*, one of the most prestigious medical journals, suggested that fiber did nothing to prevent cancer.[43] The 16-year trial of almost 90,000 nurses—called the Nurses' Health Study—found that nurses who ate low-fiber diets (less than 10 grams daily) were no more likely to develop colon cancer than those eating higher levels of fiber (about 25 grams daily). As a result, the researchers concluded that the study provided no support for the theory that fiber could reduce risk of colon cancer.

The news was surprising and reinforces the need for research studies to be duplicated. All studies have some limitations, and a number of questions can be raised regarding conclusions of the Nurses' Health Study. For example, the study relied on participants to recall their eating habits accurately.

Is this type of self-reported dietary information reliable? Back in 1980, the nurses were asked for information about their intakes of "dark" breads. However, food labels at that time did not list the fiber content of breads, and some wheat breads on the shelf had amounts of fiber similar to those found in white bread. Did the nurses mistakenly consider "dark" bread to be the same as 100 percent whole-wheat bread?

Another question is, "What are the optimal levels of fiber intake for colon cancer protection?" Some experts believe it may take more than 25 grams of fiber a day to show cancer-protective effects, which might explain the lack of effect noted in the Nurses' Health Study.

This fiber story illustrates how news reports based on only one study can leave the public with the impression that scientists can't make up their minds. It seems as if one week scientists are saying fiber is good, and the next week the word is that fiber doesn't do any good at all. The truth is that health experts and major health organizations continue to urge adults to get 21 to 38 grams of fiber a day. (The average fiber intake today is about 15 grams.)

Contrary to what some headlines imply, reputable scientists do not base their dietary recommendations for the public on findings of one or two studies. Scientists are still conducting research to determine whether fiber does in fact help prevent colon cancer, and, if so, what types of fiber and in what amount. Scientists design their research to test theories, such as the notion that eating a high-fiber diet is associated with lower risk of cancer. Other factors, however, often complicate the matter at hand. The study of fiber and colon cancer is complicated by many other factors linked with colon cancer development, including inactivity, obesity, saturated fat intake, low calcium or folate intakes, and others.

So, should you hang on to your high-fiber cereals and vegetables? Yes, according to the American Institute for Cancer Research. Its analysis of more than 7,000 research studies provides evidence that increased intake of fiber may be associated with decreased incidence of colon cancer.[44] Most important, however, the health benefits of fruits, vegetables, legumes, and whole grains go beyond possible protection from colon cancer.[45] Diets rich in fiber from these foods are also strongly associated with reduced risks of heart disease, high blood pressure, type 2 diabetes, and diverticular disease, a condition that can lead to painful inflammation of the large intestine. Fiber also promotes a feeling of fullness after you eat, which can help with weight control.

How can I tell if a nutrition news story is noteworthy and a source of credible nutrition information?

You can critique nutrition news you read by asking a series of questions (see Table 1-8). Consider the following points as a checklist for separating the bogus news stories from those worth your attention:

- Where is the study published? The study described in the news story should have been published in a journal that uses experts in the field to review research results (called *peer review*). These reviewers can point out flaws in research design and may challenge researchers' conclusions before the study is published.

- How recent is the study? Be sure that the report is about recent research. The science of nutrition continues to develop from the results of new studies that employ state-of-the-art methods and technology, and benefits from the scrutiny of experts current in the particular field of study.

- What research methods were used to obtain the data? Are the reported results from an **epidemiological study** or an **intervention study**? Epidemiological studies examine populations

TABLE 1-8	Questions to Ask about a Research Report

- Was the research done by a credible institution? A qualified researcher?
- Is this a preliminary study? Have other studies reached the same conclusions?
- Was the study done with animals or humans?
- Was the research population large enough? Was the study long enough?
- Who paid for the study? Might that affect the findings? Is the science valid despite the funding source?
- Was the report reviewed by peers?
- Do the report findings and claims avoid absolutes, such as "proves" or "causes"?
- Does the report reflect appropriate context? For example, how does the research fit into a broader picture of scientific evidence and consumer lifestyles?
- Do the results apply to a certain group of people? Do they apply to someone of your age, gender, and health condition?
- What do follow-up reports from qualified nutrition experts say?

SOURCE: From "Position of the American Dietetic Association: Food and Nutrition Misinformation," *Journal of the American Dietetic Association* 106: 605. Copyright © 2006 by Elsevier. Reprinted with permission from Elsevier.

(Continued)

to determine food patterns and health status over time. These population studies are useful in uncovering **correlations** between two factors (for example, whether a high calcium intake early in life reduces the incidence of bone fractures later in life). However, they are not considered as conclusive as intervention studies. A correlation between two factors may *suggest* a cause-and-effect relationship between the factors, but it *does not prove* a causal relationship.

Intervention studies examine the effects of a specific treatment or intervention on a particular group of subjects and compare the results to those of a similar group of people not receiving the treatment. An example is a cholesterol-monitoring study in which half of the subjects follow dietary advice to lower their blood cholesterol and half do not. Ideally, intervention studies should be randomized and controlled—that is, subjects are assigned to either an **experimental group** or a **control group** using a random selection process. Each subject has an equal chance of being assigned to either group. The experimental group receives the "treatment" being tested; the control group receives a **placebo** or neutral substance. If possible, neither the researcher nor the participants should know which subjects have been assigned to which group until the end of the experiment. A randomized, controlled study helps ensure that the results were actually due to the treatment, and it minimizes chances that the results are due to a placebo effect or bias on the part of the researcher.

- What was the size of the study? To achieve *validity*—accuracy in results—studies generally must include a sufficiently large number of people (for example, intervention studies of 50 or more persons). This reduces the chances that the results are simply a coincidence and helps researchers generalize the conclusions of the study to a wider population.

- Who were the subjects? Look for similarities between the subjects in the study and yourself. The more you have in common with the participants (age, diet pattern, gender, and so forth), the more pertinent the study results may be for you.

- Does a consensus of published studies support the results reported in the news? Even if an experiment is carefully designed and carried out perfectly, its findings cannot be considered definitive until they have been confirmed by other research. Testing and retesting reduce the possibility that the outcome was simply the result of chance, error, or oversight on the part of the experimenter. Every study should be viewed as preliminary until it becomes just one addition to a significant body of evidence pointing in the same direction.

When making dietary recommendations for the public, experts pool the results of different types of studies, such as analyses of food patterns of groups of people and carefully controlled studies on people in hospitals or clinics. Before drawing any conclusions,

they then consider the evidence from all the research. The bottom line is that if you read a report in the newspaper or watch one on television that advises making a dramatic change in your diet or lifestyle based on the results of one study, don't take it to heart. The findings may make for a good story, but they're not worth taking too seriously.

Why doesn't the government do something to prevent the media from delivering misleading nutrition information?

The **First Amendment** guarantees freedom of the press. Accordingly, people may express whatever views they like in the media, whether sound, unsound, or even dangerous. This freedom is a cornerstone of the U.S. Constitution, and to deny it would be to deny democracy. By law, writers cannot be punished for publishing misinformation unless it can be proved in court that the information has caused a reader bodily harm.

Fortunately, most professional health groups maintain committees to combat the spread of health and nutrition misinformation. Many professional organizations have banded together to form the National Council against Health Fraud (NCAHF). The NCAHF monitors radio, television, and other advertising, and it investigates complaints. For more information on separating nutrition fact from nutrition fiction, visit these websites:

- www.ncahf.org
- www.quackwatch.org
- http://www.fda.gov/Food/default.htm
- http://www.nlm.nih.gov/medlineplus/healthfraud.html

Is the Internet a reliable source of nutrition and health information?

Information is rampant on the Internet.[46] In a sense, the Internet is information, and the information is continually being revised and *created*. Internet information exists in many forms (facts, statistics, stories, opinions), is created for many purposes (to entertain, to inform, to persuade, to sell, to influence), and varies in quality from good to bad.

The information you find on the Internet is only as good as its source. One method for determining whether information found on the Internet is reliable and of good quality is the CARS checklist. The acronym *CARS* stands for *credibility*, *accuracy*, *reasonableness*, and *support*.[47]

- *Credibility*. Check credentials of the author (if there is one!) or sponsoring organization. Is the author or organization respected and well known as a source of sound, *scientific* information? Evidence of a lack of credibility includes no posted author and even the presence of misspelled words or bad grammar. A credible sponsor will use a professional approach to designing the website.

- *Accuracy.* Check to ensure that the information is current, factual, and comprehensive. If important facts, consequences, or other information are missing, the website may not be presenting a complete story. Evidence of a lack of accuracy includes no date on the document, the use of sweeping generalizations, and the presence of outdated information. How often is the site updated? Reliable sites are updated regularly and post those dates on the site. Watch out for testimonials masquerading as scientific evidence—a common method for promoting questionable products on the Internet.

- *Reasonableness.* Evaluate the information for fairness, balance, and consistency. Does the author present a fair, balanced argument supporting his or her ideas? Are the author's arguments rational? Has he or she maintained objectivity in discussing the topic? Does the author have an obvious—or somewhat hidden—conflict of interest? Evidence of a lack of reasonableness includes gross generalizations ("Foods not grown organically are all toxic and shouldn't be eaten") and outlandish claims ("Kombucha tea will cure cancer and diabetes").

- *Support.* Check to see whether supporting documentation is cited for scientific statements. Does the website refer to legitimate scientific journals and publications? Is the source of information clearly presented? An Internet document that fails to show its sources is suspect.

Many reputable health and professional organizations and government agencies host websites on the Internet. To help you explore the World Wide Web, the publishers of this book maintain a website with links to reliable nutrition and health-related information at **www.cengage.com/nutrition**.

How can I tell whether a product is bogus?

It's not always easy. Given that many misleading claims are supposedly backed by scientific-sounding statements, it is difficult for even informed consumers to separate fact from fiction. The general rule of thumb is: If it sounds too good to be true, it probably is. The following red flags can help you spot a quack:

- *The promoter claims that the medical establishment is against him or her and that the government won't accept this new "alternative" treatment.* If the government or medical community doesn't accept a treatment, it's because the treatment hasn't been proven to work. Reputable professionals don't suppress knowledge about fighting disease. On the contrary, they welcome new remedies for illness, provided the treatments have been carefully tested.

- *The promoter uses testimonials and anecdotes from satisfied customers to support claims.* Valid nutrition information comes from careful experimental research, not from random tales. A few people's reports that the product in question "works every time" are never acceptable as sound scientific evidence.

- *The promoter uses a computer-scored questionnaire for diagnosing "nutrient deficiencies."* Those computers are programmed to suggest that just about everyone has a deficiency that can be reversed with supplements the promoter just happens to be selling, regardless of the consumer's symptoms or health.

- *The promoter claims that the product will make weight loss easy.* Unfortunately, there is no simple way to lose weight. Again, if a claim sounds too good to be true, it probably is.

- *The promoter promises that the product is made with a "secret formula" available only from this one company.* Legitimate health professionals share their knowledge of proven treatments so that others can benefit from it.

- *The treatment is available only through the back pages of magazines, over the phone, or by mail-order ads in the form of news stories or 30-minute commercials (known as infomercials) in talk-show format.* Results of studies on credible treatments are reported first in medical journals and then administered by a doctor or other health professional. If information about a treatment appears only elsewhere, it probably cannot withstand scientific scrutiny.

If I do buy a product, say, to help me lose weight, but I still need some advice about dieting, should I check with a nutritionist?

To answer that question, first consider the following. About 25 years ago, Charlie Herbert became a professional of the International Academy of Nutrition Consultants. Another member of the household, Sassafras Herbert, met all the requirements for membership in the American Association of Nutrition and Dietary Consultants, a "professional association dedicated to maintaining ethical standards in nutritional and dietary consulting." The only qualification for membership is a $50 fee, regardless of your background (or even your species). Charlie Herbert is a cat, and Sassafras is a poodle. The two obtained their "credentials" with the help of the late Victor Herbert, M.D., former professor of medicine, at Mount Sinai School of Medicine, New York City, and a leader in combating nutrition fraud. Dr. Herbert had his pets added to the membership rosters of those organizations to demonstrate how easy it is for anyone to get fake nutrition credentials. This is because in some states, the term **nutritionist** is not legally defined at present.

Before you pay a fee or follow a nutritionist's advice, inquire about the person's credentials. Some "nutritionists" obtain their diplomas and titles without undergoing the rigorous training required to obtain a legitimate degree in nutrition. Lax state laws make it possible for irresponsible **correspondence schools**—also called **diploma mills**—to grant degrees to unqualified individuals for nothing more than a fee.

(Continued)

How can I check a nutritionist's credentials?

You can call the institution the person claims has awarded the degree. To find out about the existence or reputation of an institution of higher learning, you can check the directory of colleges and universities called *Accredited Institutions of Postsecondary Education*, published by the American Council on Education (www.acenet .edu). Be suspicious of diplomas or degrees issued by institutions that cannot prove that they have **accreditation** from the Council on Education.

Another option is to find out whether the person is a **registered dietitian (RD)**. The RD, an especially meaningful credential, has a standard definition: a professional who has fulfilled coursework required by the American Dietetic Association (ADA), including courses in nutrition, food science, and other related subjects; has completed an internship that includes on-the-job training for counseling people about diet; and has passed a national registration exam. All registered dietitians must keep their credentials current by completing regular continuing education requirements. In some instances, a **dietetic technician** assists dietitians in helping people make healthy food choices. Like dietitians, dietetic technicians maintain their national registration credential through continuing professional education.

You can check on any registered dietitian (RD) or dietetic technician, registered (DTR) by asking for that person's registration number and calling the Commission on Dietetic Registration (312-899-0040). If you'd like to contact an RD but don't know where to find one, call the ADA's Consumer Nutrition Hot Line at (800) 366-1655 for a referral or visit the ADA website (www.eatright.org).

© Courtesy of Marilyn Herbert

Charlie and Sassafras display their professional credentials.

ANSWERS TO SPOTLIGHT NUTRITION HEALTH I.Q. TEST: All are false except numbers 4, 8, and 10. Seven correct answers suggest that you are fairly well informed about nutrition information. Ten correct answers suggest that you are very well informed.

Miniglossary

accreditation approval; in the case of hospitals or university departments, approval by a professional organization of the educational program offered. There are phony accrediting agencies; the genuine ones are listed in a directory called Accredited Institutions of Postsecondary Education.

control group a group of individuals with characteristics that match those of the group being treated in an intervention study but who receive a sham treatment or no treatment at all.

correlation a simultaneous change in two factors, such as a decrease in blood pressure with regular aerobic activity (a direct or positive correlation) or the decrease in incidence of bone fractures with increasing calcium intakes (an inverse or negative correlation).

correspondence school a school from which courses can be taken and degrees granted by mail (or online). Schools that are accredited offer respectable courses and degrees.

dietetic technician, registered (DTR) nutrition professionals who work as part of health care and food service management teams. DTRs have completed at least a two-year associate's degree at a U.S. regionally accredited college or university and an accredited dietetic technician program, and passed a national examination.

diploma mill a correspondence school that grinds out degrees (sometimes worth no more than the paper they are printed on) the way a grain mill grinds out flour.

epidemiological study a study of a population that searches for possible correlations between

nutrition factors and health patterns over time.

experimental group the participants in a study who receive the real treatment or intervention under investigation.

First Amendment the amendment to the U.S. Constitution that guarantees freedom of the press.

intervention study a population study examining the effects of a treatment on experimental subjects compared to a control group.

nutritionist a person who claims to be capable of advising people about their diets. Some nutritionists are registered dietitians, whereas others are self-described experts whose training is questionable.

placebo a sham or neutral treatment given to a control group; an inert, harmless "treatment" that the group's members cannot recognize as different from the real thing.

registered dietitian (RD) a professional who has graduated from a program of dietetics accredited by the Commission on Accreditation for Dietetics Education (CADE) of the American Dietetic Association (ADA), has completed an internship program or the equivalent to gain practical skills, has passed a registration examination, and maintains competencies through continuing education. Some states require licensing for dietitians, thereby requiring anyone who wants to use the title "dietitian" to receive permission to do so by passing a state examination.

Nutrition on the Web

www.cengage.com/nutrition/boyle/personalnutrition8e

Go to the *Personal Nutrition* site to check for the latest updates to chapter topics or to access links to related websites.

www.healthfinder.gov

Scroll down to *Personal Health Tools* for online screening checkups; check the Health Library for more information on healthy lifestyles; this site offers many links to other reliable health-related sites.

www.cdc.gov/healthyliving

The Centers for Disease Control (CDC) website provides links for healthy living topics.

www.eatright.org/Public/

The American Dietetic Association home page includes many nutrition resources and links. Search the Food & Nutrition Topics tab for tips on eating out, food shopping, and other topics.

www.fruitsandveggiesmorematters.org/

Provides healthy resources and encourages fruit and vegetable consumption.

www.hc-sc.gc.ca/

A Canadian health and nutrition information home page.

www.healthypeople.gov

Provides updates on the *Healthy People 2020* initiative.

www.nutrition.gov

This resource provided by the United States Department of Agriculture (USDA) provides easy access to all online government information on nutrition and dietary guidance.

www.nal.usda.gov/fnic

The Food and Nutrition Information Center (FNIC) provides credible, accurate, and practical information about nutrition and health. Search under Dietary Guidance and then scroll down to Fraud and Nutrition Misinformation.

www.pueblo.gsa.gov

This resource developed by the Federal Citizen Information Center (FCIC) provides access to consumer issues including food and health information.

www.health.nih.gov

A search engine from the National Institutes of Health (NIH) with access to health information; many database links are available, such as PubMed and clinical trials.

www.nlm.nih.gov/medlineplus/healthfraud.html

This government site provides health fraud resources including how to report fraud cases.

Menu of Online Study Tools

A variety of study tools for this chapter are available at our website to deepen your understanding of chapter concepts. Go to

www.cengage.com/sso

to find

- Practice tests
- Flashcards
- Glossary
- Web links
- Animations
- Chapter summaries, learning objectives, and crossword puzzles.

Diet Analysis+ in Action

Go to Diet Analysis+ and create your profile, following the directions given in the program. Track your diet for three days, then answer the following questions: What are your three favorite foods? Why are these your favorites? Using food records for the three days you tracked your diet, write down how often you consumed your favorite foods (or record why you *didn't* consume these foods during this period). What factors influenced your food choices during the three-day period?

Look at the Intake vs. Goals report. Are any of the areas low or missing during this three-day period? How might you improve your eating habits to correct these deficiencies?

The Pursuit of a Healthy Diet

Ask Yourself . . .

Which of the following statements about nutrition are true, and which are false? For each false statement, what is true?

1. It is wise to eat the same foods every day.
2. Milk is such a perfect food that it alone can provide all the nutrients a person needs.
3. Cookies cannot be included in a healthful diet.
4. When it comes to nutrients, more is always better.
5. A person's energy needs are based on his or her age, gender, and physical activity levels.
6. From a nutritional standpoint, there is nothing wrong with grazing on snacks all day, provided the snacks meet nutrient needs without supplying too many calories.
7. If you don't meet your recommended intake for a nutrient every day, you will end up with a deficiency of that nutrient.
8. If a food label claims that a product is low in fat, you can believe it.
9. Most dietitians encourage people to think of their diets in terms of the four basic food groups.
10. According to the government, people should try to eat at least 2 cups of fruit and 2½ cups of vegetables—totaling 9 servings—a day.

Answers found on the following page.

For most people, eating is so habitual that they give hardly any thought to the foods they choose to eat. Yet, as Chapter 1 emphasized, the foods you select can have a profound effect on the quality, and possibly even the length, of your life. Given all the statistics and government mandates presented so far, however, designing a healthful diet may seem like a complicated matter involving a rigid regimen that excludes certain foods from the diet. Fortunately, that's not the case. The government, as well as many major health organizations, has devised dietary guidelines and tools (such as food labels) to help you choose the most healthful diet. This chapter provides an overview of some of the best guides and tools and shows you how to use them.

As you read the following pages, keep in mind that one of the biggest misconceptions about planning a healthful diet is believing that some foods (say, carrots and celery sticks) are "good," whereas others, like cookies and candy, are "bad." People who categorize foods this way often feel guilty every time they "splurge" on a so-called bad food.

Chapter Objectives

- List and describe five characteristics of planning a healthy diet.
- Briefly describe the various guidelines that can be used to plan a healthy diet:
  Dietary Reference Intakes (DRI)
  Dietary Guidelines for Americans
  MyPlate
  Food Labels

Contents

If the doctors of today will not become the nutritionists of tomorrow, the nutritionists of today will become the doctors of tomorrow.

THOMAS EDISON (1847–1931, AMERICAN INVENTOR)

© David Solzberg/Getty Images

The overall diet is what really counts. A diet consisting of nothing but carrot sticks is just as unhealthful as one made up of only candy bars. The trick is choosing a healthful balance of foods. The ideal diet contains primarily foods that supply adequate nutrients, fiber, and calories without an excess of fat, sugar, sodium, or alcohol.

The ABCs of Eating for Health

When you plan a diet for yourself, try to make sure it has the following characteristics:

- **Adequacy** (to provide enough of the essential nutrients, fiber, and energy—in the form of calories)

- **Balance** (to avoid overemphasis on any food type or nutrient at the expense of another)

- **Calorie control** (to supply the amount of energy you need to maintain a healthy weight—not more, not less)

- **Moderation** (to avoid excess amounts of unwanted constituents, such as solid fats and added sugars)

- **Variety** (to consume different foods rather than eating the same meals day after day)

Equally important, be sure that your diet suits you—that it consists of foods that fit your personality, family and cultural traditions, lifestyle, and budget. At best, your diet can be a source of both pleasure and good health.

Diet-planning principles:

- Adequacy
- Balance
- Calorie control
- Nutrient density
- Moderation
- Variety

adequacy characterizes a diet that provides all of the essential nutrients, fiber, and energy (calories) in amounts sufficient to maintain health.

balance a feature of a diet that provides a number of types of foods in harmony with one another, such that foods rich in one nutrient do not crowd out of the diet foods that are rich in another nutrient.

calorie control control of consumption of energy (calories); a feature of a sound diet plan.

moderation the attribute of a diet that provides no unwanted constituent in excess.

variety a feature of a diet in which different foods are used for the same purposes on different occasions; the opposite of *monotony*.

Adequacy Any nutrient can be used to demonstrate the importance of dietary adequacy. For example, iron is an essential nutrient that your body loses daily; it must be replaced continually by iron-rich foods. If your diet does not provide adequate iron—that is, if it lacks food sources of the mineral—you can develop a condition known as *iron-deficiency anemia*. If you add iron-rich foods such as meat, fish, poultry, and legumes to your diet, the condition is likely to disappear soon. (More information about iron appears in Chapter 8.)

Balance To appreciate the importance of balance, consider a second essential nutrient. Calcium plays a vital role in building a strong frame that can withstand the gradual bone loss that occurs with age. Thus, adults are advised to consume three cups of milk or milk products (the best sources of this bone-building mineral) daily to meet their calcium needs. Foods that are rich in calcium typically lack iron, however, and vice versa; so you have to balance the two in your diet.

ASK YOURSELF ANSWERS: 1. False. It is unwise to eat the same foods day in and day out; your diet will lack variety and probably will not supply all the nutrients your body needs. **2.** False. Milk rates as an excellent source of nutrients such as calcium and protein, but it contains only very small amounts of iron and several other essential nutrients. **3.** False. Any food can fit into a healthful eating plan. It's the total diet, not individual foods, that can be either good or bad for health. **4.** False. Too much or too little of a nutrient is often equally harmful. **5.** True. **6.** True. **7.** False. Even if you don't meet your recommended intake for a nutrient, you still may be consuming a sufficient amount of the nutrient. **8.** True. **9.** False. Dietitians today encourage people to think about eating from the five food groups in the MyPlate Food Guide. **10.** True.

Balancing the whole diet is a juggling act that, if successful, provides enough, but not too much, of each of the 40-odd nutrients the body needs for good health. As you will see later in the chapter, you can design a diet that is both adequate and balanced by using food group plans that help you choose from various groups the specific amounts of foods that should be eaten each day.

Calorie Control To maintain a healthy weight, energy intakes should not exceed energy needs. Calorie control helps ensure a balance between energy we take in from food and energy we expend in activity. A cup and a half of ice cream, for example, has about the same amount of calcium as a cup of milk, but the ice cream may contain more than 500 calories, whereas the milk may supply only 90. When it comes to iron, a 3-ounce serving of beef pot roast provides about twice the amount of the mineral as a 3-ounce serving of canned tuna. But whereas the beef contains 280 calories, the tuna adds only about 100 calories to the diet. The choice of which one to eat depends on personal preferences as well as the nutrient and calorie content of the other foods in the total diet.

Those who are trying to ensure optimal intakes of nutrients without excess calories should be sure to include foods that are rich in nutrients (protein, vitamins, and minerals) but relatively low in calories and fat. Such foods are referred to as **nutrient dense**. A baked potato, for example, contains more iron and vitamin C for its calories than French fries. Hence, it is more nutrient dense. Figure 2-1 compares the nutrient density of selected beverages.

Variety fosters good nutrition.

Michael Rosenfeld/Getty Images

nutrient dense refers to a food that supplies large amounts of nutrients relative to the number of calories it contains. The higher the level of nutrients and the fewer the number of calories, the more nutrient dense the food is.

1% low-fat milk (1 c, 100 cal) — Whole milk (3/4 c, 100 cal) — Cola beverage (1 c, 100 cal)

Nutrient	1% low-fat milk	Whole milk	Cola beverage
Calories	5%	5%	5%
Protein	16%[b]	11%[b]	
Vitamin A	16%[b]	5%	
Vitamin C	<2%	<1%	
Folate	3%	2%	
Riboflavin	24%[b]	17%[b]	
Calcium	30%[b]	16%[b]	
Potassium	11%[b]	6%	

Percentage of Daily Values[a]

FIGURE 2-1

Nutrient Density of Selected Beverages

Understanding the concept of nutrient density will help you locate foods that are high in the amount of nutrients provided per calorie. Compare the nutrient contributions made by the three beverages shown here. The figures show that 1% low-fat milk and whole milk are good sources of several essential nutrients, whereas the cola beverage simply provides calories (from sugar). Note that the low-fat milk is more nutrient dense per 100 calories than the whole milk.

[a] The Daily Values, used on food labels, are based on 2,000 calories a day for adults and children over 4 years old and are useful for making comparisons among foods in terms of their nutrient composition.
[b] A good source of this nutrient (more than 10 percent of the recommended intake).

Moderation   Another characteristic of a healthy diet is moderation. In other words, try to eat meals that do not contain excessive amounts of any one nutrient, such as solid fats, which are major sources of saturated and *trans* fatty acids (discussed in Chapter 5), the culprits linked to a number of chronic diseases. That's not to say that you should choose only foods that supply little or no fat. Such an approach is unrealistic and will only lead to frustration. Practice moderation by choosing such foods only on occasion (milkshake versus fat-free milk) and regularly choosing the most nutrient-dense foods from all food groups without exceeding overall calorie needs.

Variety   Aside from avoiding the monotony of eating the same foods day after day, we need variety in our diet for several reasons. First, some foods are better sources of nutrients needed in such small amounts that we don't consciously plan diets around them. Second, variety improves nutrient adequacy, as foods within a specific food group may contain different amounts of nutrients. For example, when selecting fruits, oranges are an excellent source of vitamin C, but a slice of cantaloupe would be a better choice for vitamin A. Finally, a limited diet can supply excess amounts of undesirable substances such as chemical contaminants. Eating many different foods, in contrast, greatly reduces the likelihood that large amounts of a potential toxin will be consumed.

Research underscores that variety is one of the hallmarks of a healthful diet.[1] If your diet lacks variety, chances are good that you're missing out on many nutrients necessary for optimal health. The Japanese, incidentally, recognize variety as such an important part of healthful eating that their dietary guidelines recommend consuming 30 or more different kinds of food *every* day to achieve a balance of essential nutrients.[2]

Nutrient Recommendations

At this point, knowing that foods are made of so many different combinations of nutrients, you may be wondering how to determine whether you are eating the right balance of nutrients. Obviously, if your diet lacks any of the essential nutrients, you may develop deficiencies. Even if you don't develop a full-blown deficiency disease, when you are less than optimally nourished you may get sick more easily and suffer other health problems. To help prevent such problems and provide a benchmark for people's nutrient needs, experts in Canada and the United States devised the **Dietary Reference Intakes (DRI)** for planning and assessing diets of healthy people in both countries.[3]

The Dietary Reference Intakes (DRI)   A committee of nutrition experts selected by the National Academy of Sciences (NAS) sets forth the DRI—a set of daily nutrient standards based on the latest scientific evidence regarding diet and health. The DRI for all age groups are listed on Study Cards pages 27–29 at the back of this book.

Although the DRI are widely used, many people have misconceptions about their meaning and intent. To get a proper perspective about the DRI, consider the following facts:

- The DRI estimate the energy and nutrient needs of *healthy* people. People with certain medical problems often have different nutritional needs.

- Separate recommendations are made for different groups of people. For instance, the DRI committee issues one set of recommendations for children ages 4 through 8, another set for adult men, another for pregnant women, and so on. The DRI may evolve over time because of new scientific evidence indicating a need for reevaluation of current concepts and recommended intakes.[4]

Most fats with a high percentage of saturated and/or *trans* fatty acids are solid at room temperature and are referred to as solid fats (see Chapter 5). Common solid fats include butter, beef fat, chicken fat, pork fat (lard), stick margarine, and shortening. The fat in whole milk is also considered to be solid fat since it is solid at room temperature but suspended in milk by the process of homogenization.

Nutrient Intake Standards

Dietary Reference Intakes (DRI) a set of reference values for energy and nutrients that can be used for planning and assessing diets for healthy people.

FIGURE 2-2

The Correct View of the DRI

This figure shows that people with intakes below the EAR are likely to have dietary adequacy of 50 percent or less. People with intakes between the EAR and RDA are likely to have adequacy between 50 and 97–98 percent. Intakes between the RDA or AI and the UL are likely to be adequate. At intakes above the UL, the risk of harm increases.

People often think that more is better when it comes to nutrients. Too much of a good thing can be dangerous, however, so the RDA, AI, and AMDR fall within an optimal margin of safety, and the UL help people avoid harmful excesses of nutrients.

SOURCE: Adapted from Institute of Medicine, *Dietary Intakes for Energy, Carbohydrate, Fiber, Fat, Fatty Acids, Cholesterol, Protein, and Amino Acids* (Washington, DC: National Academy Press, 2002).

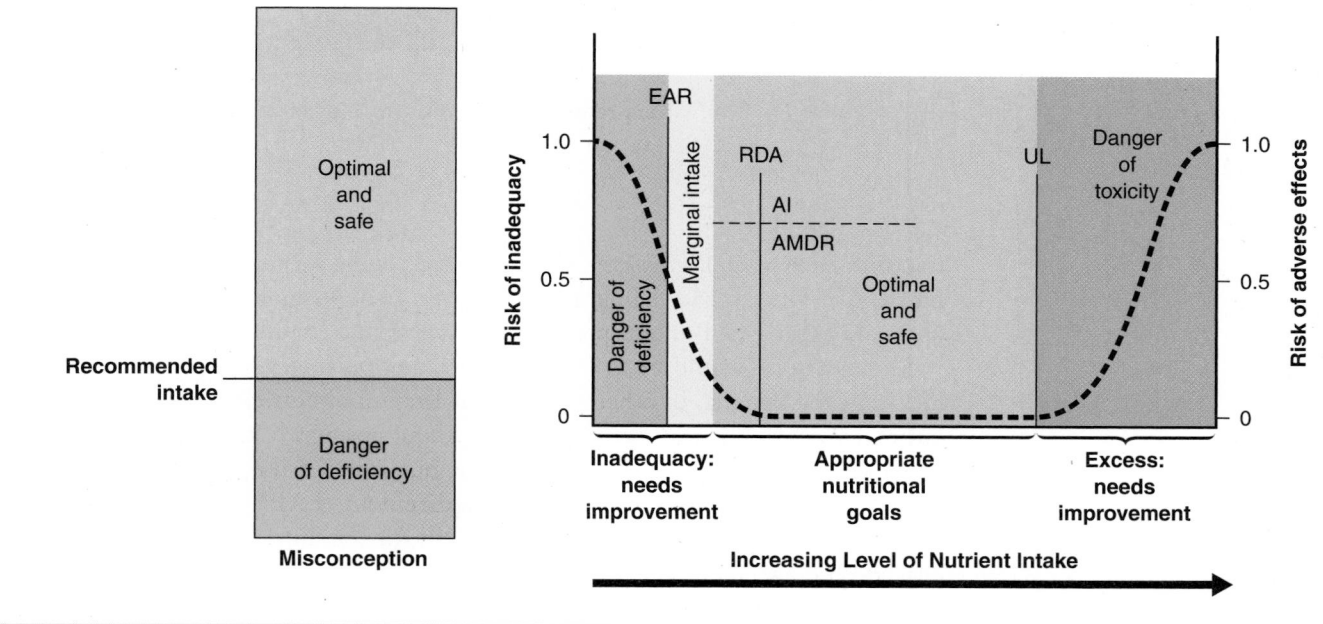

- The DRI take into account differences among individuals and establish a range within which the nutritional needs of virtually all healthy people in a particular age and gender group will be covered (see Figure 2-2).

- The DRI are illustrated in Figure 2-2 and include:
  - Estimated Average Requirement (EAR)
  - Recommended Dietary Allowances (RDA)
  - Adequate Intakes (AI)
  - Tolerable Upper Intake Levels (UL)
  - Estimated Energy Requirement (EER)
  - Acceptable Macronutrient Distribution Ranges (AMDR)

The DRI for Nutrients   The DRI aim to prevent nutrient deficiencies in a population as well as reduce risk for chronic diseases such as heart disease, cancer, and osteoporosis. To understand the process of developing the new DRI, consider the following discussion. Suppose we were the committee members, and we were called upon to set the DRI for nutrient X. First, we would determine the **requirement**— how much of that nutrient the average healthy person needs to prevent a deficiency. To do so, we would review scientific research and explore how the body stores the nutrient, what the consequence of a deficiency might be, what causes depletion of the nutrient, and what other factors affect a person's need for nutrient X. We would also consider current concepts regarding the amount of the nutrient needed for reducing risk of chronic disease (for example, optimal calcium intake and reduced risk of bone

requirement the minimum amount of a nutrient that will prevent the development of deficiency symptoms. Requirements differ from the RDA and AI, which include a substantial margin of safety to cover the requirements of different individuals.

FIGURE 2-3

Nutrient Requirements Vary from Person to Person

Each square represents a person with different nutrient needs. For example, the squares labeled A, B, and C represent Mr. A, Ms. B, and Mr. C. The EAR is the amount of a nutrient estimated to meet the needs of half of the people in a particular age and gender group. The RDA for the nutrient is the amount that covers about 98 percent of the population.

fractures later in life). However, we know that people vary in the amount of a given nutrient they need. Mr. A might need an average of 40 units of nutrient X daily for optimal health; Ms. B might need 35; Mr. C, 60; and so on.

Our next task would be to determine what amount of nutrient X to recommend for the general public. In other words, what amount of nutrient X would we consider optimal for covering most people's needs without posing a hazard? One option would be to set it at the average requirement for nutrient X (shown in Figure 2-3 at 45 units). This is the **Estimated Average Requirement (EAR)**: the amount of a nutrient that is estimated to meet the requirement for the nutrient in half of the people of a specific age and gender. But if we did so and people took us literally, half the population—Mr. C and everyone else whose requirement is greater than 45—would eventually develop deficiencies.

Instead, we use the EAR to set the **Recommended Dietary Allowance (RDA)**. To benefit the most people, the RDA is set at a point high enough to cover most healthy people without creating an excess in people who do not require as much nutrient X. In this case, a reasonable choice might be 63 units, which would add a generous safety margin to ensure that just about all healthy people are covered (see Figure 2-3).

If sufficient scientific evidence is not available for us to set an EAR—the requirement for nutrient X—that is needed to establish an RDA, an **Adequate Intake (AI)** is provided instead of an RDA.* Because sufficient scientific evidence is lacking, the AI is based on our best estimate of the need for nutrient X in practically all apparently healthy individuals of a specific age and gender. Individuals may use both the RDA and the AI as goals for nutrient intake.

For healthy individuals, no established benefits result from consuming amounts of a nutrient that exceed the recommended intake (the RDA or AI). However, the DRI include a **Tolerable Upper Intake Level (UL)**, which is the maximum daily intake of a nutrient that is unlikely to pose risk of *adverse* effects in healthy people. The UL exceeds the RDA and is not intended to be a recommended level of intake (refer to Figure 2-2). The need to set ULs is the result of more and more people using large doses of nutrient supplements and the increasing availability of **fortified foods**. Individuals can use the UL to determine whether their levels of nutrient intakes may pose risk of adverse effects over time.

fortified foods foods to which nutrients have been added, either because they were not already present or because they were present in insignificant amounts; foods to which manufacturers have added 10 percent or more of the Daily Value for a particular nutrient. Examples: margarine with added vitamin A, milk with added vitamin D, certain brands of orange juice with added calcium, and breakfast cereals with added nutrients and nonnutrients.

*This text refers to the nutrient recommendations (for example, RDA and AI) as DRI recommended intakes.

Miniglossary of DRI Terms

EAR, Estimated Average Requirement the amount of a nutrient that is estimated to meet the requirement for the nutrient in half of the people of a specific age and gender. The EAR is used in setting the RDA.

RDA, Recommended Dietary Allowance the average daily amount of a nutrient that is sufficient to meet the nutrient needs of nearly all (97–98 percent) healthy individuals of a specific age and gender.

AI, Adequate Intake the average amount of a nutrient that appears to be adequate for individuals when there is not sufficient scientific research to calculate an RDA. The AI exceeds the EAR and possibly the RDA.

UL, Tolerable Upper Intake Level the maximum amount of a nutrient that is unlikely to pose any risk of adverse health effects to most healthy people. The UL is not intended to be a recommended level of intake.

EER, Estimated Energy Requirement the average calorie intake that is predicted to maintain energy balance in a healthy adult of a defined age, gender, weight, height, and level of physical activity, consistent with good health.

AMDR, Acceptable Macronutrient Distribution Range a range of intakes for a particular energy source (carbohydrates, fat, protein) that is associated with a reduced risk of chronic disease while providing adequate intakes of essential nutrients.

Dietary Reference Intakes (DRIs)

DRI for Nutrients:
- EAR
- RDA
- AI
- UL

DRI for Energy & Energy Nutrients:
- EER
- AMDR

The DRI represent a major shift in thinking about nutrient requirements for humans, from prevention of nutrient deficiencies to prevention of chronic disease. They also herald a new way of thinking about the role of dietary supplements in achieving good health—such that, for some individuals at higher risk, use of nutrient supplements may be desirable so as to meet recommended intakes. For example, older women who are at risk of osteoporosis may benefit from dietary calcium supplements to help maintain bone mineral mass. In addition, the development of the UL signals the widespread recognition that high intakes of nutrients can create a degree of risk.

People vary in the amount of a given nutrient they need. The challenge of the DRI is to determine the best amount to recommend for everybody.

© Bob Thomas/Getty Images

The DRI for Energy and the Energy Nutrients The DRI report on energy and the energy nutrients provides guidelines for the United States and Canada regarding the consumption of energy, carbohydrates, fiber, fat, fatty acids, cholesterol, and protein; it includes guidelines for physical activity as well. To meet the body's daily energy and nutritional needs while minimizing risk for chronic disease, consumption of the energy nutrients should resemble the pattern shown in Figure 2-4.

To reduce the risk of chronic disease, the DRI report stresses the need to balance caloric intake with physical activity, recommending total calories to be consumed by individuals of given heights, weights, and genders for each of four different levels of physical activity (from sedentary to very active). For more about weight management and specific physical activity guidelines, see Chapters 10 and 11 as well as Study Card page 31 at the back of this book.

The Challenge of Dietary Guidelines

As Chapter 1 pointed out, health authorities are as concerned today about widespread nutrient excesses among Americans as they used to be about nutrient deficiencies. This is where dietary guidelines come into play. Among the most widely used of the guidelines are the *Dietary Guidelines for Americans*. The Eating Well with Canada's Food Guide recommendations presented in Appendix B are also commonly used.[5]

FIGURE 2-4

Recommended Dietary Intake Ranges for Energy Nutrients

A balanced diet is composed of approximately 10–35 percent protein, 45–65 percent carbohydrates (with no more than 10 percent of this amount from added sugars or caloric sweeteners), and 20–35 percent from fat, but derived mostly from oils rather than solid fats. Currently, most Americans consume about 15 percent of their calories from protein, 49 percent from carbohydrates (with 16 percent of this amount coming from simple sugars), and 34 percent of their total calories from fat.

Dietary Reference Intakes (DRI) for Carbohydrates, Fats, and Protein

Carbohydrates	45 to 65 percent of total calories
Fats	20 to 35 percent of total calories
Protein	10 to 35 percent of total calories

© Felicia Martinez/PhotoEdit

The *Dietary Guidelines for Americans* provide science-based advice to promote health and to reduce risk for chronic diseases through diet and physical activity. They emphasize variety, calorie control, moderation, and nutrient density. The guidelines also emphasize physical activity because it increases energy expenditure and helps in maintaining a healthy weight.

Table 2-1 presents the *Dietary Guidelines for Americans, 2010* grouped into four general topics, along with their key recommendations. The new guidelines incorporate two general themes: maintaining calorie balance over time to achieve and maintain a healthy weight, and consuming more nutrient-dense foods and beverages. Americans currently consume too much sodium and too many calories from solid fats, added sugars, and refined grains. These replace nutrient-dense foods and beverages and make it difficult for people to achieve recommended nutrient intakes without exceeding their energy needs. For this reason, the *Dietary Guidelines* emphasize three major goals for building healthy eating patterns:

- Balance calories with physical activity to manage weight.

- Consume more nutrient-dense foods such as fruits, vegetables, whole grains, fat-free and low-fat (1%) milk products, seafood, lean meats and poultry, eggs, beans and peas, and unsalted nuts and seeds.

USDA

TABLE 2-1

Key Recommendations of the *Dietary Guidelines for Americans, 2010*

Balancing Calories to Manage Weight
- Prevent and/or reduce overweight and obesity through improved eating and physical activity behaviors.
- Control total calorie intake to manage body weight. For people who are overweight or obese, this will mean consuming fewer calories from foods and beverages.
- Increase physical activity and reduce time spent in sedentary behaviors.
- Maintain appropriate calorie balance during each stage of life—childhood, adolescence, adulthood, pregnancy and breastfeeding, and older age.

Foods and Food Components to Reduce
- Reduce daily sodium intake to less than 2,300 milligrams (mg) and further reduce sodium intake to 1,500 mg among persons who are 51 and older and those of any age who are African American or have hypertension, diabetes, or chronic kidney disease.
- Consume less than 10 percent of calories from saturated fat by replacing it with monounsaturated and polyunsaturated fats.
- Consume less than 300 mg per day of dietary cholesterol.
- Keep trans fat consumption as low as possible by limiting foods such as partially hydrogenated oils, and by limiting other solid fats.
- Reduce the intake of calories from solid fats and added sugars.
- Limit the consumption of foods that contain refined grains, especially refined grain foods that contain solid fats, added sugars, and sodium.
- If alcohol is consumed, it should be consumed in moderation—up to one drink per day for women and two drinks per day for men.

Foods and Nutrients to Increase
Individuals should meet the following recommendations as part of a healthy eating pattern while staying within their calorie needs.
- Increase vegetable and fruit intake.
- Eat a variety of vegetables, especially dark-green and red and orange vegetables and beans and peas.
- Consume at least half of all grains as whole grains. Increase whole-grain intake by replacing refined grains with whole grains.
- Increase intake of fat-free or low-fat milk and milk products, such as milk, yogurt, cheese, or fortified soy beverages.
- Choose a variety of protein foods, which include seafood, lean meat and poultry, eggs, beans and peas, soy products, and unsalted nuts and seeds.
- Increase the amount and variety of seafood consumed by choosing seafood in place of some meat and poultry.
- Replace protein foods that are higher in solid fats with choices that are lower in solid fats and calories and/or are sources of oils.
- Use oils to replace solid fats where possible.
- Choose foods that provide more potassium, dietary fiber, calcium, and vitamin D, which are nutrients of concern in American diets. These foods include vegetables, fruits, whole grains, and milk and milk products.

Building Healthy Eating Patterns
- Select an eating pattern that meets nutrient needs over time at an appropriate calorie level.
- Account for all foods and beverages consumed and assess how they fit within a total healthy eating pattern.
- Follow food safety recommendations when preparing and eating foods to reduce the risk of foodborne illnesses.

Note: These guidelines are intended for adults and children ages 2 years and older, including those who are at increased risk of chronic disease. The *Dietary Guidelines for Americans, 2010* contains additional recommendations for specific populations. The full document is available at http://www.cnpp.usda.gov/Publications/DietaryGuidelines/2010/PolicyDoc/PolicyDoc.pdf

SOURCE: U.S. Department of Agriculture and U.S. Department of Health and Human Services, *Dietary Guidelines for Americans, 2010*. 7th ed. (Washington, DC: U.S. Government Printing Office, 2010).

- Consume fewer foods with sodium (salt), saturated fats, trans fats, cholesterol, added sugars, and refined grains.

Healthful eating and regular physical activity enable people of all ages to work productively, enjoy life, and feel their best. They also help children grow, develop, and do well in school. The goal of the *Dietary Guidelines* is to help people decrease their risk of some forms of cancer, heart disease, obesity, diabetes, high blood pressure, stroke, osteoporosis, and liver disease—the so-called **lifestyle diseases**. Following such recommendations certainly makes sense, given the considerable potential health benefits they confer.

Introducing the MyPlate Diet-Planning Tool

Although the DRI and the *Dietary Guidelines* provide good frameworks for healthful eating, planning daily menus requires use of other, more specific tools. Dietitians and other nutrition experts often rely on a number of tools that you, too, can use to assess and plan your own diet.

lifestyle diseases conditions that may be aggravated by modern lifestyles that include too little exercise, poor diets, and excessive drinking and smoking. Lifestyle diseases are also referred to as *diseases of affluence*.

It's 3:30 in the afternoon, and that sound you just heard is coming from your stomach. Physiologically speaking, the human digestive system is customized for us to eat about every four hours to maintain our energy level. So, it's not unreasonable to get a little hungry between meals or before we go to bed. In fact, it is now clear that healthy snacking can fit into any eating plan and is important to everyone's health.[6]

© PhotoDisc/Getty Images

Still, although **grazing** is "in," many people feel a twinge of guilt now and then about between-meal munching. Perhaps parental warnings from childhood that snacks can spoil a meal linger in the back of many minds. Nevertheless, nutritious nibbling can make it easier for many people to eat healthfully.

Snacks can supply essential vitamins, minerals, and calories to diets of young children, who often cannot eat large portions at mealtime because of their small stomachs and appetites. Teenagers typically don't seem to have the time (or inclination) to eat regularly, due to busy school schedules, sports practices, music lessons, or other activities. Snacks account for approximately 25 percent of the calories they eat and are often more energy dense than their meals.[7]

Snacks also contribute to nutritional needs of adults, who may find that fitting meals into a busy schedule is difficult. Even older adults, whose lifestyles tend to be less hectic, can benefit from

grazing eating small amounts of food at intervals throughout the day rather than or in addition to eating regular meals.

grazing. That's because lack of activity, certain medications, and isolation can blunt a formerly hearty appetite, making frequent, small meals more desirable than large breakfasts, lunches, and dinners. And, of course, a college student's busy life makes it difficult, if not impossible, to plan meals and snacks in advance, so snacking is a factor at any age.

The key to healthful snacking for everyone is to choose healthful, high-fiber, nutrient-dense foods instead of snacks that add unhealthful fats and calories to the diet and little else in the way of nutrients. A snack with a balance of carbohydrate, some fat, and some protein will satisfy hunger for a longer period of time than food with only carbohydrate or sugars (such as candy or soft drinks). How do you achieve this balance? Choosing snacks that contain at least two food groups can provide yummy and healthful snack possibilities.

Some snacks that appear nutritious may be deceiving. For example, fruit drinks, mixes, and punches are loaded with sugar (usually in the form of high-fructose corn syrup) and are more similar to soft drinks than to fruit juice. Fruit rolls and bars are nutritionally similar to jams and jelly, not fresh fruit. For these snacks, sugar is added and most of the nutrients are lost when the fruit is processed with heat. Energy and protein bars can also be deceiving because, like candy bars, many are loaded with sugar and fat. Even some varieties of microwave popcorn don't rate well as snacks, because they contain so much oil and salt. (You can easily make your own popcorn in the microwave oven without adding excessive extras.) When you feel like snacking, try some of the alternatives offered in Table 2-2. These snacks provide at least two food groups and 100–250 calories per serving.[8] Use Table 2-3 as a guide for choosing healthful snacks instead of calorie-dense snacks with little nutritional value. In addition, consider the following tips next time you're in the mood to grab a snack:

● Stock your refrigerator and kitchen cupboards with healthful foods like fruit juices, low-fat yogurt, fresh fruits and vegetables, plain popcorn, whole-grain crackers, and low-fat cheeses so that they are close at hand. Nibblers often reach for a snack

TABLE 2-2	Smart Snacking

- Low-fat yogurt with ½ cup cereal
- Fruit and yogurt smoothies
- 1-ounce serving trail mix (nuts, seeds, dried fruits)
- 1 cup cereal with low-fat milk or soy milk
- A *small* handful of tortilla chips with salsa or low-fat bean dip
- Fruit and cheese: apples or grapes with 1 ounce low-fat mozzarella string cheese
- 1–2 tablespoons peanut butter on an apple, celery, or carrots

just to have something to munch on rather than because of a desire for the food itself. If nutritious choices are easy to get to, chances are that's what you'll eat.

- Carry fresh fruit and crackers and cheese in your backpack or book bag so you won't have to resort to buying candy from a vending machine when you get the urge to munch.
- Create your own snacks. Mix together one cup each whole-grain cereal, almonds, raisins, and sunflower seeds to take with you on your next bicycle trip or hike.
- Make new versions of old favorites, such as *Chili Popcorn* and *Mexican Snack Pizzas*.[9]
- Snack with a friend. If you're craving a candy bar or chips and nothing else will do, try splitting a bar or a bag with a friend. That way, you'll satisfy your craving without going too far overboard on calories, fat, or salt.
- Try to brush your teeth—or at least rinse your mouth thoroughly—after snacking to prevent tooth decay. (See pp. 119–120 in Chapter 4 for a detailed explanation of the role of diet in dental health.)

Chili Popcorn

Mix 1 quart popped popcorn with 1 tablespoon melted margarine. In a separate bowl, mix 1¼ teaspoons chili powder, ¼ teaspoon cumin, and a dash of garlic powder. Sprinkle seasonings over popcorn and mix well. Makes about four 1-cup servings, each of which contains approximately 50 calories, 3 grams of fat, and 42 milligrams of sodium.

Mexican Snack Pizzas

Split two whole-grain English muffins and toast lightly. Mix ¼ cup tomato paste; ¼ cup canned, drained, chopped kidney beans; 1 tablespoon each chopped onion and chopped green pepper; and ½ teaspoon oregano. Spread mixture on muffin halves. Top with ¼ cup shredded part-skim mozzarella cheese and broil until cheese is bubbly (about 2 minutes). Garnish with ¼ cup shredded lettuce. Makes four servings, each of which contains 95 calories, 2 grams of fat, and 300 milligrams of sodium.

TABLE 2-3	What's in a Muncher's Healthy Snacking Menu?				
100 Calories of Grains					
1½ slices whole wheat bread	or	⅔ Lender's Original® wheat bagel	or	⅔ Dunkin' Donuts® plain bagel	
1 c Kellogg's Corn Flakes®	or	scant ½ c Kellogg's Just Right® with fruit and nuts	or	¼ c Kellogg's Healthy Choice Low Fat Granola® (without raisins)	
100 Calories of Fruit					
1¼ fresh apple	or	1 c unsweetened applesauce	or	½ c sweetened applesauce	
2 heaping cups fresh strawberries	or	½ c frozen, sweetened strawberries	or	2 tbsp strawberry jam	
100 Calories of Dairy					
⅗ c 1% cottage cheese	or	4⅓ tbsp grated Parmesan cheese	or	2 tbsp cream cheese	
100 Calories of Potato					
½ plain baked potato with skin	or	½ c mashed potato	or	10 French fries	
100 Calories of Protein Foods (meats, poultry, fish, beans, nuts)					
2 oz roasted tip round beef	or	1.4 oz extra lean broiled ground beef	or	1.2 oz regular broiled ground beef	
2.1 oz roasted chicken breast without skin	or	1.4 oz KFC Original Recipe® chicken breast with skin	or	6 macadamia nuts	
3 oz baked/broiled haddock	or	⅛ oz Mrs. Paul's Select Cut® frozen haddock fillets (breaded)			
⅜ c chickpeas	or	¼ c hummus (made with chickpeas)			
29 pistachios	or	15 almonds			
100 Calories of "Eat Sparingly" Foods					
Scant ½ cup soft-serve vanilla frozen yogurt	or	⅜ c regular vanilla ice cream	or	³⁄₁₆ c Haagen-Dazs® vanilla ice cream	
			or	¹⁰⁄₁₁ fudge-covered Oreo®	
2⅓ reduced-fat Oreos®	or	1⅖ regular Oreos®	or	⁵⁄₁₄ Snickers® bar	
⅘ Reese's Peanut Butter Cup®	or	½ Chunky® bar			

SOURCE: Adapted from "For 100 Calories, You Can Have . . . ," *Tufts University Health & Nutrition Letter* (July 2000): 8.

One of the most helpful, easy-to-use diet-planning tools is the **food group plan**, which separates foods into specific groups and then specifies the number of **servings** from each group to eat each day. The MyPlate food guide is presented in detail in Figure 2-5. This tool includes five food groups and presents tips for choosing foods from within each group.

The MyPlate Food Guidance System provides food-based guidance to help implement the recommendations of the *Dietary Guidelines for Americans* and provides a visual aid to assist in improving diet and lifestyle. MyPlate is designed to help you:

- Make smart choices from every food group;
- Find your balance between food and physical activity; and
- Get the most nutrition out of your calories by focusing on nutrient-dense foods in sensible portion sizes.

Using MyPlate to Achieve a Healthy Lifestyle   MyPlate is designed to help consumers choose foods that supply a good balance of nutrients, and it aims to moderate or limit dietary components often consumed in excess, such as saturated fat, *trans* fat, sugar, sodium, and alcohol, in keeping with the U.S. government's *Dietary Guidelines for Americans*. MyPlate includes six key components that can help you achieve a healthy lifestyle, including:

1. **Activity.** The MyPlate website promotes regular physical activity and reduced sedentary activities. As part of an overall healthful diet, balancing caloric intake with energy needs can help reduce the risk of chronic disease, help prevent weight gain, and help sustain weight loss. Adults should do at least 2 hours and 30 minutes each week of aerobic physical activity at a moderate level or 1 hour and 15 minutes each week of aerobic physical activity at a vigorous level.
2. **Variety.** MyPlate creates a foundation for good nutrition and health by guiding us to make food selections from all food groups and subgroups. Different foods contain different nutrients and other substances known to be protective against chronic diseases. No one food or no single food group provides all essential nutrients in amounts necessary for good health.
3. **Proportionality.** MyPlate shows the proportions of foods that should make up a healthful diet. MyPlate illustrates the five food groups using a familiar mealtime visual, a place setting. The plate is split into four sections: red for fruits, green for vegetables, purple for protein, and orange for grains (refer to Figure 2-5). A separate blue section, shaped like a drinking glass, represents dairy foods. MyPlate provides a visual reminder of a key nutrition principle: Fruits and

---

In June 2011, MyPlate replaced MyPyramid (released in April 2005), which replaced the Food Guide Pyramid (1992). The older food guide materials are archived at www.ChooseMyPlate.gov.

Some examples of moderate physical activity are walking briskly, mowing the lawn, dancing, swimming, and bicycling on level terrain. Some examples of vigorous physical activity are jogging, high-impact aerobic dancing, swimming continuous laps, and bicycling uphill.

food group plan   a diet-planning tool, such as MyPlate, that groups foods according to similar origin and nutrient content and then specifies the amount of food a person should eat from each group.

serving   the standard amount of food used as a reference to give advice regarding how much to eat (such as a 1-cup serving of milk).

The MyPlate Food Guidance System calls for eating a variety of foods to get the nutrients you need and at the same time the right amount of calories to maintain a healthy weight. Remember to balance the energy consumed with the energy expended in play.

FIGURE 2-5

Choose MyPlate: Steps to a Healthier You

SOURCE: U.S. Department of Agriculture, 2011.

Take Action: Ten Tips to a Great Plate

Use the ideas in this list to balance your calories, to choose foods to eat more often, and to cut back on foods to eat less often.

1. **Balance your calories.** Find out how many calories you need for a day as a first step in managing your weight. Go to www.ChooseMyPlate.gov to find your calorie level. Being physically active also helps you balance calories.

2. **Enjoy your food, but eat less.** Take the time to fully enjoy your food as you eat it. Pay attention to hunger and fullness cues before, during, and after meals. Use them to recognize when to eat and when you've had enough.

3. **Avoid oversized portions.** Use a smaller plate, bowl, and glass. Portion out foods before you eat. When eating out, choose a smaller size option, share a dish, or take home part of your meal.

4. **Eat some foods more often.** Eat more vegetables, fruits, whole grains, and fat-free or 1% dairy products. Make them the basis for meals and snacks.

5. **Make half your plate fruits and vegetables.** Choose red, orange, and dark-green vegetables like tomatoes, sweet potatoes, and broccoli, along with other vegetables for your meals. Add fruit to meals as part of main or side dishes or as dessert.

6. **Switch to fat-free or low-fat (1%) milk.** They have the same amount of calcium and other essential nutrients as whole milk, but fewer calories and less saturated fat.

7. **Make half your grains whole grains.** To eat more whole grains, substitute a whole-grain product for a refined product—such as eating whole-wheat bread instead of white bread or brown rice instead of white rice.

8. **Eat certain foods less often.** Cut back on foods high in solid fats, added sugars, and salt. They include cakes, cookies, ice cream, candies, sweetened drinks, pizza, and fatty meats like ribs, sausages, bacon, and hot dogs. Use these foods as occasional treats, not everyday foods.

9. **Compare sodium in foods.** Use the Nutrition Facts label to choose lower sodium versions of foods like soup, bread, and frozen meals. Select canned foods labeled "low sodium," "reduced sodium," or "no salt added."

10. **Drink water instead of sugary drinks.** Cut calories by drinking water or unsweetened beverages. Soda, energy drinks, and sports drinks are a major source of added sugar and calories in many diets.

Source: Adapted from Center for Nutrition Policy and Promotion, Tip Sheet No. 1. June 2011. Go to www.ChooseMyPlate.gov

GRAINS Make half your grains whole	VEGETABLES Vary your veggies	FRUITS Focus on fruits	DAIRY Get your calcium-rich foods	PROTEIN FOODS Go lean with protein
Eat at least 3 oz of whole grain cereals, breads, crackers, rice, or pasta every day. 1 oz is about 1 slice of bread, about 1 cup of breakfast cereal or 1/2 cup of cooked rice, cereal, or pasta.	Eat more dark-green veggies like broccoli, spinach, and other dark leafy greens. Eat more orange vegetables like carrots and sweet potatoes. Eat more dry beans and peas like pinto beans, kidney beans, and lentils.	Eat a variety of fruit. Choose fresh, frozen, canned, or dried fruit. Go easy on fruit juices.	Go low-fat or fat-free when you choose milk, yogurt, and other milk products. If you don't or can't consume milk, choose lactose-free products or other calcium sources such as fortified foods and beverages. 1 cup = 1 1/2 oz natural cheese, or 2 oz processed cheese.	Choose low-fat or lean meats and poultry. Bake it, broil it, or grill it. Vary your protein routine—choose more fish, beans, peas, nuts, and seeds. 1 oz = 1 oz meat, poultry, or fish; 1/4 cup cooked dry beans; 1 egg; 1 tbsp peanut butter; 1/2 oz nuts/seeds.
For a 2,000-calorie diet, you need the amounts below from each food group. To find the amounts that are right for you, go to www.ChooseMyPlate.gov.				
Eat 6 oz every day.	Eat 2 1/2 cups every day.	Eat 2 cups every day.	Get 3 cups every day; for kids aged 2 to 8, it's 2.	Eat 5 1/2 oz every day.

*Make most of your fat choices from fish, nuts, and vegetable oils. Find your allowance for oils at www.ChooseMyPlate.gov.

vegetables form the foundation of a healthy diet and should fill at least half of a plate at every meal. The MyPlate food guide advises you to eat more of some foods (fruits, vegetables, whole grains, and fat-free or low-fat milk products), and eat less of others (such as foods high in refined grains and added sugars).

4. **Moderation.** Currently, many of the foods and beverages Americans eat and drink contain *empty calories*—calories from **solid fats** and/or **added sugars**. Solid fats and added sugars add calories to the food but few or no nutrients and are thus called *empty calories*. A small amount of empty calories is fine, but most

solid fats fats that are solid at room temperature, such as butter, lard, and shortening. These fats may be visible or may be a constituent of foods such as milk, cheese, meats, or baked products.

added sugars sugars and other caloric sweeteners that are added to foods during processing or preparation. Added sugars do not include naturally occurring sugars such as those found in milk and fruits.

people eat far more than is healthy. It is important to emphasize nutrient-rich foods in the diet and limit empty calories to the amount that fits your calorie and nutrient needs. Making better choices, like unsweetened applesauce or extra-lean ground beef, can help keep your intake of added sugars and solid fats low. To find your personal total calorie needs and empty calories limit, enter your information into the Daily Food Plan section of www.ChooseMyPlate.gov.

5. **Personalization.** One size doesn't fit all. MyPlate can help you choose the foods and amounts of food that are right for you. For a quick estimate of what and how much you need to eat, enter your age, gender, and physical activity level in the Daily Food Plan box at www.ChooseMyPlate.gov. You can also print out a worksheet based on your personal food plan to help you track your progress.

6. **Gradual improvement.** Individuals can benefit from taking small steps to improve their diet and lifestyle every day (for example, taking the stairs instead of the elevator or escalator, or ordering a green salad instead of fries).

Use the Simplicity of MyPlate to Build a Healthful Diet
You can determine the right amount of foods to eat to meet your personal energy needs and promote a healthy weight in three easy steps.

Step 1: **Estimate Your Daily Energy Needs.** Use Table 2-4 to find your estimated daily energy needs, or go to www.ChooseMyPlate.gov and enter your age, gender, and usual activity level. If you are moderately physically active, you need fewer calories than those who are more active. The calorie levels for the food intake patterns in MyPlate were matched to age and gender groups using the **Estimated Energy Requirement (EER)** for a person of average height, healthy weight, and sedentary activity level in each age and gender group. The sedentary level was selected as a way to help people avoid overestimating their calorie needs.

Step 2: **Build Your Daily Eating Plan.** Use your estimated number of calories to build a daily eating plan that incorporates all five food groups. How much food do you need to eat each day? MyPlate provides guidance by indicating an appropriate amount of food to eat from each food group that will supply not only necessary nutrients but also the adequate amounts of calories. Table 2-5 specifies the amounts of foods from each group that are needed daily to create a healthful diet at several calorie levels.

Step 3: **Let MyPlate Guide Your Food Choices.** Eating is one of life's great pleasures. Different people like different foods and like to prepare the same foods in different ways. Because you can build a healthful menu using many foods in many ways, you have lots of room for choice. MyPlate can provide a starting point to develop healthful eating patterns while still allowing for personal preferences in food choices. Make choices from each major food group and combine them however you like.

The important thing to remember is that you can enjoy all foods as part of a healthy diet as long as you don't overdo it on solid fats (especially saturated fat and *trans* fat), refined carbohydrates, added sugars, sodium, and alcohol. Read labels to identify (and avoid) foods that are higher in saturated fats, *trans* fats, added sugars, and sodium. Certain items in each food group may be difficult to fit into a healthful eating plan on a regular basis.

Using MyPlate to assess and plan your own diet requires an understanding of how much food to consume from the various food groups and how much of a food counts as a serving. For instance, one slice of bread, half a bagel, and ½ cup of pasta each count as 1 ounce from the Grains group. When it comes to vegetables, ½ cup of raw or cooked vegetables or 1 cup of leafy raw vegetables chalks up one serving. Table 2-6 shows the portions that count as a serving from each of the various food groups.

For a hundred more tips, visit www.smallstep.gov.

MyPlate isn't the only food group plan. Canada has one of its own—Canada's Food Guide (shown in Appendix B).

Some popular foods don't fit neatly into one MyPlate food group. For example, a slice of cheese pizza includes servings from several groups: the crust from the grains group, the tomato sauce from the vegetable group, and the cheese from the dairy group. See "Tips and Resources" at www.ChooseMyPlate.gov for a chart showing how to count mixed dishes in a healthy diet.

The Spotlight feature at the end of this chapter shows you how to sort ethnic foods into the MyPlate food groups.

For more information on menu planning with the MyPlate tool, go to www.ChooseMyPlate.gov and click on "Interactive Tools"; also click on "Food Groups" and scroll down to see the Food Gallery photos for each food group for examples of sensible-size portions of the foods you eat.

TABLE 2-4   Estimated Daily Calorie Needs for Adults[a]

Find your gender and age. Then select the activity level that best describes your lifestyle (sedentary, moderately active, or active).

Activity Level:	Sedentary[b]	Moderate[c]	Active[d]
Females			
19–30	1,800–2,000	2,000–2,200	2,400
31–50	1,800	2,000	2,200
51+	1,600	1,800	2,000–2,200
Males			
19–30	2,400–2,600	2,600–2,800	3,000
31–50	2,200–2,400	2,400–2,600	2,800–3,000
51+	2,000–2,200	2,200–2,400	2,400–2,800

[a] The calorie levels in each gender and age group are based on persons of average height and at a healthy weight. If you are overweight, your calorie needs may be higher to maintain your weight. To lose weight, you can follow the calorie level in the chart, depending on your body weight. A range of calories is given to meet needs of different ages within the group. For adults, fewer calories are needed at older ages.
[b] **Sedentary:** less than 30 minutes a day of moderate physical activity in addition to daily activities.
[c] **Moderately active:** at least 30 minutes and up to 60 minutes a day of moderate physical activity in addition to daily activities. Examples of moderate physical activity include walking briskly (about 3-1/2 mph), mowing the lawn, dancing, swimming, or bicycling on level terrain. A person should feel some exertion but should be able to carry on a conversation comfortably during the activity.
[d] **Active:** 60 or more minutes a day of moderate physical activity in addition to daily activities.

NOTE: For more information, visit www.ChooseMyPlate.gov.

SOURCE: U.S. Department of Agriculture, 2005

TABLE 2-5   Build Your Eating Plan with Recommended Daily Amounts from Each Food Group

Find your calorie level at the top of the chart. Follow the column below your estimated calorie level to see how much food to eat from each of the food groups. (Estimated daily energy needs are shown in Table 2-4.) Recommended amounts for fruits, vegetables, and dairy are measured in cups and those for grains and meats, in ounces.

Calorie Level:	1,600	1,800	2,000	2,200	2,400	2,600	2,800	3,000
Grains	5 oz	6 oz	6 oz	7 oz	8 oz	9 oz	10 oz	10 oz
Vegetables	2 cups	2½ cups	2½ cups	3 cups	3 cups	3½ cups	3½ cups	4 cups
Fruits	1½ cups	1½ cups	2 cups	2 cups	2 cups	2 cups	2½ cups	2½ cups
Dairy	3 cups	3 cups	3 cups	3 cups	3 cups	3 cups	3 cups	3 cups
Protein Foods	5 oz	5 oz	5½ oz	6 oz	6½ oz	6½ oz	7 oz	7 oz
Oils	5 tsp	5 tsp	6 tsp	6 tsp	7 tsp	8 tsp	8 tsp	10 tsp
Daily Limit for Empty Calories*	120 cal	160 cal	260 cal	265 cal	330 cal	360 cal	400 cal	460 cal

*The *Daily Limit for Empty Calories* (extra fats and added sugars) includes the balance of calories remaining in a person's estimated energy allowance after accounting for the number of calories needed to meet recommended nutrient intakes through consumption of foods in low-fat or no-added-sugar forms. The calories assigned to empty calories may be used to increase intake from the basic food groups; to select foods from these groups that are higher in fat or with added sugars; to add oils, solid fats, or sugars to foods or beverages; or to consume alcohol.

Besides showing what foods are the equivalent of one serving from each food group, Table 2-6 also lists the nutrients supplied by each of the five food groups. To get an idea of how your current diet compares with the recommendations of the interactive MyPlate website, try the Rate Your Plate Scorecard on page 51.

Using MyPlate to Meet Nutrient Needs
MyPlate can help you get the nutrients your body needs each day. The grains, fruits, vegetables, and dairy groups serve as the foundation of a healthful diet because they supply the vitamins, minerals, and fiber lacking in many people's diets. When it comes to the dairy group, low-fat (1%) and fat-free dairy products make the best nutrient-dense choices. Although foods from the Protein foods group provide protein, iron, zinc, and other nutrients, some items can contain large amounts of saturated fat and *trans* fat. Thus, choosing wisely from this group goes a long way in limiting the more unhealthy fat content

Making Use of MyPlate
- Determine calorie needs.
- Note amount of food to eat from each food group.
- Plan meals and snacks.
- Balance food intake with physical activity.

TABLE 2-6   Are You Savvy about Recommended Amounts of Food to Eat?[a]

Food Group	Goal[b]	Nutrients Supplied	Equivalent Amounts
Grain Group	6 oz	Protein, complex carbohydrate and fiber, folate, niacin, riboflavin, thiamin, magnesium, and iron	**1 ounce is equivalent to:** 1 slice bread, 1 mini bagel, 6" tortilla, ½ English muffin 1 cup dry cereal ½ cup cooked rice, pasta, cereal 1 packet instant oatmeal 3 cups popcorn
Whole grains	3 oz		
Other grains	3 oz		
Vegetable Group	2½ cups	Vitamin C, beta-carotene, fiber, potassium, folate, calcium, magnesium	**½ cup is equivalent to:** ½ cup cut-up raw or cooked vegetables 1 cup raw leafy vegetable ½ cup vegetable juice
Dark-green vegetables	3 cups/week		
Orange vegetables	2 cups/week		
Legumes (dry beans)	3 cups/week		
Starchy vegetables	3 cups/week		
Other vegetables	6½ cups/week		
Fruit Group	2 cups	Vitamin C, beta-carotene, fiber, potassium, folate, magnesium	**½ cup is equivalent to:** ½ cup fresh, frozen, or canned fruit 1 medium fruit ¼ cup dried fruit ½ cup fruit juice
Dairy Group	3 cups	Protein, calcium, riboflavin, vitamin B_{12}, vitamins A and D	**1 cup is equivalent to:** 1 cup low-fat/fat-free milk, yogurt 1½ oz low-fat or fat-free natural cheese 2 oz low-fat or fat-free processed cheese ⅓ cup shredded cheese
Protein Foods Group	5½ oz	Protein, zinc, iron, niacin, thiamin, phosphorus, vitamin B_6, vitamin B_{12}, vitamin E, fiber (legumes)	**1 ounce is equivalent to:** 1 oz cooked lean meats, poultry, fish 1 egg ¼ cup cooked dry beans or tofu 1 tbsp peanut butter ½ oz nuts or seeds ½ cup split-pea soup
Oils[c]	6 tsp (27 grams)	Vitamin E, essential fatty acids (see Chapter 5)	**1 tsp equivalent is:** 1 tsp soft margarine 1 tbsp low-fat mayonnaise 2 tbsp light salad dressing 1 tsp vegetable oil 1 oz peanuts, nuts, or sunflower seeds has 3 teaspoons of oil
Daily limit for empty calories[d]	260 calories		

[a] All servings are per day unless otherwise noted. Vegetable subgroup amounts are per week. To follow this eating pattern, food choices over time should provide these amounts of food from each group on average.

[b] Recommended amounts to eat are based on a 2,000-calorie diet. See Table 2-4 for information about gender/age/activity levels and appropriate calorie intakes. See Table 2-5 for more information on the food groups, amounts, and food intake patterns at other calorie levels, or go to www.ChooseMyPlate.gov.

[c] The oils listed in this table are not considered to be empty calories because they are a major source of the vitamin E and polyunsaturated fatty acids, including the essential fatty acids, in the food pattern. In contrast, solid fats (i.e., saturated and *trans* fats) are included as sources of empty calories.

[d] The empty calorie allowance is the remaining amount of calories in each calorie level after nutrient-dense forms of foods in each food group are selected.

SOURCE: U.S. Department of Agriculture, 2005

of your diet. Because lean cuts of meat, skinless poultry, and fish rank lower in saturated fat than ground beef and chicken with skin, the leaner items should be chosen more often than the high-fat selections. Dry beans, which contain only a trace of fat, are good choices to add to the diet even more often.

In addition to the five food groups, we need a small amount of oil in our diet because oils provide vitamin E and essential fats. The Oils category includes vegetable oils and soft margarines containing no *trans* fat. Heart-healthy oils like olive, canola, and peanut oil are also good choices. (For more about fats, see Chapter 5.)

Using MyPlate to Moderate Energy Intakes You can follow the MyPlate guidelines to get enough nutrients without overdoing calories. The way to balance your energy needs is to make your food choices count. By incorporating the tips and strategies for healthful eating found in Table 2-7 into your eating plan, you will get the nutrients you need for good health. Choose the most nutrient-dense foods from each of the food groups. This means choosing more whole foods that are naturally high in nutrients, such as whole grains; fruits and vegetables, including legumes; and fat-free and low-fat milk and milk products. This can help you maintain a healthy weight and may lower your risk for developing chronic health problems such as diabetes, heart disease, or cancer. Be flexible and adventurous! Try new food choices in place of some of the less nutritious or higher-calorie foods you usually eat. A sample eating pattern is given in Table 2-8, and Table 2-9 translates that eating pattern into a nutritious menu. The Eat Well Be Well feature on page 54 provides additional tips for including a variety of fruits and vegetables in your diet for maximum health benefits.

Gaining Calorie Control: The Daily Limit for Empty Calories If you consistently build your diet by choosing mostly nutrient-dense foods that are low in solid fat and added sugars, you may be able to meet your nutrient needs without using your full calorie allowance. If so, you may have what is called a **daily limit for empty calories** for use in meeting the rest of your calorie needs (as illustrated in the margin).

Most empty calorie allowances are very small, between 100 and 300 calories, especially for those who are not physically active. How do we track these extra calories? One example is a regular 12-ounce soda that contains 155 calories—but all 155 of these calories are from added sugars and, thus, are considered "empty" calories. Keep in mind that, for many people, the empty calorie limit is totally used up by the foods they choose with solid fats and added sugars in each food group, such as higher-fat meats, higher-fat cheeses, whole milk, and sweetened bakery products.

Your empty calorie limit can be used to:

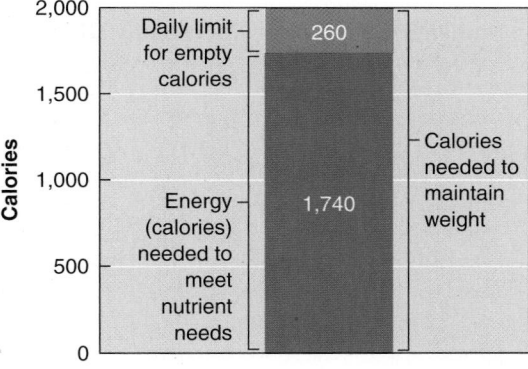

- Eat additional nutrient-dense foods from each of the food groups, such as an extra container of fat-free yogurt or an extra piece of fruit

- Select limited amounts of foods that are not in their most nutrient-dense form and/or contain solid fats or added sugars, such as whole milk, full-fat cheese, biscuits, and sweetened cereal

- Add fats or sweeteners to foods, such as sauces, gravies, sugar, syrup, butter, and jelly

- Eat or drink items that contain only fats, caloric sweeteners, and/or alcohol, such as candy, soda, wine, and beer

daily limit for empty calories the balance of calories remaining in a person's energy allowance, after accounting for the number of calories needed to meet recommended nutrient intakes through consumption of nutrient-dense foods in low-fat or no-added-sugar forms. These calories may be used to increase intake from the basic food groups; to select foods from these groups that are higher in fat or with added sugars; to add oils, solid fats, or sugars to foods or beverages; or to consume alcohol.

GRAIN GROUP

Tip: Make at least half of your grain selections whole grains.

Strategies:

- Buy products that list a whole grain or whole wheat or other whole-grain flour first on the label's ingredient list.
- Check the Nutrition Facts label for the fiber content of food products. Fiber content is a good clue to the amount of whole grain in the product.
- Choose 100% whole-grain breads, preferably, or mixed flour breads such as multi-grain or cracked wheat.
- To eat more whole grains, substitute a whole-grain product for a refined product—such as eating brown rice instead of white rice.
- Add whole grains to mixed dishes such as soups, stews, and casseroles.

Nutrient-Dense Choices (Choose Most Often)

Whole grains: Barley, brown and wild rice, bulgur, millet, oats, popcorn, quinoa, rye, wheat; whole-grain low-fat breads, cereals, crackers, and pastas

Enriched grain products: breads, bagels, cereals, grits, pastas, rice, rolls, and tortillas

Less Nutrient-Dense Choices (Limit Choices)

Biscuits, cakes, cookies, crackers, cornbread, croissants, Danish, doughnuts, fried rice, French toast, granola, muffins, pancakes, pies, presweetened cereals, taco shells, waffles

VEGETABLE GROUP

Tip: Choose from all five vegetable subgroups several times a week. Eat more dark-green vegetables, orange vegetables, and dry beans and peas.

Strategies:

- Buy fresh vegetables in season. They cost less and are likely to be at their peak flavor.
- Stock up on frozen vegetables for quick and easy cooking.
- Prepare main dishes, side dishes, and salads that include vegetables.
- Keep a bowl of cut-up vegetables in a see-through container in the refrigerator. Carrot and celery sticks are traditional, but consider broccoli, cucumber slices, or red or green pepper strips.
- Add dark-green or orange vegetables to soups, stews, casseroles, and stir-fries.
- Use romaine, spinach, or other dark leafy greens as salad greens, and eat green salads often.
- Choose main dishes, side dishes, and salads that include cooked dry beans or peas.

Nutrient-Dense Choices (Choose Most Often)

Dark-green vegetables: Broccoli and leafy greens such as arugula, kale, green-leaf and romaine lettuce; spinach, beet, collard, mustard, and turnip greens

Orange vegetables: Carrots, pumpkin, sweet potatoes, winter squash

Legumes: Black beans, black-eyed peas, chickpeas (garbanzo beans), kidney beans, pinto beans, soybeans, split peas, and lentils

Starchy vegetables: Cassava, corn, green peas, and white potatoes

Other vegetables: Artichokes, asparagus, bamboo shoots, beets, bok choy, Brussels sprouts, cabbage, cauliflower, celery, cucumbers, eggplant, green beans, iceberg lettuce, mushrooms, okra, onions, peppers, snow peas, tomatoes, vegetable juices, zucchini

Less Nutrient-Dense Choices (Limit Choices)

Baked beans, candied sweet potatoes, coleslaw, French fries, potato salad, refried beans, scalloped potatoes

FRUIT GROUP

Tip: Choose a variety of whole or cut-up fruits more often as snacks or with meals, instead of juice.

Strategies:

- Use fruit in salads, toppings, desserts, and/or snacks regularly.
- Use fruit as a topping on cereal, pancakes, and other foods rather than sugars, syrups, or other sweet toppings.
- Buy fresh fruits in season when they may be less expensive and at their peak flavor.
- Buy fruits that are dried, frozen, and canned (in water or juice) as well as fresh, so that you always have a supply on hand.

Nutrient-Dense Choices (Choose Most Often)

Apples, apricots, avocados, bananas, blueberries, cantaloupe, grapefruit, grapes, guava, kiwi, mango, oranges, papaya, peaches, pears, pineapple, plums, raspberries, strawberries, watermelon, dried fruit, unsweetened fruit juices

Less Nutrient-Dense Choices (Limit Choices)

Canned or frozen fruit in syrup; juices, punches, and fruit drinks with added sugars

TABLE 2-7

DAIRY GROUP

Tip: Make most milk group choices fat-free or low-fat (1%). Consume other calcium-rich foods or fortified cereals or beverages if milk products are not consumed.

Strategies:
- Drink fat-free or low-fat (1%) milk as a beverage.
- Use fat-free or low-fat milk or yogurt on cereal or as a snack.
- If you drink cappuccinos or lattes, ask for them with fat-free milk.
- Add fat-free or low-fat milk instead of water to oatmeal and hot cereals.
- Choose low-fat cheeses.

Nutrient-Dense Choices (Choose Most Often)

Fat-free milk and fat-free milk products, buttermilk, cheeses, cottage cheese, yogurt, fat-free fortified soy milk

Less Nutrient-Dense Choices (Limit Choices)

Reduced-fat milk, whole milk, reduced-fat milk and whole milk products such as cheeses, cottage cheese, yogurt, chocolate milk, fortified soy milk, custards and puddings, ice cream, ice milk, and frozen yogurt

PROTEIN FOODS GROUP

Tip: Make most choices lean or low-fat. Include fish, dry beans and peas, nuts, and seeds as well as meats, poultry, and eggs.

Strategies:
- Select meat cuts that are low in fat and ground beef that is extra lean (at least 90% lean).
- Trim fat from meat and remove poultry skin before cooking or eating.
- Drain fat from ground meats after cooking.
- Use preparation methods that do not add fat, such as grilling, broiling, poaching, or roasting.
- Choose lean turkey, roast beef, ham, or low-fat luncheon meats for sandwiches instead of fatty luncheon meats such as regular bologna or salami.
- Select fish as a choice from this group more often, especially fish rich in omega-3 fatty acids, such as salmon, trout, and herring.
- Choose dry beans or peas as a main dish often.
- Choose unsalted nuts as a snack, on salads, or in main dishes, to replace meat or poultry.

Nutrient-Dense Choices (Choose Most Often)

Lean meat, fish, shellfish, poultry (no skin), eggs, legumes, tofu, tempeh, peanut butter, nuts, and seeds

Less Nutrient-Dense Choices (Limit Choices)

Luncheon meats, ground beef, hot dogs, fried meats, fried fish, fried poultry, or fried eggs, poultry with skin, refried beans, sausages

OILS

Tip: Choose most fats from sources of monounsaturated and polyunsaturated fatty acids, such as fish, nuts, seeds, and vegetable oils.

Strategies:
- Substitute vegetable oils for solid fats like butter, stick margarine, shortening, or lard.
- Substitute nuts for meat or cheese as a snack or as part of a meal.

Nutrient-Dense Choices (Choose Most Often)

Liquid vegetable oils, such as canola, corn, flaxseed, olive, peanut, safflower, sesame, soybean oil, and sunflower oils; mayonnaise, oil-based salad dressings, and soft margarine with no *trans* fats; foods naturally high in oils, such as avocados, nuts, seeds, olives, fatty fish, and shellfish

SOLID FATS & ADDED SUGARS

Tip: Limit intakes of foods and beverages with solid fats or added sugars. Choose and prepare foods and beverages with little added sugars or caloric sweeteners.

Strategies:
- Limit products containing saturated fats, such as ground and processed meats, full-fat cheese, cream, ice cream, and fried foods.
- Limit foods containing partially hydrogenated vegetable oils, which contain *trans* fats, such as some commercially fried foods and some bakery goods.
- Select baked, steamed, or broiled rather than fried foods most often.
- Choose water, fat-free milk, or unsweetened tea or coffee as a beverage most often.
- Limit sweet snacks and desserts. Choose canned fruits in 100% fruit juice.
- Select unsweetened cereals.

TABLE 2-8   A Sample Eating Pattern for 2,000 Calories

Food Group	Recommended Amounts	Breakfast	Lunch	Snack	Dinner	Snack
Grains	6 oz	1½ oz	2 oz		2 oz	½ oz
Vegetables	2½ c		1 c		1½ c	
Fruits	2 c	½ c		1 c	½ c	
Dairy	3 c	1 c			1 c	1 c
Protein Foods	5½ oz		2 oz	1 oz	2½ oz	
Oils	6 tsp		2 tsp		4 tsp	
Empty Calories	260 cal					

TABLE 2-9   A Sample Menu for 2,000 Calories

The menu below provides about 1,830 calories, leaving about 170 extra calories to spend on additional nutrient-dense foods, or on foods with added sugars or fats.

Amounts	Menu Item	Calories
	Breakfast	
1½ oz whole grains	1½ c whole-grain flakes cereal	162
1 c dairy	1 c fat-free milk	90
1 c fruit	1 c sliced banana	105
	Lunch	
2 oz whole grains	2 slices rye bread	160
2 oz protein foods	1½ oz tuna, packed in water, drained	55
1 c vegetables	2 tbsp kidney beans	25
	6 baby carrots, shredded	25
	¼ c shredded Romaine lettuce	2
	2 slices tomato	10
2 tsp oils	1 tbsp low-fat mayonnaise	35
	1 tsp Dijon mustard	5
8 oz unsweetened beverage	1 c iced tea, unsweetened	0
	Snack	
½ c fruit	¼ c dried apricots	80
1 oz protein foods	½ oz almonds (12 almonds)	80
	Dinner	
	Spinach salad:	
½ c vegetables	1 c raw baby spinach leaves	8
½ oz protein foods	2 tbsp chickpeas	34
½ c fruit	½ c tangerine slices	30
2 tsp oils	2 tsp oil and vinegar dressing	60
	Pasta with meat sauce:	
2 oz enriched grains	1 c rigatoni pasta	197
1 c vegetables	½ c tomato sauce with	25
	½ c mixed frozen vegetables	25
2 oz protein foods	2 oz cooked ground turkey breast,	130
2 tsp oils	sautéed in 2 tsp oil	90
1 c dairy	1½ oz Parmesan cheese	190
	Snack	
1 c dairy	1 c fat-free fruited yogurt	120
½ oz grains	3 graham crackers	90

How Well Do We Eat?

How are consumers doing in following the nutrition recommendations presented in this chapter? The answer to this question is, "Needs improvement." Undoubtedly, trends such as eating away from home and increased portion sizes in restaurant and carry-out items negatively affect American eating habits.

Figure 2-6 provides insights into how the typical American diet compares to recommended intakes or limits.[10] In general, we eat too many calories and too much solid fat, added sugars, refined grains, and sodium. Solid fats and added sugars should make up only a small proportion of total calories in the diet. However, as illustrated in the margin, Americans consume about 35 percent of total calories from foods high in solid fats and added sugars—soft drinks, desserts, fatty meats, French fries, pizza, and whole milk products, in contrast to a recommended limit of no more than 5 to 15 percent of total calories for most individuals.

We also consume too few vegetables, fruits, whole grains, and fat-free or low-fat milk and milk products. As shown in Figure 2-6, consumers eat only 15 percent of

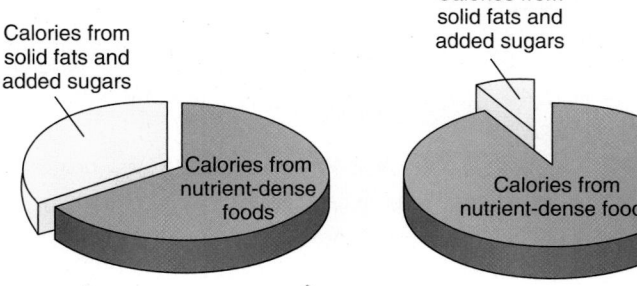

What We Eat versus Recommended Limits

FIGURE 2-6

How the U.S. Diet Measures Up to Recommended Dietary Guidelines

The bars show average intakes for all individuals (2 years or older) as a percent of the recommended intake level or recommended limit.[a]

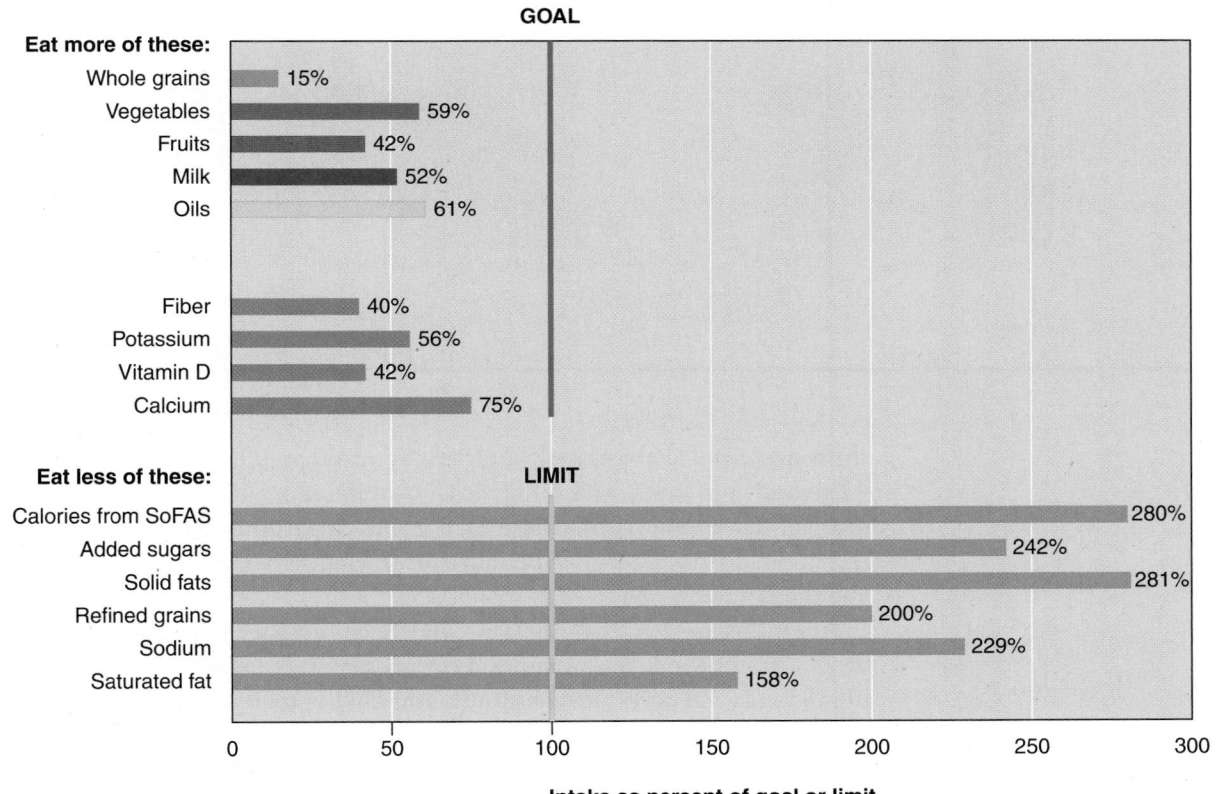

GOAL

Eat more of these:
- Whole grains: 15%
- Vegetables: 59%
- Fruits: 42%
- Milk: 52%
- Oils: 61%
- Fiber: 40%
- Potassium: 56%
- Vitamin D: 42%
- Calcium: 75%

LIMIT

Eat less of these:
- Calories from SoFAS: 280%
- Added sugars: 242%
- Solid fats: 281%
- Refined grains: 200%
- Sodium: 229%
- Saturated fat: 158%

Intake as percent of goal or limit

[a] Recommended intakes for food groups and limits for refined grains, SoFAS (solid fats and added sugars) are based on a 2,000-calorie diet.
* Recommended intakes for fiber, potassium, vitamin D, and calcium are based on the highest Adequate Intakes (AI) for ages 14 to 70 years. Limits for sodium are based on the AI and for saturated fat on 7 percent of calories.

SOURCE: U.S. Department of Agriculture and U.S. Department of Health and Human Services, *Dietary Guidelines for Americans, 2010*. 7th ed. (Washington, DC: U.S. Government Printing Office, 2010). Data source: What We Eat in America, National Health and Nutrition Examination Survey (WWEIA, NHANES) 2001–2004 or 2005–2006.

Rules of Thumb for Portion Sizes— It's All in Your Hands

In the "more must be better" viewpoint of eating in America today, it can be difficult to balance the energy we take in through the food we eat with the energy we expend on a daily basis. This is evidenced by the larger and larger portion sizes we find in restaurants, on the grocery shelves, and in almost every aspect of eating in the United States today.

What's the difference between a serving and a portion? A *serving* is a standard amount of food used as a reference to give advice regarding how much food to eat (such as 1 cup of milk or its equivalent). The amount of a food that is considered a serving usually stays the same. A *portion*, in contrast, is the amount of food *you* choose to eat, and it may vary from one meal or snack to the next based on your appetite or hunger. How do you determine how much food is in your portion? You can use the following images as a visual reference to "right-size" your portions and avoid consuming extra calories.

One fist, clenched = 8 fl oz

Two hands, cupped = 1 cup

One hand, cupped = 1/2 cup

Palm of hand = 3 oz

Two thumbs together = 1 tbsp

SOURCE: Adapted from USDA, *Dietary Guidelines for Americans,* 5th ed., 2000.

the recommended amount of whole grains, less than 60 percent for vegetables, about 42 percent for fruits, and 50 percent for milk and milk products. By eating fewer nutrient-dense foods from these food groups, consumers are at risk for lower than recommended levels of specific nutrients, including vitamin D, calcium, potassium, and dietary fiber.

According to the 2010 Dietary Guidelines Advisory Committee, Americans can benefit from the practice of mindful eating, that is, "consciously" and carefully choosing all foods and beverages consumed. You can improve your health with a diet that is:

- Energy balanced, limited in total calories, and portion controlled

- Nutrient-dense and includes:
  - Vegetables, fruits, high-fiber whole grains
  - Fat-free or low-fat dairy products
  - Seafood, lean meat and poultry, eggs, soy products, nuts, seeds, and oils

- Very low in solid fats and added sugars and low in sodium

To see how your diet measures up to the recommendations in the MyPlate food plan, follow these steps.

Step 1: Write down everything you ate yesterday, including meals *and* snacks. Make note of portion sizes as well. (Refer to Table 2-6 for help with conversions.)

Step 2: Identify the food group for each item you ate. (Refer to Figure 2-5 for help.)

Step 3: Using the five food groups, determine the amounts that are right for you. Go to "Interactive Tools" at www .ChooseMyPlate.gov and type in your age, sex, and activity level under the "Daily Food Plan" and click on "submit." This will give you a plan with a good estimate of the amount of -servings you need from each food group. In addition, it will give you the approximate number of calories you require.

Step 4: Circle the estimated amounts you should eat from the middle column. In the right column, write down the amounts you ate yesterday. (Refer to Table 2-6 and the Savvy Diner feature in this chapter for help with conversions.) Compare the two columns to see how your diet rates.

	Amount You Should Eat	Amount You Ate
Grain group servings (ounces)	5 6 7 8 9 10	_____
Vegetable group servings (cups)	2 2½ 3 4	_____
Fruit group servings (cups)	2 2½ 3	_____
Dairy group servings (cups)	3	_____
Protein Foods group (ounces)	5 5½ 6 7	_____
Solid Fats and Added Sugars	Use Empty Calorie Limit	_____

Step 5: Decide what changes in your eating habits will make your diet more healthful. If your diet is "un-proportional," with too many foods coming from one food group, make gradual changes to develop a diet with more variety. The chapters that follow offer tips on how to do so.

More Tools for Diet Planning

In 1990, Congress passed one of the most important pieces of legislation of the 20th century. Known as the Nutrition Labeling and Education Act, the law called for sweeping changes in the way foods are labeled in the United States.[11] For consumers, the law ensures that food companies provide the kind of nutrition information that best allows people to select foods that fit into a healthful eating plan.

Food Labels   Considering the great variety of packaged foods available, using the food label to understand the nutrients a food supplies or lacks is essential (see Figure 2-7). The label is one of the most important tools you can use to eat healthfully.

By law, all labels must contain the following five components:

- The name of the food, also known as the statement of identity.

- The name of the manufacturer, packer, or distributor, as well as the firm's city, state, and ZIP code.

- The net contents of the package, which tells you the quantity of the food product that is in the container and helps you compare prices.

- The **ingredients list**, with items listed in descending order by weight. The first ingredient listed makes up the largest proportion of all the ingredients in the food; the second, the second largest amount; and so on. If the first ingredient

ingredients list a listing of the ingredients in a food, with items listed in descending order of predominance by weight. All food labels are required to bear an ingredients list.

FIGURE 2-7

How to Read a Food Label

You can use food labels to help you make informed food choices for healthy eating practices. The food label allows you to compare similar products, determine the nutritional value of the foods you choose, and can increase your awareness of the links between good nutrition and reduced risk of chronic, diet-related diseases.

SOURCE: Food and Drug Administration

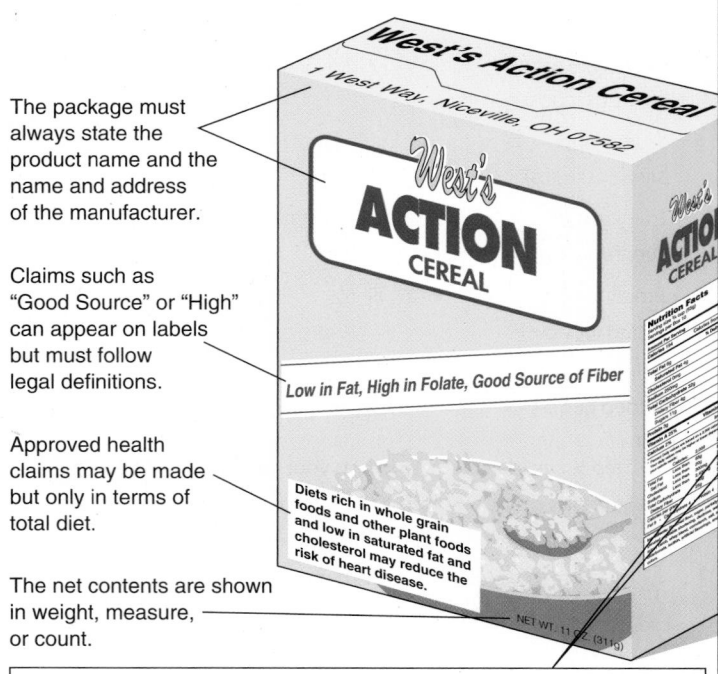

The package must always state the product name and the name and address of the manufacturer.

Claims such as "Good Source" or "High" can appear on labels but must follow legal definitions.

Approved health claims may be made but only in terms of total diet.

The net contents are shown in weight, measure, or count.

The only vitamins required to appear on the Nutrition Facts panel are vitamins A and C. If a manufacturer makes a nutrition claim about another vitamin, however, the amount of that nutrient in a serving of the product must also be stated on the panel. For instance, the cereal shown is touted as "High in Folate," so the percentage of the recommended intake for folate in a serving of the cereal (25%) is stated on the label. The manufacturer has the option of listing any other nutrients as well.

Start here

Serving size, number of servings per container, and calorie information

Limit these nutrients.

Information on sodium is required on food labels.

Get enough of these nutrients.

Guide to the % Daily Value: Quantities of nutrients as a percentage of Daily Value for nutrients based on a 2,000-calorie energy intake.
  5% or less is low.
  10% or more is good.
  20% or more is high.

Calcium and iron are required on the label. Look for foods that provide 10% or more of these minerals.

Reference values

This allows comparison of some values for nutrients in a serving of the food with the needs of a person requiring 2,000 or 2,500 calories per day.

Calorie/gram reminder

Ingredients in descending order of predominance by weight

---

in the list is sugar, for example, you know the food contains more sugar than anything else. The list is especially useful in helping people identify ingredients they avoid for health, religious, or other reasons (see Figure 2-8).

- The **Nutrition Facts panel**, unless the package is small—no larger than 12 square inches of surface area (about the size of a small candy bar or a roll of breath mints).* Small packages must carry a telephone number or address that consumers can use to obtain nutrition information.

Nutrition Facts panel a detailed breakdown of the nutritional content of a serving of a food that must appear on virtually all packaged foods sold in the United States.

*A container with less than 40 square inches of surface area for nutrition labeling (e.g., a can of tuna) is allowed to present fewer facts.

FIGURE 2-8

Using the Ingredients List on Food Labels

COURTESY: Modern Maturity Magazine.

Oats 'N' More	Morning Krisps
Ingredients: whole grain oats, (includes the oat bran), modified corn starch, wheat starch, (sugar) salt, oat fiber, trisodium phosphate, calcium carbonate, vitamin E (mixed tocopherols) added to preserve freshness. **Vitamins and Minerals:** iron and zinc (mineral nutrients), vitamin C (sodium ascorbate), vitamin B$_6$ (pyridoxine hydrochloride), riboflavin, thiamin mononitrate, niacinamide, folic acid, vitamin A (palmitate), vitamin B$_{12}$, vitamin D.	**Ingredients:** (Sugar,) wheat, (corn syrup,) (honey,) hydrogenated soybean oil, salt, caramel color, soy lecithin. **Vitamins and Iron:** sodium ascorbate (vitamin C), ferric phosphate (iron), niacinamide, pyridoxine hydrochloride (vitamin B$_6$), riboflavin, vitamin A palmitate, thiamin hydrochloride, BHT (preservative), folic acid, vitamin B$_{12}$, and vitamin D.

The cereals shown contain sugars. Learn to read the ingredients list. Labels list ingredients in order of amount by weight with the greatest amount of an ingredient present in the food listed first. Check labels for sugar terms, in addition to sugar, such as: brown sugar, corn syrup, dextrose, fructose, glucose, high-fructose corn syrup, honey, invert sugar, levulose, mannitol, molasses, sorbitol, and sucrose.

Using the Nutrition Facts Panel   The Nutrition Facts panel must indicate the amount of certain mandatory nutrients that one serving of the food contains. You can use the Nutrition Facts panel to help you follow a healthful diet by considering the nutrition information that it provides.

1. Keep serving sizes in mind. Based on the amount of food most people eat at one time, the Food and Drug Administration (FDA) has set forth a list of serving sizes for more than 100 food categories. Manufacturers must use these recommended serving sizes on food labels. For instance, the FDA might say that the serving size for product X must always be 8 ounces. This procedure ensures that consumers can easily compare one brand of a product to another without having to make difficult calculations to compare nutrient quantities for different serving sizes.
2. Notice the total calories and calories from fat per serving.
3. Check out the nutrient information that must appear on the Nutrition Facts panel, including total fat, saturated fat, *trans* fat, cholesterol, sodium, total carbohydrate, dietary fiber, sugars, protein, vitamin A, vitamin C, calcium, and iron (in that order). These nutrients were chosen to appear on the Nutrition Facts panel because they address today's health concerns.

    The FDA spelled out the ranking of the required nutrients to ensure that the label reflects the government's dietary priorities for the public. For example, fat falls near the top of the list because most consumers need to pay closer attention to the amount and type of fat in their diet. Most Americans eat too much saturated fat and trans fat, which raises the risk of developing heart disease. Protein, in contrast, appears near the bottom of the label because the amount of protein most Americans eat does not rate as a major health concern. Note that only vitamins A and C, iron, calcium, and sodium appear on the nutrition panel. Those are the only vitamins and minerals required to be on food labels, unless a manufacturer chooses to make a nutrition claim about another one. For instance, if a manufacturer says that a cereal is *high* in folate or fortified with niacin, the amount of folate or niacin in the product must appear on the label (see Figure 2-7).
4. Look for the percent Daily Values. The **Daily Values** for fats, protein, carbohydrates, and fiber are calculated according to what experts deem a healthful diet for adults (as noted in the Calculation Practice box on page 56). For instance, since the Daily Value for fat recommends that no more than

Daily Value the amount of fat, sodium, fiber, and other nutrients health experts say should make up a healthful diet. The "% Daily Values" that appear on food labels tell you the percentage of a nutrient that a serving of the food contributes to a healthful diet.

Color Your Plate for Health with a Variety of Fruits and Vegetables

The message is simple: *Eat plenty of fruits and vegetables every day for better health.*[12] Most Americans, however, find this simple bit of nutritional advice challenging to put into practice. Overwhelming evidence points to the health benefits available from diets rich in fruits and vegetables because of the vitamins, minerals, and **phytochemicals** (*phyto* is the Greek word for plant) found in plant foods.

In the classic sense, the naturally occurring phytochemicals—for example, lycopene, a pigment that makes tomatoes red and watermelon pink—are not vitamins, minerals, or nutrients because they do not provide energy or building materials. However, scientists who study them say that phytochemicals may have the potential to slow the aging process; boost immune function; decrease blood pressure and cholesterol; prevent cataracts; prevent, slow, or even reverse certain cancers; and strengthen our hearts and circulatory systems. (More about the food sources and beneficial effects of phytochemicals appears in Chapter 7.)

How can you incorporate fruits and vegetables into your diet to achieve the maximum health benefits? Aim for the amounts of fruits and vegetables recommended in MyPlate (about 2 cups of fruits and 2½ cups of vegetables on a 2,000-calorie diet).* Go to www.ChooseMyPlate.gov to determine your own recommended intake of fruits and vegetables. Also consider the following three steps in your daily food planning.

 Color Your Plate with Health-Protective Foods.

Consume many differently colored fruits and vegetables. Select at least three differently colored fruits and vegetables a day. The red pigment in tomatoes has different bioactive properties than the orange pigments in carrots, sweet potatoes, melon, or winter squash. Add dark-green leafy vegetables such as kale, mustard greens, collards, and spinach, along with broccoli and Brussels sprouts. Fruits are a vital powerhouse of the **antioxidant nutrients** that act to squelch cell-damaging molecules in the body. Blueberries top the list for being one of the highest in antioxidants. Be sure to include citrus fruits such as oranges and grapefruit regularly, as they contain many compounds that have antioxidant and other vital health benefits.

Eat Your Colors Every Day To Stay Healthy & Fit

*Low-fat diets rich in fruits and vegetables and low in saturated fat and cholesterol may reduce the risk of heart disease and some types of cancer—diseases associated with many factors.

SOURCE: © Produce for Better Health Foundation.

 Be Adventurous: Select from As Wide a Variety of Fruits and Vegetables As Possible.

Foods within each group of the MyPlate Food Guidance System have similar nutrient content. Choosing foods from each group offers a variety of the nutrients necessary for health, but variety within food groups is valuable, too. Even fruits that may seem like comparable foods often provide a number of different nutrients; no food has them all. For example, oranges are an excellent source of vitamin C and folate but do not supply beta-carotene (a precursor of vitamin A); cantaloupe is high in beta-carotene.

Make It Easy on Yourself!

Keep fruit and sliced vegetables in easy-to-reach places, such as sliced vegetables in the refrigerator and fresh fruit on the table. For convenience, keep a variety of fresh, frozen, and canned fruits and vegetables on hand to add to soups, salads, rice dishes, and other menu items. Frozen fruits and vegetables are processed right after harvesting and contain nutrient profiles similar to fresh produce. If using canned items, choose fruits packed in fruit juice (not in syrup) and remove excess salt from vegetables by rinsing before using them.

*In general, 1 cup of fruit or 100% fruit juice, or ½ cup of dried fruit is considered 1 cup from the fruit group; 1 cup of raw or cooked vegetables or vegetable juice, or 2 cups of raw leafy vegetables are considered as 1 cup from the vegetable group.

---

phytochemicals nonnutritive substances in plants that possess health-protective benefits.

antioxidant nutrients vitamins and minerals that protect other compounds from damaging reactions involving oxygen by themselves reacting with oxygen. The antioxidant nutrients are vitamin C, vitamin E, and beta-carotene. The mineral selenium also has a role in antioxidant reactions in the body.

30 percent of total calories should come from fat, the % Daily Value tells you the percentage of fat that a serving of the food contributes to a 2,000-calorie eater's fat "allowance." A 2,000-calorie diet was chosen as a good point of reference because that's about the amount recommended for most moderately active women, teenage girls, and sedentary men. Of course, more calories may be appropriate for many men, teenage boys, and active women. This is why the nutrition panel also shows Daily Values for a 2,500-calorie diet (refer to the bottom of the Nutrition Facts panel in Figure 2-7).

Using the Percent Daily Values (% DV)

To understand how the Daily Values for fats, sodium, carbohydrates, and fiber are calculated, let's go through an example. First, look for the grams of total fat and the % Daily Value for the cereal shown in Figure 2-7. The label shows that a serving supplies 3 grams of fat, with a Daily Value of 5 percent. This means that a serving of the cereal contributes 5 percent of the total fat that a person eating 2,000 calories a day should consume.

Now look at the bottom of the Nutrition Facts panel, which indicates that someone eating 2,000 calories a day should take in no more than 65 grams of fat a day. Divide 3 (the number of fat grams in a serving of the cereal) by 65. Multiply that number by 100 to obtain a percentage. The answer is 5—that is, 5 percent of the total fat.

You can use the % Daily Values to get a good idea of how various foods fit into a healthful diet, regardless of the number of calories you eat. Consider a student who eats only 1,800 calories a day. If she snacks on two servings of potato chips with a Daily Value of 15 percent fat per serving, she has already taken in 30 percent of the fat that someone eating 2,000 calories should have in an entire day. Because she eats less than 2,000 calories, the potato chips contribute slightly more than 30 percent. Thus, the 30 percent Daily Value shows that potato chips add up to a lot of fat for one snack. In short, the % Daily Value column can give you a good idea of how different foods fit into your overall diet.

Some people find it easiest to bypass the % Daily Values and simply check the grams of total fat a serving of food supplies to see how much it adds to a daily fat tally. Let's say a man eats 2,000 calories a day and therefore should consume no more than 65 grams of fat a day. If he eats a muffin (15 grams of fat) and coffee with cream (10 grams) in the morning, he's already up to 25 grams of fat. That means he can have about 40 more grams during the rest of the day to stay within his fat "budget." Once you determine the maximum number of fat grams you should have in a day, you can use the food label to get a good idea of how many grams of total fat the items you buy add to your daily tally. To figure out your fat allowance, check Table 5-4 in Chapter 5.

You can also use the Daily Values to comparison-shop. For example, if you're looking for a high-fiber cereal to increase the amount of fiber in your diet, you can check the % Daily Value for fiber on the labels of several brands of cereal. If a serving of Brand X's cereal has a Daily Value of 20 percent for fiber and Brand Y supplies only 5 percent, you know that Brand X is a better source of fiber, whereas Brand Y is low in fiber.

The % Daily Values for vitamins and minerals are calculated using standard values designed specifically for use on food labels. These values are listed on Study Card page 30 at the back of this book. These standard values for nutrients were created to help manufacturers avoid a stumbling block they face as they label foods. Because manufacturers don't know whether you're an 18-year-old woman or a 30-year-old man, they don't know exactly what your nutritional needs are. You may recall that the DRI include a different set of vitamin and mineral recommendations for each gender and age group.

To help get around this problem, the nutrient recommendations used for vitamins and minerals on labels represent the highest of all the values to ensure that virtually everyone in the population is covered. For most nutrients, the highest recommendation is for an adult man. When it comes to iron, however, the DRI for women is the highest (women require more iron than men), so the women's DRI is used as the standard value on labels.

Nutrient Content Claims

By law, foods carrying terms called **nutrient content claims**—low-fat, low-calorie, light, and so forth—must adhere to specific definitions spelled out by the Food and Drug Administration. For instance, a serving of

Learn to read food labels to help you achieve a healthful diet.

Noel Hendrickson/Blend Images LLC

"Henry likes nothing more than to curl up with a good label."

Courtesy: Modern Maturity Magazine

nutrient content claims claims such as "low-fat" and "low-calorie" used on food labels to help consumers who don't want to scrutinize the Nutrition Facts panel get an idea of a food's nutritional profile. These claims must adhere to specific definitions set forth by the Food and Drug Administration.

The Daily Values (DVs) and % Daily Values (% DVs) can help you choose foods and comparison-shop. As noted earlier, the DVs are designed for persons requiring 2,000 or 2,500 calories to meet their energy needs. The Daily Values for carbohydrates, protein, and fat are calculated using "calculation factors" set forth by the FDA for food labeling purposes as follows:*

Calculation factors:

- Fat: 30% of calories

- Saturated Fat: 10% of calories

- Protein: 10% of calories

- Carbohydrate: 60% of calories

- Fiber: 11.5 grams of fiber per 1,000 calories

To understand how the Daily Values for the macronutrients are calculated if your energy needs differ from 2,000 or 2,500 calories, let's go through an example using the calculation factors. If a person's energy requirement (EER) is 1,800 calories, determine this person's DVs for carbohydrate, protein, and fat.

Carbohydrate:
60% of 1,800 calories = 0.60 × 1,800 = 1,080 calories
1,080 calories of carbohydrate = 270 grams of carbohydrate (1,080 calories ÷ 4 calories/gram)
DV for carbohydrate = 270 grams

Protein:
10% of 1,800 calories = 0.10 × 1,800 = 180 calories
180 calories of protein = 45 grams of protein (180 calories ÷ 4 calories/gram)
DV for protein = 45 grams

Fat:
30% of 1,800 calories = 0.30 × 1,800 = 540 calories
540 calories of fat = 60 grams of fat (540 calories ÷ 9 calories/gram)
DV for fat = 60 grams

*THE FDA may update the DV calculation values soon to reflect the current DRI and AMDR values.

Nutrient content claims are strictly defined by the FDA.

a food dubbed low-fat must contain no more than 3 grams of fat. An item touted as low-calorie may provide no more than 40 calories per serving. Table 2-10 lists the claims commonly used on food labels and their legal definitions.

Health Claims

A statement linking the nutritional profile of a food to a reduced risk of a particular disease is known as a **health claim**. The FDA has set forth very strict rules governing the use of such health claims (see Table 2-10).[13] For example, if a food label bears health claims regarding calcium, a serving of the product must contain at least 20 percent of the Daily Value for calcium, among other restrictions. What's more, the manufacturers are allowed to imply only that the food "may" or "might" reduce risk of disease. They must also note the other factors, such as exercise, that play a role in prevention of the disease. Finally, they must phrase the claim so that the consumer can understand the relationship between the nutrient and the disease. For example, a health claim on a food low in fat, saturated fat, and cholesterol might read, "Whereas many factors affect heart disease, diets low in saturated fat and cholesterol may reduce the risk of this disease." The nutrient–disease relationships about which health claims can be made are listed in Table 2-11.[14]

Exchange Lists

Although food group plans provide sufficient detail to help most healthy people plan a good diet, exchange lists take meal planning a step further. As their name implies, **exchange lists** are simply lists of categories of foods, such as fruit, with portions specified in a way that allows the foods to be mixed and

health claim a statement on the food label linking the nutritional profile of a food to a reduced risk of a particular disease, such as osteoporosis or cancer. Manufacturers must adhere to strict government guidelines when making such claims.

exchange lists lists of foods with portion sizes specified. The foods on a single list are similar with respect to nutrient and calorie content and thus can be mixed and matched in the diet.

TABLE 2-10	Definitions of Nutrient Content Claims

General Terms
- Fresh, Raw: Unprocessed, or minimally processed with no added preservatives.
- Good Source of: A serving of the food supplies 10 to 19% of the Daily Value for a particular nutrient.
- Healthy: A food is low in fat, saturated fat, *trans* fat, cholesterol, and sodium, and provides at least 10% of the Daily Value for vitamin A, vitamin C, protein, calcium, iron, or fiber.
- High, Rich in, Excellent Source of: Contains 20% or more of the Daily Value for a particular nutrient.
- Less, Fewer, Reduced: A serving provides at least 25% less of a nutrient or calories than the comparison food.
- Light or Lite: A serving provides one-third fewer calories than or half the fat of the regular product. If fat supplies 50% or more of the calories to begin with, it must be reduced by half to be called "light." The sodium content of a serving of a low-calorie, low-fat food has been reduced by 50%.
- More, Extra: A serving of the food contains at least 10% more of the Daily Value of a particular nutrient than the regular food.
- Organic: At least 95% of the product's ingredients have been grown and processed according to USDA regulations regarding the use of fertilizers, herbicides, fungicides, insecticides, preservatives, and other substances (see Chapter 13).

Energy Terms
- Calorie-Free: Less than 5 calories per serving.
- Low-Calorie: No more than 40 calories per serving.
- Reduced Calories: Contains at least 25% fewer calories than the regular, unaltered product.

Carboydrates: Sugar and Fiber
- Sugar-Free: Less than 0.5 g of sugar per serving.
- High Fiber: 5 g or more fiber per serving.
- Good Source of Fiber: 2.5 g to 4.9 g per serving.

Fat and Cholesterol Terms
- Fat-Free: Less than 0.5 g of fat per serving.
- Low-Fat: 3 grams of fat or less per serving.
- Saturated Fat-Free: Less than 0.5 g of saturated fat and less than 0.5 g *trans* fat per serving.
- Low Saturated Fat: 1 gram or less saturated fat per serving and less than 0.5 g *trans* fat per serving.
- Less Saturated Fat: at least 25 percent less saturated fat and trans fat combined than the comparison food.
- *Trans* Fat-Free: Less than 0.5 g of *trans* fat *and* less than 0.5 g of saturated fat per serving.
- Cholesterol-Free: Less than 2 mg cholesterol and 2 g or less saturated fat and *trans* fat combined per serving.
- Low in Cholesterol: 20 mg or less of cholesterol and 2 g or less of saturated fat and *trans* fat combined per serving.
- Reduced or Less Cholesterol: at least 25 percent less cholesterol than the comparison food and 2 g or less saturated fat and trans fat combined per serving.
- Percent Fat-Free: An indication of the amount of a food's weight that is fat-free; can be used only on foods that are low-fat or fat-free to begin with, and must reflect the amount of fat in 100 g. For instance, a food that weighs 100 g with 3 g from fat can be labeled "97 percent fat-free." Note that this term refers to the amount of the food that is fat-free by weight, not calories. If that same food supplies 100 calories, the 3 g of fat contribute 27 of them (1 g of fat contains 9 calories). This means that 27 of the 100 calories, or 27% of the total calories, come from fat.
- "Lean" and "Extra Lean" describe the fat content of meat and poultry products:
  Lean:
  Less than 10 g of fat and
  Less than 4.5 g of saturated fat and *trans* fat combined, and
  Less than 95 mg of cholesterol per serving.
  Extra Lean:
  Less than 5 g of fat and
  Less than 2 g of saturated fat and *trans* fat combined, and
  Less than 95 mg of cholesterol per serving.

Sodium Terms
- Sodium-Free: Less than 5 mg of sodium per serving.
- Very Low Sodium: 35 mg or less sodium per serving.
- Low in Sodium: 140 mg sodium or less per serving.
- Reduced Sodium: At least 25% lower in sodium than the regular product.

SOURCE: Food and Drug Administration

matched or exchanged with one another in the diet. For instance, you might strive to eat four servings of fruit each day, and the exchange list shows you that ½ cup of orange juice, a small banana, or a small apple each counts as a fruit.

Portion sizes within groups are determined by considering the calorie, protein, carbohydrate, and fat content of the food. For example, one fruit contains about 60

Appendix A provides more details regarding Choose Your Foods: Exchange Lists for Diabetes (formerly U.S. Food Exchange System).

TABLE 2-11	Health Claims on Food Labels

Approved health claims with significant scientific agreement:[a]
- Diets adequate in calcium and vitamin D, as part of a well-balanced diet, may reduce the risk of osteoporosis.
- Eating a healthful diet low in total fat may help reduce the risk of some types of cancers.
- Low-fat diets rich in fiber-containing grain products, fruits, and vegetables may reduce the risk of some types of cancer.
- Low-fat diets rich in fruits and vegetables may reduce the risk of some types of cancer.
- Diets low in sodium may reduce the risk of high blood pressure.
- Diets low in saturated fat and cholesterol, and as low as possible in trans fat, may reduce the risk of heart disease.
- Diets low in saturated fat and cholesterol and high in fruits, vegetables, and grain products that contain fiber may lower blood cholesterol levels and reduce your risk of heart disease.
- Diets low in saturated fat and cholesterol and rich in fruits, vegetables, and grain products that contain dietary fiber, particularly soluble fiber, may reduce the risk of heart disease.
- Diets that are low in saturated fat and cholesterol and that include soluble fiber in foods such as whole oats may reduce the risk of heart disease.
- Adequate folate in healthful diets may reduce a woman's risk of having a child with a brain or spinal cord birth defect.
- Sugar alcohols may reduce the risk of tooth decay.
- Drinking fluoridated water may reduce the risk of tooth decay.
- Diets low in saturated fat and cholesterol that include 25 grams of soy protein a day may reduce the risk of heart disease.
- Diets rich in whole-grain foods and other plant foods and low in total fat, saturated fat, and cholesterol may reduce the risk of heart disease and certain cancers.
- Diets low in saturated fat and cholesterol that include 2 grams per day of phytosterols [plant sterols, plant stanols, or plant sterols and stanols] eaten with meals or snacks may reduce the risk of heart disease.
- Diets containing foods that are good sources of potassium and low in sodium may reduce the risk of high blood pressure and stroke.

A sampling of qualified health claims:[b]
- Omega-3 fatty acids and reduced risk of heart disease
- Nuts such as walnuts and reduced risk of heart disease
- Calcium and reduced risk of high blood pressure
- Calcium and reduced risk of colorectal cancer
- Monounsaturated fat from olive oil and reduced risk of heart disease
- B vitamins and reduced risk of heart disease

[a] The FDA does not require these claims to carry disclaimers.
[b] FDA requires label disclaimers when these claims are used; for example, "Health claim: Although there is scientific evidence supporting the claim, the evidence is not conclusive."

SOURCE: Food and Drug Administration

calories and 15 grams of carbohydrate. One starch, in contrast, provides about 80 calories, 15 grams of carbohydrate, 3 grams of protein, and a trace of fat. This breakdown makes exchange lists especially useful tools for people who follow carefully planned diets as a result of a health problem such as diabetes.

Exchange lists are also useful for people who are following calorie-controlled diets to lose weight. Dietitians sometimes give clients tailor-made diets centered on the exchange lists—say, a 1,500-calorie daily diet that might include six starches, five vegetables, two fruits, and so forth. A person can take such a framework and use the exchange lists to choose a wide variety of foods that fit into the basic eating plan.

American eating habits have become as diverse as the various ethnic and cultural groups that make up America's people. Throughout American history, immigrant groups—from Poles to Jews to Italians to Irish to Germans to Hispanics to African Americans to Asians—have had and continue to have profound effects on the collective American palate. As food writer and critic John Mariani has pointed out:

> [T]he United States—a stewpot of cultures—has developed a gastronomy more varied . . . than that of any other country in the world. . . . In any major American city one will find restaurants representing a dozen national cuisines, including northern Italian trattorias, bourgeois French bistros, Portuguese seafood houses, Vietnamese and Thai eateries, Chinese dim sum parlors, Japanese sushi bars, and German rathskellers.[15]

This Spotlight examines some of the more prevalent ethnic and regional food practices to see how they originated and how they fit into a healthful eating plan.

I love to eat Mexican food. Is it true that it is loaded with fat? Do the Mexican people eat a lot of high-fat food?

Although it's true that many menu selections in Mexican restaurants are loaded with high-fat ingredients such as cheese, ground beef, sour cream, guacamole, and fried tortillas, most dishes eaten regularly in Mexican American homes are much simpler. Breakfast, for example, might be tortillas (flat, thin corn or wheat pancakes) served with refried beans, eggs, or cereal, and a beverage. Lunches and dinners often consist of beans and rice, bread or tortillas, meat/sausage (often as part of a stew), a vegetable or lettuce and tomato, and a beverage.

The traditional Mexican diet drawing largely from Spanish and Indian influences was even simpler, containing mostly vegetables, including beans, squash, and maize (corn). In addition, it often included cactus parts, agave (a plant with spiny-margined leaves and flowers), chili peppers, amaranth (a grain), avocado, and guava (a sweet, juicy fruit with green or yellow skin and red or yellow flesh).

This type of diet is high in complex carbohydrates such as rice and vitamin A- and C-rich fruits and vegetables, making it a particularly healthful aspect of the traditional Mexican way of eating. Consider that no Mexican meal is complete without salsa, a low-fat condiment consisting of vitamin-rich tomatoes, chilies, and onions. Another plus of the Mexican diet is frequent use of beans, mostly the pinto variety, which rank as a particularly good source of fiber.

There are downsides to the Mexican diet, however. Most foods, even beans and rice, are typically fried rather than baked or broiled. For example, *frijoles refritos* (refried beans) are usually fried in lard and contain about 270 calories and 3 grams of fat per cup. Most flour tortillas are also made with lard, sometimes 1 or 2 teaspoons per tortilla. (Corn tortillas, in contrast, typically contain very little fat.) Another drawback of the Mexican diet is the frequent consumption of high-fat meats, such as *chorizo* (spicy pork or beef sausage) and eggs, which are used in dishes such as *chiles rellenos* (roasted mild green chili pepper stuffed with cheese, dipped in egg batter, and fried), *burritos* (warm flour tortillas stuffed with a mixture of egg, meat, beans, and/or avocado), and *chilaquiles* (tortilla casserole often made with eggs or meat).

Fortunately, health-conscious Mexican-food lovers can take advantage of the popular cuisine with a little know-how. Instead of the usual American versions of Mexican fare, such as fried tortillas packed with ground beef and smothered in cheese and sour cream, opt for corn tortillas filled with, say, regular, unfried pinto beans mixed with chopped onion and topped with a sprinkle of shredded cheese and a generous portion of lettuce, salsa, a dollop of fat-free plain yogurt or fat-free sour cream, and a garnish of sliced avocado. Or, instead of serving high-fat commercial tortilla chips with salsa, try making "no-fry" chips: Immerse several tortillas in warm water, drain quickly, cut into six to eight wedges, and place on a nonstick pan; bake in a 500-degree Fahrenheit oven for 3 to 4 minutes, flip, and continue to bake for another minute or two until golden brown.

Adventurous eaters who have access to a wide variety of exotic produce might want to try some of the different fruits and vegetables the Mexican people have enjoyed for decades, such as *jicama* (HE-cah-mah), a yam bean root vegetable that is tan outside and white inside, has a mild chestnut flavor, and is always eaten raw.[16] Figure 2-9 shows produce and other foods commonly eaten by Mexican Americans.

Are foods served in Chinese restaurants in the United States, such as chop suey, egg rolls, and fortune cookies, traditional Chinese foods? A Chinese American friend of mine says they aren't. Is she right?

Yes, it's true that chop suey and the like are American inventions. The traditional Chinese diet consists of much simpler, lower-fat dishes.

(Continued)

© Michael Newman/PhotoEdit

GRAINS	VEGETABLES	FRUITS	DAIRY	PROTEIN FOODS
bolillo[b], bread, cake, cereal, corn tortilla, crackers, flour tortilla, fried flour tortilla, graham crackers, macaroni, oatmeal, pastry, rice, sopa[b], spaghetti, sweet bread, taco shell	agave[b], beets, cabbage, carrots, cassava[b], chilis, corn, green tomatoes, iceberg lettuce, jicama[b], onion, peas, potato, prickly pear cactus leaves, purslane[b], squash, sweet potatoes, tomato, turnips	apple, avocado, banana, cherimoya[b], guava[b], mango, orange, papaya, pineapple, plantain[b], zapote[b]	cheddar cheese, custard, evaporated milk, ice cream, jack cheese, powdered milk, queso blanco, fresco, or mexicano[b]	beef, black beans, chicken, chorizo[b], eggs, fish, garbanzo beans, kidney beans, lamb, nuts, peanut butter, pinto beans, pork, refried beans

Overall, about 80 percent of the calories in traditional Chinese fare comes from grains, legumes, and vegetables while the other 20 percent comes from animal meats, fruits, and fat. In southern China, rice can be easily produced and provides the bulk of the complex carbohydrate in the diet. In northern areas, where wheat grows readily, noodles, dumplings, and steamed buns are staples. In all regions of the country, fruits and vegetables are eaten in abundance. Typically, they are not eaten raw, but rather are steamed, added to soups, or stir-fried in peanut or corn oil or (less frequently) lard. In addition, vegetables are often salted, pickled, and dried, a practice resulting from lack of the facilities needed to refrigerate and transport fresh produce across the country. As for meat and other protein foods, pork is considered the staple meat, though poultry, eggs, lamb, and fish are eaten when available. Tofu, or soybean curd, is another staple often added to stir-fries or fermented into sauces. Dairy foods, such as milk and cheese, have never been part of the Chinese diet.

Traditional Chinese meals follow basically the same pattern. Breakfast might be rice congee (rice gruel containing bits of meat), a salty side dish such as pickles, and tea. Lunch typically consists of soup, rice, and mixed dishes made with vegetables and fish, meat, or poultry; dinner is usually a larger version of lunch, occasionally followed by fresh fruit.

Four schools of cooking have developed within China as a result of differences in climate, food production, religion, and custom: Peking, Shanghai, Szechwan or Hunan, and Cantonese. *Peking cooking* comes from the north and northeast part of China and is distinguished by use of garlic, leeks, and scallions. It has given rise to familiar dishes such as Peking duck and spring rolls. *Shanghai cuisine*, from the east and coastal areas, is characterized by "red cooking"—braising foods with large amounts of soy sauce and sugar. Pickled and salted vegetables are also frequently served with meat. In the west and central part of China, people favor the *Szechwan or Hunan cooking* style, in which chili peppers and hot pepper sauces are added liberally to dishes, and the food tends to be spicy and oily. The cuisine of southern China, known as *Cantonese cooking*, is the style Americans know best. Because foods are usually steamed or stir-fried and chicken broth is often used as a cooking medium, Cantonese food tends to be the least fatty. One popular example of Cantonese food is *dim sum*, steamed or fried dumplings stuffed with pork, shrimp, beef, sweet paste, or preserves and steamed or fried.[17]

The regional differences in cooking styles and eating habits among people living in rural China make the country fertile ground for scientific research into the effects of diet on disease. Another reason rural China is such an ideal place to conduct research is that most people residing in rural China today spend their entire lives within the vicinity of the community where they were born. In addition, eating habits are dictated by climate and environment because the country has little or no means of transporting foods from one region to another. Thus, scientists have been able to carefully study eating patterns and disease rates in an attempt to see how the two are related. One of the largest such studies was

© Tony Freeman/PhotoEdit

FIGURE 2-10

A Chinese American MyPlate Food Guide[a]

[a] Empty calorie limit can be used for bacon fat, butter, coconut milk, duck sauce, honey, lard, maltose syrup, sesame paste, suet, and sugar.
[b] These foods are described in the Miniglossary.

SOURCE: Adapted from www.ChooseMyPlate.gov, 2011 and the Pyramid Packet © 1993, Penn State Nutrition Center, 417 East Calder Way, University Park, PA 16801.

ChooseMyPlate.gov

GRAINS	VEGETABLES	FRUITS	DAIRY	PROTEIN FOODS
amaranth[b], barley, bing[b], dumplings, fried rice, glutinous rice[b], mantou[b], noodles (including cellophane noodles[b], rice sticks[b], rice vermicelli[b]), rice congee, rice flour, steamed rice, sorghum, wonton wrappers	bamboo shoots, black mushroom, bok choy[b], cabbage, celery, chiles, Chinese broccoli[b], choy sum[b], eggplant, garlic, ginger, green beans, leek, mustard greens, okra, onions, Oriental radish[b], peas, pickled cucumber, potatoes, scallions, spinach, sprouts, taro[b], tomatoes, turnip, water chestnut, watercress, yard-long beans[b]	carambola, guava, jujube[b], litchi[b], longan[b], mango, orange, papaya, persimmon, pummelo, watermelon	buffalo milk, cow's milk, fish bones, soybean milk, yogurt	bean paste, beef, chestnuts, chicken, duck, eggs, fish (e.g., carp, catfish) lamb, legumes (e.g., mung beans, soybeans), shellfish and other seafood (e.g., shrimp, squid), squab

led by T. Colin Campbell, PhD, of Cornell University. Begun in 1983, the research was carried out in 130 villages located in 65 counties of rural China and included 6,500 adults. Findings suggest that the traditional, plant-based Chinese diet is associated with low rates of many of the chronic diseases that plague Americans, such as heart disease and some types of cancer.[18]

To be sure, Chinese food served in American Chinese restaurants is a far cry from the type eaten daily by rural Chinese people. Many Chinese restaurant meals are swimming in oil and contain much more meat and poultry and fewer vegetables than "real" Chinese food. Consider that a typical American Chinese meal might include wonton soup, barbecued spareribs, chicken lo mein, and fried rice—chalking up some 1,400 calories and more than 80 grams of fat. A typical meal eaten in rural China, in contrast, would likely contain a heaping portion of rice, along with fiber- and nutrient-rich vegetables and less than an ounce of meat or fish, and would thereby contain only a fraction of the fat.[19] Americans who want to enjoy both the flavor and health benefits of traditional Chinese cuisine can "stretch" one of the many delicious vegetable-based dishes with relatively large portions of rice and go easy on deep-fried appetizers such as egg rolls. People who enjoy cooking can also follow the Chinese people's lead and make low-fat, high-carbohydrate stir-fries with lots of vegetables, little oil, and small amounts of meat, poultry, or seafood. People who are sodium conscious may also want to go easy on soy sauce, which contains large amounts of the mineral, or try some of the reduced-sodium soy sauces on the market. Figure 2-10 shows Chinese foods placed into the five food groups of MyPlate.

How does Italian food rate? I've heard that olive oil in Italian pasta dishes is good for health.

Much of the Italian food served in the United States runs very high in fat and calories. Rich cream sauces in dishes such as fettuccine Alfredo; an abundance of cheese and sausage in entrees, including meat-topped pizza and lasagna; and liberal use of olive oil to prepare all manner of pasta and other Italian fare can chalk up extraordinary amounts of fat and calories.

With a little modification, however, the Italian food so many Americans love can easily fit into a healthy diet. For example, substituting vegetables for sausage and pepperoni on pizza and in pasta sauces and lasagna reduces saturated fat content considerably. Using reduced-fat cheeses, such as part-skim mozzarella and ricotta and low-fat cottage cheese, as substitutes for the high-fat versions called for in many Italian recipes also helps skim some of the fat. When it comes to olive oil, a staple in traditional Italian cuisine, it's true that using it instead of butter is preferable from a health standpoint. Figure 2-11 shows a food guide for common Italian American foods.

© Felicia Martinez/PhotoEdit

(Continued)

An Italian American MyPlate Food Guide[a]

[a] Empty calorie limit can be used for butter, olives, olive oil, and heavy cream.

SOURCE: Adapted from www.ChooseMyPlate.gov, 2011 and the Pyramid Packet © 1993, Penn State Nutrition Center, 417 East Calder Way, University Park, PA 16801.

GRAINS	VEGETABLES	FRUITS	DAIRY	PROTEIN FOODS
Rice, linguini, macaroni, spaghetti, focaccia, polenta, Italian bread	Asparagus, artichokes, broccoli, cabbage, capers, carrots, cauliflower, celery, eggplant, mushrooms, onions, peppers, tomatoes, zucchini	Apples, figs, grapes, oranges, peaches, pears, strawberries	Cheeses: mozzarella, parmesan, provolone, ricotta	Beef, calamari, chicken, fava beans, lobster, mussels, lentils, shrimp, perch, sausage

ChooseMyPlate.gov

For reasons that are explained in detail in Chapter 5, olive oil and other vegetable oils are less likely than certain other types of fat, such as butter and lard, to boost blood cholesterol levels. Some scientists even believe that people can eat large amounts of fat—in the neighborhood of 35 to 40 percent of total calories—without raising their risk of heart disease, as long as the predominant fat is olive oil. This line of thinking stems from research conducted in southern Italy and the Greek island of Crete, as well as southern regions of other European nations, including France, North African areas such as Morocco and Tunisia, and Middle Eastern countries such as Israel and Syria. Collectively known as the *Mediterranean region*, these countries share an overall dietary pattern that includes an abundance of fruits and vegetables; breads and other grains; beans, nuts, and seeds; low to moderate amounts of cheese, yogurt, fish, and poultry; small amounts of red meat; moderate consumption of wine; and liberal use of olive oil. Scientists are fascinated with this region because people living there historically have enjoyed long lives and low rates of chronic diseases. That was particularly true in Crete during the 1950s and 1960s. At that time, researchers found that residents of the island shared one of the lowest rates of heart disease and cancer ever recorded and one of the longest life expectancies. That held true despite the fact that their diets derived nearly 40 percent of calories from fat—well above the amount recommended by major U.S. health organizations.[20]

With this historical perspective in mind, some scientists recommend that Americans adopt a "Mediterranean diet" that includes lots of fruits, vegetables, and grains; little meat and other flesh food; moderate amounts of wine; and generous amounts of olive oil (see Figure 2-12). This advice has sparked a good deal of controversy within the scientific community, however. Many nutrition experts question the wisdom of recommending such a diet in America, where excess fat from olive oil and other sources might contribute to the epidemic of obesity in this country. In addition, many experts

point out that diet is not the only lifestyle factor that may have promoted the long, healthy lives of the people of Crete and Italy several decades ago. For example, the residents of this region farmed the land and thereby engaged in far more physical activity than the average American, a factor that probably played a role in their health and longevity. In addition, unlike many modern-day Americans, they enjoyed social support of extended networks of family and friends, another part of their lifestyle that probably contributed to their well-being. Given these caveats, major U.S. health organizations, including the American Dietetic Association, the American Heart Association, and the U.S. Department of Health and Human Services, maintain that Americans should keep the amount of total fat—whether in the form of olive oil or anything else—to no more than 20 to 35 percent of total calories.

Our advice is that you consider adopting the aspects of the Mediterranean diet that virtually all nutrition experts advocate: Eat an abundance of produce, grains, and legumes combined with moderate amounts of dairy products and relatively smaller amounts of meats, poultry, and fish.

I have heard that the traditional Indian diet is vegetarian; is that true?

"Traditional Indian diet" can mean many different things. Other than the heavy use of vegetables and spices throughout India, Indian cuisines share few other characteristics. Geographic location, religious beliefs, and availability all contribute to the varied nature of the Indian diet. Furthering the diversity, Indian cooking rarely follows recipes, relying more on the skill of the cook and the availability of ingredients. These phenomena have resulted in a mix of cuisines that are both tasty and various.

The history of vegetarianism in India dates back thousands of years. Many believe it started with the introduction of Buddhism and Jainism in the sixth century B.C. These religions, which are closely related, are strong advocates of the sanctity of all forms of

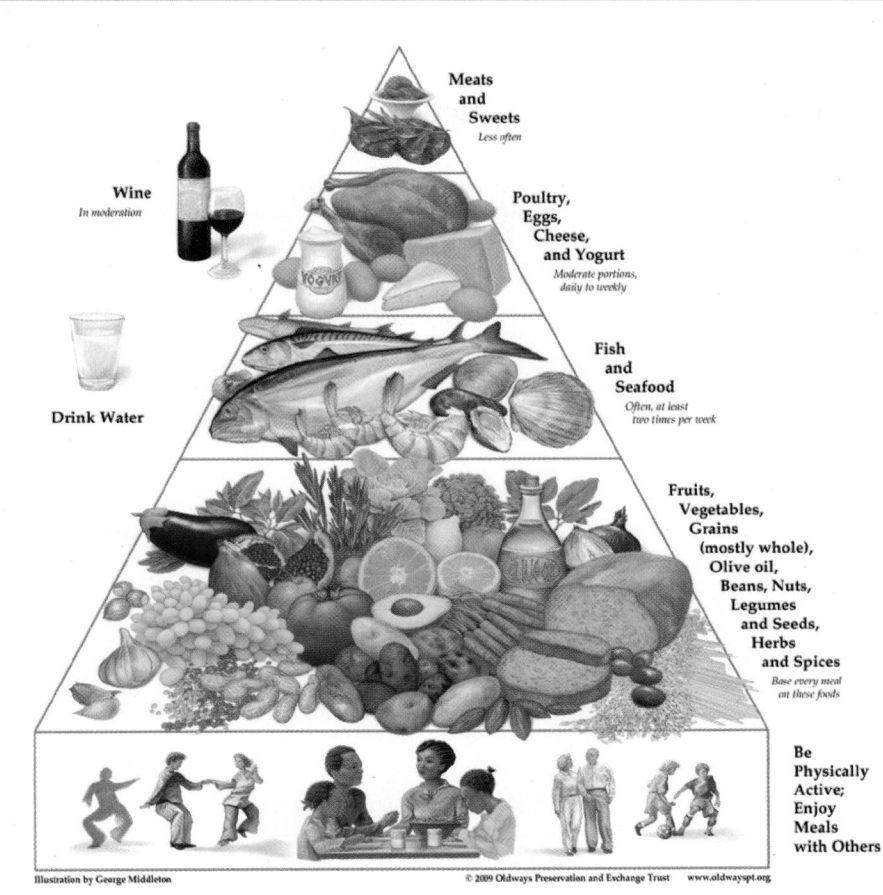

FIGURE 2-12

Mediterranean Food Guide Pyramid

*Other oils rich in monounsaturated fats such as canola or peanut oil can be substituted for olive oil. People who are watching their weight should limit their oil consumption.

SOURCE: Adapted with permission. Copyright © 2009 Oldways Preservation & Exchange Trust, http://oldwayspt.org.

Meats and Sweets
Less often

Wine
In moderation

Poultry, Eggs, Cheese, and Yogurt
Moderate portions, daily to weekly

Drink Water

Fish and Seafood
Often, at least two times per week

Fruits, Vegetables, Grains (mostly whole), Olive oil, Beans, Nuts, Legumes and Seeds, Herbs and Spices
Base every meal on these foods

Be Physically Active; Enjoy Meals with Others

life and are therefore opposed to the slaughter of animals for consumption. However, it is not uncommon for those who live along the coast to incorporate fish into their diet. Likewise, the more northern territories are known to eat much more meat than their southern counterparts. These food choices are most likely due to availability and climate differences.

Nonetheless, the idea that the Indian diet is vegetarian is not far from the truth. Anthropology sources clearly show the predominance of vegetarianism, and even those who do partake in meat consumption do so in moderation, usually no more than one day a week. More often, meat consumption is reserved for special occasions like weddings and celebrations, averaging about four times a year. Another interesting fact is that women are less likely to eat meat than men. This practice stems from the belief that meat causes masculine traits such as aggressiveness.

This feature cannot possibly encompass the vast culture and history that influences Indian cuisine. With the second largest population in the world, India is a multifaceted land with traditions and beliefs that would take books upon books to explain. Even so, its cuisines are a good illustration of the history of India, as there are so

many distinctive characteristics. Overall, the diet is based on complex carbohydrates and rich in fiber-full vegetables and fruits. In any Indian restaurant, you are bound to find a flavorful and healthy dish that will satisfy you! Some traditional Indian foods are shown in Figure 2-13.

As with most ethnic cuisines, the Americanization of Indian ingredients has not been very positive from a health perspective. With the addition of more animal-based protein sources and fat, and a decrease in fiber content, many of the advantages of the traditional Indian diet have been lost. However, those who prepare

© Bonnie Kamin/PhotoEdit

(Continued)

FIGURE 2-13

Fitting Indian Foods into the MyPlate Food Guide[a]

[a] Empty calorie limitcan be used for ghee,[b] coconut oil, and similar items.
[b] These foods are described in the Miniglossary.

SOURCE: Adapted from www.ChooseMyPlate.gov, 2011 and the Pyramid Packet (1993) Penn State Nutrition Center, 417 East Calder Way, University Park, PA 16801.

GRAINS	VEGETABLES	FRUITS	DAIRY	PROTEIN FOODS
basmati rice, bulgur wheat, chapati[b], millet, rice, roti[b]	broccoli, cabbage, carrots, cauliflower, cucumber, eggplant, onions, plantains, potato, squash, tomatoes, turnips	bananas, chutney, dates, figs, grapes, mango, melon, papaya	lassi[b], milk, yogurt	almonds, beef, cashews, chicken, chickpeas, lamb, lentils, shrimp

Indian cuisine in a traditional manner will receive the benefits of a high-carbohydrate, low-fat, low-cost diet that is as tasty as it is healthy.

What about "soul food" or "southern" cooking? Do most African Americans eat lots of it?

The term *soul food* was coined in the mid-1960s to promote ethnic pride and solidarity among African Americans, but the origins of soul food date back to a much earlier time in history. Black-eyed peas, grits (coarsely ground cornmeal), collard greens, okra, and other soul foods evolved from the traditional diet of West African slaves living in the South. When West Africans were brought to the United States to work the fields, their dietary habits revolved around foods provided by slave owners. Corn was commonly given as a staple, and it was prepared in many forms, such as grits, cornmeal pudding, and hominy (hulled, dried corn kernels with certain parts removed). Also, because salt pork was frequently a staple supplied to slaves, pork fat was used to fry and flavor greens,

breads, stews, and other foods. In addition, some owners allowed their slaves to grow vegetables in small plots. Slaves who farmed such plots grew American vegetables, including cabbage, collard and mustard greens, sweet potatoes, and turnips, and the slaves introduced okra and black-eyed peas, two West African favorites, to the United States. The slaves who worked as cooks in the homes of slave owners made popular other southern favorites, such as fried chicken and fried catfish.

After emancipation, eating patterns of African Americans did not change significantly, and they represent much of what we think of as southern cuisine today. The underpinnings of this eating style, which is very high in fat, remain corn-based dishes, greens, pork, and pork products such as chitterlings (chitlins—pig intestines) and ham hocks. Food habits of African Americans around the country tend to be influenced more by economic status and geographic location than by heritage, although soul food remains a symbol of identity and heritage for many African Americans.[21] Figure 2-14 shows traditional African American foods in the five food groups.

FIGURE 2-14

Traditional African American Foods and the MyPlate Food Guide[a]

[a] Empty calorie limit can be used for butter, candy, fruit drinks, lard, meat drippings, chitterlings,[b] soft drinks, and vegetable shortening.
[b] These foods are described in the Miniglossary.

SOURCE: Adapted from www.ChooseMyPlate.gov, 2011 and the Pyramid Packet © 1993, Penn State Nutrition Center, 417 East Calder Way, University Park, PA 16801.

GRAINS	VEGETABLES	FRUITS	DAIRY	PROTEIN FOODS
biscuits, cookies, cornbread, grits[b], pasta, rice	beets, broccoli, cabbage, corn, green peas, greens, hominy[b], okra, potatoes, spinach, squash, sweet potatoes, tomatoes, yams	apples, bananas, berries, fruit juice, peaches, watermelon	buttermilk, cheese, milk	blackeyed peas, beef, catfish, chicken, crab, crayfish, eggs, kidney beans, peanuts, perch, pinto beans, pork, red beans, red snapper, salmon, sardines, shrimp, tuna, turkey

© Bill Aron/PhotoEdit

I see many foods marked "kosher" in the supermarket, especially during Jewish holidays such as Passover. Are kosher foods better for health than regular food items? Do other religious groups eat special foods?

As pointed out in Chapter 1 (refer to Table 1-7), food is part of the symbolism and traditions of many major religions. In the predominantly Christian United States, however, most people's eating habits are not dictated by religion to a large extent.

Nevertheless, during the past few decades, "Jewish" foods have been growing in popularity among Americans of various religious backgrounds. For instance, bagels are now staple breakfast menu items in the United States.

Most traditional Jewish foods eaten in America come from a particular group known as *Ashkenazic Jews*—Jews from central and eastern European countries such as Russia, Germany, Poland, and Romania. (The other major group is Sephardic Jews, who come mainly from Spain and Portugal.) Although few Jewish people in the United States strictly abide by all dietary laws Judaism prescribes, many adhere to at least some rules of *kashrut,* biblical ordinances specifying which foods are *kosher,* or fit to eat.

Most people assume that the laws of kashrut were set forth to protect the health of the Jewish people. For example, a popular misconception is that kashrut forbids consumption of pork products because eating undercooked pork can cause serious illness. In truth, however, Jewish dietary laws are considered divine commandments set forth to maintain spiritual, not physical, health. Foods labeled as kosher are not necessarily more healthful than their unmarked counterparts. Instead, the designation "kosher" indicates that a food has been prepared in accordance with the basic tenets of kashrut. For instance, one principle of kashrut is separation of milk and meat products, meaning that an item containing, say, both ground beef and cheese would not be kosher.

Another tenet is selection of appropriate meat, poultry, and seafood items. Only animals with cloven hooves that chew their cud are allowed to become food: cattle, sheep, goats, and deer. Chicken, turkey, goose, pheasant, and duck can be kosher, but birds of prey cannot. Seafood with both fins and scales can be kosher, whereas shellfish is forbidden.

Finally, as a result of the salting process used to prepare kosher animal foods, many traditional Jewish foods are high in sodium. Herring, smoked fish, canned beef, tongue, corned beef, and other deli-style meats are examples. Other traditional Jewish foods, many of which are high in fat, include schmaltz (chicken fat), knishes (potato pastry filled with ground meat or potato), cream cheese, and chopped liver.[22]

Figure 2-15 shows how some traditional Jewish foods fit into the MyPlate food guide.

FIGURE 2-15

Fitting Jewish American Foods into the MyPlate Food Guide[a]

[a] Empty calorie limit can be used for cream cheese, honey, jelly, hard margarine, marmalade, preserves, schmaltz,[b] sherbet, sour cream, and sugar.
[b] These foods are described in the Miniglossary.

SOURCE: Adapted from www.ChooseMyPlate.gov, 2011 and the Pyramid Packet © 1993, Penn State Nutrition Center, 417 East Calder Way, University Park, PA 16801.

ChooseMyPlate.gov

GRAINS	VEGETABLES	FRUITS	DAIRY	PROTEIN FOODS
bagel, barley, bialy[b], blintz, bulgur, challah[b], crepe, dumplings, hard rolls, honey cake, kasha[b], matzoh[b], noodle pudding, pastry, pita bread, pumpernickel bread, rye bread	artichokes, asparagus, beets/borscht, broccoli, Brussels sprouts, cabbage, carrot, cauliflower, corn, garlic, green beans, greens, leeks, olives, onion, peas, peppers, pickles, potatoes, sorrel, spinach, squash, sweet potatoes, tomatoes, turnips, yams	bananas, citrus fruits, dates, dried apples, dried apricots, dried pears, figs, grapes, melons, prunes, raisins	cottage cheese, edam cheese, farmer's cheese, gouda cheese, milk, Swiss cheese, yogurt	almonds, beef, beef tongue, brisket, chickpeas, chopped liver, corned beef, dry beans, eggs, gefilte fish[b], herring, lentils, lox[b], pastrami, poultry, salmon, sardines, smelt, smoked fish, split peas, tripe, veal

(Continued)

Miniglossary of Foods

fusion cuisine a term used to describe food that combines the elements of two or more cuisines—say, European and Oriental—to create a new one

MEXICAN AMERICAN

amaranth a golden-colored grain

agave (uh-GAH-vee) a plant with spiny-margined leaves and flower

bolillo (buh-LEE-oh) a roll-like bread often used instead of tortillas or to make sandwiches

burritos warm flour tortillas stuffed with a mixture of egg, meat, beans, and/or avocado

cassava a starchy root that is never eaten raw because it must be cooked to eliminate its bitter smell

cherimoya a fruit with a rough green outer skin and sherbet-like flesh

chilaquiles (chee-lah-KEE-lehs) tortilla casserole often made with eggs or meat

chiles relleños (CHEE-lehs Ray-EN-yose) roasted mild green chili peppers stuffed with cheese, dipped in egg batter, and fried

chorizo (chuh-REE-zoh) spicy beef or pork sausage

guava a sweet, juicy fruit with green or yellow skin and red or yellow flesh

jicama (HE-cah-mah) a crisp root vegetable of the bean family that is tan outside and white inside and is always eaten raw; as popular in Mexico as the potato is in the United States

plantain a greenish, starchy banana; because it is starchy even when ripe, it is never eaten raw and is usually pan-fried

purslane a leafy vegetable that can be used in salads or cooked like spinach

queso blanco, fresco, or Mexicano soft white cheese made of part-skim milk

sopa rice or pasta that is fried and cooked in consommé

zapote (sah-POH-tay) an apple-sized fruit with green skin and black flesh

CHINESE AMERICAN

bing thin pancakes

bok choy a vegetable with broad, white or greenish-white stalks and dark-green leaves; also called *Chinese chard*

cellophane noodles thin, translucent noodles made from mung beans

Chinese broccoli a green leafy vegetable often stir-fried; also called *Chinese kale*

choy sum a bright green vegetable commonly stir fried; also called *field mustard* or *Chinese flowering cabbage*

dim sum dumplings stuffed with pork, shrimp, beef, sweet paste, or preserves and steamed or fried

glutinous rice short-grained, opaque white rice that turns sticky when cooked

jujube (JOO-joob) Chinese date

litchi (LEE-chee) small, round fruits with orange-red skin and opaque white flesh; also called *litchee* or *lychee*

longan a small round fruit with smooth brown skin and clear pulp

mantou steamed bread

Oriental radish a large, cylindrically shaped vegetable with smooth skin; also called *daikon*

rice sticks flat, opaque, wide noodles made from rice flour

rice vermicelli (vur-mi-SEL-ee) thin white noodles made from rice flour

taro a starchy vegetable with brown, hairy skin and a pink-purple interior

yard-long beans thin, tender string beans that grow to as long as 18 inches

INDIAN

chapatti a flat, unleavened bread made with finely milled whole-meal flour

ghee a clarified butter without any milk solids or water

lassi a milkshake-like drink made from yogurt

roti an unleavened flatbread, resembling a tortilla, usually made from whole-wheat flour

AFRICAN AMERICAN

chitterlings (chitlins) pig intestine

grits coarsely ground cornmeal

hominy hulled, dried corn kernels

JEWISH AMERICAN

bialy a flat breakfast roll that is softer than a bagel

challah (KAH-luh) an egg-containing yeast bread, often braided, typically served on the Sabbath and holidays

gefilte fish (guh-FIL-tuh) a chopped fish mixture often made with pike and whitefish as well as matzoh crumbs, eggs, and seasonings

kasha cracked buckwheat, barley, millet, or wheat that is served as a cooked cereal or potato substitute

knish a potato pastry filled with ground meat, potato, or kasha

kosher fit, proper, or in accordance with religious law

lox smoked salmon

matzoh a cracker-like bread eaten most often at Passover

schmaltz chicken fat

Nutrition on the Web

www.cengage.com/nutrition/boyle/personalnutrition8e

Go to the *Personal Nutrition* section to check for the latest updates to chapter topics or to access links to related websites.

www.nap.edu

The National Academy Press site contains the latest Dietary Reference Intakes (DRI) reports.

www.nal.usda.gov/fnic

The Food and Nutrition Information Center (FNIC) provides credible, accurate, and practical information about nutrition and health. Browse by Subject to access links:

- Click on Dietary Guidance and then select Dietary Guidelines.
- Click on Dietary Guidance and then select MyPlate. Scroll down to access Ethnic/Cultural and Special Audience Food Guides.
- Click on Dietary Guidance and then select Dietary Reference Intakes.
- Click on Food Composition to access Food Composition resources.
- Click on Food Labeling for links to Food Labeling resources.

www.ChooseMyPlate.gov

This interactive website provides tips and resources for planning a healthy diet using MyPlate. Browse Interactive Tools to access links to the following:

- MyPlate Food Planner—an online menu-planning tool.
- MyPlate Tracker—an online dietary and physical activity assessment tool.

www.ars.usda.gov/ba/bhnrc/ndl

Click on Search for free food analyses. Just type in the food you want to analyze, and get a breakdown of its calories, fat, fiber, protein, vitamins, and minerals.

www.fao.org/ag/AGN/nutrition/education_guidelines_country_en.stm

Dietary guidelines and food guides from around the world.

www.cdc.gov/nutrition/index.html

The Centers for Disease Control and Prevention (CDC) maintains a nutrition home page that includes many links.

www.health.gov

This website sponsored by the U.S. Department of Health and Human Services (HHS) is a portal to websites of a number of multi-agency government health initiatives pertaining to health and nutrition.

http://nat.illinois.edu/

A free diet analysis program developed at the University of Illinois–Urbana/Champaign. This site allows anyone to analyze the various nutrients in the foods they eat.

www.fda.gov/Food/LabelingNutrition/default.htm

Useful facts about food labels; updates on label health claims.

Menu of Online Study Tools

A variety of study tools for this chapter are available at our website to deepen your understanding of chapter concepts. Go to

www.cengage.com/sso

to find

- Practice tests
- Flashcards
- Glossary
- Web links
- Animations
- Chapter summaries, learning objectives, and crossword puzzles.

Diet Analysis+ in Action

Go to the Profile report in Diet Analysis+ and record the daily recommended intakes for the following nutrients:

- calories
- protein
- fat
- fiber
- water
- sodium
- vitamin C

Using the three-day diet tracking you created in Chapter 1's assignment, go again to the food guide report and record which food groups were in the preferred range, which food groups were in the less-than-preferred-range, and which food groups exceeded the range.

Make a copy of your Profile (the foods you have recorded will also be copied). Now add one fruit to each meal for all three days. Once you have done that, look again at the food guide report. Have any of the ranges improved?

Try other changes—for example, substitute at least one whole-grain food in each meal for a less-nutritious choice (whole-grain toast instead of a bagel at breakfast; brown rice instead of French fries at lunch, etc.). What difference does that change make in the ranges? Based on this assignment, are there any changes you want to make in your diet?

Anatomy for Nutrition's Sake

Ask Yourself . . .

Which of the following statements about nutrition are true, and which are false? For each false statement, what is true?

1. Essential nutrients are those nutrients that can be made by the body.
2. The body's cells need nutrients 24 hours a day.
3. Each part of the digestive tract has specific functions.
4. Digestion begins in the mouth.
5. It takes approximately two days for a meal to empty from the stomach.
6. Nutrients are absorbed in the small intestine.
7. The small intestine is longer than the large intestine.
8. The small intestine is lined with finger-like projections (called villi) that increase the absorptive area so that if the lining of the small intestine could be spread out flat, it would cover an area the size of a tennis court.
9. The liver determines the metabolic fate of every nutrient absorbed.
10. The pancreas plays no role in the digestive process.

Answers found on the following page.

The body is composed of millions of **cells**, and not one of them knows anything about food.* Although you get hungry for meat, milk, or bread, each cell of your body sits in its place waiting until the nutrients it needs pass by. Each cell is a self-contained, living entity (Figure 3-1); however, each depends on the rest of the body to supply its needs. The body cells' most basic needs, always, are for energy (or fuel), oxygen, and water.

After energy, cells need a constant supply of nutrients in order to synthesize new cells and new cell parts. These necessary nutrients cannot be made by the cells and are, therefore, **essential nutrients**. Essential nutrients must be supplied by the foods we eat. Need for these nutrients is a limitation of our heredity that we

Every cell in our body was once in the food we ate.

HELEN GUTHRIE

cells the smallest units in which independent life can exist. All living things are single cells or organisms made of cells.

essential nutrients compounds that must be obtained from food because they cannot be synthesized by the body in amounts sufficient to meet physiological needs.

---

*Material in this chapter was drawn generally from A. C. Guyton and J. E. Hall, *Textbook of medical physiology,* 11th ed. (Philadelphia: Elsevier/Saunders, 2006); and L. Sherwood, *Human physiology: From cells to symptoms,* 6th ed. (Belmont, CA: Thomson Brooks/Cole, 2007).

FIGURE 3-1

A Typical Cell

A membrane encloses each cell's contents.

These fingerlike projections are typical of cells that absorb nutrients in the intestines.

A separate inner membrane encloses the cell's nucleus.

Inside the nucleus, the genetic material, DNA, contains the genes. The genes control the inheritance of the cell's characteristics and its day-to-day workings. The DNA is faithfully copied each time the cell duplicates itself.

On these membranes, instructions from the genes are translated into proteins that perform functions in the body.

Many other structures are present. This is a mitochondrion, a structure that takes in nutrients and releases energy from them.

cannot change, and this need underlies the first principle of nutrition: Whatever foods we choose, they must provide energy, water, and essential nutrients. In a sense, the body is merely a system organized to provide for these cellular needs.

Cells are organized into *tissues* designed to perform specialized tasks. For example, some cells are organized to form muscle tissue that helps move our bodies from place to place. Tissues can be organized to form *organs*. One organ, the heart, relies on muscle tissues, nerve tissues, and connective tissues to pump blood throughout the body. When several organs work together cooperatively, these organs are considered parts of a body *system*. One such system, the digestive system, consists of organs and tissues working together to supply necessary energy, water, and essential nutrients to every cell in our bodies.

The Digestive System

You may eat meals only two or three times a day, but your body's cells need nutrients 24 hours a day. Providing the needed nutrients requires cooperation of millions of specialized cells. When the body's cells are deprived of fuel, specific nerve cells in the **hypothalamus** (part of the brain) detect this condition and generate nerve impulses that signal hunger to the conscious part of the brain. Hunger and appetite are experienced in the **cortex** of the brain—the thinking, outer layer (Figure 3-2). Hunger

hypothalamus (high-poh-THALL-uh-mus) a part of the brain that senses a variety of conditions in the blood, such as temperature, salt content, glucose content, and others, and signals other parts of the brain or body to change those conditions when necessary.

cortex an outer covering; in the brain, that part in which conscious thought takes place.

ASK YOURSELF ANSWERS: 1. False: Essential nutrients cannot be made by the body and must be obtained from food. **2.** True. **3.** True. **4.** True. **5.** False: It takes approximately four hours for a meal to empty from the stomach. **6.** True. **7.** True. **8.** True. **9.** True. **10.** False: The pancreas produces digestive enzymes and bicarbonate and secretes them into the small intestine.

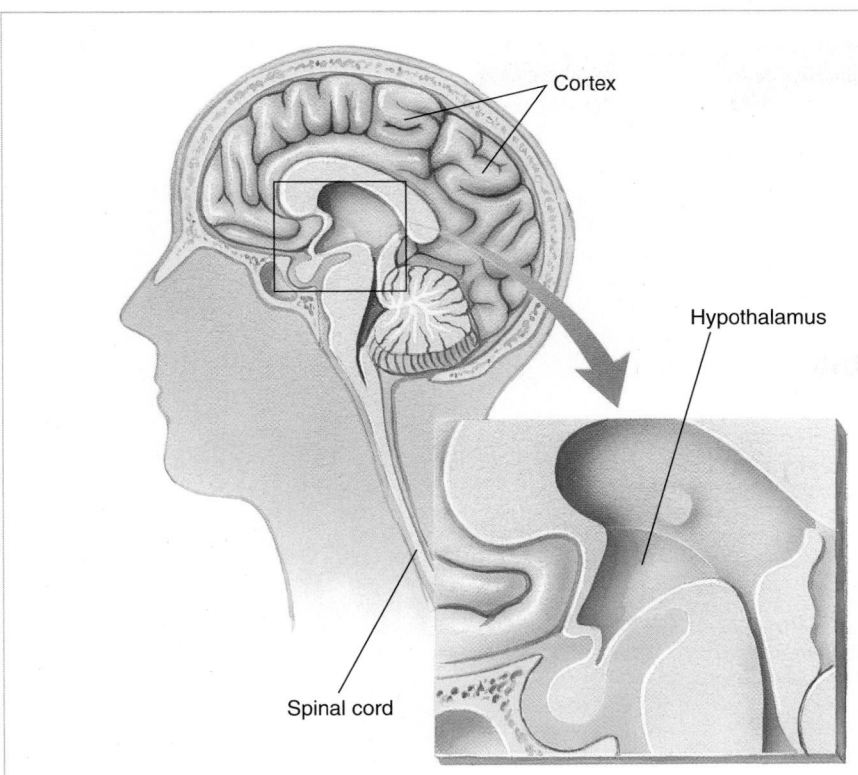

FIGURE 3-2

The Hypothalamus

The hypothalamus monitors the body's conditions and sends signals to the brain's thinking portion, the cortex, which decides on actions.

Cortex

Hypothalamus

Spinal cord

(involuntary action) has primary control over how much we eat, while the type of food we look for and eat is determined by appetite (voluntary action).

The gastrointestinal (GI) tract provides the body with a constant supply of water and nutrients. Accomplishing this seemingly simple task entails:

- Passage of food through the GI tract

- Secretion of digestive juices and enzymes

- Digestion of the food

- Absorption of water and nutrients

- Circulation of blood through the body to distribute absorbed substances

- Control of these functions

The Gastrointestinal Tract   Each part of the digestive tract has specific functions, as detailed in Figure 3-3. Some parts and functions involve basic passage of food; other parts serve as temporary storage units, and still other parts digest food and absorb the resulting nutrients.

The outer layer of the GI tract wall (called the *mesentery*) protects blood vessels, nerves, and lymphatic vessels. It also attaches the lower parts of the small intestine (jejunum and ileum) and part of the large intestine (transverse colon) to the abdominal wall to prevent the intestines from twisting.

Figure 3-4 shows a typical cross section of the intestinal wall. Movement of food through the GI tract is enabled by the different layers of smooth muscle. These muscles in the GI tract are longitudinal (up and down the length of the GI tract) and circular (around the GI tract). These two sets of muscles work together to spontaneously

FIGURE 3-3

The Digestive System

Fiber	Carbohydrate
Mouth The mechanical action of the mouth and teeth crushes and tears fiber in food and mixes it with saliva to moisten it for swallowing.	The salivary glands secrete a watery fluid into the mouth to moisten the food. The salivary enzyme amylase begins digestion: $$\text{Starch} \xrightarrow{\text{amylase}} \text{small polysaccharides, maltose. . .}$$
Esophagus Fiber is unchanged.	Digestion of starch continues as swallowed food moves down the esophagus.
Stomach Fiber is unchanged.	Stomach acid and enzymes start to digest salivary enzymes, halting starch digestion. To a small extent, stomach acid hydrolyzes maltose and sucrose.
Small intestine Fiber is unchanged.	The pancreas produces enzymes and releases them through the pancreatic duct into the small intestine: $$\text{Polysaccharides} \xrightarrow[\text{amylase}]{\text{pancreatic}} \text{disaccharides.}$$ Then enzymes on the surfaces of the small intestinal cells break disaccharides into monosaccharides, and the cells absorb them: $$\text{Maltose} \xrightarrow{\text{maltase}} \text{glucose + glucose.}$$ $$\text{Sucrose} \xrightarrow{\text{sucrase}} \text{fructose + glucose.}$$ $$\text{Lactose} \xrightarrow{\text{lactase}} \text{galactose + glucose.}$$

Diagram labels: Salivary glands, Mouth, Tongue, Airway to lungs, Esophagus, Stomach, Liver, Gallbladder, Pancreas, Pancreatic duct, Pyloric sphincter, Bile duct, Colon (large intestine), Appendix, Small intestine, Rectum, Anus

Colon (large intestine)
Most fiber passes intact through the digestive tract to the colon. Here, bacterial enzymes digest some fiber:

$$\text{Some fiber} \xrightarrow[\text{enzymes}]{\text{bacterial}} \text{fatty acids, gas.}$$

Fiber holds water; regulates bowel activity; and binds cholesterol and some minerals, carrying them out of the body as it is excreted with feces.

Fat	Protein	Vitamins	Minerals and Water
Mouth Glands in the base of the tongue secrete a fat-digesting enzyme known as lingual lipase. Some hard fats begin to melt as they reach body temperature.	In the mouth, chewing crushes and softens protein-rich foods and mixes them with saliva to be swallowed.	No action.	The salivary glands add water to disperse and carry food.
Esophagus Fat is unchanged.	No action.	No action.	No action.
Stomach The degree of hydrolysis is slight for most fats but may be appreciable for milk fats. The stomach's churning action mixes fat with water and acid. A gastric enzyme accesses and hydrolyzes a small percentage of fat.	Stomach acid works to uncoil protein strands and activate stomach enzymes. Then the enzymes break the strands into smaller fragments: $$Protein \xrightarrow[\text{HCl}]{\text{pepsin}} \text{smaller polypeptides}$$	Water-soluble vitamins need little action by the digestive organs except absorption in the small intestine. However, vitamin B_{12} requires "intrinsic factor" produced by the stomach in order to be absorbed.	The stomach secretes enough watery fluid to turn a moist, chewed mass of swallowed food into a liquid. Stomach acid acts on iron to make it more absorbable. Vitamin C and a factor in meat also increase iron absorption.
Small intestine The liver secretes bile; the gallbladder stores it and releases it through the common bile duct into the small intestine when fat arrives there. The bile emulsifies the fat, making it ready for enzyme action. The pancreas produces fat-digesting enzymes and releases them through the common bile duct into the small intestine. These enzymes split triglycerides into monoglycerides, free fatty acids, and glycerol, which are absorbed.	In the small intestine, the fragments of protein are split into free amino acids, dipeptides, and tripeptides with the help of enzymes from the pancreas and small intestine. Enzymes on the surface of the small intestinal cells break these peptides into amino acids, and they are absorbed through the cells into the blood. The large intestine carries any undigested protein residue out of the body. Normally, practically all the protein is digested and absorbed.	Bile emulsifies fat-soluble vitamins and aids in their absorption with other fats. Water-soluble vitamins are absorbed.	The small intestine, pancreas, and liver add enough fluid so that approximately 2 gallons are secreted into the intestine in a day. Many minerals are absorbed. Vitamin D aids in the absorption of calcium.
Colon Some fat and cholesterol, trapped in fiber, exit in feces.		Bacteria produce vitamin K, which is absorbed.	More minerals and most of the water are absorbed.

FIGURE 3-4

Layers of the GI Tract

Body wall

Peritoneum

Mesentery

Outer longitudinal muscle

Inner circular muscle

Mucosa

Submucosa

Duct of large
accessory digestive
gland (i.e., liver
or pancreas) emptying
into digestive tract
lumen

move food along the GI tract (Figure 3-5). As food works its way along the GI tract, it is mixed with digestive juices and enzymes in preparation for the digestive process. Optimal food processing depends on the amount of time the food remains in each part of the GI tract. These functions are controlled by nervous and hormonal feedback systems, which are discussed later in this chapter.

The innermost layer of the GI tract wall, called the *mucosa*, is made up of cells that serve vital functions in the digestive process. Some cells secrete gastric juices, digestive enzymes, and **mucus** to protect the lining of the GI tract. These cells and their functions are also discussed later.

The Mouth Digestive processes begin in the mouth. The mouth is designed for physical breakdown of foods (mechanical digestion). Teeth cut off a bite-sized portion

mucus (MYOO-kuss) a thick, slippery coating of the intestinal tract lining (and other body linings) that protects the cells from exposure to digestive juices. The adjective form is *mucous*, and the coating is often called the *mucous membrane*.

FIGURE 3-5

Peristalsis

As the wave of peristaltic contraction sweeps down the esophagus, it pushes the food bolus (a mixture of food particles and saliva) into the relaxed area ahead of it, propelling the bolus toward the stomach.

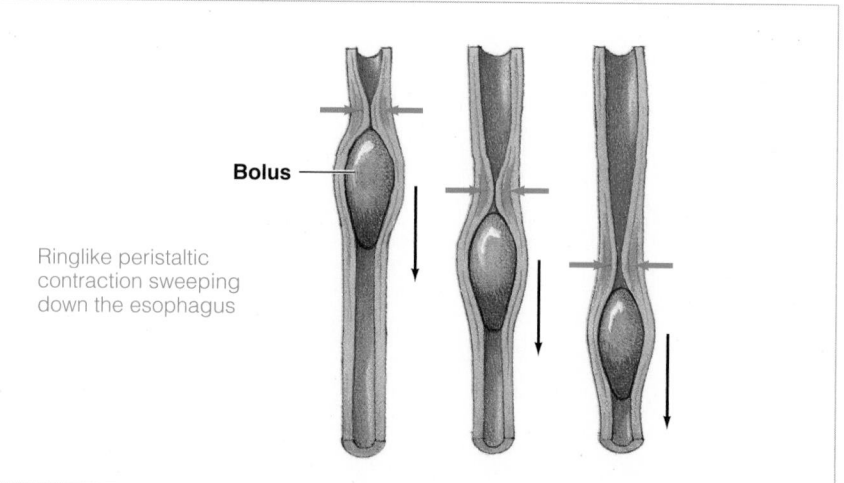

Bolus

Ringlike peristaltic
contraction sweeping
down the esophagus

and then, aided by the tongue, grind it finely enough to be mixed with saliva and swallowed. The significance of chewing becomes apparent when eating raw fruits and vegetables. These foods have indigestible membranes covering the nutrient-containing portions, and these membranes must be broken down before the food can be digested. Chewing also exposes the surface area of food to digestive secretions. This is important because digestive enzymes act only on the surface of food particles. The mouth also contains three salivary glands that secrete salivary amylase (called *ptyalin*), which moistens food during chewing and makes it easier to swallow. This salivary amylase also begins the breakdown (chemical digestion) of complex carbohydrates (starch). This action is stopped when the carbohydrate units reach the stomach (as discussed later).

When food is chewed, the result is a mixture of food particles and saliva called a *bolus*. The food bolus must move from the mouth to the esophagus. This occurs through the process of swallowing, a seemingly simple act that is in actuality a very intricate process (Figure 3-6) because the back of the throat (*pharynx*) is also a part of the respiratory system. The first part of the swallowing process is voluntary because we choose to chew and then, with the help of the tongue, move the bolus toward the pharynx. From there on, the process is pretty much automatic (involuntary) and as a rule cannot be stopped. The trachea (windpipe) closes and the pharynx propels the bolus toward the esophagus. This complete process occurs in less than two seconds!

The Esophagus No digestion takes place in the esophagus. Its purpose is strictly to serve as a vehicle for the food bolus as the bolus travels from the mouth to the stomach. During the swallowing process, a *sphincter* (circular band of muscle fibers that

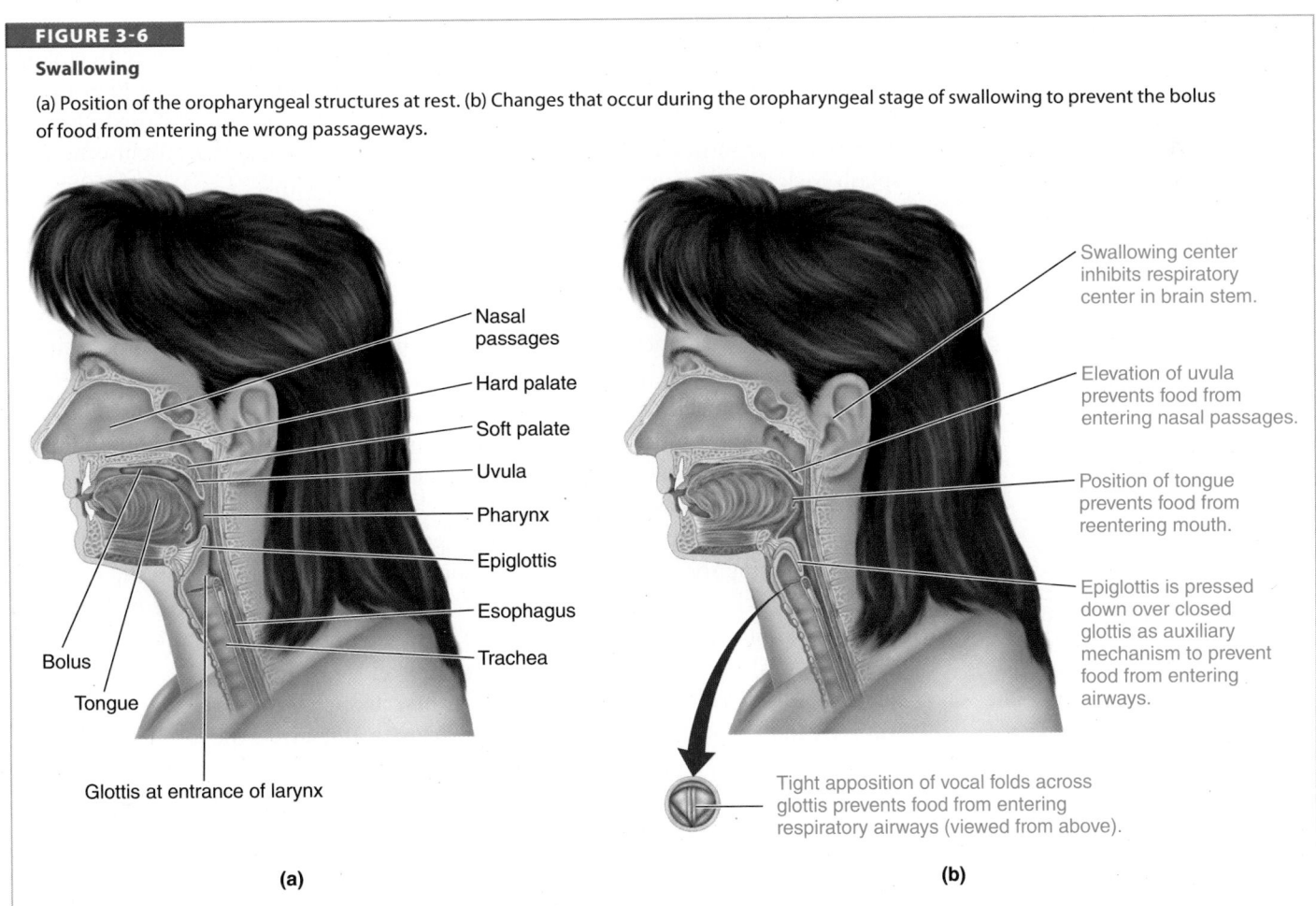

FIGURE 3-6

Swallowing

(a) Position of the oropharyngeal structures at rest. (b) Changes that occur during the oropharyngeal stage of swallowing to prevent the bolus of food from entering the wrong passageways.

Nasal passages

Hard palate

Soft palate

Uvula

Pharynx

Epiglottis

Esophagus

Trachea

Bolus

Tongue

Glottis at entrance of larynx

Swallowing center inhibits respiratory center in brain stem.

Elevation of uvula prevents food from entering nasal passages.

Position of tongue prevents food from reentering mouth.

Epiglottis is pressed down over closed glottis as auxiliary mechanism to prevent food from entering airways.

Tight apposition of vocal folds across glottis prevents food from entering respiratory airways (viewed from above).

(a)

(b)

You've just taken a bite of your lunch, and your roommate says something that is hilarious. Unfortunately, you start laughing at the same time you are trying to swallow that last bite of your cannoli. You can guess what happens next . . . it's not pretty. Pieces of food or small objects can cause choking if they get caught in your throat and block your airway. Oxygen cannot get to your lungs and brain if your airway is blocked. And, if your brain goes without oxygen for more than four minutes, you could experience brain damage or die. When you or someone else is choking, quick action can be lifesaving. How can you prevent yourself or someone else from choking?

1. Be extra careful when eating foods that are easy to choke on. This includes foods like hot dogs, nuts, grapes, raw carrots, popcorn, and hard or sticky candy.
2. Sit down while eating, take small bites, and don't talk or laugh with your mouth full. Does this sound like something your mom would say? Well, mom was right! In addition to describing good manners, this advice will help prevent choking.
3. If babies or toddlers can pick something up, they will probably put it in their mouth, so help keep them safe by picking up anything off the floor that might be dangerous to swallow. This includes items like pen caps, coins, beads, batteries, and toys with small parts. Keep hazards out of their reach, and supervise them when they eat.
4. Learn the Heimlich maneuver. Check for classes on your campus.

constricts a passage or closes a natural opening in the body) at the top of the esophagus relaxes to permit the bolus to enter. Longitudinal and circular muscle layers of the esophagus rhythmically push the bolus down the length of the esophagus to the stomach. This movement is called **peristalsis** (see Figure 3-5). A bolus passing through the esophagus is coated with mucus to make it slippery and easy to swallow. The mucus also protects the lining of the esophagus from being damaged by inadequately chewed foods.

At the lower end of the esophagus (between the esophagus and stomach), circular muscles form another sphincter called the *lower esophageal sphincter (LES)* or *gastroesophageal sphincter*. This sphincter, which is normally constricted or closed, plays an important role in protecting the esophagus from stomach acid that might otherwise splash back into the esophagus.

The Stomach   The stomach carries out three jobs during digestion. It stores a sizeable amount of food until the food can be processed; forms a mixture of food and gastric secretions, which becomes a semiliquid blend called *chyme* (KIME); and controls movement of the chyme into the **small intestine** at a rate suitable for proper digestion and absorption by the small intestine.

An empty stomach has a volume of about $1^1/_2$ ounces or 3 tablespoons. Eat some food, drink some fluids, and the stomach capacity can increase to about 1 liter (a little over 1 quart). How, you ask? The stomach is lined with deep folds (Figure 3-7) that relax and flatten as we eat. Eating more than 1 liter (overeating) causes the stomach to stretch just a little too much and triggers discomfort.

Food entering the stomach from the esophagus triggers peristalsis in the stomach, which causes the stomach to "churn," thereby mixing the food with hydrochloric acid (HCl) and other gastric secretions to form chyme. Each peristaltic wave pushes chyme toward the pyloric sphincter (Figure 3-8) at the lower end of the stomach. The *pyloric sphincter* or **pylorus** is a circular muscle surrounding the lower end of the stomach and is normally almost completely closed. It controls the exodus of stomach contents, allowing only a little at a time to be squirted forcefully into the small intestine. Gradually, the stomach empties itself by this method. The complicated chemical breakdown that takes place beyond the stomach requires that only small amounts be processed at a time.

peristalsis (pair-ih-STALL-sis) the wavelike squeezing motions of the stomach and intestines that push their contents along the digestive tract.

small intestine a long, tubular organ of digestion that is the site of nutrient absorption.

pylorus (pie-LORE-us) muscle that regulates the opening of the bottom of the stomach.

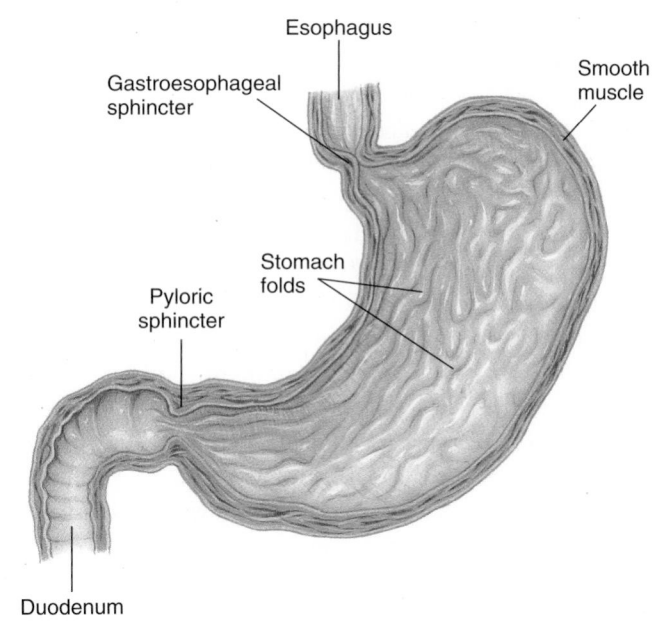

FIGURE 3-7

Simplified Anatomy of the Stomach

Esophagus

Gastroesophageal sphincter

Smooth muscle

Stomach folds

Pyloric sphincter

Duodenum

FIGURE 3-8

Mixing of Stomach Contents and Emptying of the Stomach

Stomach

Esophagus

Gastroesophageal sphincter

Pyloric sphincter

Duodenum

Direction of movement of peristaltic contraction

Movement of chyme

Peristaltic contraction

Gastric emptying

Peristaltic contraction

Gastric mixing

(1) Peristaltic contractions begin in the upper portion of the stomach and sweep down toward the pyloric sphincter.

(2) Contractions become more vigorous as they reach the lower part of the stomach.

(3) These strong contractions propel chyme forward.

(4) A small portion of chyme is pushed through the partially open sphincter into the small intestine. The stronger the contractions, the more chyme is emptied with each contraction.

(5) When chyme that was being propelled forward hits the closed sphincter, it is tossed back into the stomach. Mixing of chyme is accomplished as by propelling chyme forward then backward with each peristaltic contraction.

(6) When the peristaltic contractions reach the pyloric sphincter, the sphincter is tightly closed and prevents emptying.

Several digestive secretions are produced in the stomach. One such secretion is mucus. The stomach produces two types: one is thick and forms a protective barrier for the stomach lining, and the other is thin and watery to mix with food particles forming part of the chyme. Intrinsic factor is also produced by specialized cells in the stomach. It is necessary for absorption of vitamin B_{12} in the small intestine. Even though HCl does not really digest anything, it facilitates in digestion several ways (Table 3-1). HCl activates pepsinogen, which is the inactive form of the enzyme pepsin. Pepsin can function only in this acidic environment, where it begins the chemical breakdown of proteins. Large food particles are digested into smaller particles by breaking down muscle fiber and connective tissues. HCl also kills bacteria. Carbohydrate digestion that began in the mouth stops because the action of salivary amylase is halted by the stomach's acidic environment. Carbohydrates are chemically digested in the small intestine.

It takes approximately four hours for a meal to empty from the stomach. Liquids pass through the stomach quickly, while solids stay in the stomach until they are mixed thoroughly with stomach secretions. Carbohydrates tend to move through the stomach more quickly than proteins, and fatty foods may stay in the stomach for as long as six hours. The rate of stomach emptying (also controlled by the central nervous system and hormonal mechanisms) plays an important role in digestion and absorption of nutrients from foods. If the stomach is too slow to empty, digestion and absorption are reduced. Likewise, if the stomach empties too quickly, before food particles are thoroughly mixed with stomach secretions, very few nutrients are digested and the acidic chyme may damage the lining of the small intestine.

The Small Intestine The small intestine is the organ of digestion and absorption; it finishes the job that the mouth and stomach started. It is actually about 20 feet long, but it is called *small* because its diameter is small compared with that of the large intestine.

Chyme must touch the walls of the small intestine to make contact with the secretions and to be absorbed at the proper places. At the end of the small intestine, a circular muscle (similar in function to the pylorus at the end of the stomach) controls

TABLE 3-1	Primary Functions of Selected Digestive Enzymes	
Enzyme	**Enzyme Function**	**Enzyme Source**
A. Carbohydrate Digestion		
Amylase	Breaks down starch into smaller chains of glucose molecules	Produced in the salivary glands (salivary amylase) and the pancreas (pancreatic amylase)
Sucrase	Splits the disaccharide sucrose into glucose and fructose	Produced in the small intestine
Lactase	Splits the disaccharide lactose into glucose and galactose	Produced in the small intestine
Maltase	Splits the disaccharide maltose into two molecules of glucose	Produced in the small intestine
B. Fat Digestion		
Lipase	Breaks down fats into fragments of fatty acids and glycerol	Produced in salivary glands (lingual lipase) and the pancreas (pancreatic lipase) The action of lipase is enhanced by bile
C. Protein Digestion		
Pepsin	Separates protein into shorter chains of amino acids	Produced by the stomach
Trypsin	Splits short chains of amino acids into molecules containing one, two, or three amino acids	Produced by the pancreas

Does Heartburn Mean That Your Heart Is on Fire?

Heartburn by any other name is called pyrosis or acid indigestion. It is not a heart problem, but a gastrointestinal condition that gets its name from the painful burning sensation that occurs in the esophagus just below or behind the sternum or breastbone. Almost everyone has occasional heartburn. But if heartburn becomes frequent and ongoing, a more serious problem may be at hand . . . *gastroesophageal reflux disease* (GERD). GERD occurs when the muscles (sphincter) between the esophagus and stomach do not close tightly enough, allowing acidic contents of the stomach to back up (or reflux) into the esophagus and irritate it. Other symptoms of heartburn or GERD are:

- Burning in the throat—or hot, sour, acidic, or salty-tasting fluid at the back of the throat
- Difficulty swallowing
- Feeling of food "sticking" in the middle of the chest or throat
- Chronic cough, *sore throat*, or chronic hoarseness

Some foods we eat or drink can initiate heartburn, and so can some over-the-counter and prescription medications. Examples of these irritants include:

- Certain foods such as fatty foods, spicy foods, chocolate, caffeine, onions, tomato sauce, carbonated beverages, and mints
- Alcohol
- Large meals
- Lying down too soon after eating
- Certain medications including sedatives, antidepressants, and calcium channel blockers for high blood pressure
- Cigarette smoking
- Aspirin
- Other anti-inflammatory medications like ibuprofen (*Motrin*®, Advil®, Nuprin, etc.)

Smoking and consumption of high-fat-content foods tend to cause the lower esophageal sphincter (LES) to relax and allow acid to reflux into the esophagus. Pregnancy, obesity, and wearing clothing too tight around the waist can also cause increased pressure in the abdomen, leading to reflux. The best thing to do if you experience any of these symptoms is to see your health care provider, who may recommend over-the-counter remedies or prescribe medication.

the flow of the contents going into the large intestine (colon). It is called the *ileocecal valve.*

The small intestine works with the precision of a laboratory chemist. As the thoroughly liquefied and partially digested nutrient mixture (chyme) arrives there, hormonal messages tell the gallbladder to send its **emulsifier**, called **bile**, in amounts matched to the amount of fat present. Other **hormones** notify the pancreas to release **bicarbonate** in amounts precisely adjusted to neutralize stomach acid as well as **enzymes** of the appropriate kinds and quantities to continue dismantling whatever large molecules remain. Such messages also maintain peristalsis so the contents will be passed along to the next region. Peristalsis is stimulated by the presence of roughage or fiber and is quieted by the presence of fat, which requires a longer time for digestion.

Meanwhile, as the pancreatic and intestinal enzymes act on bonds that hold the large nutrients together, smaller and smaller units make their appearance in the intestinal fluids. Finally, units that cells can use—glucose, glycerol, fatty acids, and amino acids, among others—are released.

Once the digestive system has broken food down to its nutrient components, it must deliver those nutrients to the rest of the body. The cells lining the small intestine absorb nutrients from the mixture and deposit them in the blood and **lymph**. Every molecule of nutrient must traverse one of these cells if it is to enter body fluids.

emulsifier (ee-MULL-sih-fire) a compound with both water-soluble and fat-soluble portions that can attract lipids into water to form an emulsion.

bile a compound made from cholesterol by the liver, stored in the gallbladder, and secreted into the small intestine. It emulsifies lipids to ready them for enzymatic digestion.

hormone a chemical messenger, secreted by one organ (a gland) in response to a condition in the body, that acts on another organ or organs to change that condition.

bicarbonate a chemical that neutralizes acid; a secretion of the pancreas.

enzyme a protein catalyst. A *catalyst* is a compound that facilitates (speeds up the rate of) a chemical reaction without itself being altered in the process.

lymph (LIMF) the fluid outside the circulatory system that bathes the cells; derived from the blood by being pressed through the capillary walls; similar to the blood in composition but without red blood cells.

The cells are selective: They can recognize nutrients needed by the body. The cells are also extraordinarily efficient: They absorb enough nutrients to nourish all the body's other cells.

The intestinal tract lining is composed of a single sheet of cells, and the sheet pokes out into millions of finger-shaped projections called **villi**. Each villus has its own capillary network and a lymph vessel so that as nutrients move across the cells, they can immediately mingle into the body fluids. Every villus cell has a brush-like covering of tiny hairs, called **microvilli**, that can trap nutrient particles. Figure 3-9 provides a close look at these details.

The lining of the small intestine, villi and all, is wrinkled into thousands of folds, making its absorbent surface area enormous. If the folds and the villi that cover them were spread out flat, the total area would equal the size of a tennis court. The billions of cells of that surface, although they weigh only 4 to 5 pounds total, absorb enough nutrients in a few hours a day to nourish the other 150 or so pounds of body tissues.

Nutrients released early in the digestive process, such as simple sugars, and those requiring no special handling, such as the water-soluble vitamins, are absorbed high in the small intestine. Nutrients released more slowly are absorbed further down. The lymphatic and circulatory systems then take over the job of transporting them to the cell consumers. The lymph at first carries most of the products of fat digestion and the fat-soluble vitamins, later delivering them to the blood. The blood carries products of carbohydrate and protein digestion, water-soluble vitamins, and minerals. By the time the remaining mixture reaches the end of the small intestine, little is left but water, indigestible residue (mostly fiber), and dissolved minerals.

villi (VILL-ee, VILL-eye) fingerlike projections of the sheet of cells that line the GI tract; the villi make the surface area much greater than it would otherwise be (singular: *villus*).

microvilli (MY-croh-VILL-ee, MY-croh-VILL-eye) tiny hairlike projections on each cell of the intestinal tract lining that can trap nutrient particles and translocate them into the cells (singular: *microvillus*).

FIGURE 3-9

Details of the Lining of the Small Intestine

Stomach

Small intestine

Folds with villi on them

The wall of the small intestine is wrinkled into thousands of folds and is carpeted with villi.

Muscle layers beneath folds

A villus

Capillaries

Lymphatic vessel

Between the villi, tubular glands secrete enzyme-containing intestinal juice.

Artery

Vein

Lymphatic vessel

Microvilli

This photograph shows part of a human intestinal cell with microvilli.

Three cells of a villus. Each cell is covered with microvilli.

The Large Intestine (Colon) The small intestine is so named because of its small diameter, and the same rationale applies to the large intestine (colon). Although the diameter is larger, the large intestine is only about 5 feet long (Figure 3-10). Under normal circumstances, all chemical digestion has been completed by the time chyme reaches the large intestine; therefore, no digestive enzymes are secreted here. However, naturally occurring bacteria that reside in the large intestine break down and digest some of the fiber left over from carbohydrate digestion. Other colonic bacteria synthesize absorbable vitamin K.

Because the primary functions of the large intestine are to absorb dissolved minerals and water and eliminate waste products, cells lining the colon are specialized for absorbing these minerals and retrieving water for recycling. The final waste product is mixed with mucus produced in the large intestine to protect the innermost lining against abrasion and irritation from the chyme. Fecal material, a smooth paste of a consistency suitable for excretion, is stored in the colon until excretion. This system can adjust to whatever mixture of foods is presented. Fecal material normally consists of water, undigested fiber, bile pigments, bacteria, and a small amount of salt.

The colon has the same type of mixing and peristaltic movements as in other sections of the digestive tract, but they are more sluggish and occur less frequently. Commonly occurring after a meal, as the rectum fills with feces, the defecation reflex is set in motion and waste products are eliminated.

In addition to feces being eliminated via the colon, sometimes intestinal gas (*flatus*) passes out. Most intestinal gas is produced by bacterial fermentation. The quantity and quality of the gas depends on the types of foods eaten and characteristics of intestinal bacteria.

Accessory Organs of Digestion Organs outside the digestive tract—the liver, gallbladder, and pancreas—also contribute digestive juices through a common duct

FIGURE 3-10

Large Intestine

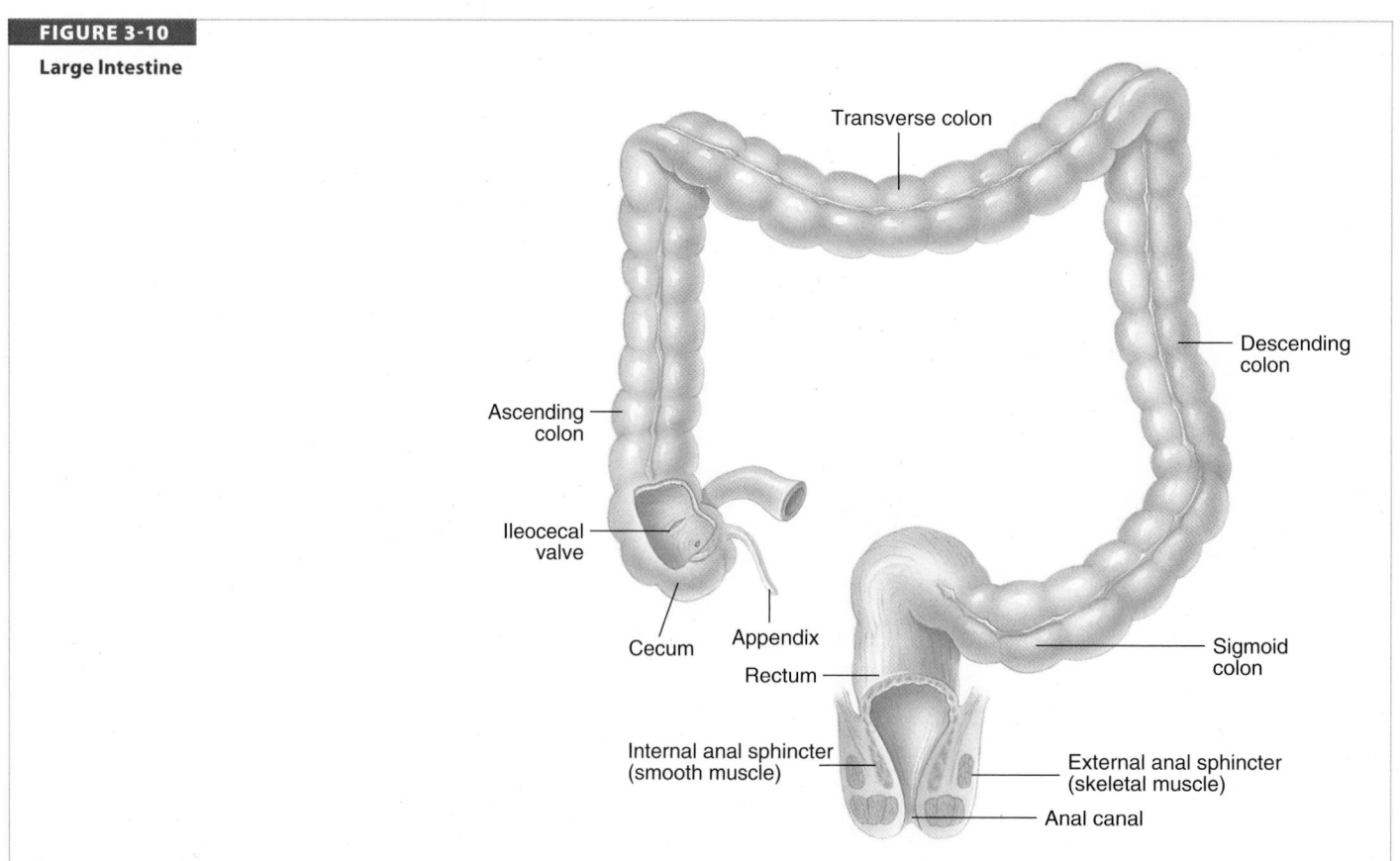

into the small intestine (Figure 3-11). Presence of these digestive juices and enzymes requires ancillary cells that specialize in protecting the digestive system. They do so by secreting a thick, viscous substance known as *mucus*, or the *mucous membrane*, which coats the intestinal tract lining and ensures that it will not itself be digested.

Liver The **liver** determines the metabolic fate of every nutrient we digest and absorb, in addition to performing a wide variety of other functions, most of them essential for life (Table 3-2). The liver produces about one liter of bile each day. *Bile*, made up of water, bile salts, bile pigments, and cholesterol, is slightly alkaline, which helps neutralize acidic chyme in the intestinal tract. Bile salts are not digestive enzymes, but do play a role in the digestion of fats. Bile salts are emulsifying agents, which break down large globules of fat into tiny globules of fat. This improves the efficiency of fat digestion by increasing the surface area of fat. Bile salts also play a role in absorption of fat-soluble vitamins. The bile pigments and cholesterol are waste products and are eventually eliminated from the body.

Gallbladder The gallbladder is a sac, attached to the liver (see Figure 3-3), where bile is stored and concentrated. Bile empties from the gallbladder into the uppermost part of the small intestine (*duodenum*) via the common bile duct. Entry of bile into the small intestine is controlled by the sphincter of Oddi. When the small intestine is empty, the sphincter is closed and bile is backed up into the gallbladder for concentration and storage until it is needed. When chyme with fatty content enters the

liver the large, lobed organ that lies under the ribs and filters the blood, in the process removing, processing, and readying for redistribution many of its materials.

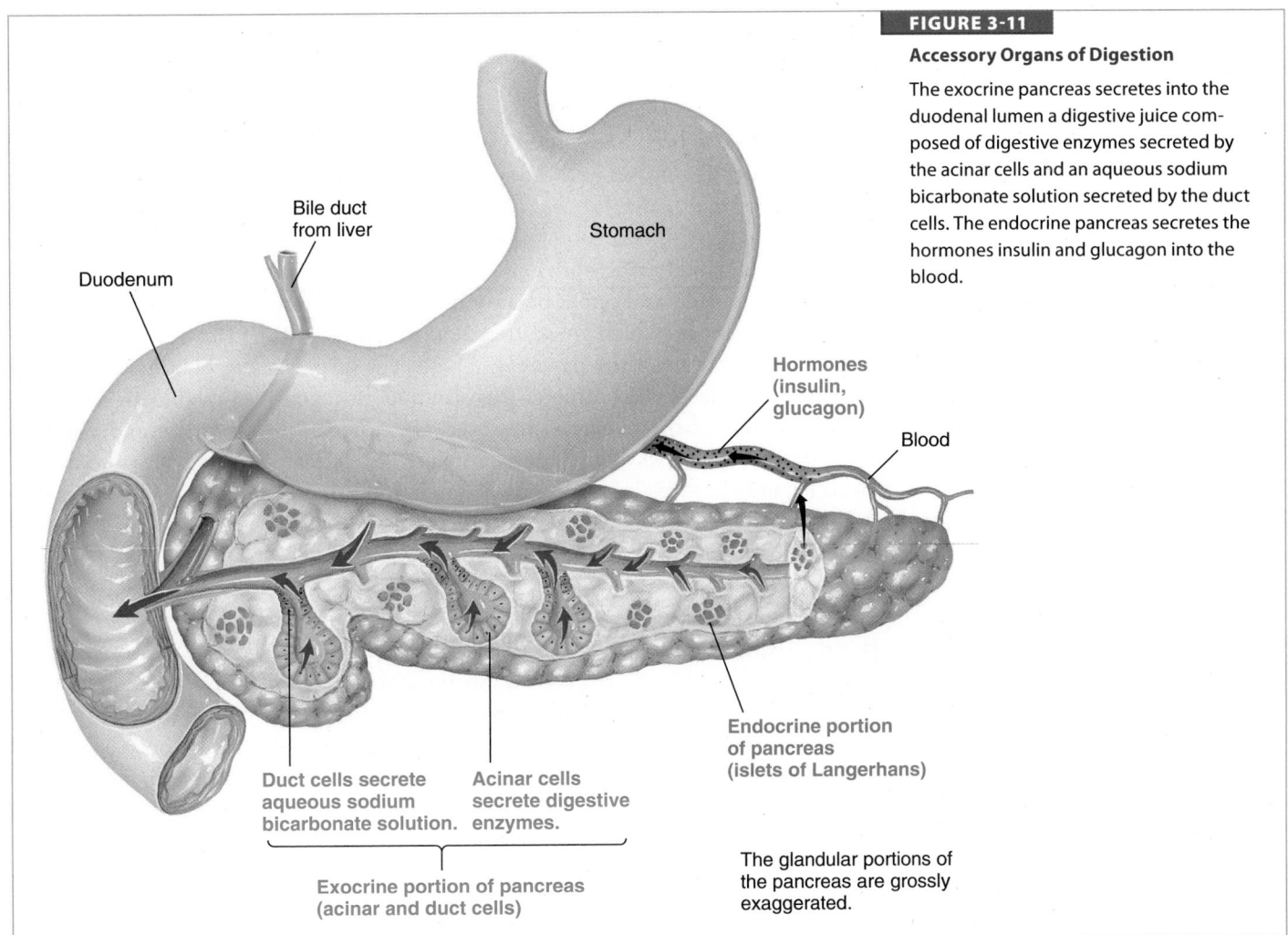

FIGURE 3-11

Accessory Organs of Digestion

The exocrine pancreas secretes into the duodenal lumen a digestive juice composed of digestive enzymes secreted by the acinar cells and an aqueous sodium bicarbonate solution secreted by the duct cells. The endocrine pancreas secretes the hormones insulin and glucagon into the blood.

Bile duct from liver

Stomach

Duodenum

Hormones (insulin, glucagon)

Blood

Duct cells secrete aqueous sodium bicarbonate solution.

Acinar cells secrete digestive enzymes.

Endocrine portion of pancreas (islets of Langerhans)

Exocrine portion of pancreas (acinar and duct cells)

The glandular portions of the pancreas are grossly exaggerated.

TABLE 3-2 Functions of the Liver

Digestive Functions

Secretion	Produces and secretes bile.
Synthesis of bile salts	Bile salts facilitate fat digestion, absorption of fats and fat-soluble vitamins.
Storage	Stores glucose in the form of glycogen. Stores iron and vitamins A, B_{12}, D, E, and K.
Detoxification	Alters chemical composition of toxic compounds and drugs such as alcohol to make them less harmful.
Excretion	Cholesterol and bile pigments from the breakdown of hemoglobin are excreted in bile.
Carbohydrate metabolism	Removes excess glucose from blood, converts it to glycogen for storage. Breaks down glycogen into glucose when more is needed. Converts noncarbohydrate molecules (such as amino acids) into glucose.
Fat metabolism	Functions in breakdown of fatty acids, synthesis of cholesterol and phospholipids. Functions in conversion of excess carbohydrates and proteins into fats.
Protein metabolism	Sends amino acids to body cells for protein synthesis. Converts certain amino acids into different amino acids as needed for protein synthesis. Converts ammonia from the breakdown of proteins into urea to be excreted in bile.

Nondigestive Functions

Synthesis of plasma proteins	Synthesizes albumin, fibrinogen, globulins (except immunoglobulin), and clotting factors.
Detoxification	Changes configuration of certain drugs (such as penicillin) and excretes them in bile to remove them from the body.
Excretion	Hormones and drugs are excreted in bile.
Filtration	Removes bacteria, damaged red blood cells, and other particles from the blood.

small intestine, a hormone (cholecystokinin [koh-lee-sis-toh-KYE-nin]) stimulates the gallbladder to contract and the sphincter to open, thus allowing bile to flow into the small intestine.

Pancreas A gland having both endocrine and exocrine functions, the **pancreas** lies behind the stomach and upper portion of the small intestine (see Figure 3-11). The islets of Langerhans (which are scattered on the surface of the pancreas) perform the **endocrine** functions of the pancreas. These cells help regulate blood glucose levels through secretion of the hormones **insulin** and glucagon into the blood. The principal functions of the pancreas, however, are **exocrine**. Bicarbonate and digestive enzymes that act on carbohydrates, proteins, and fats are secreted into the duodenum. Pancreatic amylase breaks down starch and other complex carbohydrates into simpler sugars called *disaccharides* (see Chapter 4).

Metabolism

Breaking Down Nutrients for Energy
All nutrients leaving the digestive system by way of blood are collected in thousands of **capillaries** in the membrane that supports the intestine. These converge into **veins** and then into a single large vein. This vein conveys its contents to the liver and there breaks up once again into a vast network of capillaries that weave among liver cells, allowing them access to newly arriving nutrients.

Liver cells process these nutrients. They convert sugars from carbohydrate mostly into the body's primary sugar, *glucose*. If there is a surplus, they store some as glyco-

pancreas a gland that secretes the endocrine hormone insulin and also produces the exocrine secretions that aid digestion in the small intestine. (See also *endocrine; exocrine*.)

endocrine (EN-doh-krin) a term to describe a gland secreting or a hormone being secreted into the blood.

insulin a hormone from the pancreas that helps glucose get into cells.

exocrine (EX-oh-krin) glands that secrete chemical messengers through a duct into a body cavity or onto the surface of the skin; *exo* means "out."

capillaries minute, weblike blood vessels that connect arteries to veins and permit transfer of materials between blood and tissues.

veins blood vessels that carry used blood from the tissues back to the heart.

gen (discussed in Chapter 4) and convert the remainder to fat. They reassemble fatty acids and glycerol from dietary fat into larger fats and package them with protein for transport to other parts of the body. As for amino acids from protein, liver cells alter these as needed, making glucose from some if necessary and fat from others if there is an excess, or converting one amino acid into another for use in making proteins.

Nutrients that leave the digestive tract by way of lymph are packaged as **chylomicrons** that circulate throughout the body, allowing cells to withdraw fats from them. Some also find their way into blood and circulate through the liver, which removes them, alters their components, and releases new products, including other **lipoproteins**.

The new products of liver **metabolism**—glucose, fat packaged with protein (lipoproteins—see Chapter 5), and amino acids—are released into the bloodstream again and circulated to all other cells of the body. Surplus fat is then removed by cells specialized for its storage; these fat cells are located in deposits all over the body.

The glycogen stored in the liver provides a reserve supply of the body's sugar, glucose, and thus can sustain cell activities if intervals between meals become so long that glucose absorbed from ingested foods is used up. When the body is depending solely on liver glycogen, however, the supply is used up within three to six hours. Similarly, fat cells store reserves of fat, the body's other principal energy nutrient.

Unlike the liver, however, fat cells have virtually infinite storage capacity and can continue to supply fat for days, weeks, or even months when no food is eaten. (Chapter 10 contains more about fasting, and Chapter 11 contains more about energy production.)

These storage systems for glucose and fat ensure cells will not go without energy nutrients even if the body is hungry for food, except under extreme conditions. Body stores also exist for many other nutrients, each with a characteristic capacity. For example, the third energy nutrient, protein, is held in an available pool (amino acids in the liver and blood) that is rather rapidly depleted during protein deficiency conditions (see Chapter 6). Liver and fat cells store many vitamins, and bones provide reserves of calcium, sodium, and other minerals that can be drawn on to keep blood levels constant and to meet cellular demands.

The breakdown of body compounds is known as *catabolism*. These reactions usually release energy. Glycogen can be broken down to glucose, triglycerides to fatty acids and glycerol, and protein to amino acids. When the body needs energy, it breaks any or all of the four basic units—glucose, fatty acids, glycerol, and amino acids—into even smaller units. When the body does not require energy, the end products of digestion (glucose, amino acids, glycerol, and fatty acids) are used to build body compounds in a process called *anabolism*. Anabolic reactions involve conversion of glucose to glycogen or fat, conversion of amino acids to body proteins or fat, and synthesis of body fat from glycerol and fatty acids. Catabolism and anabolism are examples of **energy metabolism**.

Other Systems

In addition to the digestive system, the body has many more systems: bones, muscles, nerves, lungs, reproductive organs, and others. All of these systems cooperate so each cell can carry on its own life. Each assures, through hormonal or nerve-mediated messages, that its needs will be met by the others, and each contributes to the welfare of the whole by doing its specialized work. Most of the body's work is done automatically and is finely regulated to achieve a state of well-being. We would do well to "listen" to our bodies and to cultivate an understanding and appreciation of the body's needs. Then we can make decisions to promote health.

chylomicron (KIGH-loh-MY-cron) a type of lipoprotein that transports newly digested fat—mostly triglyceride—from the intestine through lymph and blood.

lipoproteins (LIP-oh-PRO-teens) clusters of lipids associated with protein that serve as transport vehicles for lipids in blood and lymph (see Chapter 5).

metabolism (meh-TAB-o-lism) total of all chemical reactions that go on in living cells.

energy metabolism all reactions by which the body obtains and expends the energy from food or body stores.

Lycopene in your tomato sauce? Beta-carotene in your soup? Today, one of the hottest areas in food science and nutrition policy is *functional foods*. Nutritionists, food scientists, food marketers, and others are exploring how today's traditional foods, and perhaps new formulations, may open doors to a healthier tomorrow.[1] The fact that food is intimately linked to optimal health is not a novel concept. "Let food be your medicine and medicine be your food" was a tenet espoused by Hippocrates in approximately 400 B.C.E.[2] Almost 2,500 years later, this philosophy is once more of utmost importance, as it is the "food as medicine" philosophy that underpins the paradigm of functional foods. In this feature, we look at the promising benefits of functional foods in disease prevention, as well as how we can incorporate functional foods into our daily eating plan so we can reap their benefits.

You can visit your grocery store for a wide assortment of functional foods.

Visit the Local Grocery Store.

Functional foods are not really a new concept; grocery stores are already filled with numerous items that meet the definition of functional foods (see Figure 3-12).[3] One of the first to appear was calcium-fortified orange juice, which gave people who don't like milk as much bone-healthy calcium per glass as milk. We now have grain products fortified with folate to prevent birth defects and possibly heart disease. Many cereal grains, fruits, vegetables, nuts, and dairy products are touted for their potential cancer prevention benefits as well as cardiovascular protection.

The Quaker Oats man continues to smile as the cholesterol-lowering effects of the soluble fiber beta-glucan in oats are well documented. Omega-3 fatty acids in flaxseed oil and fatty fish such as salmon and halibut appear to reduce the risk of heart disease, as well as having anti-inflammatory properties. Research indicates

that the organosulfur compounds in garlic may have it all—cancer-fighting, cholesterol-lowering, antibiotic, and antihypertensive properties—which far outweigh the herb's ability to cause bad breath. *Probiotics* (friendly bacteria) in fermented milk products such as yogurt, and *prebiotics* (fermentable dietary fiber) in oatmeal, flax, and legumes also appear to have multiple effects, especially in promoting a healthy gastrointestinal tract. Probiotics are also associated with reducing the risk of colon cancer, lowering cholesterol, and outcompeting potentially disease-causing bacteria in the gastrointestinal tract. Soy protein has been the focus of intense research efforts because of its ability to lower cholesterol and possibly reduce the risk of cancer, osteoporosis, and symptoms associated with menopause.[4] Current research is investigating a certain type of fat called conjugated linoleic acid (CLA), found primarily in dairy products, for numerous potential health benefits, including improvements in immune function, body composition, and bone mineralization. Identification of CLA underscores the importance of meeting nutrient needs through consumption of foods from all the major food groups.[5]

Follow the Research.

So much is still unknown when it comes to functional foods that additional research is needed to clarify the role of phytochemicals in health promotion. Such research will also provide plant physiologists and human nutritionists with the vital information they need to develop an enhanced food supply and better target dietary recommendations for the general public.[6] For years, the food industry focused on "taking out the bad stuff" (fat, sodium, cholesterol, calories) and restoring or enhancing favorable nutrients (calcium, folate, fiber). We may soon see the day where we have tomatoes with extra lycopene, sulforaphane-enriched broccoli, and sweet potatoes with extra carotene.

Incorporate Functional Foods into Your Daily Diet.

You can start by incorporating the MyPlate food guide into your daily food planning if you are not already using it (see Figure 3-12 and the following box). Begin with the Grains group, and consume the recommended amount of whole-grain breads and cereals, especially oatmeal. Grind flaxseed in a coffee grinder and add it to your cereal.

Consume at least nine servings of vegetables and fruits daily. Be sure to include a variety of vegetables and fruits, including orange vegetables and fruits such as sweet potatoes, carrots, winter squash, apricots, cantaloupe, and peaches. Add dark-green leafy vegetables such as kale, mustard greens, collards, and spinach, along with broccoli and Brussels sprouts. Include tomatoes as often as possible, and also add garlic and onions to your diet. Eat plenty of purple vegetables such as eggplant and red cabbage. Try

FIGURE 3-12

Functional Food Guide

SOURCE: Adapted from University of Illinois, Functional Foods for Health.

GRAINS	VEGETABLES	FRUITS	DAIRY	PROTEIN FOODS
Psyllium-containing bread and cereal, Corn products, Flaxseed, Oat products, Rye products, Wheat bran products	Broccoli, Brussels sprouts, Cabbage, Carrots, Cauliflower, Celery, Garlic, Horseradish, Jerusalem artichokes, Leeks, Onions, Scallions/Shallots, Soybeans, Tomatoes, Watercress	Apples, Bananas, Blueberries, Cranberries, Grapefruit, Grapes/Juice, Lemons, Limes, Oranges, Raspberries	Cheese, Milk, Milk products, Soy milk products, Yogurt	Beans, Beef, Eggs, Mackerel, Salmon, Soy nuts, Soy protein, Sardines, Tuna, Walnuts

Here are examples of some foods that are potentially associated with maintaining health.

Antioxidant
Broccoli family
Carrots
Citrus fruit
Cocoa/chocolate
Flaxseed
Grapes/juice
Honey
Horseradish
Raspberries
Tomatoes

Improves Gastrointestinal Health
Bananas   Onion family
Honey   Yogurt
Horseradish
Jerusalem artichokes
Milk

Maintains Vision
Blueberries
Broccoli
Carrots
Corn products
Eggs
Leafy greens

Reduces Cancer Risk
Apples
Beans
Berries
Broccoli family
Citrus fruit
Corn products
Flaxseed
Garlic
Grapes/juice
Green or black tea
Milk products
Onion family
Rye products
Salmon
Soy products
Tomatoes
Wheat bran products

Improves Heart Health
Apples
Beans
Berries
Fish (salmon, tuna, mackerel, sardines)
Flaxseed
Garlic
Grapes/juice
Green or black tea
Milk products
Oat products
Onion family
Psyllium-containing bread and cereals
Soy products
Walnuts

Maintains Urinary Tract Health
Blueberries
Cranberries

Reduces Blood Pressure
Bananas
Celery
Cheese
Garlic
Milk
Soy products

Improves Bone Health
Cheese
Milk
Milk products
Soy milk products
Soy nuts

experimenting with soy milk, tofu, miso, and tempeh, and include them regularly in your diet.

Salmon and other fatty fish (sardines, mackerel, herring, tuna) contain omega-3 fatty acids, which are shown to offer heart protection. Try selecting two fish meals per week.

Yogurt, cheese, and milk products go a long way toward providing the calcium needed for healthy bones; be sure to have at least three servings daily. Many dairy products are functional foods. The live bacteria used to ferment milk products such as yogurts and yogurt-based drinks enhance intestinal flora.

Last but not least, top off your meal with a cup of green or black tea, a source of phenolic compounds, which many researchers are convinced protect against cancer, heart disease, and stroke. As you can see, the best advice is to consume a variety of foods that contain both known beneficial compounds and those still awaiting discovery.

(Continued)

NOTES:
1. Hasler, C. M. Functional foods: Benefits, concerns, and challenges—A position paper from the American Council on Science and Health. *Journal of Nutrition, 132 (2002)*: 3772–81. Parts of this discussion are adapted from Functional foods: Opening the door to better health, *Food Insight* (November/December 1995).
2. W. H. S. Jones, trans., *Hippocrates* (New York: Heinemann, 1932), 351.
3. Position of the American Dietetic Association: Functional foods, *Journal of the American Dietetic Association, 104 (2004)*: 814–26; Klotzbach-Shimomura, K., Functional foods: The role of physiologically active compounds in relation to disease. *Topics in Clinical Nutrition, 16 (2001)*: 68–78.
4. Anderson, J. W. et al., Meta-analysis of the effects of soy protein intake on serum lipids. *New England Journal of Medicine, 333 (1995)*: 276–82; Messina, M., and Barnes, S., The role of soy products in reducing risk of cancer. *Journal of the National Cancer Institute, 83 (1991)*: 541–46; Arjmandi, B. H., et al., Dietary soybean protein prevents bone loss in an ovariectomized rat model of osteoporosis, *Journal of Nutrition, 126 (1996)*: 162–67; Albertazzi, P., et al., The effect of dietary soy supplementation on hot flushes, *Obstetrics and Gynecology, 91 (1997)*: 6–11.
5. Raloff, J., The good *trans* fat: Will one family of animal fats become a medicine? *Science News, 159 (2001)*: 136.
6. Taylor, C. L., Regulatory frameworks for functional foods and dietary supplements. *Nutrition Reviews, 62 (2004)*: 55–59.

Tips for Adding Functional Foods to Your Daily Diet

At breakfast:

- Mix oatmeal with blueberries.
- Top whole-grain dry or hot cereal with yogurt.
- Add fresh fruits and berries and/or whole-grain cereal to yogurt.
- Spread soy nut butter on whole-grain toast.
- Drink a glass of sparkling purple grape juice with breakfast.
- Blend soy milk with fresh pineapple or frozen berries.

Healthy meal ideas:

- Mix tuna salad with grated carrots, red peppers, onions, and garlic.
- Serve whole-grain pasta with tomato sauce and fresh herbs.
- Cook leeks and onions with tomatoes as a side dish.
- Grill salmon and serve with fresh greens and yogurt salad dressing.
- Try low-fat cream of carrot, spinach, and broccoli soups.
- Enjoy a cup of green tea with a marinated tofu sandwich.
- Stir-fry fresh vegetables with extra garlic.

Snacks on the go:

- Grab a piece of fresh fruit.
- Mix soy nuts, whole-grain cereal, and dried fruit together and hit the trail.
- Grab a glass of tomato, cranberry, or orange juice.
- Try fresh broccoli, cauliflower, and carrots with tofu dip.
- Mix bananas with fresh raspberries.

Easy phytochemical-rich minestrone soup:

Four antioxidant-rich veggies make this soup a real winner.

- one 16-oz package frozen broccoli, cauliflower, and carrot blend
- two 15-oz cans stewed tomatoes
- two 14½-oz cans broth (beef, vegetable, or poultry)
- one 15-oz can Great Northern beans
- 2 oz uncooked whole-wheat vermicelli (break into 2-inch pieces)
- Grated Parmesan cheese

In a large saucepan, combine vegetables, tomatoes, broth, beans, and pasta and bring to a boil. Reduce heat, cover, and simmer 6 to 8 minutes or until vegetables and pasta are tender. Sprinkle with Parmesan cheese.

Makes four to six (1½ c) servings.

Nutrition information per serving: 210 calories and 2 grams fat

SOURCE: University of Illinois Functional Foods for Health Program: www.ag.uiuc.edu.

As you study the material in this text, you will learn that our bodies need protein to build and repair cells, that fats supply more than twice as much energy as carbohydrates, and that the "typical American" diet (Western diet) can be wholly or partially linked to chronic diseases such as heart disease (the leading cause of death in the United States), some cancers (second leading cause of death in the United States), osteoporosis, over-weight, obesity, and type 2 diabetes. But you may not know why.

As humans, 99 percent of our genetic physiology and metabolic heritage dates back roughly 40,000 years! What does this mean to present-day humans? Why is this important? And most impor-tantly, why should you care? Think about this: If our bodies are "programmed" the way our distant hunter-gatherer ancestors' were, nearly all of our metabolism and physiology are fine-tuned to conditions that existed more than 10,000 years ago, before the development of agriculture and the domestication of animals we currently use for food. Our physiology most certainly predates fast foods, convenience foods, and sedentary workplaces and lifestyles. Our **gene** pool has been programmed to perform in ancient surroundings significantly unlike those in which we now live.

If you're still thinking, "So what?" take a look at this comparison between the eating patterns of our ancient ancestors and those of today.

Food/Nutrient	Ancient Ancestors' Eating Habits	The Way We Eat Today
Grains (wheat, rice, corn, oats, barley)	Rarely used	40–90% of total caloric intake
Meat	30% total calories (undomesticated wild game, fish)	12% total calories (domesticated animals)
Fruits	50% total calories	6% total calories
Vegetables	(> 100 different species of uncultivated plant foods, roots, tubers)	5% total calories
Fiber	> 100 g/d	< 20 g/day
Milk & milk products	Nonexistent after weaning	1.8 c/day (or 14% total calories)
Fats	~ 35% total calories (varied with latitude)	20–35% total calories
Saturated fats	5–7.5% total calorie intake	11–15% total calorie intake
Sugar	Honey when available	20 to 30 tsp/day
Sodium	600 mg/day	~ 4,000 mg/day
Potassium	7,000 mg/day	~ 3,000 mg/day
Energy intake	Feast–famine	Food abundance (food available 24 hr/day)
Energy expenditure	Obligatory for survival (probably ≥ 3,000 cal/day)	Sedentary (< 30 min/d of moderate-intensity physical activity, ~ 2,000 cal/day)

What does all of this information mean? You will learn what it means as you read this text for your class, but to give you a sneak peek of information to come, consider the impact that our ances-tors' lifestyles and eating habits might have on the chronic degen-erative diseases that are the most common ailments of today.

Some of you may be thinking, "Yeah, but Stone Age humans didn't live as long as we do!" Though this is true, the major contrib-utors to our increased life expectancy are better sanitation (purer water, adequate sewage disposal), safer food, and an effective system of quarantine and immunization. Chronic degenerative dis-eases do cause death later in life, but they begin much earlier in life, usually in our late teens and early 20s. We do not have to expend energy (as compared to our ancestors) to obtain our food, making calories accessible at the lowest cost in human history without reciprocal energy expenditure.

gene a unit of a cell's inheritance; made of a chemical, DNA, that is copied faithfully so that every time the cell divides, both its offspring get identical copies. Genes direct the cells' machinery to make the proteins that form each cell's structures and enable the cell to do its work.

(Continued)

Disease	Ancient Habits and Eating Patterns
Heart disease	High intake of very lean animal protein. Habitual physical activity reduces incidence of atherosclerosis. Fiber from fruits and vegetables is predominantly soluble fiber, which protects against heart disease. Saturated fat intake was negligible and *trans* fats nonexistent.
Hypertension	High dietary potassium intake along with low dietary sodium intake. Habitual physical activity decreases blood pressure. Decreased blood pressure reduces risk of heart disease.
Cancer	Humans are the only free-living primates that consume cereal grains. Fruits and vegetables provide more potential cancer prevention (phytochemical protection from free radicals), but consumption of fruits and vegetables has largely been replaced by consumption of cereal grains. Habitual physical activity reduces risk of some cancers (especially breast cancer).
Osteoporosis	Wild plant foods (fruits and vegetables) contain considerable calcium and do not contain chemicals that bind calcium and keep it from being absorbed in the small intestine. Habitual physical activity increases bone density and muscular strength, and reduces incidence of age-related fractures. High intake of dietary potassium (fruits and vegetables) and low intake of dietary sodium counteract the effects of high animal protein intake that increases urinary calcium loss (thus weakening bones).
Diabetes/insulin resistance	Habitual physical activity can decrease serum glucose levels. Higher-fiber diets can help control blood glucose levels.
Obesity	Our ancestral genes were not selected for a sedentary lifestyle. Inactivity has a greater impact on weight gain than consuming large amounts of calories alone.

Certainly, the foods we consume today are very different from the foods of the ancient past. As Dr. Michael Eades, founder of the Arkansas Center for Health and Weight Control in Little Rock, observed, "The same thing that caused our ancestors to survive is killing us now."

SOURCES:

M. V. Chakravarthy and F. W. Booth, Eating, exercise, and "thrifty" genotypes: Connecting the dots toward an evolutionary understanding of modern chronic diseases, *Journal of Applied Physiology, 96 (2004):* 3–10.

L. Cordain et al., Origins and evolution of the Western diet: Health implications for the 21st century, *American Journal of Clinical Nutrition, 81 (2005):* 341–54.

S. B. Eaton, The ancestral human diet: What was it and should it be a paradigm for contemporary nutrition? *Proceedings of the Nutrition Society, 65 (2006):* 1–6.

S. B. Eaton and S. B. Eaton III, Paleolithic vs. modern diets: Selected pathophysiological implications, *European Journal of Nutrition, 39 (2000):* 67–70.

S. B. Eaton et al., Review: Evolutionary health promotion, Preventive Medicine, 34 (2002): 109–18.

Foods Found in Modern Diets Generally Unavailable to Our Prehistoric Ancestors

Food Group	Foods
Dairy	Whole milk Low-fat milk Cheese Butter
Cereal grains	Whole grains Refined grains Refined sugars High-fructose corn syrup (used in soft drinks, etc.) Syrups
Refined vegetable oils	Salad and cooking oils Shortening Margarine
Alcohol	Alcoholic beverages

SOURCE: Adapted from L. Cordain et al., Origins and evolution of the Western diet: Health implications for the 21st century, *American Journal of Clinical Nutrition, 81* (2005): 41–54.

You're sitting in class or church . . . everything is relatively quiet and you are very tired from studying late into the wee hours of the morning . . . or for other reasons. So you close your eyes (for just a second, you think) just to give them a little rest, and the next thing you know you are rudely awakened by a loud "ripping" noise somewhere in the vicinity of where you are sitting. You look around only to find all eyes on you! Yes, it has happened. The humiliation of the public passing of gas, sometimes called a fart. Admit it. It's happened to you. Possibly more than once.

Americans have come a long way since Hippocrates proclaimed, "Passing gas is necessary to well-being," or the Emperor Claudius decreed, "All Roman citizens shall be allowed to pass gas whenever necessary." Face it. Americans are embarrassed by all natural bodily functions (except when it comes to discussing them in television commercials): hiccupping, burping/belching, flatulence (farting), constipation, diarrhea (there are especially obnoxious TV commercials regarding this phenomenon), and hemorrhoids. Digestive disorders are common, yet most people don't know what causes them or how to avoid them. It is important to understand how your GI tract works (and you should from reading the rest of this chapter) and what you can do to keep it healthy. You don't want your GI tract to be a pain in the butt (pun intended . . . but more on hemorrhoids later), so make yourself aware of what's going on in there.

What are hiccups?

Hiccups occur when the muscle at the base of the lungs (diaphragm) spasms. This causes a person to take a short, quick breath that is interrupted by the closing of the epiglottis (Figure 3-6). The sound we hear when hiccupping is the vocal cords (Figure 3-13) slamming closed after the spasm.

What causes hiccups?

Hiccups can be associated with a variety of ailments, but there is often no obvious cause.

Some common causes of hiccups include:

- Hot and spicy foods and liquids
- Noxious fumes
- Pneumonia
- Swallowing air
- Excess smoking and drinking alcohol
- Emotions such as shock, excitement, and stress
- Nerve irritation
- Surgery, especially abdominal surgery
- Mental health issues

FIGURE 3-13

Vocal Cords

Chet Childs/Custom Medical Stock

Is there a "cure"? How do I get rid of them?

There is no one sure way to get rid of hiccups, but there are several common suggestions (home remedies) that may or may not be effective:

- Hold your breath
- Breathe repeatedly into a paper bag
- Drink a glass of cold water
- Eat a teaspoon of sugar

Usually hiccups disappear after several minutes, but a health care provider should be consulted if hiccups continue for more than a few days.

How does gas get into my digestive tract?

Everyone has gas. We know we've got it when it's time to eliminate it (and all too often everyone around us knows too). Belching or burping is nature's way of getting rid of air we've swallowed when eating or drinking. If we eat or drink too rapidly, chew gum, smoke, or wear loose-fitting dentures, even more air may get into the stomach. And just for the scientifically curious, swallowed air that is burped out of the stomach contains nitrogen, oxygen, and carbon dioxide. The average adult passes gas 14 to 23 times a day.

Any gas remaining in the stomach moves into the small intestine, where two things take place: It is absorbed in the small intestine; and a small amount finds its way into the large intestine, where it joins up with undigested foods and/or the gas produced by the friendly bacteria that live in our large intestines. All of this gas is released . . . you guessed it . . . through the rectum. Gas from

(Continued)

Miroku/Getty Images

GeoM/Shutterstock.com

Frank Siteman/PhotoEdit

Sandy Jones/Photodisc/Getty Images

Home remedies for hiccups may or may not be effective.

swallowed air is odorless, but the friendly, helpful bacteria that live in our large intestine to make vitamin K and biotin, and break down fibers in the carbohydrates we eat, sometimes produce small amounts of sulfur that does have an unpleasant odor.

I've heard that intestinal gas is caused by the types of food I eat. Is this true?

It's true and false. As you've learned from reading this chapter (and will learn in Chapter 4), humans do not digest and absorb some types of carbohydrates (sugar, starches, and fiber). We simply do not make the necessary digestive enzymes to break down fiber. Another interesting note is foods that produce gas in one person may not cause gas in another, and the amount of gas caused by certain foods varies from person to person. Trial and error is the only way to learn how specific foods affect you and your gas production.

My roommate says she has gas and diarrhea when she drinks milk. How can that happen? Isn't milk supposed to be good for us?

What is most likely happening is your roommate cannot digest the disaccharide lactose, the natural sugar in milk and some milk products such as cheese and ice cream. This occurs because a person has low to no levels of lactase, the enzyme necessary to break down lactose into its two absorbable components, the monosaccharides galactose and glucose. Nutrients are commonly broken down and absorbed in the small intestine. When a nutrient has been broken down only partially (maldigestion) or hasn't been broken down completely (intolerance) and absorbed, it causes all kinds of havoc when it reaches the large intestine. Fluid is pulled from surrounding tissues to "water down" or dilute the intact or partially intact lactose in the colon, thus causing a bloated feeling and diarrhea, along with gas and abdominal cramping caused from intestinal bacterial that digest lactose.

Humans are the only mammals that regularly drink milk past weaning. And, for many years, scientists believed that people who experienced lactose maldigestion or intolerance were the "exceptions," while in truth they were the "rule" since this included about 70 to 90 percent of the world's population (mostly those of African or Asian decent). The more recent genetic theory is that people who tolerate lactose do so because their prehistoric ancestors migrated to parts of the world where dairy cattle could be raised safely and economically, primarily northern Europe and Scandinavia.

Foods That May Cause Gas Include:

- Beans
- Broccoli
- Cabbage
- Brussels sprouts
- Onions
- Artichokes
- Asparagus
- Pears
- Apples
- Peaches
- Whole grains
- Bran
- Soft drinks
- Fruit drinks
- Milk and milk products
- Sorbitol (in sugar-free candies and gum)

Linux Patrol/Used under license from Shutterstock

Tischenko Irina/Shutterstock.com

Purestock/Getty Images

Elena Schweitzer/iStockphoto.com

Photolink/Getty Images

Margaryta Vakhterova/iStockphoto.com

Carterphoto/Dreamstime LLC

When I was growing up, my grandmother would ask me every day, "Have you gone to the bathroom?" What she wanted to know was if I had experienced a bowel movement. She thought if I didn't go every day, I was constipated and she would give me a laxative. Should I take a laxative if I don't have daily bowel movements?

Some people do believe they are constipated if they do not experience a bowel movement every day; however, "normal" stool elimination depends on what is normal for each person . . . it may be three times a day, it may be three times a week. Constipation is defined as experiencing bowel movements three times per week or less. Stools are usually hard, dry, small in size, and difficult to eliminate.

What causes constipation?

Constipation is the most common GI complaint in the United States. More than 4 million people in the United States report having frequent constipation, and over half believe it serious enough to visit their physician. To understand what causes constipation, it will help to review the section of this chapter on the large intestine (Figure 3-10). As what's left over after digestion and absorption passes through the colon, water is absorbed, forming stool. As muscle contractions push the stool toward the rectum, it becomes more and more solid in consistency. When the colon absorbs too much water or if the muscle contractions are slow-moving, stools can become hard and dry. Frequent causes of constipation include:

- Lack of fiber in the diet
- Lack of physical activity (particularly in the elderly)
- Not enough liquids
- Medications (e.g., pain medications, antacids, blood pressure medications, antidepressants, iron supplements, diuretics, antispasmodics, anticonvulsants)
- Milk
- Changes in life or routine (e.g., pregnancy, aging, travel)

(Continued)

- Over-the-counter laxative abuse
- Ignoring the urge to have a bowel movement

How is constipation treated?

Constipation is a symptom, not a disease. Most constipation is temporary and not serious. Treatment depends on the cause, severity, and duration of the constipation, but in most cases dietary and lifestyle changes will help relieve symptoms and assist in preventing them from recurring.

- Diet: 20 to 35 grams of fiber each day helps the body form soft, bulky stools. See Chapter 4 to find out what foods are high in fiber.
- Lifestyle changes: Increasing water consumption and other liquids, daily exercise, and not ignoring the urge to have a bowel movement will help prevent constipation.
- Laxatives: Most cases of mild constipation do not require use of laxatives. But if the constipation is of long duration, a physician should be consulted for the best treatment to retrain a persistently lethargic bowel.

What if constipation becomes serious?

One serious complication of constipation is hemorrhoids. Hemorrhoids are swollen and inflamed veins around the anus or lower rectum. They are either internal (inside the anus) or external (under the skin around the anus). Hemorrhoids may result from straining to have bowel movements, pregnancy, aging, chronic constipation or diarrhea, and anal intercourse. They are usually not dangerous or life-threatening, but the itching and irritation can be annoying. About half the U.S. population develops hemorrhoids by the time they are 50, but not everyone experiences symptoms. The most common symptom of internal hemorrhoids is bright red blood covering the stool, on toilet paper, or in the toilet bowl. Hemor-

rhoids are common in pregnant women. Hormonal changes and pressure of the fetus on the abdomen cause hemorrhoidal vessels to enlarge.

How are hemorrhoids treated?

Initial treatment is to relieve symptoms:

- Tub baths several times a day in plain, warm water for about 10 minutes
- Application of hemorrhoidal cream or a suppository to the affected area for a limited time

Preventing hemorrhoids from occurring or recurring requires relieving the pressure and straining of constipation. Increasing fiber and fluids (not alcohol) in the diet will result in softer, bulkier, easier-to-pass stools.

SOURCES:

G. Bloom and P. W. Sherman, Dairying barriers affect the distribution of lactose malabsorption, *Evolution and Human Behavior* 26 (2005): 301.e1-301.e33.

Constipation. National Digestive Diseases Information Clearinghouse, http://digestive.niddk.nih.gov/ddiseases/pubs/constipation/index.htm, updated July 2007.

Dehydration. National Digestive Diseases Information Clearinghouse, http://digestive.niddk.nih.gov/ddiseases/pubs/diarrhea/index.htm, updated January 2011.

Gas in the digestive tract. National Digestive Diseases Information Clearinghouse, http://digestive.niddk.nih.gov/ddiseases/pubs/gas/index.htm, updated January 2008.

Hemorrhoids. National Digestive Diseases Information Clearinghouse, http://digestive.niddk.nih.gov/ddiseases/pubs/hemorrhoids/index.htm, updated November 2010.

Hiccups. eMedicineHealth, http://www.emedicinehealth.com/hiccups/article_em.htm, updated August 2005.

Jefferies M. How hiccups work, Howstuffworks. http://health.howstuffworks.com/hiccup.htm, updated February 2011.

Medical encyclopedia: hiccups, MedlinePlus, http://www.nlm.nih.gov/medlineplus/ency/article/003068.htm, updated February 2009.

Lactose intolerances. National Digestive Diseases Information Clearinghouse, http://digestive.niddk.nih.gov/ddiseases/pubs/lactoseintolerance/index.htm, updated June 2009.

Nutrition on the Web

www.cengage.com/nutrition/boyle/personalnutrition8e
Go to the Personal Nutrition site to check for the latest updates to chapter topics or to access links to related websites.

www.healthfinder.gov
Search for information on diverticular disease, heartburn, indigestion, constipation, diarrhea, and intestinal gas.

www.nal.usda.gov/fnic
Under Browse by Subject, select Diet and Disease and click on Digestive Diseases and Disorders.

www.digestive.niddk.nih.gov
This government website is the National Digestive Diseases Information Clearinghouse (NDDIC), and it includes many links.

www.digestive.niddk.nih.gov/ddiseases/a-z.asp
Search for information about digestive diseases. A description of digestion and absorption is provided under Your Digestive System and How It Works.

www.mayoclinic.com
The Mayo Clinic maintains this website, and it provides a variety of health information. Click on the Diseases & Conditions tab, and look up digestive diseases. The Health Tools tab provides an area for self-assessment of disease risk.

www.foodinsight.org/Content/6/
functionalfoodsbackgrounder.pdf
The International Food Information Council (IFIC) provides a Backgrounder on Functional Foods.

http://nccam.nih.gov/health/atoz.htm
Search for functional food components at the National Center for Complementary and Alternative Medicine (NCCAM) website.

www.nlm.nih.gov
Free access to National Library of Medicine's Medline for information searches on a variety of digestive disease topics.

www.cdc.gov/genomics/public/index.htm
The Centers for Disease Control and Prevention (CDC) maintains this home page for Public Health Genomics.

www.ghr.nlm.nih.gov
The U.S. National Library Service maintains the Genetics Home Reference home page.

Menu
of Online Study Tools

A variety of study tools for this chapter are available at our website to deepen your understanding of chapter concepts. Go to

www.cengage.com/sso

to find

- Practice tests
- Flashcards
- Glossary
- Web links
- Animations
- Chapter summaries, learning objectives, and crossword puzzles.

Diet Analysis+ in Action

Go to the Profile report in Diet Analysis+ and record the recommended daily intake of fiber. Then go to the Intake vs. DRI Goals report and record the actual amount of fiber in the foods you ate during the three-day period you recorded. Look at both your original profile and the one you created for Chapter 2, to which you added fruit and whole grains.

How does your intake of dietary fiber compare with the goal in your original profile? In the second profile? Based on what you have read so far, what foods might you add or substitute for foods you regularly eat that would increase your daily intake of fiber? Are there any changes you want to make in your diet based on this investigation?

The Carbohydrates: Sugar, Starch, and Fiber

Ask Yourself . . .

Which of the following statements about nutrition are true, and which are false? For each false statement, what is true?

1. Fruit sugar (fructose) is less fattening than table sugar (sucrose).
2. Foods high in complex carbohydrate (starch and fiber) are good choices when you are trying to lose weight.
3. People with diabetes should never eat sugar.
4. The primary role of dietary fiber is to provide energy.
5. The brain demands the sugar glucose to fuel its activities.
6. Honey and refined sugar are the same as far as the body is concerned.
7. Of all the components of foods that increase one's risk of diseases, sugars are probably the biggest troublemakers.
8. Breads that are brown in color have more fiber than white bread.
9. Some foods labeled sugar-free actually contain calorie-bearing sugars.
10. Artificial sweeteners are safe to use in moderation.

Answers found on the following page.

Once upon a time, bread, potatoes, pasta, and other starchy foods were placed on the dieter's list of most-fattening or "illegal" foods. Today, carbohydrate bashing is still a popular pastime. This unattractive image undoubtedly comes from the practice of serving many carbohydrate-rich foods laden with fat: potatoes with sour cream and butter, vegetables or pasta with rich cream sauces, toast with butter, and salads with fat-rich dressings. People who need to lose weight must limit high-calorie foods, but they are ill advised to try to avoid all **carbohydrates**. Excess calories *are* fattening, and the fat, not the carbohydrate, raises the calorie count the most.

This chapter invites you to learn to distinguish between certain carbohydrates, such as the refined carbohydrates found in many snack crackers and baked goods, versus the complex carbohydrates found in whole grains, fruits, vegetables, and legumes, and others, such as the concentrated sugars of sweetened beverages, candies, and cookies. Since some foods are healthier carbohydrate choices than others, you will learn to choose your carbohydrates by the company they keep (see Table 4-1).

Rabbit said, "Honey or condensed milk with your bread?" [Pooh] was so excited that he said, "Both," and then, so as not to seem greedy, he added, "But don't bother about the bread, please."

FROM *WINNIE THE POOH,*
BY A. A. MILNE (1882–1956)

carbohydrates compounds made of single sugars or multiple sugars and composed of carbon, hydrogen, and oxygen atoms (*carbo* = carbon (C); *hydrate* = water (H_2O)).

It's not the potatoes that are fattening; it's the butter, sour cream, or gravy that they put on us!

BUTTER

SOUR CREAM

TABLE 4-1	Choosing Carbohydrates by the Company They Keep		
Food		**Calories**	**Grams of Fat**
Toast (2 slices)			
with margarine (2 tsp)		188	9
with low-sugar jelly/fruit spread (2 tsp)		139	2
Potato (medium)			
with margarine (1 tsp) and sour cream (1 tbsp)		287	10
with fat-free sour cream (2 tbsp)		223	0
Bagel (medium)			
with regular cream cheese (2 tbsp)		263	11
with fat-free cream cheese (2 tbsp)		188	1
Pasta (1 c)			
with Alfredo sauce (⅓ c)		390	20
with tomato and mushroom sauce (⅓ c)		232	4

SOURCE: Adapted from *Environmental Nutrition* 17:2 (February 1994).

The Body's Need for Carbohydrates

The primary role of carbohydrates is to provide the body with energy (calories). For certain body systems (for example, the brain and the nervous system), carbohydrates are the preferred energy source. Carbohydrates are the ideal fuel for the body. There are only two alternative calorie sources: protein and fat. Protein-rich foods are usually more expensive and provide no advantage over carbohydrates when used to provide fuel for the body. Fat-rich foods might be less expensive, but fat cannot be used efficiently as fuel by the brain and nerves, and diets high in solid fats are associated with many chronic diseases. Thus, of all the possible food-energy sources, carbohydrates are preferred; they provide most of the day's energy for most of the world's people.

The brain and nervous system are sensitive to the concentration of glucose in the blood. Normal blood glucose levels are important for a feeling of well-being. When your blood glucose level becomes too high, you get sleepy; when the concentration falls too low, you get weak and shaky. Only when blood glucose is within the normal range can you feel energetic and alert. The person who wants to feel energetic and alert all day should make the effort to eat so as to maintain blood glucose levels in the normal range to fuel the critical work of the brain and nervous system. Table 4-2 lists a few of the vital functions of carbohydrate in the body and in foods you eat.

Carbohydrate Basics

Carbohydrate-rich foods are obtained almost exclusively from plants. Milk is the only animal-derived food that contains significant amounts of carbohydrate. Carbohydrates are divided into two categories: complex carbohydrates and simple carbohydrates. **Complex carbohydrates** include starch and fiber. Starches make up a

complex carbohydrates long chains of sugars (glucose) arranged as starch or fiber; also called *polysaccharides* (*poly* = many; *saccharide* = sugar unit).

ASK YOURSELF ANSWERS: **1.** False. Fructose and sucrose are equally fattening because they have the same number of calories per gram. **2.** True. **3.** False. People with diabetes need to watch the total carbohydrate in their diets, but they can choose foods with sugar as a small portion of that total. **4.** False. Although certain fibers may provide negligible calories to the diet, fiber's primary role is to provide bulk for the digestive tract. **5.** True. **6.** True. **7.** False. Of all the things in foods associated with risk of diseases, fat is by far the biggest troublemaker. **8.** Not always. A brown color does not always mean the bread is high in fiber. The color can come from molasses or caramel. To be high in fiber, the label must say *whole-grain* or *whole-wheat,* not simply *wheat* bread. *Whole-grain* flour should be listed first in the ingredients list. **9.** True. **10.** True.

Lactose-reduced milk, enzyme solutions containing lactase for treating dairy products, and lactase tablets can help reduce the symptoms of lactose intolerance.

large part of the world's food supply, mostly as grains. For example, such staples as wheat, rice, and corn are rich sources of starch. Fiber is found abundantly in plants, especially in the outer portions of cereal grains, and in fruits, legumes, and most vegetables. **Simple carbohydrates** include naturally occurring sugars in fresh fruits, in some vegetables, in milk and milk products, and as added sugars in concentrated form, such as in honey, corn syrup, or sugar in the sugar bowl. All of these carbohydrates have characteristics in common, but they are of different merit nutritionally. Table 4-3 introduces the different types of carbohydrates.

The Simple Carbohydrates: Monosaccharides and Disaccharides

All carbohydrates are composed of single sugars—known as *monosaccharides*—alone or in various combinations, and all carbohydrates but fiber can be converted to **glucose** in the body. Green plants make glucose from carbon dioxide and water through a process known as *photosynthesis* in the presence of chlorophyll and sunlight, as illustrated in the margin on page 100.

simple carbohydrates (sugars) the single sugars (*monosaccharides*) and the pairs of sugars (*disaccharides*) linked together.

glucose (GLOO-koce) the building block of carbohydrate; a single sugar used in both plant and animal tissues as quick energy. A single sugar is known as a *monosaccharide* (*mono* = one).

TABLE 4-3 **Categories and Sources of Carbohydrate**

Carbohydrate Type	Common Names	Examples of Food Sources
Monosaccharides		
Glucose ●	Dextrose, blood sugar	Fruits, sweeteners
Fructose ■	Fruit sugar, levulose	Fruits, honey, high-fructose corn syrup
Galactose ▲	—	Part of lactose, found in milk
Disaccharides		
Sucrose (glucose + fructose) ●—■ = ● + ■	Table sugar	Beet and cane sugar, fruit, most sweets
Lactose (glucose + galactose) ●—▲ = ● + ▲	Milk sugar	Milk and milk products
Maltose (glucose + glucose) ●—● = ● + ●	Malt sugar	Sprouted seeds
Polysaccharides		
Starches ●—●—●—●—●—●—●	Dextrins*	Potatoes, legumes, corn, wheat, rye, and other grains
Dietary fiber ●—●—●—●—●	Roughage, bulk:	Whole grains, legumes, fruits, vegetables
Insoluble fibers	cellulose, hemicellulose	Wheat products, brown rice, vegetables, legumes, seeds
Soluble fibers	pectins, gums, mucilages, some hemicelluloses	Oat products, barley, legumes, fruits, vegetables, seeds

*Starch can be broken down during food processing to shorter chains of glucose units known as *dextrins*. The word *dextrins* sometimes appears on food labels, because dextrins can be used as thickening agents in foods.

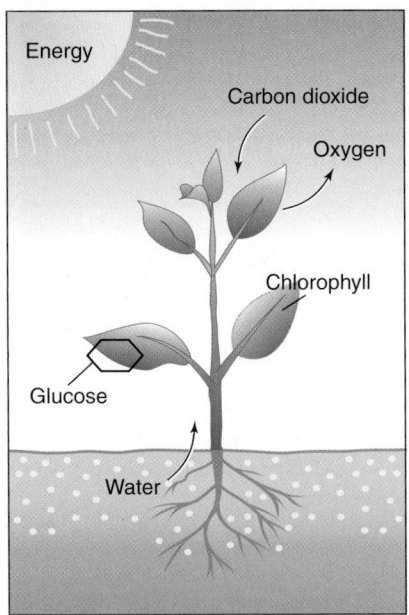

Glucose is the preferred energy source for the brain and nervous system. Plants use chlorophyll and the sun's energy to make glucose through a process called photosynthesis. Water, absorbed by the plant's roots, and carbon dioxide, absorbed through the plant's leaves, combine to form a glucose molecule.

fructose (FROOK-toce) fruit sugar; the sweetest of the single sugars.

sucrose (SOO-krose) a double sugar composed of glucose and fructose. A double sugar is known as a *disaccharide* (*di* = two).

maltose a double sugar composed of two glucose units.

enzymes proteins that facilitate chemical reactions without being changed in the process; protein catalysts.

lactose a double sugar composed of glucose and galactose; commonly known as *milk sugar.*

galactose (guh-LACK-toce) a single sugar that occurs bonded to glucose in the sugar of milk.

lactose intolerance inability to digest lactose as a result of a lack of the necessary enzyme lactase.

kefir a fermented milk made by adding *Lactobacillus acidophilus* and other bacteria that break down lactose.

starch a plant polysaccharide composed of hundreds of glucose molecules, digestible by human beings.

polysaccharide a long chain of 10 or more glucose molecules linked together in straight or branched chains; another term for *complex carbohydrates.*

staple grain a grain used frequently in the diet; for example, corn (in Mexico) or rice (in Asia).

Glucose is not a very sweet sugar, but plants can rearrange its atoms to form another sugar, **fructose,** which is sweet to the taste. Fructose is found mostly in fruits, in honey, and as part of another sugar—table sugar. Glucose and fructose are the most common single sugars in nature.

Some sugars are double sugars—known as *disaccharides*—made by bonding two single sugars together. When glucose and fructose are bonded together, they form **sucrose,** or table sugar, the product most people refer to when they use the term *sugar.* The sweet taste of sucrose comes primarily from the fructose in its structure. It occurs naturally in many fruits and vegetables. Sugar cane and sugar beets are two sources from which sucrose is purified and granulated to various extents to provide the brown, white, and powdered sugars available in the supermarket. Sucrose is one of the two most caloric ingredients of candy, cakes, pastries, frostings, cookies, presweetened ready-to-eat cereals, and other concentrated sweets. (The other major calorie contributor is fat, discussed in Chapter 5.)

Another double sugar, **maltose,** consists of two glucose units. It occurs in sprouting seeds and is produced during the digestion of starch in the human body. The malt found in beer contains maltose. **Enzymes** used in the brewing process break down the long chains of starch in barley and wheat into maltose units.

Finally, **lactose,** the major sugar in milk, is a double sugar made by mammals from galactose and glucose units. A human baby is born with the digestive enzymes necessary to split lactose into its two simple sugars—glucose and **galactose**—so that they can be absorbed by the body. Lactose also facilitates the absorption of calcium and promotes the growth of beneficial bacteria in the intestines. Breast milk and infant formula, which contain lactose, are ideal foods for babies because they provide a simple, easily digested carbohydrate to meet an infant's energy needs.

When you eat a food containing lactose, the enzyme *lactase* in your small intestine first splits the double sugar into single sugars so that they can enter your bloodstream. Many people lose the ability to digest lactose during or after childhood. Thereafter, these people may experience nausea, bloating, abdominal pain or cramping, diarrhea, or excessive gas after drinking milk or eating lactose-containing products. These conditions are created because when the intestinal bacteria use the lactose for energy, they produce gas and other products that irritate the intestine. This condition, called **lactose intolerance,** is inherited by about 70 to 90 percent of the world's people. It is most common in African Americans, Mediterranean peoples, Native Americans, and Asians, and less common in people of northern European origin.[1] It also can develop temporarily in people who are malnourished or sick, making it necessary for them to choose lactose-reduced milk products until they return to a healthy state.

Many people with lactose intolerance are able to consume small amounts of lactose without symptoms.[2] For them, fermented milk products, such as yogurt and **kefir,** aged cheeses, cottage cheese, or specially prepared milk products that have been treated with an enzyme to reduce lactose, may be tolerated. Table 4-4 shows the lactose contents of fermented dairy products compared with the lactose content of milk.

The Complex Carbohydrates: Starch

Complex carbohydrates include starch and fiber. All starchy foods are plant foods. **Starch** is a **polysaccharide** made up of many glucose units bonded together—3,000 or so in each molecule of starch. Shorter carbohydrate chains composed of 3 to 10 glucose molecules are called *oligosaccharides.*

Seeds such as grains, peas, and beans are the richest starch sources. Most societies have a primary or **staple grain** that provides most of the people's food energy. In many Asian nations, the staple grain is rice. In Canada, the United States, and Europe, the staple grain is wheat. If you consider all the food products made from wheat—bread (and other baked goods made from wheat flour), cereals, and pasta—

you will realize how pervasive this grain is in the food supply. The staple grains of other peoples include corn, millet, rye, barley, and oats.

A second important source of starch is the legume family, including dried beans and peas such as butter beans, kidney beans, pinto beans, navy beans, black-eyed peas, chickpeas, lentils, and soybeans. These vegetables are about 40 percent starch by weight and contain abundant protein. Root vegetables (such as yams) and tubers (such as potatoes) are other sources of starch that are important in many societies. Table 4-5 shows the types of carbohydrates found in foods from the USDA MyPlate food guidance system.

The Bread Box: Refined, Enriched, and Whole-Grain Breads

For many people, grains supply much of the carbohydrate, or at least most of the starch, in a day's meals. Because grains have such a primary place in the diet, be sure that the grains you choose—wheat, rice, oats, or corn—contribute the nutrients you need. Also, be sure to learn the meanings of the words associated with the flours that make up the grain products you use: **refined, enriched, fortified,** and **whole grain.** This discussion of the nutritional differences between different breads provides an example of an important principle of nutrition: Foods that are far removed from their original state of wholeness may lack significant nutrients.

The part of the wheat plant that is made into flour and then into bread and other baked goods is the *kernel.* The wheat kernel (a whole grain) has four main parts. The *germ* is the part that grows into a wheat plant, and it contains concentrated food to support the new life. It is especially rich in protein, vitamins, and minerals. The *endosperm* is the soft, white inside portion of the kernel containing starch and protein. The *bran,* a protective coating around the kernel (which is similar in function to the shell of a nut), is also rich in nutrients and fiber. The *husk,* commonly called *chaff,* is unusable for most purposes except for animal feed.

In earlier times, people milled wheat by grinding it between two stones, sifting out the inedible chaff, but retaining the nutrient-rich bran and germ as well as the

Foods such as potatoes, dried beans and peas, rice and whole-grain breads, cereals, and pastas are especially nutritious because of their starch, fiber, vitamin, and mineral content—and because they are virtually fat- and cholesterol-free.

Fresh Food Images/PhotoLibrary

TABLE 4-4	Lactose in Selected Dairy Products	
Dairy Product		**Lactose (grams)**
Cheese (1 oz)		
Parmesan or cream		0.7–0.8
Cheddar or American		0.5
Whole milk (1 c)		11.0
2% low-fat milk (1 c)		9.0–13.0
Fat-free milk (1 c)		12.0–14.0
Chocolate milk (1 c)		10.0–12.0
Lactose-reduced, low-fat milk (1 c)		3.3
Buttermilk (1 c)		9.0–11.0
Yogurt, low-fat (1 c)*		11.0–15.0
Cottage cheese, low-fat (1 c)		7.0–8.0
Ice cream/ice milk (1 c)		9.0–10.0

*Choose yogurt with "active cultures"—the bacteria in these products digest the lactose found in yogurt for their own use.

TABLE 4-5	Types of Carbohydrates Found in Selected Foods in the MyPlate Food Guide			
		Sugar	Starch	Fiber
Grains Group				
Bread, cooked grains		✓	✓	✓
Vegetable Group				
Corn, peas		✓	✓	✓
Green beans		✓		✓
Potato		✓	✓	✓
Tomato		✓		✓
Fruit Group				
Apple, banana		✓ (mostly fructose)		✓
Orange juice		✓ (mostly fructose)		
Dairy Group				
Milk		✓ (lactose)		
Protein Group				
Meat, poultry, fish, eggs				
Legumes		✓	✓	✓

refined grains grains that have been milled to remove the chaff, bran, and germ from the grain—leaving only the endosperm—in order to yield a finer texture and improved shelf life. The process also removes dietary fiber, iron and other minerals, and many B vitamins. Many refined grain products can be high in solid fats and added sugars (e.g., cookies and cakes). The recommended amount of refined grains is no more than 3 ounce-equivalents per day.

enriched refers to the process by which the B vitamins thiamin, riboflavin, niacin, folic acid, and the mineral iron are added to refined grains and grain products at levels specified by law.

fortified term for foods to which nutrients have been added. Typically, commonly eaten foods are chosen for fortification with added nutrients to help prevent a deficiency (iodized salt, milk with vitamin D) or to reduce the risk of chronic disease (juices with added calcium).

whole grain refers to a grain that is milled in its entirety (all but the husk), not refined. Whole grains include wheat, corn, rice, rye, oats, amaranth, barley, buckwheat, sorghum, and millet. Two others, bulgur and couscous, are processed from wheat grains.

FIGURE 4-1

Nutrients in Whole-Grain, Enriched White, and Unenriched White Breads

Iron — 83%, 21%, 100%
Thiamin — 26%, 100%, >100%
Riboflavin — 28%, 100%, >100%
Niacin — 2%, 100%, 98%
Folate — 64%, 100%, >100%
Vitamin B$_6$ — 18%, 18%, 100%
Fiber — 24%, 24%, 100%
Magnesium — 23%, 23%, 100%
Zinc — 36%, 36%, 100%

0 10 20 30 40 50 60 70 80 90 100

Percentage of nutrients in bread
(100% represents nutrient levels of whole-grain bread.)

Key:
= Whole-grain bread
= Enriched white bread
= Unenriched white bread

The nutrients present in the wheat plant at harvest are not always present in the wheat products you eat.

© Juan-Pablo Lira/Getty Images

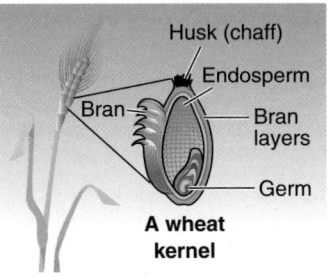

Husk (chaff)
Endosperm
Bran
Bran layers
Germ

A wheat kernel

fibers the indigestible residues of food, composed mostly of polysaccharides. The best known of the fibers are cellulose, hemicellulose, pectin, and gums.

endosperm. Improved milling machinery made it possible to remove the dark, heavy germ and bran as well, leaving a whiter, smoother-textured flour. People came to look on this flour as more desirable than the crunchy, dark-brown, "old-fashioned" flour, but initially were unaware of the nutrition implications. Bread eaters suffered a tragic loss of needed nutrients when they turned to white bread. A 1936 U.S. survey revealed that many people were suffering from deficiencies of the nutrients iron, thiamin, riboflavin, and niacin, which they had formerly received from bread. The Enrichment Act of 1942 standardized the return of these four lost nutrients to commercial flour. This legislation was amended in 1996 to include folic acid, a form of the B vitamin folate, which is considered important in the prevention of certain birth defects. Today, you can assume that almost all breads, grains such as rice, wheat products such as macaroni and spaghetti, and cereals (both cooked and ready to eat), have been enriched. Figure 4-1 shows that although enrichment makes refined bread comparable to whole-grain bread with respect to the enrichment nutrients, it does not do so with respect to other important nutrients. Therefore, although the enrichment of flour and other cereal products does improve them, it doesn't improve them enough: Whole-grain products are still preferable to enriched products. If bread is a staple food in your diet—that is, if you eat it every day—you are well advised to learn to like the hearty flavor of whole-grain breads.

The Complex Carbohydrates: Fiber

The **fibers** of a plant form the supporting structures of its leaves, stems, and seeds. Most fibers are polysaccharides, just as starch is, but with different bonds between the glucose units—bonds that cannot be broken by human digestive enzymes. The term *fiber* is used by almost everyone as though it represents a single entity. Fiber was known generations ago as "roughage." However, many compounds, mostly carbohydrates, make up fiber.* Such compounds are familiar to us as the strings of celery, the skins of corn kernels, and the membranes separating the segments in citrus fruits. When isolated from plants, fiber may be used to thicken jelly (citrus pectin), to keep salad dressing from separating (guar gum), to provide bulk (wheat and other brans), and to create other effects on food texture and consistency.

The bonds that hold the units of fiber together cannot be broken by human digestive enzymes, but some can be broken by the bacteria that reside in the human digestive tract. However, very few (if any) calories are absorbed during this process. Fiber has two forms, **insoluble fiber** and **soluble fiber,** each of which exerts important effects on people's health.

The Health Effects of Fiber

Many health experts encourage consumers to eat more fiber. According to recent evidence, inadequate levels of fiber in the diet are associated with several diseases (see the Miniglossary), whereas the consumption of recommended levels of fiber offers many health benefits.[3] Recall that dietary fiber is found only in plant foods, such as fruits, vegetables, legumes, and whole grains, and that it is the part of plant foods that human enzymes cannot digest.

Table 4-6 shows the potential health benefits from consuming these two types of fiber. As you can see from the table, not all fibers are created equal. Because insoluble and soluble fibers have different effects in the body, it is important to eat a variety of high-fiber foods to get both types.

Both types of fiber, soluble and insoluble, can help with weight control. In the stomach, they convey a feeling of fullness because they absorb water, and some of them delay the emptying of the stomach so that you feel full longer. Also, if you eat many high-fiber foods, you are likely to eat fewer empty-calorie foods such as solid fats and sweets. In fact, producers of some diet aids base the success of their products on the ability of certain fibers in the products to provide bulk and make you feel full.

Foods Rich in Insoluble Fiber

- bran
- brown rice
- green beans
- green peas
- many vegetables
- nuts
- rice
- seeds
- skins/peels of fruits and vegetables
- wheat bran
- whole-grain products

Foods Rich in Soluble Fiber

- barley
- broccoli
- carrots
- corn
- fruits (apples, citrus)
- legumes
- oat bran
- oats
- potatoes
- rye
- seeds

TABLE 4-6		Health Benefits of Dietary Fiber
Health Problem	**Fiber Type**	**Possible Health Benefits**
Obesity	Insoluble/ soluble	Replaces calories from fat, provides satiety, and prolongs eating time because of chewiness of food
Digestive tract disorders: Constipation Diverticulosis Hemorrhoids	Insoluble	Provides bulk and aids intestinal motility
Colon cancer	Insoluble	Speeds transit time through intestines and may protect against prolonged exposure to carcinogens*
Diabetes	Soluble	May improve blood sugar tolerance by delaying glucose absorption
Heart disease	Soluble	May lower blood cholesterol by slowing absorption of cholesterol and binding bile

*This effect is based on epidemiologic studies and is usually observed along with a reduced fat intake. It is unclear whether the protection comes from fiber or from other components that accompany fiber in foods, such as vitamins (for example, folate), minerals (for instance, calcium), or phytochemicals.

insoluble fiber includes the fiber types called cellulose, hemicellulose, and lignin. Insoluble fibers do not dissolve in water.

soluble fiber includes the fiber types called pectin, gums, mucilages, some hemicelluloses, and algal substances (for example, carrageenan). Soluble fibers either dissolve or swell when placed in water.

---

*The woody material of heavy stems and bark is the noncarbohydrate lignin, which is classed by some as a fiber.

appendicitis inflammation and/or infection of the appendix, a sac protruding from the large intestine.

atherosclerosis (ATH-er-oh-skler-OH-sis) a type of cardiovascular disease characterized by the formation of fatty deposits, or plaques, in the inner walls of the arteries.

colon cancer cancer of the large intestine (colon), the terminal portion of the digestive tract.

constipation hardness and dryness of bowel movements associated with discomfort in passing them.

diabetes a disease characterized by insufficiency or relative ineffectiveness of insulin that renders a person unable to regulate the blood glucose level normally.

diverticulosis (dye-ver-tik-you-LOH-siss) outpocketings of weakened areas of the intestinal wall that can rupture, like blowouts in a tire, causing dangerous infections.

hemorrhoids (HEM-or-oyds) swollen, hardened (varicose) veins in the rectum, usually caused by the pressure resulting from constipation.

obesity body weight that is high enough above normal weight to constitute a health hazard (see also Chapter 10).

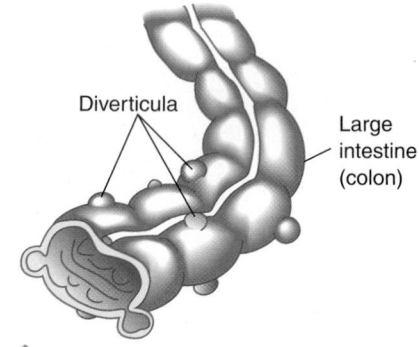

Diverticulosis: The outpocketings of intestinal linings that balloon through the weakened intestinal wall muscles are known as diverticula.

Insoluble fiber (the type in wheat bran, for example) holds water in the colon, thereby increasing bulk and stimulating the muscles of the digestive tract so that they retain their health and tone. The toned muscles can more easily move waste products through the colon for excretion. This prevents constipation, hemorrhoids (in which veins in the rectum swell, bulge out, become weak, and bleed), and diverticulosis (in which the intestinal walls become weak and bulge out in places in response to the pressure needed to excrete waste when bulk is inadequate). These types of fiber may also speed up the passage of food through the digestive tract, thus shortening the time of exposure of the tissue to agents in food that might cause certain cancers, such as colon cancer.[4]

Soluble fiber (the type in legumes, barley, and oats, for example) is credited with reducing the risks of heart and artery disease—atherosclerosis—by lowering the level of cholesterol in the blood. Certain soluble fibers can bind bile (a compound made from cholesterol) in the digestive tract and cause it to be excreted, thus lowering the body's total cholesterol level. As a result, the liver uses its own cholesterol to make new bile for digestion. It also appears that the products of bacterial digestion of soluble fiber in the colon are absorbed into the body and may inhibit the liver's production of cholesterol.[5] Cholesterol levels may also decrease if food sources of soluble fiber (for example, barley, lentils, peas, beans, oat bran, or **psyllium**-enriched cereal) are used as part of a heart-healthy, low-cholesterol diet.[6]

Soluble fibers also improve the body's handling of glucose, even for people with diabetes, perhaps by slowing the digestion or absorption rate of carbohydrates.[7] Blood glucose levels therefore stay moderate, helping to prevent symptoms of diabetes or hypoglycemia. The list of fiber's contributions to human health is impressive!

When people choose high-fiber foods in hope of receiving some of these benefits, they must choose with care. Wheat bran, which is composed mostly of cellulose, has no cholesterol-lowering effect, whereas oat bran and the fibers found in legumes, carrots, apples, and grapefruits do lower blood cholesterol. In contrast, the fiber of wheat bran in whole-wheat bread is one of the most effective stool-softening fibers, and thus can help to prevent constipation and hemorrhoids. If one practical conclusion were to be drawn from what we know about fiber, it would have to be that all whole-plant foods seem to contain many kinds of fibers and thus can be expected to have the entire range of these beneficial effects. To obtain the greatest benefits from fiber, therefore, you must eat a variety of fiber-rich foods rather than taking doses of purified fiber, such as bran, from a single source.

Guidelines for Choosing Carbohydrates

As it happens, most of the recent nutrition goals and guidelines have had one point in common: They tend to favor a return to a more **whole-food,** plant-based diet.

psyllium seed husk, an ingredient in certain cereals and bulk-forming laxatives; contains both soluble and insoluble fibers.

whole food a food that is altered as little as possible from the plant or animal tissue from which it was taken; examples are milk, oats, potatoes, and apples.

We are advised to increase our intakes of complex carbohydrates. Choose plenty of whole foods like these . . .

. . . and fewer foods like these—foods that no longer resemble their original farm-grown products.

Generally speaking, the more a food resembles the original, farm-grown product, the more nutritious it is likely to be. During processing, some nutrients may be lost, and food producers often add nutrient-poor ingredients such as sugar, salt, and fat. For example, a potato contains 20 milligrams of vitamin C, but the same number of calories in French fries contains only about 7 milligrams, and the same number of calories in potato chips contains only 2 milligrams of vitamin C. People who change their diet and begin to eat fewer processed foods and more whole-food, plant-based products can expect many health benefits.

Complex Carbohydrates in the Diet   Complex carbohydrates are thought to be our most valuable energy nutrient. The 2010 *Dietary Guidelines for Americans* includes the following recommendation concerning intakes of complex carbohydrates in the diet:[8]

- Increase vegetable and fruit intake.

- Eat a variety of vegetables, especially dark-green and red and orange vegetables and beans and peas.

- Consume at least half of all grains as whole grains. Increase whole-grain intake by replacing refined grains with whole grains.

A whole food has the nutritional advantage over its processed forms: large whole apple with peel—4 g fiber; ½ c applesauce—1.5 g fiber; ¾ c apple juice—0.2 g fiber.

	Carbohydrate	Added Sugars	Fiber		
Dietary Guidelines	45%–65% of total calories	< 10% of total calories	14 g/1,000 calories consumed		
DRI Committee	45%–65% of total calories	Maximum upper limit: 25% or less of total calories	Adequate Intake (AI)		
				Men	**Women**
			Adults under 50 years	38 g	25 g
			Adults over 50 years	30 g	21 g
Current Intake	49%	16%	14–15 g		

Likewise, the MyPlate food guide illustrates the goal of healthy eating as a shift away from a diet based on high-protein, higher-fat foods to one that uses more of the beneficial complex carbohydrates. Thus, the new MyPlate graphic reflects this shift visually and encourages you to eat proportionally more servings of whole grain products, vegetables, and fruit (see Figure 2-5 in Chapter 2). Table 4-7 lists the current recommendations for choosing carbohydrates for a healthful diet.

Grain products are low in fat, unless fat is added in processing, in food preparation, or at the table. For maximum benefit from grains, choose most often foods that contain little fat (less than 3 grams of fat per serving). According to the *Dietary Guidelines*, at least half of the recommended grain servings consumed should be whole grains, particularly because of their fiber content. Today, whole grains make up only 13 percent of the grains eaten by Americans (see Figure 4-2).[9] See the Eat Well Be Well feature that follows for more benefits and tips for making half your grains whole.

Fiber in the Diet   Undoubtedly, including fiber in a daily meal plan has benefits—but how much is enough? How much is too much? Even fiber has potential to cause harm if taken in excess.

Because fiber carries water out of the body, too much fiber can cause dehydration and intestinal discomfort. Also, iron is mainly absorbed early in the digestive process; thus, because fiber speeds the movement of foods through the digestive system, it may limit the opportunity for absorption of iron and other nutrients. Binders in some fibers link chemically with minerals such as calcium and zinc, making them unavailable for absorption and carrying them out of the body (more about this in Chapter 8). Too much bulk from the diet could reduce the total amount of food consumed and cause deficiencies of both nutrients and energy. The malnourished, some older adults, and children, because they eat small amounts of food anyway, are especially vulnerable to these concerns.

The *Dietary Guidelines* recommend increasing intakes of fiber-rich whole foods and limiting the consumption of foods that contain refined grains, especially refined grain foods that contain solid fats and added sugars.[10] Table 4-7 presents current fiber recommendations.[11] The diet can supply adequate fiber if it is high in fruits, vegetables, legumes, and whole grains—the same recommendations made in the *Dietary Guidelines*.

The wholesale addition of purified fiber to foods is ill advised because it can be taken so easily to extremes. People whose fiber intake consists of only one isolated type of fiber are deprived of the benefits the other types of fiber provide. In contrast, if you include a variety of whole grains, legumes, nuts, fruits, and vegetables in your diet, you get the various types of fiber you need, together with a package of other benefits as well: water, minerals, vitamins, the energy nutrients, and other beneficial substances.

FIGURE 4-2

Whole-Grain Consumption in the Average American Diet

SOURCE: *Healthy People 2010* Midcourse Review (Washington, D.C.: Department of Health and Human Services, Office of Disease Prevention and Health Promotion, 2007).

Proportion of grain servings

Whole grain 13%

All other grains 87%

Adults 20 years and over
Target = 1/2 whole grain

Fruits, vegetables, and foods made from whole grains should form the foundation of a nutritious diet. Why the emphasis on whole-grain foods? Vitamins, minerals, fiber, and other protective substances, including hundreds of phytochemicals, contribute to the health benefits of whole-grain foods. Refined grains are low in fiber and in the protective substances that accompany the fiber in whole grains. A diet rich in whole grains may help protect you against many chronic diseases, including diabetes, heart disease, and certain cancers, and it can also be an effective tool for managing your weight.

Evidence suggests that people consuming diets rich in whole-grain foods, including fiber-rich cereals, have improved insulin sensitivity and are less likely to develop **metabolic syndrome,** a clustering of risk factors associated with the development of type 2 diabetes and heart disease.[12] These risk factors include excess body weight (especially around the abdomen), insulin resistance (reduced ability of the hormone insulin to regulate blood glucose), high levels of fat particles in the blood (known as *triglycerides*), low levels of the body's good form of cholesterol (known as *HDL cholesterol*), and increased blood glucose levels. Recent data show that an estimated 24 percent of adults in the United States have metabolic syndrome. Metabolic syndrome can also be found among children, especially in those who are overweight.[13]

metabolic syndrome a cluster of interrelated symptoms, including obesity, high blood pressure, abnormal blood lipids, and insulin resistance; highly associated with development of type 2 diabetes and heart disease.

© Mark Antman/The Image Works

Enjoy the hearty flavor of whole grains.

How can you incorporate whole grains into your diet to achieve the maximum health benefits and protection from chronic diseases? Very easily! Consider the following three tips:

1 Count to Three.

Consume three or more servings of whole-grain products per day, with the rest of the recommended grains coming from enriched or whole-grain products.* Go to www.ChooseMyPlate.gov to determine your own recommended intake of whole grains. In general, at least half the grains you eat should come from whole grains.

2 Keep It Varied.

Whole grains differ from refined grains in the amount of fiber and nutrients they provide, and different whole-grain foods differ in the amount and type of fiber and nutrients they contain. Therefore, choose a variety of whole grains. Experiment with new or unfamiliar grains like quinoa, bulgur (cracked wheat), barley, buckwheat, and amaranth. The following tips can get you started on building a healthy foundation for your diet by selecting a variety of whole grains daily.

- Start your day with a high-fiber selection—a warm bowl of oatmeal with fresh fruit, bran cereals or shredded wheat with sliced fruit, low-fat apple bran muffin, buckwheat pancakes, or whole-grain English muffin.

- Whole grains are naturally low in fat and added sugars. Keep them that way by going easy on the fats and sugars you add as spreads, seasonings, or toppings, such as butter, margarine, oil, gravy, sauces, sugar, jams, and jellies.

- Whenever possible, substitute whole-grain flour for at least one-fourth of the all-purpose flour called for in baking.

- Make a fiber-rich snack mix from ready-to-eat whole-grain cereals, fat-free popcorn, and nuts, or enjoy air-popped or light microwaved popcorn.

- Try whole-wheat pasta or brown rice, and order sandwiches on whole-grain breads or whole-wheat tortillas for a change.

- Combine whole grains with other tasty, nutritious foods in mixed dishes. Try brown rice stuffing (cooked brown rice, onion, celery, and seasonings) in baked green peppers or tomatoes; pearl barley in vegetable soup; bulgur in vegetarian chili.

(Continued)

*One serving of whole grains from the Grains group equals the following: 1 slice whole-grain bread; ½ whole-grain English muffin, bun, or bagel; 1 ounce whole-grain dinner roll; 1 cup (1 ounce) ready-to-eat whole-grain cereal; ½ cup cooked cereal (oatmeal); ½ cup cooked brown rice or whole-wheat pasta; 5–6 whole-grain crackers; 3 cups popped popcorn.

Choose whole grains and whole-grain foods over highly processed grains and cereals by doing some investigative shopping in the grocery store. Compare products by examining the ingredients list and Nutrition Facts panel (also refer to Figures 2-7 and 2-8 in Chapter 2). Buy products that list a whole-grain or whole-wheat or other whole-grain flour *first* on the label's ingredient list. Look for brown rice, oatmeal, whole oats, bulgur, popcorn, whole rye, graham flour, pearl barley, whole wheat, whole-grain corn. If the label states "wheat flour," "enriched flour," or "degerminated cornmeal," the product is not whole grain. Believe it or not, food products labeled with the words *multigrain, stone-ground, 100% wheat, seven-grain,* or *bran* are usually *not* whole-grain products! Read the ingredient list to see if it is a whole grain.

Check the Daily Value listed in the Nutrition Facts panel to find out if a food is high or low in fiber. The "% Daily Value" for fiber provides a good clue about the amount of whole grain in the product.

A percent Daily Value (% DV) of 5 percent for fiber is low, whereas 20 percent or more is high. Use the Nutrition Facts panel to help you choose products with a higher % Daily Value.

The level of whole grains in ready-to-eat cereal also depends on the sugar content. The more sugar, the less grain and fiber. For example, an ounce of unsweetened oat cereal may contain 3 grams of fiber, but the same size serving of its sweetened counterpart may contain 13 grams of sugar and only 1 gram of fiber. Read the food label's ingredient list, looking for terms that indicate added sugars (sucrose, high-fructose corn syrup, honey, and molasses) that add extra calories. Choose grain products with fewer solid fats and added sugars.

> ### Sample Ingredients List for a Whole-grain Food
>
> **Ingredients:** Whole-wheat flour, water, wheat gluten, soybean and/or canola oil, yeast, salt, honey

A sampling of foods providing added sugars to the diet.

Small amounts of added sugars can be included as part of a nutrient-dense diet and be within your MyPlate calorie allowance:

- 3 tsp for 1,600 calories
- 5 tsp for 1,800 calories
- 8 tsp for 2,000 calories
- 9 tsp for 2,200 calories
- 12 tsp for 2,400 calories

added sugars sugars and other caloric sweeteners that are added to foods during processing or preparation. Added sugars do not include naturally occurring sugars such as those found in milk and fruits.

Added Sugars: Use Discretion   The *Dietary Guidelines for Americans* includes the following recommendation concerning intakes of **added sugars** in the diet:[14]

- Reduce the intake of calories from added sugars.

The MyPlate food guide distinguishes between naturally occurring sugars, such as those found in fruits, and concentrated added sugars containing *empty calories*, as found in soft drinks or other calorically sweetened foods.

The World Health Organization recommends that we limit our intake of added sugars to 10 percent of calories.[15] The Dietary Reference Intakes (DRI) Committee recommends that we choose most often the naturally occurring sugars present in nutrient-dense dairy products and fruits and minimize our intake of added sugars.[16] For people who otherwise meet their nutrient needs, maintain a healthy body weight, and still need additional calories to meet energy needs, the DRI committee sets the *maximum* intake for added sugars at 25 percent of total calories or less, because diets high in added sugars tend to displace intakes of vitamin A, calcium, iron, zinc, and other essential vitamins and minerals. For most people, 3 to 12 teaspoons of sugar per day represent the maximum allowance for one's intake of empty calories. Table 4-8 shows the amounts of sugar in some common products.

Currently, adults in the United States consume an average of 21 teaspoons of sugar a day, or about 16 percent of total calories from sugar.[17] Added sugar intakes are highest in young adults, particularly men aged 19 to 30, with some individuals having added sugar intakes as high as 55 teaspoons of sugar a day. The major sources of added sugars in the American diet include (in decreasing order): regular soft drinks; sugars and candy; cakes, cookies, pies; fruit drinks; dairy products (such as ice cream, sweetened yogurt, and sweetened milk); and sweetened grains (such as waffles and cinnamon toast).[18] A later section of this chapter offers pointers on using discretion in "Keeping Sweetness in the Diet."

How the Body Handles Carbohydrates

Just as glucose is the original unit from which the many different carbohydrate foods are made, so is glucose the basic carbohydrate unit that each cell of the body uses

TABLE 4-8	Sugar in Selected Foods			
Food	Teaspoons of Sugar per Serving (1 tsp = 4 grams)	Food	Teaspoons of Sugar per Serving (1 tsp = 4 grams)
Fruit drink, ade (12 oz)	12	Cereal, Sugar Pops (¾ c)	4
Chocolate shake (10 oz)	9	Shredded wheat (¾ c)	0
Chocolate (2 oz)	8	Fruit, canned in heavy syrup (½ c)	4
Cola (12 oz)	8	Gelatin dessert (½ c)	4
Jellybeans (10)	7	Chocolate milk, 2% (1 c)	3
Yogurt, fruit flavored (1 c)	7	Corn, canned (½ c)	3
Apple pie (⅙ of pie)	6	Ice cream, ice milk, or frozen yogurt (½ c)	3
Cake, frosted (1/16 of cake)	6	Syrup or honey (1 tbsp)	3
Angel food cake (1/12 of cake)	5	Dairy creamer (1 tbsp)	2
Applesauce, sweetened (½ c)	5	Doughnuts, plain (1)	2
Fig bars (1)	5	Catsup (1 tbsp)	1
Fudge (1 oz)	5	Chewing gum (2 sticks)	1
Sherbet (½ c)	5	Cookie, Oreo type (1)	1
Doughnuts, glazed (1)	4	Sugar, jam, or jelly (1 tsp)	1

for energy. Cells cannot use lactose, sucrose, or starch—they require glucose. The task of the **digestive system,** then, is to disassemble the double sugars and starch into single sugars so that these monosaccharides can be absorbed into the blood. Following absorption, the liver converts to glucose any carbohydrates that were not absorbed in the form of glucose, so they can be used by the cells. The cells can then store the glucose, use it for current energy needs, or convert it to fat.

The first digestive enzymes to work on starch are those in the saliva; they begin taking the starch apart, and the enzymes in the stomach and intestines continue digestive action (see Figure 4-3). The enzymes release the individual glucose units, which are absorbed across the intestinal wall into the blood. Cooking facilitates the digestive process by spreading out the tightly packaged chains of glucose so that during **digestion** the digestive enzymes can break the chains down into glucose units for **absorption.** One to four hours after a meal, all the starch has been digested and absorbed and is circulating to the cells as glucose.

If the blood delivers more glucose than the cells need, the liver and muscles take up the surplus to build the polysaccharide **glycogen** (see Figure 4-4). The muscles hold two-thirds of the body's total store of this carbohydrate and use it during exercise. (Chapter 11 explores the relationship between glycogen and exercise and offers tips on how to make the most of glycogen stores.) The liver stores the remaining one-third of the body's glycogen, making it available as needed to help maintain the blood glucose level.

Maintaining the Blood Glucose Level After the glycogen stores are full and the cells' immediate energy needs are met, the body takes a third path for using carbohydrates. For example, perhaps you have eaten a dinner that includes enough carbohydrates to fill your glycogen stores. Now, you watch a movie, eat popcorn, and drink a cola. Your digestive tract delivers glucose from the popcorn and soda to your liver; your liver breaks these extra energy compounds into small fragments and puts them together into the more permanent energy-storage compound: fat (see Figure 4-4). The fat is then released from the liver, carried to the fatty tissues of the body, and deposited there. (Fat cells, too, can utilize excess glucose to make fat for storage.) Unlike the liver cells, however, which can store only about half a day's glucose needs as glycogen, the fat cells can store unlimited quantities of fat.

Refer to Chapter 3 for a review of the structure and function of the many body organs discussed in this section.

digestive system the body system composed of organs and glands associated with the ingestion and processing of food for absorption of nutrients into the body.

digestion the process by which foods are broken down into smaller, absorbable products.

absorption the passage of nutrients or substances into cells or tissues; nutrients pass into intestinal cells after digestion and then into the circulatory system (for example, into the bloodstream).

glycogen (GLEYE-ko-jen) a polysaccharide composed of chains of glucose, manufactured in the body and stored in liver and muscle. As a storage form of glucose, liver glycogen can be broken down by the liver to maintain a constant blood glucose level when carbohydrate intake is inadequate.

FIGURE 4-3

A Summary of Carbohydrate Digestion and Absorption

Salivary glands

Liver

Stomach

Pancreas

Gallbladder

Small intestine

Large intestine

1. Carbohydrate digestion begins in the mouth. The **salivary glands** secrete a watery fluid into the mouth to moisten the food. An enzyme begins digestion by splitting starch into smaller polysaccharides and maltose. This digestion continues after the food is swallowed until stomach acid and enzymes start to digest the salivary enzymes.

2. The **pancreas** produces carbohydrate-digesting enzymes and releases them through the common bile duct into the small intestine. These enzymes split polysaccharides into disaccharides. Then enzymes on the surface of the cells of the **small intestine** break these into simple sugars (monosaccharides). Absorption of the monosaccharides takes place in the small intestine.

3. These simple sugars are then absorbed into the blood and travel to the **liver**; the liver regulates the amount of glucose circulating in the blood in response to the hormones insulin and glucagon.

4. Most fiber passes intact through the digestive tract through the **large intestine** and is eventually excreted with the feces. Some fiber is digested by bacteria in the large intestine.

FIGURE 4-4

Feasting and Fasting

In A, the person is storing energy. In B, the person is drawing on stored energy.

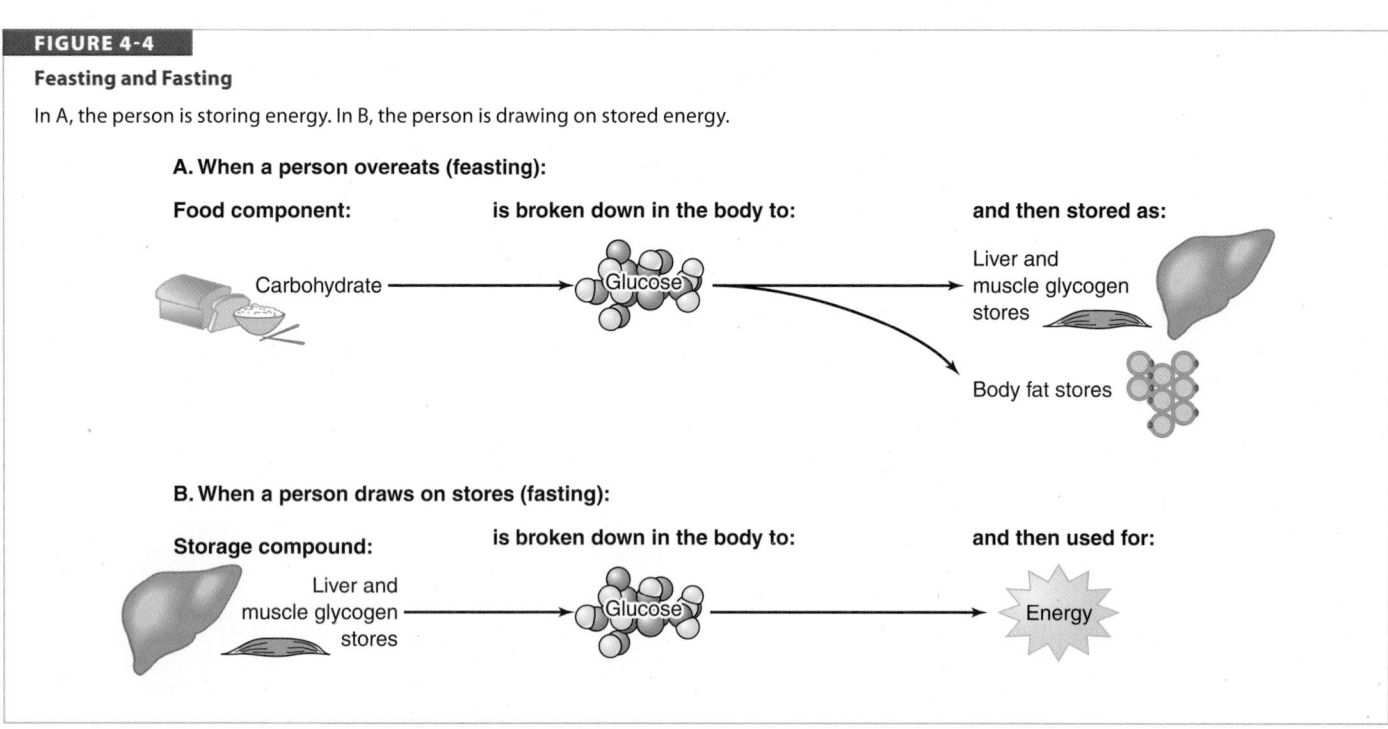

A. When a person overeats (feasting):

Food component:	is broken down in the body to:	and then stored as:
Carbohydrate	Glucose	Liver and muscle glycogen stores
		Body fat stores

B. When a person draws on stores (fasting):

Storage compound:	is broken down in the body to:	and then used for:
Liver and muscle glycogen stores	Glucose	Energy

Maintaining a normal blood glucose level depends on two safeguards, as shown in Figure 4-5. When it gets too high (for example, immediately after a meal), the blood glucose level can be corrected by siphoning off the excess glucose and converting it into liver glycogen, muscle glycogen, or body fat. When it gets too low (for example, following an overnight fast), it can be replenished by drawing on liver glycogen stores.

When the blood glucose level rises, the body adjusts by storing the excess. The first organ to detect the excess glucose is the pancreas, which releases the hormone **insulin** in response. Most of the body's cells respond to insulin by taking up glucose from the blood to make glycogen or fat. Thus, the blood glucose level is quickly brought back down to normal as the body stores the excess. Insulin's opposing hormone, released by the pancreas when blood glucose is too low, is **glucagon,** which draws forth glucose from storage, making it available to supply energy. Insulin and glucagon both work to maintain the concentration of glucose in the blood within the normal range—neither too high nor too low.

insulin a hormone secreted by the pancreas in response to high blood glucose levels; it assists cells in drawing glucose from the blood.

glucagon (GLOO-kuh-gahn) a hormone released by the pancreas that signals the liver to release glucose into the bloodstream.

Blood Glucose Regulation

A. When a person eats, blood glucose rises. High blood glucose stimulates the pancreas to release insulin. Insulin serves as a key for entrance of blood glucose into cells. Liver and muscle cells store the glucose as glycogen. Excess glucose can also be stored as fat.

B. Later, when blood glucose is low, the pancreas releases glucagon, which serves as the key for the liver to break down stored glycogen into glucose and release it into the blood to raise blood glucose levels.

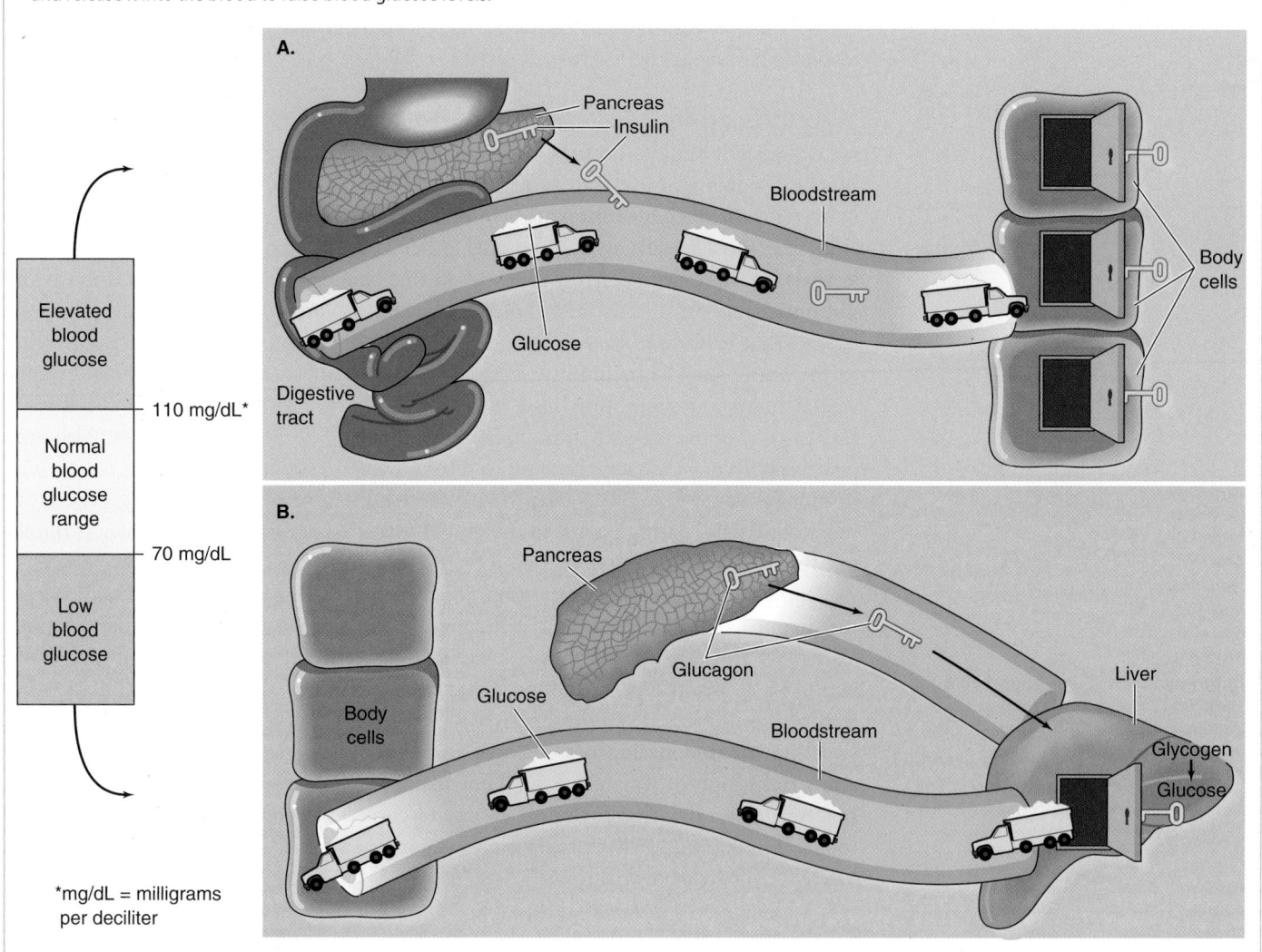

Obviously, when the blood glucose level falls and stores are depleted, a meal or a snack can replenish the supply. An appropriate choice is to eat a balanced meal containing foods that offer carbohydrates (including fiber), protein, and fat, for the following reasons:

- The carbohydrates in the meal provide a quick source of glucose.
- The protein in the meal stimulates glucagon secretion, which opposes insulin and prevents it from storing glucose too quickly.
- The soluble fibers and fat in the meal slow down digestion so that a steady stream of glucose is received rather than a sudden flood.[19]

By these standards, a bowl of whole-grain cereal and fresh fruit with low-fat yogurt for breakfast (offering complex carbohydrate, protein, and fat) is a good choice. Eating well-spaced, carefully chosen meals that provide the balance of protein, carbohydrates, and fat recommended in the *Dietary Guidelines* can prevent rapid rises and falls in the blood glucose level.

Hypoglycemia and Diabetes

In some people, conditions develop in which the body is unable to maintain normal blood glucose levels without careful medical and dietary interventions. The following discussions help you appreciate the difficulties of the person with **hypoglycemia,** whose insulin response is excessive, or the person with **diabetes,** whose insulin response is slow or ineffective.

Hypoglycemia   Suppose that your blood glucose falls, your glycogen reserves are exhausted, but you do *not* eat. Gradually, your body will shift into a fasting state, breaking down its muscle to provide amino acids to the liver. In this state, the liver converts some of these amino acids into glucose to fuel the brain, and the fat released from the muscle cells is used to fuel other cells. (Chapter 9 describes this state, **ketosis,** in more detail.) Most times when this happens, the transition is smooth and not noticeable. At other times, though, when your blood glucose level falls rapidly or below what is normal for you, you may experience symptoms of glucose deprivation to the brain, such as irritability, weakness, and dizziness. Your muscles may become weak, shaky, and trembling, and your heart may race in an attempt to speed more fuel to your brain. These symptoms disappear once calories are consumed. Although *true* hypoglycemia is rare, the symptoms of low blood sugar may occur in people who do not eat often enough to keep their blood sugar levels from dropping. The treatment consists of eating balanced meals and learning to eat more frequently throughout the day (see the sample menu on page 116).

People who suspect they might have *true* hypoglycemia should consult a physician for diagnosis and treatment. It is a serious condition in which abnormal amounts of insulin are secreted, perhaps because of a pancreatic tumor or other health problem. As a result, the person's blood glucose level is constantly too low. This type of hypoglycemia causes symptoms such as headache, confusion, fatigue, nervousness, sweating, or unconsciousness. People with hypoglycemia are told to avoid alcohol and snacks high in simple sugars. They usually benefit from eating six small, evenly spaced meals a day that include a variety of complex carbohydrates and lean protein foods. These foods will increase blood sugar levels slowly and more gradually than highly refined carbohydrates.

Diabetes   Because you know how the blood glucose level is maintained, you can appreciate the problem of people with diabetes whose insulin response is slow or ineffective. Most adults with diabetes fall into this category (see Table 4-9).[20] Even if the

hypoglycemia (HIGH-po-gligh-SEEM-ee-uh) an abnormally low blood glucose concentration, below 60 to 70 mg/100 mL.

diabetes (dye-uh-BEET-eez) a disorder (technically termed *diabetes mellitus*) characterized by insufficiency or relative ineffectiveness of insulin, which renders a person unable to regulate the blood glucose level normally.

ketosis abnormal amounts of ketone bodies in the blood and urine; ketone bodies are produced from the incomplete breakdown of fat when glucose is unavailable for the brain and nerve cells.

TABLE 4-9	Distinctions between the Two Major Types of Diabetes	
	Type 1 Diabetes	**Type 2 Diabetes**
Incidence	5%–10% of cases	90%–95% of cases
Age of Onset	< 30	> 40*
Insulin Deficiency	Yes; pancreas unable to make insulin to meet needs.	In some cases there may be insufficient insulin, or cells may be unresponsive to insulin.
Risk Factors	Genetic predisposition plus environmental factor (for example, viral infection)	Genetic predisposition plus obesity, family history, aging
Treatment	Insulin injections, diet, and physical activity	Weight loss, diet, and physical activity; oral drugs or insulin sometimes needed.

*The incidence of type 2 diabetes is increasing among children. In the majority of cases, the children are overweight or obese and have a family history of type 2 diabetes.

blood glucose level rises too high (**hyperglycemia**), glucose still fails to get into cells and the blood glucose level stays too high for an abnormally long time. The kidneys may respond by shifting some glucose into the urine so that it can be excreted. Short-term effects of hyperglycemia may include thirst, frequent urination, weakness, lack of ability to concentrate, hunger, and blurred vision. People with diabetes must be careful to eat regularly scheduled, balanced meals—providing a constant, steady, moderate flow of glucose to the bloodstream—so the body's insulin response can keep up with the need to process the glucose.

The most common type of diabetes, type 2 diabetes, is usually diagnosed in people over the age of 40 and results from genetic factors, too little insulin, or cellular resistance to insulin.[21] As noted in Figure 4-6, the incidence of type 2 diabetes is rising, mostly because the U.S. population is aging, sedentary, and gaining excess weight.[22] Obesity is considered a major risk factor for this type of diabetes, especially when the excess fat is carried around the abdomen. Usually, eating a healthful diet, exercising, and losing weight can normalize blood glucose levels for people with type 2 diabetes. For others, oral medications and sometimes insulin shots are necessary.

Recent studies show that type 2 diabetes, which normally occurs only in adults, is now affecting an increasing number of children and adolescents. Obesity in children and teens seems to play a major role in the early development of type 2 diabetes.[23]

hyperglycemia an abnormally high blood glucose concentration, often a symptom of diabetes.

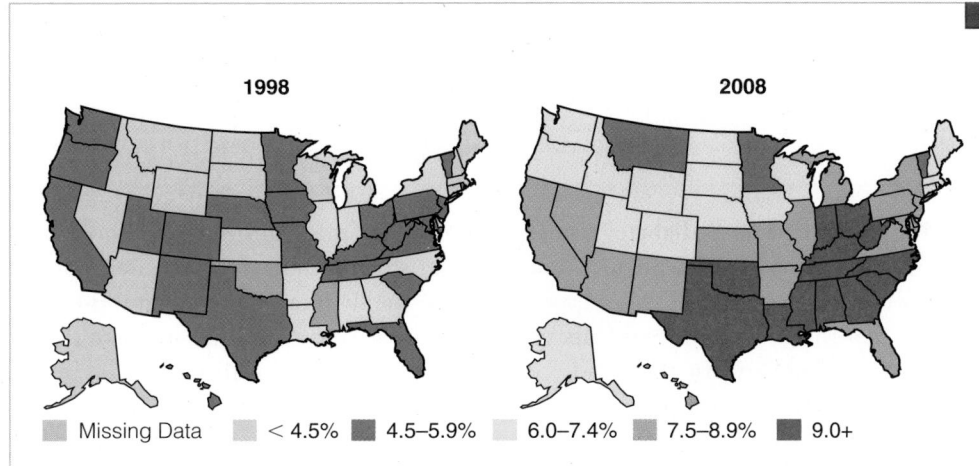

1998   2008

▨ Missing Data	▨ < 4.5%	▨ 4.5–5.9%	▨ 6.0–7.4%	▨ 7.5–8.9%	▨ 9.0+

FIGURE 4-6

Prevalence of Diagnosed Diabetes by State, United States, 1998 and 2008

This series of maps displays the increasing prevalence (existing cases) of diagnosed diabetes among the adult populations of states between 1998 and 2008. In 1998, 15 states had a prevalence of less than 4.5% and 11 states had a prevalence of 6% or greater. However, by 2008, no state had a prevalence of diagnosed diabetes of less than 4.5%, and all but 3 states had a prevalence of 6% or greater, with 13 of those states 9% or higher.

SOURCE: Centers for Disease Control and Prevention.

*W*ith the low-carbohydrate trend now on its way out, many people are wondering, "What are they going to tell us next about carbohydrates?" The answer to that question might lie in a decade-old theory concerning the glycemic effect of carbohydrates, which is making its way back into the spotlight. The **glycemic effect** refers to the effect of food on a person's blood glucose and insulin response: how fast and how high the blood glucose rises and how quickly the body responds by bringing it back to normal. This phenomenon is the basis of many diets, including the "South Beach" and "Nutri System" plans, and it relates to the effect particular carbohydrates have on raising or lowering blood glucose and the time it takes to do so.

© Buntrock/StockFood Munich/Stockfood America

A Closer Look at the Glycemic Effect of Foods

Researchers rank carbohydrate foods according to how fast each is digested into glucose and how much the food causes blood glucose to rise (see Figure 4-7). In general, legumes produce the most even blood glucose response, dairy products next, and fruits and cereals next. Pure sugar produces the greatest rise in the blood glucose level.[24]

In rating this effect with a scale called the **glycemic index (GI)**, foods that quickly raise blood glucose are given a higher score, and lower scores are assigned to those that don't. Some believe this can help people pick "good carbohydrates" that are digested more slowly and therefore increase **satiety**. Nevertheless, many

---

FIGURE 4-7

Glycemic Effect of Carbohydrates

Researchers rank foods according to how fast each is digested into glucose and how much the food causes blood glucose to rise following ingestion. In general, legumes produce the most even blood glucose response, whereas pure sugar produces the greatest rise in the blood glucose level.

others are quick to point out that there are well-known flaws in this system. For example, foods of different glycemic index scores when eaten together behave much differently than when eaten alone, therefore voiding practical use of these scores.

A glycemic index ranks foods on the basis of the extent to which the foods raise the blood glucose level as compared with pure glucose. The blood glucose response to foods with a *high* glycemic index (such as white bread, cornflakes, waffles, and jelly beans) is fast and high. As shown in the margin on page 115, foods with a *low* glycemic index *tend* to be wholesome, fiber-rich foods (such as fruits, legumes, whole oats, bran cereals) and dairy products. Because these foods are more slowly digested, they release glucose gradually into the blood.

The effect of a food on the blood glucose level is important to people with abnormalities of blood glucose regulation, notably diabetes and hypoglycemia. Rapid swings in blood glucose can also affect the performance of an athlete during an endurance event or an office employee following lunch.[25] However, a problem with using the glycemic index as a tool for food selection is that the actual glycemic index of a food can vary from person to person, as well as from meal to meal, depending on the overall composition of the meal. Due to the high variability of the index, it may take

People with type 2 diabetes (those whose bodies make insulin but whose cells resist responding to it) tend to become obese, storing more fat than normal and being constantly hungry. This happens because the liver takes the glucose from the blood that the cells cannot take up, converts it to fat, and then ships it out to the fat cells for storage. Because their body cells are resistant to the insulin produced by the pancreas, insulin does not move glucose into cells effectively, and the message that energy fuels are coming in from food is delayed, causing the person with diabetes to be constantly hungry. Unfortunately, the larger the fat cells become, the more resistant they may be to insulin, thus making the diabetes worse. People with diabetes in some experimenting on an individual's part to learn what combinations of foods yield the most satisfying effects.

However, many nutrition professionals agree about the theory of so-called good carbohydrates. These "good carbohydrates" are usually thought of as containing little added sugar and having adequate amounts of complex carbohydrates and fiber—which often create a better GI score. In fact, this belief is a major underpinning of the new dietary guidelines, which recommend increasing intakes of fruits, vegetables, and whole grains.

Here is a comparison of glycemic index, carbohydrate, fiber, fat, and calorie content of two similar meals:

Meal 1	Meal 2
White bread	Wheat bread
Carrots	Black-eyed peas
Baked white potato	Sweet potato
Soda	Skim milk
Chicken breast	Chicken breast
Watermelon	Peaches
Total: (GI) 492, (fat) 4g, (fiber) 9g, (carb.) 77g, (calories) 620	**Total: (GI) 341, (fat) 3g, (fiber) 11g, (carb.) 84g, (calories) 663**

As shown by the second meal, a lower glycemic index does not necessarily mean that the meal will be lower in calories and/or higher in fiber. This illustrates why use of the GI as a "dieting" tool might be up for debate.

Overall, people are wise to choose diets rich in fruits, vegetables, legumes, whole grains, low-fat dairy, nuts, fish, and lean meats. Quite often, such a diet will have a low glycemic index, plus an array of other healthful attributes. Here are some nutrition "do's" that ensure a healthy diet:

- Choose foods minimally processed and with fiber intact.
- Forgo items with sugar added.
- Look for "whole grain" as the main ingredient.

Whether it is the glycemic index or any other way of dealing with carbohydrates, these principles will not steer you and your diet wrong no matter what the current fad happens to be!

High Glycemic Index Foods

- French, white, and other soft-textured breads or bagels
- rice (medium-grain white or brown)
- certain cereals (Cheerios, Corn Flakes, Rice Krispies)
- waffles, doughnut
- mashed potatoes
- honey, regular soft drinks, sports drinks, jelly beans
- pretzels

Intermediate Glycemic Index Foods

- Watermelon
- Cream of Wheat, instant oatmeal, Shredded Wheat
- sourdough and rye breads
- banana, pineapple, orange juice
- ice cream
- popcorn
- raisins

Low Glycemic Index Foods

- whole-grain, heavy-textured breads
- long-grain brown rice, barley
- bran cereals, whole oats
- apples, oranges, pears, cherries
- carrots, sweet potatoes
- milk, yogurt, soy milk
- cashews, peanuts, soybeans
- baked beans, lentils, other legumes, tomato juice

glycemic effect (also called *glycemic response*) the effect of food on a person's blood glucose and insulin response; how fast and how high the blood glucose rises and how quickly the body responds by bringing it back to normal.

glycemic index (GI) a ranking of foods based on their potential to raise blood glucose levels.

satiety (sah-TIE-eh-tee) the feeling of fullness and satisfaction that occurs after a meal.

their family are urged not to gain excess weight, for it is likely to precipitate the onset of the disease.

Persons with the less common type of diabetes, type 1, produce no (or very little) insulin and require insulin injections to control their blood sugar levels. Type 1 diabetes is more likely to be diagnosed during childhood and has been linked to a possible viral or allergic reaction in susceptible persons that causes the destruction of pancreatic cells that produce insulin. The person who does not produce any insulin is likely to experience a sudden onset of the disease. Such a person may lose weight rapidly because, without insulin, the cells cannot store glucose or fat. Thus, the two

See Chapter 12 for a discussion of *gestational diabetes*—a form of diabetes seen only in pregnancy.

types of diabetes have opposite effects on body weight: one makes the person fat and the other makes the person thin. To determine your own risk for diabetes, take the test in the box on the bottom of this page.

The American Diabetes Association eased its restriction on sugar use in the diets of people with diabetes because sucrose-containing foods do not seem to exert any higher glycemic response than do some starchy foods.[26] However, according to their guidelines, any foods containing sucrose must be used in place of other carbohydrate-containing foods and should not exceed 5 to 10 percent of the total carbohydrate calories. Many people with diabetes learn to count carbohydrates and estimate the caloric content of foods by using the exchange system presented in Appendix A.

A Sample Meal Plan for Keeping Blood Sugar Levels on an Even Keel

Breakfast
Spread ¼ cup low-fat ricotta cheese on 2 halves of a small cinnamon-raisin bagel. Top with 2 tbsp raisins and sprinkle with cinnamon. Heat under broiler for 1 to 2 minutes.

Midmorning Snack
Mix 1 sliced fresh peach with 1 cup low-fat plain yogurt and top with ¼ cup low-fat granola.

Lunch
Stuff a whole-wheat pita with lettuce, ¾ cup Zesty Vegetable Tuna (see recipe), and top with shredded lettuce. Enjoy 1 cup fresh melon cubes as a side dish.

Midafternoon Snack
Spread 2 tbsp of hummus over 4 slices of whole-grain melba toast.

Dinner
¾ cup Quick Kidney Bean Salad (see recipe)
3 oz grilled chicken breast
1 cup steamed broccoli florets
1 medium baked sweet potato
2 tsp butter

Evening Snack
Combine ¾ cup fat-free milk, 1 frozen peeled banana, and 1 tbsp peanut butter and blend for 1 to 2 minutes until smooth.

Zesty Vegetable Tuna
one 6-oz can water-packed tuna
½ cup chopped green pepper
½ cup chopped red pepper
½ cup shredded carrots
3 tbsp light mayonnaise
2 tbsp Dijon mustard
3 tbsp balsamic vinegar

Combine last three ingredients, then add remaining ingredients.
Makes 3 servings.

Quick Kidney Bean Salad
one 16-oz can kidney beans, rinsed and drained
½ cup chopped celery
½ cup chopped onion
⅓ cup orange juice
2 tbsp cider vinegar

Mix all ingredients in a bowl and refrigerate until chilled.
Makes 4 servings.

SOURCE: Adapted from "Getting Even with Low Blood Sugar," *Health Wise* 5 (1999): 6.

What Is Your Risk for Diabetes?

1. I have given birth to a baby weighing more than nine pounds.
   Yes = 1   No = 0   _____

2. I have a sister or a brother with diabetes.
   Yes = 1   No = 0   _____

3. I have a parent with diabetes.
   Yes = 1   No = 0   _____

4. My weight is equal to or above that listed in the At-Risk Weight Chart shown at right.
   Yes = 1   No = 0   _____

5. I am under 65 years of age and I get little or no exercise during a typical day.
   Yes = 1   No = 0   _____

6. I am between 45 and 64 years of age.
   Yes = 1   No = 0   _____

7. I am 65 years or older.
   Yes = 9   No = 0   _____

   Total: _____

At-Risk Weight Chart

If you weigh the same or more than the weight listed here for your height (BMI > 27), you may be at risk for diabetes.

Height, without shoes (feet/inches)	Weight, without clothing (pounds)	Height, without shoes (feet/inches)	Weight, without clothes (pounds)
4'10"	128	5'8"	177
4'11"	133	5'9"	183
5'0"	138	5'10"	188
5'1"	143	5'11"	193
5'2"	147	6'0"	198
5'3"	152	6'1"	204
5'4"	157	6'2"	210
5'5"	162	6'3"	218
5'6"	167	6'4"	221
5'7"	172		

SCORING

3–9 points: Chances are you are at low risk for diabetes now. Maintain a healthy weight, eat nutritious low-fat meals, and exercise regularly to keep risk low.

10 or more points: You are at high risk for diabetes. See your doctor to be tested.

SOURCE: Reprinted with permission, © 2000, American Diabetes Association. For more information about diabetes, call 1-800-DIABETES.

Sugar and Health

Sugar is often in the headlines and has been accused of contributing to a host of human ills, such as tooth decay, obesity, diabetes, heart disease, hyperactive behavior in children, and even criminal behavior. Nevertheless, research studies have not shown a direct link between sugar and any of these conditions, except tooth decay.[27] However, eating too many high-sugar foods could also mean that you are eating inadequate amounts of foods containing essential nutrients. Also, if you eat a lot of high-sugar foods without eating less of other foods, you might be getting too many calories. Excess calories from any energy nutrient, even protein, are stored as body fat. Evidence from population studies in many countries shows that obesity rates increase as sugar consumption increases. One reason may be that many sugary foods, such as candy bars, are also high in fat.

The dietary guideline to limit added sugar intake does not apply to *all* sugars in the diet. The diluted *naturally occurring* sugars found in milk and fruits should not be confused with concentrated refined sugars, such as table sugar, honey, and high-fructose corn syrup. These concentrated sweets should be used in moderation, so that they do not displace needed nutrients.

When people learn that fruit's energy comes from simple sugars, they may think that eating fruit is the same as consuming concentrated sweets such as candy or soft drinks. However, fruits differ from candy and soft drinks in important ways. Their sugars are diluted in large volumes of water, packaged in fiber, and mixed with many vitamins and minerals needed by the body (see Table 4-10). In contrast, concentrated sweets such as honey and table sugar are merely—as the popular phrase calls them—**empty-calorie foods**. How much sugar *do* you eat? Try the Carbohydrate Consumption Scorecard, which gives you the opportunity to check your diet for sugar and fiber and find out.

Keeping Sweetness in the Diet

The taste of sweetness is a pleasure; the liking for it is innate. The *Dietary Guidelines for Americans* recommend that you "reduce the intake of calories from added sugars."[28] To help with this task while still catering to the sweet tooth, consider the tips listed on page 119.

The honey and table sugar (sucrose) are concentrated sweeteners, each containing the simple sugars fructose and glucose. In the orange, these sugars are naturally occurring sugars. Compared with honey or sugar, the orange is more nutrient dense—providing vitamins, minerals, and fiber along with its sugars.

empty-calorie foods a phrase used to indicate that a food supplies calories but negligible nutrients.

TABLE 4-10	Sample Nutrients in Fruits and Sugars						
	Size of 100-Calorie Portion	Carbohydrate (g)	Fiber (g)	Vitamin A (µg)	Vitamin C (mg)	Folate (µg)	Potassium (mg)
Fruits							
Apricots	6	24	6	137	20	18	622
Apple	1 large	25	5	4	12	6	244
Cantaloupe	½ 5-in. melon	23	2	444	116	47	853
Orange	1¼ c sections	26	5	24	120	68	408
Pineapple	1¼ c chunks	24	2.5	2.5	30	20	219
Strawberries	2 c	20	6	4	164	50	478
Watermelon	1¼ c chunks	22	2	56	30	6	352
Sugars							
Cola beverage	8 oz	26	0	0	0	0	3
Honey	1½ tbsp	26	0	0	0	< 1	22
Jelly	2 tbsp	26	0	0	< 1	0	24
Sugar, white	2 tbsp	24	0	0	0	0	0
Sugar, brown	3 tbsp	24	0	0	0	0	90
Adult Dietary Reference Intake (DRI)		**> 130**	**21–38**	**700–900**	**75–90**	**400**	**4,700**

Scorecard Carbohydrate Consumption—Rating Your Diet: How Sweet Is It?

Now that you are aware of some of the sources of added sugars, let's take a look at *your* diet. Check the box that most closely describes your eating habits to see how the foods you choose affect the amount of added sugars in your diet.

How often do you	Seldom or Never	1 or 2 Times a Week	3 to 5 Times a Week	Almost Daily
1. Drink soft drinks, sweetened fruit drinks, or punches?	☐	☐	☐	☐
2. Choose sweet desserts and snacks, such as cakes, pies, cookies, and ice cream?	☐	☐	☐	☐
3. Use canned or frozen fruits packed in heavy syrup or add sugar to fresh fruit?	☐	☐	☐	☐
4. Eat candy?	☐	☐	☐	☐
5. Add sugar to coffee or tea?	☐	☐	☐	☐
6. Use jam, jelly, or honey on bread or rolls?	☐	☐	☐	☐

How Did You Do?

The more often you choose the items listed above, the higher your diet is likely to be in sugars. You may need to cut back on sugar-containing foods, especially those you checked as "3 to 5 times a week" or more. This does not mean eliminating these foods from your diet. You can moderate your intake of sugars by choosing foods that are high in sugar less often and by eating smaller portions.

Check Your Diet for Fiber

To reap the benefits of fiber, healthy adults need between 21 and 40 grams of fiber a day. Most Americans get about 15 grams a day. To figure how much fiber you consume in a day, write down the number of servings you eat in a typical day for each of the food categories below. Next, multiply by the factor shown. (This number represents the average amount of fiber from a serving in that food group.) Then add up the total and determine how you compare to the recommended goal.

Food Group	Number of Servings	Amount (g)
Vegetables (½ c cooked; 1 c raw)	_____ × 2 g	= _____
Fruits (1 medium; ½ c cut; ¼ c dried)	_____ × 2 g	= _____
Dried beans, lentils, split peas (½ c cooked)	_____ × 8 g	= _____
Nuts, seeds (¼ c; 2 tbsp peanut butter)	_____ × 2 g	= _____
Whole grains (1 slice bread; ½ c rice, pasta; ½ bun, bagel, muffin)	_____ × 2 g	= _____
Refined grains[a] (1 slice bread; ½ c rice, pasta; ½ bun, bagel, muffin)	_____ × 1 g	= _____
Breakfast cereals[b] (1 oz)	_____ × g	= _____
	Total = _____	

[a] *Refined grains* refers to products made with white or wheat flour (not whole-wheat flour); white rice, or other processed grains.
[b] Check the Nutrition Facts panel on the cereal you eat to determine the number of grams of fiber in a 1-ounce serving.

Scoring

21–40: Good news! Your fiber intake meets the recommended goal.

15–20: You consume more fiber than the average American. Another serving from the fruit, vegetable, and whole-grain categories would help bring your fiber intake into the recommended goal area.

10–15: Your intake is similar to the average American intake. Consider adding a serving of legumes and additional servings of fruits, vegetables, and whole grains to your daily diet.

0–10: Your fiber intake is too low. Try getting at least five servings of fruits and vegetables, along with three servings of whole-grain foods each day. Add a high-fiber cereal to your morning routine, and consider adding legumes to some of your weekly meals.

- Use less of all sugars, including white sugar, brown sugar, honey, jelly, and syrups.

- Choose sensibly to limit your intake of beverages and foods that are high in added sugars, such as soft drinks, energy drinks, fruit drinks, candy, ice cream, cakes, cookies, and pies.

- Select fresh fruits to satisfy your urge for sweets.

- Learn to read the ingredients list. Check food labels for clues about added sugar content. If any of the following sugars appears first or second in the ingredients list, or if several names are listed, the food is likely to be high in sugar (see Figure 2-8 in Chapter 2). Examples of added sugars include brown sugar, confectioner's sugar, corn syrup, dextrose, fructose, high-fructose corn syrup, fruit juice concentrate, honey, malt syrup, molasses, pancake syrup, powdered sugar, raw sugar, and white sugar.

- For dental health, remember that how frequently you eat sugar is as important as, and perhaps more important than, how much sugar you eat at one time (see the section that follows).

- Alternatives to sweet desserts might be whole-grain crackers, low-fat cheese, and fresh fruit with unsweetened yogurt. Snacks for children could include fruits, vegetables, string cheese, popcorn, homemade fruit juice pops, and other wholesome foods.

- Substitute fruit *juices* or water for fruit drinks, regular soft drinks, and punches that contain considerable amounts of sugar.

- Buy *unsweetened* cereals so that you can control the amount of sugar added. Many cereals are presweetened. Check the Nutrition Facts panel for the grams of sugar present. Many cereals list sugar first or second among their ingredients, indicating a high amount of added sugar.

- Experiment with reducing the sugar in your favorite recipes. Some recipes taste just the same even after a 50 percent reduction in sugar content.

- The sweet spices—allspice, anise, cardamom, cinnamon, ginger, and nutmeg—can replace substantial sugar in recipes. Use half as much sugar and increase one and a half times the amount of spice the recipe calls for. Increasing the amount of extracts like vanilla can enhance sweetness, too. Experiment with other extracts like maple, coconut, banana, and chocolate, or try adding dried fruit to baked goods for extra sweetness and nutrients as you decrease sugar.

Still another alternative for reducing sugar intake is to use sugary foods that convey nutrients as well as calories. For example, rather than using table sugar as a topping, add raisins and banana slices, which are really very sweet. The person with nutrition sense and a taste for sweets can artfully combine the two by using sugar with creative imagination to enhance the flavors of nutritious foods.

You may wonder whether using artificial sweeteners to reduce some of the total sugar in your diet is a safe and recommended strategy. The Spotlight feature "Sweet Talk" at the end of this chapter will help you decide.

Keeping a Healthy Smile   For years, conventional wisdom has held that staying away from candy, cookies, sugary soda pop, and the like is the most important line of dietary defense against cavities. Even Aristotle asked, in about 350 B.C., "Why do figs when they are sweet, produce damage to the teeth?"[29]

These days, dentists advise that cutting down on the amount of sugar eaten is not the only way to prevent **dental caries.** Every bit as important (if not more) as when and how often a food is eaten is whether that food clings to the teeth and lingers in the mouth. It has to do with the mouth's level of acid. Each time you bite into a food, the bacteria that live in your mouth feed on the sugar in it and release an acid that

Naturally sweet foods such as fruit can satisfy your sweet tooth.

© Tom McCarthy Photography

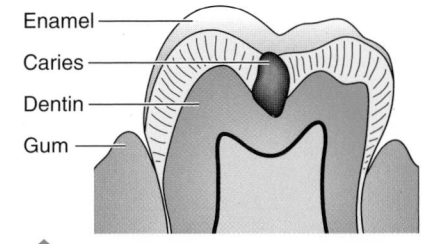

Enamel
Caries
Dentin
Gum

Bacteria living in the mouth feed on sugar found in foods and release an acid that can eat away at tooth enamel and result in a cavity.

dental caries   decay of the teeth; tooth cavities (*caries* = rottenness).

"Just pull my sweet tooth."

For optimal dental health, the American Dental Association (ADA) recommends the following:[32]

- Eat a balanced diet.
- Keep snacking to a minimum, if possible. The ADA recognizes that some people, such as people with diabetes, may require snacks. For others, however, the ADA suggests limiting the number of snacks if brushing the teeth is not possible shortly after eating them.
- Eat sweets with meals rather than between them.
- Brush and floss thoroughly each day to remove dental plaque.
- Use an ADA-accepted fluoride toothpaste and mouth rinse, and talk to your dentist about the need for supplemental fluoride.
- Visit a dentist regularly.
- Do not allow an infant to sleep with a bottle in the mouth that contains sweetened liquids, fruit juices, milk, or formula.

dental plaque a colorless film, consisting of bacteria and their byproducts, that is constantly forming on the teeth.

periodontal disease inflammation or degeneration of the tissues that surround and support the teeth.

nursing bottle syndrome (also called *baby bottle tooth decay*) decay of all the upper and sometimes the back lower teeth that occurs in infants given carbohydrate-containing liquids when they sleep. The syndrome can also develop in toddlers given bottles of liquid to carry around and sip all day.

eats away at tooth enamel. When a large number of bacteria living in a film referred to as **dental plaque** produce enough acid to dissolve a "hole" in the enamel over a period of time, the result is a cavity.

Although consuming sugary items such as soft drinks boosts acid production in the mouth, so does munching on starchy foods such as crackers or pretzels. This happens because enzymes in the saliva can break the carbohydrate in the cracker or pretzel into the simple sugars that bacteria feast on. If a crumb or two from a starchy food gets caught between the teeth, it might provide enough carbohydrate for bacteria to feed on for hours, thereby prolonging the teeth's exposure to acid. In fact, some research indicates that even high-sugar foods such as chocolate bars and hot fudge sundaes are less likely to contribute to cavity formation than stickier, starchier items that are likely to linger in the mouth, such as potato chips and crackers.[30]

To be sure, the carbohydrate content and stickiness of a food are only two of the many factors that influence the food's effect on teeth. Another is how often you eat the food. Each time you eat a carbohydrate-containing food, your teeth are bathed in acids for about 20 minutes. Thus, the more often you eat, the longer your teeth are exposed to harmful acids.[31]

Yet another factor to consider is the combination of foods that you eat. Starchy, sugary foods tend to be less harmful to the teeth when eaten with a meal than when consumed alone. One reason may be that the mouth makes more saliva during a full meal. Saliva production is crucial, because even though it helps make the sugar that encourages bacteria to thrive, it also clears food particles from the mouth and neutralizes destructive acids before they can dissolve the teeth. One study at the University of Iowa's College of Dentistry studied people who nibbled on foods that have a strong tendency to produce acid, such as raisins and chocolate bars, and then chewed sugarless gum. They found that within 10 minutes, the gum helped stimulate the release of enough saliva to neutralize the acid flow that the sweets had caused.[33] In addition, some research suggests that, like sugarless chewing gum, foods such as cheese and peanuts help fight acid attacks that are stimulated by eating carbohydrate-rich foods.[34]

Of course, although saliva helps fight cavities, the best way to ensure good dental health is to brush your teeth as soon as possible after eating, or at least swish the mouth with water after a meal to help rinse the teeth and dislodge stuck particles. Flossing daily is also important because flossing rids the mouth not only of food but also of plaque—before it becomes so widespread that it produces tooth-threatening levels of acid. Regular visits to the dentist also play a role in maintaining dental health. In addition, drinking water containing fluoride, a mineral that helps strengthen tooth enamel, can prevent cavities. People whose drinking water does not supply fluoride should check with a dentist about the need for fluoride supplements.

Along with practicing good dental hygiene, eating a balanced diet helps keep the mouth healthy. One reason is that if the diet lacks essential nutrients, mouth tissues can become compromised, leaving them particularly vulnerable to infection. In fact, some experts believe that **periodontal disease** is especially severe among people who have poor diets.[35] Another reason to eat an adequate diet, particularly for children, is that it helps ensure proper tooth development.[36] Similarly, eating well should rank as a high priority for a pregnant woman, whose unborn baby's teeth, among other things, start forming after just six weeks and begin to harden between the first and second trimester of pregnancy.

Incidentally, parents should never allow an infant to sleep with a bottle filled with sweetened liquids, fruit juices, milk, or formula. Little ones allowed to do so often develop what is known as **nursing bottle syndrome.** As a baby sucks on a bottle, the tongue pushes outward slightly and covers the lower teeth. If the infant falls asleep with the bottle in the mouth, the liquid bathes the teeth, particularly the upper teeth not protected by the tongue, thereby literally soaking them in cavity-causing carbohydrates for hours at a time.

It all started back in 1879—the year that a substance called saccharin was first discovered and found to be able to sweeten foods without adding calories. Since that time, the sale of **artificial sweeteners** for use as tabletop sweeteners—such as Sweet'N Low and Equal—and in artificially sweetened soft drinks, yogurt, and numerous other products has soared.[37] The most common and popular artificial sweeteners approved for use in the United States today are **acesulfame-K, aspartame, neotame, saccharin, stevia,** and **sucralose.** Despite the ever-growing popularity of such sweeteners, doubts about their safety have stirred a good deal of controversy over the years. In addition, confusion about the characteristics of the various alternatives to sugar (see Table 4-11) and their role in managing problems such as obesity, diabetes, and tooth decay prevails among the millions of Americans who use them. The following discussion about sugar substitutes will help put matters into perspective.

Aspartame, sucralose, and other sugar substitutes are used to sweeten a wide variety of products.

© Polara Studios

Is it true that small amounts of saccharin cause cancer?

This assertion has never been proven. It's true that some studies of saccharin (the sweetener in Sweet'N Low) have found that it can cause bladder cancer in laboratory rats. A human, however, would have to drink 850 cans of soft drinks a day to take in a dose equivalent to what the rats in those studies were given. Moreover, some research suggests that although very high doses of saccharin may promote bladder cancer in rats, the sweetener does not have the same effect on mice, hamsters, monkeys, or humans.[38] In addition, investigations of large groups of people have yet to establish any clear-cut link between saccharin consumption and the risk of cancer. One study, conducted by the National Cancer Institute and involving more than 9,000 men and women, showed no association between saccharin use and bladder cancer.[39] Thus, it appears that the health risks, if any, posed by saccharin are minuscule at most. People who fear getting cancer would be better off quitting smoking or reducing the amount of fat in their diet than worrying about putting Sweet'N Low in their coffee now and then. As the American Medical Association's Council on Scientific Affairs has stated, "In humans, available evidence indicates that the use of artificial sweeteners, including saccharin, is not associated with an increased risk of bladder cancer."[40] The government removed saccharin from a list of potential cancer-causing agents, and in December 2000, federal legislation was signed to remove the saccharin warning label that had been required on saccharin-sweetened foods and beverages in the United States since 1977.[41]

Does aspartame cause headaches?

Not in most people. Although the U.S. Food and Drug Administration has received numerous complaints from consumers who claim to suffer from headaches, nausea, anxiety, and other symptoms after consuming aspartame-containing foods or beverages, scientific studies have never confirmed that the sweetener is truly the culprit. Scientists at Duke University, who conducted carefully controlled research into the matter, concluded that tablets containing the amount of aspartame in about 4 liters of diet soft drinks were no more likely to prompt headaches in people than **placebo** pills administered for the sake of comparison.[42] Furthermore, numerous careful scientific investigations carried out both before and after aspartame first appeared on the market have indicated that the product brings about adverse health effects in only a small group of people with a rare metabolic disorder known as **phenylketonuria,** or PKU. People born with PKU must carefully control their consumption of phenylalanine (one of the two amino acids that make up aspartame) to prevent health problems as severe as mental retardation. The American Medical Association's Council on Scientific Affairs, the Centers for Disease Control and Prevention, the Food and Drug Administration, the American Dietetic Association, and numerous other health organizations all consider the sweetener safe for use except by people with PKU.[43]

I have heard a lot about Splenda as a sweetener. Why is it so popular?

Splenda (sucralose) is billed as the sweetener that is "made from sugar so it tastes like sugar." Although taste is purely subjective, it is true that Splenda is a derivative of sugar (sucrose). Splenda is the result of a multi-step process that selectively substitutes three atoms of chlorine for three hydroxyl groups (OH groups) on the sugar molecule, resulting in a sweetener 600 times sweeter than regular sugar. Splenda provides no calories. Because Splenda is

(Continued)

stable, even when used at extremely high or low temperatures, it can be used in a variety of frozen and cooked foods that some of its competitor sweeteners cannot.

Splenda's marketing has gone well beyond simply claiming sugar-like qualities. Its manufacturer has produced "Splenda Granular" that mimics the characteristics and quality of regular sugar; it is intended for use in cooking and specifically baking. Because a molecule of sucralose is 600 times sweeter than a molecule of sucrose, the actual amount of Splenda you would need is much smaller. Therefore, bulking agents are added to allow use of Splenda cup for cup in place of sugar when cooking. A similar product that combines half regular sugar with half "Splenda Granular" is available for baking. Although this product contains calories, the baking results are claimed to be superior, and it also can be measured cup for cup like regular sugar.

Products that use Splenda claim that the taste is superior to other artificial sweeteners, but that is for the consumer to decide. The real advantage to using Splenda is its packaging and its quality of being highly interchangeable with regular sugar.

Can the use of artificial sweeteners bring about weight loss?

If only it were that simple! Obviously, because artificially sweetened foods typically contain fewer calories than their sugar-sweetened counterparts, substituting a low-calorie alternative for a sugar-laden food can allow a person to enjoy sweet-tasting foods and still save calories. But simply *adding* foods sweetened with sugar substitutes to the diet will not do the trick. Moreover, eating artificially sweetened foods to create an excuse to splurge on a different high-calorie food defeats the purpose. A person who drinks diet soda to justify eating an ice cream sundae later is reaping little, if any, benefit. In other words, it's *how* you fit artificial sweeteners into the rest of your diet that really counts. The American Dietetic Association takes the position that individuals who desire to lose weight may choose to use artificial sweeteners, but they should do so as part of a sensible weight management program that includes a nutritious diet and regular physical activity.[44] (See Chapter 10 for healthy weight-management guidelines.)

On the flip side, some consumers may be under the impression that artificial sweeteners bolster the appetite. A report that aspartame sends mixed signals to the brain and thereby increases appetite as well as food consumption spawned the notion that artificial sweeteners can interfere with weight loss.[45] However, that report was based on comparisons of the effects of drinking plain water and drinking aspartame- and sugar-sweetened water, rather than consuming actual foods or beverages that are typically sweetened with aspartame. Moreover, the researchers did not measure how much food people ate after drinking the liquids. Thus, the study did not show how consuming aspartame-sweetened foods influences appetite and eating behavior in the real world.

Since that time, about a dozen investigations into the relationship between appetite and foods and beverages sweetened with the substance have indicated that aspartame either lessens or doesn't affect feelings of hunger or the amount of food ultimately eaten.[46] So for now, at least, it appears that aspartame's impact on appetite need not concern people who are trying to lose weight.

I have heard that foods sweetened with sugar substitutes do not contribute to tooth decay. Is this true?

Not necessarily. As pointed out earlier, any carbohydrate-containing food, be it sugar sweetened or otherwise, can promote tooth decay. When you eat a sugary food, the millions of bacteria lurking on the surfaces of your teeth and between them feast on the sugar and, in the process, release an acid that eats away at tooth enamel. Once inside the mouth, carbohydrates can be devoured by cavity-causing bacteria because enzymes in the saliva break the complex carbohydrates into simple sugars—on which bacteria thrive.

Should people with diabetes eat foods sweetened only with sugar substitutes?

That depends. It is often assumed that because one of the hallmarks of diabetes is high blood sugar (glucose), sugar from foods is the major culprit behind high blood sugar and should therefore be off limits for people with diabetes. But it's not that simple. The total amount of carbohydrate, including both simple and complex, exerts the most influence on blood glucose levels. What's more, many other factors, including the amount of fat and fiber in a food, affect the body's blood glucose response to it. That's why people with diabetes do not have to limit themselves to only artificially sweetened foods. Sugar-containing items can be incorporated into a carefully designed eating plan.

Another reason blanket statements about the use of sugar substitutes are not made for people with diabetes is that the various types of sweeteners behave differently in the body. A group of sugar substitutes known as **alternative sweeteners** (fructose, sorbitol, mannitol, and xylitol), for instance, contain calories but were formerly recommended for people with diabetes because the body generally absorbs them more slowly than it does table sugar. Most of these alternative sweeteners, however, have side effects that detract from their desirability. Some research suggests, for example, that large amounts of fructose may contribute to increases in blood cholesterol levels, making fructose a poor choice for the many people whose diabetes goes hand in hand with heart disease. By the same token, some people experience diarrhea after consuming large amounts of sorbitol or mannitol, and, according to the American Diabetes Association, those sweeteners' overall effect on blood glucose control is insignificant. Products with sorbitol and mannitol may carry the following warning on the label because high intakes increase the risk of malabsorption: "Excess consumption may have a laxative effect." Thus, the role sugar substitutes play

TABLE 4-11	How Sweet It Is		
Name (Trade Name)	**Sweeteners Compared with Sucrose**	**Typical Uses**	**Characteristics[a]**
Sweeteners approved for use in the United States:			
Acesulfame-K (Sunett/Sweet One)	200 times sweeter	Tabletop sweetener, puddings, gelatins, chewing gum, candies, baked goods, desserts, diet drink mixes	Stable in high temperatures; soluble in water. A new blend of Ace-K and aspartame synergizes their sweetening power.
Aspartame[b] (Equal/Nutra Sweet/others)	200 times sweeter	General-purpose sweetener in foods, beverages, chewing gum; tabletop sweetener	Loses sweetness at high temperatures and may lose sweetness over time
Neotame	8,000 times sweeter	General-purpose sweetener in beverages, dairy products, frozen desserts, baked goods, and gums	Stable at high temperatures
Saccharin (Sweet'N Low/ Sweet Twin/Sugar Twin)	450 times sweeter	Soft drinks, tabletop sweetener, baked goods, candy	Stable at high temperatures
Stevia (Truvia/ PureVia/ Sun Crystals)	200 to 300 times sweeter	General-purpose sweetener in foods and beverages, tabletop sweetener	Stable at various temperatures
Sucralose (Splenda)	600 times sweeter	Soft drinks, jams, frozen desserts, dairy products, baked goods, chewing gum, salad dressings, syrups, tabletop sweetener	Stable at various temperatures

[a] In addition to the characteristics listed, all of the artificial sweeteners have a synergistic effect when combined with other sweeteners. In other words, together they enhance each other's sweetness, yielding a combined sweetness greater than the sum of each individual substance's sweetness.

[b] The Food and Drug Administration has set forth an "acceptable daily intake" of 50 milligrams of aspartame per kilogram (2.2 pounds) of body weight. Most people, however, take in much less—on the order of fewer than 5 milligrams per kilogram daily. To reach the FDA's limit, consider that a 150-pound adult would have to consume about nineteen 12-ounce cans of diet soda or 97 packets of Equal, and a 40-pound child would have to consume four 12-ounce cans of diet soft drinks or 24 packets of Equal.

SOURCE: Adapted from "Position of the American Dietetic Association: Use of Nutritive and Nonnutritive Sweeteners," *Journal of the American Dietetic Association* 104 (2004): 255–268.

in the overall diet is a decision that should be made individually with the help of a dietitian and a physician.[47]

Isn't it true that sugar-free chewing gum doesn't contain any calories?

No. Some sugar-free chewing gum is sweetened with certain alternative sweeteners, such as xylitol and sorbitol, also known as **sugar alcohols.** Although sugar alcohols impart a sweet taste and supply calories (about 8 per stick), unlike sucrose, they do not promote tooth decay. Xylitol may actually inhibit the production of tooth-damaging acid by the caries-producing bacteria in the mouth and prevent them from adhering to the teeth. For this reason, the FDA authorizes use of the health claim on food labels that sugar alcohols do not promote tooth decay.

Some manufacturers sweeten chewing gum with artificial sweeteners or a combination of sweeteners (for example, both xylitol and aspartame). Chewing gum sweetened with artificial sweeteners contains negligible calories and also does not promote tooth decay.

I've never heard of neotame. What is it?

Neotame is a derivative of two amino acids, aspartic acid and phenylalanine, and is approximately 7,000 to 13,000 times sweeter than

table sugar. It was approved for use in the United States by the FDA in July 2002. And while neotame contains phenylalanine, it is considered safe for use by individuals with PKU (see page 121). It was approved as a tabletop sweetener and for use in soft drinks, juices, dairy foods, yogurts, jellies, chewing gum, confections, baked goods, sauces, and cereals.[48]

My roommate's parents own a health food store and told me that stevia is better to use instead of artificial sweeteners or sugar since it is a natural product. I've never heard of it. What is it?

Stevia (STEE-vee-uh) is a plant that belongs to the daisy family and is native to Brazil and Paraguay. It has long been used in South America and Asia as a tabletop sweetener and in beverages, candy, gum, baked goods, dairy products, and frozen desserts. The main ingredient in stevia is stevioside, which is 200 to 300 times sweeter than sugar and virtually calorie free. The Food and Drug Administration approved stevia as a tabletop sweetener and food additive in foods and beverages in 2008.[49]

Miniglossary of Sweeteners

acesulfame-K (AY-see-sul-fame) a derivative of acetoacetic acid approved for use in the United States in 1988. Because it is not metabolized by the body, acesulfame-K does not contribute calories and is excreted from the body unchanged. It is currently approved for use in more than 70 countries and is found in more than 100 international products, including chewing gum, gelatins, nondairy creamers, powdered drink mixes, and puddings.

alternative sweeteners nutritive (calorie-containing) sweeteners such as fructose, sorbitol, mannitol, and xylitol.

artificial sweeteners nonnutritive sugar replacements such as acesulfame-K, aspartame, neotame, saccharin, and sucralose.

aspartame a dipeptide (see Chapter 6) containing the amino acids aspartic acid and phenylalanine and used in the United States and Canada since 1981. Although it is digested as protein and supplies calories, it is so sweet that only small amounts, which contribute negligible calories, are needed to sweeten foods. Thus, it is classified as a nonnutritive sweetener. Aspartame is often sold under the trade name NutraSweet, and it is also blended with lactose and an anticaking agent and sold commercially as Equal.

neotame a zero-calorie, heat-stable sweetener approved in 2002 that is a derivative of the dipeptide composed of aspartic acid and phenylalanine. It is 7,000 to 13,000 times as sweet as sugar, depending on how it is used. Unlike aspartame, it is not metabolized to phenylalanine, and thus no special labeling for people with PKU is required.

phenylketonuria (FEN-il-KEY-toe-NYOO-ree-ah) or PKU an inborn error of metabolism, detectable at birth, in which the body lacks the enzyme needed to convert the amino acid phenylalanine to the amino acid tyrosine. If not detected and treated, derivatives of phenylalanine accumulate in the blood and tissues, where they can cause severe damage, including mental retardation.

placebo (plah-SEE-bo) a sham treatment given to a control group; an inert, harmless "treatment" that the group's members cannot recognize as different from the real thing. Using a placebo minimizes the chance that an effect of the treatment will appear to have occurred due to the healing effect of the *belief* in the treatment (known as the *placebo effect*), rather than the treatment itself.

saccharin a zero-calorie sweetener discovered in 1879 and used in the United States since the turn of the century. A possible link to bladder cancer led to saccharin being banned as a food additive in Canada, although it is available there as a tabletop sweetener. Saccharin is the sweetening agent in Sweet'N Low.

stevia (STEE-vee-uh) a small, shrubby perennial with small white flowers; native to Paraguay and Brazil. The plant's sweetness appears to be concentrated in the leaves. Historically, it has been used as a natural sweetener in South America, Japan, Taiwan, China, Malaysia, and South Korea.

sucralose a nonnutritive sweetener approved in 1998 as a tabletop sweetener and for use in a variety of desserts, confections, and nonalcoholic beverages. Sucralose is a noncaloric, heat-stable sweetener derived from a chlorinated form of sugar. Although sucralose is made from sugar, the body does not recognize it as sugar, and the sucralose molecule is excreted in the urine essentially unchanged.

sugar alcohols (mannitol, sorbitol, isomalt, xylitol) can be derived from fruits or commercially produced from dextrose and are absorbed more slowly and metabolized differently than other sugars in the human body. The sugar alcohols are not readily used by bacteria ordinarily found in the mouth and therefore are associated with less cavity formation. Although the sugar alcohols are used as sugar substitutes, they do add calories (about 1.5 to 3 calories per gram) to a food product. They are found in a wide variety of chewing gums, candies, and dietetic foods. Sorbitol and mannitol can have a laxative effect in some people.

Nutrition on the Web

www.cengage.com/nutrition/boyle/personalnutrition8e

Go to the Personal Nutrition site to check for the latest updates to chapter topics or to access links to related websites.

www.healthfinder.gov

Search for information on lactose intolerance, tooth decay, diabetes, and artificial sweeteners.

www.eatright.org

The American Dietetic Association home page includes many nutrition resources and links.

www.ada.org

For information about dental health, search A–Z Oral Health Topics on the American Dental Association's website.

www.foodinsight.org

Search the International Food Information Council (IFIC) site for information on carbohydrates and sugar, sugar and health, sugar substitutes, and fiber.

www.wheatfoods.org

The Wheat Foods Council provides information about wheat and other grains.

www.wholegrainscouncil.org

The Whole Grains Council provides information to help identify whole-grain foods.

www.flaxcouncil.ca

Look for information about the benefits of fiber from flaxseed.

www.cdc.gov/diabetes/index.htm

This government website provides information about diabetes including research, statistics, and educational publications.

www.diabetes.org

Search the American Diabetes Association website for consumer information about diabetes including research, prevention, and living with diabetes.

www.ndep.nih.gov

This website accesses the National Diabetes Education Program (NDEP), a public and private partnership sponsored to improve prevention, diagnosis, and treatment of diabetes.

www.diabetes.niddk.nih.gov

This government website is the National Diabetes Information Clearinghouse (NDIC), and it includes many links.

www.mayoclinic.com

The Mayo Clinic maintains this website, and it provides a variety of health information. Click on the Diseases & Conditions tab, and look up diabetes.

Menu of Online Study Tools

A variety of study tools for this chapter are available at our website to deepen your understanding of chapter concepts. Go to

www.cengage.com/sso

to find

- Practice tests
- Flashcards
- Glossary
- Web links
- Animations
- Chapter summaries, learning objectives, and crossword puzzles.

Diet Analysis+ in Action

Go to the Intake Spreadsheet report in Diet Analysis+ that shows the results of your three-day tracking. List the three foods that contribute the most carbohydrates to your daily intake, the three foods that contributed the most fiber to your daily intake, and the three foods that were either low in fiber or had no fiber. What percentage of your calories came from carbohydrates? How does your intake compare with the recommendation that 45 to 65 percent of your calories should come from carbohydrates? How many grams of fiber did you consume? How does this amount compare with the recommendations of 25 grams per day for women and 38 grams per day for men? What changes could you make in your diet to increase your fiber intake?

The Lipids: Fats and Oils

Ask Yourself . . .

Which of the following statements about nutrition are true, and which are false? For each false statement, what is true?

1. The body can store fat in virtually unlimited amounts.
2. Dietary cholesterol is found only in animal foods.
3. A person's blood level of cholesterol is a predictor of that person's risk of having a heart attack.
4. For the health of your heart, the fat you should avoid eating, most of all, is cholesterol.
5. The more monounsaturated fats you consume, the better it is for your health.
6. Fruits are essentially fat-free.
7. In general, the softest margarines are the most polyunsaturated.
8. Polyunsaturated fat has the same number of calories as saturated fat.
9. All the foods you eat should contain less than 10 percent of calories from saturated fat.
10. No one is free of atherosclerosis.

Answers found on the following page.

*W*hile you and your friend are visiting a health fair at a local shopping mall, you learn that your blood cholesterol level is high. Your friend, a registered dietitian, urges you to see your health care provider to request a blood test to determine the level of triglycerides and the ratio of "good" to "bad" cholesterol in your blood. As you leave the mall, you notice a bookstore display with one diet book that urges you to cut your fat intake to virtually nothing to reduce cancer risk and another book touting the benefits of a high-fat diet for weight loss. Then you stop at the window of a health food store. Your friend points to the freshly ground peanut butter, mentioning to you that it is not hydrogenated; meanwhile, you stare at the bottles of antioxidant and fish oil supplements that line the wall. Later, during the evening news, you see a television commercial featuring a margarine that promises to reduce your blood cholesterol level. "What does all of this mean?" you ask.

Actually, you may know more than you think you know about **lipids**, more commonly called **fats** and **oils**. The most obvious dietary sources of fat are oil, butter, margarine, and shortening. Other food sources that supply fat to the diet are meats, nuts, mayonnaise, salad dressings, eggs, bacon, gravy, cheese, ice cream, and whole milk. You may also know that egg yolk and liver are high in cholesterol,

Fresh Food Images/PhotoLibrary

Chapter Objectives

- *Describe the functions of fat in the body.*
- *Describe the difference between saturated fat, polyunsaturated fat, and monounsaturated fat.*
- *Describe how fat affects heart health.*

Contents

The ultimate reality of nutrition rests with the chemistry of the food we eat and its effects on the processes of life.

R. M. DEUTSCH (1928–1988, NUTRITION AUTHOR AND EDUCATOR)

lipids a family of compounds that includes triglycerides (fats and oils), phospholipids (lecithin), and sterols (cholesterol).

fats lipids that are solid at normal room temperature.

oils lipids that are liquid at normal room temperature.

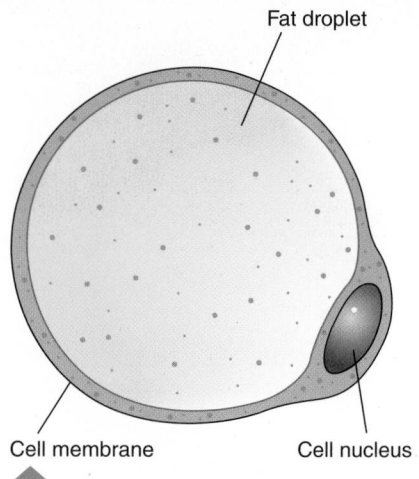

Fat droplet

Cell membrane   Cell nucleus

A Fat Cell

Within the fat cell, lipid is stored in a droplet. This droplet can enlarge, and the cell membrane will grow to accommodate its swollen contents.

and you probably know that the cholesterol in the body is in some way related to heart disease. However, you may be confused by the many terms related to the fat in your diet. This chapter explores the terminology of fat and describes how fats can both contribute to health and detract from it.

A Primer on Fats

Because most people have the impression that fat is bad for them, it may come as a surprise to learn that fats are valuable. More than valuable, some are absolutely essential, and some fats *must* be present in the diet for you to maintain good health. Even if you wanted to, it would be impossible to remove all the fat from your diet, because at least a trace of fat is found in almost all foods.

The Functions of Fats in the Body   Fat is the body's chief storage form for the energy (or calories) from food eaten in excess of immediate need. Fats provide about 60 percent of the energy needed to perform much of the body's work during rest and slightly more during extended bouts of light to moderately intense exercise.

Fat serves as an energy reserve. Whenever you eat, you store some fat, and within a few hours after a meal, you take the fat out of storage and use it for energy until the next meal. Both glucose and fat are stored after meals, and both are released later when needed as energy to fuel the cells' work. However, whereas excess carbohydrate and protein can be converted to fat, the process does not work in reverse. Fat cannot be converted back into protein and carbohydrate. Fat can serve only as an energy fuel for cells equipped to use it.

The body has scanty reserves of carbohydrate and virtually no protein to spare, but it can store fat in practically unlimited amounts. A pound of body fat is worth 3,500 calories, and a person's body can easily carry 30 to 50 pounds of fat without appearing fat at all.

Both fats and oils are found in your body, and both help to keep your body healthy. Table 5-1 summarizes the major functions of fats in the body. Fat is important to all your body's cells as a major component of cell membranes. Natural oils in the skin provide a radiant complexion; in the scalp they help to nourish the hair and make it glossy. Fat insulates the body and cushions the vital organs. It serves as a shock absorber. The fat under the skin also provides insulation from extremes of temperature, thus achieving internal climate control.

TABLE 5-1	The Functions of Fats

Fats in the Body
- Provide a concentrated source of energy to fuel the body's work
- Serve as the body's chief energy reserve
- Form the major components of cell membranes
- Nourish skin and hair
- Insulate the body from extremes of temperature
- Cushion the vital organs to protect them from shock
- Provide raw materials for body compounds such as bile and vitamin D

Fats in Foods
- Provide calories
- Provide satiety
- Carry fat-soluble vitamins and essential fatty acids
- Contribute aroma and flavor

Muscles derive fuel from fat.

Pete Saloutos/Shutterstock.com

ASK YOURSELF ANSWERS: **1.** True. **2.** True. **3.** True. **4.** False. For the health of your heart, the fat you should avoid eating, most of all, is saturated fat. **5.** False. Monounsaturated fats such as olive or canola oil are good for the health of your heart, but like all fats, they should be consumed in moderation. **6.** True. **7.** True. **8.** True. **9.** False. Your *total* diet, not individual foods, should contain less than 10 percent of calories from saturated fat. **10.** True.

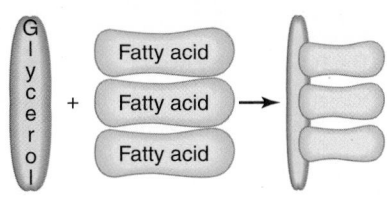

Glycerol + 3 Fatty acids ⟶ **Triglyceride**

▲ Triglyceride

The Functions of Fats in Foods

Fat is a nutrient found in many foods. As the most concentrated source of calories, fat contains more than twice as many calories, ounce for ounce, as protein or carbohydrate. High-fat foods may therefore deliver many *unneeded* calories in only a few bites to the person who is not expending much physical energy. Fats in foods also provide **satiety** by slowing the rate at which the stomach empties. This is the reason you feel fuller longer after eating meals that include fat.

Fat is important for another reason. Some essential nutrients are soluble in fat and therefore are found mainly in foods that contain it. These nutrients are the essential fatty acids (to be described shortly) and the fat-soluble vitamins—A, D, E, and K (described in Chapter 7). Fat also carries many dissolved compounds that give foods their aroma and flavor. This accounts for the aromatic smells associated with foods that are being fried, such as onions or French fries. It also helps explain why a plain doughnut is more flavorful than a plain roll—it is higher in fat. Table 5-1 summarizes the functions of fats in foods.

A Closer View of Fats

When the energy from any of the energy-yielding nutrients is to be stored as fat, it is first broken into fragments: small molecules made of carbon, hydrogen, and oxygen. These fragments are then linked together into chains known as **fatty acids,** the major building blocks of triglycerides, which are the chief form of fat found in the body. **Triglycerides** are composed of three fatty acids that are attached to the compound **glycerol,** which is why glycerol is known as the "backbone" of triglycerides.

About 95 percent of the lipids in foods and in the human body are triglycerides. Other members of the lipid family are the **phospholipids** (of which **lecithin** is one) and the **sterols** (**cholesterol** is the best known of these).

Saturated versus Unsaturated Fats

Fatty acids differ from one another in two ways: in chain length and in degree of saturation. *Chain length* refers to the number of carbon atoms that are hooked together in the fatty acid.* Chain length is significant because it affects the solubility of fat in water; short-chain fatty acids are somewhat soluble in water. Milk, butter, and cheese are rich in short-chain fatty acids; in contrast, vegetable oils and red meat contain triglycerides with long-chain fatty acids, which are insoluble in water.

More significant than chain length, the other way fatty acids differ from one another is in their degree of saturation. *Saturation* refers to the chemical structure: specifically, to the number of hydrogen atoms the fatty acid chain is holding. If every available bond from the carbons is holding (or attached to) a hydrogen atom, we say the chain is a **saturated fatty acid**—filled to capacity, or *saturated*, with hydrogen (see Figure 5-1).

Sometimes, especially in the fatty acids in plants and fish, there is a place in the chain where hydrogens are missing and can easily be added. This is known as an "empty spot," or point of unsaturation (see Figure 5-1). A chain that possesses a point of unsaturation is an **unsaturated fatty acid.** If there is one point of unsaturation, it is a **monounsaturated fatty acid.** If there are two or more points of unsaturation, it is a **polyunsaturated fatty acid.**

satiety the feeling of fullness or satisfaction that people experience after meals.

fatty acids basic units of fat composed of chains of carbon atoms with an acid group at one end and hydrogen atoms attached all along their length.

triglycerides (try-GLISS-er-ides) the major class of dietary lipids, including fats and oils. A triglyceride is made up of three units known as *fatty acids* and one unit called *glycerol.*

glycerol (GLISS-er-all) an organic compound that serves as the backbone for triglycerides.

phospholipid (FOSS-foe-LIP-id) a lipid similar to a triglyceride but containing phosphorus; one of the three main classes of lipids.

lecithin (LESS-ih-thin) a phospholipid, a major constituent of cell membranes; manufactured by the liver and also found in many foods.

sterols (STEER-alls) lipids with a structure similar to that of cholesterol; one of the three main classes of lipids.

cholesterol (koh-LESS-ter-all) one of the sterols, manufactured in the body for a variety of purposes.

saturated fatty acid a fatty acid carrying the maximum possible number of hydrogen atoms (having no points of unsaturation). Saturated fats are found in animal foods such as meat, poultry, and full-fat dairy products as well as in tropical oils such as palm and coconut.

unsaturated fatty acid a fatty acid with one or more points of unsaturation. Unsaturated fats are found in foods from both plant and animal sources. Unsaturated fatty acids are further divided into monounsaturated fatty acids and polyunsaturated fatty acids.

monounsaturated fatty acid a fatty acid containing one point of unsaturation; found mostly in vegetable oils such as olive, canola, and peanut.

polyunsaturated fatty acid (sometimes abbreviated PUFA) a fatty acid in which two or more points of unsaturation occur; found in nuts and vegetable oils such as safflower, sunflower, and soybean as well as in fatty fish.

---

*Short-chain fatty acids contain 4 to 6 carbons, medium-chain fatty acids contain 8 to 10 carbons, and long-chain fatty acids contain 12 or more carbons.

FIGURE 5-1

The Types of Fatty Acids

*Omega, the last letter of the Greek alphabet (ω), is used by chemists to refer to the position of the endmost double bond in a fatty acid counting from the far end (the methyl [CH$_3$] end). The omega-6 fatty acids have their endmost double bonds after the sixth carbon in the chain; the omega-3 acids, after the third. Linoleic acid and linolenic acid are essential fatty acids.

Saturated fatty acid: palmitic acid

Monounsaturated fatty acid: oleic acid (omega-9)*

Polyunsaturated fatty acid: linoleic acid (omega-6)*

Polyunsaturated fatty acid: linolenic acid (omega-3)*

C is a carbon atom.
H is a hydrogen atom.
— is a single bond.
= is a double bond.
C=C is a point of unsaturation.

The Essential Fatty Acids
(Polyunsaturated)

Omega-6

Linoleic acid

Arachidonic acid

Omega-3

Linolenic acid

(DHA)   (EPA)

The Essential Fatty Acids

Linoleic acid and linolenic acid are essential fatty acids and must be supplied by the diet. The fatty acids in fish oils include eicosapentaenoic (EYE-kossa-PENTA-ee-NOH-ic) acid (EPA) and docosahexaenoic (DOE-cosa-HEXA-ee-NOH-ic) acid (DHA), both of which are omega-3 fatty acids.

linoleic (lin-oh-LAY-ik) acid, linolenic (lin-oh-LEN-ik) acid polyunsaturated fatty acids, essential for human beings.

essential fatty acid a fatty acid that cannot be synthesized in the body in amounts sufficient to meet physiological need.

omega-6 fatty acids polyunsaturated fatty acids that have their endmost double bonds after the sixth carbon in the chain.

omega-3 fatty acids polyunsaturated fatty acids that have their endmost double bonds after the third carbon in the chain.

The Essential Fatty Acids

The human body can synthesize all the fatty acids it needs from carbohydrate, fat, or protein, except two—**linoleic acid** and **linolenic acid.** These two cannot be made from other substances in the body or from each other, and they must be supplied by the diet; they are, therefore, **essential fatty acids.** Linoleic acid and linolenic acid are polyunsaturated fatty acids, widely distributed in the food supply, especially in plant and fish oils. Fortunately, they are readily stored in the adult body, making deficiencies unlikely. Still, deficiency symptoms can appear in the person deprived of these acids; these symptoms include a characteristic skin rash and, in children, poor growth.

Omega-6 versus Omega-3 Fatty Acids

One further classification system for unsaturated fatty acids identifies the fatty acids as either **omega-6 fatty acids** or **omega-3 fatty acids** (see Figure 5-1). Of the two essential fatty acids, linoleic acid is an omega-6 fatty acid, related to a whole series of others. Linolenic acid is an omega-3 fatty acid, with a similar family of its own.

Of interest in relation to dietary fat are findings about the omega-3 fatty acids found in fish oils, which offer a protective effect on health. Interest in fish oils was first kindled when someone thought to ask why the Eskimos of Greenland, who eat a diet very high in fat, have such a low rate of heart disease. The trail led to the abundance of fish they eat, then to the oils in those fish, and finally to the omega-3 fatty acids—EPA and DHA—in the oils. Now scientists are unraveling the mystery of what those fatty acids do.[1]

People who eat two to three meals of oily coldwater fish (such as salmon or bluefish) a week tend to have lower blood cholesterol and triglyceride levels and slower clot-forming rates.[2] More than 70 studies have documented other connections as well. Diets rich in omega-3 fatty acids seem to bring about enhanced defenses against cancer (via the immune response) and reduced inflammation in arthritis and asthma sufferers.[3] The Eskimos apparently did well using fish for food. See the Spotlight feature later in this chapter for more about the benefits of omega-3 fats and fish in the diet.

The more unsaturated a fat, the more liquid it is at room temperature. The more polyunsaturated the fat is, the sooner it melts.

Characteristics of Fats in Foods

The amount of unsaturated fatty acids in a fat affects the temperature at which the fat melts. The more unsaturated a fat, the more liquid it is at room temperature. In contrast, the more saturated a fat (the more hydrogen it has), the firmer it is. Thus, of three fats—beef fat, chicken fat, and corn oil—beef fat is the most saturated and the hardest; chicken fat is less saturated and somewhat soft; and corn oil, which is the most unsaturated, is a liquid at room temperature. If your health care provider tells you to use monounsaturated fats or polyunsaturated fats, you can usually judge which ones to choose by their hardness at room temperature. Fats and oils contain mixtures of saturated, monounsaturated, and polyunsaturated fatty acids; the fatty acids that predominate determine whether the fat is solid or liquid at room temperature. Most fats with a high percentage of saturated or *trans* fatty acids are solid at room temperature and are referred to as **solid fats**, while those with more unsaturated fatty acids are usually liquid at room temperature and are referred to as oils.[4] Solid fats are found in most animal foods but also can be made from vegetable oils through the process of hydrogenation, as described below. Figure 5-2 compares the most common fats and oils with regard to their various types of fat. As shown in Figure 5-2, animal fats tend to have a higher proportion of saturated fatty acids (seafood being the major exception), and plant foods tend to have a higher proportion of monounsaturated and/or polyunsaturated fatty acids (coconut oil, palm kernel oil, and palm oil being the exceptions).

Fats in common use. The vegetable oil is polyunsaturated; the olive oil is monounsaturated; the butter is largely saturated; and the margarine, no doubt, is partially hydrogenated.

Because fats differ chemically, they behave differently in foods. To control their characteristics, food manufacturers sometimes alter them—and sometimes the alterations have health consequences, as described later in this chapter.

Points of unsaturation in fatty acids are like weak spots, in that they are vulnerable to attack by oxygen. When the unsaturated points react with oxygen, the oils become rancid. This is why unprocessed oils should be stored in tightly covered containers. If they are to be stored for long periods, they must be refrigerated to prevent spoilage.

One way to prevent spoilage of oils containing unsaturated fatty acids is to change them chemically by **hydrogenation**, but this causes them to lose their unsaturated character and the health benefits that go with it. When food producers want to use a polyunsaturated oil such as corn oil to make a spreadable margarine, they hydrogenate the oil. Hydrogen is forced into the oil, some of the unsaturated fatty acids accept the hydrogen, and the oil becomes harder. The spreadable margarine that results is more saturated than the original oil but not as saturated as butter.

A second way to prevent spoilage of oils is to add a chemical that will compete for the oxygen and thus protect the oil. Such an additive is called an **antioxidant**. Examples are the well-known additives BHA and BHT,* which are listed on the labels of many processed foods, such as breakfast cereals and snack foods, and the natural antioxidants vitamin C and vitamin E. A third alternative, as previously mentioned, is to keep the product refrigerated.

solid fats fats that are usually solid at room temperature. Most solid fats contain saturated fats, cholesterol, and/or *trans* fats. Examples include butter, beef fat, chicken fat, pork fat (lard), stick margarine, and shortening.

hydrogenation (high-drah-jen-AY-shun) the process of adding hydrogen to unsaturated fat to make it more solid and more resistant to chemical change.

antioxidant (an-tee-OKS-ih-dant) a compound that protects other compounds from oxygen by itself reacting with oxygen.

*BHA and BHT are butylated hydroxyanisole and butylated hydroxytoluene.

A Comparison of Saturated and Unsaturated Fatty Acids in Dietary Fats and Oils

All fats are a mixture of saturated, monounsaturated, and polyunsaturated fatty acids (though we usually call them by the name of the fatty acid they have the most of). Look for the least saturated (red), and a good mixture of everything else (for example, canola oil). Polys (yellow and green) and monos (blue) lower cholesterol if you eat them in place of saturated fats. Alpha-linolenic acid (green) is an omega-3 polyunsaturated fat that may protect the heart. Canola, soy, and flaxseed oil are good sources.

SOURCE: USDA Nutrient Database for Standard Reference (Release 14), the National Sunflower Association, and the Flax Council of Canada. Reprinted with permission from *Nutrition Action Healthletter* (July/August 2002), p. 7.

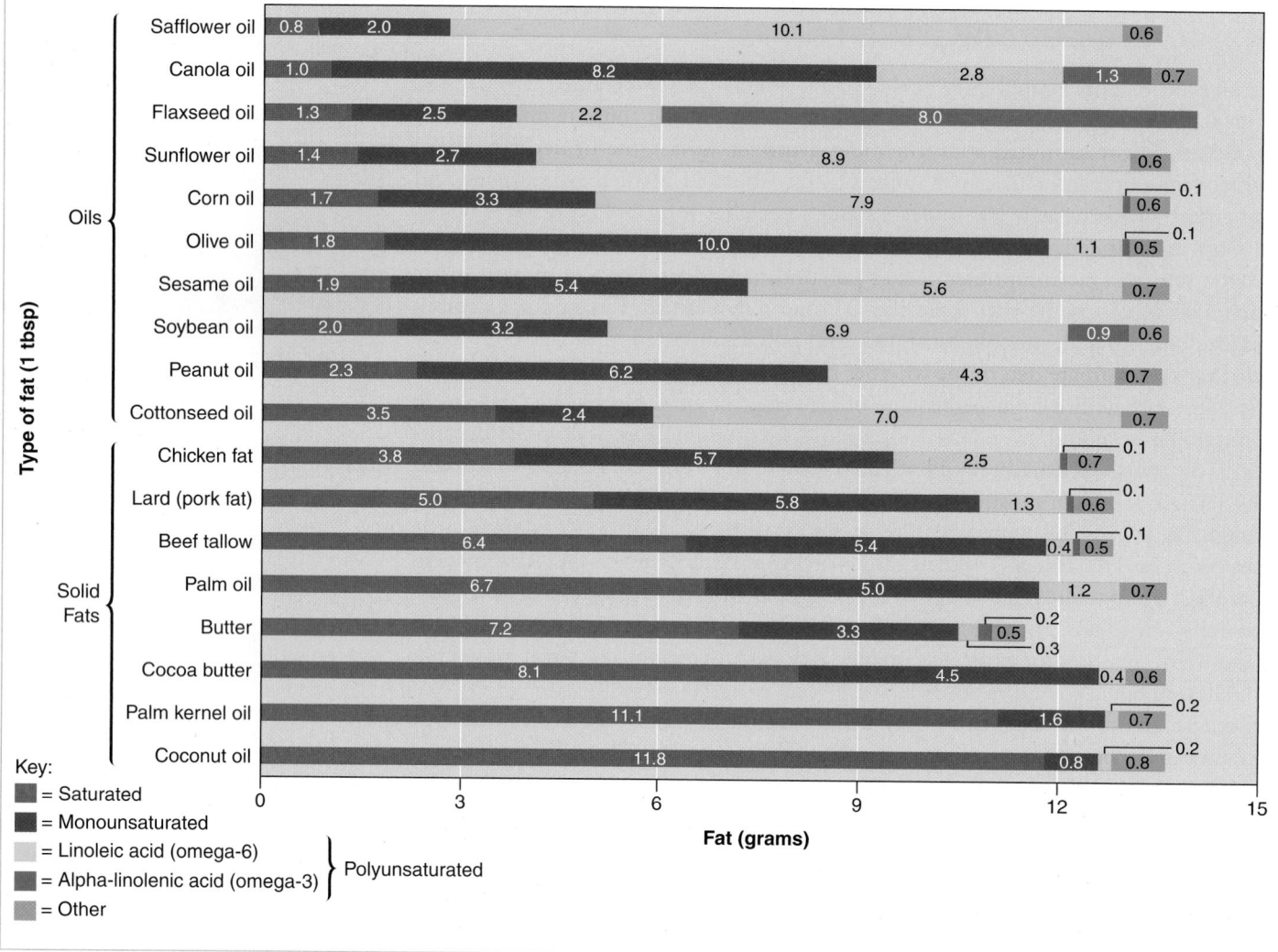

Another way in which the food industry alters natural fats and oils is by adding an **emulsifier** to a food product to allow fats and water to mix and remain mixed in that product. Mayonnaise, margarines, salad dressings, and cake mixes often list an emulsifier on their labels. Monoglycerides and diglycerides are good emulsifiers, as is lecithin, which is an emulsifier found in egg yolk.

The Other Members of the Lipid Family: Phospholipids and Sterols

Lecithin and other phospholipids are important components of cell membranes. Because of the way the phospholipids are constructed, with a water-soluble head and a fat-soluble tail, they serve as emulsifiers in the body by joining with both water and fat. Thus, they help fats travel back and forth across the lipid-containing membranes of cells into the watery fluids on both sides. Because it functions as an emulsifier,

emulsifier a substance that mixes with both fat and water and can break fat globules into small droplets, thereby suspending fat in water.

lecithin is often used as a food additive and thus appears in the ingredients list on food labels. Food manufacturers add lecithin to foods such as mayonnaise, margarine, chocolate, salad dressings, and frozen desserts to keep the fats evenly dispersed with the other ingredients. Lecithin is widespread in food and is also made by the liver in abundant quantities; therefore, most people's diets contain adequate amounts.

Sterols, such as cholesterol, are large molecules with a multiple-ring structure, as shown in the margin on this page. Cholesterol is found only in animal foods and is also made in the body, where it is an important compound with many functions. It is a part of **bile**, which is necessary in the digestion of fats, and it is the starting material from which the sex hormones and many other hormones are made. In the skin, one of cholesterol's derivatives is made into vitamin D with the help of sunlight. It is also an important lipid in the structure of brain and nerve cells. In fact, cholesterol is a part of every cell. Still, although it is widespread in the body and necessary to the body's function, it also is the major component of the plaque that narrows the arteries in the killer disease atherosclerosis. (See the Spotlight at the end of this chapter for more on heart disease.) Figure 5-3 shows the cholesterol content of selected foods.

How the Body Handles Fat

A summary of lipid digestion is shown in Figure 5-4. After digestion in the upper small intestine, the products of fat digestion—fatty acids, glycerol, and **monoglycerides**—must enter the bloodstream to be of use to the body's cells. The shortest free fatty

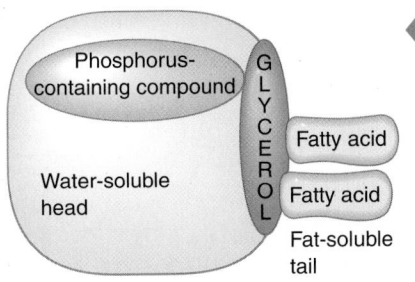

A Phospholipid: Lecithin

This phospholipid (lecithin) consists of a water-soluble head and a fat-soluble tail. Thus, lecithin can travel back and forth across lipid-containing cell membranes to the watery fluids on both sides. A phospholipid is a phosphorus-containing fat.

Cholesterol

Sterols such as cholesterol have a multiple-ring structure. Cholesterol can be:

- Incorporated as an integral part of the structure of cell membranes
- Used to make bile for digestion
- Used to make sex hormones (estrogen and testosterone)
- Made into vitamin D
- Deposited in the artery walls, leading to plaque buildup and heart disease

---

FIGURE 5-3

Cholesterol in Foods

NOTE: Five foods contribute up to 70% of the cholesterol in the U.S. diet: eggs (30%), beef (16%), poultry (12%), cheese (6%), and milk (5%).

Food	Serving Size	Cholesterol (mg)
Beef liver	3 oz	410
Egg (egg yolk)	1	185
Shrimp	3 oz	166
Cheeseburger	double patty	100
Ground beef (lean)	3 oz	85
Pork chop (lean)	3 oz	78
Chicken breast/turkey (skinless)	3 oz	70
Hot dog	1	45
Bologna (beef)	2 slices	40
Tuna (canned, in water)	3 oz	30
Ice cream (10% fat)	1/2 c	30
Cheddar cheese	1 oz	27
Milk (whole)	1 c	25
Fish (baked cod)	3 oz	24
Milk (reduced fat, 2%)	1 c	17
Yogurt (plain, low-fat)	1 c	15
Butter	1 tsp	11
Cottage cheese (reduced fat)	1/2 c	9
Milk (skim)	1 c	5

Daily Value

0 50 100 150 200 250 300 350 400 450

bile a mixture of compounds, including cholesterol, made by the liver, stored in the gallbladder, and secreted into the small intestine. Bile emulsifies lipids to ready them for enzymatic digestion and helps transport them into the intestinal wall cells.

monoglyceride (mahn-oh-GLISS-er-ide) a glycerol molecule with one fatty acid attached to it. A *diglyceride* is a glycerol molecule with two fatty acids attached to it.

FIGURE 5-4

A Summary of Lipid Digestion and Absorption

A. Digestion of Fat

1. Mouth
Some hard fats begin to melt as they reach body temperature.

2. Stomach
The stomach's churning action mixes fat with water and acid. A stomach enzyme accesses and breaks apart a small amount of fat. Fat is last to leave the stomach.

3. Liver, Gallbladder, and Small Intestine
Once in the small intestine, fat encounters **bile**, an emulsifier made in the liver (see Part B). The gallbladder, a storage organ, squirts bile into the contents of the small intestine to blend the fat with the watery digestive secretions.

4. Pancreas
Fat-digesting enzymes from the pancreas (pancreatic lipase) enter the small intestine. The enzymes can attack fat only after emulsification by bile. They break down the triglycerides to fatty acids, glycerol, and monoglycerides.

5. Large Intestine
Some fat and cholesterol, trapped in fiber, is carried out of the body with other wastes.

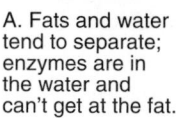

Mouth
Salivary glands
Liver
Stomach
Pancreas
Gallbladder
Small intestine
Large intestine

B. Emulsification of Fat by Bile

Fat
Water
Bile
Enzymes
Enzyme
Emulsified fat

A. Fats and water tend to separate; enzymes are in the water and can't get at the fat.

B. Bile (emulsifier) has affinity for fats and for water so it can bring them together.

C. Small droplets of emulsified fat. The enzymes now have access to the fat, which is mixed in the water solution.

C. Absorption of Fat: The Chylomicron

Most of the newly digested fats are absorbed into lymph as part of a special package—the chylomicron. A chylomicron (lipoprotein) contains an interior of triglycerides and cholesterol surrounded by phospholipids. Proteins cover the structure. Such an arrangement of hydrophobic (water-fearing) molecules (the fatty acids) on the inside and hydrophilic (water-loving) molecules (proteins) on the outside allows lipids to travel through the watery fluids of the body.

Cholesterol
Phospholipid
Protein
Triglyceride

---

acids pass into the cells that line the intestine by simple diffusion. Because these short-chain fatty acids are somewhat water-soluble, they can, without any further processing, enter the body's capillaries. Like the products of carbohydrate digestion, short-chain fatty acids are transported from these capillaries through collecting veins to the capillaries of the liver. The liver cells pick them up and convert them to other substances the body needs. The glycerol follows the same path as the short-chain fatty acids because it, too, is water-soluble.

The larger products of fat digestion (long-chain fatty acids, cholesterol, and phospholipids) are insoluble in water, a difficulty that must be overcome. The body's fluids—**lymph** and blood—are watery and will not accept these larger molecules as they are. The longer-chain fatty acids do pass into the intestinal cells, but there they reconnect with glycerol or with monoglycerides, forming new triglycerides. Then the cells package them for transport before releasing them into the lymph system.

The cells allow triglycerides and other lipids to form and combine with special proteins to make **chylomicrons,** one of the four types of **lipoproteins** found in the blood (see Figure 5-4, Part C). Within the body, the larger fats always travel in the form of lipoproteins. In this ingenious configuration, the water-soluble proteins enable the fats to travel in the watery body fluids. That way, when the tissues of the body need energy from fat, they can extract the energy they need from the lipoproteins in the blood and lymph. The lipoprotein remnants that remain are picked up

lymph (LIMF) the body fluid that transports the products of fat digestion toward the heart and eventually drains back into the bloodstream; lymph consists of the same components as blood with the exception of red blood cells.

chylomicron (KIGH-low-MY-krahn) a type of lipoprotein that transports newly digested fat (mostly triglyceride) from the intestine through lymph and blood.

lipoproteins (LYE-poh-PRO-teenz) clusters of lipids associated with protein that serve as transport vehicles for lipids in blood and lymph. The four main types of lipoproteins are chylomicrons, VLDL, LDL, and HDL.

FIGURE 5-5

The Lipoproteins

A. Functions and Interactions of Lipoproteins

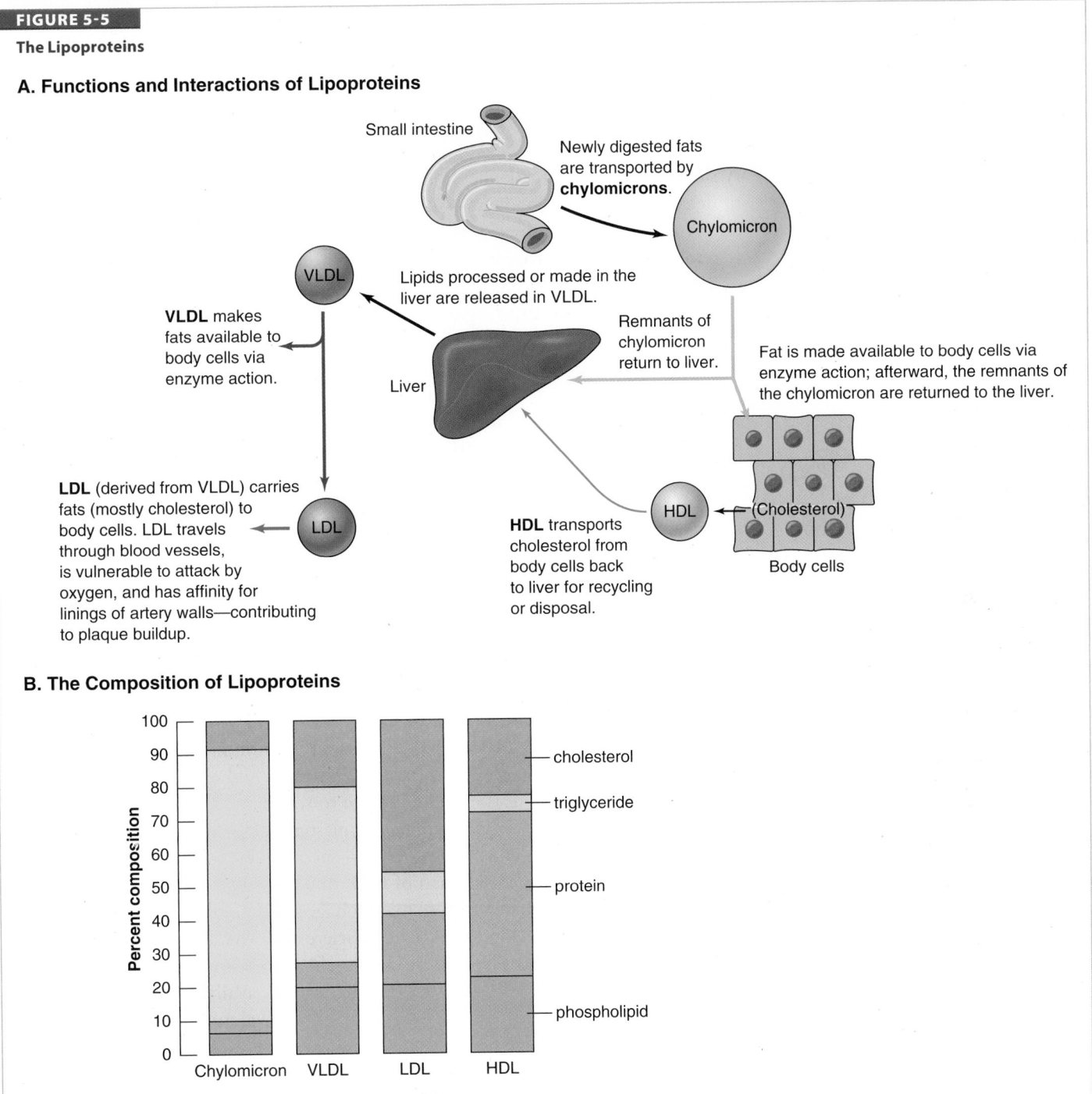

B. The Composition of Lipoproteins

by the liver, which dismantles them and reuses their parts. The characteristics of the four types of lipoproteins circulating in the blood are shown in Figure 5-5.

Lipoproteins are very much in the news these days. In fact, the health care provider who measures your blood lipid profile is interested not only in the types of fats in your blood (triglycerides and cholesterol) but also in the lipoproteins that carry them. One distinction among types of lipoproteins is very important because it has implications for the health of the heart and blood vessels: that distinction is between low-density lipoproteins (**LDL**) and high-density lipoproteins (**HDL**). The more protein in the lipoprotein molecule, the higher its density.

Of course, the intestine also processes the few other dietary fats, such as phospholipids and cholesterol, that may have entered the body in food. These fats enter

VLDL (very-low-density lipoprotein) carries fats packaged or made by the liver to various tissues in the body.

LDL (low-density lipoprotein) carries cholesterol (much of it synthesized in the liver) to body cells. A high blood cholesterol level usually reflects high LDL.

HDL (high-density lipoprotein) carries cholesterol in the blood back to the liver for recycling or disposal.

- High LDL blood cholesterol level
- Low HDL blood cholesterol level
- High blood pressure
- Cigarette smoking
- Obesity
- Physical inactivity
- Diabetes
- An "atherogenic" diet (high in saturated and *trans* fats, and low in fruits, vegetables, legumes, and whole grains)

Other risk factors cannot be changed: advanced age, male gender, and family history

LDL, the "bad" cholesterol (L = Lousy)

HDL, the "good" cholesterol (H = Healthy)

trans fat a type of fatty acid created when an unsaturated fat is hydrogenated. Found primarily in margarines, shortenings, commercial frying fats, and baked goods, *trans* fatty acids have been implicated by research as culprits in heart disease.

oxidized LDL-cholesterol (o-LDL) the cholesterol in LDLs that is attacked by reactive oxygen molecules inside the walls of the arteries; o-LDL is taken up by scavenger cells and deposited in plaque.

foam cells cells from the immune system containing scavenged oxidized LDL-cholesterol; thought to initiate arterial plaque formation.

the circulation the same way the triglycerides do, and they also travel packaged in chylomicrons. After transport, they end up in the liver as part of the chylomicron remnants.

"Good" versus "Bad" Cholesterol

A silent, symptomless risk factor for heart disease is much talked about but little understood: elevated blood cholesterol levels. Blood cholesterol levels may be high for any of a number of reasons. Some people inherit tendencies to make too much cholesterol or to fail to destroy it on schedule. Others have high blood cholesterol for any or all of the following lifestyle reasons: eating too much saturated fat and **trans** fat, exercising too little, or carrying too much weight.[5] The blood lipid profile mentioned earlier can give you an idea of your standing as to this risk factor. (Table 5-8 in the Spotlight at the end of this chapter shows how to interpret your blood lipid profile.)

The underlying cause of most heart disease is atherosclerosis: a narrowing of the arteries caused by a buildup of cholesterol-containing plaque in the arterial walls (this chapter's Spotlight further discusses atherosclerosis and heart disease). The initiating step in the process of atherosclerosis is some form of injury to or inflammation of the artery wall.[6] High blood pressure, high blood cholesterol levels, and cigarette smoke are potential sources of injury, as are the other causes listed in the margin (see also Table 5-7 later in this chapter). Raised LDL concentrations in the blood are a sign of high heart attack risk because LDLs in the blood tend to deposit cholesterol in the arteries.

Researchers now theorize that LDL-cholesterol is damaging to the artery walls once it has been oxidized. Circulating LDL-cholesterol is more likely to settle along the linings of the artery walls after it first reacts with an unstable form of oxygen to become **oxidized LDL-cholesterol (o-LDL)** (see Figure 5-6).

Researchers believe that scavenger cells from the immune system known as *macrophages* ingest more and more of the o-LDL particles and eventually become **foam cells**—so called because of their resemblance to sea foam. These foam cells eventually burst and deposit their accumulated cholesterol as debris on the arterial wall, leading to the development of fatty streaks—the precursors of plaque. Thus, scientists speculate that the oxidized form of LDL catalyzes the process of atherosclerosis in the artery walls by attracting these macrophages to the arterial area.

A key step in the development of atherosclerosis is the accumulation of o-LDL in the walls of the artery. The more LDLs in the circulating blood, the greater the chance for oxidation to occur. This leaves more o-LDL available for ingestion by the scavenger cells and more debris left behind in the arterial wall. In theory then, steps to reduce the total amount of LDL circulating in the blood (reducing saturated fat and *trans* fat in the diet) or steps to prevent the oxidation of LDL-cholesterol (ample antioxidants in the body) would reduce the formation of foam cells and cause less injury to arterial walls. Thus, the process of atherosclerosis could be slowed or possibly prevented. See Table 5-2 for the effects of various types of fat on blood lipids.

Most blood cholesterol is carried in LDL and correlates *directly* with heart disease risk, but some is carried in HDL and correlates *inversely* with risk. In fact, the most potent single predictor of heart attack risk may be the HDL level. Research indicates that an acceptable total blood cholesterol reading of 200 milligrams per deciliter (mg/dL) or below may not be protective against heart disease if the HDL level is low. Raised HDL concentrations relative to LDL represent cholesterol on its way out of the arteries back to the liver—and a reduced risk of heart attack.

Some cases of elevated blood cholesterol do not respond to changes in lifestyle. In such cases, cholesterol-lowering drugs might be prescribed. However, for many people, a few simple changes in diet can improve cholesterol readings, as discussed in the Eat Well Be Well feature later in this chapter.

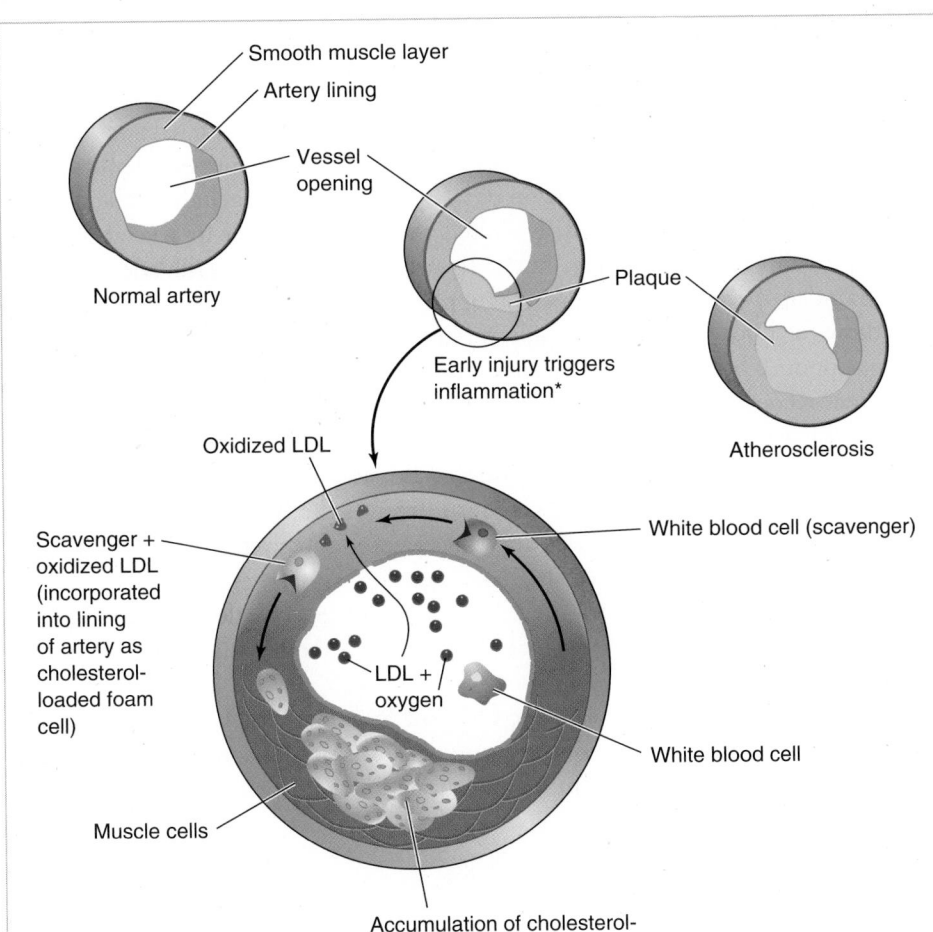

FIGURE 5-6

Atherosclerosis

As LDL particles penetrate the walls of the arteries, they become oxidized LDL and next are scavenged by the body's white blood cells. These foam cells are then deposited into the lining of the artery wall. This process, known as atherosclerosis, causes plaque deposits to enlarge, artery walls to lose elasticity, and the passage through the artery to narrow.

*Early injury may be the result of smoking, high blood pressure, elevated blood cholesterol, elevated homocysteine level, diabetes, genetics, or possibly a bacterial or viral infection.

TABLE 5-2	The Effects of Various Kinds of Fat on Blood Lipids

Type of Fat[a]	Dietary Sources	Effects on Blood Lipids
Saturated Fat	All animal meats, beef tallow, butter, cheese, coconut, cocoa butter, coconut oil, cream, cream cheese, hydrogenated oils, lard, palm oil, stick margarine, shortening, sour cream, whole milk, and ice cream	Increases total cholesterol Increases LDL-cholesterol

Polyunsaturated Fat	Almonds, corn oil, cottonseed oil, fish, liquid/soft margarine, mayonnaise, pumpkin seeds, safflower oil, salad dressing, sesame oil, soybean oil, sunflower oil, sunflower seeds, walnuts	If used to *replace* saturated fat in the diet, polyunsaturated fat may:
		   Decrease total cholesterol
		   Decrease LDL-cholesterol
		   Decrease HDL-cholesterol[b]

Monounsaturated Fat	Almonds, avocados, canola oil, cashews, filberts, hazelnuts, macadamia nuts, olive oil, olives, peanut butter, peanut oil, peanuts, pecans, pistachios, sesame seeds	If used to *replace* saturated fat in the diet, monounsaturated fat may:
		   Decrease total cholesterol
		   Decrease LDL-cholesterol without decreasing HDL-cholesterol

Omega-3 Fat	Canola and soybean oils, ground flaxseed, flaxseed oil, oily coldwater fish (salmon, mackerel, tuna, herring), soy foods, walnuts, wheat germ	If used to *replace* saturated fat in the diet, omega-3 fat may:
		   Decrease total cholesterol
		   Decrease LDL-cholesterol
		   Increase HDL-cholesterol
		   Decrease triglycerides

***Trans* Fat**	Margarine (hydrogenated), cake, cookies, doughnuts, crackers, chips, meat and dairy products, hydrogenated peanut butter, nondairy creamer, shortening, many fast foods, foods fried in hydrogenated oils or shortening	Increases total cholesterol Increases LDL-cholesterol

[a] All fats, whether classified as mainly saturated fat, monounsaturated fat, or polyunsaturated fat, contain mixtures of saturated and unsaturated fats and provide the same number of calories: 9 calories per gram.

[b] If consumed in large amounts (> 10% of total calories).

SOURCE: USDA Nutrient Database for Standard Reference (Release 14), the National Sunflower Association, and the Flax Council of Canada. Reprinted with permission from *Nutrition Action Healthletter* (July/August 2002), p. 7.

Fat in the Diet

"Excuse me, but which is it that practically kills you—polysaturated or polyunsaturated!?"

The type of fat consumed may be more important in influencing the risk of heart disease than is the total amount of fat in the diet.[7] Some researchers warn, however, that diets rich in any type of fat can be calorie-dense and could worsen the global problem of obesity. The question of what kind of fat to include in the diet can be puzzling. Research has potentially linked the fat in the diet to several diseases, including certain types of cancer, heart disease, arthritis, and gallbladder disease. The remainder of this chapter helps you apply what you have learned about fats— that is, how to choose foods that supply enough but not too much of the right kinds of fat to support a healthy weight and optimal health and provide pleasure in eating. The current recommendations for dietary fat intake are listed in Table 5-3.

Today, the average American diet includes about 34 percent of its calories from fat, about 11 percent of calories from saturated fat, and about 2.6 percent of calories from *trans* fat. The Institute of Medicine recommends that *total* fat not exceed 20 to 35 percent of the day's total calories and that it come mostly from polyunsaturated and monounsaturated fat sources such as fish, nuts, and vegetable oils.[8] Saturated fat should be kept low and contribute less than 10 percent of calories. *Trans* fat consumption should be kept as low as possible. Table 5-4 shows you how to determine your daily fat allowance.

To reduce the intake of both saturated fat and *trans* fat, the *Dietary Guidelines for Americans* recommend limiting consumption of foods high in solid fat and replacing them with foods that are rich in monounsaturated fat and polyunsaturated fat. For

---

TABLE 5-3	Dietary Recommendations for Fats in the Diet*

From the *Dietary Guidelines for Americans*:
- Keep total fat intake between 20 to 35 percent of calories, with most fats coming from sources of polyunsaturated and monounsaturated fatty acids, such as fish, nuts, and vegetable oils.
- When selecting and preparing meat, poultry, dry beans, and milk or milk products, make choices that are lean, low-fat, or fat-free.
- Consume less than 10 percent of calories as saturated fat with an ultimate goal of limiting saturated fat intake to less than 7 percent of calories, replacing these calories with those from monounsaturated or polyunsaturated fatty acids.
- Limit dietary cholesterol to less than 300 mg per day (and less than 200 mg per day in persons at high risk for heart disease or diabetes).
- Keep *trans* fatty acid consumption as low as possible, especially by limiting foods that contain synthetic sources of *trans* fats, such as partially hydrogenated oils.
- Reduce intake of solid fats (major sources of saturated and *trans* fatty acids).
- Consume two 4-ounce servings of seafood per week as a source of omega-3 fatty acids.

From the American Heart Association:
- Saturated fat: less than 7% of total calories
- *Trans* fat: less than 1% of total calories
- Dietary cholesterol: less than 300 mg/day on average

From the DRI Committee:
An *Acceptable Macronutrient Distribution Range (AMDR)* for:
- Total fat: 20%–35% of total calorie intake
- Polyunsaturated fats:
  - Omega-6 fats: 5%–10% of total calorie intake
    Adequate intake of linoleic acid (essential omega-6 fatty acid):

Men	Women
17 g/day	12 g/day

  - Omega-3 fats: 0.6%–1.2% of total calorie intake
    Adequate intake of linolenic acid (essential omega-3 fatty acid):

Men	Women
1.6 g/day	1.1 g/day

On Food Labels: Daily Values (DV) for a 2,000-Calorie Diet:

Total Fat	Saturated Fat	Cholesterol
65 g	20 g	300 mg

* Refer to Table 5-2 for dietary sources of various types of fat.

TABLE 5-4	How to Determine Daily Fat Allowances

We are advised to choose a diet that is low in saturated fat and cholesterol and moderate in total fat. The DRI Committee has set an acceptable range of from 20 to 35 percent of total calories from fat. Recommended intake for saturated fat is less than 10 percent of total daily calories. Every single food you eat does not need to conform to these allowances if you balance higher-fat items with lower-fat foods throughout the day.

If you usually eat this many calories:	Your total fat intake should be no more than 20 to 35 percent of calories:			Your saturated fat intake should be no more than:
	20% of Total Calories from Fat[a] (g/day)	30% of Total Calories from Fat[b] (g/day)	35% of Total Calories from Fat[a] (g/day)	10% of Total Calories (g/day)
1,200	27	40	47	13 or less
1,500	33	50	58	17 or less
1,800	40	60	70	20 or less
2,000[c,d]	40	65	75	20 or less
2,400[e]	53	80	93	27 or less
2,500[d]	55	80	95	28 or less
2,800	62	93	109	31 or less
3,000[f]	67	100	117	33 or less

[a] The DRI committee set an Acceptable Macronutrient Distribution Range for fat intake at 20% to 35% of total calories. For more discussion about AMDR, see Chapter 2.
[b] The American Heart Association determines fat allowances based on a diet with no more than 30% of total calorie intake from fat.
[c] Percent Daily Values on Nutrition Facts labels are based on a 2,000-calorie diet.
[d] Values for 2,000 and 2,500 calories are rounded to the nearest 5 grams to be consistent with the Nutrition Facts label.
[e] Estimated Energy Requirement (EER) for 19-year-old women; subtract 7 calories per day for females for each year of age above 19. EER values are determined at four physical activity levels; the values shown here are for the "active" female.
[f] Estimated Energy Requirement (EER) for 19-year-old men; subtract 10 calories per day for males for each year of age above 19. EER values are determined at four physical activity levels; the values shown here are for the "active" male.

Calculation Practice

Use Table 5-4 to find your total daily fat allowance, based on your daily calorie intake, to keep your fat grams within the acceptable range. To find a reasonable calorie intake for yourself, see the DRI table on Study Card page 27 at the back of this book. Or, use the following quick method to determine your calorie allowance: Multiply your body weight in pounds by 15 (if you're active). If you weigh 150 pounds, you expend about 2,250 calories (150 × 15) in a typical day. If you're sedentary, multiply your weight by 13.

Example: If you eat 1,800 calories per day, multiply 1,800 by 10 percent *to determine the maximum number of calories that should come from saturated fat* in one day (1,800 × 0.1 = 180 calories from saturated fat). Since 1 gram of fat provides 9 calories, divide the calories from saturated fat by 9 to see the maximum amount of saturated fat you can have per day (180 ÷ 9 = 20 grams of saturated fat per day):

Total daily calories × 0.10 = total saturated fat calories
Total saturated fat calories ÷ 9 = total saturated fat grams

Likewise, *to determine an acceptable range (20% to 35% of calories) for total fat intake:*

Total daily calories × 0.20 = total fat calories
Total fat calories ÷ 9 = total fat grams

To determine maximum amount of calories from fat:

Total daily calories × 0.35 = total fat calories
Total fat calories ÷ 9 = total fat grams

Identify the amount of calories from added sugars and solid fats contained in the foods and beverages you consume at www .myfoodapedia.gov.

A 3-ounce portion of lean beef, chicken, or fish is roughly the size of a deck of playing cards or the palm of the average woman's hand.

phytochemicals (FIGHT-oh-KEM-ih-kulz) physiologically active compounds found in plants that appear to help promote health and reduce risk for cancer, heart disease, and other conditions. Also called *phytonutrients (phyto = plant)*.

example, when preparing foods at home, solid fats such as butter and lard can be replaced with vegetable oils such as canola or olive oil (see Figure 5-2). Foods high in solid fats include many cheeses, creams, whole milk, ice creams, well-marbled cuts of meats, regular ground beef, bacon, sausages, poultry skin, and many baked goods such as cookies, crackers, doughnuts, pastries, and croissants.

"Fat on the plate" includes visible fats and oils, such as butter, the oil in salad dressing, and the fat you trim from a steak. It also refers to some you cannot see, such as the fat that marbles a steak or that is hidden in such foods as nuts, cheese, biscuits, crackers, doughnuts, cookies, muffins, avocados, olives, fried foods, and chocolate. Most of the fat, especially saturated fat, in our diet comes from animal products (see Figure 5-7). Meats probably conceal most of the fat that people unwittingly consume. Many people, when choosing a serving of meat, don't realize that they are electing to eat a large amount of fat. When selecting beef or pork, look for the words *loin* or *round* on the label—these words represent lean cuts from which the fat can be trimmed. Table 5-5 shows a few examples of the differences in the saturated fat content of different forms of commonly eaten foods. Note that choices for lower saturated fat can be made within the same food group (for example, low-fat milk versus whole milk).

Fat Can Be Healthy

Think that all fat is evil? Think again. Researchers are finding that people living in Greece, France, Italy, and other countries around the Mediterranean Sea have only a fraction of the heart disease and related deaths experienced by people living in Western, industrial countries like the United States.

What causes this difference? Genetic differences aside, other factors suggest that nutritional differences may play a role because dietary patterns in countries around the Mediterranean Sea vary significantly from those in industrialized countries like the United States, Canada, and northern European countries. What do people who live around the Mediterranean Sea eat that protects them from coronary heart disease? The "Mediterranean diet" is rich in fruits, vegetables, grains, fish, and beans, whereas most American diets include much more meat and other animal products known to be high in saturated fat. Thus, the Mediterranean diet is high in the vitamins, minerals, fiber, and **phytochemicals** that keep bodies healthy and low in saturated fat, because Mediterranean people consume so little meat and butter (see

FIGURE 5-7

Major Sources of Saturated Fat in the U.S. Diet

The current American diet contains about 11 percent of total calories as saturated fat.

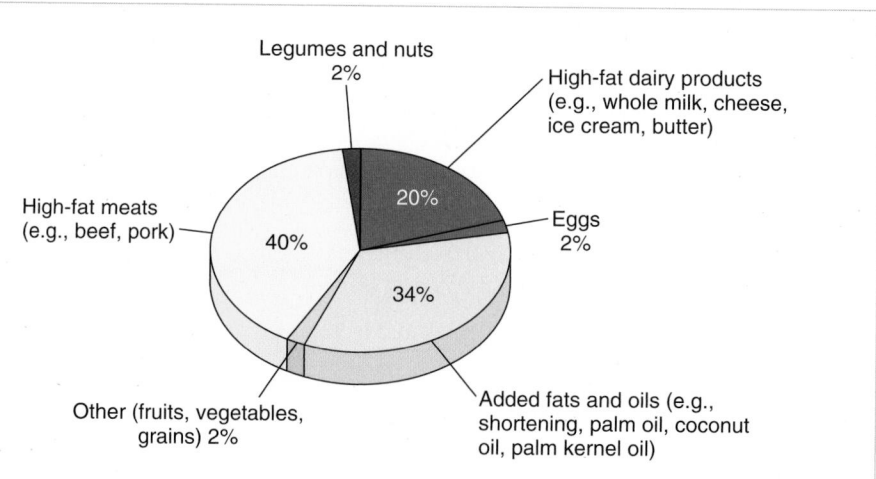

- Legumes and nuts 2%
- High-fat dairy products (e.g., whole milk, cheese, ice cream, butter) 20%
- Eggs 2%
- Added fats and oils (e.g., shortening, palm oil, coconut oil, palm kernel oil) 34%
- Other (fruits, vegetables, grains) 2%
- High-fat meats (e.g., beef, pork) 40%

		Saturated Fat	
Food Category	Portion	(grams)	Calories
Added fat			
Butter	1 tbsp	7	100
Olive oil	1 tbsp	2	119
Cheese			
Regular cheddar cheese	1 oz	6.0	114
Low-fat cheddar cheese	1 oz	1.2	49
Milk			
Whole milk	1 cup	4.6	146
Low-fat (1%) milk	1 cup	1.5	102
Fat-free (skim) milk	1 cup	0	90
Frozen desserts			
Regular ice cream	1/2 cup	4.9	145
Low-fat frozen yogurt	1/2 cup	2.0	110
Meat, Poultry, Fish			
Regular ground beef (25% fat)	3 oz (cooked)	6.1	236
Extra lean ground beef (5% fat)	3 oz (cooked)	2.6	148
Spareribs	3 oz (cooked)	9.5	337
Ham, extra lean	3 oz (cooked)	1.5	123
Fried chicken (leg with skin)	3 oz (cooked)	3.3	212
Roasted chicken (breast no skin)	3 oz (cooked)	0.9	140
Fish fillet, breaded, fried	3 oz (cooked)	2.4	197
Salmon	3 oz (cooked)	1.2	125
Snack Items			
Potato chips	10 chips	3.1	155
Mixed nuts	2 tbsp	1	100

TABLE 5-5 Compare the Saturated Fat and Calorie Content of Selected Foods

SOURCE: U.S. Department of Agriculture, Agricultural Research Service, Nutrient Data Laboratory, 2010, USDA National Nutrient Database for Standard Reference, Release 23.

Figure 5-8). Most Americans don't eat enough whole-grain products, fruits, and vegetables to get the necessary vitamins, minerals, fiber, and phytochemicals that will protect their bodies from coronary heart disease and other chronic diseases.

Study the pyramid in Figure 5-8, and you will see that the Mediterranean diet is not fat-free. About 30 to 35 percent of total calories are provided by fat. The key is the *type* of fat eaten. Only about 7 to 8 percent of total calories come from saturated fat, whereas average Americans consume about 11 percent of their calories from saturated fat. As you already know, saturated fat is the major dietary risk factor for heart disease.[9] What do these Mediterranean people use instead of animal fat (butter, meats, ice cream)? They use olive oil and other fats from plant sources that are higher in monounsaturated and polyunsaturated fats.[10]

Olive oil is high in oleic acid, a monounsaturated fatty acid that seems to keep the heart healthy. Oleic acid helps keep "good" HDL-cholesterol (the artery cleaner) high while lowering levels of "bad" LDL-cholesterol (the artery clogger). It also contains a good ratio of omega-3 fatty acids to omega-6 fatty acids. This is a good thing. Keep in mind, though, that olive oil is still a fat. It provides as many calories as any other pure fat. Therefore, olive oil can promote overweight and obesity if too much is consumed. Other vegetable oils high in monounsaturated fats are peanut and canola oils.

UNDERSTANDING THE PYRAMID

The Mediterranean Diet Pyramid depicts the traditional foods and drinks that make up the healthy, balanced Mediterranean Diet. It contains many of the foods you will find in other dietary pyramids. The principal difference is in the frequency that some foods are eaten. Almost all foods can be part of a balanced and healthy diet – but your overall health and well-being can be greatly affected by *how often* you eat different foods, and the *portion size* you choose.

MEATS AND SWEETS These are "sometimes" foods to eat less often. If you eat meat, choose small portions of lean cuts, such as round, shoulder, tenderloin, strip, T-bone, and flank. Enjoy sweets at a celebration or as a treat.

YOGURT, CHEESE, POULTRY, AND EGGS form a central part of the Mediterranean Diet and are eaten in moderate portion sizes several times a week. Cheese, for example, is eaten regularly but in small amounts.

WINE AND WATER – Wine can be consumed regularly but moderately: up to one glass per day for women, two for men. Water is essential for proper hydration, and contributes to health, well-being, and energy.

FISH AND SEAFOOD occupy their own section, since they are important sources of protein. Fish such as tuna, herring, salmon, and sardines are rich in heart-healthy omega-3 fatty acids, and shellfish including mussels, oysters, shrimp, and clams have similar benefits. Enjoy at least twice a week.

WHOLE GRAINS, FRUITS, VEGETABLES, BEANS, HERBS AND HEALTHY FATS, SUCH AS THOSE FOUND IN OLIVE OIL represent the core of the diet. Base every meal on fruits, vegetables, whole grains, legumes, and herbs and spices. Olive oil, the main source of dietary fat, is used for almost all cooking and baking, and for dressing salads and vegetables.

DAILY PHYSICAL ACTIVITY, which is important for overall good health, includes strenuous exercise such as running and aerobics, more leisurely activities such as walking and house-or-yard work, and simple changes, such as taking the stairs instead of the elevator. Add physical activity to each day.

Other Sources of Omega-3 Fatty Acids Another place to find omega-3 fats is fish, especially oily coldwater fish like salmon, cod, farmed catfish, lake trout, herring, bluefish, sardines, albacore tuna, mackerel, and shellfish (see Table 5-6). Eating these fish at least two times a week as part of a balanced diet can help reduce blood clot formation (which can lead to heart attack and stroke), decrease risk of arrhythmias (which can lead to sudden cardiac death), decrease triglyceride levels, decrease growth rate of atherosclerotic plaque, improve the health of the arteries, and slightly lower blood pressure (another risk factor for heart disease).[11]

The FDA approved a claim for a reduced risk of heart disease for foods, such as fish, that contain the EPA and DHA omega-3 fatty acids: "Supportive, but not conclusive research shows that consumption of EPA and DHA omega-3 fatty acids may reduce the risk of coronary heart disease."[12] So, you don't like fish and think you can get the healthy fats you need from fish oil capsules? Think again. The American Heart Association recommends more research to confirm the health benefits of omega-3 fatty acid supplements.[13]

TABLE 5-6	Sources of Omega-3 Fats for the Diet[a]			

Common Seafood Varieties[a,b]	EPA + DHA[c] mg/4 oz	Other Sources of Omega-3 Fats	Omega-3 Content (linolenic acid)[e] grams/tbsp
Salmon: Atlantic, Chinook, Coho	1,200–2,400	Flaxseed oil	8.0
Anchovies, herring	2,300–2,400	Walnuts, ¼ c	2.7
Mackerel: Atlantic and Pacific (not king)	1,350–2,100	Flaxseeds, ground	1.9
Tuna: bluefin and albacore	1,700	Walnut oil	1.4
Sardines: Atlantic and Pacific	1,100–1,600	Canola oil	1.3
Oysters[d]: Pacific	1,550	Canola oil mayonnaise	1.0
Trout: freshwater	1,000–1,100	Soybeans, cooked, ½ c	1.0
Tuna: white (albacore) canned	1,000	Soybean oil	0.9
Mussels[d]: blue	900	Tofu, ½ c	0.7
Salmon: pink and sockeye	700–900	Olive oil	0.1
Squid	750		
Pollock: Atlantic and walleye	600		
Crab[d]: blue, king, snow	200–550		
Tuna: skipjack and yellowfin	150–350		
Flounder and sole	350		
Clams[d]	200–300		
Tuna: light canned	150–300		
Catfish	100–250		
Cod: Atlantic and Pacific	200		
Scallops[d]: bay and sea	200		
Haddock and hake	200		
Lobster[d]: Northern	200		
Crayfish[d]	200		
Tilapia	150		
Shrimp[d]	100		

[a] EPA and DHA values are for cooked, edible portion rounded to the nearest 50 mg. Ranges are provided when values are comparable. Values are estimates.

[b] Methylmercury is a heavy metal toxin found in varying levels in nearly all fish and shellfish. However, some fish and shellfish contain higher levels of mercury that may harm the developing nervous system of an unborn baby or young child. The Food and Drug Administration (FDA) and the Environmental Protection Agency (EPA) advise women who may become pregnant, pregnant women, lactating women, and young children to avoid some types of fish, including shark, swordfish, king mackerel, or tilefish, because they contain high levels of mercury. Women who are pregnant or breastfeeding should also limit white (albacore) tuna to 6 ounces per week. Instead, choose up to 12 ounces a week of a variety of fish and shellfish that are lower in mercury, such as shrimp, canned light tuna, salmon, flounder, pollock, or catfish. For updates on the mercury status of fish and shellfish, go to the Center for Food Safety at www.cfsan.fda.gov.

[c] A total of 1,750 mg of EPA and DHA per week (an average of 250 mg per day) is the goal amount for consumers to achieve at the recommended two 4-ounce seafood meals per week.

[d] Cooked by moist heat.

[e] Dietary Reference Intake (DRI) for omega-3 fats (linolenic acid): Men: 1.6 g/day
Women: 1.1 g/day

SOURCE: USDA, Nutrient Data Laboratory, 2010, USDA National Nutrient Database for Standard Reference, Release 23; P. M. Kris-Etherton, W. S. Harris, & L. J. Appel, Fish consumption, fish oil, omega-3 fatty acids, and cardiovascular disease, *Circulation* 106 (2002): 2747–57; and Institute of Medicine, *Seafood Choices* (Washington, DC: National Academies Press, 2006).

Other places to find omega-3 fatty acids are ground flaxseed, flaxseed oil, canola oil, soy products, pecans, and walnuts. The type of omega-3 fatty acid in these plant foods is less potent than that found in fish oils, but they are good sources nonetheless.[14] Research has demonstrated that people who eat nuts tend to eat more dietary fat, but the fat is primarily monounsaturated. Also, nuts contain insoluble fiber, another dietary component that protects us from coronary heart disease.[15]

Current Western intakes of omega-3 fats are well below the levels that experts consider optimal.[16] Health experts now advise us to decrease consumption of foods rich in omega-6 fatty acids, such as vegetable oils (corn, safflower, sesame, sunflower), and increase the intake of omega-3 fatty acids (those found in fish, canola and soybean oils, flaxseed, walnuts, other nuts, and soy foods) to achieve a more healthful balance between the two types of fats.

Choose Fats Sensibly

It is not the food but how you prepare it that often determines the total fat (and calories) in a food. Compare the fat in 3 ounces of broiled chicken (3 grams fat, 141 calories) versus 3 ounces of fried chicken (18 grams fat, 364 calories). What makes the difference in calories? Fat. This feature suggests a variety of ways to reduce added fats in your diet when preparing home-cooked meals.

- Use nonstick sprays rather than fat to coat pans.

- Try reducing the fat in recipes a little at a time, and notice that you can do so and still have a tasty meal. If you reduce fat by one tablespoon of butter or oil, you lower the total fat in the product by about 12 grams and at least 100 calories. Or, to reduce the fat even more, try substituting unsweetened apple-sauce or fruit purees for the oil called for in your favorite cake, quick bread, or cookie recipe.

Season with herbs and spices.

Bake, broil, poach, or steam.

- When you sauté a vegetable such as onion in butter, margarine, or oil, try reducing the amount used and substituting water, fat-free vegetable or chicken broth, or wine in its place.

- Prepare broths, soups, and stews ahead of time, refrigerate them, and then skim off hardened fat from the surface.

- Prepare lean cuts of meat, remove visible fat from meat, remove skin from poultry, and cook meats on a rack so that the invisible fat can drain off.

- For flavor in sauces and dressings, experiment with herbs and spices, onions or garlic, salsa, ginger, lemon juice, plain fat-free yogurt with lemon juice, mustard, or butter-flavored granules as replacements for butter, margarine, or oil.

- Try using fruit jam instead of butter on your bagel or toast. Most fruit jams contain half the calories per teaspoon of butter. Or, try the whipped varieties of butter, margarine, or cream cheese, which contain about half the calories of the regular types.

What's Your Meal Mentality?   Another difference between how Americans and Mediterranean people eat is the way meals are eaten. The Mediterranean diet involves a wide variety of foods served in several small courses at a slower pace that tends to prevent overeating. Note that one of the dietary guidelines in Greece advises everyone to "eat slowly, preferably at regular times of the day, and in a pleasant environment." Overall, the Mediterranean diet emphasizes olive oil (but no other vegetable oils, hydrogenated oils, or tropical oils); less animal protein (particularly red meats); poultry, eggs, and sweets only a few times a week; and daily use of cheese, yogurt, fruits, vegetables, and whole-grain products. Still, other factors undoubtedly contribute to the lower incidence of heart disease in the Mediterranean region.

In understanding the paradox of how a moderate-fat diet (about 30 to 35 percent of calories from mostly monounsaturated fats) can coexist with a low incidence of heart disease, it is important to note that people from the Mediterranean region in general are more physically active, have the social and emotional support found in extended networks of family and friends, and consume more of each day's calories earlier in the day than people in the United States and Canada. Any one of these factors may also be significant in maintaining lower incidence of heart disease in the region. In addition to these possible lifestyle changes, a healthful change in eating patterns (with a Mediterranean twist) for most people in the United States and Canada would be to adopt a diet that is lower in saturated and *trans* fat and higher in complex carbohydrates and fiber (fruits, vegetables, whole-grain breads and other grains, and legumes).

The *Trans* Fatty Acid Controversy— Is Butter Better?

Conventional nutrition wisdom has long held that margarine is better than butter to use on food, including bagels and other breads. In recent years, though, rumors have been spreading that margarine may not be the most healthful choice after all. The heart of the controversy is a growing body of research suggesting that a certain type of fat found in margarine may be as likely to boost blood levels of "bad" LDL-cholesterol as the saturated fat found in butter.[17] What's more, the research suggests that in large amounts, *trans* fatty acids lower "good" HDL-cholesterol in the blood.[18]

The alleged culprit, *trans* fatty acid, is formed when margarine is processed. To make margarine, manufacturers take a highly unsaturated vegetable oil and partially hydrogenate (add hydrogen to) it. Hydrogenation gives the spread its relatively solid consistency and helps protect against rancidity. During the hydrogenation process, however, a chemical "fluke" occurs. Hydrogenating the oil creates a new chemical configuration, known as a *trans* fatty acid, in which hydrogen atoms of the fat lie on opposite sides of the point of unsaturation in the carbon chain instead of side by side (see Figure 5-9). (Recall that the body of a fatty acid molecule consists of a chain of carbon atoms to which hydrogen is attached, as shown in Figure 5-1.)

Because *trans* fatty acids are formed whenever oils are hydrogenated, margarine is only one of the foods on the market that contain *trans* fatty acids (see Figure 5-10). Shortenings, baked goods, certain brands of peanut butter, the commercial frying fats used in many fast-food outlets to cook French fries and other fried items, and many other foods that list hydrogenated or partially hydrogenated vegetable oil on the label are among the foods containing *trans* fatty acids. Fueling the controversy even further, an analysis by researchers at Harvard University suggested that more than 30,000 American deaths each year are attributable to *trans* fatty acids. The authors of the report went so far as to call for government legislation mandating that manufacturers include *trans* fatty acid amounts on food labels and phase out the use of partially hydrogenated oils in the United States.[19]

FIGURE 5-9

Types of Unsaturated Fatty Acids: Cis versus *Trans*

Unsaturated fatty acids are either in *cis* form or in *trans* form, depending on the way in which the hydrogen atoms are attached to the points of unsaturation in the carbon chain. If the hydrogen atoms are attached to the same side of the points of unsaturation, the arrangement is called *cis*. If the hydrogen atoms are attached to different sides, the arrangement is called *trans*.

cis   trans

cis = same
trans = across

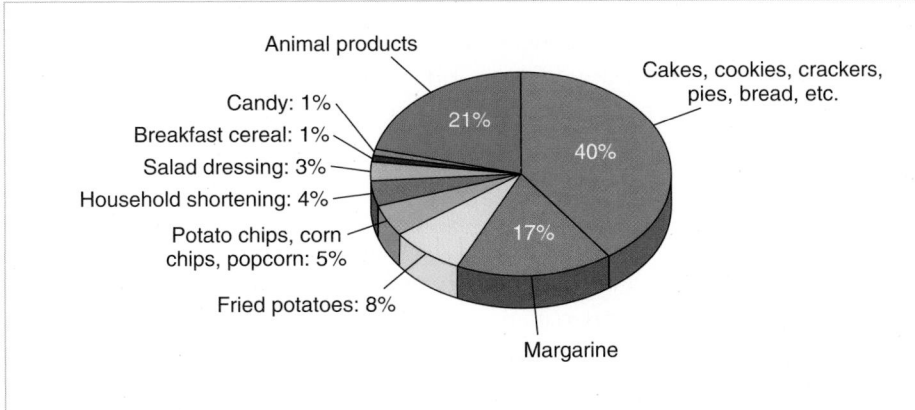

Animal products

Candy: 1%
Breakfast cereal: 1%
Salad dressing: 3%
Household shortening: 4%
Potato chips, corn chips, popcorn: 5%
Fried potatoes: 8%

21%

40% Cakes, cookies, crackers, pies, bread, etc.

17%

Margarine

FIGURE 5-10

Major Sources of *Trans* Fatty Acids in the U.S. Diet

Trans fatty acids occur naturally in meat, poultry, and dairy products. They show up in packaged and fast foods to which hydrogenated vegetable oils have been added. Note that high levels of *trans* fatty acids go hand in hand with foods containing solid fats. Thus, if you choose foods prepared with little or no solid fats, you will consume minimal amounts of *trans* fatty acids.

SOURCE: FDA Consumer, 2003.

FIGURE 5-11

Compare Spreads

Read manufacturers' labels to determine both the amounts and the types of fat contained in foods. Shop for margarine containing no more than 2 grams of saturated fat per tablespoon and with liquid vegetable oil as the first ingredient. Choose soft tub or liquid forms over stick; "hard" stick margarines typically contain the most *trans* fatty acids. Look for the canola- or olive-oil-based margarines containing zero *trans* fat.

SOURCE: Adapted from Center for Food Safety and Applied Nutrition, Office of Nutritional Products, Labeling, and Dietary Supplements, U.S. Food and Drug Administration, January 16, 2004.

Butter	Margarine, stick	Margarine, tub
Nutrition Facts Serving Size 1 Tbsp (14 g) Servings per Container 32	**Nutrition Facts** Serving Size 1 Tbsp (14 g) Servings per Container 32	**Nutrition Facts** Serving Size 1 Tbsp (14 g) Servings per Container 32
Amount Per Serving	**Amount Per Serving**	**Amount Per Serving**
Calories 100 — Calories from Fat 100	**Calories** 100 — Calories from Fat 100	**Calories** 60 — Calories from Fat 60
% Daily Value	% Daily Value	% Daily Value
Total Fat 11g — 17%	**Total Fat** 11g — 17%	**Total Fat** 7g — 11%
Saturated Fat 7g — 35%	Saturated Fat 2g — 10%	Saturated Fat 1g — 5%
Trans Fat 0g	*Trans* Fat 3g	*Trans* Fat 0.5g
Cholesterol 30mg — 10%	**Cholesterol** 0mg — 0%	**Cholesterol** 0mg — 0%
Saturated Fat: 7g + *Trans* Fat: 0g Combined Amt.: 7g	Saturated Fat: 2g + *Trans* Fat: 3g Combined Amt.: 5g	Saturated Fat: 1g + *Trans* Fat: 0.5g Combined Amt.: 1.5g

Types and Uses of Fat-Replacers

Carbohydrate based: Carrageenan (a seaweed derivative), fruit purees, gelatin, gels derived from cellulose or starch, guar gum, xanthan gum, maltodextrins made from corn, cornstarch (Stellar), polydextrose, Oatrim (made from oat fiber), and Z-trim (a modified form of insoluble fiber)

Uses: Dairy products, frozen desserts, salad dressings, baked goods, cake and cookie mixes, frostings

Protein based: Whey protein concentrate (Dairy-Lo), microparticulated protein products (Simplesse K-Blazer) made from whey, or milk and egg-white protein

Uses: Cheese, butter, salad dressings, mayonnaise, sour cream, dairy products, ice cream, baked goods

Fat based: Mono- and diglycerides; Caprenin (a substitute for cocoa butter in candy) and Salatrim (found in reduced-fat baking chips) both contain long-chain fatty acids, which are partially absorbed, and short-chain fatty acids (providing 5 calories per gram); Olestra (noncaloric artificial fat made from fatty acids and sucrose)

Uses: Savory snacks, chocolate confections, bakery products

In January 2006, the Food and Drug Administration began requiring that manufacturers list the *trans* fat content on a separate line within the Nutrition Facts panel on all food products (as shown in Figure 5-11). Manufacturers can label a product "*trans* fat free" if the food contains less than 0.5 gram of *trans* fat and less than 0.5 gram of saturated fat per serving.[20] Canada also requires labeling of *trans* fat content of foods.

Many major health organizations point out that the *trans* fatty acid research does not indicate that consumers should switch back from eating margarine to eating butter, which contains an excess of saturated fat. That's because *trans* fatty acids account for only about 2.6 percent of the total calories in a typical American diet, whereas saturated fat constitutes some 11 percent of calories.[21]

Although the evidence that has come to light doesn't warrant going back to butter, it does underscore the value of keeping the intake of *total* saturated and *trans* fat to a minimum. The foods that contain large amounts of *trans* fatty acids—French fries, potato chips, deep-fat-fried doughnuts—are also high in total fat. With the addition of *trans* fat to the Nutrition Facts panel, you can now evaluate the total fat, saturated fat, and *trans* fat content per serving for various food products (see Figure 5-11). Thus, health-conscious eaters who want to hedge their bets against developing high blood cholesterol levels should stick with the principles of heart-healthy eating outlined in the Savvy Diner feature on page 144. The Scorecard feature later in this chapter will help you rate your own dietary selections for fat.

Understanding Fat Replacers

Fat-free ice cream, cookies, cakes, and salad dressings have long been the stuff that dieters' dreams are made of. The food industry makes such fatless fare a reality. With more and more low-fat and fat-free items introduced daily into supermarkets across the country, sales of fat-free products are soaring.

The boom in both low-fat and fat-free products can be accounted for by the country's expanding health consciousness. Although most Americans have heard the warnings that fat can contribute to heart disease, cancer, and obesity, many people find low-fat diets particularly unpalatable because fat adds a desirable flavor and texture—known as "mouth feel"—to foods. With innovations in food chemistry, however, the food industry can now concoct new recipes or ingredients that yield low-fat or nonfat products that retain the characteristic flavor and mouth feel of fat. The types of fat-reduction ingredients currently in use are listed in the margin on page 146.

Some manufacturers of fat-free cakes and cookies, for example, make those products by substituting egg whites for whole eggs and fat-free milk for whole milk, as well as by removing the butter from their recipes. The end products are sweets that contain fewer calories and fewer grams of fat per serving than traditional desserts.[22]

Another technique manufacturers use to replace fat is to add starches, gums, and gels to their products. For example, Kraft uses cellulose gel, a complex carbohydrate, as a filler to make fat-free salad dressings.[23] Other companies add starches and gums—also complex carbohydrates—to items such as sauces and yogurts to skim fat from those foods. That's because starches and gums hold water and impart a smooth, creamy texture similar to that of fat, and they add form and structure to foods. These substitutes cannot, however, replace the fat used for cooking and frying.[24]

Yet another, more innovative approach the food industry has taken to provide fat-free fare is the development of fat replacers. In 1990, a substance called **Simplesse** became the first such product to gain the approval of the Food and Drug Administration (FDA). Six years in the making, Simplesse is a mixture of food proteins such as egg white, whey, and milk protein that are cooked and blended to form tiny round particles that trap water. Inside the mouth, the particles roll over one another, and the tongue perceives them as a creamy, smooth liquid similar to fat.

The FDA allows use of Simplesse in foods including cheese, baked goods, ice creams, frozen desserts, mayonnaise, salad dressings, yogurts, sour cream, and butter. Foods made with this fat replacer contain considerably less fat and fewer calories than their traditional fat-containing counterparts.[25] The reason is that whereas fat contains 9 calories per gram, Simplesse supplies only 1 to 2. Simplesse cannot be used for frying or baking, because heat causes it to gel and lose its creaminess.

An artificial fat approved for use in salty snacks and crackers is **olestra** (sold under the brand name Olean). Created by Procter & Gamble, the olestra molecule resembles a triglyceride but is structured in a way that prevents it from being broken down by digestive enzymes in the body, thereby allowing it to pass through the digestive tract completely unabsorbed (see Figure 5-12). Olestra interferes with the absorption of the fat-soluble vitamins (A, D, E, and K), as well as beta-carotene. To compensate for this effect, olestra must be fortified with the fat-soluble vitamins. Consumption of large amounts of olestra can cause nutrient losses, abdominal cramping, and loose stools.[26]

Certainly, the growing number of fat-free foods and fat replacers provides consumers with viable alternatives to fattier fare. Nevertheless, many experts view the fat-free boom with skepticism. Although reduced-fat foods can help lower the overall fat content of the diet, they *do* contain calories, and they are *not* a replacement for a healthful diet rich in whole grains, fresh fruits, and vegetables. Nor are they likely to become the panacea that will prevent problems such as heart disease and obesity. When included in a heart-healthy diet, however, they can help consumers reach and adhere to the goal of taking in no more than 20 to 35 percent of total calories as fat.[27]

FIGURE 5-12

Olestra: A Fat-Free Fat

Compare the structure of olestra to that of a triglyceride. Whereas triglycerides have 3 fatty acids attached to a glycerol core, olestra has from 6 to 8 fatty acids on a sucrose core. The extra fatty acids make it too large to be digested or absorbed by the body, which is why it adds no fat or calories to the diet. Thus a 1-ounce bag of potato chips, which typically has 10 grams of fat and 150 calories, has 0 grams of fat and 60 calories when Olean is added in place of fat.

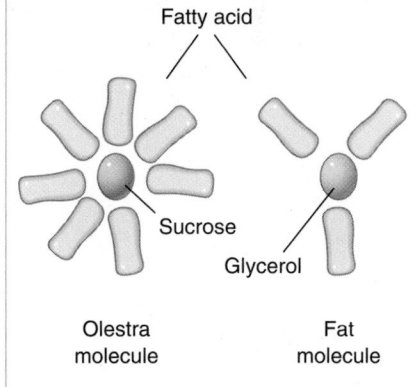

Fatty acid

Sucrose

Glycerol

Olestra molecule

Fat molecule

Simplesse the trade name for a protein-based, low-calorie fat replacer, approved by the FDA for use in foods such as frozen desserts; cannot be used for frying or baking.

olestra an artificial fat derived from sucrose and fatty acids combined in such a way that the body cannot break them down. Sold under the brand name Olean, olestra does not contribute calories to food.

Heart disease rarely develops from a single risk factor. Clusters of risk factors usually occur, and a small increase in one risk factor, such as blood pressure, becomes more critical when combined with other risk factors, such as obesity. Luckily, a similar pattern happens in reverse: Even moderate changes in one risk factor can decrease several others at the same time. For example, a weight loss of just 5 to 10 pounds can reduce blood pressure in overweight people. Likewise, becoming physically active can lower blood pressure, increase HDL ("good" cholesterol), and help control weight.

In the past few years, attention has focused on reducing heart disease risk by both *reducing* contributory factors in the diet—notably, saturated fat and *trans* fat—and by *increasing* the intake of protective factors, such as fiber and the antioxidant vitamins. How does this translate into a heart-healthy lifestyle? What foods should you eat, for example, to achieve the goal of reducing your saturated fat intake to less than 10 percent of total calories, while also increasing your intake of the protective fibers and antioxidant nutrients? This feature offers some general tips that will help you make the heart-healthy food choices needed to achieve these goals. See this chapter's Spotlight for more tips on how to raise your HDL ("good" cholesterol) level and lower your LDL ("bad" cholesterol) level.

micro10x/Shutterstock.com

Be a savvy supermarket shopper. Read food labels to compare the total fat, saturated fat, *trans* fat, and cholesterol content of similar foods, in addition to fat calories per serving.

1 Become a Savvy Supermarket Shopper.

Learn to read food labels to help you choose nutritious, heart-healthy food products. Read manufacturers' labels to determine both the amounts and the types of fat contained in foods. When comparing foods, look at the Nutrition Facts panel, and choose the food with the lower amounts of saturated fat, *trans* fat, and cholesterol (see Figure 5-13). For saturated fat and cholesterol, keep in mind that 5 percent of the Daily Value (% DV) or less is low and 20 percent or more is high. There is no % DV for *trans* fat, but we are advised to keep our intake of *trans* fat as low as possible while consuming a nutritionally adequate diet.

2 Keep Blood Cholesterol at or Below the Recommended Levels.

Among the most influential diet-related factors that raise blood cholesterol levels are saturated fat and *trans* fat intakes. For every 1 percent reduction in a high blood cholesterol level, there is a 2 to 3 percent reduction in the risk of heart attack. As it turns out, the changes in diet that reduce blood cholesterol concentrations

© Polara Studios

Enjoy a variety of nutritious foods for the health of your heart.

FIGURE 5-13

Checking Out the Food Label for Fat Information

Papa Solo's French Bread Pizza
French Bread Pizza with Tomato Sauce and Mozzarella Cheese

Nutrition Facts
Serving Size 1 pizza (158 g)
Servings per Container 1

Amount Per Serving

Calories 310   Calories from Fat 45

	% Daily Value*
Total Fat 5g	**8%**
Saturated Fat 2g	10%
Trans Fat 1.5g	
Polyunsaturated Fat 0.5g	
Monounsaturated Fat 1g	
Cholesterol 10mg	**3%**
Sodium 470mg	**20%**
Total Carbohydrate 49g	**16%**
Dietary Fiber 6g	24%
Sugars 3g	
Protein 20g	

Vitamin A 4%	•	**Vitamin C** 0%
Calcium 30%	•	**Iron** 15%

* Percent Daily Values are based on a 2,000 calorie diet. Your Daily Values may be higher or lower depending on your calorie needs:

	Calories:	2,000	2,500
Total Fat	Less than	65g	80g
Sat Fat	Less than	20g	25g
Cholesterol	Less than	300mg	300mg
Sodium	Less than	2,400mg	2,400mg
Total Carbohydrate		300g	375g
Dietary Fiber		25g	30g

Calories per gram
Fat 9 • Carbohydrate 4 • Protein 4

Ingredients: French bread (enriched unbleached wheat flour), water, wheat gluten, oat fiber, sugar, soybean oil, nonfat dry milk, mozzarella cheese, tomatoes. Contains 2% or less of each of the following: pizza spice mix (spices, salt, onion and garlic), sugar, coloring (dextrose, cabbage extract, annatto).

MAMA MIA's
Pepperoni Pizza-for-One

Nutrition Facts
Serving Size 1 pizza (191g)
Servings per Container 1

Amount Per Serving

Calories 520   Calories from Fat 243

	% Daily Value*
Total Fat 27g	**42%**
Trans Fat 8g	
Saturated Fat 10g	50%
Cholesterol 25mg	**8%**
Sodium 1,280mg	**53%**
Total Carbohydrate 53g	**18%**
Dietary Fiber 4g	16%
Sugars 6g	
Protein 19g	

Vitamin A 45%	•	**Vitamin C** 0%
Calcium 30%	•	**Iron** 10%

* Percent Daily Values are based on a 2,000 calorie diet. Your Daily Values may be higher or lower depending on your calorie needs:

	Calories:	2,000	2,800
Total Fat	Less than	65g	80g
Sat Fat	Less than	20g	25g
Cholesterol	Less than	300mg	300mg
Sodium	Less than	2,400mg	2,400mg
Total Carbohydrate		300g	375g
Dietary Fiber		25g	30g

Calories per gram
Fat 9 • Carbohydrate 4 • Protein 4

Ingredients: CRUST: Wheat flour with malted barley flour, water, partially hydrogenated vegetable oil (soybean and/or cottonseed oil) with soy lecithin, soybean oil, yeast, high fructose corn syrup, salt; TOPPING: Part-skim mozzarella cheese substitute, pepperoni (pork and beef, salt, water, dextrose, spices, sodium nitrite), part-skim mozzarella cheese; SAUCE: Tomato puree, water, green peppers, salt, lactose and flavoring, spices, corn oil, xanthum gum.

Total fat refers to all the fat in the food: saturated, monounsaturated, polyunsaturated, and *trans* fat. Total fat, saturated fat, *trans* fat, and cholesterol information is required on the label. Listing the amount of monounsaturated and polyunsaturated fats in the food is voluntary.

Ingredients list—For sources of *total* fat, look for terms such as coconut oil, diglycerides, lard, monoglycerides, oil, palmitate, palm oil, stearate, triglycerides, and vegetable shortening.

Also check ingredients lists for *hydrogenated* or *partially hydrogenated oil*. The higher on the list it appears, the more of both total fat and *trans* fat the product probably contains, and the more difficult it is to fit into a healthful diet. As an alternative, look for similar products made with no hydrogenated oils and smaller amounts of total fat.

mostly do so by reducing LDL-cholesterol. Dietary modifications that help lower LDL-cholesterol include:

- Substitute highly monounsaturated fats (canola or olive oils) and highly polyunsaturated fats (vegetable oils and fish oils) for saturated fats in the diet, because both lower LDL-cholesterol levels in the blood; refer to Table 5-2 on page 137. Choose fats and oils with 2 grams or less of saturated fat per tablespoon, such as liquid and tub margarines and olive or canola oils.[28] Shop for margarine with liquid vegetable oil as the first ingredient. Look for the canola- or olive-oil-based margarines containing zero *trans* fat.

- Replace saturated fats and *trans* fats with unsaturated fats from vegetables, legumes (beans), and nuts. Choose healthful portions of skinless poultry, lean meat, and fish, especially fish rich in omega-3 fatty acids (like salmon). Limit foods high in saturated fat, *trans* fat, and/or cholesterol, such as full-fat milk products, fatty meats, tropical oils (coconut or palm), partially hydrogenated vegetable oils, and egg yolks. Eat fewer high-fat desserts (such as cheesecake, ice cream, brownies, pies, pastries, and cakes with butter cream frosting). Choose fruits as desserts most often. Use the tips in the Recipe Modification box to make your own favorite recipes more healthful.

(Continued)

Recipe Modification

Use this substitution information to modify your own favorite recipes. They'll be more healthful but will still look and taste as good as your originals!

Instead of:	Substitute:
1 whole egg	2 whipped egg whites or ¼ cup egg substitute
Whole milk	Fat-free or 1% milk
Evaporated milk	Evaporated nonfat or low-fat milk
Heavy cream	Evaporated nonfat or low-fat milk
Sour cream	Reduced-fat or fat-free sour cream or plain yogurt
Solid shortening (baking)	Vegetable oil, margarine
Solid shortening (stir-frying)	Peanut or olive oil
Mayonnaise	Reduced-fat or fat-free mayonnaise
Cream cheese	Reduced-fat or fat-free cream cheese
Regular cheese	Cheese made from part-skim milk (mozzarella, Swiss lace, farmer) or reduced-fat cheeses
Salad dressing	Fat-free salad dressings or reduced-calorie dressings
White sauce (made with cream and butter)	Use low-fat milk and cornstarch or flour by blending the starch into cold liquid to eliminate the need for fat.

- Add bulk. People on high-fiber diets have been shown to excrete more cholesterol and fat than those on low-fiber diets. A 5 percent decrease in total cholesterol can be achieved with a 5- to 10-gram increase in soluble fiber intake.[29] One reason for this is that the high-fiber diet decreases food's transit time through the digestive tract, allowing less time for cholesterol to be absorbed. When cholesterol from the diet is thus reduced, the body must turn to its own supply for making necessary body compounds. Diets high in fiber are typically low in fat and cholesterol—another advantage of emphasizing fiber. The various dietary fibers have varying effects on blood cholesterol. Soluble fiber, such as that found in rolled oats, oat bran, and psyllium-fortified cereals, have favorable effects on blood cholesterol, whereas wheat bran does not appear to be effective.[30] Apples, pears, peaches, oranges, and grapes are good sources of pectin, another type of cholesterol-lowering fiber.

- Eat generous amounts of fruits and vegetables—at least five to nine servings a day—as a source of antioxidants and other protective compounds in the diet. Antioxidants, such as beta-carotene, vitamins C and E, and the mineral selenium, strengthen the body's natural defenses against cell damage

free radicals highly toxic compounds created in the body as a result of chemical reactions that involve oxygen. Environmental pollutants such as cigarette smoke and ozone also prompt the formation of free radicals.

by blocking the potentially damaging free radicals that arise as a part of numerous normal cell activities. **Free radicals**, the chemical compounds that oxidize LDL-cholesterol, become a problem when they are too plentiful. Antioxidants in the body may reduce the amount of LDL-cholesterol that becomes oxidized because the antioxidants (particularly vitamin E) can neutralize the highly reactive oxygen before it gets a chance to oxidize the LDL particle, thereby lessening the buildup of plaque in the artery walls.[31] Scientists are now exploring the potential abilities of all the antioxidant substances found in foods to protect against heart disease.[32]

③ Balance Energy Intake with Energy Needs.

For heart health, maintain a healthy body weight and a level of physical activity that keeps you fit and matches the number of calories you eat.[33] Walk, or do other activities on most days. Moderate exercise, such as brisk walking, may both lower LDL levels and raise HDL levels if the activity is consistently pursued for long enough periods.

It seems that the factors affecting the health of the heart are all tangled together, but all evidence points to the same general recommendations. For good health and to avoid heart disease, do not smoke; reduce blood pressure and weight, if necessary; exercise regularly; and eat a heart-healthy diet.

Rate Your Fats and Health IQ

Do You?	Often	Sometimes	Rarely
Trim or drain the fat from meats, remove the skin from chicken, and serve fish-based meals?	10	5	1
Eat a variety of fresh, frozen, or canned fruits and vegetables?	10	5	1
Eat high-fat foods such as bacon, sausage, regular franks, and luncheon meat several times a week?	1	5	10
Limit whole eggs or egg yolks to seven per week?	10	5	1
Read food labels when shopping?	10	5	1
Choose low-fat or fat-free milk, yogurt, cheese, and sour cream?	10	5	1
Bake, rather than fry, foods?	10	5	1
Maintain a healthy weight?	10	5	1
Think "eating right" and make trade-offs when eating out?	10	5	1
Choose doughnuts, croissants, or sweet rolls for breakfast?	1	5	10
Choose reduced-fat or fat-free products when available?	10	5	1
Routinely add margarine, butter, salad dressing, and sauces to foods?	1	5	10
Balance a high-fat dinner by choosing low-fat foods for breakfast and lunch?	10	5	1
Plan physical activity (walking, running, swimming, bike riding) into your schedule on most days of the week?	10	5	1
TOTAL	_____	_____	_____

Scoring

113–150: You practice heart-healthy habits. Keep up the good work.

75–112: Not a bad score, but there's room for improvement.

5–74: Too low a score! Learn more about the relationships between fats and health.

Large potato with 1 tablespoon butter and 1 tablespoon sour cream (14 grams fat, 350 calories).

Large potato with 2 tablespoons fat-free sour cream or yogurt seasoned with chives (less than 1 gram fat, 235 calories).

SOURCE: Adapted from American Dietetic Association and Mosby Great Performance, *Healthy Eating for the Whole Family* (St. Louis, MO: Mosby-Year Book, 1995), 5.

Spotlight Diet and Heart Disease

More than one of every four people who die in the United States each year die of heart and blood vessel disease. The underlying condition that contributes to most of these deaths is atherosclerosis, which leads to closure of the arteries that feed the heart and brain and thus to heart attacks and strokes. In terms of direct health care costs, lost wages, and lost productivity, heart disease costs the United States more than $315 billion a year. There is little wonder, then, that much effort has been focused on preventing it.

The twin demons that lead to most forms of heart disease are atherosclerosis and hypertension. Atherosclerosis, the subject of this Spotlight, is the narrowing of the arteries caused by a buildup of cholesterol-containing plaque in the arterial walls. Hypertension (which is discussed in the Nutrition Action feature of Chapter 8) is high blood pressure, and each of these conditions aggravates the other.

How can I know whether I have atherosclerosis?

No one is free of atherosclerosis. The question is not whether you have it, but how far advanced it is and what you can do to retard or reverse it. As already mentioned, atherosclerosis usually begins with the accumulation of soft mounds of lipids, known as *plaques,* along the inner walls of the arteries, especially at the branch points. These plaques gradually enlarge, making the artery walls lose their elasticity and narrowing the passage through them. Most people have well-developed plaques by the time they are 30.

Normally, the arteries expand with each heartbeat to accommodate the pulses of blood that flow through them. Because arteries that are hardened and narrowed by plaques cannot expand, the blood pressure rises. The increased pressure puts a strain on the heart and further damages the artery wall. At damaged points, plaques are especially likely to form; thus, the development of atherosclerosis is a self-accelerating process.

Hypertension makes atherosclerosis worse. A stiffened artery, already strained by each pulse of blood surging through it, is stressed even more if the internal pressure is high. Hypertension causes injured places within the blood vessels to develop more frequently, and plaques grow faster. Hardened arteries also fail to let blood flow freely through the body's blood-pressure-sensing organs, the kidneys, which respond as if the blood pressure were too low and raise it further.

How can I slow the process down?

Learn your risk factors, and control the ones you can. Among the many factors linked to heart disease are smoking, gender (being male), age (men older than 45 and women older than 55), postmenopausal status in women, heredity, diabetes, high blood pressure, lack of exercise, obesity, high blood cholesterol level, and low

A normal artery provides open passage for blood to circulate.

Jean-Claude Revy/ISM/Phototake

Plaques along an artery narrow the passage and obstruct blood flow.

Zephyr/Photo Researchers, Inc.

HDL-cholesterol level (see Table 5-7). Some of the risk factors are also powerful *predictors* of heart disease. If you have none of the risk factors, the statistical likelihood of your developing heart disease may be only 1 in 100. If you have three major risk factors, your chance may rise to more than 1 in 20. Some of the factors that have emerged as powerful predictors of risk are high LDL-cholesterol, low HDL-cholesterol, high blood pressure, smoking, obesity, and physical inactivity.[34]

What can I do to reduce my risk?

Obviously, some risk factors cannot be altered. Being born male or inheriting a predisposition to develop high blood cholesterol or high blood pressure are factors beyond your control. Even so, you can

TABLE 5-7	Leading Risk Factors for Heart Disease

Heart disease rarely develops from a single risk factor. Clusters of risk factors usually occur, and a small increase in one risk factor, such as blood pressure, becomes more critical when combined with other risk factors. Luckily, a similar pattern happens in reverse: Even moderate changes in one risk factor can decrease several others at the same time. For example, a weight loss of just 5 to 10 pounds can reduce blood pressure in overweight people. Or, becoming physically active can lower blood pressure, increase HDL ("good" cholesterol), and help control weight.

Risk Factors You Can Change

Risk Factor	How to Minimize the Risk
High LDL-cholesterol	Increase your intake of soluble fiber, soy foods, and omega-3 fats.
Low HDL-cholesterol	Maintain a physically active lifestyle. Control high blood pressure with medication and a heart-healthy diet. Maintain a healthy weight. Losing just 5 to 10 pounds may lower your blood pressure. Compare sodium in soups, frozen meals, and other foods—and choose the foods with less sodium.
Cigarette smoking	Stop smoking. Nicotine constricts blood vessels and forces your heart to work harder. Carbon monoxide reduces oxygen in blood and damages the lining of blood vessels.
Diabetes	Maintain proper weight. Losing excess weight helps control blood sugar level. Eat high-fiber foods. Limit solid fats and added sugars. Get regular physical activity.
Physical inactivity	Increase physical activity by seeking ways to be more physically active every day. Aim for the equivalent of 150 minutes (2 hours and 30 minutes) of moderate-intensity aerobic activity (such as brisk walking) each week. See Chapter 11 for more physical activity guidelines.
Obesity	Maintain a healthy weight and be physically active. Being only 10% overweight increases heart disease risk. See Chapter 10 for more about weight management.
"Atherogenic" diet	Keep saturated fat to less than 7% of daily calories. Substitute olive and canola oils for saturated fat. Keep *trans* fat to less than 1% of calories (as low as possible). Limit consumption of foods with solid fats and added sugars. Increase soluble fiber intake by eating whole grains, legumes, fruits, and vegetables. Eat a variety of fruits and vegetables a day to receive the beneficial antioxidants, vitamins, minerals, and phytochemicals they contain.
Stress	Get regular physical activity. Avoid excessive caffeine and alcohol. Practice relaxation techniques. Maintain good social relationships.

Risk Factors You Can't Change

Age	Men over age 45 and women over age 55 are at increased risk.
Gender	Men are at higher risk. Estrogen may protect women before menopause.
Genetics	Risk is increased if you have a father or brother under age 55 or a mother or sister under 65 who had heart disease.

make conscious choices that may reduce your risk of developing heart disease. Let's examine one of the two major risk factors related to diet—high blood cholesterol—with an eye toward learning which dietary and lifestyle changes will help reduce your heart disease risk.

To what extent does a high blood cholesterol level raise the risk of developing heart disease?

The likelihood of a person developing or dying from heart disease increases as total blood cholesterol level rises. Figure 5-14 presents this relationship graphically, showing that the number of deaths from heart disease increases steadily among those with elevated blood cholesterol levels, particularly when the level rises above 200 milligrams per deciliter. Individuals with a blood cholesterol level in the neighborhood of 300 milligrams per deciliter run four times the risk of dying from heart disease compared with those whose cholesterol level is lower than 200 mg/dL. Having a low HDL-cholesterol value is now also recognized as being a risk factor

(Continued)

FIGURE 5-14

Relationship between Blood Cholesterol Level and Death Rate from Heart Disease

Persons with blood cholesterol levels of 300 mg/dL are four times as likely to die from heart disease as those with blood cholesterol levels below 200 mg/dL.

SOURCE: Data from 361,662 men screened for the Multiple Risk Factor Intervention Trial (MRFIT). Adapted from National Cholesterol Education Program, *Report of the Expert Panel on Population Strategies for Blood Cholesterol Reduction* (Bethesda, MD: U.S. Department of Health and Human Services, Public Health Service, National Institutes of Health, National Heart, Lung, and Blood Institute, NIH Publication No. 90-3046, November 1990), 7.

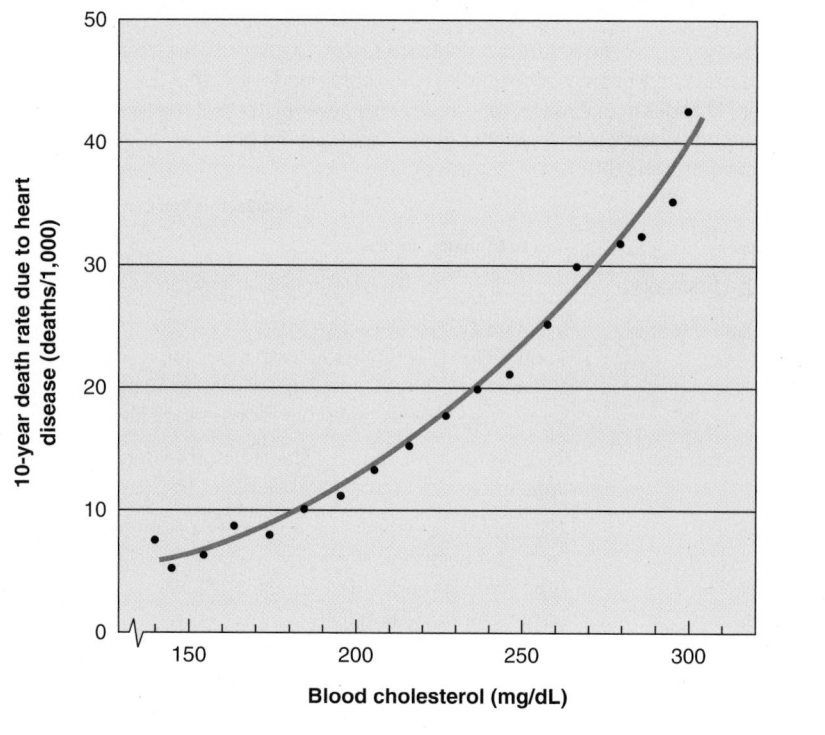

for heart disease.[35] A low HDL value is defined as one below 40 mg/dL. A combined effort to lower LDL-cholesterol and raise HDL-cholesterol delivers a double punch in the fight against heart disease.

Reducing high blood cholesterol levels, particularly the "bad" LDL-cholesterol level, is thus an important strategy for lowering the risk of heart disease. This is especially true for persons having one or more of the other major heart disease risk factors, such as smoking. Table 5-8 provides information for determining what to do if your blood cholesterol level is high and whether you might need to seek treatment.

What sorts of dietary changes should I make to reduce my total blood cholesterol and LDL-cholesterol levels?

Probably the most significant dietary changes you can make are to reduce the amounts of saturated fat and *trans* fat that you eat; high intakes of these fats are related to high blood levels of LDL-cholesterol.[36] The goals to work toward are to reduce your intake of saturated fat to less than 10 percent of total calories, and to keep *trans* fatty acid consumption as low as possible. People with elevated levels of LDL-cholesterol, existing heart disease, diabetes, or combinations of risk factors are advised to reduce their intake of saturated fat even further, to less than 7 percent of total calories, and their cholesterol intakes should be less than 200 milligrams per day.[37] See this chapter's Eat Well Be Well feature for more specifics on reducing your LDL-cholesterol level.

Can the addition of plant sterols to the diet help reduce blood cholesterol levels?

Certain brands of margarine contain phytosterols (plant stanols or plant sterols), which are compounds derived from plant oils that may actually reduce blood cholesterol when consumed as part of a low-fat diet.[38] A *phytosterol* is a phytochemical that is structurally similar to the steroid hormones (for example, estrogen). Examples of margarines containing phytosterols include Benecol and Take Control. Benecol uses a stanol ester derived from pine trees, and Take Control uses one made from soy. These compounds block the absorption of dietary cholesterol from the intestine. The FDA has given the manufacturers approval to use labels stating that the products "promote healthy cholesterol levels." Margarines containing these compounds, however, typically cost more than regular margarine, and the safety of their long-term daily use is not known.

Should I also reduce my cholesterol intake?

Dietary cholesterol can raise LDL levels, depending on the amount consumed and on the body's ability to compensate by making less. The *Dietary Guidelines* recommend that we limit the cholesterol in our daily diets to 300 milligrams. Consuming less than 200 mg per day is recommended for persons at high risk of heart disease.[39] On average, men consume 350 milligrams of cholesterol and women consume about 240 milligrams daily.

TABLE 5-8 Standards for Blood Cholesterol Levels and Risk of Heart Disease

Evaluating Blood Lipid Levels

	Desirable	Borderline High	High Risk
Total blood cholesterol (mg/dL)	< 200	200–239	≥ 240
LDL-cholesterol (mg/dL)	< 100[a]	130–159	160–189[b]
HDL-cholesterol (mg/dL)	≥ 60	40–59	< 40
Triglycerides, fasting (mg/dL)	< 150	150–199	200–499[c]

Treating LDL-Cholesterol Levels

Risk Status	Target/Recommendations for LDL-Cholesterol
Low-risk individuals[d]	Less than 160 mg/dL Begin dietary therapy at 160 mg/dL; consider drug therapy at 190 mg/dL.
High-risk individuals[e]	Less than 130 mg/dL Begin dietary therapy at 130 mg/dL; consider drug therapy at 160 mg/dL.
Very high-risk individuals[f]	Less than 100 mg/dL Begin dietary therapy at 100 mg/dL; consider drug therapy at 130 mg/dL.

[a] 100–129 mg/dL LDL indicates a near- or above-optimal level.

[b] ≥ 190 mg/dL LDL indicates a very high risk.

[c] ≥ 500 mg/dL triglycerides indicates very high risk.

[d] Low-risk individuals are those with one or no risk factors for heart disease. Major risk factors (other than LDL-cholesterol) include cigarette smoking, hypertension or on antihypertensive medication, low HDL-cholesterol (< 40 mg/dL), family history of premature heart disease, age (men ≥ 45 years; women ≥ 55 years).

[e] High-risk individuals are those who have at least two risk factors for coronary heart disease. (Men over age 45 and postmenopausal women are considered to have one risk factor.)

[f] Very high-risk individuals are those with known heart disease (history of heart attack or previous bypass surgery or angioplasty) and diabetes.

What factors determine my HDL-cholesterol level?

One is gender. Women have higher HDL-cholesterol levels than men. However, heart disease is the major cause of death among women after menopause, when low levels of estrogen lead to a decrease in HDL levels along with an increase in LDL levels. Another factor, interestingly, seems to be smoking habits. Nonsmokers have uniformly higher HDL levels than do smokers. Still another factor is weight reduction for those who are overweight.[40]

If any dietary factors are significant, one important one may be to include some fish in the diet.[41] Eat fish such as salmon, trout, and herring in place of meat or poultry twice a week. Another may be including soy foods and foods containing soluble fibers that lower LDL levels.*

By far the most powerful influence on HDL levels is not a dietary factor at all—it is *regular physical activity*. The discovery that physical activity raises HDL levels gave impetus to the physical fitness movement, and especially to the popularity of running and walking. The earliest reports indicated increased HDL levels in long-distance run-

ners, and continuing study of this elite group has repeatedly demonstrated that running does indeed elevate HDL. People do not, however, need to become competitive athletes to raise their HDL. Moderate exercise, such as walking, may both lower LDL levels and raise HDL levels if the activity is consistently pursued for long enough periods.

What is homocysteine, and how does it contribute to heart disease?

Homocysteine (pronounced ho-mo-SIS-teen) has been linked to heart disease in recent years. The body naturally produces homocysteine when it breaks down protein. The B vitamins (folate, B_6, and B_{12}) then convert homocysteine into other amino acids that the body requires. However, when dietary intakes of these B vitamins are low, blood levels of homocysteine rise. Scientists believe that high levels of homocysteine cause damage to the linings of arteries and accelerate the formation of blood clots. Research shows that people with high blood levels of homocysteine have significantly more heart attacks and strokes.[42]

* For a discussion of the benefits of soy in prevention of heart disease, see the Spotlight in Chapter 6.

(Continued)

The good news is that as a risk factor for heart disease, high homocysteine levels may be preventable and reversible by consuming generous amounts of the B vitamins. Good sources of folate include orange juice; enriched breads, cereals, and other grain products; green leafy vegetables; and legumes. Vitamin B_6 is found in meat, chicken, fish, whole grains, and legumes. Vitamin B_{12} is found in fortified cereals, fortified soy milk, and all animal foods. (See Chapter 7 for more information about these B vitamins.)

What about alcohol? I've heard that moderate alcohol intake can be beneficial, at least with respect to heart disease risk.

The evidence with regard to alcohol and blood lipids points to a possible protective mechanism between moderate alcohol consumption and heart disease risk factors.[43] Investigators have reported that the consumption of moderate amounts of alcohol appears to raise HDL levels.[44] *Moderate drinking* is defined as no more than one drink a day for most women, and no more than two drinks a day for most men. A strong association between moderate alcohol consumption and low rates of heart disease was first observed in France and then in other wine-drinking areas of the Mediterranean. Researchers have since identified antioxidants and other compounds in red wine that decrease blood clotting, and this may help explain part of the "French paradox," or how a region with high fat intakes could have such low rates of heart disease. Researchers are quick to point out, however, that other factors certainly may contribute to the paradox. For example, the French consume about 57 percent of their day's calories before 2 p.m., whereas most Americans have consumed only about 38 percent of their calories by that time.[45] Also, the French mostly consume their wine with meals, and they eat more fruits and vegetables and leaner cuts of meat.

Scientists warn that caution is needed in recommending moderate alcohol consumption to the public. Many people need to refrain from alcohol consumption altogether (pregnant women, recovering alcoholics, people under age 21, persons on certain medications, and those intending to drive a vehicle).[46] Keep in mind that alcohol can have profoundly negative effects on the body and is associated with many disease states (see Chapter 9).

What about garlic? I've heard that garlic pills and whole cloves help lower blood cholesterol.

A sizeable body of evidence from around the world suggests that garlic may help lower blood cholesterol, reduce blood pressure, and help arteries remain elastic.[47] Although the research is promising, some studies have shown conflicting results regarding the effect of garlic on blood cholesterol. Further studies are needed before scientists can come to any conclusions about the usefulness of fresh garlic or garlic pills (which vary a great deal in composition) in lowering blood cholesterol levels. See Chapter 7 for more about the phytochemical benefits derived from garlic and its relatives (leeks, onions, chives, scallions, and shallots).

What about eating chocolate? Are there any health benefits derived from chocolate?

Some evidence suggests that modest consumption of dark chocolate or cocoa is associated with health benefits in the form of reduced risk for heart disease. However, chocolate is high in calories (a 3-ounce chocolate bar can contain more than 400 calories), so the benefits need to be balanced with caloric intake to prevent weight gain. The predominant fat in chocolate is stearic acid, a saturated fat which has been shown to not raise blood cholesterol. The beneficial effects from dark chocolate are attributed to its phytochemicals, in particular the flavonoids, which act as antioxidants. The benefits most frequently associated with consumption of dark chocolate include reducing inflammation; decreasing total cholesterol; decreasing LDL-cholesterol; protecting against blood clot formation; and decreasing blood pressure.[48] Many foods and beverages that are less calorie-dense (berries, citrus fruit, grapes, soy foods, and tea) contain flavonoids, and chocolate is considered a minor source. See Chapter 7 for more about phytochemicals, and see Chapter 9 for more about the flavonoids found in both tea and cocoa.

How do I translate these recommendations into a heart-healthy lifestyle?

Most people have a difficult time translating the dietary recommendations into actual meal patterns. This chapter's Eat Well Be Well feature provided many tips for reducing your LDL-cholesterol level. Consider the following suggestions as well:[49]

- Eat a variety of antioxidant-rich fruits and vegetables. Choose nine servings per day.

- Choose fat-free or low-fat dairy products, such as fat-free milk or low-fat or fat-free yogurt. Choose milk or yogurt more often than cheese; they are better sources of potassium and are lower in sodium than most cheeses.

- Consume abundant legumes of many varieties, including soybeans, kidney beans, and lentils.

- Eat a variety of whole grains and other foods high in soluble fibers and eat fewer refined grain products.

- Choose soy foods, skinless poultry, lean meat, and fish, especially fish such as salmon that is rich in omega-3 fatty acids, in place of foods high in saturated fat.

- Limit your intake of foods that are high in saturated fat and *trans* fat, such as full-fat milk products, fatty meats, tropical oils, and partially hydrogenated vegetable oils.

- Limit your intake of foods that are high in calories from solid fats and added sugars, such as cakes, cookies, candy, and ice cream; pizza; cheese; processed and fatty meats; and sweetened beverages.

- Adopt heart-healthy cooking methods, such as broiling, baking, steaming, and stir-frying.

- Reduce daily sodium intake to less than 2,300 milligrams and further reduce intake to 1,500 milligrams if you are 51 or older, African American, or have high blood pressure or diabetes.

- Get more potassium in your diet from foods such as cantaloupe, bananas, beans, and yogurt.

- Consume alcohol only in moderation, if at all. Have no more than one alcoholic drink per day if you're a woman and no more than two if you're a man. "One drink" means it has no more than 1/2 ounce of pure alcohol. Examples of one drink are 12 ounces of beer, 5 ounces of wine, 1½ ounces of 80-proof spirits, or 1 ounce of 100-proof spirits.

Attention to emotional health is also important in reducing the risk of heart disease. Both love and affection seem to affect the heart. People with many social ties appear to develop less heart disease than do people with few or none. Similarly, pet owners, even owners of pet fish, have lower blood pressure than do people without pets. Clearly, the mystery of heart disease, like all the great human mysteries, involves the mind and spirit as well as the body. So nourish yourself in all ways—not just physically.

Should children, like their parents, eat low-fat diets for heart health?

The current *Dietary Guidelines* and experts from major health and professional organizations recommend that children aged 2 and older eat diets containing no more than 10 percent of total calories from saturated fat and no more than 300 milligrams of cholesterol daily. The reason is that hardening of the arteries often begins in childhood. However, from birth to 2 years of age, a child's fat consumption should not be restricted, because fat is a concentrated source of the calories needed to ensure proper physical development.

The panel also advised that children or teens should get their blood cholesterol measured if they have one parent with a high blood cholesterol level. For children and adolescents, a total of 200 milligrams per deciliter or more is considered high, 170 to 199 is borderline high, and less than 170 is acceptable. In addition, children born to families with a history of premature heart disease should have both total blood cholesterol and HDL-cholesterol checked.[50]

Nutrition on the Web

www.cengage.com/nutrition/boyle/personalnutrition8e

Go to the Personal Nutrition site to check for the latest updates to chapter topics or to access links to related websites.

www.healthfinder.gov

Search for information on atherosclerosis, heart disease, and stroke.

www.eatright.org

The American Dietetic Association home page includes many nutrition resources and links.

www.nal.usda.gov/fnic

Under Browse by Subject, select Diet and Disease and click on Heart Health.

www.mayoclinic.com

The Mayo Clinic maintains this website that provides a variety of health information. Click on the Diseases & Conditions tab, and look up atherosclerosis, heart disease, or stroke.

www.fda.gov

Click on Food and search for information on Olestra and other fat substitutes and labeling of *trans* fatty acids in foods.

www.nhlbi.nih.gov/index.htm

This government website is a clearinghouse for the National Heart Lung and Blood Institute (NHLBI). Under Public, click Heart and Vascular Diseases. Click on the health assessment tools, recipes, and educational campaigns links for more information.

www.heart.org/HEARTORG

The American Heart Association provides practical information about diet, physical activity, and cardiovascular risk reduction under the Getting Healthy tab.

www.caloriecontrol.org

Under the Sweeteners and Lite tab, click on Fat Replacers.

www.flaxcouncil.ca

The Flax Council of Canada gives useful information on the health benefits of flaxseed. Click on Nutrition, Recipes, or Food Products tabs.

www.healthyfridge.org

This site provides information on heart disease, practical shopping tips for stocking a healthy refrigerator, heart-healthy recipes, and a quiz to test your saturated fat IQ.

www.foodinsight.org

Search the International Food Information Council (IFIC) for information on managing fats and fat replacers.

www.ncbi.nlm.nih.gov/PubMed

Use this search engine to help locate information from current scientific articles on any topic related to fat.

Menu of Online Study Tools

A variety of study tools for this chapter are available at our website to deepen your understanding of chapter concepts. Go to

www.cengage.com/sso

to find

- Practice tests
- Flashcards
- Glossary
- Web links
- Animations
- Chapter summaries, learning objectives, and crossword puzzles.

Diet Analysis+ in Action

Go to the Intake vs. Goals report in Diet Analysis+ that shows the results of your three-day tracking. How many grams of total fat did you consume? Now go to the Intake Spreadsheet report and list the three foods that contributed the most total fat to your intake. Now list the top three foods that contributed the most saturated fat to your daily intake. What foods might you substitute for those high in saturated fat? If your overall intake of fat exceeded the recommended amount (20 to 35 percent of total calories), what changes could you make to improve your fat intake?

The Proteins and Amino Acids

Ask Yourself . . .

Which of the following statements about nutrition are true, and which are false? For each false statement, what is true?

1. When more protein is eaten than the body needs, it is stored intact in the body (the way fat is stored) so that it can be used when a person's diet falls short of supplying the day's need for essential proteins.
2. No new living tissue can be built without protein.
3. Whenever cells are lost, protein is lost.
4. All enzymes and hormones are made of protein.
5. When antibodies enter the body, they produce illness.
6. When a person doesn't eat enough food to meet the body's energy needs, the body devours its own protein tissue.
7. Once the body has assembled its proteins into body structures, it never lets go of them.
8. Milk protein is the standard against which the quality of other proteins is usually measured.
9. It is impossible to consume too much protein.
10. People who eat no meat need to eat a lot of special foods to get enough protein.

Answers found on the following page.

The **proteins** are perhaps the most highly respected of the three energy nutrients, and the roles they play in the body are far more varied than those of carbohydrate or fat. First named 150 years ago after the Greek word *proteios* ("of prime importance"), proteins have revealed countless secrets about how living processes take place, and they account for many nutrition concerns. How do we grow? How do our bodies replace the materials they lose? How does blood clot? What makes us able to become immune to diseases we have been exposed to? To a great extent, the answers to these and many other such questions are found in an understanding of the nature of the proteins.

The amino acids of proteins are the raw materials of heredity, the keys to life chemistry, handed from generation to generation.

R. M. DEUTSCH (1928–1988, NUTRITION AUTHOR AND EDUCATOR)

proteins compounds composed of atoms of carbon, hydrogen, oxygen, and nitrogen and arranged as strands of amino acids. Some amino acids also contain atoms of sulfur.

© Michael Cogliantry/Stockfood America

An Amino Acid: glycine

Amine group — Acid group

Side group

An Amino Acid: phenylalanine

What Proteins Are Made Of

To appreciate the many vital functions of proteins, we must understand their structure. One key difference from carbohydrate and fat, which contain only carbon, hydrogen, and oxygen atoms, is that proteins contain nitrogen atoms. These nitrogen atoms give the name *amino* ("nitrogen containing") to the **amino acids** of which protein is made. Another key difference is that in contrast to the carbohydrates—whose repeating units, glucose molecules, are identical—the amino acids in a strand of protein are different from one another.

All amino acids have the same, simple chemical backbone, with an **amine group** (the nitrogen-containing part) at one end and an acid group at the other end. The differences among the various amino acids are due to the varying structures of the chemical side chains that are attached to the backbone. Twenty amino acids with 20 different side chains make up most of the proteins of living tissue.

The side chains vary in complexity from a single hydrogen atom, like that on glycine, to a complex ring structure, like that on phenylalanine. Not only do these structures differ in composition, size, and shape, but they also differ in electrical charge. Some are negative, some are positive, and some have no charge. These side chains help to determine the shapes and behaviors of the larger protein molecules that the amino acids make up.

Essential and Nonessential Amino Acids The body can make about half of the amino acids (known as *nonessential amino acids*) for itself, when it has the needed parts: nitrogen to form the amine group and backbone fragments, which are derived from carbohydrate or fat. But even the healthy body cannot make some amino acids. These are known as **essential amino acids.** * If the diet does not supply them, the body cannot make the proteins it needs to do its work. The indispensability of the essential amino acids makes it necessary for people to eat protein food sources every day.

Proteins as the Source of Life's Variety In the first step of **protein synthesis**, each amino acid is hooked to the next. A bond, called a **peptide bond**, is formed between the amino end of one and the acid end of the next. Proteins are made of many amino acid units, from several dozen to many hundred.

A strand of protein is not straight; it is more like a tangled chain. The amino acids at different places along the strand are attracted to one another, and this attraction causes the strand to coil into a shape similar to that of a metal spring. Not only does the strand of amino acids form a long coil, but the coil tangles, forming a globular structure.

The charged amino acids are attracted to water, and in the body fluids they orient themselves on the outside of the globular structure. The neutral amino acids are

---

The nine essential amino acids for human adults that must be obtained from the diet:

histidine	phenylalanine
isoleucine	threonine
leucine	tryptophan
lysine	valine
methionine	

The nonessential amino acids— also important in nutrition:

alanine	glutamine
arginine	glycine
asparagine	proline
aspartic acid	serine
cysteine	tyrosine
glutamic acid	

amino (uh-MEEN-oh) acids building blocks of protein; each is a compound with an amine group at one end, an acid group at the other, and a distinctive side chain.

amine (a-MEEN) group the nitrogen-containing portion of an amino acid.

essential amino acids amino acids that cannot be synthesized by the body or that cannot be synthesized in amounts sufficient to meet physiological need.

protein synthesis the process by which cells assemble amino acids into proteins. All individuals are unique because of minute differences in the ways their body proteins are made. The instructions for making all the proteins in our bodies are transmitted in the genetic information we receive at conception.

peptide bond a bond that connects one amino acid with another.

ASK YOURSELF ANSWERS: 1. False. When more protein is eaten than the body needs, it is *not* stored in the body (the way fat is stored), so it must be eaten every day to avoid protein depletion. **2.** True. **3.** True. **4.** False. All enzymes, but not all hormones, are made of protein. **5.** False. Antibodies protect the body from illness caused by antigens. **6.** True. **7.** False. Your body loses protein every day. **8.** False. Egg-white protein, not milk protein, is the standard against which the quality of other proteins is usually measured. **9.** False. It is possible to consume too much protein. **10.** False. People who eat no meat can easily get enough protein without eating a lot of special foods.

---

*The distinction between essential and nonessential amino acids is not quite as clear-cut as the list in the margin makes it appear. For example, cysteine and tyrosine normally are not essential, because the body makes them from methionine and phenylalanine. However, if there are not enough of these precursors from which to make cysteine and tyrosine, they must be supplied by the diet.

repelled by water and are attracted to one another; they tuck themselves into the center, away from the body fluids. All these interactions among the amino acids and the surrounding fluid result in the unique architecture of each type of protein. Additional steps may be needed for the protein to become functional. A mineral or a vitamin may be needed to complete the unit and activate it, or several proteins may gather to form a functioning group.

The differing shapes of proteins enable them to perform different tasks in the body. In proteins that give strength and elasticity to body parts, several springs of amino acids coil together and form rope-like fibers. Other proteins, like those in the blood, do not have such structural strength but are water-soluble, with a globular shape like a ball of steel wool. Some are shaped like hollow balls that can carry and store minerals in their interiors. Still others provide support to tissues. Some—the enzymes—act on other substances to change them chemically.

Denaturation of Proteins   Proteins can undergo **denaturation**, resulting in distortion of shape by heat, alcohol, acids, bases, or the salts of heavy metals. The denaturation of a protein is the first step in breakdown of the protein. Denaturation is useful to the body in digestion. During the digestion of a food protein, an early step is denaturation by the stomach acid, which opens up the protein's structure, permitting digestive enzymes to break the peptide bonds (see Figure 6-1 later in this chapter). Denaturation can also occur during food preparation. For example, cooking an egg denatures the proteins of the egg and makes the egg firmer. Perhaps more important, cooking denatures two raw-egg proteins that bind the B vitamin biotin and the mineral iron, as well as a protein that slows the digestion of other proteins. Thus, cooking eggs liberates biotin and iron and aids in protein digestion.

The Functions of Body Proteins

No new living tissue can be built without protein, for protein is part of every cell. About 20 percent of our total body weight is protein. Proteins come in many forms: enzymes, antibodies, hormones, transport vehicles, oxygen carriers, tendons and ligaments, scars, the cores of bones and teeth, the filaments of hair, the materials of nails, and more (see Table 6-1). A few of the many vital functions of proteins are

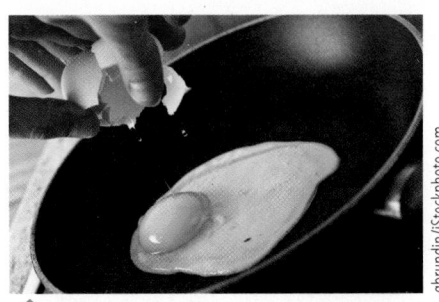

Cooking an egg denatures its proteins.

Peptide bond

A Molecule of Insulin

Amino acids are linked together with peptide bonds to form strands of protein. The sulfur groups (S) on two cysteine (cys) molecules can bond together, creating a "sulfur bridge" between the two protein strands.

TABLE 6-1	The Functions of Body Proteins

1. Growth and Maintenance
- Proteins provide *building materials*—amino acids—for growth and repair of body tissues.
- *Body structures*. Proteins form vital parts of most body structures, such as skin, nails, hair, membranes, muscles, teeth, bones, organs, ligaments, and tendons.

2. Regulatory Roles
- *Enzymes*. Proteins facilitate numerous chemical reactions in the body; all enzymes are proteins.
- *Hormones*. Some proteins act as chemical messengers, regulating body processes; not all hormones are proteins.
- *Antibodies*. Proteins assist the body in maintaining its resistance to disease by acting against foreign disease-causing substances.
- *Fluid balance*. Proteins help regulate the quantity of fluids in body compartments.
- *Acid–base balance*. Proteins act as buffers to maintain the normal acid and base concentrations in body fluids.
- *Transportation*. Proteins move needed nutrients and other substances into and out of cells and around the body.

3. Energy Production
- *Energy*. Protein can be used to provide calories (4 calories per gram) to help meet the body's energy needs.

denaturation a change in the shape of a protein brought about by heat, alcohol, acids, bases, salts of heavy metals, or other agents.

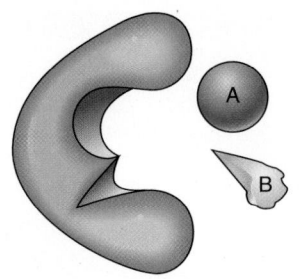

Enzyme plus two compounds, A and B

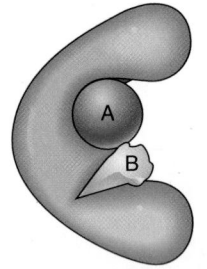

Enzyme complexed with A and B

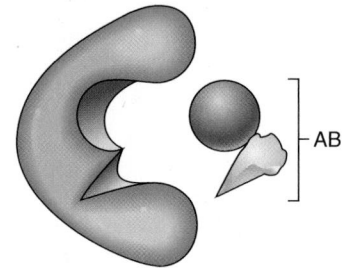

Enzyme plus new compound AB

▲
Enzyme Action

Each enzyme facilitates a specific chemical reaction.

hormones chemical messengers. Hormones are secreted by a variety of glands in the body in response to altered conditions. Each affects one or more target tissues or organs and elicits specific responses to restore normal conditions.

antibodies large proteins of the blood and body fluids that are produced by one type of immune cell in response to invasion of the body by unfamiliar molecules (mostly foreign proteins). Antibodies inactivate the foreign substances and so protect the body. The foreign substances are called *antigens.*

immunity specific disease resistance derived from the immune system's memory of prior exposure to specific disease agents and its ability to mount a swift response against them.

described here to show why they have rightfully earned their position of importance in nutrition.

Growth and Maintenance

One function of dietary protein is to ensure the availability of amino acids for building the proteins of new tissue. New tissue is needed in an embryo; in a growing child; in the blood that replaces that which has been lost in burns, hemorrhage, or surgery; in the scar tissue that heals wounds; and in new hair and nails. Not so obvious, but equally important, is the protein that helps replace worn-out cells. The cells that line the digestive tract live for only about three days, and they are constantly being shed and excreted. You have probably observed that the cells of your skin die, rub off, and are replaced from underneath. For this new growth, amino acids must constantly be resupplied by food.

Enzymes

All enzymes are proteins, and they are among the most important proteins formed in living cells. Enzymes are *catalysts*—biological spark plugs—that help chemical reactions take place. There are thousands of enzymes inside a single cell, each type facilitating a specific chemical reaction. Enzymes are involved in such processes as the digestion of food, the release of energy from the body's stored energy supplies, and the growth and repair of tissue.

The mystery of how an enzyme can be specific for a particular reaction is only partially understood. Scientists believe that the surface of the enzyme is contoured so that it can recognize only the substances it works on, and ignore others. The surface provides a site that attracts one or more specific chemical compounds and promotes a specific chemical reaction. For example, two substances might first become attached to the enzyme and then to each other. The newly formed product is then expelled by the enzyme into the fluid of the cell. Enzymes are the hands-on workers in the production and processing of all substances needed by the body.

Hormones

Hormones are similar to enzymes in creating profound effects. However, hormone molecules differ from enzyme molecules. For one thing, not all of them are made of protein. For another, hormones do not catalyze chemical reactions directly; rather, they act as messengers that elicit the appropriate responses to maintain a normal environment in the body. Hormones regulate overall body conditions, such as the blood glucose level (the hormones insulin and glucagon) and the metabolic rate (thyroid hormone).

Antibodies

Of all the great variety of proteins in living organisms, **antibodies** best demonstrate that proteins are specific to individual organisms. Antibodies form in response to the presence of *antigens* (foreign proteins or other large molecules) that invade the body. The foreign protein may be part of a bacterium, a virus, or a toxin, or it may be something present in food that causes a reaction we call an *allergy*. The body, after recognizing that it has been invaded, manufactures antibodies that deactivate the foreign substance. Without sufficient protein to make antibodies, the body cannot maintain its resistance to disease.

One of the most fascinating aspects of the antibody response to foreign substances is that each antibody is uniquely designed to destroy a specific foreign substance. An antibody that has been manufactured to combat one strain of flu virus is of no help in protecting a person against another strain. Once the body has learned to make a particular antibody, it never forgets, and the next time it encounters that same foreign substance, it will be equipped to destroy it even more rapidly. In other words, the body develops an **immunity**. This is the principle underlying the effectiveness of the vaccines that have nearly eradicated most childhood diseases in the Western world.

Clearly, malnutrition injures the immune system. Without adequate protein in the diet, the immune system will not be able to make its specialized cells and the other tools required to function optimally. Often, protein deficiency and immune incompetence appear together. For this reason, measles can be fatal to a malnourished child.

Many nutrients other than proteins participate in conferring immunity, and many factors other than antibodies are involved.[1] Nevertheless, the immune system is extraordinarily sensitive to nutrition, and almost any nutrient deficit can impair its efficiency and reduce resistance to disease.

Fluid Balance

Proteins help regulate the quantity of fluids in the compartments of the body. To remain alive, a cell must contain a constant amount of fluid; this is known as the **fluid balance**. Too much fluid might cause a cell to rupture, and too little would make it unable to function. Although water can diffuse freely into and out of the cell, proteins cannot—and proteins attract water. By maintaining a store of internal proteins, the cell retains the fluid it needs. (It also uses minerals this way.) Similarly, the cells secrete proteins (and minerals) into the spaces between them to keep the fluid volume constant in those spaces. Also, because the proteins secreted into the blood cannot cross the blood vessel walls, they help to maintain the blood volume in the same way.

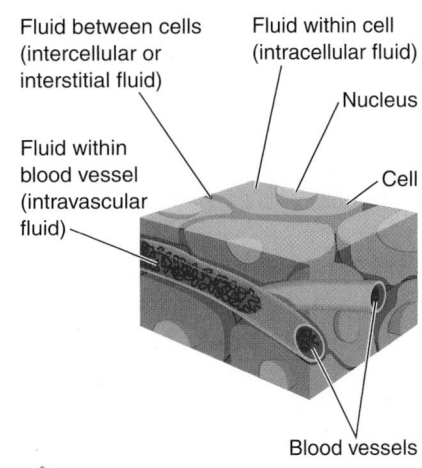

Fluid between cells (intercellular or interstitial fluid)
Fluid within cell (intracellular fluid)
Nucleus
Fluid within blood vessel (intravascular fluid)
Cell
Blood vessels

Shown here are the fluids within and surrounding a cell. Body proteins help hold fluid within cells, tissues, and blood vessels.

pH Values of Selected Fluids

A fluid's acidity or alkalinity is measured in pH units.

Basic

14 — Concentrated lye
13
12 — Household ammonia
11
10
9 — Baking soda
— Bile
8 — Pancreatic juice
— Blood
pH neutral 7 — Water
— Saliva
6 — Urine
5 — Coffee
4 — Orange juice
— Soda water
3 — Vinegar
2 — Lemon juice
— Gastric juice
1
Acidic 0 — Battery acid

Acid–Base Balance

Normal processes of the body continually produce **acids** and their opposite, **bases**, which must be carried by the blood to the organs of excretion. The blood must do this without allowing its own **acid–base balance** to be affected. To accomplish this, some proteins act as **buffers** to maintain the blood's normal **pH**. They pick up hydrogen **ions** when there are too many in the blood (the more hydrogen, the more concentrated the acid). Likewise, protein buffers release hydrogens again when there are too few in the blood. The secret is that the negatively charged side chains of the amino acids can accommodate additional hydrogens (which are positively charged) when necessary.

The acid–base balance of the blood is one of the most accurately controlled conditions in the body. If it changes too much, either the dangerous acidic condition called **acidosis** or the opposite basic condition, **alkalosis**, can cause coma or death. The hazards of these conditions are a result of their effect on proteins. When the proteins' buffering capacity is exceeded—for example, when proteins have taken on board or released all the acid hydrogens they can—additional acid or base deranges protein structures by pulling them out of shape; that is, it denatures the proteins—rendering them useless. Knowing how indispensable the structures of proteins are to their functions and how vital protein functions are to life, you can imagine how many body processes would be halted by such a disturbance.

Transport Proteins

A specific group of the body's proteins specializes in moving nutrients and other molecules into and out of cells. Some of these **transport proteins** act as pumps, picking up compounds on one side of the membrane and depositing them on the other, and thereby regulating what substances the cell will

fluid balance distribution of fluid among body compartments.

acids compounds that release hydrogens in a watery solution; acids have a low pH.

bases compounds that accept hydrogens from a watery solution; bases have a high pH.

acid–base balance equilibrium between acid and base concentrations in the body fluids.

buffers compounds that help keep the acidity (amount of acid) or alkalinity (amount of base) of a solution constant.

pH the concentration of hydrogen ions. The lower the pH, the stronger the acid: pH 2 is a strong acid, pH 7 is neutral, and a pH above 7 is alkaline (base).

ions (EYE-onz) electrically charged particles, such as sodium (positively charged) and chloride (negatively charged).

acidosis (a-sih-DOSE-is) blood acidity above normal, indicating excess acid.

alkalosis (al-kuh-LOW-sis) blood alkalinity above normal.

transport proteins proteins that carry nutrients and other molecules in body fluids. Some transport proteins reside in cell membranes and act as "pumps" by picking up compounds on one side of the membrane and releasing them on the other side as needed.

take up or release. One such pump is the "sodium–potassium pump," which resides in the cell membrane and acts as a revolving door, picking up potassium from outside the cell and depositing it inside the cell, and picking up sodium from within the cell and depositing it outside the cell as necessary. The protein machinery of cell membranes can be switched on or off in response to the body's needs. Often hormones do the switching with marvelous precision.

Other transport proteins move about in the body fluids, carrying nutrients and other molecules from one organ to another. Those that carry lipids in the lipoproteins are examples. Special proteins also can carry fat-soluble vitamins, water-soluble vitamins, and minerals. As a result, a protein deficiency can cause a vitamin A deficiency or a deficiency of whatever other nutrient needs a transport protein to reach its destination in the body.

This sampling of the major roles proteins play in the body illustrates their versatility, uniqueness, and importance. All the body's tissues and organs—muscles, bones, blood, skin, and nerves—are made largely of proteins. No wonder proteins are said to be the primary material of life.

Protein as Energy Only protein can perform all the functions previously described, but it will be sacrificed to provide needed energy if insufficient fat and carbohydrate are available from the diet. The body's number one need is for energy. All other needs have a lower priority.

When amino acids are degraded so they can be used for energy, their amine groups are usually incorporated by the liver into **urea** and sent to the kidney for excretion in the urine. The remaining components are carbon, hydrogen, and oxygen, which are available for immediate energy use by the body.

Only when the **protein-sparing** calories from carbohydrate and fat are sufficient to power the cells are the amino acids used for their most important function: making proteins. Thus, energy deficiency (starvation) is always accompanied by the symptoms of protein deficiency.

When amino acids are oversupplied, the body has no place to store them. It removes them, excretes their amine groups, and converts the remaining fragments into glucose and glycogen or into fat for energy storage. Amino acids are not stored in the body except in the sense that they are present in proteins in all the tissues. When there is a great shortage of amino acids, the body must break down such tissues as blood, muscle, and skin so that it can use their amino acids to maintain the heart, lungs, and brain.

How the Body Handles Protein

When a person eats a food protein, whether from cereals, vegetables, meats, or dairy products, the digestive system breaks down the protein and delivers the separated amino acids to the body cells. The cells then put the amino acids together in the order necessary to produce the particular proteins they need. (To review the digestive and absorptive systems relevant to the body's handling of protein, turn to Chapter 3.)

The stomach initiates protein digestion (see Figure 6-1). By the time proteins move into the small intestine, they are already broken into different-sized pieces: some single amino acids and many strands of two amino acids, **dipeptides**; some strands of three amino acids, **tripeptides**; and some longer chains. Digestion continues until almost all pieces of protein are broken into dipeptides, tripeptides, and free amino acids. Absorption of amino acids takes place all along the small intestine. Dipeptides and tripeptides are captured on the surface of the cells that line the small intestine. These cells split the peptides into amino acids, absorb them, and then release them into the bloodstream.

urea (yoo-REE-uh) the principal nitrogen excretion product of metabolism, generated mostly by the removal of amine groups from unneeded amino acids or from those amino acids being sacrificed to a need for energy.

protein-sparing a description of the effect of carbohydrate and fat, which, by being available to yield energy, allow amino acids to be used to build body proteins.

dipeptides (dye-PEP-tides) protein fragments two amino acids long. A *peptide* is a strand of amino acids.

tripeptides (try-PEP-tides) protein fragments three amino acids long.

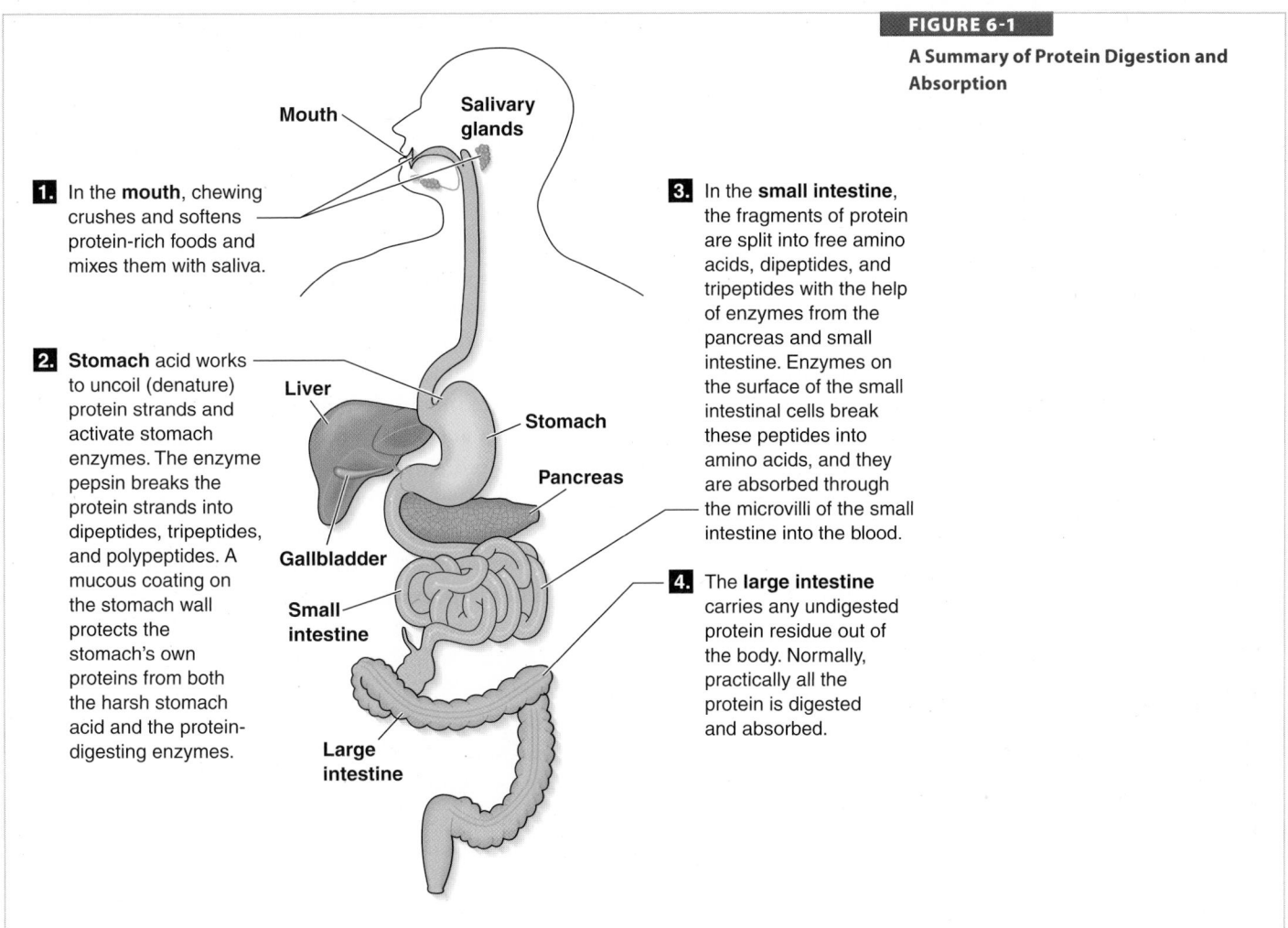

FIGURE 6-1

A Summary of Protein Digestion and Absorption

1. In the **mouth**, chewing crushes and softens protein-rich foods and mixes them with saliva.

2. **Stomach** acid works to uncoil (denature) protein strands and activate stomach enzymes. The enzyme pepsin breaks the protein strands into dipeptides, tripeptides, and polypeptides. A mucous coating on the stomach wall protects the stomach's own proteins from both the harsh stomach acid and the protein-digesting enzymes.

3. In the **small intestine**, the fragments of protein are split into free amino acids, dipeptides, and tripeptides with the help of enzymes from the pancreas and small intestine. Enzymes on the surface of the small intestinal cells break these peptides into amino acids, and they are absorbed through the microvilli of the small intestine into the blood.

4. The **large intestine** carries any undigested protein residue out of the body. Normally, practically all the protein is digested and absorbed.

Mouth
Salivary glands
Liver
Stomach
Pancreas
Gallbladder
Small intestine
Large intestine

Once they are circulating in the bloodstream, the amino acids are available to be taken up by any cell of the body. The cells can use them to make proteins, either for the cell's own use or for secretion into the circulatory system for other uses.

If a *non*essential amino acid (that is, one the body can make for itself) is unavailable for a growing protein strand, the cell will make one and will continue attaching amino acids to the strand. If, however, an essential amino acid (one the body cannot make) is missing, the building of the protein will halt. The cell cannot hold partially completed proteins to complete them later, for example, the next day. Instead, it must dismantle the partial structures and return surplus amino acids to the circulation, making them available to other cells. If other cells do not soon pick up these amino acids and insert them into protein, the liver will remove their amine groups for the kidney to excrete. Other cells will then use the remaining fragments for other purposes. The nutritional need calling for the production of that particular protein will not be met.

Protein Quality of Foods

The role of protein in food, as already mentioned, is not to provide body proteins directly but to supply the amino acids from which the body can make its own proteins. Because body cells cannot store amino acids for future use, it follows that all the essential amino acids must be eaten as part of a balanced diet. To manufacture

FIGURE 6-2

How Two Plant Proteins Combine to Yield a Complete Protein

Two incomplete proteins (for example, legumes plus grains) can be combined to equal a complete protein (peanut butter sandwich). In this example, the peanut butter provides adequate amounts of the amino acid lysine but is lacking in methionine. The bread "complements" the peanut butter because it contains adequate methionine but is lacking in lysine. When combined as a sandwich, all essential amino acids are present.

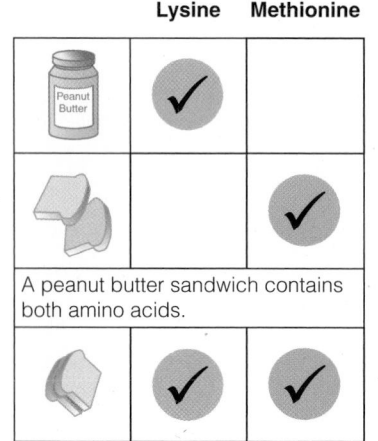

	Lysine	Methionine
Peanut Butter	✔	
Bread		✔
A peanut butter sandwich contains both amino acids.	✔	✔

complete proteins proteins containing all the essential amino acids in the right proportion relative to need. The *quality* of a food protein is judged by the proportions of essential amino acids that it contains relative to our needs. Animal and soy proteins are the highest in quality.

incomplete protein a protein lacking or low in one or more of the essential amino acids.

limiting amino acid a term given to the essential amino acid in shortest supply (relative to the body's need) in a food protein; it therefore *limits* the body's ability to make its own proteins.

complementary proteins two or more food proteins whose amino acid assortments complement each other in such a way that the essential amino acids limited in or missing from each are supplied by the others.

protein quality a measure of the essential amino acid content of a protein relative to the essential amino acid needs of the body.

biological value (BV) a measure of protein quality, assessed by determining how well a given food or food mixture supports nitrogen retention.

reference protein egg-white protein, the standard to which other proteins are compared to determine protein quality.

body proteins, then, all the needed amino acids must be available to the cells. Three important characteristics of dietary protein, therefore, are (1) that it should supply at least the nine essential amino acids, (2) that it should supply enough other amino acids to make nitrogen available for the synthesis of whatever nonessential amino acids the cell may need to make, and (3) that it should be accompanied by enough food energy (preferably from carbohydrate and fat) to prevent sacrifice of its own amino acids for energy.

This presents no problem to people who regularly eat **complete proteins**, such as those of meat, fish, poultry, cheese, eggs, milk, or many soybean products, as part of balanced meals.[2] The proteins of these foods contain ample amounts of all the essential amino acids relative to our bodies' need for them, and the rest of the diet provides protein-sparing energy and needed vitamins and minerals. An equally sound diet choice is to eat two or more **incomplete protein** foods from plants, each of which supplies the **limiting amino acid** in the other—also, of course, as part of a balanced diet. The *quality* of plant proteins (legumes, grains, and vegetables) having different limiting amino acids can therefore be balanced by combining different sources of plant proteins, either during a meal or over the course of a day, making sufficient amounts of all the essential amino acids available for protein synthesis. This strategy of using **complementary proteins** is shown in Figure 6-2. Note that by combining a grain (whole-wheat bread) that is low in lysine but high in methionine with a legume (peanut butter) that is low in methionine but high in lysine, all the essential amino acids are provided.

A person in good health can be expected to use dietary protein efficiently. However, malnutrition or infection can seriously impair digestion (by reducing enzyme secretion), absorption (by causing degeneration of the absorptive surface of the small intestine or losses from diarrhea), and the cells' use of protein (by forcing amino acids to meet other needs). In addition, infections cause increased production of antibodies, which are made of protein. Thus, malnutrition or infection can greatly increase protein needs while making it hard to meet them.[3]

People usually eat many foods containing protein. Each food has its own characteristic amino acid balance, and a mixture of foods almost invariably supplies plenty of each individual amino acid. However, when food energy intake is limited, this is not the case (as discussed in the section "Protein-Energy Malnutrition," later in the chapter). Also, even when food-energy intake is abundant, if the selection of foods available is severely limited, protein intake may not be adequate. The primary food source of protein must be taken into account because its quality is of great importance.

Researchers have studied many different individual foods as protein sources and have developed many different methods of evaluating the **protein quality** of foods. Protein quality is determined by two factors—the protein's digestibility and its amino acid composition. One method of evaluating protein quality judges how easily the body can absorb the protein. In general, amino acids from animal and soy proteins are the most easily absorbed (90 to 100 percent). Amino acids from other legumes are next best (80 to 90 percent); the absorption rates for those from grains and other plant foods vary (70 to 90 percent).

When amino acids are wasted, their amine groups (which contain their nitrogen) cannot be stored. Therefore, the efficiency of a protein can be assessed experimentally by measuring the net loss of nitrogen from the body. The higher the amount of nitrogen retained, the higher the quality of the protein. This is the basis for determining the **biological value (BV)** of proteins. A high-quality protein by this standard is egg-white protein, which has been designated the **reference protein** and given a score of 100. Other proteins are compared with it. The best guarantee of amino acid adequacy is to eat a variety of foods containing protein in the presence of adequate amounts of vitamins, minerals, and energy from carbohydrate and fat.

Recommended Protein Intakes

Recommended protein intakes can be stated in two ways: as a percentage of total calories, or as an absolute number (grams per day). The DRI committee recommends that protein provide 10 to 35 percent of total caloric intake.[4] The recommended protein allowance for a healthy adult is 0.8 gram per kilogram (or 2.2 pounds) of desirable body weight per day.

The recommendation for protein uses the desirable, not the actual, weight for a given height, because the desirable weight is proportional to the *lean* body mass of the average person. Lean body mass, not total weight, determines protein need. This is because fat tissue is composed largely of fat, which does not require much protein for maintenance.

The recommendations for protein intake are based on the assumption that the protein source will be a combination of plant and animal proteins, that it will be consumed with adequate calories from carbohydrate and fat, and that other nutrients in the diet will be adequate. These protein recommendations apply only to healthy individuals with no unusual metabolic need for protein.

Protein and Health

Protein deficiency effects are well known because, together with energy deficiency, they are the world's main form of malnutrition. In contrast, the health effects of too much protein—and particularly the effects of proteins of different kinds—are far less well known. The following sections discuss protein deficiency, excess protein, and types of protein. The Nutrition Action feature in this chapter discusses the problem of protein-related food allergies in children and adults.

Protein-Energy Malnutrition Protein deficiency and energy deficiency very often go hand in hand and are called **protein-energy malnutrition (PEM)**. The two diseases and their symptoms overlap all along the spectrum, but the extremes have names of their own. Protein deficiency is **kwashiorkor**, and energy deficiency is **marasmus**.[5]

Kwashiorkor is the Ghanaian name for "the evil spirit that infects the first child when the second child is born." In countries where kwashiorkor is prevalent, parents customarily give their newly weaned children watery cereal rather than the food eaten by the rest of the family. The child has been receiving the mother's breast milk, which contains high-quality protein designed to support growth. However, when a new baby is born, the first child is weaned and suddenly is fed only a weak drink with scant protein of very low quality. It is not surprising that the just-weaned child becomes sick when the new baby arrives.

The child who has been banished from its mother's breast faces this threat to life by engaging in as little activity as possible. Apathy is one of the earliest signs of protein deprivation. The body is collecting all its forces to meet the crisis and so cuts down on any expenditure of protein not needed for the heart, lungs, and brain. As the apathy increases, the child doesn't even cry for food. All growth ceases, and the child is no larger at age 4 than at age 2. New hair grows without the protein pigment that gives hair its color. The skin also loses its color, and open sores fail to heal. Digestive enzymes are in short supply, the digestive tract lining deteriorates, and absorption

Calculation Practice

To calculate the percentage of calories you derive from protein:

1. Use your total calories as the denominator (example: 1,900 cal).
2. Multiply your total protein intake in *grams* by 4 cal/g to obtain calories from protein as the numerator (example: 70 g protein + 4 cal/g = 280 cal).
3. Divide to obtain a decimal, multiply by 100, and round off (example: 280/1,900 × 100 = 15% cal from protein).

To calculate your recommended protein intake (RDA):

1. Find the desirable weight for a person your height (see Appendix A). Assume that this weight is appropriate for you.
2. Change pounds to kilograms (divide pounds by 2.2; 1 kilogram = 2.2 pounds).
3. Multiply kilograms by 0.8 g/kg.

Example (for a 5'8" male):

1. Desirable weight: about 150 lb.
2. 150 lb ÷ 2.2 lb = 68 kg (rounded off).
3. 68 kg × 0.8 g/kg = 54 g protein (rounded off).

Chapter 11 discusses the protein needs of athletes, and the Spotlight in Chapter 11 presents the pros and cons of using popular amino acid and other supplements.

protein-energy malnutrition (PEM) also called protein-calorie malnutrition (PCM); the world's most widespread malnutrition problem; includes both kwashiorkor and marasmus as well as the states in which they overlap.

kwashiorkor (kwash-ee-OR-core) a deficiency disease caused by inadequate protein in the presence of adequate food energy.

marasmus (muh-RAZ-mus) an energy deficiency disease; starvation.

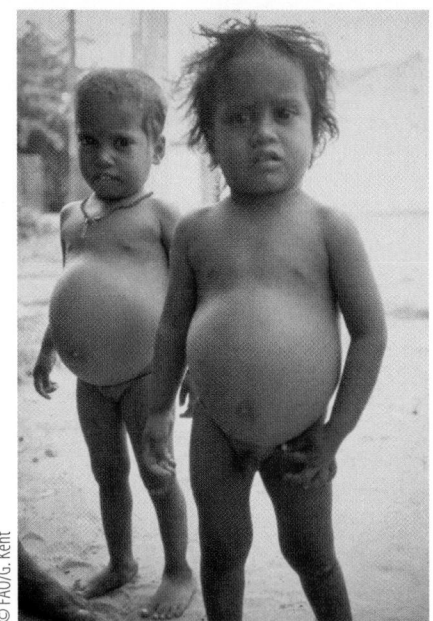

Kwashiorkor. These children have the characteristic edema and swollen belly often seen with kwashiorkor.

Marasmus. This child is suffering from the extreme emaciation of marasmus.

fails. The child can't assimilate what little food is eaten. Proteins and hormones that previously kept the fluids correctly distributed among the compartments of the body now are diminished, so that fluid leaks out of the blood (**edema**) and accumulates in the belly and legs. Blood proteins, including hemoglobin, are not synthesized, so the child becomes anemic, which increases the child's weakness and apathy. The kwashiorkor victim often develops a fatty liver, caused by a lack of the protein carriers that transport fat out of the liver. Antibodies to fight off invading bacteria are degraded to provide amino acids for other uses; hence, the child becomes an easy target for any infection. Then **dysentery**, an infection of the digestive tract that causes diarrhea, further depletes the body of nutrients, especially minerals. Measles, which might make a healthy child sick for a week or two, kills the child victim of kwashiorkor within 2 or 3 days. If the condition is caught in time, the starving child's life may be saved by careful nutrition therapy.

Children with *marasmus* suffer symptoms similar to those of children with kwashiorkor because both conditions cause loss of body protein tissue. However, there are differences between the two conditions. A marasmic child looks like a wizened little old person—just skin and bones. The child is often sick because his or her resistance to disease is low. All the muscles are wasted, including the heart muscle, and the heart is weak. Metabolism is so slow that body temperature is subnormal. There is little or no fat under the skin to insulate against cold. The experience of hospital workers with victims of this disease is that the victims' primary need is to be wrapped up and kept warm. The disease occurs most commonly in children from 6 months to 18 months of age. Because the brain normally grows to almost its full adult size within the first two years of life, marasmus impairs brain development and thus may have a permanent effect on a child's learning ability.

PEM is prevalent in Africa, Central America, South America, and Asia.[6] PEM has also been recognized in many undernourished hospital patients, including those with anorexia nervosa, **AIDS**, cancer, and other wasting conditions. The extent and severity of malnutrition worldwide is a political and economic problem. It is discussed further in the Spotlight feature in Chapter 13.

Too Much Protein   Many of the world's people struggle to obtain enough food and enough protein to keep themselves alive, but in the developed countries, where protein is abundant, the problems of protein excess can be seen. Animals fed high-protein diets experience a protein overload effect, most notable in the enlargement of their livers and kidneys. In human beings, when protein intake is high, calcium excretion increases.[7] The higher a person's intake of animal-protein sources such as meat, the more likely it is that fruits, vegetables, and whole grains will be crowded out of the diet, creating deficiencies of other nutrients.

Although protein is essential to health, the body converts extra protein to energy (glucose), which is stored as body fat when energy needs are met. Despite the flood of new protein-packed snack bars and other products in the marketplace, there are no

edema (eh-DEEM-uh) swelling of body tissue caused by leakage of fluid from the blood vessels, seen in (among other conditions) protein deficiency.

dysentery (DISS-en-terry) an infection of the digestive tract that causes diarrhea.

AIDS (acquired immune deficiency syndrome) an immune system disorder caused by the human immunodeficiency virus (HIV).

known benefits from consuming excess protein. The recommended upper range for protein intake applies *when calorie intake is adequate*. Note the qualification "when calorie intake is adequate" in the preceding statement. Remember that your recommended protein intake can be stated as a percentage of calories in the diet or as a specific number of grams of dietary protein. The recommended protein intake for a 150-pound person is roughly 55 grams, or about 12 percent of their daily caloric intake. Fifty-five grams of protein is equal to 220 calories and equals 11 percent of a 2,000-calorie intake, which is reasonable for a 150-pound active person. If this person were to drastically reduce his or her caloric intake to, say, 800 calories a day, then 220 calories from protein is suddenly 28 percent of the total. However, it is still this person's recommended intake for protein, and a reasonable intake. It is the caloric intake that is unreasonable in this example. Similarly, if the person eats too many calories—say, 4,000—this protein intake represents only 6 percent of the total caloric intake, yet it is still a reasonable intake. It is the caloric intake that may be unreasonable.

Thus, it is important to be careful when judging protein intakes as a percentage of calories. Always ask what the absolute number of grams is, too, and compare it with the recommended protein intake in grams. As calorie intake decreases, it is necessary to increase the percentage of calories from protein so as to consume the RDA for protein. Recommendations stated as a percentage of calories are useful only when food energy intakes (calories) are within reason.

Protein in the Diet   Misconceived notions abound regarding protein in the diet; the most obvious of these is that more is better. American women eat about 60 to 65 grams of protein a day, notably higher than the recommended 46 grams a day. Young men average about 100 grams a day and drop to about 75 to 85 grams as older adults, still considerably higher than their recommended intake of 56 grams. Moreover, more than 65 percent of this protein comes from animal and dairy products (see Figure 6-3). Saturated fats supply half or more of the calories in some animal-protein foods. You can better balance your food choices by selecting one-third or less of your protein from animal sources and the rest from plants. (See the Eat Well Be Well feature later in this chapter). To limit your intake of saturated fat, consider the following tips when shopping for protein foods:

Both meals shown here supply an adequate assortment of amino acids needed for health.

- Dairy foods are excellent sources of protein, calcium, and other important nutrients. Look for fat-free and low-fat varieties for the recommended three servings each day. Look for low-fat cheeses that have less than 5 grams of fat per ounce, such as part-skim or fat-free ricotta or mozzarella, feta cheese, string cheese, or other reduced-calorie cheeses. These choices are lower in saturated fat and calories than their full-fat counterparts.

- Fish and shellfish—fresh, frozen, or canned in water—make excellent protein choices and are low in fat. Experts recommend that you eat at least two fish meals per week.

- Meat, chicken, and fish all provide excellent protein, as well as iron, zinc, and vitamin B_{12}. We are advised to choose low-fat varieties. Look for the leanest meats:

  - Flank steak, round steak, sirloin, tenderloin, or extra lean ground beef

  - Lean ham, Canadian bacon, pork tenderloin, and center-loin pork chops

  - Chicken, turkey, or game hens without the skin; fresh ground turkey breast or chicken breast meat

FIGURE 6-3

Protein Contributed by Food Groups in the Average American Diet

These foods provide 70% of the protein in the U.S. diet. The following foods also contribute at least 1% (in descending order): cold cuts (excluding ham), ready-to-eat cereal, white potatoes, sausage, flour and baking ingredients, ice cream, sherbet and frozen yogurt, nuts and seeds, cooked rice and other grains, and canned tuna.

SOURCE: Adapted from P. A. Cotton and coauthors, Dietary sources of nutrients among U.S. adults, 1994–1996, *Journal of the American Dietetic Association* 104 (2004): 921–930.

Foods that supply protein in abundance are shown here in the Dairy group and the Protein Foods group of the MyPlate food guide (top two photos). Servings of foods from the Vegetable group and the Grains group can also contribute protein to the diet (bottom two photos).

legumes (LEG-yoomz or leh-GYOOMZ)
plants of the bean and pea family having roots with nodules containing bacteria that can trap nitrogen from the air in the soil and make it into compounds that become part of the seed. The seeds are rich in high-quality protein compared with those of most other plant foods.

- At the deli counter, select items with less than 1 gram of fat per ounce, such as lean ham, turkey or chicken breast, and lean roast beef.

- Nuts and nut butters are good sources of protein but are also high in fat. Look for nonhydrogenated varieties. A 1-tablespoon serving of peanut butter counts as 1 ounce of meat in the MyPlate food guide.

- Eggs are another excellent source of protein. Also look for egg substitutes, and substitute ¼ cup for one whole egg in recipes. One egg counts as 1 ounce of meat.

- Many interesting sources of protein are available. Try adding **legumes** to your meals using the tips in the Savvy Diner feature that follows. A ¼-cup serving counts as 1 ounce of meat in the MyPlate food guide.

As the vegetarian knows, one can easily design a perfectly acceptable diet around plant foods alone by choosing an appropriate variety. This chapter's Spotlight reviews the benefits of soy products, an excellent source of protein in the diet.

Adequate protein in the diet is easy to obtain. A breakfast of one egg, two slices of wheat bread, and a glass of milk provide nearly 20 grams of protein. This meets about 35 percent of an average man's recommended protein intake of 56 grams and about 43 percent of a woman's recommended protein intake of 46 grams. The Scorecard feature in this chapter shows you how to estimate your own protein intake.

The Vegetarian Diet

More and more people are following vegetarian diets. The reasons for becoming vegetarian vary widely.[8] Some people have health reasons, whereas others have religious, philosophical, or ethical reasons. Some believe that vegetarianism is ecologically sound, and others believe that it is less costly than the meat-eating alternative.

Legumes—dried beans, peas, and lentils—have been very highly praised in recent years.[9] The *Dietary Guidelines* recommend that Americans eat more beans. On the MyPlate food guide, legumes are the only food featured in two different categories, the Vegetables group and the Protein Foods group. Legumes are rich in the B vitamins and fiber. They are good sources of protein, iron, and zinc and are naturally low in fat and cholesterol-free. The Food and Drug Administration approved a health claim for bean packages and cans: "Diets including beans may reduce your risk of heart disease and certain cancers."[10] However, Americans typically eat only about half the recommended amount of legumes: one cup instead of the recommended 1.5 to 2 cups per week for a 2,000-calorie diet (see bean recommendations for other calorie levels at www.ChooseMyPlate.gov).

Here are three simple ways to increase your intake of this low-cost, easy-to-include group of foods:

1. Enjoy adding more legumes to your weekly meals. Include legumes in entrees (tacos), side dishes (baked beans), soups (split pea or lentil), and salads (chickpeas from the salad bar, or three-bean combos). Recipes can be prepared from dried beans (soaked, rinsed, and cooked) or more quickly from canned beans (rinsed to remove excess sodium). Thorough rinsing is recommended to remove the gas-causing sugar, known as *raffinose*, found in beans.

2. Explore the many varieties of legumes used in cooking—as noted in the Miniglossary.

3. Enjoy learning more about these nutritional powerhouses online. The following three sites can help:

   ● *American Dry Bean Board:* www.americanbean.org/. Includes bean basics, photos, health tips, and 1,001 recipes.

   ● *U.S. Dry Bean Council:* www.beansforhealth.com/. A more global focus, with many links and a useful bibliography.

   ● *USA Dry Pea & Lentil Council:* www.pea-lentil.com/. Nutrition information, cooking instructions, and recipes.

Legumes include such plant foods as the soybean, kidney bean, garbanzo bean, black bean, lentil, garden pea, black-eyed pea, and lima bean.

© Tom McCarthy Photography

Miniglossary of Legumes

black, Cuban, or turtle beans medium-size, black-skinned oval beans that have a rich, sweet taste. They are best served in Mexican and Latin American dishes or thick soups and stews.

black-eyed peas small, oval-shaped, creamy-white peas with a black spot. They have a vegetable flavor with mealy texture. Use in salads with rice and greens.

garbanzo beans or chickpeas large, round, tan-colored legumes that have a nutty flavor and crunchy texture. Use in soups and stews and puréed for dips.

Great Northern beans a medium-white, kidney-shaped bean. Enjoy the delicate flavor and firm texture in salads, soups, and main dishes.

kidney beans these familiar beans are large, red, and kidney shaped (the white variety is called *cannellini*). They have a bland taste and soft texture but tough skins. Use in chili, bean stews, and Mexican dishes (for red) or Italian dishes (for white).

lentils small, flat, and round legumes. Usually brown, lentils can also be green, pink, or red. They have a mild taste with firm texture. Best used when combined with grains or vegetables in salads, soups, or stews.

lima or butter beans limas are soft and mealy in texture. They are flat, oval shaped, and white tinged with green. The smaller variety has a milder taste. Use in soups and stews.

pinto beans medium-size, oval beans that are mottled beige and brown and have an earthy flavor. They are most often used in Mexican dishes, such as refried beans, stews, or dips.

red beans this versatile bean is a medium-size, dark-red oval. The taste and texture are similar to kidney beans. Use in soups and stews, and serve with rice.

soybeans you can find these creamy white ovals in numerous food products, such as tofu, flour, grits, and milk. They have a firm texture and bland flavor. See the Spotlight feature later in this chapter for a discussion of the possible health benefits derived from soy foods.

split peas green or yellow, these small, halved peas supply an earthy flavor with a mealy texture. They are best used in soups and with rice or grains.

white navy beans small, white, oval beans that are best used in soups and stews and as baked beans.

SOURCE: Adapted from K. Mangum, *Life's Simple Pleasures: Fine Vegetarian Cooking for Sharing and Celebration* (Boise, ID: Pacific Press, 1990), 149.

The ancient inhabitants of South America liked to eat a kind of paste made from peanuts. But modern peanut butter came into being around 1890 as the bright idea of a St. Louis physician, who thought it would be a good health food for elderly people. It was not linked with jelly until the 1920s.[11]

Many health organizations now recommend a diet that emphasizes vegetables, fruits, legumes, and whole grains to protect against cancer, heart disease, stroke, diabetes, and obesity.[12] The key to getting enough, but not too much, protein seems to be to use a variety of plant-based foods and to de-emphasize meats.[13]

1 In the Kitchen.

- Small meat portions tend to work best when mixed into dishes with lots of vegetables and grains; try stir-fries, pastas, soups and stews, burritos, and main-dish salads. For example, cook a large pot of soup, stew, or chili. Minimize the amount of meat you use, and load up the pot with vegetables (fresh or frozen) and cooked beans.

© PhotoDisc/Getty Images

- Go meatless one or more days each week. Experiment with recipes from health-minded cookbooks.
- Take a fresh look at your favorite recipes. Try to use less meat and add more vegetables. Instead of chicken and broccoli stir-fry, try stir-fried veggies with a little chicken.
- For quick, colorful, meals rich in nutrients and flavor, try various combinations of stir-fried vegetables on beds of steamed brown rice, whole-grain bulgur, or couscous.

2 In the Lunch Box.

- Get out of the peanut butter and jelly rut by filling sandwiches with water-packed tuna mixed with mandarin oranges, bean sprouts, and a bit of plain low-fat yogurt; chopped, cooked, skinless chicken combined with raw sliced vegetables and a little French dressing; cooked, mashed dried beans seasoned with chopped onion, garlic powder, rosemary, thyme, and pepper; or low-fat cottage cheese flavored with drained, chopped pineapple.
- Take a thermos filled with chili, vegetable soup, or a milk-based soup, such as cream of tomato prepared with nonfat milk, instead of a sandwich. Try cold lunches such as low-fat yogurt and fruit, brown rice with cubes of skinless poultry, or cooked pasta tossed with raw vegetables, low-fat cheese, and a bit of Italian dressing.

3 At the Table.

- When dining out, choose an ethnic restaurant with plant-based entrées on the menu. Consider Spanish paella, Asian stir-fries, Moroccan stew, Indian curries, or French ratatouille as your entrée. Or try Chinese, Vietnamese, or Thai take-out with lots of rice and vegetables.
- Twice a week, make seafood the protein on your plate.
- Eat more legumes as well as red, orange, and dark green vegetables, such as tomatoes, sweet potatoes, and broccoli.
- Make whole grains, vegetables, and legumes the main event of your meals. At least two-thirds of your meal should come from these plant-based foods and one-third or less from lean meat, poultry, fish, or low-fat dairy products.

Scorecard Estimate Your Protein Intake

The average American consumes much more than his or her recommended protein intake. How do you compare? First, figure your recommended protein intake (divide your weight in pounds by 2.2 and then multiply by 0.8). Next, write down everything you ate and drank yesterday. Using the values given below, estimate the grams of protein you ate from both animal and plant sources. If an item is not listed here, use Appendix D to determine the amount of protein it contains. How close are you to your recommended protein intake? What percentage of your protein comes from animal versus plant sources?

Recommended Protein Intake: _____ **grams**

Protein Foods	Amount	Grams of Protein in 1 Serving	Grams of Protein in Your Typical Diet
Animal Sources			
Hard cheese (e.g., cheddar)	1 oz	7	_____
Cottage cheese	½ c	14	_____
Milk	1 c	9	_____
Yogurt	1 c	12	_____
Egg	1 large	7	_____
Poultry	3 oz	21	_____
Ground beef, lean	3 oz	24	_____
Beef steak, lean	3 oz	26	_____
Pork chop, lean	3 oz	20	_____
Other	_____	_____	
		Animal Proteins Subtotal (grams):	_____
Plant Sources			
Vegetables	½ c	2	_____
Legumes, cooked	½ c	8	_____
Tofu	4 oz	9	_____
Cereals	1 c	2–6	_____
Bread	1 slice	2	_____
Tortilla	1	2	_____
Rice	½ c	3	_____
Pasta	½ c	3	_____
Peanut butter	1 tbsp	4	_____
Nuts	2 tbsp	3	_____
Seeds	2 tbsp	3	_____
Other	_____	_____	
		Plant Proteins Subtotal (grams):	_____

Day's Total Protein: _____ **grams**

In addition to the traditional types of vegetarians (see Table 6-2), some people eat seafood but not other meats, and some include chicken and other poultry but not red meat. Whatever the particular reasons for choosing a vegetarian diet, the vegetarian needs to be aware of its implications for nutrition and health.

Important goals for *any* diet planner include the following:

- Obtain neither too few nor too many calories—that is, to maintain a healthy weight

- Obtain adequate quantities of complete protein

TABLE 6-2	Types of Vegetarians

Semivegetarian Some but not all groups of animal-derived products, such as meat, poultry, fish, seafood, eggs, milk, and milk products, are included in this diet.

Lacto-vegetarian Milk and milk products are included in this diet, but meat, poultry, fish, seafood, and eggs are excluded.
possible limiting nutrient: iron

Lacto-ovovegetarian Milk and milk products and eggs are included in this diet, but meat, poultry, fish, and seafood are excluded.
possible limiting nutrient: iron

Ovovegetarian Eggs are included in this diet, but milk and milk products, meat, poultry, fish, and seafood are excluded.
possible limiting nutrients: iron, vitamin D, calcium, riboflavin

Strict vegetarian/vegan All animal-derived foods, including meat, poultry, fish, seafood, eggs, milk, and milk products are excluded from this diet.
possible limiting nutrients: iron, vitamin D, calcium, riboflavin, vitamin B_{12}, omega-3 fatty acids, high-quality protein

Macrobiotic diet Extremely restrictive diet based on metaphysical beliefs and consisting mostly of legumes, whole grains, and certain vegetables.
when taken to extremes, includes only brown rice and water or herbal teas and can cause malnutrition and death

- Obtain the needed vitamins and minerals

- Limit intake of foods with added sugars or solid fats, especially saturated fat and *trans*-fat, and foods that are high in sodium

With guidance in meal planning, a vegetarian diet can be a healthful choice for individuals during all stages of the life cycle, including pregnancy, lactation, infancy, childhood, and adolescence.[14] An increasing variety of fortified foods such as soy and rice beverages, meat replacements, juices, and breakfast cereals are available that can add substantially to vegetarians' intakes of key nutrients such as calcium, iron, zinc, vitamin B_{12}, vitamin D, riboflavin, and omega-3 fatty acids.[15] In addition, vegetarians can use the special Vegetarian Food Pyramid (see Figure 6-4) to balance their diets.

© Polara Studios

Well-planned, plant-based meals consisting of a variety of whole grains, legumes, nuts, vegetables, fruits—and for some vegetarians, eggs and dairy products—can offer sound nutrition and health benefits to vegetarians and nonvegetarians alike.

Proteins The vegetarian needs adequate amounts of all the essential amino acids. Because proteins from animals contain ample amounts of the essential amino acids, the lacto-ovovegetarian can get a head start on meeting protein needs by drinking recommended amounts of milk daily or by consuming the equivalent in milk products in the day's diet.

Adequate amounts of amino acids can be obtained from a plant-based diet when a varied diet is routinely consumed on a daily basis. Mixtures of proteins from whole grains, vegetables, beans, nuts, seeds, nut butters, peas, and soy products (tofu, tempeh, veggie burgers) eaten over the course of a day complement one another's amino acid profiles so that deficits in one are made up by the assets of another. Table 6-3 gives examples of how such mixtures of foods can be combined to form complete proteins.

Fats Vegetarian diets are typically adequate in omega-6 fatty acids, but they may be marginal in omega-3 fatty acids.[16] As discussed in Chapter 5, the omega-3 fatty acids that are most important nutritionally are alpha-linolenic acid, docosahexaenoic acid (DHA), and eicosapentaenoic acid (EPA). Fish, especially

FIGURE 6-4

A Vegetarian Food Pyramid and Meal-Planning Tips

SOURCE: V. Messina, V. Melina, and A. R. Mangels, "A new food guide for North American vegetarians," *Journal of the American Dietetic Association* 103 (2003). Reprinted by permission of Elsevier.

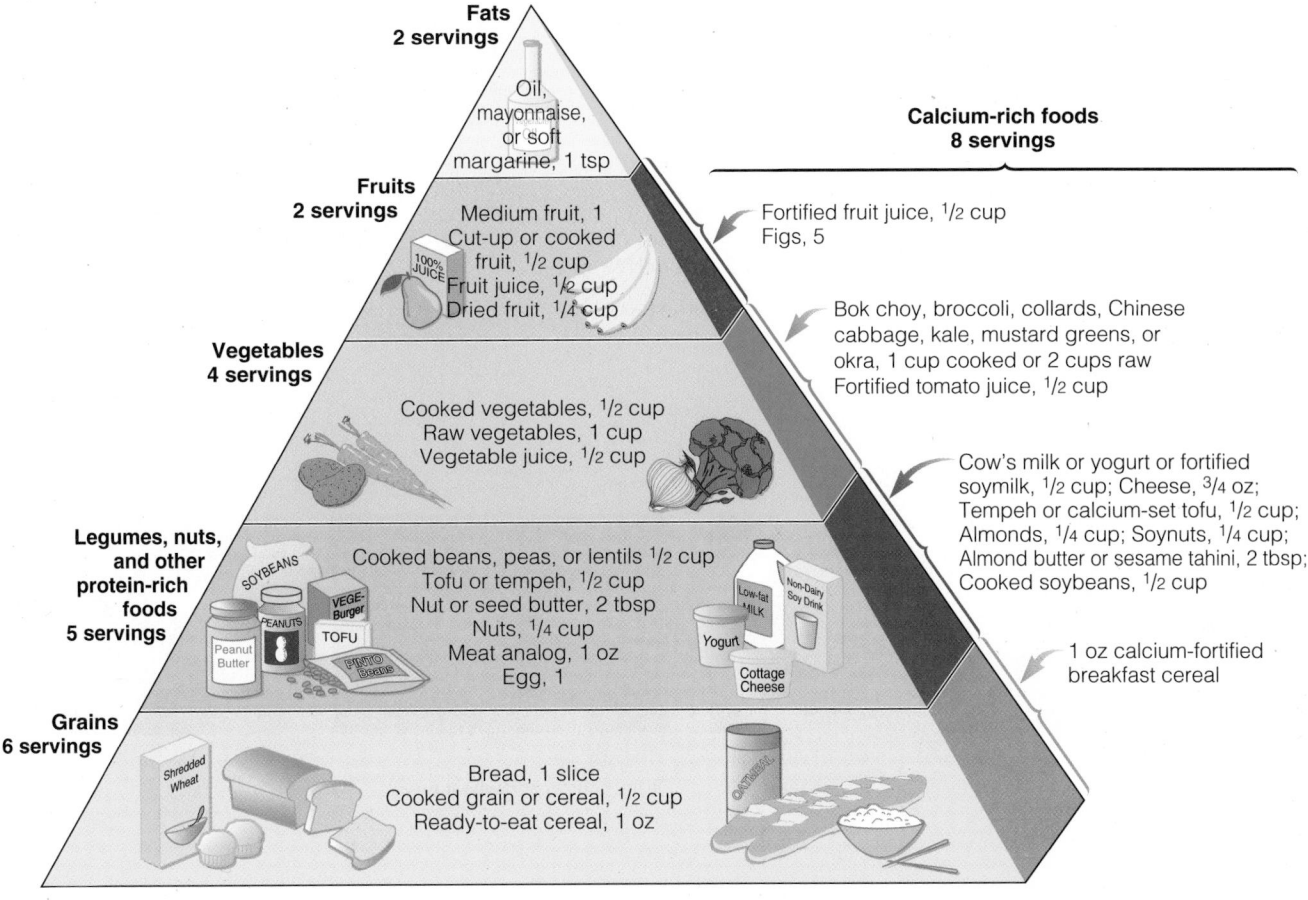

Meal Planning Tips

1. Choose a variety of foods.
2. The number of servings in each group is for minimum daily intakes.
3. A serving from the calcium-rich food group provides approximately 10% of adult daily requirements. Choose 8 or more servings per day. These also count towards servings from the other food groups in the guide. For example, 1/2 cup of fortified fruit juice counts as a calcium-rich food and also counts towards servings from the fruit group.
4. Include 2 servings every day of foods that supply omega-3 fats. Foods rich in omega-3 fat are found in the legumes/nuts group and in the fats group. A serving is 1 teaspoon of flaxseed oil, 3 teaspoons of canola or soybean oil, 1 tablespoon of ground flaxseed, or 1/4 cup walnuts. For the best balance of fats in your diet, olive and canola oils are the best choices for cooking.
5. Servings of nuts and seeds may be used in place of servings from the fats group.
6. Be sure to get adequate vitamin D from daily sun exposure or through fortified foods or supplements. Cow's milk and some brands of soy milk and breakfast cereals are fortified with vitamin D.
7. Include at least 3 good food sources of vitamin B_{12} in your diet every day. These include 1 tablespoon of *Red Star* Vegetarian Support Formula nutritional yeast, 1 cup fortified soy milk, 1/2 cup cow's milk, 3/4 cup yogurt, 1 large egg, 1 ounce fortified breakfast cereal, 1 1/2 oz fortified meat analog. If you don't eat these foods regularly (at least 3 servings per day), take a daily vitamin B_{12} supplement of 5 to 10 μg or a weekly B_{12} supplement of 2000 μg.
8. If you include sweets or alcohol in your diet, consume these foods in moderation. Get most of your daily calories from the foods in the Vegetarian Food Guide.

fatty fish, are good sources of DHA and EPA, because the fish eat marine algae containing DHA and EPA. Some vegetarian diets may contain low levels of EPA and DHA from dairy products and DHA-fortified eggs, but vegan diets do not. Some brands of soy milk and breakfast bars are now fortified with omega-3 fatty acids. All vegetarians should include sources of omega-3 fats from ground flaxseed, flaxseed oil, canola oil, soy products, and walnuts.

Vitamins   The lacto-ovovegetarian diet can be adequate in all vitamins, but several vitamins may be a problem for the vegan. One such vitamin is vitamin B_{12}, which does not occur naturally in plant foods but is available in fortified foods, such

TABLE 6-3   Complementary Protein Combinations That Provide High-Quality Protein

	Combine		Examples
Cereal Grains	+	**Legumes**	
Barley		Dried beans	Bean taco
Bulgur		Dried lentils	Chili and corn bread
Oats		Dried peas	Lentils or beans and rice
Rice		Peanuts	Peanut butter sandwich
Whole-grain breads			
Pasta			
Cornmeal			
Legumes (or grains)	+	**Seeds and Nuts**	
Dried beans		Sesame seeds	Hummus (chickpea and sesame paste)
Dried lentils		Sunflower seeds	Split-pea soup and sesame crackers
Dried peas		Walnuts	Noodles with sesame seeds
Peanuts		Cashews	
		Nut butters	

Examples

Hummus and bread   Corn and black-eyed peas   Peanut butter and wheat bread   Tofu and rice

as breakfast cereals or **nutritional yeast** that is grown in a vitamin B_{12}-enriched environment. The vegan needs a reliable B_{12} source, such as vitamin B_{12}-fortified soy and rice beverages, breakfast cereals, or **meat replacements**. Some vegetarians use seaweeds, fermented soy, and other products in the belief that they provide vitamin B_{12} in adequate amounts, but these products are not currently recommended as reliable sources.

Another vitamin of concern to the vegan is vitamin D. The milk drinker is protected if the milk is fortified with vitamin D, but there is no practical source of vitamin D in plant foods. Fortified margarines, soy milk, rice milk, orange juice, and breakfast cereals can supply some vitamin D. Regular exposure to the sun can help prevent a deficiency, too.

Riboflavin, another B vitamin obtained from milk, is present in the diet of the vegan who eats ample servings of dark greens, whole and enriched grains, mushrooms, legumes, nuts, and seeds. The vegan who doesn't eat these foods, however, may not meet the body's riboflavin needs.

Minerals   Iron and zinc need special attention in the diets of all vegetarians.[17] Whole-grain products, soy foods, other legumes, dried fruit, nuts, and seeds are important sources of iron in the vegetarian diet. The iron in these foods, however, is not as easily absorbed by the body as that in meat. Because the vitamin C in fruits and vegetables can triple the absorption of the iron provided by other foods eaten at the same meal, vegetarian meals should be rich in foods offering vitamin C.

Zinc is widespread in plant foods, but its availability may be reduced by the fibers and other binders found in fruits, vegetables, and whole grains. Vegetarians

nutritional yeast a fortified food supplement containing B vitamins, iron, and protein; can be used to improve the quality of a vegetarian diet.

meat replacements textured vegetable protein products formulated to look and taste like meat, fish, or poultry; also called meat analogs. Many of these are designed to match the known nutrient contents of animal protein foods.

are advised to eat varied diets that include wheat germ, legumes, nuts, pumpkin seeds, and whole-grain products. Milk, yogurt, and cheese provide zinc to the lacto-vegetarian as well.

Special efforts are necessary to meet the vegan's calcium needs.[18] Whereas the milk-drinking vegetarian is protected from calcium deficiency, the vegan must find other sources of calcium. Some good sources of calcium are *regular* servings of calcium-fortified breakfast cereals and juices; legumes; firm-style tofu; other soy foods, including calcium-fortified soy milk; dried figs, some nuts, such as almonds; certain seeds, such as sesame seeds; and some vegetables, such as broccoli, collard greens, kale, mustard greens, turnip greens, okra, rutabaga, and Chinese cabbage (bok choy). The choices should be varied, because the absorption of the calcium provided by some of these foods is hindered by binders in them. The strict vegetarian is urged to use *calcium-fortified* soy or rice beverages, juices, or cereals in *ample quantities, regularly.*[19]

Health Benefits   Vegetarian protein foods are higher in fiber, richer in certain vitamins and minerals, and lower in fat than meats.[20] Vegetarians can enjoy a heart-healthy diet, provided that they eat high-fat foods such as margarine, oil, cheese, and sour cream in moderation. Table 6-4 offers tips for nutritious, easy-to-fix vegetarian meals and snacks.

Studies have found that people with vegetarian or near-vegetarian traditions, such as the Seventh-Day Adventists and the Chinese, have lower rates of heart disease, cancer, diabetes, and obesity than those consuming the typical North American diet.[21] Informed vegetarians are more likely to be at a healthy weight and to have lower blood cholesterol levels, lower blood pressure, lower rates of certain types of cancer, better digestive function, and better health in other ways.[22] Even compared with people who are health conscious, vegetarians experience fewer deaths from cardiovascular disease. Often vegetarianism goes with a healthful lifestyle (no smoking, lower alcohol intakes, emphasis on supportive family life, and so forth), so it is unlikely that dietary practices *alone* account for all the aspects of improved health. However, they may contribute to it.

TABLE 6-4	Easy-to-Prepare Vegetarian Meals and Snacks

Breakfast
- Fortified breakfast cereal; eat with fat-free or low-fat milk, yogurt, or soy milk.
- Hot cereals: add fresh fruit slices and yogurt and sprinkle with cinnamon.
- Toast, bagels: top with low-fat cheese, low-fat cottage cheese, or 1 to 2 tbsp of peanut butter.

Snacks
- Assorted fresh fruits and vegetables with yogurt dip.
- Low-fat cheese or peanut butter on rice cakes or crackers.
- Fat-free or low-fat yogurt.
- English muffin pizza with part-skim mozzarella cheese.
- Hummus with pita bread wedges and crisp vegetables.*

Lunch and Dinner
- Salads: add tofu, chickpeas, three-bean salad, kidney beans, low-fat cottage cheese, sunflower seeds, and hard-cooked egg.
- Salad dressings: add salad seasonings to plain yogurt or blenderized tofu.
- Pasta: add diced tofu and/or canned kidney beans to tomato sauce; top with grated, part-skim mozzarella.
- Baked potato: top it with canned beans, steamed vegetables, or low-fat cheese.
- Hearty soups: enjoy lentil, split pea, bean, and minestrone soups, either homemade or canned.
- Vegetarian pizza: top with nonfat or low-fat cheeses and lots of vegetables.

*To make hummus: Blend 1 cup cooked chickpeas, 2 tablespoons tahini (sesame seed paste), 2 tablespoons lemon juice, 1 clove minced garlic, and ¼ cup chopped fresh parsley in food processor until smooth. Chill and serve with pita bread wedges, crackers, or crisp vegetables. Makes two servings.

In December 1995, a 33-year-old woman with a peanut allergy read the contents of the label of a container of split-pea soup. Peanuts were not listed. She ate only part of the soup, but within minutes experienced a life-threatening reaction and was rushed to the emergency room for treatment. Following the incident, the split-pea soup was analyzed and found to contain peanut flour as a component of the "flavoring" ingredients. The soup manufacturer subsequently discontinued using peanut flour in the product.[23]

The woman in the story belongs to the estimated 3 to 4 percent of the adult population in the United States suffering from food allergies. Interestingly, although a relatively small percentage actually have food allergies, nearly 25 percent of people think they do and thus develop **food aversions**.[24] What is the explanation? The answer may be in understanding the difference between **food allergy** and **food intolerance**. Although the physical responses to a food allergy and food intolerance may be very similar, the difference between the two is whether or not the immune system is involved in the reaction.

© PhotoDisc/Getty Images

Food Intolerance versus Food Allergy

Food intolerance is far more common than food allergy. A food intolerance is an **adverse reaction** to food that *does not* involve the immune system. Lactose intolerance is one example of a food intolerance. A person with lactose intolerance lacks the enzyme needed to digest lactose. When that person eats milk products, symptoms such as gas, cramps, and bloating can occur. Gluten intolerance is caused by an intolerance to gluten, a protein found in wheat, oats, barley, and rye. People with gluten intolerance must avoid products prepared with these grains. Some people may develop sensitivity to various other agents in a food. Tyramine, found in cheese or red wine, can induce a headache in some people. Others may have sensitivity to certain food additives, such

as monosodium glutamate (MSG), sulfites, or coloring agents. The physical reactions to these agents can include hives, rashes, nasal congestion, or asthma.

A food allergy, in contrast, is an abnormal response to a food triggered by the immune system. The allergic reaction involves three main components: **food allergens**, immunoglobulin E (IgE), and mast cells. Food allergens are the fragments of food that are responsible for the allergic reaction. They consist of proteins from the food that are not broken down during the digestive process, which then cross the gastrointestinal lining to enter the bloodstream. IgE is a type of protein called an *antibody* that circulates through the blood. When allergic people eat certain foods, their immune system reacts to the food allergen by making IgE that is specific to that food. Once released, the IgE antibody attaches to a cell called the *mast cell*. Mast cells are specialized cells of the immune system that serve as the storehouse for various chemical substances, including **histamine**. Mast cells are found in all body tissues, but they are especially common in the areas of the body that are typical sites of allergic reaction: the nose and throat, lungs, skin, and gastrointestinal tract. When an allergic reaction occurs, the food allergen interacts with the IgE on the surface of the mast cells, which triggers those cells to release histamine. Depending on the tissue in which the histamine is released, these chemicals will cause a person to have various symptoms of a food allergy (see Figure 6-5). The most severe allergic reaction is **anaphylaxis**. This potentially fatal condition occurs when several parts of the body experience food allergy reactions at the same time. Signs of anaphylaxis include difficulty breathing, swelling of the mouth and throat, a drop in blood pressure, and loss of consciousness. The reaction can occur in a few seconds or minutes, and without

food aversion a strong desire to avoid a particular food.

food allergy an adverse reaction to an otherwise harmless substance that involves the body's immune system.

food intolerance a general term for any adverse reaction to a food or food component that does not involve the body's immune system.

adverse reaction an unusual response to food, including food allergies and food intolerances.

food allergen a substance in food (usually a protein) that is seen by the body as harmful and causes the immune system to mount an allergic reaction.

histamine a substance released by cells of the immune system during an allergic reaction to an antigen; causes inflammation, itching, hives, dilation of blood vessels, and a drop in blood pressure.

anaphylaxis (an-uh-fuh-LACKS-is) a potentially fatal reaction to a food allergen that causes reduced oxygen supply to the heart and other body tissues. Symptoms include difficulty breathing; low blood pressure; pale skin; a weak, rapid pulse; and loss of consciousness.

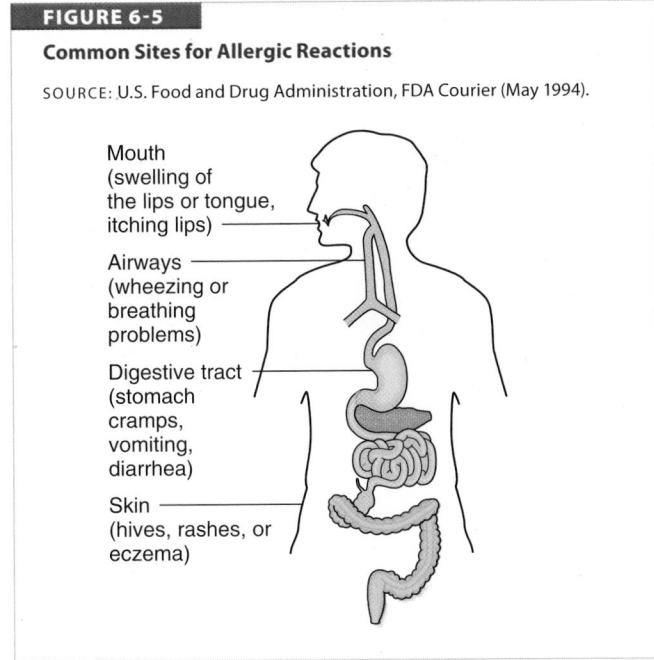

FIGURE 6-5

Common Sites for Allergic Reactions

SOURCE: U.S. Food and Drug Administration, FDA Courier (May 1994).

Mouth
(swelling of
the lips or tongue,
itching lips)

Airways
(wheezing or
breathing
problems)

Digestive tract
(stomach
cramps,
vomiting,
diarrhea)

Skin
(hives, rashes, or
eczema)

immediate medical attention, death may result. The foods most associated with anaphylactic reactions include peanuts, tree nuts (for example, walnuts, cashews), eggs, and shellfish.

According to the Food Allergy Network, eight foods cause 90 percent of all allergic reactions: egg, fish, milk, peanuts, shellfish, soy, tree nuts, and wheat.[25] In adults, the foods that most commonly cause allergic reactions include shellfish, peanuts, tree nuts, fish, and egg. In children, the problem foods are typically egg, milk, and peanuts. **Cross-reaction** is a concern for someone diagnosed with a food allergy. For instance, if someone has a history of allergic reaction to shrimp, testing may show that person is also allergic to other shellfish, such as crab and lobster (see Table 6-5).

Children and Food Allergy

The prevalence of food allergy is greatest in the first few years of life, with 6 to 8 percent of children younger than 4 years experiencing allergic responses.[26] Increased susceptibility of infants to food allergic reactions is believed to be the result of an immature immune system. The immature digestive system may also allow more intact allergenic proteins to enter the bloodstream. Cow's milk and soy are the most common allergens for infants, producing reactions such as hives, bloating, and diarrhea. Breastfeeding is recommended for the first 12 months to avoid early exposure to cow's milk or soy, thus avoiding allergy. Breast milk contains agents that stimulate the development and maintenance of the digestive tract,

cross-reaction the reaction of one antigen with antibodies developed against another antigen.

which further reduces the risk of food allergy. Early exposure to certain foods can also play a role in the development of food allergy in children. Delaying introduction of common allergenic foods may delay the onset of food allergies by allowing the digestive system and immune system to further develop before exposure to these known allergens. Young children often lose their sensitivity to most of the common allergenic foods in a few years. Unfortunately, sensitivity to certain foods, such as peanuts, tree nuts, fish, and shellfish, is rarely lost, and sensitivity persists into adulthood.

Diagnosis and Management of Food Allergy

Because both food intolerance and food allergy present very similar symptoms, diagnosing a food allergy involves determining if the reaction is mediated by the immune system. A complete physical examination can rule out the possibility that an underlying physical condition may be causing the symptoms. The physician takes a detailed case history and considers the type and timing of symptoms as well as the suspected offending food. Individuals may be asked to keep a food diary and record all physical symptoms.

Once food allergy is identified as a likely cause of symptoms, confirmation of the diagnosis and positive identification of the allergen involve various tests. A prick skin test (PST) is a simple test done in the doctor's office. The doctor puts a drop of the substance being tested on the forearm of the patient and pricks it with a needle, which allows a tiny amount of the substance to enter the skin. If the patient is allergic, a small bump will occur at the site in approximately 15 minutes. The radioallergosorbent test (RAST) requires a blood sample. Laboratory tests are conducted using specific foods to find out if the patient has IgE antibodies to those foods. Also, the physician may recommend an elimination diet, in which the patient

TABLE 6-5	Foods That Commonly Provoke Food Allergy
Food	**Cross-Reacting Foods**
Cows' milk*	Goats' milk, ewes' milk
Hens' eggs*	Eggs from other birds
Peanuts*	Soybeans, green beans, green peas
Soybeans	Peanuts, green beans, green peas
Cod	Mackerel, herring
Shrimp	Other crustaceans
Wheat	Other grains, mostly rye
Patients allergic to pollen may have cross-reactions with hazelnuts, green apples, peaches, almonds, kiwis, tomatoes, and potatoes (birch pollen) or wheat, rye, and corn (grass pollen).	

*Most common allergenic foods for children.

SOURCE: Adapted from C. Bindslev-Jensen, ABCs of allergies: Food allergy, *British Medical Journal* 316 (1998): 7140.

does not eat the food suspected of causing the allergy for a period of about 2 to 4 weeks. If the allergic symptoms improve, a diagnosis can be made.

The only treatment for food allergy is to avoid the offending food. People with food allergies must develop a skill for reading food labels. Effective January 2006, the Food and Drug Administration requires food labels to clearly state if food products include any ingredients that contain protein derived from the eight major allergenic foods.[27] Manufacturers are required to identify in plain English the presence of ingredients that contain protein derived from milk, eggs, fish, crustacean shellfish, tree nuts, peanuts, wheat, or soybeans in the list of ingredients or to say "contains" followed by name of the source of the food allergen after or adjacent to the list of ingredients.

The possibility of cross-contamination is also a problem for people with food allergies. When one food comes in contact with another food, trace amounts of each food mix with the other. Cross-contamination can occur in processing plants when different foods are processed on the same equipment. It can also occur at home—a knife inserted into the peanut butter jar and then used in the jelly will contaminate the jelly with peanut protein. The emerging field of biotechnology presents an additional challenge through genetic engineering (see Chapter 13). In genetic engineering, a gene from one species can be "spliced" into another species. The FDA requires that any food product of biotechnology that contains protein from a food known to be a common allergen must be properly labeled.

Currently, no drugs are available to cure food allergy. However, some medications, such as antihistamines and corticosteroids, are used to treat the symptoms. Individuals with a history of anaphylactic reactions to food allergens typically carry self-injecting syringes of epinephrine to use in case of accidental exposure.

Many questions remain to be answered in the field of food allergy. As scientists gain insight into the reactions of the immune system to various components of allergens, we can expect advances in the diagnosis and treatment of this disorder.[28] As the mystery of DNA is unraveled, it is conceivable that food allergy can be prevented. Until that time, however, many organizations offer additional information on food allergy and food intolerance, as listed on this page.

© Scott Hirko/Hespenheide Design

The top eight foods causing adverse reactions in some individuals: milk, eggs, peanuts, nuts, fish, shellfish, soy, and wheat.

The Food Allergy Network

www.foodallergy.org

American Academy of Allergy, Asthma, and Immunology

www.aaaai.org

American Dietetic Association

www.eatright.org

Asthma and Allergy Foundation of America

www.aafa.org

Soybeans have a long, rich history in the cuisine of the Eastern world. The early Chinese recognized the importance of this food, and they called it *Ta Tau,* which means "greater bean." According to Chinese tradition, soybeans were named as one of the five most sacred crops by the emperor who reigned 5,000 years ago.

Not too long ago in the Western world, soybeans were fodder for livestock. During the 1970s, many people turned to a vegetarian style of eating, often as a form of protest, but more frequently as a way to adopt a healthier lifestyle. Soy protein became the meat substitute of choice. Twenty-five years later, soybeans are recognized as the ideal functional food—a food that has the potential to reduce the risk of disease.

Soy foods are currently a hot area of research. Recent findings show that substances such as phytoestrogens and isoflavones, which are found in soybeans, can lower cholesterol and help prevent disease. Numerous studies attest to the role soy foods may play in reducing the risk of certain forms of cancer, heart disease, and osteoporosis, as well as in controlling diabetes and easing a woman's transition through menopause. Is it any wonder that more than 5,000 years ago, the soybean was called the greater bean?

What Is Soy?

Soybeans are legumes, members of the same plant family that includes other beans, peas, and lentils. Among edible legumes, the soybean is somewhat unusual because it is relatively low in carbohydrates. Soybeans are, however, high in fiber.

Among plant foods, legumes are high in protein. What distinguishes the soybean from its cousins, however, is the nature of the protein: Soybeans supply all of the essential amino acids needed for health. The amino acid pattern of soy protein is essentially the same quality as that found in meat, milk, and egg protein. Soybeans

TABLE 6-6	Protein and Isoflavone Content of Selected Soy Foods		
	Serving Size	Protein[a] (g)	Isoflavones[b] (mg)
Green soy beans, edamame	4 oz	14	50
Soy nuts	3 oz	20	90
Miso	2 tbsp	4	15
Soy milk	1 c	10	24
Soy flour, roasted	¼ c	7	42
Tempeh	4 oz	19	36
Tofu, firm	4 oz	13	24
Texturized soy protein, dry	¼ c	6	29
Soy burger (check label)	1	10–12	38–55
Soy protein isolate, dry	1 oz	23	28
Soy protein concentrate (alcohol extracted)	1 oz	17	4[c]

[a] Soy protein data (rounded to whole numbers) from USDA Nutrient Database for Standard Reference, Release 1–3.
[b] Isoflavone data (rounded to whole numbers) from USDA–Iowa State University database on the Isoflavone Content of Foods, 2002.
[c] To isolate soy protein from defatted soybeans, the carbohydrates must be removed by using a solvent, which can remove some or all of the isoflavone content. Because isoflavones are soluble in alcohol, much of the isoflavone content is lost if alcohol or repeated water washings are used in the extraction process. Soy milk, soy flour, tofu, tempeh, and soy protein isolate are not prepared with alcohol or repeated water extraction and therefore have a higher isoflavone content than soy protein concentrate.

are the only vegetable food that contains complete protein. What emerges from this nutritional analysis of the soybean is the image of "balance." Soybeans as a food can basically stand alone, give or take a few vitamins and minerals. Add some vegetables to the beans, and you have created a high-quality, nutrient-dense meal!

You have mentioned isoflavones. What are they?

Isoflavones are a type of *phytoestrogen*—compounds that have weak estrogenic activity. Many types of phytoestrogenic compounds are available in edible plants. Foods made from soybeans have varying amounts of isoflavones (see Table 6-6), depending on how the foods are processed. Foods such as tofu, soy milk, soy flour, and soy nuts have higher isoflavone concentrations than foods made with a combination of soy and grains. Soy sauce and soybean oil have virtually no isoflavones.

Research in several areas of health care has shown that consumption of soy foods may play a role in lowering risk of disease.[29] Soy isoflavones are being studied intensively to clarify their

© Scott Hirko/Hespenheide Design

physiological effects. In some cases, research has shown that isoflavones may be a key factor in the disease-fighting potential of soybeans.

It is important to keep in mind that our knowledge of the long-term effects of isoflavones is based on their content in soy foods. These foods have been consumed for hundreds of years, and are known to be safe. It is still best to obtain isoflavones by enjoying a variety of soy foods.

The new information about soy seems promising. What are the potential health benefits of adding soy foods to my diet?

Soy and Heart Disease

High blood cholesterol is a major risk factor for heart disease. A great deal of evidence indicates that soy protein helps lower blood cholesterol levels. Replacing animal protein with soy protein in the diet lowers total and LDL cholesterol levels in people with high cholesterol.[30] A meta-analysis of 38 research studies concluded that soy protein lowers total and LDL cholesterol and triglycerides, without lowering HDL-cholesterol, in people with high cholesterol.[31]

In these studies, the average consumption of soy protein was 47 grams per day. The greatest decreases in blood cholesterol were seen in those with the highest starting levels. Even adding soy protein to an omnivorous diet has been shown to produce this effect.[32] The newly approved health claim for food labels states that as little as 25 grams of soy protein per day may be enough to lower cholesterol levels. Refer to Table 6-6 for the protein content of selected soy foods.

Soy and Osteoporosis

Soybeans and soy foods may help prevent and treat osteoporosis, a disease that weakens bones and often results in bone fractures. As women age, it becomes more important than ever to maintain adequate levels of calcium in bones. Soy foods such as fortified soy milk, texturized soy protein, and tofu made with calcium salt are all good calcium sources.

Isoflavones found in soy protein may also play an important role in protecting bones.[33] A breakthrough study at the University of Illinois at Urbana concluded that consuming soybean isoflavones can increase bone mineral content and bone density.

Miniglossary—What Foods Contain Soy?[34]

green soybeans (edamame) these large soybeans are harvested when the beans are still green and sweet and can be served as a snack or a main vegetable dish, after boiling in water for 15 to 20 minutes. They are high in protein and fiber and contain no cholesterol.

hydrolyzed vegetable protein (HVP) a protein obtained from any vegetable, including soybeans. The protein is broken down into amino acids by a chemical process called *acid hydrolysis*. HVP is a flavor enhancer that can be used in soups, broths, sauces, gravies, flavoring and spice blends, canned and frozen vegetables, and meats and poultry.

meat alternatives meat alternatives made from soybeans contain soy protein or tofu and other ingredients mixed together to simulate various kinds of meat. These meat alternatives are sold as frozen, canned, or dried foods.

miso a rich, salty condiment that characterizes the essence of Japanese cooking. The Japanese make miso soup and use miso to flavor a variety of foods. A smooth paste, miso is made from soybeans and a grain such as rice, plus salt and a mold culture, and then aged

in cedar vats for 1 to 3 years. Miso should be refrigerated. Use miso to flavor soups, sauces, dressings, marinades, and pâtés.

nondairy soy frozen dessert nondairy frozen desserts are made from soy milk or soy yogurt. Soy ice cream is one of the most popular desserts made from soybeans and can be found in many grocery stores.

soy cheese cheese made from soy milk. Its creamy texture makes it an easy substitute for sour cream or cream cheese. It can be found in a variety of flavors in many food stores.

soy flour flour made from roasted soybeans ground into a fine powder. Soy flour gives a protein boost to recipes. Soy flour is gluten-free, so yeast-raised breads made with soy flour are denser in texture. Replace one-quarter to one-third wheat flour with soy flour in recipes for muffins, cakes, cookies, pancakes, and quick breads.

soy protein, texturized *texturized soy protein* usually refers to products made from texturized soy flour. Texturized soy flour is made by running defatted soy flour through an extrusion cooker, which allows the creation of many

different forms and sizes. When hydrated, it has a chewy texture. It is widely used as a meat extender.

soy yogurt yogurt made from soy milk. Its creamy texture makes it an easy substitute for sour cream or cream cheese. Soy yogurt can be found in a variety of flavors in many food stores.

soybeans as soybeans mature in the pod, they ripen into a hard, dry bean. Most soybeans are yellow. However, brown and black varieties are also available. Whole soybeans can be cooked and used in sauces, stews, and soups.

soy milk, soy beverages soybeans that are soaked, ground fine, and strained produce a fluid called *soybean milk*. Soy milk is an excellent source of high-quality protein and B vitamins. Look for calcium-fortified varieties.

soy-nut butter made from roasted, whole soy nuts, which are then crushed and blended with soy oil and other ingredients. Soy-nut butter has a slightly nutty taste, contains significantly less fat than peanut butter, and provides many other nutritional benefits.

soy nuts whole soybeans that have been soaked in water and then baked until browned. Soy

nuts can be found in a variety of flavors, including chocolate covered. High in protein and isoflavones, soy nuts are similar in texture and flavor to peanuts. Try sprinkling some on salads.

tempeh a chunky, tender soybean cake; a traditional Indonesian food. Whole soybeans, sometimes mixed with another grain such as rice or millet, are fermented into a rich cake of soybeans with a smoky, nutty flavor. Tempeh can be marinated and grilled and added to soups, casseroles, or chili.

tofu and tofu products *tofu*, also known as *soybean curd*, is a soft, cheese-like food made by curdling fresh, hot soy milk with a coagulant. Tofu is a bland product that easily absorbs the flavors of other ingredients with which it is cooked. Tofu is rich in high-quality protein and B vitamins and is low in sodium. Firm tofu is dense and solid and can be cubed and served in soups, stir-fried, or grilled. Firm tofu is higher in protein, fat, and calcium than other forms of tofu. Silken tofu is a creamy product and can be used as a replacement for sour cream in many dip recipes.

As little as 40 grams of soy protein, consumed each day for 6 months, led to positive results in a test group of postmenopausal women.[35] Forty grams of soy protein can be found in 2 ounces of soy protein isolate (see Table 6-6).

Soy and Menopause

The hormonal changes that occur during menopause can cause a variety of symptoms and increase risk for heart disease and osteo-porosis.[36] Soy foods that contain phytoestrogens are being studied for their possible efficacy in decreasing the negative effects of menopause. Fluctuating levels of estrogen can cause hot flashes, night sweats, insomnia, vaginal dryness, or headaches. Hormone replacement therapy (HRT) had been commonly prescribed to help prevent the negative health effects of menopause. However, findings from the Women's Health Initiative indicated possible increased risk for breast cancer, stroke, and heart disease from HRT.[37]* Scientists are now investigating the question: Can soy foods provide the same kinds of health benefits as HRT, without the risks?

In women who are producing little estrogen, phytoestrogens may produce enough estrogenic activity to relieve symptoms such as hot flashes. A recent study found that women who were fed 45 grams of soy flour per day had a 40 percent reduction in the incidence of hot flashes.[38] From an epidemiological point of view, it is interesting to note that in Japan, where soy consumption is high, menopause symptoms of any kind are rarely reported. In addition, bones tend to be stronger in Asia, and broken hips and spinal fractures are less common.

Soy contains phytoestrogens in the form of isoflavones, genistein, and daidzein. These are known to have weak estrogenic effects when consumed by animals and humans.[39] Researchers continue to study the physiological effects of the isoflavones to find out whether they can serve some of the same functions as estrogen, and thereby decrease the health risks associated with menopause.

Soy and Cancer

One out of every four deaths in the United States is due to cancer. Epidemiological studies show that populations that consume a typical Asian diet have lower incidences of breast, prostate, and colon cancers than those consuming a Western diet.[40] The Asian diet includes mostly plant foods, including legumes, fruits, and vegetables, and is low in fat. The Japanese have the highest consumption of soy foods. Japan has a very low incidence of hormone-dependent cancers. The mortality rate from breast and prostate cancers in Japan is about one-fourth that of the United States.[41] Some evidence suggests that the difference in cancer rates is due not to genetics, but to diet. Migration studies have shown

© Polara Studios

that when Asians move to the United States and adopt a Western diet, they ultimately have the same cancer incidence as Americans.[42] Other long-term studies have noted an inverse association between regular consumption of miso soup and breast cancer risk in premenopausal women.[43] In Hawaii, a long-term study of 8,000 men of Japanese ancestry showed that men who ate tofu daily were only one-third as likely to get prostate cancer as those who ate tofu once a week or less.[44]

Soybeans contain five classes of compounds that have been identified as anticarcinogens.[45] Most of these compounds can be found in many different plant foods, but soy is the only significant dietary source of isoflavones. Soy isoflavones, especially genistein, have been the subject of a tremendous amount of cancer research.

Conflicting data abound in almost all nutritional research, and soy is no exception. Some studies have actually shown a correlation between soy consumption and the development of breast cancer, and still others associate the time in life when soy is actually consumed with the extent of its effects.[46] Whatever the case, it could be years before scientists reach definite conclusions. The field of soy research is still relatively new.[47] As this area of nutritional research evolves, more conflicting information will no doubt be published. Thus far, however, the potential benefits of soy seem to outweigh the risks.

Soy and Diabetes

Another interesting benefit of soy foods is their effect on glucose control. Recently, scientists have become interested in the role of soy foods in regulating diabetes.[48] Emerging evidence indicates that the phytoestrogens, soy protein, and soy polysaccharides found in soy foods might play a beneficial role in diabetes. Although further long-term investigation is needed, this emerging area of research will be important in years to come. In addition, because soybeans are a complex carbohydrate and also have a low glycemic index, they are an ideal food to help regulate blood glucose levels in people with diabetes. Overall, it seems that soy foods can have a positive synergistic effect on diabetes.

How much soy should I eat on a daily basis to receive these health benefits?

Science has not yet established a recommended daily amount of soy consumption to achieve all of the health benefits mentioned

*The National Institutes of Health (NIH) began the Women's Health Initiative in 1991 to study strategies for preventing the major causes of death and disability in postmenopausal women, including heart disease, breast and colon cancer, and osteoporosis.

here. In many cases, just one serving of soy per day may help improve your health. Although the new health claim for food labels states that 25 grams of soy protein each day may lower risk of heart disease, no recommendation for daily isoflavone intake has yet been made. In Asian countries, where people typically consume 25 to 40 milligrams of isoflavones a day, the incidence of osteoporosis, heart disease, and certain cancers is low. Still, many questions remain about how soy acts in the body and how much is needed for benefits at various stages of the life cycle. A balanced diet that includes soy is recommended, but loading up on one food, nutrient, or phytochemical is not advisable. Here are some quick tips for adding some soy to your diet:

- Pour soy milk over breakfast cereal once or twice a week.

- Cook oatmeal or cream of wheat in soy milk. You might try vanilla-flavored soy milk.

- Make pancakes or French toast with vanilla soy milk.

- Use soy milk to make hot chocolate.

- Add firm tofu chunks in place of meat or poultry in stir-fries, fajitas, or shish kebabs.

- Try a "garden" burger or soy burger. Check food labels; not all are made with soy.

- Replace ground beef on nachos, pizzas, or in spaghetti sauce with crumbled soy burgers.

- Add tofu to the blender with your favorite seasoning packet, such as ranch, onion soup mix, or taco seasoning, and serve with tortilla chips or fresh vegetables.

- Add soy flour to quick bread recipes. It adds moisture and a soy protein boost. Just replace one-fourth of the total flour with soy flour in recipes for quick breads, muffins, or cakes.

- Enjoy soy-nut butter on bagels, breads, and English muffins, or carrot and celery sticks.

- Make your own breakfast sandwich with a whole-wheat English muffin, slice of soy cheese, and soy sausage-style patty.

- After a workout, whip up a chocolate banana soy smoothie made from vanilla soy yogurt, chocolate soy milk, and a frozen banana.

Quick and Easy Soy Recipes

Strawberry-Banana Frosty

3 c plain or vanilla soy milk
1 c strawberries
1 ripe banana
Blend in blender until smooth.
Makes 6 servings.

Multigrain Apple Pancakes

1¼ c yellow cornmeal
1½ c rolled oats
1½ c unbleached flour
1½ c soy flour
1 tsp cinnamon
1 tbsp baking powder
1½ c plain soy milk
1½ c applesauce

In a large bowl, combine the rolled oats, cornmeal, unbleached flour, soy flour, cinnamon, and baking powder. Add the soy milk, and blend with a few swift strokes. Fold in the applesauce. Pour 1¼ cup of the batter on a hot nonstick griddle or pan. Cook for about 2 minutes or until bubbles appear on the surface. Flip the pancake and cook for another minute or until heated through. Serve the pancakes with maple syrup, fruit spread, or applesauce.
Makes 12 pancakes.

Soy Nut Trail Mix

2 c roasted soy nuts
1 c raisins
1 c oat-ring cereal
1 c dried cranberries
2 c mini-wheat squares cereal
½ c dried cherries

Mix all ingredients in large bowl or container. Keep tightly closed in container or zippered plastic bag.
Yield: 7 cups

SOURCE: *Soyfoods Cookbook*, www.soyfoods.com/recipes. © 1999 Indiana Soybean Board. Reprinted by permission.

Nutrition on the Web

www.cengage.com/nutrition/boyle/personalnutrition8e
Go to the Personal Nutrition site to check for the latest updates to chapter topics or to access links to related websites.

www.healthfinder.gov
Search for information on vegetarianism and food allergies.

www.eatright.org
The American Dietetic Association home page includes many nutrition resources and links.

www.fda.gov
Go to Food and search for information on food allergens and vegetarian diets.

www.who.int/nutrition/en/
Search this site for more information about protein-energy malnutrition worldwide.

www.nal.usda.gov/fnic

- Under Browse by Subject, select Lifecycle Nutrition, scroll down, and click on Vegetarian Nutrition.
- Under Browse by Subject, select Diet and Disease and click on Allergies and Food Sensitivities.
- Under Browse by Subject, select Food Composition and click on Macronutrients to access protein requirements.

www.vrg.org
The Vegetarian Resource Group provides information on vegetarianism.

http://vegetariannutrition.net
Vegetarian Nutrition Dietetic Practice Group

www.choosemyplate.gov/tipsresources/vegetarian_diets.html
Center for Nutrition Policy and Promotion

www.nal.usda.gov/fnic/pubs/bibs/gen/vegetarian.pdf
Food and Nutrition Information Center

www.mayoclinic.com/health/vegetarian-diet/HQ01596
Mayo Clinic

www.nlm.nih.gov/medlineplus/vegetariandiet.html
Medline Plus, Vegetarian Diet

www.ncbi.nlm.nih.gov/PubMed
This search engine helps you locate information from current scientific articles on any topic related to protein.

www.soyfoods.com
This U.S. Soy Foods Directory website includes a searchable database, recipes, and research information about the health benefits of soy foods.

www.soyconnection.com
The United Soybean Board provides information and answers questions about soy foods.

www.foodallergy.org
The Food Allergy and Anaphylaxis Network (FAAN) website provides information about food allergies and allergens.

www.nlm.nih.gov/medlineplus/foodallergy.html
This government website provides information and links about food allergies.

Menu of Online Study Tools

A variety of study tools for this chapter are available at our website to deepen your understanding of chapter concepts. Go to

www.cengage.com/sso

to find

- Practice tests
- Flashcards
- Glossary
- Web links
- Animations
- Chapter summaries, learning objectives, and crossword puzzles.

Diet Analysis+ in Action

Go to the Intake vs. Goals report in Diet Analysis+ that shows the results of your three-day tracking. Compare your daily recommended intake of protein (in grams) to your actual intake. What percent of your Dietary Reference Intake did you consume in protein? Now go to the Intake Spreadsheet report and identify the two sources of animal protein that contributed the most to your total intake. Identify the two sources of plant protein that contributed the most to your total intake. Looking at the MyPlate report, how many servings from each of the protein food groups did you consume? Were they within the recommended number of servings? How can you improve your intake?

The Vitamins: A Functional Approach

Ask Yourself . . .

Which of the following statements about nutrition are true, and which are false? For each false statement, what is true?

1. The most important role that vitamins play is providing energy.
2. You can't overdose on vitamins, because the body excretes them in the urine.
3. Serving for serving, fruits and vegetables tend to be the richest sources of vitamins.
4. Vitamin D is also known as the "sunshine vitamin" because the body can make vitamin D with the help of sunlight.
5. Vitamin C supplements prevent colds.
6. Several major public health associations recommend that all adults take antioxidant supplements.
7. Fresh vegetables contain more vitamins than frozen vegetables.
8. Phytochemicals are beneficial nonnutrient substances found in fruits, vegetables, and whole grains.
9. In general, nutrients are absorbed equally as well from foods as from supplements.
10. Maintaining a healthy weight may be one of the most important ways to protect against cancer.

Answers found on the following page.

Contents

About a century ago, scientists ushered in a new era in the science of nutrition—the discovery of vitamins. They quickly realized that these substances, found in minute amounts in foods, were just as essential to health as fats, carbohydrates, and proteins. A diet lacking in one vitamin could cause a barrage of symptoms and, ultimately, death. Knowledge of the vital roles played by vitamins quickly advanced, and today life-threatening vitamin deficiencies are rare in developed countries such as the United States and Canada.

Still, the vitamin research that has been conducted during the past decade or so has marked the beginning of yet another chapter in the annals of nutrition. Throughout the past 10 years, more and more scientists have investigated the possibility that large doses of certain vitamins will help stave off chronic diseases such as cancer and heart disease, which rank as major killers today. In fact, the study of vitamins, particularly a class known as the *antioxidant vitamins*, is one of the hottest, most widely publicized areas in current nutrition research. In addition,

In France old Crainquebille sold leeks from a cart, leeks called "the asparagus of the poor." Now asparagus sells for the asking, almost, in California markets, and broccoli, that strong age-old green, leaps from its lowly pot to the Ritz's copper saucepan.

Who determines, and for what strange reasons, the social status of a vegetable?

M. F. K. FISHER (1908–1992, U.S. FOOD WRITER)

The more varied the kinds of fruits and vegetables you eat, the better nourished you are likely to be.

The typical dry, crusty dermatitis of pellagra is found only on areas of the body exposed to sunlight.

scurvy the vitamin C-deficiency disease characterized by bleeding gums, tooth loss, and, in severe cases, death.

rickets a disease that occurs in children as a result of vitamin D deficiency; characterized by abnormal growth of bone, which in turn leads to bowed legs and an outward-bowed chest.

pellagra (pell-AY-gruh) niacin deficiency characterized by diarrhea, inflammation of the skin, and, in severe cases, mental disorders and death.

the pros and cons of taking vitamin supplements are the subject of heated debate within the scientific community. To help you sort through the steady stream of controversial and often contradictory claims regarding vitamins, this chapter explores the history, roles, and current thrust of research on the various vitamins and offers practical advice on how to incorporate the information into decisions about your own lifestyle.

Turning Back the Clock

Many of the vitamin deficiency diseases that have been virtually eliminated today were first recognized in Greek and Roman times, and ultimately led to the discovery of vitamins centuries later. One of the most prevalent was **scurvy**, a disease caused by lack of vitamin C and characterized by bleeding gums, tooth loss, and even death. The scourge of armies, sailors, and other travelers forced to do without vitamin C-rich foods for weeks on end, scurvy was recognized by Hippocrates, a Greek physician heralded today as the father of medicine.[1]

A cure for scurvy was not recorded until the 16th century, however, when a beverage made of spruce needles or oranges and lemons was recommended. In 1753, a British physician named James Lind published a famous report recommending consumption of herbs, lettuce, endive, watercress, and summer fruits to prevent scurvy. By the early 1800s, sailors in the British navy had been dubbed "limeys" because they were required to drink lemon or lime juice daily.[2] Although they still didn't know that vitamin C was the actual antidote, they did recognize that certain foods prevented and cured the illness.

Similarly, a deficiency disease called **rickets** dates back to Roman times, when children frequently suffered skeletal deformities as a result of a lack of vitamin D. By the 1600s, rickets was known as the "English disease" because it afflicted so many English children. Some 200 years later, cod liver oil was finally recognized as a cure for the disease; no one knew at the time, however, that the "magic" ingredient in the oil was vitamin D.

Another deficiency disease, called **pellagra**, was not recognized until 1730, when a Spanish physician named Gaspar Casal first described the crusty, dry, scabby, blackish patches of skin symptomatic of the disease. In Italy, the disease was named pellagra, from the Italian *pelle agra*, meaning "sour skin." Called *mal de la rosa* in Spanish, pellagra was thought to be incurable until Dr. Casal noticed that the people who developed the disease were typically poor and had inadequate diets made up of mostly corn and little meat.

By the 19th century, physicians had recognized that certain foods prevented or cured pellagra and other deficiency diseases, although they still hadn't determined exactly what it was in these various foods that worked as remedies. By the middle of the 19th century, however, the science of chemistry had advanced to a point at which foods could be analyzed. Chemists had determined that foods consisted of fats, proteins, and carbohydrates, along with minerals and water, and they assumed that they had identified all the nutritionally significant compounds.

Then, in the early 20th century, scientists detected minute amounts of other substances that they found to be essential in preventing disease and maintaining health.

ASK YOURSELF ANSWERS: 1. False. Vitamins do not provide energy, though they do play roles in energy-yielding reactions in the body. **2.** False. Excess doses of any of the vitamins can be toxic. **3.** True. **4.** True. **5.** False. Vitamin C has never been proved to prevent colds; at best, it may reduce the severity of cold symptoms. **6.** False. No major health organization recommends that all adults take antioxidant pills. **7.** False. Fresh vegetables do not necessarily contain more vitamins than their frozen counterparts; vitamin content depends on such factors as how the fresh vegetable has been stored and how long it has been since the fresh produce was harvested. **8.** True. **9.** False. In general, nutrients are absorbed best from foods because they are accompanied by other ingredients that facilitate their absorption. **10.** True.

The substances were dubbed *vitamines*, a term coined in 1912 by a scientist named Dr. Casimir Funk to indicate that these substances were *vital* for survival and that they contained nitrogen—that is, they were *amines*. (The *e* was later dropped when scientists discovered that some vitamins were not amines.)

Over the next few decades, scientists identified the various **vitamins**, established their chemical formulas, and determined their functions in the body. They also measured the amount of vitamins in various foods and determined human and animal requirements for the compounds. Knowledge of vitamins constantly evolves as scientists continue to study the actions of these substances in the human body.

Today, scientists recognize vitamins as potent compounds that perform many tasks in the body to promote growth and reproduction and maintain health and life. Vitamins constantly work to keep your nerves and skin healthy; build bone, teeth, and blood; and heal wounds, among many other things. Although they do not provide calories, vitamins are essential in helping the body make use of the calories consumed via foods.

The Two Classifications of Vitamins

Vitamins fall into two categories: those that dissolve in water (water-soluble vitamins) and those that dissolve in fat (fat-soluble vitamins). To date, scientists have identified 13 vitamins, each with its own special roles to play. As Figure 7-1 shows, each of the food groups in the MyPlate Food Guide supplies a number of vitamins. Eating plans that exclude entire food groups or fail to include the minimum number of servings from each group may lead to vitamin deficiencies over time.

The nine water-soluble vitamins—eight B vitamins and vitamin C—are found in the watery components of foods, such as the juice of an orange. Table 7-1 identifies major functions and good food sources of the water-soluble vitamins. These vitamins are distributed into water-filled compartments of the body, including the fluid that surrounds the spinal cord. The body excretes excess water-soluble vitamins if blood levels rise too high. As a result, these vitamins rarely reach toxic levels in the body. This is not to say, however, that excess levels cannot cause problems, at least in some people.

In contrast, the four fat-soluble vitamins—A, D, E, and K—are generally found in the fats and oils of foods. Table 7-2 identifies major functions and good food sources of the fat-soluble vitamins. Because they are stored in the liver and in body fat, it is possible for megadoses of fat-soluble vitamins to build up to toxic levels in the body and cause undesirable side effects. Table 7-3 shows the contrasts between the two classes of vitamins.

Water-Soluble Vitamins include:

- B vitamins
  - Thiamin
  - Riboflavin
  - Niacin
  - Vitamin B_6
  - Folate
  - Vitamin B_{12}
  - Pantothenic acid
  - Biotin
- Vitamin C

Fat-Soluble Vitamins include:

- Vitamin A
- Vitamin D
- Vitamin E
- Vitamin K

vitamin a potent, indispensable compound that performs various bodily functions that promote growth and reproduction and maintain health. Vitamins are *essential* nutrients required in minute amounts and *organic*, meaning that they contain or are related to carbon compounds. Contrary to popular belief, vitamins do not supply calories.

FIGURE 7-1

Good Sources of Vitamins in the USDA MyPlate Food Guide*

*Serving sizes shown here are based on a 2,000-calorie diet. Go to www.ChooseMyPlate.gov for a quick estimate of how much food you should eat from the different food groups based on your age, gender, and activity level.

FRUITS
2 cups every day
Vitamin A (as beta-carotene)
Vitamin C
Folate

VEGETABLES
2¹/2 cups every day
Vitamin A (as beta-carotene),
Vitamin C, Vitamin K,
Riboflavin, Folate

DAIRY 3 cups every day
Riboflavin, Vitamin A, Vitamin D,
Vitamin B_{12}

GRAINS 6 oz every day
Niacin, Riboflavin,
Thiamin, Folate

PROTEIN FOODS 5¹/2 oz every day
Vitamin B_6, Vitamin B_{12}, Niacin, Biotin,
Thiamin, Vitamin D

TABLE 7-1   A Guide to the Water-Soluble Vitamins

Water-Soluble Vitamins

(Chemical Name)	Best Sources[a]	Chief Roles	Deficiency Symptoms	Toxicity Symptoms[b]
Thiamin	Meat, pork, liver, fish, poultry, whole-grain and enriched breads, cereals and grain products, nuts, legumes	Helps enzymes release energy from carbohydrate; supports normal appetite and nervous system function	Beriberi: edema, heart irregularity, mental confusion, muscle weakness, apathy, impaired growth	None reported
Riboflavin	Milk, leafy green vegetables, yogurt, cottage cheese, liver, meat, whole-grain or enriched breads, cereals and grain products	Helps enzymes release energy from carbohydrate, fat, and protein; promotes healthy skin and normal vision	Eye problems, skin disorders around nose and mouth, magenta tongue, hypersensitivity to light	None reported
Niacin	Meat, eggs, poultry, fish, milk, whole-grain and enriched breads, cereals and grain products, nuts, legumes, peanuts	Helps enzymes release energy from energy nutrients; promotes health of skin, nerves, and digestive system	Pellagra: flaky skin rash on parts exposed to sun, loss of appetite, dizziness, weakness, irritability, fatigue, mental confusion, indigestion, delirium	Flushing, nausea, headaches, cramps, ulcer irritation, heartburn, abnormal liver function, rapid heartbeat with doses above 500 mg per day
Vitamin B$_6$ (pyridoxine)	Meat, poultry, fish, shellfish, legumes, fruits, soy products, whole-grain products, green leafy vegetables	Protein and fat metabolism; formation of antibodies and red blood cells; helps convert tryptophan to niacin	Nervous disorders, skin rash, muscle weakness, anemia, convulsions, kidney stones	Depression, fatigue, irritability, headaches, numbness, damage to nerves, difficulty walking
Folate (folacin, folic acid)	Green leafy vegetables, liver, legumes, seeds, citrus fruits, melons, enriched breads, cereals, and grain products	Red blood cell formation; protein metabolism; new cell division	Anemia, heartburn, diarrhea, smooth red tongue, depression, poor growth, neural tube defects, increased risk of heart disease, stroke, and certain cancers	Diarrhea, insomnia, irritability, may mask a vitamin B$_{12}$ deficiency
Vitamin B$_{12}$ (cobalamin)	Animal products: meat, fish, poultry, shellfish, milk, cheese, eggs; fortified cereals	Helps maintain nerve cells; red blood cell formation; synthesis of genetic material	Anemia, smooth red tongue, fatigue, nerve degeneration progressing to paralysis	None reported
Pantothenic acid	Widespread in foods	Coenzyme in energy metabolism	Rare; sleep disturbances, nausea, fatigue	None reported
Biotin	Widespread in foods	Coenzyme in energy metabolism; fat synthesis;	Loss of appetite, nausea, depression, muscle pain	None reported
Vitamin C (ascorbic acid)	Citrus fruits, cabbage-type vegetables, tomatoes, potatoes, dark green vegetables, peppers, lettuce, cantaloupe, strawberries, mangoes, papayas	Synthesis of collagen (helps heal wounds, maintains bone and teeth, strengthens blood vessel walls); antioxidant; strengthens resistance to infection; helps body absorb iron	Scurvy: anemia, depression, frequent infections, bleeding gums, loosened teeth, pinpoint hemorrhages, muscle degeneration, rough skin, bone fragility, poor wound healing, hysteria	Intakes of more than 1 g per day may cause nausea, abdominal cramps, diarrhea, and increased risk for kidney stones

[a] The recommended intakes for the vitamins are listed on Study Card page 28 at the back of this book.
[b] The Tolerable Upper Intake Levels (UL) for the vitamins are listed on Study Card page 29 at the back of this book.

TABLE 7-2	A Guide to the Fat-Soluble Vitamins			
Fat-Soluble Vitamins				
(Chemical Name)	**Best Sources[a]**	**Chief Roles**	**Deficiency Symptoms**	**Toxicity Symptoms[b]**
Vitamin A	*Retinol:* fortified milk and margarine, cream, cheese, butter, eggs, liver *Beta-carotene:* Spinach and other dark leafy greens, broccoli, deep orange fruits (apricots, peaches, cantaloupe), and vegetables (squash, carrots, sweet potatoes, pumpkin)	Vision; growth and repair of body tissues; maintenance of mucous membranes; reproduction; bone and tooth formation; immunity; hormone synthesis; antioxidant (in the form of beta-carotene only)	Night blindness, rough skin, susceptibility to infection, impaired bone growth, abnormal tooth and jaw alignment, eye problems leading to blindness, impaired growth	Red blood cell breakage, nosebleeds, abdominal cramps, nausea, diarrhea, weight loss, blurred vision, irritability, loss of appetite, bone pain, dry skin, rashes, hair loss, cessation of menstruation, liver disease, birth defects
Vitamin D (cholecalciferol)	Self-synthesis with sunlight; fortified milk, fortified margarine, eggs, liver, fish	Calcium and phosphorus metabolism (bone and tooth formation); aids body's absorption of calcium	Rickets in children; osteomalacia in adults; abnormal growth, joint pain, soft bones	Deposits of calcium in organs such as the kidneys, liver, or heart, mental retardation, abnormal bone growth
Vitamin E	Vegetable oils, green leafy vegetables, wheat germ, whole-grain products, liver, egg yolk, salad dressings, mayonnaise, margarine, nuts, seeds	Protects red blood cells; antioxidant (protects fat-soluble vitamins); stabilization of cell membranes	Muscle wasting, weakness, red blood cell breakage, anemia, hemorrhaging	Doses over 800 IU/day may increase bleeding (blood clotting time)
Vitamin K	Bacterial synthesis in digestive tract, liver, green leafy and cabbage-type vegetables, soybeans, milk, vegetable oils	Synthesis of blood-clotting proteins and a blood protein that regulates blood calcium	Hemorrhaging, decreased calcium in bones	Interference with anticlotting medication; synthetic forms may cause jaundice

[a] The recommended intakes for the vitamins are listed on Study Card page 28 at the back of this book.
[b] The Tolerable Upper Intake Levels (UL) for the vitamins are listed on Study Card page 29 at the back of this book.

TABLE 7-3	General Characteristics of Water-Soluble and Fat-Soluble Vitamins	
Characteristic	**Water-Soluble Vitamins (B vitamins and vitamin C)**	**Fat-Soluble Vitamins (A, D, E, and K)**
Dietary intake	Excess intake usually detected and excreted by the kidneys	Excess intake tends to be stored in fat-storage sites
Body stores	Only a short-term supply available, with little storage; daily intake recommended	Long-term storage available in body tissues; regular intake recommended
Deficiency	Deficiency symptoms appear relatively quickly	Deficiency symptoms are slow to develop
Toxicity	Lower risk of toxicity	Higher risk of toxicity
Absorption and transport	Easily absorbed into blood; travel freely in blood	Like lipids, absorbed into lymph; many require protein carriers if they are to travel in the blood
Solubility	Dissolves in water	Dissolves in lipid
Stability in food handling and processing	Less stable	Stable
Role(s) in the body	Most B vitamins share similar roles; vitamin C serves many different roles in the body.	Each has unique roles in the body.

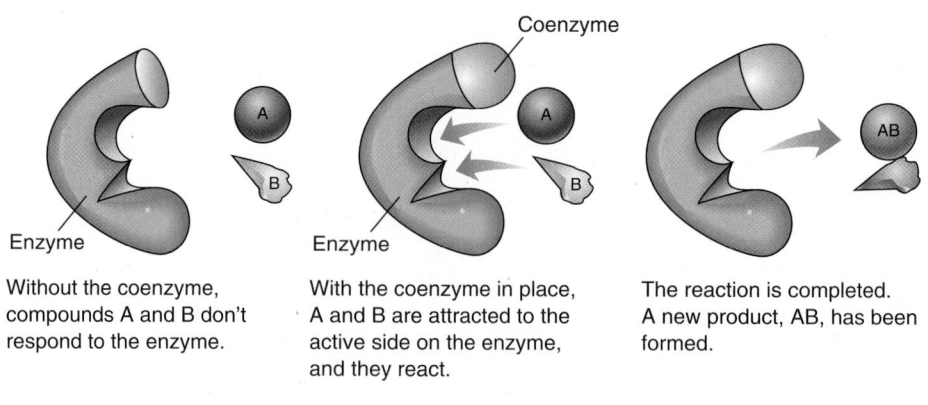

FIGURE 7-2

How a Coenzyme Works

Coenzyme

Enzyme

Without the coenzyme, compounds A and B don't respond to the enzyme.

Enzyme

With the coenzyme in place, A and B are attracted to the active side on the enzyme, and they react.

The reaction is completed. A new product, AB, has been formed.

Water-Soluble Vitamins In the body, water-soluble vitamins act as **coenzymes**—that is, they assist enzymes in doing their metabolic work within the body (see Figure 7-2). You may recall from Chapter 6 that enzymes are proteins that act as catalysts to help boost chemical reactions in the body.

In foods, the water-soluble vitamins are relatively fragile. Although large amounts of them are naturally present in many foods, they can be washed out or destroyed during food storage, processing, and preparation. These effects are described in detail in the Savvy Diner feature in this chapter.

Fat-Soluble Vitamins The four fat-soluble vitamins—A, D, E, and K—are absorbed from the digestive tract with the aid of fats in the diet and bile produced by the liver. Any disorder that interferes with fat digestion or absorption can precipitate a deficiency of the fat-soluble vitamins. Once in the bloodstream, these vitamins are escorted by protein carriers because they are *insoluble* in water. Because they are stored in the liver and in body fat, you need not consume them daily unless your intakes are usually marginal.

This chapter discusses the vitamins known to be important in human nutrition and provides information about food sources and human requirements. The discussions about the vitamins are organized according to the biological roles these substances play in the body.

The Antioxidant Vitamins

As their name suggests, **antioxidants** are "antioxygen": they fight oxygen, in a manner of speaking. Some chemical reactions that occur in the body involve the use of oxygen. Although these reactions are essential to the body's ability to function, they also lead to the creation of highly toxic compounds called **free radicals** (see Figure 7-3). Environmental pollutants such as cigarette smoke and ozone also prompt the formation of free radicals. Left unchecked, these compounds can cause severe cell injury and ultimately may contribute to the development of chronic diseases such as cancer and heart disease.

Fortunately, the body has a built-in defense system to protect against potential damage from free radicals. That defense system makes use of the antioxidant nutrients: vitamin C, vitamin E, and the carotenoid called **beta-carotene**. In addition, the body manufactures certain enzymes, one of which contains the mineral selenium—discussed in Chapter 8—that help fight free radicals.

coenzymes enzyme helpers; small molecules that interact with enzymes and enable them to do their work. Many coenzymes are made from water-soluble vitamins.

antioxidant a substance, such as a vitamin, that is "anti-oxygen"; that is, it helps prevent damage done to the body as a result of chemical reactions involving oxygen.

free radicals highly toxic compounds created in the body as a result of chemical reactions that involve oxygen. Environmental pollutants such as cigarette smoke and ozone also prompt the formation of free radicals.

beta-carotene an orange pigment found in plants that is converted into vitamin A inside the body. Beta-carotene is also an antioxidant.

FIGURE 7-3

A.

The Antioxidants versus Free Radicals in the Body

A. Free radicals—unstable oxygen molecules—can be formed from sunlight, in cigarette smoke and environmental pollution, and as a result of many normal chemical reactions involving oxygen in the body. These free radicals attack healthy molecules in the body in hopes of stealing an electron to help stabilize themselves, which in turn can cause cell and tissue damage to the body. Free-radical activity can cause damage to the body's enzymes, cell membranes, and nuclear DNA, or it can result in the formation of oxidized LDL-cholesterol in the arteries.

B. The antioxidant team includes vitamin C, vitamin E, the carotenoids (for example, beta-carotene, lutein, zeaxanthin, lycopene), selenium (a trace mineral), and many naturally occurring nonnutrients—called phytochemicals—found in fruits, vegetables, legumes, and whole grains. The antioxidants prevent free radicals from attacking cells and causing damage by neutralizing the free radicals and converting them back into stable oxygen molecules.

SOURCE: Adapted from *Antioxidant Nutrients* (Mount Olive, NJ: BASF Corporation, 1997).

B.

Vitamin C

The antioxidants work to squelch free radicals before they injure the body, as explained in Figure 7-3. Vitamin C helps stop free radicals in their tracks, working with vitamin E to block damaging chain reactions that appear to promote heart disease and cancer. In addition, vitamin C is a powerful scavenger of environmental air pollutants. In fact, the National Academy of Sciences advises smokers to consume an additional 35 milligrams of vitamin C a day as compared to nonsmokers. The more smoke a person inhales, the more free radicals are produced, and the more vitamin C is needed to fight them.

Other Important Roles for Vitamin C

Vitamin C is required for the production and maintenance of **collagen**, the proteinaceous foundation material for the body's connective tissue, including bones, teeth, skin, and tendons. Vitamin C has also been touted as a nutrient that can help fight stress. In times of stress, the body

Antioxidant Vitamins:

- Enhance the immune system
- Limit free-radical formation
- Destroy free radicals and prevent oxidative damage to cells
- Reduce inflammation in blood vessel walls
  - Vitamin C defends body cells and tissues such as skin, lungs, and blood.
  - Vitamin E and beta-carotene defend the body's lipids such as cell membranes and lipoproteins.

collagen (KAHL-uh-jen) the characteristic protein of connective tissue (*kolla* = glue; *gennan* = to produce).

"A little vitamin C ought to clear that up in no time."

Used with permission © V. G. Myers

TABLE 7-4	Vitamin C in Foods
(mg)*	**Sources**
187	Papaya (1)
113	Cantaloupe (½)
75	Orange (1)
74	Kiwi (1)
71	Grapefruit juice (¾ c)
62	Orange juice, fresh (½ c)
60	Green pepper (½ c)
57	Mango (1)
51	Broccoli, cooked (½ c)
48	Brussels sprouts, cooked (½ c)
43	Strawberries (½ c)
36	Pink/red grapefruit (½)
27	Cauliflower, cooked (½ c)
26	Baked potato (1)
22	Bok choy, cooked (½ c)
22	Cabbage, raw (1 c)
22	Tomato, fresh (1)
16	Raspberries (½ c)
10	Asparagus, cooked (½ c)
9	Spinach, cooked (½ c)

*DRI adult: men, 90 mg/day; women, 75 mg/day.

does use more vitamin C than usual because the vitamin is involved in the release of stress hormones. However, the amount of extra vitamin C used as a result of, say, work-related stress or the stress of ending a significant relationship is minuscule and is more than accounted for by a diet that regularly includes vitamin C-rich foods such as oranges and broccoli (see Table 7-4).

Vitamin C also boosts the body's ability to fight infection, and a growing body of research suggests that it may protect against heart disease and certain types of cancer. Vitamin C's potential role as a chronic disease fighter stems from its function as an antioxidant.[3]

Of course, vitamin C is most famous for its long-standing notoriety as a cure for the common cold. Ever since the publication of the controversial book *Vitamin C and the Common Cold*, by the award-winning scientist Linus Pauling, millions of Americans have followed Dr. Pauling's advice and swallowed megadoses of vitamin C. Despite the popularity of vitamin C as a cold remedy, however, many carefully controlled studies have shown that it plays an insignificant role—if any at all—in preventing colds. At best, the nutrient in amounts up to 1 gram per day may shorten the duration of a cold by one day and *slightly* reduce the severity of cold symptoms in some people.[4] To be sure, many people swear by vitamin C, and their belief in the nutrient may be so strong that they experience a placebo effect as a result of their faith in its curative powers.

Sources of Vitamin C in Foods   Vitamin C-rich foods are widely available in the United States and include not only oranges and other citrus fruits but also broccoli, Brussels sprouts, cantaloupe, and strawberries. A single serving of any of those foods provides more than half of the DRI for the vitamin. Potatoes also contribute significant amounts of vitamin C to the American diet because they are eaten so often.

Vitamin C is widespread in the food supply. Still, deficiencies arise in infants who are not given a source of vitamin C, as well as in children and older adults who do not consume adequate amounts of fruits and vegetables.

Vitamin E   Vitamin E performs a key role as a fat-soluble antioxidant in the body. Vitamin E is involved in protecting red blood cells, nerve cells, and the membranes of the lungs, brain, and other organs against damage from pollutants and other environmental hazards. Vitamin E may also protect against heart disease because it can neutralize the free radicals that might otherwise damage the walls of blood vessels and contribute to coronary artery disease.[5] Scientists believe that the vitamin E residing in the fatty cell membranes that surround cells acts as a scavenger of free radicals that enter the area.[6] When vitamin E is absent, the free radicals can attack the vulnerable polyunsaturated fatty acids found within the cell membrane and start a chemical chain reaction that damages the cell membrane and ultimately causes it to break down completely.

Vitamin E helps maintain a healthy immune system by protecting white blood cells. Vitamin E also protects other lipids, such as LDLs, throughout the body and fat-soluble compounds, including Vitamin A. Additionally, vitamin E is important throughout life for normal neuromuscular function.

The cell membranes harbor vitamin E, and a deficiency of the nutrient causes those membranes, particularly those in red blood cells, to rupture and cause a type of

SOURCES OF VITAMIN C
Adult DRI* (RDA) is 75 to 90 mg.

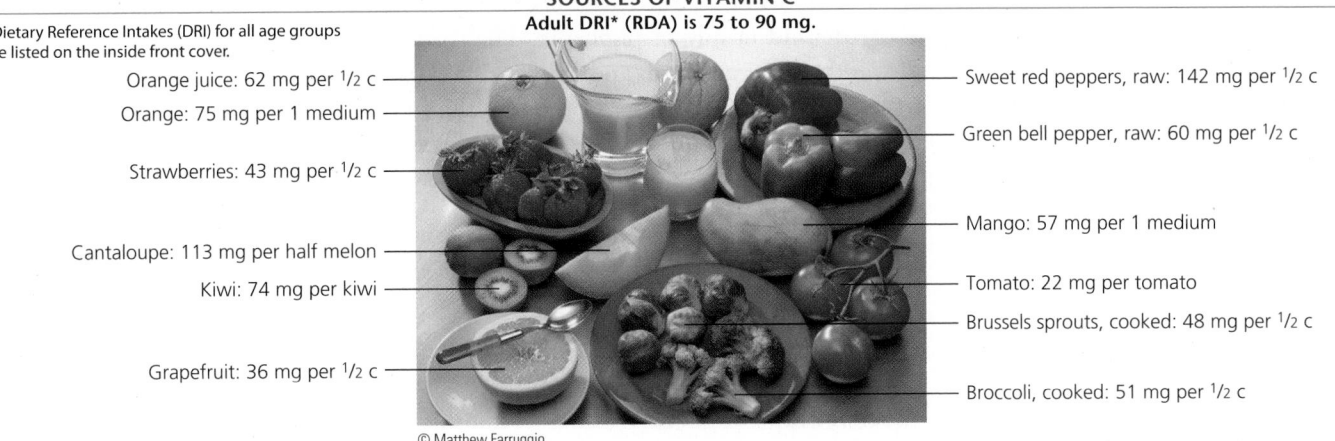

*Dietary Reference Intakes (DRI) for all age groups are listed on the inside front cover.

Orange juice: 62 mg per ½ c
Orange: 75 mg per 1 medium
Strawberries: 43 mg per ½ c
Cantaloupe: 113 mg per half melon
Kiwi: 74 mg per kiwi
Grapefruit: 36 mg per ½ c

Sweet red peppers, raw: 142 mg per ½ c
Green bell pepper, raw: 60 mg per ½ c
Mango: 57 mg per 1 medium
Tomato: 22 mg per tomato
Brussels sprouts, cooked: 48 mg per ½ c
Broccoli, cooked: 51 mg per ½ c

© Matthew Farruggio

anemia. Although extremely rare in healthy adults, this problem sometimes occurs in premature infants who are born before enough vitamin E is transferred to them from their mothers in the last few weeks of pregnancy. Other groups of people who run the risk of deficiency include those who cannot absorb fats as a result of diseases and those with certain blood disorders.

Vitamin E toxicity appears to be rare, occurring only in people who take extremely high doses. Suspected symptoms include alteration of the body's blood-clotting mechanisms and interference with the function of vitamin K.

Sources of Vitamin E in Foods
Vitamin E is widespread in plant foods. The great majority of vitamin E in the diet comes from vegetable oils and products made from them, such as margarine and salad dressings. Soybean, sunflower, canola, and safflower oils contain generous amounts of vitamin E, as do nuts, seeds, and wheat germ. Smaller amounts come from fruits and vegetables. Table 7-5 lists sources of vitamin E.

TABLE 7-5	Vitamin E in Foods
(mg)*	**Sources**
8.9	Almonds (¼ c)
8.5	Sunflower seeds (2 tbsp)
3.0	Peanut butter (2 tbsp)
3.0	Peanuts (1 oz)
2.8	Avocado (1)
2.6	Wheat germ (2 tbsp)
2.3	Mango (1)
1.0	Shrimp (3 oz)
Fats & Oils	
6.0	Safflower oil (1 tbsp)
3.0	Corn oil (1 tbsp)
2.9	Canola oil (1 tbsp)
1.7	Mayonnaise (1 tbsp)
1.7	Olive oil (1 tbsp)
1.6	Peanut oil (1 tbsp)

*DRI adult: 15 mg/day.

SOURCES OF VITAMIN E
Adult DRI (RDA) is 15 mg.

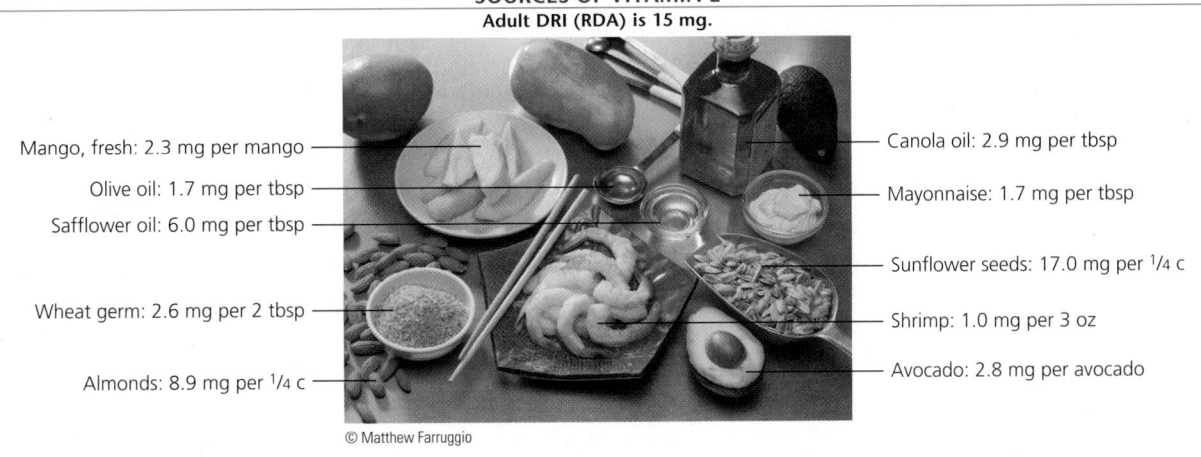

Mango, fresh: 2.3 mg per mango
Olive oil: 1.7 mg per tbsp
Safflower oil: 6.0 mg per tbsp
Wheat germ: 2.6 mg per 2 tbsp
Almonds: 8.9 mg per ¼ c

Canola oil: 2.9 mg per tbsp
Mayonnaise: 1.7 mg per tbsp
Sunflower seeds: 17.0 mg per ¼ c
Shrimp: 1.0 mg per 3 oz
Avocado: 2.8 mg per avocado

© Matthew Farruggio

See this chapter's Spotlight feature for more information about the role of the carotenoids in disease prevention.

For additional perspectives on the use of vitamin and mineral supplements, see the Eat Well Be Well feature later in this chapter on page 214.

carotenoids (kah-ROW-teh-noyds) a group of pigments (yellow, orange, and red) found in foods. (See the discussion on phytochemicals later in the chapter for more about this family of compounds.)

age-related macular degeneration (AMD) oxidative damage to the central portion of the eye (the *macula*) that allows you to focus and see details clearly (peripheral vision remains unimpaired). The carotenoids lutein and zeaxanthin (found in broccoli, Brussels sprouts, kale, spinach, corn, lettuce, and peas) may protect normal macular function.

precursor a compound that can be converted into another compound. For example, beta-carotene is a precursor of vitamin A (*pre* = before; *cursor* = runner, forerunner).

pigment a molecule capable of absorbing certain wavelengths of light. Pigments in the eye permit us to perceive different colors.

retina (RET-in-uh) the paper-thin layer of light-sensitive cells lining the back of the inside of the eye.

retinal (RET-in-al) one of the active forms of vitamin A that functions in the pigments of the eye. Other active forms of vitamin A include retinol and retinoic acid.

night blindness slow recovery of vision following flashes of bright light at night; an early symptom of vitamin A deficiency.

epithelial (ep-ih-THEE-lee-ul) tissue the cells that form the outer surface of the body and line the body cavities and the principal passageways leading to the exterior. Examples include the cornea, digestive tract lining, respiratory tract lining, and skin. The epithelial cells produce mucus to protect these tissues from bacteria and other potentially harmful substances. Without this mucus, infections become more likely.

preformed vitamin A vitamin A in its active form.

retinol one of the active forms of vitamin A.

retinol activity equivalents (RAE) a measure of the amount of retinol the body will derive from a food containing preformed vitamin A or beta-carotene and other vitamin A precursors. Note that some tables list vitamin A in terms of *International Units (IU)*. (See Appendix A for methods of converting from one measure to another.)

The Vitamin A Precursor: Beta Carotene

Beta-carotene is a member of the **carotenoid** family of pigments. The carotenoids possess antioxidant properties and work with vitamins C and E in the body to protect against free-radical damage that leads to chronic diseases of the respiratory tract. Certain carotenoids with antioxidant properties, found in dark-green leafy vegetables such as spinach, kale, collard greens, and Swiss chard, may help prevent **age-related macular degeneration**, as well as lower the risk of cataracts.[7] These carotenoids work by filtering out harmful light rays that could cause free-radical damage to the eye.

Beta-carotene is an orange plant pigment that is a vitamin A **precursor**. Inside the body, beta-carotene is converted into vitamin A; however, this happens so slowly that excess amounts are not stored as vitamin A, but are stored in fat deposits instead.

Other Roles for Vitamin A

Vitamin A is one of the most versatile vitamins, playing roles in several important body processes. The best-known function of vitamin A relates to vision. For a person to see, light reaching the eye must be transformed into nerve impulses that the brain interprets to produce visual images. The *transformers* are molecules of **pigment** in the cells of the **retina**, a paper-thin tissue lining the back of the eye. A portion of each pigment molecule is **retinal**, a compound the body can synthesize only if vitamin A is supplied by the diet in some form. Thus, when vitamin A is deficient, vision is impaired. Specifically, the eye has difficulty adapting to changing light levels. For a person deficient in vitamin A, a flash of bright light at night (after the eye has adapted to darkness) is followed by a prolonged spell of **night blindness**. Because night blindness is easy to diagnose, it aids in the diagnosis of vitamin A deficiency. (Night blindness is only a symptom, however, and it may indicate a condition other than vitamin A deficiency.)

Vitamin A also helps to maintain healthy **epithelial tissue**: skin and the cells (called *epithelial cells*) lining such body cavities as the small intestine and lungs. Vitamin A is also involved in the production of sperm, the normal development of fetuses, the immune response, hearing, taste, and growth.

As much as a year's supply of vitamin A can be stored in the body—90 percent of it in the liver. If you stop eating good food sources of vitamin A, deficiency symptoms will not begin to appear until your stores are depleted. Then, however, the consequences are profound and include blindness and reduced resistance to infection. Although vitamin A deficiency is rarely seen in developed countries such as the United States and Canada, it is a serious public health problem in developing countries, where millions of children suffer from blindness, infections, and the other consequences of vitamin A deficiencies.

Vitamin A toxicity, in contrast, is not nearly as widespread as deficiency. Nevertheless, it can also lead to severe health consequences, including joint pain, dryness of skin, hair loss, irritability, fatigue, headaches, weakness, nausea, and liver damage. Thus, it's especially important not to take megadoses of this nutrient.

Although toxicity poses a hazard to people who take supplements of **preformed vitamin A**, toxicity poses virtually no risk to people who obtain vitamin A from foods in the form of beta-carotene.

Sources of Vitamin A and Beta-Carotene in Foods

Because the body uses both the preformed vitamin A and the beta-carotene in foods to make **retinol**, the amount of vitamin A that comes from foods is usually expressed in **retinol activity equivalents (RAE)**—a measure of the amount of retinol the body will derive from the food. Table 7-6 shows the amount of vitamin A, expressed in RAE, that comes from various foods.

The major sources of vitamin A (in the form of beta-carotene) are almost all brightly colored, in hues of green, yellow, orange, and red. Any plant food with significant vitamin A activity must have some color, because beta-carotene is a rich,

SOURCES OF VITAMIN A AND BETA CAROTENE
Adult DRI (RDA) is 700 to 900 µg RAE.

*Preformed vitamin A
†Beta-carotene

Carrots, cooked: 671 µg† per ½ c

Sweet potato, cooked: 961 µg† per ½ c

Cantaloupe: 444 µg† per half melon

Fortified milk: 150 µg* per c

Mango: 78 µg† per mango

Spinach, cooked: 472 µg† per ½ c

Apricots, dried: 60 µg† per ¼ c

© Matthew Farruggio

TABLE 7-6	Vitamin A in Foods		
(µg RAE)[a]	**Sources**	**(µg RAE)[a]**	**Sources**
6,582	Beef liver (3 oz)[b]	85	Egg, boiled (1)[b]
971	Carrot, fresh (½ c)	78	Mango (1)
961	Sweet potato (½ c)	72	Romaine lettuce (1 c)
472	Spinach, cooked (½ c)	60	Apricot halves, dried (¼ c)
444	Cantaloupe (½)	60	Broccoli, cooked (½ c)
361	Butternut squash (½ c)	53	Watermelon (1 slice)
198	Turnip greens, cooked (½ c)	52	Tomato, fresh (1)
150	Fortified milk (1 c)[b]	29	Tomatoes, cooked (½ c)
109	Bok choy, cooked (½ c)	28	Tomato juice (½ c)

[a] DRI adult: men, 900 µg RAE/day; women, 700 µg RAE/day.
[b] Preformed vitamin A. The rest of the items on the chart derive vitamin A from beta-carotene.

deep-yellow, almost orange color. (Preformed vitamin A is pale yellow.) The dark-green leafy vegetables contain large amounts of the green pigment **chlorophyll**, which masks the carotene in them.

In the United States, about half of the vitamin A consumed in foods comes from fruits and vegetables, and about half of that comes from *dark* leafy greens, such as broccoli and spinach and *rich* yellow or *deep*-orange vegetables, such as winter squash, carrots, and sweet potatoes. The other half of the vitamin A comes from milk, cheese, butter, and other dairy products, eggs, and a few meats, such as liver. When whole milk is processed to produce fat-free milk, the vitamin is lost along with the fat that is skimmed. (Remember that vitamin A is found in the fat.) In the United States and Canada, fat-free milk is fortified with vitamin A to compensate for this loss. Likewise, margarine is fortified to provide the amount of vitamin A typically found in butter. (Milk and margarine are also fortified with vitamin D.)

Vitamins for Healthy Blood

Every cell in the body needs a continuous supply of water, oxygen, energy, and building materials. As the blood, pumped by the heart, travels through the circulatory system, it picks up and delivers materials as needed. Every cell continuously uses up oxygen (producing carbon dioxide) and nutrients (producing waste products). The blood transports oxygen and nutrients to the cells and carries carbon dioxide and waste away from them.

Many vitamins serve blood-related functions in the body. Folate and vitamin B_{12} assist with the formation of new blood cells. Vitamin B_6, is associated with

chlorophyll the green pigment of plants that traps energy from sunlight and uses this energy in photosynthesis (the synthesis of carbohydrate by green plants).

Vitamins Essential for Healthy Blood:

- Folate is needed for formation and maintenance of healthy red blood cells.
- Vitamin B$_{12}$ is needed for formation of healthy red blood cells.
- Vitamin B$_6$ is needed for synthesis of hemoglobin, which carries oxygen in the blood.
- Vitamin K is needed for blood clotting.

This child has spina bifida—a birth defect characterized by the incomplete closing of the casing around the spinal cord. Afflicting some 3,000 U.S. infants born each year, the problem causes partial paralysis.

hemoglobin, the compound that carries oxygen in the blood. Vitamin K is needed for blood clotting.

Folate Folate (also called *folic acid* or *folacin*) is a coenzyme with many functions in the body. It is particularly important in the synthesis of DNA and the formation of red blood cells. A folate deficiency creates misshapen red blood cells that are unable to carry sufficient oxygen to the body's other cells, thereby causing a certain kind of **anemia**. Thus, folate deficiency results in a generalized malaise with many symptoms, including fatigue, diarrhea, irritability, forgetfulness, lack of appetite, and headache. Folate deficiency can easily be confused with general ill health, depressed mood, and dementia in older adults. Folate deficiency may also elevate a person's risk for certain cancers—notably cervical cancer in women and colon cancer.[8]

Folate occurs naturally in fresh green, leafy vegetables (hence its derivation from the word *foliage*), but it is easily lost when foods are overcooked, canned, dehydrated, or otherwise processed. People who are growing rapidly run a high risk of folate deficiency because folate is needed to promote the rapid multiplication of cells that occurs during growth. That's why, for example, the need for folate increases during pregnancy, when the vitamin is needed to support the growth of the fetus.

Folate plays a crucial role in a healthy pregnancy. A growing body of evidence indicates that consuming the recommended amount of folate reduces the risk of bearing a baby with a type of birth defect called **neural tube defect**. Afflicting some 3,000 infants born in the United States each year, neural tube defects include *spina bifida*—the incomplete closing of the bony casing around the spinal cord that causes partial paralysis—and *anencephaly*—a condition in which major parts of the brain are missing. All women of childbearing age are advised to consume the recommended amount of folate, because adequate levels of the nutrient must be ingested before and during the first few weeks of pregnancy—the period during which the neural tube of the embryo is closing, but when most women are not even aware that they are pregnant.

anemia any condition in which the blood is unable to deliver oxygen to the cells of the body. Examples include a shortage or abnormality of the red blood cells. Many nutrient deficiencies and diseases can cause anemia.

neural tube defects malformations of the brain and/or spinal cord during embryonic development.

SOURCES OF FOLATE
Adult DRI (RDA) is 400 µg.

Spinach, fresh: 58 µg per cup
Broccoli, cooked: 104 µg per cup
Lentils, cooked: 179 µg per ½ c
Soy nuts, dry-roasted: 44 µg per 2 tbsp
Black-eyed peas: 179 µg per ½ c

Romaine lettuce: 76 µg per cup
Enriched cereal: 97 µg per cup
Orange juice: 75 µg per cup
Asparagus: 146 µg per ½ c
Cantaloupe: 47 µg per half melon

© Matthew Farruggio

(µg)*	Sources
179	Black-eyed peas (½ c)
179	Lentils, cooked (½ c)
146	Asparagus, cooked (½ c)
113	Spinach, cooked (½ c)
104	Broccoli, cooked (1 c)
100	Oatmeal, instant (½ c)
≥97	Fortified cereal (1 c)
85	Turnip greens, cooked (½ c)
76	Romaine lettuce (1 c)
71	Peanuts (½ c)
65	Kidney beans (½ c)
58	Spinach, fresh (1 c)
55	Lima beans (½ c)
47	Cantaloupe (½)
41	Sunflower seeds (2 tbsp)
40	Orange (1)
27	Cauliflower, cooked (½ c)
19	Tofu (soybean curd) (½ c)
14	Whole-wheat bread (1 slice)
13	Fat-free milk (1 c)
13	Strawberries (½ c)
8	Sirloin steak (3 oz)

TABLE 7-7   Folate in Foods

*DRI adult: 400 µg/day.

When you think of the B vitamin folate, think of fresh produce. Folate derives its name from the word *foliage*, and it is naturally abundant in fresh dark-green vegetables (such as spinach and broccoli), some fresh fruits (oranges and melons), and legumes (lentils and black-eyed peas). Other sources include foods made from grains enriched with folic acid such as breakfast cereals, breads, rice, or pasta.

Sources of Folate in Foods   Although folate is abundant in vegetables, legumes, and seeds, as shown in Table 7-7, most women typically consume foods that provide only about half the 400-microgram amount recommended. For this reason, the Food and Drug Administration has mandated that all enriched grain products be fortified with folic acid, an easily absorbed synthetic form of folate, to improve intakes in the United States.[9]* As a result, the United States has recorded significant declines in neural tube defects since folic acid fortification of all enriched cereal-grain products began. Labels on fortified products may make the health claim "adequate intake of folate has been shown to reduce the risk of neural tube defects." A person can obtain the recommended 400 micrograms of folate by increasing consumption of foods naturally rich in folate (orange juice and green vegetables) and foods fortified with folic acid (breakfast cereals, breads, rice, or pasta), or by taking a folic acid supplement daily.[10] Because high levels of blood folate can mask a true vitamin B_{12} deficiency, total folate intake should not exceed 1 milligram daily.[11]

B Vitamins and Heart Disease   Low intakes of three B vitamins—folate, vitamin B_{12}, and vitamin B_6—are linked with increased risk of fatal heart disease in both men and women.[12] People with low blood levels of these B vitamins tend to have high blood levels of the protein-related compound **homocysteine**.[13] High levels

---

*All enriched cereal-grain products are fortified with 1.4 milligrams of folic acid per 100 grams of food. Recommended folate intakes are stated in micrograms (µg) DFE. A microgram is a thousandth of a milligram. *DFE* stands for Dietary Folate Equivalent, a unit of measure expressing the amount of folate available to the body from naturally occurring food sources. The measure accounts for the differences in absorption between food folate and the more absorbable synthetic folic acid added to foods and supplements. Appendix A offers a conversion factor for calculating DFE.

of homocysteine seem to enhance blood clot formation and damage to arterial walls, as well as raising the risk of suffering a heart attack or stroke as much as fourfold.[14] In addition, scientists suspect that high homocysteine levels may not only be damaging to the cardiovascular system, but may also be toxic for brain tissue and impair cognitive ability.[15] The B vitamins clear homocysteine from the blood and prevent its toxic buildup. The research linking a vitamin B-poor diet with increased homocysteine levels highlights the importance of consuming generous amounts of these nutrients.

Vitamin B₁₂ Vitamin B_{12} maintains the sheaths that surround and protect nerve fibers. The nutrient also works closely with folate, enabling it to manufacture red blood cells. When a B_{12} deficiency is present, folate is unable to do its work of building red blood cells. As a result, a person suffering a lack of vitamin B_{12} ends up with the same sort of anemia seen in people with a folate deficiency, characterized by large, immature red blood cells. Although extra folate will clear up the anemia, it does not take care of the other problems resulting from a B_{12} deficiency: namely, a creeping paralysis of the nerves and muscles that can cause permanent nerve damage if left untreated. Thus, because excess folate can clear up the blood problems that signal an otherwise hard-to-diagnose vitamin B_{12} deficiency, the amount of folate that can be added to enriched foods is limited by law.

To be sure, dietary deficiencies of vitamin B_{12} are unlikely for people who eat animal foods such as meat, milk, cheese, and eggs, all of which supply generous amounts of the nutrient (see Table 7-8). Strict vegetarians who avoid meat, eggs, and dairy products, however, need to find alternative sources of the nutrient, such as vitamin B_{12}-fortified soy beverages, fortified cereals, or B_{12} supplements.

Several other groups of people are also at high risk for vitamin B_{12} deficiency, not because of a lack of the vitamin in their diets, but because of physical conditions that hamper the body's ability to make use of the nutrient. One such group is people who inherit a genetic defect that leaves the body

TABLE 7-8	Vitamin B₁₂ in Foods
(µg)*	**Sources**
14.00	Chicken liver (3 oz)
7.50	Sardines (3 oz)
3.00	Tuna (3 oz)
2.01	Ground beef (3 oz)
2.00	Cottage cheese (1 c)
1.40	Plain nonfat yogurt (1 c)
1.27	Shrimp (3 oz)
1.18	Haddock (3 oz)
0.93	Fat-free milk (1 c)
0.50	Egg (1)
0.36	Cheddar cheese (1.5 oz)
0.29	Chicken breast (½)

*DRI adult: 2.4 µg/day.

homocysteine (hoe-moe-SIS-teen) a chemical that appears to be toxic to the blood vessels of the heart. High blood levels of homocysteine have been associated with low blood levels of vitamin B_{12}, vitamin B_6, and folate.

SOURCES OF VITAMIN B₁₂
Adult DRI (RDA) is 2.4 µg.

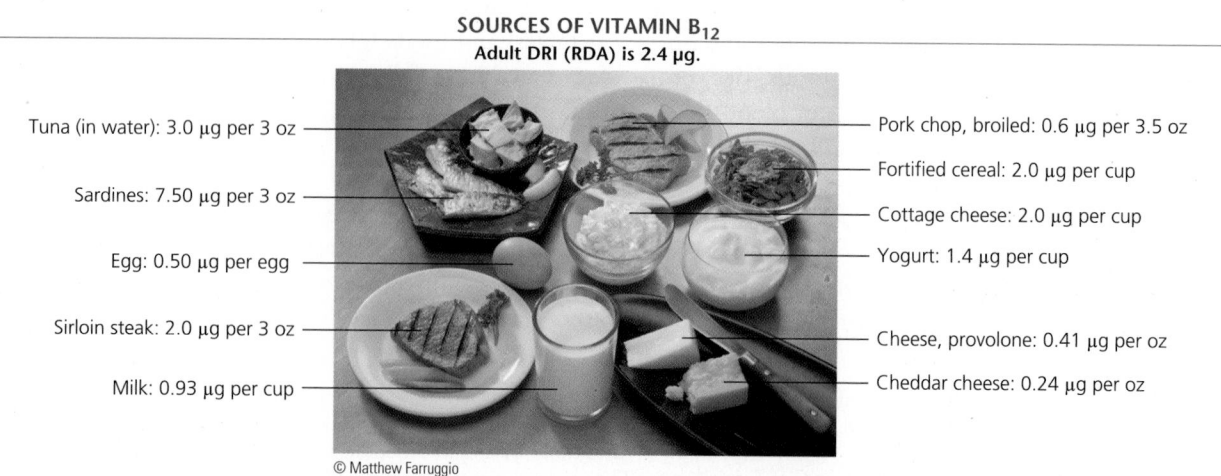

Tuna (in water): 3.0 µg per 3 oz

Sardines: 7.50 µg per 3 oz

Egg: 0.50 µg per egg

Sirloin steak: 2.0 µg per 3 oz

Milk: 0.93 µg per cup

Pork chop, broiled: 0.6 µg per 3.5 oz

Fortified cereal: 2.0 µg per cup

Cottage cheese: 2.0 µg per cup

Yogurt: 1.4 µg per cup

Cheese, provolone: 0.41 µg per oz

Cheddar cheese: 0.24 µg per oz

© Matthew Farruggio

Buying a variety of vitamin-rich fruits and vegetables at the market is one of the first steps to obtaining plenty of those all-important nutrients in your diet. The next step is storing and cooking those foods in ways that minimize the loss of vitamins that can occur as a result of improper storage and preparation. To help get the most from the produce you buy, use the following tips on fruit and vegetable storage and preparation.

- Shop for produce at least once a week. The longer fruits and vegetables stay in your refrigerator before being eaten, the more nutrients are likely to be lost.

- Store fruits and vegetables (other than bananas, tomatoes, and potatoes) in your refrigerator rather than in a fruit bowl or on the kitchen counter. Chilling slows the metabolic rate of the cells of a fruit or vegetable, which in turn causes the cells to use less of their own nutrients. Thus, chilling prevents nutrient depletion.

- Store fruits and vegetables whole, peeling and cutting only what you need immediately before cooking or eating. Once you cut into the skin of an item and expose it to air, vitamin loss begins. After slicing, the vitamin C content of oranges, grapefruits, tomatoes, and strawberries begins to decline. If you do have leftover cut produce, wrap it tightly in airtight plastic or store it in an airtight container inside the refrigerator.

- Try to eat frozen vegetables within a month or two of purchase, because nutrient losses occur over time in stored vegetables.

- Cook vegetables in the least amount of water and for the shortest period of time possible. Water-soluble vitamins readily dissolve into cooking water, and heat destroys some as well. To minimize such losses, steam vegetables over water, or cook vegetables in a microwave oven.

Once you've taken the time to select fresh, vitamin-rich fruits and vegetables, be sure to handle them with care when you get them home.

unable to make a compound known as **intrinsic factor**. Produced in the stomach, intrinsic factor enables the body to absorb and make use of vitamin B_{12}; without the compound, vitamin B_{12} deficiency develops. In this instance, or in the case of stomach damage that interferes with the production of intrinsic factor, people must get vitamin B_{12} injections.

Older adults are also likely to experience vitamin B_{12} deficiencies. An estimated 20 percent of older adults in their 60s and 40 percent of those in their 80s develop **atrophic gastritis**, an age-related condition characterized by the stomach's inability to produce enough acid, which in turn hampers the body's ability to use vitamin B_{12}. In severe cases, the condition also limits the stomach's ability to make intrinsic factor. Vitamin B_{12} deficiencies resulting from atrophic gastritis appear to be easily treated with vitamin B_{12} supplements or injections.[16]

Vitamin B_6   Like the other B vitamins, vitamin B_6 functions as an indispensable coenzyme. For example, vitamin B_6 helps make **hemoglobin** for red blood cells. Vitamin B_6 also plays many roles in protein metabolism. In fact, a person's requirement for vitamin B_6 is proportional to protein intakes. Because vitamin B_6 performs this and so many other tasks, a deficiency causes a multitude of symptoms, including weakness, irritability, and insomnia. Low levels of vitamin B_6 may also weaken the

intrinsic factor a compound made in the stomach that is necessary for the body's absorption of vitamin B_{12}.

atrophic gastritis an age-related condition characterized by the stomach's inability to produce enough acid, which in turn leads to vitamin B_{12} deficiencies.

hemoglobin (HEEM-uh-glow-bin) the oxygen-carrying protein of the blood; found in the red blood cells.

SOURCES OF VITAMIN B₆
Adult DRI (RDA) is 1.3 mg.

Acorn squash, baked: 0.20 mg per ½ c

Watermelon: 0.22 mg per cup

Avocado, California: 0.48 mg per avocado

Banana: 0.66 mg per banana

Tomato juice: 0.34 mg per cup

Salmon: 0.65 mg per 3.5 oz

Potato, baked: 0.70 mg per potato

Oatmeal, cooked: 0.74 mg per cup

Chicken breast: 0.51 mg per 3.5 oz

© Matthew Farruggio

TABLE 7-9	Vitamin B₆ in Foods		
(mg)*	Sources	(mg)*	Sources
0.70	Baked potato (1)	0.30	Tuna (3 oz)
0.66	Banana (1)	0.26	Ground beef (3 oz)
0.65	Salmon (3.5 oz)	0.22	Spinach, cooked (½ c)
0.51	Chicken breast (3.5 oz)	0.22	Watermelon (1 c)
0.42	Figs, dried (10)	0.20	Flounder/sole (3 oz)
0.35	Pork chop (3 oz)	0.20	Soybeans (½ c)
0.31	Cantaloupe (½)	0.10	Fat-free milk (1 c)

*DRI adult (19–50 years): 1.3 mg/day.

body's immune response and increase a person's risk of heart disease. Vitamin B₆ is found in meats, vegetables, and whole-grain cereals, and true vitamin B₆ deficiencies are rare—occurring in some people who eat inadequate diets and whose nutrient needs are higher than usual because of pregnancy, alcohol abuse, some diseases, use of certain prescription drugs, and other unusual circumstances. Table 7-9 shows the vitamin B₆ content of various foods.

Vitamin K   The *K* in vitamin K stands for the Danish word *koagulation* (coagulation or clotting). The key function of vitamin K is its role in the blood-clotting system of the body, where its presence can mean the difference between life and death. It is essential for the synthesis of at least 4 of the 13 proteins involved, along with cal-

SOURCES OF VITAMIN K
Adult DRI is 90 to 120 µg.

Broccoli: 248 µg per ½ c

Cauliflower, raw: 96 µg per ½ c

Strawberries: 23 µg per cup

Chickpeas: 52 µg per ½ c

Soy nuts: 20 µg per ¼ c

Cabbage, raw: 104 µg per cup

Brussels sprouts: 460 µg per ½ c

Milk, whole, 10 µg per cup

Spinach, cooked: 380 µg per ½ c

Eggs: 25 µg per egg

© Matthew Farruggio

TABLE 7-10   Vitamin K in Foods

(μg)*	Sources	(μg)*	Sources
520	Kale, cooked (½ c)	58	Broccoli, raw (½ c)
460	Brussels sprouts (½ c)	52	Chickpeas (½ c)
380	Spinach, cooked (½ c)	30	Sauerkraut (½ c)
364	Turnip greens, cooked (½ c)	25	Soybean oil (1 tbsp)
248	Broccoli, cooked (½ c)	25	Egg (1)
140	Spinach, raw (1 c)	20	Soybeans, dry roasted (¼ c)
104	Cabbage, raw (1 c)	19	Canola oil (1 tbsp)
96	Cauliflower, raw (½ c)	10	Strawberries (½ c)
72	Asparagus, cooked (½ c)	10	Milk, whole (1 c)
60	Lettuce (1 c)		

*DRI adult: 90 to 120 μg/day.

cium, in making the blood clot. When *any* of these blood-clotting factors is absent, blood cannot clot, leaving a person vulnerable to excessive bleeding upon injury.

Vitamin K can be synthesized by the **intestinal flora**—the bacteria that reside in the digestive tract. In addition, as Table 7-10 shows, many foods supply ample amounts of the vitamin; particular standouts are green leafy vegetables and members of the cabbage family.

Because vitamin K is obtained both in the diet and via the intestinal bacteria, deficiencies are rare and occur only under unusual circumstances. Taking antibiotics for an extended period of time, for instance, could kill some of the intestinal bacteria and thereby prompt a deficiency.

Newborn babies are the one group that is commonly susceptible to vitamin K deficiency, because a baby's digestive tract contains no bacteria before birth. After birth, the infant's intestinal tract gradually becomes populated with bacteria, but this happens over time. What's more, baby formula or breast milk generally doesn't contain adequate amounts of vitamin K. Thus, newborns are given a dose of vitamin K to prevent the possibility of a life-threatening hemorrhage in the case of injury.

Vitamin K toxicity is rare, but it can occur when supplemental doses are taken. In particular, adults who must pay attention to the amount of vitamin K in their diet are those who take anticoagulant drugs designed to prevent the blood from clotting and possibly causing a stroke or heart attack. People taking such medications are advised to keep their consumption of vitamin K fairly constant from day to day, because large fluctuations can limit the effectiveness of the anticlotting drugs.[17]

Vitamins for Healthy Bones

Bones are made up of a complex matrix of living tissue based on the protein collagen, into which the crystals of the bone minerals discussed in the next chapter—principally calcium and phosphorus—are deposited. When calcium intakes are low, hormones and vitamin D call upon the bones to release calcium into the blood for use by the rest of the body whenever levels fall too low.* Over time, the lacy network of bones becomes less dense and more fragile as calcium deposits are withdrawn.

Vitamins Essential for Healthy Bones:

- Vitamin D
- Vitamin K
- Vitamin C

---

*Vitamin D and parathyroid hormone circulate in the blood whenever the calcium concentration in the blood falls too low. These hormones cause the release of calcium from the bones, decreased excretion of calcium by the kidneys, and increased absorption of dietary calcium by the intestines. The hormone calcitonin is released from the thyroid gland when levels of blood calcium get too high; it acts to stop the release of calcium from bone and to slow intestinal absorption of the mineral.

intestinal flora the normal bacterial inhabitants of the digestive tract (*flora* = plant inhabitants).

Electron micrograph of healthy trabecular bone.

Electron micrograph of trabecular bone affected by osteoporosis.

Several vitamins are needed for the growth and maintenance of a healthy skeleton. Vitamin D directs a large bone-making and bone-maintenance team composed of several nutrients and other compounds, including vitamin C and vitamin K; hormones; the protein collagen; and several minerals.

Vitamin D

Vitamin D's special role in bone health involves assisting in the absorption of dietary calcium and helping to make calcium and phosphorus available in the blood that bathes the bones, so that these minerals can be deposited as the bones harden. In addition, vitamin D acts very much like a hormone—a compound manufactured by one organ of the body that affects another. Indeed, vitamin D exerts an influence on a number of organs, including the kidneys and the intestines.

Another unique feature of vitamin D is that the body can synthesize it with the help of sunlight, regardless of dietary consumption. Vitamin D is commonly called the "sunshine vitamin" because sunlight can transform a cholesterol compound in skin into a vitamin D precursor, which is absorbed directly into the blood. The liver alters the molecule, and the kidney alters it further to produce the active form of the vitamin. This is why diseases affecting either the liver or the kidneys, which in turn upset vitamin production, may ultimately lead to bone deterioration.

Because the body can make vitamin D with the help of sunlight, you can meet your needs for the nutrient either via sun exposure or through diet.* However, significant amounts of the nutrient come from only a few animal foods—notably eggs, liver, and some fish (see Table 7-11). And, even in these foods, the vitamin D content varies greatly.

Although vitamin D is not prevalent in the diet, adults living in sunny, southern regions may not need to worry about the vitamin D content of the foods they eat, because their bodies are getting plenty of the nutrient as a result of sun exposure. Sun exposure of the face, hands, and arms for just 5 to 15 minutes several times a week is usually all it takes to meet vitamin D needs. However, during the winter (from November through February) people who live in most parts of the country, north of an imaginary line drawn at latitudes above 35 degrees (above northern borders

This child has the bowed legs characteristic of rickets. Worldwide, rickets affects many children who live in poverty and do not have access to sunlight or adequate foods containing vitamin D.

---

*To protect against skin cancer, avoid prolonged exposure to the sun.

SOURCES OF VITAMIN D
Adult DRI (RDA) is 15 to 20 µg (600 to 800 IU).

Sunshine: Sunlight promotes vitamin synthesis in the skin.

Margarine: 0.5 µg per tsp

Eggs: 0.6 µg per egg

Salmon: 4.3 µg per 3 oz

Fortified cereal: ≥ 0.5 µg per cup

Fortified milk: 2.5 µg per cup

Shrimp: 3.0 µg per 3 oz

© Matthew Farruggio

of Alabama and Georgia), are not exposed to enough ultraviolet rays from the sun to synthesize vitamin D. The same holds true for those with very dark skin and housebound or institutionalized older adults—who not only get outside less often than younger people but also tend to be much less efficient at producing vitamin D via the skin/sun synthesis pathway. Others at risk for vitamin D deficiency include people with **osteoporosis** or conditions that interfere with vitamin D absorption. For these people, eating vitamin D-rich foods, such as fortified milk and yogurt, fatty fish (including sardines, herring, mackerel, trout, and salmon), eggs, fortified orange juice, and some fortified cereals, is particularly important.

Some evidence suggests an association between low blood levels of vitamin D and increased risk for certain cancers, inflammation, high blood pressure, and other conditions, but thus far the evidence is inconclusive.[18] However, in an effort to improve bone health, the Institute of Medicine recently issued new recommendations regarding vitamin D intakes for adults and older adults—now recommending 15 micrograms for adults between ages 18 and 70, and 20 micrograms for adults age 71 or older (the 1997 DRIs ranged from 5 to 15 micrograms).[19]

Children who fail to get enough vitamin D characteristically develop bowed legs, which are often the most obvious sign of the deficiency disease rickets. In adults,

TABLE 7-11	Vitamin D in Foods
(µg)*	**Sources**
4.3	Salmon (3 oz)
3.0	Shrimp (3 oz)
2.5	Fat-free milk (1 c)
0.9	Cod liver oil (1 tbsp)
0.6	Egg (1)
0.5	Margarine (1 tsp)
0.5	Fortified cereals

*DRI adult, 18–70 years: 15 µg/day; ≥ 70 years: 20 µg/day.

osteoporosis (OSS-tee-oh-pore-OH-sis) also known as *adult bone loss*; a disease in which the bones become porous and fragile (*osteo* = bones; *poros* = porous).

Brian Goodman/Shutterstock.com

Vitamin D is the sunshine vitamin. A 15-minute walk, on a clear summer day, two to three times per week, can help supply much of your needed vitamin D. Note that although many people can meet at least some of their vitamin D needs through exposure to sunlight, factors such as season, time of day, cloud cover, smog, skin melanin content, and sunscreen can affect ultraviolet radiation exposure and vitamin D synthesis.

vitamin D deficiency causes **osteomalacia**, most often in women whose diets lack calcium, who get little exposure to the sun, and who go through several closely spaced pregnancies and prolonged periods of breastfeeding. Osteomalacia causes the bones, particularly the leg bones and spine, to become soft, porous, and weak.

Although vitamin D deficiency depresses calcium absorption, resulting in low blood calcium levels and abnormal bone development, an excess of vitamin D does just the opposite. It increases calcium absorption, causing abnormally high concentration of the mineral in the blood, which then tends to be deposited in the soft tissues. This is especially likely to happen in the kidneys, resulting in the formation of calcium-containing stones called *kidney stones*. The Tolerable Upper Intake Level for vitamin D is set at 50 micrograms per day.

Vitamin K   Accumulating evidence also supports an active role for vitamin K in the maintenance of bone health. Vitamin K works in conjunction with vitamin D to synthesize a bone protein that helps regulate the calcium levels in the blood. Low levels of vitamin K in the blood have been associated with low bone mineral density, and researchers have noted a lower risk of hip fracture in older women who have optimal intakes of vitamin K than in those who have low intakes of the vitamin (see Table 7-10).[20]

Vitamins and Energy Metabolism

As described in Chapter 3, glycogen can be broken down in the body to glucose, triglycerides to fatty acids and glycerol, and protein to amino acids. When the body needs energy, it breaks any or all of the four basic units—glucose, fatty acids, glycerol, and amino acids—into even smaller units. When the body does not require energy, the end products of digestion (glucose, amino acids, glycerol, and fatty acids) are used to build body compounds such as glycogen, body proteins, or body fat. Many vitamins (as well as minerals) are essential for **energy metabolism**, as described in the following sections.

Thiamin   One of the B vitamins, *thiamin*, acts primarily as a coenzyme in reactions that release energy from carbohydrates. Thiamin also plays a crucial role in processes involving the nerves. Thiamin is so vital to the functioning of the entire body that a deficiency affects the nerves, muscles, heart, and other organs. A severe deficiency, called **beriberi**, causes extreme wasting and loss of muscle tissue, swelling all over the body, enlargement of the heart, irregular heartbeat, and paralysis. Ultimately, the person with beriberi dies from heart failure. A mild thiamin deficiency, in contrast, often mimics other conditions and typically manifests itself as vague, general

As mentioned earlier, vitamin C is required for the production and maintenance of collagen, the proteinaceous foundation material for the body's connective tissue, including bones and teeth.

Vitamins Essential for Producing Energy:

- Thiamin
- Riboflavin
- Niacin
- Folate
- Vitamin B$_6$
- Vitamin B$_{12}$
- Pantothenic acid
- Biotin

osteomalacia (os-tee-oh-muh-LAY-shuh) the disease resulting from vitamin D deficiency in adults and characterized by softening of the bones. (Its counterpart in children is called rickets.) Symptoms include bowed legs and a curved spine.

energy metabolism all the reactions by which the body obtains and expends the energy from food.

beriberi the thiamin deficiency disease, characterized by irregular heartbeat, paralysis, and extreme wasting of muscle tissue.

SOURCES OF THIAMIN
Adult DRI (RDA) is 1.1 to 1.2 mg.

Green peas: 0.23 mg per ½ c

Enriched cereal: 0.36 mg per cup

Sunflower seeds: 0.41 mg per 2 tbsp

Watermelon: 0.23 mg per 1" by 10" wedge

Black beans: 0.21 mg per ½ c

Pork chop, broiled: 0.97 mg per 3 oz

Flour tortilla: 0.38 mg per 10" tortilla

© Matthew Farruggio

symptoms such as stomachaches, headaches, fatigue, restlessness, sleep disturbances, chest pains, fevers, personality changes (aggressiveness and hostility), and neurosis.

Thiamin is found in a wide variety of foods, and virtually no single food will supply your daily needs in a single serving (in the same way that, say, an orange provides a plentiful supply of vitamin C). However, people who eat a balanced diet following the framework of the MyPlate food guide typically consume plenty of thiamin. As Table 7-12 shows, thiamin is found in a variety of meats, legumes, fruits, and vegetables, as well as in all enriched and whole-grain products.

Riboflavin   Like thiamin, the B vitamin *riboflavin* acts as a coenzyme in energy-releasing reactions in the body. In addition, riboflavin helps to prepare fatty acids and amino acids for breakdown. Deficiencies of the vitamin, which are rare, are characterized by severe skin problems, including painful cracks at the corners of the mouth; a red, swollen tongue; and teary or bloodshot eyes.

Table 7-13 shows the riboflavin content of foods. Milk and dairy products contribute a good deal of the riboflavin in many people's diets. Meats are another good source, as are dark-green vegetables such as broccoli. Leafy green vegetables and whole-grain or enriched bread and cereal products also supply a generous amount of riboflavin in most people's diets. Because riboflavin can be destroyed by the ultraviolet rays of the sun or by fluorescent lamps, milk is usually sold in protective cardboard or opaque plastic containers rather than in transparent glass bottles.

Niacin   Like thiamin and riboflavin, the B vitamin *niacin* is part of a coenzyme that is vital to producing energy. Without niacin to form this coenzyme, energy-yielding reactions come to a halt. Over time, a deficiency of niacin leads to the disease pellagra, characterized by diarrhea, dermatitis, and, in severe cases, dementia—a progressive mental deterioration resulting in delirium, mania or depression, and eventually death.

TABLE 7-12	Thiamin in Foods
(mg)*	Sources
0.97	Pork chop (3 oz)
0.41	Sunflower seeds (2 tbsp)
> 0.35	Fortified cereal (1 c)
0.24	Salmon (4 oz)
0.23	Watermelon (1 slice)
0.23	Green peas (½ c)
0.22	Baked potato (1)
0.22	Enriched pasta (½ c)
0.21	Black beans (½ c)
0.21	Peanuts (⅓ c)
0.17	Black-eyed peas (½ c)
0.13	Oatmeal, cooked (½ c)
0.11	Sirloin steak (3 oz)
0.11	Orange (1)
0.11	Wheat bread (1 slice)
0.09	Milk, fat-free (1 c)

*DRI adult: men, 1.2 mg/day; women, 1.1 mg/day.

TABLE 7-13	Riboflavin in Foods
(mg)*	Sources
0.60	Low-fat yogurt (1 c)
0.45	Fat-free milk (1 c)
0.37	Almonds (⅓ c)
> 0.35	Fortified cereal (1 c)
0.24	Pork chop (3 oz)
0.23	Ricotta cheese (½ c)
0.23	Sirloin steak (3 oz)
0.21	Beet greens, cooked (½ c)
0.21	Egg, cooked (1)
0.21	Spinach, cooked (½ c)
0.20	Ground beef (3 oz)
0.17	Cheddar cheese (1.5 oz)
0.16	Turkey (3 oz)
0.11	Asparagus, cooked (½ c)
0.10	Strawberries (1 c)
0.08	Wheat bread (1 slice)

*DRI adult: men, 1.3 mg/day; women; 1.1 mg/day.

SOURCES OF RIBOFLAVIN
Adult DRI (RDA) is 1.1 to 1.3 mg.

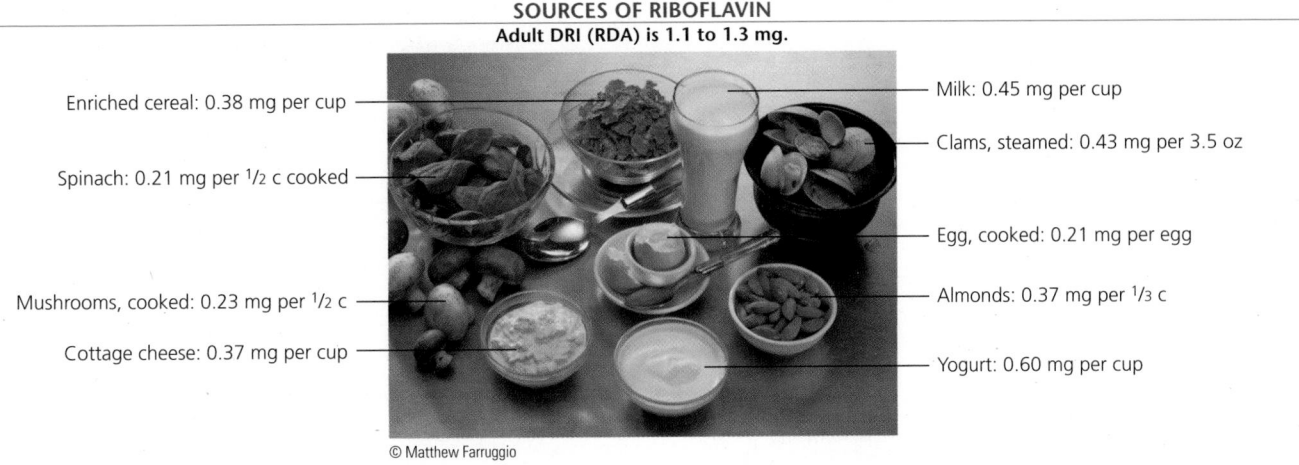

Enriched cereal: 0.38 mg per cup

Spinach: 0.21 mg per ½ c cooked

Mushrooms, cooked: 0.23 mg per ½ c

Cottage cheese: 0.37 mg per cup

Milk: 0.45 mg per cup

Clams, steamed: 0.43 mg per 3.5 oz

Egg, cooked: 0.21 mg per egg

Almonds: 0.37 mg per ⅓ c

Yogurt: 0.60 mg per cup

© Matthew Farruggio

SOURCES OF NIACIN
Adult DRI (RDA) is 14 to 16 mg NE.

Mango: 1.5 mg per 1 medium

Bagel, plain: 3.3 mg per 2.5 oz

Peanuts: 4.93 mg per ¼ c

Tuna (in water): 11.3 mg per 3 oz

Potato, baked: 3.3 mg per potato

Enriched cereal: 4.8 mg per cup

Flour tortilla: 2.6 mg per 10" tortilla

Salmon: 8.0 mg per 3.5 oz

Chicken breast: 8.9 mg per half breast

© Matthew Farruggio

TABLE 7-14	Niacin in Foods
(mg NE)*	**Sources**
11.30	Tuna (3 oz)
8.90	Chicken breast (½)
6.05	Halibut (3 oz)
5.08	Ground beef (3 oz)
4.63	Turkey (3 oz)
4.29	Peanut butter (2 tbsp)
> 4.00	Fortified cereal (1 c)
3.31	Baked potato (1)
3.29	Sirloin steak (3 oz)
1.85	Flounder/sole (3 oz)
1.53	Cantaloupe (½)
1.49	Brown rice, cooked (½ c)
1.13	Wheat bread (1 slice)
0.97	Asparagus, cooked (½ c)
0.89	Broccoli, cooked (½ c)
0.86	Peach (1)

*DRI adult: men, 16 mg NE/day; women, 14 mg NE/day.

Although we can prevent niacin deficiency by eating a diet rich in niacin itself, consuming plenty of protein also staves off the problem. That's because the essential amino acid tryptophan, which is a component of protein, can be converted to niacin in the body. In fact, 60 milligrams of tryptophan yield 1 milligram of niacin. Thus, the DRI for niacin is expressed in niacin equivalents (NEs)—that is, the amount of niacin present in food, including the amount that can theoretically be made from the tryptophan in the food. Milk, eggs, meat, poultry, and fish contribute the bulk of the niacin equivalents consumed by most people, followed by enriched breads and cereals. Table 7-14 shows the niacin content of some common foods.

Diet aside, in recent years, niacin has been increasingly used as a drug-like supplement to help lower cholesterol. Doses ranging from 10 to 15 times the RDA have been shown to reduce "bad" LDL-cholesterol and raise "good" HDL-cholesterol.[21] The hitch, however, is that such high doses of niacin can lead to side effects such as nausea, flushing of the skin, rash, fatigue, and liver damage.

Pantothenic Acid and Biotin   Two other B vitamins—*pantothenic acid* and *biotin*—are needed to synthesize coenzymes that are active in energy metabolism and in a multitude of body systems. Biotin is also required for cell growth, synthesis of DNA (the genetic "blueprint" present in every cell), and maintenance of blood glucose levels. Because both pantothenic acid and biotin are widespread in foods, people who eat a varied diet are not at risk for deficiencies.

SOURCES OF BIOTIN AND PANTOTHENIC ACID*

*Information concerning biotin and pantothenic acid in foods is incomplete: Deficiencies are rare.

Adult DRI (AI):   Biotin: 30 µg
Pantothenic acid: 5 mg

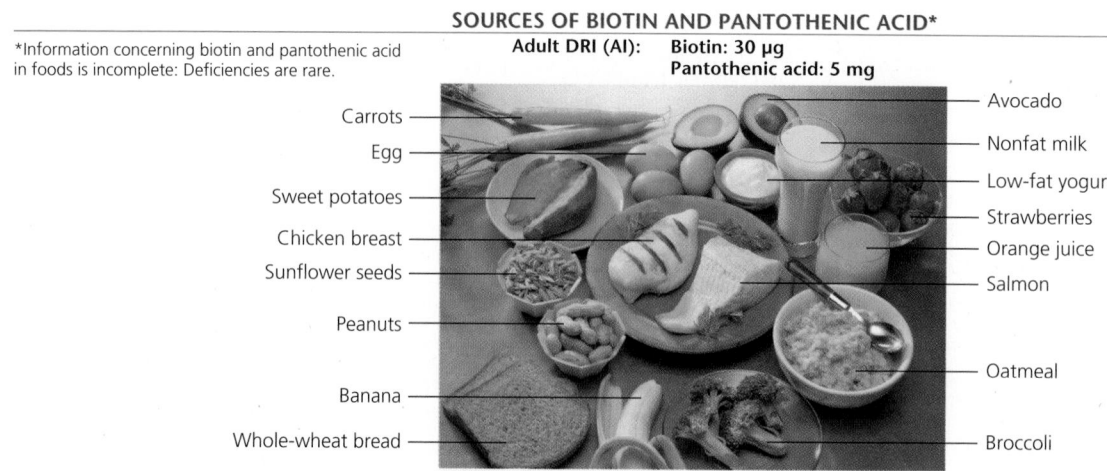

Carrots

Egg

Sweet potatoes

Chicken breast

Sunflower seeds

Peanuts

Banana

Whole-wheat bread

Avocado

Nonfat milk

Low-fat yogurt

Strawberries

Orange juice

Salmon

Oatmeal

Broccoli

© Matthew Farruggio

Vitamin-Like Compounds

In addition to B vitamins, **choline**—a vitamin-like substance—is needed by the body to metabolize homocysteine. Choline is also needed to make lecithin and other molecules—and is considered a "conditionally" essential nutrient because the body becomes unable to make sufficient amounts of choline when fed a choline-free diet.[22] However, because choline is found in milk, eggs, peanuts, and many other foods, deficiencies are rare. The recommended intake for choline is listed in the DRI table on Study Card page 28 at the back of this book.

Nonvitamins   Various other substances have been mistaken as essential nutrients for humans because bacteria or animals need them to live. Sometimes called nonvitamins, some of the substances are important to cell membranes (inositol) and cellular activities (carnitine) but are not essential in human diets because our bodies can make these compounds as needed. Moreover, these substances are abundant in common foods. Research about the role of these substances for humans is ongoing.

Other substances that are sometimes added to vitamin supplements because they are essential for growth in certain nonhuman species are truly nonessential in human diets. These include PABA (para-aminobenzoic acid), the bioflavonoids ("vitamin P" or hesperidin) and ubiquinone (coenzyme Q_{10}), vitamin B_{15} (a hoax), and vitamin B_{17} (laetrile, a falsely touted cancer cure).

Phytonutrients in Foods: The Phytochemical Superstars

One of the newest and most exciting areas of nutrition research today focuses on a class of substances in plant foods called **phytochemicals**, nonnutritive substances in plants that possess health-protective benefits. Phytochemicals are the compounds that give plants their brilliant colors (for example, lycopene, a pigment that makes tomatoes red and watermelon pink) and distinctive aromas (for example, the allium compounds that give us "garlic breath"). These natural compounds also protect plants from the ravages of overexposure to sunlight and other environmental threats and insects.

Most recently, the term *phytochemicals* has been popularized to refer in particular to plant chemicals that may affect health and prevent disease. Of particular interest are phytochemicals in edible plants, including fruits and vegetables, whole grains, legumes, herbs, and seeds. See Table 7-15 for a sampling of the phytochemicals that have been identified to date, as well as common food sources and beneficial effects and attributes of each phytochemical group. In the classic sense, the naturally occurring phytochemicals are not nutrients, for they do not provide energy or building materials. However, ongoing research shows that phytochemicals might perform important functions by acting as powerful antioxidants, decreasing blood pressure and cholesterol, preventing cataracts, reducing menopause symptoms and risk for osteoporosis, and slowing or even reversing certain cancers.

An individual fruit or vegetable may contain many (50 or more) of the thousands of known phytochemicals, but one, or only a select few, are usually present in a large amount. For example, garlic contains more than 160 identified compounds. When a clove of garlic is cut or crushed, it produces sulfur compounds, such as *allicin*, which scientists believe is partly responsible for the health benefits attributed to garlic. Finding the specific chemical in a food that offers the disease-protecting potential is not easy, however. Clinical studies in which people eat foods rich in certain phytochemicals are currently under way. If health benefits, such as protection against cancer, are observed, then the active ingredient in the food must be determined. Is

choline a nonessential nutrient used by the body to synthesize various compounds, including the phospholipid lecithin; the body can make choline from the amino acid methionine.

phytochemicals (FIGHT-oh-KEM-ih-kulz) physiologically active compounds found in plants that are not essential nutrients but appear to help promote health and reduce risk for cancer, heart disease, and other conditions. Also called *phytonutrients* (*phyto* = plant).

TABLE 7-15 **A Sampling of Phytochemicals: Classifications, Food Sources, and Possible Health Benefits**

Phytochemical	Food Sources	Possible Health Benefits
Beta glucan	Oat bran, rolled oats, oat flour	May reduce risk of coronary heart disease
Capsaicin	Hot peppers	May reduce risk of fatal clotting in heart disease
Carotenoids	Deeply colored fruits and vegetables	Act as antioxidants
Beta-carotene	Orange fruits and vegetables (apricots, cantaloupe, carrots, mango, pumpkin, sweet potato, winter squash); dark-green vegetables (spinach, kale, turnip greens)	May help reduce risk of many cancers and strengthen the immune system
Lutein and zeaxanthin	Pumpkin, summer squash, corn, eggs, broccoli; dark-green leafy vegetables, such as kale, collard greens, and spinach	May reduce risk of age-related eye disorders by protecting the retina from harmful ultraviolet radiation and by neutralizing free radicals[a]
Lycopene	Tomatoes, tomato products, watermelon, red grapefruit, red peppers	May help reduce risk of prostate and other cancers
Curcumin	Turmeric, a spice	May inhibit enzymes that activate carcinogens
Flavonoids and other phenols[b]		
Anthocyanin Ellagic acid Resveratrol	Fruits (apples, berries, cherries, citrus, grapes, pears, pomegranates, prunes); whole grains, nuts, chocolate, black or green tea, eggplant, potato peel, red cabbage, beets, celery, peppers, onions, soy, red wine	May act as antioxidants, decrease inflammation, reduce plaque buildup in arteries, increase HDL-cholesterol levels, deactivate carcinogens, and inhibit cancer development
Indoles	Cruciferous vegetables such as broccoli, kale, cauliflower, cabbage, turnip, and Brussels sprouts	May stimulate enzymes that make the hormone estrogen less effective, possibly reducing breast cancer risk
Isothiocyanates (most notably, sulforaphane)	Broccoli, kale, and other cruciferous vegetables; horseradish	May help stimulate protective enzymes that detoxify carcinogens, bolstering the body's natural ability to ward off cancer
Monoterpenes (such as limonene)	Citrus (fruits, juices, peels, oils)	May act as antioxidants and increase production of enzymes that may help the body dispose of carcinogens
Organosulfur compounds (such as allicin)	Onions, garlic, chives, leeks	May block the action of cancer-causing chemicals; may offer heart protection by decreasing production of cholesterol by the liver
Phytic acid	Whole grains	May prevent free-radical formation by binding to minerals and thereby reduce cancer risk
Phytoestrogens (part of flavonoid family)		
Isoflavones (genistein and daidzein)	Soy foods and other legumes	May protect against heart disease by lowering blood cholesterol; may lower risk of breast, ovarian, and other cancers by blocking the action of the hormone estrogen
Lignans	Flaxseed, whole grains	Exhibits estrogen-blocking activity and may lower risk of breast, ovarian, colon, and prostate cancer
Saponins	Sprouts, potatoes, green vegetables, tomatoes, nuts, whole grains, soy foods, legumes	May strengthen the immune system and interfere with DNA replication, preventing cancer cells from multiplying
Tannins	Grapes, red and white wine, tea	Act as antioxidants; may inhibit enzymes that activate carcinogens

[a] Disorders causing vision loss in older adults, including age-related macular degeneration (AMD) and cataracts.
[b] Flavonoids are a subset of phenols. Flavonoids of interest include anthocyanins, catechins, ferulic acid, flavones, glycosides, quercetin, resveratrol, rutin, tangeretin, and nobiletin. The larger class of phenolic phytochemicals includes phenols, ellagic acid, capsaicin (in chili peppers), coumarin, curcumin, and others.

the phytochemical, a vitamin, fiber, or the low-fat content of the increased fruit or vegetable diet responsible for the benefit? Based on current research, any one or a combination of these factors might be responsible.

Mechanisms of Actions of Phytochemicals   Many foods contain numerous phytochemicals, each one acting with or through one or several mechanisms. Some have antioxidant properties (protecting against harmful cell damage), others have anticancer properties (preventing initiation and promotion of cancer), and some have anti-estrogen properties (blocking the action of estrogen and lowering the risk of some cancers).

Different phytochemicals have different modes of action, and one individual phytochemical may exhibit more than one mechanism of action. A phytochemical may influence one or more stages of cancer, from initiation to promotion and progression to a malignant tumor. Phytochemicals act by both *direct* and *indirect* mechanisms. They may act directly to *inhibit* enzymes that activate carcinogens or to *induce* enzymes that detoxify carcinogens. They may act indirectly by stimulating the immune response or scavenging free radicals to prevent DNA damage. See the Spotlight feature in this chapter for a discussion of the role of nutrition in cancer.

Although many phytochemicals act on cells to suppress cancer development, they may also help protect against other diseases, notably heart disease. Some phytochemicals may influence blood pressure and blood clotting, whereas others reduce the synthesis and absorption of cholesterol. Certain pigments (especially carotenoids) in plant foods may protect the eye against free-radical damage and thus prevent or postpone macular degeneration, which can lead to blindness in older adults. One of the most widely studied groups of phytochemicals, the isoflavones found in soy products, is currently being researched for its potential role in preventing diseases such as osteoporosis, heart disease, and various cancers (mainly prostate, breast, and ovarian), and for alleviating symptoms associated with menopause. See the Spotlight feature in Chapter 6 for more about the benefits of soy.

How to Optimize Phytochemicals in a Daily Eating Plan   Research indicates that pure extracts of phytochemicals in supplements are less effective than phytochemicals in whole foods. Some phytochemicals might not be metabolized in pure form, and some might not function by themselves. Thus, phytochemicals have less protective power when ingested as concentrated extracts, such as in pills.

Visit your local farmers' market or grocery store produce aisle for a cornucopia of health-promoting compounds: carotenoids in carrots, spinach, and tomatoes; sulforaphane in broccoli and bok choy, capsaicin in chili peppers. To find the farmers' market nearest you, log on to the USDA website at www.ams.usda.gov/farmersmarkets.

We know that the absorption, metabolism, and distribution of some nutrients are dependent upon the presence of other nutrients. Likewise, it appears that the absorption, metabolism, distribution, and function of phytochemicals also depend on the combination of one phytochemical with other phytochemicals or other substances that occur naturally in food. An individual phytochemical does not necessarily provide protection against disease; rather, the combination of the phytochemical with other phytochemicals or food components provides the protection.

Until more is known about phytochemicals and how they function, it is best to follow the recommendations of the MyPlate food guide and consume the recommended amounts of vegetables and fruits per day along with a variety of whole grains, soy foods, and other legumes, nuts, and seeds. Use the Scorecard feature at the end of the chapter to evaluate your current intake of fruits and vegetables.

For years, health experts have been saying that most healthy people can meet their vitamin and mineral needs with a balanced diet. Nevertheless, almost half of the U.S. population, including about half of young adults and college students, pop vitamin and mineral pills. Americans spend more than $20 billion annually on pills, powders, liquids, and other vitamin and mineral supplements, often in the mistaken belief that such preparations will ensure proper nutrition, help reduce stress, decrease fatigue, and increase pep and energy.

Before you buy a supplement, remember that most major health organizations—from the American Dietetic Association to the American Medical Association to the American Academy of Pediatrics—essentially agree that healthy children and adults should be able to get all the nutrients they need by eating a variety of foods. However, those organizations and other experts say that taking a multivitamin-mineral supplement, under the guidance of a physician or dietitian, may be in order for these particular groups of people:

- People following very low-calorie diets
- People with certain diseases or those taking medications that interfere with appetite, absorption, or excretion of nutrients
- Strict vegetarians, whose diets may fall short in vitamin B_{12}, vitamin D, calcium, iron, and zinc
- Women who are pregnant or breastfeeding, phases that bolster the need for nutrients, including iron and folate
- Women with excessive menstrual bleeding, who may need iron supplements
- Women during their childbearing years who do not consume folate-rich foods or foods fortified with folic acid may need more folate in their diets to prevent neural tube defects in infants
- Anyone with lactose intolerance, or who does not consume milk or other dairy products, needs a source of calcium; those with inadequate exposure to sunlight may also need vitamin D
- Older adults who may have difficulty choosing an adequate diet, chewing problems, or a reduced ability to absorb and metabolize certain nutrients
- People who are recovering from surgery, burn injuries, or other illnesses that increase nutrient needs
- People with heart disease or who are at risk for heart disease and consume diets inadequate in the B vitamins (folate, vitamin B_6, and vitamin B_{12})
- People with chronic diseases of the digestive tract or other conditions that lead to poor intake or deplete nutrient stores
- People with alcohol or other drug addictions are likely to have a shortage of vitamins and minerals in their diets

This symbol on a dietary supplement label helps you know that the supplement actually contains the listed ingredients in the declared amount, does not contain harmful levels of contaminants, will break down and release its ingredients in the body, and has been made under good manufacturing practices. See also http://www.uspverified.org.

If you decide to start taking a supplement, keep the following points in mind when choosing one:

1. Remember that price is not an indication of quality. Many products sold at major retail chains and drugstores are just as high in quality as pricier versions sold in health food stores. Look for a product that meets high standards for manufacturing. One way to do this is to check the label to see whether the product meets USP standards—manufacturing practices set forth by the U.S. Pharmacopeia, the organization that establishes drug standards. The organization's standards require that a supplement be able to disintegrate and dissolve thoroughly in the stomach within a certain period of time, thereby increasing the chances that the nutrients inside will be absorbed and used by the body. Also look for a bottle or package that carries an expiration date. If it doesn't, you run the risk of buying a product that has been sitting on a shelf for a long time. After a while, the product may lose its potency.

2. Look for a supplement that contains both vitamins and minerals, with no more than 100 percent to 150 percent of the recommended Daily Values for each. For the most part, nutrients work in concert with one another, promoting the body's ability to make use of them. Products that include a balanced mix of vitamins and minerals are the best bet for most people. Steer clear of products containing extraneous substances such as PABA, hesperidin, inositol, and bee pollen. These nonvitamin substances have never been proved essential to humans and only add to the price of the supplement. Be wary of taking multivitamins that also contain herbs. Although herbal products are considered to be dietary supplements, the unregulated herbal industry of today is a buyer-beware market.

3. Buy products sold in childproof bottles or packages if you have children around. Vitamins and minerals, especially iron, can be highly toxic to children. Every year, tens of thousands of children swallow excess vitamin-mineral supplements, and iron-tablet overdoses alone are one of the top causes of accidental death in youngsters.

Throughout human history,

people have relied on herbal medicines, and the use of herbs and medicinal plants for any number of ailments is a universal phenomenon. The World Health Organization estimates that about 80 percent of the world's population depends on traditional herbal medicine for primary health care.[23] Only with the development of 20th-century Western medicine have synthetic chemicals found their place in the medical system. Yet, even in a modern pharmacy in the United States, more than 25 percent of medicines are extracted from plants or are synthetic copies or derivatives of plant chemicals.[24]

In the United States, plant medicines composed of whole plants (crude drugs) or complex extracts are sold as dietary supplements because natural (or herbal) medicines are not economically viable candidates for drug research and development. Most botanicals contain one or several relatively dilute compounds, and thus they tend to have milder actions than the more concentrated chemicals found in most drugs. Therefore, herbal medicines usually take longer to act than regular medicinal products, and few herbs have the potency of a prescription drug.

Pharmaceutical companies are less willing to spend the millions of dollars needed to fund research on plants that grow in the wild (and therefore cannot be patented), and most herb manufacturers

© Steve Terrill/Corbis

don't have the funds to support large research studies. Despite this, herbal products are becoming increasingly popular in the United States, and natural food stores are reporting annual growth rates as high as 60 to 80 percent for medicinal herbs in bulk, capsules, extracts, tinctures, tablets, and teas.[25] Since 1990, the use of herbal medicines in the United States has increased by 380 percent, and currently, sales of herbal products are in the range of $25 billion annually.[26] See Table 7-16 on page 217 for a list of herbs thought to be effective and those that should be avoided.

What is driving this trend toward increased use of herbal products? Primarily, the growth is consumer driven. Most consumers learn of herbal products through the media, either in magazines, television, or radio commercials, or by word of mouth from others. Other factors responsible for this trend are an interest in returning to a more natural lifestyle; dissatisfaction with the current state of Western health care; the unwanted side effects of prescription drugs; the spiraling cost and disarray of managed health care; aging baby boomers who want a better quality of health; a strong interest in alternative and complementary therapies; and, finally, a large arena of sales and marketing campaigns, often making use of famous personalities to market herbal products to consumers.

Many herbal product users believe that herbal medicines are the "natural" way to good health. However, *natural* is not always synonymous with *safe*—there are no regulations or oversight agencies for the manufacture and marketing of herbal supplements. Also, when it comes to botanicals, several plant species may look identical, but one may in fact be toxic. If the person collecting the herbs is not entirely knowledgeable, there is a danger that the toxic herb may be mixed with the medicinal herb; such mistakes have been made. A further concern is that consumers have no way of knowing if the product they purchase has an "effective" amount of the active compound.

(Continued)

© Jessica Wecker/Photo Researchers, Inc.

Research is currently under way in the United States to test the safety and efficacy of a few of the most popular herbs on the market today.

In 1994, Congress passed the Dietary Supplement Health Education Act (DSHEA), which severely restricted the FDA's authority over virtually any product labeled "supplement" as long as the product made no claim to affect a disease. The DSHEA allowed herbal medicines to be marketed without prior approval from the FDA. What the DSHEA does allow manufacturers to state on a label is how the product affects a structure or function of the body, such as the claim that the herbal product can "support," "promote," or "maintain" health. The DSHEA states that a product cannot claim that it affects disease, and a manufacturer cannot state on the label that the herbal product will "prevent," "treat," "diagnose," "mitigate," or "cure" disease. A disclaimer must always be included on the label, stating: "This product has not been evaluated by the Food and Drug Administration. This product is not intended to diagnose, treat, cure, or prevent any disease." Therefore, herbal products are not obliged to meet any standards of effectiveness or safety that have been established for other medicines, which require extensive laboratory and clinical trials before the FDA grants approval. Today, a supplement is presumed safe until the FDA receives well-documented reports of adverse reactions.

In October 1998, Congress established the National Center for Complementary and Alternative Medicine (NCCAM), a division of the National Institutes of Health. The center is devoted to conducting and supporting basic and applied research and training, and it disseminates information on complementary and alternative medicine to practitioners and the public. Some of the herbs that are currently undergoing research here in the United States are garlic, St. John's wort, ginkgo biloba, saw palmetto, echinacea, hawthorn, and cranberry.

Until we have further research studies with which to evaluate the safety and efficacy of herbal medicines, physicians, health professionals, and consumers must continue to seek valid information and further education from reliable sources. Use the following guidelines for choosing and using herbal medicines:

- Be informed; seek out unbiased, scientific sources. Inform your physician, especially if taking prescribed medications.

- Do not exceed recommended doses or use herbal medicines for prolonged periods. Call your physician or the FDA Med Alert hotline at 800-332-1088 if you experience adverse effects.

- To find out more about herbal medicines, try the following recommended publications and websites:

  - American Botanical Council, The Complete German Commission E Monographs: Therapeutic Guide to Herbal Medicines

  - S. Foster and V. E. Tyler, Tyler's Honest Herbal: A Sensible Guide to the Use of Herbs and Related Remedies (2000)

  - American Botanical Council, HerbalGram, a peer-reviewed journal

  - American Botanical Council: www.herbalgram.org

  - Herb Research Foundation: www.herbs.org

  - U.S. Food and Drug Administration: www.fda.gov

  - National Institutes of Health/NCCAM: nccam.nih.gov

  - U.S. Pharmacopeia: www.usp.org

TABLE 7-16 The Garden of Herbal Remedies

These herbs are thought to be effective:*

Herb	Why It Is Used	How It Works	Cautions
Black cohosh	Reduces symptoms of premenstrual syndrome, painful menstruation, and the hot flashes associated with menopause.	Appears to function as an estrogen substitute and a suppressor of luteinizing hormone.	Occasional stomach pain or intestinal discomfort. Because no long-term studies have been done, use of black cohosh should be limited to 6 months.
Capsicum	Best known as the hot red peppers cayenne and chili. Applied topically for chronic pain from conditions such as shingles and trigeminal neuralgia.	The active ingredient, capsaicin, works as a counterirritant and decreases sensitivity to pain by depleting substance P, a neurotransmitter that facilitates the transmission of pain impulses to the spinal cord.	Overuse can result in a prolonged insensitivity to pain. More concentrated products can cause a burning sensation. Users must avoid contact with eyes, genitals, and other mucous membranes.
Echinacea	These species of purple coneflower appear to shorten the intensity and duration of colds and flus; may help to control urinary tract infections; when applied topically, may speed the healing of wounds.	Though echinacea lacks direct antibiotic activity, it helps the body muster its own defenses against invading microorganisms.	Experts warn against using echinacea for more than 8 weeks at a time and against its use by people with autoimmune diseases such as multiple sclerosis and rheumatoid arthritis, or by those who are infected with HIV.
Feverfew	The dried leaves of this plant have been shown to reduce the frequency and severity of migraine, as well as its frequently associated symptoms of nausea and vomiting.	The active ingredient, parthenolide, appears to act on the blood vessels of the brain, making them less reactive to certain compounds.	Most commercial preparations recommend doses that are much too high; 250 micrograms a day of parthenolide—or 125 milligrams of the herb—is an adequate dose.
Garlic	Active against viruses, fungi, and parasites. It may also lower cholesterol and inhibit the formation of blood clots, actions that might help to prevent heart attacks.	When fresh garlic is crushed, enzymes convert alliin to allicin, a potent antibiotic. Garlic tablets and capsules containing alliin and the enzyme can be absorbed when dissolved in the intestines rather than the stomach.	More than five cloves of garlic a day can result in heartburn, flatulence, and other gastrointestinal problems. People taking anticoagulants should be cautious about taking garlic.
Ginger	A time-honored remedy for settling an upset stomach, ginger has been shown in clinical studies to prevent motion sickness and nausea following surgery.	Components in the aromatic oil and resin of ginger have been found to strengthen the heart and to promote secretion of saliva and gastric juices.	May prolong postoperative bleeding, aggravate gallstones, and cause heartburn. There is also debate about its safety when used to treat morning sickness.
Ginkgo	Used medicinally in China for hundreds of years, ginkgo biloba was recently reported to improve short-term memory and concentration in people with early Alzheimer's disease.	Appears to work by increasing the brain's tolerance for low levels of oxygen and by enhancing blood flow to the brain and extremities.	Possible side effects include indigestion, headache, and allergic skin reactions.
Milk Thistle	One of the few herbs that has been demonstrated to protect the liver against toxins. Also encourages regeneration of new liver cells.	The seeds contain a compound called silymarin, which helps liver cells keep out toxins and may promote formation of new liver cells.	When used as capsules containing 200 milligrams of concentrated extract (140 milligrams of silymarin), no harmful effects have been reported.

(Continued)

TABLE 7-16   The Garden of Herbal Remedies *(continued)*

Herb	Why It Is Used	How It Works	Cautions
Psyllium	A laxative that doesn't undermine the natural action of the gut. It may reduce the risk of colorectal cancer and reduce blood levels of cholesterol.	The seeds of this herb have husks that are filled with mucilage, a fiber that swells with water in the intestines to add bulk and lubrication to the stool.	Increase in flatulence in some people, especially if a great deal is consumed.
Saw palmetto	Studies on patients with an enlarged prostate have shown that extracts of this palm tree can reduce urinary symptoms even though the gland may not shrink.	Nonhormonal chemicals in saw palmetto appear to work through antiandrogenic and anti-inflammatory activity.	Some experts are concerned that those taking the herb may have inaccurate PSA readings, used as an early warning sign of prostate cancer.
St. John's Wort	Some reports attest to this herb's ability to relieve mild depression. It may also have sedating and antianxiety activity.	Inhibits uptake of serotonin by nerve cells. Though products are standardized for hypericin, hyperforin now appears to be a more potent antidepressant than hypericin.	Based on the sun-induced toxicity of hypericin in animals, people are advised to avoid exposure to bright sunlight.
Valerian	Perhaps best characterized as a mild tranquilizer.	Parts of this plant have antianxiety effects, making it potentially useful in treating nervousness and insomnia.	Long-term use can cause headache, restlessness, sleeplessness, and heart function disorders.

These herbs should be avoided:

Herb	Why It Is Used	Reasons for Caution
Aconite	Pain, rheumatism, headaches	Numerous poisonings in China
Belladonna	Spasms, gastrointestinal pain	Contains three toxic alkaloids, including atropine
Blue Cohosh	Menstrual ailments, worms	Can induce labor
Borage	Coughs, diuretic, mood booster	May contain liver toxins and carcinogens
Broom	Intoxicant, diuretic, heart problems	May slow heart rhythm; contains toxic alkaloids
Chaparral	Arthritis, cancer, pain, colds	Can cause severe hepatitis and liver failure
Comfrey	Cuts, bruises, ulcers	Contains toxins linked to liver disease and death
Ephedra	Stimulant, decongestant	Contains cardiac toxins that have resulted in dozens of deaths
Germander	Digestion, fever	Stimulant can cause heart problem
Kava	Sedative, anxiety, insomnia, restlessness	Possible liver damage; should not be used with alcohol or other depressants, or by pregnant women
Kombucha Tea	AIDS, insomnia, acne	Can cause liver damage, intestinal problems, and death
Lobelia	Mood booster	Can cause rapid heartbeat, coma, and death
Pennyroyal	Stimulant, gastric distress	Liver damage, convulsions, and deaths
Poke Root	Emetic, rheumatism	Extremely toxic; can cause low blood pressure, respiratory depression
Sassafras	Stimulant, sweat producer, syphilis	Contains the carcinogen safrole; banned from use in food
Skullcap	Tranquilizer	Can cause liver damage
Wormwood	Tonic, digestion	Can cause convulsions, loss of consciousness, and hallucinations

*In general, much more controlled research is needed to examine the potential active components, efficacy, and long-term safety of the herbs listed here.

SOURCES: *The Physician's Desk Reference for Herbal Medicines*; V. E. Tyler with S. Foster, Tyler's Honest Herbals; Tyler's Herbs of Choice; *Journal of the American Dietetic Association* (October 1997); The Mayo Clinic Food and Nutrition Center @ www.mayoclinic.com. Copyright © 1999 by The New York Times Co. Reprinted with permission.

Are You Reaping the Power of Produce?

Overwhelming evidence points to the health benefits derived from diets rich in fruits and vegetables, including an enhanced immune system and reduced risk of many chronic conditions, such as heart disease, high blood pressure, certain types of cancer, and age-related vision loss. To achieve the maximum benefits from the vitamins, minerals, antioxidants, and phytochemicals available in produce, aim for a *minimum* of five servings every day. Be adventurous! Select from as wide a variety of fruits and vegetables as possible. Be sure to frequently include members of the "superstar" categories of fruits and vegetables listed here. To evaluate your intake of fruits and vegetables, take the "Fruits and Veggies: More Matters" challenge by answering the following questions and checking the appropriate boxes.

A. Evaluate Your Intake of the Superstar Fruits and Vegetables

	Daily	Occasionally	Seldom	Never
I eat citrus fruits (oranges, grapefruits, etc.).	☐	☐	☐	☐
I eat berries (strawberries, blueberries, etc.), kiwi, or grapes.	☐	☐	☐	☐
I eat cruciferous vegetables such as cabbage, broccoli, cauliflower, Brussels sprouts, and kale.	☐	☐	☐	☐
I eat dark-green leafy vegetables such as spinach, turnip greens, collard greens, mustard greens, or beet greens.	☐	☐	☐	☐
I eat deep-yellow, orange, and red fruits and vegetables such as sweet potatoes, carrots, winter squash, pumpkin, cantaloupe, nectarines, peaches, apricots, mangos, papayas, tomatoes, red peppers, and watermelon.	☐	☐	☐	☐

B. Take the "Fruits and Veggies: More Matters" Challenge

Number of Servings
(each box represents one serving)*

I start the day with a serving of fruit at breakfast.	☐	☐	☐	☐
I snack on fruits or vegetables during the day.	☐	☐	☐	☐
I include a serving of fruit or a vegetable at lunch.	☐	☐	☐	☐
I eat one or more vegetable servings at dinner.	☐	☐	☐	☐
I choose berries, melon, pears, apples, or other fruit for a sweet dessert.	☐	☐	☐	☐

Total Boxes Checked: _____

Scoring for the "Fruits and Veggies: More Matters" Challenge

Fewer than five boxes checked:

Ouch! You're missing out on the health benefits from produce. Visit the *FruitsandVeggiesMatter.gov* website to get tools and information to help you eat more fruits and vegetables each day!

At least five boxes checked:

Good! You're getting the *minimum* number of servings of fruits and vegetables for a healthy diet. Be sure to select a variety of fruits and vegetables and to include servings from the superstar groups as well.

Nine boxes or more checked:

Excellent! You're getting the recommended number of servings for fruits and vegetables in a healthy diet. Be sure to select a variety of fruits and vegetables and to include servings from the superstar groups as well.

*A serving of fruit equals one medium piece of fruit, ½ cup cut or cooked fruit, ¼ cup dried fruit, or ½ cup fruit juice. A serving of vegetables equals 1 cup leafy vegetables, ½ cup cooked or cut vegetables, or ½ cup vegetable juice.

You probably know someone who has cancer, who has recovered from it, or who has died of it. After heart disease, cancer is our most prevalent disease and can be expected, in one form or another, to affect one out of every three Americans living today. Given what is known now about the link between diet and cancer, you are well advised to learn about this connection. Eating pattern is only one of many factors that contribute to the development of cancer (see Figure 7-4), but unlike so many factors in our environment, the food you choose to eat is a factor you can control to a great extent. This discussion attempts to answer questions about the connection between diet and cancer.

How is diet associated with cancer?

Numerous studies, conducted in both laboratory animals and humans, over the past two decades have shown that many connections exist between diet and cancer. Constituents in foods may be responsible for starting the cancer (a process called *initiation*) or for speeding its development (a two-step process that includes tumor *promotion* and *progression*)—or they may protect against cancer (see Figure 7-5). Also, for the person who has cancer, diet can make a crucial difference in recovery by helping to restore body weight and improve nutritional status.

Not all studies have shown a firm relationship between cancer and food and nutrient intake. In particular, some epidemiologic studies—that is, studies of disease rates and food patterns of groups of people—have failed to demonstrate such a link. Even when a positive association seems likely, interpreting data that may link dietary components with cancer or other chronic diseases requires a great deal of caution. Remember that an increase in one component of the diet can cause increases or decreases in others. For example, if a close correlation is shown between cancer and

the consumption of animal protein by a human population, can we be sure that the critical factor is the animal protein? The real factor could be increased fat consumption, because fat is consumed with animal protein. Or the cancer might be caused by what is crowded out of the diet: vitamins, minerals, fiber, or nonnutrients that would otherwise have been supplied by the missing fruits and vegetables.

These issues must be considered when examining the results of studies describing a connection between diet and cancer. Our diets are complex and diverse, making it difficult to separate the effect of a single dietary component from the hundreds of other constituents in foods. In addition, the difficulty of evaluating the diet–cancer link is compounded by the fact that many cancers take up to 20 years to develop. Thus, assessing a cancer patient's diet today is not as helpful as knowing how that patient ate 10, 20, or even 30 years before the cancer was diagnosed.

Is nutrition related to causing cancer the way other environmental factors are, such as smoking or air pollution?

Yes. The National Cancer Institute estimates that more than 85 percent of all cancers are associated with lifestyle and environmental factors, including nutrition. Lifestyle factors include tobacco use, diet, exercise, alcohol consumption, exposure to sunlight, patterns of sexual behavior, and personal hygiene; these can usually be controlled by the individual. In contrast, individuals typically have little control over environmental factors, such as exposure to carcinogens in the workplace, radiation during medical and dental procedures, and contaminants (either naturally occurring or artificially created) present in the soil, air, and water. We also have no control over the genetic factors that contribute to the development of cancer.

Nutrition is a lifestyle factor that may account for as much as 35 percent of all cancer deaths.[27] Thus, researchers take the study

---

FIGURE 7-4

The Impact of Genetic, Environmental, and Lifestyle Risk Factors on Cancer

SOURCES: American Institute for Cancer Research, *Stopping Cancer Before It Starts: The American Institute for Cancer Research's Program for Cancer Prevention* (New York: Golden Books, 1999), 10; *Food, Nutrition, Physical Activity, and the Prevention of Cancer: A Global Perspective* (Washington, DC: World Cancer Research Fund/American Institute for Cancer Research), 2007.

Genetic and related risk factors (Total) **14%**
Family history of cancer 5%
Perinatal factors/growth 5%
Reproductive factors 3%
Prescription drugs/medical procedures 1%

Lifestyle risk factors (Total) **77%**
Tobacco use 30%
Adult diet/obesity 30%
Sedentary lifestyle 5%
Occupation/job-related factors 5%
Alcohol 3%
Socioeconomic status 3%
Salt/food additives, and other preservatives/contaminants 1%

Environmental risk factors (Total) **9%**
Viruses/other biologic agents 5%
Pollution 2%
Ionizing/ultraviolet radiation 2%

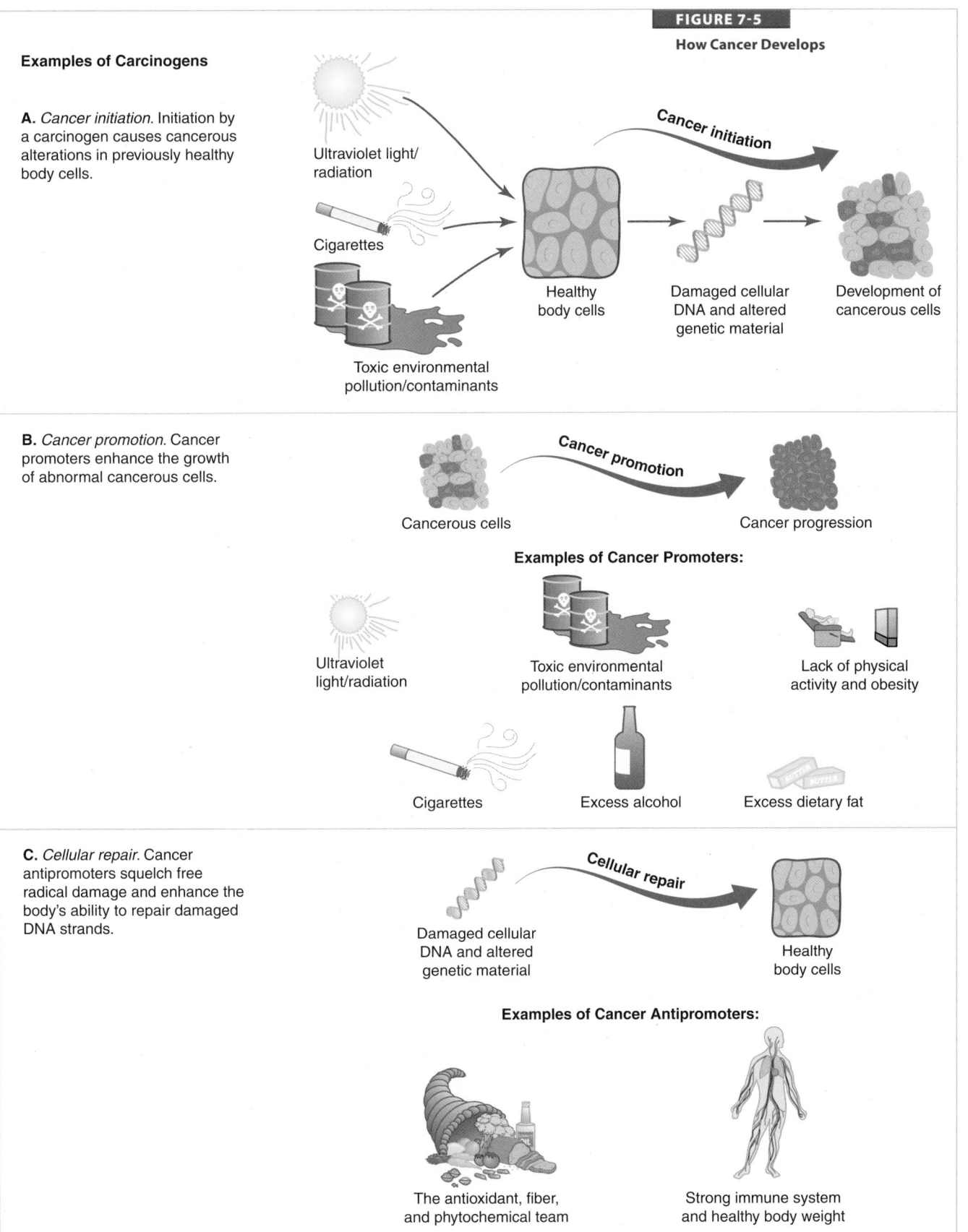

FIGURE 7-5

How Cancer Develops

Examples of Carcinogens

A. *Cancer initiation.* Initiation by a carcinogen causes cancerous alterations in previously healthy body cells.

Ultraviolet light/radiation

Cigarettes

Toxic environmental pollution/contaminants

Cancer initiation

Healthy body cells

Damaged cellular DNA and altered genetic material

Development of cancerous cells

B. *Cancer promotion.* Cancer promoters enhance the growth of abnormal cancerous cells.

Cancer promotion

Cancerous cells

Cancer progression

Examples of Cancer Promoters:

Ultraviolet light/radiation

Toxic environmental pollution/contaminants

Lack of physical activity and obesity

Cigarettes

Excess alcohol

Excess dietary fat

C. *Cellular repair.* Cancer antipromoters squelch free radical damage and enhance the body's ability to repair damaged DNA strands.

Cellular repair

Damaged cellular DNA and altered genetic material

Healthy body cells

Examples of Cancer Antipromoters:

The antioxidant, fiber, and phytochemical team

Strong immune system and healthy body weight

(Continued)

TABLE 7-17	Diet and Cancer
Cancer Site	**Associated with These Diet and Exercise Risk Factors**
Breast	High intakes of calories and alcohol; obesity, low fruit and vegetable intake; low level of physical activity
Colon or Rectum	High red meat and processed meat intake; excessive alcohol; low-fiber diets; low calcium and vitamin D intake; obesity; low level of physical activity
Esophagus	Excessive alcohol intakes; low intakes of vitamins and minerals; obesity
Lung	Low fruit and vegetable intake; low level of physical activity
Ovary and Endometrium	High-fat diets; obesity; low fruit and vegetable intake; low level of physical activity
Prostate	Diets high in saturated fat may promote tumor growth; obesity; low level of physical activity
Pancreas	Low fruit and vegetable intake
Stomach	Regular consumption of smoked foods and foods cured with salt or nitrite compounds; low fruit and vegetable intake
Liver	Ingestion of aflatoxin-contaminated grains; regular consumption of smoked foods and foods cured with salt or nitrite compounds; alcohol abuse
Mouth and Throat	Excessive alcohol intake; low fruit and vegetable intake

SOURCES: *Food, Nutrition, Physical Activity, and the Prevention of Cancer: A Global Perspective* (Washington, DC: World Cancer Research Fund/American Institute for Cancer Research, 2007); and Proceedings of the International Research Conference on Food, Nutrition, and Cancer, *Journal of Nutrition* 132 (2002): 3449S–3534S.

of nutrition and cancer seriously. They are attempting to discover what dietary differences exist between people who do and do not get cancer. In the process, they are trying to identify the various dietary factors that may contribute to or protect against many different types of cancer (see Table 7-17).

What is some of the evidence linking diet and cancer?

Studies of the eating habits of different population groups provide some of the knowledge we already have about the diet–cancer connection. Research involving Seventh-Day Adventists, a group of people with a remarkably low death rate from cancers of all kinds, is particularly telling. This religious group has rules against smoking and using alcohol, discourages the use of hot condiments and spices, and encourages a meatless diet. After cancers linked to smoking and alcohol are discounted, these people still have a cancer mortality rate about one-half to two-thirds that of the rest of the population. This may be due to a low consumption of meat (and therefore *fat*) and to a high consumption of fruits, vegetables, and cereal grains.

When it comes to colon cancer, studies lend weight to the theory that this particular cancer is associated with indicators of affluence, such as a high-fat diet rich in animal protein.[28] In one study, people with colon cancer were compared with a carefully matched set of people without cancer. Those with cancer were found to have a strikingly higher consumption of meat, especially red meat.[29] Another study showed fiber consumption to be lower in people with colon cancer than in comparable people who did

not have cancer. However, another study showed people with colon cancer to be eating about the same amount of fiber as people without cancer, indicating that the exact role of fiber in colon cancer prevention is yet to be determined.[30] Furthermore, among U.S. women between the ages of 34 and 59, the risk of colon cancer was related to their intake of animal (but not vegetable) fat, according to researchers at Harvard Medical School.[31] These examples are just a few of the many studies that have led researchers to the view that a diet high in total fat, especially animal fat, and possibly low in fiber is associated with cancer of the colon.[32] Other results suggest that beta-carotene, vitamin C, calcium, and folate, or possibly other components of the foods that contain them, can help reduce the risk of colon cancer.[33]

So, diet is associated with the development of colon cancer. What about breast cancer?

The case of the connection between diet and breast cancer is a good example of the difficulty of interpreting study results. Laboratory animals, particularly rodents, consistently develop mammary or breast tumors when fed diets high in either vegetable oils (omega-6 fatty acids) or animal fats.[34] Despite the strength of this link in the animal model, similar results have not been seen consistently in human studies.[35]

A group of researchers in Athens, Greece, who studied 120 women with breast cancer and 120 women without cancer, reported no association between breast cancer and the consumption of fats and oils.[36] By comparison, a research group in China

Cruciferous vegetables, such as cauliflower, broccoli, and Brussels sprouts, contain nutrients and nonnutrients that protect against cancer.

reported that women with a high intake of fat and calories and a low intake of vegetables and dietary fiber had an increased risk of breast cancer.[37] As you can see from these apparently conflicting results, the role of dietary fat in breast cancer development remains unclear. Most studies of human populations fail to confirm an association between dietary fat and breast cancer. Laboratory studies, however, show that eating a low-fat diet can reduce estrogen levels and, theoretically, may reduce the risk of breast cancer. Many ongoing studies continue to examine the possible role of dietary fat in breast cancer.

Of course, excess fat in the diet can contribute to overweight, and recent studies have indicated that women who gain excess weight as adults may have an increased risk of breast cancer.[38] Health experts advise us to limit weight gain during adulthood to no more than 10 pounds and to get regular physical activity. In fact, regular physical activity throughout the reproductive years may significantly lower breast cancer risk. Body fat is involved in the production of estrogen. Therefore, by discouraging the accumulation of excess body fat (especially around the abdomen), exercise may decrease the total amount of estrogen a woman is exposed to over her lifetime.

What should consumers do?

The American Cancer Society and the National Cancer Institute have reviewed the evidence independently and have pointed out the following specific concerns.[39]

- *Total calorie intake.* Studies on animals show that reduced food intake reduces cancer incidence at any age, but the evidence is less clear for human beings. Obesity, however, does increase risks for some cancers in both animals and humans.[40]

- *Fat.* Both animal studies and population studies support the view that high fat intake increases the incidence of cancers of the ovaries, colon, and prostate.

- *Protein.* High protein intake may be associated with increased risks of certain kinds of cancer, but the evidence is not yet firm enough to permit a definitive statement.

- *Carbohydrate.* There is little evidence that carbohydrates as such play a role in cancer development.

- *Beta-carotene.* Inadequate intakes of beta-carotene correlate with a high incidence of cancers of the lung, bladder, and larynx; by inference, adequate intakes may help protect against these cancers.[41]

- *Vitamin C.* Vitamin C may help prevent the formation of cancer-causing agents and thereby protect against cancers of the esophagus and stomach.[42]

- *Cruciferous vegetables.* The consumption of cauliflower, cabbage, and broccoli is associated with a reduced incidence of cancer at several sites.[43] *Cruciferous vegetables,* a group of vegetables named for their cross-shaped blossoms, have been shown to protect against cancer in laboratory animals.

- *Other vitamins and minerals.* Bits of evidence suggest that other nutrients may protect against certain types of cancer, but no firm conclusions can yet be made. The effect of the antioxidant nutrients—vitamin C, beta-carotene, vitamin E, and selenium—on cancer risk is an area of active research.[44] The trace mineral selenium, used in the body's production of its own antioxidants, is believed to protect against cancer of the esophagus, stomach, colon, and rectum.[45] Likewise, vitamin E may protect against cancer, particularly cancer of the gastrointestinal tract.[46]

- *Calcium.* Low calcium intake has been associated with increased incidence of colon cancer. Conversely, people who consume more calcium tend to develop less colon cancer.[47]

- *Fiber.* Fiber may help protect against some cancers by, for example, speeding up the passage of all materials through the colon so that its walls are not exposed for long to cancer-causing substances.

Fat is linked to certain cancers, and fiber is associated with cancer prevention. Do vegetarians have a lower incidence of those cancers?

Yes, they do. The Seventh-Day Adventists have already been mentioned. Vegetarian women also have less breast cancer than do women who eat meat.

A number of studies have examined the relationship between cancer and vegetable consumption. Many of them have shown

(Continued)

that people with colon cancer eat vegetables less frequently than do others. One study revealed that people with colon cancer specifically consumed less cabbage, broccoli, and Brussels sprouts than did people free of cancer. Similarly, comparisons of the diets of people who have stomach cancer with those of carefully matched people without cancer showed lower consumption of vegetables in the cancer group: in one case, vegetables in general; in another case, fresh vegetables; in still others, lettuce and other fresh greens or vegetables containing vitamin C. Some of the suspects for causing stomach cancer are chemicals known as nitrosamines, which are produced in the stomach and intestines from nitrites found in foods. Vegetables may help in cancer prevention by contributing vitamin C, which inhibits the conversion of nitrites to nitrosamines.[48]

Another healthy aspect of many plant-based diets is the use of soy foods. A number of studies conducted in China and Japan have shown that consumers of soy foods have lower cancer rates than those who rarely consume these foods (see the Spotlight feature in Chapter 6). The data suggest that consuming one or two servings of soy a day (for example, 1 cup of soy milk or 4 ounces of tofu) may reduce risk of breast, prostate, lung, and colon cancers.[49]

What about alcohol? Is it connected with cancer?

Yes. Environmental causes of head, neck, and esophageal cancer have been studied, and the major factor appears to be the combination of alcohol and tobacco use. However, dietary factors have turned up, pointing to a low intake of fruits and raw vegetables, specifically of the fruits and vegetables that contribute the orange pigment beta-carotene (which converts to vitamin A in the body) and the vitamin riboflavin. Beta-carotene—noted for giving carrots, winter squash, sweet potatoes, apricots, cantaloupe, and other

The National Cancer Institute's National Fruit and Vegetable program urges consumers to eat five to nine servings (or more) of fruits and vegetables rich in fiber, vitamins, minerals, antioxidants, and phytochemicals every day to help protect against cancer.

fruits and vegetables their familiar colors—and its relatives may also be important in reducing the risk of skin cancer.

Researchers are probing the possible association of alcohol consumption with increased risk of breast cancer.[50] Although some studies have shown a modest but significant increase in risk of breast cancer for women who were classified as heavy drinkers (more than four drinks per day), other studies have found no relationship.[51] More research is needed to clarify this issue.

What is the association between beta-carotene and other carotenoids and cancer?

Among the known actions of vitamin A (beta-carotene) are its important roles in maintaining immune function. A strong immune system may be able to prevent cancers from gaining control even after they have been initiated in the body.

The carotenoids are a family of powerful antioxidants found in dark-yellow, orange, and red fruits and vegetables and in dark-green vegetables. They are potent free-radical fighters that may play important roles in preventing cancer. The family includes alpha- and beta-carotene, lycopene, lutein, zeaxanthin, and others. Alpha- and beta-carotene appear to protect against the progression of cancer, whereas other carotenoids may offer more protection at earlier stages of cancer development. Carotenoids may work against cancer by boosting the immune system and supporting the enzymes that detoxify carcinogens. Beta-carotene and lycopene may be effective in preventing damaged cells from proliferating and becoming malignant. Some scientists now believe that lycopene, which is found in red fruits and vegetables such as tomatoes, tomato products, red peppers, and watermelon, is the most powerful of the antioxidant carotenoids. Research is currently under way to examine a link between lycopene and lowered risk

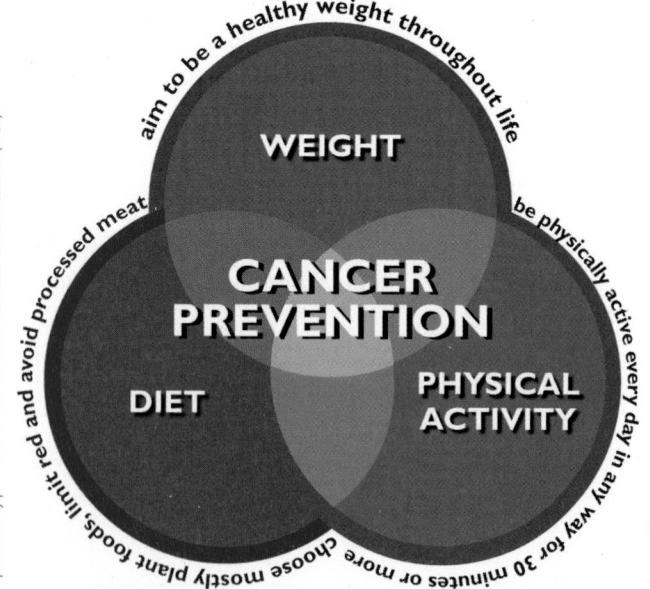

aim to be a healthy weight throughout life

WEIGHT

be physically active every day in any way for 30 minutes or more

CANCER PREVENTION

DIET

PHYSICAL ACTIVITY

choose mostly plant foods, limit red and avoid processed meat of prostate and other cancers.[52] Meanwhile, we are advised to consume five to nine servings of a variety of fruits and vegetables each day. Be sure to include deep-green and brightly colored fruits and vegetables every day. By doing so, your ability to repair free-radical damage in the body is likely to be enhanced by the carotenoids (or possibly by something else) in these plant foods.

Would that "something" in the vegetables be a vitamin?

Not necessarily. Both fruits and vegetables appear to have a protective effect beyond those already discussed for beta-carotene, vitamin C, and fiber. Researchers have identified substances known as *phytochemicals*—naturally occurring plant compounds—in fruits and vegetables that may play a role in decreasing cancer risk and strengthening the immune system.[53] (Refer to Table 7-15 for a sampling of these compounds.) These vegetables also contain folate, a vitamin, which is involved in cell multiplication, and may prove to play an important role in prevention of colon and cervical cancers.[54] The effects of the members of the cabbage family may be due to substances known as *indoles*, a type of phytochemical, which may act by inducing an enzyme in the host that destroys cancer-causing agents.

Do these findings have any implications for the way a person should eat right now, today?

Although clearly there is still much to learn, many experts believe that enough is known to take the first preventive steps. An exhaustive review of some 7,000 previous studies on cancer has led scientists at the American Institute for Cancer Research to release a comprehensive set of recommendations for cancer prevention.[55] To put these recommendations into action, consider these simple steps:

1. Be as lean as possible without becoming underweight. Maintaining a healthy weight is one of the most important things you can do to reduce your risk of cancer. Aim to be at the lower end of the healthy weight range (see Chapter 10).
2. Be physically active for at least 30 minutes every day. Physical activity in any form helps lower cancer risk. Aim to build more activity, like brisk walking, into your daily routine.
3. Avoid sugary drinks. Limit consumption of energy-dense foods (particularly processed foods high in added sugar, low in fiber, or high in solid fat). Choosing healthy foods and drinks instead of those that are high in refined carbohydrates and often in added sugar and fat (energy-dense foods) can help us avoid overweight and obesity and thereby reduce our cancer risk.
4. Eat more of a variety of vegetables, fruits, whole grains, and legumes such as beans. Basing our diets on plant foods, which contain fiber and other nutrients, can reduce our risk of cancer.
5. Limit consumption of red meats (such as beef, pork, and lamb) and avoid processed meats. To reduce your cancer risk, eat no more than 18 ounces (cooked weight) per week of red meats, like beef, pork, and lamb, and avoid processed meat such as ham, bacon, salami, hot dogs, and sausages.
6. If you do drink alcohol, limit your consumption to no more than two drinks a day for men and one drink a day for women.
7. Limit consumption of salty foods and foods processed with salt (sodium). Consuming too much salt can be harmful to our health, increasing our risk of stomach cancer as well as high blood pressure.
8. Don't use supplements to protect against cancer. To reduce your risk of cancer, choose a balanced diet with a variety of foods rather than taking supplements.
9. And always remember—*do not smoke or chew tobacco.*

Your lifestyle choices regarding diet, physical activity, and smoking can be powerful tools in reducing your risk for cancer (see Figure 7-6). Taken together, this advice agrees with most recommendations made to the public for helping prevent heart disease, diabetes, and many other ills, as well as cancer.

Public efforts are now under way to reduce the major risk factors for cancer.[56] The National Fruit and Vegetable Program is a national partnership to increase consumption of fruits and vegetables by all Americans. The *Fruits & Veggies—More Matters* initiative reflects the new dietary guideline to eat more servings of fruits and vegetables a day.[57] Go to www.fruitsandveggiesmatter.gov to find the amount of fruits and vegetables that is right for you each day!

Does it make a difference if I choose to take supplements instead of eating vegetables?

Yes. Fiber, vitamin C, beta-carotene, riboflavin, indoles, and other phytochemicals found in *foods* appear to have preventive effects on cancer development. And because it is obvious that researchers do not have "the answer"—just many partial answers—it is best to stick with foods.[58] Supplements may not contain some as yet unidentified components found in foods that may help to protect against cancer. Also, with vitamin-mineral supplementation, there is always the risk of an excessively high intake. (Recall that vitamin A, selenium, and other vitamins and minerals can be toxic at high doses.) *It is best to rely on foods.* Fruits and vegetables add vitamins, minerals, and fiber to your diet—as well as color and flavor.

Are any anticancer benefits derived from drinking tea?

Scientists note that people in Asia, where large quantities of tea—both green and black—are consumed, have a lower incidence of esophageal, colon, stomach, and other cancers. Laboratory studies suggest that *polyphenols*—phytochemicals having potent antioxidant properties—found in tea may help protect against these

(Continued)

cancers by blocking the formation of carcinogenic compounds in the body. However, the implications for humans are not yet known.

Does cooking food at high temperatures increase the risk of cancer?

Some research suggests that frying or charcoal-broiling meats—especially fatty meats—at very high temperatures creates chemicals on the surface of the food that may increase cancer risk. Preserving meats by methods involving smoke also increases their content of potentially carcinogenic chemicals. We are advised to eat grilled, smoked, or cured foods only occasionally. When you do, consume them with fruits and vegetables that contain protective antioxidant and phytochemical factors. Choose techniques such as braising, poaching, stewing, baking, or roasting instead.

Why should I vary my diet?

The recommendation to eat a varied diet is based on an important cancer-prevention strategy: dilution. The standard advice to eat a variety of foods takes on new meaning in this quote: "The wider the variety of food intake, the greater the number of different chemical substances consumed, and the less is the chance that any one chemical will reach a hazardous level in the diet."[59] In other words, when you add new foods to the diet, you dilute what is in one food with what is in another.

The variety principle traditionally means to eat foods from each of the various food groups. This principle must also be applied within each of the food groups. Don't alternate just between corn and potatoes. Select different vegetables each time you go to the store—broccoli, peas, green beans, carrots, squash, and many others.

Although you cannot control many cancer-causing factors, you can decide which food habits you will keep and which ones you will change. By using these guidelines in making your choices, you will have every reason to feel confident that you are providing your body with the best nutrition at the lowest possible risk. Remember that in the final analysis, your risk of developing cancer can be reduced significantly by not smoking, by consuming alcohol in moderation if at all, and by adopting healthful eating and physical activity habits.

FIGURE 7-6

Factors That Decrease or Increase Cancer Risk*

*From: Cancer Facts and Figures, 2011; available at www.cancer.org.

SOURCE: Adapted from *Food, Nutrition, Physical Activity, and the Prevention of Cancer: A Global Perspective* (Washington, DC: World Cancer Research Fund/American Institute for Cancer Research), 2007.

Cancer Type	Estimated Annual Deaths (2011)*	Vegetables	Fruits	Fiber	Physical activity	Alcohol	Salt and salting	Red meat	Grilled or barbecued (animal foods)	Total and saturated animal fats	Obesity	Smoking tobacco
Lung	156,940											
Colon and rectum	49,380											
Breast	39,970											
Pancreas	37,660											
Prostate	33,720											
Liver	19,590											
Esophagus	14,710											
Kidney	13,120											
Stomach	10,340											
Oral cavity and pharynx	7,900											
Uterus (endometrium)	8,120											

Legend:
- ■ Decreases risk convincingly
- ■ Decreases risk probable
- ■ Decreases risk possible
- ■ Increases risk convincingly
- ■ Increases risk probable
- ■ Increases risk possible

Nutrition on the Web

www.cengage.com/nutrition/boyle/personalnutrition8e

Go to the *Personal Nutrition* site to check for the latest updates to chapter topics or to access links to related websites.

www.nal.usda.gov/fnic

The Food and Nutrition Information Center (FNIC) provides credible and accurate information about vitamins. Browse by Subject to access links:

- Click on Dietary Guidance and then select Dietary Reference Intakes (DRIs).
- Click on Dietary Supplements to access dietary supplement resources.
- Click on Food Composition and then select Resources on Individual Vitamins and Phytonutrients.
- Click on Diet and Disease and then select Cancer.

www.nutrition.gov

This government website provides information and links. Under Browse by Subject, select Dietary Supplements.

www.nap.edu

The National Academies Press contains the latest Dietary Reference Intakes (DRI) reports.

www.healthfinder.gov

Search for reliable consumer information on individual vitamins.

www.iom.edu/fnb

Search for updates regarding new Dietary Reference Intakes at the website for the Institute of Medicine. Search for updates on vitamins.

www.eatright.org

The American Dietetic Association home page includes many nutrition resources and links.

www.fruitsandveggiesmorematters.org

Information about the Fruits & Veggies: More Matters program.

www.fda.gov

The Food and Drug Administration provides information about Dietary Supplements. Search for individual vitamins and minerals.

www.nlm.nih.gov/medlineplus

Search for information about herbal dietary supplements, vitamins, and minerals at this government website.

www.nccam.nih.gov

Search for accurate and reliable information about vitamins, minerals, and dietary supplements at the National Center for Complementary and Alternative Medicine (NCCAM).

www.ods.od.nih.gov/Health_Information/IBIDS.aspx

Search the International Bibliographic Information on Dietary Supplements (IBIDS) database for vitamin, mineral, and dietary supplement information.

www.cancer.gov

Information about cancer, research, and statistics is found at this government website.

www.cancer.org

The American Cancer Society provides information on this website.

www.aicr.org

The American Institute for Cancer Research (AICR) provides information about cancer.

Menu of Online Study Tools

A variety of study tools for this chapter are available at our website to deepen your understanding of chapter concepts. Go to

www.cengage.com/sso

to find

- Practice tests
- Flashcards
- Glossary
- Web links
- Animations
- Chapter summaries, learning objectives, and crossword puzzles.

Diet Analysis+ in Action

Go to the Intake vs. Goals report that shows the results of your three-day tracking and record the vitamins that were less than 75 percent of your recommended nutrient intake. What foods could you add to your diet to increase your intake of these vitamins? Is there a food you could remove to balance out the calorie intake? Now list the vitamins that exceeded 200 percent of your recommended nutrient intake. Go to the Intake Spreadsheet and list the three foods that contributed most to your intake of vitamins. What health effects might your current intake have? Can you adjust your diet to create a better balance of vitamins?

The Minerals and Water: A Functional Approach

Ask Yourself . . .

Which of the following statements about nutrition are true, and which are false? For each false statement, what is true?

1. Water accounts for about 60 percent of an adult's body weight.
2. Milk is necessary for children, but adults can find replacements for it.
3. It is generally harder for women than for men to obtain diets that are adequate in iron.
4. Sodium is bad for the body and should be avoided.
5. When a person becomes deficient in iron, the very first symptom to appear is anemia.
6. Zinc is toxic in excess.
7. Both too little and too much iodine in the diet can cause swelling of the thyroid gland, known as goiter.
8. A diet high in salt is associated with high blood pressure in some individuals.
9. Osteoporosis is a disease that can affect men and women at any age.
10. Bone damage is associated with smoking cigarettes.

Answers found on the following page.

For hundreds, if not thousands, of years, the physical and chemical properties of minerals such as gold, silver, lead, and copper have been known to metallurgists and alchemists. Even so, the role of some minerals in biological processes has been recognized only within the past few hundred years. The discovery of iron in blood, for example, occurred in 1713. The identification of calcium in bone was made in 1771.[1] The role of **minerals** in human nutrition was not fully appreciated until the late 19th century.

The previous chapter described the water-soluble and fat-soluble vitamins, their biological roles, food sources, and human requirements. This chapter discusses the minerals known to be important in human nutrition. In some respects, minerals are similar to vitamins. Like the vitamins, minerals themselves do not contribute energy (calories) to the diet. Of the minerals important in human nutrition, most have diverse functions within the body and work with enzymes to facilitate chemical reactions. Also like the vitamins, most minerals are required in the diet in very small amounts.

In other respects, minerals are different from vitamins. Whereas vitamins are organic compounds, minerals are **inorganic** compounds that occur naturally

The pleasure of eating . . . is of all times, all ages, all conditions. . . . Because it may be enjoyed with other enjoyments, and even console us for their absence. . . . Because its impressions are more durable and more dependent on our will. . . . Because in eating we experience a certain indescribably keen sensation of pleasure, by what we eat we repair the losses we have sustained, and prolong life.

ANTHELME BRILLAT-SAVARIN (1755–1826, FRENCH POLITICIAN AND GOURMET; AUTHOR OF *PHYSIOLOGY OF TASTE*)

minerals naturally occurring, inorganic, chemical elements; serve as structural components of the body and are used in many vital body processes.

inorganic being or composed of matter other than plant or animal.

TABLE 8-1   A Guide to the Minerals

A Guide to the Minerals

Mineral	Best Sources	Chief Roles	Deficiency Symptoms	Toxicity Symptoms
Major Minerals				
Calcium	Milk and milk products, small fish (with bones), tofu, certain green vegetables, legumes, fortified juices	Principal mineral of bones and teeth; involved in muscle contraction and relaxation, nerve function, blood clotting, blood pressure	Stunted growth in children; bone loss (osteoporosis) in adults	Excess calcium is usually excreted except in hormonal imbalance states
Phosphorus	Meat, poultry, fish, dairy products, soft drinks, processed foods	Part of every cell; involved in acid–base balance and energy transfer	Muscle weakness and bone pain (rarely seen)	May cause calcium excretion
Magnesium	Nuts, legumes, whole grains, dark green vegetables, seafood, chocolate, cocoa	Involved in bone mineralization, protein synthesis, enzyme action, normal muscular contraction, nerve transmission	Weakness, confusion, depressed pancreatic hormone secretion, growth failure, hallucinations, muscle spasms	Excess intakes (from overuse of laxatives) has caused low blood pressure, lack of coordination, coma, and death
Sodium	Salt, soy sauce; processed foods such as cured, canned, pickled, and many boxed foods	Helps maintain normal fluid and acid–base balance; nerve impulse transmission	Muscle cramps, mental apathy, loss of appetite	High blood pressure
Chloride	Salt, soy sauce; processed foods	Part of hydrochloric acid found in the stomach, necessary for proper digestion, fluid balance	Growth failure in children, muscle cramps, mental apathy, loss of appetite	Normally harmless (the gas chlorine is a poison but evaporates from water); vomiting
Potassium	All whole foods: meats, milk, fruits, vegetables, grains, legumes	Facilitates many reactions, including protein synthesis, fluid balance, nerve transmission, and contraction of muscles	Muscle weakness, paralysis, confusion; can cause death; accompanies dehydration	Causes muscular weakness; triggers vomiting; if given into a vein, can stop the heart
Sulfur	All protein-containing foods	Component of certain amino acids; part of biotin, thiamin, and insulin	None known; protein deficiency would occur first	Would occur only if sulfur amino acids were eaten in excess; this (in animals) depresses growth

The Major Minerals

Calcium
Phosphorus
Magnesium
Sodium
Chloride
Potassium
Sulfur

The Trace Minerals

Iodine
Iron
Zinc
Copper
Fluoride
Selenium
Chromium
Molybdenum
Manganese

major mineral an essential mineral nutrient found in the human body in amounts greater than 5 grams.

trace mineral an essential mineral nutrient found in the human body in amounts less than 5 grams.

in the earth's crust. Unlike vitamins, some minerals (such as calcium) contribute directly to the building of body structures (such as bone).

As with other areas of research in human nutrition, many questions related to mineral metabolism remain unanswered. Scientists continue to study the biochemical functions of minerals, the mechanisms by which they activate enzymes, the factors that control their levels in blood and tissues, and the ways in which the composition of the diet affects the body's ability to make use of these dietary components. The many complex metabolic interactions of minerals make this a challenging area of research. The discussions about the minerals that follow are organized according to the biological roles these substances play in the body.

The Two Classifications of Minerals

Nutritionists traditionally divide minerals into two large classes: the **major minerals** and the **trace minerals**. The distinction between them is that the major minerals occur in relatively large quantities in the body and are needed in the daily diet in relatively large amounts—on the order of a gram or so each. The trace minerals occur

ASK YOURSELF ANSWERS: **1.** True. **2.** False. Strictly speaking, milk is not absolutely necessary in anyone's diet, but its nutrients are hard to obtain from other foods, and it is recommended for both children and adults. **3.** True. **4.** False. Sodium is an essential nutrient, but excess sodium should be avoided. **5.** False. When a person becomes deficient in iron, one of the last symptoms to appear is anemia; fatigue and weakness appear first. **6.** True. **7.** True. **8.** True. **9.** True. **10.** True.

| TABLE 8-1 | A Guide to the Minerals *(continued)* |

A Guide to the Minerals

Mineral	Best Sources	Chief Roles	Deficiency Symptoms	Toxicity Symptoms
Trace Minerals				
Iodine	Iodized salt, seafood, bread	Part of thyroxine, which regulates metabolism	Goiter, cretinism	Depressed thyroid activity
Iron	Red meats, fish, poultry, shellfish, eggs, legumes, dried fruits, fortified cereals	Hemoglobin formation; part of myoglobin; energy utilization	Anemia: weakness, pallor, headaches, reduced immunity, inability to concentrate, cold intolerance	Iron overload: infections, liver injury, acidosis, shock
Zinc	Protein-containing foods: meats, fish, shellfish, poultry, grains, vegetables	Part of insulin and many enzymes; involved in making genetic material and proteins, immunity, vitamin A transport, taste, wound healing, making sperm, fetal development	Growth failure in children, delayed development of sexual organs, loss of taste, poor wound healing	Fever, nausea, vomiting, diarrhea, kidney failure
Copper	Meats, seafood, nuts, drinking water	Helps make hemoglobin; part of several enzymes	Anemia, bone changes (rare in human beings)	Nausea, vomiting, diarrhea
Fluoride	Drinking water (if fluoride containing or fluoridated), tea, seafood	Formation of bones and teeth; helps make teeth resistant to decay	Susceptibility to tooth decay	Fluorosis (discoloration of teeth); nausea, vomiting, diarrhea
Selenium	Seafood, meats, grains, vegetables (depending on soil conditions)	Helps protect body compounds from oxidation; works with vitamin E	Fragile red blood cells, cataracts, growth failure, heart damage	Nausea, abdominal pain; nail and hair changes; liver and nerve damage
Chromium	Meats, unrefined foods, vegetable oils	Associated with insulin needed for release of energy from glucose	Abnormal glucose metabolism	Occupational exposures damage skin and kidneys
Molybdenum	Legumes, cereals, organ meats	Facilitates, with enzymes, many cell processes	Unknown	Enzyme inhibition
Manganese	Widely distributed in foods	Facilitates, with enzymes, many cell processes	In animals: poor growth, nervous system disorders, abnormal reproduction	Poisoning, nervous system disorders

in the body in minute quantities and are needed in smaller amounts in the daily diet. Table 8-1 lists the minerals known to be essential in human nutrition, and Figure 8-1 offers tips for locating the minerals in the MyPlate food guide.

If you could extract all of the trace minerals from the body, you would obtain only a bit of dust, hardly enough to fill a teaspoon. As tiny as their quantities are, though, each of the trace minerals performs several vital roles for which no substitute will do. A deficiency of any of them may be fatal, and an excess of many can be equally deadly.

FIGURE 8-1

Good Sources of Minerals in the USDA MyPlate food guide[a]

[a] Serving sizes shown here are based on a 2,000-calorie diet. Go to www.ChooseMyPlate.gov for a quick estimate of how much food you should eat from the different food groups based on your age, gender, and activity level.
[b] In whole-grain choices.

FRUITS
2 cups
potassium, magnesium, iron

VEGETABLES
2½ cups
iron, magnesium, selenium, calcium, potassium, sodium

DAIRY 3 cups
calcium, phosphorus, magnesium, sodium, zinc, potassium, iodine

GRAINS 6 oz
chromium[b], iron, zinc, magnesium[b], sodium, sulfur, chloride, iodine, copper[b], potassium[b], phosphorus

PROTEIN FOODS 5½ oz
chromium, iron, copper, sulfur, iodine, zinc, phosphorus, magnesium, sodium, potassium, fluoride

Minerals for Healthy Bones

Bones are made up of a complex matrix of living tissue based on the protein collagen, into which the crystals of the bone minerals—principally calcium and phosphorus—are deposited. Bone tissue has two forms: **trabecular bone** and **cortical bone**. The thick, ivory-like outer portion of a bone is the cortical bone. Cortical bone provides a covering for the inner trabecular bone, which is a lacy network of calcium-containing crystals that are almost spongelike in appearance (see Figure 8-2). When calcium intakes are low, hormones and vitamin D call first upon the trabecular bone to release calcium into the blood for use by the rest of the body whenever levels fall too low.* Over time, the lacy network of bones becomes less dense and more fragile as calcium deposits are withdrawn (see the photos on page 206).

trabecular (tra-BECK-you-lar) bone the lacy inner network of calcium-containing crystals; spongelike in appearance, it supports the bone's structure.

cortical bone the dense outer ivory-like layer of bone that provides an exterior shell over trabecular bone.

FIGURE 8-2

A Bone's Life

A. Bone is living tissue that continuously remodels itself. Bone contains cells embedded in a hard, mineralized matrix. The osteoblast cells make new bone; the osteoclast cells break down old or damaged bone.

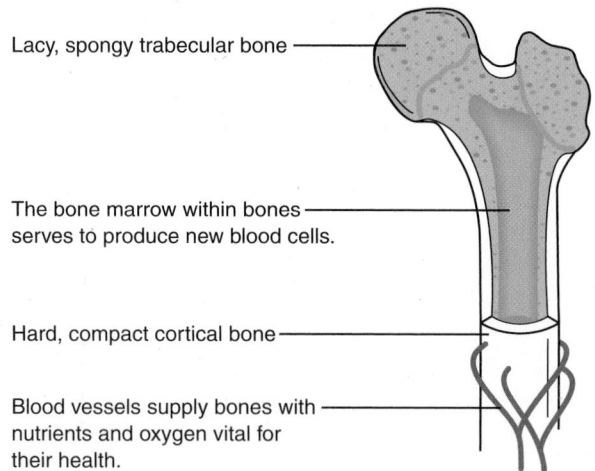

Lacy, spongy trabecular bone

The bone marrow within bones serves to produce new blood cells.

Hard, compact cortical bone

Blood vessels supply bones with nutrients and oxygen vital for their health.

Cross section of bone. The lacy structural elements are trabeculae (tra-BECK-you-le), which can be drawn on to replenish blood calcium.

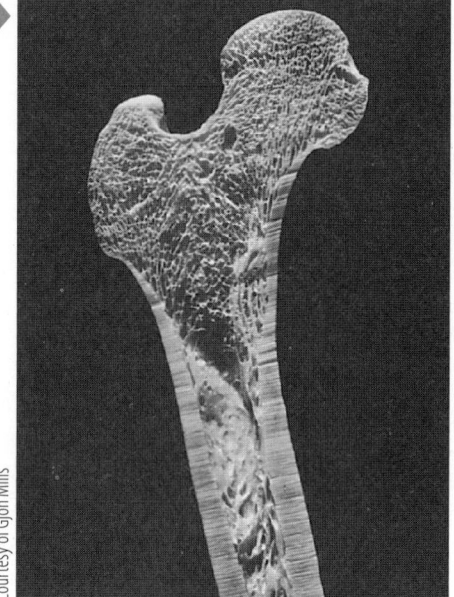

Courtesy of Gjon Mills

B. Peak bone mass occurs at approximately 30 years of age. Afterward, bone loss starts to outpace bone deposition; at menopause there is a surge of calcium out of the bones.

SOURCE: Adapted from The Stay Well Company, San Bruno, CA. © 1998.

Bone mass

Bone mass increases rapidly during the growth years.

Peak bone mass

Menopause
Menopause leads to increased bone loss due to a lack of estrogen.

When bone mass stays high, bones resist breaking.

As bone mass decreases, risk of fracture increases.

Age 5 10 15 20 25 30 35 40 45 50 55 60 65 70 75 80 85 90

---

*As mentioned in Chapter 7, vitamin D and parathyroid hormone circulate in the blood whenever the calcium concentration in the blood falls too low. These hormones cause the release of calcium from the bones, decreased excretion of calcium by the kidneys, and increased absorption of dietary calcium by the intestines. The hormone calcitonin is released from the thyroid gland when levels of blood calcium get too high; it acts to stop the release of calcium from bone and to slow intestinal absorption of the mineral.

Although bones undergo remodeling throughout life—adding and losing bone minerals—the total amount of bone mass in the human body reaches a peak by about age 30.[2] Afterward, bones lose strength and density as bone minerals are lost. The principal determinant of bone health is *peak bone mass.* Think of bone mass as money in the bank: The larger the savings account, the easier it will be to withstand some loss of bone as one ages. To attain a healthful peak bone mass, it is necessary to have an optimal intake of calcium during the years of bone growth, all through the growing years and on into the years of young adulthood.[3]

Besides calcium, several other vitamins and minerals are needed for the growth and maintenance of a healthy skeleton (see the margin list). As discussed in Chapter 7, vitamin D directs a large bone-making and bone-maintenance team composed of several nutrients and other compounds, including vitamin C and vitamin K; hormones; the protein collagen; and the minerals calcium, phosphorus, magnesium, and fluoride.

Calcium

Calcium is the most abundant mineral in the body. Ninety-nine percent of the body's calcium is found in the bones, which play two important roles. First, the bones support and protect the body's soft tissues. Second, they serve as a calcium bank, providing calcium to the body fluids whenever the supply runs low.

Although only a small part (about 1 percent) of the body's calcium is in its fluids, circulating calcium is vital to life. Calcium is required for the transmission of nerve impulses. It is essential for muscle contraction and thus helps maintain the heartbeat. It appears to be essential for the integrity of cell membranes and for the maintenance of normal blood pressure. Calcium must also be present if blood clotting is to occur, and it is a **cofactor** for several enzymes.

Everyone knows that children need calcium daily to support the growth of their bones and teeth, but not everyone is aware of adults' needs for daily intakes of calcium. Abundant evidence now supports the importance of calcium for adults, especially women, who need about as much calcium in their later years as they did when they were adolescents.[4] A calcium deficit during the growing years and in adulthood contributes to gradual bone loss—**osteoporosis**, which can totally cripple a person in later life. This chapter's Spotlight feature describes the many factors that lead to osteoporosis.

Other nutrients are also important to bone growth and maintenance. Fluoride and vitamin D deficiencies, like calcium deficiencies, can cause loss of bone density. So can heredity, abnormal hormone levels, alcohol, prescription medications, other drugs, and lack of exercise (especially weight-bearing exercises)—but dietary calcium is one of the most important factors.

Table 8-2 shows that calcium appears almost exclusively in three classes of foods: milk and milk products; green vegetables such as broccoli, kale, bok choy, collards, and turnip greens; and a few fish and shellfish. Milk and milk products typically contain the most calcium per serving. Many greens are also good choices, but a complication enters in—absorption. It is not clear to what extent calcium is absorbed from certain green vegetables (notably, spinach and Swiss chard), whereas calcium is known to be very well absorbed from milk. Milk contains both vitamin D and lactose, both of which enhance calcium absorption and promote bone health. Milk and milk products also normally supply about 40 percent of people's intake of riboflavin. Figure 8-3 shows the DRI committee's current recommendations for calcium intakes.

Some food components interfere with calcium absorption. Phytate is one example; it occurs in some fruits, vegetables, and whole grains, as well as in nonherbal tea. Oxalate is found in beet greens, rhubarb, and spinach, among other foods. These **binders** seem to depress absorption of the calcium present in the *same* food as the binder but do not depress the absorption of calcium from other foods consumed

Minerals for Healthy Bones

- Calcium
- Phosphorus
- Magnesium
- Fluoride

TABLE 8-2	Calcium in Foods
(mg)[a]	Sources
448	Yogurt, plain, low-fat (1 c)
408	Swiss cheese (½ oz)
350	Orange juice, calcium-fortified (1 c)
348	American cheese (2 oz)
345	Yogurt with fruit (1 c)
325	Sardines (with bones) (3 oz)
316	Fat-free milk (1 c)
306	Cheddar cheese (1½ oz)
300	Parmesan cheese (1 oz)
300	Rice drink, calcium-fortified (1 c)
275	Shrimp (3 oz)
250	Pizza (1 slice)
248	Frozen yogurt (1 c)
> 200	Soy milk, calcium-fortified (1 c)
> 200	Cereal and snack bars, calcium-fortified (1 bar)
197	Turnip greens, cooked (1 c)
186	Cream soup (1 c)
182	Salmon (with bones) (3 oz)
180	Kale, cooked (1 c)
179	Collard greens, cooked (½ c)
165	Instant oatmeal (1 packet)
160	Bok choy (1 c)
154	Cottage cheese (1 c)
144	Pudding, chocolate (½ c)
130[b]	Tofu (½ c)
126	Almonds (⅓ c)
122	Waffle (1)
94	Broccoli (1 c)
87	Ice cream (½ c)
52	Tortilla, corn (1)
50	Dried beans, cooked (½ c)

[a] DRI adult: 1,000 mg/day; > 50 years (women): 1,200 mg/day; > 70 years: 1,200 mg/day.
[b] The calcium content of tofu varies depending on processing methods. Look for tofu processed with calcium salts.

cofactor a mineral element that, like a coenzyme, works with an enzyme to facilitate a chemical reaction.

osteoporosis (OSS-tee-oh-pore-OH-sis) also known as *adult bone loss;* a disease in which the bones become porous and fragile (*osteo* = bones; *poros* = porous).

binders in foods, chemical compounds that can combine with nutrients (especially minerals) to form complexes the body cannot absorb. For example, *phytates* (FYE-tates) in whole grains and *oxalates* (in vegetables) may bind iron and calcium and interfere with their absorption to a minor extent.

FIGURE 8-3

Calcium Recommendations*

*In general, 1 cup of milk or yogurt, 1½ ounces of natural cheese, or 2 ounces of processed cheese can be considered as 1 cup from the milk group. Equivalent choices include nondairy servings of beverages or food such as calcium-fortified juices that provide a similar amount of calcium as found in a 1 cup serving of milk.

†The USDA MyPlate food guide sets a minimum of 3 cups of milk each day for a 2,000-calorie diet and 2 cups per day for children under 8 years of age.

Age (years)	Calcium needed (mg) per day	Number of milk or milk product servings per day (or the equivalent)†
1–3	700	
4–8	1,000	
9–18	1,300	
19–50	1,000	
51+	1,200	

at the same time. Because fiber in general seems to hinder calcium absorption, the higher your diet is in fiber, the higher it should also be in calcium. This fact doesn't detract from the overall value of high-fiber foods, which are important and nutritious for many reasons.

Protein also affects calcium status by affecting excretion. The higher the diet is in protein, the greater the amount of calcium excreted. This is why people in the United States and Canada are told to ingest more calcium than people in countries whose protein intakes are lower.

Alternative Sources of Calcium   Milk need not be your only source of calcium. Yogurt is an acceptable substitute for regular milk, and puddings, custards, and baked goods can be prepared in such a way that they also contain appreciable amounts of milk. Powdered nonfat milk, which is an excellent and inexpensive source of protein, calcium, and other nutrients, can be added to many foods (such as cookies, soups, casseroles, and meatloaf) during preparation. Nonfat yogurt fortified with extra milk solids is another excellent calcium source. Also, small fish, such as Atlantic sardines or canned pink salmon with soft edible bones, are similar to milk and milk products in calcium richness.

A number of calcium-fortified foods are available, including calcium-fortified juices, fruit drinks, soy drinks, breads, cereals, waffles, snack bars, and hot cocoa mixes. Many times, these foods are fortified to match or exceed the amount of calcium in a cup of milk. For example, an 8-ounce serving of fortified orange juice provides about 350 milligrams of calcium. Remember to check the Nutrition Facts panel on food labels for the amount of calcium contained in the foods you eat.

Got milk? If not, be sure to consume the recommended amount of calcium—the amount found in three to four glasses of milk—from the many alternate sources of calcium available on the market today (see Table 8-2).

Advertising Archives

Adults need between 1,000 and 1,200 milligrams of calcium daily—the amount found in 3 to 4 cups of milk or the equivalent. To estimate your calcium intake, answer these questions by considering what you ate yesterday.

1. Did you eat plain yogurt (low-fat or fat-free)? Give yourself *4 points* for each 1-cup serving. _____

2. Did you eat or drink any of these items? If so, give yourself *3 points* for each serving.
- Calcium-fortified juice, 1 cup _____
- Calcium-fortified soy or rice beverage, 1 cup _____
- Firm cheese (cheddar, Swiss, mozzarella), $1\frac{1}{2}$ ounces _____
- Fruit-flavored regular or custard-style low-fat yogurt, 1 cup _____
- Milk (non-fat, 1%, 2%, whole, chocolate), 1 cup _____
- Nonfat milk powder, 1/3 cup _____
- Ricotta cheese, part skim, 1/2 cup _____
- Yogurt beverage or yogurt smoothie, 1 cup _____

3. Did you eat or drink any of these items? If so, give yourself *2.5 points* for each serving.
- Calcium-fortified cereal (1 cup) with $\frac{1}{2}$ cup of milk _____
- Canned fish with bones (salmon, mackerel), 4 ounces _____
- Lattes or mochas, made with milk, 12 ounces ($1\frac{1}{2}$ cup) _____
- Milkshake, any flavor, 8 ounces (1 cup) _____
- Sardines, 3 ounces ($\frac{1}{2}$ can) _____

4. Did you eat any of these items? If so, give yourself *1.5 points* for each serving.
- Baked beans, white beans, soybeans, 1 cup _____
- Bok choy Chinese cabbage, cooked, 1 cup _____
- Cereal (1 cup) with $\frac{1}{2}$ cup of milk _____
- Collard greens, spinach, $\frac{1}{2}$ cup, cooked _____
- Custard, flan, pudding made with milk, $\frac{1}{2}$ cup _____
- Processed cheese, string cheese, 1 ounce _____
- Soft cheeses (feta, Brie, Camembert), $1\frac{1}{2}$ ounces _____
- Soup made with milk, 1 cup _____
- Tofu processed with calcium, 3 ounces _____

5. Did you eat any of these items? If so, give yourself *1 point* for each serving.
- Cottage cheese, 1% or 2%, $\frac{3}{4}$ cup _____
- Ice cream, ice milk, or frozen yogurt, $\frac{1}{2}$ cup _____

6. Did you eat any of these items? If so, give yourself $\frac{1}{2}$ *point* for each serving.
- Almonds, $\frac{1}{4}$ cup _____
- Broccoli or kale, 1 cup _____
- Hummus, $\frac{1}{2}$ cup _____
- Kidney beans, black beans, lima beans, 1 cup _____
- Orange, 1 _____
- Parmesan cheese, 1 tablespoon _____
- Soynuts, $\frac{1}{4}$ cup _____

Now add up all your points. *Total Points:* _____ Multiply your total points × 100: _____ This gives you an idea of how many milligrams of calcium you are getting each day.

SOURCE: Adapted from Dairy Council of California, Online Calcium Quiz, 2009, available at www.dairycouncilofca.org/Tools/CalciumQuiz/Default.aspx.

SOURCES OF CALCIUM
Adult DRI (RDA): Men: 19 to 70 years, 1,000 mg; 70 years and older, 1,200 mg
Women: 19 to 50 years, 1,000 mg; 51 years and older, 1,200 mg

Orange juice, fortified: 350 mg per cup

Chinese cabbage: 160 mg per cup

Almonds: 50 mg per 2 tbsp

Tofu, firm: 130 mg per ½ c

Cheddar cheese: 306 mg per 1½ oz

Milk: 316 mg per cup

Navy beans: 71 mg per ½ c

Yogurt, plain: 448 mg per cup

Salmon, canned with bones: 182 mg per 3 oz

Soy nuts: 120 mg per ¼ c

© Matthew Farruggio

The word *daily* must be stressed with respect to calcium intake. Because of the body's limited ability to absorb calcium, it cannot handle massive doses periodically; it needs frequent opportunities to take in small amounts.

Phosphorus Phosphorus is second to calcium in mineral abundance in the body. About 85 percent of it is found combined with calcium in the crystals of the bones and teeth as calcium phosphate, the chief compound that gives them strength and rigidity. Phosphorus is also a part of DNA and RNA, the genetic code material present in every cell. Phosphorus is thus necessary for all growth, because DNA and RNA provide the instructions for new cells to be formed.

Phosphorus plays many key roles in the cells' use of energy nutrients. Many enzymes and the B vitamins become active only when a phosphate group is attached. Phosphorus is critical in energy exchange and in maintaining fluid balance in the body.

Some lipids (phospholipids) contain phosphorus as part of their structure. They help transport other lipids in the blood; they also form part of the structure of cell membranes, where they affect the transport of nutrients and wastes into and out of the cells.

Animal protein is the best source of phosphorus, because phosphorus is so abundant in the energetic cells of animals. People who eat large amounts of animal protein have high phosphorus intakes. People who regularly consume carbonated beverages also have high phosphorus intakes, because those drinks contain phosphoric acid. When soft drinks replace milk in the diet, fracture risks increase, especially for girls and women.[5]

SOURCES OF PHOSPHORUS
Adult DRI (RDA) is 700 mg.

Navy beans: 143 mg per ½ c

Milk: 235 mg per cup

Split pea soup: 213 mg per cup

Sirloin steak: 200 mg per 3 oz

Yogurt: 325 mg per cup

Trout: 260 mg per 3 oz

Sunflower seeds: 377 mg per ¼ c

Almonds: 215 mg per ¼ c

© Matthew Farruggio

The recommended intake of phosphorus is lower than that for calcium. It is believed to provide sufficient phosphorus to ensure adequate absorption and retention of calcium. Higher intakes of phosphorus can interfere with calcium absorption. The good news is that people do not need to make a special effort to eat foods containing phosphorus, because it is present in virtually all foods (see Table 8-3).

Magnesium   Magnesium helps relax muscles after contraction and promotes resistance to tooth decay by helping to hold calcium in tooth enamel. Bone magnesium seems to be a reservoir to ensure that some of this mineral will be on hand for vital reactions regardless of recent dietary intake. Magnesium also acts in all the cells of the muscles, heart, liver, and other soft tissues, where it forms part of the protein-making machinery and is necessary for the release of energy. Areas of the country that have hard water—higher in magnesium and calcium—have lower rates of death from cardiovascular disease. A deficiency of magnesium may be related to sudden death from heart failure and to high blood pressure.[6] A dietary deficiency of magnesium is not likely, but may occur as a result of vomiting, diarrhea, alcohol abuse, or protein malnutrition. Good food sources of magnesium include nuts, legumes, cereal grains, dark-green vegetables, seafoods, chocolate, and cocoa (see Table 8-4).

TABLE 8-3	Phosphorus in Foods
(mg)*	**Sources**
422	American cheese (2 oz)
341	Cottage cheese (1 c)
325	Yogurt (1 c)
280	Salmon (canned) (3 oz)
242	Pork (3 oz)
235	Fat-free milk (1 c)
200	Sirloin steak (3 oz)
189	Sunflower seeds (2 tbsp)
186	Turkey (3 oz)
174	Peanuts (⅓ c)
161	Hamburger (3 oz)
149	Shredded wheat (1 c)
143	Navy beans (½ c)
139	Tuna (3 oz)
115	Potato (1)
104	Peanut butter (2 tbsp)
67	Corn (½ c)
51	Cola (12 oz)
46	Broccoli (½ c)
46	Wheat bread (1 slice)
45	Diet cola (12 oz)

*DRI adult: 700 mg/day.

TABLE 8-4	Magnesium in Foods
(mg)*	**Sources**
140	Almonds (⅓ c)
119	Cashews (⅓ c)
95	Raisin bran (1 c)
85	Peanuts (⅓ c)
78	Spinach, cooked (½ c)
70	Tofu (½ c)
60	Black beans (½ c)
55	Oysters (steamed) (3 oz)
54	Baked potato (1)
46	Soy milk (1 c)
45	Avocado (½ c)
45	Black-eyed peas (½ c)
43	Yogurt, plain (1 c)
42	Brown rice, cooked (½ c)
40	Lima beans (½ c)
33	Dried figs (¼ c)
28	Fat-free milk (1 c)
27	Chicken (3 oz)
21	Sunflower seeds (2 tbsp)
21	Hamburger (3 oz)
20	Pork (3 oz)
17	Milk chocolate (1 oz)

*DRI adult: men (19–30 years), 400 mg/day; men (> 31 years), 420 mg/day; women (19–30 years), 310 mg/day; women (> 31 years), 320 mg/day.

Fluoride   Only a trace of fluoride occurs in the human body, but studies have demonstrated that for people who live where diets are high in fluoride, the crystalline deposits in their bones and teeth are larger and more perfectly formed than in people who live where diets are low in fluoride. The trace mineral fluoride not only protects children's teeth from decay but also makes the bones of older people resistant to adult bone loss (osteoporosis). Thus, its continuous presence in body fluids is desirable.

SOURCES OF MAGNESIUM
Adult DRI (RDA) is 310 to 420 mg.

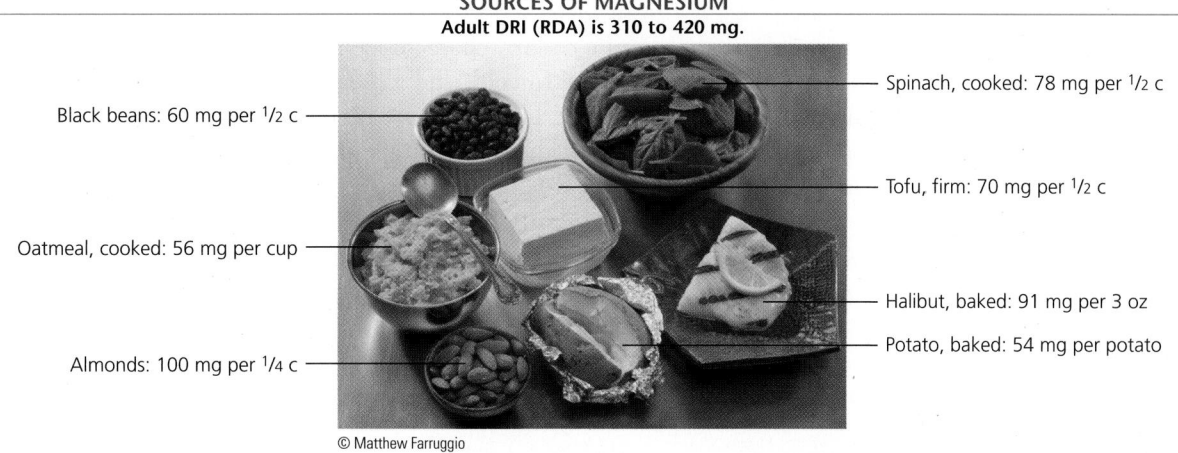

Black beans: 60 mg per ½ c

Oatmeal, cooked: 56 mg per cup

Almonds: 100 mg per ¼ c

Spinach, cooked: 78 mg per ½ c

Tofu, firm: 70 mg per ½ c

Halibut, baked: 91 mg per 3 oz

Potato, baked: 54 mg per potato

© Matthew Farruggio

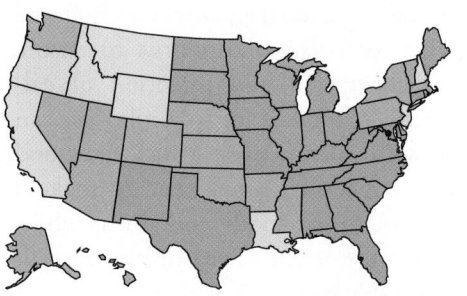
Alternatives to water fluoridation:

- Fluoride toothpastes
- Fluoride treatments for teeth
- Fluoride tablets and drops

Drinking water is the usual source of fluoride. Where fluoride is lacking in the water supply, the incidence of dental decay is very high. Fluoridation of community water, where needed to raise its fluoride concentration to one part per million (ppm), is thus an important public health measure (see Figure 8-4).

In some communities, the natural fluoride concentration in water is high (2 to 8 ppm), and children's teeth develop with mottled enamel—a condition called **fluorosis**. True toxicity from fluoride overdoses can occur, but usually only after years of chronic daily intakes of 20 to 80 times the amounts normally consumed from fluoridated water.

Minerals for Healthy Blood

Minerals for Healthy Blood

- Iron
- Zinc
- Copper

All body cells require a continuous supply of water, oxygen, energy, and building materials. As the blood circulates through the body, it picks up and delivers materials as needed. The blood transports oxygen and nutrients to the cells and carries cellular wastes away from them. Along with the vitamins discussed in Chapter 7, three of the *trace* minerals serve blood-related functions in the body (see margin list). Iron, zinc, and copper are associated with **hemoglobin**, the compound that carries oxygen in the blood.

fluorosis (floor-OH-sis) discoloration of the teeth from ingestion of too much fluoride during tooth development.

hemoglobin (HEEM-uh-glow-bin) the oxygen-carrying protein of the blood; found in the red blood cells.

iron-deficiency anemia a reduction of the number and size of red blood cells and a loss of their color because of iron deficiency.

anemia any condition in which the blood is unable to deliver oxygen to the cells of the body. Examples include a shortage or abnormality of the red blood cells. Many nutrient deficiencies and diseases can cause anemia.

Iron Every living cell contains iron. Bound into the protein hemoglobin in the red blood cells, it helps transport oxygen from lungs to tissues and thus permits the release of energy from fuels to do the cells' work. When the iron supply is too low, **iron-deficiency anemia** occurs, characterized by weakness, tiredness, apathy, headaches, increased sensitivity to cold, and a paleness that reflects the reduction in the number and size of the red blood cells. A person with this **anemia** can do very little muscular work without disabling fatigue, but can replenish iron status by eating iron-rich foods (see Table 8-5).

Iron deficiency is the most common and widespread nutritional disorder in the world.[7] People begin to feel the impact of iron deficiency, without knowing it, long before anemia is diagnosed. They don't appear to have an obvious deficiency disease; they just appear unmotivated and apathetic. Because they work and play less, they are less physically fit. In the United States, iron deficiency remains relatively prevalent among toddlers, adolescent girls, and women of childbearing age.[8]

The cause of iron deficiency is usually malnutrition—that is, inadequate intake, either from limited access to food or from high consumption of foods low in iron. Among causes other than malnutrition, blood loss is the primary one, caused in many countries by parasitic infections of the digestive tract.

The typical Western mixed diet provides only about 5 to 6 milligrams of iron in every 1,000 calories. The recommended daily intake for an adult man is 8 milligrams. Because most men easily eat more than 2,000 calories, they can meet their iron needs without special effort.

The situation for women is different. Women may have normal blood cell counts or hemoglobin levels and yet may need more iron because their body stores are depleted, a factor that doesn't show up in standard tests. Because most women typically eat less food than men, their iron intakes are lower. Also, because women menstruate, their iron losses are greater. These two factors may put women much closer to the borderline of deficiency.

The recommended intake for a woman before menopause is 18 milligrams per day; pregnant women need 30 milligrams daily. Because women typically consume fewer than 2,000 calories per day, they understandably have trouble achieving adequate iron intakes. A woman who wants to meet her iron needs from foods must increase the iron-to-calorie ratio of her diet so that she will receive about double the average amount of iron: about 10 milligrams per 1,000 calories. This means she must emphasize iron-rich foods in her daily diet.

Table 8-5 shows the amounts of iron in foods. Meats, fish, and poultry are superior sources on a per serving basis. People must select foods carefully to obtain enough iron, because it is present in such small quantities in most foods. The best meat sources are liver, red meats, poultry, fish, oysters, and clams. Among the grains, whole grains and enriched and fortified breads and cereals are best, and dried beans are a good source. Some fruits and vegetables contain appreciable amounts of iron, as shown in the table. Foods in the milk group are notoriously poor iron sources.

TABLE 8-5	Iron in Foods
(mg)*	**Sources**
23.80	Clams, steamed (3 oz)
8.20	Instant Cream of Wheat (¾ c)
8.00	Enriched cereal (1 c)
3.20	Spinach, cooked (½ c)
3.13	Beef pot roast (3 oz)
2.90	Sirloin steak (3 oz)
2.75	Baked potato (1)
2.65	Shrimp (3 oz)
2.48	Sardines (3 oz)
2.35	Hamburger (3 oz)
2.30	Navy beans (½ c)
2.25	Lima beans (½ c)
2.25	Prune juice (¾ c)
2.15	Black-eyed peas (½ c)
2.00	Swiss chard (½ c)
1.61	Kidney beans (½ c)
1.59	Oatmeal, cooked (1 c)
1.40	Tofu (½ c)
1.30	Tuna (3 oz)
1.30	Dried figs (¼ c)
1.26	Green peas (½ c)
0.94	Wheat bread (1 slice)
0.82	Apricot halves, dried (5)
0.76	Raisins (¼ c)
0.65	Broccoli, cooked (½ c)

*DRI adult: men, 8 mg/day; women (19–50 years), 18 mg/day; women (> 50 years), 8 mg/day.

Ways to Enhance Iron Absorption   Iron occurs in two forms in foods: as **heme iron**, bound into the iron-carrying proteins such as hemoglobin in meats, poultry, and fish; and as **nonheme iron**, in both plant and animal foods. Heme iron is much more reliably absorbed than is nonheme iron. Absorption of nonheme iron is affected by many factors, including the amount of vitamin C consumed with meals.[9] Vitamin C promotes iron absorption and can triple the amount of nonheme iron absorbed from foods eaten at the same meal.

heme (HEEM) iron   the iron-holding part of the hemoglobin protein, found in meat, fish, and poultry. About 40 percent of the iron in meat, fish, and poultry is bound into heme. Meat, fish, and poultry also contain a factor (MFP factor) other than heme that promotes the absorption of iron, even of the iron from other foods eaten at the same time as the meat.

nonheme iron   the iron found in plant foods.

SOURCES OF IRON
Adult DRI (RDA) is 8 to 18 mg.

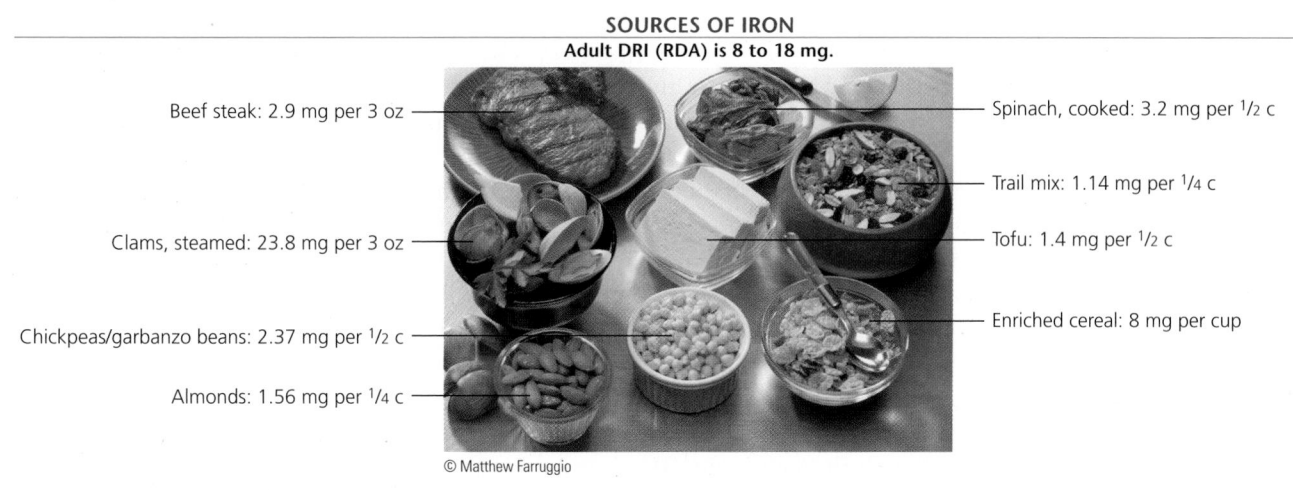

Beef steak: 2.9 mg per 3 oz

Clams, steamed: 23.8 mg per 3 oz

Chickpeas/garbanzo beans: 2.37 mg per ½ c

Almonds: 1.56 mg per ¼ c

Spinach, cooked: 3.2 mg per ½ c

Trail mix: 1.14 mg per ¼ c

Tofu: 1.4 mg per ½ c

Enriched cereal: 8 mg per cup

© Matthew Farruggio

You can combine foods to achieve maximum iron absorption—the heme iron in the meat and the vitamin C in the tomatoes in this chili help you absorb the nonheme iron from the beans.

Iron cookware adds supplemental iron to foods.

contamination iron iron found in foods as the result of contamination by inorganic iron salts from iron cookware, iron-containing soils, and the like.

iron overload a condition in which the body contains more iron than it needs or can handle; excess iron is toxic and can damage the liver. The most common cause of iron overload is the genetic disorder hemochromatosis.

Contamination iron—that is, iron obtained from cookware or soil—can also increase iron intake significantly.[10] Consumers who cook their foods in iron cookware thereby contribute to their iron intake. For example, the iron content of a half-cup of spaghetti sauce simmered in a glass dish is 3 milligrams, but it is 87 milligrams when the sauce is cooked in an iron skillet. Similarly, dried apricots and raisins contain more iron than the fresh fruit because they are dried in iron pans. Admittedly, this form of iron is not as well absorbed as the iron from meat, but every little bit helps.

As mentioned earlier, some foods contain binders that combine chemically with iron and other minerals, such as calcium and zinc, thus preventing absorption of those substances and carrying them out of the body with other wastes. For example, *phytates* render the iron, calcium, zinc, and magnesium in certain foods less available than they might otherwise be; *oxalate* binds iron and calcium. Phytate is found in oatmeal and other whole grains. Other binders are the tannins, which occur in black teas, coffee, cola drinks, chocolate, and red wines. Fiber also can reduce iron absorption because it speeds up the transit of materials through the intestines.

Iron Toxicity   Large amounts of iron can be toxic to the body.[11] **Iron overload** is a condition in which the body absorbs excessive amounts of iron. As a result, tissue damage occurs, especially in organs such as the liver that store iron. Infections are more likely to occur when this condition exists, because bacteria thrive on iron-rich blood. This condition is more common in men than in women.

Although iron is an important essential nutrient in the body, it can also act as a powerful oxidizing agent in reactions that produce free radicals in the body. Therefore, avoid taking extra supplemental iron unless you have been diagnosed as deficient. Be sure to keep iron supplements out of the reach of children, too—they can fatally overdose on such tablets.

Zinc   Zinc is found in every cell of the body and plays a major role along with more than 50 enzymes that regulate cell multiplication and growth; normal metabolism of protein, carbohydrate, fat, and alcohol; and disposal of damaging free radicals. Zinc is associated with the hormone insulin, which regulates the body's fuel supply. It is involved in the utilization of vitamin A, taste perception, thyroid function, wound healing, the synthesis of sperm, and the development of sexual organs and bone.

Zinc is also known for its role in promoting a healthy immune system.[12] A flurry of zinc lozenges has appeared on the market following the results of a study seeking an antidote for the common cold.[13] People who used zinc gluconate lozenges every 2 hours, starting within 24 hours of the onset of a cold, reduced the length of their colds by three days. However, other studies have not reported similar results, and whether zinc lozenges are any more effective than placebos remains to be determined. Furthermore, the long-term effects of such supplemental use of zinc are unknown.

Zinc deficiencies were first reported in the 1960s in the Middle East, where studies on adolescent boys revealed severe growth retardation and delayed sexual maturation—symptoms responsive to zinc supplementation. The native diets were typically low in animal protein and zinc, but they were high in fiber and other compounds that bind minerals. The researchers learned that the binders were carrying zinc out of the boys' bodies, thus causing the deficiency.

Since then, cases of zinc deficiency have been discovered closer to home.[14] Zinc deficiency can cause night blindness, hair loss, poor appetite, susceptibility to infection, delayed healings of cuts or abrasions, decreased taste and smell sensitivity, and poor growth in children. Because zinc is lost from the body daily in much the same way as protein is, it must be replenished daily.

The highest zinc needs are those for people who are building new tissue: infants, children, teenagers, and pregnant women. The pregnant teenager is at particular risk because she needs zinc for her own growth as well as her fetus's growth. Pregnant vegetarians are at risk, too, because their diets are high in fiber and zinc-binding factors. Dieters also need to be reminded that very low-calorie diets cause not only low zinc intake but also a loss of zinc from body tissues as they break down to release fuel.

Zinc is a relatively nontoxic element. However, it can be toxic if consumed in large enough quantities. Consumption of high levels of zinc can cause a host of symptoms, including vomiting, diarrhea, fever, and exhaustion.[15] The hazards of overconsumption are greatest when consumers dose themselves with supplements. Excess supplemental zinc can cause imbalances of both copper and iron in the body. Chronic consumption of zinc exceeding 15 milligrams per day is not recommended without close medical supervision.

Table 8-6 shows the zinc amounts in foods. An average 1,500-calorie diet provides about 6.3 milligrams of zinc per day, or about 60 to 80 percent of the recommended intake. Zinc is highest in foods of high protein content, such as shellfish (especially oysters), meats, and liver. As a rule of thumb, two servings a day of animal protein will provide most of the zinc a healthy person needs. Whole-grain products are good sources of zinc if large quantities are eaten (the phytate in grains does not inhibit the absorption of zinc in people consuming ordinary diets). Cow's milk protein (casein) binds zinc avidly and seems to prevent its absorption somewhat; infants absorb zinc better from human breast milk. Fresh and canned vegetables vary in zinc content, depending on the soil in which they are grown. The zinc content of cooking water varies from region to region as well.

Copper Copper is a trace mineral and a component of several enzymes with important roles in a variety of metabolic and physiological processes. For example,

TABLE 8-6	Zinc in Foods
(mg)*	**Sources**
28.00	Oysters, cooked (3 oz)
6.50	Crabmeat (3 oz)
5.27	Ground beef (3 oz)
4.90	Sirloin steak (3 oz)
4.65	Beef pot roast (3 oz)
4.00	Soybeans, dry-roasted (½ c)
3.10	Enriched cereal (1 c)
2.20	Yogurt (1 c)
1.73	Turkey (3 oz)
1.70	Peanuts (¼ c)
1.66	Swiss cheese (½ oz)
1.50	Shrimp (3 oz)
1.32	Cheddar cheese (½ oz)
1.10	Black-eyed peas (½ c)
1.00	Black beans (½ c)
1.00	Tofu (½ c)
0.98	Fat-free milk (1 c)
0.75	Green peas (½ c)
0.70	Kidney beans (½ c)
0.68	Spinach, cooked (½ c)
0.55	Whole-wheat bread (1 slice)

* DRI adult: men, 11mg/day; women, 8 mg/day.

The Egyptian boy in this picture is 17 years old but is only 4 feet tall, the height of an average 7-year-old in the United States. His genitalia are like those of a 6-year-old. The retardation, known as dwarfism, is rightly ascribed to zinc deficiency because it is partially reversible when zinc is restored to the diet.

© Harold Sanstead, M.D.

SOURCES OF ZINC
Adult DRI (RDA) is 8 to 11 mg.

Black beans: 1.0 mg per ½ c

Shrimp: 1.5 mg per 3 oz

Soy nuts: 2.1 mg per ¼ c

Chickpeas/garbanzo beans: 1.25 mg per ½ c

Oysters, steamed: 28 mg per 3 oz

Beef steak: 4.9 mg per 3 oz

Yogurt: 2.2 mg per cup

Tofu: 1.0 mg per ½ c

Fortified cereal: 3.1 mg per cup

Peanuts: 1.7 mg per ¼ c

© Matthew Farruggio

copper is involved in making red blood cells, manufacturing collagen, healing wounds, and maintaining the sheaths around nerve fibers. Good food sources of copper include whole grains, nuts, seeds, legumes, and shellfish.

Minerals and Energy Metabolism

As described in Chapter 3, the carbohydrate glycogen can be broken down in the body to glucose, triglycerides to fatty acids and glycerol, and protein to amino acids. When the body needs energy, it breaks any or all of the four basic units—glucose, fatty acids, glycerol, and amino acids—into even smaller units. When the body does not require energy, the end products of digestion (glucose, amino acids, glycerol, and fatty acids) are used to build body compounds such as glycogen, body proteins, or body fat. Along with several vitamins (discussed in the previous chapter), several minerals are essential for **energy metabolism**, as described in the following sections.

Chromium

Chromium is a trace mineral involved in carbohydrate and lipid metabolism. Chromium works closely with the hormone insulin by enhancing its ability to help the cells take up glucose and break it down for energy.[16] Chromium is found in a wide variety of unrefined foods. Good sources include nuts, mushrooms, asparagus, prunes, meats (especially liver), brewer's yeast, dark chocolate, and whole grains.

Sulfur

Sulfur is a major mineral and a component of the hormone insulin. It is also a part of the B vitamins—thiamin and biotin—which have important roles in helping enzymes convert nutrients into energy for the body's use. Sulfur is present in some amino acids and in all dietary proteins. It helps strands of protein assume and hold a particular shape, thus enabling them to do their specific jobs, such as enzyme work. Skin, hair, and nails contain some of the body's more rigid proteins, and they have a high sulfur content. There is no recommended intake for sulfur, and no deficiencies are known. Only a person who lacks dietary protein to the point of severe protein deficiency will lack the sulfur-containing amino acids.

Iodine

Iodine is a trace mineral that occurs in the body in an infinitesimal quantity, but its principal role in human nutrition is well known. It is an integral part of the thyroid hormones, which regulate body temperature, metabolic rate, reproduction, and growth. The hormones enter every cell of the body to control the rate at which the cells use oxygen and release energy.

When the iodine level of the blood is low, the thyroid gland may enlarge in an attempt to trap more iodine, until it causes swelling in the throat area—a condition called **goiter**. Goiter is estimated to affect about 200 million people worldwide.

In addition to causing sluggishness and weight gain, an iodine deficiency can have serious effects on fetal development. Severe thyroid undersecretion by a woman during pregnancy causes extreme and irreversible mental and physical retardation of the child; this condition is known as **cretinism**. A person with cretinism has mental retardation and a face and body with many abnormalities. Much of the mental retardation associated with cretinism can be averted by early diagnosis and treatment of the mother's iodine deficiency.

The amount of iodine in foods reflects the amount present in the soil in which plants are grown or on which animals graze. Soil iodine is greatest along the coastal regions. In the inland areas of the United States, where the soil is low in iodine, widespread goiter appeared in the local inhabitants during the 1930s. Iodized salt was introduced as a preventive measure, and iodine deficiency disappeared.

The average American dietary iodine intake is above the recommended intake, but falls below toxic levels. Excessive intakes of iodine can cause an enlargement of the thyroid gland resembling a goiter; in infants, the enlargement can be so severe

Lester V. Bergman/CORBIS

In iodine deficiency, the thyroid gland enlarges—a condition known as simple goiter.

energy metabolism all the reactions by which the body obtains and expends the energy from food.

goiter (GOY-tur) enlargement of the thyroid gland caused by iodine deficiency.

cretinism (CREE-tin-izm) severe mental and physical retardation of an infant caused by iodine deficiency during pregnancy.

as to block the airways and cause suffocation. Some of the excess iodine may come from salted fast foods, bakery products, and from milk. Iodates (iodine salts) are used as dough conditioners in the baking industry; and many dairy producers give cows iodine-containing medications, and use iodine-containing disinfectants during the milk treatment process. Today, the baking and dairy industries have reduced their use of iodine-containing compounds, and there is an increasing use of noniodized salt in manufactured foods.[17] The sudden emergence of this problem points to a need for continued surveillance of the food supply to prevent the effects associated with nutrient toxicities.

People sometimes ask if they should be sure to buy *iodized* salt in the grocery store to ensure adequate iodine intake. Although most consumers now have access to fruits and vegetables grown in coastal areas rich in iodine, health experts state the importance of using iodized salt to maintain an adequate iodine intake. Check the label, because only "iodized salt" has iodine added. A half-teaspoon of iodized salt meets the entire adult RDA for iodine. However, seafood, bread, dairy products, and fruits and vegetables grown in iodine-rich soil can also provide significant amounts of iodine to the diet.

Selenium   *Selenium* is a trace mineral found in the soil in varying amounts. Like iodine, selenium is necessary for the synthesis of the active form of the thyroid hormone that regulates the body's metabolic processes. Selenium is also an antioxidant nutrient; it functions as part of an antioxidant enzyme system that defends the body by preventing the formation of free radicals. Selenium can also substitute for vitamin E in some of that vitamin's antioxidant activities. Research is currently under way to investigate a possible role for selenium in protecting against the development of some forms of cancer.[18] A deficiency of selenium is associated with a form of heart disease prevalent in some areas of China with selenium-poor soil. Good sources of selenium include whole grains, meat, seafood, and fruits or vegetables (depending on the soil content).

Water, Minerals, and Fluid Balance

Throughout history, humans have known that water is indispensable to life itself as well as essential for health. Although we often take it for granted, water is by far the nutrient most needed by the body. Consisting entirely of a combination of hydrogen and oxygen atoms, water makes up part of every cell, tissue, and organ in the body and accounts for about 60 percent of body weight (see Figure 8-5), even contributing to body parts thought of as "dry." For example, bone is more than 20 percent water, muscle is 75 percent water, and teeth are about 10 percent water.

Inside the body, water performs many tasks vital to life (see Table 8-7), such as helping to transport the nutrients needed to nourish the cells. The blood is a river of water that flows through the arteries, capillaries, and veins, bringing each cell the exact substances and particles it requires. This same river carries away waste products formed during the reactions that take place in the cells. In addition, water acts as a shock absorber in joints and around the spinal cord; lubricates the digestive tract, as well as all the tissues moistened with mucus; surrounds and cushions an unborn child; and plays a key role in maintaining body temperature.

About 40 percent of the body's water weight is inside the cells, and about 15 percent bathes the outsides of the cells. The remainder is in the blood vessels. Special conditions are needed to regulate the amount of water inside and outside the cells, so that the cells do not collapse from water leaving them or swell up under the stress of too much water entering them. The cells cannot manage this by pumping water across their membranes, because water slips back and forth freely. However, they can pump minerals across their membranes, and these minerals attract the water to come along with them. This is how the cells maintain water balance. Minerals used for this

The body's antioxidant nutrients include:

- Vitamin C
- Vitamin E
- Beta-carotene
- Selenium

See Chapter 7 for details regarding the antioxidant nutrients.

Keshan (KEE-shan) disease is the form of heart disease associated with selenium deficiency and named after the province in China where it was studied. This form of heart disease is characterized by enlargement of the heart with fibrous tissue.

See Chapter 9 for a discussion of water balance in the body and the sources of water in our diet.

Life begins in water.

FIGURE 8-5

Water—The Number One Nutrient

SOURCE: C. Lecos, Water: The Number One Nutrient, FDA Consumer.

30 lb — { Protein, Carbohydrate, Vitamins, Minerals }

30 lb — Fat

90 lb — Water

150-lb Man

TABLE 8-7	The Functions of Water in the Body

- Transports nutrients throughout the body
- Carries away waste from the blood and body tissues
- Moistens eyes, mouth, and nose
- Hydrates skin
- Ensures adequate blood volume
- Forms main component of body fluids
- Participates in many chemical reactions
- Helps maintain normal body temperature
- Acts as a lubricant around joints
- Serves as a shock absorber inside the spinal cord and in the amniotic sac surrounding a fetus

A Guide to the Minerals for Fluid and Electrolyte Balance

- Sodium
- Chloride
- Potassium
- Phosphorus

purpose take the special form of **ions** or **electrolytes**. In this form, as single, electrically charged particles, the minerals play many roles, including helping to maintain water balance and acid–base balance.

Sodium, potassium, phosphorus, and chloride are examples of *electrolytes*— dissolved substances in blood and body fluids that carry electric charges. Sodium is the chief positively charged ion needed to maintain the volume of fluid outside cells; potassium is the chief positively charged ion inside the cells. Chloride is the major negatively charged ion in the fluids outside the cells, where it is found mostly in association with sodium. Phosphorus is the major negatively charged ion found inside cells. Electrolytes influence the distribution of fluids among the various body compartments. Specifically, when present in healthful amounts, electrolytes serve to keep the fluids inside and outside the cells in balance so that the cells can function properly and allow body fluids to bring cells the needed nutrients and remove waste products from the cells. With a normal diet, many factors in addition to the intake of sodium and chloride found in salt work together to keep the fluid volume fairly constant inside and outside of cells.

Electrolytes also help create the environment in which the cells' work takes place—work such as nerve-to-nerve communication, heartbeats, and contraction of muscles. When a person's body loses fluid, whether as sweat, blood, or urine, the person also loses electrolytes. Maintaining healthful concentrations of electrolytes is crucial to the life-sustaining activities of the vital organs. When large amounts of body fluid

Fluid inside cells is rich in potassium (K^+).

Na^+ Cl^-

Fluid outside cells is rich in sodium (Na^+) and chloride (Cl^-).

Electrolytes

Sodium, potassium, and chloride are examples of body electrolytes. Potassium, which is usually found in the fluids inside the cells, carries a positive charge. Sodium and chloride are usually found in the fluids outside the cells. Sodium carries a positive charge, whereas chloride carries a negative charge.

ions (EYE-onz) electrically charged particles, such as sodium (positively charged) and chloride (negatively charged).

electrolytes compounds that partially dissociate in water to form ions; examples are sodium, potassium, and chloride.

are lost, as in **heat stroke**, infant diarrhea, or injury, replacing them is a critical medical task. People who exercise also lose fluids, and they must replace those fluids to avoid dehydration. Chapter 11 discusses fluid needs during and after exercise, as well as popular sports drinks that can help replace electrolytes that may have been lost along with the fluids.

Sodium, Potassium, and Chloride

Sodium Sodium is part of sodium chloride, ordinary table **salt**, a food seasoning and preservative. The recommended adequate intake (AI) for sodium is set at 1,500 milligrams for young adults, 1,300 milligrams for adults aged 51 through 70, and 1,200 milligrams for older adults. Because average sodium intakes are about 3,400 milligrams per day, substantially higher than recommended, the *Dietary Guidelines for Americans* recommends consuming little sodium and salt and staying below the upper limit of 2,300 milligrams of sodium per day (approximately 1 teaspoon of salt). Reducing sodium intake to 1,500 milligrams is recommended for adults who are 51 and older, and for those of any age who are African American, or have high blood pressure, diabetes, or chronic kidney disease. The use of highly salted foods can contribute to high blood pressure (**hypertension**) in those who are genetically susceptible. (A discussion of the causes and prevention of hypertension appears in the Nutrition Action feature later in this chapter.)

If any members of your family have high blood pressure, you are advised to curtail your sodium consumption and make sure that your potassium and calcium intakes are ample. Table 8-8 shows a sampling of the sodium content of common foods and

TABLE 8-8	Where's the Sodium?
Food Groups	**Sodium (mg)**
Grains and Grain Products	
Cooked cereal, rice, pasta, unsalted, ½ c	0–5
Ready-to-eat cereal, 1 c	100–360
Bread, 1 slice	110–175
Vegetables	
Fresh or frozen, cooked without salt, ½ c	1–70
Canned or frozen with sauce, ½ c	140–460
Tomato juice, canned, ¾ c	820
Fruit	
Fresh, frozen, canned, ½ c	0–5
Low-Fat or Fat-Free Dairy Foods	
Milk, 1 c	120
Yogurt, 8 oz	160
Natural cheeses, 1½ oz	110–450
Processed cheeses, 1½ oz	600
Nuts, Seeds, and Dry Beans	
Peanuts, salted, ⅓ c	120
Peanuts, unsalted, ⅓ c	0–5
Beans, cooked from dried or frozen, without salt, ½ c	0–5
Beans, canned, ½ c	400
Meats, Fish, and Poultry	
Fresh meat, fish, poultry, 3 oz	30–90
Tuna, canned, water pack, no salt added, 3 oz	35–45
Tuna, canned, water pack, 3 oz	250–350
Ham, lean, roasted, 3 oz	1,020

SOURCE: Adapted from the National Heart, Lung, and Blood Institute, "The DASH Eating Plan," 2006.

heat stroke an acute and dangerous reaction to heat buildup in the body, requiring emergency medical attention; also called *sun stroke*.

salt a pair of charged mineral particles, such as sodium (Na^+) and chloride (Cl^-), that associate together. In water, they dissociate and help carry electric current; that is, they become *electrolytes*.

hypertension sustained high blood pressure (*hyper* = too much; *tension* = pressure).

Choose and Prepare Foods with Less Salt

You've decided to cut your salt intake. How do you know which foods to buy? How can you cook foods with less salt and good flavor? Fortunately, cutting some of the salt out of your diet isn't difficult. Here are a few basic principles.

At the Supermarket

- Read the Nutrition Facts label to compare the amount of sodium in processed foods, such as frozen dinners, cereals, soups, salad dressings, and sauces. The amount in different types and brands often varies widely. Look for labels that say "low sodium." They contain 140 mg (about 5 percent of the Daily Value) or less sodium per serving. Look for the key words *salt* and *sodium* in the ingredients list.

- Limit your intake of cured foods (such as bacon and ham), foods packed in brine (such as pickles, pickled vegetables, olives, and sauerkraut), and condiments (such as MSG, mustard, horseradish, catsup, soy sauce, and barbecue sauce).

- Cut back on your intake of frozen dinners, mixed dishes such as pizza, packaged mixes, canned soups or broths, and salad dressings; these products often contain a lot of sodium. Choose reduced-sodium or no-salt-added products when possible. Many commercially prepared products are available in sodium-reduced versions. For example, most supermarkets carry reduced-sodium soy sauce, canned tuna, soups, tomato sauce, nuts, canned vegetables, crackers, and pretzels.

- Buy fresh, natural foods more frequently than processed foods, which tend to be high in salt.

At Home

- Explore what herbs and spices complement specific foods. Add flavor with herbs, spices, lemon, lime, vinegar, or salt-free seasoning blends.

- Cook rice, pasta, and hot cereals without salt. Cut back on instant or flavored rice, pasta, and cereal mixes, which usually have added salt, and use only half of the seasoning packet provided.

- Rinse canned foods, such as black beans or tuna, to remove some sodium.

- Consult the *DASH Eating Plan* for these and more sodium-reducing tips at www.nhlbi.nih.gov.

When Eating Out

- Ask how foods are prepared. Ask that they be prepared without added salt or salt-containing ingredients.

- Move the salt shaker away.

- Limit condiments, such as mustard, catsup, and pickles.

- Choose fruit or vegetables, instead of salty snack foods.

Read the Nutrition Facts label to compare the amount of sodium in processed foods.

Rinse canned foods, such as black beans, to remove some sodium.

*Tolerable Upper Intake Level is no more that 2,300 mg/day.

Adult DRI (AI):*   19 to 50 years: 1,500 mg
51 to 70 years: 1,300 mg
71+ years: 1,200 mg

10 mg
(1 ear)

20 mg
(1 small potato)

68 mg
(1 tsp)

167 mg
(1 tbsp)

830 mg
(1 pickle)

1,210 mg
(1 dinner)

400 mg
(1/2 c)

120 mg
(1 c)

50 mg
(3 oz)

50 mg
(1/2 c)

300 mg
(1 tsp)

2,000 mg
(1 tsp)

56 mg
(1 tsp)

169 mg
(20 chips)

1,470 mg
(1 fast-food
breakfast biscuit)

400 mg
(1-oz slice)

© Matthew Farruggio

reveals that, in general, the more processed a food is, the more sodium it contains. As the box below shows, however, whole, unprocessed foods tend to be high in potassium and low in sodium.

People who wish to choose and prepare foods with less salt need to know that what they pour from the salt shaker may be only a sixth of the total salt they consume. Up to 75 percent of the salt in the diet has been added to foods by food processors. Because processed foods don't always taste salty, eating can sometimes be a guessing game. Remember to check food labels for sodium content. See the Savvy Diner feature for suggested ways to reduce salt intake.

Potassium   Potassium is critical to maintaining the heartbeat. The sudden deaths that occur during fasting, severe diarrhea, or severe vomiting are thought to be due to heart failure caused by potassium loss. As the principal positively charged ion inside body cells, potassium plays a major role in maintaining water balance and cell integrity.

In dehydration, sodium is lost along with water that is lost from the body, but more critical damage comes because potassium is lost from inside cells at the same time. Potassium deficiency is especially dangerous because it affects the brain cells, making the victim unaware of the need for water. Adults are warned not to take **diuretics** except under the direction of a physician, because some of them cause potassium excretion. Physicians prescribing such diuretics tell their patients to eat potassium-rich foods to compensate for the losses and, depending on the diuretic, may also advise a lowered sodium intake.

The relationship of potassium and sodium in maintaining the blood pressure is not entirely clear. Abundant evidence supports the simple view that the two minerals have opposite effects. In any case, it is clear that increasing the potassium in the

diuretics (dye-you-RET-iks) medications and other substances causing increased water excretion (*dia* = through; *ouron* = urine).

Whole Foods versus Processed Foods			
Food	Potassium (mg)	Sodium (mg)	Potassium-to-Sodium Ratio
Corn (cooked), 1 c	242	8	30:1
Corn flakes, 1 c	25	300	1:12
Peaches (fresh), 1	193	0	171:1
Peach pie, 1 piece	131	253	1:2

SOURCES OF POTASSIUM
Adult DRI (AI) is 4,700 mg.

Banana: 422 mg per banana

Orange juice: 496 mg per cup

Raisins: 256 mg per ¼ c

Lima beans, cooked: 485 mg per ½ c

Cantaloupe: 213 mg per quarter melon

Milk: 407 mg per cup

Fish, baked: 405 mg per 3 oz

Potato, baked, with skin: 844 mg

© Matthew Farruggio

TABLE 8-9	Potassium in Foods
(mg)*	**Sources**
844	Baked potato (1)
529	Yogurt (1 c)
496	Orange juice (1 c)
485	Lima beans (½ c)
422	Banana (1 medium)
419	Spinach, cooked (½ c)
407	Fat-free milk (1 c)
400	Pinto beans (½ c)
329	Kidney beans (½ c)
319	Salmon (3 oz)
315	Bok choy, cooked (½ c)
299	Pork (3 oz)
297	Hamburger (3 oz)
273	Tomato (1 medium)
256	Raisins (¼ c)
254	Raisin bran (1 c)
237	Chicken (3 oz)
232	Carrots (½ c)
228	Broccoli (½ c)
213	Cantaloupe (¼ melon)

*DRI adult: 4,700 mg/day.

diet can promote sodium excretion under most circumstances and thereby lower the blood pressure.[19] A lifelong intake of foods that are low in sodium and high in potassium protects against hypertension and is thought to play a role in the low blood pressure seen in vegetarians.[20]

A dietary deficiency of potassium is unlikely, but high-sodium diets that are also high in processed foods and low in fresh fruits and vegetables can make it a possibility (see the box on page 247). Whole foods of all kinds, including fruits, vegetables, grains, meats, fish, and poultry, are among the richest sources of potassium. Potassium is also abundant in milk. Table 8-9 shows the potassium content of foods.

Some people have medical reasons for needing potassium supplementation, but these people must be medically supervised. For example, people on medically supervised, very low-calorie weight-loss diets may be advised to take a potassium supplement. A physician must monitor the potassium status of these people and order supplements commensurate with the degree of depletion.

Potassium supplements should never be self-prescribed. Toxicity from potassium in supplement form is a greater concern than potassium deficiency. The body protects itself from this eventuality as best it can. If you consume more than you need, the kidneys accelerate their excretion and so maintain control. Should their limit be exceeded (if you ingest too much potassium too fast), a vomiting reflex is triggered. However, if the digestive tract is bypassed and potassium is injected into a vein, the heart can stop.

Chloride The negative ion chloride accompanies sodium in the fluids outside the cells. Because chloride can move freely across membranes, it is also found inside the cells in association with potassium. In the blood, chloride helps in maintaining the acid–base balance. In the stomach, the chloride ion is part of hydrochloric acid, which maintains the strong acidity of the stomach needed for protein digestion. Nearly all dietary chloride comes from sodium chloride or salt.

Nutrition Action

Diet and Blood Pressure—Beyond Just the Salt Shaker

Think "diet" and "blood pressure," and the first thing that comes to mind is salt. Small wonder, given that experts have long emphasized that a diet high in salt or, more specifically, the sodium it contains, contributes to a public health problem known as *hypertension* (or high blood pressure), as shown in Figure 8-6.[21] Nevertheless, a number of other, albeit less notorious, dietary factors play a role in the unhealthful rise in blood pressure that afflicts some 50 million Americans.

Obesity, for instance, inarguably ranks as the number one dietary culprit linked to hypertension.[22] That's because excess weight forces the heart to work harder to supply blood to the extra pounds of fat tissue. High blood pressure occurs about twice as often among obese people as among thinner people. Fortunately, losing as little as 5 percent of the excess weight has been shown, in some cases, to lower blood pressure to the point that antihypertensive medications may become unnecessary.[23] In other words, a little weight loss, regardless of sodium intake, can be enough to get high blood pressure back under control. With increasing obesity, age, and blood pressure levels comes an increasing probability that a person will become *salt sensitive,* meaning that the blood pressure rises in proportion to the amount of sodium the person consumes.[24] For these people, of course, keeping dietary sodium to a minimum can help lower blood pressure.

© PhotoDisc/GettyImages

Drinking excessive amounts of alcohol is another dietary factor that may lead to high blood pressure (not to mention poor compliance with drug regimens and other treatments for the problem). In fact, consuming more than a couple of ounces of alcohol daily has

(Continued)

FIGURE 8-6

Are Your Numbers Up?

	Systolic		Diastolic
High blood pressure	>140	or	>90
Blood pressure that increases risk	120 to 139 or 80 to 89 (Prehypertension)		
Optimal blood pressure	<120	and	<80

Hypertension has been called a "silent killer" that cannot be felt and may go undetected for years. That's why it is crucial to have your blood pressure checked regularly. Diagnosis of hypertension requires at least two elevated readings. The first of the two numbers in a blood pressure reading, systolic pressure, represents the force exerted by the heart as it contracts to pump blood throughout the body, as measured by the number of millimeters that the pressure pushes a column of mercury up a tube. The second number, diastolic pressure, is a measure of the pressure the blood exerts on artery walls between heartbeats. A reading of below 120 over 80 millimeters of mercury is considered "optimal" for adults 18 years and older. Measurements between 120 over 80 and 139 over 89 are classified as prehypertension. Individuals with prehypertension are at increased risk for progression to hypertension and are encouraged to make lifestyle changes to prevent hypertension. Individuals in the 130 over 80 to 139 over 89 range are at twice the risk to develop hypertension as those with lower values. Beyond those levels, the risks of heart attacks and strokes rise in direct proportion to increasing blood pressure.*

*The categories are for adults not taking high blood pressure medications. If your systolic and diastolic numbers fall into different categories, your overall status depends on the higher category.

SOURCE: National Institutes of Health, National Heart, Lung, and Blood Institute, *The Seventh Report of the Joint National Committee on Prevention, Detection, Evaluation, and Treatment of High Blood Pressure* (Washington, DC: U.S. Department of Health and Human Services, May 2003).

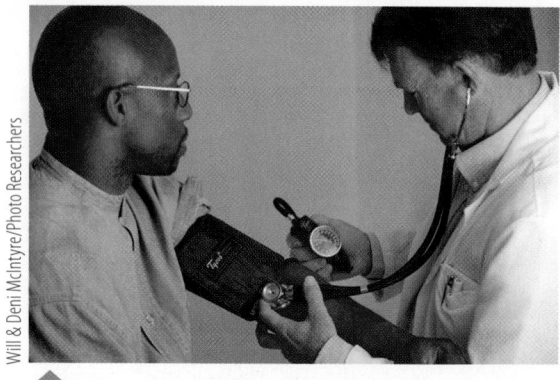

The most effective single measure you can take to protect yourself against high blood pressure is to know what your blood pressure is.

Will & Deni McIntyre/Photo Researchers

been blamed for leading to 5 to 11 percent of cases of hypertension among men.[25] The National High Blood Pressure Education Program thus recommends that to control high blood pressure, those who drink should do so only in moderation.[26]

In addition to recommending that hypertensive people lose weight and drink alcohol in moderation, it is clear that modest changes in lifestyle can add up to significant gains in controlling high blood pressure. It appears that attention to a healthful lifestyle can also help prevent the problem from arising in the first place. For example, researchers at Northwestern University Medical School in Chicago put one group of adults prone to high blood pressure on a preventive program that included losing weight, lowering sodium intake, cutting down on alcohol, and exercising more. They then compared them to a similar group whose members were left to their own devices. The researchers found that over the course of five years, only 1 in 11 persons who changed their lifestyle ended up with hypertension. Of those who didn't alter their habits, though, one in five became hypertensive.[27]

Along with the factors just explained, scientists are exploring other nutrients that may influence blood pressure.[28] Some researchers believe, for instance, that it is not dietary sodium per se that alters blood pressure, but rather the ratio between it and potassium, sodium's partner in regulating the body's water balance.

Risk factors for high blood pressure:

- Obesity
- Family history of high blood pressure
- Race (African Americans are more likely than whites to develop hypertension.)
- Age (In the United States, blood pressure tends to rise with age.)
- Excess alcohol consumption
- Sedentary lifestyle

In other words, the more potassium and less sodium in the diet, the greater the likelihood that the body will maintain a normal blood pressure. Keep in mind that diets rich in high-potassium foods such as fresh fruits and vegetables tend to go hand in hand with low sodium consumption.[29]

In addition, some studies indicate that people who eat high-calcium diets are less likely to have high blood pressure; conversely, people with inadequate calcium intakes are more likely to have hypertension.[30] It is recommended, therefore, that people eat a diet including two to three servings daily of calcium-rich foods, such as milk or calcium-fortified orange juice.

Finally, some evidence suggests that eating a diet that includes monounsaturated fat and is low in saturated fat helps lower blood pressure. It certainly makes sense for those with high blood pressure to eat such a diet, if for no other reason than to help prevent heart disease, a condition that persons with hypertension run a high risk of developing (refer to the Spotlight feature in Chapter 5).[31]

An Eating Plan to Reduce High Blood Pressure

Researchers have shown that following a particular eating plan—called the *DASH diet*—and reducing sodium intake lowers blood pressure. The first study, Dietary Approaches to Stop Hypertension (DASH), examined the effects of overall diet on 459 adult participants with normal to high blood pressures. After just eight weeks, those people with mild hypertension who were on the DASH diet—a diet high in fruits, vegetables, whole grains, and low-fat dairy products and low in total fat, saturated fat, and cholesterol—saw their blood pressures decrease.[32] A second landmark study, DASH-Sodium, has shown that a combination of the DASH diet and sodium reduction can lower blood pressure even more. The DASH-Sodium study looked at the effect on blood pressure of a reduced dietary sodium intake as 412 participants followed *either* the DASH diet or an eating pattern resembling the typical American diet. Participants were followed for a month at each of three sodium levels: 3,300 milligrams per day (a level consumed by many Americans), 2,300 milligrams per day, and the lower-intake DASH diet of 1,500 milligrams per day. Results showed that reducing dietary sodium reduced blood pressure in participants following the DASH diet and those on the typical American diet, but the biggest reductions in blood pressure were seen for those following the DASH diet. Although those with hypertension saw the biggest reductions, those without the disease also saw declines in blood pressure. Figure 8-7 illustrates the DASH diet.

Overall, the results of the DASH studies echo the recommendations for an overall healthful diet: Choose low-fat dairy products, smaller portions of meat, plenty of fruits and vegetables, and ample servings of high-fiber, whole-grain products. This eating pattern may reduce the risk not only for hypertension, but also for heart disease (low in saturated fat and cholesterol), diabetes (high in com-

FIGURE 8-7

The DASH Diet and MyPlate Food Guide

Sample DASH Menu: The DASH eating pattern shown on the next page is lower in fat, saturated fat, cholesterol, and sodium, and higher in complex carbohydrates, potassium, magnesium, and calcium than the typical American diet. The sample menu provides 2,000 calories, 5 servings fruits, 7 servings vegetables, 3 servings dairy foods, 59 grams fat, 121 milligrams cholesterol, and 1,356 milligrams sodium.

*Sweets should be low in fat (for example, jelly, jam, sugar, maple syrup, fruit-flavored gelatin, jelly beans, hard candy, sorbet, fruit ices).

SOURCE: NHLBI, *Facts About the DASH Eating Plan*; available at www.nhlbi.nih.gov.

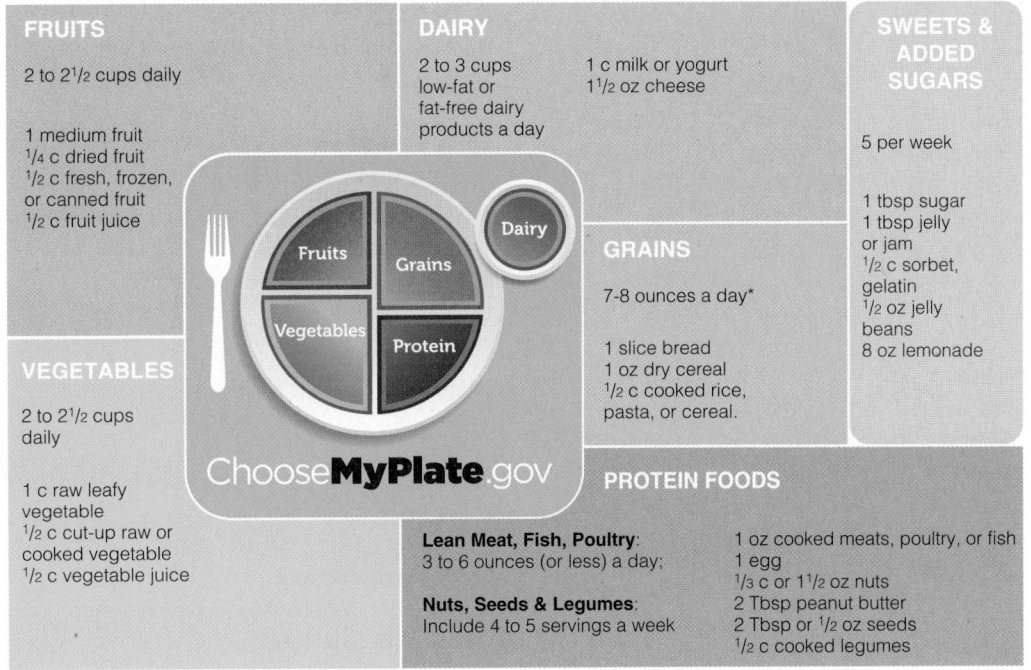

FRUITS

2 to 2½ cups daily

1 medium fruit
¼ c dried fruit
½ c fresh, frozen, or canned fruit
½ c fruit juice

VEGETABLES

2 to 2½ cups daily

1 c raw leafy vegetable
½ c cut-up raw or cooked vegetable
½ c vegetable juice

DAIRY

2 to 3 cups low-fat or fat-free dairy products a day

1 c milk or yogurt
1½ oz cheese

GRAINS

7-8 ounces a day*

1 slice bread
1 oz dry cereal
½ c cooked rice, pasta, or cereal.

PROTEIN FOODS

Lean Meat, Fish, Poultry:
3 to 6 ounces (or less) a day;

Nuts, Seeds & Legumes:
Include 4 to 5 servings a week

1 oz cooked meats, poultry, or fish
1 egg
⅓ c or 1½ oz nuts
2 Tbsp peanut butter
2 Tbsp or ½ oz seeds
½ c cooked legumes

SWEETS & ADDED SUGARS

5 per week

1 tbsp sugar
1 tbsp jelly or jam
½ c sorbet, gelatin
½ oz jelly beans
8 oz lemonade

ChooseMyPlate.gov

*Whole grains are recommended for most grain servings as a good source of fiber and nutrients

plex carbohydrates and fiber), osteoporosis (adequate in calcium), and cancer (lots of fruits and vegetables). The take-home message seems to be that if you want to help prevent a rise in blood pressure as you age and reduce your risk of developing high blood pressure, you can do so by following these six tips:

1. Adopt an eating pattern rich in fruits, vegetables, legumes, and low-fat dairy products (similar to the DASH diet) with reduced saturated-fat content.

2. Maintain a normal weight. Lose weight if you're overweight; even losing just a few pounds can reduce blood pressure if you're overweight.

3. Keep your sodium intake at *or below* recommended levels: not more than 2,300 milligrams a day, or not more than 1,500 mil-

ligrams a day if you are age 51 or older, or African American, or have diabetes, high blood pressure, or chronic kidney disease.

4. Pursue an active lifestyle: Walk briskly, swim, jog, cycle, or do other moderately paced aerobic activities. Aim for at least 30 minutes of activities daily.

5. If you drink, do so in moderation: no more than one drink a day for women, and no more than two drinks a day for men.

6. Don't smoke. Cigarette smoking raises blood pressure and seriously increases risk for heart disease.

The Savvy Diner feature on page 246 offers tips for designing your own healthful diet with a focus on seasoning foods without excess salt.

(Continued)

Sample DASH Menu	Amount	Amount of Sodium (mg)
Breakfast:		
Cereal, shredded wheat	½ c	2
Skim milk	1 c	126
Orange juice	1 c	5
Banana, raw	1 medium	1
Bread, 100% whole wheat	1 slice	149
Lunch:		
Chicken salad*	¾ c	151
Whole-wheat bread	2 slices	298
Dijon mustard	1 tsp	125
Tomatoes, raw, fresh	2 large slices	6
Mixed cooked vegetables	1 c	25
Fruit cocktail, juice pack	½ c	5
Dinner:		
Spicy baked cod*	3 oz	93
Beans, snap, green, cooked from frozen, without salt	1 c	11
Potato, baked, with skin	1 large	15
Sour cream, low fat	2 tbsp	30
Chives (or scallions)	1 tbsp	0
Fat-free natural cheddar cheese	3 tbsp	169
Tossed salad with mixed greens	1½ c	25
Olive oil and vinegar dressing	2 tbsp	0
Snacks:		
Orange juice	½ c	3
Almonds	⅓ c	3
Raisins	¼ c	5
Yogurt, blended, fat-free	1 c	103

*Recipes

Chicken Salad

(makes 5 servings)

3¼ c chicken breast, cooked, cubed
3 tbsp light mayonnaise
¼ c celery, chopped
1 tbsp lemon juice
½ tsp onion powder

Steps: Mix ingredients in a large bowl and serve.
Serving size: ¾ cup

Spicy Baked Cod

(makes 4 servings)

1 lb cod or other fish fillet, fresh or thawed from frozen
1 tbsp olive oil
1 tsp spicy seasoning mix (see below)

Steps:

1. Preheat oven to 350°F. Coat a small baking dish with cooking oil spray.
2. Wash and dry cod. Place in dish and drizzle with oil–seasoning mix.
3. Bake uncovered for 12 minutes or until fish flakes with fork.
4. Cut into 4 pieces and serve.

Spicy seasoning mix

Mix together the following ingredients and store in airtight container for other recipes:

1½ tsp white pepper
½ tsp cayenne pepper
½ tsp black pepper
1 tsp onion powder
1¼ tsp garlic powder
1 tbsp dried basil
1½ tsp dried thyme

SOURCE: NHLBI, *Facts About the DASH Eating Plan*; available at www.nhlbi.nih.gov.

- As she waited in a grocery checkout line, Nancy Miller Friesen felt a stab of pain in her left hip. The 40-year-old Fort Worth woman had always been healthy and active—suddenly she could barely stand. When she went to her family doctor, she learned that several of her vertebrae had fractured. The radiologist studying her X-rays thought he was looking at the bones of a 70-year-old woman.

- Lila Rubin had just teed off at a New Jersey golf course. Putting down her club, the robust 63-year-old felt a searing back pain. When Rubin met with her orthopedic surgeon, she was told she'd suffered a serious compression fracture.

- Debra Epstein of New Haven, Connecticut, wasn't worried when her physician diagnosed a mild curvature of her upper spine 7 years ago. She was only 31 and in otherwise excellent health. Then last year, her right wrist began aching and her back tired easily. After tests, she learned that at age 37, she had the brittle bones of a woman twice her age.

The three women just described share the potentially crippling and sometimes deadly condition *osteoporosis*.[33] Osteoporosis threatens the integrity of the skeleton in some 44 million adults in the United States. One out of two women and one out of four men over the age of 50 will suffer an osteoporosis-related fracture in their lifetime. Osteoporosis is more prevalent in women than in men after age 50, for several reasons:[34]

- Women generally have less bone mass than men.

- Women typically have lower calcium intakes than men.

Osteoporosis strikes the bones. This 75-year-old woman has lost height—not because her legs have grown shorter, but because vertebrae in her back have collapsed. These fractures occur on the front side of the spinal bones, resulting in a hunched-over posture.

© Larry Mulvehill/Photo Researchers, Inc.

- Women more often use weight-loss diets, which tend to be low in calcium and lead to bone loss.[35]

- Bone loss begins earlier in women because of women's different hormonal makeup, and the loss is accelerated at **menopause**, when their protective **estrogen** secretion declines.

- Pregnancy and lactation decrease the calcium reserves in bones whenever calcium intake is inadequate.

- Women live longer than men, and bone loss continues with aging.

What is bone loss?

Adult bone loss affects the entire skeleton, but it occurs first in the pelvis and the spine. Unfortunately, osteoporosis establishes itself silently in its victims' lives. Symptoms do not usually occur until late in life. The disease often results in fractures of the hip, spine, and wrists—more than 1.5 million such fractures a year—that occur with very little or no stress or pressure (see Figure 8-8).[36] Often, osteoporosis first becomes apparent when someone's hip suddenly gives way. Fewer than 50 percent of all women with hip fractures return to previous activity levels, and many are confined to wheelchairs for the remainder of their lives. Approximately 20 percent of these women die of complications within the first year after hip fracture.

Can osteoporosis be prevented?

Although there is no cure for osteoporosis, the good news is that you can take steps to minimize your risk. First, determine your risk by considering the following list of factors that play a role in osteoporosis:

- *Age.* During childhood, adolescence, and the young adult years, the cells that build bone (**osteoblasts**) form more bone than the bone-dismantling cells (**osteoclasts**) take away. With aging, the bone-building cells become less active, but the bone dismantlers continue to work, causing bone tissue and strength to decline.

- *Gender.* Osteoporosis is four times more common in women. Men achieve a higher **peak bone mass** than women do, and the rate of age-related bone loss is lower in men than in women (see Figure 8-9).

- *Age-related decline in hormones.* Hormones are important in bone health for both men and women. Menopause deprives women of the protective effects of estrogen. Estrogen improves calcium absorption from the intestines and reduces excretion of the mineral by the kidneys.[37] Bone loss accelerates in women for

(Continued)

FIGURE 8-8

Bone Loss and Most Common Types of Bone Fractures in Women

A total of 1.5 million fractures occur each year, including 250,000 fractures at sites not shown here.

Spinal Vertebrae Fracture
• More likely at ages 55 to 75 years.
• Fracture can occur from bending, lifting, or spontaneously.
• Bone weakens and collapses, leading to loss of height and chronic back pain.
• Fractures occur on front side of vertebrae, and a hunched-over posture results.

Hip Fracture
• Most serious type of fracture.
• Most occur at 70 years or older.
• May lead to serious complications.
• 50 percent of victims become institutionalized.

Wrist or Forearm Fracture
• Most occur at 50 years or older.
• Fracture often occurs when hands are used to break a fall.
• Can be early warning sign for osteoporosis.

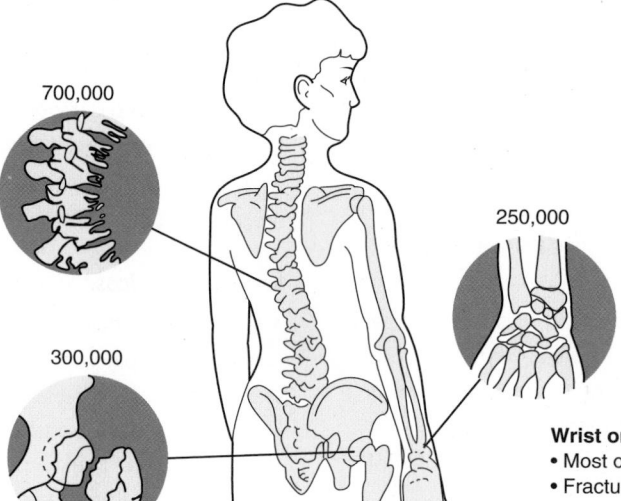

700,000

250,000

300,000

FIGURE 8-9

Bone Loss That Occurs with Age

SOURCE: Data from National Osteoporosis Foundation; www.nof.org.

(Graph: Bone mass (High to Low) vs. Age (years) 0–100, showing curves labeled "Men" and "Women")

the 5 to 10 years following menopause. The earlier menopause occurs, the greater a woman's risk of osteoporosis. Women who have ceased menstruating (a reliable indicator of reduced estrogen levels) comprise the largest segment of people with osteoporosis. Menopause occurs naturally in women typically between the ages of 45 and 55, but it can occur earlier in life with the removal of diseased ovaries. Risk of osteoporosis increases in men with an age-related decline in testosterone; low testosterone levels increase risk of fractures in men.[38]

● *Abnormal absence of menstrual periods (estrogen deficiency).* Menstruation can also cease in women who overexercise and are underweight, a condition called *athletic amenorrhea* (discussed in Chapter 11). For the same reason, eating disorders (anorexia nervosa and bulimia) increase the risk of osteoporosis later in life.[39]

● *Family history.* The greater the history of skeletal fractures of the hips and vertebrae among elderly relatives, the greater your risk for osteoporosis.

● *Race and ethnic background.* Those of British, northern European, Chinese, Japanese, or Mexican American background, or Hispanic people from Central and South America, are at the highest risk. African American women tend to have denser bones and are at lower risk.[40] Undoubtedly, environmental factors—such as calcium intakes, physical activity, smoking, body weight, and alcohol intake—can influence the ultimate outcome of one's genetic heritage.

● *Body build.* The smaller the frame and the thinner the person (under 127 pounds), the greater the risk. Petite women have less bone to lose than larger-boned women.

Soft drinks often replace milk as a beverage in the diet of teenagers, accounting for about 10 percent of their caloric intake.

- *Sedentary lifestyle.* Inactivity leads to bone loss.[41]

- *Smoking and alcohol.* Smoking and excessive alcohol intake increase the risk. People who smoke tend to have lower body weights than nonsmokers. Some women smokers have lower levels of estrogen in their blood and experience an earlier onset of menopause compared to nonsmokers.[42] Alcohol decreases the activity of the bone-building cells, interferes with the absorption of calcium, and may increase excretion of calcium from the kidneys. People who abuse alcohol also tend to have poor nutrition status, including low intakes of calcium.

- *Medical conditions.* Certain medical conditions, including type 1 diabetes, thyroid disorders, rheumatoid arthritis, asthma, seizures, and organ transplants, are associated with increased risk, primarily because of the drugs used to treat them. Prolonged use of medications such as thyroid hormone, antiseizure drugs, and the anti-inflammatory drugs (such as prednisone) used to treat asthma, arthritis, and some cancers may reduce calcium absorption, impair bone formation, and accelerate bone loss.

- *A bone-healthy diet that includes calcium, vitamin D, and other nutrients.* Calcium intake early in life affects the attainment of peak bone mass, achieved by about age 30, which may be the most important determinant of the risk of a fracture in later life.[43] Hence, ensuring an adequate calcium intake throughout life appears to be a sensible strategy. Adequate intake of vitamin D is required for the absorption of calcium. Many older adults, who typically have lower intakes of vitamin D and reduced ability to synthesize the vitamin from sunshine, absorb less calcium as they age.

The calcium intake of a sizeable portion of the U.S. population, especially female adolescents and female adults, is estimated to be below the recommended intake (see Figure 8-10).[44] Between the ages of 18 and 30 years, the time of the attainment of peak bone mass, more than two-thirds of all American women consume less calcium than they need. One explanation is the obsession (especially among adolescent girls) with dieting. Consumption of diet sodas has increased dramatically over the past three decades, often displacing milk as an accompanying beverage to meals.[45]

Other nutrients also play a role in bone health.[46] For example, high intakes of phosphorus, protein, vitamin A (as retinol), and sodium may adversely affect calcium in the body.[47] Adequate intakes of vitamins C and K are needed to maintain healthy bone. Low vitamin K intakes have recently been associated with increased risk of hip fractures in older adults.[48] Potassium, magnesium, and fruit and vegetable intakes are also associated with greater bone density in older adults.[49] The many phytochemicals found in fruits, vegetables, and legumes also offer protective benefits to bone (as discussed in Chapter 7).

Should I consider taking calcium supplements if I am unable to meet the recommended intake for calcium using foods? If so, what should I consider when choosing a supplement?

Although it is preferable to meet calcium needs by consuming calcium-rich foods, calcium supplements are an alternate way to obtain calcium.[50] Supplements may be needed by those who either cannot or will not adjust their diets to get enough calcium from food. Supplements have the advantage of being easy to take, but they have the disadvantage of not offering other nutrients, such as the thiamin, riboflavin, niacin, potassium, phosphorus, magnesium, zinc, vitamin A, and vitamin D found naturally in milk products.

Choosing a calcium supplement can be confusing. Calcium comes in a number of different salts, and the cost of a one-month supply of 1,000 milligrams per day can vary widely. The organic salts include calcium citrate, calcium citrate-malate, and calcium lactate; the inorganic salts include calcium carbonate and calcium phosphate. Calcium carbonate is the least expensive and contains the most calcium per pill, but an organic salt, such as calcium citrate, is better absorbed, especially by older people whose secretion of stomach acid—a factor known to enhance calcium absorption—tends to be reduced.[51] A strategy to overcome this problem is to take calcium carbonate two or three times a day in divided doses, with meals, to improve absorption. Need other alternatives? Try a chewable form, or take the calcium carbonate pill with a glass of orange juice to help it dissolve. Because regular vitamin-mineral pills contain only small amounts of calcium, read the label carefully. Note, too, that many calcium supplements come with added nutrients, such as vitamin D and magnesium. You may not need these additional nutrients if you are consuming a nutritious diet, taking a multivitamin-mineral pill, or eating highly fortified cereals.

(Continued)

Keep in mind that a cup of milk contains about 300 milligrams of calcium. If you are using supplements to replace the calcium found in milk, choose a supplement that contains at least 300 milligrams of calcium. Also, because calcium is best absorbed in doses of 500 milligrams or less, consuming supplements with more than 500 milligrams of calcium per pill is not advantageous. For example, if you need to take 1,000 milligrams of calcium in supplement form, choose a 500-milligram tablet and take one in the morning and one in the evening. A generic brand is fine to use—just be sure to check the label to see how much elemental calcium and which form of calcium it contains. Avoid calcium supplements derived from dolomite, oyster shell, or bone meal, because some have been found to be contaminated with lead.[52] Table 8-10 offers tips for selecting from among the many popular calcium supplements available today. See also the Eat Well Be Well feature in Chapter 7 for more about choosing a vitamin-mineral supplement.

Is there a way to detect bone loss before the onset of a fracture?

A **bone-density** test can detect low bone density before a fracture occurs, and it can predict your chances of fracturing a bone in the future. Since 1998, bone-density screening has been covered by Medicare for all women over the age of 65. The National Osteoporosis Foundation encourages women to discuss with their physician the possibility of having their bone density measured at the onset of menopause, when bone loss sharply accelerates, or if they have other medical conditions that increase their risk for osteoporosis. The more risk factors you have for osteoporosis, the more you need to evaluate your bone density as you age.

Are any drug treatments available to reverse bone loss?

Yes. Among the bone-conserving drugs currently approved by the Food and Drug Administration are **estrogen replacement therapy**, **designer estrogens**, and **bisphosphonates** for osteoporosis prevention. These options and the hormones **calcitonin** and **parathyroid hormone** are also available for the treatment of osteoporosis.[53] These drugs work by slowing bone loss, allowing bones to slowly rebuild and increase bone density, and by reducing the risk of fractures by as much as 50 percent. To be effective, these treatments should be accompanied by the recommended intakes of calcium (1,200 mg for women over 50 years of age) and vitamin D (15 micrograms for adults 51 to 70 years; 20 micrograms for adults 71 years and older), as well as regular weight-bearing exercise.[54]

What can be done to prevent the debilitating effects of osteoporosis?

Not surprisingly, the same recommendations have been made before for lowering risks for other chronic diseases. Eat a nutritious diet, and exercise *throughout life*. Consider the following steps for building bone mass and lowering your risk for osteoporosis:[55]

- Maximize peak bone mass and be vigilant about keeping the bones well supplied with calcium. Consume the recommended amount of calcium (and vitamin D) for your age group (refer to Figure 8-3). According to researcher Robert Heaney, only a few simple changes can add an extra 1,000 milligrams of calcium to an ordinary diet. Put a couple of heaping teaspoons of nonfat dried milk powder in each cup of coffee or tea, drink calcium-fortified orange juice, and eat 1 cup of low-fat yogurt, a dark-green leafy vegetable, and a 1-inch cube of hard cheese.[56]

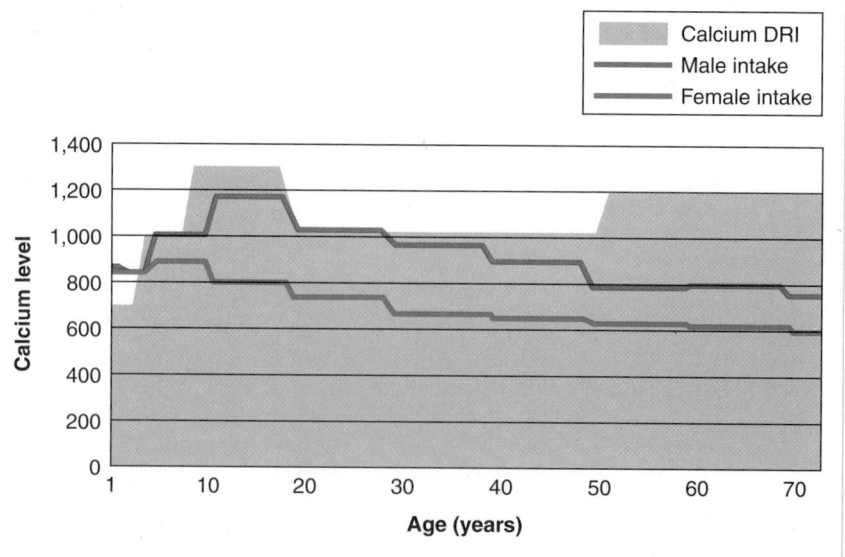

FIGURE 8-10

The Calcium Gap: Daily Calcium Intakes in the United States Compared to Recommended Intakes

Maintaining adequate calcium intake during childhood and adolescence is necessary for the attainment of peak bone mass, which may be important in reducing the risk of fractures and osteoporosis later in life. For the most part, very young children meet their recommended calcium intakes. In contrast, according to national survey data, most children older than 8 years fail to achieve the recommended intake of calcium. Some 81 percent of girls and 48 percent of boys fail to get enough calcium in their diets after age 11—including 7 out of 10 teen boys and 9 out of 10 teen girls. Likewise, most adult women of any age and adult men after age 50 do not meet their recommended calcium intakes.

TABLE 8-10	Tips for Selecting and Using a Calcium Supplement			
Calcium Supplement[a]	**(milligrams/pill)**	**Advantages**	**How to Take**[b]	
Calcium Carbonate		Least expensive and most available	Take with meals	
Caltrate 600 Plus (tablets or chewables)	600			
Os-Cal Ultra	600			
Tums Ultra[d]	400			
VIACTIV[d]	500			
Solgar[c]	600			
Calcium Citrate		Easily absorbed, regardless of the amount of stomach acid	Take on an empty stomach	
Bluebonnet Liquid Calcium Magnesium Citrate	500 per tbsp			
Citracal Caplets + D	315			
Solgar Calcium Citrate with Vitamin D	250			
Twin Labs Calcium Citrate Wafers	250			
Calcium Phosphate		May be less likely to cause constipation	Take with meals	
Posture-D	600			

[a] Calcium with other supplements:
- If you also take an iron supplement, take it at a different time of day. Each mineral is better absorbed alone.
- Very high calcium intakes may limit zinc absorption—a concern for the many Americans who already get too little zinc. High intakes may also increase the risk of kidney stones, especially when taken on an empty stomach. The safe, upper limit for total calcium (from foods and supplements) is 2,000 milligrams per day (adults > 50 years).
- Although vitamin D is necessary for calcium absorption, excessive amounts may be harmful. Supplemental vitamin D in the amount of 600 to 800 IU per day should be sufficient for most individuals.

[b] Calcium with medications:
- Calcium should be taken separately from some medications, such as Fosamax, or the drug's effectiveness can be diminished. Ask your doctor or pharmacist if you have any questions about food–drug interactions.

[c] Calcium source is oyster shells.

[d] Chewable.

SOURCE: Adapted in part from How to choose and use a calcium supplement, *Nutrition in Clinical Care 1*, no. 2 (1998): 100.

- Consume alcohol only in moderation, if at all, and avoid cigarettes altogether.

- Exercise regularly, because exercise can reduce the risk of developing osteoporosis by making bones stronger and increasing their ability to absorb calcium.[57] Incorporate regular sessions of weight-bearing exercise, such as brisk walking, weight training, stair climbing, rope jumping, dancing, hiking, tennis, or aerobic dance classes, into your weekly schedule to slow and possibly even prevent bone loss.[58] Swimming and cycling are less effective at building bone mass, because they put less weight on the bones. Exercises designed to strengthen the back muscles and improve posture are also recommended, because they help the skeleton bear its burden of weight. Increased muscle strength and flexibility improve balance and help prevent debilitating falls.[59]

- For women at or nearing menopause, talk to your health care provider about the need for bone-density testing, medications, and nondrug treatments to slow bone loss after menopause.

Taking these steps to lower your risk of osteoporosis or to treat it if you already have it improves the chances of enjoying a quality life in your later years. Determine your own risk for osteoporosis by taking the quiz on the following page, which examines your current lifestyle for osteoporosis risk factors.

© PCN Photography/Alamy

Incorporate regular sessions of weight-bearing exercise, such as brisk walking, into your weekly schedule to slow and possibly even prevent bone loss (for more about exercise, see Chapter 11).

Do You Have a Bone-Healthy Lifestyle?

No matter what your age, your food and lifestyle choices make a difference to your bone health. How are you "banking" on bones? Give yourself 10 points for each "always," 5 points for each "sometimes," and zero for each "never."

Do You . . .	Always	Sometimes	Never
1. Eat three or more servings of milk, yogurt, cheese, or other calcium-rich foods?	☐	☐	☐
2. Drink vitamin D-fortified milk or get moderate exposure to sunlight?	☐	☐	☐
3. Keep your body moving—with at least 30 minutes of weight-bearing activity, at least 3 times weekly?	☐	☐	☐
4. Consume enough calories to maintain a healthy weight (not too thin)?	☐	☐	☐
5. Avoid cigarettes?	☐	☐	☐
6. Go easy on alcoholic beverages?	☐	☐	☐
Score	____ +	____ +	____ = ____

Scoring

A perfect 60: You are banking on beautiful bones as a lifelong investment. Toast to a perfect score with an ice-cold glass of milk!

35–55 points: For healthier bones, you'd be wise to improve your investment strategies. Otherwise, there may be future penalties.

30 or less points: Your calcium withdrawals may exceed your deposits. Make changes now before your window of opportunity closes.

SOURCE: *Banking on Beautiful Bones: A Lifelong Commitment to Calcium* (Rosemont, IL: National Dairy Council, 1998). Reprinted by permission.

Miniglossary

bisphosphonates drugs that decrease the risk of fractures by acting on the bone-dismantling cells (osteoclasts) and inhibiting their resorption of bone tissue; examples are Fosamax, Actonel, and Boniva. The side effects of long-term use of these drugs have yet to be determined.

bone density a measure of bone strength that reflects the degree of bone mineralization. The DEXA (dual-energy X-ray absorptiometry) bone-density test compares your bone density to that of a healthy young adult. Bone-density tests—using a dual beam of low-level X-rays—take a snapshot of bone density in the spine, wrist, and hip. Simpler, less precise tests use ultrasound to measure bone density in the wrist or heel, and they can be done in a physician's office to identify individuals who might need more precise testing.

calcitonin a hormone used as a drug to decrease the rate of bone loss in osteoporosis. Administered as a nasal spray (Miacalcin) or by injection, calcitonin works by inhibiting the bone resorption activity of osteoclasts.

designer estrogens (selective estrogen receptor modulators—SERMs) drugs that act on *estrogen receptors* in osteoblasts to promote an increase in bone mass; an example is raloxifene (Evista). Unlike estrogen, SERMs have little effect on reproductive tissues of the breast or uterus. *Estrogen receptors* are cellular molecules that bind to estrogen, selective estrogen receptor modulators, or phytoestrogens and deliver these compounds to the nucleus of the cell. Phytoestrogens, estrogen-like compounds found in soy foods, may act as SERMs (see the Spotlight feature in Chapter 6).

estrogen a major female hormone; important in connection with nutrition because it maintains calcium balance and because its secretion abruptly declines at menopause.

estrogen replacement therapy (ERT) administration of estrogen to replace the natural hormone that declines with menopause. Because ERT may increase the risk of uterine and breast cancer, *hormone replacement therapy* (the administration of a combination of estrogen with the hormone progesterone) is often used. However, the FDA recommends that HRT and ERT be considered only for women with significant risk of osteoporosis that outweighs the risk of the drugs (increased risk for breast cancer, stroke, and heart disease).

menopause the time of life at which a woman's menstrual cycle ceases, usually at about 45 to 50 years of age.

osteoblast a bone-building cell; responsible for formation of bone.

osteoclast a bone-destroying cell; responsible for resorption and removal of bone.

parathyroid hormone a hormone used as a drug (teriparatide) to stimulate new bone formation; administered by injection once a day in the thigh or abdomen. Daily injections of teriparatide (brand name Forteo) lead to increased bone mineral density.

peak bone mass the highest bone density achieved for an individual; accumulated over the first three decades of life; typically occurs by 30 years of age. After age 30, bone resorption slowly begins to exceed bone formation.

Nutrition on the Web

www.cengage.com/nutrition/boyle/personalnutrition8e
Go to the *Personal Nutrition* site to check for the latest updates to chapter topics or to access links to related websites.

www.nal.usda.gov/fnic
The Food and Nutrition Information Center (FNIC) provides credible and accurate information about vitamins and minerals. Browse by Subject to access links:

- Click on Dietary Guidance and then select Dietary Reference Intakes (DRIs).
- Click on Food Composition and then select Resources on Individual Minerals.
- Click on Diet and Disease and then select Osteoporosis.

www.nap.edu
The National Academies Press contains the latest Dietary Reference Intakes (DRI) reports.

www.healthfinder.gov
Search for reliable consumer information on individual minerals.

www.iom.edu/fnb
Search for updates regarding new Dietary Reference Intakes at the website for the Institute of Medicine. Search for updates on minerals.

www.eatright.org
The American Dietetic Association home page includes many nutrition resources and links.

www.fda.gov
The Food and Drug Administration provides information about Dietary Supplements. Search for individual minerals.

www.nlm.nih.gov/medlineplus
Search for information about minerals at this government website.

www.nationaldairycouncil.org
Search here for the latest in calcium and nutrition research.

www.nhlbi.nih.gov/hbp
The National Heart Lung and Blood Institute (NHLBI) provides up-to-date and practical information for preventing and controlling high blood pressure.

www.nhlbi.nih.gov/health/public/heart/hbp/dash
Access the DASH Eating Plan—"Lowering Your Blood Pressure with DASH"—from this government website.

www.niams.nih.gov/Health_Info/Bone
Search for information about bone health at this government website.

www.4women.gov
Search The National Women's Health Information Center for information about bone health.

Menu
of Online Study Tools

A variety of study tools for this chapter are available at our website to deepen your understanding of chapter concepts. Go to

www.cengage.com/sso

to find

- Practice tests
- Flashcards
- Glossary
- Web links
- Animations
- Chapter summaries, learning objectives, and crossword puzzles.

Diet Analysis+ in Action

Go to the Intake vs. Goals report that shows the results of your three-day tracking, and record the minerals that were less than 75 percent of your recommended nutrient intake. What foods could you add to your diet to increase your intake of these minerals? Is there a food you could remove to balance out the calorie intake? Now list the minerals that exceeded 200 percent of your recommended nutrient intake. Go to the Intake Spreadsheet, and list the three foods that contributed most to your intake of minerals. What health effects might your current intake have? Can you adjust your diet to create a better balance of minerals?

Impact of Fluids and Beverages on Nutritional Health

Objectives

- Describe the functions of water (fluid) in the body.
- Describe how alcohol is metabolized in the body.
- Identify factors that influence absorption and metabolism of alcohol.

Ask Yourself . . .

Which of the following statements about nutrition are true, and which are false? For each false statement, what is true?

1. "Soft" water is healthier than "hard" water.
2. Bottled water sold in the United States is always cleaner and safer than most tap water.
3. Caffeine can cross the placenta and enter the fetus.
4. Even moderate caffeine users can suffer from withdrawal symptoms if they stop using caffeine cold turkey.
5. Soft drinks sweetened with sugar contribute to empty calorie intake.
6. Energy drinks are regulated and therefore always safe to consume.
7. Enhanced waters provide an important source of nutrients.
8. A 12-ounce beer, a 5-ounce glass of wine, and a 1-ounce shot of tequila all contain the same amount of alcohol.
9. Reflexes are not impaired if your blood alcohol concentration is below the legal limits of intoxication.
10. Drinking alcohol may be associated with an increased risk of breast cancer.

Answers found on the following page.

Contents

From the time humans first appeared on Earth, we have consumed a plethora of various beverages: breast milk, water, plant juices, animal fluids, fermented plant and animal products, and various mixtures. Beverages are used to quench thirst or provide recreation. Some beverages, such as wine, possess cultural and religious significance, whereas others are consumed for medicinal value. Beverages have even been specially formulated for use in outer space.[1]

What do Americans drink now? This chapter discusses selected beverages and evaluates their impact on nutrition and health.

In wine there is wisdom, in beer there is strength, in water there is bacteria.

DAVID AUERBACK, 2002

Water

Adults consume and excrete some 1½ to 3 quarts of water a day (see Figure 9-1). Although most of the water we take in comes from juice, milk, soft drinks, and other beverages, including tap water, foods also add considerable amounts of water to the diet. For example, water makes up as much as 85 to 95 percent of fruits and vegetables. Adults are advised to consume 1.0 to 1.5 milliliters of water from all sources for each calorie expended. For instance, if you expend 2,000 calories a day, you need approximately 2 quarts of fluid.[2]

Just as the sources of water in our diet vary, so too does the water we drink "straight." The makeup of water differs depending on where it comes from and how it is processed—variations that can have significant health implications. One of the most basic distinctions, hard versus soft water, is based on concentrations of three minerals: calcium, magnesium, and sodium. **Hard water** usually comes from shallow ground and contains relatively high levels of minerals, primarily calcium and magnesium. **Soft water**, in contrast, generally flows from deep in the earth and has a higher concentration of sodium.

Although your water utility company can tell you whether your water is soft or hard, you can probably distinguish between the two based on your own experience. Soft water helps soap lather better than hard water and leaves less of a ring on the bathtub. Hard water, to the contrary, doesn't clean clothes as thoroughly as soft water and leaves a residue of rocklike crystals on the inside of the teakettle over time. That's why many consumers prefer soft water.

From a health standpoint, however, hard water seems to be the better alternative.[3] One reason is that the excess sodium carried in soft water, even in small amounts,

---

FIGURE 9-1

Water Balance in the Body

Water enters the body in liquids and foods, and some water is created in the body as a by-product of metabolic processes. Water leaves the body through the evaporation of sweat, in the moisture of exhaled breath, in the urine, and in the feces.

[a] This amount equals 1½ to 3 quarts (1 oz equals approximately 30 mL).
[b] Adults are advised to consume 1.0 to 1.5 mL of water from all sources for each calorie expended. For instance, if you require 2,000 calories a day, you need approximately 2 quarts of fluid.

$$2,000 \text{ cal} \times 1 \text{ mL/cal} = 2,000 \text{ mL}$$

$$2,000 \text{ mL} \div 30 \text{ mL/1 oz} = 67 \text{ oz}$$

$$67 \text{ oz} \div 32 \text{ oz/qt} = 2 \text{ qt}$$

SOURCE: Adapted from F. Sizer and E. Whitney, *Nutrition: Concepts and Controversies*, 11th ed. (Belmont, CA: Thomson Wadsworth, 2008), 273. Information on fluid requirements from S. Kleiner, Water: An essential but overlooked nutrient, *Journal of the American Dietetic Association* 99 (1999): 200.

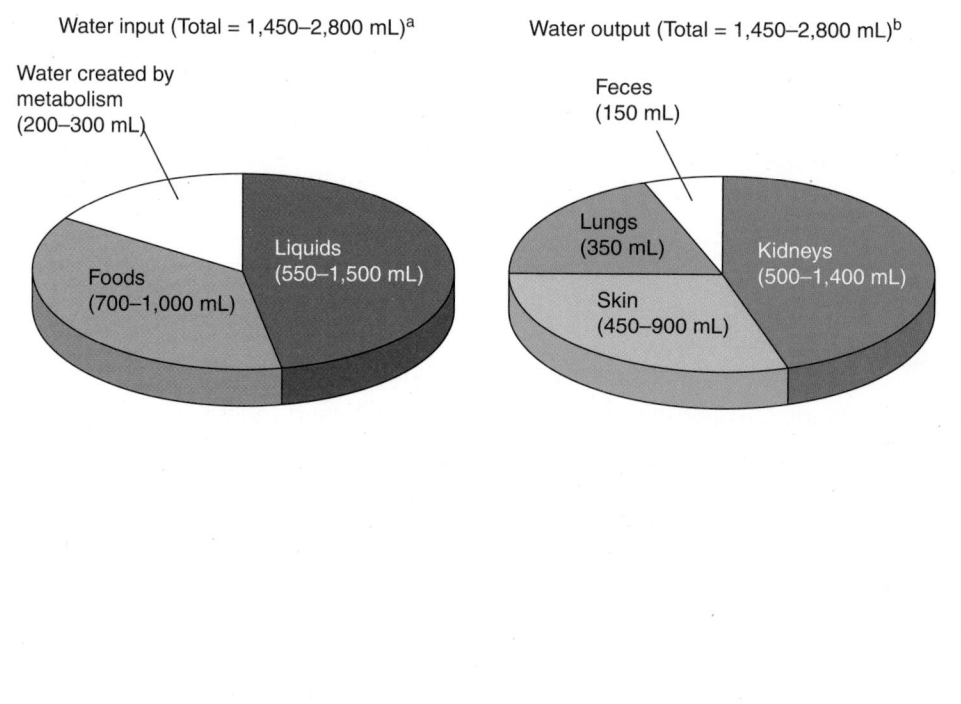

Water input (Total = 1,450–2,800 mL)[a]

Water created by metabolism (200–300 mL)
Foods (700–1,000 mL)
Liquids (550–1,500 mL)

Water output (Total = 1,450–2,800 mL)[b]

Feces (150 mL)
Lungs (350 mL)
Skin (450–900 mL)
Kidneys (500–1,400 mL)

---

hard water water that has a high mineral (calcium and magnesium) content.

soft water water containing a high sodium concentration and little or no calcium or magnesium.

ASK YOURSELF ANSWERS: **1.** False. Soft water is higher in sodium and may increase risk for high blood pressure. **2.** False. Depending on the brand, bottled water may itself be tap water. **3.** True. **4.** True. **5.** True. **6.** False. Energy drinks are not regulated; therefore, manufacturers do not have to prove the safety or efficacy of these products. **7.** False. A balanced diet is usually the best choice. **8.** True. **9.** False. Having a measurable amount of alcohol in the blood can mean that judgment and reflexes are impaired. **10.** True.

adds more of that mineral to our already sodium-laden diets. More important, soft water dissolves potentially toxic substances, such as lead, from pipes. Therefore, people who install water softeners in their homes for the purpose of getting cleaner laundry and better mileage from soap would do well to connect them only to their hot-water lines. That way, they can use hot, soft water for washing and bathing and use cold, hard water for drinking and cooking.

Keeping Water Safe   In addition to having varying concentrations of minerals, water taken from the earth contains different levels of bacteria, microorganisms, and heavy metals such as lead. To ensure that water flowing from the tap is safe to drink, the Environmental Protection Agency (EPA), the arm of the U.S. government responsible for monitoring municipal water supplies, sets limits for potential contaminants such as mercury, nitrate, and silver in drinking water. By law, the public must be notified by a water utility company within 24 hours of the discovery of any potentially dangerous contaminants in drinking water. The EPA also mandates that tap water be disinfected if bacteria levels run high, to prevent the spread of waterborne diseases such as typhoid and dysentery. Such precautionary measures go a long way toward keeping the U.S. water supply one of the safest in the world.

One potential contaminant is a parasite called *Cryptosporidium*. Found in lakes and rivers that have come into contact with sewage or animal waste, *Cryptosporidium* has emerged as a health threat to vulnerable people because it is highly resistant to the chlorine and other disinfectants used in municipal water supplies. In healthy people, the parasite can cause diarrhea and other flu-like symptoms that typically subside within a week to 10 days. In people with weak immune systems, however, this parasite can lead to severe, long-lasting gastrointestinal problems that can even cause death.

Because of the *Cryptosporidium* hazard, the EPA and the Centers for Disease Control and Prevention recommend special precautions for people with severely weakened immune systems (those with HIV infection, cancer, transplant patients taking immunosuppressive drugs, and people born with a weakened immune system). Public health groups advise such individuals to talk to a health care provider about boiling tap water for 1 minute (if it is meant to be used for drinking water) to destroy any *Cryptosporidium* present, or about buying a special water filter system.[4]

Another potential health threat over which the EPA has little control is the level of lead that comes out of your faucet. Although the EPA has put a ceiling on the concentration of lead that may be present in public water supplies, once the water leaves the reservoir, unhealthy levels of the metal may dissolve into it. Before the water comes out of your faucet, it may flow through pipes made of lead or joined by lead solder; the metal can leach into the water as it passes through. Even though the U.S. government banned the use of lead-containing plumbing systems back in 1986, dwellings (and portions of public water systems) built before that time may not have lead-free pipes.

The issue of possible lead contamination in water has generated a good deal of publicity and concern. Certainly, the attention is warranted, given that when lead accumulates in the body, it begins to damage the nerves, kidneys, and liver along with the cardiovascular, reproductive, immunologic, and gastrointestinal systems. Lead is especially toxic to children and fetuses, in whom it can cause neurologic problems as severe as brain damage.[5]

Luckily, lead levels can usually be kept to a minimum simply by "flushing" the tap—that is, letting the water run until it becomes as cold as possible, thereby ridding it of water that has been sitting in pipes and dissolving lead for any length of time. Pipes should be flushed with cold water only, because cold water is less likely than warm or hot water to pick up lead as it flows through them. In some instances, however, flushing is not enough to reduce harmful lead levels. Some homes or apartments may require special water treatment systems, the necessity of which can be determined only by subjecting tap water to a lead test. Names of certified laboratories that analyze water can be obtained from a local branch of the EPA, or by calling the EPA's Safe Drinking Water Hotline (800-426-4791).

Bottled Water   Bottled water has become one of America's most popular beverages today, at a total cost of more than $11 billion a year.[6] Although reasons for this trend are many, the perceived health benefits of bottled water fall near the top of the list. Surveys have found that about 25 percent of bottled-water drinkers choose the

Miniglossary for Bottled Water

Michael A. Keller

The Food and Drug Administration classifies some bottled water according to its origin (Figure 9-2).

artesian water or artesian well water   water drawn from a well that taps a confined water-bearing rock or rock formation.

ground water   water that comes from an underground body of water that does not come into contact with any surface water.

mineral water   water that is drawn from an underground source and contains at least 250 parts per million of dissolved solids. If the water contains between 250 and 500 parts per million total dissolved solids, the statement "low mineral content" must appear. If it contains more than 1,500 parts per million, the statement "high mineral content" must appear. If a cup of the water contains at least 20 milligrams of calcium, 0.36 milligram of iron, or 5 milligrams of sodium, the product must carry nutrition labeling.

purified water (also known as demineralized water, distilled water, deionized water, or reverse osmosis water)   water from which all minerals have been removed, thereby eliminating the possibility that the minerals might corrode pipes or appliances (such as a steam iron).

sparkling bottled water   water in which carbon dioxide (the ingredient that makes soda pop bubbly) is naturally present; that is, carbonation is not added from an outside source.

spring water   water derived from an underground formation from which water flows naturally to the surface of the earth and to which minerals have not been added or taken away. It may be collected either at the spring itself or through a hole tapping the underground formation feeding the spring.

well water   ground water derived from a rock formation by way of a hole bored, drilled, or otherwise constructed in the ground.

from a community water system or from a municipal source   statement that must appear on bottles containing water derived from a municipal water supply. The phrase must conspicuously precede or follow the name of the brand.

seltzer*   tap water injected with carbon dioxide and containing no added salts.

club soda*   artificially carbonated water containing added salts and minerals.

tonic water*   artificially carbonated water with added sugar and/or high-fructose corn syrup, sodium, and quinine.

*These items are not considered bottled water in government parlance. The FDA defines bottled water as water that is sealed in bottles or other containers and is intended for human consumption, excluding soda, seltzer, flavored, and vended water products.

The Origins of Bottled Water

FIGURE 9-2

Origins of Bottled Water

SOURCE: A. C. Bullers, "Bottled water: Better Than the Tap?" *FDA Consumer Magazine*, July–August 2002.

beverage for health and safety reasons, and another 25 percent believe that bottled water is pure and free of contaminants.[7]

Regardless of its pristine image, bottled water is not necessarily any more pure or healthful than water that flows out of the tap. The Food and Drug Administration (FDA), the watchdog of the bottled-water industry, does not require that bottled water meet higher standards for quality, such as the maximum level of contaminants, than public water supplies regulated by the EPA. For the most part, the FDA simply follows EPA's regulatory lead. Typically, bottled water is filtered to remove chemicals such as chlorine that may impart a certain taste, but that doesn't make it any safer. In fact, about 25 to 40 percent of bottled water comes from the same municipal water supplies as tap water.[8] Furthermore, some bottled waters do not contain any or enough of the fluoride needed to fight cavities.[9] The only way to determine whether certain water contains this mineral is to check with the company that bottles it. If you use bottled water, look for the trademark of the International Bottled Water Association (IBWA)—a trade organization that supports the FDA regulations.

This is not to say that bottled water is necessarily any better or worse, from a health standpoint, than tap water. It's certainly preferable to tap water for those who like its taste. The problem is that the demand for bottled water increases the use of non-renewable resources to make the plastic bottles, which in turn creates an energy demand for recycling the used bottles. Additionally, many consumers pay 250 to 10,000 times more per gallon for bottled water than for tap water because they think bottled water is more healthful. Bottlers add to the confusion by sprinkling terms such as *pure*, *crystal pure*, and *premium* on labels illustrated with pictures of glaciers, mountain streams, and waterfalls, even when the water inside comes from a public reservoir. However, the FDA has set forth regulations mandating clear labeling of bottled waters. The miniglossary for bottled water explains what some of the terms used on bottles actually mean.

Juices

It is hard to believe that once upon a time, the only choices of juice in supermarkets were orange, grapefruit, grape, and cranberry. Now there is an apparently infinite assortment of not only juices, but also blended fruit and vegetable juices. You've already learned that fruit and vegetable juices are great sources of antioxidants and phytochemicals that help fight diseases like heart disease, cancer, and diabetes. You've also learned that any fruit or vegetable juice you enjoy can count toward the MyPlate recommended number of servings.

Is there a downside to fruit and vegetable juices? There can be. It is very important to be a conscientious label reader when selecting juices. Juices can add extra calories, and not all juices are uniformly beneficial. Fruit/vegetable flavor does not necessarily equal fruit/vegetable nutrition.[10] The best nutritional choice is 100 percent fruit/vegetable juice. "Juice drinks," "fruit-flavored drinks," or "juice blends" contain little real juice, but serve up quite a bit of added sugar. Remember, too, that even 100 percent fruit/vegetable juice lacks the fiber of the whole fruit or vegetable. Additionally, some juices use concentrated white grape juice as a sweetener, making the beverage higher in simple sugar and calories even though the label can truthfully claim that the beverage is 100 percent fruit juice.[11]

Valentyn Volkov/Shutterstock.com

Juices with Benefits   After making sure your juice is 100 percent pure juice, look at the color of the juice. As a general rule, the more colorful the juice is, or the darker its color, the more nutritious the juice is. One example is pink grapefruit juice, which contains two chemicals that function as antioxidants and phytochemicals not found in regular grapefruit juice: lycopene and beta-carotene.[12] Cranberry juice products available commercially vary extensively, but 10–12 ounces of 27 percent cranberry juice per day is protective against urinary tract infections. Cranberry juice also contains the phytochemical polyphenol, which is believed to protect against heart disease and cancer.[13]

Citrus fruits and juices are rich in vitamins (vitamin C, folate) and flavonoids (phytochemicals) that help reduce overall rate of cancer occurrence. Ounce for ounce, orange juice is the most nutrient-dense juice consumed in the United States. Additionally, citrus juices are rich in potassium and low in sodium content, which may reduce risk of high blood pressure and stroke.[14] The best orange juice, of course, is freshly squeezed; the packaged orange juice you purchase in stores loses vitamin C content daily. Compared to orange juice purchased in gable-top cartons, orange juice sold in plastic jugs retains vitamin C content the best, losing only 2 percent of its vitamin C daily. Orange juice stored in gable-top cartons will have lost most of its vitamin C content by its expiration date.[15]

Orange juice fortified with calcium is also available. Each ounce of calcium-fortified orange juice provides the same amount of absorbable calcium as an ounce of milk. This is an easy way to get calcium or increase calcium intake if you don't like or can't consume dairy foods.

Tomatoes and tomato products are rich in nutrients and the phytochemical lycopene. Lycopene has been found to be protective against prostate cancer and cardiovascular disease. One word of caution: Salt is often added to tomato products, especially tomato juice. Those with high blood pressure or at risk for high blood pressure may prefer to use low-sodium versions of these products.[16]

Milk

"Drink milk, for it wipes away heat from the heart, strengthens the back, increases the brain, augments the intelligence, renews vision, and drives away forgetfulness."[17] Humans are the only mammals that consume milk past infancy. Milk is a white, opaque liquid made by mammary glands, but the term *milk* is also applied to non-animal beverages such as soy milk (see discussion in Chapter 6). Milk is the infant's principal source of nutrition before maturation of the gastrointestinal tract and ability to digest other types of foods. Nutritional composition of milk differs widely between mammal species. For example, cow's milk contains a relatively high percentage of protein compared to human milk due to the much larger size of calves as compared to human infants. This and other differences make human milk an unsuitable substitute for cow's milk for the calf since it would not support the growth of such a large animal. Likewise, cow's milk is an unsuitable substitute for human infants because the high protein content damages the human infant's immature gastrointestinal tract and puts stress on the kidneys to excrete the excessive amount of protein.

Being a nutrient-dense food (providing a high concentration of nutrients in relation to caloric content), milk supplies significant amounts of protein, calcium, phosphorus, riboflavin, vitamin A, vitamin D, potassium, magnesium, vitamin B_6, vitamin B_{12}, and zinc to the human diet.[18] Additionally, milk/calcium intake has been related to weight control.[19] There appears to be an inverse relationship between intakes of calcium/dairy foods and obesity.[20] Lower dairy intake seems to equal higher amounts of body fat, and higher intakes of milk and dairy foods seem to promote greater weight reduction.[21, 22]

With all it has going for it (and us!), milk should top the list of beverage consumption in the United States. Right? Wrong! Fewer and fewer Americans "got milk." Throughout the last half of the 20th century, total per capita milk consumption has declined. This is attributed to the displacement of milk by carbonated beverages, fruit juices, fruit drinks, and flavored teas. The upshot is that innovative advertising campaigns are slowly reversing this trend and resulting in increased milk consumption, especially of flavored milks.[23]

Chepko Danil Vitalevich/Shutterstock.com

Caffeinated Beverages: Coffee, Tea, Cocoa, and Soft Drinks

Caffeine is a chemical compound containing four of the most common elements on earth: carbon, hydrogen, nitrogen, and oxygen. At room temperature, caffeine resembles the powdery consistency of cornstarch, is odorless, is slightly bitter to the taste, and is highly soluble in hot water. Caffeine is one of a group of chemicals called *xanthines*.[24] In the United States, most caffeine is consumed in coffee, but significant amounts are also consumed in teas, carbonated soft drinks, and energy drinks.

In 1657, when merchants first introduced Londoners to a Middle Eastern brew known as *coffee*, they boasted it as a "wholesome and physical drink," an elixir of health suitable for treating colds, coughs, gout, and many other ills.[25] Today, at least 8 out of 10 Americans consume caffeine, the most widely used behaviorally active drug in the world. Modern-day coffee drinkers have been subjected to a barrage of reports linking coffee and other caffeine-containing products with more than 100 diseases. Fingers have repeatedly pointed at caffeine as the culprit behind breast disease, cancer, heart disease, birth defects, and high blood pressure, to name just a few.

Despite the brouhaha, the jury is still out as to whether caffeine is truly to blame. Scientists have yet to confirm long-standing suspicions that caffeine contributes to any health problems other than jitteriness. One reason for all the controversy is that much of the evidence linking the substance with different diseases has been clouded by a number of methodological issues in the studies. Some studies do not measure caffeine sources except for coffee and tea, excluding soft drinks, chocolate, and certain medications. In addition, most studies fail to account for the fact that the amount of caffeine and other substances in coffee or tea can vary considerably depending on how the beverage is brewed.

For example, consider the widely debated question of whether drinking coffee raises blood cholesterol. A great deal of strong evidence suggests that the beverage does contribute to high blood cholesterol and therefore to heart disease. The hitch is that most of the evidence comes from Scandinavia, where coffee is boiled rather than

caffeine a type of compound, called a methylxanthine, found in coffee beans, cola nuts, cocoa beans, and tea leaves. A central nervous system stimulant, caffeine's effects include increasing the heart rate, boosting urine production, and raising the metabolic rate.

brewed in automatic drip coffeemakers or electric percolators. Subsequent research has found that although boiled coffee appears to boost blood cholesterol levels, filtered coffee does not, most likely because substances in boiled coffee other than caffeine may be the cholesterol-raising culprits.[26]

Another issue that most research has not filtered out is that coffee drinking itself may not contribute to ill health; however, it seems to be part of a lifestyle that does. After questioning some 2,600 men and women about their health habits, a group of Boston-based researchers found that women who opted for decaffeinated coffee exclusively were more likely than regular coffee drinkers to eat vegetables frequently and exercise regularly (among other things). Male decaf drinkers also tended to have adopted more healthful habits, such as eating low-fat diets, than did their counterparts who drank regular coffee. Thus it may not be the coffee or caffeine that contributes to health problems, but rather the poor health habits that often go along with a coffee-drinking lifestyle.[27]

None of this is meant to say that caffeine is necessarily good for people. As anyone who can't get going in the morning without a cup of coffee or can of caffeinated soda is well aware, consuming caffeine day in and day out can be habit forming. In fact, some researchers have identified a condition called **caffeine dependence syndrome**, characterized by at least three of the four following criteria: withdrawal symptoms such as headache and fatigue; caffeine consumption despite knowledge that it may be causing harm; repeated, unsuccessful attempts to cut back on caffeine; and tolerance to caffeine.[28]

Along with people who are dependent on caffeine, people with certain medical conditions would do well to consume caffeine in moderation or avoid it completely. Pregnant women, for example, should limit their caffeine intake. Although it has never been proven to cause birth defects, caffeine does cross the placenta and enter the fetus, where large amounts can affect the unborn baby's heart rate and breathing.[29] Some research also suggests that the amount of caffeine in three or four cups of coffee could raise the risk of suffering a miscarriage, perhaps by decreasing blood flow through the placenta.[30] Women who drink caffeine during pregnancy are generally advised to limit consumption to less than 150 milligrams per day—the equivalent of about 16 ounces of coffee.[31] In addition, anyone with ulcers should steer clear of both caffeinated *and* decaffeinated coffee, because both stimulate the secretion of acid, which can irritate the stomach lining. Chapter 11 discusses the effect of caffeine on physical performance.

For a healthy person, drinking one or two cups of coffee, tea, or cola a day does not seem to pose any hazard. Only those who are particularly sensitive to caffeine and suffer symptoms such as headaches, nervousness, and insomnia after consuming it really need to consider avoiding it—or at least cutting back (see Table 9-1 to figure out how much caffeine you're taking in each day).[32]

If you drink coffee or a can or two of cola every day and decide to quit, do it gradually. Even moderate caffeine users who try to stop cold turkey often suffer from withdrawal symptoms, such as splitting headaches, fatigue, moodiness, and nausea. Try instead to cut back gradually by, say, drinking no more than one cup every two or three days. Another way to cut back on caffeine is to mix some decaffeinated grounds in with your regular coffee and gradually use more and more decaf and less regular. Likewise, you can drink a can of decaffeinated soda now and then until you've weaned yourself off the caffeine.

Coffee   Coffee is one of the most widely consumed beverages in the world. Can it also be a beneficial drink healthwise? According to recent research, the answer may be yes! Habitual coffee consumption appears to be associated with a lower risk of type 2 diabetes. It is not known exactly how this apparent protective mechanism

caffeine dependence syndrome
dependence on caffeine characterized by at least three of the four following criteria: withdrawal symptoms such as headache and fatigue; caffeine consumption despite knowledge that it may be causing harm; repeated, unsuccessful attempts to cut back on caffeine; and tolerance to caffeine.

works. It may well be that the antioxidants found in coffee help control cell damage that can be a factor in development of type 2 diabetes. Whatever the mechanism, it is not the caffeine in coffee affording the benefits: Decaffeinated coffee alone offers the same degree of protection.[33]

The antioxidants found in coffee have also been found to inhibit inflammation and thereby reduce risk of cardiovascular and other inflammatory diseases in post-menopausal women. The most protective effect was found at a dose of 1 to 3 cups of coffee (caffeinated or decaffeinated) per day.[34] In males, consumption of 8 or more cups of coffee a day reduced the risk of cerebral infarction (stroke) when compared to non-coffee drinkers.[35] These anti-inflammatory properties of coffee may explain why coffee appears to reduce the risk of alcohol-related cirrhosis and liver cancer.[36] In fact, regular coffee consumption appears to provide a modest protective benefit from all causes of death in both men and women.[37]

To be sure, the antioxidants in coffee may protect us from anti-inflammatory diseases, but it remains prudent to remember that the caffeine in coffee can have negative effects on some people (review the preceding section). For adults consuming 3 to 4 cups of coffee a day, there is little indication of health risks and some suggestion of health benefits. Nevertheless, coffee consumption is linked with increases in cardiovascular disease risk factors such as blood pressure. This finding implies that individuals with high blood pressure may be more vulnerable to the adverse effects of caffeine.[38]

Tea "Of all the beverages consumed today, tea is undoubtedly one of the oldest, most widely known, and most widely consumed. Consumption was introduced to the world by traders and travelers."[39] The method by which tea is processed determines whether green, black, or oolong is produced. Green tea is made by steaming fresh leaves at high temperatures, followed by a sequence of drying and rolling steps, resulting in tea that is for the most part the same chemical composition as the fresh

Hugh Threlfall/Alamy Limited

leaves. Black tea is fermented. The extent of the fermentation period determines the chemical composition of the tea; hence, the chemical composition of most black teas differs.[40] Oolong is produced by "solar withering" of tea leaves followed by partial fermentation.[41]

Tea leaves are rich in phytochemicals, antioxidants, and caffeine, among a myriad of other chemicals. The amount of caffeine in a serving of tea is usually less than half the caffeine in a serving of coffee, but actual caffeine content can vary widely depending on the specific blend of tea and the brew strength.[42]

Health benefits most frequently associated with tea include:[43]

- Lowers risk of cardiovascular disease by
  - reducing inflammation
  - decreasing total cholesterol
  - decreasing LDL-cholesterol
  - protecting against blood clot formation
  - decreasing blood pressure

- Decreases risk of cancer by
  - Inhibiting carcinogenesis
  - Suppressing formation of carcinogenic substances
  - Increasing the effectiveness of a number of chemotherapeutic agents

- Protects against neurodegenerative diseases by
  - Decreasing plaque formation associated with Alzheimer's disease
  - Raising dopamine levels to ward off Parkinson's disease
  - Protecting against deterioration of brain function

The health benefits of tea appear to be dose dependent, meaning that drinking more tea provides the greatest effect.[44]

Maceofoto/istockphoto.com

Cocoa   Cocoa is indigenous to equatorial rain forests of Central and South America, where it was utilized as a beverage, food, and drug by the Olmec, Mayan, and Aztec empires preceding its introduction to Europe in the 1500s. According to Mayan and Aztec religions, cocoa was discovered by the gods and was prepared as a beverage. Perceived as an intoxicating food, the cocoa beverage was reserved for use only by adult males, specifically priests, highest government officials, military officers, distinguished warriors, and occasionally sacrificial victims for ritual purposes.[45]

Chocolate was introduced to Europe in 1544 by Mayan nobles who were taken to Spain by Dominican friars. Within a century, culinary and medicinal use of chocolate expanded to France, England, and other countries in Western Europe.[46] Although the confectionery history of chocolate is familiar to almost everyone, only recently, in the late 20th century, have the health benefits of chocolate returned to the forefront.

Cocoa contains several minerals (potassium, magnesium, phosphorus, and calcium among others), although mineral composition varies among different cocoa varieties and even within the same variety depending on how the cocoa is processed. Other nutrients found in cocoa are fats and proteins, in addition to antioxidants, phytochemicals, and caffeine. As with coffee and tea, cocoa and chocolate (more specifically, dark chocolate) may positively influence cardiovascular health by some of the same mechanisms.[47]

FIGURE 9-3

Generic Names for Soft Drinks

Soft Drinks Whether you call it soda, pop, soda pop, coke (Figure 9-3), or fizzy drinks (United Kingdom), you are probably talking about a carbonated soft drink. To be precise, a *soft drink* is a beverage that does not contain alcohol; for the purposes of this chapter, though, *soft drink* refers to a cold, carbonated, sweetened (sugar or noncaloric) beverage.

The history of soft drinks dates back to the medieval Middle East. The earliest soft drinks were made of fruit juices or extracts of flowers or herbs combined with sugar and water to form a syrup that was later added to water, ice, or even snow.[48] The first noncarbonated soft drinks appeared in the 17th century, and the process of carbonating water was soon mastered in the 18th century. Carbonated water soon led to flavored carbonated water: American pharmacists began adding herbs and chemicals for flavoring to carbonated water in the 19th century. Root beer, Coca-Cola™, Pepsi-Cola™, and Dr. Pepper™ were conceived and sold in pharmacies as health tonics.[49]

Carbonated soft drinks are no longer the unique products of pharmacies—and soft drinks are definitely no longer marketed as health tonics. Until the 1980s, soft drinks were sweetened with refined cane sugar or corn syrup. Today, regular (nondiet) soft drinks get their sweet taste from high-fructose corn syrup (HFCS). The biggest controversy regarding soft drinks in the diet today concerns their contribution to empty-calorie intake in the U.S. population, its replacement of nutrient-dense foods (such as milk), and potential health problems associated with their consumption.

A review of available scientific research regarding the effect of (regular) soft-drink consumption and nutrition and health issues found the following:[50]

- Increased energy consumption

- Increased weight

- Lower consumption of calcium, protein, riboflavin, fiber

Miniglossary of Soft-Drink Ingredients

Ever wonder about the ingredients used to make soft drinks? (As with any other food, ingredients used in soft drinks are approved and regulated by the U.S. Food and Drug Administration.) Depending on the product, ingredients may include:

filtered water regular soft drinks are 90 percent water, and diet soft drinks can contain up to 95 percent water.

carbon dioxide provides a unique taste (when mixed with water) and the characteristic "pop" you hear and "fizz" you see when you open a bottle.

flavors most soft drinks contain a combination of many individual flavors (some are natural, some are fruit-flavored extracts, others are herbs, spices, or artificial flavors).

colors are obtained from natural and synthetic sources.

acidulants add tartness and act as a preservative; usually in the form of phosphoric acid and citric acid, although sometimes malic acid or tartaric acid are used.

preservatives the acidity and carbonation of soft drinks usually keep the drinks from spoiling, but some soft drinks contain small amounts of preservatives to improve their shelf life.

potassium is found naturally in the water used to make soft drinks and in some of the flavoring agents and other ingredients used.

sodium is present in small quantities.

regular sweeteners most soft drinks are sweetened with high-fructose corn syrup (HFCS).

low-calorie sweeteners aspartame, saccharin, sucralose, and acesulfame K are approved for use in soft drinks.

SOURCE: The American Beverage Association. www.ameribev.org/

- Lower intake of milk, fruit, fruit juices

- Increased risk of several medical problems
  - Type 2 diabetes (associated with as little as 1 serving per day)
  - Osteoporosis and bone fracture (due to increased calcium excretion in the urine)
  - Dental caries
  - Increased blood pressure

- Research results funded by the food industry found no relationship (or only weak relationships) between soft-drink consumption and increased calorie intake.

- When regular soft drinks were replaced with diet soft drinks, many of the negative health effects (previously listed) were not seen—but the risk of obesity appears to be even higher in individuals who consume diet soft drinks.[51]

There are more than a few health risks and indications of harm, and *no* indisputable positive health benefits associated with soft-drink consumption. Soft drinks now provide more than one-third of total dietary intake of refined sugar, especially in children. Soft-drink consumption increases caloric intake but decreases total nutrient intake; in particular, soft drinks replace milk in the diet. Increased soft-drink consumption is to a certain extent to blame for the increased rates of obesity, osteoporosis, dental caries, and other health problems.[52]

Functional Beverages

"Want some whiskey in your water? Sugar in your tea? What's all these crazy questions they're askin' me?"*

These were a few of the simple questions asked by college students in the 1970s. Today, college students are asking:

Mama Told Me (Not to Come), Three Dog Night, 1970. Lyrics by R. Newman.

Increase in serving sizes of soft drinks

Standard serving size

1950s	6.5 oz
1960s	12 oz
1990s	20 oz
2000+	32 oz & 64 oz

"Do I want vitamins or minerals in my water?"

"Do I want amino acids in my water?"

"Do I want fiber in my water?"

"Do I want a beverage to enhance my sports performance?"

"Do I want a beverage to help me stay alert and provide energy so I can study?" Beverage choices have exploded! You can find a beverage for just about any need you can ponder. But are the claims true? What is beneficial? What may be harmful?

Rachel Epstein/PhotoEdit, Inc.

Energy Drinks Blue Ox™. Venom™. Full Throttle™. Monster Energy™. Rage™. Cocaine™. Atomic Energy™. Red Bull™. The names say it all. And they promise even more—unrestrained energy in a can. Need a pick-me-up? Get an energy drink! Want to enhance your workout? Get an energy drink! Need to study all night for that midterm exam? Get an energy drink! Apparently, the ads are working: More than $2 billion a year is spent on energy drinks.

Energy drinks are nonalcoholic beverages containing purportedly energy-enhancing ingredients. The primary ingredients are sugar and caffeine. Other ingredients may include vitamins, amino acids, guarana, carnitine, inositol, ginseng, glutamic acid, ginkgo biloba, royal jelly, and yohimbe, among others.[53]

There are several nutrition and health concerns regarding use of energy drinks:[54]

- Energy drinks are not regulated by the Food and Drug Administration. Any amount and any mixture of ingredients can be used without being proven safe or effective.

- The level of caffeine found in most energy drinks is not suitable for children or anyone sensitive to caffeine.

Miniglossary of Energy Drink Ingredients

What's in my energy drink?

caffeine central nervous system stimulant.

carnitine [KAR-nih-teen] synthesized from the amino acids lysine and methionine in sufficient quantities to meet the needs of most people; it helps the body convert fatty acids into energy.

creatine [KREE-uh-teen] an amino acid synthesized in the body; used as a dietary supplement to increase lean body mass and strength; the amount added to most energy drinks is too small to be of any benefit.

ginkgo biloba [GING-coe bye-LOW-buh] an herb used to treat any number of conditions, including fatigue and failing memory.

ginseng [gin-SING] an herb used to support overall health and boost the immune system; research has not been conclusive enough to prove health claims.

glutamic acid [gloo-TAM-ik] a nonessential amino acid found in plant and animal tissue; used as a flavor-intensifying seasoning.

guarana [gwuh-RAH-nuh] an herb chemically equivalent to caffeine.

inositol [ih-NAH-sih-tall] once identified in association with the B vitamins; a type of sugar (related to glucose) found synthesized in the body and many foods.

kola nut an herb chemically equivalent to caffeine.

pyruvate [pie-ROO-vate] fatigue fighter and fat burner.

royal jelly substance secreted from the saliva glands of honeybees; provides food for all young larvae and is the only food for larvae that will develop into queen bees. It has been incorrectly claimed to impart extra energy and to have restorative properties; can be potentially hazardous for individuals who may suffer allergic reactions.

yerba maté an herb chemically equivalent to caffeine.

yohimbe [yoh-HIM-bee] touted as an aphrodisiac, a cure-all for sexual dysfunction; also claimed to be able to alter body mass, muscle mass, or exercise performance when combined with resistance training. None of these claims have been proven by research.

SOURCE: L. Bonci, "Energy" drinks: Help, harm, or hype? www.gssiweb.com/Article_Detail.aspx?articleid=310.

- Mixing energy drinks with alcohol increases the risk of heart arrhythmias.

- Using energy drinks before or during exercise may cause nervousness, light-headedness, and nausea.

- Caffeine is a diuretic and may be counterproductive for replacing fluids lost during exercise.

- Mixing several different stimulants (caffeine, guarana, kola nuts, yerba maté) can intensify their effects and become unsafe.

- Sudden cessation of use of heavily caffeinated energy drinks may cause caffeine withdrawal. Withdrawal symptoms include headaches, mood swings, and trouble concentrating.

Energy drinks tend not to live up to most of the claims they make. Don't take the ads at face value. It's also a good idea to question the marketing ethics behind energy drinks. These beverages should be used with caution, and they certainly should not be used by everyone.[55] Want more energy? Want to enhance your workout? Want to be at your best studying for that midterm? Eat a healthful diet, drink plenty of water, and get some sleep!

Sports Drinks   The first sports drink was invented by Dr. Robert Cade and a team of scientists at the University of Florida (UF) in 1965. They were attempting to design a beverage to alleviate dehydration and muscle cramps encountered by their football players. To do this, their beverage had to replace fluids and electrolytes lost due to heat and humidity during practices and games. They named their drink after the UF mascot, and thus Gatorade™ was born.[56] The sports beverages market exploded in the 1990s, and today it is a $2.6 billion business.[57]

Sports drinks are formulated to replace fluids and electrolytes (minerals) lost through sweat and to provide energy for muscles.[58] They may help delay fatigue and help with postexercise recovery of electrolytes and glycogen synthesis (see Chapter 11). Flavorings are generally added to enhance fluid intake.[59]

What should you look for in a sports drink?[60]

- Six percent carbohydrate in the form of glucose, sucrose, and fructose

- Sodium

- No carbonation

- No caffeine

See Chapter 11 for more about water, fluid-replacement drinks, and exercise.

Sports drinks containing only fructose should be avoided. It slows fluid absorption and causes abdominal cramps. Carbonation can also cause stomach discomfort and therefore should be avoided. Caffeine, as you know, can cause dehydration and should not be a component of a sports drink.[61]

"Enhanced" Waters   Enhanced water? If you think plain water is boring, you might be tempted to try some of the newest "waters" to hit the market. You might think they taste better, and they certainly are prettier. As you gaze at the display shelf, how do you decide which one to choose? The pink one? The purple one? The peach-colored one? And you just might be convinced that you actually need water with added vitamins, minerals, amino acids, herbs, antioxidants, and fiber.

Let's begin with waters that have added vitamins. Water-soluble vitamins (taken in excess of amounts we need) added to water will flush out of the body along with the water. Fat-soluble vitamins are almost impossible to absorb without the presence of fat (see Chapter 5). The only ingredient in "vitamin waters" that your body just might use is the sugar. If the calories are in excess of what you need for energy, your body will use it to make fat.

How about the water with fiber? Everyone knows Americans don't get the recommended amount of fiber each day. The fiber added to water is maltodextrin. There is no research to demonstrate that maltodextrin works in the same way as fiber from whole grains, vegetables, and beans. Also, because maltodextrin is a soluble fiber (Chapter 4), it won't help with "regularity."[62]

Although no research has ever shown that antioxidant *supplements* help prevent disease, "antioxidant water" is in stores, claiming to "help protect your body." What makes "invigorating" or "energizing" water so invigorating or energizing? It just might be the 50 milligrams of caffeine per bottle (about the same as a cup of coffee). Herbs added to enhanced water are not added in amounts that might be therapeutic.[63]

Most enhanced waters are no more healthful than other beverages containing added sugar. Adding vitamins and minerals to calorie-dense foods does not make them nutrient dense. Some of the ingredients added to these specialty waters may not have proven health benefits. If you really need enhancement, a vitamin-mineral supplement is a better choice.

© Polara Studios, Inc.

Alcohol

Is alcohol a nutrient? Or is it a drug? **Alcohol** is *not* a nutrient; actually, it is a general term used to describe a group of organic chemicals with common properties. Members of this group include ethanol, methanol, and isopropanol, among others. This chapter discusses the most commonly ingested substance in this group: ethyl alcohol or ethanol (EtOH). Therefore, for the remainder of this chapter, *alcohol* refers to ethanol.

Alcohol is a clear, volatile liquid that burns easily. In fact, it is so flammable that it can easily be used for fuel. It has a slight, characteristic odor and is very soluble in water. Physiologically, alcohol is a sedative and central nervous system depressant. Just as with other drugs, the extent to which central nervous system function is impaired by alcohol consumption is directly related to the amount of alcohol in the blood. Similarly, as with other drugs, any health benefits resulting from alcohol consumption depend on the amount of alcohol consumed.

It's difficult to classify alcohol in terms of nutrition. Like the energy nutrients (carbohydrates, proteins, and fats), alcohol supplies energy (7 calories/gram). Unlike the energy nutrients, alcohol is not an essential nutrient, nor is it stored in the body.

Absorption and Metabolism of Alcohol
When alcohol is consumed, it passes down the esophagus, through the stomach, and into the small intestine. About 20 percent of the alcohol is absorbed in the stomach, and about 80 percent is absorbed in the small intestine (see Figure 9-4).[64] Alcohol absorbed from the small intestine passes into the portal vein and is transported directly to the liver, where a portion of the alcohol is then metabolized by enzymes. One enzyme in particular, **alcohol dehydrogenase**, facilitates conversion of alcohol into **acetaldehyde** and water. Acetaldehyde is quickly transformed to acetate by other enzymes and then ultimately metabolized to form carbon dioxide and water. Until all the alcohol consumed has been metabolized, it is disseminated throughout the body via the bloodstream and so affects the brain and other tissues.[65]

Nearly all alcohol ingested is metabolized in the liver, but a small amount remains unmetabolized and can be measured in breath (lungs exhale about 5 percent of alcohol, which can be detected by devices such as the Breathalyzer) and urine (kidneys eliminate approximately 5 percent of alcohol in the urine).[66] Alcohol is also eliminated through sweat, feces, breast milk, and saliva.

alcohol clear, colorless, volatile liquid; the most commonly ingested form is ethyl alcohol or ethanol (EtOH).

alcohol dehydrogenase a liver enzyme that mediates the metabolism of alcohol.

acetaldehyde (ass-et-AL-duh-hide) a substance into which drinking alcohol (ethanol) is metabolized.

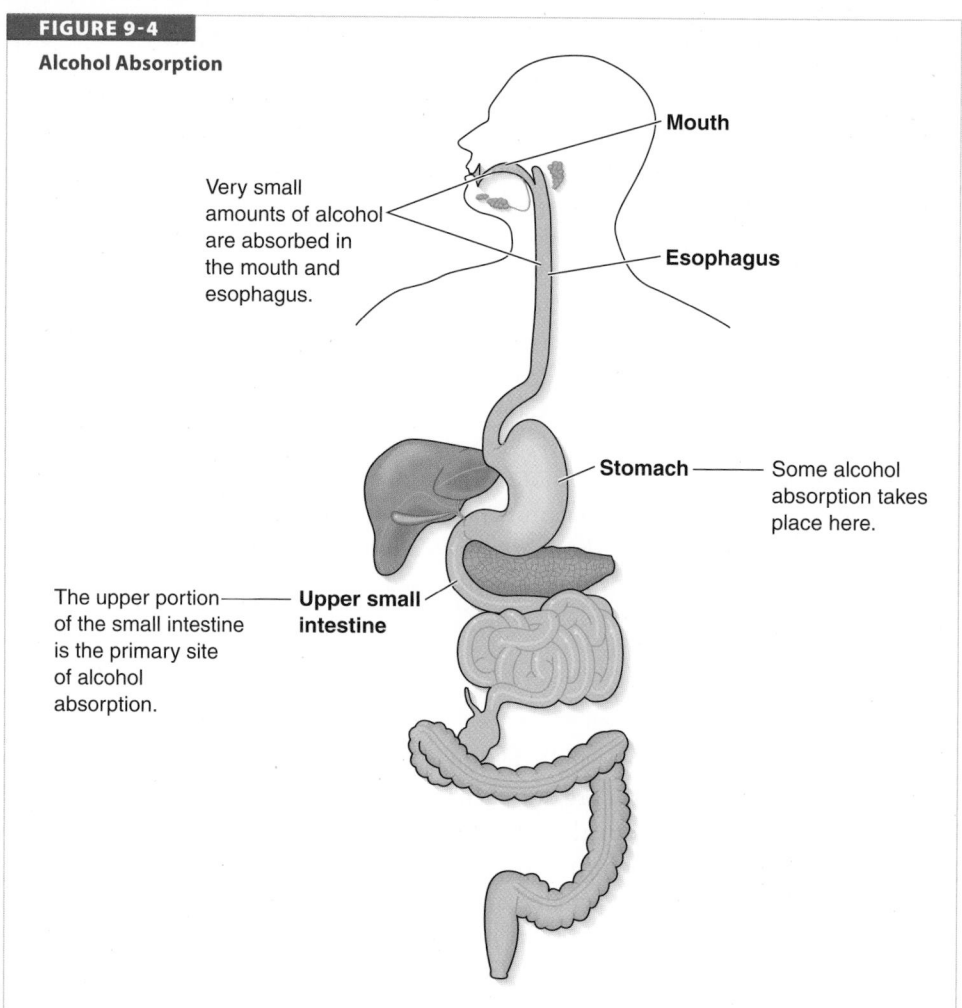

FIGURE 9-4

Alcohol Absorption

Mouth

Very small amounts of alcohol are absorbed in the mouth and esophagus.

Esophagus

Stomach — Some alcohol absorption takes place here.

The upper portion of the small intestine is the primary site of alcohol absorption.

Upper small intestine

Even though this man and woman are the same height and weight, the woman has a lower capacity for metabolizing alcohol.

- Body composition: Because women have less water in their bodies, the concentration of alcohol in their bodies is higher (drink for drink, more alcohol gets into women's bodies).
- Enzymes: Alcohol dehydrogenase is about 40 percent less active in a woman's stomach, allowing more undiluted alcohol to enter the bloodstream.
- Hormones: Alcohol can change estrogen levels during a woman's menstrual cycle, increasing the risk for breast cancer.

The liver can metabolize only a limited amount of alcohol per hour, no matter how much is consumed. A healthy average person can eliminate ½ ounce (15 milliliters) of alcohol per hour. The rate of alcohol metabolism depends, in part, on the amount of alcohol dehydrogenase formed in the liver. The amount of alcohol dehydrogenase produced in the liver varies among individuals and appears to depend on a person's genetic makeup.[67] Nonetheless, alcohol is metabolized more slowly than it is absorbed,[68] and it takes approximately one hour to metabolize one standard drink[69] (see the Savvy Diner feature later in this chapter). Controlled consumption can prevent accumulation of alcohol in the body and, therefore, can prevent intoxication as well.[70]

Factors Influencing Absorption and Metabolism

Food Presence of food in the stomach slows absorption of alcohol.[71] The rate at which alcohol is absorbed depends on how quickly or slowly the stomach empties its contents into the small intestine. Dietary fat delays emptying time of the stomach and consequently slows absorption of alcohol.

Gender Men and women absorb and metabolize alcohol differently. Women have higher blood alcohol content (BAC) after consuming the same amount of alcohol as men. A woman absorbs 30 percent more alcohol into her bloodstream than does a man of the same weight. One drink for a woman could potentially have the same effect as two drinks for a man. Additionally, women are more susceptible to alcoholic liver disease, heart muscle damage,[72] and brain damage.[73] These metabolic differences can be attributed to smaller amounts of body water[74] in women's bodies and a lower activity of alcohol dehydrogenase in the stomach, causing a larger proportion of ingested alcohol to reach the blood.[75]

Ethnicity Native Americans have higher rates of liver damage due to alcohol consumption than do members of other ethnic groups.

Effects of Alcohol

Because it is distributed so quickly and thoroughly in the body, alcohol rapidly affects the central nervous system. Even small amounts of alcohol in the blood can slow reactions (see Figure 9-5). The body responds to alcohol in stages; therefore, higher BAC leads to increased loss of mental and physical control (see Table 9-2). If a large amount of alcohol is consumed over a short period of time, a person may lose consciousness or even die.[76]

Alcohol and Medications Use of prescription or over-the-counter medications can increase the effects of alcohol. Chronic, heavy drinking (more than two drinks a day for women or more than four drinks a day for men over a couple of years) appears to activate an enzyme that may be responsible for changing the over-the-counter pain reliever acetaminophen (for example, Tylenol) and many others into chemicals that can produce liver damage, even when acetaminophen is taken in recommended doses[77] (four to five "extra-strength" pills taken over the course of the day).[78] This interaction is more likely to occur when acetaminophen is taken after, rather than before, the alcohol has been metabolized.

Alcohol and Sex Hormones Alcohol metabolism alters the balance of sex hormones in men and women.[79] In men, alcohol metabolism contributes to testicular injury and impairs testosterone synthesis and sperm production.[80] Prolonged testosterone deficiency may contribute to feminization, such as breast enlargement, in males.[81]

Alcohol and Urine Output Although some people believe it is simply due to the amount of extra liquid they are ingesting, excessive urination when drinking alcohol really has more to do with the alcohol itself. In fact, an ounce of hard liquor will hypothetically cause the same amount of urination as a 12-ounce beer. Alcohol blocks the antidiuretic hormone (ADH), which is secreted from the pituitary gland and is responsible for regulating urine production. ADH works by causing the kidneys to conserve fluids. Thus, when this hormone is blocked, urine output increases considerably because the kidneys are not properly absorbing fluids. This leads to significant loss of water and eventually to dehydration.

Alcohol and Hangovers A *hangover* actually consists of a group of ailments, usually including but not limited to headache, nausea, vomiting, sensitivity to light and sound, dry mouth, and irritability. As explained earlier, dehydration results from excessive consumption of alcohol, and it is thought to be the main factor contributing to what is referred to as a hangover. In addition, alcohol affects the stomach lining, causing considerable distention. As the depressive effects of alcohol wear off, the person's sensitivity to noise and sound are amplified. Although theories (and myths) abound about how and why hangovers occur and how to subdue their effects, the best way to avoid these unwanted side effects is to drink in moderation.

FIGURE 9-5

Alcohol's Effects on the Brain

Alcohol is rightly termed an anesthetic because it puts brain centers to sleep in this order: first the cortex, then the emotion-governing centers, then the centers that govern muscular control, and finally the deep centers that control respiration and heartbeat.

Next most sensitive: voluntary muscle control and emotion-governing centers (sensory area)

Most sensitive: judgment and reasoning (motor and speech area)

Last to be affected: respiration and heart action

TABLE 9-2   Effects of Drinking Alcohol

Approximate BAC[a]	Stage	Recognizable Consequence[b]
0.01–0.02	Subclinical	Behavior nearly normal by ordinary observation
0.03–0.12	Euphoria	Increased self-confidence or daring
0.08–0.10 or greater (legal limit defining intoxication in most states)		Attention span shortens Skin flushes Judgment diminishes; may say first thought that comes to mind rather than appropriate comment for a given situation Has trouble with fine motor movements, such as writing or signing name
0.09–0.25	Excitement	Becomes sleepy Has trouble understanding or remembering things (even recent events); impaired perception and comprehension Decreased coordination, increased reaction times Blurred vision, reduced peripheral vision and glare recovery Decreased sensation (hearing, tasting, feeling, etc.)
0.18–0.30	Confusion	Disorientation, mental confusion (persons might not know where they are or what they are doing) Dizziness and staggering Exaggerated emotional states (aggressive, withdrawn, overly affectionate) Vision disturbances Sleepiness Slurred speech Uncoordinated movements (trouble catching a thrown object) Increased pain threshold (may not feel pain as readily as a sober person)
0.25–0.4	Stupor	Can barely move at all Cannot respond to stimuli Cannot stand or walk Vomiting, incontinence Lapses in and out of consciousness
0.35–0.50	Coma	Unconscious Reflexes depressed (pupils do not respond appropriately to light) Body temperature lowers Breathing slows and becomes shallower Heart rate slows Possible death
>0.50	Death	Death from respiratory arrest

[a] Grams per 100 milliliters of blood or grams per 210 liters of breath.

[b] [Blood alcohol content] BAC and effects of drinking alcohol vary from person to person and depend on body weight, amount of food eaten while drinking, and the person's ability to tolerate alcohol.

SOURCE: Adapted from C. C. Freudenrich, *How Alcohol Works*, available online at www.howstuffworks.com/alcohol2.htm.

Alcohol and Blood Alcohol Level Blood alcohol level (BAL) indicates the amount of alcohol in the bloodstream and is measured in milligrams of alcohol per 100 milliliters of blood. This system does not directly relate to how many drinks you consume; instead, it indicates the amount of alcohol actually circulating in the blood at a given time. BAL is affected by the amount of alcohol ingested and how quickly it is consumed. Because the liver can metabolize only one drink per hour, any amount beyond that will hypothetically enter the bloodstream at full strength. As mentioned already, many variables determine the effects of alcohol on a person. Height, weight, and gender all have strong associations with what BAL will be after a given amount of alcohol is consumed. Although a Breathalyzer or urinalysis provides a far more accurate reading, Table 9-3 provides an estimate of BAL for different people at different levels of alcohol consumption (see Table 9-2 for a reminder of what these levels represent).

Alcohol and Driving The best advice is *never* to drink and drive. Studies have shown that even one drink can considerably impair a driver's response time. Although driv-

TABLE 9-3	Estimate of Blood Alcohol Level (BAL) at Different Body Weights and Levels of Alcohol Consumption							
Drinks Consumed/Sex		Weight (in pounds)							
		100	120	140	160	180	200	220	240
1	Male	.04	.04	.03	.03	.02	.02	.02	.02
	Female	.05	.04	.04	.03	.03	.03	.02	.02
2	Male	.09	.07	.06	.05	.05	.04	.04	.04
	Female	.10	.08	.07	.06	.06	.05	.05	.04
3	Male	.13	.11	.09	.08	.07	.07	.06	.05
	Female	.15	.13	.11	.10	.08	.08	.07	.06
4	Male	.17	.15	.13	.11	.10	.09	.08	.07
	Female	.20	.17	.15	.13	.11	.10	.09	.09
5	Male	.22	.18	.16	.14	.12	.11	.10	.09
	Female	.25	.21	.18	.16	.14	.13	.12	.11
6	Male	.26	.22	.19	.16	.15	.13	.12	.11
	Female	.30	.26	.22	.19	.17	.15	.14	.13
7	Male	.30	.25	.22	.19	.17	.15	.14	.13
	Female	.36	.30	.26	.22	.20	.18	.16	.15
8	Male	.35	.29	.25	.22	.19	.17	.16	.15
	Female	.41	.33	.29	.26	.23	.20	.18	.16
9	Male	.39	.35	.28	.25	.22	.20	.18	.16
	Female	.46	.38	.33	.29	.26	.23	.21	.19
10	Male	.39	.35	.28	.25	.22	.20	.18	.16
	Female	.51	.42	.36	.32	.28	.25	.23	.21
11	Male	.48	.40	.34	.30	.26	.24	.22	.20
	Female	.56	.46	.40	.35	.31	.27	.25	.23
12	Male	.53	.43	.37	.32	.29	.26	.24	.21
	Female	.61	.50	.43	.37	.33	.30	.28	.25
13	Male	.57	.47	.40	.35	.31	.29	.26	.23
	Female	.66	.55	.47	.40	.36	.32	.30	.27
14	Male	.62	.50	.43	.37	.34	.31	.28	.25
	Female	.71	.59	.51	.43	.39	.35	.32	.29
15	Male	.66	.54	.47	.40	.36	.34	.30	.27
	Female	.76	.63	.55	.46	.42	.37	.35	.32

SOURCE: From California State University Bakersfield Counseling Center, Blood alcohol level. www.csub.edu/counselingcenter/mentalhealth/bloodalcohol.shtml.

ing is a very normal task, drivers must remember that behind the wheel of a car, they can do substantial damage to property and can even take people's lives, including their own. Drinking and driving is simply not worth the risk. In most states, the legal BAC limit is 0.08; for those under the age of 21, there is zero tolerance for any trace of alcohol. If this limit is exceeded, the driver could receive a citation for DUI (driving under the influence) or DWI (driving while intoxicated), which in some states is a felony.

Alcohol and Tolerance **Tolerance** is the decrease in effectiveness of a drug after a period of prolonged or heavy use. Continued exposure to alcohol causes physiological adaptations that increase tolerance. These adaptations change a person's behavior. Two types of tolerance are at work with alcohol adaptation. **Metabolic tolerance** occurs because, with continued exposure, alcohol is metabolized at a higher rate (up to 72 percent faster). Long-term exposure to alcohol increases levels of alcohol dehydrogenase in the liver, resulting in lower peak blood alcohol concentrations. This means that the body becomes more effective at removing high levels of alcohol

tolerance decrease in effectiveness of a drug after a period of prolonged or heavy use.

metabolic tolerance increased efficiency of removing high levels of alcohol from the blood due to long-term exposure; may lead to more drinking and possible addiction.

in the blood. Even so, it also means the person must drink more alcohol to experience effects comparable to those felt before, which leads to more drinking and can contribute to addiction.[82]

Functional tolerance results from an actual change in sensitivity to a drug. Compared to the average person, chronic alcohol users can have twice the tolerance for alcohol, since the normal chemical and electrical functions of their nerve cells increase to counteract the inhibitory effects of alcohol exposure. The increased nerve activity helps chronic alcohol users function normally when they have a higher BAC, but it also tends to make them irritable when they are not drinking. Furthermore, the increased nerve activity may make them crave alcohol. Undoubtedly, the increased nerve activity contributes to hallucinations and convulsions (such as delirium tremens) that may occur when alcohol is removed, making it challenging to overcome **alcohol abuse** and **alcohol dependency**.[83]

Impact of Alcohol on Nutrition   If you are a light drinker, in good health, and otherwise well nourished, the occasional consumption of alcohol will probably have little effect on your nutritional status. The biggest risk to you will most likely come from the additional calories provided by alcohol, which may contribute to unwanted weight gain. See Table 9-4 for the calorie content of commonly consumed alcoholic beverages. This is not to say that alcohol has *no effect* on your nutritional status, because it does. Fundamental changes in metabolism occur whenever you consume alcohol. The extent to which alcohol affects your nutritional status depends on how much alcohol you consume and your current nutritional and health status.

If you drink excessively on a regular basis, your nutritional status will become compromised. Protein deficiency can develop, both from the depression of protein synthesis in the cells and, in the drinker who substitutes alcohol for food, from poor diet. Alcohol affects every tissue's nutrient metabolism in different ways. Stomach cells become inflamed and vulnerable to ulcer formation. Intestinal cells fail to absorb thiamin, folate, and vitamin B_{12}. Liver cells lose efficiency in activating vitamin D, and they alter their production and excretion of bile. Rod cells in the retina, which normally process vitamin A alcohol (*retinol*) into the form needed for vision (*retinal*), end up processing drinking alcohol (*ethanol*) instead. The kidneys excrete increased quantities of magnesium, calcium, potassium, zinc, and folate. Acetaldehyde interferes with metabolism, too. It dislodges vitamin B_6 from its protective binding protein so that it is destroyed, causing a vitamin B_6 deficiency and thereby lower production of red blood cells.

functional tolerance actual change in sensitivity to a drug; often results in hallucinations and convulsions when alcohol is withheld.

alcohol abuse continued use of alcohol despite negative psychological, social, family, employment, or school problems caused by the alcohol use.

alcohol dependency (alcoholism) a condition marked by compulsive, uncontrollable drinking that has negative effects on the affected person's physical health, family relationships, and social health.

TABLE 9-4	Calories in Alcoholic Beverages and Mixers		
Beverage		**Amount (oz)**	**Calories**
Beer		12	150
Light beer		12	100
Gin, rum, vodka, whiskey (80 proof)		1½	100
Dessert wine		3½	140
Table wine (red, rosé, white)		5	100
Wine cooler		12	170
Liqueurs		1½	155–185
Tonic, ginger ale		8	80
Cola		8	100
Fruit juice		8	110
Club soda, plain seltzer, diet soda		8	0

Health Benefits of Alcohol   Drinking moderate amounts (see the Savvy Diner feature later in this chapter) of alcohol appears to be safe for healthy people who do not have alcohol abuse or dependency problems.[84] People who consume one to two drinks daily have lower mortality rates than nondrinkers.[85] Alcohol, like any other drug, has a beneficial dose and a level (dose) that will cause harm. Whether alcohol consumption will produce beneficial or adverse effects depends on a person's alcohol consumption, age, and background for cardiovascular risk. Rates of death from all causes are lowest among those who report consuming one drink per day.[86] Although most research indicates that wine consumption is the most beneficial, it appears that the benefits are from the alcohol itself, not the other components of each type of alcoholic beverage.[87]

Alcohol and Cardiovascular Health   Cardiovascular disease is the leading cause of death worldwide.[88] More than a dozen research studies have demonstrated a consistent, positive correlation between moderate alcohol consumption and decreased incidence of heart disease.[89] The protective effect of alcohol is the result of increased levels of high-density lipoprotein (HDL) cholesterol.[90] HDL-cholesterol removes cholesterol from the lining of artery walls and carries it back to the liver for excretion. Alcohol also inhibits blood from forming clots, reducing risk of death from heart attack.[91] This anticlotting effect of moderate drinking reduces the risk of thrombotic or ischemic stroke (blockage of blood vessel in the brain) but increases the risk of hemorrhagic stroke (rupture of a blood vessel within the brain).

Health Risks of Alcohol   Moderate drinking is not risk-free. Deaths reduced by moderate alcohol consumption are generally found in age groups with high rates of coronary heart disease—in other words, in people 45 years and older. However, most deaths due to alcohol consumption occur in people younger than 45 years.[92] Among young adults, the risks of alcohol consumption (alcohol abuse and dependence, alcohol-related violent behavior and injuries) far outweigh any benefits that may accrue later in life.[93]

Accidents   Alcohol affects judgment and slows reflexes, which can lead to falls or accidents with vehicles or other heavy machinery. It can also increase the likelihood of homicide and suicide. Alcohol makes men less able to control impulses toward violence and makes women less capable of recognizing cues to potentially violent behavior.

Drug Interactions   Like alcohol, many other **drugs** are metabolized in the liver. The liver has limited processing capacity, and alcohol and drugs may compete with each other for metabolism. Alcohol may hinder the metabolism of another drug, keeping the medication in the system longer than intended and possibly increasing risk of side effects. See Table 9-5 for specific alcohol–drug interactions.

Other Risks

Night Blindness   Alcohol blocks formation of retinal, a compound in the eye responsible for vision in low light.

Breast Cancer   Even when consumed in moderate amounts (one drink per day for women), alcohol may increase the risk of breast cancer in women.

Other Cancers   Heavy alcohol use, especially when combined with smoking, appears to increase risk of cancer of the throat and esophagus. Risk of cancer increases for people who have hepatitis and cirrhosis. Risk of colon cancer also increases with alcohol use.

drugs substances that can modify one or more of the body's functions.

TABLE 9-5	Specific Alcohol–Drug Interactions
Drug	**Interaction with Alcohol**
Antibiotics	Nausea, vomiting, headache, and possible convulsions
Antidepressants	Increases sedative effect, impairing mental skills used for driving; other interactions can produce a dangerous increase in blood pressure
Oral hypoglycemic drugs (used to treat type 2 diabetes)	Can prolong effects, causing nausea, headache, or a dangerous decrease in blood glucose levels; chronic alcohol consumption can decrease the drug's effect
Over-the-counter antihistamines (such as Benadryl) used to treat symptoms of allergy and insomnia	May intensify the sedation, causing dizziness or drowsiness
Barbiturates	Prolongs sedative effect; chronic alcohol use decreases sedative effect
Cardiovascular medications	Can cause dizziness or fainting upon standing up; chronic alcohol use may decrease therapeutic effect of some cardiovascular drugs
Cocaine/crack	Increases heart rate three to five times as much as when either drug is used alone (cocaine increases metabolism, which in turn increases blood alcohol levels)
Ecstasy/MDMA	Increases dehydration to dangerous levels
GHB	Slows rate of breathing to dangerous levels
Heroin	Can induce heroin overdose, leading to cessation of breathing
Marijuana	Increases suppression of gag reflex (inhibiting vomiting), which can increase risk of alcohol poisoning
Narcotic pain relievers (opiates, morphine, codeine, Darvon, Demerol)	Enhances sedative effect of both substances, increasing risk of death by overdose
Nonnarcotic pain relievers (aspirin and similar nonprescription pain relievers; acetaminophen [Tylenol])	Increases stomach bleeding and inhibition of blood clotting; increases risk of liver damage
Oral contraceptives	May hinder absorption of alcohol into bloodstream, delaying onset of intoxication
PCP/Special K	Compounds central nervous system (CNS) depression
Sedatives and hypnotics (sleeping pills, Valium)	Severe drowsiness, increased risk of household and automotive accidents
Tobacco	Doubles risk of heart attack; increases risk for lung, throat, mouth, and bladder cancers, stroke, high blood pressure, miscarriage, and low birth-weight infants

SOURCE: *Drinking: A Student's Guide*, "Alcohol and Legal Drugs," available at www.mcneese.edu/community/alcohol/index.html.

Liver Damage **Alcoholic hepatitis** (inflammation of the liver) and **cirrhosis** (scarring of liver tissue that interferes with blood flow and liver function) are two common consequences of heavy drinking. Hepatitis can lead to permanent liver damage if alcohol consumption continues.

High Blood Pressure and Stroke Heavy drinking causes high blood pressure (hypertension) as the heart compensates for the initial reduction in blood pressure caused by alcohol consumption. Hypertension is a risk factor for stroke and heart disease.

Pancreatitis Long-term alcohol consumption can result in recurrent attacks of severe pain caused by inflammation of the pancreas. The pancreas is responsible for production of many digestive enzymes. Continued use of alcohol can ultimately cause permanent damage to the pancreas.

Gastrointestinal Symptoms Alcohol irritates the lining of the stomach and impairs intestinal enzymes and transport systems. Thus alcohol consumption can cause a wide range of common, uncomfortable, but reversible problems, including gastritis

alcoholic hepatitis inflammation and injury to the liver due to excess alcohol consumption.

cirrhosis a chronic, degenerative disease of the liver in which the liver cells become infiltrated with fibrous tissues; blood flow through the liver is obstructed, causing back pressure and eventually leading to coma and death unless the cause of the disease is removed; the most common cause of cirrhosis is chronic alcohol abuse.

The Savvy Diner — What Is a Drink?

Ancient Persians ate five almonds to prevent hangovers. For Romans and Greeks, celery was thought to cure a hangover. By the late 1800s, celery still had a medicinal use: Sears, Roebuck & Company advertised a celery tonic to calm the nerves.[94]

According to the *Dietary Guidelines for Americans* (2005), "If you drink alcoholic beverages, do so in moderation." But what is drinking in moderation? If you decide to drink alcohol, you should have as much information as possible so that you can drink responsibly.

Moderation is defined as

- Men: no more than two drinks per day
- Women: no more than one drink per day

These limits are based on the disparity between men and women in both weight and alcohol metabolism. Whereas drinking in moderation provides little, if any, health benefits for young adults, moderate drinking for men over the age of 45 and women over the age of 55 may lower their risk for coronary heart disease.

More than one drink a day for women or two drinks a day for men can increase risk for motor vehicle accidents, other injuries, high blood pressure, stroke, violence, suicide, and certain types of cancer. Even one drink per day can slightly increase the risk of breast cancer. Alcohol consumption during pregnancy (see the Spotlight feature in this chapter) increases the risk of birth defects.

What Constitutes a Drink?

Alcohol is alcohol is alcohol. It does not matter if the beverage of choice is beer, wine, a wine cooler, a cocktail, or a mixed drink. A standard serving of beer, distilled spirits, and wine each contain the same amount of alcohol (see the photo). A *standard serving* is:

- 12 ounces of regular beer (150 calories)
- 5 ounces of wine (100 calories)
- 1½ ounces of 80-proof distilled spirits (100 calories)
- 12 ounces of wine/malt or spirit-based cooler (210–230 calories)
- 3 ounces of sherry or port (130 calories)
- 9¾ ounces of malt liquor (135 calories)

When it comes to drinking alcohol, the old adage is true: It doesn't matter what you drink, it's really *how much* you drink that counts.

Sometimes responsible drinking means not drinking at all. Some people who should not drink alcoholic beverages include:

- Children and adolescents
- Individuals of any age who cannot restrict their drinking to moderate levels
- Women who *may* become pregnant or who are pregnant. A safe alcohol intake has not been established for women at any time during pregnancy, including the first few weeks (see the Spotlight feature later in this chapter).
- Individuals who plan to drive, operate machinery, or take part in any activities that require attention, skill, or coordination. Most people retain some alcohol in their blood for up to two to three hours after a single drink.
- Individuals who are taking prescription or over-the-counter medications that can interact with alcohol

SOURCES: American Dietetic Association, Alcohol beverages: Making responsible drinking choices (available online at www.eatright.org); *Dietary Guidelines for Americans* (available online at www.health.gov/dietaryguidelines); National Consumers League, Alcohol: How it all adds up (available online at www.nclnet .org).

12 oz beer — 10 oz wine cooler — 1½ oz hard liquor (80 proof whiskey, gin, brandy, rum, vodka) — 5 oz wine

© Polara Studios

(inflammation of the lining of the stomach), stomach and intestinal ulcers, diarrhea, and weight loss.

Brain Damage Brain cells in various parts of the brain die, reducing total brain mass.

Decreased Sex Hormone Production Alcohol causes decreased testosterone secretion from the hypothalamus/pituitary and testes, resulting in decreased sperm production. It can cause infertility in women by disrupting or changing the menstrual cycle.

Anemia Poor nutrition as a result of excessive alcohol intake decreases levels of iron and B vitamins, leading to anemia.

Emotional and Social Problems Alcohol affects emotional centers in the limbic system, causing anxiety and depression. Emotional and physical effects of alcohol can contribute to marital and family problems, including domestic violence, as well as to work-related problems such as excessive absences and poor work performance.

Weighing the Pros and Cons of Alcohol Consumption

How do you decide if low to moderate alcohol consumption will provide health benefits or health risks for you? One way is to compare your age and gender to the leading causes of death for those of similar ages and gender. The leading causes of death for men under the age of 40 and women under the age of 50 (premenopausal) are accidents and breast cancer, respectively. In these cases, the risks of low to moderate alcohol consumption outweigh the benefits. The leading cause of death for men over the age of 40 and women over the age of 50 is heart disease. Thus, for these people, the benefits of low to moderate alcohol consumption outweigh the risks.

What Is Alcohol Abuse or Alcoholism?

Alcoholism is a dependency on alcohol characterized by craving (a strong need to drink), loss of control (being unable to stop drinking despite a desire to do so), physical dependence and withdrawal symptoms, and tolerance (increasing difficulty in becoming drunk).

Although many people stereotype persons with alcoholism as those who spend their lives at a bar or as skid-row bums, in fact anyone can be an alcoholic. Previous notions of alcoholism as a moral flaw have basically been replaced with the theory that alcoholism is a disease with genetic, psychosocial, and environmental factors and influences. Successful people such as lawyers, doctors, professors, and even college students can suffer from what is commonly known as *alcoholism*.

The following is a scale of alcohol addiction from Narcotics Anonymous (NA) and Alcoholics Anonymous (AA):[95]

Use: Alcohol "use" is considered to be the ingestion of alcohol or other drugs without experiencing any negative consequences. If a college student drinks a beer at a fraternity party, she has used alcohol. (This can also apply to any drug.)

Misuse: When a person experiences negative consequences from his or her use of alcohol or other drugs, it is "misuse." A large percentage of the population misuses drugs or alcohol at some point in their lives. However, this does not imply that the negative consequences are minor. For example, the same college student continues to use alcohol on an infrequent basis. Her friends throw her a surprise party and she drinks more than usual. On the way home, she is arrested for DUI. Although she does not really have a problem with alcohol, in this instance the consequences are not minor.

Abuse: *Abuse* consists of continued use of alcohol or other drugs in spite of negative consequences. Let's continue to observe the college student who was arrested for DUI. If she does *not* have a substance abuse problem, she will probably cut way back on her alcohol consumption, even if she does not give it up altogether. Getting a DUI should be enough of a deterrent. However, if shortly thereafter she goes to another party, drinks in excess, and then drives her car, it is alcohol abuse.

Dependency/Addiction: Compulsive use of alcohol or other drugs, regardless of adverse or negative consequences, presents the problem of dependency or addiction. For example, suppose that our college student has received three

DUI citations in one year. She is now on probation and will be sentenced to one year in prison if caught using alcohol again. However, she continues to drink at parties and at local establishments. This young woman is clearly addicted to alcohol, because the negative consequences do not affect her decisions or deter her use.

When an individual is clearly in the first or second stage of alcoholism (use or misuse), there are no indications as to whether he or she will naturally progress to the final stages. However, once an individual reaches the abuse stage, a high probability exists that he or she will progress into dependency on the drugs or alcohol. Such individuals should seek professional help to stop the progression.

What Is AA? Alcoholics Anonymous (AA) is perhaps the best-known method for dealing with alcohol addiction. Alcoholics Anonymous defines itself as "a fellowship of men and women who share their experience, strength, and hope with each other that they may solve their common problem and help others to recover from alcoholism." This not-for-profit organization, established in 1935, has helped millions of people deal with this life-altering disease. Through group meetings, self-proclaimed alcoholics come together for counseling, guidance, and support with the primary purpose of staying sober and helping other alcoholics achieve sobriety. Their well-known "12 Step Program" focuses on the physical act of abstaining from alcohol as well as on the psychosocial aspects. To preserve anonymity, membership records do not exist, and people are welcome from all walks of life and at all levels of alcoholism.

This screening test is designed to help you determine if you have a problem with alcohol. Read through the following questions about your use of alcoholic beverages during the past year. In the questions, a "drink" is defined as 12 ounces of beer, 5 ounces of wine, or 1 to 1½ ounces of 80-proof liquor. Select the most appropriate answer for each question by checking yes or no.

Early Warning Signs of Alcohol Abuse

	Yes	No
1. Have you missed classes more than once due to a hangover?	☐	☐
2. Have you thought that you should cut down on your drinking?	☐	☐
3. Have you decided to cut down on your drinking and found out you could not?	☐	☐
4. Have you been angered by the criticism of others about your drinking?	☐	☐
5. Have you been in fights while drinking?	☐	☐
6. Is excessive alcohol use a significant part of your weekly social and recreational activities?	☐	☐
7. Have you had problems with resident assistants or campus police because of your drinking?	☐	☐
8. Do you routinely "binge" drink? (Binge drinking for women is defined as drinking four or more drinks during an episode of drinking. For men, five or more drinks is considered binge drinking.)	☐	☐
9. Have you ever had periods of time you cannot account for while you were drinking or after drinking occurrences?	☐	☐
10. Have you had sexual experiences after drinking that you felt bad about later?	☐	☐

Give yourself 1 point for each Yes response.

0 Congratulations! Your response on this test suggests that your use of alcohol is not causing you any ongoing negative experiences that indicate early warning signs of alcohol abuse or dependence.

1 Now is the time to evaluate how much you are drinking, how often, and the impact your alcohol consumption is having on you. A score of 1 also indicates that you should probably reduce the quantity of alcohol you consume.

2 or greater More than one yes response indicates the definite need for you to limit your alcohol use by either abstaining or limiting your use to responsible levels of consumption. If you are unable to control your use, then it's time to abstain completely.

What Is Binge Drinking?

What would you think if you were at a party or restaurant and saw someone nearby drink an entire six-pack of a soft drink in one sitting? That's 72 ounces of pop. Strange, right? What will all that sugar do to the person's teeth? What will all of those calories do to his or her waistline? What about all of that caffeine at one time?

Now let's say that same person is drinking a six-pack of beer instead of soft drinks. Sound weird, too? Probably, but many college students do it all the time. It's called *binge drinking*—drinking at least five drinks at one time if you are a man or four drinks at one time if you are a woman, in one sitting. The short-term reactions caused by consuming large amounts of alcohol in a brief period of time are serious: vomiting, dizziness, impaired mental capabilities, and hangover. Other effects can include risky sexual behavior, alcohol-related injuries, or death.

Think it doesn't happen all that often? About 50 percent of college men and 37 percent of college women report drinking more than five drinks at one time in one sitting.

SOURCES: *Drinking: A Student's Guide*, "Binge Drinking," www.mcneese.edu/community/alcohol/binge.html. Reprinted from L. Hickman, *Self Tests*, Early signs of alcohol abuse. ucc.nd.edu/self-help/alcohol-addiction/binge-drinking/

Scientists first created the term *fetal alcohol syndrome* (FAS) more than 30 years ago to describe a pattern of birth defects found in children of mothers who drank alcohol during pregnancy.[96] In the United States, FAS is one of the primary causes of birth defects and is considered to be the most common cause of preventable mental retardation. It is defined by four criteria:[97]

- Maternal drinking during pregnancy
- Characteristic pattern of facial abnormalities (see Table 9-6)
- Growth retardation
- Brain damage, often manifested by intellectual difficulties or behavioral problems (see Table 9-6). Studies have demonstrated abnormalities of certain brain regions in babies exposed to alcohol during their mothers' pregnancy (see Figure 9-6).

As you can see, the effects of fetal exposure to alcohol last throughout the child's life.

How does alcohol get into the baby's body?

Alcohol consumed by a pregnant woman travels through her bloodstream and across the placenta to her baby. The unborn

TABLE 9-6	Signs of Fetal Alcohol Syndrome

Characteristic Facial Features of FAS
- Small eyes with drooping upper lids
- Short, upturned nose
- Flattened cheeks
- Small jaw
- Thin upper lip
- Flattened groove in middle of upper lip

Central Nervous System Problems
- Mental retardation
- Hyperactivity
- Delayed development of gross motor skills (rolling over, crawling, walking)
- Delayed development of fine motor skills (grasping objects with the thumb and index finger, transferring objects from one hand to another)
- Impaired language development
- Memory problems, poor judgment, distractibility, impulsiveness
- Learning problems
- Seizures

SOURCE: National Institutes of Health, "Highlights from the 10th Special Report to Congress," *Alcohol Research and Health* 24, no. 1 (2000).

FIGURE 9-6

Brain Regions Affected by FAS

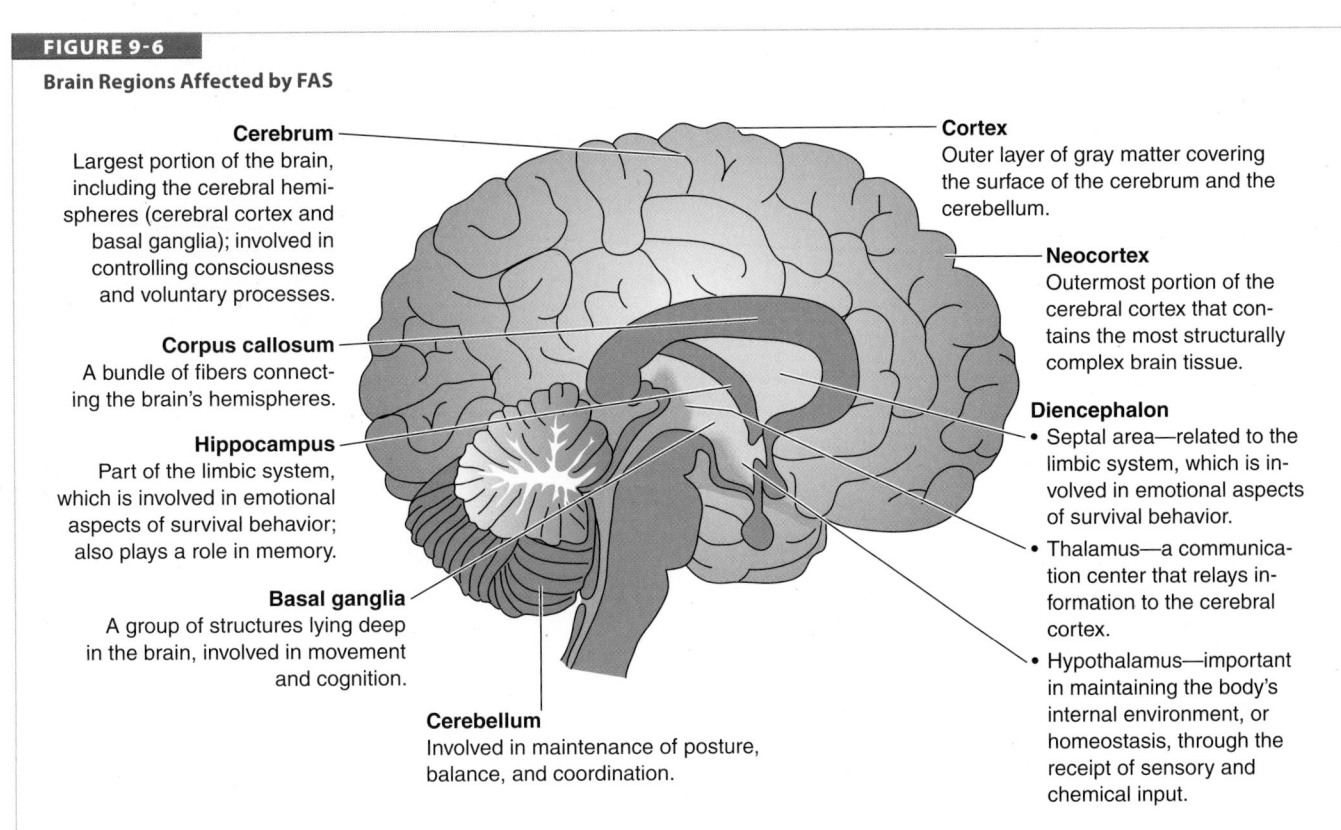

Cerebrum
Largest portion of the brain, including the cerebral hemispheres (cerebral cortex and basal ganglia); involved in controlling consciousness and voluntary processes.

Corpus callosum
A bundle of fibers connecting the brain's hemispheres.

Hippocampus
Part of the limbic system, which is involved in emotional aspects of survival behavior; also plays a role in memory.

Basal ganglia
A group of structures lying deep in the brain, involved in movement and cognition.

Cerebellum
Involved in maintenance of posture, balance, and coordination.

Cortex
Outer layer of gray matter covering the surface of the cerebrum and the cerebellum.

Neocortex
Outermost portion of the cerebral cortex that contains the most structurally complex brain tissue.

Diencephalon
- Septal area—related to the limbic system, which is involved in emotional aspects of survival behavior.
- Thalamus—a communication center that relays information to the cerebral cortex.
- Hypothalamus—important in maintaining the body's internal environment, or homeostasis, through the receipt of sensory and chemical input.

(Continued)

These facial traits (low nasal bridge, short eyelid opening, small head circumference, undeveloped groove in center of upper lip) are typical of fetal alcohol syndrome, caused by drinking alcohol during pregnancy. Irreversible abnormalities of the brain and other organs accompany these facial features.

baby's body can metabolize the alcohol, but it does so at a much slower rate than the adult body. As a result, the alcohol level in the baby's blood is higher than in the mother's, and the alcohol remains in the baby's blood longer.

How much alcohol does a woman have to drink to cause FAS?

Women who drink frequently (more than four alcoholic beverages a day) seriously increase the likelihood that their babies will have FAS. Binge drinking (four or more drinks per occasion) is an especially hazardous drinking pattern in terms of FAS risk.[98] However, no quantity of alcohol use during pregnancy has been established to be safe. Effects of FAS have been seen in children whose mothers drank only moderately or lightly during pregnancy. An average of one drink a day increases a baby's risk of FAS. Children with FAS do not outgrow the signs listed in Table 9-6. These problems are lifelong.

How can FAS be prevented?

Unlike most birth defects, FAS is completely preventable, because its direct cause—maternal drinking—is a controllable behavior.[99] Simply put, pregnant women can prevent FAS by *not consuming alcohol*. No amount of alcohol use during pregnancy has been proven to be safe. For that reason, any woman who suspects she might be pregnant should *stop drinking immediately*. Women who are attempting to get pregnant *should not drink* alcohol. Because many women of childbearing age drink regularly, it's likely that their babies will be exposed to alcohol before pregnancy is detected. It is common for a woman to be pregnant for four to six weeks before she knows she is pregnant. Alcohol can hurt a baby even during the first one to two months of pregnancy, and no type of alcoholic beverage—beer, wine, wine coolers, and liquor (whiskey, vodka, tequila, gin, and rum)—is exempt.

The bottom line? Everything a woman eats or drinks affects her baby. Fetal alcohol syndrome could be completely eliminated if pregnant women did not consume alcohol.[100]

Targeted media campaigns can help increase public awareness of the adverse effects of alcohol use during pregnancy. Because levels of binge and frequent drinking among nonpregnant women have not declined, all women of childbearing age should be warned about the adverse effects of alcohol use to avert early prenatal exposure before women become aware of pregnancy. Additional information about CDC activities to prevent alcohol-exposed pregnancies is available at www.cdc.gov/ncbddd/fas.

Nutrition on the Web

www.cengage.com/nutrition/boyle/personalnutrition8e
Go to the *Personal Nutrition* site to check for the latest updates to chapter topics or to access links to related websites.

www.nal.usda.gov/fnic

- The Food and Nutrition Information Center (FNIC) provides credible and accurate information.
- Click on Consumer Corner, select Eating for Health, and then select Water and Fluid Needs.
- Under Browse by Subject, select Dietary Supplements and then select Ergogenic Aids.

www.nutrition.gov
This government website provides information and links. Under Browse by Subject, select Dietary Supplements/Dietary Supplements for Athletes to find individual ingredients in ergogenic aids.

www.ods.od.nih.gov/Health_Information/IBIDS.aspx
Search the International Bibliographic Information on Dietary Supplements (IBIDS) database for ingredients found in ergogenic aids and functional food beverages.

www.cdc.gov
Search the Centers for Disease Control (CDC) website for information about meeting your daily fluid needs, water quality, water safety, alcohol and public health, binge drinking, and fetal alcohol syndrome.

www.nlm.nih.gov/medlineplus
Search for information about drinking water, caffeine, tea, and alcohol at this government website.

www.foodinsight.org
Search the International Food Information Council (IFIC) site for information about caffeine, tea, alcohol, and functional food ingredients.

www.eatright.org
The American Dietetic Association home page includes many nutrition resources and links.

www.fda.gov
Search the Food and Drug Administration (FDA) website for information about water, caffeine, and dietary supplements found in beverages.

www.beverageguidancepanel.org
Compare the relative health benefits for a variety of beverages at this website.

www.epa.gov
Search the Environmental Protection Agency (EPA) website for information about water quality and water safety.

www.nida.nih.gov
Search the National Institute on Drug Abuse (NIDA) website for information about alcohol abuse.

www.aa.org
Search the website for Alcoholics Anonymous (AA) for information about alcohol abuse.

www.niaaa.nih.gov
Search the National Institute on Alcohol Abuse and Alcoholism (NIAAA) website to learn more about alcohol abuse and alcoholism research.

www.intox.com/about_alcohol.asp
Learn about alcohol absorption and measuring blood alcohol content (BAC).

www.marchofdimes.com
March of Dimes (MOD) provides a health library with fact sheets on a variety of subjects, including fetal alcohol syndrome.

www.higheredcenter .org/services/ pubs
Browse the Higher Education Center for Alcohol and Other Drug Abuse and Violence Prevention website for information about binge drinking on campus and alcohol abuse prevention.

Menu of Online Study Tools

A variety of study tools for this chapter are available at our website to deepen your understanding of chapter concepts. Go to

www.cengage.com/sso

to find

- Practice tests
- Flashcards
- Glossary
- Web links
- Animations
- Chapter summaries, learning objectives, and crossword puzzles.

Diet Analysis+ in Action

Go to the Intake Spreadsheet report that shows the results of your three-day tracking. Does your intake include sports drinks or fortified (vitamin) waters? Do you consume either of these types of beverages? If so, please go back to your Track Diet and include your average consumption on at least one of the days. You may need to create a food label for your fortified beverage of choice. Once you have included your sports drink or vitamin water, take another look at the Intake vs. Goals report. Have the percentages of vitamins and minerals changed? Does your intake exceed the recommended intake for vitamins and minerals? Do you also take a vitamin and mineral supplement? What are some consequences of regularly consuming beverages that have been fortified with vitamins and minerals? What changes, if any, might you make as a result of this investigation?

Weight Management

Ask Yourself . . .

Which of the following statements about nutrition are true, and which are false? For each false statement, what is true?

1. The less you weigh, the better it is for your health.
2. Obese people pay higher insurance premiums than thin people.
3. If you weigh too much according to the scales, you are too fat.
4. If you are too fat, it is because you eat too much.
5. Basal metabolism contributes only a small percentage of a person's daily energy output.
6. Probably the most important contributor to the obesity problem in our country is underactivity.
7. Any food can make you fat, even carrot sticks, if you eat enough of it.
8. You can lose weight faster on a properly designed diet and exercise program than on a total fast.
9. Fad diets are popular because their followers achieve quick and permanent weight loss.
10. Anorexia nervosa is a disease in which a person has no appetite.

Answers found on the following page.

Say the word *diet*, and most people think "starvation," "deprivation," "hunger," and something to go "on" and "off." As this chapter shows, such thinking is unrealistic and self-defeating. Strict, temporary diets—and the attitudes that go with them—don't work for the great majority of overweight people. In fact, although Americans spend more than $40 billion a year on programs and products aimed at weight loss, obesity rates are higher than ever.[1] Currently, 68 percent of adults and approximately 17 percent of children and adolescents in the United States are either overweight or obese; that is, exceeding their healthy weight range (see Figure 10-1). The annual costs of overweight and obesity are staggering: more than $117 billion, including $61 billion in direct costs (treatment of related disease) and $56 billion in indirect costs (lost productivity due to disability, morbidity, and mortality).

Before you begin a thing remind yourself that difficulties and delays quite impossible to foresee are ahead. . . . You can see only one thing clearly, and that is your goal. Form a mental vision of that and cling to it through thick and thin.

K. NORRIS

FIGURE 10-1

Trends in Prevalence of Overweight and Obesity among Children (1988–2008) and Adults (1988–2008), United States

*See Figure 10-3 for how to calculate and interpret body mass index (BMI).

SOURCE: Adapted from C. L. Ogden and coauthors, Prevalence of high body mass index in U.S. children and adolescents, 2007–2008, *Journal of the American Medical Association* 303 (2010): 242–49; and C. L. Ogden & coauthors, Prevalence and trends in obesity among U.S. adults, 1999–2008, *Journal of the American Medical Association* 303 (2010): 23541.

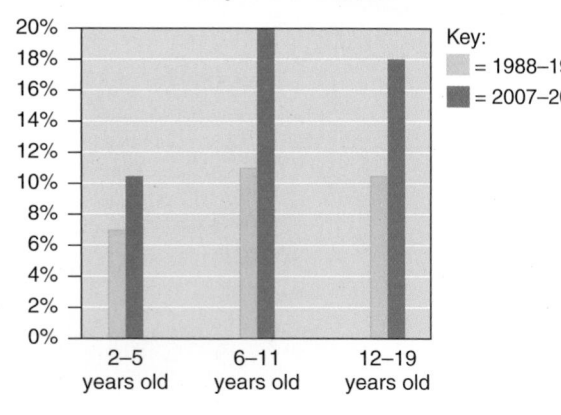

Trends in Prevalence of Overweight and Obesity in U.S. Children

Key:
 = 1988–1994
 = 2007–2008

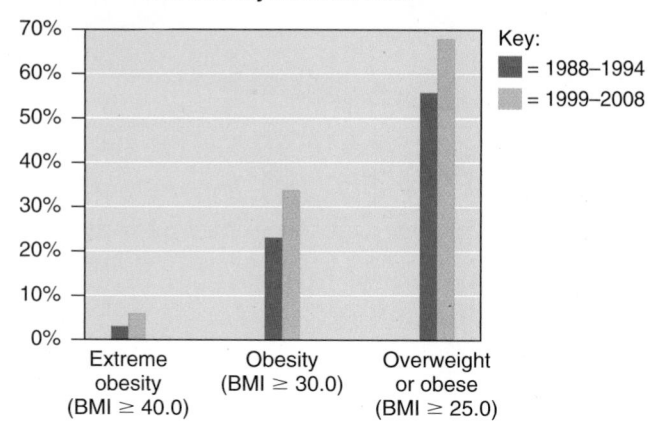

Trends in Prevalence of Overweight and Obesity in U.S. Adults*

Key:
 = 1988–1994
 = 1999–2008

Bodies come in many shapes and sizes. Which are healthy?

A Closer Look at Obesity

The World Health Organization describes obesity as "an escalating epidemic" and one of the greatest neglected public health problems of our time (see Figure 10-2). Because obesity is a disease with multiple health risks, such as type 2 diabetes, heart disease, and some forms of cancer, the increasing rates of obesity are projected to result in increased rates of disability and preventable deaths.[2] The *Healthy People* national health initiative includes objectives for increasing the proportion of adults who are at a healthy weight and reducing the number of adults and children who are obese by the year 2020.[3]

Although many factors (including genetics) influence body weight, excess energy intake and physical inactivity are the leading causes of **overweight** and **obesity**, and they represent the best opportunities for prevention and treatment.[4] Consider how the following trends in our society have increased opportunities for poor nutrition (particularly excess calories) and decreased opportunities for physical activity:

- Food portion sizes and obesity rates have grown at the same rate. In the 1960s, an average fast-food meal of a hamburger, fries, and a 12-ounce cola provided 590 calories; today, supersized fast-food meals deliver 1,500 calories or more.[5] Even an extra 150 calories a day—the calorie cost of supersizing a soft drink—can convert to about 16 extra pounds of body fat every year.

overweight conventionally defined as weight between 10 and 20 percent above the desirable weight for height, or a body mass index (BMI) of 25.0 through 29.9.

obesity conventionally defined as weight 20 percent or more above the desirable weight for height, or a BMI of 30 or greater.

ASK YOURSELF ANSWERS: **1.** False. Being thin is good for health only to a point; being too thin is as risky as being too fat. **2.** True. **3.** False. A high weight according to the scales may reflect heavy bones and muscles rather than excess fatness. **4.** False. If you are too fat, it could be because you exercise too little. **5.** False. Basal metabolism contributes about 60 percent or more of the average person's daily energy output. **6.** True. **7.** True. **8.** True. **9.** False. Although you may lose weight quickly on a fad diet, much of the weight you lose may be muscle or water and the weight may soon be regained when the diet ends. **10.** False. People with anorexia nervosa are constantly hungry but control their eating.

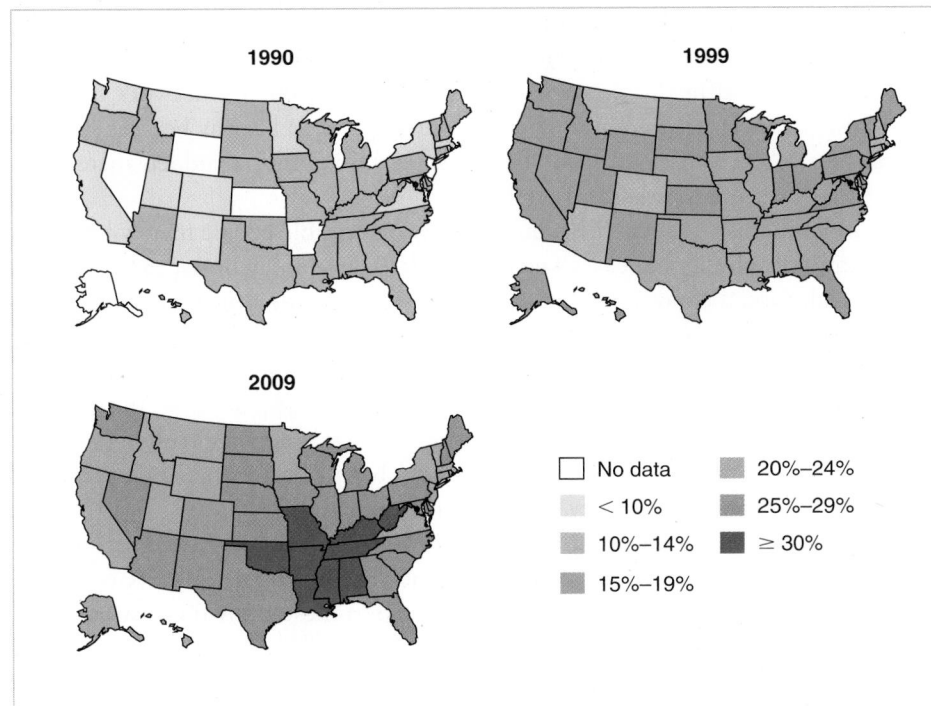

FIGURE 10-2

The Epidemic of Obesity among U.S. Adults

Note the increasingly upward trend of obesity among the 50 states. In 1990, ten states had a prevalence of obesity less than 10% and no states had prevalence equal to or greater than 15%. By 1999, no state had prevalence of less than 10%, 18 states had a prevalence of obesity between 20 and 24%, and no state had prevalence equal to or greater than 25%. In 2009, only one state (Colorado) and the District of Columbia had a prevalence of obesity less than 20%. Thirty-three states had prevalence equal to or greater than 25%; nine of these states (Alabama, Arkansas, Kentucky, Louisiana, Mississippi, Missouri, Oklahoma, Tennessee, and West Virginia) had a prevalence of obesity equal to or greater than 30%.

SOURCE: U.S. Obesity Trends 1985–2009, Centers for Disease Control and Prevention. www.cdc .gov.

- Vending machines selling soft drinks, high-fat snacks, and highly sugared snacks are common in schools and workplaces. Milk, juices, water, and healthful snacks are far less accessible.

- Adults spend more time in sedentary activities, such as watching television, working on the computer, or commuting to and from work and school.

- Children watch 12 to 14 hours of television a week and spend 7 hours playing video games.[6]

- Schools offer fewer physical education classes for children.

- More families live in communities that are designed for car use but are unsuitable (lack of green space for recreation) and often unsafe (lack of sidewalks, inadequate street lighting) for activities such as walking, biking, and running.

Problems Associated with Weight

Clearly, each individual can weigh too much or too little to be healthy. The **underweight** person has minimal body fat stores and is at a disadvantage in situations where energy reserves are needed, such as a prolonged period of physiological stress or injury. Other problems for the underweight person may include menstrual irregularity, infertility, and osteoporosis.

The physical risks of overweight and obesity are greater for some people than for others, depending on inherited susceptibilities to conditions such as high blood pressure, high blood cholesterol, and diabetes. High blood pressure is made worse by weight gain and can often be normalized merely by weight loss. Diabetes can be precipitated, in genetically susceptible people, by becoming overweight.

Obesity also increases the risk of heart disease, partly because excess fat pads crowd the heart muscle and the lungs within the body cavity. These fat pads encumber the heart as it beats, requiring it to work harder to deliver oxygen and nutrients to the rest of the body. Also, the lungs cannot expand fully, thus limiting the oxygen intake of

underweight weight 10 percent or more below the desirable weight for height, or a BMI of less than 18.5.

TABLE 10-1	Problems Associated with Obesity

- Abdominal hernias
- Accidents
- Certain cancers: colon, rectal, prostate, breast, uterus, cervical, ovarian
- Complications during pregnancy
- Complications with surgical procedures
- Decreased longevity
- Decreased quality of life
- Depression
- Diabetes (type 2)
- Fertility problems
- Gallbladder and liver disease
- Gout
- Heart disease
- High blood cholesterol levels
- Hormonal imbalances
- Hypertension
- Injury to weight-bearing joints
- Kidney abnormalities
- Metabolic syndrome*
- Osteoarthritis (knees, hips, lower spine)
- Poor self-esteem
- Respiratory problems
- Sleep disturbances
- Varicose veins

*Metabolic syndrome is a clustering of risk factors associated with the development of type 2 diabetes and heart disease. These risk factors include excess body weight (especially around the abdomen), insulin resistance, high blood pressure, high levels of triglycerides, low levels of HDL-cholesterol, and increased blood glucose levels.

underwater weighing a measure of density and volume; the less a person weighs underwater compared to the person's out-of-water weight, the greater the proportion of body fat (fat is less dense or more buoyant than lean tissue).

skinfold test a method for measuring body fat in which the thickness of a fold of skin on the back of the arm (the triceps), below the shoulder blade (subscapular), or in other areas is measured with an instrument called a caliper.

bioelectrical impedance estimation of body fat content made by measuring how quickly electrical current is conducted through the body.

central obesity excess fat on the abdomen and around the trunk. *Peripheral obesity* is excess fat on the arms, thighs, hips, and buttocks.

each breath and causing the heart to work even harder. Furthermore, because each extra pound of fat tissue demands to be fed by way of miles of capillaries, the heart must work extra hard in obese people to pump blood through a network of blood vessels that is vastly larger than that of a thin person. Even a healthy heart is strained by excess fat. When a diseased heart finds itself in this bind, a sudden increase in workload may be more than it can handle.

Gallbladder disease, too, can be brought on in susceptible people merely by excess weight.[7] Similarly, obesity increases a woman's risk of developing breast cancer.[8] Table 10-1 shows additional conditions that can be caused, or made worse, by obesity.

In addition to being at risk for these health hazards, millions of obese people incur risks from ill-advised, misguided diet programs. Some fad diets are actually more hazardous to health than obesity! Many claims, treatments, devices, and gadgets for losing weight are, at best, simply ineffective, whereas others are truly dangerous. Over the centuries, "magic" weight-loss plans have been offered time and again, and their success lies in their popularity, not in their results.

Although some obese people suffer none of these physical health hazards, no one in our society quite escapes one disadvantage of obesity. In many parts of North America, obesity is a social and economic handicap.[9] Obese individuals suffer from discrimination in many areas, including social relationships and the job market. Psychologically, a body size that embarrasses or shames a person can be a private anguish. For people who perceive themselves as fat in a society that prizes thinness, real or imagined obesity can thrust them into withdrawal, shame, humiliation, and isolation.

How thin, then, is too thin—and how fat is too fat? The following section discusses the concept of a healthy weight.

What Is a Healthy Weight?

The question of what weight is healthy is harder to answer than you might at first think. Although Table 10-2 provides healthy weight ranges, to think merely in terms of weight oversimplifies the issue of body fatness and health. This chapter first asks what range of weights is compatible with wellness and long life and then suggests that personal preferences dictate the choice of a weight within that range.

Body Weight versus Body Fat   Two people of the same sex, age, and height may both weigh the same, yet one may be too fat and the other too thin. The difference lies in their body composition. One may have small, light bones and minimally developed muscles, while the other has big, heavy bones and well-developed muscles. The first person could have too much body fat and the second person too little. For example, football players, bodybuilders, and other athletes may technically be overweight for their height according to the weight tables. However, they typically have less body fat than would pose a health risk. Likewise, many sedentary persons who weigh in as "normal" may actually be too fat. Thus, obesity must be defined by amount of body fat rather than by weight.

The health risks of obesity refer to people who have too much body fat. On the average, men having more than 25 percent body fat and women having more than 33 percent body fat are considered obese.[10] Desirable measures are 12–20 percent body fat for most men and 20–30 percent body fat for most women.

Measuring Body Fat   Body fatness is hard to measure. One very accurate way is to obtain a measure of the body's density—that is, weight divided by volume. Lean tissue is denser than fat tissue. Weight is easy enough; just step on an accurate scale. To calculate body density, though, you have to immerse the whole body in

Hydrostatic (underwater) weighing.

© David Young-Wolff/PhotoEdit

The fatfold test gives a fair approximation of total body fat.

Copyright © David Young-Wolff/Photo Edit

TABLE 10-2	Weight for Height Standards[a]	

Healthy Weight Ranges[b]

Height	Weight (lb)[c]	
	Midpoint	Range
4'10"	105	91–119
4'11"	109	94–124
5'0"	112	97–128
5'1"	116	101–132
5'2"	120	104–137
5'3"	124	107–141
5'4"	128	111–146
5'5"	132	114–150
5'6"	136	118–155
5'7"	140	121–160
5'8"	144	125–164
5'9"	149	129–169
5'10"	153	132–174
5'11"	157	136–179
6'0"	162	140–184
6'1"	166	144–189
6'2"	171	148–195
6'3"	176	152–200
6'4"	180	156–205

BMI Cutoff Weights for Overweight and Obesity

Height	BMI of 25 overweight (lb)	BMI of 30 obese (lb)
4'11"	124	148
5'0"	128	153
5'1"	132	158
5'2"	136	164
5'3"	141	169
5'4"	145	174
5'5"	150	180
5'6"	155	186
5'7"	159	191
5'8"	164	197
5'9"	169	203
5'10"	174	207
5'11"	179	215
6'0"	184	221
6'1"	189	227
6'2"	194	233
6'3"	200	240
6'4"	205	246

[a] See Appendix A for a quick estimation of desirable body weight and a method for determining frame size.
[b] The higher weights in the ranges generally apply to men, who tend to have more muscle and bone; the lower weights more often apply to women, who have less muscle and bone.
[c] Without shoes or clothes.

a tank of water and measure the amount of water displaced. However, not many health professionals have space in their offices for the equipment needed for this procedure, known as *hydrostatic weighing* or **underwater weighing**. Body density can more easily be determined by air-displacement methods* such as the BOD POD, which measures the volume of air displaced by a person seated in a sealed device of known volume.[11] Most professionals employ the **skinfold test**, using a caliper—a pinching device that measures the thickness of a fold of fat in such areas as the back of the arm, below the shoulder blade, and the side of the waist. About 50 percent of the body's fat lies beneath the skin, and its thickness at these locations can be compared with standard tables to give a fair approximation of total body fat, at least for most people.

Although costly, the same technology used to measure bone density—*dual-energy X-ray absorptiometry* (the DEXA test)—can yield an accurate image of the body's fat-free tissue and total fat content. An alternate method for assessing body fatness is **bioelectrical impedance**, in which electrodes are attached to a person's hand and foot. This method measures a person's fat by measuring how fast a slight electrical current is conducted through the body (from the ankle to the wrist). Because fat is a poor conductor of electricity, the more fat one has, the more resistance this current encounters in the body.[12]

Distribution of Body Fat   Not everyone carries body fat in the same places. To complicate matters further, how body fat is distributed has health implications. Excess fat around the middle—**central obesity**—is associated with increased health hazards.[13] Body types are described as being either apple shaped or pear shaped. Apple-shaped people, those who store most of their excess fat around the abdomen (typically men), are at a greater risk for developing diabetes, hypertension, elevated levels of blood cholesterol, and heart disease than pear-shaped people, who store excess fat on the hips, thighs, and buttocks (typically women). A simple calculation of a person's waist circumference can be used to assess abdominal fat.

Weighing in for Health   Obesity presents one of the most serious health risks that people face—and one that people should, theoretically, be able to control. In response to the rising epidemic of excessive weight, guidelines for the assessment and

---

*The air-displacement technique used by the BOD POD relies on the physics of Boyle's law, which states that pressure and volume vary inversely with one another (for example, as pressure goes up, volume goes down, and vice versa). Measuring air displacement (pressure changes) caused by a person seated in the sealed BOD POD chamber of known volume provides the data needed to calculate the person's body density.

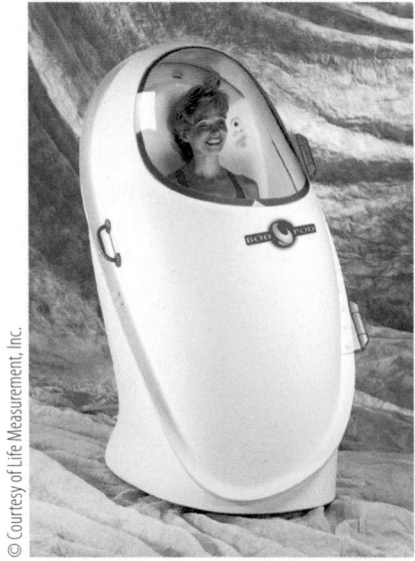

The BOD POD measures body density.

Dual-energy X-ray absorptiometry (DEXA).

Bioelectrical impedance measures body fat.

body mass index (BMI) an index of a person's weight in relation to height that correlates with total body fat content.

waist circumference a measure used to assess abdominal (visceral) fat; excess fat in the abdomen increases a person's risk of health problems.

treatment of overweight and obesity have been developed.[14] A person's health risk due to weight depends on three factors: body weight, amount and location of body fat, and current health status.

Body weight is assessed using the **body mass index (BMI)**, an index of your weight in relation to your height. *Overweight* is defined as a BMI of 25–29.9, and *obesity* is defined as a BMI of 30 or above (refer to Table 10-2 for body weights that equate with these BMI values).[15] Keep in mind that BMI does not account for location of fat in the body, and a muscular person with a low percentage of body fat may have a high BMI (see Figure 10-3). In consideration of this shortcoming of the BMI, the other two factors are taken into account.

The **waist circumference** measurement provides information about the distribution of fat in the abdomen. Excess abdominal fat poses a greater health risk than excess fat in the hips and thighs. The extra abdominal fat crowds the organs and its proximity to the liver means that, when metabolized, abdominal fat can raise blood

---

Understanding Body Mass Index (BMI)

SOURCE: Adapted from G. Bray, *Contemporary Diagnosis and Management of Obesity*, © 1998 Handbooks in Healthcare, Co., a division of AAM Co., Inc.

Benefits of Using BMI
BMI correlates strongly with body fatness and risk of disease and death.

Very low BMI (<18) increases risk

18 to 27: Low risk

27 to 30: Moderate risk

30 to 35: High risk

35 and above: Very high risk

Cautions in Using BMI
May overestimate body fat in athletes and underestimate body fat in adults over 65.

A muscular person, such as this bodybuilder, often has a low percentage of body fat but a high BMI.

Losevsky Pavel/Shutterstock.com

Body Mass Index (BMI) Can Be Calculated As Follows:

BMI = [weight (pounds) ÷ height (inches)^2] × 703

For example, the BMI for a woman who is 5'5" tall and weighs 145 lb would be:

[145 ÷ (65 × 65)] × 703 = 24

- BMI < 18.5 = underweight
- BMI 18.5 to 24.9 = normal weight
- BMI 25 to 29.9 = overweight
- BMI ≥ 30 = obese
- BMI ≥ 40 = severely obese

A wide range of weights is compatible with good health. Within this range, the definition of desirable or healthful weight is up to the individual, depending on such factors as family history, occupation, physical and recreational activities, and personal preferences. To determine if your weight is a healthful weight for you:

1. *Calculate your body mass index (BMI)*. Find your height and weight in the following chart.* Your BMI is at the top of the column that contains your weight. Record your BMI here: _____

BMI:	19	20	21	22	23	24	25	26	27	28	29	30	35	40
Height							Weight (pounds)							
5'0"	97	102	107	112	118	123	128	133	138	143	148	153	179	204
5'1"	100	106	111	116	122	127	132	137	143	148	153	158	185	211
5'2"	104	109	115	120	126	131	136	142	147	153	158	164	191	218
5'3"	107	113	118	124	130	135	141	146	152	158	163	169	197	225
5'4"	110	116	122	128	134	140	145	151	157	163	169	174	204	232
5'5"	114	120	126	132	138	144	150	156	162	168	174	180	210	240
5'6"	118	124	130	136	142	148	155	161	167	173	179	186	216	247
5'7"	121	127	134	140	146	153	159	166	172	178	185	191	223	255
5'8"	125	131	138	144	151	158	164	171	177	184	190	197	230	262
5'9"	128	135	142	149	155	162	169	176	182	189	196	203	236	270
5'10"	132	139	146	153	160	167	174	181	188	195	202	209	243	278
5'11"	136	143	150	157	165	172	179	186	193	200	208	215	250	286
6'0"	140	147	154	162	169	177	184	191	199	206	213	221	258	294
6'1"	144	151	159	166	174	182	189	197	204	212	219	227	265	302
6'2"	148	155	163	171	179	186	194	202	210	218	225	233	272	311
	Healthy Weight						Overweight					Obese		

SOURCE: National Heart, Lung and Blood Institute.

2. Determine if your fat distribution is associated with health risks. Measure your waist circumference by placing a tape measure around your waist just above your belly button. Record your waist in inches here. _____

3. Is your weight affecting your health? Do you have any of these weight-related health problems or risk factors?
   - Heart disease
   - High LDL-cholesterol
   - Osteoarthritis
   - Cigarette smoking
   - Type 2 diabetes
   - Low HDL-cholesterol
   - Recurrent gallstones
   - Sedentary lifestyle
   - High blood pressure
   - High triglycerides
   - Sleep disturbances
   - Male > 45 years or postmenopausal female

4. How does your current weight measure up to these considerations?
   - If your BMI is acceptable for good health and if your waist measurement is not high, you will want to maintain this weight. If you need to lose weight or gain weight, consider the tips offered throughout this chapter for healthfully changing your weight.
   - You should consider losing weight if:
     Your BMI is 30 or greater.
     Your BMI is 25 to 29 and you have two or more of the weight-related health problems or risk factors listed above.
     Your waist circumference exceeds 40 inches (for men) or 35 inches (for women) and you have two or more weight-related health problems or risk factors.
   - Weight loss is optional for you if your BMI is 25 to 29 and you do not have two or more weight-related health problems (particularly if your BMI is under 27 or you have large muscles and bones).

*Additional heights and weights are listed on Study Card page 31 at the back of this book, or you can use this formula to calculate your BMI:

$$\frac{\text{weight (in pounds)}}{\text{height}^2 \text{ (in inches)}} \times 703$$

Example: A 5'8" person weighing 145 pounds has a BMI of 22.

$$\text{BMI} = \frac{145}{68^2} \times 703$$

$$\text{BMI} = \frac{145}{4624} \times 703$$

BMI = 22 (rounded)

Central obesity, characterized by an "apple-shaped" body with large abdominal-type fat stores, is a strong risk factor for type 2 diabetes, heart disease, and other problems.

FIGURE 10-4

How the Body Expends Energy

The body expends most of its energy on *basal metabolism*—maintaining basic physiological processes such as breathing, heartbeat, and other involuntary activities. The second largest amount of energy is expended for voluntary *physical activities*—an amount that will vary by activity level. A minor amount of energy is also used for the *thermic effect of food*—the energy needed to digest, absorb, and process the food you eat.

Thermic effect of food
5%–10%

Basal metabolism
60%–65%

Physical activity
25%–35%

basal metabolism the sum total of all the chemical activities of the cells necessary to sustain life, exclusive of voluntary activities; that is, the ongoing activities of the cells when the body is at rest.

basal metabolic rate (BMR) the rate at which the body spends energy to support its basal metabolism. The BMR accounts for the largest component of a person's daily energy (calorie) needs.

cholesterol levels and lower the body's sensitivity to insulin. Disease risk rises significantly with a waist circumference of more than 35 inches in women and more than 40 inches in men.[16]

The third factor to consider is current health status: the presence of weight-related health problems and risk factors for diseases. These may include family health history, heart disease, type 2 diabetes, high blood cholesterol, high blood pressure, cigarette smoking, osteoarthritis, gallstones, or sleep apnea (irregular breathing during sleep). According to the treatment guidelines, an initial goal for treatment of overweight and obese people with risk factors is to reduce body weight by about 10 percent at a rate of about 1 to 2 pounds per week.[17] For overweight individuals, losing as little as 5 to 10 percent of their body weight may improve or even resolve many of the problems linked to overweight, such as diabetes and high blood pressure. The Scorecard feature on page 297 helps you make sense of these guidelines and evaluate your own weight status.

Energy Balance

Suppose you decide that you are too fat or too thin and that you got that way by having an unbalanced energy budget—that is, by eating either more or less food energy than you spent. Fatness and thinness are reflections of excessive or deficient energy stores. You store extra energy as fat only if you eat *more* food energy in a day than you use to fuel your metabolic and other activities. Similarly, you lose stored fat only if you eat *less* food energy in a day than you use as fuel. A day's energy balance can be stated like this:

$$\text{Change in energy stores} = \text{Energy in} - \text{Energy out}$$

Calories in

Calories out

You already know about the "energy in" side of this equation. An apple brings you 80 calories; a candy bar, 290 calories. You may also know that for each 3,500 *excess* calories consumed ("excess" meaning more than your body needs), you store 1 pound of body fat.

For the "energy out" side of the equation, the body spends energy in two major ways: to fuel its metabolic activities and to fuel its muscle activities. As shown in Figure 10-4, you also spend a small amount of energy to digest, absorb, transport, process, and store the food you eat, which is known as the *thermic effect of food*. The following three sections discuss the body's needs for energy.

Basal Metabolism   About 60 percent or more of the energy the average person spends goes to support the ongoing metabolic work of the body's cells, the **basal metabolism**.[18] This is the work that goes on all the time, without conscious awareness. The beating of the heart, the inhaling and exhaling of air, the maintenance of body temperature, and the sending of nerve and hormonal messages to direct these activities are the basal processes that maintain life. Basal metabolic needs are surprisingly large. A person whose total energy expenditure amounts to 2,000 calories per day spends as many as 1,200 to 1,400 of them to support basal metabolism.

The **basal metabolic rate** (often abbreviated **BMR**) is influenced by a number of factors (see Table 10-3). In general, the younger a person is, the higher the basal metabolic rate, partly because of the increased activity of cells undergoing division. The BMR is most pronounced during the growth spurts that take place during infancy, puberty, and pregnancy. Body composition also influences metabolic rate. Muscle tissue is highly active even when it is resting, whereas fat tissue is comparatively inactive. Therefore, when comparing two people of the same body weight, the person with more lean muscle tissue will have a higher BMR, and the person with more fat tissue will have a lower BMR. Lean body mass decreases with age, but consistent physical activity—especially endurance and strength-building activities—may prevent some of the decline.

Gender correlates roughly with body composition. Men generally have a faster metabolic rate than women, and researchers believe that this is because of men's greater percentage of lean tissue. (A female athlete has a greater percentage of lean tissue than a sedentary man of the same weight and so would have a higher metabolic rate.) Fever increases the energy needs of cells, as their increased activities to fight off infection and generate heat speed up the metabolic rate.

Fasting and constant malnutrition lower the metabolic rate because of the loss of lean tissue and the slowdown of activities the body can't afford to support fully. This slowing of metabolism seems to be a protective mechanism to conserve energy when there is a shortage, and it hampers weight loss in a person who fasts or undertakes a very low-calorie diet.

Some hormones—the stress hormones, for example—influence metabolism. They increase the energy demands of every cell and thus raise the metabolic rate. This raised metabolism partly accounts for the weight loss sometimes seen in people experiencing extreme stress in their life, although other factors, such as upset digestion and loss of appetite, also enter into this picture. The activity of the thyroid gland also influences the basal metabolic rate. The less thyroid hormone secreted, the lower the energy requirement for maintenance of basal functions.

Voluntary Activities

Muscular activity does not contribute as much to most people's energy outputs as basal metabolism does. On the average, it amounts to only about 30 percent of the total. Unlike basal metabolism, though, which cannot be changed immediately, physical activity can be changed at will. If you want to tinker with your energy balance, this is the component—on the output side—that you can alter significantly in the short term. If you increase physical activity consistently, you will also ultimately increase the energy your body spends on metabolic activity, because you will have an increase in lean body mass.

The energy spent on physical activity is the energy spent moving the body's skeletal muscles—the muscles of the arms, back, abdomen, legs, and so forth—and the extra energy needed to speed up the heartbeat and respiration rate. The number of calories spent depends on three factors: (1) the amount of muscle mass required, (2) the amount of weight being moved, and (3) the amount of time the activity takes. Thus, an activity involving both the arms and the legs requires more calories than an activity of the same intensity involving only the legs; an activity performed by a heavier person requires more energy than the same activity performed by a lighter person; and an activity performed for 40 minutes requires twice as much energy as that same activity performed for 20 minutes.

It may be disheartening for a college student to discover that mental activity requires little energy, even though it may be tiring. Studying for an exam may be hard work, but it doesn't

© David Madison/Getty Images

Body composition influences metabolic rate. Weight training can help shift your body composition toward more lean tissue, thereby speeding up your metabolism. See Chapter 11 for guidelines on how to get started.

TABLE 10-3	Factors That Influence the Basal Metabolic Rate

Factors That Increase BMR:
- Caffeine
- Fever
- Growth (higher in children and pregnant women)
- Height (higher in tall, thin people)
- High thyroid hormone
- Male gender (more lean tissue)
- Muscle mass (the more lean tissue, the higher the BMR)
- Smoking (nicotine)
- Stress

Factors That Decrease BMR:
- Age (slows down with age)
- Low thyroid hormone
- Reduced energy intake (fasting, starvation, low-calorie diets)
- Sleep (BMR is lowest when sleeping)

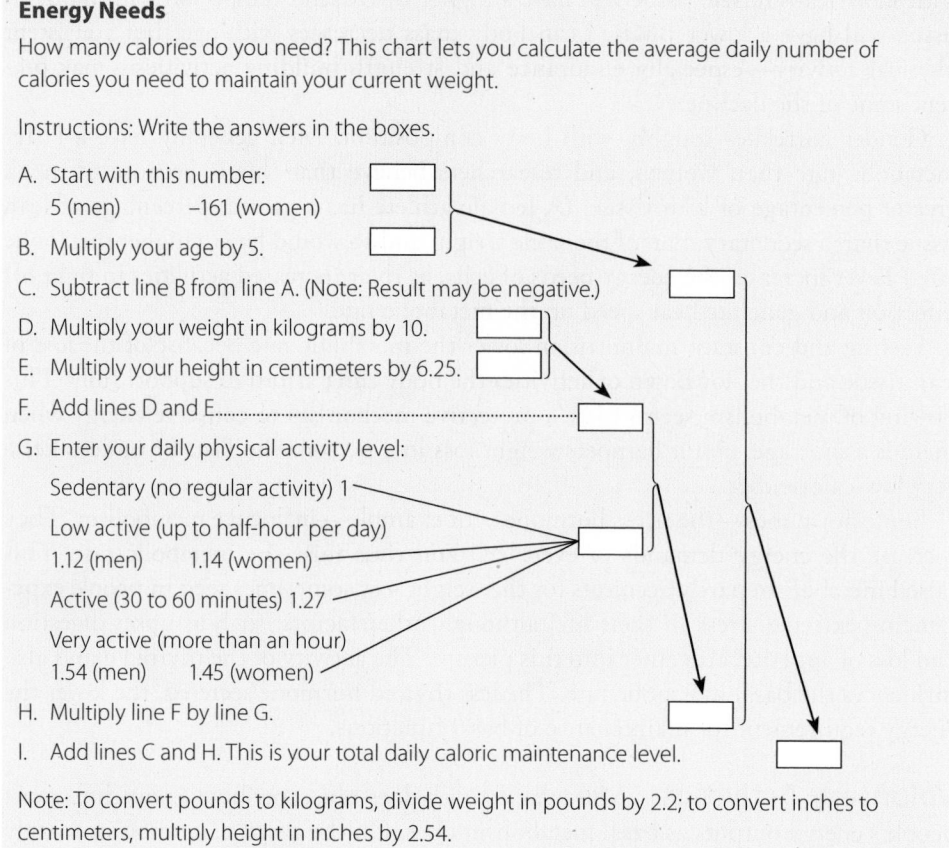

Energy Needs

How many calories do you need? This chart lets you calculate the average daily number of calories you need to maintain your current weight.

Instructions: Write the answers in the boxes.

A. Start with this number:

  5 (men)   −161 (women)

B. Multiply your age by 5.

C. Subtract line B from line A. (Note: Result may be negative.)

D. Multiply your weight in kilograms by 10.

E. Multiply your height in centimeters by 6.25.

F. Add lines D and E.

G. Enter your daily physical activity level:

  Sedentary (no regular activity) 1

  Low active (up to half-hour per day)
  1.12 (men)   1.14 (women)

  Active (30 to 60 minutes) 1.27

  Very active (more than an hour)
  1.54 (men)   1.45 (women)

H. Multiply line F by line G.

I. Add lines C and H. This is your total daily caloric maintenance level.

Note: To convert pounds to kilograms, divide weight in pounds by 2.2; to convert inches to centimeters, multiply height in inches by 2.54.

SOURCES: Adapted from *Consumer Reports on Health* (January 2003), p. 5; M. D. Mifflin, S. T. St. Jeor et al. A new predictive equation for resting energy expenditure in healthy individuals. *Journal of the American Dietetic Association* 51 (2005):241–247.

burn body fat. People who are very, very busy—surfing the Internet, making phone calls, riding in their cars—may wonder why they gain weight. Even though they are socially or intellectually active, such activities involve few muscles and therefore little energy expenditure.

Total Energy Needs   A typical breakdown of the total energy spent by a lightly active person (for example, a student who walks back and forth to classes) might look like this:

Energy for basal metabolism:	1,400 calories
Energy for physical activity:	560 calories
Total:	1,960 calories

The first component (basal metabolism) is larger, and you cannot change it much. You can, however, change the second component—physical activity—and use more calories. To increase your basal metabolic output, make exercise a daily habit. Your body composition will gradually change, and your basal energy output will pick up the pace as well. You can figure out roughly how much energy you need in a day by using the method illustrated in the box at the top of this page.

In summary, the amount of fat stored in a person's body depends on the balance between the total food energy the person has taken in and the total energy the person has expended. But why do so many people have excessive fat stores?

Causes of Obesity

Some people eat more than they need or exercise less than they should to maintain their body weight, and they get fat. Some eat less or exercise more, and they get thin. Amazingly, many people eat exactly what they need and stay at the same weight year after year. A single extra pat of butter each day would make them gain 5 pounds in a year, but if these people overeat by that much on one day, they apparently undereat by the same amount the next day. How do they do this, and (in contrast) why do some people fail to maintain their weight? In general, most experts believe that obesity is associated with both inside-the-body causes and environmental factors.[19]

Genetics   The theory that a hereditary, inside-the-body factor for obesity may exist is supported by the existence of animal strains that are genetically fat. Such animals tend to be fat in any environment; that is, they are fat regardless of the kind or variety of food that is available. In humans, studies have shown that identical twins—whether raised together or apart—tend to have similar weight-gain patterns. Also, twins raised by adoptive parents tend to have body shapes similar to those of their biological parents. Not all studies confirm this, however.[20] Moreover, pairs of twins purposefully overfed in clinical experiments tend to respond similarly to the extra calories. Some sets of twins gain considerable weight when fed a certain number of calories, whereas other pairs put on relatively few pounds even on the same diet.[21]

Set-Point Theory   One popular inside-the-body theory is the so-called **set-point theory**. Noting that many people who lose weight on reducing diets subsequently return to their original weight, some researchers have suggested that the body "wants" to maintain a certain amount of fat and regulates eating behaviors and hormonal actions to defend its "set point." The theory implies that science should search inside obese people to find the causes of their problems—perhaps in their hunger-regulating mechanisms.

Leptin   Researchers have identified a gene—named *ob* (for *obese*)—that appears to produce a hormone called *leptin*, after the Greek word for "slender." When it is released from fat cells, leptin seems to tell the body to stop eating.[22] Researchers report that as body-fat stores increase, blood leptin also increases. The brain responds by decreasing appetite and increasing energy expenditure. Likewise, when body-fat stores decrease, blood leptin decreases, and the brain responds by stimulating appetite and decreasing energy expenditure. Mice who have a defective form of the gene fail to produce leptin and can weigh as much as three times more than normal mice. Overweight people, too, may have a defective form of this gene (or may be unresponsive to leptin).[23] More research is needed to clarify this mechanism.

Fat Cell Theory   Some overweight infants become overweight adults, but most grow out of their obesity in childhood. An overweight child, however, is more likely to remain overweight into adulthood.[24] Some researchers propose the **fat cell theory**, which holds that childhood obesity is persistent because early overfeeding (during the growing years) may cause fat cells to increase abnormally in *number*. According to this theory, a person's *number* of fat cells is relatively fixed by the time he or she reaches adulthood; after that, a gain or loss of weight either increases or diminishes the size of the fat cells. Unfortunately, persons with greater numbers of fat cells are less likely to lose weight successfully. Some researchers suggest that the body triggers hunger signals when the fat stored in these cells begins to decrease. Because fat cells increase in number during childhood, prevention of obesity is critical during the growing years.

set-point theory theory that the body tends to maintain a certain weight by adjusting hunger, appetite, and food energy intake on the one hand and metabolism (energy output) on the other so that a person's conscious efforts to alter weight may be foiled.

fat cell theory states that during the growing years, fat cells respond to overfeeding by producing additional fat cells; the number of fat cells eventually becomes relatively fixed, and overfeeding from this point on causes the body to enlarge existing fat cells.

Additionally, fat cells of obese people contain higher levels of the enzyme **lipoprotein lipase (LPL)**, which determines the rate at which adipose cells store fat. The larger the fat cell (and the greater the number of fat cells), the more LPL and the more easily the body can pull triglycerides into fat cells for storage. Unfortunately, LPL activity rises further with weight loss, enhancing the body's ability to regain the lost weight.[25] A question still to be answered is whether some people develop obesity because their fat cells contain an abnormal amount of LPL from birth.

Environment   Environmental factors linked with obesity suggest that people overeat or underexercise because they are pushed to do so by factors in their surroundings: foremost among them, the availability of a multitude of delectable foods and a lack of opportunity for vigorous physical activity.

Some people seem to have inherited or learned a way of resisting external stimuli to eat, but others have not. In a classic experiment with "cafeteria rats," ordinary rats who were fed regular rat chow maintained normal weight (for rats), but when those very same rats were offered free access to a wide variety of tempting, rich, highly palatable foods, they greatly overate and became obese. Similarly, one study found a positive correlation between overfatness and a diet offering a wide variety of snacks and sweets.[26] This is the basis of the **external cue theory**, which posits that, at least in some people, the internal regulatory systems are easily overridden by environmental influences. Does this mean obesity is hereditary, environmental, or both?

It seems likely that both hereditary and environmental factors influence obesity in human beings. The tendency to obesity is probably inherited, but the environment is probably influential in the sense that it can prevent or permit the development of obesity when the potential is there.

A Closer Look at Eating Behavior   In human beings, learning plays an important role. Although we have genetically inborn instincts, things learned during early childhood experiences combined with our current environments may conflict with these instincts. Thus, the **hunger** drive is programmed by heredity, but **appetite** is influenced by learned responses that may cause us to ignore or over-respond to our hunger. Another way to say this is to say that hunger is physiological, whereas appetite is psychological, and the two don't always coincide. We have all experienced appetite without hunger: "I'm not hungry, but I'd love to have some." We also often experience the reverse: "I know I'm hungry, but I don't feel like eating."

The ways people respond to hunger and appetite determine whether they eat too much, too little, or just enough to maintain their weight. A third factor, **satiety**, which signals that it is time to stop eating, also affects our eating patterns and weight issues.

In human physiology, research is beginning to find possible answers to what regulates eating behavior. The stomach's nerves perceive stretching, and you stop eating when your stomach feels stretched full. Blood glucose level is also thought to be involved. You get hungry when your blood glucose level falls—or perhaps when your liver glycogen is beginning to be exhausted. Blood lipids, and possibly amino acids and other molecules, also play a role. When you eat, you secrete hormones to regulate digestive activity; these hormones may also convey the message to the brain that it is time to start or stop eating.*

Where in the brain are these messages received (whatever they are)? One brain area stands out as a regulator for food behavior: the **hypothalamus**. The hypothala-

CAFETERIA

lipoprotein lipase (LPL) an enzyme located on the surfaces of fat cells that enables the cell to convert blood triglycerides into fatty acids and glycerol to be pulled into the cell for reassembly and storage as body fat.

external cue theory the theory that some people eat in response to such external factors as the presence of food or the time of day rather than to such internal factors as hunger.

hunger the physiological drive to find and eat food, experienced as an unpleasant sensation.

appetite the psychological desire to find and eat food, experienced as a pleasant sensation, often in the absence of hunger.

satiety the feeling of fullness or satisfaction experienced following a meal.

hypothalamus (high-poh-THALL-ah-mus) a part of the brain that senses a variety of conditions in the blood, such as temperature, salt content, and glucose content, and then signals other parts of the brain or body to change those conditions when necessary.

---

*Researchers have discovered *ghrelin* (GRELL-in), a protein produced by the stomach cells that acts as a hormone affecting the hypothalamus and stimulating appetite. Ghrelin levels are high before a meal (reflecting hunger) and fall rapidly after a meal (reflecting satiety). Researchers continue to investigate ghrelin's exact role in food intake regulation.

mus communicates with the hormone system and the nervous system. It integrates many signals received from the rest of the body, including information about the blood's temperature, sodium content, and glucose content. We know it is important in regulating eating because damage to the hypothalamus produces derangements in eating behavior and body weight. In some cases it can cause severe weight loss, and in others it can cause extreme overeating. In the person with a normal hypothalamus, however, appropriate eating behavior seems to be a response to a whole host of signals rather than a response to a single signal arriving at some one location in the hypothalamus. Somehow these many inputs become integrated into a final common path: the act of eating.

A person who eats inappropriately may have established a habitual behavior pattern that inappropriately links certain stimuli to the act of eating. The study of behavior offers insight into the causes of overeating by viewing it as a conditioned response. Sometimes eating behavior is initiated by the wrong triggers. For example, a crying child with a bruised knee who is offered a lollipop may learn to associate food with comfort and inappropriately feel the need for food when experiencing emotional pain later in life.

Eating behavior, then, may be a response not only to hunger or appetite but also to complex human sensations such as yearning, craving, addiction, or compulsion. Often, eating is used to relieve boredom or to ward off depression. Some people respond to anxiety—or, in fact, to any kind of **arousal**—by eating. Significantly, however, when these people are able to name their aroused condition, they often gain a feeling that they have some control over it and are not as likely to overeat.

Stress may also directly promote the accumulation of body fat. The stress hormones favor the breakdown of energy stores (glycogen and fat) into glucose and fatty acids, which can be used to fuel the human fight-or-flight response. If a person doesn't use the fuel in physical exertion, however, the body cannot turn these fragments back into glycogen. Thus, its only alternative is to convert them to fat. Each time glucose is pulled out of storage in response to stress and then transformed into fat, the lowered glucose level or exhausted glycogen will signal hunger, and the person will eat again soon after.

Stress eating may appear in different patterns. Some people eat excessively at night, whereas others binge during emotional crises. Some people react oppositely and reject food when stressed. We do not know why these behaviors occur, but research continues.

The many possible causes of obesity mentioned so far all relate to the input side of the energy equation. What about output? Probably the most important contributor to the obesity problem in our country is underactivity.[27] Some obese people eat less than lean people, but due to extraordinary inactivity, they still manage to store surplus calories. Some people move more efficiently than others, too. Two people of the same age, height, and weight might use different amounts of calories walking five miles because of the different ways in which they move their muscles.

No two people are alike, either physically or psychologically, and the causes of obesity are no doubt as varied as the obese people themselves. Many causes may contribute to the problem in a single person. Given this complexity, it is obvious that there is no panacea. The top priority should be prevention, but where prevention has failed, the treatment of obesity must involve a simultaneous attack on many fronts.[28]

Weight Gain and Loss

When you step on the scale and see that you weigh a pound more or less than you did the last time, you haven't necessarily gained or lost body fat. Changes in body weight reflect shifts in many different materials—not only fat but also fluid, bone minerals, and lean tissues such as muscles. The gain of a pound, for example, may reflect

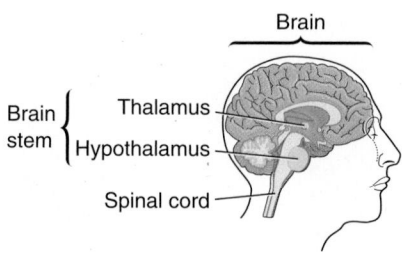

Researchers believe that the hypothalamus controls the sensations of hunger and satiety.

arousal as used in this context, heightened activity of certain brain centers associated with excitement and anxiety.

The loss (or gain) of a pound does not always reflect the loss (or gain) of body fat.

gained muscle and bone and an overall shift toward a leaner body type. Because it is so important for people concerned with weight control to realize this, this section discusses the various changes that take place with gains and losses of weight.

A healthy man or woman about 5 feet, 10 inches tall who weighs 150 pounds carries about 90 of those pounds as water and 30 as fat. The other 30 pounds are the so-called *lean tissues*: muscles; organs such as the heart, brain, and liver; and the bones of the skeleton.* Stripped of water and fat, then, the person weighs only 30 pounds. This lean tissue is the body's vital machinery that maintains health and life. When a person who is too fat seeks to lose weight, it should be fat, not this precious lean tissue, that is lost. For someone who wants to gain weight, it is desirable to gain lean *and* fat, not just fat.

Weight is gained or lost in different body tissues, depending on how a person goes about it. Most quick-weight-loss diet schemes promote large losses of fluid that create large, temporary changes in a person's scale weight with little or no real loss of body fat. The rest of this chapter underscores this distinction, and a later section on weight-loss strategy emphasizes exercise as a means of supporting lean tissue during weight loss.

Weight Gain   When you eat more calories than you need, where does the excess go? The energy nutrients—carbohydrate, fat, and protein—contribute to body stores as follows:

- Carbohydrate is broken down to glucose for absorption. Inside the body, glucose may be built up to glycogen or converted to and stored as fat.

- Fat is broken down to its component parts (including fatty acids) for absorption. Inside the body, these components are easily converted to storage fat.

- Protein, too, is broken down to its basic units (amino acids) for absorption. Inside the body, these units may be used to replace body proteins. The amino acids that are not used lose their nitrogen and are converted to fat.

Notice in Figure 10-5 that although three kinds of materials enter the body, they are stored for later use in only two forms, glycogen and fat. Also notice that when protein is stored in the form of fat, it cannot be converted back into protein later. The amino acids lose their nitrogen, which is then excreted in the urine. It does not matter whether you are eating hamburgers, brownies, or carrot sticks: if you eat enough of any food, the excess will be turned to fat within hours.

Weight Loss and Fasting   When the tables are turned and you stop eating altogether, your body must draw on its stored supplies of nutrients to keep going. Nothing is wrong with this; in fact, it is a great advantage that you can eat periodically, store fuel, and then use up that fuel between meals. The between-meal interval is ideally about 4 to 6 hours (about the length of time it takes to use up most of the available liver glycogen) or 12 to 14 hours at night, when body systems are slowed down and the need for energy is lower. If a person doesn't eat for, say, three days or even a week, the body makes one adjustment after another.

The body's first adjustment to fasting is to use the liver's glycogen for needed fuel. (The glycogen in muscles is reserved for the muscles' own use—and they do use it.) The liver's glycogen, remember, is the body's source of blood glucose to fuel brain and nerve activities. Ordinarily, the brain and nerves can use no other fuel, but after about a day without food, the primary supply is gone. Where, then, does the body

_____

*For a healthy woman or man 5 feet tall who weighs 100 pounds, the comparable figures would be 60 pounds of water, 20 pounds of fat, and 20 pounds of lean tissue.

turn to keep the nervous system going? Whatever it has to do, it will do, for the nervous system runs the body, and when it stops, the body dies.

An obvious alternative source of energy is the abundant fat stores that most people carry. At first, this fat is of no use to the nervous system. The muscles and other organs can use fat as fuel, but the nervous system ordinarily cannot. Nor can the body convert this fat to glucose, because it possesses no enzymes to do so. It does, however, possess enzymes to convert protein to glucose.

When the fast continues, the body turns to its own lean tissues to provide the necessary supply of glucose (see Figure 10-5). One reason people lose weight so dramatically within the first three days of a fast is that they are devouring their own protein tissues as fuel. Because protein contains only half as many calories per pound as fat, it disappears twice as fast. Also, with each pound of body protein, three or four pounds of associated water are lost. This same process accounts for the rapid weight loss seen in the early stages of a low-carbohydrate diet.

[Text continued on page 308.]

FIGURE 10-5

Feasting and Fasting

In A, the person is storing energy. In B, the person is drawing on stored energy. In C, the person is in ketosis.

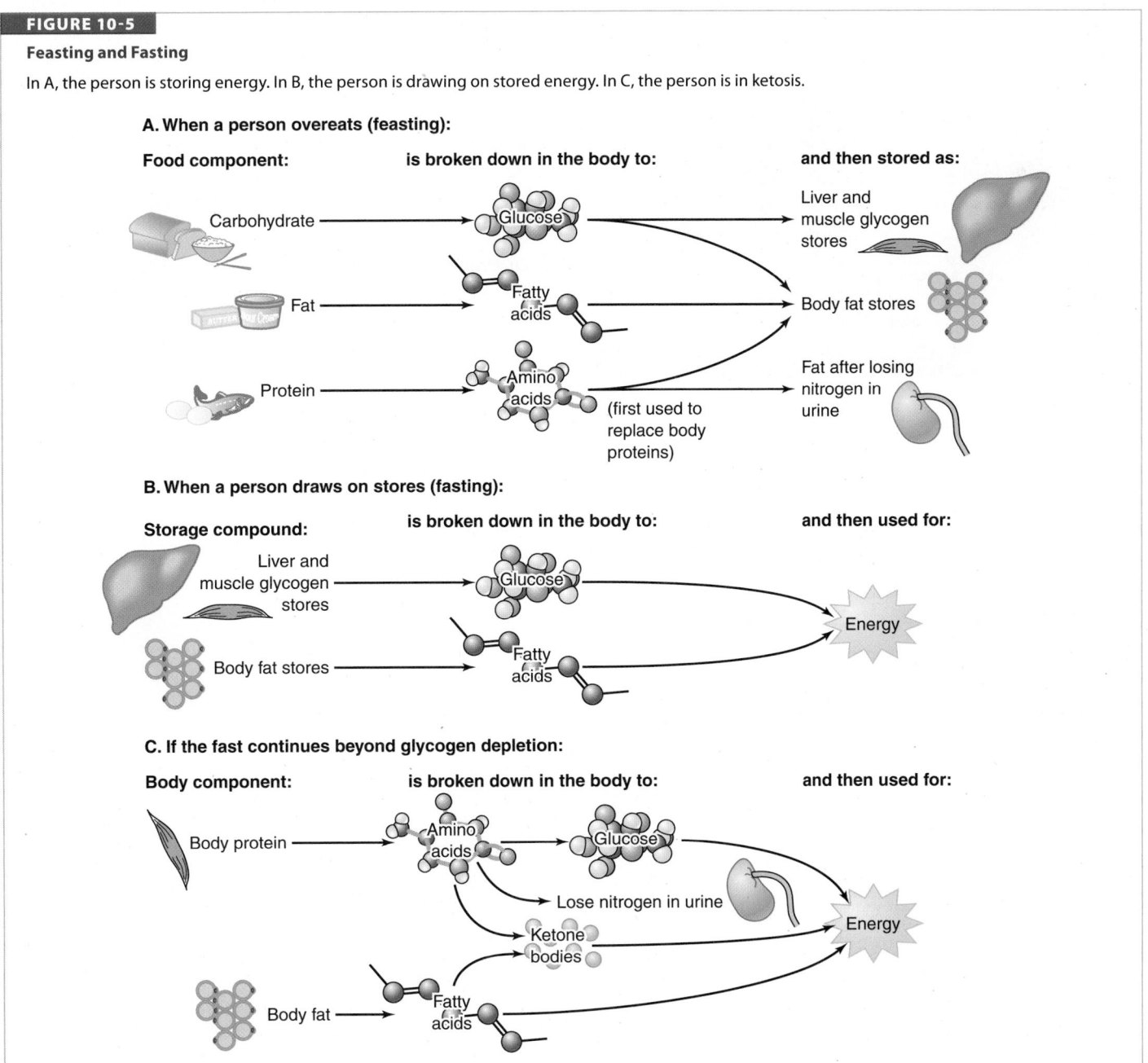

Lose weight while you sleep! Lose 30 pounds in just 20 days! Eat the foods you love and lose weight! You will never be hungry! Do these claims sound familiar? With the recent focus on the increased rates of obesity in the United States and the world, diet books, products, and programs abound. The truth is that although most diets enable a short-term weight loss, few people can lose weight and keep it off permanently, and some of these diets may actually be harmful. Nevertheless, dieting is big business in the United States. One national survey found that 33 percent of American adults are on a diet—an increase from 24 percent in 2000.[29]

Tony Freeman/PhotoEdit, Inc.

How Do Diets Work?

Diets work because people limit their food consumption. Excess weight is the consequence of an energy imbalance, caused by overconsumption of food or decreased physical activity relative to individual requirements. Limiting dietary intake can take several forms:

- Elimination or restriction of certain food groups, such as carbohydrates
- Portion control through prepackaged meals, snacks, or drinks
- Alteration of meal patterns or content
- Control of food intake through point systems or monitoring

A comparison of the approximate caloric content and macronutrient distribution of several types of diets is provided in Table 10-4.

What Are Common Diets?

Although diet fads change quickly, certain types of diets have appeared during the past few years. Here are some of the most common:[30]

- **Dr. Atkins New Diet Revolution.** In this diet, consumption of high-fat meats, cheeses, and fats is encouraged and consumption of carbohydrates (such as fruit, breads, and cereals) is

severely limited. The premise of the diet is that elimination of these foods will produce a "benign dietary ketoacidosis," which leads to decreased hunger and slows excessive food consumption. Ketosis can be accompanied by bad breath, nausea, headaches, and fatigue. High protein intake may exacerbate gout and kidney disease, and high saturated fat intake can increase blood cholesterol levels.

- **The Zone Diet.** This rigid eating plan separates foods into "macronutrient blocks."
- **The South Beach Diet.** This regimen is a more healthful version of the Atkins high-protein, low-carbohydrate diet, incorporating lower-fat protein sources such as chicken and fish, whole grains, and vegetables and fruits. The plan does limit some foods, such as carrots, bananas, pineapple, and watermelon, and the first phase of the diet is more restrictive than later phases.
- **Weight Watchers.** Dieters use a point system to select and eat foods to reduce caloric intake and lose weight. In Weight Watchers, no food is forbidden, but all foods must be balanced with other choices.
- **Dr. Ornish Eat More, Weigh Less.** Weight loss is based on consuming a very low-fat diet (10 percent of calories from fat), with little meat, oils, nuts, butter, dairy (except nonfat), sweets, or alcohol. The original Ornish plan included diet together with exercise and stress reduction.
- **Eat Right for Your Blood Type.** This diet is based on the claim that your blood type determines the types of foods you should eat and how your body absorbs nutrients. For example, people with type O blood should consume meat, seafood, fruits, and vegetables, but less wheat and beans. There is no scientific basis for this claim.
- **Dr. Phil's Ultimate Weight Solution.** The book describing this diet focuses on "Keys to Weight-loss Freedom," but these concepts do not include defined meal plans or recipes. The diet promotes seafood, poultry, meat, low-fat dairy, whole grains, fruits, vegetables, and some oils. Supplements, weight-loss bars, and shakes are also promoted.
- **The New Glucose Revolution.** This eating plan encourages consumption of low-glycemic foods, such as beans, pasta, most fruits, vegetables, low-fat dairy, and meats. Unfortunately, the glycemic index is not always a reliable measure of increases in blood glucose, which can vary with the food itself and with other foods consumed at the same time. In addition, changes in blood glucose level depend on the amount of food consumed.

A study published in the *Journal of the American Medical Association* evaluated four of these diets (the Atkins, Ornish, Weight Watchers, and Zone diets) and found that, after one year on the

TABLE 10-4

TABLE 10-4 Comparison of Diet Programs/Eating Plans to Typical American Diet

Type of Diet	Example	General Dietary Characteristics	Comments
Typical American diet		Carb.: 50% Protein: 15% Fat: 35% Average of 2,200 cal/day	• Low in fruits and vegetables, dairy and whole grains • High in saturated fat and refined carbohydrates
Balanced nutrient, moderate-calorie approach	DASH Diet or Diet based on MyPlate food guide; commercial plans such as Diet Center, Jenny Craig, Nutrisystem, Physician's Weight Loss, Shapedown Pediatric Program, Weight Watchers, Setpoint Diet, Sonoma Diet, TOPS, Volumetrics	Carb.: 55–60% Protein: 15–20% Fat: 20–30% Usually 1,200 to 1,700 cal/day	• Based on set pattern of selections from food lists using regular grocery store foods or prepackaged foods supplemented by fresh food items • Low in saturated fat and ample in fruits, vegetables, and fiber • Recommend reasonable weight-loss goal of 0.5 to 2.0 lb/week • Prepackaged plans may limit food choices • Most recommend exercise plan • Many encourage dietary record-keeping • Some offer weight-maintenance plans/support
Very low-fat, high-carbohydrate approach	Ornish Diet (Eat More, Weigh Less), Pritikin Diet, T-Factor Diet, Choose to Lose, Fit or Fat	Carb.: 65% Protein: 10–20% Fat: ≤ 10–19% Limited intake of animal protein, nuts, seeds, other fats	• Long-term compliance with some plans may be difficult because of low level of fat • Can be low in calcium • Some plans restrict healthful foods (seafood, low-fat dairy, poultry) • Some encourage exercise and stress-management techniques
Low-carbohydrate, high-protein, high-fat approach	Atkins New Diet Revolution, Protein Power, Stillman Diet (The Doctor's Quick Weight Loss Diet), the Carbohydrate Addict's Diet, Scarsdale Diet	Carb.: ≤ 20% Protein: 25–40% Fat: ≥ 55–65% Strictly limits Carb. to less than 100–125 g/day	• Promote quick weight loss (much is water loss rather than fat loss) • Ketosis causes loss of appetite • Can be too high in saturated fat • Low in carbohydrates, vitamins, minerals, and fiber • Not practical for long-term because of rigid diet or restricted food choices
Moderate-carbohydrate, high-protein, moderate-fat approach	The Zone Diet, Sugar Busters, South Beach Diet	Carb.: 40–50% Protein: 25–40% Fat: 30–40%	• Some diets rigid and difficult to maintain • Enough carbohydrates to avoid ketosis • Can be low in carbohydrates; can be low in vitamins and minerals
Novelty diets	Immune Power Diet, Rotation Diet, Cabbage Soup Diet, Beverly Hills Diet, Dr. Phil	Most promote certain foods, or combinations of foods, or nutrients as having unique (magical) qualities	• No scientific basis for recommendations
Very low-calorie diets	Health Management Resources (HMR), Medifast, Optifast	Less than 800 cal/day	• Requires medical supervision • For clients with BMI ≥ 30 or BMI ≥ 27 with other risk factors; may be difficult to transition to regular meals
Weight-loss online diets	Cyberdiet, DietWatch, eDiets, Nutrio.com	Meal plans and other tools available online	• Recommend reasonable weight-loss goal of 0.5 to 2.0 lb/week • Most encourage exercise • Some offer weight-maintenance plans/support

SOURCE: Adapted from Weighing the diet books, *Nutrition Action Newsletter* (January–February 2004): 3–8; M. Freedman and coauthors, Popular diets: A scientific review, *Obesity Research* 9 (2001): 1S–39S; and A guide to rating the weight-loss websites, *Tufts University Health and Nutrition Letter*, May 2001, pp. 1–4.

(Continued)

diet, all four modestly reduced body weight and some cardiac risk factors.[31] Adherence to each diet for the 12-month period varied, ranging from 50 percent for the Ornish diet and 53 percent for the Atkins diet to 65 percent for both the Weight Watchers and Zone diets. The subjects who had the best adherence to the diets had the best results, and cardiac risk factors were more closely associated with weight loss than with diet type. In general, the subjects had more difficulty following the more restrictive diets (the Ornish and Atkins diets). Although this is just one study with small sample sizes, it does suggest that there are many ways to lose weight, that people find it difficult to adhere to very restrictive diets for a long time, and that we need to find methods of keeping people motivated to stay on any new eating plan.

How Can You Determine If a Diet Is Healthful?

What can you do to determine whether a particular diet plan is useful *and* healthful? Use the following checklist:[32]

1. *Does the weight-loss program systematically eliminate one group of foods from a person's eating pattern?* For example, are all carbohydrates systematically eliminated from the diet? Are dairy products eliminated? In general, a diet that eliminates a certain food group is probably lacking in important nutrients and dietary variety, and it will be difficult for a person to adhere to that eating plan.

2. *Does the weight-loss program encourage specific supplements or foods that can be purchased only from selected distributors?* These supplements or foods often contain ingredients that may be harmful or unproven.

3. *Does the weight-loss program tout magic or miracle foods or products that burn fat?* The only way to burn fat is to increase your physical activity levels or decrease the amount of total food that you consume. You cannot "burn" fat with sauna belts, body wraps, thigh-reducing creams, or similar products. If you consume more than you expend, or if you lower your physical activity level and keep your food intake the same, your body will store the extra calories as fat.

4. *Does the weight-loss program promote bizarre quantities of only one food or one type of food?* Some diets require that you eat only one food each day or eat unlimited amounts of certain foods, such as grapefruit or cabbage soup. Such advice runs counter to everything we know about the broad spectrum of human nutritional needs.

5. *Does the weight-loss program have rigid menus?* If a diet has specific meal plans and times to eat, it will be difficult to incorporate individual preferences. People are unique, so no one diet plan will work for everyone. A person who loves Thai food will not succeed on a diet if there is no way to incorporate Thai food into his or her eating plan.

6. *Does the weight-loss program promote specific food combinations?* Some diets include combinations of foods that should or should not be eaten at the same time. These food combinations have no basis in fact and needlessly restrict the dieter's options for reasonable food choices.

7. *Does the weight-loss program promise a weight loss of more than 2 pounds per week for an extended period of time?* If so, the initial weight loss will probably be due to water loss. A more realistic diet plan will aim for a weight loss of 0.5 to 2.0 pounds per week.

8. *Does the weight-loss program provide a warning to people with diabetes, high blood pressure, or other health conditions?* People with preexisting health conditions should consult a physician or other health care provider before beginning any diet. Elimination of certain food groups or eating excessive amounts of certain foods can exacerbate these problems and may interfere with the effectiveness of certain medications.

9. *Does the weight-loss program encourage or promote increased physical activity?* Although people can lose weight by limiting food intake alone, research has shown that the most success-

ketosis (key-TOE-siss) an adaptation of the body to prolonged (several days) fasting or carbohydrate restriction; body fat is converted to ketones, which can be used as fuel for some brain cells.

If the body were to continue to consume itself at this rate, death would ensue within about 10 days. After all, the liver, the heart and skeletal muscles, the lung tissue, and the blood—all vital tissues—are being burned as fuel. (In fact, fasting or starving people remain alive only until their body fat is gone or until half their lean tissue is gone, whichever comes first.) But now the body plays its last trump card: It begins converting fat stores into a form it can use to help feed the nervous system and so forestall the end. This is known as **ketosis**.

TABLE 10-5	Risks Associated with the Very Low-Calorie Diets

- Blood sugar imbalance
- Cold intolerance
- Constipation
- Decreased basal metabolic rate
- Dehydration
- Diarrhea
- Emotional problems
- Fatigue/weakness
- Gallstones and kidney stones
- Headaches
- Heart irregularity
- Ketosis
- Loss of lean body tissue
- Kidney infection
- Menstrual irregularity
- Mineral and electrolyte imbalances
- Sleeplessness
- Sudden death

SOURCE: Adapted from Position of the American Dietetic Association: Very low-calorie weight loss diets, *Journal of the American Dietetic Association* (May 1990), p. 722.

ful weight-loss plans include lifestyle changes, such as increasing exercise.

10. *Does the weight-loss program encourage an intake that is very low in calories (below 800 calories/day) without supervision of medical experts?* Very low-calorie diets should be used only by persons with severe obesity or obesity with other health-related problems. Because the calorie intake is so low, the diet must be supplemented with vitamins and minerals. In addition, the dieter must be strictly observed by a physician for any of the adverse health effects listed in Table 10-5. Finally, the person needs dietary counseling to handle "real" food choices before the end of the diet, or weight gain can quickly ensue.

Don't Give Up!

For some folks, weight loss may seem like a lost cause, but don't give up! Several strategies and diets have been proven successful. The strategies supported by the most evidence are detailed in a recent analysis by the USDA and backed up by data from the National Weight Control Registry, a study that examines people who have lost at least 30 pounds and have maintained that loss for at least a year.[33]

You can take several steps to help you make wise decisions about dieting:

1. Be familiar with the current fad diets, and study any diet that interests you. These food plans often include scientifically based statements intermingled with inaccuracies, so you have to know the literature to refute any incorrect claims.

2. Seek out appropriate weight-loss strategies and programs. A recent evidence-based review indicates that most weight loss is associated with diets that include about 1,400 to 1,500 calories per day. It is essential to control energy intake for any weight-loss plan.[34] Weight Watchers has been cited as a good option in many recent studies, because of the variety of foods offered and because its principles are based on scientific evidence. Internet-based programs are available for people who like to keep records and need support but cannot attend group sessions. The DASH diet has been found to significantly reduce high blood pressure and improve other chronic disease outcomes, and it is free on the National Institutes of Health (NIH) website. About half of the people in the National Weight Control Registry (NWCR) lost weight without any formal program, indicating that the more individualized a program, the more likely it is that people will adhere to it for longer periods of time. Finally, most successful attempts to lose weight include some type of regular exercise.

3. The following websites contain good information or handouts that you can use to determine whether following a particular diet will be harmful or not: (1) the Federal Trade Commission (FTC) website, www.ftc.gov, which includes "Weighing the Evidence in Diet Ads"; and (2) the American Heart Association's *Fad Diets*, available at www.americanheart.org.

4. Report fraudulent or deceptive weight-loss claims. Any weight-loss claims that are distributed via the Internet, television, or print media can be reported at www.ftc.gov or by calling 1-877-FTC-HELP (1-877-382-4357).

SOURCE: Adapted from D. M. Hoelscher & C. McCullum-Gomez, Addressing the obesity epidemic: An issue for public health policy, in M. Boyle & D. Holben, *Community Nutrition in Action: An Entrepreneurial Approach*, 5th ed. (Belmont, CA: Wadsworth/Cengage Learning, 2010).

Ketosis is an adaptation to fasting or carbohydrate deprivation. Instead of breaking fat molecules all the way down to carbon dioxide and water as it normally does, the body takes partially broken-down fat fragments, combines them into ketone bodies (compounds that are normally rare in the blood), and lets them circulate in the bloodstream. The advantage for the body is that about half of the brain cells can use these compounds for energy. Thus, indirectly, the nervous system begins to feed on the body's fat stores. This reduces the nervous system's need for glucose, spares the muscle and other lean tissue from being devoured so quickly, and prolongs the starving person's life. Because of ketosis, an initially healthy person who is totally deprived of food can live for as long as six to eight weeks.

Fasting has been practiced as a periodic discipline by respected, wise people in many cultures. However, ketosis may be harmful to the body by upsetting the acid–base balance of the blood. For the person who merely wants to lose weight, then, fasting is not the best way. For one thing, even in ketosis, the body's lean tissue continues to be lost at a rapid rate to supply glucose to those nervous system cells that cannot use ketones as fuel. For another, the body becomes conservative during a fast and slows its metabolism so as to lose as little energy as it possibly can. A well-designed, low-calorie diet, accompanied by the appropriate physical activity program, has actually been observed to promote the same rate of *weight* loss as, and a faster rate of *fat* loss than, a total fast. The Nutrition Action feature that follows discusses what to look for in a weight-loss diet, including tips for how *not* to design a diet.

Drugs and Weight Loss   The search is on to find a safe and effective drug solution to the problem of obesity. The ideal drug will have to be safe, free of undesirable side effects and abuse potential, and effective at reducing body fat. Additionally, as with drug treatments for other chronic disorders (for example, high blood pressure), the ideal drug should be safe and effective for long-term use. To foster long-term success, any such drug treatment should be combined with lifestyle changes, including physical activity and a healthy diet.[35] Until recently, most of the available drugs were appetite suppressants, which were available as prescriptions or over the counter (see Table 10-6).

Numerous other diet aids on the market include products with mysterious-sounding ingredients, such as spirulina, chitosan, chromium picolinate, bitter orange extract, "carb blockers," "fat blockers," ginseng, and many others. Manufacturers typically claim or suggest that such products can aid in weight loss, but their ingredients often serve as little more than fillers. To date, none have proven effective in aiding weight loss. For example, manufacturers of products that include chromium picolinate praise its drug-like abilities to reduce fat, build lean tissue, suppress appetite, and increase metabolism, and they imply that our diets lack chromium. However, chromium picolinate has not been approved for weight loss by the FDA, and claims regarding it are not backed by scientific data.[36] The FDA and the Federal

TABLE 10-6	Drugs for Weight Loss	
	Action in the Body	**Potential Side Effects and/or Comments**
Prescription Drugs (Trade Name)		
orlistat (Xenical)[a]	A lipase inhibitor; acts by inhibiting the enzymes—gastric and pancreatic lipase—needed for fat digestion; binds to the enzymes, making them unavailable to digest dietary fat. Because fat absorption is reduced by as much as 30%, people often lose weight.	Bloating, gas, frequent bowel movements, and anal leakage of undigested fats in people who do not adhere to a low-fat diet. Because Xenical may block the absorption of fat-soluble vitamins, users are advised to take a multivitamin supplement. The drug is intended only for the obese person—with a BMI greater than 30.
Phentermine (Adipex-P or Fastin)	Raises the level of norepinephrine, signals satiety and acts as a stimulant to increase the rate at which calories are burned.	Increased heart rate and blood pressure, dry mouth, insomnia, nausea, agitation, diarrhea, and constipation.
Over-the-Counter Drugs		
benzocaine	An anesthetic found in gum or candy form that numbs the taste buds and reduces the desire for food.	Trade names: Diet Ayds candy or Slim Mint gum; the only over-the-counter appetite-suppressant ingredient currently approved by the FDA without prescription.

[a]An over-the-counter low-dose form of orlistat is also available (trade name alli).

Trade Commission have begun taking action against diet aids and products touting false claims for weight loss.[37]

The popularity of herbal supplements for weight loss is on the rise, partly because consumers have the mistaken notion that herbs are a "natural" alternative to prescription drugs. Thus far, herbal preparations in the United States are produced and marketed free of regulation to ensure their safety or effectiveness, and some have been linked to side effects ranging from nausea and headaches to heart attacks and death. For example, the Chinese herb *ma huang*—commonly called ephedra—contains ephedrine, a stimulant that mimics the action of the drug phentermine. Ephedrine can cause tremors, insomnia, severe headaches, high blood pressure, heart attacks, and stroke. Some herbal preparations combine ephedra with other substances (for example, caffeine-containing guarana), to enhance the ephedrine's effects. More than 800 reports of illness and dozens of deaths are linked to ephedra, and both the FDA and Canada prohibit its use.[38]

Other herbal preparations sold as dieter's tea contain senna, rhubarb root, aloe, cascara, or buckthorn. The "tea" has a laxative effect that can cause dehydration, diarrhea, nausea, fainting, and in some cases, even death.[39] For more about herbs thought to be effective and those that should be avoided, see the Nutrition Action feature in Chapter 7.

Surgery and Weight Loss   Sheer desperation prompts some obese people to request weight-loss surgery. Surgery may be an option for people who cannot lose weight by traditional means and are severely obese (BMI ≥ 40) or have a BMI of at least 35 with one or more severe health problems.[40] Two common types of surgery are gastric bypass and gastric banding (see Figure 10-6). Common side effects include nausea, vomiting, diarrhea, heartburn, abdominal pain, and band slippage or pouch enlargement. Other more severe effects, including complications during surgery, infections, and death, may occur as well.

Gastric bypass is the most common surgical procedure for obesity in the United States. First, the surgeon creates a small pouch in the stomach with staples or a plastic band. Next, a Y-shaped section of the small intestine is attached to the pouch to allow food to bypass the lower stomach and the upper portion of the small intestine. This

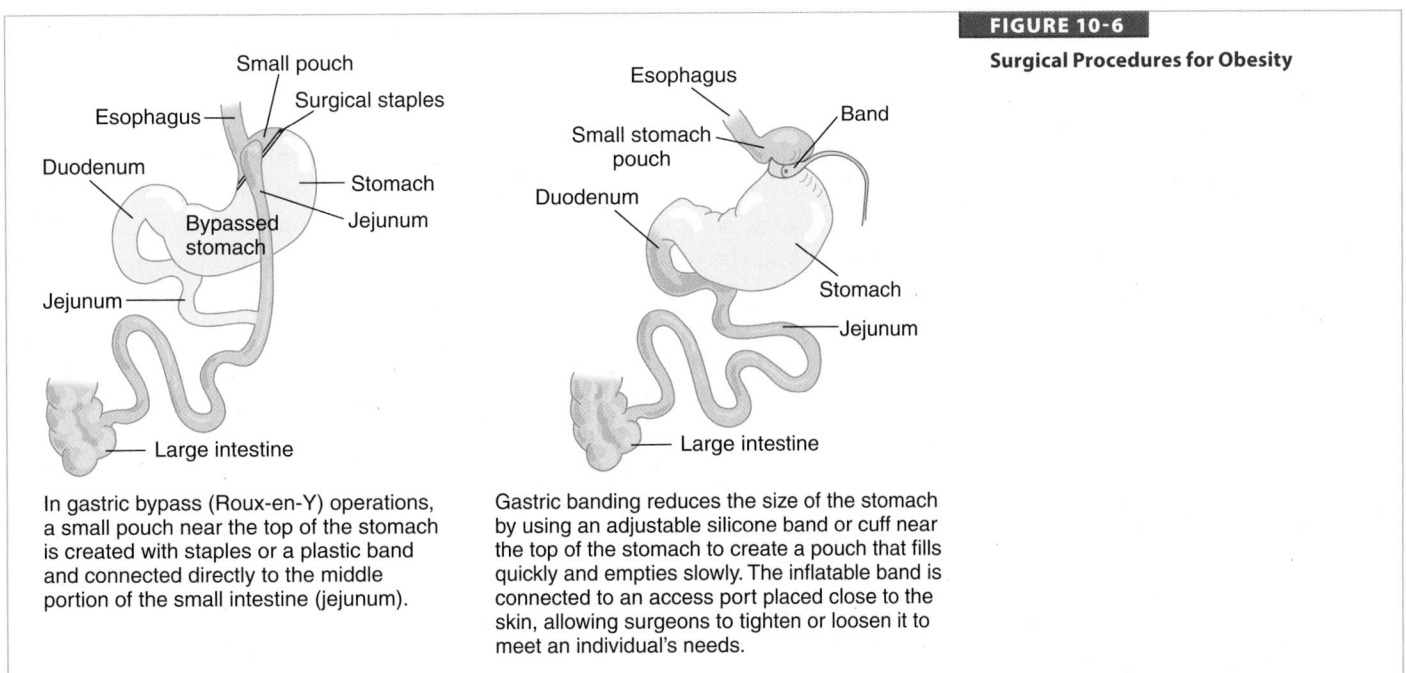

FIGURE 10-6

Surgical Procedures for Obesity

In gastric bypass (Roux-en-Y) operations, a small pouch near the top of the stomach is created with staples or a plastic band and connected directly to the middle portion of the small intestine (jejunum).

Gastric banding reduces the size of the stomach by using an adjustable silicone band or cuff near the top of the stomach to create a pouch that fills quickly and empties slowly. The inflatable band is connected to an access port placed close to the skin, allowing surgeons to tighten or loosen it to meet an individual's needs.

The problem with going on a "crash" diet with a goal of, say, a 15-pound weight loss in 3 weeks is that it's a quick fix—a temporary solution to what is typically a chronic problem. Often, the dieter tries to rigidly restrict eating by, for example, skipping meals or eating salads all day. But trying not to eat is like trying not to breathe, as one group of psychologists put it.[41] The body and mind soon rebel and, like a person gasping for air, the dieter loses control and binges. In fact, experts believe that rigid diets play a strong role in the development of binge-eating problems.[42] After the binge, the person feels that he has failed, gives up the diet altogether, eats more to make himself feel better, puts on some weight, feels even worse, gains even more weight, decides to try another restrictive diet, and begins the whole cycle all over again. The solution, say many experts, is *not* to focus on losing a certain amount of weight within a set period of time. This feature shows you how.

1 Adopt a Nondiet Approach to Weight Loss.

A more healthful, nondiet approach to weight loss is to gradually develop habits that you can live with permanently and that will help you shed pounds and keep them off over the long run. Instead of measuring your success by the needle on the scale, gauge your progress by the strides you make in adopting good eating and exercise habits as well as healthful attitudes about yourself and your body.

2 Set Achievable Goals.

Most people recognize the importance of the eating and exercise habits discussed in this chapter, but few dieters recognize that the attitude problems that often accompany restrictive dieting can

stand in the way of long-term weight loss. One of the most common problems is "all-or-nothing" thinking in which people view the world as black or white, right or wrong, good or bad, and so forth. When it comes to diets, this attitude translates into "good or bad" foods and "on or off" limits to diet food or junk food.[43] Such thinking sets the stage for failure as the dieter tries to live up to extremely rigid, unrealistic goals: "Ice cream is bad, so I must never eat it," "I will jog three miles every day," or "I will never order anything but a salad when I go out to eat." Typically, a person who has an all-or-nothing attitude might go several days without, say, ice cream, which she views as "bad junk food," and then splurge on a double-dip cone. Instead of recognizing that ice cream isn't "bad" and that enjoying some every now and then won't undo all her previous efforts, she tells herself, "I've blown my diet completely— I have no willpower." Consequently, feeling guilty and depressed because she's "blown it" anyway, she eats another pint of ice cream and some cookies to make herself feel better. If she had a different attitude—"I've been eating lots of fruits and vegetables and go bicycling several times a week, so one ice cream cone won't hurt me"—she might be able to avoid this bout of despair and binging.

3 Focus on Health Rather than Appearance.

Another attitude that can thwart efforts to achieve a healthy weight is the "lookist" attitude: the notion that weight and appearance determine a person's worth and happiness.[44] Nearly everyone holds lookist attitudes in one form or another and wish that they could change something about their appearance in the hopes that it would make life more enjoyable. In our "thin-is-in" society, though, overweight people suffer tremendous prejudice and tend to be painfully self-conscious about their bodies. As a result, they

---

procedure restricts both food intake and the amount of calories and nutrients the body absorbs. The second most common operation in the United States for weight loss is called *gastric banding*. The procedure involves placement of an adjustable silicone band around the upper part of the stomach to create a small pouch; this limits food consumption and creates an earlier feeling of fullness.

At first, the small stomach pouch that results from weight-loss surgery holds about 1 ounce of food, though it may later stretch to hold 2 to 3 ounces. Nausea and vomiting occur if the person continues to overeat following surgery. Most people can eat about ½ to 1 cup of food without discomfort or nausea, but the food must be soft, moist, and well chewed. Most people who undergo bariatric surgery lose more than 50 percent of their excess weight and keep that weight off long term. As with other treatments for obesity, the best results are achieved by healthful eating behaviors and regular physical activity.

Another weight-loss approach involves cosmetic surgery. One such procedure is **liposuction**, in which the surgeon uses a small hollow tube to suction out fatty tissue from beneath the skin. People who wish to remove the fat from a particular area can

liposuction a type of surgery (also called lipectomy) that vacuums out fat cells that have accumulated, typically in the buttocks and thighs. If the person continues to eat more calories than are expended through physical activity, fat will return to the fat cells that remain in those regions.

TABLE 10-7   The Obese Self versus the Healthy Self

The Obese Self	The Healthy Self
Uses appearance as measure of self-worth	Uses caring for others/self as measure of self-worth
Is socially isolated	Is socially involved
Rejects self	Accepts self
Goes on and off extreme diets	Eats healthfully
Rarely exercises (or overdoes it and quits)	Exercises regularly and sensibly
Focuses on mistakes; feels like a failure	Views mistakes as learning experiences and recognizes that nobody is perfect
Uses obesity as excuse for failures in life	Doesn't allow obesity to interfere with life
Focuses on past	Focuses on present
Allows negative emotions to reduce self-control	Uses positive attitude to enhance self-control
Focuses on appearance	Focuses on health
Wants to improve appearance to get love/affection	Wants to be healthy to participate fully in loving relationships

SOURCE: Adapted from J. P. Foreyt & G. K. Goodrick, *Living without Dieting* (New York: Warner Books, 1992), 55.

may mistakenly believe that all their dreams will come true once they lose weight and that all their problems will be lost along with the pounds. By pinning all their hopes for happiness on losing weight, they put themselves under enormous pressure to shed pounds. In addition, people with this attitude often do not live in the present, because they are so caught up in fantasizing about what they think life will be like once they lose weight.

Using weight and appearance as a measure of self-worth and happiness can be extremely destructive. The person who does so may desperately try to lose weight with restrictive, unrealistic diets or unhealthful, strenuous exercise regimens. Each slip-up whittles away at the person's self-esteem, which in turn may lead to feel-ings of rejection, depression, and social isolation, which in turn may prompt a binge, and so forth. Even the person who loses the desired number of pounds may find that he or she still has many of the same life problems as before, and this realization can lead to depression and loss of self-esteem.

The way around lookist thinking is to focus on health—both mental and physical—rather than appearance. Table 10-7 lists some common characteristics of the obese self (which holds lookist atti-tudes) and the healthy self. Although it can be difficult to overcome society's prejudices about weight, striving to adopt the ideals of the healthy self, regardless of your weight, can be a major step in help-ing you take care of your mental and physical health.

elect this procedure, which sometimes brings pleasing results but sometimes produces a figure in which one part of the body is disproportionately thin relative to the others.

Surgery is appropriate in some instances. Cosmetic surgery can minimize disfig-urements, improve self-confidence, and ease the way toward concentration on life issues that are more important than external appearance. After surgery, however, the same person resides within the skin as before. A changed appearance does not guar-antee changed eating habits, a better personality, reduced interpersonal conflicts, or any other improvements in the quality of one's life.

Successful Weight-Loss Strategies

Given that so many approaches are likely to fail, what weight-loss strategies work? How can a person lose weight safely and permanently? The secret is a sensible (not to say *easy*) three-pronged approach involving healthful eating habits, exercise, and behavior change. Such an approach takes tremendous dedication, especially at first, as a person whose habits have promoted obesity makes new habits out of the hundred

Weight-Loss Readiness Quiz

Are you ready to lose weight? Your attitude affects your ability to succeed. Take this readiness quiz to see if you're mentally ready before you begin.

Mark each statement as true or false. Be honest with yourself! The answers should reflect the way you really think, not how you'd like to be or think you "should" be!

1. I have thought a lot about my eating habits and physical activities, and I know what I might change.
2. I know that I need to make permanent, not temporary, changes in my eating and activity patterns.
3. I will feel successful only if I lose a lot of weight.
4. I know that it's best if I lose weight slowly.

Now score yourself:

Look at your answers for items 1, 2, 4, 5, 7, and 9. Score "1" if you answered true and "0" if you answered false. For items 3, 6, 8, 10, and 11, score "0" for each true answer and "1" for each false answer.

No single item indicates if you're ready to start losing weight. But the higher your total score, the more likely you are to be successful.

If you scored 8 or higher, you probably have good reasons to lose weight now, and you know some of the steps that can help you succeed.

Interpret Your Score

Your answers can be clues to some stumbling blocks in your weight-management success. Any item you scored as "0" suggests a misconception about weight loss, or a problem area for you. So let's look at each item a bit more closely.

1. You can't change what you don't understand, and that includes your eating habits and activity pattern. Keep records for a week to pinpoint when, what, why, and how much you eat, as well as patterns and obstacles to regular physical activity.
2. You may be able to lose weight in the short run with drastic or highly restrictive changes in your eating habits or activity pattern—but those changes may be hard to live with permanently. Your food and activity plans should be healthful ones that you can enjoy and sustain.
3. Many people fantasize about reaching a weight goal that's unrealistically low. If that sounds like you, rethink your meaning of success. A reasonable goal takes body type into consideration and sets smaller, achievable "mile markers" along the way.
4. If you equate success with fast weight loss, you'll have problems keeping weight off. This "quick fix" attitude can backfire when you face the challenges of weight maintenance. The best and healthiest approach is to lose weight slowly while learning strategies to keep weight off permanently.
5. To be successful, the desire for and commitment to weight loss must come from you, not your best friend or a family member. People who lose weight and keep it off take responsibility for their weight goals and choose their own approach.

5. I'm thinking about losing weight now because I really want to, not because someone else thinks I should.
6. I think losing weight would solve other problems in my life.
7. I am willing and able to increase my regular physical activity.
8. I can lose weight successfully if I have no slip-ups.
9. I am willing to commit time and effort each week to organize and plan my food and activity choices.
10. Once I lose a few pounds but reach a plateau (and can't seem to lose more), I usually lose the motivation to keep going toward my weight goal.
11. I want to start a weight-loss program, even though my life is unusually stressful right now.

If you scored 5 to 7 points, you may need to reevaluate your reasons for losing weight and the strategies you'd follow.

If you scored 4 or less, now may not be the right time for you to lose weight. You may be successful initially, but you may not be able to sustain the effort to reach or maintain your weight goal. Reconsider your reasons and approach.

6. Being overweight may contribute to some social problems, but it's rarely the single cause. Although body image and self-esteem are strongly linked, thinking that you can solve all your problems by losing weight isn't realistic. Furthermore, it may set you up for disappointment.
7. A habit of regular, moderate physical activity is a key factor in successfully losing weight and keeping it off. For weight control, physical activity doesn't have to be strenuous to be effective. Any moderate physical activity that you enjoy and will do regularly counts.
8. Most people don't expect perfection in their daily lives, yet they often feel they must stick to a weight-loss program perfectly. Perfection at weight loss isn't realistic. Rather than viewing lapses as catastrophes, see them as opportunities to discover what triggers your problems and to develop strategies for the future.
9. To successfully lose weight, you must take time to assess your problem areas, then develop the approach that's best for you. Success requires planning, commitment, and time.
10. First, a plateau in an ongoing weight loss program is perfectly normal, so don't give up too soon! Before you lose your motivation, think about any past efforts that have failed, then identify strategies that can help you overcome those hurdles.
11. Weight loss itself can be a source of stress, so if you're already under stress, you may find a weight-loss program somewhat difficult to implement right now. Try to resolve other stressors in your life before starting your weight-loss effort.

SOURCE: From ADA, *The American Dietetic Association's Complete Food and Nutrition Guide*, pp. 32–33, Copyright © 2002 John Wiley & Sons, Inc. Reproduced with permission of John Wiley & Sons Ltd.

or so behaviors necessary to promote a healthful weight. Even the most effective weight-loss programs reveal a grim pattern: Many people complete the program, lose about 10 percent of their body weight, regain two-thirds of it within 1 year, and regain almost all of it within 5 years.[45] Still, many others who resolve to lose weight do so, and they manage to keep it off. The University of Pittsburgh School of Medicine maintains a National Weight Control Registry of more than 3,000 persons who have successfully achieved and maintained weight loss for a number of years. A recent survey of dieters found that 20 percent of respondents had lost up to 42 pounds and kept it off for an average of 7 years.[46] Those who succeed typically do so because they have employed many of the techniques described in this chapter, as well as those in the Eat Well Be Well feature on page 312.

Personalize Your Weight-Loss Plan

The way a particular person loses weight is a highly individual matter. Two different weight-loss plans may be equally successful and yet have little or nothing in common. You have to find a plan that's right for you. To emphasize the personal nature of weight-loss plans, the following sections are written as advice to "you," even though you may not need to lose weight at all. To help you understand the issues of weight loss, you can pretend that you have 10 to 200 pounds to lose and are being competently counseled by someone familiar with the techniques known to be effective. Notes in the margin highlight the principles involved.

No particular diet is magical, and no particular food must be either included or avoided. Because you are the one who will have to live with the eating plan, you had better be involved in creating it. Don't think of it as going "on" a diet—because then you may be tempted to go "off" it. Lifestyle changes are successful only if the pounds do not return. Think of it as an eating plan that you will adopt for life. The diet must consist of foods that you like or can learn to like, that are available to you, and that are within your means.

If you want to lose weight, a deficit of 500 calories a day for 7 days (3,500 calories a week) is enough to lose a pound of body fat each week. If you choose to spend an extra 250 calories a day in some form of exercise, you could increase this energy deficit.

Choose a calorie level you can live with. The "10-calorie rule" will enable you to lose a pound or two a week while still supporting your basal metabolism: Allow 10 calories a day for each pound of your present body weight. As you lose weight, you can gradually adjust calories downward to keep losing at this rate. Thus, a person who weighs 220 pounds should begin with 2,200 calories a day; one who weighs 150 pounds should begin with 1,500 calories a day.

Put nutritional adequacy high on your list of priorities. This is a way of putting yourself first: "I like me, and I'm going to take good care of me." This means including foods that are rich in valuable nutrients, such as vegetables, fruits, whole-grain breads and cereals, and a reasonable amount of protein-rich foods, such as lean meats, skinless poultry, fish, eggs, legumes, low-fat cheeses, and fat-free milk. Within these categories, learn what foods you like, and eat them often. If you resolve to include a certain number of servings of food from each of these groups each day, you may be so busy making sure you get what you need that you will have little time or appetite left for high-calorie or empty-calorie foods. Researchers have shown that reducing the intake of fat alone can promote significant weight loss, especially when coupled with a diet high in complex carbohydrates.[47]

A small amount of fat should be included in each meal to make it satisfying and keep you from getting hungry again too soon. You don't have to use pure fat, such as butter, margarine, or oil. Rather, most of the fat should come from protein-rich foods, such as lean meats, eggs, poultry, fish, and low-fat cheeses. Add any pure fat with extra caution. A slip of the butter knife adds more calories than a slip of the

Eating Plan Strategies

1. Get personally involved.
2. Adopt a realistic plan, and then keep track of calories.
3. Make the eating plan adequate.
4. Emphasize high nutrient density.
5. Individualize. Eat foods you like.
6. Stress "dos," not "don'ts."
7. Eat regular meals.
8. Take a positive view of yourself.
9. Visualize a changed future self.
10. Take well-spaced weighings to avoid discouragement.

FIGURE 10-7

Sample Balanced Weight-Loss Diets Using the MyPlate Food Guide[a]

[a] Assumes no alcohol intake.

[b] Not recommended for pregnant or lactating women, children (depending on age), or those who have special dietary needs. At or below this low level of calorie intake, it may not be possible to obtain recommended amounts of all nutrients from foods; therefore, it is important to make careful food choices, and the need for dietary supplements should be evaluated.

[c] The 1,200 and 1,400 plans include a 170 empty calorie allowance and 4 teaspoons of fats/oils. The 1,800 plan includes 195 empty calories and 5 teaspoons of fats/oils.

[d] For maximum nutritional value, make whole-grain, high-fiber choices.

[e] Choose fat-free or low-fat (1%) dairy products.

[f] Select lean meat and use cooking methods that do not require added fat.

Daily recommendations for a
1,200-calorie diet[b,c]

GRAINS	VEGETABLES	FRUITS
4 oz[d]	1.5 cups	1 cup

DAIRY	PROTEIN FOODS
2 cups[e]	3 oz[f]

Daily recommendations for a
1,400-calorie diet[c]

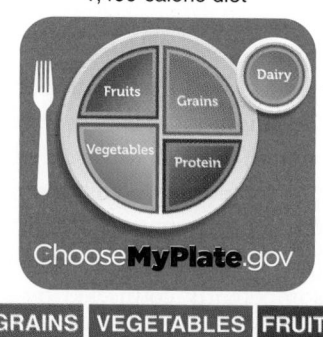

GRAINS	VEGETABLES	FRUITS
5 oz[d]	1.5 cups	1.5 cups

DAIRY	PROTEIN FOODS
2 cups[e]	4 oz[f]

Daily recommendations for a
1,800-calorie diet[c]

GRAINS	VEGETABLES	FRUITS
6 oz[d]	2.5 cups	1.5 cups

DAIRY	PROTEIN FOODS
3 cups[e]	5 oz[f]

sugar spoon. Keep concentrated sweets to a minimum, and let your carbohydrate come from fruits, vegetables, and whole-grain foods. Figure 10-7 shows three suitable patterns for weight-loss eating plans. Table 10-8 offers more tips for cutting back on fat and calories, and the Savvy Diner feature later in this chapter gives suggestions for defensive dining.

If you include alcohol or other empty-calorie items in your eating plan, limit them to no more than 150 calories a day. Budget this amount into your chosen calorie level, and reconcile yourself to a slower rate of weight loss.

TABLE 10-8	Healthful Tips for Weight Loss

- Eat breakfast.
- Watch out for second helpings of higher-calorie foods.
- Choose low-fat and low-calorie versions of foods you like.
- Go easy on foods that are high in fat or sugar.
- Limit alcoholic beverages.
- Roast, broil, boil, steam, or poach foods rather than fry them.
- Select lean cuts of meat, and trim visible fat.
- Plan ahead for snacks of fruit, vegetables, or low-fat yogurt.
- Use spices and herbs instead of sauces, butter, or other fats.
- Consume low-fat or fat-free dairy products.
- Try fresh fruit for dessert or baked products made with less fat and sugar (e.g., angel food cake).
- Use alternatives to foods as rewards (such as long walks, relaxing baths, a visit with a friend, a hobby, gardening, a good book).
- Keep a food diary.
- Aim for 60 minutes of exercise on most days of the week; try wearing a pedometer and walking 10,000 to 12,000 steps a day.

Three meals a day is standard for our society, but no law says you shouldn't have four or five meals—just be sure that they are smaller, of course. What is most important is to eat regularly and, if at all possible, to eat before you become very hungry.

Keep a record of what you have eaten each day for at least a week or two until your habits are automatic. Resume recordkeeping whenever you need to.

At first it may seem as if you are spending all your waking hours thinking about and planning your meals. Such a massive effort is always required when you are learning a new skill. After about 3 weeks, however, it will be much easier as your new eating pattern becomes a habit. Some characteristics of successful dieters are listed in the margin on this page.[48]

Aim for Gradual Weight Loss
Do not weigh yourself more than once a week. Although 3,500 calories roughly equals a pound of body fat, there is no simple relationship between calorie balance and weight loss over short intervals. A gain or loss of a pound or more in a few days can reverse quickly; the smoothed-out average is what is real. Don't expect to lose continuously as fast as you did at first. A sizable water loss is common during the first week, but it will not happen again.

Expect to Reach a Plateau
Many dieters experience a temporary plateau after about three weeks—not because they are slipping, but because they have gained water weight temporarily while they are still losing body fat. The fat they are hoping to lose must be used for energy. To use it, the body must combine it with oxygen (oxidize it) to make carbon dioxide and water. These compounds are heavier than the fat they are made from because oxygen has been added to them.* The carbon dioxide will be exhaled quickly, but the water takes longer to leave the cells. First, it makes its way into the spaces between the cells and finally enters the bloodstream. Only after the water has arrived in the blood do the kidneys remove it and send it to the bladder for excretion. Meanwhile, this excess water creates a temporary weight gain, but one day the plateau will break. The signal that this is happening is frequent urination.

Aim for a Positive Gain in Lean Body Mass
If you have been working out lately, successive weighings may show a slight occasional gain when you expect a loss. This may reflect a welcome development: the gain of lean body mass, just what you want to be healthy. In fact, weight loss without exercise can have a negative effect on body composition. No doubt you've heard someone say as a joke, "I've lost 200 pounds, but I've never been more than 20 pounds overweight." Typically, this is the person who diets, loses weight, regains the weight, and diets again throughout life—*without* exercising. When this pattern becomes a lifetime of weight fluctuations, it can have a lasting impact on the makeup of the body and the way it burns calories. Dubbed the "yo-yo effect," this pattern of losing, gaining, and losing weight again may actually cause the body to accumulate a greater percentage of fat and less lean muscle with each round of dieting.[49]

If you cut back drastically the number of calories you consume without exercising for, say, a month or two, you'll probably lose a great deal of weight each week, in the form of not only fat but also lean muscle. Then, if you later put the weight back on, the regained pounds may be primarily fat, *not* muscle. Thus, you may weigh the same as you did when you began the diet, but your body is now composed of more fat and less muscle. Because muscle burns more calories just to sustain itself than fat does, your body needs fewer calories to maintain its weight than it did before. If you go back to eating the way you used to, chances are you will gain even more weight.

---

*Water weight accumulates during fat oxidation because one fatty acid weighing 284 units leaves behind water weighing 324 units—14 percent more.

They know their weight. (They check their scale weight weekly.)

They are motivated from within.

They know what they eat. (They keep records.)

They engage in regular exercise (60 to 90 minutes of moderate exercise on most days).

They have social support (relationships/groups).

They control their intake of alcohol, fat, and sugar.

They follow a personalized diet plan—one that they can enjoy permanently.

They lose no more than a pound or two a week.

They set reasonable goals, such as losing just 10 percent of their weight, as a start.

They accept an occasional lapse rather than aiming for perfection.

They view weight management as a long-term commitment.

Exercise is essential for weight control. ►

Kzenon/Shutterstock.com

Maintain Bone Health by Increasing Calcium Intake   Weight loss—even modest weight loss—can also result in a depletion of bone minerals and hence a loss of bone density.[50] Fat loss may lower the levels of estrogen, a hormone necessary for healthy bone maintenance. Recent research emphasizes the importance of including both adequate calcium intakes and exercise in any weight-loss regimen, particularly weight-bearing exercise such as walking, jogging, stair climbing, and weight lifting. Researchers at the University of Tennessee provide additional evidence for the benefits of dietary calcium in weight-loss regimens. New research suggests that low calcium intakes favor increased deposition of body fat, whereas higher calcium intakes (at levels generally recommended for health) may not only inhibit storage of body fat but also stimulate the use of body fat for energy.[51]

In addition to the risk of altering the body's overall composition during repeated bouts of weight loss, consider that each unsuccessful attempt to keep weight off is often viewed by the dieter as a personal failure, which can erode the person's self-esteem and trigger painful feelings of guilt and depression.[52] A more healthful alternative to crash dieting is to gradually develop healthful, permanent lifestyle habits that will help you to lose weight and keep it off.

Adopt a Physically Active Lifestyle   Physical activity makes three contributions to a weight-management program. It increases calorie expenditure, alters body composition in a desirable direction, and alters metabolism.[53] Physical activity also offers the psychological benefits of looking and feeling healthy, and the increased self-esteem that accompanies these benefits can enhance the motivation to maintain a healthful lifestyle for the long run.

Compared with lean tissue, fat tissue is relatively inactive metabolically. Metabolic activity burns calories—lots of them. Thus, the more lean tissue you develop, the faster your metabolism becomes, the more calories you spend, and the more you can afford to eat. This brings you both pleasure and nutrients. Physical activity, by shifting body composition toward more lean tissue, speeds up the metabolism *permanently*—that is, for as long as you keep your body conditioned. Furthermore, the more muscle and lean tissue you have, the more fat you will burn—all day long, even when you are resting.

The next chapter offers many pointers about becoming fit, but a few notes on strategy are in order here. For one thing, keep in mind that if physical activity is to help with weight loss, it must be *active* exercise of moderate intensity—voluntary moving of muscles. Being moved passively, such as by a machine at a health spa or by a massage, does not increase calorie expenditure. The more muscles you move, the more calories you spend (see Table 10-9). You can tell that the physical activity is moderate if you are breathing a little faster than normal but can still easily carry on a conversation.

People sometimes ask about spot reducing. Can you lose fat in particular locations? Unfortunately, muscles don't "own" the fat that surrounds them. Because all body fat is shared by all the muscles and organs, spot-reducing exercises that work only the flabby parts won't help reduce the fat located there. The good news is, however, that tightening muscles in trouble spots by way of a balanced, all-over exercise program may improve the appearance of the fatty areas.

Another thing to keep in mind is that the number of calories spent in an activity depends more on how much a person weighs than on how fast the person can do the exercise (see Table 10-9). For example, a person who weighs 125 pounds burns off 78 calories by running a 9-minute mile. That same person, walking a mile in 15 minutes, burns almost the same amount: 82 calories. Similarly, a 200-pound person spends 125 calories on the 9-minute mile, and a similar amount—131 calories—on the 15-minute walk. The rule seems to be that you don't have to work fast to use up calories effectively. If you choose to walk rather than run the distance, you will use up about the same energy; it will just take you longer.

A good suggestion that you may have heard already is to incorporate more physical activity into your daily schedule in many simple, small-scale ways. Park the car at the far end of the parking lot; use the stairs instead of the elevator; do a round of sit-ups before you get up in the morning. If you do these types of things and also incorporate regular aerobic exercise and strength training into your schedule, your heart and lungs as well as your skeletal muscles will be fit (see Chapter 11).

Weight-Gain Strategies

It is as hard for a person who tends to be thin to gain a pound as it is for a person who tends to be fat to lose one. The person who wants to gain weight is faced with some of the same challenges as the one who wants to lose weight: learning new habits, learning to like new foods, and establishing discipline related to meals and mealtimes. But there are major differences.

TABLE 10-9	Calories Spent during Various Activities	
	Calories Expended per Hour[a]	
Activity	**Man**[b]	**Woman**[b]
Sitting quietly	100	80
Standing quietly	120	95
Light activity:	300	240
Cleaning house		
Office work		
Playing baseball		
Playing golf		
Moderate activity:	460	370
Walking briskly (4 mph)		
Gardening		
Cycling (5.5 mph)		
Dancing		
Playing basketball		
Strenuous activity:	730	580
Jogging (9 min/mile)		
Playing football		
Swimming		
Very strenuous activity:	920	740
Running (7 min/mile)		
Racquetball		
Skiing		

[a] May vary depending on environmental conditions.
[b] Healthy man, 175 lb; healthy woman, 140 lb. (To determine caloric expenditure at a different weight, both men and women should divide their body weight by 175, then multiply by the man's caloric expenditure for the activity.)

SOURCE: U.S. Department of Agriculture, U.S. Department of Health and Human Services, *Dietary Guidelines for Americans*.

Add extra calories to the milk or juice you drink by tossing in a frozen banana or other fruit and whipping up a smoothie in the blender.

Try eating more food at each meal. If you normally eat one sandwich for lunch, try eating an extra sandwich between meals.

Knowing that vigorous physical activity costs calories, an active person may wonder whether it is advisable to curtail activity. The answer is no, not unless the underweight condition is so extreme as to be endangering your health. The healthful way to gain weight is to build yourself up with patient and consistent training while eating nutritious foods containing enough extra calories to support the weight gain. If you add extra snacks of high-calorie nutritious foods, such as a peanut butter sandwich with a glass of milk before bedtime, and a healthful midafternoon fruit smoothie or milkshake, you can add 700 to 800 extra calories a day and achieve a healthful weight gain of 1 to 1½ pounds per week. Choose calorie-dense snacks, such as fruit yogurt, granola, dried fruits, apple or grape juice, nuts, sunflower seeds, fig bars, and baked potatoes topped with low-fat cheese.

A person wanting to gain weight often has to learn to eat different foods. No matter how many helpings of carrots you consume, you won't gain weight very fast because carrots simply don't offer enough calories. Thus, the person who can't consume much food volume is encouraged to eat calorie-dense foods at meals (the very ones the dieter tries to avoid). For example, if you are trying to gain weight, you should choose a milkshake instead of milk, peanut butter instead of lean meat, an avocado instead of a cucumber, and a blueberry bran muffin instead of whole-wheat bread. When you do eat carrots, dip them in hummus and use creamy dressings on salads, yogurt on fruit, cottage cheese on potatoes, and so forth. Because fat contains twice as many calories per teaspoon as sugar, it adds calories without adding bulk. For heart health, be sure to choose the "good" fats, found in foods such as avocados, olives, peanuts, pistachios, and other nuts and seeds, and limit your intake of saturated fats.

Also, eat more frequently. Make three sandwiches in the morning, and eat them between classes in addition to the day's three regular meals. Spend more time eating each meal. If you fill up fast, eat the highest-calorie items first. Don't start with soup or salad; start with the main course. Finish with dessert. Many underweight people have simply been too busy (for months) to eat enough to gain or maintain weight. These strategies will help you change this behavior pattern.

Whether you need to gain, lose, or maintain weight, paying attention to what you eat will pay off in long-term wellness benefits. This chapter emphasizes the relationship of food and eating to body weight and has the same message as other parts of the book: To support wellness, eat regular, balanced meals composed of a wide variety of foods you enjoy.

Breaking Old Habits

"Just do it," urge the makers of Nike sportswear on billboards, in magazines, and wherever else they promote their products. This motto implies that adhering to a fitness program is just a matter of donning sneakers and heading for the track or gym. But, as most of us know from experience, sticking to a regular physical activity program for the first time or changing eating habits isn't always easy. Most people who successfully quit smoking make three or four attempts before they finally kick the habit; most people make the same New Year's resolution at least five years in a row before they stick to it permanently.[54] Breaking old patterns of behavior and developing new ones involves a number of stages before reaching the point at which you're able to change for good.

First, you must accept the belief that you *can* change: "I can lose some weight and try to become more physically fit," you might tell yourself. This might inspire you to *want* to change, which in turn will motivate you to find ways to actually make the change (for example, talking to a registered dietitian about diet strategies and reading about various forms of physical activity). After you've gone through those steps,

Elements of Behavior Change[55]

1. *Precontemplation:* You need to change, but you're not yet ready to accept that fact.

2. *Contemplation:* You want to change, but you're not sure how.

3. *Preparation:* You gain knowledge to set up a plan of action for change.

4. *Action:* You jump in and "just do it."

5. *Maintenance:* You work on sticking to your plan of action.

6. *Termination:* You have achieved lasting change and experience few, if any, temptations or relapses.

Pizza Yankee . . . The story goes that a pizza-parlor waitress asked Yogi Berra whether he wanted his pizza cut into six or eight slices. "Six, please," he said. "No way am I hungry enough to eat eight."[56]

Dining out can be a pleasant time to relax, socialize, taste the foods of different cultures, and provide a break from your usual schedule—and it does *not* necessarily mean that you overindulge in high-calorie foods. With practice, you can maximize the benefits of eating out by adopting some of the following strategies:

- Examine your options before selecting foods at buffets, cafeterias, or food courts. Otherwise, you may find yourself halfway through the line with a full plate and end up with a mountain of food on your plate when you reach the end of the line.

- Take the edge off hunger by starting with a broth-based soup, small salad, fruit, or light appetizer.

- Make specific requests. Ask for salad dressings, sauces, or gravies *on the side*. This puts you in control of the portions. Ask if reduced-fat or fat-free salad dressings are available.

- Ask for entrées to be broiled, baked, grilled, steamed, or roasted. See "Eating Well on the Run" for tips on ordering at quick-service restaurants.

- Request fresh fruit, sorbet, or low-fat frozen yogurt for dessert. It will give you the sweetness you want without excess calories from fat. If you choose to splurge on your favorite chocolate dessert, enjoy it by sharing it with a friend. After the dinner out, go back to your healthful eating habits and exercise program.

- Be a menu sleuth. The following terms typically describe high-fat items: au gratin, Alfredo, batter-dipped, breaded, béarnaise, carbonara, creamy, crispy, croquette, deep-fried, flaky, fritters, parmigiana, tempura, with gravy, or with hollandaise sauce. Order red pasta sauces rather than white.

- Downsize your order. Many restaurants serve portions that are two or more times the recommended serving size. Request a doggie bag so you can take half the amount home. Or, consider sharing an entrée with a friend and ordering a salad to go with it. You can also order à la carte: one or two nutritious, low-fat appetizers such as steamed shrimp or raw vegetables and a low-fat dip along with a salad instead of a huge main meal.

- Finally, eat slowly and enjoy your meal, along with the company and conversation. You'll be giving your body time to digest the food, you'll feel satisfied, and you'll have enjoyed yourself at the same time.

Eating Well on the Run

Here are some smart choices for those with little time:

Sandwich Shop: Fresh sliced veggies in a pita with low-fat dressing, cup of minestrone soup, turkey breast sandwich with mustard, lettuce, tomato, fresh fruit.

Rotisserie Chicken: Chicken breast (remove skin), steamed vegetables, mashed sweet potatoes, tossed salad, fruit salad. Select plain rolls instead of cornbread or biscuits.

Fast Food: Grilled chicken breast sandwich (no sauce), single hamburger without cheese, grilled chicken salad, garden salad, low-fat or nonfat yogurt, fat-free muffin, cereal, low-fat milk.

Salad Bars: Broth-based soups, fresh bread or bread sticks, fresh greens, chopped veggies, beans, low-fat dressing, fresh fruit salad. Avoid marinated beans and oily pasta salads.

Asian Take-Out: Wonton soup, pho (Vietnamese noodle soup), hot and sour soup, steamed vegetable dumplings, steamed vegetable mixtures over rice or noodles.

Pizza Night: Choose flavorful, low-fat toppings such as peppers, onions, sliced tomatoes, spinach, broccoli, or mushrooms. Ask for your pizza with less cheese.

SOURCE: Adapted from *Healthy Eating Away from Home* (Washington, DC: American Institute for Cancer Research, www.aicr.org).

---

you will be ready to take action—that is, to "just do it." Finally, to maintain your behavior change, you can create a game plan to help you handle the inevitable lapses that will occur over time. When you follow these steps and then stick with your plan for a good deal of time, you're home free.

Behavior Modification Making lasting behavioral changes poses ongoing challenges. A process known as **behavior modification** can help. Developed by psychologists to help people change their habits, behavior modification techniques are

behavior modification a process developed by psychologists for helping people make lasting behavior changes.

Behavior Modification Steps

Step 1: Identify the goal.

Step 2: Record your present behavior pattern. Identify the reasons you practice these behaviors.

Step 3: Identify the behaviors that will lead to the goal and the rewards of those behaviors.

Step 4: Commit yourself to changing. Face what you'll have to give up or change to make the desired behavior a reality. Envision your changed future self.

Step 5: Plan. Divide the behavior into manageable portions. Set small, achievable goals and plan periodic rewards.

Step 6: Try out the plan. Modify the plan, if necessary, in ways that will help you succeed.

Step 7: Evaluate your progress on a regular basis.

similar to mapping out a game plan by following the specific steps listed in the margin on this page.

Set a Goal and Get to Know Yourself For *Step 1*, identify your goals. You might decide, for example, that your goal is to lower the fat and calorie content of your diet. Next, in *Step 2*, you record your present behavior pattern and think about the reasons for those behaviors. In this case, for example, you might keep a food diary, writing down everything you eat for five days and describing how you feel each time you eat (see Table 10-10 for a sample food diary). This technique will help you understand why you behave the way you do in certain situations. You might find, for instance, that every time you are angry at your significant other or get a bad grade, you eat a candy bar or two to comfort yourself. Or, perhaps without even thinking about it, you nibble on whatever happens to be handy while watching television.

Change Behaviors and Reward Yourself Having identified your goal and current patterns of behavior, you can now move to *Step 3*, which is to determine strategies to help you meet your goal and to choose some rewards for when you follow these strategies. An example of a possible behavior-change strategy might be that you decide to talk to a friend instead of heading to the kitchen cupboard or vending machine when you're angry. You could save hundreds of calories and probably experience emotional relief by getting your feelings off your chest.

This is the time to remind yourself of the progress you're making and the benefits you're gaining: weight loss, increased energy, better health. It's also the time to give yourself a tangible reward, such as a night out at the movies or a new item of clothing. This step is key because behavior research indicates that once you have instituted a change in your life, you will maintain it only by positive reinforcement. Rewards don't have to be expensive. For example, you can take time for a long, candlelit bath, or celebrate with a leisurely Sunday morning curled up with a book you've been wanting to read.

In *Step 4*, after you've pinpointed behaviors you want to alter, you must *commit* yourself to make the changes and take the time to envision the healthier person you'll become as a result.

Plan to Take Small Steps *Step 5* is planning, and the way to do this is to take small steps. Divide your primary goal into smaller goals that can be achieved one at a time. Although you might have identified 10 behaviors that you want to change to meet your overall goal, trying to accomplish all 10 in the first week or two is probably very unrealistic, and it may simply set the stage for disappointment and failure (and cause

TABLE 10-10	Sample Entries in Food Diary				
Food Eaten	**Time and Place**	**With Whom**	**Mood**		**Other Activity**
1 large blueberry muffin	7:30 A.M., doughnut shop	Alone	Stressed—late for class		Skimming notes for upcoming class
1 cup coffee with 1 oz cream & 1 tsp sugar					
1 hot dog with 2 tbsp ketchup	12:45 P.M., cafeteria	2 friends	Relaxed		Talking with friends
1 small bag potato chips					
12-oz can diet soda					
1 candy bar	2:35 P.M., bedroom	Alone	Angry about argument with significant other		Watching TV

cathy®

by Cathy Guisewite

you to give up completely). To give yourself a more realistic chance for success, choose only one or two of the smaller goals, do your best to stick to them for, say, two weeks, and then reward yourself for doing so before going on to work on your next goals.

Give It a Try and Adapt Your Plan! In *Step 6* of behavior modification, you put your plan into action. As you meet your goals and reward yourself for doing so, also be aware of when you are not meeting them. If something is not working, don't give up. Instead, go on to *Step 7* and feel free to modify your goals—perhaps those that you initially set were too difficult or too simple to achieve. You may, for example, decide to choose the smaller goal of cutting some of the fat out of your diet for the first two weeks and wait for a later time to work on reducing the calorie count. As you can see, when you evaluate your progress on a regular basis, you offer yourself the flexibility to adapt your plan so that you can continue to move forward to meeting your goal.

Be Prepared for a Relapse While you're working on modifying your behavior, it is critical to keep in mind that you're only human, and there will be days when you slip and revert to an old pattern of behavior. Instead of dwelling on it or berating yourself, realize that relapses are par for the course, forgive yourself, and simply move on. Keep in mind that even though you may have had a setback, you're still moving forward if you learn from your mistakes and get right back on track again. That's why some experts prefer the term *recycle* to the term *relapse*; people who make a mistake and then learn from it can reuse, or recycle, their new knowledge.

Set Priorities Setting priorities helps ensure that you adopt one or two behaviors that you stick to all the time. Also, pay attention to how other areas of your life may be affecting your readiness to make changes and meet your goals. Many people waste a great deal of time and energy trying to make changes at difficult times. Someone who is going through the breakup of a long-term relationship or having difficulties at work or school, for example, may be under too much stress to deal with making a behavior change on top of everything else.

Consider the "Rule of Three" Keep in mind that you might need to try new behaviors more than once before you begin to like them. Dietitians sometimes use a rule of three in counseling. Try fat-free milk once, and you may not like it. Try it a second time, and you may be able to tolerate drinking it, even though it doesn't taste as good as whole milk. Try it a third time, and you may conclude that although it's not as good as whole milk, you don't really mind drinking it. Then you'll be able to maintain your new habit more easily.

Continued Motivation

- Persist long enough to experience the rewards, such as improved self-image and enhanced self-esteem.
- Remember the price of the old behavior.
- Keep in mind where you started.
- Tune in to the benefits of the new behavior.

A Recipe for Success

Be realistic.
Make small changes over time in what you eat and your activity level. Aim for losing no more than 1 to 2 pounds per week. Even a small amount of weight loss can bring major health benefits.

Be adventurous.
Expand your tastes to enjoy a wide variety of foods and activities. Get fewer calories by going "large" on flavor and small on portions.

Be flexible.
You don't have to give up your favorite foods to manage your weight. Just balance what you eat and what you do over several days.

Be sensible.
Enjoy all foods, just don't overdo any of them. Slow down when you're eating and listen to your body's signals to know when you're hungry or full.

Be active.
It's not *what* activity you choose, it's whether you enjoy doing it. If you enjoy it, you are likely to do it more often!

SOURCE: Adapted from Weighing In on Health, *Food and Nutrition News* 70 (1998): 6.

Strategies for Changing the Way You Eat   The steps in this section are designed specifically to help with weight control. (Chapter 11 includes strategies for modifying exercise behavior.) First, begin by keeping a record of your present eating behavior that you can compare with your future progress. Your eating records will eventually help you see how far you've come in changing your eating habits.

This record will also help you with the second step, which is to identify cues and situations that trigger your behavior of reaching for food. For example, some cues, such as watching television or walking by a vending machine, may prompt you to eat when you're not hungry. Once the cues are identified, you can resolve to stop responding to them and try to eat only in one place in a certain room, such as at the table in the dining room. In addition, try the following tips to eliminate the temptation to eat when you really don't want to:

- Don't buy hard-to-resist foods such as cakes, cookies, and ice cream.

- Don't shop when you're hungry and thereby more likely to buy tempting foods.

- Don't leave large amounts of food within easy reach. Keep serving dishes off the table, for instance, and put the cookie jar out of sight.

- Make small portions of food look large by spreading food out and putting it on smaller plates. Garnish empty space with low-calorie vegetables.

- Try to eat regular meals and snacks instead of skipping them, thereby reducing the likelihood of becoming uncomfortably hungry and then overeating later.

- Ask your family and friends to encourage you to eat a healthy diet and not to criticize you if you splurge occasionally.

Third, make it easy for yourself to eat the way you want to eat.

- Keep a variety of nutritious foods, such as fruits and vegetables, readily available.

- If you like to eat between meals, plan snacks that fit easily into your diet, such as low-fat yogurt and fruit.

Fourth, take a look at and then, if necessary, alter the manner in which you eat. If you eat quickly, slow down by chewing food thoroughly, pausing between bites, putting down your utensils, and swallowing before reloading your fork or spoon. Eating slowly will give your body a chance to feel full and satisfied.

Finally, reward yourself for positive behaviors. Plan to buy something new to wear or do something you enjoy, such as attending a sporting event or a concert, each time you meet your goals.

The relentless pursuit of thinness and fear of being fat are a haunting nightmare that drives millions of American teens and adults to starve, vomit, or purge.[57] The illness of **anorexia nervosa**—self-starvation—has been recognized as a psychiatric syndrome since the 1870s. Its companion disease **bulimia nervosa**—gorging on food and then purging—was not recognized as a separate eating disorder until the 1960s and 1970s. Some researchers suspect that a complex interplay between environmental, social, and perhaps genetic factors triggers an eating disorder in victims, who are mostly but not always women. Others question whether the fear of fatness isn't a mask for underlying emotional problems. Experts speculate that focusing on the body diverts these people's attention from and suppresses the painful emotion of anger, feelings of low self-esteem, the inability to express their feelings, or poor family relationships. The acute focus on the body develops into an intense fear of fatness, a characteristic intrinsically linked with food.[58] As a result, for many people, food is seen only as a source of body fat and so becomes carefully controlled. But why food? Why not use some other method of coping with stress?

Enter the societal link. In a society where thinness is equated with material success and even self-worth, especially for women, becoming thin appears to be the yellow brick road that leads to happiness. Unfortunately, as victims of eating disorders come

Looking in the mirror, a woman with anorexia nervosa distorts her body image and overestimates her body fatness.

© Tony Freeman/PhotoEdit

to learn, practicing self-starvation or gorging and purging leads instead to physical and emotional pain.

What is considered an eating disorder?

The term **eating disorder** comprises a wide spectrum of conditions, including anorexia nervosa, bulimia nervosa, and **binge-eating disorder**.[59] Although the various conditions differ in origin and consequences, they appear to have similarities: all of the conditions involve excessive preoccupation with weight, a fear of fatness, and a distorted body image. Many times the person with an eating disorder falls short of the diagnostic criteria shown in Table 10-11 for anorexia nervosa or bulimia nervosa. Some of these people, described as having an **unspecified eating disorder**, include those who:[60]

- Meet all the criteria for anorexia nervosa except irregular menses.
- Meet all the criteria for anorexia nervosa except that their weight remains within a normal range.
- Meet all the criteria for bulimia nervosa except that their binges are less frequent.
- Have recurrent episodes of binge eating but do not compensate using the methods of those with bulimia nervosa, a condition known as binge-eating disorder (see Table 10-12).[61]

What is the difference between an eating disorder and disordered eating?

Disordered eating occurs when a person eats (or doesn't eat) because of an external stimulus rather than an internal one. For example, an external stimulus might be, "I'm lonely," "I'm angry," or "I'm anxious," and "these chocolate cookies will distract me and help me feel better." An internal stimulus is a physical sensation: a message from your brain about your current state of hunger or satiety, such as "my stomach is growling," or "I feel full." People with eating disorders spend more time in "external eating" than in responding to internal hunger cues.[62] Many people eat for reasons other than hunger from time to time, but if this type of behavior controls someone's thoughts and interferes with their everyday routines, it becomes an eating disorder.

What are the symptoms of anorexia nervosa?

Anorexics deprive themselves of food except for controlled amounts of very low-calorie foods such as unbuttered toast or popcorn, apples, and green beans. Even intake of these foods is painstakingly limited. After three to four days of eating very small amounts of food, hunger pangs subside. Once their appetite is suppressed, anorexics report feeling quite energetic, as if on a high, making the strict fast easier to stick to—but the body needs fuel to

(Continued)

TABLE 10-11	Diagnosis of Eating Disorders

Anorexia Nervosa

A person with anorexia nervosa demonstrates the following:

1. Refusal to maintain body weight at or above a minimal normal weight for age and height

2. Intense fear of weight gain or becoming fat, even though underweight

3. Distorted body image, undue influence of body weight or shape on self-evaluation, or denial of the seriousness of the current low body weight

4. In females past puberty, amenorrhea (the absence of at least three consecutive menstrual cycles)

Two types:

Restricting type: During the episode of anorexia nervosa, the person does not regularly engage in binge eating or purging behavior (that is, self-induced vomiting or the misuse of laxatives, diuretics, or enemas).

Binge-eating/purging type: During the episode of anorexia nervosa, the person regularly engages in binge eating or purging behavior (that is, self-induced vomiting or the misuse of laxatives, diuretics, or enemas).

Bulimia Nervosa

A person with bulimia nervosa demonstrates the following:

1. Recurrent episodes of binge eating characterized by:

   a. Eating in a discrete period of time an amount of food that is definitely larger than most people would eat during a similar period of time and under similar circumstances

   b. A sense of lack of control over eating during the episode

2. Recurrent compensatory behavior to prevent weight gain, such as self-induced vomiting; misuse of laxatives, diuretics, enemas, or other medications; fasting; or excessive exercise

3. Binge eating and compensatory behaviors that occur, on an average, at least twice a week for three months

4. Self-evaluation unduly influenced by body shape and weight

5. Does not occur exclusively during episodes of anorexia nervosa.

Two types:

Purging type: The person regularly engages in self-induced vomiting or the misuse of laxatives, diuretics, or enemas.

Nonpurging type: The person uses other behaviors, such as fasting or excessive exercise, but does not regularly engage in self-induced vomiting or the misuse of laxatives, diuretics, or enemas.

SOURCE: Adapted from American Psychiatric Association, *Diagnostic and Statistical Manual of Mental Disorders*, 4th ed. (Washington, DC: American Psychiatric Association, 2000).

run. To compensate for the lack of fuel from food, the body turns inward for its fuel and begins to slowly destroy muscle and fat tissue for energy. The following are some of the physical symptoms associated with the starvation seen in anorexics:[63]

- Wasting of the whole body, including muscle tissue and bones

- Arrested sexual development and stopping of menstruation due to loss of body fat

- Drying and yellowing of the skin from an accumulation of a stored vitamin A compound released from body fat

- Intolerance to cold weather due to loss of subcutaneous fat

- Growth of hair on the body, perhaps in response to a decrease in body temperature

- Loss of health and texture of hair

- Pain on touch

- Lowered blood pressure and metabolic rate

- Anemia

- Severe sleep disturbance[64]

- Depression, possibly related to changes in neurotransmitter function in the brain[65]

At the same time the physical symptoms appear, distorted psychological symptoms develop. When looking in the mirror, anorexics do not see the emaciated body others see, but continue to see themselves as being too fat.[66] A preoccupation with death develops, accompanied by a frantic pursuit of physical fitness by means of stringent exercise routines. The anorexic deals with parents and family in a manipulative way so as to become the center of attention. Diet becomes so totally engrossing that the anorexic may be quite socially isolated except for friends who stick close by and worry without knowing how to help.

By this time, the anorexic has reached absolute minimum body weight (for example, 65 to 70 pounds for a woman of average height). The person is on the verge of incurring permanent brain damage and chronic debilitation or death. The National Association of Anorexia Nervosa and Associated Disorders (ANAD) estimates that of those with severe eating disorders, 6 percent die, usually because major organs—heart and kidneys—fail.

What are the symptoms of bulimia?

Unlike anorexics, bulimics don't shrink to skeleton-like proportions. They usually have a healthful body weight or may even be slightly overweight. Bulimics also follow rigid rules of dietary restraint, but their routine is not as rigorous as the anorexic's starvation routine, and they occasionally break their own rigid rules.

Quickly, and usually privately, bulimics gorge on foods that are often sweet, starchy, and high in fat or calories and require little chewing. The binge ends when it would hurt to eat any more, when they are interrupted, or when they go to sleep or induce vomiting to expel the food just eaten.

Bulimics often feel controlled by the vicious circle that develops: anxiety about being "fat" leads to rigid dietary restraint. Mounting hunger from not eating and an increased preoccupation with food cause them to break their rules, and binging results. After gorging, an intense fear of fatness overtakes the person, who then vomits to get rid of the food and release the fear of becoming "fatter." Feelings of guilt and shame follow the purge, building a new level of anxiety concerning the body, and the cycle begins again. Each binge reinforces the idea that additional rigidity of dietary restraint is required to prevent weight gain. Excessive use of laxatives or diuretics or bouts of vigorous exercise replace vomiting in some cases.

Binge eating is seldom life threatening, but at the extreme, it can be physically damaging, causing lacerations of the stomach, tearing or irritation of the esophagus (in those who vomit frequently), dental caries (from acidic vomit attacking the teeth), electrolyte imbalances, and malnutrition (in those who vomit and those who take laxatives).

Bulimics also suffer from a distorted body image.[67] They see themselves as fat and needing to restrict food, even though they usually have a healthful body weight. Bulimics prefer a body size somewhat smaller than normal.

How are anorexia nervosa and bulimia treated?

There are several philosophies regarding the treatment of eating disorders. The four major approaches include individual psychotherapy, hospitalization, family therapy, and behavior modification therapy.[68] Some therapists use more than one approach. Most treatment methods focus on identifying the societal and environmental pressures that triggered the eating disorder and on exposing the emotions masked by it. An interdisciplinary team made up of a psychologist, a social worker, a family therapy counselor, and a dietitian work with the family and the patient to reestablish emotional and nutritional health.[69] Length of treatment varies from two months to two to three years, depending on the patient's readiness for change and the type of treatment. American researcher Hilde Bruche, MD, focuses on the problems of low self-esteem, guilt, anxiety, depression, and a sense of helplessness. In addition, family therapy focuses on changing patterns of family interaction.

Normal nutrition must be restored in the anorexic. After some progress is made in counseling, the dietitian can help the patient gain a new understanding of a healthful eating pattern and clear up misconceptions about food and nutrition.[70]

(Continued)

TABLE 10-12	Criteria for Diagnosis of Binge-Eating Disorder

A person with a binge-eating disorder demonstrates the following:

1. Recurrent episodes of binge eating. An episode of binge eating is characterized by both of the following:

    a. Eating, in a discrete period of time (for example, within any 2-hour period), an amount of food that is definitely larger than most people would eat in a similar period of time under similar circumstances.

    b. A sense of lack of control over eating during the episode (such as a feeling that one cannot stop eating or control what or how much one is eating).

2. Binge-eating episodes are associated with at least three of the following:

    a. Eating much more rapidly than normal.

    b. Eating until feeling uncomfortably full.

    c. Eating large amounts of food when not feeling physically hungry.

    d. Eating alone because of being embarrassed by how much one is eating.

    e. Feeling disgusted with oneself, depressed, or very guilty after overeating.

3. The binge eating causes marked distress.

4. The binge eating occurs, on average, at least twice a week for six months.

5. The binge eating is not associated with the regular use of inappropriate compensatory behaviors (purging, fasting, excessive exercise) and does not occur exclusively during the course of anorexia nervosa or bulimia nervosa.

SOURCE: Adapted from American Psychiatric Association, *Diagnostic and Statistical Manual of Mental Disorders*, 4th ed. (Washington, DC: American Psychiatric Association, 2000).

For many people with anorexia nervosa, a full day's diet may consist of no more than three or four items.

For many people with bulimia, guilt, depression, and self-condemnation follow a binge-eating episode.

usually causes sufficient weight loss to stop menstrual cycles in women.

People with eating disorders were usually good children who did not indulge in rebellion. Not all anorexics and bulimics exhibit all symptoms. Early detection is vital. Use the Eating Attitudes Quiz on page 330 to see if your own eating attitudes and behaviors are within a normal range.

What are the early warning signs of eating disorders?

Table 10-13 shows the similarities and differences among the eating disorders. Families and friends can be alerted to several possible warning signs of eating disorders (see Table 10-14). Severe dieting often precedes the illness. Anorexics develop an exaggerated interest in food but at the same time deny their hunger and stop eating. Their distorted body image makes them feel fat even as weight loss continues. The anorexic begins to have sleep problems, shows unusual devotion to schoolwork, and often undertakes a program of unrelenting exercise. Bulimics may binge and self-induce vomiting or use excessive amounts of laxatives. Reduced food intake

What can be done to prevent eating disorders?

Consider the following suggestions for helping people to remain healthy and for preventing the occurrence of eating disorders altogether:[71]

- Discourage restrictive dieting and meal skipping. Severe dieting may lead to eating disorders, especially in adolescent girls.

- Model a lifestyle of healthful eating and exercise, and never encourage unhealthful "quick-fix" weight loss.

- Promote fitness and a healthy body rather than thinness and numbers on a scale.

- Help teens understand and accept the normal physiological changes in body composition and weight that occur with puberty—changes that may be interpreted as "getting fat."

TABLE 10-13	How the Eating Disorders Compare		
	Anorexia Nervosa	**Bulimia Nervosa**	**Binge-Eating Disorder**
Estimated Prevalence*	Up to 0.5%–1.0%	Up to 1.0%	Up to 2.0%
Male versus Female Incidence	5%–10% male versus 90%–95% female	5%–10% male versus 90%–95% female	Unknown
Typical Age of Onset	Early to middle adolescence	Late teens to early twenties	Any age, but not usually recognized until adulthood
Weight	Extremely thin and emaciated; < 85% of desirable body weight	Near-desirable body weight, but often has weight fluctuations	Usually overweight or obese
Self-Esteem	Low	Low	Low
Depression	Common	Common	Common
Substance Abuse	Rare	Common	Rare
Rate of Weight Loss	Rapid	Repeatedly loses and gains weight or chronically diets without losing weight	Repeatedly loses and gains weight or chronically diets without losing weight
Past Dieting	Yes	Yes	Yes

* Determining accurate statistics is difficult because physicians are not required to report eating disorders to a health agency and because people who have eating disorders tend to deny that they have a problem and are very secretive about their behaviors.

SOURCE: J. Kirby, for the American Dietetic Association, *Dieting for Dummies* (Foster City, CA: IDG Books Worldwide, Copyright © 2003), 67–68. Reproduced by permission of John Wiley & Sons, Inc.

TABLE 10-14	Warning Signs of an Eating Disorder

- Preoccupation with weight, food, calories, and dieting, to the extent that it consistently intrudes on conversations and interferes with other activities
- Excessive exercise—despite weather, fatigue, illness, and injury
- Dramatic weight loss; arrested growth in children
- Anxiety about being fat that does not diminish with weight loss
- Reluctance to be weighed
- Inability to gain weight
- Difficulty eating in social situations
- Deceptive or secretive behavior
- Alternating periods of severely restrictive dieting and overeating
- Evidence of binge eating or consumption of large quantities of food inconsistent with the person's weight
- Evidence of use of laxatives, diuretics, or agents that cause vomiting (emetics)
- Extreme concern about appearance as a defining feature of self-esteem (for example, I am "thin and good," or "gross and bad")
- Social withdrawal or avoidance of activities because of weight and shape concerns
- Absence from school or work
- Depression
- Disrupted menstruation
- Constipation or diarrhea
- Paleness or lightheadedness not accounted for by other medical conditions
- Susceptibility to fractures

SOURCES: Adapted from Eating Disorders Awareness and Prevention, Inc. materials; and D. Rosen, Dieting disorder, *New England Journal of Medicine* (April 8, 1999).

- Don't plant unhealthy seeds. Avoid assigning "good" or "bad" labels to food. Don't offer food as a reward. Don't imply that the shape or size of your body, or someone else's, has anything to do with self-worth (for example, "I'd be happy if I were thin").

- Be sensitive and careful about making weight-related comments or recommendations. Even playful teasing may trigger long-term struggles with disordered eating or eating disorders.

- Encourage normal expression of the basic emotions: joy, anger, fear, sadness, loneliness, guilt, and shame.

What can I do if I am worried that someone I know may have an eating disorder?[72]

If you or others observe behaviors in a friend, roommate, or family member that suggest the possibility of an eating disorder, consider the following:

- Make a plan to approach the person in a private place when there is no immediate stress and you have time to talk.

- Present in a caring but straightforward way what you have observed and what your concerns are. Tell the person that you are worried and want to help.

- Give the person time to talk, and encourage her or him to verbalize feelings. Listen carefully and accept what is said in a nonjudgmental manner.

- Do not argue about whether there is or is not a problem—power struggles are not helpful. Perhaps you can say, "I hear what you are saying, and I hope you are right that this is not a problem. But I am still very worried about you and your health, and that is not going to go away."

- Provide information about resources for treatment. Offer to go with the person and wait while she or he has a first appointment with a counselor, physician, or dietitian.

- If you are concerned that the eating disorder is severe or life threatening, enlist help from the counseling center, a relative, a friend, or the person's roommate *before* you intervene. Present a united and supportive front together.

- If the person denies the problem, becomes angry, or refuses treatment, understand that this is often part of the illness. Besides, they have a right to refuse treatment (unless their life is in danger). You may feel helpless, angry, and frustrated with the person. You might say, "I know you can refuse to go for help, but I am still your friend, and I am worried about you. I may bring

(Continued)

Miniglossary

anorexia nervosa literally, "nervous lack of appetite," a disorder (usually seen in teenage girls) involving self-starvation to the extreme (*an* = without; *orexis* = appetite).

bulimia nervosa, bulimarexia (byoo-LEE-me-uh, byoo-lee-ma-REX-ee-uh) binge eating (literally, "eating like an ox"), combined with an intense fear of becoming fat and usually followed by self-induced vomiting or taking laxatives (*buli* = ox).

eating disorder general term for several conditions (anorexia nervosa, bulimia nervosa, binge-eating disorder) that exhibit an excessive preoccupation with body weight, a fear of body fatness, and a distorted body image.

binge-eating disorder an eating disorder characterized by uncontrolled chronic episodes of overeating (compulsive overeating) without other symptoms of eating disorders. Typically, the episodes of binge eating occur at least twice a week for a period of six months or more.

unspecified eating disorders condition in which a person exhibits some but not all of the criteria for specific eating disorders.

disordered eating eating food as an outlet for emotional stress rather than in response to internal physiological cues.

this up again later, and maybe we can talk more about it then." Follow through on that—and on any other promise you make.

- Do not try to be a hero or a rescuer; you will probably be resented. Eating disorders are stubborn problems, and treatment is most effective when the person is truly ready for it. You may have planted a seed that helps the person get ready.

For more information about anorexia nervosa, bulimia, and associated disorders, log on to the following websites:

- Anorexia Nervosa and Related Eating Disorders: www.anred .com

- *National Eating Disorders Association*: www.nationaleating disorders.org
- *National Institute of Mental Health*: www.nimh.nih.gov
- *Gurze Books' Eating Disorders Resource Catalog*: http://www .bulimia.com/index.cfm
- *Eating Disorder Referral and Information Center*: www.edreferral .com
- *Canadian National Eating Disorder Information Centre*: www.nedic.ca

---

Eating Attitudes Quiz

Answer the following questions to evaluate your own eating attitudes and behaviors. Do they fall within a normal range? Use these responses:

A = always
U = usually
O = often
S = sometimes
R = rarely
N = never

1. I am terrified about being overweight.
2. I avoid eating when I am hungry.
3. I find myself preoccupied with food.
4. I have gone on eating binges where I feel that I may not be able to stop.
5. I cut my food into very small pieces.
6. I am aware of the calorie content of the foods I eat.
7. I particularly avoid foods with high carbohydrate content.
8. I feel that others would prefer that I eat more.
9. I vomit after I eat.
10. I feel extremely guilty after eating.
11. I am preoccupied with a desire to be thinner.
12. I think about burning up calories when I exercise.
13. Other people think I am too thin.

14. I am preoccupied with the thought of having fat on my body.
15. I take longer than other people to eat my meals.
16. I avoid foods with sugar in them.
17. I eat diet foods.
18. I feel that food controls my life.
19. I display self-control around food.
20. I feel that others pressure me to eat.
21. I give too much time and thought to food.
22. I feel uncomfortable after eating sweets.
23. I engage in dieting behavior.
24. I like my stomach to be empty.
25. I enjoy trying new rich foods.
26. I have the impulse to vomit after meals.

SCORING

Never = 3
Rarely = 2
Sometimes = 1
Always, usually, and often = 0

A total score under 20 points may indicate abnormal eating behavior and the risk of having or developing an eating disorder. Share your results with a health professional for further evaluation.

SOURCE: J. A. McSherry, Progress in the diagnosis of anorexia nervosa, *Journal of the Royal Society of Health* 106 (1986): 8–9; *Eating Attitudes Test* and scoring developed by Dr. P. Garfinkel.

Nutrition on the Web

www.cengage.com/nutrition/boyle/personalnutrition8e
Go to the *Personal Nutrition* site to check for the latest updates to chapter topics or to access links to related websites.

www.nal.usda.gov/fnic
The Food and Nutrition Information Center (FNIC) provides credible and accurate information.

- Under Browse by Subject, select Weight and Obesity.
- Under Browse by Subject, select Diet and Disease and then Eating Disorders.
- Click on Consumer Corner, select Eating for Health, and then select Weight Management.

www.ChooseMyPlate.gov
Search under Subjects, and select Steps to a Healthier Weight.

www.smallstep.gov
Search for information about portion control and improving health of adults and teens on this government website.

www.nutrition.gov
This government website provides information and links. Under Browse by Subject, select Weight Management.

www.nhlbi.nih.gov/health/public/heart/obesity/lose_wt/index.htm
The National Heart, Lung and Blood Institute website provides Aim for a Healthy Weight.

www.cdc.gov/nccdphp/dnpa
The Centers for Disease Control (CDC) website contains many topics related to nutrition and weight management.

www.nlm.nih.gov/medlineplus
Search for information about overweight, obesity, underweight, and eating disorders at this government website.

www.ncbi.nlm.nih.gov/PubMed
Use this search engine to locate information on obesity and related topics.

www.eatright.org
The American Dietetic Association home page includes many nutrition resources and links.

www.fda.gov/cder
Search the Center for Drug Evaluation and Research website for information about weight-loss drugs.

www.healthfinder.gov
Search for information on overweight, obesity, underweight, and eating disorders on this U.S. government site.

www.ars.usda.gov
Look up calories in foods and beverages at this government website.

win.niddk.nih.gov/publications/choosing.htm
The Weight-control Information Network website provides safe and reliable information about successful weight loss programs.

www.nwcr.ws
The National Weight Control Registry is provided at this website.

www.intelihealth.com
This website provides information about weight management, obesity, "the freshman 15," and eating disorders and numerous interactive tools.

www.mayoclinic.com
Visit the Mayo Clinic's Food and Nutrition Healthy Living Center; check out the virtual cookbook for healthy recipes and find tools to help with weight management.

www.caloriecontrol.org
The Calorie Control Council provides healthy recipes, an online calorie counter, and virtual calculators for figuring body mass index and calories expended during various exercises.

www.national eatingdisorders .org
The National Eating Disorders Association (NEDA) website provides information about prevention and treatment of eating disorders.

Diet Analysis+ in Action

Go to your Personal Profile and record your body mass index (BMI). Look at the textbook to see how you are classified based on this BMI. Now look at the Energy Balance report that shows the results of your three-day tracking, and record your average calorie intake for this period. Were you over, in line with, or under the recommended intake during this period? If you continued to consume calories at this level for 30 days, would you gain, lose, or maintain your current weight? If you are over- or underweight, create a second profile that includes your desirable weight (be sure to copy and then edit the profile so that your three-day diet tracking is included in the new profile). What changes will you need to make to attain—or work toward—your desirable weight? What one or two changes could you make each day that would help you reach your goal? Make these changes in the diet tracking to see the result.

Nutrition, Physical Activity, and Fitness

Ask Yourself . . .

Which of the following statements about nutrition are true, and which are false? For each false statement, what is true?

1. Regular physical activity can help people increase their lean body mass and reduce their fat tissue.
2. Less than 34 percent of U.S. adults exercise adequately.
3. People who fail to exercise regularly are more likely to fall prey to degenerative diseases such as heart disease, osteoporosis, and diabetes.
4. Essentially, to be fit means to be at desirable weight and to have strong muscles.
5. People should never push themselves to exercise longer or harder than they can easily manage to do.
6. Of all the components of fitness, cardiovascular endurance has the most impact on health and longevity.
7. If you run out of breath, it is a sign that your heart and lungs are not strong enough to perform the desired tasks.
8. When a muscular athlete stops exercising, much of his or her muscle tissue turns to fat.
9. The use of steroid hormones can cause a disfiguring disease.
10. Athletes can lose two or more quarts of fluid during every hour of heavy physical activity and must rehydrate before, during, and after exercising.

Answers found on the following page.

How physically active are you? The benefits of regular physical activity make up an impressive list, and the list keeps growing longer as new discoveries are made. People who begin a fitness program to trim fat or add muscle are soon pleasantly surprised to find that they have more energy, feel less tense, sleep better, and feel healthier.[1] And so they keep it up! Most people do not realize the many benefits of regular physical activity. Read the list in Table 11-1 and imagine how being more physically active might change your life for the better.

In addition, mounting evidence suggests that our bodies need regular, moderate physical activity that gets our hearts beating and forces our muscles to work harder than they usually do to stay healthy. Physiologically speaking, overall fitness is a balance between different body systems. With respect to the joints, flexibility is important. With respect to muscles, strength and endurance are important.

Get health. No labor, effort, nor exercise that can gain it must be grudged.

R. W. EMERSON (1803–1882, AMERICAN ESSAYIST AND POET)

Being fit is more than being free of disease; it is feeling full of vitality and enthusiasm for life.

This chapter illustrates the effects of nutrition on fitness—and vice versa, for a two-way relationship exists. Optimal nutrition contributes to athletic performance, and, conversely, regular physical activity contributes to a person's ability to use and store nutrients optimally. Together, the two are indispensable to a high quality of life. **Fitness**, like good nutrition, is an essential component of good health. In fact, in 2010, in conjunction with the First Lady's *Let's Move!* Initiative that addresses the epidemic of child obesity, President Obama launched the President's Council on Fitness, Sports, and Nutrition (formerly known as the President's Council on Physical Fitness and Sports). Both initiatives include a focus on healthy eating as well as physically active lifestyles.

No one can promise that you will receive all of these benefits if you exercise, but almost everyone who exercises reaps at least some of them. Physical activity need not be strenuous to yield health benefits. Moderate amounts of physical activity are recommended for people of all ages. However, at present, only about 33 percent of adults engage regularly in sustained physical activity of any intensity during leisure time. About 38 percent of adults report no physical activity at all.

TABLE 11-1	Benefits of Physical Activity

- Increased self-confidence
- Easier weight control
- More energy
- Less stress and anxiety
- Improved sleep
- Enhanced immunity
- Lowered risk of heart disease
- Lowered risk of certain cancers
- Stronger bones
- Lowered risk of diabetes
- Lowered risk of high blood pressure
- Increased quality of life
- Increased independence in life's later years

Major risk factors for heart disease:

- Sedentary lifestyle
- Family history of heart disease
- Cigarette smoking
- High blood pressure
- High blood cholesterol (> 200 mg/dL)
- Obesity (BMI ≥ 30)
- Diabetes

Getting Started on Lifetime Fitness

The notion that a certain *minimum* daily average amount of activity is indispensable to health is just now reaching public consciousness. The *Physical Activity Guidelines for Americans* report that health benefits occur with at least 2 hours and 30 minutes a week of moderate-intensity physical activity, such as brisk walking, or any one of numerous forms of physical activities (see Figure 11-1).[2] Additional benefits occur with more physical activity. Just mix and match your preferred activities in periods as short as 10 minutes. Alternatively, adults can choose a minimum of 1 hour and 15 minutes a week of vigorous-intensity aerobic physical activity by doing activities like jogging, aerobic dancing, and jumping rope. The new guidelines stress the value of moderate activity and suggest that the total *amount* of activity is more important than the manner in which it is carried out. For total fitness, a physical activity program that incorporates aerobic activity, strength training, and stretching is best. The more active you are, the more fit you are likely to be.

In proceeding with a fitness program, keep in mind that fitness builds slowly, so activity should increase gradually. For beginners, consistency is very important. Establish a regular pattern of physical activity first (for example, 30 minutes cumulative to start) and plan to increase that amount over time. View your physical activity time as a lifelong commitment.

If you are just starting on a fitness program, a few precautions are important. For most apparently healthy people, moderate physical activity such as walking should not pose any problem. However, medical advice concerning suitable type of activity may be necessary for anyone with any of the risk factors shown in the margin or for anyone diagnosed with cardiac or other known diseases.[3]

ASK YOURSELF ANSWERS: 1. True. 2. True. 3. True. 4. False. To be fit means not only to be at desirable weight and to have strong muscles, but also to be flexible and, most importantly, to have muscular and cardiovascular endurance. 5. False. The overload principle states that people should push themselves to exercise longer or harder than they can easily manage to do—although not, of course, to the point of strain. 6. True. 7. False. If you run out of breath, it is not a sign that your heart and lungs are weak but a sign that you are going into oxygen debt. 8. False. Muscle tissue does not turn to fat, but for a muscular athlete who stops exercising, muscle tissue is lost and fat is gained. 9. True. 10. True.

fitness the body's ability to meet physical demands; composed of four components: flexibility, strength, muscle endurance, and cardiovascular endurance.

FIGURE 11-1

Physical Activity Guidelines for Adults

Physical activity helps you feel and perform at your best. The key to sticking with a regular physical activity program is to choose activities that you enjoy. Be flexible. Remember that you don't have to spend 60 consecutive minutes engaged in physical activity—a few minutes here and a few minutes there add to the benefits of moderate activity. Keep in mind that *any* activity is good (for example, "just do something"), but *more* is better. The goal is to have a *physically active* lifestyle.*

SOURCE: Adapted from USDA's MyPyramid and from the U.S. Department of Health and Human Services, 2008 Physical Activity Guidelines for Americans, by University of Missouri Cooperative extension, 2009.

*More tips for increasing daily physical activity can be found at www.ChooseMyPlate.org.

LIFESTYLE ACTIVITIES	AEROBIC ACTIVITY*	STRENGTH & FLEXIBILITY	INACTIVITY
As often as possible.	At least 150 minutes (2 hours and 30 min.) of moderate intensity each week OR At least 75 minutes (1 hour and 15 min.) of vigorous intensity each week	At least 2 times each week	Limit
Pick an activity you like and one that fits your life. Some physical activity is better than none.	What is a moderate-intensity activity? • You can talk while you do it, but you can't sing. What is a vigorous-intensity activity? • You can only say a few words without stopping to catch your breath	Strength • Include all major muscle groups. • Perform 8–12 repetitions per set (1–3 sets) Flexibility • Perform flexibility activities at least two days each week for at least 10 minutes each day	Screen time (television, computer, video games). Sitting longer than 60 minutes.

*Most health benefits occur with at least 150 minutes (2 hours and 30 min.) a week of moderate-intensity physical activity.

For additional health benefits, adults should increase their aerobic physical activity to 300 minutes (5 hours) a week of moderate-intensity, or 150 minutes a week of vigorous-intensity aerobic physical activity, or a combination of both. Additional benefits include lower risk of colon and breast cancer and prevention of unhealthy weight gain.

To learn more about these guidelines visit: www.health.gov/paguidelines

The term *fitness* is not restricted to the seasoned athlete. With a basic understanding of the concept of total fitness and a personal commitment to a physically active lifestyle, anyone can become fit. To be fit, you don't have to be able to finish the local marathon, nor do you have to develop the muscles of a Mr. Universe or Miss Olympia. Rather, what you need is a healthy body weight (refer to Chapter 10) and enough flexibility, muscle strength, muscle endurance, and cardiovascular endurance to meet the everyday demands that life places on you, plus some to spare. The Scorecard feature later in this chapter helps you evaluate your own level of physical activity.

For fitness, remember the FIT principle:

F—frequency: number of exercise sessions per week; at least three to five sessions per week are recommended.

I—intensity: how hard you exercise (for example, the degree of exertion while exercising); it is recommended that you exercise at 55 to 90 percent of your maximum heart rate per minute, known as your target heart rate.

T—time: duration or length of time that you exercise with your heart rate elevated into your target heart rate zone (the minimum amount is thought to be 20 to 30 minutes per session).

The Components of Fitness

True fitness consists of several components: strength, flexibility, muscle endurance, and cardiovascular endurance. Although you may choose to emphasize one or two aspects, each component is equally important to overall fitness and optimum health. The way to get there is through physical conditioning.

Physical Conditioning

Physical conditioning refers to a planned program of physical activity directed toward improving the function of a particular body system. Placing regular, physical demand on your body and forcing it to do more than it usually does will cause it to adapt and function more efficiently. In terms of exercise, this is called **overload**. Muscles respond to an overload (more than normal amount) of exercise by gaining strength and ability to endure. The overload principle applies equally to all aspects of fitness: flexibility, muscle strength, muscle endurance, and cardiovascular endurance.

You can apply overload to your physical activity program in several ways:

- You can do the activity more often—that is, increase its *frequency*.

- You can do the activity more strenuously—that is, increase its *intensity*.

- Or you can do it for longer periods of *time*—that is, increase its *duration*.

People's bodies are shaped by what they do.

All three strategies work well, and you can pick one or a combination of the three, depending on your fitness goals. Strength gain may not be perceptible in all cases. For some, though, such as male bodybuilders, muscles increase in strength and size. This response is called **hypertrophy**. The converse is also true. If muscles are not called on to perform, they decrease in size. This response is called **atrophy**.

Strength

Strength is the ability of the muscles to work against resistance, such as pulling yourself out of a swimming pool, carrying a backpack full of books, or opening a jar of pickles. The purpose of resistance training is to build well-toned muscles that let you accomplish daily activities at work and during recreation, as well as to prevent injury. As muscles get stronger, individual fibers thicken and enlarge. Our ability to respond to resistance training continues to a very old age.[4] The connective tissues making up muscles, tendons, and ligaments also strengthen and become more efficient at using energy. The benefit is that strong muscles, tendons, and ligaments play a key role in preventing injury. For example, strong quadriceps—the muscles on the front of the thigh—stabilize your knee as you bike. Strong calf muscles and ankle ligaments decrease the risk of an ankle sprain when you walk briskly or jog. Resistance training also helps with weight loss by increasing lean muscle mass and thus increasing your basal metabolic rate.

Flexibility

Keeping your muscles and joints pliable is critical for developing a fit body. A flexible body can move as it was designed to move and will bend rather than tear or break in response to sudden stress. **Flexibility** (range of motion) depends on the condition and interrelationships of bones, ligaments, muscles, and tendons.

overload an extra physical demand placed on the body. A principle of training is that for a body system to improve, its workload must be increased by increments over time.

hypertrophy an increase in size in response to use.

atrophy a decrease in size in response to disuse.

strength the ability of muscles to work against resistance.

flexibility the ability to bend or extend without injury; depends on the elasticity of the muscles, tendons, and ligaments and on the condition of the joints.

Nutrition and Fitness—Forever Young— or You're Not Going to Take Aging Lying Down, Are You?

No matter how old you are, you have the body of a modern human being—and that body (and how it works) was designed more than 100,000 years ago. One could even conclude that we humans have inherited genes that have been fine-tuned to support a physically active lifestyle. In fact, scientists from the University of Missouri, University of Pennsylvania, and East Carolina University not only hypothesize this to be true, but go on to say "that physical inactivity (defined as the activity equivalent of less than 30 minutes of brisk walking each day) in sedentary societies directly contributes to multiple chronic health disorders."[5] They conclude that physical inactivity is an abnormal state because our bodies have been "programmed" to expect physical activity; this is why inactivity causes the metabolic dysfunctions that lead to a host of chronic health conditions, such as heart disease, cancer, obesity, diabetes, and high blood pressure.

Axel Leschinski/Alamy

YESTERDAY'S GENES, TODAY'S LIFESTYLE

Nearly all of your biochemistry and physiology (in other words, how you process foods and metabolize energy) was fine-tuned by and for conditions of life that existed more than 10,000 years ago. However, what we eat has changed more in the past 40 years than in the previous 40,000 years. So what's the big problem? Our genes don't know it! We still "process" food the same way our ancestors did—very efficiently (sometimes called the "thrifty gene" hypothesis). To understand this concept, you need only compare "yesterday's" diet and physical activity patterns to "today's" diet and physical activity habits (see the photos at the top of page 338).

CHOOSE YOUR WEAPON

You have your choice of weapons against the life-threatening diseases associated with sedentary aging, as listed in Table 11-2. Recommendations for physical activity include 2 hours and 30 minutes of moderate-intensity activity (brisk walking between classes,

bicycling from your dorm to classes, climbing the stairs instead of taking the elevators) a week and resistance exercise (using free weights, exercise bands, weight machines, or push-ups and squats). Again, it doesn't matter if the activity is accumulated in small increments or at one time.[6] Resistance exercise should average three sessions (20 to 45 minutes per session) per week, working one to two muscle groups during each session. The different muscle groups are the chest, shoulders, arms, back, abdominals, and legs. Each session should include one to three sets (8 to 12 repetitions per set), resting 2 to 3 minutes between sets.[7] The photos on the following page clearly demonstrate the need for physical activity throughout life.

So you don't have free weights? Don't have any of that fancy exercise equipment advertised on TV? Not to worry. For no-frills weight equipment, simply take two empty 1-gallon milk containers and add 1 pint (2 cups) of water (1 pint = 1 pound). Increase the amount of water in the milk containers to increase the amount of weight you lift. For a makeshift barbell, hang the milk containers on each end of a broomstick. Yes, it's a little hokey. But it's cheap!

GET OFF YOUR "BUTS"

Perhaps you are thinking, "I can't afford membership to a gym." Or "I don't have any exercise equipment." Or maybe your first thought is, "I go to school full-time and work part-time, and I have to study when I'm not working or going to class, and I have friends and a family. . . . I don't have time to exercise." Don't let those thoughts in! Remember, instead, that you don't need special equipment, special clothes, or membership to a fitness club. All you need to do is what we are (physiologically) supposed to be doing: short bouts of activity throughout the day.

2-28

"Now all we need is fake exercise."

(Continued)

Preagricultural Hunter-Gatherers

Burned ~3,000 calories per day

Moderate physical activity > 30 minutes per day to provide basic necessities (food, water, shelter, gathering materials for warmth, etc.)

Feast or famine

Lean wild game or fish

Uncultivated fruits and vegetables

Industrialized Modern Humans

Burn ~1,800 calories per day

Sedentary (desk jobs, computer games, TV, dishwashers, drive-through windows, escalators and elevators, shopping by phone, remote controls, e-mail/Internet, lawn services, etc.)

Abundance of food (continuous feeding)

Grain-fattened meats

Refined sugar

The photo at the left shows cross sections of the thighs of an active 20-year-old woman. The small white circle is the femur or thighbone. The photo at the right shows the thigh muscles of an older, sedentary woman in her late 60s. The older woman's thighs have lost muscle and gained fat. Strength-training activities such as walking, stair climbing, knee extensions, and leg curls can help prevent as well as reverse at least some of the muscle loss that accompanies aging and a sedentary lifestyle.[8] In other words, such a loss of muscle mass and the accompanying decrease in basal metabolism are not inevitable changes that occur with healthy aging. However, as the photos illustrate, these changes will accompany sedentary aging.

Flexibility tends to decrease with age, but improves in response to stretching, and it can be maintained in most people by frequent stretching. Stretching activities improve flexibility by increasing muscle and tendon elasticity and length. Stretching should be done slowly. When you feel a slight strain in the muscle, hold the position for 20 to 30 seconds. Bouncy, rapid stretches can cause minute tears in the muscle and often do more damage than good.

Stretching routines are commonly part of a warm-up routine before exercise. Low-intensity preliminary exercise allows your heart—also a muscle—to slowly accelerate and make necessary adjustments in blood flow and oxygen supply, preparing the heart for the work it is about to perform. Using calisthenics as a warm-up, such as walking, marching in place, or doing some other moderate rhythmic activity, prepares your heart muscle for action. After your light warm-up, stretch the muscles that you will be using in your main physical activity. Waiting to stretch until after

For example, walk to your classes instead of driving. When you do have to drive to campus or the mall, park at the far end of the parking lot. Take stairs instead of elevators. Is the weather bad? Go out to the mall and walk. You will be active and will get to window-shop at the same time. Remember, our ancient ancestors didn't have exercise equipment or fitness clubs, either. But the required activities of their lives—hunting food, building shelters, chopping wood—kept them alive, strong, and healthy. The need for physical activity to stay alive hasn't changed today.

TABLE 11-2	Health Consequences of Physical Inactivity
Cardiovascular Diseases	Coronary artery disease Congestive heart failure Hypertension Stroke Peripheral artery disease
Metabolic Diseases	Type 2 diabetes Obesity Low levels of HDL-cholesterol High levels of triglycerides, LDL-cholesterol, and total cholesterol Gallbladder disease
Cancers	Breast cancer Colon cancer Prostate cancer Pancreatic cancer Melanoma
Pulmonary Diseases	Asthma
Immune Dysfunction	Susceptibility to viral infections
Musculoskeletal Disorders	Osteoporosis Physical frailty
Neurological Disorders	Cognitive dysfunction

Bill Bachman/Alamy Limited

Strength training promotes healthy bones and muscles at any age.

your warm-up allows blood to move into the muscles, making them easier to stretch. Doing stretches after your activity session gives your heart a chance to gradually slow its pace. It also allows you to lengthen any muscles that have become tight and tense from the activity. You can make greater gains in flexibility by stretching after your workout because your muscles are warm and easier to stretch.

Muscle Endurance
Muscle endurance, the third component of fitness, is the ability of a muscle to keep working for long periods. Your muscle endurance influences your performance in the last set of a tennis match, your swing on the 18th hole of a golf game, or your ability to pedal during the last 10 miles of a 100-mile bike tour. Endurance of certain muscles can be tested by the number of sit-ups or push-ups you can accomplish in a certain period of time. Remember, though, that these tests evaluate only the abdominal and upper arm muscles.

endurance the ability to sustain an effort for a long time. One type, **muscle endurance**, is the ability of a muscle to contract repeatedly within a given time without becoming exhausted. Another type, **cardiovascular endurance**, is the ability of the cardiovascular system to sustain effort over a period of time.

To develop cardiovascular fitness, choose an aerobic activity such as:

- Aerobic dancing
- Bench stepping
- Bicycling
- Cross-country skiing
- Fast walking
- Jogging
- Roller blading
- Roller skating
- Rope jumping
- Rowing
- Running
- Speed skating
- Stair climbing
- Swimming
- Treadmill walking or running

Cardiovascular Endurance   Another important aspect of endurance is the length of time that you can keep going with an elevated heart rate—that is, how long your heart can endure a given demand. This kind of endurance is called **cardiovascular endurance**. The heart is a muscle, and, like your other muscles, it can respond to repeated demands by becoming larger and stronger.

Activities that promote cardiovascular endurance are the best for making short-term fitness gains and for long-term health, as well as for weight control. The best activities to develop cardiovascular endurance are those that repetitively use large muscle groups—arms and legs—and that last for a continuous 20 to 60 minutes. Examples include brisk walking, aerobic dance, running, cycling, cross-country skiing, and rowing. The American College of Sports Medicine recommends that people participate in cardiovascular conditioning activities at least five times a week for a continuous 30 to 60 minutes.[9]

Energy for Physical Activity

Your body runs on water, oxygen, and food—primarily carbohydrate and fat. The chemical process that converts these substances to energy is called *metabolism*. Two interrelated energy-producing systems are at work in your body. One system, which depends on oxygen, is called **aerobic** metabolism. The other system functions without oxygen and is called **anaerobic** metabolism. Understanding how the two systems work is important because it explains why you would choose certain activities over others to strengthen your heart, why you eat what you do, and what factors influence your performance during sporting events.

Aerobic and Anaerobic Metabolism   At rest, your muscles burn mostly fat and some carbohydrate for energy. During exercise, though, the amount of energy the muscles use depends on the interplay between fuel availability and oxygen availability. To an exercising muscle, oxygen is everything. With ample oxygen, muscles can extract all available energy from carbohydrate and fat by means of aerobic metabolism. During moderate activity, your lungs and circulatory system have no trouble keeping up with the muscles' need for oxygen. You breathe deeply and easily, and your heart beats steadily—the activity is aerobic. Still, the heart and lungs can supply only so much oxygen only so fast.

When the muscles' exertion becomes great enough that their energy demand outstrips the oxygen supply, they must also rely on anaerobic metabolism for energy. Because the anaerobic metabolic pathway can only burn carbohydrate for fuel, it draws heavily on your limited body stores of carbohydrate. Nevertheless, this system provides an immediate energy source without requiring oxygen. Thanks to anaerobic metabolism, you can dash out of the way of an oncoming car or sprint ahead of your

aerobic *requiring oxygen.*

anaerobic *not requiring oxygen.*

competitor at the finish line. Unfortunately, however, this energy-yielding system is extremely inefficient. Only 5 percent of carbohydrate's energy-producing potential is harnessed by this pathway.[10]

Because the anaerobic metabolic pathway only partially burns your carbohydrate, it also litters your muscle with lactic acid—partly broken-down portions of glucose. When lactic acid builds up in the muscles, it causes burning pain and can lead to muscle exhaustion within seconds if it is not drained away. An effective strategy for dealing with lactic acid buildup is to relax the muscles at every opportunity, allowing the circulating blood to carry it away and bring in oxygen to support aerobic metabolism. Fortunately, lactic acid is not a waste product. The blood delivers it to your liver, where it is converted back into glucose.

Neither the aerobic nor the anaerobic metabolic pathway functions exclusively to supply energy to your body. The two work together, complementing and supporting each other. Keep in mind, however, that carbohydrate is absolutely essential for exercise. Without it, your muscles can't perform. When you exercise aerobically, muscles burn fat and extract energy from carbohydrate more efficiently in the presence of oxygen, thereby conserving your body's limited store of carbohydrate. Thus, you want to exercise at an intensity that allows your heart and lungs to keep pace with the oxygen needs of your working muscles.

Aerobic Exercise—Exercise for the Heart To meet your body's increased oxygen needs during aerobic exercise, your heart must pump oxygen-rich blood to muscles at a faster pace than normal. This increased demand on the heart makes the heart stronger and increases its endurance. In addition, aerobic exercise improves the endurance of the lungs and the muscles along the arteries and in the walls of the digestive tract and, of course, the muscles directly involved in the activity. These all-over improvements are called **cardiovascular conditioning** or the **training effect**. In cardiovascular conditioning, the total blood volume increases so that the blood can carry more oxygen. The heart muscle becomes stronger and larger. Therefore, because each beat of the heart pumps more blood, it needs to pump less often. The muscles that work the lungs gain strength and endurance, and breathing becomes more efficient. Circulation through the body's arteries and veins improves. Blood moves easily, and the blood pressure falls. Muscles throughout the body become firmer. Figure 11-2 shows the major relationships among the heart, circulatory system, and lungs.

To make these gains in cardiovascular conditioning, you must elevate your heart rate (pulse). This elevated heart rate—called your **target heart rate**—must be considerably faster than the resting rate to push (overload) the heart but not so fast as to strain it. To achieve this goal, you must work up to the point at which you can exercise aerobically for at least 20 minutes or longer.

An informal pulse check can give you some indication of how conditioned your heart is to start with. As a rule of thumb, the average resting pulse rate for adults is around 70 beats per minute, but the rate can be higher or lower. Active people may have resting pulse rates of 50 or even lower.

For cardiovascular conditioning, you can calculate your target heart rate using your age as the starting point. The older you are, the lower your maximum target heart rate. As your heart becomes stronger, more intense activity will be required to reach the same target rate. For example, at first, walking at a pace of 3 miles per hour may cause you to reach your target heart rate. After 6 to 8 weeks of walking at this pace, you may notice that you no longer reach your target heart rate. That's because your heart is stronger. It now needs more of a challenge to beat faster. Increasing the intensity of your workout by walking faster can provide this challenge.

cardiovascular conditioning or **training effect** the effect of regular exercise on the cardiovascular system; includes improvements in heart, lung, and muscle function and increased blood volume; also called *cardiorespiratory endurance or cardiorespiratory training.*

target heart rate the heartbeat rate that will achieve a cardiovascular conditioning effect for a given person: fast enough to push the heart, but not so fast as to strain it.

To calculate your target heart rate range, take the following steps:

1. *Estimate your maximum heart rate (MHR).* Subtract your age from 220. This provides an estimate of the absolute maximum heart rate possible for a person your age. You should never exercise at this rate, of course.
2. *Determine your target heart rate range.* Multiply your MHR by 55 percent and 90 percent to find your upper and lower limits (see margin example on page 345).

When you can work out at your target heart rate for 20 to 60 minutes, you know that you have arrived at your cardiovascular fitness goal.

Fuels for Physical Activity

Your energy-producing pathways require oxygen and the two muscle fuels, glucose and fatty acids. As Figure 11-2 shows, the oxygen comes from the lungs, which pass it to the blood, which carries it to the muscles. Your muscles, and to some extent your liver, supply carbohydrate to your muscles from their carbohydrate supply (see Figure 11-3). The fatty acids come mostly from fat inside the muscles but partly from fat that is released from the body's fat stores, and the blood delivers these fatty acids to the muscles.

FIGURE 11-2

Delivery of Oxygen by the Heart and Lungs to the Muscles

The more fit a muscle is, the more oxygen it draws from the blood. That oxygen is drawn from the lungs, so the person with more fit muscles extracts more oxygen from the inhaled air than a person with less fit muscles. The cardiovascular system responds to the demand for oxygen by building up its capacity to deliver oxygen. Researchers can measure cardiovascular fitness by measuring the amount of oxygen a person consumes per minute while working out, a measure called VO_2 max.

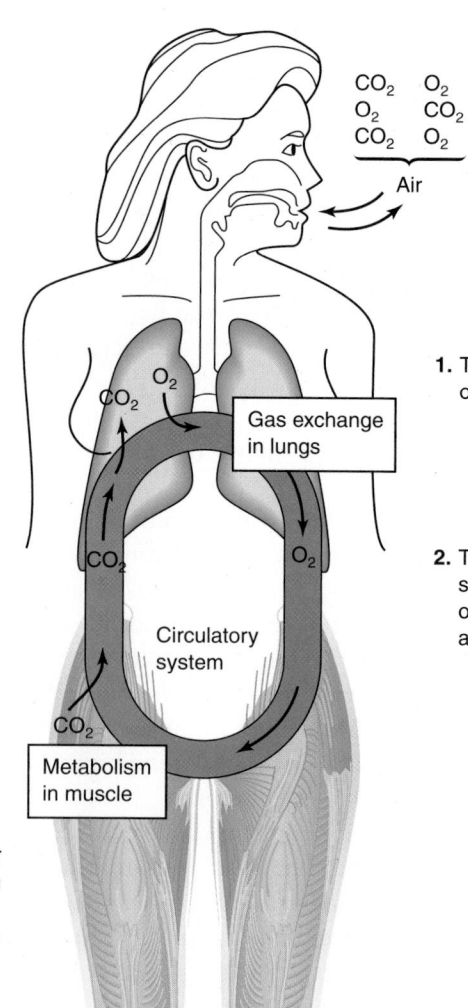

5. The lungs remove carbon dioxide, making the blood ready to be reloaded with oxygen.

4. The blood carries the carbon dioxide back to the lungs.

3. The muscles and other tissues remove oxygen from the blood and release carbon dioxide into it.

CO_2 O_2
O_2 CO_2
CO_2 O_2
_____
Air

CO_2 O_2

Gas exchange in lungs

CO_2 O_2

Circulatory system

Metabolism in muscle

CO_2

1. The lungs deliver oxygen to the blood.

2. The circulatory system carries the oxygenated blood around the body.

How physically active are you?
For each question answered yes, give yourself the number of points indicated. Then total your points to determine your score.

A. VIGOROUS PHYSICAL ACTIVITY ROUTINES

1. I participate in active recreational sports such as tennis or racquetball for an hour or more:
   a. about once a week (2 points)
   b. about twice a week (4 points)
   c. three times a week (6 points)
   d. four times a week (8 points)
   e. not at all (0 points)

2. I participate in vigorous fitness activities like aerobic dancing, roller blading, jogging, or swimming (at least 20 minutes each session):
   a. about once a week (3 points)
   b. about twice a week (6 points)
   c. three times a week (9 points)
   d. four times a week (12 points)
   e. not at all (0 points)

B. OTHER PHYSICAL ACTIVITY ROUTINES

3. At least two times a week, I work out with weights for at least 10 minutes:
   a. two sessions a week (2 points)
   b. three sessions a week (3 points)
   c. four or more sessions a week (4 points)
   d. not at all (0 points)

4. At least two times a week, I perform floor workouts (sit-ups, push-ups) for at least 10 minutes:
   a. two sessions a week (2 points)
   b. three sessions a week (3 points)
   c. four or more sessions a week (4 points)
   d. not at all (0 points)

5. At least two times a week, I participate in yoga or perform stretching exercises for at least 10 minutes:
   a. two sessions a week (2 points)
   b. three sessions a week (3 points)
   c. four or more sessions a week (4 points)
   d. not at all (0 points)

C. OCCUPATION AND DAILY ACTIVITIES

6. I walk to and from school, work, and shopping (1/2 mile or more each way), two or three times a week or more. (1 point)

7. I climb stairs rather than using elevators or escalators, every other day or more. (1 point)

8. My school, job, or household routine involves physical activity that fits the following description:
   a. Most of my day is spent in desk work or light physical activity. (0 points)
   b. Most of my day is spent in farm activities, moderate physical activity, brisk walking, or comparable activities. (4 points)
   c. My typical day includes several hours of heavy physical activity (shoveling, lifting, etc.). (2 points per day)

D. LEISURE ACTIVITIES

9. I do several hours of gardening, lawn work, or similar hobby work each week. (1 point)

10. At least once a week, I dance vigorously (folk or line dancing) for an hour or more. (1 point)

11. In season, I play 9 to 18 holes of golf at least once a week, and I do not use a power cart. (2 points)

12. I walk for physical activity or recreation:
    a. 1–2 hours a week (1 point)
    b. 3–4 hours a week (2 points)
    c. 5 hours or more a week (3 points)
    d. not at all (0 points)

13. In addition to the above, I engage in other forms of physical activity:
    a. 1–2 hours a week (1 point)
    b. 3–4 hours a week (2 points)
    c. 5 hours or more a week (3 points)

SCORING

Record your point scores here.

Category	Score
A. Vigorous physical activity routines	_____
B. Other physical activity routines	_____
C. Occupation and daily activities	_____
D. Leisure activities	_____
Total:	_____

Evaluation of total score (circle one):
- Inactive (0–5 points)
- Moderately active (6–11 points)
- Active (12–20 points)
- Very active (21 points or over)

If your score categorizes you as inactive or only moderately active, think of activities that you could realistically engage in on a regular basis to raise your score to "active" (12 points).

SOURCE: Adapted with permission from Russell Pate (University of South Carolina, Human Performance Laboratory).

FIGURE 11-3

The Use of Glycogen and Body Fat for Energy* during Physical Activity

Training can increase the amount of glycogen a muscle can conserve during exercise. The more fit a muscle is, the more fat it can burn for energy when oxygen is present—sparing the valuable glycogen.

*In order to perform physical activity, the body generates ATP (adenosine triphosphate) from the breakdown of the energy nutrients (primarily carbohydrate and fat). ATP is the "energy molecule" used by all cells of the body. ATP is composed of the compound adenosine plus three phosphate groups attached to each other by high-energy bonds. Energy is released each time a phosphate group is split off from the ATP molecule. For example, ATP converts to ADP (adenosine diphosphate) + P when a phosphate bond is broken to release energy.

4. The muscle can convert its own limited supply of glycogen to glucose for use as energy. Muscle triglycerides can also be converted to fatty acids and used for energy.

Muscle

Muscle Triglyceride
↓
Fatty acids

Glycogen
↓
Glucose

Liver

Blood

Glucose

Muscle Glycogen
↓
Glucose

1. The liver can convert its limited store of glycogen to glucose to help meet the energy demands of the working muscles.

Triglycerides
↓
Fatty acids

Body fat

Fatty acids

2. The body can also help meet the energy demands of the working muscles by breaking down its supply of body fat (triglycerides) to fatty acids.

3. The circulatory system carries fuel (glucose and fatty acids) to the muscle.

Energy

5. The working muscles can pick up circulating glucose and fatty acids from the blood and metabolize them for energy. Since the trained muscle is better equipped to use fat for energy, it can use more fat for energy than the untrained muscle and can thereby conserve its limited glycogen supply for a longer period of time.

FIGURE 11-4

Fuel Use and Duration/Intensity of Physical Activity

For most people, fat isn't used much as a fuel for exercise until you've been working out aerobically for at least 20 minutes, and it is not used as a primary fuel until after 2 hours.

[a] The more moderate the intensity of activity (brisk walking, jogging, aerobic dancing), and the longer the duration, the greater the use of fat for fuel.
[b] The higher the intensity of activity (sprinting, hurdles, rowing), the greater the use of carbohydrate for fuel.

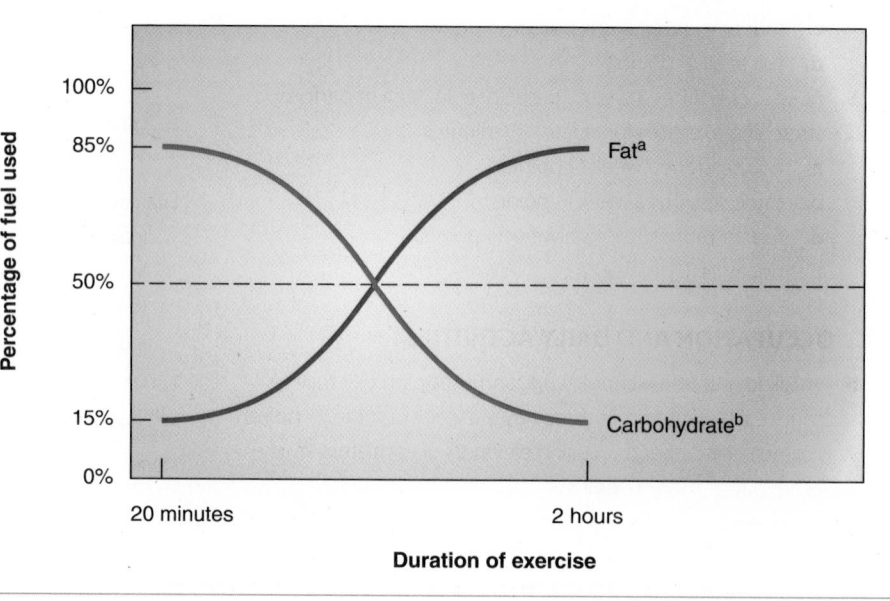

Percentage of fuel used

100%

85%   Fat[a]

50%

15%   Carbohydrate[b]

0%

20 minutes   2 hours

Duration of exercise

Glucose Use during Physical Activity   Glucose comes from carbohydrate-rich foods: breads, pasta, rice, legumes, fruits, vegetables, milk, and yogurt. Your body stores glucose in your liver and muscles in the form of glycogen, a long chain of glucose molecules linked together.

During physical activity, the body supplies glucose to the muscles from the stores of glycogen in the liver and in the muscles themselves. The longer the activity lasts or the more intense it is, the more glucose a person uses. Recall that activities done at an intensity that outstrips the ability of the heart and lungs to supply oxygen to working muscles relies primarily on glucose for fuel. Thus, activities such as sprinting quickly deplete the body's stores of glycogen. Other activities, such as jogging or brisk walking, in which the body can meet the muscles' oxygen demands, use glycogen more conservatively. Nonetheless, joggers and walkers still use glycogen, and eventually they can run out of it.

When a person begins exercising, for the first 20 minutes or so, about one-fifth of the body's total glycogen store is rapidly used.[11] If exercise continues beyond 20 minutes, glycogen use slows down (see Figure 11-4). To conserve the remaining glycogen supply, the body begins to rely more on fat for fuel. At some point, if exercise continues long enough, glycogen will run out almost completely. People who run out of muscle glycogen during an event (for example, before the finish line in a marathon) "hit the wall." They must slow down their pace because muscle glycogen is no longer available as fuel. Exercise can continue for a short time after that, only because the liver scrambles to produce the minimum amount of glucose needed to briefly forestall body shutdown. When blood sugars dip too low, the nervous system function comes almost to a halt, making exercise difficult, if not impossible, even though there is still plenty of fat left to burn.

Another factor that influences glycogen use during exercise is how well trained the person is. In the beginning stages of building a physical activity program, a person uses more glucose than a trained athlete. This is because their "untrained" muscles can quickly and easily extract energy from glucose. However, for muscles to easily extract energy from fat, the muscle cells must contain abundant fat-burning enzymes. With training, the muscles adapt and pack their cells with more fat-burning enzymes. Thus, trained muscles are able to use more fat and conserve their glucose.

The amount of glycogen that is present in the muscles before exercise also influences glycogen use. By following the diet prescription in Figure 11-5, athletes can provide their muscles with enough glycogen to support exercise. In low- to moderate-intensity activities (walking, bicycling, dancing), casual exercisers rarely deplete the glycogen in their stores. Carbohydrate loading—a practice that endurance athletes use to trick their muscles into storing extra glycogen—may not be beneficial for people who exercise less than 90 minutes per workout at a low intensity, although competitive athletes who exercise at a high intensity for more than 90 minutes at a time may benefit from carbohydrate loading. Muscles typically have enough glycogen to fuel 1½- to 2-hour bouts of activity.

An athlete who follows the glycogen-loading technique in preparation for an upcoming event will first exercise intensely without restricting carbohydrates and then gradually cut back on exercise the week

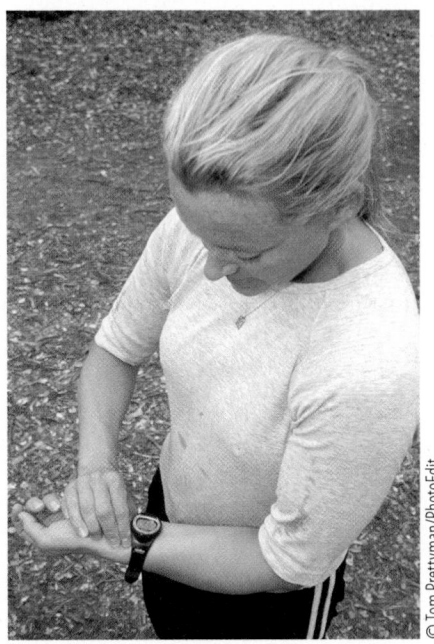

To take your pulse and monitor your heart rate during exercise, lightly press your middle and index fingers on the radial artery (on the thumb side of the wrist), as shown here. Count your pulse for 10 seconds, and then multiply by 6 to give beats per minute.

Example: Jennifer, age 25
Maximum heart rate: 220 − 25 = 195
Lower limit (55%) of target heart rate range:
0.55 × 195 = 107
Upper limit (90%) of target heart rate range:
0.90 × 195 = 176

Target heart rate range: 107 to 176 beats per minute. Therefore, when Jennifer exercises aerobically, her heart should beat at least as fast as 107 beats per minute but no faster than 176 beats per minute.

© Tom Prettyman/PhotoEdit

Pete Saloutos/Alamy Limited

Anaerobic exercise. Glucose is the principal source of energy for activities of high intensity.

FIGURE 11-5

The Effect of Diet on Physical Endurance

A high-carbohydrate diet can increase an athlete's endurance. In this study, the fat and protein diet provided 94 percent of calories from fat and 6 percent from protein; the normal mixed diet provided 55 percent of calories from carbohydrate; and the high-carbohydrate diet provided 83 percent of calories from carbohydrate.

Fat and protein diet

Normal mixed diet

High-carbohydrate diet

© Cory Sorensen/Corbis

Maximum endurance time:

57 min

114 min

167 min

People who participate in endurance events know to build up their reserves of muscle glycogen before an event, so that they do not run out or "hit the wall" before the finish line.

© Tony Freeman/PhotoEdit

Aerobic activity. Fats are the main source of energy for activities of low to moderate intensity.

Tyler Olson/Shutterstock.com

before the competition, rest completely the day before, and eat a very high-carbohydrate diet. Endurance athletes who follow this plan can keep going longer than their competitors without ill effects.[12] In a hot climate, extra glycogen offers an additional advantage. As glycogen breaks down, it releases water, which helps meet the athlete's fluid needs.

Fat Use during Physical Activity

When you exercise, the fat your muscles burn comes from the fatty deposits all over the body, especially from those with the greatest amounts of fat to spare. That is why physically fit people look trim all over: they reduce their fat stores all over the body, not just those overlaying the working muscles.

A person with a desirable body weight may store 25 to 30 pounds of body fat but only about 1 pound of carbohydrate. Although your supply of fat is almost unlimited, the ability of your muscles to use fat for energy is not.

Recall that for a working muscle to burn fat, oxygen must be present. If you work out at a rate that allows your heart to supply ample oxygen to working muscles, the muscles will draw heavily on fat stores for fuel. When exercise intensity outstrips your oxygen supply, fat still contributes as much energy as ever, but glucose pitches in and is burned by the anaerobic pathway. Thus, the percentage of the total energy supplied by fat declines. Your breathing rate can signal which fuel is providing most of the energy. A rule of thumb for gauging exercise intensity for aerobic workouts is this: If you can't talk normally, you are incurring oxygen debt and are burning more glucose than fat; if you can sing, you aren't getting a cardiovascular workout or burning much of anything (so speed up).

Athletic training also controls the amount of fat used during an exercise period. Exercise training improves the body's ability to deliver fat to working muscles, and trained muscles have an increased ability to use the fat.

Much attention has been focused on the type of fuel used for varying exercise intensities and duration. Research shows that when athletes exercise at a moderate

The word routine almost always accompanies *exercise plan* when the goal is to enhance health through physical activity. It should be no surprise that positive health effects cannot be realized with inconsistent physical activity. Sporadic binges of activity are of little or no benefit and often do more harm than good. But how do you find time and motivation to get fit and stay fit in today's hectic lifestyle? Many people are dying to tell you how, and they claim to have just the plan for you! Look at any late-night infomercial or shopping channel to see the many exercise gurus with videos, equipment, and supplements to allegedly keep you motivated and provide great results in a short amount of time. But just as fad diets come and go, the promised success with these gizmos and gadgets is usually unsustainable, and the equipment that works for one person may not work for another.

What's usually missing from these infomercials is the whole story. Although many products and programs promise to burn loads of fat in minimal time, they typically fail to mention that success with their product *must* be combined with a total fitness routine—which would mean that their product wasn't necessary in the first place! Don't be swayed by commercials that liberally use the phrase "scientific research." If their "scientific research" is not peer reviewed and published in a respected scientific journal, the results are most likely invalid or, at the very least, biased. Those who market these programs and products are skilled with their presentations, and their real goal is to separate you from your hard-earned money. If their product sounds too good to be true, it probably is!

Instead, pay attention to the guidelines in this text, do additional reading, consult experts, talk to your physician, and let education and common sense guide your fitness routine. Even though physical activity is not always easy, it is worthwhile and can change your life! In your physical activity program, consistency is the key, and no amount of gizmos or gadgets can replace that. Here are some tips for sustaining a physical activity program:

Make physical activity a habit. Choose an activity you enjoy.

1. Check with your physician before starting any physical activity program or weight loss plan. Your doctor can assess your ability for a given level of activity and refer you to qualified professionals for assistance, such as physical therapists, exercise physiologists, dietitians, and/or certified personal trainers.
2. Find an exercise buddy. You can motivate each other. Use a gradual approach and set realistic goals. Don't jump out of bed tomorrow morning and expect to be able to run 10 miles.

Set a realistic smaller first goal and use it as motivation to set your next goal. Experiment a little to find the time that is right for you to exercise. Many people like to exercise in the morning before their day begins. If you try that and it doesn't work, don't give up. You might have more luck with a lunchtime routine or an after-work exercise schedule.
3. Don't overdo it, especially in the beginning! Listen to your body and let it set the pace while still challenging yourself. "No pain, no gain" is a myth that can be dangerous.
4. Don't focus on weight loss. Focus on your new energy level and how much better you feel as your clothes start to fit again!

intensity, they initially use more carbohydrate than fat for fuel.[13] Gradually, as exercise continues for more than 20 minutes, the fuel ratio shifts, and the athletes use more fat. For athletes participating in endurance sports, such as marathon runners and long-distance cyclists, who want to conserve their limited supply of carbohydrate, switching to a fat-burning energy system is crucial.

Protein Needs for Fitness

Fit people have more muscle than fat; exercise involves muscles; muscles are made largely of protein. It seems logical, then, that to become or stay fit, an athlete might need more protein. Although it's true that fat and glucose are the primary fuels for working muscles, 5 to 10 percent of energy needs for weight lifters and endurance-sport athletes comes from muscle protein. So, do athletes need more protein?

The athlete's body may use slightly more protein, especially during the early stages of training. Initial increases in muscle mass, numbers of red blood cells to carry oxygen, and amounts of aerobic enzymes in muscles to use fuel efficiently may elevate an athlete's protein needs. In addition, hormonal changes during exercise can temporarily slow the amount of protein the muscle makes and can encourage the muscle to break down its protein stores.[14] How much protein an athlete uses for fuel during hard exercise (endurance exercise and heavy weight lifting) depends on exercise intensity and duration, the athlete's fitness level, and the glycogen stores in the athlete's muscles. When glycogen stores are well stocked, however, protein contributes only 5 percent of fuel needs.

The important factor here is that, although muscle protein breakdown dominates during heavy exercise, muscle *growth* escalates after exercise. The muscles use the available amino acids to repair and build, and the net effect of these changes is muscle protein buildup. Consistent training enhances muscle protein buildup after exercise.

The American College of Sports Medicine recommends that endurance athletes consume 1.2 to 1.6 grams of protein per kilogram of desirable body weight.[15] Athletes who are involved in prolonged heavy resistance training may need even more protein (see Table 11-3).

Many people wonder if eating even more protein will help build muscles. Unfortunately, muscles don't respond to excess protein by simply accepting it. Instead, they respond to the hormones that regulate them and to the demands put upon them. Thus, the way to make muscle cells grow is to put a demand on them—that is, to make them work. They will respond by taking up nutrients—amino acids included—so that they can grow.

TABLE 11-3	Protein Recommendations for Athletes
	Recommendations (g/kg/day)
RDA for adults	0.8
Endurance athletes	1.2–1.4
Resistance training (bodybuilders, strength athletes)	1.2–1.7

SOURCE: Position of the American Dietetic Association, Dietitians of Canada, and the American College of Sports Medicine: Nutrition and athletic performance, *Journal of the American Dietetic Association* 109 (2009): 509–27.

Fluids and Physical Activity

The water in your blood—known as *plasma volume*, or just *plasma*—serves a function similar to the function of the water in your car's radiator. This particular function of water is critical to humans during physical activity because, as the blood continually circulates throughout your body, it picks up the tremendous amount of heat generated by your working muscles. The plasma then transports this heat to your skin, through which it is expelled from the body primarily by evaporation of sweat. You can think of sweating as your body's air-conditioning system. As sweat evaporates from your skin, it expels large amounts of heat, helping to keep your body cool. However, sweating cools your body only when the sweat evaporates from your skin. If the sweat simply rolls down your face or down your back, body heat is not released. On a humid day, for example, the air is already saturated with water, impairing the evaporation of sweat. Hot, humid days, then, are doubly dangerous because you continue to sweat and lose precious body water, but your body temperature doesn't fall.

When exercising on humid days, it is extremely important to pay particular attention to your fluid needs.

Thus, as your body heats up because of the energy released during exercise, it loses water by sweating. How much sweat you lose depends on the intensity and duration of the activity. The more intense the activity, the more heat you generate and the more you sweat. If you don't replace this lost water, your plasma volume will decrease. In its attempt to maintain plasma volume, your body will pull water from your muscles and organs. As water is pulled from muscles, cramps may occur, along with premature fatigue and a noticeable decline in performance.

When plasma levels are low (meaning that blood volume is lower), your heart is forced to beat faster to supply sufficient oxygen to your muscles. Finally, because less plasma is circulating to transport the heat to your skin, the heat builds up, and your body's internal temperature continues to rise. All these changes force your body to work at a higher intensity, leading to early exhaustion. A water loss equal to 2 percent of body weight can reduce muscular work capacity by 20 to 30 percent.

Replenishing fluid lost during exercise is easy, yet many athletes and fitness enthusiasts either don't drink enough or don't drink at all.[16] Ignoring body fluid needs can hinder performance and increase risk of heat-related injury.[17]

The amount of fluid needed to prevent dehydration and **heat stroke** can be surprising. Athletes can lose two or more quarts of fluid during every hour of heavy exercise, and they must rehydrate *before*, *during*, and *after* exercise to replace the lost fluid. Even casual exercisers must drink some fluids while exercising. Thirst is an unreliable indicator of how much to drink: It signals too late, after fluid stores are depleted.[18] Table 11-4 presents one schedule of hydration. To know how much water is needed to replenish fluid losses after a workout, weigh yourself before and after: The difference is all water. One pound equals roughly 2 cups of fluid.

Endurance athletes need to be sure to replace both the water *and* sodium lost in heavy sweating during prolonged events (e.g., marathons) in order to avoid the dangerous condition known as **hyponatremia**. The symptoms of hyponatremia resemble dehydration to some extent and are listed in the margin.

Water and Fluid Replacement Drinks

For fitness enthusiasts, the choice between water and a sports drink is primarily a matter of personal taste and desired performance abilities. In contrast, for endurance events (continuous exercise for longer than 60 minutes), mounting evidence indicates that consuming a properly balanced sports drink during exercise will enhance energy status and endurance and help maintain plasma volume levels better than drinking water does.[19]

How the body manages water and carbohydrate use during exercise determines how well it performs. Sports drinks are designed to enhance the body's use of carbohydrate and water (see Table 11-5). The carbohydrate in a sports beverage serves three

Plan to drink fluids before, during, and after exercise.

Signs of hyponatremia:

- Severe headache
- Nausea, vomiting
- Muscle cramps
- Bloating
- Confusion
- Seizure

Signs of dehydration and heat stroke include:

- Very high body temperature (104° Fahrenheit or higher)
- Hot, dry, red skin
- Sudden cessation of sweating
- Deep breathing and fast pulse
- Blurred vision
- Confusion, delirium, hallucinations
- Convulsions
- Loss of consciousness

To prevent heat stroke, drink plenty of fluid before, during, and after physical activity; avoid overexercising in hot weather; and stop exercising at any sign of heat exhaustion.

Signs of heat exhaustion include:

- Cool, clammy, pale skin
- Dizziness
- Dry mouth
- Fatigue/weakness
- Headache
- Muscle cramps
- Nausea
- Sweating
- Weak and rapid pulse

TABLE 11-4	Schedule of Hydration Before, During, and After Exercise*
When to Drink	**Amount of Fluid**
2 hours before exercise	2–3 cups
15 minutes before exercise	1–2 cups
Every 15 minutes during exercise	½–2 cups
After exercise	2 cups fluid for each pound of body weight lost

*These guidelines are for exercise lasting less than 1 hour. During intense exercise lasting more than 1 hour, the consumption of approximately 1 liter of sports drink per hour (containing 4–8 percent carbohydrate per liter) is recommended to maintain oxidation of carbohydrates and delay fatigue.

SOURCE: American College of Sports Medicine, Position Stand, Exercise and Fluid Replacement, *Medicine & Science in Sports & Exercise* 39 (2007): 377–90.

heat stroke an acute and dangerous reaction to heat buildup in the body, requiring emergency medical attention; also called sun stroke.

hyponatremia (HIGH-poe-na-TREE-mee-ah) a decreased concentration of sodium in the blood.

TABLE 11-5	Fluid Replacement Drinks		
Sports Drink	Calories/Cup	Carbohydrate Percentage	Sodium (mg)
All Sport	70	9.0	55
Gatorade Thirst Quencher	50	6.0	110
Isostar	70	8.0	150
Met-Rx	75	8.0	125
Accelerade	60	7.0	127
PowerAde	72	8.0	55

Sports drinks can enhance fluid and energy status during endurance events.

purposes during exercise: (1) It becomes an energy source for working muscles, (2) it helps maintain blood glucose at an optimum level, and (3) it helps increase the rate of water absorption from the small intestine, helping better maintain plasma volume. In addition, the drink can supply water and minerals lost through sweating.[20]

There are many factors to consider when choosing a sports drink. The ideal beverage should leave the digestive tract rapidly and enter circulation, where it is needed. Carbohydrate solutions don't all empty from the stomach at the same rate. The drink should contain at least 4 percent but no more than 8 percent carbohydrate by volume. Drinks containing more than 10 percent carbohydrate, such as sodas, fruit juice, Kool-Aid types of drinks, and some sports drinks, take longer to absorb. Some can cause cramps, nausea, bloating, and diarrhea. Drinks with less than 4 percent carbohydrate may not offer an endurance-enhancing effect. Drinks using a blend of glucose polymers—short chains of carbohydrate—and fructose leave the stomach at the same rate as water, speeding the availability of the carbohydrate and water to working muscles.[21]

Sodium is another important factor to consider when exercising.[22] Because most people consume enough salt in their regular diet to replace the sodium they lose during exercise, it's not essential for a fluid replacement drink to provide large amounts of sodium. In fact, too much sodium can delay muscles' receipt of water.

Research shows that about 50 milligrams of sodium per cup will help stimulate water absorption from your gut. Other studies have found that people who drink a beverage with some sodium tend to drink more of it. If the drink tastes good, athletes and exercisers will want to drink it and thereby meet their fluid needs.

For people who are exercising to lose weight, however, drinking a full quart of a sports drink may in fact simply resupply the amount of calories they expended during a 40-minute aerobic class or 30 minutes of swimming or biking. For these people, plain water is a better choice.

Vitamins and Minerals for Physical Activity

Your muscles burn food and oxygen to make energy. How well they burn these fuels depends, however, on your supply of vitamins and minerals. Without small amounts of these potent substances, your muscles' ability to work is compromised.

The Vitamins
Vitamins are the links and regulators of energy-producing and muscle-building pathways. Without them, your muscles' ability to convert food energy to body energy is hindered, and muscle protein formation is slowed. Table 11-6 lists a few vitamins and minerals and their exercise-supporting functions.

The B vitamins are of special interest to athletes and exercisers because they govern the energy-producing reactions of metabolism. Needs for these vitamins increase proportionally with energy expenditure. A person who expends 4,000 calories per day needs twice as much of the B vitamins as someone who expends 2,000 calories.

TABLE 11-6 Exercise-Related Functions of Vitamins and Minerals

Vitamin or Mineral	Function
Thiamin, riboflavin, pantothenic acid, niacin, magnesium	Energy-releasing reactions
Vitamin B_6, zinc	Building of muscle protein
Folate, vitamin B_{12}, copper	Building of red blood cells to carry oxygen
Biotin	Fat and glycogen synthesis
Vitamin C	Collagen formation for joint and other tissue integrity; antioxidant ability may reduce oxidative tissue damage
Vitamin E	Protect cell membranes from oxidative damage
Iron	Transport of oxygen in blood and in muscle tissue
Calcium, vitamin D, vitamin A, phosphorus	Building of bone structure; muscle contractions; nerve transmissions
Sodium, potassium, chloride	Maintenance of fluid balance; transmission of nerve impulses for muscle contraction
Chromium	Assistance in insulin's glucose-storage function
Magnesium	Cardiac and other muscle contraction

Note: This is just a sampling. All vitamins and minerals play indispensable roles in exercise.

A well-balanced diet that meets athletes' energy needs and that features complex carbohydrate-rich foods will ensure B vitamin intakes proportional to energy intake.

Researchers are presently studying the protective effects of antioxidants on recovery from exercise and performance. Because the body uses oxygen at a higher rate during exercise, the generation of free radicals and the potential for exercise-induced tissue damage increase in the body. Although more research is needed, preliminary studies support a role for the antioxidant nutrients, which may enhance recovery from exercise by reducing exercise-induced oxidative injury.[23] Meeting the recommendation of eating five or more fruits and vegetables per day will help athletes meet recommended intakes for the antioxidant nutrients.

The Minerals Iron is a core component of the body's oxygen taxi service: hemoglobin and myoglobin. A lack of oxygen compromises the muscles' ability to perform. Iron deficiency has not been reported as a problem for fitness enthusiasts who exercise moderately.[24] Male and female endurance athletes, though, may be prone to developing mild iron deficiency, diagnosed by low blood ferritin levels (a measure of the body's store of iron). Menstruating female athletes are at particular risk: growth and menstruation combined with strenuous training can take a toll on a woman's iron stores.[25]

A combination of factors increases an athlete's chances of depleting his or her iron stores. Inadequate dietary intakes of iron-rich foods combined with iron losses due to physical activity can compromise iron status. Physical activity may cause increased iron losses in sweat, feces, and urine, plus increased destruction of red blood cells that occurs during exercise. (Chapter 8 contains numerous suggestions for obtaining sufficient iron from foods.)

Sometimes iron deficiencies can be corrected only with iron supplements. If you are concerned about your iron level, see a physician. Iron supplements should not be taken without medical supervision. High iron intakes can induce deficiencies of trace minerals, such as copper and zinc, and can produce an iron overload in some people.

An apparent anemia—sometimes called **sports anemia**—also can occur in athletes. This condition does not reflect a reduction in the blood's iron supply, but

Muscles grow in response to work, not to eating protein. Several vitamins regulate the muscle-building pathways.

Eliza Snow/iStock Photo

sports anemia a temporary condition of low blood hemoglobin level, associated with the early stage of athletic training.

The condition characterized by the potentially fatal combination of disordered eating, amenorrhea, and low bone density is referred to as the female athlete triad.

stress fracture bone damage or breakage caused by stress on bone surfaces during exercise.

amenorrhea cessation of menstruation associated with strenuous athletic training.

rather indicates an increase in the blood plasma volume.[26] This occurs because athletic training causes the kidneys to conserve sodium and water. In other words, the extra blood volume dilutes the concentration of iron, thereby making it seem as if the blood does not contain enough of the mineral. Sports anemia is considered a temporary state and probably reflects a normal adaptation to physical training.

The Bones and Exercise   Bones absorb great stresses during exercise, and like the muscles, they respond by growing thicker and stronger. Weight-bearing activities—running, walking, dancing, rope skipping, or activities such as strength training in which significant muscular force can be generated against the long bones of the body—encourage bone development. A bone that is not strong enough to withstand the strain of an athletic exertion can break, causing what has become known as a **stress fracture**. When a person suffers such a break, there are three probable causes. The first is unbalanced muscle development, which allows strong muscles to pull against the bone opposed only by weaker, undeveloped muscles, thereby leaving the bone susceptible to fractures. The second cause of stress fractures is bone weakness caused by inadequate calcium intake. A possible third cause, which occurs in women who have ceased menstruating, is when reduced estrogen concentration leads to bone mineral loss and therefore to fragile bones.

Balanced muscle development can protect the bones from undue stresses. Each set of muscles pulling against bone should be kept in check by an equally strong set of opposing muscles. Thus, when you work one set of muscles in training, you should also work the opposing muscles. For example, if you work your back and leg muscles (by jogging or walking, for example), also work your abdominal muscles, too (do sit-ups). Bones, like muscles, take time to develop strength. To avoid the likelihood of stress fractures, give your bones and muscles plenty of time to build up to one level of performance before moving up to the next level. Maintaining adequate calcium intake throughout life may be one of the primary defenses against developing weak bones.

Some women who exercise strenuously cease to menstruate, a condition called **amenorrhea**. Such women have lower than normal amounts of estrogen, a hormone essential for maintaining the integrity of the bones. With low estrogen levels, the mineral structures of the bones are rapidly dismantled, weakening the skeleton. Women who have athletic amenorrhea are at risk for stress fractures now and adult bone loss later in life.[27] To reverse the condition, they should not stop exercising altogether, because reasonable amounts of exercise may be a key defense against bone depletion. They should, however, seek evaluation from a health care provider who specializes in sports medicine to find the cause of and receive treatment for their amenorrhea.

Eating disorders are sometimes related to athletic amenorrhea, and a logical part of diagnosis is to look carefully at the woman's diet. It could be that a diet too low in calories, coupled with low body fat stores and strenuous activity, sets the stage for amenorrhea to develop. In such cases, calcium intakes between 1,000 and 1,300 milligrams per day may help protect the bones.

The best nutrition prescription for peak performance is a well-balanced diet. Although no eating plan meets every athlete's needs, certain components are common to all well-balanced diets. The athlete's diet should account for increased energy needs, vitamin and mineral needs, the relative efficiency of various foods as fuels, and current knowledge about long-term health. An eating plan that supplies 60 percent of calories from complex carbohydrate, 15 percent of calories from protein, and 25 percent of calories from fat supplies a proper fuel mix to the muscles of athletes and fitness enthusiasts and will also maintain health.[28] Two critical nutrition periods for the athlete are the training diet and the pre-competition diet.

Planning the Diet

- A diet rich in complex carbohydrate and low in fat provides the best balance of nutrients for health and also the best support for physical activity. The following table shows some sample balanced eating plans for athletes who wish to increase their carbohydrate intake along with their calories.

- Choose foods to provide nutrients as well as calories—extra milk for calcium and riboflavin, many vegetables for B vitamins, meat or alternates for iron and other vitamins and minerals, and whole grains for magnesium and chromium. The photos on page 354 show examples of high-carbohydrate meals for the athlete.

- An athlete may be able to eat more food by consuming it in six or eight meals each day rather than in three or four meals. Large snacks of milkshakes, dried fruits, peanut butter sandwiches, or cheese and crackers can add substantial calories and nutrients.

The Pre-Game Meal

- The best choices for the meal before a competitive event are foods that are high in carbohydrate and low in fat, protein, and fiber. Fat and protein slow the stomach's emptying, and the waste products generated during protein metabolism require that too much water be excreted with them.

- Fiber intake is not desirable immediately before physical exertion, because it stays in the digestive tract too long and draws water out of the blood.

- A high-carbohydrate meal supports blood glucose levels during competition. Olympic training tables are laden with foods such as breads, whole-grain cereals, pasta, rice, potatoes, and fruit juices.

- For pre-game meals and snacks, choose foods such as grape juice, apricot nectar, pineapple juice, Jell-O, sherbet, popsicles, raisins, apricots, figs, dates, jams and toast, pancakes with syrup, honey, pasta, baked white or sweet potatoes, steamed vegetables, low-fat frozen yogurt, angel food or sponge cake with fresh fruit, and graham crackers.

- Stay away from higher-fat foods such as meats, cheese, nuts, gravies, cream, French fries, muffins, croissants, biscuits, butter, potato chips, pies, and ice cream.

- Include plenty of fluids—two or more 8-ounce glasses of water or juice per meal—to ensure adequate hydration.

- Any meal should be finished at least 2 to 4 hours before the event, because digestion requires routing the blood supply to the digestive tract to pick up nutrients. By the time the contest begins, the circulating blood should be freed from that task and, instead, should be available for carrying oxygen and fuel to the muscles.

(Continued)

High-Carbohydrate Eating Patterns for Various Energy Levels

Use the number of servings indicated to arrive at the specified energy levels.[a]

Food Group	Calorie Level					
	1,500	2,000	2,500	3,000	3,500	4,000
Dairy	3	3	4	4	4	4
Fruits	5	6	7	9	10	12
Vegetables	3	3	3	5	6	7
Grains	7	11	16	18	20	24
Oils[b]	2	3	5	6	8	10
Protein Foods[c]	5	5	5	5	6	6
Percentage of carbohydrate	58	58	63	64	60	62

[a] Refer to Table 2-6 in Chapter 2 for serving sizes.
[b] A serving of fat is equivalent to 1 tsp butter, margarine, or oil.
[c] Meat servings are given as total ounces of meat; a typical serving includes 2 to 3 ounces.

Sample Meals for High-Carbohydrate Intakes for the Exercise Enthusiast at Two Calorie Levels

Breakfast
1 c coffee
8 oz low-fat milk
2 pieces whole-wheat toast
4 tsp jelly
½ c strawberries
½ c orange juice
1 c oatmeal and raisins with 2 tsp brown sugar

Breakfast
1 c coffee
½ c strawberries
8 oz nonfat milk
1 c oatmeal and raisins

Morning Snack
4 tsp trail mix

Morning Snack
4 tsp trail mix

Lunch
12 oz iced tea with sugar
1 orange
1 banana
2 beef and bean burritos

Lunch
12 oz iced tea with sugar
1 orange
1 beef and bean burrito

Afternoon Snack
A smoothie made from 12 oz nonfat milk, 1 frozen banana
1 apple
4 rye wafers with 1 oz low-fat cheese

Dinner
½ c sherbet
1 c spinach salad with 1 tbsp dressing
8 oz nonfat milk
¼ tomato
1 c broccoli
½ c noodles with parsley and 2 tsp butter
4 oz salmon

All © Felicia Martinez/PhotoEdit

Dinner
8 oz low-fat milk
1 c sherbet
1 c spinach salad with 1 tbsp dressing
1 dinner roll with 2 tsp butter
¼ tomato
1 c broccoli
4 oz salmon
¾ c noodles with parsley and 2 tsp butter

Total Calories: 3,119
61% cal from carbohydrates, 24% cal from fat, 15% cal from protein

Total Calories: 1,759
57% cal from carbohydrates, 24% cal from fat, 19% cal from protein

Competitors in the ancient Greek Olympiad reportedly used mushrooms and herbs.[29] Since then, virtually every food has at one time or another been touted as the "magic bullet" that will enhance performance. Athletes have been known to swallow everything from bee pollen to brewer's yeast to kelp to wheat germ in their quest to gain the competitive edge. Although the idea of using pills and potions to achieve peak performance (to run faster or jump higher) may be seductive, scientific evidence supporting such claims is sorely lacking.[30]

Can nutritional supplements enhance the benefits I achieve from my everyday workouts?

Most so-called **ergogenic aids**—that is, substances that increase the ability to exercise harder—are costly versions of vitamins, minerals, sugar, and other substances that can be provided easily by a balanced diet.[31] Table 11-7 describes some of the many substances currently promoted as ergogenic aids.

For example, bee pollen is simply a mixture of protein, carbohydrate, a bit of fat, and a few vitamins and minerals.[32] Though touted by one manufacturer as a "natural and balanced source of extra energy" appreciated by "athletes worldwide," it has been tested at Louisiana State University among both runners and swimmers and found to confer no benefit whatsoever on an athlete's training or performance abilities.[33] The same goes for chromium picolinate, promoted by some to increase lean body mass and delay fatigue due to its role in glucose utilization. Although chromium is necessary for muscle function by transferring glucose from the blood to the muscle cells, true chromium deficiencies are rare to nonexistent. To date, there is no evidence that chromium supplements improve athletic performance in healthy individuals who do not have a chromium deficiency.[34]

Athletes rank as prime targets of amino acid supplement manufacturers, whose products promote false hopes.

© Felicia Martinez/PhotoEdit

Surveys have found that more than 50 percent of athletes use a vitamin or mineral supplement, although no evidence exists that doing so improves performance.[35] More than 40 years of research has provided no strong evidence that popping vitamins and minerals increases energy or athletic prowess of adequately nourished people. Except for iron, vitamin and mineral deficiencies are rare among athletes. Because most athletes eat more food than non-athletes eat to meet their increased energy demands, the extra food intake usually provides the additional vitamins, minerals, and other beneficial substances they need, if they choose a well-balanced diet.

Of course, when athletes firmly hold that one or another ergogenic aid does indeed improve performance, convincing them otherwise can be extremely difficult. One reason for this is their profound belief that a particular substance can actually produce a psychological benefit. This phenomenon is known as the **placebo effect**.

I see many claims about amino acids and athletic ability. Can I improve my performance by using amino acid supplements?

Athletes rank as prime targets for amino acid supplement manufacturers. Pick up any copy of one of the bodybuilding magazines, and you'll probably see ads for supplements packed with "free-form," "predigested," and "peptide-bond" amino acids touted as optimum sources of protein for athletes. Scientific-sounding names notwithstanding, such products have never been proven to increase muscle size or enhance athletic prowess. Consider that one comparison of U.S. Marine officer candidates given protein supplements with another set of trainees who received a placebo indicated that the groups performed equally well before, during, and after the program.[36] (Refer to the Spotlight feature in Chapter 1 for some tips on how to spot fraudulent nutritional products.)

An additional important fact is that your body can't store extra amino acids, whether they come from food you eat or from supplements. Your body converts the excess into fat. This conversion generates urea, which increases your body's need for water. Increased urination of urea can lead to dehydration, which impedes training and performance.

What risks are associated with amino acid supplements?

The FDA recently asked a panel of experts to review the safety of amino acids currently available in the marketplace. The panel's conclusions underscore the need for better regulation of these supplements. For instance, the panel found that the labels of most amino acid supplements failed to carry vital information such as suggested doses, shelf life, and contraindications for use of the product. In addition, the panel identified certain groups of people who may be at particularly high risk for suffering health problems as a result of swallowing amino acid supplements. Children and teenagers, for example, may not grow properly if they take amino acid pills or powders. That's because young, underdeveloped

(Continued)

TABLE 11-7   A Sampling of Popular Ergogenic Aids[a,b]

Ergogenic Aids with Unproven Claims

Amino acids (for example, arginine, ornithine, glycine) nonessential amino acids falsely promoted to increase muscle mass and strength by stimulating growth hormone and insulin. Individual amino acids do not significantly increase muscle mass or growth hormone. Weight lifting and endurance training do.

Anabolic steroids synthetic male hormones (related to testosterone) that stimulate growth of body tissues, with many adverse effects (see the footnote on page 357).

Bee pollen mixture of bee saliva, plant nectar, and pollen touted falsely to enhance athletic performance. May cause allergic reactions in people with sensitivity to bee stings and honey allergies.

Carnitine a compound synthesized in the body from two amino acids (lysine and methionine) and required in fat metabolism. Falsely claimed to increase the use of fatty acids and spare glycogen during exercise, delay fatigue, and decrease body fat. The body produces sufficient amounts on its own. No evidence that supplementation in healthy people improves energy or enhances fat loss.

Chromium picolinate Chromium is an essential component of the glucose tolerance factor, which facilitates the action of insulin in the body. Picolinate is a natural derivative of the amino acid tryptophan. Falsely promoted to increase muscle mass, decrease body fat, enhance energy, and promote weight loss. Choose instead a diet rich in whole, unprocessed foods.

Coenzyme Q10 a lipid made by the body and used by cells in energy metabolism; falsely touted to increase exercise performance and stamina in athletes; potential antioxidant role. May increase oxygen use and stamina in patients with heart disease, but no significant effect is seen in healthy athletes.

DHEA (dehydroepiandrosterone) a precursor of the hormones testosterone and estrogen; falsely promoted to increase production of testosterone, build muscle, burn fat, and delay the effects of aging. Long-term effects unknown; self-supplementation not recommended.

Ginseng a collective term used to describe several species of plants, belonging to the genus Panax, that contain bioactive compounds in their roots. Falsely touted to enhance exercise endurance and boost energy. There is a lack of well-controlled research; what little has been done has yielded inconclusive evidence for the benefits of ginseng. The potential for adverse drug–herb or herb–herb interactions with ginseng exists.[a]

Pyruvate a three-carbon compound derived from the breakdown of glucose for energy in the body. Falsely promoted to increase fat burning and endurance. Side effects include intestinal gas and diarrhea, which could interfere with performance.

Ergogenic Aids with Some (Not All) Scientific Support for Claims

Creatine A nitrogen-containing substance made by the body that combines with phosphate to form the high-energy compound creatine phosphate (CP). CP is stored and used by muscle for ATP production. Some (not all) studies show that creatine may increase CP content in muscles and improve short-term (< 30 seconds) strenuous exercise performance (for instance, sprinting, weight lifting). Long-term effects are unknown.

Caffeine a stimulant that increases blood levels of epinephrine. Caffeine is promoted for improved endurance and utilization of fatty acids during exercise. Consuming 2 to 3 cups of coffee (equal to 3 to 6 milligrams of caffeine per kilogram body weight) 1 hour before exercise may improve endurance performance. High caffeine consumption may cause dehydration, headache, nausea, muscle tremors, and fast heart rate.

HMB (beta-hydroxy-beta-methylbutyrate) a metabolite of the branched-chain amino acid leucine that is promoted to increase muscle mass and strength by preventing muscle damage or speeding up muscle repair during resistance training. More research on long-term safety and effectiveness is needed.

Phosphate salt a salt with claims for improved endurance. Found to increase a substance in red blood cells (diphosphoglycerate) and enhance the cell's ability to deliver oxygen to muscle cells and reduce levels of disabling lactic acid in elite athletes. However, more research on safety and efficacy of phosphate loading is needed. Excess can cause loss of bone calcium.

Sodium bicarbonate baking soda is touted to buffer lactic acid in the body and thereby reduce pain and improve maximal-level anaerobic performance. High sodium load may cause diarrhea in users. Effects of repeated ingestion are unknown.

[a] See also Table 7-16 in Chapter 7 for information on herbal supplements.

[b] For more information, see M. Dunford and M. Smith, Dietary supplements and ergogenic aids, in M. Dunford (ed.), *Sports Nutrition: A Practice Manual for Professionals*, 4th ed. (Chicago: American Dietetic Association, 2006), pp. 116–41; S. A. Sarubin, *The Health Professional's Guide to Popular Dietary Supplements*, 3rd ed. (Chicago: American Dietetic Association, 2006).

bodies may metabolize amino acids differently than adult bodies, possibly leaving young people more vulnerable to harmful effects of excess amino acids.

The panel also found that much of the advertising and product label information about the effectiveness of amino acid supplements is based on anecdotes rather than grounded in careful, scientific research. In fact, the lack of data to support the usefulness of many amino acid supplements has been a concern for years.

Health risks aside, consumers should also note that special protein supplements are typically very expensive. Foods, in contrast, supply ample amounts of protein at a fraction of the cost. One glass of milk, a serving of rice and beans, or a 3-ounce portion of chicken, for example, provides a generous helping of all nine essential amino acids for less than half the price of a dose of most amino acid tablets, liquids, or powders.

What about arginine and other amino acid products advertised for weight control? Do these products work?

Arginine is an amino acid that has been promoted as "causing weight loss overnight" by stimulating secretion of a substance called *human growth hormone*, which in turn supposedly promotes weight loss. Although it's true that arginine can prompt the release of the hormone, it does so only when people take whopping doses that are unlikely to be found in supplements. Furthermore, even if a person were to take enough arginine to prompt a surge of the hormone in the body, he or she wouldn't automatically shed pounds. Human growth hormone has not been found to cause weight loss. Thus, claims that arginine "burns fat" are spurious at best. An FDA advisory panel on over-the-counter weight loss products investigated arginine along with 11 other amino acids touted as diet aids—cystine, histidine, isoleucine, leucine, L-lysine, methionine, phenylalanine, threonine, tryptophan, tyrosine, and valine—and found no basis for the claims about the effectiveness of any of these products in controlling weight.

Can anabolic steroids help me increase the size and strength of my muscles?

Psychological impact aside, pill popping may be harmful in some cases. Swallowing supplements known as **anabolic steroids**—synthetic hormones that appear to help build muscle—can be dangerous.

A popular practice particularly among weight lifters and bodybuilders, steroid abuse often begins around the age of 18 years.[37] Although steroids may help increase muscular size and strength in some people, they can also bring about numerous side effects, including acne, liver abnormalities, temporary infertility, and offensive outbursts often referred to as "roid rages."*

*Side effects of steroids include acne, anxiety, blood clots, blood poisoning, cancer, diarrhea, dizziness, fatigue, heart disease, hypertension, irreversible baldness in women, jaundice, kidney damage, liver damage, male pattern baldness, mood swings, nausea, oily skin, prostate enlargement, psychotic depression, shrunken testicles, sterility (reversible), stroke, stunted growth in adolescents, swelling of feet or lower legs, and yellowing of the eyes or skin.

Among adults, many effects of steroid use are reversible. Unfortunately, adolescents aren't so lucky. Several studies show that adolescent steroid users may suffer the serious consequences of premature skeletal maturation, decreased spermatogenesis, and elevated risk of injury.[38]

Along with being unhealthful, steroid use is considered unethical by domestic and international sports organizations such as the American College of Sports Medicine and the International Olympic Committee. As a case in point, track star Ben Johnson lost his Olympic gold medal for the 100-meter sprint in 1988 after officials discovered he had been using steroids.[39]

What about energy drinks? They seem perfect for athletic performance!

Carbohydrates, vitamins, minerals, and hydration all in one bottle . . . sounds too good to be true! Well, for the athlete who uses "energy drinks" for an extra edge, it probably is.[40] For the past 10 years, energy drinks have increased in variety and market share of beverage sales across the United States. Their formulation of high amounts of caffeine, simple carbohydrates, and mixtures of vitamins and minerals has caused many people to use them as a "pick me up" to get through their day. In particular, manufacturers target athletes because of their desire to gain an advantage over the competition. Unfortunately, however, energy drinks are not properly formulated to increase athletic performance.

Caffeine   Thought of as the central ingredient in almost all "energy drinks," caffeine is a central nervous system stimulant that has been shown to increase athletic ability slightly when consumed in moderate amounts. However, higher doses are considered "doping" in most athletic organizations, and they cause unwanted side effects such as light-headedness and twitchiness, along with laxative and diuretic effects. Even though energy drinks do not all have exactly the same ingredients, nearly all contain high concentrations of caffeine.[41]

Carbohydrates   Although carbohydrates are the preferred source of fuel for athletes, high levels of carbohydrates in an energy drink pose undesirable side effects. One problem is that absorption of water is slowed considerably, putting the athlete's hydration status at risk. Furthermore, like caffeine, simple carbohydrate ingestion immediately before physical exertion might cause gastrointestinal distress and create a laxative effect.

Vitamins and minerals   There is little chance that an athlete who consumes a normal diet will be at risk for vitamin or mineral deficiency. Accordingly, adding nutrients to beverages has never been proven to markedly increase athletic performance. A daily multivitamin and a balanced diet are far less costly than relying on fortified energy drinks as a source of vitamins and minerals.

Other ingredients   It is important to realize that these drinks sometimes include additional ingredients such as amino acids and herbal supplements. Even though they may be catchy items to list on the label, few, if any, have been proven to increase athletic

(Continued)

performance. In fact, herbal ingredients may increase the risk of serious drug–nutrient interactions if the athlete is also taking medication.[42]

What questions should I ask if I'm considering consuming energy drinks?[43]

- Does the drink contain herbal ingredients that will affect a medication I am currently taking?

- Does any research back up the claimed or proposed benefits of this beverage?

- Does the beverage label include a Nutrition Facts panel? If not, avoid this product!

- Can I do anything else to positively affect my energy level when training? (For example, am I eating a balanced diet, getting enough rest and hydration, and training for the proper length of time?)

- Is the cost of the product really worth what I am getting?

- How high is the caffeine level? Will it cause me to fail a doping test?

Some athletes use caffeine or alcohol instead of popping pills to promote athletic prowess. Can these substances improve my athletic performance?

Many exercisers drink caffeine-containing beverages such as coffee, tea, or cola to enhance performance and endurance. Caffeine apparently stimulates the release of fats into the blood that the body can then use instead of glycogen as a source of energy. Thus, the glycogen is "spared," or saved, for later use, and the amount of time an exerciser can endure physical activity before running out of fuel is prolonged.

The glycogen-sparing effect of caffeine, however, is beneficial only for athletes who exercise for more than 1½ to 2 hours at a time. As noted earlier, the muscles generally store enough glycogen to fuel as much as 90 minutes of activity. Moreover, even endurance athletes can experience certain downsides to consuming caffeine. Because caffeine is a diuretic, it promotes frequent urination and fluid loss that can lead to dehydration. In addition, caffeine can induce rapid heart rate and jitters, which can interfere with performance. Athletes would also do well to remember that caffeine is a drug that neither the American College of Sports Medicine nor the International Olympic Committee condones for use among athletes.

Along with caffeinated beverages, alcoholic drinks are often touted as choice fluids for athletes. Beer, for example, is sometimes portrayed as the perfect carbohydrate-containing complement to both before- and after-competition meals. Despite such images, alcoholic drinks rank as poor sources of fluid and energy, for several reasons. For one, alcohol is a diuretic that can bring about fluid loss and dehydration. More importantly, the amount of alcohol in just one beer or glass of wine depresses the nervous system, thereby slowing an athlete's reaction time and interfering with reflexes and coordination. Also, one can of beer provides only 50 carbohydrate calories. The rest of the calories come from alcohol, which must be metabolized by your liver, not your muscles. The American College

An endless array of ergogenic aids are marketed to athletes and other sports enthusiasts. Although big on claims, few are based on scientific evidence.

of Sports Medicine and the American Dietetic Association both conclude that use of alcohol hinders performance. (Chapter 9 presents a detailed explanation of how alcohol affects the body.)

The special supplements discussed here are just a sampling of the many "magic" pills and potions promoted to athletes. A nutritious diet and regular physical activity enhance performance far better than these products that are supposed to help you gain the competitive edge. If you're in doubt about a particular product you see touted as an ergogenic aid, ask yourself some of the following questions:

1. Is the promised action of the product based on magical thinking? ("Develop a trim body with no exercise.")
2. Does the promotion claim that "doctors agree" or "research has determined," without clarification? (Which doctors? What research?)
3. Does the promoter use scare tactics to pressure you into buying the product? ("It's the only one available without poisons.")
4. Is the product advertised as having a multitude of different beneficial effects? ("Makes bigger muscles; gives that pumped-up feeling; improves digestion, coordination, and breathing.")
5. Is the product available only from the sponsor by mail order and with payment in advance?
6. Does the promoter use many case histories or testimonials from grateful users?

Every yes answer is a point against the claimant—a warning signal that you are dealing with misinformation. Three or more points is a sure sign of quackery.

Miniglossary

ergogenic aids anything that helps increase the capacity to work or exercise (ergo = work; genic = give rise to).

placebo effect an improvement in a person's sense of well-being or physical health in response to the use of a placebo (a substance having no medicinal properties or medicinal effects).

anabolic steroids synthetic male hormones with a chemical structure similar to that of cholesterol; such hormones have wide-ranging effects on body functioning.

Nutrition on the Web

www.cengage.com/nutrition/boyle/personalnutrition8e

Go to the site to check for the latest updates to chapter topics or to access links to related websites.

www.nal.usda.gov/fnic

The Food and Nutrition Information Center (FNIC) provides credible and accurate information. Under Browse by Subject:

- Select Lifecycle Nutrition, scroll down, and click on Fitness, Sports and Sports Nutrition.
- Select Dietary Supplements, scroll down, and click on Ergogenic Aids.

www.cdc.gov/nccdphp/sgr/sgr.htm

The Centers for Disease Control website contains the Surgeon General's Report on Physical Activity and Health and other links related to health and physical activity.

www.fitness.gov

The President's Council on Fitness, Sports, and Nutrition website contains information about healthy eating, fitness, physical activity, and sports.

www.smallstep.gov

Select the Get Active tab at this government website to access practical fitness resources including an interactive activity tracker.

www.sportsci.org

This website for *Sportscience* allows access to information about sport research.

www.acsm.org

Click on the News tab to access research and publications from the American College of Sports Medicine (ACSM) website.

www.shapeup.org/fitness

This website provides practical advice on fitness and nutrition.

www.cdc.gov/nccdphp/dnpa

The Centers for Disease Control (CDC) website contains information on topics pertaining to fitness.

www.nhlbi.nih.gov/health/public/heart/obesity/lose_wt/index.htm

The National Heart Lung and Blood Institute website provides Aim for a Healthy Weight.

www.gssiweb.com

The Gatorade Sports Science Institute provides updates on exercise science, dietary supplements, sports drinks, and eating disorders in athletes.

www.ods.od.nih.gov/Health_Information/IBIDS.aspx

Search the International Bibliographic Information on Dietary Supplements (IBIDS) database for individual ingredients found in ergogenic aids.

www.ncahf.org

The National Council Against Health Fraud website offers current information on ergogenic aids and nutrition fads.

www.fda.gov/ForConsumers/ConsumerUpdates/ucm244206.htm

The Food and Drug Administration website provides resources to protect against health fraud.

www.mayoclinic.com

The Mayo Clinic maintains this website, and it provides a variety of health information. Click on the Healthy Living tab and then select Fitness.

www.acefitness.org

The American Council on Exercise provides fact sheets, an online newsletter, and an information resource center.

www.kidshealth.org

This website provides information about nutrition and fitness for children and teens. Choose the Parents Site (for Parents) and then click on the Nutrition and Fitness link.

Menu of Online Study Tools

A variety of study tools for this chapter are available at our website to deepen your understanding of chapter concepts. Go to

www.cengage.com/sso

to find

- Practice tests
- Flashcards
- Glossary
- Web links
- Animations
- Chapter summaries, learning objectives, and crossword puzzles.

Diet Analysis+ in Action

Go to the Track Activity portion of your Diet Analysis+ profile. If you haven't already done so, fill in at least one day's activity. Now go to the Energy Balance report for the day on which you tracked your actual activity. Did you consume more, fewer, or the same number of calories as you burned? If you want to burn more calories, look at the Scorecard activity in this chapter (you will find it on p. 343) for ideas about various types of activities. Choose one of those activities and include it in your activity profile for a second day of tracking. Look again at the Energy Balance report to see how your calorie intake matched up with your calorie expenditure. After doing this exercise, what changes might you make in your weekly activity to expend more calories? What changes would you have to make in your schedule to accommodate this level of activity?

The Life Cycle: Conception through the Later Years

Ask Yourself . . .

Which of the following statements about nutrition are true, and which are false? For each false statement, what is true?

1. The poor nutrition of a pregnant woman can impair the health of her grandchild, even after that child has grown up.
2. A woman needs twice as many calories per day in late pregnancy as she did before she was pregnant.
3. Even one alcoholic beverage, if taken at the wrong time during pregnancy, can damage the development of the nervous system in the unborn infant.
4. A woman who craves a food during pregnancy instinctively knows that she needs the nutrients in that food.
5. Substances in a mother's milk can protect the infant against certain diseases to which the mother has been exposed.
6. If a child loses his or her appetite, the caretaker must insist that the child eat his or her meals anyway.
7. School lunches provide all of the nutrients children need in a day.
8. For older adults, age-related weight gain is inevitable.
9. The number of older people is declining.
10. As you grow older, you need more calories, but fewer vitamins and minerals, to stay healthy.

Answers found on the following page.

Nutrition shares the responsibility for maintaining good health with other lifestyle factors. The complete prescription for good health presented in Chapter 1 reads as follows: Avoid excess alcohol, don't smoke, maintain a healthy weight, exercise regularly, get regular sleep, and eat nutritious, regular meals. A person who practices good health habits can expect to delay the onset of even minimal disability by several years, compared with a person who practices few or none of them.[1] If you believe in accepting the things you can't control and controlling the things you can, you are in luck. Your nutritional health can be controlled, and it deserves your conscientious attention. This chapter follows people through the life cycle, considering their special nutritional needs at each stage.

How far you go in life depends on your being tender with the young, compassionate with the aged, sympathetic with the striving, and tolerant of the weak and the strong. Because someday in life you will have been all of these.

GEORGE WASHINGTON CARVER
(1864–1943, AMERICAN BOTANIST)

- Age 15 or under
- Unwanted pregnancy
- Many pregnancies close together (depletes nutrient stores)
- History of poor pregnancy outcome
- Poverty
- Lack of access to health care
- Low education level
- Inadequate diet (such as that due to food faddism or dieting)
- Iron-deficiency anemia early in pregnancy
- Cigarette smoking
- Alcohol or drug abuse
- Chronic disease requiring special diet (for example, diabetes)
- Underweight or overweight
- Insufficient or excessive weight gain in pregnancy
- Carrying twins or triplets

Notice the umbilical cord connecting this 16-week-old fetus with the placenta. The placenta is the organ inside the uterus in which maternal blood vessels lie side by side with fetal blood vessels entering it through the umbilical cord. This close association between the two circulatory systems permits the mother's bloodstream to deliver nutrients and oxygen to the fetus and to carry away fetal waste products.

placenta (pluh-SEN-tuh) the organ inside the uterus in which the mother's and fetus's circulatory systems intertwine and in which exchange of materials between maternal and fetal blood takes place. The fetus receives nutrients and oxygen across the placenta; the mother's blood picks up carbon dioxide and other waste materials to be excreted via her lungs and kidneys.

prenatal prior to birth.

postnatal after birth.

trimester one-third of the normal duration of pregnancy; the first trimester is 0 to 13 weeks, the second is 13 to 26 weeks, and the third trimester is 26 to 40 weeks.

Pregnancy: Nutrition for the Future

The only way nutrients can reach the developing fetus in the uterus is through the **placenta**, the special organ that grows inside the uterus to support the new life. If the mother's nutrient stores are inadequate early in pregnancy when the placenta is developing, the fetus will develop poorly, no matter how well the mother eats later. After getting such a poor start on life, a female child may grow up poorly equipped to support a normal pregnancy, and she, too, may bear a poorly developed infant. Thus, a woman who has poor nutrition habits during her early pregnancy can even impair the health of her *grandchild*.

Infants born to malnourished mothers are more likely to become ill, to have birth defects, and to suffer retarded mental or physical development than infants who are born to healthy women. Malnutrition in the **prenatal** and early **postnatal** periods also affects learning ability and behavior. According to the fetal origins hypothesis, if a woman's nutrient intake is under- or oversupplied—particularly at critical phases of fetal development—long-term alterations in tissue function may occur. For example, if a woman's energy intake is low during the third trimester of pregnancy, pancreatic cell development may be hindered, resulting in impaired glucose tolerance for the baby and increased risk of developing diabetes later in life.[2] Clearly, it is critical to provide the best nutrition at the early stages of life. Ideally, a woman will start pregnancy at a healthful weight, with filled nutrient stores, and with the firmly established habit of eating a balanced and varied diet. The Scorecard feature in this chapter helps women evaluate their nutritional readiness for pregnancy and identify which of their eating habits might need improvement.

Nutritional Needs of Pregnant Women
For pregnant and lactating women, nutrient needs are higher than at any other time in their adult life, and they have greater need for certain nutrients than for others (see Figure 12-1). Notice that although their nutrient needs are much higher than usual, their energy needs are not.

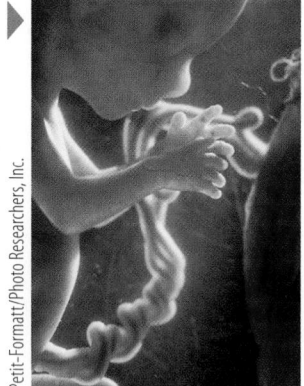

© Petit-Formatt/Photo Researchers, Inc.

To support the metabolic demands of pregnancy and fetal development, the recommended average increase is only about 17 percent higher than recommended maintenance calories. An additional 340 calories during the second **trimester** and an additional 450 calories per day during the third trimester are recommended.[3]

Nearly all nutrients are recommended in increased amounts during pregnancy and lactation. The nutrient needs of pregnancy are best met by the routine intake of a variety of foods (see Figure 12-2). The nutrients deserving special attention in the diets of pregnant women include protein, folate, iron, zinc, and calcium as well as vitamins known to be toxic in excess amounts.[4]

ASK YOURSELF ANSWERS: 1. True. 2. False. A woman needs only 17 percent more calories per day during pregnancy than she did before. 3. True. 4. False. A woman's cravings during pregnancy do not seem to reflect real physiological needs. 5. True. 6. False. A child's appetite regulates food intake to meet need; caretakers should not force food on children because this will only create conflict. 7. False. School lunch is designed to provide one-third of the nutrients schoolchildren need in a day. 8. False. Although age-related weight gain is a fact of life for many people, it is not inevitable. Consuming low-fat meals within your calorie allowance and making physical activity part of your daily routine can help. 9. False. Today, 12.4 percent of the U.S. population is 65 or older. By the year 2030, 20 percent of the population will be more than 65 years of age. The fastest-growing group is those over age 85. 10. False. Older people need fewer calories (due to lower basal metabolic rates) but the same amounts of most vitamins and minerals as younger people. Some authorities recommend higher intakes of certain nutrients (calcium, vitamin D, antioxidants, B vitamins) for older adults.

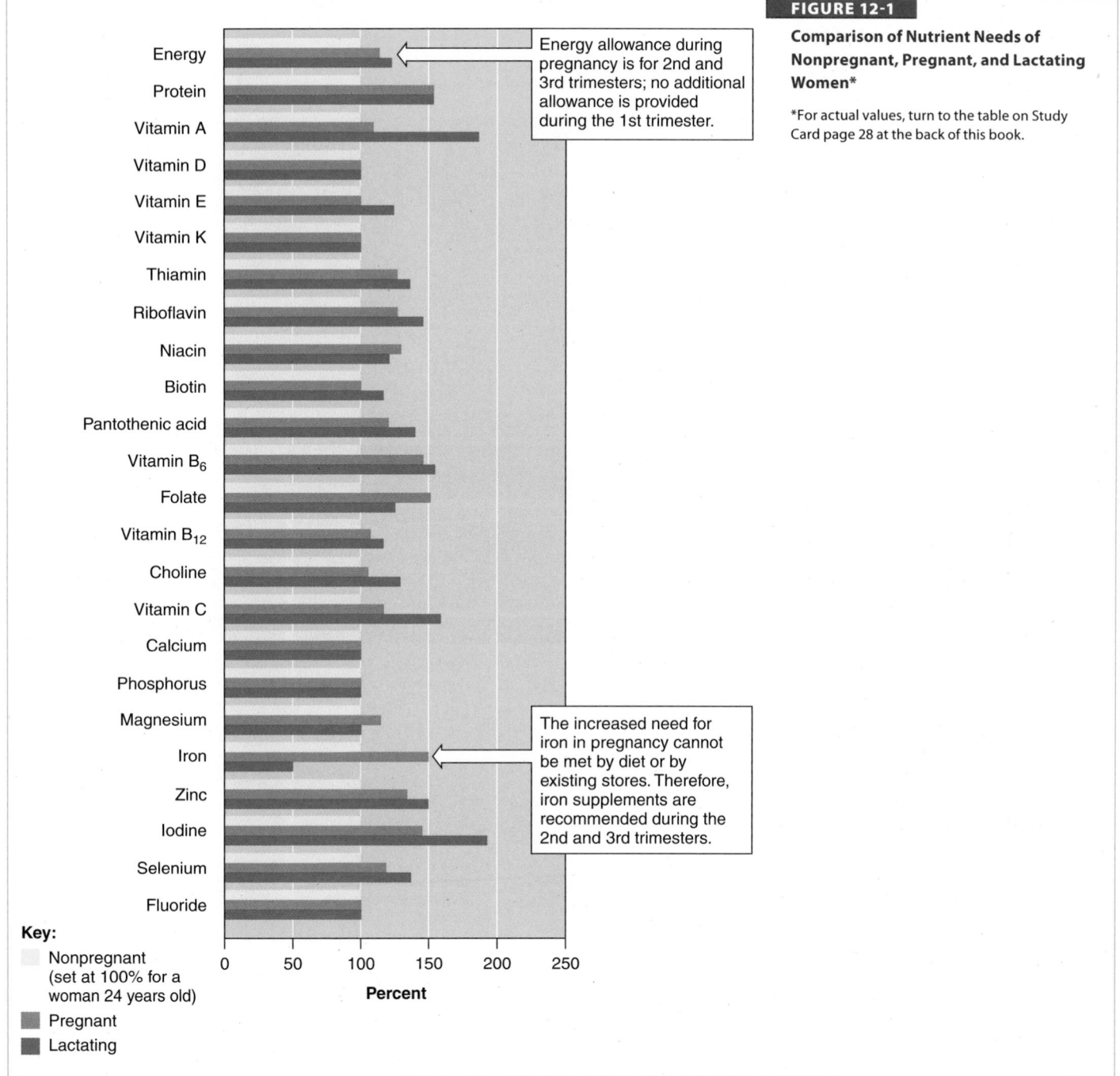

FIGURE 12-1

Comparison of Nutrient Needs of Nonpregnant, Pregnant, and Lactating Women*

*For actual values, turn to the table on Study Card page 28 at the back of this book.

Energy allowance during pregnancy is for 2nd and 3rd trimesters; no additional allowance is provided during the 1st trimester.

The increased need for iron in pregnancy cannot be met by diet or by existing stores. Therefore, iron supplements are recommended during the 2nd and 3rd trimesters.

Key:

Nonpregnant (set at 100% for a woman 24 years old)

Pregnant

Lactating

The recommended protein intake is about an additional 20 grams per day more than nonpregnant requirements. Many women already eat enough protein to cover the increased demand of pregnancy.

The pregnant woman's recommended folate intake is 50 percent greater than the normal requirement due to her large increase in blood volume and the rapid growth of the fetus. Certain studies have shown that folate supplements given around the time of conception reduce the recurrence of **neural tube defects**, such as spina bifida, in the infants of women who previously have had such births.[5] To lower the risk of neural tube birth defects, women are advised to get the recommended amounts of folate—especially *before* becoming pregnant (400 micrograms) and during the first trimester of pregnancy (600 micrograms).[6]

All refined grain products (bread, cereal, cornmeal, farina, flour, grits, pasta, and rice) are fortified with folate. The folate contained in fortified foods and supplements

neural tube defects any of a number of birth defects in the orderly formation of the neural tube during early gestation. Both the brain and the spinal cord develop from the neural tube, and defects cause various central nervous system disorders. The two main types are spina bifida (incomplete closure of the bony casing around the spinal cord) and anencephaly (a partially or completely missing brain). Because the neural tube closes before the sixth week of pregnancy, women are advised to consume adequate folate from as early as 3 months before conception.

FIGURE 12-2

A Food Guide for Pregnant and Lactating Women

^a Refer to Table 2-6 and Figure 2-5 in Chapter 2 for a summary of foods in each group and serving sizes.
^b Most pregnant and lactating women need a total of 2,200 to 2,900 calories per day; see "MyPlate for Moms" at www.ChooseMyPlate.gov.

What Should I Eat?

When you are pregnant or breastfeeding, you have special nutritional needs. Follow the plan below to help you and your baby stay healthy.

Food Group	1st Trimester	2nd and 3rd Trimesters	Breast-feeding only	Breast-feeding plus formula	What counts as 1 cup or 1 ounce?	Remember to...
	Eat this amount from each group daily.*					
Fruits	2 cups	2 cups	2 cups	2 cups	1 cup fruit or juice ½ cup dried fruit	*Focus on fruits—* Eat a variety of fruits.
Vegetables	2½ cups	3 cups	3 cups	3 cups	1 cup raw or cooked vegetables or juice 2 cups raw leafy vegetables	*Vary your veggies—* Eat more dark-green and orange vegetables and cooked dry beans.
Grains	6 ounces	8 ounces	8 ounces	7 ounces	1 slice bread 1 ounce ready-to-eat cereal ½ cup cooked pasta, rice, or cereal	*Make half your grains whole—*Choose whole instead of refined grains.
Protein Foods	5½ ounces	6½ ounces	6½ ounces	6 ounces	1 ounce lean meat, poultry, or fish ¼ cup cooked dry beans ½ ounce nuts or 1 egg 1 tablespoon peanut butter	*Go lean with protein—* Choose low-fat or lean meats and poultry.
Dairy	3 cups	3 cups	3 cups	3 cups	1 cup milk 8 ounces yogurt 1½ ounces cheese 2 ounces processed cheese	*Get your calcium-rich foods—*Go low-fat or fat-free when you choose milk, yogurt, and cheese.

*These amounts are for an average pregnant woman. You may need more or less than the average. Check with your doctor to make sure you are gaining weight as you should.

In each food group, choose foods that are low in "extras"—solid fats and added sugars.

Pregnant women and women who may become pregnant should not drink alcohol. Any amount of alcohol during pregnancy could cause problems for your baby.

Most doctors recommend that pregnant women take a prenatal vitamin and mineral supplement every day **in addition to** eating a healthy diet. This is so you and your baby get enough folic acid, iron, and other nutrients. But don't overdo it. Taking too much can be harmful.

Get a Food Guide Plan for Moms designed just for you.
Go to www.ChooseMyPlate.gov for your Plan and more.
Click on "Pregnancy and Breastfeeding."

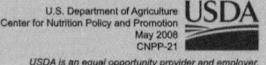
U.S. Department of Agriculture
Center for Nutrition Policy and Promotion
May 2008
CNPP-21
USDA is an equal opportunity provider and employer.

is almost twice as well absorbed as the folate that is naturally available in such foods as green leafy vegetables, citrus fruits, whole-grain breads, or legumes. Women are advised to choose a variety of foods naturally high in folate as well as foods that have been fortified with folate. Because high folate intake can mask a vitamin B_{12} deficiency, folate intake should not exceed 1 milligram per day. If the woman's dietary intake of folate is low, a 300-microgram folate supplement is recommended.[7]

The body conserves iron even more than usual during pregnancy. Menstruation ceases, and absorption of iron increases up to threefold. However, the developing fetus draws on its mother's iron stores to create stores of its own to carry it through the first 3 to 6 months of life. This drain on the mother's iron supply can precipitate a deficiency. Furthermore, she will lose blood when she gives birth.

The recommended iron intake during pregnancy is 27 milligrams per day—an increase of 50 percent above standard recommendations—to meet maternal and fetal needs. Because iron deficiency is a common problem among nonpregnant women, many women begin pregnancy with diminished iron stores. For this reason, an iron supplement of 30 milligrams ferrous iron daily during the second and third trimesters is recommended. To facilitate absorption from the supplement, iron should be taken between meals with vitamin C-rich fruit juices or at bedtime.

The Dietary Reference Intakes (DRI) for calcium during pregnancy are 1,300 milligrams for teens and 1,000 milligrams for adults over 18 years of age. Intestinal absorption of calcium doubles early in pregnancy, and the mineral is stored in the mother's bones. Later, during the last trimester of pregnancy, when fetal skeletal

The neural tube (outlined by the delicate red arteries) has successfully closed after only 6 weeks of pregnancy.

Foods containing calcium:

- Milk
- Other dairy products (yogurt, cheese)
- Green leafy vegetables
- Legumes
- Fortified juice or soy milk
- Certain brands of tofu

Foods containing iron:

- Red meat, fish, and other meat
- Dried fruits
- Legumes
- Whole-grain and fortified breads and cereals
- Dark-green vegetables

TABLE 12-1	Recommended Weight Gain for Pregnant Women	
Weight Category*		**Recommended Gain (lb)**
Underweight		28–40
Normal weight		25–35
Overweight		15–25
Obesity		11–20

*Underweight is defined as BMI < 18.5; normal weight as BMI 18.5 to 24.9; overweight as BMI 25.0 to 29.9; and obesity as BMI ≥ 30.0.

SOURCE: Adapted from Institute of Medicine and National Research Council of the National Academies, *Weight Gain During Pregnancy: Reexamining the Guidelines* (Washington, D.C.: The National Academies Press, 2009).

low birthweight (LBW) a birthweight of 5½ lb (2,500 g) or less; used as a predictor of poor health in the newborn and as a probable indicator of poor nutrition status of the mother during and/or before pregnancy. Normal birthweight for a full-term baby is 6½ to 8¾ lb (about 3,000 to 4,000 g). LBW infants are either born premature or have suffered growth failure in the uterus; they may or may not be born early, but they are small.

growth is at maximum and teeth are being formed, the fetus draws approximately 300 milligrams of calcium per day from the maternal blood supply.[8] Dairy products are a recommended calcium source because they include valuable vitamin D and riboflavin as well. This increased calcium intake is particularly important for women under 25 years of age because their bone mineral density is still increasing. For the woman who normally consumes less than 600 milligrams of calcium a day, a 600-milligram supplement of calcium per day during pregnancy is recommended.

Routine use of vitamin supplements during pregnancy is not advised, and excess intakes of certain vitamins, notably vitamins A and D, can cause fetal malformations.[9] Supplements should not contain more than one to two times the recommended levels. For example, taking more than three times the recommended intake of vitamin A per day early in pregnancy may cause malformations of the newborn.[10]

Nutrient supplements may be appropriate in certain circumstances, however. For example, a multivitamin-mineral supplement (providing 100 percent of the DRI) beginning in the second trimester may be recommended for women who do not ordinarily consume an adequate diet or who are in high-risk categories (such as women carrying more than one fetus, heavy cigarette smokers, and alcohol and other drug abusers).[11]

Because energy needs increase less than nutrient needs, the pregnant woman must select foods of high nutrient density. A woman who already eats well can simply increase her servings of nutritious foods to meet her increasing nutrient needs (refer to Figure 12-2).

Maternal Weight Gain   Normal weight gain and adequate nutrition support the health of the mother and the development of the fetus. The recommendations for weight gain take into account a mother's pre-pregnancy weight for height or body mass index (BMI), as shown in Table 12-1. A woman who begins pregnancy at a healthful weight should gain between 25 and 35 pounds. A woman pregnant with twins needs to gain 35 to 45 pounds. An underweight woman needs to gain between 28 and 40 pounds; an obese woman, between 11 and 20 pounds. Weight gains at the upper end of the range are recommended for pregnant teenagers because of their increased risk of low weight gains and delivery of low-birthweight infants.

Low weight gain in pregnancy is associated with increased risk of delivering a **low-birthweight (LBW)** infant. Not all small babies are unhealthy, but birthweight and length of gestation are the primary indicators of an infant's future health status. A low-birthweight baby is more likely than a normal-weight baby to experience complications during delivery and has a statistically greater chance of having physical and mental birth defects, developing diseases, and dying early in life.

Excessive weight gain during pregnancy increases the risk of complications during labor and delivery as well as of postpartum obesity.[12] Obese women also have an increased risk for complications during pregnancy, including hypertension and gestational diabetes.

An infant at birth weighs only about 6½ to 8 pounds, but the mother's built-up body tissues (blood, blood vessels, muscle, fat stores, and others) that provided a healthful environment for the developing fetus weigh more than 20 pounds (see Table 12-2). Weight gain should be lowest during the first trimester—3 to 4 pounds for the entire trimester—followed by a steady gain of about 1 pound per week thereafter. If a woman gains more than the recommended amount of weight early in pregnancy, however, she should not try to diet in the last weeks. Weight-loss dieting during pregnancy is not recommended. A *sudden* large weight gain may indicate the onset of pregnancy-induced hypertension (discussed later). A woman who experiences this type of weight gain should see her health care provider.

Some of the weight a woman gains in pregnancy is lost at delivery. For the woman who has gained only the recommended 25 to 35 pounds, the remainder is generally lost within a few months as her blood volume returns to normal and she loses the fluids she has accumulated.

Practices to Avoid   Optimal pregnancy outcome is influenced by maternal nutrient intake, but it can also be affected by maternal use of nonfood substances, excess caffeine, low-calorie diets, megadoses of certain vitamins, tobacco, alcohol, and illicit drugs. See Chapter 9 for recommendations regarding caffeine and pregnancy.

The term **pica** refers to the craving of nonfood items that have little or no nutritional value. When pica occurs during pregnancy, it typically involves the consumption of dirt, clay, or laundry starch; but episodes of pica have included compulsive ingestion of such things as ice, paper, and coffee grounds.[13] The medical consequences of pica can include malnutrition (as nonfood items replace nutritious foods in the diet), obesity (from overconsumption of items such as starch), poisoning (from ingestion of toxic compounds), or intestinal obstruction (due to consuming large amounts of clay or starch).

Some other practices are also truly harmful, and their potential impact on pregnancy outcome is too great to risk. For example, low-carbohydrate or low-calorie diets that cause ketosis can deprive the fetus's brain of needed glucose and cause congenital deformity. Protein deprivation can cause children's height and head circumference to diminish markedly and irreversibly.

Another harmful maternal practice is smoking, which restricts the blood supply to the growing fetus, thereby limiting the delivery of nutrients and removal of wastes. Smoking stunts fetal growth, thus increasing the risk of premature delivery, low birthweight, retarded development, and spontaneous abortions. Smoking is responsible for 20 to 30 percent of all low-birth-weight deliveries in the United States.[14] Sudden infant death syndrome (SIDS) has been linked to smoking during pregnancy and to postnatal exposure to secondhand smoke.[15]

As discussed in Chapter 9, research confirms that alcohol consumption adversely affects fetal development. Even as few as one or two drinks daily can cause **fetal alcohol syndrome (FAS)**—irreversible brain damage and mental and physical retardation in the fetus. The most severe impact of maternal drinking is likely to occur in the first month, before the woman even is sure she is pregnant. This preventable condition (FAS) is estimated to occur in approximately 1 to 2 infants per 1,000 live births and is the leading known cause of mental retardation in the United States.[16] Birth defects, low birth-weight, and spontaneous abortions occur more often in pregnancies of women who drink even as little as 2 ounces of alcohol daily during pregnancy. Accumulating evidence that even one drink may be too much has led the American Academy of Pediatrics

For most women, the surest way to have a healthy baby is to follow a healthy lifestyle: Get early prenatal care, eat a well-balanced diet, exercise regularly with your doctor's permission, and avoid cigarettes, alcohol, and other drugs.

Ken Hurst/Shutterstock.com

TABLE 12-2	An Example of the Pregnant Woman's Weight Gain	
Development		**Weight Gain (lb)**
Infant at birth		7–8
Placenta		1
Increase in mother's blood volume to supply placenta		4
Increase in mother's fluid volume		4
Increase in size of mother's uterus and the muscles to support it		2
Increase in size of mother's breasts		2
Fluid to surround infant in amniotic sac		2
Mother's fat stores (varies)		3–12
Total		25–35

Note: The pattern of gain should be about 1 pound a month for the first 3 months and 1 pound a week thereafter. Different patterns of weight gain are suggested for underweight, normal-weight, and overweight women.

pica the craving of nonfood items such as clay, ice, and laundry starch; does not appear to be limited to any particular geographic area, race, sex, culture, or social status.

fetal alcohol syndrome (FAS) the cluster of symptoms seen in an infant or child whose mother consumed excess alcohol during pregnancy; includes retarded growth, impaired development of the central nervous system, and facial malformations. A lesser condition—called *fetal alcohol effect (FAE)*—causes learning impairment and other more subtle abnormalities in infants exposed to alcohol during pregnancy.

to take the position that women should stop drinking as soon as they *plan* to become pregnant.[17]

Drugs other than alcohol taken during pregnancy can also cause birth defects. A particularly dramatic example is the acne medication Accutane (isotretinoin), which causes major deformities during fetal development. Pregnant women should avoid taking all drugs except on the advice of their physician.

Although many pregnant women view herbal products as safe and natural alternatives to over-the-counter medications, most herbal products have not been formally evaluated in terms of how they might affect pregnancy. Therefore, pregnant women are advised to avoid using herbal supplements during pregnancy.[18] See the Nutrition Action feature in Chapter 7 for more information on the use of herbal remedies, including safety and regulatory issues.

Pregnant women, as well as young children, are particularly vulnerable to the effects of environmental contaminants, notably lead and mercury. These substances can severely impair an unborn child's developing nervous system. To reduce exposure to lead during pregnancy, the Food and Drug Administration (FDA) advises pregnant women to avoid frequent use of ceramic mugs for hot beverages such as coffee or tea and not to use lead crystal glassware daily.

To reduce exposure to mercury, the FDA advises that children under 12 years of age, pregnant or lactating women, and women planning pregnancy within a year should do the following:

- Consume 8 to 12 ounces of seafood per week from a variety of seafood types. Due to their methyl mercury content, limit white (albacore) tuna to 6 ounces per week and avoid eating large ocean fish such as shark, swordfish, king mackerel, and tilefish. These long-lived predatory fish accumulate the highest levels of mercury and pose the greatest risk to people who eat them regularly.

Common Nutrition-Related Problems of Pregnancy   Common physical problems in pregnancy include morning sickness and, later, constipation. The nausea of morning sickness seems unavoidable because it arises from the hormonal changes taking place early in pregnancy, but it can sometimes be alleviated. Suggested strategies to alleviate the nausea and vomiting of morning sickness include eating soda crackers, hard candies, or other dry, starchy foods before getting up in the morning; eating small, frequent meals as soon as you feel hungry; and avoiding any specific food (especially highly seasoned foods or foods with strong odors) that triggers nausea or vomiting.[19]

Later, as the hormones of pregnancy alter her muscle tone and the growing fetus crowds her intestinal organs, an expectant mother may complain of constipation. A high-fiber diet, plentiful fluid intake, and regular exercise will help relieve this condition.

Women's cravings during pregnancy reflect alterations in taste and smell sensitivities, rather than real physiological needs. If a woman craves pickles and chocolate sauce at 2 o'clock in the morning, for example, it is probably not because she lacks a combination of nutrients uniquely supplied by these foods. The woman is, however, expressing a need as real and as important as her need for nutrients: the need for support, understanding, and love. More serious problems requiring control during pregnancy include hypertension and diabetes.

Hypertension in Pregnancy   Ideally, a woman with preexisting hypertension has her blood pressure under control before becoming pregnant. Otherwise, she may have an increased risk of delivering a low-birthweight baby. Some women develop a *transient hypertension of pregnancy* during the second half of their pregnancy. Usually, this is a mild form of hypertension with no adverse effects on pregnancy outcome, and the mother's blood pressure returns to normal shortly after the baby is born. Sometimes, however, high blood pressure in a pregnant woman signals the onset of **pregnancy-induced hypertension (PIH)**. Preeclampsia and eclampsia are hypertensive conditions induced by pregnancy. **Preeclampsia** is characterized by high blood pressure, protein in the urine, and generalized edema that may cause sudden, large weight gain from retained water. Fluid retention alone, which is quite common in pregnant women, is not sufficient for a diagnosis of preeclampsia.[20] Warning signs of preeclampsia include severe and constant headaches; sudden weight gain (1 lb/day); swelling of face, hands, and feet; dizziness; and blurred vision. **Eclampsia**, the most severe form of pregnancy-induced hypertension, is characterized by convulsions that may lead to coma. PIH can retard fetal growth and cause the placenta to separate from the uterus, resulting in stillbirth. Both conditions present serious health risks to mother and fetus and demand careful medical treatment.

Diabetes   Infants born to women with diabetes are at greater risk for premature birth, congenital defects, excessively high birthweight, and respiratory distress syndrome.[21] Metabolic control of diabetes before and throughout pregnancy is critical. In some women, pregnancy can alter carbohydrate metabolism and precipitate a condition known as **gestational diabetes**. The abnormal blood glucose levels usually occur during the second half of pregnancy and return to normal after pregnancy for most women with the condition. However, some women with gestational diabetes are at increased risk of developing type 2 diabetes after pregnancy, especially if too much weight is gained. Risk factors include a previous history of gestational diabetes, glucose in the urine, obesity, and a strong family history of diabetes. Some women with gestational diabetes have the classic symptoms of diabetes—increased thirst, hunger, urination, weakness—but other women have no warning signs. For this reason, the American Diabetes Association recommends that all women be assessed for risk of gestational diabetes at their first prenatal examination.

Risk factors for gestational diabetes:

- Glucose in the urine
- Family history of diabetes
- Member of an ethnic or racial group with a high prevalence of diabetes (Hispanic, Native American, Asian, African American, or Pacific Islander)
- Obesity
- Personal history of gestational diabetes

Adolescent Pregnancy

More than 757,000 teenagers become pregnant in the United States each year—one out of every eight babies is born to a teenager—and more than a tenth of these mothers are under the age of 15.[22] The complexity of social, emotional, and physical factors makes teen pregnancy one of the most challenging situations for meeting nutritional needs. According to a position paper from the American Dietetic Association, pregnant adolescents are nutritionally at risk and require early intervention and special care throughout pregnancy.[23] Medical and nutritional risks are particularly high when the teenager is within 2 years of menarche (usually 15 years of age or younger). Risks for pregnant teens include higher rates of pregnancy-induced hypertension, iron-deficiency anemia, premature birth, stillbirths, low-birthweight infants, and prolonged labor.

Pregnancy places adolescent girls, who are already at risk for nutrition problems, at even greater risk because of the increased energy and nutrient demands of pregnancy. To support the needs of both mother and infant, adolescents are encouraged to strive for pregnancy weight gains at the upper end of the ranges recommended for pregnant women (refer to Table 12-1). Those who gain between 30 and 35 pounds during pregnancy have lower risks of delivering low-birthweight infants.[24] Adequate nutrition can substantially improve the course and outcome of adolescent pregnancy.[25] A model program for providing nutritional help to teenage mothers is, among others, the Women's, Infants' and Children's (WIC) program, a federally funded program that provides nutrition education and low-cost, nutritious foods to low-income pregnant women, mothers, and their children.

pregnancy-induced hypertension (PIH) high blood pressure that develops during the second half of pregnancy.

preeclampsia a condition characterized by hypertension, fluid retention, and protein in the urine.

eclampsia a severe extension of preeclampsia characterized by convulsions; may lead to coma.

gestational diabetes the appearance of abnormal glucose tolerance during pregnancy.

Nutrition of the Breastfeeding Mother Adequate nutrition for the mother makes a highly significant contribution to successful lactation. A nursing mother produces 30 ounces of milk a day, on the average, with wide variations possible. Current recommendations suggest that 400 of the calories needed to support this milk production come from added food and that the rest come from the stores of fat the mother's body has accumulated during pregnancy for this purpose. (Figure 12-2 shows a food pattern that will meet the lactating woman's nutrient needs.)

The period of lactation is the natural time for a woman to lose the extra body fat she has accumulated during pregnancy. Once lactation is established, if her food choices are judicious, a nursing mother can tolerate a calorie deficit and a gradual loss of weight (1 pound per week) without any effect on her milk output. Fat is mobilized slowly, however, and an energy deficit that is too large will inhibit lactation.

Healthy Infants

The growth of infants directly reflects their nutritional well-being and is the major indicator of their nutrition status. A baby grows faster during the first year of life than it ever will again, doubling its birthweight during the first 4 to 6 months and tripling its birthweight by the end of the first year. Adequate nutrition during infancy is critical to support this rapid rate of growth and development. Clearly, from the viewpoint of nutrition, the first year is the most important year of a person's life. This section provides an overview of nutrient requirements, current recommendations for feeding healthy infants, and the relationship between infant feeding and selected pediatric nutrition issues.

Milk for the Infant: Breastfeeding Breastfeeding has both emotional and physical health advantages. Emotional bonding is facilitated by many events and behaviors of mother and infant during the early months and years; one of the first can be breastfeeding.

During the first 2 or 3 days of lactation, the breasts produce **colostrum**, a premilk substance containing antibodies and white cells from the mother's blood. Because it contains immunity factors, colostrum helps protect the newborn infant from those infections against which the mother has developed an immunity—precisely those in the environment against which the infant needs protection. Upon entering the infant's body with the milk, these antibodies inactivate bacteria within the digestive tract, where they could otherwise cause intestinal infections. Breast milk also contains antibodies, although not as much as colostrum contains.

Colostrum and breast milk both contain the **bifidus factor** that favors growth of "friendly" probiotic bacteria, such as *Lactobacillus bifidus*, in the infant's digestive tract so that other, harmful bacteria cannot grow there. (Probiotics are discussed in Chapter 3.) Breast milk also contains the powerful antibacterial agent **lactoferrin**, as well as other factors, including several enzymes, several hormones, and lipids that help protect the infant against infection. Breast-fed infants have lower rates of hospital admissions, ear infections, diarrhea, rashes, allergies, asthmatic disease, and other health problems than bottle-fed infants.[26]

Breast milk is tailor-made to meet the nutrient needs of the young infant. It offers its carbohydrate in the easy-to-assimilate form of lactose; its fat contains a generous proportion of the essential omega-6 fatty acid linoleic acid; and its protein, alpha-lactalbumin, is one that the infant can easily digest. Depending on the mother's diet, breast milk can also deliver beneficial omega-3 fatty acids to the infant. Except for vitamin D, its vitamin contents are ample. As for minerals, calcium, phosphorus, and magnesium are present in amounts appropriate for the rate of growth expected in a human infant, and breast milk is low in sodium. Its iron is highly absorbable, and the presence of a zinc-binding protein facilitates absorption of the zinc it contains.

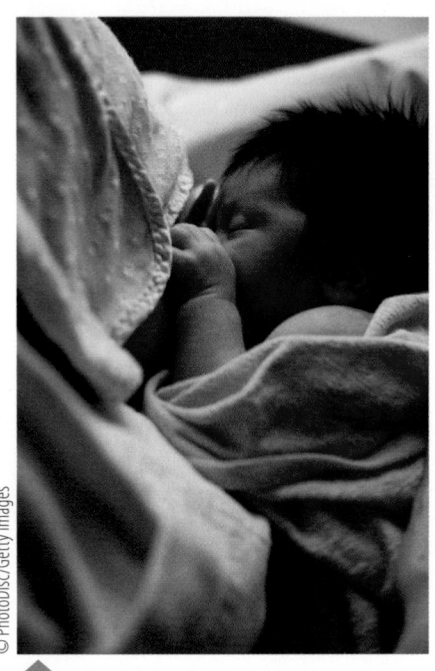

Breast milk is a very special substance.

colostrum (kuh-LAHS-trum) a milk-like secretion from the breast, rich in protective factors, present during the first day or so after delivery and before milk appears.

bifidus factor (BIFF-id-us) a factor in colostrum and breast milk that favors growth in the infant's intestinal tract of the "friendly" bacteria *Lactobacillus bifidus*; colonization with friendly bacteria means that other, less desirable intestinal inhabitants will not flourish.

lactoferrin (lak-toe-FERR-in) a factor in breast milk that binds and helps absorb iron and keeps it from supporting the growth of the infant's intestinal bacteria.

The American Academy of Pediatrics recommends that infants receive exclusive breastfeeding for the first 6 months of life, and breastfeeding with complementary foods for at least 12 months.[27] Despite the health benefits, however, the incidence of breastfeeding in the United States declined from the mid-1980s until the early 1990s.[28] Today, approximately 69 percent of mothers currently initiate breastfeeding—lower than the *Healthy People 2020* goal, which is "to increase to at least 82 percent the proportion of mothers who breastfeed their infants." Only 44 percent of mothers, however, are still breastfeeding at 6 months. Analysis of data from a survey of mothers indicates that breastfeeding rates continue to be the highest among women who are older, well educated, relatively affluent, and/or live in the western United States. Among those least likely to breastfeed are women who are in low-income groups, black, under age 20, and/or live in the southeastern United States.[29] Table 12-3 lists 10 steps hospitals can take to promote breastfeeding.

A number of barriers to achieving the nation's health objective for increasing the incidence of breastfeeding have been identified. They include lack of knowledge, an absence of work policies and facilities that support lactating women (for example, extended maternity leave, part-time employment, facilities for pumping breast milk or breastfeeding, and on-site child care), and the portrayal of bottle-feeding rather than breastfeeding as the norm in the American society.

Contraindications to Breastfeeding   Breastfeeding is not recommended in certain circumstances. A woman must not breastfeed if she has a communicable disease such as active, untreated tuberculosis, or if she takes a medication that is secreted in breast milk and is known to affect the infant. Drug addicts—including alcohol abusers—are capable of ingesting such high doses of their drug that their infants can become addicts by way of breast milk. In such cases, breastfeeding is also contraindicated.

Because the human immunodeficiency virus (HIV), which is responsible for AIDS, can be passed to an infant through breast milk, HIV-infected mothers should not breastfeed. Unfortunately, sometimes the nutritional and immunologic benefits of breast milk outweigh the risks of HIV transmission through breastfeeding. Such is the case in many developing countries, where lack of safe drinking water to make formula increases the risk of diarrhea and disease for the infant. However, the World Health Organization encourages HIV-positive mothers in developing countries to feed formula whenever their circumstances assure a regular supply of infant formula and safe drinking water.[30]

TABLE 12-3	Baby-Friendly Hospitals: 10 Steps to Successful Breastfeeding

To promote breastfeeding, every maternity facility should:

- Develop a written breastfeeding policy that is routinely communicated to all health care staff.
- Train all health care staff in the skills necessary to implement the breastfeeding policy.
- Inform all pregnant women about the benefits and management of breastfeeding.
- Help mothers initiate breastfeeding within ½ hour of birth.
- Show mothers how to breastfeed and how to maintain lactation, even if they need to be separated from their infants.
- Give newborn infants no food or drink other than breast milk, unless medically indicated.
- Practice rooming-in, allowing mothers and infants to remain together 24 hours a day.
- Encourage breastfeeding on demand.
- Give no artificial nipples or pacifiers to breastfeeding infants.*
- Foster the establishment of breastfeeding support groups and refer mothers to them at discharge from the facility.

*Compared with nonusers, infants who use pacifiers breastfeed less frequently and stop breastfeeding at a younger age.

SOURCE: United Nations Children's Fund and World Health Organization.

Most prescription drugs taken by mothers do not reach nursing infants in sufficiently large quantities to affect them adversely. As a precaution, however, a nursing mother should consult with her physician before taking any drug. *Minimal* use of alcohol between feedings is compatible with breastfeeding, but both alcohol and nicotine can enter breast milk. These substances also adversely affect maternal production and composition of breast milk. It is important not to expose the infant to secondhand smoke, due to the risk of health problems such as impaired growth, respiratory problems, and sudden infant death. Coffee drinking is fine in moderation (two to three cups a day), as is eating foods such as garlic and spices. Sometimes, a particular food may affect the baby's liking for the mother's milk, but this matter requires individual detective work. (Examples are chocolate for some babies, excess caffeine for others, and foods that cause gas in the mother for still others.) If a woman has an ordinary cold, she can continue nursing without worry. The infant may catch it from her anyway, but may actually be less susceptible than a bottle-fed baby would be, thanks to the immunologic protection offered by breast milk.

Feeding Formula   Like the breastfeeding mother, the mother who offers formula to her baby has reasons for making her choice, and her feelings should be honored. Infant formulas are manufactured to approximate the nutrient composition of breast milk. National and international standards have been set for the nutrient content of infant formulas. The immunologic protection of breast milk, however, cannot be duplicated. One major advantage of formula feeding is the relief it provides to the mother whose attempts at breastfeeding have met with frustration. Formula provides adequate nourishment for the infant, and a mother can choose this alternative with confidence.

Many mothers breastfeed at first and then wean the baby within the first 1 to 6 months. When a woman chooses to wean her infant during the first 12 months of life, it is imperative that she shift to *infant formula*. Cow's milk, whether whole or reduced-fat, is not recommended during the first year of life, according to the American Academy of Pediatrics (AAP).[31] Cow's milk is an inappropriate replacement for breast milk or infant formula because it provides insufficient vitamin C and iron and excessive sodium and protein. Feeding cow's milk to infants may increase the risk of iron-deficiency anemia and cow's milk protein allergy.

For infants with special problems, many variations of infant formulas are available. Special formulas based on soy protein are available for infants allergic to milk protein, and formulas with the lactose replaced can be used for infants with lactose intolerance.

Supplements for the Infant   Breast milk or formula and the infant's own internal stores will meet most nutrient needs for the first 6 months. Thereafter, the introduction of properly chosen juices and foods will normally keep up with the infant's changing requirements. At 6 months, infants require additional iron, preferably in the form of iron-fortified cereal.

Breast milk does not provide enough vitamin D for the infant, and vitamin D deficiency causes impaired bone mineralization in children. Manufacturers fortify infant formulas with vitamin D, but due to the low concentration of vitamin D in breast milk, pediatricians routinely prescribe a vitamin D supplement for breast-fed infants whose mothers are vitamin D deficient.

If the water supply is severely deficient in fluoride, the infant may also need supplemental fluoride after 6 months of age. The pediatrician should prescribe it, if appropriate.

Food for the Infant   The infant's rapid growth and metabolism demand an adequate supply of all essential nutrients.[32] Because of their small size, infants need smaller total amounts of the nutrients than adults do; but when comparisons are

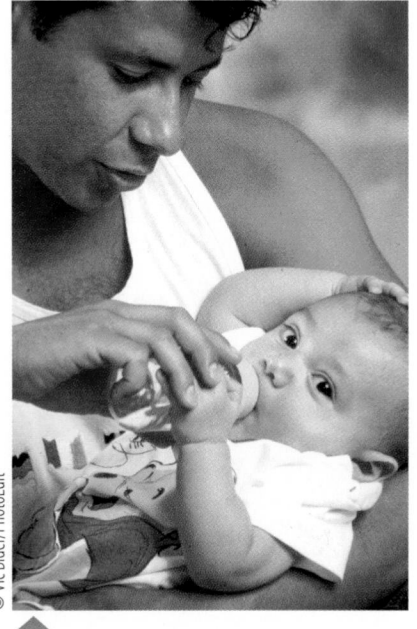

Using formula feeding or pumping breast milk into a bottle allows other family members to enjoy feeding the infant.

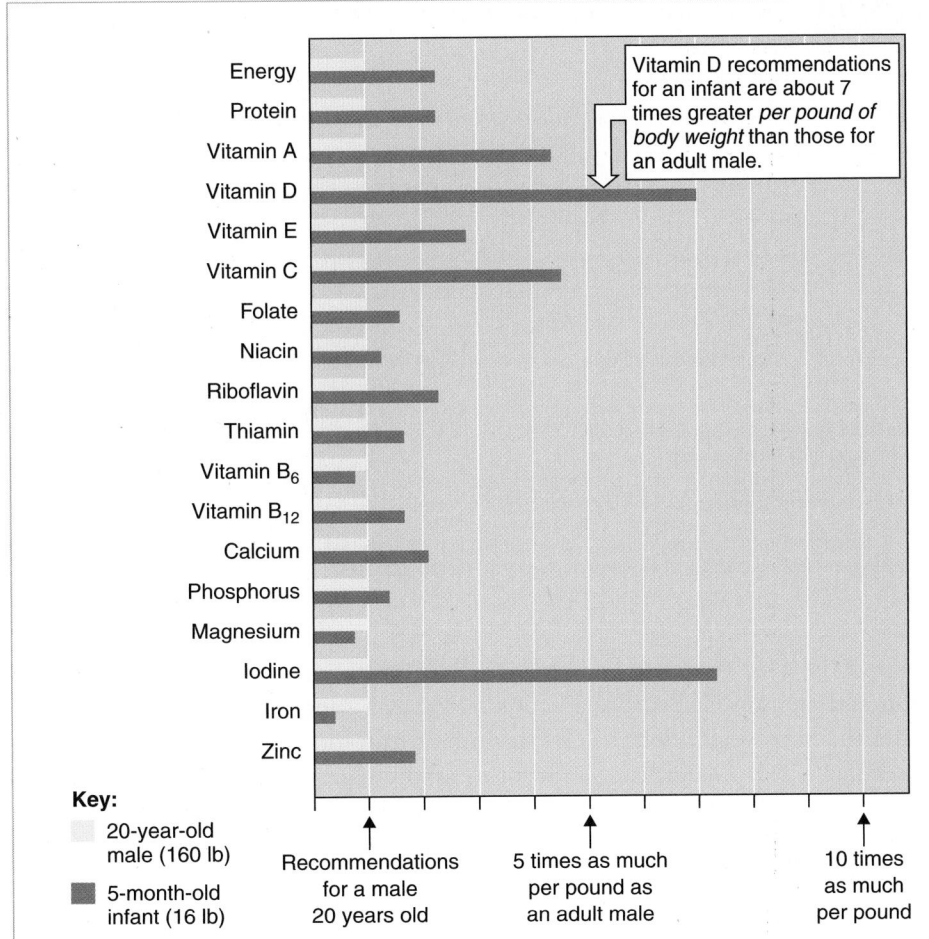

FIGURE 12-3

Recommended Intakes of Nutrients for an Infant and an Adult Compared on the Basis of Body Weight

SOURCE: Adapted with permission from E. Whitney and S. Rolfes, *Understanding Nutrition*, 12th ed. Copyright © 2011 by Wadsworth, Cengage Learning. All rights reserved.

Vitamin D recommendations for an infant are about 7 times greater *per pound of body weight* than those for an adult male.

Key:
20-year-old male (160 lb)
5-month-old infant (16 lb)

Recommendations for a male 20 years old

5 times as much per pound as an adult male

10 times as much per pound

made based on body weight, infants need more than twice as much of many of the nutrients. Figure 12-3 compares a 5-month-old baby's needs with those of an adult man. As you can see, some of the differences are extraordinary. After 6 months, calorie needs increase less rapidly as the growth rate begins to slow down, but some of the energy saved by slower growth is spent on increased activity.

The most important nutrient of all—for infants as for everyone—is the one easiest to forget: water. The younger the child, the greater the percentage of the body weight is water and the more inefficient the kidneys are at concentrating waste, making water easy to lose. Conditions that cause fluid loss, such as vomiting, diarrhea, sweating, or normal urination, can rapidly propel an infant into life-threatening dehydration unless fluid is replaced. Fluid and electrolyte imbalances caused by diarrhea and infection kill more of the world's children than any disease or disaster. Because infants can only cry, and cannot tell you what they are crying for, it is important to remember that they may need fluid and to let them drink plain water until their thirst is quenched.

Solid foods may normally be added to a baby's diet when the baby is about 6 months old, depending on readiness. The infant is developmentally ready when he or she can sit upright with support and can control head movements.

Solids should not be introduced too early, because infants are more likely to develop allergies to them in the early months. Nonetheless, all babies are different, and the program of food additions should depend on the individual baby, not on any rigid schedule. Table 12-4 presents a suggested sequence for feeding infants.

The addition of foods to a baby's diet should be governed by three considerations: the baby's nutrient needs, the baby's physical readiness to handle different forms of

Energy saved by slower growth is spent in increased activity.

Bananastock/Jupiter Images

TABLE 12-4 First Foods for the Infant

Age (months)	Food Additions
4 to 6[a]	Iron-fortified rice cereal, followed by other cereals (baby can swallow nonliquid foods now);[b] pureed fruits and vegetables
6 to 8	Mashed vegetables and fruits, infant breads and crackers; unsweetened fruit juices[c]
8 to 10	Protein foods (soft cheeses, yogurt, tofu, mashed cooked beans, finely chopped meat, fish, chicken, egg yolk), toast, teething crackers (for emerging teeth), soft-cooked vegetables, and fruit
10 to 12	Whole egg (allergies are less likely now), whole milk (at 1 year)

[a] The AAP supports exclusive breastfeeding for 6 months but acknowledges that infants may be developmentally ready for certain foods between 4 and 6 months of age.
[b] Mix with breast milk, formula, or water.
[c] All baby juices are fortified with vitamin C. Orange juice may cause allergies; apple juice may be a better juice to feed first. Offer juices in a cup to prevent nursing bottle syndrome.

SOURCE: Adapted from Committee on Nutrition, American Academy of Pediatrics, *Pediatric Nutrition Handbook* (6th ed.), ed. R. E. Kleinman (Elk Grove Village, IL: American Academy of Pediatrics, 2009).

foods, and the need to detect and control allergic reactions. Nutrients needed early are iron and vitamin C. Because a baby's stored iron supply from before birth runs out after the birthweight doubles, breast milk or iron-fortified formula, followed by iron-fortified cereals—and, later, meat or meat alternates such as legumes—are recommended. Fruits that contain vitamin C can be introduced to enhance the absorption of iron. Fruit juices can be introduced at 6 months of age and should be served in a cup rather than a bottle.

There is no room in a baby's diet for empty-calorie foods such as sweets (soda pop, baby-food desserts, candy, rich pies, and cakes). In contrast, naturally sweetened fruits and juices supply not only calories but also needed nutrients that support normal growth and development. However, the AAP recommends limiting juice consumption by infants and young children to 4 to 6 ounces per day, to avoid the displacement of other important nutrients from their diets.[33]

A baby's physical readiness to handle foods develops in many small steps. For example, the ability to swallow solid food develops at around 4 to 6 months, and experience with solid food at 6 months helps develop swallowing ability by desensitizing the gag reflex. Later still, when a baby can sit up, can handle finger foods, and is teething, hard crackers and other hard finger foods may be introduced. Such foods promote the development of manual dexterity and control of the jaw muscles. (An infant can choke on these foods, however, so an adult should keep a watchful eye during the learning process.)

An alternative for parents who want the baby to have family foods is to "blenderize" a small portion of the table food at each meal. This requires cooking without salt, however, because foods that adults prepare for themselves often contain much more salt than do commercial baby foods. Canned vegetables are inappropriate for infants because their sodium content is often too high. It is also important to take precautions against food poisoning. Honey should *never* be fed to infants because of the risk of botulism. Babies and even young children have difficulty swallowing certain foods: popcorn, whole grapes, hard candies, bite-sized hot dogs, nuts, and spoonfuls of peanut butter, for instance. An infant can easily choke on these foods, and it is not worth the risk to give such foods to infants.

At 1 year of age, the obvious food to supply most of the nutrients the baby needs is still milk; 2 to 3 cups a day are now sufficient. Infants less than 2 years old should drink whole milk rather than low-fat or fat-free milk. They need the fat and vitamins A and D of fortified whole milk until 2 years of age. The other foods—meat, iron-fortified cereal, enriched or whole-grain bread, fruits, and vegetables—should be supplied in variety and in sufficient amounts to round out total calorie needs. Ideally, the 1-year-old is sitting at the table and eating many of the same foods everyone else eats. A meal plan that meets the requirements for the 1-year-old is shown in Table 12-5.

The wise parent of a 1-year-old offers nutrition and love together. Both promote growth. It is literally true that "feeding with love" produces better growth in both weight and height of children than feeding the same food in an emotionally negative climate. It also promotes better brain development. The formation of nerve-to-nerve connections in the brain depends both on nutrients and on environmental stimulation.

The person feeding a 1-year-old should keep in mind that the baby is also developing eating habits that will persist throughout life.[34] Mealtimes should be relaxed and leisurely. Children should learn to eat slowly, pause and enjoy their table companions, and stop eating when they are full. Physical activity should be encouraged on a daily basis to promote strong skeletal and muscular development and to establish habits that will promote good health throughout life.

Nutrition-Related Problems of Infancy
Iron deficiency and food allergies are two of the most significant nutrition-related problems of infants.

Iron Deficiency Iron deficiency remains a prevalent nutritional problem for infants, although it has declined in recent years largely due to the increasing use of iron-fortified formulas. Feeding cow's milk earlier than recommended in infancy can cause iron deficiency because of its poor iron content and the potential to cause gastrointestinal blood loss in susceptible infants.[35] Other factors contributing to iron deficiency in infancy include breastfeeding for more than 6 months without providing supplemental iron, feeding infant formula that is not fortified with iron, the infant's rapid rate of growth, low birthweight, and low socioeconomic status. To prevent iron deficiency, the American Academy of Pediatrics recommends that infants be fed breast milk or iron-fortified formula for the first year of life, with appropriate foods added as shown in Table 12-4.

Food Allergies Genetics is probably the most significant factor affecting an infant's susceptibility to food allergies. At-risk infants can be identified by means of careful skin testing and by a family history. Breast milk is recommended for those infants allergic to cow's milk protein and is preferable to soy or goat's milk formulas because infants are sometimes allergic to these proteins as well. To reduce the risk of food sensitivity or allergic reactions to other foods, new foods should be introduced one at a time to facilitate prompt detection of allergies. For example, when cereals are introduced, try rice cereal first for 5 to 7 days; it causes allergy least often. Try wheat cereal last; it is the most common offender. If a cereal causes irritability from skin rash, digestive upset, or respiratory discomfort, discontinue its use before going on to the next food. Allergies aren't easy to detect. About 9 times out of 10, the allergy won't be evident immediately but will manifest itself in vague symptoms occurring up to 5 days after the offending food is eaten. Allowing several days to elapse between the introduction of each new food will provide time for the clinical symptoms to appear so that the offending food may be identified. The Nutrition Action feature in Chapter 6 provides more information about food allergies and intolerances in children and adults.

TABLE 12-5	Meal Plan for a 1-Year-Old

Breakfast
¼ c whole milk
½ c iron-fortified cereal
1–2 tbsp fruit

Snack
½ c yogurt
Teething crackers
½ c fruit

Lunch
½ c whole milk
½ c vegetables
1 egg or 1 oz chopped meat or well-cooked mashed legumes
½ c noodles

Snack
½ c whole milk
½ slice toast
1 tbsp apple butter

Supper
½ c whole milk
1 oz chopped meat or well-cooked mashed legumes
¼ c potato or rice
½ c vegetables
2 to 3 tbsp fruit

Michelle D. Milliman/Shutterstock.com

Let infants handle food as they become ready.

Early and Middle Childhood

Childhood is a critical time in human development. Children typically grow taller by 2 to 3 inches and heavier by 5 or more pounds each year between the age of 1 and adolescence. They master fine motor skills (including those related to eating and drinking), become increasingly independent, and learn to express themselves appropriately. This section describes the nutrient requirements of children and the primary nutritional problems of this population.

Growth and Nutrient Needs of Children   After age 1, a child's growth rate slows, but the body continues to change dramatically. At age 1, most babies have just learned to stand and toddle; by age 2, they can take long strides with solid confidence and are learning to run, jump, and climb. The internal change that makes these new accomplishments possible is the accumulation of a larger body mass and greater density of bone and muscle tissue. These changes are obvious in the photos at the bottom of the page.

Children generally become leaner between the ages of 6 months and 6 years, after which time there is a gradual increase in fat thickness in both males and females until puberty. Females have a greater body fat content than males at all stages of development. The energy requirements of children are determined by their individual basal metabolic rates, activity patterns, and rates of growth. Toddlers (ages 1 to 3 years) need about 1,000 calories per day. By the age of 10, children need about 2,000 calories per day. Appetite decreases markedly around the age of 1 year, in line with the great reduction in growth rate. Thereafter, the appetite fluctuates; a child will need and demand much more food during periods of rapid growth than during slow periods.

To provide the gradually increasing needs for all nutrients during the growing years, the MyPlate food guide recommends a balance among milk and dairy products, meats and meat alternates, fruits, vegetables, and grains. The interactive *Food Guide for Preschoolers* and *Food Guide for Kids* present the recommended proportion of food from each food group and focus on the importance of making healthy food choices every day (see Figure 12-4). They also provide a general guide regarding which foods should be offered on a daily basis. In addition, the www .ChooseMyPlate.gov website offers interactive games and guides to help children and parents along the way.

After the crucial first year, a parent can still do a great deal to foster the development of healthful eating habits.[36] Table 12-6 offers tips on making feeding times

 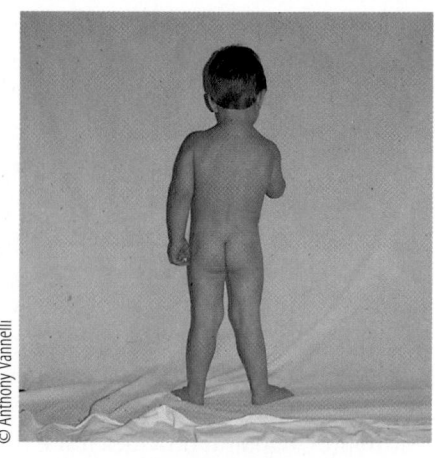

▲
1-Year-Old and 2-Year-Old
The 2-year-old has lost much of the baby fat; the muscles (especially in the back, buttocks, and legs) have firmed and strengthened, and the leg bones have lengthened and increased in density.

FIGURE 12-4

A Food Guide for Young Children

SOURCE: United States Department of Agriculture, Food and Nutrition Service, available at www.ChooseMyPlate.gov

Grains Make half your grains whole	**Vegetables** Vary your veggies	**Fruits** Focus on fruits	**Milk** Get your calcium-rich foods	**Meat & Beans** Go lean with protein
Start smart with breakfast. Look for whole-grain cereals.	Color your plate with all kinds of great-tasting veggies.	Fruits are nature's treats—sweet and delicious.	Move to the milk group to get your calcium. Calcium builds strong bones.	Eat lean or low-fat meat, chicken, turkey, and fish. Ask for it baked, broiled, or grilled—not fried.
Just because bread is brown doesn't mean it's whole grain. Search the ingredients list to make sure the first word is "whole" (like "whole wheat").	What's green and orange and tastes good? Veggies! Go dark green with broccoli and spinach, or try orange ones like carrots and sweet potatoes.	Go easy on juice and make sure it's 100%.	Look at the carton or container to make sure your milk, yogurt, or cheese is low fat or fat-free.	It's nutty, but true. Nuts, seeds, peas, and beans are all great sources of protein, too.

For a 1200-kcalorie diet (suitable for many preschoolers ages 2 to 5), include the amounts below from each food group.

Eat 4 oz. every day; at least half should be whole	**Eat 1 ½ cups every day**	**Eat 1 cup every day**	**Get 2 cups every day**	**Eat 3 oz. every day**

For a 1800-kcalorie diet (suitable for many children ages 6 to 11), include the amounts below from each food group.

Eat 6 oz. every day; at least half should be whole	**Eat 2 ½ cups every day**	**Eat 1 ½ cups every day**	**Get 3 cups every day;** for kids ages 6 to 8, it's 2 cups	**Eat 5 oz. every day**

Oils   Oils are not a food group, but you need some for good health. Get your oils from fish, nuts, and liquid oils such as corn oil, soybean oil, and canola oil.

Find your balance between food and fun
- Move more. Aim for at least 60 minutes every day, or most days.
- Walk, dance, rollerblade—it all counts. How great is that!

Fats and sugars—know your limits
- Get your fat facts and sugar smarts from the Nutrition Facts label.
- Limit solid fats as well as foods that contain them.
- Choose food and beverages low in added sugars and other kcaloric sweeteners.

enjoyable for both parent and child. The goal is to teach children to like nutritious foods in all food categories. (Nutrient intake recommendations for children are given on Study Card page 28 at the back of this book.)

Candy, cola, and other concentrated sweets must be limited in a child's diet to help ensure that the needed nutrients will be supplied by other foods. A child can't be expected to choose nutritious foods on the basis of taste alone, because our human

TABLE 12-6   Strategies to Foster Healthful Eating Habits and Happy Mealtimes

These tips may make feeding time easier and more relaxing for both parent and child:

- Schedule regular meals and snacks for toddlers; they require frequent feeding to ensure adequate intake of calories and nutrients.
- Offer a variety of foods that includes at least one that the child likes.
- Remain calm if the child leaves an entire meal untouched.
- Do not be concerned about short food jags, stretches of time when the child wants the same food over and over.
- Allow the child to eat slowly.
- Offer healthful food in a relaxed manner, and children will eat what they need. Try these suggestions for healthful snacking:
  - Keep plenty of washed and cut raw vegetables in the refrigerator. Team them up with a yogurt or bean dip.
  - Top frozen waffles and pancakes with fresh fruit for a tasty and refreshing snack.
  - Make your own frozen juice pops in an ice-cube tray.
  - Put together a large batch of cereal, pretzel, and nut mix. Divide into individual plastic bags.

SOURCE: Adapted from M. G. Hermann, *The ABCs of Children's Nutrition* (Chicago: American Dietetic Association, 1991), 3–7.

preference for sweets is innate. In contrast, an active child can enjoy the higher-calorie nutritious foods in each category: ice cream or pudding in the milk group and whole-grain crackers in the grains group. These foods, made from milk and grain, carry valuable nutrients and encourage a child to learn, appropriately, that eating is fun.

Other Factors That Influence Childhood Nutrition   Although parents do what they can to establish favorable eating behaviors during the transition from infancy to childhood, when their children enter preschool or grade school they encounter foods prepared and served by outsiders. The U.S. government funds several programs to provide nutritious, high-quality meals for children at school. School lunches are designed to meet certain requirements. As shown in Table 12-7, they must include specified servings of milk, protein-rich foods (meat, cheese, eggs, legumes, or

TABLE 12-7   School Lunch Patterns for Different Ages

Food Group	Preschool (age)		Grade School through High School (grade)		
	1 to 2	**3 to 4**	**K to 3**	**4 to 6**	**7 to 12**
Meat or meat alternate (1 serving)					
Lean meat, poultry, or fish	1 oz	1½ oz	1½ oz	2 oz	3 oz
Cheese	1 oz	1½ oz	1½ oz	2 oz	3 oz
Large egg(s)	½	¾	¾	1	1½
Cooked dry beans or peas	¼ C	⅜ C	⅜ C	½ C	¾ C
Peanut butter or other nut or seed butters	2 tbsp	3 tbsp	3 tbsp	4 tbsp	6 tbsp
Peanuts, soy nuts, tree nuts, seeds[a]	½ oz	¾ oz	¾ oz	1 oz	1½ oz
Yogurt, plain or flavored	4 oz	6 oz	6 oz	8 oz	12 oz
Vegetable and/or fruit					
2 or more servings, both to total	½ C	½ C	½ C	¾ C	¾ C
Bread or bread alternate					
Servings[b]	5 per week	8 per week	8 per week	8 per week	10 per week
Milk					
1 serving of fluid milk	¾ C	¾ C	1 C	1 C	1 C

[a] Can be used as up to ½ serving of meat, but must be accompanied by other meat/meat alternate in the meal.
[b] A serving is 1 slice bread; 1 biscuit, roll, or muffin; ½ cup cooked rice, pasta, or cereal grain, minimum of ½ serving per day for ages 1–2 and 1 serving per day for all others.

SOURCE: U.S. Department of Agriculture, *Food Program Facts—National School Lunch Program*; www.fns.usda.gov/cnd/governance/regulations.htm.

peanut butter), vegetables, fruits, and bread or other grain foods. These programs are designed to follow the recommendations of the *Dietary Guidelines for Americans* and to provide at least one-third of the recommended intakes for protein, vitamins A and C, iron, calcium, and calories.[37]

Children growing up today not only need to be fed well in the interest of growth and development but also need to learn enough about nutrition to make healthy food choices when the choices become theirs to make. Thus, it is desirable for children to learn to like nutritious foods in all of the food groups. With one exception, this liking usually develops naturally. The exception is vegetables, which young children sometimes dislike and refuse. Children prefer vegetables that are slightly undercooked and crunchy, attractive in color and shape, and easy to eat. Vegetables should be served warm, not hot, because a child's mouth is much more sensitive than an adult's. Children tend to prefer mild flavors, and they want smooth foods like mashed potatoes or pea soup to have *no lumps*. (A child wonders, with some disgust, what the lumps might be.)

Little children like to eat at little tables and to be served little portions of food. They also love to eat with other children and have been observed to stay at the table longer and eat much more when in the company of their peers. A bright, unhurried atmosphere free of conflict is also conducive to good appetite.

Ideally, each meal is preceded, not followed, by the activity the child looks forward to the most. A number of schools have discovered that children eat a much better lunch if recess occurs before rather than after the meal. When recess follows the meal, children are likely to hurry out to play, leaving food on their plates that they were hungry for and would otherwise have eaten. Before sitting down to eat, small children should be helped to wash their hands and faces to decrease the likelihood of contaminating the food with bacteria.

Many little children, both boys and girls, enjoy helping in the kitchen. Their participation provides many opportunities to encourage good food habits. Vegetables are pretty, especially when fresh, and provide opportunities to learn about color, about growing things and their seeds, about shapes and textures—all of which are fascinating to young children. Measuring, stirring, decorating, cutting, and arranging vegetables are skills even a very small child can practice with enjoyment and pride.

When introducing new foods at the table, parents are advised to offer them one at a time—and only a small amount at first. Whenever possible, the new food should be presented at the beginning of the meal, when the child is hungry. If the child is irritable, or feeling sick, don't insist; withdraw the new food and try it again a few days later. Remember, parents have inclinations and dislikes to which they feel entitled; children should be accorded the same privilege.

Nutrition-Related Problems of Childhood   Nutrition plays a critical role in the development and growth of children. However, as shown in Figure 12-5, the diet quality of most children aged 2 to 9 is less than optimal.[38] Children aged 7 to 9 have a lower diet quality than younger children, and the lower quality is associated with low fruit and poor sodium scores—perhaps because as children get older, they consume more fast food and salty snacks. Dietary quality continues to decline from childhood to adolescence, especially with the decreased consumption of vegetables, fruits, and milk, and the increased intake of soft drinks. More than two-thirds of children consume well above the recommended levels of total fat and saturated fat.[39] For all children, the percent of total fat from milk and eggs has decreased, whereas the percentage of fat from cheese and snacks has increased.[40]

National surveys show that beverage choices for children of all ages has changed from whole milk to lower-fat milk, soft drinks, and fruit and fruit-flavored drinks.[41] These changes are especially pronounced for adolescents. On any given day, a majority of all children consume soft drinks, which are the second leading energy source

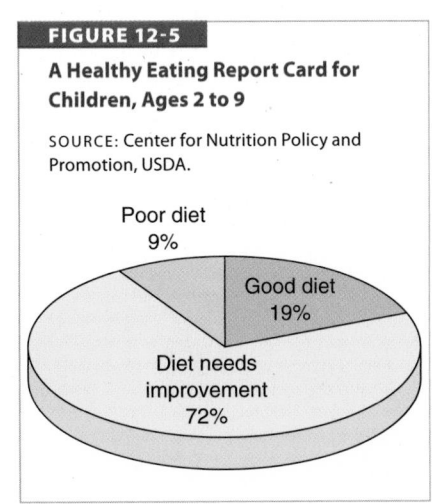

FIGURE 12-5

A Healthy Eating Report Card for Children, Ages 2 to 9

SOURCE: Center for Nutrition Policy and Promotion, USDA.

Poor diet 9%

Good diet 19%

Diet needs improvement 72%

for children ages 2 to 18. The primary energy sources for children and teens comes from sweetened beverages and a high consumption of grain products, found in dough-based dishes such as pizza, pastas, and Mexican food—often high in calories, saturated fat, and sodium and low in fiber.[42]

The most common nutrition-related problems among U.S. children include overweight and obesity, iron-deficiency anemia, and high blood cholesterol levels. The epidemic of obesity among U.S. children requires a multifaceted strategy to address the health, social, physical, and environmental issues that may prevent and treat this escalating public health problem.[43] See the Spotlight feature later in this chapter for more about the problem of obesity in children and adolescents.

Iron-Deficiency Anemia Of all nutritional disorders, other than obesity, in U.S. children, the most common is iron-deficiency anemia. It is most prevalent in low-birthweight infants, babies from 6 months to 2 years of age, and in children and adolescents from low-income families.[44] To ensure adequate iron nutrition, parents should offer an abundance of iron-rich foods such as lean meats, fish, poultry, eggs, and legumes. Grain products should be whole-grain or enriched only. Milk, beneficial as it is, is a poor iron source, and dairy products should be consumed only in the amounts needed to ensure optimal calcium intakes.

High Blood Cholesterol Considerable evidence exists that atherosclerosis begins in childhood and that this process is related to high blood cholesterol levels.[45] Children and adolescents in the United States have higher blood cholesterol levels and higher dietary intakes of saturated fat and cholesterol than children in other countries.[46] A panel of experts from major U.S. health and professional organizations has recommended that children aged 2 and older eat diets containing no more than 30 percent of total calories from fat, less than 10 percent of total calories from saturated fat, and less than 300 milligrams of cholesterol daily. However, from birth to 2 years of age, a child's fat consumption should *not* be restricted, because fat is a concentrated source of the calories needed to ensure proper physical development.

The panel also advised that youngsters and teens should have their blood cholesterol measured if they have one parent with a high blood cholesterol level.[47] For children and adolescents, a total of 200 milligrams per deciliter of blood or more is considered high; 170 to 199 is borderline high; and less than 170 is acceptable.

The Importance of Teen Nutrition

Adolescence is a time of change. Between the ages of about 10 and 18 years in girls and 12 and 20 years in boys, marked changes take place in physical, intellectual, and emotional growth and development. The maturation process is initiated and controlled by a variety of hormones, including, among others, growth hormone, prolactin, estrogen, testosterone, and the thyroid hormones. Many aspects of the maturation process are influenced by dietary intake and nutritional status. This section reviews the nutrient requirements and special nutrition-related problems of teenagers.

Nutrient Needs of Adolescents
The dramatic changes in body composition and the rate of growth that occur during early adolescence give rise to the familiar phrase "the adolescent growth spurt." The magnitude of these changes is such that the linear growth increments during adolescence can contribute about 15 to 25 percent of adult stature. The rate of weight gain can contribute anywhere from 40 to 50 percent of the adult body mass. This remarkable growth rate requires adequate intake of energy and nutrients.[48] The individual energy needs for teenagers are influenced by body size, activity levels, and biological factors affecting growth. Consult

the DRI table shown on Study Cards pages 27–29 at the back of this book for the energy and nutrient intake recommendations for adolescents.

The rates and patterns of growth for individual teenagers varies widely.[49] The only way to be sure teenagers are growing normally is to compare their heights and weights with previous measures taken at intervals and note whether reasonable progress is being made. Physical growth charts for children and adolescents and information on assessing children's growth status can be accessed at www.cdc.gov/growthcharts.

Nutritious snacks can supply added nutrients to the active teenager's diet.

Nutrition-Related Problems of Adolescents   Most adolescents in the United States are perceived as "healthy." However, many U.S. adolescents experience a variety of health and nutrition problems, some related to their risk-seeking behaviors and an inability to deal with abstract notions such as "good health" and the link between current behaviors and long-term health. Nutrition-related problems among U.S. adolescents include overweight and obesity (discussed later in this chapter's Spotlight feature), undernutrition, iron-deficiency anemia, low dietary calcium intakes, high blood cholesterol levels, dental caries, and eating disorders.

Undernutrition   Some groups of adolescents are at risk for reduced energy and food intakes, such as adolescents from low-income families and those who have run away from home or abuse alcohol or other drugs. Approximately 17 percent of youth aged 18 or younger are estimated to be living below the poverty line. African American and Hispanic teenagers are nearly twice as likely to live in poverty as are white youth. Chronic dieters are also at risk. Thirteen percent of the 17,354 females in grades 7 through 12 who were interviewed in the Minnesota Adolescent Survey reported being chronic dieters, defined as always being on a diet or having been on a diet for 10 of the previous 12 months.[50]

Iron-Deficiency Anemia   Iron needs increase during adolescence, especially for females as they start to menstruate. In boys, the requirements for absorbed iron increase because of an expanding blood volume and rise in hemoglobin concentration that accompanies the development of larger muscles and sexual maturation. (After the adolescent growth spurt, a male's need for iron falls off.) Whereas most males have an adequate iron intake during adolescence, many females between 12 and 22 years of age have iron intakes below the current recommended intake. Girls typically consume fewer total calories and less meat than boys do.

Low Calcium Intakes   Another problem nutrient during the teen years is calcium. Low intakes of calcium-rich foods during adolescence compromise peak bone mass development and increase the risk of osteoporosis later in life. Adolescents need a minimum of 1,300 milligrams of calcium each day to achieve an optimal peak bone mass and healthy bones. However, the actual calcium intake by many teenagers, especially girls, is below the recommended amount.[51] Adolescent girls on low-calorie diets who drink diet soft drinks instead of milk are at particular risk for age-related bone loss and subsequent fractures. See the Spotlight feature in Chapter 8 for more about calcium and osteoporosis.

High Blood Cholesterol   Teenagers have many of the same risk factors for high blood cholesterol as adults: family history of coronary heart disease; diets high in total fat, saturated fat, and cholesterol; hypertension; low activity levels; and smoking. Although the process of atherosclerosis is not completely understood, it is believed that the fatty streaks in blood vessels that develop in young people progress to the fibrous plaques of adulthood.[52] For more information about diet and heart disease, see the Spotlight feature in Chapter 5.

Dental Caries Although dental caries are largely preventable, this remains the most common chronic disease of children aged 5 to 17 years.[53] By age five, 60 percent of all children have had tooth decay, and more than 90 percent of 18-year-olds have experienced decay.[54] Children in low-income households, and especially those who are American Indian, African American, or Hispanic, have three times the risk of tooth decay because they lack access to or encounter barriers to accessing dental services. Fortunately, the incidence of dental caries has decreased by as much as 30 to 50 percent over the past two decades, partly because of fluoridation of public drinking water, improved dental hygiene, and the use of fluoride in toothpastes and mouthwashes.

Eating Disorders Eating disorders have become a serious health problem in recent years. The most common eating disorders are anorexia nervosa and bulimia nervosa. A constellation of individual, familial, sociocultural, and biological factors contribute to these disorders, which threaten physical health and psychological well-being. See the Spotlight feature in Chapter 10 for more information about eating disorders.

Some individuals are more predisposed to developing an eating disorder than others. For example, about 90 percent of people with eating disorders are female. Most are Caucasian, with few cases seen among blacks and other minority groups. Most individuals who develop eating disorders are adolescents or young adults who typically began experiencing food-related and self-image problems between the ages of 14 and 30 years. Because these syndromes are surrounded by secrecy, their prevalence is not known with certainty, although it has increased dramatically within the past three decades.[55] Estimates in the general population range from 1 percent for anorexia nervosa to 1 to 5 percent for bulimia nervosa. These two eating disorders are also sometimes seen in adolescent athletes, many of whom compete in sports such as gymnastics, wrestling, distance running, diving, horse racing, and swimming, which demand a rigid control of body weight.

Nutrition in Later Life

In the United States, the life expectancy is now 78 years (75 years for men and 80 years for women).[56] To what extent is aging inevitable? Apparently, aging is an inevitable, natural process programmed into our genes at conception. Nevertheless, we can adopt lifestyle habits, such as consuming a healthful diet, exercising, and paying attention to our work and recreational environments, that will slow the aging process—within the natural limits set by heredity.[57] Clearly, good nutrition can slow down and ease the aging process in many significant ways. However, no potions, foods, or pills will prolong youth.

One approach to the prevention of aging has been to study other cultures in the hope of finding an extremely long-lived race of people and then learning their secrets. (See the Eat Well Be Well feature in Chapter 1 for more insights regarding nutrition and longevity.) The views of the experts are summed up best by simply saying that disease can *shorten* people's lives, and poor nutrition practices make diseases more likely to occur. Thus, by postponing and slowing disease processes, optimal nutrition can help prolong life and can promote health to make later years more enjoyable.[58] This section focuses on the nutrient requirements of older adults, the diseases that seem to come with age, their risk factors, and the relevance of nutrition to these diseases.

Demographic Trends and Aging
The number of older adults (persons aged 65 years and older) in the United States will double by 2030 to more than 70 million people. In 2000, people aged 65 and over accounted for 12.4 percent of

the population, and this proportion is expected to increase to nearly 20 percent by 2030. Nearly 12 percent of the population will be over the age of 74 by 2030. The increased growth in the older adult population in the United States is illustrated in Figure 12-6.[59]

Improved life expectancy and the baby boom that took place between 1946 and 1964 are important contributors to the growing older adult population in the United States. Baby boomers increased the numbers of the older middle-aged (ages 46 through 63) until 2010; now they will begin to swell the ranks of the retired population. Improved life expectancy has resulted from better prenatal and postnatal care and more effective means of combating disease in older adults. For example, the death rate from heart disease began to decline in the 1960s and continues to fall today. More than half of the drop is attributed to a decline in smoking and fewer people with high blood pressure or high blood cholesterol.

Healthy Adults

An individual's current health profile is substantially determined by behavioral risk factors. Among the leading causes of death for adults ages 25 and older are heart disease, cancer, stroke, injuries, chronic lung disease, diabetes, and liver disease, all of which have been associated with behavioral risk factors. To what extent can nutrition prevent or retard the development of these diseases?

Many of the health problems associated with the later years are preventable or can be controlled.[60] For example, eliminating certain risk behaviors and adopting healthy ones can improve the quality of life for older persons and reduce their risk of disability. For example, incorporating physical activity and a healthful diet into one's lifestyle can contribute to weight loss and to controlling three important risk factors for heart disease: high saturated fat intake, overweight, and a sedentary lifestyle. Figure 12-7 illustrates how the same set of dietary recommendations can promote general health and help prevent a broad spectrum of chronic diseases.[61]

Aging and Nutrition Status

Growing old is often associated with frailty, sickness, and a loss of vitality. Although the aging members of our society do experience chronic illness and the associated disabilities, there is great heterogeneity among this population. That is, older people vary greatly in their social, economic, and lifestyle situations as well as their capacity to function and their physical condition.[62] Each person ages at a different rate, sometimes making chronological age different from biological age. Most older persons live at home, are fully independent, and have lives of good quality.[63] Older persons sometimes have problems with the **activities of daily living (ADL)**. Because they depend on others to perform these essential activities, they are more likely to be at risk for malnutrition.

Nutritional Needs and Intakes

Many of the nutrient needs of older adults are the same as for younger persons, but some special considerations deserve emphasis. Calorie needs decline with age because of a decrease in basal metabolism related

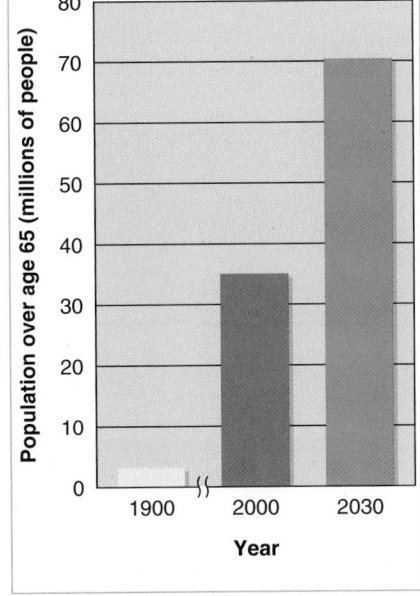

FIGURE 12-6

The Aging of the Population

In 1900, 4 percent of the U.S. population (3.1 million people) were more than 65 years of age; in 2000, 12.4 percent (35 million people) were more than 65; by 2030, 20 percent (about 70 million people) will have reached age 65.

SOURCE: Data from *A Profile of Older Americans, 2008;* available at www.aoa.gov.

activities of daily living (ADL) include bathing, dressing, grooming, transferring from bed to chair, going to the bathroom, and feeding oneself.

FIGURE 12-7

Consistency of Recommendations for Reducing the Risk of Chronic Diseases and Their Complications

[a] Starch refers to complex carbohydrates provided by fruits, vegetables, and whole-grain products.
[b] Aim for a minimum of 5 to 9 servings of fruits and vegetables daily.
[c] Gastrointestinal diseases affected by dietary factors are primarily gallbladder disease (fat and calories), diverticular disease (fiber), and cirrhosis of the liver (alcohol).

SOURCE: J. M. McGinnis and M. Nestle, The Surgeon General's Report on Nutrition and Health: Policy implications and implementation strategies, *American Journal of Clinical Nutrition* 49 (1989): 26. © American Society for Clinical Nutrition; AHA Conference Proceedings, Summary of a Scientific Conference on Preventive Nutrition: Pediatrics to geriatrics, *Circulation* 100 (1999): 450–56.

Change diet → / Reduce risk ↓	Reduce fats	Control calories	Increase starch[a] and fiber	Reduce sodium	Control alcohol	Increase antioxidants[b]
Heart disease	🍎	🍎	🍎	🍎		🍎
Cancer	🍎	🍎	🍎		🍎	🍎
Stroke	🍎	🍎	🍎	🍎	🍎	🍎
Diabetes	🍎	🍎	🍎			🍎
Gastrointestinal diseases[c]	🍎	🍎	🍎		🍎	

to loss of lean tissue and a decrease in physical activity. The recommended energy intake decreases by 10 calories per day for males and 7 calories per day for females for each year of age above 19.[64] Given their lower energy allowances, older adults are advised to select mostly nutrient-dense foods. General nutrition guidelines for older adults are shown in Figure 12-8.[65] The recommended eating pattern reflects the lower caloric needs of most healthy older adults and emphasizes the need for adequate fluid intake. The recommended energy and nutrient intakes for older adults are shown inside the front cover.

On one side of the energy budget, energy is taken in; and on the other side, it is spent. If you are motivated to maintain your good health into the later years, you should plan regular physical activity into your days. Ideally, the activity should be intense enough to increase the heartbeat and respiration rate and to prevent the atrophy of all muscles (not only the heart) that otherwise takes place. Many older people believe that they can't participate in strenuous activity, but studies have shown that they can do more than they think they can. Any exercise—even a 10-minute walk each day—is better than none. With persistence, great improvement is possible. Modest endurance training can improve cardiovascular and respiratory function and promote good muscle tone as well as control the accumulation of body fat.

Although caloric needs may decrease with age, the need for certain nutrients—such as calcium, vitamin D, vitamin C, vitamin B_{12}, and vitamin B_6—may actually increase with the effects of aging (see Table 12-8).[66] For example, as many as 30 percent of persons older than 50 years may experience reduced stomach acidity, which can interfere with their ability to absorb vitamin B_{12}, calcium, and iron from foods. In addition, older adults who have low intakes of fortified dairy products may have increased needs for vitamin D because their skin has become less efficient at making vitamin D when exposed to sunlight.[67]

Assessing the diet quality of older adults is important for identifying issues relevant to their health and nutrition status. According to the Healthy Eating Index, a composite measure of overall dietary quality based on the *Dietary Guidelines for Americans*, only 13 percent of the population aged 45 to 64 years consumes a

Any form of physical activity—even a 10-minute walk around the block—is better than nothing.

© Michael Newman/PhotoEdit

FIGURE 12-8

A Food Guide for Older Adults

Vit. B12
Vit. D
Calcium

Older adults often find it difficult to get adequate amounts of some nutrients from foods alone and may need supplements.

To prevent dehydration, older adults need to drink plenty of water and other beverages.

Regular physical activity helps older adults maintain independence and enjoy life more fully.

SOURCE: Copyright © 2007 Tufts University. Reprinted with permission from Lichtenstein AH, Rasmussen H, Yu WW, Epstein SR, Russell RM. Modified MyPyramid for Older Adults. *J Nutr.* 2008;138:78–82.

The food guide for older adults highlights several major features, including good choices within each food category; a foundation depicting a row of water glasses and physical activities, emphasizing the increased importance of both fluid intake and regular physical activity in older adults; and a flag on the top to suggest that some older adults, due to biological changes, may need supplemental vitamin B_{12}, vitamin D, and calcium. The food guide emphasizes variety, nutrient density, and fiber-rich foods from all food groups; whole grains, vegetables, and fruits (including easy-to-use frozen foods); low-fat and fat-free dairy products, including reduced lactose alternatives in the dairy group; low-saturated-fat and *trans*-fat choices in the oils group; and low-saturated-fat and plant-based options in the protein foods group. To meet nutritional needs, older adults should use the www.ChooseMyPlate .gov site to find their recommended amounts from each food group.
A sample diet might include:

- At least 3 servings of low-fat dairy products or suitable alternatives.
- 5 ounces from the protein foods group featuring dried beans, fish, eggs, and lean cuts of meat and poultry.
- A variety of fruits (2 cups) and vegetables (3 cups) featuring foods that are richly colored (dark green, orange, red, or yellow).
- 6 ounces of nutrient-dense, fiber-rich whole grains and fortified cereals.
- At least 8 glasses of water per day to prevent dehydration.

"good" diet, with the dairy and fruit groups needing the most improvement. Figure 12-9 on page 387 summarizes the overall dietary quality of independent, free-living, middle-aged and older adults.

Some health practitioners recommend nutritional supplements for older adults, particularly for chronic conditions such as osteoporosis, arthritis, or anemia. However, advertisers often target older adults by recommending supplements to reduce the effects of aging and increase longevity. Such advertising makes the older adult vulnerable to exploitation by quacks, when their money is better spent on nutritious foods. Surveys designed to find out what kinds of nutrient supplements older people are using tend to support the view that in many cases, older people are wasting their money.[68]

Nutrition-Related Problems of Older Adults

Although aging is not completely understood, we know that it involves progressive changes in every body

The evidence for using dietary supplements to slow the aging process is not as strong as the evidence supporting a diet rich in fruits, vegetables, and other nutrient-dense foods along with a physically active lifestyle.

David Young-Wolff/Getty Images

TABLE 12-8   The Impact of Aging on Nutrient Needs

Nutritional Concern	Age-Related Finding	Implications for Healthy Aging
Energy	Energy needs decrease due to decline in amount of lean tissue.	Get regular physical activity, including strength-training exercises (see Chapter 11).
Water	Reduced sense of thirst; increased urine output.	To prevent dehydration, drink 6 to 8 glasses of fluid a day.
Fiber	Constipation is a common complaint due to low fiber and fluid intakes and lack of exercise.	Consume 21 to 38 grams of fiber from a variety of sources such as fresh fruits, vegetables, legumes, and whole-grain products; get adequate amounts of fluid and exercise.
Vitamin D	Intakes may be low; skin becomes less able to synthesize vitamin from sunlight; less exposure to sunlight.	Choose fortified milk and cereals, salmon, and other fatty fish; get daily exposure to sunlight if possible; or use a supplement (600 to 800 IU per day).
Calcium	Intakes typically low; reduced gastric acidity impairs absorption; reduced bone mass.	To offset bone loss, choose calcium-rich foods (low-fat or fat-free milk, yogurt, cheese), fortified sources of calcium (orange and other juices), or take a calcium supplement; get regular weight-bearing exercise.
Iron	Anemia less common after menopause; reduced gastric acidity may impair absorption; chronic blood loss or low energy intakes may increase risk of deficiency.	Consume adequate calories from a variety of nutrient-dense foods with a source of vitamin C to enhance absorption of dietary iron.
Protein, vitamin E, vitamin B_6, zinc	Intakes may be low; decline in immune function.	Meet nutrient needs from low-fat servings of meat, fish, poultry, and legumes; use nuts in moderation; choose at least 6 servings of whole-grain products.
Antioxidants: vitamin C, vitamin E, carotenoids (beta-carotene, lutein)	Decline in vision (cataracts, macular degeneration); increased oxidative stress to body tissues.	Eat at least 5 to 9 servings of fruits and vegetables a day; choose regular servings of green leafy vegetables, brightly colored fruits and vegetables, and citrus fruits; choose whole-grain products; use nuts in moderation.
Vitamin A	Decline in liver's uptake of vitamin A; increased absorption may lower the aging body's requirement.	Avoid supplements of vitamin A.
Vitamin B_{12}	Intakes may be low; reduced gastric acidity impairs absorption of food-bound vitamin B_{12}.	Choose foods fortified with vitamin B_{12} (e.g., breakfast cereals), or take a supplement that contains vitamin B_{12}.
Folate, vitamin B_6, vitamin B_{12}	Intakes may be low, causing elevated blood levels of homocysteine and increased risk of heart disease; use of certain medications may impair B-vitamin status.	Consume citrus and other fresh fruits and green leafy vegetables, legumes, whole grains, and foods fortified with folate and vitamin B_{12} (such as breakfast cereals).

Changes in biological function between the ages of 30 and 70:

- Cardiac output: ↓ 30%
- Maximum heart rate: ↓ 25%
- Vital capacity: ↓ 40%
- Maximum O_2 uptake: ↓ 60%
- Muscle mass: ↓ 30%
- Hand grip, flexibility: ↓ 30%
- Bone mineralization: ↓ 20–30%
- Renal function: ↓ 40%
- Taste and smell: ↓ 90%
- Basal metabolic rate: ↓ 15%

tissue and organ: the brain, heart, lungs, digestive tract, and bones. After age 35, functional capacity declines in almost every organ system. Some changes, including oral problems, interfere with nutrient intake; others affect absorption, storage, and utilization of nutrients; still others increase the excretion of and need for specific nutrients. Examples of various changes associated with aging that can affect nutritional status include sensory impairments, altered endocrine, gastrointestinal, and cardiovascular functions, as well as changes in the renal and musculoskeletal systems (see the list in the margin). Both genetic and environmental factors contribute to these declines. Many of the changes are inevitable, but a healthful lifestyle that combines regular physical activity with adequate intakes of all essential nutrients can forestall degeneration and improve the quality of life into the later years.

As a person gets older, the chances of suffering a chronic illness or functional impairment become greater. In the United States, some type of disability (e.g., difficulty in hearing, vision, cognition, ambulation, self-care, or independent living) was reported by 37% of older persons in 2009.[69] A majority of older persons have at least one chronic condition, and many have multiple conditions. Among the most frequently occurring conditions in older persons are hypertension (49%), diagnosed arthritis (36%), all types of heart disease (31%), diabetes (16%), and sinusitis (15%).

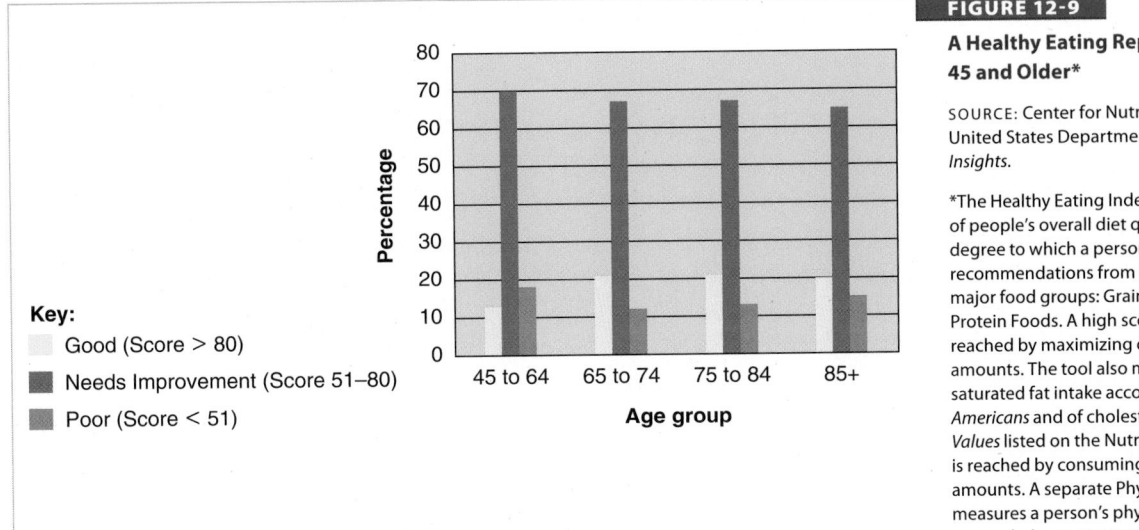

FIGURE 12-9

A Healthy Eating Report Card for Adults Age 45 and Older*

SOURCE: Center for Nutrition Policy and Promotion, United States Department of Agriculture, *Nutrition Insights.*

*The Healthy Eating Index gives a summary measure of people's overall diet quality. The tool measures the degree to which a person's diet conforms to the serving recommendations from the USDA Food Guide's five major food groups: Grains, Vegetables, Fruits, Dairy, and Protein Foods. A high score for these components is reached by maximizing consumption of recommended amounts. The tool also measures compliance of saturated fat intake according to the *Dietary Guidelines for Americans* and of cholesterol and sodium from the *Daily Values* listed on the Nutrition Facts label. A high score is reached by consuming at or below recommended amounts. A separate Physical Activity assessment tool measures a person's physical activity status and overall energy balance at www.ChooseMyPlate.gov.

Other chronic conditions contributing to disability include strokes and disorders of vision and hearing (see Figure 12-10). Dementia (especially Alzheimer's disease) is a major contributor to disability and nursing home placement for those over the age of 75. As noted in Table 12-9, malnutrition can occur secondary to these conditions, many of which require special diets that can further compromise nutrition status in the older adult.

FIGURE 12-10

Common Causes of Disability in Older Persons

SOURCE: National Center for Health Statistics, 2008

TABLE 12-9	Malnutrition That Is Secondary to Disease or Physiological State
Disease or Condition	**Effects on Nutrition Status**
Atherosclerosis	May increase difficulties in regulating fluid balances if caused by congestive heart failure.
Cancer	Weight loss, lack of appetite, and secondary malnutrition are common.
Dental and oral disease	May alter the ability to chew and thus reduce dietary intake. Increased likelihood of choking and aspiration.
Depression and dementia	Increased or decreased food intakes are common. A person with dementia may have decreased ability to get food, or the appetite may be very small or very great. Judgment and balance in meal planning are generally absent.
Diabetes mellitus	Increased risk of other diet-related diseases. Weight loss is needed if obesity is present.
Emphysema	May be difficult to eat because of breathing problems.
End-stage kidney disease	Alters fluid and electrolyte needs. Infections and low-grade fever may increase energy output and weight loss.
Gastrointestinal disorders	Increased risk of malabsorption of nutrients and consequent undernutrition.
High blood pressure	Weight gain may exacerbate high blood pressure.
Osteoarthritis	Makes motion difficult, including those activities related to purchasing, serving, eating, and cleaning up after meals. Predisposes people to a sedentary lifestyle and may give rise to obesity.
Osteoporosis	Limits the ability to purchase and prepare food if mobility is affected. If severe scoliosis is present, the appetite may be altered.
Stroke	May alter abilities in the cognitive and motor realms related to food and eating.

SOURCE: Adapted from Institute of Medicine, *The Second Fifty Years: Promoting Health and Preventing Disability* (Washington, DC: National Academy Press, 1992), pp. 168–69.

Factors that contribute to malnutrition in older adults:

- Disease
- Eating poorly
- Tooth loss or oral pain
- Economic hardship
- Reduced social contact
- Multiple medications
- Involuntary weight loss or gain
- Need of assistance with self-care
- Advanced age (older than 80 years)

Up to one-quarter of all older adults and one-half of all hospitalized older adults may be suffering from malnutrition.[70] About 10 percent of the older population lives in poverty and 26 percent of the older population is classified as "low income." Older women are more likely than older men (12 percent versus 7 percent) to live in poverty. In addition, older women are more than twice as likely as older men to live alone (40 percent and 19 percent, respectively). About 45 percent of older adults take multiple prescription drugs that can interfere with appetite and nutrient absorption, and 42 percent have difficulties with basic activities (for example, preparing meals and taking medications). All of these factors place older persons at risk for malnutrition.[71]

Financial resources, living arrangements, and a social support network, including availability of caregivers, can also directly affect a person's nutrition status. Poverty and social isolation in particular impair the nutrition status of many older adults, as noted perceptively by a professor of psychiatry:

> It is not what the older person eats but with whom that will be the deciding factor in proper care for him. The oft-repeated complaint of the older patient that he has little incentive to prepare food for only himself is not merely a statement of fact but also a rebuke to the questioner for failing to perceive his isolation and aloneness and to realize that food . . . for one's self lacks the condiment of another's presence which can transform the simplest fare to the ceremonial act with all its shared meaning.[72]

Sources of Nutritional Assistance Socioeconomic problems—low income, inadequate facilities for preparing food, lack of transportation, and inability to afford dental care, among others—trouble many older adults and may lead to malnutrition. In response, federal, state, and local agencies have mandated nutrition programs for older adults.

The Supplemental Nutrition Assistance Program (SNAP)—formerly the Food Stamp program—enables people who qualify to obtain electronic benefits in the

Eating a food you haven't tasted before prolongs your life by 75 days, according to an old Japanese saying.[73] Perhaps finding someone to share that food with might extend your life even more.

Planning nourishing meals for one person that offer variety and optimal nutrition can be challenging. The following tips can help singles of any age buy and prepare foods.

- Keep cupboards and refrigerator stocked with milk, eggs, bread, tortillas, pita bread, canned beans, jars of spaghetti sauce, rice, pasta or noodles, potatoes, onions, canned soups and broth, margarine, cooking oil, and frozen vegetables.

- Keep fresh fruits or vegetables, low-fat yogurt or cheese, and popcorn on hand for easy-to-grab snacking.

- Buy large bags of frozen vegetables if you have sufficient freezer space. Take out the exact amount you need at mealtime.

- Keep an assortment of whole-grain breads, bagels, and muffins in the freezer. Take out individual servings as you need them.

- Buy large packages of meat and poultry when they are on sale. Divide the package into individual servings and freeze them separately.

- Buy fruits and vegetables in season; they will be cheaper and most flavorful at these times.[74]

  - *Winter:* oranges, grapefruits, sweet potatoes, rutabagas, greens

  - *Spring:* asparagus, green beans, sweet peas, rhubarb

  - *Summer:* berries, peaches, zucchini, melons

  - *Fall:* apples, pears, acorn and butternut squash

- Simplify cooking and clean-up.[75] Prepare a one-dish meal that includes foods from several food groups, such as meats, whole grains, legumes, and vegetables (for example, chicken, vegetable, and rice casserole, or vegetarian chili with chopped vegetables).

- Downsize your meal. If you're short on time or energy, make a healthy snack rather than a full meal. For example, spread a brown rice cake with low-fat ricotta cheese and fresh fruit; serve corn muffins with apple and cheese slices; or try fat-free refried beans mixed with salsa, a small amount of low-fat sour cream, and baked tortilla chips.

- Recruit friends and neighbors for a cooking club—a great way to share new recipes and make meal preparation more enjoyable.

- Keep it simple and full of vegetables. A heaping bowl of hot soup or stew can make a delightfully healthy and satisfying meal on a cold night. Search the Internet for healthy recipes such as Corn and Shrimp Chowder, Sweet Potato and Peanut Soup, or Autumn Harvest Stew.

- Cook for several meals at a time. Roast a turkey breast, or skinless, boneless chicken breasts, and use half for dinner and the rest for lunches—in sandwiches, tortillas, stir-fry meals, or salads.

- Double a favorite recipe. Label and store the extra servings in the freezer. Date these so that you will use the oldest first.

- Add your own steamed vegetables to a frozen entrée. Add a tossed salad, whole-grain roll, and a naturally sweet fruit.

- Choose frozen entrees that contain no more than 10 grams of fat per 300 calories and fewer than 800 milligrams of sodium per serving.

- When you cook for yourself, imagine that you are cooking for special guests. Or, make your meal preparation time more meaningful by asking friends or relatives to join you for a home-cooked meal.

Bochkarev Photography/Shutterstock.com

form of a "credit" card they can use to buy food. In many areas, food banks allow older people on limited incomes to buy good food for less money. A food bank buys industry "irregulars"—products that have been mislabeled, underweighted, redesigned, or mispackaged and would otherwise be thrown away. Nothing is wrong with this food, the industry can credit it as a donation, and the buyer (often a food-preparing site) can obtain the food for a small handling fee and make it available at a greatly reduced price.

Physiological
- Dietary intake
- Lack of appetite
- Inactivity/immobility
- Poor taste and smell acuity
- Alcohol or drug abuse
- Presence of chronic disease
- Polypharmacy
- Disability
- Oral health problems

Socioeconomic
- Cultural beliefs
- Poverty
- Limited education
- Limited access to health care
- Institutionalization

Psychological
- Loneliness
- Cognitive impairment
- Dementia
- Depression
- Loss of spouse
- Social isolation

Environmental
- Inadequate housing
- Inadequate cooking facilities
- Lack of transportation
- Lack of access to health services

The Older Americans Act Nutrition Program—also known as the Elderly Nutrition Program (ENP)—is intended to improve older people's nutrition status and help them avoid medical problems, continue living in communities of their own choice, and stay out of institutions. Its specific goals are to provide:

- Low-cost, nutritious meals

- Opportunities for social interaction

- Nutrition education and shopping assistance

- Counseling and referral to other social and rehabilitation services

- Transportation services

The program makes available at noon, five days a week, one hot meal supplying a third of the recommended nutrient intakes. There is no cost for these meals, but participants sometimes make voluntary contributions.

A part of ENP is the Congregate Meal Program. Administrators select sites for congregate meals to feed as many of the eligible older adults as possible. The congregate meal sites are often community centers, senior centers, religious facilities, schools, extended care facilities, or retirement housing complexes. Volunteers may also deliver meals to those who are homebound either permanently or temporarily; these efforts are known as the Home-Delivered Meals Program. The home delivery program ensures nutrition, but its recipients miss out on the social benefit of the congregate meal sites. Every effort is made to persuade recipients to attend the shared meals if they can.

Evaluations of the congregate and home-delivered meals programs generally show that they improve the dietary intake and nutrition status of their clients.[76] Participants generally have greater diversity in their diets and higher intakes of essential nutrients and are less likely to report food insecurity than nonparticipants. Additional benefits come as a result of screening and the referrals generated by such programs, as well as from the activities associated with the congregate meals: diet counseling, physical activity, adult education, and activities. Participants also benefit from the opportunity for improved socialization.

Looking Ahead and Aging Wisely

As a nation, we tend to value the future more than the present, putting off enjoying today so that we will have money, prestige, or time to have fun tomorrow. Older adults feel this loss of future. The present is their time for leisure and enjoyment, but they often have no experience in using leisure time.

The solution is to begin to prepare for old age early in life, both psychologically and nutritionally. Preparation for this period should, of course, include financial planning; but other lifelong habits should be developed as well (see Figure 12-11). Every adult needs to learn to reach out to others to forestall the loneliness that may otherwise ensue. Adults need to develop some skills or activities such as volunteer work, reading, games, hobbies, or intellectual pursuits that they can continue into their later years to give meaning to their lives. Every adult also needs to develop the habit of adjusting to change, especially when it comes without consent, so that it is not seen as a loss of control over his or her life. The goal is to arrive at maturity with as healthy a mind and body as possible; this means cultivating good nutrition status and maintaining a program of daily physical activity.

In general, the ability of the older adult to function well varies from person to person and depends on several factors. The "life advantages" listed on the next page seem to contribute to good physical and mental health in later years.[77]

FIGURE 12-11

The Aging Well Pyramid

The time to prepare for old age is early in life. Practice the items found at the base of the pyramid to achieve an optimal sense of well-being. Use the inner four compartments of the pyramid to create a balance among all aspects of your life: nutrition, physical activity, social health, and emotional well-being. Use the top of the pyramid to manage everyday stresses such as traffic gridlock, exams, and work deadlines.

AGING WELL

Stress Busters
- Relax
- Go for a walk
- Breathe deeply
- Think positively

Emotional Health
- Reduce stress
- Learn relaxation techniques
- Cultivate a garden
- Seek out laughter
- Take time for spiritual growth
- Adopt and love a pet
- Take time off

Social Health
- Be socially active
- Volunteer for a special cause
- Make new friends
- Enroll in lifelong learning
- Be active in your community

Nutritional Health
- Choose nutrient-dense foods
- Eat at least 5 to 9 fruits and vegetables every day
- Drink plenty of water
- Keep fat intake at a healthy level
- Get adequate fiber

Physical Health
- Be physically active
- Get adequate sleep
- Challenge your mental skills
- Do aerobic and strength-training exercises
- Stretch for flexibility

Lifelong Habits for Successful Aging
- Cherish your personal values and goals
- Develop good communication skills
- Balance diet and exercise to maintain a healthful weight
- Practice preventive health care
- Develop skills and hobbies to enjoy for a lifetime

- Manage time
- Learn from mistakes
- Nurture relationships with family and friends
- Enjoy, respect, and protect nature
- Accept change as inevitable
- Plan ahead for financial security

- Genetic potential for extended longevity. Some persons seem to have inherited a reduced susceptibility to degenerative diseases.

- A continued desire for new knowledge and new experiences. Some studies suggest that active minds, ever involved in learning new things, may be more resistant to decline.

- Socialization, intimacy, and family integrity. Older persons thrive in situations that nurture love, understanding, shared responsibility, and mutual respect.

- Adhering to a healthy diet while avoiding excess food energy, saturated fat, *trans* fat, cholesterol, and sodium. A diet with adequate intakes of all essential nutrients has a positive impact on health and weight management.

- Avoiding substance abuse.

- Acceptable living arrangements.

- Financial independence.

- Access to health care, including a family physician, health clinic, public health nursing service providing home health care, dentist, podiatrist, physical therapist, pharmacist, and registered dietitian.

Everyone knows older people who have maintained many contacts—through relatives, friends, religious institutions, or fraternal orders—and have not allowed themselves to drift into isolation. Upon analysis, you will find that their favorable environment came from a lifetime of effort. These people spent their entire lives reaching out to others and practicing the art of weaving others into their own lives. Likewise, a lifetime of effort is required for good nutrition status in the later years. Those who have eaten a wide variety of foods, stayed trim, and remained physically active are best able to withstand the assaults of change.

Aging Well Resources

http://health.howstuffworks.com/eat-right-seniors.htm
How to Eat Right as a Senior. How Stuff Works provides an easy-to-follow guide for healthful eating.

www.nlm.nih.gov/medlineplus/seniorshealth.html
MedLine Plus Seniors' Health. Includes coverage of the latest news topics on senior health issues, resource materials, and more.

http://nihseniorhealth.gov
NIH Senior Health. The National Institutes of Health provides health and wellness information on many topics, such as bone health, nutrition, and physical activity. This site allows you to adjust text size or have it read out loud to you.

www.usa.gov/Topics/Seniors.shtml
Information and services from the U.S. government on caregiver resources, health, travel, consumer protection, volunteerism, end-of-life issues, and more.

http://nutritionandaging.fiu.edu
The National Resource Center on Nutrition, Physical Activity, and Aging. Offers direct links to numerous nutrition and health sites that provide information for older adults.

www.eldercare.gov
The Eldercare Locator. A public service of the U.S. Administration on Aging to enable older persons to live independently in their communities. Helps older persons and caregivers locate local services such as meals, home care, transportation, home repair, recreation, social activities, legal, and other community services.

http://seniorhealth.about.com/
About.com. Offers many health-related articles for older adults.

http://nihseniorhealth.gov/exerciseforolderadults/toc.html
Lists the benefits derived from exercise, demonstrates how to exercise safely, and offers tips on charting your progress.

http://www.mayoclinic.com/health/healthy-aging/MY00374
Mayo Clinic Senior Health Center. Offers resources and insights into what older adults can expect as they age and tips for preventing some of the most common health problems.

Addressing Weight Problems in Children and Adolescents

The physical and emotional health of an entire generation and the economic health and *security of our nation is at stake.*
—First Lady Michelle Obama at the Let's Move! launch on February 9, 2010

For the past three decades, the health status of U.S. children and adolescents has generally improved, as seen in reduced infant mortality rates and a decline in the nutrient deficiency diseases of the past. However, approximately 9 million children over 6 years of age are overweight.[78] The increasing prevalence of overweight is now among the most important public health challenges in the United States.

Exactly how many children are overweight?

Two terms are used to refer to children and adolescents whose excess body weight could pose medical risks:

- *Overweight* is defined as a BMI at or above the 85th percentile and lower than the 95th percentile.

- *Obesity* is defined as a BMI at or above the 95th percentile for children of the same age and sex.

The number of children who are overweight has more than tripled in the United States. The prevalence of obesity has increased from 6.5 percent in 1980 to 20 percent today among children aged 6 to 11, and from 5 percent to 18 percent among adolescents.[79] These increases are greater in certain ethnic and racial groups (particularly Hispanics and African Americans) as well as in certain regions of the country (notably the South). Figure 10-1 in Chapter 10 illustrates the increasing trend toward overweight in U.S. children.

So what's the big deal?

Overweight in childhood involves significant risks to physical and emotional health. It has been estimated that one in three children born in the United States is at risk for being diagnosed with type 2 diabetes at some point in life. This risk increases for minority children. One out of two children born in 2000 will develop diabetes sometime in life.[80] The same conditions associated with overweight adults, such as type 2 diabetes, high blood lipids, high blood pressure, and gallbladder disease, are now appearing with greater frequency in young children and teens. Overweight children are also at risk for injury to weight-bearing joints, decreased quality of life, depression, poor self-esteem, respiratory problems and sleep disturbances, and other serious health problems. Because overweight children and adolescents are more likely to become overweight and obese adults, experts project that the rate of chronic diseases will skyrocket even higher in the United States if something isn't done about this epidemic.[81]

Television can have adverse effects on children by influencing their selection of foods, snacking habits, and activity levels.

What has caused this problem?

The rapid increases in overweight and obesity over the past three decades are due primarily to societal and environmental factors.[82] Factors in the environment that may contribute to overweight include increased calorie intake due to increased portion sizes; increased consumption of foods and meals away from home; excessive consumption of soft drinks; lack of positive role models; and physical inactivity. According to one recent study, the strongest predictor of overweight and overfatness in fifth graders was their prior status in third grade.[83] Second to prior overweight, the strongest predictor of subsequent overweight in children is television viewing.[84]

What are children and adolescents eating, and how is it affecting their health?

Eating practices influence a child's physical growth, cognitive development, and overall health.[85] Reports from a number of organizations have shown that children are failing to meet the recommended nutrition guidelines by not consuming enough fruits and vegetables and by eating too many foods high in solid fat and added sugars. Children's eating habits have changed over the past three decades. Dietary data, collected in large nationwide surveys to determine trends in nutrient intakes, have shown most children and adolescents have either poor diets or diets that need improvement.[86]

Over the past 25 years, the portion sizes of commonly consumed foods, such as soft drinks and hamburgers, have increased. Large food portions that provide more calories than smaller portions may be contributing to the increasing prevalence of overweight in children and teens. For example, children 3 to 5 years of age consumed 25 percent more of an entrée when they were

(Continued)

The popularity of fast-food meals poses a challenge to the nutritional quality of children's diets, because these meals are typically high in saturated fat, sodium, sugar, and calories and lacking in fruits and vegetables.

presented with portions that were double an age-appropriate standard size.[87]

Today, with the demise of traditional family meals, many busy parents rely on fast-food meals to nourish their children. Research shows that children who ate fast food consumed more total energy and had poorer diet quality (e.g., more added sugars, more sugar-sweetened beverages, less milk, and fewer fruits and nonstarchy vegetables) on days when they ate fast food, compared to days without fast food.[88]

How does the media factor into this?

Children spend a large amount of their leisure time using various media, including television, DVDs, video games, computers, and cell phones. Children hear a great deal about foods via television. Many authorities are concerned television commercials may have an undesirable impact on children. It is estimated the average child sees more than 30,000 commercials a year. The Institute of Medicine report *Preventing Childhood Obesity: Health in the Balance* noted that more than half of television advertisements directed at children promote foods and beverages such as candy, fast food, snack foods, soft drinks, and sweetened breakfast cereals that are high in calories and fat and low in fiber and other essential nutrients.[89] Billions of advertising dollars are spent each year in the effort to sell these foods to children.

The more time children spend watching television, the less time they have for physical activity. Research results show children who watch television more often had lower activity levels and were less likely to participate in organized sports or community activities. Parents can encourage physical activity in children, and thereby help prevent overweight, by participating in activities with their

children, such as cycling, swimming, rollerblading, skating, hiking, and helping them find physical activities they enjoy![90]

What are some simple tips for helping children learn good eating habits?

- Parents can set good examples with their own eating habits.
- Discourage children from eating while watching TV, working/playing on the computer, or doing homework.
- Eat meals together as a family when possible; keep fast food meals to a minimum.
- Encourage children to "listen to their bodies (stomachs)" and eat only when hungry.
- Limit consumption of high-fat or high-sugar foods, including soft drinks.
- Include children in food purchasing and preparation of meals as a way of helping them become more interested in and aware of what they eat.

Is school food to blame?

Even though school meal programs appear to have a positive effect on children's consumption of milk, fruit, vegetables, and some vitamins and minerals, there is evidence school meals contribute too much fat in menu items such as pizza, chicken sandwiches, French fries, and baked goods. In light of these findings, the USDA launched a reform of the School Lunch Program in 1994 that directed schools to upgrade the nutritional value of their meals to adhere to the *Dietary Guidelines for Americans*. To meet the nutrition standards, menus now include more fruits, vegetables, whole grains, and low-fat and reduced-fat foods. More vegetarian options are available, as well as salad bars and a variety of prepared salads.[91]

The environment in some schools discourages students from eating meals provided by the School Lunch Program and encourages food choices and eating habits that are not consistent with the *Dietary Guidelines for Americans*. Efforts to promote healthful eating habits and provide nutrition instruction may be contradicted in school settings where the sale of food and beverages, many with low nutritional appeal, is promoted in snack bars, school stores, and vending machines. Even less healthful à la carte items sold by the school cafeteria may "compete" with school meals for a student's meal money.[92]

How can school food be more healthful?

Many states, and many school districts, are developing policies that limit the sale of *competitive foods*—foods sold in competition with USDA school meals that are categorized as "foods of minimal nutritional value"—such as soft drinks, candies, and candy-coated popcorn. Some states have already implemented sweeping changes to nutrition policies in schools, including eliminating deep-fat frying; restricting portion sizes of chips and French fries, fruit drinks, and certain snacks and sweets; limiting fats and sugar; and offering fruits and vegetables daily at every point of service.[93]

The Surgeon General's call to action to prevent and decrease overweight and obesity outlines the following actions for creating healthful school environments:[94]

- Provide age-appropriate nutrition and health education to help students develop lifelong, healthful lifestyle habits.
- Adopt policies that require all foods and beverages available on school campuses and school events to contribute toward eating patterns that are consistent with the *Dietary Guidelines for Americans*.
- Ensure that healthful snacks and foods are provided in vending machines, school stores, and other venues.

How can child obesity be prevented?

Prevention of overweight and obesity involves a focus on *energy balance*—calories consumed versus calories expended—so taking action against overweight in childhood must address the factors that influence both eating and physical activity.[95] In general, most interventions for prevention of overweight in children have been conducted through the schools.[96] Effective interventions among children tend to include a component on decreasing television viewing. School-based programs that increase physical activity also appear effective in reducing body size, especially when physical activity is for 1 hour or more.[97] Most interventions involve a family component. For small children, the parents are often "gatekeepers" for diet and physical activity because they control the types of foods in the house, the access and availability of those foods, opportunities for activity or inactivity, and meal patterns. In addition, parents often serve as role models for their children. Although family is important for preschool and elementary-aged children, adolescents are often more interested in peer relationships and influences, so interventions for adolescents should include a peer component as well.[98] See the accompanying box for additional recommendations from the Institute on Medicine regarding the prevention of childhood obesity. For additional information, visit the websites listed in the margin.

Key Recommendations for the Prevention of Childhood Obesity[99]

- The food industry should make obesity prevention in children and youth a priority by developing and promoting food products that are low in dietary fat and energy.
- Nutrition labeling should be clear and useful so that parents and youth can make informed product comparisons to achieve and maintain energy balance at a healthful weight.
- The media should incorporate positive behavior-change messages into television programs and popular magazines, and they should regulate television advertising aimed at children.
- Local governments, schools, and community organizations should collaboratively develop programs that encourage healthful eating behaviors and regular physical activity.
- Local governments, private developers, and community groups should expand opportunities for physical activity, including recreational facilities, parks, playgrounds, sidewalks, and bike paths.
- Parents and schools should provide a consistent environment that is conducive to healthful eating behaviors and regular physical activity.

Websites

- *Dietary Guidelines 2010;* http://health.gov/dietaryguidelines
- *University of California Berkeley Center for Weight and Health*: www.cnr.berkeley.edu/cwh/
- *USDA Food and Nutrition Information Center, Weight Control and Obesity*: www.nal.usda.gov/fnic/etext/000060.html
- *Coordinated Approach to Child Health*: www.catchinfo.org
- *Planet Health*: www.hsph.harvard.edu/prc/proj_planet.html
- *Stanford's Student Media Awareness to Reduce Television (SMART)*: hprc.stanford.edu/pages/store/itemDetail.asp?169
- *HUGS International, Inc. for teens and adults*: www.hugs.com
- *Team Nutrition*: www.fns.usda.gov/tn

Nutrition on the Web

www.cengage.com/nutrition/boyle/personalnutrition8e

Go to the *Personal Nutrition* site to check for the latest updates to chapter topics or to access links to related websites.

www.nal.usda.gov/fnic

The Food and Nutrition Information Center (FNIC) provides credible and accurate information. Under Browse by Subject, select Lifecycle Nutrition to access a variety of nutrition topics by age group.

www.nutrition.gov

This government website provides information and links. Under Browse by Subject, select Life Stages to access a variety of nutrition topics by age group.

www.ChooseMyPlate.gov

This interactive website provides tips and resources for planning a healthy diet using MyPlate. Browse Subjects to access links to the following:

- Click on Pregnancy and Breastfeeding.
- Click on Project M.O.M.
- Click on For Kids.

www.healthfinder.gov

Search for reliable consumer information on life-cycle nutrition topics.

www.cdc.gov

Search the Centers for Disease Control (CDC) website for pregnancy information, child development, healthy youth, adolescent health, and healthy aging.

www.cdc.gov/pednss/index.htm

Search the Centers for Disease Control Pediatric and Pregnancy Nutrition Surveillance system for statistics.

www.girlshealth.gov

This government website provides information about health and nutrition for girls ages 10–16.

www.eatright.org

The American Dietetic Association home page includes many nutrition resources and links.

www.nlm.nih.gov/medlineplus

Search for information about life-cycle nutrition topics at this government website.

www.mayoclinic.com

Visit the Mayo Clinic's website, and click on the Health Information tab. Under the Healthy Living tab, scroll down and click on a variety of life-cycle nutrition topics.

www.marchofdimes.com

Information on birth defects from the March of Dimes.

www.nofas.org

The National Organization on Fetal Alcohol Syndrome website.

www.womenshealth.gov

Click on Health Topics, and search for pregnancy and breastfeeding information at this government website.

www.llli.org

The La Leche League website provides information about breastfeeding.

www.aap.org

The American Academy of Pediatrics website provides information about health and nutrition for infants, children, and teens.

www.foodinsight.org

The International Food Information Council provides information on promoting healthy lifestyles for children.

www.kidshealth.org

This website provides information on a variety of life-cycle nutrition topics for infants through teens.

www.kidnetic.com

This is an interactive website combining food, fitness, and fun for children.

www.letsmove.gov

Let's Move is dedicated to solving the problem of child obesity and provides resources for helping children get healthy by empowering parents; providing healthy food in schools; improving access to healthy, affordable foods; and increasing physical activity

www.nhlbi.nih.gov/health/public/heart/obesity/wecan/index.htm

We Can! (**W**ays to **E**nhance **C**hildren's **A**ctivity & **N**utrition) provides resources for the national initiative designed to give parents, caregivers, and communities a way to help children ages eight to thirteen years old stay at a healthy weight.

www.aarp.org

The American Association for Retired Persons (AARP) provides health and nutrition information for older people. Click on the Health tab to access a variety of topics.

www.nia.nih.gov

The National Institute on Aging (NIA) website provides resources and links.

www.fns.usda.gov

The Food and Nutrition Service (FNS) website for U.S. Food Assistance Programs provides information and links.

Diet Analysis+ in Action

Go to the Track Diet portion of your Diet Analysis+ profile and, using the ideas in the Savvy Diner feature in this chapter, plan one day's meals as if you were preparing breakfast, lunch, and dinner for yourself. Now go to the Intake Spreadsheet report and compare this day's calories and nutrient intake with those of the previous days you tracked. Did you consume more or fewer calories in this hypothetical day? Was the nutritional balance better or worse on this day than on the previous days you tracked? Does this activity give you ideas for improving your diet? If so, what are the obstacles to incorporating a healthier diet into your life? How can you overcome those obstacles?

Menu
of Online Study Tools

A variety of study tools for this chapter are available at our website to deepen your understanding of chapter concepts. Go to

www.cengage.com/sso

to find

- Practice tests
- Flashcards
- Glossary
- Web links
- Animations
- Chapter summaries, learning objectives, and crossword puzzles.

Food Safety and the Global Food Supply

Ask Yourself . . .

Which of the following statements about nutrition are true, and which are false? For each false statement, what is true?

1. Pesticides rank as the number one hazard in the U.S. food supply.
2. Two frequent causes of foodborne illness in homes and restaurants are inadequate cooking and improper cooling of foods.
3. If a food contains a toxic substance, a person who eats it will become ill.
4. Tainted mayonnaise frequently causes food poisoning.
5. Imported foods may contain residues of pesticides that are illegal in the United States.
6. A USDA rule on organic crops allows the use of genetically engineered ingredients and irradiation in organic foods.
7. Food insecurity in the United States almost exclusively afflicts unemployed homeless people.
8. Legal pesticides are poisonous only to pests, not to people.
9. Most foods that cause food poisoning are contaminated by the manufacturer or processor.
10. Lead poisoning ranks as one of the most common childhood environmental health problems.

Answers found on the following page.

So far this book has dealt primarily with nutrients and how your body handles them. This chapter takes the study of nutrition one step further and examines some nonnutrient components of food—bacteria, additives, and pesticides, to name a few—and how they affect the food supply. In addition, the chapter takes a global view of the science of nutrition, looking at how foodways in different countries influence each other and how some of our food habits affect the environment.

What additives do foods contain, and what are the effects of those additives? Are foods ever contaminated? How can you reduce your risk of food poisoning? What potential do new food technologies, such as irradiation and genetic engineering, hold for our lives and for the health of the environment? This chapter addresses these and other questions.

The role of the infinitely small is infinitely large.
LOUIS PASTEUR (1822–1895, FRENCH CHEMIST AND MICROBIOLOGIST WHO DEVELOPED THE PASTEURIZATION PROCESS)

Americans enjoy the safest, most diverse food supply in the world.

© PhotoDisc/Getty Images

Foodborne Illnesses and the Agents That Cause Them

North America has the safest, most plentiful food supply in the world, thanks to the concerted efforts of food suppliers, food processors, and federal, state, and local governments, all of which are concerned with food safety. North Americans also enjoy the most diverse food supply in the world, consisting of an incredible array of fresh and processed foods. Most American supermarkets stock a variety of foods from other countries: cookies and crackers from Denmark and the United Kingdom, Belgian chocolates, Italian cheeses, beef and veal products from New Zealand and Australia, goose liver paté from France, and specialty foods from Japan, Mexico, and even China. Our international food interests can be seen in the produce section as well, where exotic star fruit, papaya, and mango from overseas markets are widely available. Still, the diverse mix of fresh and processed foods coming from local, national, and international markets underscores the need to understand where the culprits behind foodborne illnesses originate.

Within hours of the September 11, 2001, attacks on the Pentagon and New York City, the nation's food and water supplies were identified as likely targets of terrorists. *Food safety* today refers to a food supply that is free of foodborne pathogens as well as one that is safe from bioterrorism.[1] As part of a heightened awareness of bioterrorism threats, the Food and Drug Administration (FDA) has created a special website that links to information about bioterrorism as well as other information sources. See www.fda.gov/Food/FoodDefense/default.htm.

Foodborne illness is one of the greatest concerns of public health experts and the food industry.* According to the Centers for Disease Control and Prevention, as many as 48 million Americans experience foodborne illness each year, and an estimated 3,000 deaths are linked to tainted foods.[2] Incredible as these figures are, they probably represent only a fraction of the whole picture. For a number of reasons, many mild cases of foodborne illness are never reported: the victims pass off the symptoms as flu and do not seek medical attention, the illness is misdiagnosed as another problem with similar symptoms, the victim fails to recognize food as the source of the illness, or the physician doesn't report the illness to local health agencies. Diarrhea, nausea, abdominal pain, or vomiting without fever or upper respiratory distress is often taken to be flu, but people who experience such symptoms are highly likely to be suffering from foodborne illness.

The fact that more people aren't afflicted with foodborne illness is surprising because disease-producing microorganisms proliferate in our environment. Consider that all raw foods contain microbes, and foods often pick up more microbes during production, processing, packaging, transport, storage, and preparation. For this reason, the food industry uses various control measures to limit the risk of foodborne illness. Bacteria or their spores are destroyed or inactivated by heat treatments such

---

*The FDA Food Safety Modernization Act was signed into law on January 4, 2011. It aims to ensure the U.S. food supply is safe by shifting the focus of federal regulators from responding to contamination to preventing it.

ASK YOURSELF ANSWERS: **1**. False. The greatest hazard present in the U.S. food supply today is not pesticides but bacteria, viruses, mold, and other microorganisms that cause food poisoning. **2**. True. **3**. False. If a food contains a toxic substance, a person who eats it *may* become ill, depending on whether enough of the toxin to cause illness is present. **4**. False. Mayonnaise rarely carries high levels of the bacteria that cause illness. **5**. True. **6**. False. A USDA rule on organic crops rejects the use of genetically engineered ingredients and irradiation in organic foods. **7**. False. Food insecurity in the United States affects not only the unemployed homeless but also homeowners and the working poor. **8**. False. Legal pesticides can be poisonous to people, animals, and plants, depending on the amount of the pesticide. **9**. False. Most cases of food poisoning are the result of improper handling of food *after* it leaves the processor or manufacturer. **10**. True.

foodborne illness illness occurring as a result of ingesting food or water contaminated with a poisonous substance, such as a toxin or chemical (*food intoxication*) or an infectious agent, such as bacteria, viruses, or parasites (*foodborne infection*); commonly called *food poisoning*.

TABLE 13-1	FDA's Rank of Areas of Concern in the Food Supply

Although most people think that chemicals such as pesticides and additives rank as the most dangerous, illness-producing substances in our food supply, the real hazard comes from naturally occurring bacteria and other microbes.

Most dangerous:
1. Microbial foodborne illness
2. Naturally occurring toxins in foods
3. Residues in foods, including:
   - Environmental contaminants, such as industrial chemicals
   - Pesticides
   - Animal drugs, such as hormones or antibiotics
4. Food processing and nutrients in foods

Least dangerous:
5. Intentional food additives
6. Genetic modification of foods

as **pasteurization** and canning. Freezing, dehydrating, and refrigerating food also halts or slows down bacterial growth. In addition, special packaging techniques and antimicrobial preservatives help control food-related pathogens.

What causes most cases of foodborne illness? Many people believe that chemical additives and pesticide residues added during the growing and processing of food pose the greatest risk. However, most foodborne disease is actually caused by mishandling of food, either in food service establishments such as restaurants or in homes. In fact, food from restaurants, cafeterias, and other food service establishments accounts for about two-thirds of all reported outbreaks of foodborne illness, whereas food eaten at home accounts for about one-fourth.[3] Most cases of foodborne disease are caused by faulty handling, cooking, and storing of food long after it has left the manufacturer or processor. Table 13-1 ranks the most common food hazards.

Experts classify foodborne diseases into two types: intoxications and infections. **Food intoxications** occur when a chemical or toxin transmitted by way of food causes the body to malfunction. An example of food intoxication is the vomiting, nausea, abdominal cramping, sweating, and chills that result from eating food contaminated with a strain of bacteria called *Staphylococcus aureus*. This bacterium produces what is known as an **enterotoxin**, a toxin that causes severe gastrointestinal distress. Other types of bacteria can produce **neurotoxins**, or toxins that afflict the nervous system. Food intoxication can also be caused by eating food that has been contaminated with a chemical, such as lead or some other heavy metal.

Foodborne infections, in contrast, occur as a result of eating a food that contains living microorganisms, such as bacteria, viruses, or parasites that are capable of multiplying and thriving in the body. When ingested in large amounts, such microorganisms can wreak havoc in the digestive tract or other areas of the body. An example of this type of foodborne illness is infection with *Vibrio* bacteria, which often reside in raw seafood such as oysters and clams. Inside the body, the bacteria settle in quickly and cause abrupt onset of chills, fever, or vomiting. The following section provides an overview of the various agents that can cause a foodborne illness, be it intoxication or an infection.

Microbial Agents

When we order a meal at a restaurant or reach for an egg or glass of milk from the refrigerator at home, we don't usually think about the microorganisms that might be lurking in the foods or their potential to cause illness. We tend to assume that the foods and beverages we consume are safe to eat or drink. Granted, most of the time we don't need to worry. Typically, a food must harbor thousands of microorganisms before it causes nausea, diarrhea, cramps, or other

"I need 148 get-well cards."

pasteurization the process of sterilizing food via heat treatment.

food intoxication illness caused by eating food that contains a harmful toxin or chemical.

enterotoxin a toxic compound, produced by microorganisms, that harms mucous membranes, as in the gastrointestinal tract (*entero* = intestine).

neurotoxin a poisonous compound that disrupts the nervous system (*neuro* = nerve).

foodborne infection illness caused by eating a food containing bacteria or other microorganisms capable of growing and thriving in a person's tissues.

TABLE 13-2   Microbial Food Agents: Organisms That Can Bug You

Disease and Organism That Causes It	Annual Incidence (Deaths)	Source of Illness	Usual Onset after Eating	Symptoms
Bacteria:				
Botulism (*Botulinum* toxin produced by *Clostridium botulinum* bacteria)	60 (4 deaths)	Improperly canned foods, especially home-canned vegetables, soups, beets, asparagus, mushrooms, tuna, and liver paté; also found in luncheon meats, ham, sausage, stuffed eggplant, lobster, fermented fish, bottled garlic, and herb-flavored oils.	12–72 hours	Vomiting, diarrhea, blurred vision, double vision, muscle weakness, difficulty swallowing, speech difficulty, and progressive paralysis of the respiratory system. Get medical help immediately; can be fatal.
Campylobacteriosis (*Campylobacter jejuni*)	2 million (100 deaths)	Raw (or undercooked) poultry, meat, eggs, and unpasteurized milk, contaminated water.	2–5 days	Diarrhea, abdominal cramping, fever, vomiting, and sometimes bloody stools. Lasts 7–10 days.
Listeriosis (*Listeria monocytogenes*)	2,500 (500 deaths)	Soft cheeses made with unpasteurized milk, raw meat and seafood, unpasteurized milk, ready-to-eat deli meats, imported seafood products, frozen cooked crab meat, cooked shrimp, and cooked surimi (imitation shellfish).	9–48 hours (though symptoms can strike 2–6 weeks after eating)	Fever, muscle aches, headache, nausea, diarrhea, and vomiting.
Perfringens food poisoning (*Clostridium perfringens*)	250,000 (7 deaths)	In most instances, caused by failure to keep food hot. A few organisms are often present after cooking and multiply to toxic levels during cooldown and storage of prepared foods. Meats and meat products are most frequently implicated.	8–12 hours	Abdominal pain and diarrhea and sometimes nausea and vomiting.
Salmonellosis (*Salmonella* bacteria)	1,400,000 (600 deaths)	Raw or undercooked eggs, meats, poultry, unpasteurized milk or juice, cheese, shrimp, frog legs, yeast, coconut, pasta, chocolate, contaminated raw fruits and vegetables.	6–48 hours	Nausea, abdominal cramps, diarrhea, fever, and headache.
E. coli infection (shiga toxin-producing *E. coli* [STEC] released by *Escherichia coli* O157:H7 and others)	110,000 (78 deaths)	Water or food contaminated with human feces. Undercooked beef (especially hamburger), unpasteurized milk and juice, raw apple cider, mayonnaise, raw fruits and vegetables (such as alfalfa sprouts).	1–8 days	Abdominal cramps, vomiting, nausea, watery diarrhea that often turns bloody, low-grade fever, and, in severe cases, kidney failure, strokes, and seizures.
Shigellosis (*Shigella* bacteria)	90,000 (14 deaths)	Raw produce, contaminated drinking water, uncooked foods and cooked foods that are not reheated after contact with an infected food handler.	4–7 days	Abdominal cramps, diarrhea, fever, sometimes vomiting, and blood and mucus in stools.
Staphylococcal food poisoning (staphylococcal toxin produced by *Staphylococcus aureus* bacteria)	185,000 (2 deaths)	Toxin produced when food contaminated with the bacteria is left too long at room temperature. Meats, poultry, egg products, tuna, potato and macaroni salads, and cream-filled pastries are good environments for these bacteria to produce toxin.	1–6 hours	Sudden onset of severe nausea, diarrhea, vomiting, abdominal cramps. Fever may be present.
Vibrio infection (*Vibrio vulnificus*)	47 (18 deaths)	The bacteria live in coastal waters and can infect humans either through open wounds or through consumption of contaminated undercooked or raw seafood, such as shellfish (especially oysters).	1–7 days	Chills, fever, abdominal cramps, vomiting, diarrhea. Can be fatal to people with liver disease or weakened immune systems.

Disease and Organism That Causes It	Annual Incidence (Deaths)	Source of Illness	Usual Onset after Eating	Symptoms
Parasite:				
Cryptosporidiosis (*Cryptosporidium parvum*)	30,000 (7 deaths)	Swimming in or drinking contaminated water; contaminated raw fruits and vegetables, unpasteurized milk, juices, and ciders.	2–10 days	Diarrhea, nausea, fever; can be symptomless.
Cyclosporiasis (*Cyclospora cayetanensis*)	16,000 (no deaths)	Contaminated water or produce.	1–14 days (usually 7 days)	Anorexia, diarrhea, weight loss, fatigue, vomiting, muscle aches; can be symptomless.
Trichinosis (*Trichinella spiralis*)	50 (no deaths)	Raw or undercooked pork or wild game. Worms burrow into muscle tissue.	24 hours	Nausea, vomiting, abdominal pain, diarrhea, fever; after 2 weeks, muscle pain, fluid retention, weight loss, and fever.
Giardiasis or "Traveler's Diarrhea" (*Giardia duodenalis,* formerly *G. lamblia*)	200,000 (1 death)	Contaminated water, or may be transmitted by uncooked foods that become contaminated while growing or after cooking by infected food handlers.	12 hours to several days	Sudden onset of explosive watery stools, abdominal cramps, anorexia, nausea, and vomiting.
Virus:				
Hepatitis (hepatitis A virus)	4,200 (4 deaths)	Raw produce, contaminated drinking water, shellfish from contaminated waters, uncooked foods and cooked foods that are not reheated after contact with an infected food handler.	15–50 days	Begins with malaise, appetite loss, nausea, vomiting, and fever. After 3–10 days, person develops jaundice with darkened urine. Severe cases can cause liver damage and death.
"Stomach flu"* Noroviruses (Norwalk-type viruses)	9,200,000 (124 deaths)	Raw produce, contaminated drinking water, salads, sandwiches, or other foods contaminated by infected food handlers; raw oysters and other shellfish from contaminated waters.	12–48 hours	Headache, fever, nausea, vomiting, diarrhea, abdominal pain.

*A misnomer; unrelated to influenza.

SOURCES: Adapted from A. Hecht, *The Unwelcome Dinner Guest: Preventing Foodborne Illness* (Washington, DC: U.S. Government Printing Office, DHHS Publication No. (FDA) 91-2244); and Economic Research Service, *Economics of Foodborne Disease: Food and Pathogens,* available at www.ers.usda.gov.

symptoms of foodborne illness. What's more, a healthy body can usually defend itself against small amounts of the "bad bugs."

However, when proper food-handling procedures are not followed carefully, the risk of food poisoning from bacteria or other microbial agents soars. When food items are not cooked or stored properly, they provide the perfect medium in which microorganisms can flourish. For children, the elderly, people with compromised immune systems, and other vulnerable people, it might take only small amounts of the offending microorganisms to cause trouble. Table 13-2 summarizes the common food sources of the microbial agents responsible for most outbreaks of foodborne illness.

Of the microbial pathogens, *Campylobacter jejuni* ranks as one of the most prevalent pathogens and one of the leading causes of foodborne illness in the United States. In fact, two pathogens—*C. jejuni* and *Salmonella*—account for about 70 percent of diagnosed cases of bacterial foodborne illness in the United States.[4] Symptoms of the illness caused by *C. jejuni,* called campylobacteriosis, usually begin 2 to 5 days after the contaminated food is eaten and last for 7 to 10 days. Many cases of

Covered, bottled garlic-in-oil left at room temperature provides the perfect warm, oxygen-free, low-acid environment for the toxin-releasing spores of *Clostridium botulinum*.

campylobacteriosis probably go unreported because the illness's characteristic diarrhea, abdominal pain, and fever mimic flulike gastrointestinal ills.

In the past, egg products were a major source of salmonellosis—an illness caused by *Salmonella* bacteria. However, this is no longer the case, because of the mandatory pasteurization of eggs used in the commercial preparation of ice cream, baked goods, and other egg-containing food products. Raw eggs, however, have been implicated in outbreaks of salmonellosis, as has raw, unpasteurized milk. (Because of this concern, the sale of raw milk is banned in most states.) Today, most outbreaks of salmonellosis are caused by faulty handling of raw meat and poultry, unpasteurized juice, raw sprouts, and mangos.

Staphylococcus aureus is another strain of bacteria responsible for many of the reported cases of foodborne illnesses that occur in the United States. The bacterium, found in the nose and throat and on the skin of most people, can be transmitted to food when an individual with an infected wound or boil or a respiratory infection handles food improperly. The bacterium itself isn't directly responsible for the illness, however. Rather, it produces staphylococcal enterotoxin, which causes food poisoning within ½ to 8 hours of eating a contaminated food. The foods typically implicated in *S. aureus* intoxication include meat and poultry products, egg products, tuna, potato and macaroni salads, and cream-filled pastries.

Another particularly deadly foodborne pathogen is *Clostridium botulinum*, the agent that causes the nausea, vomiting, dizziness, and muscle paralysis known as *botulism*. This severe illness results from eating food in which the bacterium has flourished and produced a neurotoxin. The toxin binds irreversibly to nerve endings and causes progressive paralysis, which makes swallowing and breathing difficult. Botulism typically develops within 4 to 36 hours of eating the contaminated food. Because it can be fatal, botulism requires immediate medical attention.

Spores of *C. botulinum* are ubiquitous, having been detected in everything from shellfish to fruits and vegetables to honey to corn syrup.* However, the *botulinum* bacteria produce the deadly neurotoxin only if they are in a warm, oxygen-free, low-acid environment. That's why improperly sterilized, low-acid canned goods are the most common culprits in cases of botulism. Note also that trendy oils flavored with garlic or herbs can harbor *C. botulinum*. Manufacturers of such mixtures must add antibacterial agents to their preparations, but people who make their own flavored oils should be sure to store them in the refrigerator.

Another especially virulent pathogen that is emerging as a major public health concern is *Escherichia coli* O157:H7.† First recognized as a cause of foodborne illness in 1982, *E. coli* garnered national attention about a decade later when it caused a major outbreak of foodborne illness in the northwest part of the United States. Undercooked, contaminated hamburgers sold at a major fast-food chain prompted the 1993 scare, which ultimately led to more than 500 reported cases of illness, some 150 hospitalizations, and 3 deaths. The outbreak sparked a national debate about the safety of the U.S. meat and poultry supply.[5]

Found in the intestinal tracts of mammals, *E. coli* is usually transmitted via animal feces. It poses special concern because it is so dangerous. Unlike most other illness-producing microorganisms, *E. coli* need not be present in large numbers to make a person sick; just a few bacteria seem to do the trick. Once ingested, the bacterium clings to the intestinal wall, where it releases an enterotoxin that causes abdominal pain and watery diarrhea that often turns bloody. In vulnerable people,

---

*Honey has been found to contain dormant bacterial spores, which can awaken in the human body to produce botulism. In adults, this is not a hazard, but infants under 1 year of age should never be fed honey. Honey has been implicated in several cases of sudden infant death.
†*Escherichia coli* O157:H7 is a particular strain of *E. coli* bacteria. When *E. coli* is mentioned throughout the rest of this chapter, it refers to the O157:H7 strain.

such as children and the elderly, *E. coli* may cause serious complications, including kidney failure and death.

Most *E. coli* outbreaks have been linked to undercooked beef, particularly hamburger. Alfalfa sprouts, raw milk, raw spinach, and fresh apple cider, presumably made from apples exposed to tainted animal manure, have also been implicated in some outbreaks.[6] As with all illness-causing microorganisms, the best defense against *E. coli* is careful food handling.

Mold is another potential microbial food contaminant. Certain molds produce poisonous compounds called **aflatoxins**. These compounds are powerful liver toxins in animals, are known to be carcinogenic in some species, and can be lethal if consumed in large doses. Aflatoxins have been found on peanuts, wheat, corn, meat pies, dry beans, and even refrigerated and frozen pastries. Molds that produce the aflatoxins typically flourish when foods such as corn and peanuts get wet and are then stored in a warm place, such as a grain silo or railroad boxcar. Control of aflatoxins in food is difficult, but it is given high priority by the food industry and regulatory agencies.

To be sure, not all microorganisms are bad. A mold called *Penicillium roquefortii* imparts the special, pungent flavor of Roquefort cheese. Another mold strain, *P. camembertii*, lends flavor to Camembert cheese. Likewise, a strain of mold called *Bacteria aceti* causes the alcohol in wine and hard cider to turn to vinegar.[7] What's more, yogurt owes its existence to active cultures of bacteria added during processing, and a new area of research—called probiotics—suggests that the "good bugs" in the yogurt may help fight "bad bugs" that cause yeast infections and other ills.[8]

Natural Toxins

Most people assume that products derived from plants are safe because they are "natural." Nevertheless, natural food **toxicants** in plants—especially herbs—are sometimes to blame for poisonings. Even familiar foods that are generally safe sometimes harbor potential toxicants. For example, potatoes contain a substance called *solanine*, a powerful inhibitor of nerve impulses. The green substance accumulates just beneath the vegetable's skin, usually in harmless amounts. When potatoes are exposed to light, however, they sometimes develop excess solanine. When this happens, the green area and about one-half inch of potato around it should be removed before cooking.

When plants are transformed into powders and potions, their components become more concentrated, as do their potentially harmful effects. Nevertheless, people often assume that because they come from plants, these substances must be safe. Unlike the chemical composition of standard pharmaceutical drugs, however, the composition of herbal products is not regulated—by the government or anyone else. This lack of safety standards for herbal pills, powders, teas, and other potions, which have become increasingly popular in recent years, ranks as a major concern among public health officials. The potential risks are amplified when an herb is mislabeled, misidentified, or mixed with another potentially toxic substance.

For example, several children and adults suffered life-threatening respiratory problems and liver malfunctions as a result of taking a Chinese herb called Jin Bu Huan. Investigators found, among other problems, that the plant from which the product was derived had been misidentified and that the product carried false and misleading medical claims.[9] Along the same lines, the FDA has banned the use of a plant derivative called ma huang, which contains ephedrine (an amphetamine-like substance often used in weight-loss products) because it has been linked to rapid heart rate, stroke, nerve damage, and several deaths.[10]

These are just a few of the herb-related problems that continue to surface. They highlight the need to be cautious about using herbs of any sort. Adults should not take large amounts of a particular herbal product or take more than one product at a time without consulting a competent professional regarding the various effects of that product. Pregnant and breastfeeding women should avoid using herbs because

Numerous herbs have been implicated in liver failure and other health problems. Here are just a few:[11]

- Chaparral
- Kava
- Jin Bu Huan
- Ma huang
- Germander
- Comfrey
- Mistletoe
- Skullcap
- Margosa oil
- Maté tea
- Gordolobo yerba tea
- Pennyroyal (squawmint) oil

See Chapter 7 for more information about herbal remedies.

aflatoxin a poisonous toxin produced by molds.

toxicants poisons; that is, agents that cause physical harm or death when present in large amounts.

they can expose a fetus or an infant to a toxic dose.[12] In addition, people with any type of liver disease or condition should be wary of herbal products; liver failure is one of the hallmarks of an herbal overdose, because toxins accumulate in that organ.

In addition to plant foods, other types of food, notably seafood, often harbor natural toxicants. For example, a type of fish called puffers, long considered a delicacy in Japan, contains a potent poison—tetrodotoxin—which doesn't harm the fish but can be lethal to people and other animals. Tetrodotoxin, which is 275 times deadlier than cyanide, works by blocking nerve impulses. Over a period of several hours, it eventually closes down a person's entire nervous system. The toxin is so deadly that sale of puffers is illegal in the United States.[13]

Puffer poisoning is just one example of a foodborne disease traced to eating a toxic sea creature. Others also exist. For instance, paralytic shellfish poisoning can occur after eating mollusks (clams, mussels, oysters, and scallops) contaminated with marine algae that produce a neurotoxin. The mollusks themselves don't become ill; but people who eat them do, experiencing such symptoms as nausea, vomiting, cramps, and muscle weakness. In each of these cases, the toxin is not destroyed by the heat of cooking. Fish can also harbor viruses, including hepatitis viruses, worms, and other parasites. For these reasons, it is especially important to buy seafood from reputable vendors and to handle it with care.

Safe Food Storage and Preparation

Commercially prepared, canned, or packaged food is usually free of harmful microbial agents when it leaves the manufacturer. When a batch of food is contaminated, batch numbering ensures that it can be recalled quickly, and the public can be forewarned. Although tampering with food products is unlikely to occur in grocery store foods, your best protection is to carefully inspect the seals and wrappers of every purchase. Jars should be firmly sealed (many have safety buttons—areas of the lid designed to pop up once opened). Packages should be free of holes or tears. A broken seal or mangled package does not provide protection against microorganisms or other contaminants. Likewise, do not purchase any canned goods that have dents, cracks, or bulges, which can indicate possible contamination with *C. botulinum*.

Most cases of food poisoning occur in the home or the restaurant and are caused by improper food storage or handling. Food poisoning caused by kitchen mistakes can be avoided by doing three simple things: keep cold food cold; keep your hands, the utensils, and the kitchen clean; and keep hot food hot. Most bacteria flourish in warm environments, but heat kills them and cold temperatures halt their growth.

Keep Cold Foods Cold The first step, keeping cold food cold, starts when you leave the grocery store. If you are running errands in a car, do the grocery shopping last so that the food does not stay in the car too long, especially in hot weather. As soon as you get home, put foods in a refrigerator set at 40 degrees Fahrenheit or a freezer kept at 0 degrees Fahrenheit, and be sure not to leave food in the refrigerator too long before eating it (see Figure 13-1). Place packages of raw meat, poultry, or fish on a plate before refrigerating, or store them in plastic storage bags to prevent bacteria-containing juices from dripping onto other foods. Thaw frozen food in the refrigerator or microwave oven—not on the kitchen counter. Because bacteria can multiply at room temperature, they can thrive in the relatively warm exterior of a food before the interior has thawed.

Wash Hands and Surfaces Often Along with keeping foods properly chilled, keeping the kitchen clean prevents contamination of otherwise wholesome foods. Before you handle food, wash your hands in warm, soapy water. In addition, be sure to wash your hands after touching meat, poultry, or fish to prevent the spread of any

Never thaw food on a kitchen counter. Bacteria can flourish at room temperature.

FIGURE 13-1

Take Control of Home Food Safety

24-hour bug? Or something you ate? Very often what seems like the flu may be foodborne illness. The following four simple actions will help you take control of food safety in your kitchen.

SOURCE: Adapted from the American Dietetic Association and the ConAgra Foundation's *Home Food Safety . . . It's In Your Hands* program. For more information, visit www.homefoodsafety.org and www.fightbac.org.

1. CLEAN: Wash Hands and Surfaces Often Using Hot Soapy Water.

Hand washing may eliminate half of all cases of foodborne illness and reduce the spread of the common cold and flu.

• Wash hands in warm, soapy water for a minimum of 20 seconds before preparing foods and after handling raw meat, poultry, and seafood.

4. CHILL: Refrigerate Promptly Below 40°F.

• Defrost foods in the refrigerator or in the microwave. Marinate foods in the refrigerator. Refrigerate foods quickly to slow the growth of bacteria and prevent foodborne illness. Set your refrigerator below 40°F to keep perishable foods out of the "danger zone"—40° to 140°F.

2. SEPARATE: Keep Raw Meats and Ready-to-Eat Foods Separate.

• Never work with raw meat, poultry, or seafood on the same surface that you use for other foods without cleaning the surface after you've finished.

• Use different cutting boards: one to cut raw meat, poultry, and seafood and another for ready-to-eat foods, such as breads and vegetables. Discard old cutting boards that have cracks, crevices, and excessive knife scars.

3. COOK: Cook to Proper Temperatures.

Harmful bacteria are destroyed when food is cooked to proper temperatures. Buy a meat thermometer and use it!

"IT'S SAFE TO BITE WHEN THE TEMPERATURE IS RIGHT!"

General Guidelines for Leftovers

Perishable food	Keeps up to
Cooked fresh vegetables	3–4 days
Cooked pasta	3–5 days
Cooked rice	1 week
Deli counter meats	5 days
Greens	1–2 days
Meat	
Ham, cooked and sliced	3–4 days
Hot dogs, opened	1 week
Lunch meats, prepackaged, opened	3–5 days
Cooked beef, pork, poultry, fish, meat casseroles	3–4 days
Cooked patties and nuggets, gravy and broth	1–2 days
Seafood, cooked	2 days
Soups and stews	3–4 days
Stuffing	1–2 days

When in doubt, throw it out!

Food Safety and Inspection Service, USDA

Safe Cooking Temperatures

140°F: Ham, fully-cooked (to reheat)

145°F: Beef, lamb & veal steaks & roasts (medium rare)

160°F: Ground beef, pork, veal & lamb; Beef, lamb & veal steaks & roasts (medium); Pork chops, ribs & roasts; Egg dishes

165°F: Ground chicken & turkey; Chicken & turkey—whole bird, breasts, legs, thighs & wings; Stuffing & casseroles; Leftovers

---

bacteria that your hands have picked up. Likewise, keep countertops and all kitchen equipment clean with soap and hot water. Because bacteria love to nestle down in the fibers of kitchen cloths, sponges, and wooden cutting boards, take particular care to keep such items clean.

Keep Hot Foods Hot Keeping hot food hot means first cooking food thoroughly to ensure that the heat destroys any bacteria that might have been present in the uncooked food. See Figure 13-1 for proper cooking temperatures. After cooking, foods must be kept hot until serving to prevent bacterial growth. Never leave perishable food at room temperature for more than 2 hours. Before refrigerating a large quantity of hot food, such as a pot of chili or a large casserole, divide it up and place it in shallow containers to allow easy cooling. Otherwise the food inside the pot may stay warm for a dangerously long time, even in the refrigerator.

Prevent Cross-Contamination Meat, poultry, and fish require special handling because they often harbor high levels of bacteria. In addition, they provide a moist, nutritious environment—just right for microbial growth. Wash anything that has come into contact with such foods to prevent **cross-contamination**. For instance, after marinating raw meat in a dish, don't put the meat back in the same dish after cooking it, and don't use the marinade unless it has been cooked thoroughly. Wash the dish in hot, soapy water before reusing it; otherwise, the bacteria inevitably left in the dish from the raw meat can contaminate and grow in the cooked product or other food—a classic example of cross-contamination. Similarly, wash a cutting board (and your hands) after, say, skinning chicken on it. If you don't, and you use the contaminated board to chop raw vegetables for a salad, the vegetables can pick up the bacteria from the poultry. Because the salad won't be heated, the bacteria won't be killed.

Food Safety for Meats Ground meat is especially susceptible to bacterial contamination. Steaks and roasts are not as risky, because bacteria usually settle on the outside of the cuts and are easily destroyed when the outside is heated. But, because the process of grinding meat spreads the bacteria throughout, foods made with ground meat, such as hamburger patties or meat loaf, must be cooked all the way through to kill the bacteria in the middle. To decrease your risk of eating contaminated ground beef, cook burgers until the juices run clear and not a trace of pink is left on the inside, and always order burgers well done at restaurants. When baking a meat loaf, use a thermometer to test the internal temperature. Be especially careful when cooking ground meat in a microwave oven, which sometimes cooks foods unevenly.

Reports of outbreaks of **mad cow disease**, primarily in the United Kingdom but in other countries as well, raise concerns among American consumers regarding the safety of consuming meat from infected animals. Agricultural officials in the United Kingdom now prohibit the addition of mammalian meat and bone meal to the feed of any food-producing animals, because these tissues are suspected to be the source of the disease. As a result, the rate of newly reported cases of the disease is decreasing. The USDA also bans the use of most mammalian protein tissues in the manufacture of animal feed given to ruminant animals (for example, cows, sheep, and goats).[14]

Food Safety for Seafood Seafood also should be handled with care, especially fish intended to be eaten raw or only lightly steamed. The dangerous foodborne diseases that may lurk in normal-appearing seafood—worms, parasites, severe viral intestinal disorders, and hepatitis—can be much worse than those resulting from normal bacterial spoilage. Raw fish dishes such as sushi and sashimi are safe for most healthy peo-

cross-contamination the inadvertent transfer of bacteria from one food to another that occurs, for instance, by chopping vegetables on the same cutting board that was used to skin poultry.

mad cow disease also called *bovine spongiform encephalopathy* (BSE); a rare and fatal degenerative disease first diagnosed in 1986 in cattle in the United Kingdom. The bovine disease may be passed to humans who eat the meat of infected animals and may lead to death due to brain and nerve damage.

The U.S. Department of Agriculture requires that all fresh meat and poultry products carry this label as a reminder to handle the products carefully.

Safe Handling Instructions

THIS PRODUCT WAS PREPARED FROM INSPECTED AND PASSED MEAT AND/OR POULTRY. SOME FOOD PRODUCTS MAY CONTAIN BACTERIA THAT CAN CAUSE ILLNESS IF THE PRODUCT IS MISHANDLED OR COOKED IMPROPERLY. FOR YOUR PROTECTION, FOLLOW THESE SAFE HANDLING INSTRUCTIONS.

KEEP REFRIGERATED OR FROZEN.
THAW IN REFRIGERATOR OR MICROWAVE.

KEEP RAW MEAT AND POULTRY SEPARATE FROM OTHER FOODS. WASH WORKING SURFACES (INCLUDING CUTTING BOARDS), UTENSILS, AND HANDS AFTER TOUCHING RAW MEAT OR POULTRY.

COOK THOROUGHLY.

KEEP HOT FOODS HOT. REFRIGERATE LEFTOVERS IMMEDIATELY OR DISCARD.

ple to eat if they are prepared with fresh fish that has been commercially frozen at temperatures lower than most home freezers can attain. This type of freezing kills any parasites that might be present. However, eating raw or undercooked oysters, clams, mussels, and whole scallops is especially risky because these types of seafood sometimes carry a strain of bacteria known as *Vibrios*, which can multiply even during refrigeration. Although *Vibrios* bacteria are destroyed by thorough cooking, they can thrive in raw shellfish and cause serious illness, sometimes even causing a deadly blood poisoning that can kill a person within a day or two.[15] Thus, the hazards of eating raw or undercooked seafood are especially serious for vulnerable people, such as those with liver disease, diabetes, gastrointestinal disorders, and HIV infection. People with these and other diseases that may compromise the body's immune system, and hence its ability to defend itself against food poisoning (see Table 13-3), need to be especially cautious when preparing and consuming seafood.

Handle raw meat and poultry with care, and cook it thoroughly to destroy any bacteria present. Place it on a clean plate when it is cooked.

Food Safety for Picnics When planning picnics, choose foods that last without refrigeration, such as fresh fruits and vegetables, breads and crackers, and canned spreads and cheeses that can be opened and used on the spot. Aged cheeses, such as cheddar and Swiss, do well for an hour or so, but they should be kept in an ice chest for longer periods. Chill dishes containing mayonnaise or other types of dressing before, during, and after the picnic. Burgers, chicken breasts, and other foods intended for grilling should not be partially cooked ahead of time. Partial cooking may not kill all the bacteria, and if these half-cooked items are not heated thoroughly later, chances of bacterial contamination run high. Partial cooking is a safe method only when, for example, you take a burger or other food *directly* from the microwave oven to the grill.[16]

In general, remember that fresh food smells fresh. Do not eat any food that carries an "off" odor, because the smell is probably the result of bacterial wastes and indicates a dangerous number of bacteria in the food. To be sure, not all types of food poisoning are detectable by odor; but if a food smells bad, chances are high that it

TABLE 13-3	Individuals at High Risk for Foodborne Illness and Foods They Should Avoid
High-Risk Patient Categories	**Foods to Avoid**
Young children	Raw fish or shellfish
Pregnant women	Raw or unpasteurized milk or cheeses
Elderly individuals	Soft, French-style cheeses and patés
Immunocompromised individuals	Raw or undercooked eggs or foods containing raw or lightly cooked eggs (e.g., certain salad dressings, cookie and cake batters, sauces, and beverages such as unpasteurized eggnog)
	Raw or undercooked meat or poultry
	Precooked processed meats that have not been reheated (e.g., deli meats, hot dogs)
	Raw sprouts (alfalfa, clover, and radish)
	Unpasteurized fruit or vegetable juices

SOURCE: Data from *Diagnosis and Management of Foodborne Illnesses: A Primer for Physicians,* available at www.ama-assn. org ama/pub/article/ 3707–3938.html; Nutrition and the MD (Hagerstown, MD: Lippincott & Wilkins, 2003), p. 8.

The tradition of tea drinking has been popular for more than 2,000 years in China. Contaminated water may have been one of the reasons. A hot drink made with boiled water was less likely to cause digestive problems than plain water.[17]

To Prevent Foodborne Illness

- When running errands, make the grocery store your last stop. When you get home, refrigerate perishables immediately.

- Refrigerate leftovers promptly, and heat them thoroughly to at least 165°F before serving. Refrigerate leftovers within 2 hours of serving (or 1 hour if it is warmer than 90° F).

- Use clean eggs with intact shells, and cook eggs before eating them (soft-boiled for 7 minutes, poached for 5, or fried for 3 minutes on each side).

- Keep hot foods hot (140°F or above). Keep cold foods cold (40°F or below).

- Do not prepare food if you have a skin infection or infectious disease. Anyone, though, may be a carrier of bacteria and should avoid coughing or sneezing over food.

- When in doubt, throw it out. Do not even taste food that is suspect. (An "off" odor, however, is not necessarily detectable in a food containing toxins.)

- Discard food from cans that leak or bulge. Dispose of the food in a manner that will protect other people and animals from accidental use.

- Cook all meat and poultry to 160°F or higher, until the flesh is no longer pink and all the juices run clear. Order hamburgers and other beef items well done when eating out.

- Wash fruits and vegetables thoroughly before eating.

To Ensure a Safe Catch: Seafood Safety Tips

- Choose fresh fish steaks and fillets that are moist, with no drying or browning around the edges.

- Buy seafood only from reputable dealers, and cook fish within 2 days of purchase. Use packaged, frozen seafood before the expiration date.

- Make sure the fish odor is fresh and mild. Fresh fish and shellfish should not have an overly "fishy" odor. Refrigerate seafood immediately below 40°F after buying.

- Cook fish to 145°F or until the flesh is opaque and separates easily with a fork.

- Refrigerate smoked, pickled, vacuum-packed, and modified-atmosphere-packed fish products.

Many of the chemicals that contaminate foods are the waste products of industry.

contaminants potentially dangerous substances, such as lead, that can accidentally get into foods.

organic halogens compounds that contain one or more of a class of atoms called *halogens*, including fluorine, chlorine, iodine, or bromine.

heavy metals any of a number of mineral ions, such as mercury and lead, so named because of their relatively high atomic weight. Many heavy metals are poisonous.

is spoiled. Refer to the Savvy Diner feature for recommendations about preventing foodborne illness. See also the Scorecard feature in this chapter to help you rate your food-safety knowledge.

Pesticides and Other Chemical Contaminants

Food producers and food processors exert major efforts to maintain a safe food supply. Some risk of consuming undesirable substances, or **contaminants**, however, is unavoidable. Our industrial society's reliance on chemical processes means that foods may become contaminated by a variety of chemicals introduced into the environment. In addition, pesticides used in agricultural techniques affect the food supply. The following section examines some of the major chemical players in the food supply and looks at some ways that scientists hope to reduce them.

Chemical Agents Some of the problem industrial chemicals prevalent in the environment and food supply are **organic halogens**, such as polychlorinated biphenyl (PCB) and polybrominated biphenyl (PBB); and **heavy metals**, such as lead and mercury. Fortunately, episodes of direct, excessive chemical contamination such as chemical spills are rare. However, when contamination *does* occur, the effects can be far-reaching. A list of several chemical contaminants of particular concern in foods is presented in Table 13-4.

© Phil Borden/PhotoEdit

TABLE 13-4	Examples of Contaminants in Foods		
Name and Description	**Sources**	**Toxic Effects**	**Typical Route to Food Chain**
Cadmium (heavy metal)	Used in industrial processes, including electroplating, plastics, batteries, alloys, pigments, smelters, and burning fuels; present in cigarette smoke	No immediately detectable symptoms; slowly and irreversibly damages kidneys and liver	Enters air in smokestack emissions, settles on ground and is absorbed into plants, consumed by farm animals, and eaten in meat and produce by people. Sewage sludge and fertilizers leave large amounts in the soil; runoff contaminates shellfish.
Lead (heavy metal)	Lead crystal, improperly manufactured and old ceramic ware, paint, old plumbing, and leaded gasoline	Displaces calcium, iron, zinc, and other minerals from their sites of action in the nervous system, bone marrow, kidneys, and liver, causing failure to function; causes breakage of red blood cells (anemia) and interferes with the immune response	Air pollution, leaded gasoline, water pipes, improperly manufactured and old ceramic ware, and lead crystal
Mercury (heavy metal)	Widely dispersed in gases from the Earth's crust; local high concentrations from industry, electrical equipment, paints, and agriculture	Poisons the nervous system, especially in fetuses	Inorganic mercury released into waterways by industry and acid rain is converted to methylmercury by bacteria and ingested by fish.*
Polychlorinated biphenyl (PCB) (organic compound)	No natural source; produced for use in electrical equipment	Causes long-lasting skin eruptions, eye irritation, growth retardation in children of exposed mothers, anorexia, fatigue, and other effects	Discarded electrical equipment, accidental industrial leakage, or reuse of PCB containers for food

*The FDA recommends that pregnant women and lactating women eliminate from their diets certain species of large, predatory fish, because these fish may contain high levels of mercury that can cause neurological damage during sensitive stages of fetal brain development. These include shark, swordfish, king mackerel, tilefish, and tuna steaks. As an extra precaution, women who plan to become pregnant should avoid those species for 6 months before conception. Visit www.cfsan.fda.gov for more information and updates.

Under normal circumstances, even though a chemical substance is toxic, it tends to pose only a small hazard because it is usually carefully controlled. However, when a chemical spill or other accident occurs, the risk can suddenly soar. That's why scientists differentiate between **toxicity** (a property of all substances) and **hazard** (the likelihood of a substance's actually causing harm). All substances are potentially toxic, but they are hazardous only if consumed in large enough quantities. In other words, the dose makes the poison. Thus, a chemical that is present in foods in minuscule amounts does not pose a significant hazard until for some reason it becomes concentrated in the food, that is, in toxic amounts.

Most experts agree that although chemicals in foods pose a small hazard, the chief concern is accidental gross contamination. In some instances, however, chemical contamination can be subtle and insidious. The problem of chronic low-level lead poisoning is a prime example. In the United States, lead poisoning ranks as one of the most common childhood environmental health problems. More than 300,000 children under the age of 6 have blood lead levels high enough to cause behavioral, mental, and other health problems.[18]

Lead contamination usually does not poison a person all at once; rather, low levels build up gradually in the soft tissues of the kidneys, bone marrow, liver, and brain. Over time, the accumulated lead can cause such health problems as diminished intelligence and impaired development.

Historically, lead has entered the food and water supply largely through leaded gasoline exhaust, which contaminates rainfall, which in turn pervades crop soil and water supplies. Lead contamination of the water supply continues to be a source of concern (refer to the discussion on water in Chapter 9). In addition, lead solder used to seal the seams of cans was a source of the contaminant for many years.

Some antique ceramic and crystal items are best left on the shelf. The older the piece, the higher the chances that it leaches lead.

toxicity the ability of a substance to harm living organisms. All substances are toxic if present in high enough concentrations.

hazard state of danger; used to refer to any circumstance in which harm is possible.

Improper storage of food not only increases the risk of food poisoning but also almost always results in a loss of nutrients and good taste. The following quiz is designed to measure your food-safety savvy.

1. If you're packing a picnic, it's okay for the cold foods to be at room temperature before packing as long as they are placed in a cooler with ice or ice packs. *True or False?*

2. Foods prepared with mayonnaise—macaroni salad, potato salad, and coleslaw—are common sources of food poisoning. *True or False?*

3. Raw ground meat or poultry can be stored in the refrigerator, but these items should be used within:
   a. 1–2 days.
   b. 2–3 days.
   c. 3–4 days.
   d. 1 week.

4. What signs indicate that canned foods may be contaminated?
   a. Bulging
   b. Leaking
   c. Spurting of liquid when opened
   d. All of the above

5. Which of the following foods have been linked to food poisoning?
   a. Cooked rice
   b. Apple cider
   c. Shellfish
   d. All of the above

6. The best place in the refrigerator to store milk is in the door. *True or False?*

7. When bringing home groceries from the market, it's a good idea to rewrap meat and poultry before placing them in the refrigerator or freezer. *True or False?*

8. Fresh fish should smell "fishy." *True or False?*

9. The best way to handle green-skinned potatoes is to
   a. throw them away.
   b. soak them in cold water.
   c. peel the skin and remove some of the flesh before cooking.
   d. remove the green section after cooking.

10. Canned foods can be stored in the pantry indefinitely. *True or False?*

11. You can tell if a food is contaminated by the way it looks, smells, or tastes. *True or False?*

12. What is the maximum time perishable foods can be kept at room temperature?
   a. ½ hour
   b. 1 hour
   c. 2 hours
   d. 24 hours

13. It's okay to thaw frozen ham on the kitchen counter because salted and smoked meats are free of bacteria. *True or False?*

14. You can reduce your exposure to chemical contaminants in whole fish by proper cleaning. *True or False?*

Fortunately, a 25-year process to phase out lead in gasoline reached its goal in 1995, and lead-soldered cans have been eliminated. Today the levels of lead in foods and beverages are the lowest in history.

Nevertheless, lead paint abounds in older housing, and new sources of lead contamination have surfaced during recent years. When it comes to lead exposure via food and beverages, scientists have identified ceramic hollowware, lead crystal, and foil capsules on wine bottles as potential sources. Lead can leach from the glaze of ceramic bowls, mugs, and pitchers that have not been properly formulated or fired—

15. Cloudy liquid around packaged hot dogs indicates that bacteria have started growing. *True or False?*

Answers

1. False. Be sure food is cold or frozen before placing it in a cooler. This minimizes the chance of microbial growth. Use ice packs between food items. Frozen juice boxes can also be packed and enjoyed later in the day after they have thawed.

2. False. Adding mayonnaise to food does not increase the chance of food poisoning. Most store-bought mayonnaise contains ingredients (vinegar, lemon juice, and salt) that actually slow bacterial growth.

3. (a) Ground meat and ground poultry are more perishable than other meats. They should be refrigerated and cooked (or frozen) within 1–2 days of purchase.

4. (d) Never buy or use products with bulging lids, leaking cans, or cracked jars. All are warning signs that a product may be contaminated with the bacteria that cause deadly botulism.

5. (d) Although the most common offenders are poultry, meat, and shellfish, any food can cause food poisoning, including cooked rice, if mishandled.

6. False. The refrigerator door does not stay as cold as the rest of the refrigerator. Highly perishable foods like milk and eggs should be stored on an inside shelf. Use the door for condiments.

7. False. Leave meat and poultry in store wrapping. The less you handle it, the better. This is especially true for ground meat and ground poultry.

8. False. Fresh fish should smell like a fresh sea breeze. If it smells "fishy," don't buy it.

9. (c) Green-skinned potatoes contain a natural toxin called solanine, which develops when potatoes are exposed to sunlight. It imparts a bitter taste and may cause stomach upset if eaten in large quantities. Before cooking, simply remove the green area and about ½ inch of flesh around it.

10. False. Canned foods have an extended but not infinite shelf life. Remember to place newly purchased cans behind older ones. To be safe, throw out beans and high-acid foods (for example, pineapple, tomatoes) after 1 year and other canned foods after 2 years. Use canned fish within 6 months.

11. False. Food spoilage may leave telltale signs such as changes in looks, smell, or taste, but contaminated food does not. Remember this rule of thumb: "When in doubt, throw it out."

12. (c) The longer a perishable food is kept at room temperature, the greater the likelihood that bacteria will multiply to dangerous levels. Food kept unrefrigerated for more than 2 hours is a prime target for bacterial growth.

13. False. Many people think that salted or smoked meats are immune to bacterial contamination. That's not so, especially since many manufacturers have gradually lowered the salt content of cured meats. With any meat or poultry, the safest way to thaw it is in the refrigerator overnight.

14. True. By removing the skin and trimming the fatty tissue along the back and belly, where chemicals tend to concentrate, you can reduce your exposure to contaminants.

15. True. Although hot dogs are processed to last longer than other meat products, *Listeria monocytogenes* can grow even under refrigeration. If you notice cloudy liquid, discard the franks. Freeze hot dogs if you don't plan to use them within 2 weeks. If opened, use within a week.

How Did You Score?

Count up the number of questions you answered correctly.

12–15   Congratulations! You're quite the food-safety scholar.

8–11   You're fairly savvy when it comes to food safety, but don't push your luck. Brush up on the questions you had trouble with.

7 or below   You're a likely target for food poisoning. Mend your ways; it's never too late.

SOURCE: Adapted with permission from A. Schepers, "What's your food safety I.Q.?" *Environmental Nutrition* 22 (1999): 1, 6. © Copyright 1999 by Environmental Nutrition, Inc., 52 Riverside Drive, New York, NY 10024. www.environmentalnutrition.com

especially when the food or beverage inside is hot and acidic, such as coffee, tea, or tomato soup. Wine and other alcoholic beverages also promote the leaching of lead, so experts advise against storing alcohol in pitchers and other containers made with lead crystal. In addition, it's a good idea to use a damp cloth to wipe the rims of wine bottles capped with foil, which has been shown to leach lead. For help on lowering your own risk, see the tips on the following page.

Although chemical contamination is not the greatest hazard posed by the food supply, there are still many unknowns. For example, what are the effects of prolonged

Getting the lead out:

To protect against overexposure to lead:

- Do not store fruit juices or other acidic foods in ceramic containers.
- Do not store beverages in lead crystal containers.
- Do not eat or drink from items that show a dusty or chalky gray residue on the glaze after they are washed.
- Do not feed babies from crystal bottles.
- Run cold water for a minute before using it for drinking or cooking.

For more information, call the National Lead Information Center at 1-800-424-LEAD.

exposure? Despite the uncertainties, keep in mind that more often than not, healthful eating habits and overall good health protect against the toxicity of food contaminants and other environmental pollutants. Eating a wide variety of food ensures an adequate supply of essential nutrients and minimizes exposure to potential environmental contaminants.

Pesticide Residues   **Pesticides** are substances used to prevent, destroy, or repel harmful pests, including insects, spiders, bacteria, weeds, molds and mildews, rodents, and other living things. Unlike many other chemicals present in the food supply, pesticides are intended specifically to poison living things. The trouble with pesticides, of course, is that they can inadvertently harm wildlife, people, and other species.

When farmers first began using pesticides, they chose potent ones: chemicals designed to keep on killing for as long as they remained in the soil. Unfortunately, after years of widespread pesticide use, some unsettling side effects of the chemicals surfaced. They washed from the soil into lakes, rivers, and oceans, contaminating water supplies; they poisoned farmworkers, who breathed the chemicals day in and day out; they endangered many species of wildlife, disrupting the animals' abilities to reproduce and wiping out entire populations; and, most disturbing, they would not go away.

For example, in the 1960s, scientists observed that a popular pesticide called DDT had begun accumulating in the body fat of animals. DDT threatened the survival of the American eagle by weakening eggshells to the point of collapse, killing the developing chicks inside. DDT also appeared in big fish, carnivorous animals, people, and human breast milk. Finally, after years of widespread agricultural use, the United States banned DDT. However, the pesticide still lingers in the environment. Many foreign countries continue to use DDT, including countries from which the United States buys produce. Regrettably, despite the U.S. ban on DDT, U.S. companies are still allowed to sell it (as well as other banned pesticides) to countries where DDT use remains legal. This practice may come back to haunt us, however, when we buy imported produce that has been exposed to DDT. In addition, the DDT situation illustrates the need to consider environmental issues from a global perspective; residues of chemicals banned in one country can travel the entire world not only through the imported and exported foods but also through wind, rain, and waterways.

Foods imported from other countries may harbor residues of pesticides that have been banned for use in the United States.

DDT taught us another lesson about pesticides: A contaminant that builds up in the body carries the greatest potential risk. In the body, a contaminant that is quickly broken down to some harmless compound poses the least risk to health. Likewise, if the body can easily and rapidly excrete a pesticide residue, the body may not be harmed by it. But if the residue enters the body and stays there, all the while interacting with the body's cells and systems, it may wreak havoc. Additional doses piled on top of the first ones compound the damaging effects. Moreover, when a substance resists breakdown either inside the body (by the body's own enzymes) or outside (by microorganisms) and furthermore accumulates from one species to the next, it builds up in the food chain (see Figure 13-2). DDT causes all these problems, which is why it is so deadly.

Even so, pesticides do have a place in the farming community. Farmers would face daunting obstacles if they tried to grow crops without them. Careful use of pesticides often boosts crop yields, which in turn helps keep the price of fruits and vegetables

pesticides chemicals intentionally applied to plants, including foods, to prevent or eliminate pest damage. *Pests* include all living organisms that destroy or spoil foods: bacteria, molds and fungi, insects, and rats and other rodents, to name a few.

FIGURE 13-2

Level 4
A 150-pound person

Level 3
100 pounds of larger fish

Level 2
A few tons of plant-eating fish

Level 1
Several tons of plants

How a Food Chain Works

A person who eats fish regularly may consume about 100 pounds of it in a year. These fish will, in turn, have eaten a few tons of small plant-eating fish during their lifetime. The little plant eaters will have ingested several tons of plants. If the plants have been contaminated with toxic chemicals, the bodies of the small fish that eat them will contain high concentrations of the chemicals; the larger fish that eat the little fish will harbor even higher amounts of the chemicals; and so on through the food chain. If none of the chemicals are lost along the way, the person at the top of the food chain ultimately eats the same amount of chemical contaminants that was contained in the original *several tons* of plants at the bottom of the food chain.

down and the availability of a wide variety of produce high. Still, as the DDT experience illustrates, pesticides must be evaluated scrupulously and used judiciously.

The ideal pesticide destroys the target pest and then breaks down quickly into other products that pose minimal hazard to people and other animals. Scientists have made many strides in developing relatively innocuous pesticides over the years, and the search for better, safer pesticides continues. What's more, since the introduction of pesticides decades ago, many national and international agencies have adopted strict **regulations** on pesticide use.

In the United States, the Environmental Protection Agency (EPA) determines whether a particular chemical may be used on U.S. crops. In deciding whether to approve a pesticide, the EPA scrutinizes dozens of studies that assess the possible effects of the substance on people, wildlife, fish, or plants, as well as its potential to cause such problems as cancer and birth defects. To do so, the EPA examines the estimated amount of a pesticide residue that a person might be exposed to during the course of a 70-year lifetime and the **risk** of harm that amount of exposure might pose. If the risk is unacceptably high, the pesticide is ruled out. This extensive research process often requires years of research and costs millions of dollars.[19]

Once the EPA approves a pesticide for use, it sets forth safety standards. For those pesticides it allows, the EPA determines which crops may be treated with it and how much may be applied. It also establishes a **tolerance**—that is, the maximum amount of a pesticide residue allowed in or on a food (see Table 13-5). The EPA identifies what is called a **reference dose** for the pesticide: the amount that could be consumed daily without posing any health risk. First, scientists use animal studies to estimate the maximum amount of the chemical that a person could take in daily without suffering harm. Next, to ensure an extra **margin of safety**, they calculate a small fraction of this amount, usually $\frac{1}{100}$, which becomes the reference dose. In other words, the reference dose is $\frac{1}{100}$ of the maximum amount of the substance that appears to be safe. Scientists factor in this large margin of safety as a precaution to help ensure

regulation a legal mandate that must be obeyed. Failure to follow a regulation brings about legal consequences.

risk for pesticide residues, the harm a substance may cause. Scientists estimate risk by assessing the amount of a chemical that each person in a population might consume over time (also called *exposure*) and by considering how toxic the substance might be (*toxicity*).

$risk = exposure \times toxicity$

$exposure = $ amount of substance in food $\times$ amount of food eaten

tolerance the maximum amount of a particular substance allowed in food.

reference dose the estimated amount of a chemical that could be consumed daily without causing harmful effects.

margin of safety from a food-safety standpoint, the *margin* is a zone between the maximum amount of a substance that appears to be safe and the amount allowed in the food supply.

TABLE 13-5	Pesticides in Perspective

When you hear about pesticide residues in food, the amounts measured are either parts per million (ppm), parts per billion (ppb), or parts per trillion (ppt). The following comparisons show just how tiny these amounts are.

1 ppm: 1 gram of residue in 1 million grams of food

1 inch in 16 miles

1 minute in 2 years

1 cent in $10,000

1 ppb: 1 gram of residue in 1 billion grams of food

1 inch in 16,000 miles

1 second in 32 years

1 cent in $10 million

1 ppt: 1 gram of residue in 1 trillion grams of food

1 inch in 16 million miles

1 second in 32,000 years

1 square foot on floor tile the size of Indiana

SOURCE: Adapted from International Food Information Council, *Pesticides and Food Safety* (Washington, DC: International Food Information Council, 1995), 2.

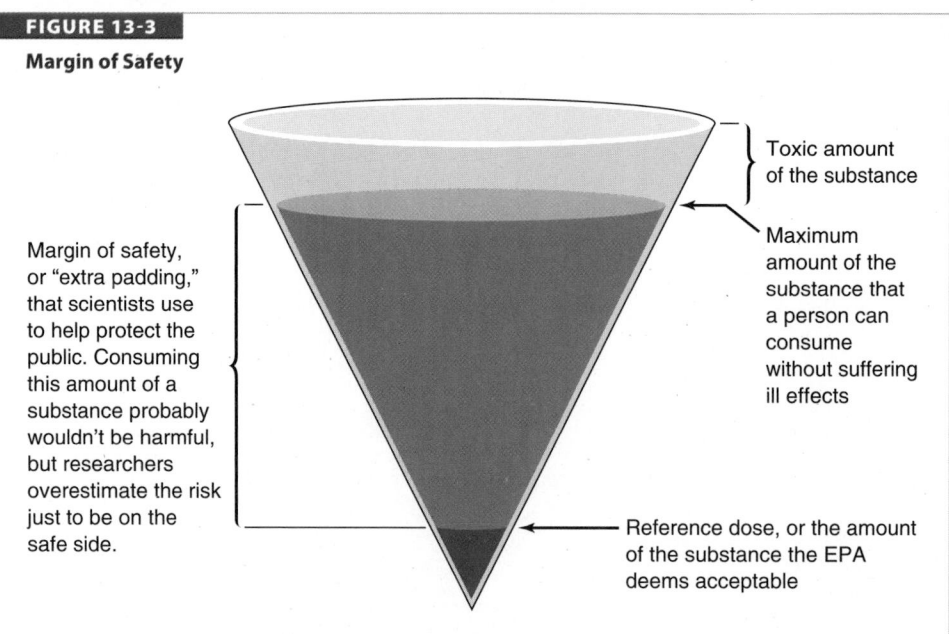

FIGURE 13-3
Margin of Safety

Margin of safety, or "extra padding," that scientists use to help protect the public. Consuming this amount of a substance probably wouldn't be harmful, but researchers overestimate the risk just to be on the safe side.

Toxic amount of the substance

Maximum amount of the substance that a person can consume without suffering ill effects

Reference dose, or the amount of the substance the EPA deems acceptable

that even highly vulnerable people won't be harmed by the substance in question (see Figure 13-3).[20]

After the EPA approves a pesticide, the FDA, in its ongoing monitoring program, begins to check for residues. Inspectors collect samples from packers, shippers, and other food handlers and test them for pesticide residues. If a food contains residues that exceed the EPA's tolerance limits, the FDA can seize the entire shipment and press criminal charges. Meanwhile, how can you protect yourself against unacceptably high levels of pesticides? The Nutrition Action feature later in this chapter offers a perspective.

Although EPA and FDA regulations go a long way in protecting the public from harmful pesticide residues, problems with the monitoring system still exist. As explained earlier, other countries may use pesticides banned in the United States, and imported foods might not be tested for the presence of those pesticides. In addition, consider that the National Academy of Sciences issued the results of a large-scale, 5-year study on the risk of pesticide residues in the diets of infants and children who may be more susceptible than adults to adverse effects from pesticides. Although it concluded that the U.S. food supply is safe for children, it called for changes in the methods used to assess health risks from pesticides to better account for differences between adults and children.[21] To keep pesticide residues from all types of produce to a minimum, use the produce-handling tips in Table 13-6.

Food Additives

food additive any substance added to food, including substances used in production, processing, treatment, packaging, transportation, or storage.

intentional food additives substances intentionally added to food; examples include nutrients, colors, spices, and herbs.

incidental food additives (indirect additives) substances that accidentally get into food as a result of contact with it during growing, processing, packaging, storing, or some other stage before the food is consumed.

From a safety standpoint, **food additives** rank among the least hazardous substances in food, although consumers tend to rank them high on their list of food-related concerns. The great majority of food additives enhance the color, flavor, texture, or stability of foods or even improve the nutritional value of certain items, as shown in Table 13-7. Many additives are common substances such as vitamins, herbs, and spices, deliberately added to foods and called *direct* or **intentional food additives**.

In contrast, **incidental food additives**, such as packaging materials or processing chemicals, often accidentally get into foods during processing. The federal government regulates all food additives and requires that food processors perform tests to determine whether additives are present in safe levels.

TABLE 13-6

TABLE 13-6   Handling Produce Properly*

To keep pesticide residues to a minimum, use the following tips:
- Rinse produce thoroughly with running water for at least 10 seconds and scrub with a vegetable brush. When present, most residues reside on the surface of a product. Peel produce to which wax has been applied.
- Discard the outer leaves of lettuce, cabbage, and other leafy vegetables. Rinse the interior leaves thoroughly to remove dirt and debris.
- Use a knife to peel an orange or grapefruit; do not bite into the peel.
- Eat a variety of fruits and vegetables. Farmers use different chemicals for different crops, so eating a wide variety of produce helps decrease exposure to any particular pesticide residue.
- Consider purchasing certified organic foods, especially for items most likely to be contaminated (such as peaches, apples, pears, bell peppers, celery, nectarines, strawberries, cherries, kale, lettuce, and imported grapes).

*To reduce pesticide residue intake from meats, trim fat from meat and remove the skin from poultry and fish (pesticide residues concentrate in the animals' fat tissues). Discard fats and oils in broths and pan drippings.

SOURCE: Adapted from *Daily Tip: Protection from Pesticides* (Chicago: American Dietetic Association, July 13, 2007).

Manufacturers must go through a lengthy, costly process to get FDA approval for a new food additive. The manufacturer must conduct extensive research to show that the additive in question does what it is supposed to do, that it can be detected and measured in foods to which it has been added, and that it is safe in the amounts in which it will be used.

As with pesticide residues, many food additives are allowed only in amounts that ensure a wide margin of safety. Most additives that pose any potential risk are allowed in foods only at levels $\frac{1}{100}$ of those at which the risk is still known to be zero. Even nutrient food additives are subject to the margin-of-safety concept. For example, even though iodine has long been added to salt to prevent iodine deficiency, the amounts added have been controlled because excess iodine can be deadly.

Food additives extend the shelf life of many commonly eaten foods.

Substances Generally Recognized As Safe (GRAS)   Attention to certain aspects of food law will help you to better understand the issues surrounding the use of additives in foods. For the first half of the 20th century, scientists evaluated the safety of food additives using a simple approach: an added substance was either "safe" and therefore permitted for use in foods or "poisonous and deleterious" and

[Text continues on page 420.]

TABLE 13-7   Major Uses of Food Additives

Function of Additive	Examples	Foods in Which Often Used
Impart or maintain consistency	Stabilizers, thickeners, anticaking agents, and emulsifiers including alginates, lecithin, mono- and diglycerides, glycerine, pectin, guar gum	Baked goods, cake mixes, salad dressings, ice cream, processed cheese, table salt, chocolate
Improve or maintain nutritional value	Vitamins A and D, thiamin, niacin, folic acid, ascorbic acid (vitamin C), calcium citrate, zinc oxide, iron, iodine	Flour, bread, biscuits, breakfast cereals, pasta, margarine, milk, iodized salt, juices
Maintain palatability and wholesomeness	Ascorbic acid, butylated hydroxyanisole (BHA), butylated hydroxytoluene (BHT), benzoates, sulfites	Bread, crackers, frozen and dried fruit, margarine, lard, potato chips, cake mixes, meat
Produce light texture; control acidity or alkalinity	Yeast, sodium bicarbonate, citric acid, lactic acid, phosphoric acid	Cakes, cookies, quick breads, crackers, butter, soft drinks
Enhance flavor or impart desired color	Cloves, ginger, fructose, aspartame, MSG, FD&C Red No. 40, caramel, turmeric, vegetable dyes	Spice cake, gingerbread, soft drinks, yogurt, soup, candy, cheese, jams, gum

SOURCE: Adapted from *Food Additives* (Rockville, MD/Washington, DC: Food and Drug Administration in cooperation with International Food Information Council, www.ific.org).

Should You Buy Organically Grown Produce or Meats?

Once sold only in health food stores, organic fruits, vegetables, meats, and other **certified organic foods** can now be found in most mainstream markets nationwide. Sales of organic food products have grown from $78 million in 1980 to more than $21 billion today. Fresh produce is the best-selling organically grown food worldwide, but organic dairy products, soy foods, eggs, meat, and poultry are also popular.

Organically grown foods have been available for more than 40 years, but they have only recently been regulated. The United States Department of Agriculture (USDA) has created a national reference standard to define what is and what is not organic. Organic crops must be raised without the use of pesticides, petroleum-based fertilizers, or sewage sludge-based fertilizers. Animals raised on organic farms must be fed organic feed and given access to the outdoors. Additionally, they may not be given antibiotics or growth hormones.[22]

These regulations were put into place in October 2002 to regulate foods labeled "organic," whether grown in the United States or imported from other countries. The regulations mandate certification of organic products by USDA-approved state or private regulating authorities and set strict labeling standards for processed organic food. For manufacturers to use the USDA organic seal on labels, raw products must be 100 percent organic, and processed foods must contain 95 percent organic ingredients. If a food contains between 70 and 95 percent organic contents, the label can read "product made with organic ingredients." Additionally, irradiated or genetically engineered products cannot be labeled as organic.[23] See Figure 13-4 for more about consumer information on food labels.

The USDA makes no claims that organically grown food is safer or more nutritious than conventionally produced foods. It is important to understand that organic agricultural practices cannot guarantee that organically grown products are entirely free of residues. Additionally, organic produce is more expensive because it is grown without synthetic pesticides or chemicals and thus is more labor intensive. Organic crop harvests are often not as high as those grown conventionally, and fewer farmers use organic methods and sustainable agriculture practices. As a result, the cost of organically grown foods reflects the greater demands placed on the farmer.[24]

© Nancy Richmond/The Image Works

FIGURE 13-4

Consumer Information about Organic Foods and Product Labels

These sample cereal boxes show the four labeling categories. Look for the name and address of the government-approved certifier on all packaged products that contain at least 70 percent organic ingredients. For more detailed information on the USDA organic standards, visit www.ams.usda.gov/nop.

SOURCE: Adapted from The National Organic Program: Labeling Packaged Products, January 9, 2003; available at www.ams.usda.gov/nop/ProdHandlers/LabelTable.htm.

Foods made with 100 percent organic ingredients may claim "100% organic" and use the USDA seal.

Foods made with at least 95 percent organic ingredients may claim "organic" and use the USDA seal.

Foods made with at least 70 percent organic ingredients may list up to three of those ingredients on the front panel.

Foods made with less than 70 percent organic ingredients may list them on the side panel but may not make any claims on the front.

However, organically grown food casts a vote in favor of agricultural techniques that promote the well-being of the environment and farming communities over the long run. Reputable organic farmers typically use more "eco-friendly" techniques, such as crop rotation, to help keep pests under control than do conventional farmers. These practices help protect the soil, the groundwater, and the farmworkers themselves against chemical contamination. The following box lists some of the pros and cons of buying organic foods.

Still, many farmers who don't meet "organic" standards per se use other techniques to keep pesticide use to a minimum. For example, more and more farmers have adopted a system called **integrated pest management**, a technique by which farmers cut back on chemical use by combining strategies such as crop rotation, genetic engineering (discussed at the end of this chapter), and biological controls such as the release of a predator insect on a crop to get rid of a pest.

The USDA organic seal is intended to provide information to consumers. It will help organic farmers and ranchers further expand their already growing markets.

certified organic foods crops or livestock grown and processed according to USDA regulations concerning use of pesticides, herbicides, fungicides, fertilizers, preservatives, other synthetic chemicals, growth hormones, antibiotics, or other drugs.

integrated pest management the use of biological controls, crop rotation, genetic engineering, and other tactics to reduce chemical use in the growing and production of crops.

Advantages of Certified Organic Foods

- Less synthetic pesticide, hormone, and other chemical fertilizer residues in foods and waterways
- Beneficial to the environment: Improved soil fertility and crop diversity; increased water and energy conservation
- Farmers may follow more humane animal welfare standards

Disadvantages of Certified Organic Foods

- More expensive
- Tend to spoil faster, because they contain no preservatives
- May be fertilized with improperly composted animal manure containing potentially harmful organisms, such as E. coli O157:H7
- May be cross-contaminated with pesticides from nearby conventionally sprayed fields

Miniglossary of Organic Buzz Words

natural most foods labeled natural, except for meat and poultry, are not subject to government regulations; this term applies broadly to foods minimally processed and free of synthetic preservatives, artificial sweeteners, colors, flavors, and other artificial additives, hydrogenated oils, stabilizers, and emulsifiers.

organic legal term that applies to a food and how the food was produced; crops must be grown without use of synthetic pesticides, bioengineered genes, petroleum-based fertilizers, and sewage sludge-based fertilizers; organic livestock must have access to outdoors and be given no antibiotics or growth hormones; organic foods may not be irradiated.

free range a method of producing livestock where animals are allowed to roam freely instead of being contained in any manner; this term is not regulated and can be used to imply the animal product has been produced more humanely than it actually has been. USDA certifies "free range" chickens as chickens raised for their meat that have access to the outdoors; there is no USDA certification for "free range" beef, pork, or other non-poultry products; free-range eggs also have no legal definition.

grass fed refers to livestock allowed to forage on pasture grass; not legally defined.

hormone free generally refers to products free from recombinant bovine somatotropin (rBST), a hormone synthetically produced from recombinant DNA administered to dairy cattle to increase milk production.

rBST free refers to milk (dairy products) produced from cattle that have not been treated with recombinant BST (bovine somatotropin hormone); all milk contains some BST as it is produced in pituitary glands of cattle.

genetically modified organisms (GMOs)/genetically engineered organisms (GEOs) organisms from bacteria, plants, or animals that have been genetically changed in a laboratory through DNA technology. Genetically engineered plants have been developed to increase nutritional value, be resistant to pests, herbicides, or harsh environmental conditions like drought, and for other purposes.

non-GMO term used to imply a food contains no ingredients made from GMO/GEO products; not legally defined and therefore cannot be enforced.

therefore banned. As the study of toxic agents advanced, scientists realized that eventually they would be able to show that virtually every substance poses a health hazard if the dose is large enough. Consequently, they recognized that simply classifying an additive as "safe" or "poisonous" was not an effective means of evaluation.

To get around the problem, Congress proposed the bill that later became the Food Additives Amendment of 1958. As the members of Congress debated the bill, a question arose as to how to deal with the additives already in use. Congress decided that a "safe" substance in use before 1958 would be deemed a "Substance Generally Recognized as Safe" (a GRAS substance) and be put on the **GRAS list**. Substances not in use before that time would be classified as food additives and subject to regulation under the Food Additives Amendment.

With the establishment of the amendment, the FDA put hundreds of substances on the GRAS list. Everything from vegetable oils, salt, pepper, sugar, caffeine, vinegar, and baking powder to meat, poultry, eggs, milk, seafood, cereals, vegetables, and fruit was—and still is—classified as GRAS substances.

The GRAS list came under scrutiny in 1969, however, when safety questions arose about the GRAS substance cyclamate, an artificial sweetener. After reviewing hundreds of studies, the FDA decided to ban cyclamate from use in foods and beverages. As a result of this incident, President Richard Nixon ordered a reevaluation of the safety of all substances on the GRAS list. The FDA conducted a sweeping review and removed about 300 substances from the list. Today, the more than 400 substances classified as GRAS are continually subject to reexamination as new facts and concerns arise.

New Technologies on the Horizon

Back in the 1860s, a French scientist named Louis Pasteur came up with a radical, "newfangled" process by which the disease-producing microorganisms in a food could be destroyed by exposing the food to heat. Dubbed *pasteurization*, after Dr. Pasteur, the process marked a major breakthrough in the science of food safety. At the time, however, Dr. Pasteur's discovery met with widespread fear and opposition. In fact, it wasn't accepted as a vital public health measure until 1909, when the city of Chicago set forth the first U.S. law requiring pasteurization of milk.

Today, of course, few consumers even question the wisdom of drinking pasteurized milk. Still, according to many experts, new technologies such as irradiation have prompted a public outcry similar to the turn-of-the-century resistance to pasteurization.[25] As public health organizations strive to feed a fast-growing world population while ensuring a safe food supply, debate about new ways of doing so is sure to heat up. The following sections explore some of the controversial food technologies under consideration as alternatives to help improve the safety of our food supply and to maintain the nutritional value of the foods available in the marketplace.

Irradiation   **Irradiation** is one of the foremost technologies earmarked by food-safety experts for increased future use in the United States. The process involves exposing food to low doses of radiation, which destroys insects and several types of bacteria, including *Salmonella*. Contrary to popular belief, irradiation does not make a food radioactive. Rather, the radiation rays pass through food, leaving behind no radioactive residues.

Aside from the misunderstanding that irradiation makes food radioactive, many of the criticisms about the process center on substances called **unique radiolytic products**: compounds that are not present in food naturally or after conventional processing. Critics charge that these products may pose health hazards to consumers who eat irradiated food. According to a comprehensive report from the World

GRAS (Generally Recognized As Safe) list a list of ingredients, established by the FDA, that had long been in use and were believed safe. The list is subject to revision as new facts become known.

irradiation the process of exposing a substance to low doses of radiation, using gamma rays, X-rays, or electricity (electron beams) to kill insects, bacteria, and other potentially harmful microorganisms.

unique radiolytic products substances unique to irradiated food and apparently created during the process of irradiation.

It's Saturday morning in Anytown, USA, and in the center of the old "downtown" a large gathering of community members bustle around, chat, and enjoy their morning out. No, it's not a flashback to the 1950s. These people are buying locally grown produce and goods from farmers in their community.

Farmers' markets have existed in different forms for many years and have enjoyed 100 percent growth over the past decade. Through these markets, farmers are able to build a local clientele and reap profits without a middleman. Buyers also benefit from consuming produce that is ultra-fresh and in season. In addition, programs like the USDA's Seniors Farmers' Market and WIC Farmers' Market provide vouchers for low-income participants to buy fresh, nutritious produce from farmers' markets. In addition, communities as a whole benefit from farmers' markets because of the small investment and potential boost to their economies.

Farmers' markets are much more than a place to buy produce, however. In an age when mega-supermarkets have overtaken the landscape, these markets provide a great locale for a community gathering place. Farmers' markets give customers the unique experience of being able to converse with the actual grower of the produce. In doing so, these consumers can make suggestions about produce they would like to buy in the future and learn how different food items are grown. Try doing that at a supermarket!

Why Choose a Farmers' Market?

1. They help develop a food culture—a variety of foods that are commonly grown in the local area can be displayed and purchased seasonally.

2. Foods are less processed, and less packaging is used. This creates less waste, which is good for the environment. Moreover, energy resources used to transport foods produced in more distant areas can be saved, thus helping to diminish the size of the region's ecological footprint (see Figure 13-5).

3. Foods are usually fresher. Most of the foods have been picked within 24 hours of purchase and have not been frozen or treated with preservatives.

4. Farmers' markets encourage healthier food choices by promoting fresh-picked seasonal fruit and vegetables.

5. Farmers' markets boost local economies because the food distributor (middleman) is eliminated. Thus, the farmer's income is more likely to be reinvested within the community.

To locate your local farmers market, visit www.ams.usda.gov/farmersmarkets for information, dates, and times of a market near you!

FIGURE 13-5

Ecological Footprints

According to the United Nations, the ecological footprint is a measure of the resources needed to support a region's consumption of food, materials, and energy—in other words, its impact on the environment. In this figure, each region is represented by a rectangle whose width is proportional to the population and whose height represents per capita resource consumption in terms of the area required to produce those natural resources. The ecological footprint accounts for both the per capita resource consumption and the population growth. Thus Asia, which has a population more than 10 times that of North America but a per capita resource consumption level only one-sixth as large, has a footprint only slightly larger than North America. Take the *Ecological Footprint Quiz* at www.myfootprint.org to estimate how much productive land and water you need to support what you use and what you discard.

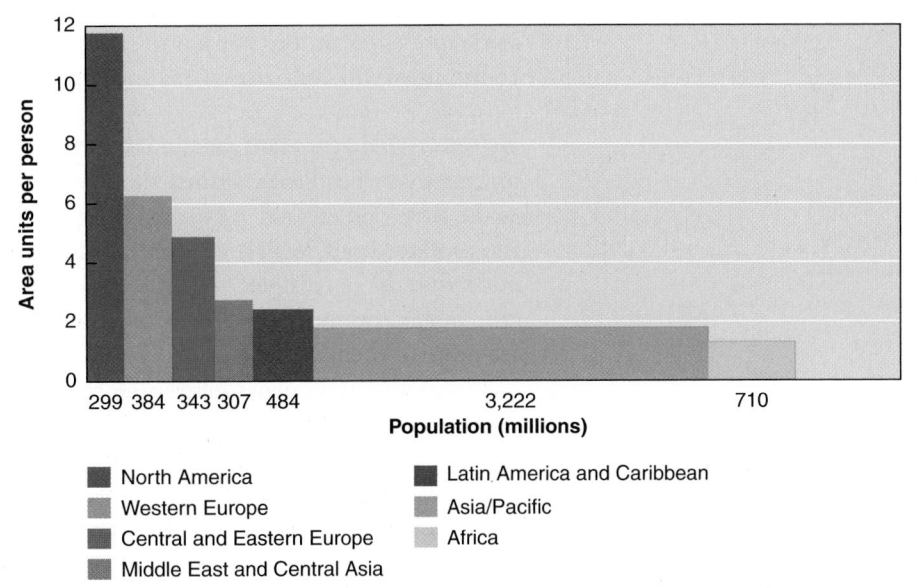

SOURCE: Data from United Nations Population Fund, www.unfpa.org/swp/2001/english/ch03.html.

The FDA requires that the labels of irradiated foods carry this internationally known radura symbol for radiation. The circle in the middle represents an energy source, the five breaks in the outer circle symbolize rays generated by the energy source, and the two petals signify food. For unpackaged irradiated meat, the statement and logo must be displayed at the point of sale for consumers. The labeling requirements do not pertain to food service establishments such as restaurants.

The FDA approves irradiation of:

- Citrus fruits
- Eggs
- Flour
- Fresh or frozen red meats
- Mushrooms
- Onions
- Potatoes
- Poultry
- Spices
- Spinach
- Strawberries
- Tomatoes
- Tropical fruits
- Wheat

genetic engineering the use of biotechnology to alter the genes of a plant in an effort to create a new plant with different traits; also called *genetic modification*. In some cases, a plant's genes may be deleted or altered, or genes may be introduced from different organisms or species. The foods or crops produced are called genetically engineered (GE) or genetically modified (GM) foods or crops.

Health Organization, however, concern about unique radiolytic products is "probably unfounded" because many of the substances found in irradiated foods are similar, if not identical, to compounds found in other foods. Even those that appear to be unique may not have been detected in conventionally processed foods, because those items haven't undergone the same extensive testing as have irradiated foods.[26]

Another common concern about irradiation is that it might alter the nutritional value of a food. Again, the World Health Organization, along with numerous other public health agencies, points out that nutrient losses, if any, are insignificant.

That's not to say, however, that irradiation poses no risks. One of the most troublesome is that the radioactive materials used in the irradiation process may put workers and communities at undue risk. However, as with any technology, the risk of irradiation must be weighed carefully against the benefits and risks of alternate technologies. Consider spices, which typically harbor high levels of pathogens. Often, manufacturers douse them with a toxic, explosive chemical called *ethylene oxide* to rid them of bacteria. Yet because of its toxic, explosive nature, ethylene oxide puts workers at risk, may pollute the air, and may leave residue on the spices. In fact, the gas, which was once used to sterilize medical supplies, has been deemed so dangerous that most manufacturers now use irradiation instead. (Cotton swabs, tampons, teething rings, and a number of other consumer goods are also sterilized via irradiation.)[27] Recently, USDA authorized irradiation of meat already inspected by the agency and approved as safe for consumption. As a result, food companies now have the option of using irradiation on raw meat and meat products (for example, ground beef, frozen hamburger patties, or frozen poultry).

The World Health Organization (WHO), the Food and Agriculture Organization of the United Nations, the Food and Drug Administration, and numerous other agencies encourage the use of irradiation in the fight against foodborne disease and food loss, and some 40 countries have approved it for use. According to WHO, each year spoilage, insect infestation, and the like lead to losses of as much as 50 percent of the world's food supply—losses that could be eliminated with irradiation. What's more, the process may help prevent deaths resulting from foodborne illness. Clearly, as the pressures to feed the world safely continue to mount, the public and scientific community will continue to examine the pros and cons of irradiation.

Biotechnology and Genetic Engineering   In May 1994, a new breed of tomato made headlines. Called the Flavr Savr, the product garnered national repute as the first food created via **genetic engineering** to hit the market. It wasn't so much the product itself, which is simply a slow-ripening tomato, that caused such a stir; but rather the opening of the regulatory door that could pave the way to a whole new crop of genetically engineered products. What are these high-tech foods, and will we see more of them in the future?

Just as a person's genes determine hair color, eye color, and other characteristics, a plant's genes dictate the plant's structure, resistance to spoilage, and other qualities. With genetic engineering, scientists can alter a plant's genes to make a particular trait more desirable. To create the Flavr Savr tomato, scientists first identified the gene that causes ripe tomatoes to soften and rot. Then they isolated the gene, reversed it (or turned it backward, so to speak), and inserted the reversed gene back into tomato plants. The backward gene in the new tomatoes suppresses, or "turns off," the rotting gene. As a result, the Flavr Savr tomato stayed ripe longer than regular tomatoes (see Figure 13-6). This allowed growers more time to ship it to the consumer without refrigeration.[28]

Although the idea of altering a plant's genes to create a different version sounds new, scientists have used a crude version of the concept for centuries. In the 1500s, farmers crossed, say, a good food crop with another plant, which was resistant to disease, to form a hybrid that contained traits of both plants. Many of the fruits and vegetables we enjoy today were produced by this sort of cross-breeding. For example, corn as we know it today came from cross-breeding it with a type of corn that has a much harder outer shell on the kernel; kiwi was originally a small, hard berry; and nectarines are modified peaches.[29] Genetic engineering is different, however, because of its speed and the precision it affords scientists. Whereas cross-breeding typically takes more than a decade, genetic engineering can yield new products in about 5 years.

In addition, genetic engineering allows scientists to "mix and match" genes from different species—say, mixing a gene from a fish with the genes of a vegetable. This type of research, called **transgenetics**, raises many ethical questions, especially among vegetarians and other people who fear that inserting an animal gene into a fruit or vegetable makes a food that is on some level both vegetable and animal. The possibilities created by this technology are seemingly endless. It has been suggested that ultimately, the world will obtain most of its food, fuel, fiber, and some of its pharmaceuticals from genetically altered vegetation and trees.[30] Major companies spend billions of dollars annually on **biotechnology** as researchers continue to develop new applications for genetically engineered products.

The U.S. Department of Agriculture reports that more than 40 new genetically engineered agricultural products have completed all the federal regulatory requirements and are now sold commercially. Recently released government data on acreage of biotechnology-derived crops indicates that genetically engineered soybean, cotton, and corn account for up to 44 percent of total acreage planted.[31] Most of these plants have been engineered with genes from bacteria that destroy pests or protect plants against herbicides. The resistant plants created through genetic engineering lower the

Farmers bred corn as we know it today from this wild corn.

transgenetics the process of transferring genes from one species to another unrelated species.

biotechnology the use of biological systems or living organisms to make or modify products. Includes traditional methods used in making products such as wine, beer, yogurt, and cheese; cross-breeding to enhance crop production; and modification of living plants, animals, and fish through the manipulation of genes (genetic engineering).

FIGURE 13-6

Creating the Flavr Savr Tomato

SOURCE: Adapted from Calgene's recipe for genetically engineered tomatoes, *FDA Consumer* (April 1995): 9.

Agrobacterium cell — Altered *PG* gene is placed in *Agrobacterium*.

1. The *PG* (polygalacturonase) gene, which causes ripe tomatoes to soften and rot, is isolated and cloned. The sequence of *PG* gene is reversed so that the gene is backward (in what scientists call the "antisense" orientation).

2. The reversed *PG* gene is put into *Agrobacterium*, which is a bacterium that infects plants and is commonly used by genetic engineers to insert modified or foreign genes into other target cells.

3. The *Agrobacteria* are then placed in a petri dish with pieces of leaf from a tomato plant. The cut edges of the leaves absorb the *Agrobacteria* and antisense *PG* gene. Antisense gene thus becomes part of the genetic material of the tomato plant cells.

4. The leaf cuttings regenerate into tomato plants that contain the reversed *PG* gene. The new plants sprout roots, are transplanted to soil, and grow to mature tomato plants. The seeds collected from these greenhouse tomatoes are planted outdoors for field trials and more seed production.

5. In the genetically engineered tomato, the natural *PG* gene's production of the fruit-rotting *PG* enzyme is repressed by the reversed gene. This gives the commercial tomato extended shelf life, allowing it to ripen more fully on the vine and still have time to get to market before it spoils.

farmers' dependence on the use of pesticides and weed killers while increasing overall production.

Genetic engineering is used to boost the nutritional value of foods. Advances in biotechnology can increase levels of the natural health-enhancing substances in plants known as *phytochemicals*. These compounds, many of which are found in soybeans, tomatoes, garlic, and other plant foods, have been shown to protect against cardiovascular disease, cancer, and free-radical damage to tissues. One such compound, beta-carotene, belongs to a class of compounds known as *carotenoids*. Scientists with the USDA have released new tomato-breeding lines that contain about 10 to 25 times more beta-carotene than typical tomatoes.[32] Also, researchers recently succeeded at manipulating a gene that improves the quality of the proteins made by sweet potato, an important food crop in poorer countries where high-quality protein foods are at a shortage. Early study results show the protein content of the genetically engineered potatoes increased as much as fivefold.[33]

Scientists are now developing plants that will produce edible vaccines. Such plants could be extremely valuable in developing countries where some environmental conditions can make conventional vaccine administration impractical. Also, an edible vaccine is being investigated that will, it is hoped, protect against diarrhea, a major cause of infant mortality in developing countries.[34]

The introduction of genetically engineered food has encountered several problems. Some people are concerned about the potential impact of an organism that

---

Risks vs. Benefits: The Debate over Genetically Engineered (GE) Foods and Crops

Some Currently Available GE Crops*

- **Maize**: Insect resistance (reduced insect damage)
- **Soybean**: Herbicide tolerance (greater weed control)
- **Cotton**: Insect resistance (reduced insect damage)
- **Rice**: Pro-Vitamin A (increase vitamin A supply)

Potential Benefits of GE Foods and Crops

- **Increased nutritional value of staple foods**: Genes are being inserted into rice to make it produce beta-carotene, which the body converts into vitamin A. This experimental transgenic "golden rice" has the potential to reduce vitamin A deficiency, a leading cause of blindness and a significant factor in many child deaths.

- **Reduced environmental impact**: Scientists are developing trees with modified cell lignin content. When used to make pulp and paper, the modified wood requires less processing with harsh chemicals.

- **Increased fish yield**: Researchers have modified the gene that governs growth hormones in tilapia, a farmed fish, offering the prospect of increased yield and greater availability of fish protein in local diets.

- **Increased nutrient absorption by livestock**: Animal feed under development will improve animals' absorption of phosphorus. This reduces the phosphorus in animal waste, which pollutes groundwater.

- **Tolerance of poor environmental conditions**: Scientists are working to produce drought-resistant or salt-tolerant transgenic crops, which can then be grown on marginal land.

Potential Risks of GE Foods and Crops

- **Inadequate controls**: Although safety regimes are being improved, control over GE crop releases is not completely effective. In 2000, for example, a maize variety cleared only for animal consumption was found in food products.

- **Transfer of allergens**: Allergens can be transferred inadvertently from an existing to a target organism, and new allergens can be created. For example, when a Brazil-nut gene was transferred to soybean, tests found that a known allergen had also been transferred. However, the danger was detected during testing, and the soybean was not released.

- **Unpredictability**: GE crops may have unforeseen effects on farming systems, for example, by taking more resources from the soil or using more water than normal crops use.

- **Undesired gene movement**: Genes brought into a species artificially may cross accidentally to an unintended species. For example, resistance to herbicide could spread from a GE crop into weeds, which could then become herbicide resistant themselves.

- **Environmental hazards**: GE fish might alter the composition of natural fish populations if they escape into the wild. For example, fish that have been genetically modified to eat more in order to grow faster might invade new territories and displace native fish populations.

*A full biotechnology glossary is available at www.fao.org/biotech/gloss.htm

SOURCE: Adapted from *World Food Summit Fact Sheet: Biotechnology and Food Security* (Rome, Italy: The Food and Agriculture Organization of the United Nations, WFS FS 01-E); available at www.fao.org/worldfoodsummit/english/fsheets/biotech.pdf.

would not have evolved under normal conditions. We cannot know the long-term environmental consequences as these genetically engineered plants multiply and mutate. Some people fear that naturally occurring cross-pollination between genetically engineered plants with nearby weeds may spread traits from plants to weeds. Potentially, this so-called superweed would be resistant to insects and herbicides.

People with food allergies express concern that new varieties of food produced by transgenetics may introduce allergens not found in the food before it was altered. Indeed, researchers have shown that allergens can be transferred through bioengineering, as in one case where an allergic protein showed up in soybeans that had been genetically altered with proteins from Brazil nuts.[35]

Another widely debated issue surrounding genetic engineering involves labeling. Many consumer groups have called for across-the-board labels on genetically engineered foods. But according to the FDA, with some exceptions, the food must carry distinct consequences to consumers who eat it before it requires special labeling. When genes from peanuts and other foods known to be common causes of allergies are put into a food, the label must indicate that the food contains an allergen, unless the manufacturer can prove that the item's potential to cause allergies has not been transferred via the gene. In addition, a food that has been genetically engineered to significantly change, say, its fiber content or nutrient composition must bear a label that states the nature of the change.[36] A number of petitions currently calling for labeling of foods with genetically engineered ingredients may lead to FDA approval of required labeling for these foods in the future.[37] The box on page 424 lists some examples of currently available genetically engineered crops and highlights some of the potential benefits and risks associated with them.

Despite the concerns, many scientists are hopeful that careful use of genetic engineering will confer long-term benefits. For instance, the development of insect- and disease-resistant plants may allow farmers to grow crops with fewer chemicals. The possibilities are enormous, and each new product considered for entry into the marketplace will require careful scrutiny.

This book emphasizes the problems of *overnutrition*: obesity, heart disease, cancer, and others—diseases of economically developed nations. However, many people in developing nations, as well as people in the less privileged parts of developed nations, suffer from problems of **undernutrition**, which is characterized by chronic debilitating hunger and malnutrition. These conditions are most visible in times of **famine**, but they are widespread and persistent even when famine does not occur.

All people need food. Regardless of our race, religion, gender, or nationality, our bodies react similarly to the effects of hunger and its companion **malnutrition**: listlessness, weakness, failure to thrive, stunted growth, mental retardation, muscle wastage, scurvy, pellagra, beriberi, anemia, rickets, osteoporosis, goiter, tooth decay, blindness, and a host of other effects, including death. Apathy and shortened attention span are only two of a number of behavioral symptoms often mistaken for laziness, lack of intelligence, or mental illness in undernourished people.

How are hunger and poverty related?

The phenomenon of hunger today is being discussed in terms of **food security** or **food insecurity**. *Food security* is defined as access by all people at all times to enough food for an active, healthy life and at a minimum includes (1) the ready availability of nutritionally adequate and safe foods and (2) the ability to acquire personally acceptable foods in a socially acceptable way.[38] Food insecurity was once viewed as a problem of overpopulation and inadequate food production, but now many people recognize it as a problem of **poverty**. Food is *available* but is not *accessible* to

© Louise Gubb/The Image Works

Children can be the first to show the signs of undernutrition due to their high nutrient needs. No famine, no flood, no earthquake, no war has ever claimed the lives of 200,000 children in a single week. Yet UNICEF estimates that malnutrition and disease claim more than that number of child victims under the age of five years every week at the rate of one every three seconds.

FIGURE 13-7

Undernourishment in 2010, by region (millions)

Developed countries 19

Near East and North Africa 37

Latin America and the Caribbean 53

Sub-Saharan Africa 239

Asia and the Pacific 578

Total = 926 million people

SOURCE: Food and Agriculture Organization (FAO), 2011.

the poor, who have neither land nor money. Poverty is much more than an economic condition; it exists for many reasons, including overpopulation, greed, unemployment, and the lack of productive resources such as land, tools, and credit. Consequently, if we are to provide adequate nutrition for all the Earth's hungry people, we must transform the economic, political, and social structures that limit food production, distribution, and consumption and create a gap between rich and poor.

Approximately how many people worldwide are affected by food insecurity?

The Food and Agriculture Organization (FAO) estimates that of the more than 6.6 billion people in the world, at least 926 million people suffer from chronic, severe undernutrition, consuming too little food each day to meet even minimum energy requirements (see Figure 13-7).[39] Three micronutrient deficiencies are of particular concern worldwide: vitamin A deficiency, the world's most common cause of preventable child blindness and vision impairment; iron-deficiency anemia; and iodine deficiency, which causes high levels of goiter and child retardation.[40]

Worldwide, 10 million people die every year of hunger and hunger-related diseases. Millions of children die each year from the diseases of poverty: parasitic and infectious diseases such as dysentery, whooping cough, measles, tuberculosis, cholera, and malaria (see Figure 13-8). These diseases interact with poor nutrition to form a vicious cycle in which the outcome for many is death.[41]

When is the risk for undernutrition highest?

Undernutrition runs high when the body's nutrient needs are high, as in times of rapid growth. If family food is limited, pregnant and lactating women, infants, and children are the first to show the signs of undernutrition. Low birthweight contributes to more than half of the deaths worldwide of children less than 5 years of age. **UNICEF**

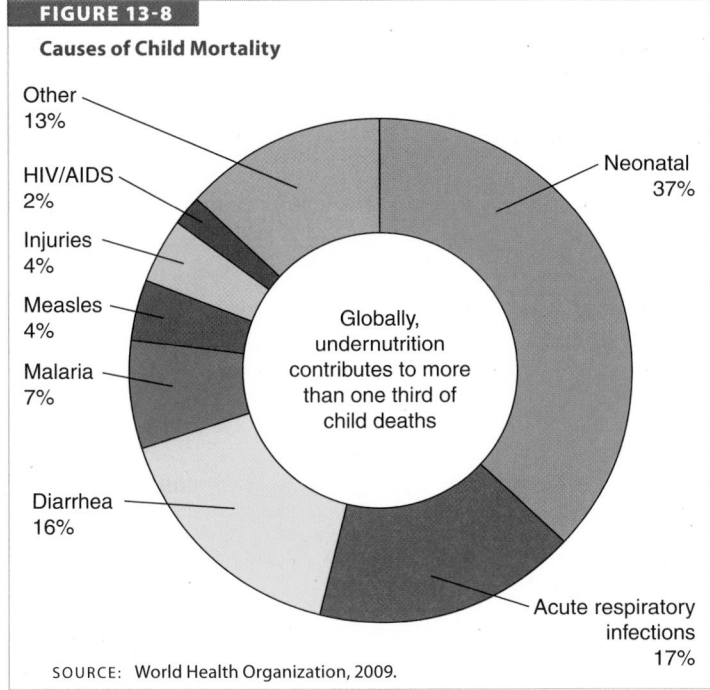

FIGURE 13-8

Causes of Child Mortality

- Other 13%
- HIV/AIDS 2%
- Injuries 4%
- Measles 4%
- Malaria 7%
- Diarrhea 16%
- Neonatal 37%
- Acute respiratory infections 17%

Globally, undernutrition contributes to more than one third of child deaths

SOURCE: World Health Organization, 2009.

refers to the **under-5 mortality rate (U5MR)** as the best indicator of children's overall health and well-being.[42] UNICEF argues that the U5MR reflects a country's overall resources directed at children:

> The U5MR reflects the nutritional health and the health knowledge of mothers; the level of immunization and use of **oral rehydration therapy**; the availability of maternal and health services (including prenatal care); income and food availability in the family; the availability of clean water and safe sanitation; and the overall safety of the child's environment.[43]

Until the middle of the 20th century in most of the developing countries, babies were breastfed for their first year of life, and supplements of other milk and cereal gruel were added to their diets after the first several months. Today, only about half of all infants are exclusively breastfed to the age of 4 months.[44] A number of factors contributed to this unfortunate decline, including the aggressive promotion and sale of infant formula to new mothers; the encouragement by health care practitioners for mothers to bottle-feed (hospitals sent home free samples after delivery of the newborn); and the global pattern of urbanization and the accompanying loss of cultural ties supporting breastfeeding combined with more women working outside the home. Overall, WHO estimates that more than 1.5 million children's lives could be saved each year if all mothers gave their babies nothing but breast milk for the first 6 months of life.[45]

Breastfeeding permits infants in many developing countries to achieve weight and height gains equal to those of children in developed countries until about 6 months of age. Even if infants are protected by breastfeeding at first, they must eventually be weaned. The weaning period is one of the most dangerous periods for children in developing countries. Newly weaned infants often receive nutrient-poor diluted cereals or starchy root crops, and the infants' foods are often prepared with contaminated water, making infection almost inevitable.

What is the status of food insecurity in the United States?

In 2010, more than 1 in 6 households experienced food insecurity. This represents 49 million people, including more than 16 million children. Food insecurity rates are higher than average in female-headed households, households with children (especially African American and Hispanic households), and households in inner-city areas.[46] A recent analysis of state hunger surveys found consistent results across all surveys and evidence for several broad conclusions:[47]

- Food insecurity is a chronic problem in the United States.
- Food insecurity is not due to food shortages. Hunger results from unequal distribution of economic resources; that is, poverty.
- People who lack access to a variety of resources—not just food—are most at risk of hunger. When income is inadequate to meet the costs of housing, utilities, health care, and other fixed expenses, these items compete with and may take precedence over food.
- Private charity cannot solve the food insecurity problem. Voluntary activities are limited in expertise, time, and resources and are likely to require government support to continue.

What are the causes of food insecurity in the United States?

The most compelling single reason is poverty. Poverty and food insecurity are interdependent. Nutrition surveys investigating people's nutritional health in the United States have demonstrated consistently that the lower a family's income, the less adequate the family's nutrition status.

(Continued)

▲ Feeding the hungry—in the United States

© Bettman/CORBIS

Although poverty is the major cause of food insecurity in the United States, other problems contribute as well, including, among others, homelessness; the reluctance of people to accept what they perceive as "welfare" or "charity"; delays in receiving public assistance benefits; health problems of older adults; lack of access to assistance programs; insufficient community food resources for the hungry; and insufficient community transportation systems to deliver food to people who have no transportation.

What is being done to help reduce problems of food insecurity?

The U.S. Department of Agriculture's (USDA) Food and Nutrition Service (FNS) implements an array of programs as a "food safety net," to provide children and low-income people with food or the means to purchase food.[48] Five programs—The Supplemental Nutrition Assistance Program (SNAP), previously called the food stamp program; National School Lunch Program; Special Supplemental Nutrition Program for Women, Infants, and Children (WIC); Child and Adult Care Food Program; and School Breakfast Program—account for 94 percent of all federal expenditures for food assistance. Other food assistance programs for seniors (Elderly Nutrition Program) are administered by the Department of Health and Human Services (DHHS).

The terms **food recovery** and **gleaning** refer to programs that collect excess wholesome food for delivery to hungry people. Approximately 96 billion pounds—more than one-quarter of the 356 billion pounds of food produced in this country for human consumption—are lost at the retail and food service levels. In an effort to reduce food waste, the National Summit on Food Recovery and Gleaning set a goal of providing an additional 500 million pounds of food a year to feeding organizations.

Additionally, concerned citizens work through community programs and churches to provide meals to the hungry. **Feeding America**, the nation's largest supplier of surplus food, distributed more than 3 billion pounds of food to nearly 200 **food banks** and some 50,0000 agencies for direct distribution around the nation in 2011.[49]

How does food insecurity in developing countries differ from food insecurity in the United States?

Global food insecurity is more extreme than domestic food insecurity. In fact, most people would find it hard to imagine the severity of poverty in the developing world:

> Many hundreds of millions of people in the poorest countries are preoccupied solely with survival and elementary needs. For them, work is frequently not available, or pay is low, and conditions barely tolerable. Homes are constructed of impermanent materials and have neither piped water nor sanitation. Electricity is a luxury. Health services are thinly spread, and

in rural areas only rarely within walking distance. Permanent insecurity is the condition of the poor . . . in the wealthy countries, ordinary men and women face genuine economic problems. . . . But they rarely face anything resembling the total deprivation found in the poor countries.[50]

Food insecurity in developing nations is a problem of supply and demand, of inappropriate technology, of environmental abuse, of demographic distribution, of unequal access to resources, of extremes in dietary patterns, and of unjust economic systems. People who are poor are often powerless to change their situation because they have less access to vital resources such as education, training, food, and health services.

How are international trade and debt connected to food insecurity?

Over the years, developing countries have seen the prices of imported fuels and manufactured items rise much faster than the prices they receive for their export goods (such as bananas, coffee, and various raw materials) on the international market. The combination of high import costs with low export profits often pushes a developing country into accelerating international debt that sometimes leads to bankruptcy.

Debt and trade are closely related to the progress a country can make toward achieving an adequate diet for its people. As import prices increase relative to export prices, more of a country's total money base moves abroad to pay for the imports. With more and more of its money abroad, the country is forced to borrow money, usually at high interest rates, to continue functioning at home. As more and more of its financial resources are used to pay off interest on the country's trade debts, less and less money is available to deal with food insecurity at home. Each year, the debt crisis worsens and leads to further problems with hunger.

What about the role of multinational corporations in this issue?

Typically, large landowners and **multinational corporations** hire indigenous people for below-subsistence wages to work in the fertile farmlands growing crops to be exported for profit—leaving little fertile land for the local farmers to use to grow their own food. The local people work hard cultivating cash crops for others, not food crops for themselves. The money they earn is not even enough to buy the products they help to produce. They do not adequately share in the profits realized from the marketing of products grown with their labor. The end results of this process are that imported foods—bananas, beef, cocoa, coconuts, coffee, pineapples, sugar, tea, winter tomatoes, and others—fill the grocery stores of developed countries, while the poor who labored to grow and produce these foods have less food and fewer resources than when they farmed the land for their own use. Additional cropland is diverted for nonfood, cash crops such as tobacco, rubber, cotton, and other agricultural products. These practices have also adversely affected the financial status of many U.S. farmers. The foreign cash crops often undersell the same U.S.-grown produce. The U.S. farmer cannot compete against these lower-priced imported foods and may be forced out of business.

How does overpopulation fit into the picture?

The current world population is approximately 6.6 billion, and the United Nations projects this figure to be approximately 9 billion by the year 2050. The Earth may not be able to adequately support this many people. The world's present population is certainly a concern, as is the projected population increase. As serious as the population problem is, it is only one cause of the world food problem. Poverty seems to be at the root of both hunger and overpopulation.

Three major factors affect population growth: birthrates, death rates, and standards of living. Low-income countries have high birthrates, high death rates, and low standards of living.

When a people's standard of living rises, giving them better access to health care, family planning, and education, the death rate falls. In time, the birth rate also falls. As the standard of living continues to improve, the family earns sufficient income to risk having smaller numbers of children. A family depends on its children to cultivate the land, secure food and water, and provide for the adults in their old age. Under conditions of ongoing poverty, parents choose to have many children to ensure that some will survive to adulthood. Children represent the "social security" of the poor. Improvements in economic status help relieve the need for this "insurance" and thus help reduce the birth rate.

Is there a better way to distribute world resources?

Land reform—for example, giving people a meaningful opportunity to produce food for local consumption—can combine with population control to increase everyone's assets. Poor nations must be allowed to increase their agricultural productivity. Much is involved, but to put it simply, poor nations must gain greater access to five things simultaneously: land, capital, water, technology, and knowledge.[51] Equally important, each nation must adopt the political priority of improving the conditions of all its people. International food aid may be required temporarily during the development period, but eventually this aid will become less and less necessary.

Governments can learn from recent history the importance of developing local agricultural technology. A major effort made in the 1960s and 1970s—the green revolution—demonstrated the potential for increased grain production in Asia. It was an effort to bring the agricultural technology of the industrial world to the developing countries, but the high-yielding strains of wheat and rice that were selected required irrigation, chemical fertilizers, and

pesticides—all costly and beyond the economic means of too many of the farmers in the developing world.

Instead of transplanting industrial technology into the developing countries, small, efficient farms and local structures for marketing, credit, transportation, food storage, and agricultural education should be developed. International research centers need to examine the conditions of tropical countries and orient their research toward **appropriate technology**: labor-intensive rather than energy-intensive agricultural methods. For example, labor-intensive technology, such as the use of manual grinders for grains, is appropriate in some places because it makes the best use of human, financial, and natural resources. A manual grinder can process 20 pounds of grain per hour, replacing the mortar and pestle, which in the same time can pound a maximum of only 3 pounds.[52] The specific technology that is appropriate varies from situation to situation.

Environmental concerns must be taken more seriously as well. As important as the amount of land available for crop production is the condition of the soil and the availability of water. Soil erosion is now accelerating on every continent at a rate that threatens the world's ability to continue feeding itself.[53] Erosion of soil has always occurred; it is a natural process. In the past, however, processes that build the soil up, such as the growth of trees, have compensated for erosion.

Is there hope for a world without hunger for women and children?

Women make up 50 percent of the world's population. Any solution to the problems of poverty and food insecurity is incomplete and even hopeless if it fails to address the role of women in developing countries, because women and their children represent the majority of those living in poverty.

Women play a vital role in the nutrition of their nation's people. Their nutrition during pregnancy and lactation determines the future health of their children. If women are weakened by malnutrition, the consequences ripple outward to affect many other individuals. The importance of the role women play in these countries is increasingly appreciated, and many countries now offer development programs with women in mind.

Seven basic strategies are at the heart of women's programs:

- Removing barriers to financial credit
- Providing access to time-saving technologies
- Providing appropriate training to promote self-reliance
- Teaching management and marketing skills
- Making health and day care services available
- Forming women's support groups[54]
- Providing information and technology to promote planned pregnancies

(Continued)

The recognition of women's needs by some development organizations is an encouraging trend in the efforts to fight the global malnutrition crisis.

There is hopeful news for children in developing countries—the group that is most strongly affected by poverty, malnutrition, and food insecurity and its relationship to the environment.[55] **GOBI**, a child survival plan set forth by UNICEF, has made outstanding progress in cutting the number of hunger-related child deaths. GOBI is an acronym formed from four simple, but profoundly important, elements of UNICEF's Child Survival campaign: **g**rowth charts, **o**ral rehydration therapy (ORT), **b**reast milk, and **i**mmunization.

In this program, a mother learns to weigh her child every month and chart the child's growth on a specially designed paper chart. She can learn to detect the early stages of hidden malnutrition.

Oral rehydration therapy (ORT) is important because most children who die of malnutrition do not starve to death; rather, they die because their health has been compromised by dehydration due to infections that cause diarrhea. The spread of ORT is preventing an estimated 1 million dehydration deaths each year.[56] Oral rehydration therapy is the administration of a simple solution that mothers can make up themselves, using locally available ingredients, that increases a body's ability to absorb fluids 25-fold.[57] International development groups also provide mothers with packets of pre-measured salt and sugar to be mixed with water in rural and urban areas. A safe and sanitary supply of drinking water is a prerequisite for the success of the ORT program.

The promotion of breastfeeding among mothers in developing countries has many benefits. Breast milk is hygienic, is readily available, is nutritionally sound, and provides infants with immunologic protection specific for their environment. In the developing world, the advantages of breastfeeding over formula feeding can mean the difference between life and death.

Immunizations (the *I* of GOBI) could prevent most of the 5 million deaths each year from measles, diphtheria, tetanus, whooping cough, poliomyelitis, and tuberculosis. The immunization achievements of the last two decades are credited with the prevention of approximately 3 million deaths a year, as well as the protection of many millions more from disease, malnutrition, blindness, deafness, and polio.[58]

The first World Summit for Children in history was convened by UNICEF in September of 1990 for the purpose of making a renewed commitment to ending child deaths and child malnutrition. Significantly, *nutrition* was mentioned for the first time in world history as an internationally recognized human right.[59] An immediate result of this summit has been an increase in the number of governments actively adopting the child survival strategies of UNICEF: universal immunization; oral rehydration therapy; a massive effort to promote breastfeeding as the ideal food for at least the first 6 months of an infant's life; an attack on malnutrition involving nutrition surveillance focusing on growth monitoring and weighing of infants at least once every month for the first 18 months of the child's life; and nutrition and literacy education that will empower women in developing countries and lead to a reduction in nutrition-related diseases among vulnerable children.[60]

What can I do to help alleviate hunger problems?

The problems of hunger can appear so great that they sometimes seem approachable only by way of worldwide political decisions. To this end, many individuals and groups are working to improve the chances of the future well-being of the world and its people through a number of national and international organizations.

Regardless of the type and level of involvement a person chooses, each person can make a difference. Individual people can do any of the following:

- Assist in government and community programs as volunteers.
- Help to increase the visibility and accessibility of existing programs and services to those who need them.
- Document the needs that exist in their own communities.
- Join with others in their community who have similar interests.
- Follow current hunger legislation; call and write legislators about hunger issues.

Individuals can also help change the world through the personal choices they make each day. Our choices affect the way the rest of the world's people live and die. People in affluent nations have the freedom and means to choose their lifestyles. We can find ways to reduce our consumption of the world's nonrenewable resources and use only what is absolutely required. As one person put it, "The widespread simplification of life is vital to the well-being of the entire human family."[61] Personal lifestyles do matter, for a society is nothing more than the sum of its individuals. As we go, so goes our world.

George McGovern, former U.S. senator and current U.S. ambassador to the UN Agencies on Food and Agriculture in Rome, calls us all to action with the following words—excerpted from his book, *The Third Freedom: Ending Hunger in Our Time*:[62]

Hunger is a political condition. The Earth has enough knowledge and resources to eradicate this ancient scourge. Hunger has plagued the world for thousands of years. But ending it is a greater moral imperative now than ever before, because for the first time humanity has the instruments in hand to defeat this cruel enemy at a very reasonable cost.

What will it cost if we don't end the hunger that now afflicts so many of our fellow humans? The World Bank has concluded that each year malnutrition causes the loss of 46 million years of productive life, at a cost of $16 billion annually, several times the cost of ending hunger and turning this loss into productive gain.

Of course it is impossible to evaluate with dollars the real cost of hunger. What is the value of a human life? The twentieth century was the most violent in human history. With nearly 150 million people killed by war. But in just the last half of that century nearly three times as many died of malnutrition or related causes. How does one put a dollar figure on this terrible toll silently collected by the Grim Reaper? What is the cost of 850 million hungry people dragging through shortened and miserable lives, unable to study, work, play, or otherwise function normally because of the ever-present drain of hunger and malnutrition on body, mind, and spirit? What is the cost of millions of young mothers breaking under the despair of watching their children waste away and die from malnutrition? This is a problem we can resolve at a fraction of the cost of ignoring it. We need to be about that task now. I give you my word that anyone who looks honestly at world hunger and measures the cost of ending it for all time will conclude that this is a bargain well worth seizing. More often than not, those who look at the problem and the cost of its solution will wonder why humanity didn't resolve it long ago.

Miniglossary

appropriate technology a technology that utilizes locally abundant resources in preference to locally scarce resources. For developing countries, which usually have a large labor force and little capital, the appropriate technology would therefore be labor intensive.

famine widespread lack of access to food caused by natural disasters, political factors, or war; characterized by a large number of deaths due to starvation and malnutrition.

Feeding America a national food-banking network to which most food banks belong.

food banks nonprofit community organizations that collect surplus commodities from the government as well as edible but often unmarketable foods from private industry for use by nonprofit charities, institutions, and feeding programs at nominal cost.

food insecurity the inability to acquire or consume an adequate quality or sufficient quantity of food in socially acceptable ways, or the uncertainty that one will be able to do so.

food recovery such activities as salvaging perishable produce from grocery stores and wholesale food markets; rescuing surplus prepared food from restaurants, corporate cafeterias, and caterers; and collecting nonperishable, canned, or boxed processed food from manufacturers, supermarkets, or people's homes. The items recovered are donated to hungry people.

food security access by all people at all times to enough food for an active and healthy life. Food security has two aspects: ensuring that adequate food supplies are available and ensuring that households whose members suffer from undernutrition have the ability to acquire food, either by producing it themselves or by being able to purchase it.

gleaning the harvesting of excess food from farms, orchards, and packing houses to feed the hungry.

GOBI an acronym formed from the elements of UNICEF's Child Survival campaign: **g**rowth charts, **o**ral rehydration therapy, **b**reast milk, and **i**mmunization.

malnutrition the impairment of health resulting from a relative deficiency or excess of food energy and specific nutrients necessary for health.

multinational corporations international companies with direct investments and/or operative facilities in more than one country. U.S. oil and food companies are examples.

oral rehydration therapy (ORT) the treatment of dehydration (usually due to diarrhea caused by infectious disease) with an oral solution; ORT as developed by UNICEF is intended to enable a mother to

mix a simple solution for her child from substances that she has at home.

poverty the state of having too little money to meet minimum needs for food, clothing, and shelter. The U.S. Department of Agriculture defined the poverty level in the United States as an annual income of $22,350 for a family of four in 2011.

under-5-mortality rate (U5MR) the number of children who die before the age of 5 for every 1,000 live births.

undernutrition (also called hunger) as used in this discussion, a term that describes the domestic and world food problem of a continuous lack of the food energy and nutrients necessary to achieve and maintain health and protection from disease.

UNICEF the United Nations International Children's Emergency Fund, now referred to as the United Nations Children's Fund.

Nutrition on the Web

www.cengage.com/nutrition/boyle/personalnutrition8e

Go to the *Personal Nutrition* site to check for the latest updates to chapter topics or to access links to related websites.

www.nal.usda.gov/fnic

The Food and Nutrition Information Center (FNIC) provides credible and accurate information. Under Browse by Subject select Food Safety.

www.nutrition.gov

This government website provides information and links. Under Browse by Subject, select Shopping, Cooking and Meal Planning. Scroll down and click on Food Storage and Preservation.

www.foodsafety.gov

This website is the gateway to government food-safety information.

www.cdc.gov/foodsafety

The Centers for Disease Control (CDC) website provides food-safety information and tips for international travelers.

www.nal.usda.gov

The National Agriculture Library provides consumer food-safety information. Under Browse by Subject, select Food and Nutrition. Scroll down and select Food Safety Information Center.

www.fda.gov/Food/default.htm

This government website provides information on food safety from the Center for Food Safety and Applied Nutrition. The toll-free information hotline is 1-888-SAFEFOOD.

www.fsis.usda.gov

This Food Safety and Inspection Service (FSIS) website provides information about ensuring the safety of the U.S. food supply.

www.pueblo.gsa.gov

The Federal Citizen Information Center website contains food-safety resources.

www.fightbac.org

The Partnership for Food Safety Education provides consumer and professional information about food safety.

www.healthfinder.gov

Use this search engine to locate home food-safety information for handling meat, poultry, and seafood.

www.epa.gov

The Environmental Protection Agency (EPA) website provides information about food additives, pesticides, and other chemicals.

www.ams.usda.gov/nop

This government website provides information on organic foods.

www.organic-center.org

Learn more about the benefits of organic farming and products to society.

www.myfootprint.org

Take the Ecological Footprint Quiz to estimate how much productive land and water you need to support what you use and what you discard.

www.hendpg.com

The Hunger and Environmental Nutrition Dietetic Practice Group of the American Dietetic Association website provides topics of interest about awareness and action on hunger and environmental issues.

www.ift.org

Search the Institute of Food Technologists site for information about biotechnology and other chapter topics.

www.ucsusa.org

The Union of Concerned Scientists website provides information regarding biotechnology and sustainable agriculture.

www.who.int/en

The World Health Organization website provides resources on world hunger, poverty, and overpopulation.

http://feedingamerica.org

This website addresses food distribution, community food banks, and food security topics.

http://frac.org

The Food Research and Action Center website provides information on U.S. food security issues and food assistance programs.

www.thehungersite.com

You can click a button that says "Click Here to Give," and a food donation is made to the United Nations World Food Program.

www.wfp.org

The United Nations World Food Programme website provides resources about world hunger.

www.worldhungeryear.org

Information about food insecurity in the United States is provided.

www.fns.usda.gov/fns

This website provides information about U.S. food assistance and food recovery programs.

www.fao.org/sd

At this website, FAO provides information about sustainable development.

www.oxfamamerica.org

This website provides information about global hunger and poverty.

www.bread.org

This website addresses domestic and world hunger issues.

www.unicef.org

Visit this website for information on child survival programs worldwide.

Menu of Online Study Tools

A variety of study tools for this chapter are available at our website to deepen your understanding of chapter concepts. Go to

www.cengage.com/sso

to find

- Practice tests
- Flashcards
- Glossary
- Web links
- Animations
- Chapter summaries, learning objectives, and crossword puzzles.

Appendix A

AIDS TO CALCULATIONS AND THE FOOD EXCHANGE LISTS FOR DIABETES

Many mathematical problems have been worked out for you as examples at appropriate places in the text. This appendix aims to help with the use of the metric system and with those problems not fully explained elsewhere. Chapter 2 introduced dietary guidelines, food group plans, and the exchange system. This appendix also provides more detail about the *Choose Your Foods: Exchange Lists for Diabetes.*

Conversion Factors*

Conversion factors are useful mathematical tools in everyday calculations, like the ones encountered in the study of nutrition. A conversion factor is a fraction in which the numerator (top) and the denominator (bottom) express the same quantity in different units. For example, 2.2 pounds and 1 kilogram are *equivalent;* they express the same weight. The conversion factor used to change pounds to kilograms or vice versa is:

$$\frac{2.2 \text{ lb}}{1 \text{ kg}} \quad \text{or} \quad \frac{1 \text{ kg}}{2.2 \text{ lb}}$$

Because either of these factors equals 1, a measurement can be multiplied by the factor without changing the value of the measurement. Thus its units can be changed.

The correct factor to use in a problem is the one with the unit you are seeking in the numerator (top) of the fraction. Following are three examples of problems commonly encountered in nutrition study; they illustrate the usefulness of conversion factors.

EXAMPLE 1
Convert $1/4$ cup to an approximate number of milliliters for use in a recipe.

1. The conversion factor is:

$$\frac{1 \text{ c}}{250 \text{ mL}} \quad \text{or} \quad \frac{250 \text{ mL}}{1 \text{ c}}$$

2. Multiply $1/4$ cup by the factor:

$$1/4 \text{ c} \times \frac{250 \text{ mL}}{1 \text{ c}} = 62.5 \text{ mL, or about } 60 \text{ mL}$$

EXAMPLE 2
Convert the weight of 130 pounds to kilograms.

1. Choose the conversion factor in which the unit you are seeking is on top:

$$\frac{1 \text{ kg}}{2.2 \text{ lb}}$$

---

*For a listing of specific conversion factors, see Table A-1 on page A-4.

2. Multiply 130 pounds by the factor:

$$130 \text{ lb} \times \frac{1 \text{ kg}}{2.2 \text{ lb}} = \frac{130 \text{ kg}}{2.2} = 59 \text{ kg (rounded off to nearest whole number)}$$

EXAMPLE 3

How many grams of saturated fat are contained in a 3-ounce hamburger? A 4-ounce hamburger contains 7 grams of saturated fat.

1. You are seeking grams of saturated fat; therefore, the conversion factor is:

$$\frac{7 \text{ g saturated fat}}{4 \text{ oz hamburger}}$$

2. Multiply 3 ounces of hamburger by the conversion factor:

$$3 \text{ oz hamburger} \times \frac{7 \text{ g saturated fat}}{4 \text{ oz hamburger}} = \frac{3 \times 7\text{g}}{4} = \frac{21}{4}$$

$$= 5 \text{ g saturated fat (rounded off to nearest whole number)}$$

Percentages

A *percentage* is a comparison between a number of items (perhaps your intake of calories) and a standard number (perhaps the number of calories recommended for your age and gender). The standard number is the number you divide by. The answer you get after the division must be multiplied by 100 to be stated as a percentage (*percent* means "per 100").

EXAMPLE 4

What percentage of the DRI for energy (EER) is your calorie intake?

1. We'll use 2,400 calories to demonstrate. (The EER are lised on the inside front cover.)
2. Total your calorie intake for a day—for example, 1,200 calories.
3. Divide your calorie intake by the EER calories:

$$1,200 \text{ cal (your intake)} \div 2,400 \text{ cal (EER)} = 0.50$$

4. Multiply your answer by 100 to state it as a percentage:

$$0.50 \times 100 = 50.0 = 50\%$$

In some problems in nutrition, the percentage may be more than 100. For example, suppose your daily intake of vitamin C is 500 mg and your RDA (male) is 90 mg. Your intake as a percentage of the RDA is more than 100 percent (that is, you consume more than 100 percent of your vitamin C RDA). The following calculations show your vitamin C intake as a percentage of the RDA:

$$500 \div 90 = 5.55$$
$$5.55 \times 100 = 555\% \text{ of RDA}$$

Sometimes the comparison is between a part of a whole (for example, your calories from protein) and the total amount (your total calories). In this case, the total number is the one you divide by as shown in Example 5.

EXAMPLE 5

What percentages of your total calories for the day come from protein, fat, and carbohydrate?

1. Using Appendix E and your diet record, find the total grams of protein, fat, and carbohydrate you consumed—for example, 60 grams protein, 80 grams fat, and 285 grams carbohydrate.
2. Multiply the number of grams by the number of calories from 1 gram of each energy nutrient (conversion factors):

$$60 \text{ g protein} \times \frac{4 \text{ cal}}{1 \text{ g protein}} = 240 \text{ cal}$$

$$80 \text{ g fat} \times \frac{9 \text{ cal}}{1 \text{ g fat}} = 720 \text{ cal}$$

$$285 \text{ g carbohydrate} \times \frac{4 \text{ cal}}{1 \text{ g carbohydrate}} = 1{,}140 \text{ cal}$$

$$240 + 720 + 1{,}140 = 2{,}100 \text{ cal}$$

3. Find the percentage of total calories from each energy nutrient (see Example 4):

Protein: $240 \div 2{,}100 = 0.114$
$0.114 \times 100 = 11.4 = 11\%$ of calories

Fat: $720 \div 2{,}100 = 0.343$
$0.343 \times 100 = 34.3 = 34\%$ of calories

Carbohydrate: $1{,}140 \div 2{,}100 = 0.543$
$0.543 \times 100 = 54.3 = 54\%$ of calories

$11\% + 34\% + 54\% = 99\%$ of calories (total)

The percentages total 99 percent rather than 100 percent because a little was lost from each number in rounding off. Either 99 or 101 is a reasonable total.

Nutrient Units and Unit Conversions

International Units (IU)

To convert IU to:
  µg vitamin D: divide by 40 or multiply by 0.025.
  mg αTE:[a] divide by 1.5

Folate

To convert micrograms of synthetic folate in supplements and enriched foods to Dietary Folate Equivalents (µg DFE):

µg synthetic folate × 1.7 = µg DFE

For naturally occurring folate, assign each microgram folate a value of 1 µg DFE:

µg folate = µg DFE

Sodium

To convert milligrams of sodium to grams of salt

mg sodium ÷ 400 = g of salt

The reverse is also true

g salt × 400 = mg sodium

Niacin

1 mg NE (niacin equivalent) = 1 mg niacin = 60 mg dietary tryptophan

Vitamin A

1 µg RAE[b] = 1 µg retinol
  = 12 µg beta-carotene
  = 24 µg other vitamin A carotenoids

To convert older RE[c] values to micrograms RAE:

1 µg RE retinol = 1 µg RAE retinol
6 µg RE beta-carotene = 12 µg RAE beta-carotene
12 µg RE other vitamin A carotenoids = 24 µg RAE other vitamin A carotenoids

1 IU = 0.3 µg retinol
  = 3.6 µg beta-carotene
  = 7.2 µg other vitamin A carotenoids

[a]Alpha-tocopherol equivalents (vitamin E).
[b]Retinol activity equivalents (vitamin A).
[c]Retinol equivalents (vitamin A).

TABLE A-1 — Conversion Factors

Length	Volume	Temperature
1 inch = 2.54 cm	1 tbsp = 3 tsp or 15 mL	Steam — 100°C 212°F — Steam
1 ft = 30.48 cm	1 oz = 2 tbsp or 30 mL	Body temperature — 37°C 98.6°F — Body temperature
1 m = 39.37 inches	½ c = about 125 mL	Ice — 0°C 32°F — Ice
	1 c = 16 tbsp or about 250 mL	Centigrade Fahrenheit
	1 qt = 4 c or .95 liter	$t_F = \frac{9}{5} t_c + 32$
	1 qt = 32 oz (fluid)	$t_c = \frac{5}{9} (t_F - 32)$
	1 L = 1.06 qt or 1000 mL	
	1 gal = 16 c or 4 qt or 3.79 liter	

Weight		Energy
1 oz = about 30 g (28.35 g)	1 tsp any powder or liquid = about 5 g or 5 mL	1 cal = 4.2 kJ = 0.004 mJ
1 lb = about 454 g	1 tbsp any powder or liquid = about 15 g or 15 mL	1 mJ = 240 cal
1 kg = 1000 g	½ c any vegetable, fruit, or fluid weighs about 100 g	1 kJ = 0.24 cal
1 kg = 2.2 lb		1 g carbohydrate = 4 cal = 17 kJ
1 g = 1000 mg		1 g fat = 9 cal = 37 kJ
1 mg = 1000 µg		1 g protein = 4 cal = 17 kJ
1 µg = $^1/_{1000}$ mg		1 g alcohol = 7 cal = 29 kJ

Body Weights: Quick Estimation of Desirable Body Weight

Men	Women
For 5 ft, consider 106 lb a reasonable weight.	For 5 ft, consider 100 lb a reasonable weight.
For each inch over 5 ft, add 6 lb.	For each inch over 5 ft, add 5 lb.
Subtract 6 lb for each inch under 5 ft. Add 10% for a large-framed individual; subtract 10% for a small-framed individual.	Subtract 5 lb for each inch under 5 ft. Add 10% for a large-framed individual; subtract 10% for a small-framed individual.
Example: A man 5 ft 8 inches tall (medium frame) would start at 106 lb, add 48, and arrive at a reasonable weight of 154 lb.	For each year under 25 (down to 18), subtract 1 lb.
	Example: A woman 21 years old, 5 ft 4 inches tall (medium-frame), would start at 100 lb, add 20, and subtract 4, arriving at a reasonable weight of 116 lb.

Table A-2 shows how to determine frame size.

TABLE A-2 — Frame Size

To make a simple approximation of your frame size:

Extend your arm and bend the forearm upward at a 90-degree angle. Keep the fingers straight and turn the inside of your wrist away from your body. Place the thumb and index finger of your other hand on the two prominent bones on either side of your elbow. Measure the space between your fingers against a ruler or a tape measure.[a] Compare the measurements with the following standards. (These standards represent the elbow measurements for medium-framed men and women of various heights. Measurements smaller than those listed indicate you have a small frame, and larger measurements indicate a large frame.[b])

Men		Women	
Height in 1-in. Heels	**Elbow Breadth**	**Height in 1-in. Heels**	**Elbow Breadth**
5 ft 2 in. to 5 ft 3 in.	2½ to 2⅞ in.	4 ft 10 in. to 4 ft 11 in.	2¼ to 2½ in.
5 ft 4 in. to 5 ft 7 in.	2⅝ to 2⅞ in.	5 ft 0 in. to 5 ft 3 in.	2¼ to 2½ in.
5 ft 8 in. to 5 ft 11 in.	2¾ to 3 in.	5 ft 4 in. to 5 ft 7 in.	2⅜ to 2⅝ in.
6 ft 0 in. to 6 ft 3 in.	2¾ to 3⅛ in.	5 ft 8 in. to 5 ft 11 in.	2⅜ to 3⅝ in.
6 ft 4 in. and over	2⅞ to 3¼ in.	6 ft 0 in. and over	2½ to 3¾ in.

[a]For the most accurate measurement, have your health care provider measure your elbow breadth with a caliper.
[b]A simple estimate of frame size can be derived by circling your wrist with the thumb and middle finger of your other hand. If the thumb and finger do not meet, you most likely have a large frame. If the thumb and finger just meet, you most likely have a medium frame, and if the fingers overlap greatly, you may have a small frame.

Choose Your Foods: Exchange List for Diabetes

The food exchange system divides the foods suitable for use in planning a healthy diet into seven lists—the starch; fruit; milk; sweets, desserts, and other carbohydrates; non-starchy vegetables; meat and meat substitutes; and fat lists.* The exchange lists are shown in Table A-3. Figure A-1 provides examples of foods, portion sizes, and energy-nutrient contributions for each of the exchange lists.

TABLE A-3	The Food Exchange System				
List	Portion Size	Carbohydrate (g)	Protein (g)	Fat (g)	Energy (cal)
Carbohydrate Group					
Starch	1 slice; ½ c	15	0–3	0–1	80
Fruit	varies	15	—	—	60
Milk	1 c				
Fat-free, low-fat, 1%		12	8	0–3	100
Reduced-fat, 2%		12	8	5	120
Whole		12	8	8	150
Sweets, desserts, and other carbohydrates	varies	15	varies	varies	varies
Nonstarchy vegetables	½ c	5	2	—	25
Meat and Meat Substitutes	1 oz				
Lean		0	7	0–3	45
Medium-fat		0	7	4–7	75
High-fat		0	7	8+	100
Plant-based proteins	varies	varies	7	varies	varies
Fat Group	1 tsp pure fat	—	—	5	45
Alcohol	varies	varies	—	—	100

*The food exchange system presented here is based on material in *Choose Your Foods: Exchange Lists for Diabetes*, 2008, prepared by committees of the American Diabetes Association and the American Dietetic Association.

FIGURE A-1

The Exchange System: Example Foods, Portion Sizes, and Energy-Nutrient Contributions

Starch

1 starch exchange is like:

1 slice bread

¾ c ready-to-eat cereal

½ c cooked grits, rice noodles, or bulgur

⅓ c cooked rice or pasta

½ c cooked beans[a]

½ c corn, peas, or yams

1 small (3 oz) potato

½ bagel, English muffin, or bun

1 tortilla, waffle, roll, taco, or matzoh

(1 starch = 15 g carbohydrate, 0–3 g
    protein, 0–1 g fat, and 80 cal.)

[a] ½ c cooked beans = 1 lean meat exchange *plus* 1 starch exchange.

Nonstarchy vegetables[b]

1 vegetable exchange is like:

½ c cooked carrots, greens, green beans,
    brussels sprouts, beets, broccoli, cauli-
    flower, or spinach

1 c raw carrots, radishes, or spinach

1 lg tomato

(1 vegetable = 5 g carbohydrate, 2 g
    protein, and 25 cal.)

[b] Salad greens are on the Free-Foods list.

Fruits

1 fruit exchange is like:

1 small banana, nectarine, apple, or orange

½ large grapefruit, pear, or papaya

½ c orange, apple, or grapefruit juice

17 small grapes

⅓ cantaloupe (or 1 c cubes)

2 tbs raisins

1½ dried figs

3 dates

1½ carambola (star fruit)

(1 fruit = 15 g carbohydrate and 60 cal.)

Lean meats and meat substitutes

1 lean meat exchange is like:

1 oz cod, flounder, or trout

1 oz clams, crab, lobster, scallops, shrimp, or
    imitation seafood

1 oz cheese with 3 g fat or less

2 egg whites (or ¼ c egg substitute)

1 oz beef or pork tenderloin trimmed of fat

¼ c cottage cheese

1 oz game meat

1 oz chicken without skin

1 oz tuna (canned in water or oil, drained)

(1 lean meat = 7 g protein, 0–3 g fat, and
    45 cal.)

Medium-fat meats and meat substitutes

1 medium-fat meat exchange is like:

1 oz ground beef

1 oz pork cutlet

1 egg

¼ c ricotta cheese

(1 medium-fat meat = 7 g protein, 4–7 g fat,
    and 75 cal.)

The Exchange System: Example Foods, Portion Sizes, and Energy-Nutrient Contributions—*Continued*

Sweets, desserts, and other carbohydrates

1 other carbohydrates exchange is like:

2 small cookies

1 small brownie or cake

5 vanilla wafers

1 granola bar

½ c ice cream

(1 other carbohydrate = 15 g carbohydrate and may be exchanged for 1 starch, 1 fruit, or 1 milk. Because many items on this list contain added sugar and fat, their fat and calorie values vary, and their portion sizes are small.)

THE FAT GROUP
Fats

1 fat exchange is like:

1 tsp butter

1 tsp margarine or mayonnaise (1 tbs reduced fat)

1 tsp any oil

1 tbs salad dressing (2 tbs reduced fat)

8 large black olives

10 large peanuts

⅛ medium avocado

1 slice bacon

2 tbs shredded coconut

1 tbs cream cheese (1½ tbs reduced fat)

(1 fat = 5 g fat and 45 cal.)

Milks (fat-free and low-fat, 1%)

1 fat-free milk exchange is like:

1 c fat-free or 1% milk

¾ c fat-free yogurt, plain

1 c fat-free or low-fat buttermilk

½ c evaporated fat-free milk

⅓ c dry fat-free milk

(1 fat-free milk = 12 g carbohydrate, 8 g protein, 0–3 g fat, and 100 cal.)

Milks (reduced-fat, 2%)

1 reduced-fat milk exchange is like:

1 c 2% milk

6 oz reduced-fat yogurt, plain

(1 reduced-fat milk = 12 g carbohydrate, 8 g protein, 5 g fat, and 120 cal.)

High-fat meats and meat substitutes

1 high-fat meat exchange is like:

2 slices pork bacon

3 slices turkey bacon

1 oz American cheese

1 hot dog[c]

1 oz pork sausage

(1 high-fat meat = 7 g protein, ≥8 g fat, and 100 cal.)

[c] A beef or pork hot dog counts as 1 high-fat meat exchange *plus* 1 fat exchange.

Plant-based proteins

1 plant-based protein exchange is like:

½ c baked beans (1 starch + 1 lean meat)

½ c edamame (½ CHO + 1 lean meat)

⅓ c hummus (1 CHO + 1 high-fat meat)

3 oz meatless burger, soy based (½ CHO + 2 lean meats)

1 tbs peanut butter (1 high-fat meat)

½ c refried beans, canned (1 CHO + 1 lean meat)

4 oz (½ c) tofu (1 medium-fat meat)

Milks (whole)

1 whole milk exchange is like:

1 c whole milk

½ c evaporated whole milk

(1 whole milk = 12 g carbohydrate, 8 g protein, 8 g fat, and 150 cal.)

CANADIAN NUTRITION AND PHYSICAL ACTIVITY GUIDELINES

Canada's Guidelines for Healthy Eating

Canada's Guidelines for Healthy Eating encourage healthy Canadians over 2 years of age to

- Enjoy a variety of foods.

- Emphasize cereals, breads, other grain products, vegetables, and fruits.

- Choose lower-fat dairy products, leaner meats, and foods prepared with little or no fat.

- Achieve and maintain a healthy body weight by enjoying regular physical activity and healthy eating.

- Limit salt, alcohol, and caffeine.

Canada's Recommended Nutrient Intakes

The Dietary Reference Intakes (DRI) reports have replaced both the 1989 RDA in the United States and the 1990 Recommended Nutrient Intakes (RNI) in Canada. The DRI are presented on Study Cards pages 27–28 at the back of this book.

Eating Well with Canada's Food Guide

Figure B-1 presents the 2007 *Eating Well with Canada's Food Guide*, which gives detailed information for selecting foods to meet *Canada's Guidelines for Healthy Eating*. The *Food Guide* was designed to meet the nutritional needs of all Canadians 2 years of age and older and takes a total diet approach. The *Food Guide* shows the four food groups with pictorial examples of foods in each group and shows the number of servings recommended for each food group.

For more detailed information and interactive tools using the *Food Guide*, go to www .healthcanada.gc.ca/foodguide. Additional publications are available from Health Canada's Office of Nutrition Policy and Promotion at www.hc-sc.gc.ca.

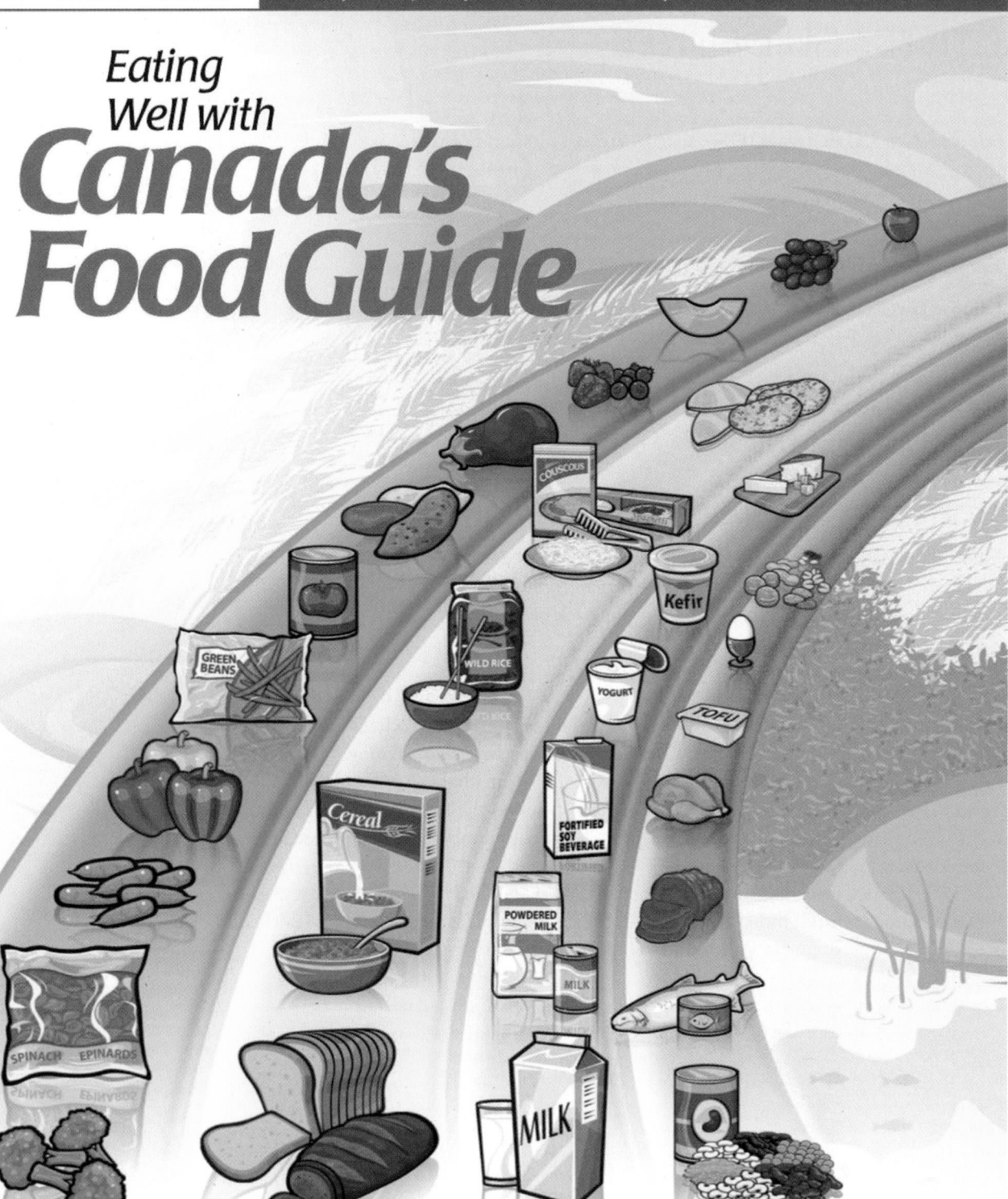

Health Canada **Santé Canada** Your health and safety... our priority. Votre santé et votre sécurité... notre priorité.

Eating Well with **Canada's Food Guide**

Canada

Recommended Number of *Food Guide Servings* per Day

	Children			Teens		Adults			
Age in Years	2-3	4-8	9-13	14-18		19-50		51+	
Sex	Girls and Boys			Females	Males	Females	Males	Females	Males
Vegetables and Fruit	4	5	6	7	8	7-8	8-10	7	7
Grain Products	3	4	6	6	7	6-7	8	6	7
Milk and Alternatives	2	2	3-4	3-4	3-4	2	2	3	3
Meat and Alternatives	1	1	1-2	2	3	2	3	2	3

The chart above shows how many Food Guide Servings you need from each of the four food groups every day.

Having the amount and type of food recommended and following the tips in *Canada's Food Guide* will help:

• Meet your needs for vitamins, minerals and other nutrients.

• Reduce your risk of obesity, type 2 diabetes, heart disease, certain types of cancer and osteoporosis.

• Contribute to your overall health and vitality.

FIGURE B-1

Eating Well with Canada's Food Guide—Continued

What is One Food Guide Serving?
Look at the examples below.

 Fresh, frozen or canned vegetables
125 mL (½ cup)

 Leafy vegetables
Cooked: 125 mL (½ cup)
Raw: 250 mL (1 cup)

 Fresh, frozen or canned fruits
1 fruit or 125 mL (½ cup)

 100% Juice
125 mL (½ cup)

 Bread
1 slice (35 g)

 Bagel
½ bagel (45 g)

 Flat breads
½ pita or ½ tortilla (35 g)

 Cooked rice, bulgur or quinoa
125 mL (½ cup)

 Cereal
Cold: 30 g
Hot: 175 mL (¾ cup)

 Cooked pasta or couscous
125 mL (½ cup)

 Milk or powdered milk (reconstituted)
250 mL (1 cup)

 Canned milk (evaporated)
125 mL (½ cup)

 Fortified soy beverage
250 mL (1 cup)

 Yogurt
175 g
(¾ cup)

 Kefir
175 g
(¾ cup)

 Cheese
50 g (1 ½ oz.)

 Cooked fish, shellfish, poultry, lean meat
75 g (2 ½ oz.)/125 mL (½ cup)

 Cooked legumes
175 mL (¾ cup)

 Tofu
150 g or
175 mL (¾ cup)

 Eggs
2 eggs

 Peanut or nut butters
30 mL (2 Tbsp)

 Shelled nuts and seeds
60 mL (¼ cup)

 ## Oils and Fats

- Include a small amount – 30 to 45 mL (2 to 3 Tbsp) – of unsaturated fat each day. This includes oil used for cooking, salad dressings, margarine and mayonnaise.
- Use vegetable oils such as canola, olive and soybean.
- Choose soft margarines that are low in saturated and trans fats.
- Limit butter, hard margarine, lard and shortening.

Make each Food Guide Serving count...
wherever you are – at home, at school, at work or when eating out!

▶ **Eat at least one dark green and one orange vegetable each day.**
- Go for dark green vegetables such as broccoli, romaine lettuce and spinach.
- Go for orange vegetables such as carrots, sweet potatoes and winter squash.

▶ **Choose vegetables and fruit prepared with little or no added fat, sugar or salt.**
- Enjoy vegetables steamed, baked or stir-fried instead of deep-fried.

▶ **Have vegetables and fruit more often than juice.**

▶ **Make at least half of your grain products whole grain each day.**
- Eat a variety of whole grains such as barley, brown rice, oats, quinoa and wild rice.
- Enjoy whole grain breads, oatmeal or whole wheat pasta.

▶ **Choose grain products that are lower in fat, sugar or salt.**
- Compare the Nutrition Facts table on labels to make wise choices.
- Enjoy the true taste of grain products. When adding sauces or spreads, use small amounts.

▶ **Drink skim, 1%, or 2% milk each day.**
- Have 500 mL (2 cups) of milk every day for adequate vitamin D.
- Drink fortified soy beverages if you do not drink milk.

▶ **Select lower fat milk alternatives.**
- Compare the Nutrition Facts table on yogurts or cheeses to make wise choices.

▶ **Have meat alternatives such as beans, lentils and tofu often.**

▶ **Eat at least two Food Guide Servings of fish each week.***
- Choose fish such as char, herring, mackerel, salmon, sardines and trout.

▶ **Select lean meat and alternatives prepared with little or no added fat or salt.**
- Trim the visible fat from meats. Remove the skin on poultry.
- Use cooking methods such as roasting, baking or poaching that require little or no added fat.
- If you eat luncheon meats, sausages or prepackaged meats, choose those lower in salt (sodium) and fat.

Enjoy a variety of foods from the four food groups.

Satisfy your thirst with water!

Drink water regularly. It's a calorie-free way to quench your thirst. Drink more water in hot weather or when you are very active.

* Health Canada provides advice for limiting exposure to mercury from certain types of fish. Refer to www.healthcanada.gc.ca for the latest information.

Advice for different ages and stages...

Children

Following *Canada's Food Guide* helps children grow and thrive.

Young children have small appetites and need calories for growth and development.

- Serve small nutritious meals and snacks each day.

- Do not restrict nutritious foods because of their fat content. Offer a variety of foods from the four food groups.

- Most of all... be a good role model.

Women of childbearing age

All women who could become pregnant and those who are pregnant or breastfeeding need a multivitamin containing **folic acid** every day. Pregnant women need to ensure that their multivitamin also contains **iron**. A health care professional can help you find the multivitamin that's right for you.

Pregnant and breastfeeding women need more calories. Include an extra 2 to 3 Food Guide Servings each day.

Here are two examples:

- Have fruit and yogurt for a snack, or

- Have an extra slice of toast at breakfast and an extra glass of milk at supper.

Men and women over 50

The need for **vitamin D** increases after the age of 50.

In addition to following *Canada's Food Guide*, everyone over the age of 50 should take a daily vitamin D supplement of 10 µg (400 IU).

How do I count Food Guide Servings in a meal?

Here is an example:

Vegetable and beef stir-fry with rice, a glass of milk and an apple for dessert		
250 mL (1 cup) mixed broccoli, carrot and sweet red pepper	=	2 **Vegetables and Fruit** Food Guide Servings
75 g (2 ½ oz.) lean beef	=	1 **Meat and Alternatives** Food Guide Serving
250 mL (1 cup) brown rice	=	2 **Grain Products** Food Guide Servings
5 mL (1 tsp) canola oil	=	part of your **Oils and Fats** intake for the day
250 mL (1 cup) 1% milk	=	1 **Milk and Alternatives** Food Guide Serving
1 apple	=	1 **Vegetables and Fruit** Food Guide Serving

Eat well and be active today and every day!

The benefits of eating well and being active include:

- Better overall health.
- Lower risk of disease.
- A healthy body weight.
- Feeling and looking better.
- More energy.
- Stronger muscles and bones.

Be active

To be active every day is a step towards better health and a healthy body weight.

Canada's Physical Activity Guide recommends building 30 to 60 minutes of moderate physical activity into daily life for adults and at least 90 minutes a day for children and youth. You don't have to do it all at once. Add it up in periods of at least 10 minutes at a time for adults and five minutes at a time for children and youth.

Start slowly and build up.

Eat well

Another important step towards better health and a healthy body weight is to follow *Canada's Food Guide* by:

- Eating the recommended amount and type of food each day.
- Limiting foods and beverages high in calories, fat, sugar or salt (sodium) such as cakes and pastries, chocolate and candies, cookies and granola bars, doughnuts and muffins, ice cream and frozen desserts, french fries, potato chips, nachos and other salty snacks, alcohol, fruit flavoured drinks, soft drinks, sports and energy drinks, and sweetened hot or cold drinks.

Read the label

- Compare the Nutrition Facts table on food labels to choose products that contain less fat, saturated fat, trans fat, sugar and sodium.
- Keep in mind that the calories and nutrients listed are for the amount of food found at the top of the Nutrition Facts table.

Limit trans fat

When a Nutrition Facts table is not available, ask for nutrition information to choose foods lower in trans and saturated fats.

Nutrition Facts
Per 0 mL (0 g)

Amount	% Daily Value
Calories 0	
Fat 0 g	0 %
Saturates 0 g	0 %
+ Trans 0 g	
Cholesterol 0 mg	
Sodium 0 mg	0 %
Carbohydrate 0 g	0 %
Fibre 0 g	0 %
Sugars 0 g	
Protein 0 g	

Vitamin A	0 %	Vitamin C	0 %
Calcium	0 %	Iron	0 %

Take a step today...

✓ Have breakfast every day. It may help control your hunger later in the day.

✓ Walk wherever you can – get off the bus early, use the stairs.

✓ Benefit from eating vegetables and fruit at all meals and as snacks.

✓ Spend less time being inactive such as watching TV or playing computer games.

✓ Request nutrition information about menu items when eating out to help you make healthier choices.

✓ Enjoy eating with family and friends!

✓ Take time to eat and savour every bite!

For more information, interactive tools, or additional copies visit Canada's Food Guide on-line at:
www.healthcanada.gc.ca/foodguide

or contact:
Publications
Health Canada
Ottawa, Ontario K1A 0K9
E-Mail: publications@hc-sc.gc.ca
Tel.: 1-866-225-0709
Fax: (613) 941-5366
TTY: 1-800-267-1245

Également disponible en français sous le titre :
Bien manger avec le Guide alimentaire canadien

This publication can be made available on request on diskette, large print, audio-cassette and braille.

CHAPTER NOTES

CHAPTER 1

1. American Society for Nutritional Sciences, *Nutrition Notes* 38 (2002): 2.
2. A. E. Sloan, Top ten global food trends, *Food Technology* 59 (2005): 31; and The Food Marketing Institute, U.S. grocery shopper trends (Washington, DC: Food Marketing Institute, 2010).
3. Position of the American Dietetic Association: Food and nutrition misinformation, *Journal of the American Dietetic Association* 106 (2006): 601–7.
4. Federal Trade Commission, Bureau of Consumer Protection, *Deception in Weight-Loss Advertising Workshop: Seizing Opportunities and Building Partnerships to Stop Weight-Loss Fraud* (Washington, DC: Author, 2003).
5. A. E. Sloan, "Top ten global food trends," *Food Technology* 59 (2005): 20–32.
6. Centers for Disease Control and Prevention, *National Vital Statistics Report, 2010* (www.cdc.gov/nchs).
7. T. L. S. Visscher et al., Obesity and unhealthy life-years in adult Finns, *Archives of Internal Medicine* 164 (2004): 1413–20; and G. A. Bray, Medical consequences of obesity, *Journal of Clinical Endocrinology and Metabolism* 89 (2004): 2583–89.
8. L. Breslow and N. Breslow, Health practices and disability: Some evidence from Alameda County, *Preventive Medicine* 22 (1993): 86–95.
9. National Cancer Institute, *5 A Day for Better Health Program Evaluation Report* (Bethesda, MD: National Institutes of Health, 2006); available at www.cancercontrol.cancer.gov/5ad_exec.html.
10. L. Breslow and N. Breslow, Health practices and disability: Some evidence from Alameda County, *Preventive Medicine* 22 (1993): 86–95; and A. J. Vita et al., Aging, health risks, and cumulative disability, *New England Journal of Medicine* 338 (1998): 1035–41.
11. Parts of the following discussion were adapted from B. J. Willcox, D. C. Willcox, and M. Suzuki, *The Okinawa Program: Learn the Secrets to Healthy Longevity* (New York: Three Rivers Press, 2001); and D. Buettner, The secrets of long life, *National Geographic* 208 (2005): 2–27.
12. *Ibid.*
13. M. Minkler, Health education, health promotion, and the open society: An historical perspective, *Health Education Quarterly* 16 (1989): 17–30; Position of the American Dietetic Association: Roles of registered dietitians and dietetic technicians, registered in health promotion and disease prevention, *Journal of the American Dietetic Association* 106 (2006): 1875–84; and Position of the American Dietetic Association: Total diet approach to communicating food and nutrition information, *Journal of the American Dietetic Association* 107 (2007): 1224–32.
14. U.S. Department of Health and Human Services, Office of Disease Prevention and Health Promotion, *Healthy People 2020* (Washington, DC: U.S. Government Printing Office, 2010).
15. *Ibid.*
16. M. Eure, Healthy eating starts with healthy food shopping, About.com *Senior Health*; available at http://seniorhealth.about.com/library/nutrition/bl_food_shop.htm.
17. U.S. Department of Agriculture, Center for Nutrition Policy and Promotion, *Recipes and Tips for Healthy, Thrifty Meals* (Washington, DC: Author, May 2000).
18. The list of barriers in the margin is from Princeton Survey Research Associates, *Shopping for Health 2002: Self Care Perspectives* (Washington, DC: Food Marketing Institute; and Emmaus, PA: Rodale, 2002), 14.
19. H. A. Raynor et al., A cost-analysis of adopting a healthful diet in a family-based obesity treatment program, *Journal of the American Dietetic Association* 102 (2002): 645–56.
20. *Ibid.*
21. K. Collins, Eating healthy on the cheap, MSNBC Health, *Nutrition Notes* (www.msnbc.msn.com/id/3035439/).
22. B. W. Hickman et al., Nutrition claims in advertising: A study of four women's magazines, *Journal of Nutrition Education* 25 (1993): 227–35.
23. K. Kotz and M. Story, Food advertisements during children's Saturday morning television programming: Are they consistent with dietary recommendations? *Journal of the American Dietetic Association* 94 (1994): 1296–1300; and A. Batada, M. Seitz, M. Wootan, and M. Story, Nine out of 10 food advertisements shown during Saturday morning children's television programming are for foods high in fat, sodium, or added sugars, or low in nutrients, *Journal of the American Dietetic Association* 108 (2008): 673–78.
24. Consumers Union, *Out of Balance* (2005); available at www.consumersunion.org/pub/core_health_care/002657.html.
25. American Dietetic Association, *Nutrition and You: Trends 2008* (Chicago: American Dietetic Association, 2008).
26. J. MacClancy, *Consuming Culture: Why You Eat What You Eat* (New York: Henry Holt, 1992), 38.
27. The discussion of sustainability in the food system is adapted from C. McCullum et al., Evidence-based strategies to build community food security, *Journal of the American Dietetic Association* 105 (2005): 278–83.
28. A. H. Harmon and B. L. Gerald, Position of the American Dietetic Association: Food and nutrition professionals can implement practices to conserve natural resources and support ecological sustainability, *Journal of the American Dietetic Association* 107 (2007): 1033–43.
29. B. Storper, Moving toward healthful sustainable diets, *Nutrition Today* 38 (2003): 57–59.
30. J. Walsh and N. de Beaufort, Dietitians and local farmers: A unique alliance to change the way people think about food, *Hunger and Environmental Nutrition Dietetic Practice Group Newsletter* 1 (Spring 2003): 3; and Position of the American Dietetic Association: Food and nutrition professionals can implement practices to conserve natural resources and support ecological sustainability, *Journal of the American Dietetic Association* 107 (2007): 1033–43.
31. W. H. Glinsmann and G. K. Beauchamp, Babies need sugars in moderation, *Pediatric Basics* 69 (1994): 19–21.
32. Fact of the Day Archive (December 13, 2002): National Restaurant Association, Meal Consumption Behavior (www.restaurant.org/research/facts/).
33. *Ibid.*
34. The food composition data in the photos on page 20 are from McDonald's USA Nutrition Facts (www.mcdonalds.com/us/en/food.html).
35. L. R. Young and M. Nestle, The contribution of expanding portion sizes to the U.S. obesity epidemic, *American Journal of Public Health* 92 (2002): 246–48; and S. J. Nielsen and B. M. Popkin, Patterns and trends in food portion sizes, 1977–1998, *Journal of the American Medical Association* 289 (2003): 450–53.
36. L. R. Young and M. Nestle, Portion sizes in dietary assessment: Issues and policy implications, *Nutrition Reviews* 53 (1995): 149–58.
37. *Ibid.*; B. J. Rolls, D. Engell, and L. L. Birch, Serving portion size influences 5-year-old but not 3-year-old children's food intake, *Journal of the American Dietetic Association* 100 (2000): 232–34; and B. J. Rolls, The supersizing of America: Portion size and the obesity epidemic, *Nutrition Today* 38 (2003): 42–53.
38. National Center for Chronic Disease Prevention and Health Promotion, Centers for Disease Control and Prevention (CDC), *Obesity and Overweight—A Public Health Epidemic*; available at www.cdc.gov/obesity.
39. National Alliance for Nutrition and Activity (NANA), *From Wallet to Waistline: The Hidden Cost of Super Sizing* (Washington, DC: NANA, June 2002).
40. Nutrition information available at www.wendys.com.

41. American Dietetic Association, Fast food: Can fast foods fit into a healthy eating plan? (www.eatright.org).

42. *Food, Nutrition, Physical Activity, and the Prevention of Cancer: A Global Perspective* (Washington, DC: World Cancer Research Fund/American Institute for Cancer Research, 2007); and Position of the American Dietetic Association: Health implications of dietary fiber, *Journal of the American Dietetic Association* 108 (2008): 1716–31.

43. C. S. Fuchs et al., Dietary fiber and the risk of colorectal cancer and adenoma in women, *New England Journal of Medicine* 340 (1999): 169–76.

44. *Food, Nutrition, Physical Activity, and the Prevention of Cancer: A Global Perspective* (Washington, DC: World Cancer Research Fund/American Institute for Cancer Research, 2007).

45. J. A. Nettleton et al., Incident heart failure is associated with lower whole-grain intake and greater high-fat dairy and egg intake in the Atherosclerosis Risk in Communities (ARIC) study, *Journal of the American Dietetic Association* 108 (2008): 1881–87.

46. The discussion about finding credible sources of information on the Internet is adapted from M. Boyle and D. Holben, *Community Nutrition in Action: An Entrepreneurial Approach*, 4th ed. (Belmont, CA: Thomson Wadsworth, 2006), 198.

47. R. Harris, Evaluating Internet research sources online (www.virtualsalt .com/evalu8it.htm).

CHAPTER 2

1. K. Kant et al., Dietary diversity and subsequent mortality in the First National Health and Nutrition Examination Survey Epidemiologic Follow-up Study, *American Journal of Clinical Nutrition* 57 (1993): 434–40.

2. S. L. Anderson, A look at the Japanese dietary guidelines, *Journal of the American Dietetic Association* 90 (1990): 1527.

3. The discussion of nutrient recommendations is adapted from Food and Nutrition Board, Institute of Medicine, *Dietary Reference Intakes for Calcium, Phosphorus, Magnesium, Vitamin D, and Fluoride* (Washington, DC: National Academy Press, 1997), S-1–S-13; Food and Nutrition Board, Institute of Medicine, *Dietary Reference Intakes for Energy, Carbohydrate, Fiber, Fat, Fatty Acids, Cholesterol, Protein, and Amino Acids* (Washington, DC: National Academy Press, 2002); and Practice Paper of the American Dietetic Association: Using the Dietary Reference Intakes, *Journal of the American Dietetic Association* 111 (2011): 762–77.

4. Food and Nutrition Board, Institute of Medicine, *Dietary Reference Intakes for Energy, Carbohydrate, Fiber, Fat, Fatty Acids, Cholesterol, Protein, and Amino Acids* (Washington, DC: National Academy Press, 2002).

5. M. A. Bush et al., Eating well with Canada's Food Guide: A tool for the times, *Canadian Journal of Dietetic Practice Research* 68 (2007): 92–96.

6. American Dietetic Association, Snacking (www.eatright.org/); and American Dietetic Association, Incorporate healthy snacking in your eating plan (www.eatright.org).

7. L. Jahns, A. M. Siega-Riz, and S. M. Popkins, The increasing prevalence of snacking among U.S. children from 1977 to 1996, *Journal of Pediatrics* 138 (2001): 493–98.

8. D. Voyatzis, Sensible snacking, *Tufts Daily*, September 30, 2002.

9. American Heart Association, *Nutritious Nibbles* (Dallas: American Heart Association, 1984), 5; and U.S. Department of Agriculture, *Making Bag Lunches, Snacks, and Desserts Using the Dietary Guidelines*, USDA HNS Home and Garden Bulletin No. 232-9 (Washington, DC: Government Printing Office), 2–25.

10. U.S. Department of Agriculture and U.S. Department of Health and Human Services, *Dietary Guidelines for Americans, 2010,* 7th ed. (Washington, DC: U.S. Government Printing Office, 2010).

11. C. L. Taylor and V. L. Wilkening, How the nutrition food label was developed, Part 1: The Nutrition Facts Panel, *Journal of the American Dietetic Association* 108 (2008): 437–42.

12. K. Baghurst, Fruits and vegetables: Why is it so hard to increase intakes? *Nutrition Today* 38 (2003): 11–20; and U.S. Department of Agriculture and U.S. Department of Health and Human Services, *Dietary Guidelines for Americans, 2010,* 7th ed. (Washington, DC: U.S. Government Printing Office, 2010).

13. C. L. Taylor and V. L. Wilkening, How the nutrition food label was developed, Part 2: The purpose and promise of nutrition claims, *Journal of the American Dietetic Association* 108 (2008): 618–23.

14. Food and Drug Administration, Center for Food Safety and Applied Nutrition, *A Food Labeling Guide*, Appendix C: Health Claims (www .cfsan.fda.gov/~dms/flg-6c.html); K. C. Ellwood, What are qualified health claims? *Nutrition Today* 41 (2006): 56–61; and M. Ternus et al., Qualified health claim for nuts and heart disease prevention: Development of consumer-friendly language, *Nutrition Today* 41 (2006): 62–66.

15. J. F. Mariani, *Dictionary of American Food and Drink* (New York: Hearst Books, 1994).

16. S. J. Algert, E. Brzenzinski, and T. H. Ellison, *Mexican American Food Practices, Customs, and Holidays*, Ethnic and Regional Food Practices: A Series (Chicago: American Dietetic Association and American Diabetes Association, 1998), 1–9, 23–26.

17. K. M. Ma, *Chinese American Food Practices, Customs, and Holidays*, Ethnic and Regional Food Practices: A Series, 1–10, 27–31.

18. T. C. Campbell and J. Chen, Diet and chronic degenerative diseases: A summary of results from an ecologic study in rural China, in N. J. Temple and D. P. Burkitt, *Western Diseases* (Totowa, NJ: Humana Press, 1994), 67–118.

19. Should Americans be eating more Chinese food? *Tufts University Diet & Nutrition Letter* 8 (1990): 5.

20. M. Nestle, Mediterranean diets: Historical and research overview, *American Journal of Clinical Nutrition* 61 (1995): 1313S–29S.

21. P. G. Kittler and K. Sucher, *Food and Culture in America*, 3rd ed. (Belmont, CA: Wadsworth Publishing, 2000).

22. C. Higgins et al., *Jewish Food Practices, Customs, and Holidays*, Ethnic and Regional Food Practices: A Series, 1–7, 18–20.

CHAPTER 4

1. M. L. G. Tolstoi, Adult-type lactase deficiency, *Nutrition Today* 35 (2000): 134–41; and A. K. Campbell, J. P. Waud, and S. B. Matthews, The molecular basis of lactose intolerance, *Science Progress* 88 (2005): 157–202.

2. B. A. Pribila et al., Improved lactose digestion and intolerance among African-American adolescent girls fed a dairy-rich diet, *Journal of the American Dietetic Association* 100 (2000): 524–28; and F. Guarner et al., Should yoghurt cultures be considered probiotic? *British Journal of Nutrition* 93 (2005): 783–86.

3. Position of the American Dietetic Association: Health implications of dietary fiber, *Journal of the American Dietetic Association* 108 (2008): 1716–31; and J. W. Anderson et al., Health benefits of dietary fibers, *Nutrition Reviews* 67 (2009): 188–205.

4. A. Schatzkin et al., Dietary fiber and whole-grain consumption in relation to colorectal cancer in the NIH-AARP Diet and Health Study, *American Journal of Clinical Nutrition* 85 (2007): 1353–60; and S. Bingham, Symposium on plant foods and public health: The fiber-folate debate in colo-rectal cancer, *Proceedings of the Nutrition Society* 65 (2006): 19–23.

5. Position of the American Dietetic Association: Health implications of dietary fiber, *Journal of the American Dietetic Association* 108 (2008): 1716–31; and M. B. Andon and J. W. Anderson, State of the art reviews: The oatmeal-cholesterol connection: 10 years later, *American Journal of Lifestyle Medicine* 2 (2008): 51–57.

6. J. A. Nettleton et al., Incident heart failure is associated with lower whole-grain intake and greater high-fat dairy and egg intake in the Atherosclerosis Risk in Communities (ARIC) study, *Journal of the American Dietetic Association* 108 (2008): 1881–87; P. B. Mellen, T. F. Walsh, and D. M. Harrington, Whole-grain intake and cardiovascular disease: A meta-analysis, *Nutrition, Metabolism, and Cardiovascular Diseases* (2007): 283–90; and R. Sola et al., Effects of soluble fiber (*Plantago ovata* husk) on plasma lipids, lipoproteins, and apolipoproteins in men with ischemic heart disease, *American Journal of Clinical Nutrition* 85 (2007): 1157–63.

7. T. T. Fung et al., Whole-grain intake and the risk of type 2 diabetes: A prospective study in men, *American Journal of Clinical Nutrition* 76 (2002): 535–40; and M. K. Jensen et al., Whole grains, bran, and germ in relation to homocysteine and markers of glycemic control, lipids, and inflammation, *American Journal of Clinical Nutrition* 83 (2006): 275–83.

8. U.S. Department of Agriculture and U.S. Department of Health and Human Services, *Dietary Guidelines for Americans, 2010,* 7th ed. (Washington, DC: U.S. Government Printing Office, 2010).

9. U.S. Department of Health and Human Services, Public Health Service, *HP 2010 Midcourse Review: Nutrition and Overweight* (Washington, DC: U.S. Department of Health and Human Services, Office of Disease Prevention and Health Promotion, 2007).

10. U.S. Department of Agriculture and U.S. Department of Health and Human Services, *Dietary Guidelines for Americans, 2010,* 7th ed. (Washington, DC: U.S. Government Printing Office, 2010).

11. Position of the American Dietetic Association: Health implications of dietary fiber, *Journal of the American Dietetic Association* 108 (2008): 1716–31.

12. A. D. Liese et al., Whole-grain intake and insulin sensitivity: The Insulin Resistance Atherosclerosis Study, *American Journal of Clinical Nutrition* 78 (2003): 965–71.

13. E. S. Ford, W. H. Giles, and W. H. Dietz, Prevalence of the metabolic syndrome among U.S. adults: Findings from the third National Health and Nutrition Examination Survey, *Journal of the American Medical Association* 287 (2002): 356–59; and S. Cook et al., Prevalence of a metabolic syndrome phenotype in adolescents: Findings from the third National Health and Nutrition Examination Survey, 1988–1994, *Archives of Pediatrics and Adolescent Medicine* 157 (2003): 821–27, as cited in M. Boyle and D. Holben, *Community Nutrition in Action: An Entrepreneurial Approach,* 4th ed. (Belmont, CA: Thomson Wadsworth, 2006), 242.

14. U.S. Department of Agriculture and U.S. Department of Health and Human Services, *Dietary Guidelines for Americans, 2010,* 7th ed. (Washington, DC: U.S. Government Printing Office, 2010).

15. Report of a Joint WHO/FAO Expert Consultation, *Diet, Nutrition, and the Prevention of Chronic Diseases,* WHO Technical Report Series 916 (Geneva: World Health Organization, 2003).

16. Food and Nutrition Board, Institute of Medicine, *Dietary Reference Intakes for Energy, Carbohydrate, Fiber, Fat, Fatty Acids, Cholesterol, Protein, and Amino Acids* (Washington, DC: National Academy Press, 2002).

17. *Ibid.*

18. U.S. Department of Agriculture and U.S. Department of Health and Human Services, *Dietary Guidelines for Americans, 2010,* 7th ed. (Washington, DC: U.S. Government Printing Office, 2010).

19. Position of the American Dietetic Association: Health implications of dietary fiber, *Journal of the American Dietetic Association* 108 (2008): 1716–31.

20. Centers for Disease Control and Prevention. National Diabetes Fact Sheet: National estimates and general information on diabetes and prediabetes in the United States, 2011 (Atlanta, GA: U.S. Department of Health and Human Services, Centers for Disease Control and Prevention, 2011).

21. *Ibid.*

22. *Ibid.*

23. E. J. Mayer-Davis, Type 2 diabetes in youth: Epidemiology and current research toward prevention and treatment, *Journal of the American Dietetic Association* 108 (2008): S45–S51; and A. H. Mokdad et al., Prevalence of obesity, diabetes, and obesity-related health risk factors, *Journal of the American Medical Association* 289 (2003): 76–79.

24. D. J. Jenkins et al., Glycemic index: Overview of implications in health and disease, *American Journal of Clinical Nutrition* 76 (2002): 266S–73S.

25. J. P. Kirwan et al., Effects of a moderate glycemic meal on exercise duration and substrate utilization, *Medicine and Science in Sports and Exercise* 33 (2001): 1517–23.

26. M. J. Franz et al., The evidence for medical nutrition therapy for type 1 and type 2 diabetes in adults, *Journal of the American Dietetic Association* 110 (2010): 1852–89; and American Diabetes Association, Nutrition recommendations and interventions for diabetes, *Diabetes Care,* 31 (2008): S61–S78.

27. A. M. Coulston and R. K. Johnson, Sugar and sugar myths: Myths and realities, *Journal of the American Dietetic Association* 102 (2002): 351–53.

28. U.S. Department of Agriculture and U.S. Department of Health and Human Services, *Dietary Guidelines for Americans, 2010,* 7th ed. (Washington, DC: U.S. Government Printing Office, 2010).

29. J. H. Shaw, Causes and control of dental caries, *New England Journal of Medicine* 317 (1987): 996–1004.

30. S. Kashket et al., Lack of correlation between food retention on the human dentition and consumer perception of food stickiness, *Journal of Dental Research* 70 (1991): 1314–19.

31. R. Harris et al., Risk factors for dental caries in young children: A systematic review of the literature, *Community Dental Health* 21 (2004): 71–85; and C. C. Mobley, Nutrition needs and oral health in children, *Topics in Clinical Nutrition* 20 (2005): 200–10.

32. American Dental Association, *Diet & Dental Health* (Chicago: American Dental Association, 1993), 2.

33. M. E. Jensen, Responses of interproximal plaque pH to snack foods and effect of chewing sorbitol-containing gum, *Journal of the American Dental Association* 113 (1986): 262–66.

34. C. A. Palmer, Important relationships between diet, nutrition, and oral health, *Nutrition in Clinical Care* 4 (2001): 4–14; and R. Touger-Decker, Nutrition and oral health in older adults, *Topics in Clinical Nutrition* 20 (2005): 211–18.

35. C. O. Enwonwu, Interface of malnutrition and periodontal diseases, *American Journal of Clinical Nutrition* 61 (1995): 430S–36S; Position of the American Dietetic Association: Oral health and nutrition, *Journal of the American Dietetic Association* 107 (2007): 1418–28.

36. J. O. Alvarez, Nutrition, tooth development, and dental caries, *American Journal of Clinical Nutrition* 61 (1995): 410S–16S.

37. Personal communication with Marketing Data Enterprises, Inc., Valley Stream, New York.

38. S. Cohen et al., Saccharin and urothelial proliferation: A threshold phenomenon, *Journal of the American Societies for Experimental Biology* 6 (1992): A-1594.

39. Position of the American Dietetic Association: Use of nutritive and nonnutritive sweeteners, *Journal of the American Dietetic Association* 104 (2004): 255–75.

40. Council on Scientific Affairs, Saccharin: Review of safety issues, *Journal of the American Medical Association* 254 (1985): 2622–44.

41. U.S. Department of Health and Human Services, National Toxicology Program, *The Twelfth Report on Carcinogens,* 12th ed. (2011); available at http://ntp.niehs.nih.gov/ntp/roc/twelfth/roc12.pdf.

42. S. S. Schiffman et al., Aspartame and susceptibility to headache, *New England Journal of Medicine* 317 (1987): 1181–85.

43. P. A. Spiers et al., Aspartame: Neuropsychologic and neurophysiologic evaluation of acute and chronic effects, *American Journal of Clinical Nutrition* 68 (1998): 531–37; American Council on Science and Health, *Low-Calorie Sweeteners* (New York: American Council on Science and Health, 1993), 7–13; and Position of the American Dietetic Association: Use of nutritive and nonnutritive sweeteners, *Journal of the American Dietetic Association* 104 (2004): 255–75.

44. Position of the American Dietetic Association: Use of nutritive and nonnutritive sweeteners, *Journal of the American Dietetic Association* 104 (2004): 255–75.

45. J. E. Blundell and A. J. Hill, Paradoxical effects of an intense sweetener (aspartame) on appetite, *Lancet* 1 (1986): 1092–93.

46. B. J. Rolls, Effects of intense sweeteners on hunger, food intake, and body weight: A review, *American Journal of Clinical Nutrition* 53 (1991): 872–78; and A. Drewnowski, Comparing the effects of aspartame and sucrose on motivational ratings, taste preferences, and energy intakes in humans, *American Journal of Clinical Nutrition* 59 (1994): 338–45.

47. Position of the American Dietetic Association: Use of nutritive and nonnutritive sweeteners, *Journal of the American Dietetic Association* 104 (2004): 255–75.

48. *Ibid.*

49. GRAS Notice Inventory, GRN No. 348, August 16, 2010.

CHAPTER 5

1. N. D. Riediger et al., A systematic review of the roles of n-3 fatty acids in health and disease, *Journal of the American Dietetic Association* 109 (2009): 668–79.

2. Position of the American Dietetic Association and Dietitians of Canada: Dietary Fatty Acids, *Journal of the American Dietetic Association* 107 (2007): 1599–1611.

3. Expert Panel on Detection, Evaluation, and Treatment of High Blood Cholesterol in Adults, Summary of the Third Report of the National Cholesterol Education Program (NCEP) Expert Panel on Detection, Evaluation, and Treatment of High Blood Cholesterol in Adults, *Journal of the American Medical Association* 285 (2001): 2486–98.

4. Position of the American Dietetic Association and Dietitians of Canada: Dietary Fatty Acids, *Journal of the American Dietetic Association* 107 (2007): 1599–1611.

5. *Ibid.*

6. G. Wolf, The role of oxidized low-density lipoprotein in the activation of peroxisome proliferator-activated receptor (gamma): Implications for atherosclerosis, *Nutrition Reviews* 57 (1999): 88–91; and R. Ross,

Atherosclerosis: An inflammatory disease, *New England Journal of Medicine* 340 (1999): 115–26.

7. K. Zelman, The great fat debate: A closer look at the controversy—questioning the validity of age-old dietary guidance, *Journal of the American Dietetic Association* 111 (2011): 655–58; and N. D. Riediger et al., A systematic review of the roles of n-3 fatty acids in health and disease, *Journal of the American Dietetic Association* 109 (2009): 668–79.

8. Food and Nutrition Board, Institute of Medicine, *Dietary Reference Intakes for Energy, Carbohydrate, Fiber, Fat, Fatty Acids, Cholesterol, Protein, and Amino Acids* (Washington, DC: National Academy Press, 2002).

9. Position of the American Dietetic Association and Dietitians of Canada: Dietary Fatty Acids, *Journal of the American Dietetic Association* 107 (2007): 1599–1611; and L. Mosca et al., Evidence-based guidelines for cardiovascular disease prevention in women: 2007 update, *Circulation* 115 (2007): 1481–1501.

10. M. Clarke, The Mediterranean food guide, *Timely Topics from the Department of Human Nutrition*, Kansas State Research and Extension (www.ksre.ksu.edu/humannutrition1/_timely/MEDRDIET.HTM).

11. U.S. Department of Agriculture and U.S. Department of Health and Human Services, *Dietary Guidelines for Americans, 2010,* 7th ed. (Washington, DC: U.S. Government Printing Office, 2010).

12. P. M. Kris-Etherton, W. S. Harris, and L. J. Appel, Fish consumption, fish oil, omega-3 fatty acids, and cardiovascular disease, *Circulation* 106 (2002): 2737–57.

13. American Heart Association, Fish oil and omega-3 fatty acids: American Heart Association Recommendation (www.americanheart.org); and P. M. Kris-Etherton and A. M. Hill, n-3 Fatty acids: Food or supplements? *Journal of the American Dietetic Association* 108 (2008): 1125–30.

14. Position of the American Dietetic Association and Dietitians of Canada: Dietary Fatty Acids, *Journal of the American Dietetic Association* 107 (2007): 1599–1611.

15. J. Sabaté, K. Oda, and E. Ros, Nut consumption and blood lipid levels: A pooled analysis of 25 intervention trials, *Archives of Internal Medicine* 170 (2010): 821–27.

16. Food and Nutrition Board, Institute of Medicine, *Dietary Reference Intakes for Energy, Carbohydrate, Fiber, Fat, Fatty Acids, Cholesterol, Protein, and Amino Acids* (Washington, DC: National Academy Press, 2002).

17. J. L. Gerberding, Safer fats for healthier hearts: The case for eliminating dietary artificial *trans* fat intake, *Annals of Internal Medicine* 151 (2009): 137–38.

18. A. H. Lichtenstein et al., Effects of different forms of dietary hydrogenated fats on serum lipoprotein cholesterol levels, *New England Journal of Medicine* 340 (1999): 1933–40.

19. D. Mozaffarian et al., Trans fatty acids and cardiovascular disease, *New England Journal of Medicine* 354 (2006): 1601–13; and K. Zelman, The great fat debate: A closer look at the controversy—questioning the validity of age-old dietary guidance, *Journal of the American Dietetic Association* 111 (2011): 655–58.

20. Revealing *trans* fats, *FDA Consumer* (September/October 2003): 20–26; M. J. Albers et al., 2006 marketplace survey of *trans*-fatty acid content of margarines and butters, cookies and snack cakes, and savory snacks, *Journal of the American Dietetic Association* 108 (2008): 367–70; S. Borra et al., An update of *trans*-fat reduction in the American diet, *Journal of the American Dietetic Association* 108 (2008): 2048–50; and R. Eckel et al., Understanding the complexity of *trans* fatty acid reduction in the American diet, *Circulation* 115 (2007): 2231–46.

21. Food and Nutrition Board, Institute of Medicine, *Dietary Reference Intakes for Energy, Carbohydrate, Fiber, Fat, Fatty Acids, Cholesterol, Protein, and Amino Acids* (Washington, DC: National Academy Press, 2002); and U.S. Department of Agriculture and U.S. Department of Health and Human Services, *Dietary Guidelines for Americans, 2010,* 7th ed. (Washington, DC: U.S. Government Printing Office, 2010).

22. Personal communication with Consumer Affairs, Entenmann's, Bay Shore, New York.

23. Personal communication with Consumer Affairs, Kraft, Glenview, Illinois.

24. G. E. Ruoff, Reducing fat intake with fat substitutes, *American Family Physician* 43 (1991): 1235–42.

25. *Simplesse All Natural Fat Substitute: A Scientific Overview* (Deerfield, IL: The Simplesse Company, 1991), 3–10.

26. U.S. Department of Health and Human Services, FDA approves fat substitute, olestra, *HHS News* (January 24, 1996); FDA changes labeling requirement for olestra, *FDA Talk Paper #T03-59* (August 2003).

27. Position paper of the American Dietetic Association: Fat replacers, *Journal of the American Dietetic Association* 105 (2005): 266–75.

28. A. H. Lichtenstein et al., AHA Dietary Guidelines, Revision 2006: A statement for healthcare professionals from the nutrition committee of the American Heart Association, *Circulation* 114 (2006): 82–96.

29. Adult Treatment Panel III, *Journal of the American Medical Association* 285 (2001): 2486–98.

30. Position of the American Dietetic Association: Health implications of dietary fiber, *Journal of the American Dietetic Association* 108 (2008): 1716–31.

31. J. Bleys et al., Vitamin-mineral supplementation and the progression of atherosclerosis: A meta-analysis of randomized controlled trials, *American Journal of Clinical Nutrition* 84 (2006): 880–87.

32. *Ibid.*

33. U.S. Department of Agriculture and U.S. Department of Health and Human Services, *Dietary Guidelines for Americans, 2010,* 7th ed. (Washington, DC: U.S. Government Printing Office, 2010).

34. Summary of the Third Report of the National Cholesterol Education Program (NCEP) Expert Panel on Detection, Evaluation, and Treatment of High Blood Cholesterol in Adults, *Journal of the American Medical Association* 285 (2001): 2486–98.

35. *Ibid.*

36. N. D. Riediger et al., A systematic review of the roles of n-3 fatty acids in health and disease, *Journal of the American Dietetic Association* 109 (2009): 668–79; and Position of the American Dietetic Association and Dietitians of Canada: Dietary fatty acids, *Journal of the American Dietetic Association* 107 (2007): 1599–1611.

37. A. H. Lichtenstein et al., AHA Dietary Guidelines, Revision 2006: A statement for healthcare professionals from the nutrition committee of the American Heart Association, *Circulation* 114 (2006): 82–96.

38. B. S. Levine and C. Cooper, Plant stanol esters: A new tool in the dietary management of cholesterol, *Nutrition Today* 35 (2000): 61–66.

39. L. Van Horn et al., The evidence for dietary prevention and treatment of cardiovascular disease, *Journal of the American Dietetic Association* 108 (2008): 287–331; and J. F. Swain et al., Characteristics of the diet patterns tested in the Optimal Macronutrient Intake Trial to Prevent Heart Disease (OmniHeart): Options for a heart-healthy diet, *Journal of the American Dietetic Association* 108 (2008): 257–65.

40. D. Canoy et al., Body fat distribution and risk of coronary heart disease in men and women in the European prospective investigation into cancer and nutrition in Norfolk cohort: A population-based prospective study, *Circulation*, 116 (2007): 2933–43.

41. U.S. Department of Agriculture and U.S. Department of Health and Human Services, *Dietary Guidelines for Americans, 2010,* 7th ed. (Washington, DC: U.S. Government Printing Office, 2010).

42. K. M. Fairfield and R. H. Fletcher, Vitamins for chronic disease prevention in adults, *Journal of the American Medical Association* 287 (2002): 3116–26; G. S. Omenn et al., Preventing coronary heart disease: B vitamins and homocysteine, *Circulation* 97 (1998): 421–24; and O. Nygard et al., Major lifestyle determinants of plasma total homocysteine distribution: The Hordaland Homocysteine Study, *American Journal of Clinical Nutrition* 67 (1998): 263–70.

43. K. Reynolds et al., Alcohol consumption and risk of stroke: A meta-analysis, *Journal of the American Medical Association* 289 (2003): 579–88.

44. J. M. Gaziano et al., Moderate alcohol intake, increased levels of high-density lipoprotein and its subfractions, and decreased risk of myocardial infarction, *New England Journal of Medicine* 329 (1993): 1829–34; and K. J. Mukamal et al., Roles of drinking pattern and type of alcohol consumed in coronary heart disease in men, *New England Journal of Medicine* 348 (2003): 109–18.

45. S. L. Englebardt, Eat, drink, go back to work, *American Health* (June 1994): 90.

46. T. A. Pearson and P. Terry, What to advise patients about drinking alcohol: The clinician's conundrum, *Journal of the American Medical Association* 272 (1994): 967–68.

47. K. Breithaupt et al., Protective effect of chronic garlic intake on elastic properties of the aorta in the elderly, *Circulation* 96 (1997): 2649–55; A. Orekhov and J. Grunwald, Effects of garlic on atherosclerosis, *Nutrition* 13 (1997): 656–63; and J. Isaacsohm et al., Garlic powder and plasma lipids and lipoproteins, *Archives of Internal Medicine* 158 (1998): 1189–94.

48. B. Buijsse et al., Chocolate consumption in relation to blood pressure and risk of cardiovascular disease in German adults, *European Heart Journal* 13 (2010): 1616–23.

49. A. H. Lichtenstein et al., AHA Dietary Guidelines, Revision 2006: A statement for healthcare professionals from the nutrition committee of the American Heart Association, *Circulation* 114 (2006): 82–96.

50. C. McMahan, S. Gidding, and H. McGill, Coronary heart disease risk factors and atherosclerosis in young people, *Journal of Clinical Lipidology* 2 (2008): 118–26; American Heart Association, Guidelines for cardiovascular risk reduction for high-risk children, *Circulation* 107 (2003): 1562–66.

CHAPTER 6

1. R. K. Chandra, Nutrition and the immune system: An introduction, *American Journal of Clinical Nutrition* 66 (1997): S460–63.

2. V. R. Young, Soy protein in relation to human protein and amino acid nutrition, *Journal of the American Dietetic Association* 91 (1991): 828–35.

3. D. G. Schroeder and R. Martorell, Enhancing child survival by preventing malnutrition, *American Journal of Clinical Nutrition* 65 (1997): 1080–81.

4. Food and Nutrition Board, Institute of Medicine, *Dietary Reference Intakes for Energy, Carbohydrate, Fiber, Fat, Fatty Acids, Cholesterol, Protein, and Amino Acids* (Washington, DC: National Academy Press, 2002).

5. B. Torún, Protein-energy malnutrition, in M. E. Shils et al., eds., *Modern Nutrition in Health and Disease*, 10th ed. (Philadelphia: Lippincott Williams & Wilkins, 2006), 881–908.

6. Position of the American Dietetic Association: Food insecurity in the United States, *Journal of the American Dietetic Association* 110 (2010): 1368–77.

7. L. K. Massey, Does excess dietary calcium adversely affect bone? Symposium overview, *Journal of Nutrition* 128 (1998): 1048–50.

8. The discussion on adding legumes to the diet is adapted from Eat Right Montana's *Celebrating Healthy Families 2005* (August 2005).

9. *Ibid.*

10. I. Chalmers, *The Great Food Almanac* (San Francisco: Collins Publishers, 1994).

11. American Institute for Cancer Research, *Stopping Cancer Before It Starts* (New York: Golden Books, 1999), 214–17.

12. C. L. Vecchia, Vegetable consumption and risk of chronic disease, *Epidemiology* 9 (1998): 208–10; American Institute for Cancer Research (AICR), *The New American Plate* (Washington, DC: AICR, 2002).

13. J. Carol and J. Sobal, Model of the process of adopting vegetarian diets: Health vegetarians and ethical vegetarians, *Journal of Nutrition Education* 30 (1998): 196–202; and E. Lea and A. Worsley, The cognitive contexts of beliefs about the healthiness of meat, *Public Health Nutrition* 5 (2002): 37–45.

14. Position of the American Dietetic Association: Vegetarian diets, *Journal of the American Dietetic Association* 109 (2009): 1266–82.

15. V. R. Young and P. L. Pellett, Plant proteins in relation to human protein and amino acid nutrition, *American Journal of Clinical Nutrition* 59 (1994): 1203S–12S; Position of the American Dietetic Association: Vegetarian diets, *Journal of the American Dietetic Association* 109 (2009): 1266–82.

16. C. Lamberg-Allardt et al., Low serum 25-hydroxyvitamin D concentrations and secondary hyperparathyroidism in middle-aged white strict vegetarians, *American Journal of Clinical Nutrition* 58 (1993): 684–89.

17. Position of the American Dietetic Association: Vegetarian diets, *Journal of the American Dietetic Association* 109 (2009): 1266–82.

18. S. I. Barr et al., Spinal bone mineral density in premenopausal vegetarian and nonvegetarian women: Cross-sectional and prospective comparisons, *Journal of the American Dietetic Association* 98 (1998): 760–65.

19. V. Messina, V. Melina, and R. Mangels, A new food guide for North American vegetarians, *Journal of the American Dietetic Association* 103 (2003): 771–75.

20. K. Burke, The use of soy foods in a vegetarian diet, *Topics in Clinical Nutrition* 10 (1995): 37–43.

21. T. J. Key et al., Dietary habits and mortality in 11,000 vegetarians and health-conscious people: Results of a 17-year follow up, *British Medical Journal* 313 (1996): 775–79; T. J. Key et al., Mortality in vegetarians and nonvegetarians: Detailed findings from a collaborative analysis of five prospective studies, *American Journal of Clinical Nutrition* 70 (1999): 516S–24S; and T. J. Key, P. N. Appleby, and M. S. Rosell, Health effects of vegetarian and vegan diets, *Proceedings of the Nutritional Society* 65 (2006): 35–41.

22. A. Dewell et al., A very-low-fat vegan diet increases intake of protective dietary factors and decreases intake of pathogenic dietary factors, *Journal of the American Dietetic Association* 108 (2008): 347–56.

23. B. Hunter, Food allergies: No trivial matter, *Consumers' Research Magazine* 82 (1999): 2.

24. Food allergies: How worried should you be? *Tufts University Health & Nutrition Letter* 17 (1999): 2.

25. Food Allergy Network, *Information About Food Allergies* (2007); www.foodallergy.org.

26. H. Sampson, Food allergy, *Journal of the American Medical Association* 278 (1997): 22–26; and S. Sicherer, Manifestations of food allergy: Evaluation and management, *American Family Physician* 59 (1999): 2–8.

27. FDA to require food manufacturers to list food allergens, *FDA News* (December 20, 2005).

28. S. Gottlieb, Scientists develop vaccine strategy for peanut allergy, *British Medical Journal* 318 (1999): 71–88.

29. United Soybean Board, Soy and health: Discovering the health benefits of isoflavones fact sheet (1999); S. M. Potter, Soy—New health benefits associated with an ancient food, *Nutrition Today* 35 (2000): 53–59; M. Friedman and D. L. Brandon, Nutritional and health benefits of soy proteins, *Journal of Agriculture and Food Chemistry* 49 (2001): 1069–1086; and M. Messina and V. Messina, Provisional recommended soy protein and isoflavone intakes for healthy adults, *Nutrition Today* 38 (2003): 100–109.

30. *U.S. Soyfoods Directory: Alphabetical List of Soyfoods* available at www.soyfoods.com.

31. J. R. Crouse et al., A randomized trial comparing the effect of casein with that of soy protein containing varying amounts of isoflavones on plasma concentrations of lipids and lipoproteins, *Archives of Internal Medicine* 159 (1999): 2070–76.

32. J. W. Anderson et al., Meta-analysis of the effects of soy protein intake on serum lipids, *New England Journal of Medicine* 333 (1995): 276–82; and S. Zhan and S. C. Ho, Meta-analysis of the effects of soy protein containing isoflavones on the lipid profile, *American Journal of Clinical Nutrition* 81 (2005): 397–408.

33. R. M. Bakhit et al., Intake of 25 g of soybean protein with or without fiber alters plasma lipids in men with elevated cholesterol concentrations, *Journal of Nutrition* 124 (1994): 213–22; F. M. Steinberg, Soybeans or soymilk: Does it make a difference for cardiovascular protection? Does it even matter? *American Journal of Clinical Nutrition* 81 (2007): 927–28; and F. M. Sacks et al., Soy protein, isoflavones, and cardiovascular health: An American Heart Association advisory for professionals from the Nutrition Committee, *Circulation* 113 (2006): 1034–44.

34. S. Kreijkamp-Kaspers et al., Effect of soy protein containing isoflavones on cognitive function, bone mineral density, and plasma lipids in postmenopausal women, *Journal of the American Medical Association* 292 (2004): 65–74; B. H. Arjmandi and B. J. Smith, Soy isoflavones' osteoprotective role in postmenopausal women: Mechanism of action, *Journal of Nutritional Biochemistry* 13 (2002): 130–137; and H. Marini et al., Effects of the phytoestrogen genistein on bone metabolism in osteopenic postmenopausal women: A randomized trial, *Annals of Internal Medicine* 146 (2007): 839–47.

35. National Osteoporosis Foundation, *Fast Facts on Osteoporosis* (Washington, DC: National Osteoporosis Foundation, 2008).

36. M. Lock, Menopause in cultural context, *Experimental Gerontology* 29 (1994): 307–17.

37. L. Bren, The estrogen and progestin dilemma: New advice, labeling guidelines, *FDA Consumer* 37 (2003): 10–11; M. Glazier, M. B. Gina, and M. A. Bowman, Review of the evidence for the use of phytoestrogens as a replacement for traditional estrogen replacement therapy, *Archives of Internal Medicine* 161 (2001): 1161–72; North American Menopause

Society (NAMS), *Report from the NAMS Advisory Panel on Postmenopausal Hormone Therapy* (October 6, 2002), available at www.menopause.org; and J. Rymer, R. Wilson, and K. Ballard, Making decisions about hormone replacement therapy, *British Journal of Medicine* 326 (2003): 322–26.

38. A. L. Murkies et al., Dietary flour supplementation decreases postmenopausal hot flushes: Effect of soy and wheat, *Maturitas* 221 (1995): 189–95.

39. D. C. Knight and J. A. Eden, A review of the clinical effects of phytoestrogens, *Obstetrics and Gynecology* 87 (1996): 897–904; M. Messina, Soy foods and soybean isoflavones and menopausal health, *Nutrition in Clinical Care* 5 (2002): 272–82; and K. K. Han et al., Benefits of soy isoflavone therapeutic regimen on menopausal symptoms, *Obstetrics and Gynecology* 99 (2002): 389–94.

40. American Cancer Society, *Cancer Facts & Figures 2007* (New York: ACS, Inc., 2007).

41. *Ibid.*

42. L. N. Kolonel, Variability in diet and its relation to risk in ethnic and migrant groups, *Basic Life Sciences* 43 (1988): 129–35.

43. A. Nomura et al., Breast cancer and diet among the Japanese in Hawaii, *American Journal of Clinical Nutrition* 31 (1978): 2020–25.

44. R. K. Severson et al., A prospective study of demographics, diet, and prostate cancer among men of Japanese ancestry in Hawaii, *Cancer Research* 49 (1989): 1857–60.

45. M. Messina and S. Barnes, The role of soy products in reducing risk of cancer, *Journal of the National Cancer Institute* 83 (1991): 541–46; and D. Li et al., Soybean isoflavones reduce experimental metastasis in mice, *Journal of Nutrition* 129 (1999): 1075–78.

46. G. Maskarinec, Soy foods for breast cancer survivors and women at high risk for breast cancer? *Journal of the American Dietetic Association* 105 (2005): 1524–28; and A. Wu et al., Adolescent and adult soy intake and risk of breast cancer in Asian-Americans, *Carcinogenesis* 23 (2003): 1491–96.

47. M. Messina, C. Gardener, and S. Barnes, Gaining insight into the health effects of soy but a long way still to go: Commentary on the Fourth International Symposium on the Role of Soy in Preventing and Treating Chronic Disease, *Journal of Nutrition* 132 (2002): 547S–51S.

48. T. J. Stephonson and J. W. Anderson, Isoflavone-rich protein healthy addition to diabetic diet, *The Soy Connection* (Summer 2002).

CHAPTER 7

1. J. Jaramillo-Arango, The conquest of nutritional diseases (vitamins), in *The British Contribution to Medicine* (Edinburgh: E. & S. Livingstone, Ltd., 1953), 140–62.

2. R. Hill et al., The discovery of vitamins, in *The Chemistry of Life* (Cambridge, UK: Cambridge University Press, 1970), 156–70.

3. R. A. Jacob and G. Sotoudeh, Vitamin C function and status in chronic disease, *Nutrition in Clinical Care* 5 (2002): 66–74; S. J. Padayatty et al., Vitamin C as an antioxidant: Evaluation of its role in disease prevention, *Journal of the American College of Nutrition* 22 (2003): 18–35; and G. Bjelakovic et al., Mortality in randomized trials of antioxidant supplements for primary and secondary prevention: Systematic review and meta-analysis, *Journal of the American Medical Association* 297 (2007): 842–57.

4. L. C. Pauling, *Vitamin C and the Common Cold* (San Francisco: W. H. Freeman, 1970); H. Hemilia, Vitamin C supplementation and common cold symptoms: Factors affecting the magnitude of the benefit, *Medical Hypotheses* 52 (1999): 171–78; and C. Audera et al., Mega-dose vitamin C in treatment of the common cold: A randomized controlled trial, *Oncology* 175 (2001): 359–62.

5. P. M. Kris-Etherton et al., Antioxidant vitamin supplements and cardiovascular disease, *Circulation* 110 (2004): 637–41.

6. G. Bjelakovic et al., Mortality in randomized trials of antioxidant supplements for primary and secondary prevention: Systematic review and meta-analysis, *Journal of the American Medical Association* 297 (2007): 842–57.

7. E. R. Miller et al., Meta-analysis: High dosage vitamin E supplementation may increase all-cause mortality, *Annals of Internal Medicine* 142 (2005): 37–46; M. G. Traber, How much vitamin E? . . . Just enough! *American Journal of Clinical Nutrition* 84 (2006): 959–60.

8. E. J. Johnson, The role of the carotenoids in human health, *Nutrition in Clinical Care* 5 (2002): 56–65; Age Related Eye Disease Research Group, A randomized, placebo-controlled clinical trial of high dose supplementation with vitamins C and E, beta carotene, and zinc for age-related macular degeneration and vision loss, AREDS Report No. 8,

Archives Ophthalmology 119 (2001): 1417–36; and Age Related Eye Disease Research Group, A randomized, placebo-controlled clinical trial of high dose supplementation with vitamins C and E and beta carotene for age-related cataract and vision loss, AREDS Report No. 9, *Archives Ophthalmology* 119 (2001): 1439–52.

9. L. Bailey, G. Rampersaud, and G. Kauwell, Folic acid supplements and fortification affect the risk for neural tube defects, vascular disease, and cancer: Evolving science, *Journal of Nutrition* 133 (2003): 1961S–68S; and G. Rampersaud, L. Bailey, and G. Kauwell, Relationship of folate to colorectal and cervical cancer: Review and recommendations for practitioners, *Journal of the American Dietetic Association* 102 (2002): 1273–82.

10. E. R. Cena et al., Folate intake and food-related behaviors in nonpregnant, low-income women of childbearing age, *Journal of the American Dietetic Association* 108 (2008): 1364–68; and Centers for Disease Control and Prevention, Spina bifida and anencephaly before and after folic acid mandate—United States, 1995–1996 and 1999–2000, *Morbidity & Mortality Weekly Report* 53 (2004): 362–65.

11. C. W. Suitor and L. B. Bailey, Dietary folate equivalents: Interpretation and application, *Journal of the American Dietetic Association* 100 (2000): 88–94.

12. Food and Nutrition Board, Institute of Medicine, *Dietary Reference Intakes for Thiamin, Riboflavin, Niacin, Vitamin B_6, Folate, Vitamin B_{12}, Pantothenic Acid, Biotin, and Choline* (Washington, DC: National Academy Press, 1998).

13. K. M. Riggs et al., Relations of vitamin B_{12}, vitamin B_6, folate, and homocysteine to cognitive performance in the Normative Aging Study, *American Journal of Clinical Nutrition* 63 (1996): 306–14; K. M. Fairfield and R. H. Fletcher, Vitamins for chronic disease prevention in adults, *Journal of the American Medical Association* 287 (2002): 3116–26; and E. Lonn et al., Homocysteine lowering with folic acid and B vitamins in vascular disease, *New England Journal of Medicine* 354 (2006): 1567–77.

14. G. S. Omenn et al., Preventing coronary heart disease: B vitamins and homocysteine, *Circulation* 97 (1998): 421–24; and O. Nygard et al., Major lifestyle determinants of plasma total homocysteine distribution: The Hordaland Homocysteine Study, *American Journal of Clinical Nutrition* 67 (1998): 263–70.

15. E. Lonn et al., Heart Outcomes Prevention Evaluation (HOPE), Homocysteine lowering with folic acid and B vitamins in vascular disease. *New England Journal of Medicine,* 354 (2006): 1567–77.

16. S. Seshadri et al., Plasma homocysteine as a risk factor for dementia and Alzheimer's disease, *New England Journal of Medicine* 346 (2002): 466; R. LeBoeuf, Homocysteine and Alzheimer's disease, *Journal of the American Dietetic Association* 103 (2003): 304–7; I. H. Rosenberg, B vitamins, homocysteine, and neurocognitive functions, *Nutrition Reviews* 59 (2001): S69–S74; and M. S. Morris, Folate, homocysteine, and neurological function, *Nutrition in Clinical Care* 5 (2002): 124–32.

17. S. P. Stabler and R. H. Allen, Vitamin B_{12} deficiency as a worldwide problem, *Annual Review of Nutrition* 24 (2004): 299–326; and P. J. Stover, Physiology of folate and vitamin B_{12} in health and disease, *Nutrition Reviews* 62 (2004): S3–S4.

18. M. A. Johnson, Influence of vitamin K on anticoagulant therapy depends on vitamin K status and the source and chemical forms of vitamin K, *Nutrition Reviews* 63 (2005): 91–100.

19. M. F. Holick, Sunlight and vitamin D for bone health and prevention of autoimmune diseases, cancers, and cardiovascular disease, *American Journal of Clinical Nutrition* 80 (2004): 1678S–88S; M. F. Holick, Vitamin D deficiency, *New England Journal of Medicine* 357 (2007): 266–81; J. N. Hathcock et al., Risk assessment for vitamin D, *American Journal of Clinical Nutrition* 85 (2007): 6–18; Committee to Review Dietary Reference Intakes for Vitamin D and Calcium, Food and Nutrition Board, A. C. Ross et al., eds., *Dietary Reference Intakes: Calcium and Vitamin D* (Washington, DC: Institute of Medicine of the National Academies, The National Academies Press, 2010); and A. C. Ross et al., The 2011 Dietary Reference Intakes for calcium and vitamin D: What dietetics practitioners need to know, *Journal of the American Dietetic Association,* 111 (2011): 524–27.

20. S. L. Booth et al., Vitamin K intake and bone mineral density in women and men, *American Journal of Clinical Nutrition* 77 (2003): 512–16; and S. Cockayn et al., Vitamin K and the prevention of fractures: Systematic review and meta-analysis of randomized control trials, *Archives of Internal Medicine* 166 (2006): 1256–61.

21. P. L. Canner, Benefits of niacin in patients with versus without the metabolic syndrome and healed myocardial infarction (from the Coronary Drug Project), *American Journal of Cardiology* 97 (2006): 477–79.

22. S. H. Zeisel, Choline: Critical role during fetal development and dietary requirements in adults, *Annual Review of Nutrition* 26 (2006): 229–50; and L. M. Sanders and S. H. Zeisel, Choline: Dietary requirements and role in brain development, *Nutrition Today* 42 (2007): 181–86.

23. N. R. Farnsworth et al., Medicinal plants in therapy, *Bulletin of the World Health Organization* 63 (1985): 965–1170.

24. N. R. Farnsworth, The role of medicinal plants in drug development, in P. Krogsgaard-Larsen, S. Brogger Christensen, and H. Kofod, eds., *Natural Products and Drug Development*, Proceedings of Alfred Benzon Symposium 20 (Copenhagen: Munksgaard, 1984), 17–30.

25. P. Goldman, Herbal medicines today and the roots of modern pharmacology, *Annals of Internal Medicine* 135 (2001): 594–600; and D. M. Marcus and A. P. Grollman, Botanical medicines: The need for new regulations, *New England Journal of Medicine* 347 (2002): 2073–76.

26. P. Brevort, *The Economics of Botanicals: The U.S. Experience.* Presentation at the NIH/OAM Conference on Botanicals: A Role in U.S. Health Care? (Washington, DC: December 16, 1994); and L. G. Tolstoi, Herbal remedies: Buyer beware! *Nutrition Today* 36 (2001): 223–30.

27. *Food, Nutrition, Physical Activity, and the Prevention of Cancer: A Global Perspective* (Washington, DC: World Cancer Research Fund/American Institute for Cancer Research, 2007); P. Lichtenstein et al., Environmental and heritable factors in the causation of cancer—Analyses of cohorts of twins from Sweden, Denmark, and Finland, *New England Journal of Medicine* 343 (2000): 78–81; National Cancer Policy Board, Institute of Medicine, *Fulfilling the Potential of Cancer Prevention and Early Detection* (Washington, DC: National Academies Press, 2003); and G. N. Wogan et al., Environmental and chemical carcinogenesis, *Seminars in Cancer Biology* 14 (2004): 473–86.

28. E. Giovannucci, Diet, body weight, and colorectal cancer: A summary of the epidemiologic evidence, *Journal of Women's Health* 12 (2003): 173–82.

29. *Food, Nutrition, Physical Activity, and the Prevention of Cancer: A Global Perspective* (Washington, DC: World Cancer Research Fund/American Institute for Cancer Research, 2007).

30. C. S. Fuchs et al., Dietary fiber and the risk of colorectal cancer and adenoma in women, *New England Journal of Medicine* 340 (1999): 169–76.

31. W. C. Willett et al., Relation of meat, fat, and fiber intake to the risk of colon cancer in a prospective study among women, *New England Journal of Medicine* 323 (1990): 1664–72.

32. American Gastroenterological Association, Medical position statement: Impact of dietary fiber on colon cancer occurrence, *Gastroenterology* 118 (2000): 1233–34; B. S. Reddy, Role of dietary fiber in colon cancer: An overview, *American Journal of Medicine* 106 (1999): S16–S19; and M. C. Jansen et al., Dietary fiber and plant foods in relation to colorectal cancer mortality: The Seven Countries Study, *International Journal of Cancer* 81 (1999): 174–79.

33. E. Giovannucci, Epidemiologic studies of folate and colo-rectal neoplasia: A review, *Journal of Nutrition* 132 (2002): 2350S–54S; M. Ferraroni et al., Selected micronutrient intake and the risk of colorectal cancer, *British Journal of Cancer* 70 (1994): 1150–55; and J. M. Lappe et al., Vitamin D and calcium supplementation reduces cancer risk: Results of a randomized trial, *American Journal of Clinical Nutrition* 85 (2007): 1586–91.

34. National Cancer Policy Board, Institute of Medicine, *Fulfilling the Potential of Cancer Prevention and Early Detection* (Washington, DC: National Academies Press, 2003).

35. P. L. Zock, Dietary fats and cancer, *Current Opinions in Lipidology* 12 (2001): 5–10.

36. K. Katsouyanni et al., The association of fat and other macronutrients with breast cancer: A case-controlled study from Greece, *British Journal of Cancer* 70 (1994): 537–41.

37. Xiu-Ying Qi et al., The association between breast cancer and diet and other factors, *Asia Pacific Journal of Public Health* 7 (1994): 98–104.

38. J. B. Magen, The relationship between obesity and breast cancer risk and mortality, *Nutrition Reviews* 61 (2003): 73–76; and J. G. Brody, Environmental pollutants, diet, physical activity, body size, and breast cancer: Where do we stand in research to identify opportunities for prevention? *Cancer* 109 (May 14, 2007): 2627–34.

39. World Cancer Research Fund/American Institute for Cancer Research, *Policy and Action for Cancer Prevention. Food, Nutrition, and Physical Activity: A Global Perspective* (Washington, DC: AICR, 2009), 12–28; and H. Kushi et al., American Cancer Society guidelines on nutrition and physical activity for cancer prevention: Reducing the risk of cancer with healthy food choices and physical activity, *CA—A Cancer Journal for Clinicians* 56 (2006): 254–81.

40. G. A. Bray, The underlying basis for obesity: Relationship to cancer, *Journal of Nutrition* 132 (2002): 3451S–55S; C. M. Friedenreich and M. R. Orenstein, Physical activity and cancer prevention: Etiologic evidence and biological mechanisms, *Journal of Nutrition* 132 (2002): 3456S–66S; and J. Verloop et al., Physical activity and breast cancer risk in women aged 20 to 54 years, *Journal of the National Cancer Institute* 92 (2000): 128–35.

41. K. A. Steinmetz and J. D. Potter, Vegetables, fruit, and cancer prevention: A review, *Journal of the American Dietetic Association* 96 (1996): 1027–39; and N. J. Temple and K. K. Gladwin, Fruit, vegetables, and the prevention of cancer: Research challenges, *Nutrition* 19 (2003): 467–70.

42. S. J. Padayatty et al., Vitamin C as an antioxidant: Evaluation of its role in disease prevention, *Journal of the American College of Nutrition* 22 (2003): 18–35.

43. P. Talalay and J. W. Fahey, Phytochemicals from cruciferous plants protect against cancer by modulating carcinogen metabolism, *Journal of Nutrition* 131 (2001): 3027S; and W. J. Craig, Phytochemicals: Guardians of our health, *Journal of the American Dietetic Association* 97 (1997): S199–S204.

44. *Food, Nutrition, Physical Activity, and the Prevention of Cancer: A Global Perspective* (Washington, DC: World Cancer Research Fund/American Institute for Cancer Research, 2007).

45. L. C. Clark et al., Reducing cancer risk with selenium, *Journal of the American Medical Association* 276 (1995): 1957–63.

46. *Food, Nutrition, Physical Activity, and the Prevention of Cancer: A Global Perspective* (Washington, DC: World Cancer Research Fund/American Institute for Cancer Research, 2007).

47. J. M. Lappe et al., Vitamin D and calcium supplementation reduces cancer risk: Results of a randomized trial, *American Journal of Clinical Nutrition* 85 (2007): 1586–91.

48. M. A. Rogers et al., Consumption of nitrate, nitrite, and nitrosodimethylamine and the risk of upper aerodigestive tract cancer, *Cancer Epidemiology, Biomarkers, and Prevention* 4 (1995): 29–36; and H. Chen et al., Dietary patterns and adenocarcinoma of the esophagus and distal stomach, *American Journal of Clinical Nutrition* 75 (2002): 137–41.

49. M. Messina and S. Barnes, The role of soy products in reducing risk of cancer, *Journal of the National Cancer Institute* 83 (1991): 541–46.

50. P. Boffetta et al., The burden of cancer attributable to alcohol drinking, *International Journal of Cancer* 119 (2006): 884–87.

51. J. L. Freudenheim et al., Lifetime alcohol consumption and risk of breast cancer, *Nutrition and Cancer* 23 (1995): 1–11.

52. E. Giovannucci, Tomatoes, tomato-based products, lycopene, and cancer: Review of the epidemiologic literature, *Journal of the National Cancer Institute* 91 (1998): 317–31.

53. Position of the American Dietetic Association: Functional foods, *Journal of the American Dietetic Association* 109 (2009): 735–46; and A. S. Bloch, Phytochemicals and functional foods for cancer risk reduction, *Topics in Clinical Nutrition* 15 (2000): 24–28.

54. J. Childers et al., Chemoprevention of cervical cancer with folic acid: A Phase III Southwest Oncology Group Intergroup Study, *Cancer Epidemiology, Biomarkers, and Prevention* 4 (1995): 155–59.

55. *Food, Nutrition, Physical Activity, and the Prevention of Cancer: A Global Perspective* (Washington, DC: World Cancer Research Fund/American Institute for Cancer Research, 2007).

56. National Cancer Policy Board, Institute of Medicine, *Fulfilling the Potential of Cancer Prevention and Early Detection* (Washington, DC: National Academies Press, 2003).

57. *Food, Nutrition, Physical Activity, and the Prevention of Cancer: A Global Perspective* (Washington, DC: World Cancer Research Fund/American Institute for Cancer Research, 2007).

58. Position of the American Dietetic Association: Nutrient supplementation, *Journal of the American Dietetic Association* 109 (2009): 2073–85; and

D. R. Jacobs, M. D. Gross, and L. C. Tapsell, Food synergy: An operational concept for understanding nutrition, *American Journal of Clinical Nutrition* 89 (2009): 1543S–48S.

59. Committee on Comparative Toxicity of Naturally Occurring Carcinogens, *Carcinogens and Anticarcinogens in the Human Diet* (Washington, DC: National Academies Press, 1996).

CHAPTER 8

1. C. M. McCay, Anorganic substances, in F. Verzar, ed., *Notes on the History of Nutrition Research* (Vienna: Hans Huber, 1973), 156–84.

2. Committee to Review Dietary Reference Intakes for Vitamin D and Calcium, Food and Nutrition Board, A. C. Ross et al., eds., *Dietary Reference Intakes: Calcium and Vitamin D* (Washington, DC: Institute of Medicine of the National Academies, The National Academies Press, 2010); A. C. Ross et al., The 2011 Dietary Reference Intakes for calcium and vitamin D: What dietetics practitioners need to know, *Journal of the American Dietetic Association,* 111 (2011): 524–27; and V. Matkovic, Nutrition, genetics, and skeletal development, *Journal of the American College of Nutrition* 15 (1996): 556–69.

3. C. Palacios, The role of nutrients in bone health, from A to Z, *Critical Reviews in Food Science and Nutrition* 46 (2006): 621–628; and J. Ma, R. A. Johns, and R. S. Stafford, Americans are not meeting current calcium recommendations, *American Journal of Clinical Nutrition* 85 (2007): 1361–66.

4. Committee to Review Dietary Reference Intakes for Vitamin D and Calcium, Food and Nutrition Board, A. C. Ross et al., eds., *Dietary Reference Intakes: Calcium and Vitamin D* (Washington, DC: Institute of Medicine of the National Academies, The National Academies Press, 2010); and A. C. Ross et al., The 2011 Dietary Reference Intakes for calcium and vitamin D: What dietetics practitioners need to know, *Journal of the American Dietetic Association,* 111 (2011): 524–27.

5. L. R. Vartanian, M. B. Schwartz, and K. D. Brownell, Effects of soft drink consumption on nutrition and health: A systematic review and meta-analysis, *American Journal of Public Health* 97, no. 4 (2007): 667–75; and C. D. Frary, R. K. Johnson, and M. Q. Wang, Children and adolescents' choice of foods and beverages high in added sugars are associated with intakes of key nutrients and food groups, *Journal of Adolescent Health* 34 (2004): 56–63.

6. S. H. Jee et al., The effect of magnesium supplementation on blood pressure: A meta-analysis of randomized clinical trials, *American Journal of Hypertension* 15 (2002): 691–96; and K. Ueshima, Magnesium and ischemic heart disease: A review of epidemiological, experimental, and clinical evidences, *Magnesium Research* 18 (2005): 275–84.

7. J. B. Mason et al., *The Micronutrient Report: Current Progress and Trends in the Control of Vitamin A, Iron, and Iodine Deficiencies* (Ottawa, Canada: The Micronutrient Initiative, 2001).

8. Centers for Disease Control and Prevention, Iron deficiency—United States, 1999–2000, *Morbidity and Mortality Weekly Report* 51 (2002): 897–99; U.S. Department of Health and Human Services, Public Health Service, *Healthy People 2020* (Washington, DC: Government Printing Office, 2010); E. M. Ross, Evaluation and treatment of iron deficiency in adults, *Nutrition in Clinical Care* 5 (2002): 220–24; and J. C. McCann and B. N. Ames, An overview of evidence for a causal relationship for iron deficiency during development and deficits in cognitive or behavioral function, *American Journal of Clinical Nutrition* 85 (2007): 931–45.

9. J. D. Cook and M. B. Reddy, Effect of ascorbic acid intake on non-heme iron absorption from a complete diet, *American Journal of Clinical Nutrition* 73 (2001): 93–98; and M. B. Reddy, R. F. Hurrell, and C. D. Cook, Estimation of nonheme iron bioavailability from meal composition, *American Journal of Clinical Nutrition* 71 (2000): 937–43.

10. E. V. M. Borigato and F. E. Martinez, Iron nutritional status is improved in Brazilian preterm infants fed food cooked in iron pots, *Journal of Nutrition* 128 (1998): 855–59.

11. J. T. Salonen et al., High stored iron levels are associated with excess risk of myocardial infarction in Eastern Finnish men, *Circulation* 86 (1992): 803–11; K. Klipstein-Grobusch et al., Dietary iron and risk of myocardial infarction in the Rotterdam Study, *American Journal of Epidemiology* 149 (1999): 421–28; Food and Nutrition Board, Institute of Medicine, *Dietary Reference Intakes for Vitamin A, Vitamin K, Arsenic, Boron, Chromium, Copper, Iodine, Iron, Manganese, Molybdenum, Nickel, Silicon, Vanadium,*

and Zinc (Washington, DC: National Academy Press, 2001), 357–78; and M. B. Reddy and L. Clark, Iron, oxidative stress, and disease risk, *Nutrition Reviews* 62 (2004): 120–24.

12. C. F. Walker and R. E. Black, Zinc and the risk for infectious disease, *Annual Review of Nutrition* 24 (2004): 255–75.

13. Zinc lozenges reduce the duration of common cold symptoms, *Nutrition Reviews* 55 (1997): 82–88.

14. Food and Nutrition Board, Institute of Medicine, *Dietary Reference Intakes for Vitamin A, Vitamin K, Arsenic, Boron, Chromium, Copper, Iodine, Iron, Manganese, Molybdenum, Nickel, Silicon, Vanadium, and Zinc* (Washington, DC: National Academy Press, 2001).

15. J. C. King and R. J. Cousins, Zinc, in M. E. Shils et al., eds., *Modern Nutrition in Health and Disease*, 10th ed. (Philadelphia: Lippincott Williams & Wilkins, 2006).

16. B. J. Stoecker, Chromium, in M. E. Shils et al., eds., *Modern Nutrition in Health and Disease*, 10th ed. (Philadelphia: Lippincott Williams & Wilkins, 2006); and Food and Nutrition Board, Institute of Medicine, *Dietary Reference Intakes for Vitamin A, Vitamin K, Arsenic, Boron, Chromium, Copper, Iodine, Iron, Manganese, Molybdenum, Nickel, Silicon, Vanadium, and Zinc* (Washington, DC: National Academy Press, 2001).

17. Food and Nutrition Board, Institute of Medicine, *Dietary Reference Intakes for Vitamin A, Vitamin K, Arsenic, Boron, Chromium, Copper, Iodine, Iron, Manganese, Molybdenum, Nickel, Silicon, Vanadium, and Zinc* (Washington, DC: National Academy Press, 2001).

18. Food and Nutrition Board, Institute of Medicine, *Dietary Reference Intakes for Vitamin C, Vitamin E, Selenium, and Carotenoids* (Washington, DC: National Academy Press, 2000); R. F. Burk, Selenium, an antioxidant nutrient, *Nutrition in Clinical Care* 5 (2002): 75–79; and R. F. Burk and K. E. Hill, Selenoprotein P: An extracellular protein with unique physical characteristics and a role in selenium homeostasis, *Annual Review of Nutrition* 25 (2005): 215–35.

19. Standing Committee on the Scientific Evaluation of Dietary Reference Intakes, Food and Nutrition Board, Institute of Medicine, *Dietary Reference Intakes: Water, Potassium, Sodium, Chloride, and Sulfate* (Washington, DC: National Academy Press, 2004), 4–26; and H. J. Androgué and N. E. Madias, Sodium and potassium in the pathogenesis of hypertension, *New England Journal of Medicine* 356 (2007): 1966–78.

20. Position of the American Dietetic Association: Vegetarian diets, *Journal of the American Dietetic Association* 109 (2009): 1266–82.

21. P. M. Suter, C. Sierro, and W. Vetter, Nutritional factors in the control of blood pressure and hypertension, *Nutrition in Clinical Care* 5 (2002): 9–19; A. Avic, Salt and hypertension, *Archives of Internal Medicine* 161 (2001): 507–9; and H. J. Androgué and N. E. Madias, Sodium and potassium in the pathogenesis of hypertension, *New England Journal of Medicine* 356 (2007): 1966–78.

22. T. A. Kotchen and J. M. Kotchen, Nutrition, diet, and hypertension, in M. E. Shils et al., eds., *Modern Nutrition in Health and Disease*, 10th ed. (Philadelphia: Lippincott Williams & Wilkins, 2006), 1095–1107; and K. Rahmouni et al., Obesity-associated hypertension: New insights into mechanisms, *Hypertension* 45 (2005): 9–14.

23. The Joint National Committee on Prevention, Detection, Evaluation, and Treatment of High Blood Pressure, The Seventh Report of the Joint National Committee on Prevention, Detection, Evaluation, and Treatment of High Blood Pressure, *Journal of the American Medical Association* 289 (2003): 2560–72.

24. T. A. Kotchen and J. M. Kotchen, Nutrition, diet, and hypertension, in M. E. Shils et al., eds., *Modern Nutrition in Health and Disease*, 10th ed. (Philadelphia: Lippincott Williams & Wilkins, 2006), 1095–1107.

25. The Joint National Committee on Prevention, Detection, Evaluation, and Treatment of High Blood Pressure, The Seventh Report of the Joint National Committee on Prevention, Detection, Evaluation, and Treatment of High Blood Pressure, *Journal of the American Medical Association* 289 (2003): 2560–72; and P. M. Suter and W. Vetter, The effect of alcohol on blood pressure, *Nutrition in Clinical Care* 3 (2000): 24–34.

26. The Joint National Committee on Prevention, Detection, Evaluation, and Treatment of High Blood Pressure, The Seventh Report of the Joint National Committee on Prevention, Detection, Evaluation, and Treatment of High Blood Pressure, *Journal of the American Medical Association* 289 (2003): 2560–72.

27. R. Stamler et al., Primary prevention of hypertension by nutritional-hygienic means, *Journal of the American Medical Association* 262 (1989): 1801–7.

28. H. J. Androgué and N. E. Madias, Sodium and potassium in the pathogenesis of hypertension, *New England Journal of Medicine* 356 (2007): 1966–78; and T. A. Kotchen and J. M. Kotchen, Dietary sodium and blood pressure: Interactions with other nutrients, *American Journal of Clinical Nutrition* 65 (1997): 708S–11S.

29. H. J. Androgué and N. E. Madias, Sodium and potassium in the pathogenesis of hypertension, *New England Journal of Medicine* 356 (2007): 1966–78.

30. L. M. Resnick, The role of dietary calcium in hypertension: A hierarchical overview, *American Journal of Hypertension* 12, pt. 1 (1999): 99–112; and D. A. McCarron and M. E. Reusser, Finding consensus in the dietary calcium blood pressure debate, *Journal American College of Nutrition* 18 (1999): 398S–405S.

31. The Joint National Committee on Prevention, Detection, Evaluation, and Treatment of High Blood Pressure, The Seventh Report of the Joint National Committee on Prevention, Detection, Evaluation, and Treatment of High Blood Pressure, *Journal of the American Medical Association* 289 (2003): 2560–72.

32. The DASH clinical trial, *New England Journal of Medicine* 336 (1997): 1117; and F. M. Sacks et al., Effects on blood pressure of reduced dietary sodium and the dietary approaches to stop hypertension (DASH) diet, *New England Journal of Medicine* 334 (2001): 3–10.

33. The opening vignettes are from P. Ola and E. D'Aulaire, The health risk women can no longer ignore, *Reader's Digest* (August 1994): 91–95.

34. NIH Consensus Development Panel, Osteoporosis prevention, diagnosis, and therapy, *Journal of the American Medical Association* 285 (2001): 785–95; L. W. Turner et al., Osteoporotic fracture among older U.S. women, *Journal of Aging and Health* 10 (1998): 372–91; and E. Siris et al., Design of NORA, the National Osteoporosis Risk Assessment Program: A longitudinal U.S. registry of postmenopausal women, *Osteoporosis International* 8 (1998): S62–S69.

35. T. A. Ricci et al., Calcium supplementation suppresses bone turnover during weight reduction in postmenopausal women, *Journal of Bone and Mineral Research* 13 (1998): 1045–50.

36. National Center for Injury Prevention and Control, Falls and hip fractures among the elderly, *Unintentional Injury Fact Sheet* (2006), available at www.cdc.gov.

37. R. L. Smith et al., Prevention of postmenopausal osteoporosis: A comparative study of exercise, calcium supplementation, and hormone replacement therapy, *New England Journal of Medicine* 325 (1991): 1189–95.

38. F. H. Anderson, Osteoporosis in men, *International Journal of Clinical Practice* 52 (1998): 176–80.

39. M. L. Rencken et al., Bone density at multiple skeletal sites in amenorrheic athletes, *Journal of the American Medical Association* 276 (1996): 238–40.

40. C. J. Strange, Boning up on osteoporosis, *FDA Consumer* [reprint] (April 2004).

41. E. L. Smith and C. Gilligan, Effects of inactivity and exercise on bone, *Physician and Sportsmedicine* 15 (1997): 91–102; and Department of Health and Human Services, *2008 Physical Activity Guidelines for Americans* (www.health.gov/PAGuidelines).

42. M. N. Hadley and S. V. Reddy, Smoking and the human vertebral column: A review of the impact of cigarette use on vertebral bone metabolism and spinal fusion, *Neurosurgery* 41 (1997): 116–24; and C. W. Slemenda, Cigarettes and the skeleton, *New England Journal of Medicine* 330 (1994): 430–31.

43. F. R. Greer and N. F. Krebs, Optimizing bone health and calcium intakes of infants, children, and adolescents, *Pediatrics* 117 (2006): 578–85; H. J. Kalkwarf, J. C. Khoury, and B. P. Lamphear, Milk intake during childhood and adolescence, adult bone density, and osteoporotic fractures in U.S. women, *American Journal of Clinical Nutrition* 77 (2003): 257–65; and R. P. Heaney et al., Dietary changes favorably affect bone remodeling in older adults, *Journal of the American Dietetic Association* 99 (1999): 1228–33.

44. F. R. Greer and N. F. Krebs, Optimizing bone health and calcium intakes of infants, children, and adolescents, *Pediatrics* 117 (2006): 578–85.

45. K. Glanz and A. Steffen, Development and reliability testing for measures of psychosocial constructs associated with adolescent girls' calcium

intake, *Journal of the American Dietetic Association* 108 (2008): 857–61; L. Harnack, J. Stang, and M. Story, Soft drink consumption among U.S. children and adolescents: Nutritional consequences, *Journal of the American Dietetic Association* 99 (1999): 436–41; and S. A. Bowman, Beverage choices of young females: Changes and impact on nutrient intakes, *Journal of the American Dietetic Association* 102 (2002): 1234–39.

46. P. Weber, The role of vitamins in the prevention of osteoporosis: A brief status report, *International Journal for Vitamin and Nutrition Research* 69 (1999): 194–97; and G. Ferland, Vitamin K-dependent proteins: An update, *Nutrition Reviews* 56 (1998): 223–30.

47. D. E. Sellmeyer et al., A high ratio of dietary animal to vegetable protein increases the rate of bone loss and the risk of fracture in postmenopausal women, *American Journal of Clinical Nutrition* 73 (2001): 118–22; U. S. Barzel and L. K. Massey, Excess dietary protein can adversely affect bone, *Journal of Nutrition* 128 (1998): 1051–53; S. J. Whiting and B. Lemke, Excess retinol intake may explain the high incidence of osteoporosis in northern Europe, *Nutrition Reviews* 57 (1999): 192–95; and F. Ginty, A. Flynn, and K. D. Cashman, The effect of dietary sodium intake on biochemical markers of bone metabolism in young women, *British Journal of Nutrition* 79 (1998): 343–50.

48. D. Feskanich et al., Vitamin K intake and hip fractures in women: A prospective study, *American Journal of Clinical Nutrition* 69 (1999): 74–79.

49. K. L. Tucker et al., Bone mineral density and dietary patterns in older adults: The Framingham Osteoporosis Study, *American Journal of Clinical Nutrition* 76 (2002): 245–52.

50. B. Dawson-Hughes et al., Effect of calcium and vitamin D supplements on bone density in men and women 65 years of age or older, *New England Journal of Medicine* 337 (1997): 670–75; K. M. Chiu, Efficacy of calcium supplements on bone mass in postmenopausal women, *Journal of Gerontology* 54 (1999): 275–80; Position of the American Dietetic Association: Nutrient supplementation, *Journal of the American Dietetic Association* 109 (2009): 2073–85; and D. R. Jacobs, M. D. Gross, and L. C. Tapsell, Food synergy: An operational concept for understanding nutrition, *American Journal of Clinical Nutrition* 89 (2009): 1543S–48S.

51. J. R. Saltzman, Nutritionally significant changes in gastrointestinal functioning with aging, *Nutrition in Clinical Care* 1 (1998): 20–29.

52. C. C. Collins and M. A. Summa, Clinical significance of lead content in dietary calcium supplements, *Nutrition in Clinical Care* 1 (1998): 156–59.

53. Food and Drug Administration, FDA approves new labels for estrogen and estrogen with progestin therapies for postmenopausal women following review of Women's Health Initiative data, *FDA News* (January 8, 2003); L. Pachucki-Hyde, Cutting your risk of osteoporosis, *Diabetes Self-Management* (May/June, 1998): 36–43; E. Barrett-Connor, Hormone replacement therapy, *British Medical Journal* 317 (1998): 457–61; R. Lindsay, The role of estrogen in the prevention of osteoporosis, *Endocrinology and Metabolism Clinics of North America* 27 (1998): 399–405; and M. McClung et al., Alendronate prevents postmenopausal bone loss in women without osteoporosis, *Annals of Internal Medicine* 128 (1998): 253–61.

54. J. W. Nieves et al., Calcium potentiates the effect of estrogen and calcitonin on bone mass: Review and analysis, *American Journal of Clinical Nutrition* 67 (1998): 18–24.

55. P. Taxel, Osteoporosis: Detection, prevention, and treatment in primary care, *Geriatrics* 53 (1998): 22–40; NIH Consensus Development Panel, Osteoporosis prevention, diagnosis, and therapy, *Journal of the American Medical Association* 285 (2001): 785–95.

56. Top ten advances of 1993, *Harvard Health Letter* 19, no. 5 (1994): 7.

57. K. Woolf et al., Physical activity is associated with risk factors for chronic disease across adult women's life cycle, *Journal of the American Dietetic Association* 108 (2008): 948–59; and T. V. Nguyen et al., Bone loss, physical activity, and weight change in elderly women: The Dubbo Osteoporosis Epidemiology Study, *Journal of Bone Mineral Research* 13 (1998): 1458–67.

58. R. L. Prince et al., Prevention of postmenopausal osteoporosis: A comparative study of exercise, calcium supplementation, and hormone replacement therapy, *New England Journal of Medicine* 325 (1991): 1189–95; and E. Ernst, Exercise for female osteoporosis: A systematic review of randomized clinical trials, *Sports Medicine* 25 (1998): 359–68.

59. N. K. Henderson et al., The roles of exercise and fall risk reduction in the prevention of osteoporosis, *Endocrinology and Metabolism Clinics of North America* 27 (1998): 369–87; and C. M. Arnold, M. M. Sran, and E. L.

Harrison, Exercise for fall risk reduction in community-dwelling older adults: A systematic review, *Physiotherapy Canada* 60 (2008): 358–72.

CHAPTER 9

1. L. E. Grivetti and T. Wilson, A brief history of human beverage consumption, in T. Wilson and N. J. Temple, eds., *Beverages in Nutrition and Health* (Totowa, NJ: Humana Press, 2004), 3–18.

2. S. M. Kleiner, Water: An essential but overlooked nutrient, *Journal of the American Dietetic Association* 99 (1999): 200–206; and Standing Committee on the Scientific Evaluation of Dietary Reference Intakes, Food and Nutrition Board, Institute of Medicine, *Dietary Reference Intakes: Water, Potassium, Sodium, Chloride, and Sulfate* (Washington, DC: National Academy Press, 2004), 4–26.

3. M. P. Sauvant and D. Pepin, Geographic variation of the mortality from cardiovascular disease and drinking water in a French small area (Puy de Dome), *Environmental Research* 84 (2000): 219–27; R. Maheswaran et al., Magnesium in drinking water supplies and mortality from acute myocardial infarction in northwest England, *Heart* 82 (1999): 455–60; and C. Y. Yang and H. F. Chiu, Calcium and magnesium in drinking water and the risk of death from hypertension, *American Journal of Hypertension* 12 (1999): 894–99.

4. U.S. Environmental Protection Agency and Centers for Disease Control and Prevention, *Guidance for People with Severely Weakened Immune Systems* (Washington, DC: U.S. Environmental Protection Agency, June 1999).

5. J. F. Rosen and P. Mushak, Primary prevention of childhood lead poisoning—The only solution, *New England Journal of Medicine* 344 (2001): 1470–71; and National Center for Environmental Health, CDC Childhood Lead Poisoning Prevention Program (www.cdc.gov/nceh/lead/lead.htm).

6. E. Gias, Rising sales of bottled water trigger strong reactions from U.S. conservationists, *International Herald Tribune* (March 19, 2008), available at www.iht.com.

7. L. M. Posnick and H. Kim, Bottled water regulation and the FDA, *Food Safety Magazine* (August/September 2002).

8. A. C. Bullers, Bottled water: Better than the tap? *FDA Consumer Magazine* (July–August 2002), 14–18.

9. J. Stannard et al., Fluoride content of some bottled waters and recommendation for fluoride supplementation, *Journal of Pedodontics* 14 (1990): 103–7; Position of the American Dietetic Association: The impact of fluoride on health, *Journal of the American Dietetic Association* 105 (2005): 1620–28; and J. Helm, Hot topic: Bottled water (American Dietetic Association, May 2008), available at www.eatright.org.

10. American Dietetic Association, Not all juices are equal (www.eatright.org/cps/rde/xchg/ada/hs.xsl/home_4639_ENU_HTML.htm).

11. K. M. Zelman, Nutrition in a glass, WebMD (www.webmd.com).

12. *Ibid.*

13. T. Wilson, Cranberry juice effects on health, in T. Wilson and N. J. Temple, eds., *Beverages in Nutrition and Health* (Totowa, NJ: Humana Press, 2004), 51–62.

14. C. R. McGill, A. M. R. Wilson, and Y. Papanikolaou, Health benefits of citrus juices, in T. Wilson and N. J. Temple, eds., *Beverages in Nutrition and Health* (Totowa, NJ: Humana Press, 2004), 63–78.

15. C. S. Johnson, Orange juice: Are the health benefits of oranges lost during processing? in T. Wilson and N. J. Temple, eds., *Beverages in Nutrition and Health* (Totowa, NJ: Humana Press, 2004), 79–91.

16. C. W. Hadley, S. J. Schwartz, and S. K. Clinton, Tomato-based beverages: Implications for the prevention of cancer and cardiovascular disease, in T. Wilson and N. J. Temple, eds., *Beverages in Nutrition and Health* (Totowa, NJ: Humana Press, 2004), 107–23.

17. Anonymous. The medicine of the prophet. Eastern Caliphate, 11th century.

18. L. D. McBean, G. D. Miller, and R. P. Heaney, Effect of cow's milk on human health, in T. Wilson and N. J. Temple, eds., *Beverages in Nutrition and Health* (Totowa, NJ: Humana Press, 2004), 205–21.

19. *Ibid.*

20. M. B. Zemel et al., Regulation of adiposity by dietary calcium, *Federation of American Societies for Experimental Biology Journal* 14 (2000): 1132–38.

21. B. R. Carruth and J. D. Skinner, The role of dietary calcium and other nutrients in moderating body fat in preschool children, *International Journal of Obesity* 25 (2001): 559–66.

22. M. B. Zemel et al., Dietary calcium and dairy products accelerate weight and fat loss during energy restriction in obese adults. *American Journal of Clinical Nutrition* 75 (2002): 342S–43S.

23. L. D. McBean, G. D. Miller, and R. P. Heaney, Effect of cow's milk on human health, in T. Wilson and N. J. Temple, eds., *Beverages in Nutrition and Health* (Totowa, NJ: Humana Press, 2004), 205–21.

24. B. A. Weinberg and B. K. Bealer, Caffeine and health, in T. Wilson and N. J. Temple, eds., *Beverages in Nutrition and Health* (Totowa, NJ: Humana Press, 2004), 171–85.

25. R. Tannahil, *Food in History* (New York: Crown, 1988), 275.

26. R. Urgert et al., Effects of cafestol and kahweol from coffee grounds on serum lipids and serum liver enzymes in humans, *American Journal of Clinical Nutrition* 61 (1995): 149–54.

27. A. Leviton and E. N. Allred, Correlates of decaffeinated coffee choice, *Epidemiology* 5 (1994): 537–40.

28. E. C. Strain et al., Caffeine dependence syndrome, *Journal of the American Medical Association* 272 (1994): 1043–48; and E. H. Hogan, B. A. Hornick, and A. Bouchoux, Communicating the message: Clarifying the controversies about caffeine, *Nutrition Today* 37 (2002): 28–35.

29. S. G. Oei et al., Fetal arrhythmia caused by excessive intake of caffeine by pregnant women, *British Medical Journal* 298 (1989): 1075–76.

30. S. Cnattingius et al., Caffeine intake and the risk of first trimester spontaneous abortion, *New England Journal of Medicine* 343 (2000): 1839–45.

31. Position of the American Dietetic Association: Nutrition and lifestyle for a healthy pregnancy outcome, *Journal of the American Dietetic Association* 108 (2008): 553–61.

32. E. H. Hogan, B. A. Hornick, and A. Bouchoux, Communicating the message: Clarifying the controversies about caffeine, *Nutrition Today* 37 (2001): 28–35.

33. R. M. van Dam and F. B. Hu, Coffee consumption and risk of type 2 diabetes, *Journal of the American Medical Association* 294, no. 1 (2005): 97–104.

34. L. F. Andersen et al., Consumption of coffee is associated with reduced risk of death attributed to inflammatory and cardiovascular diseases in the Iowa Women's Health Study, *American Journal of Clinical Nutrition* 83 (2006): 1039–46.

35. S. C. Larsson et al., Coffee and tea consumption and risk of stroke subtypes in male smokers, *Stroke* 39 (2008): 1681–87.

36. A. L. Klatsky et al., Coffee, cirrhosis, and transaminase enzymes, *Archives of Internal Medicine* 116 (2006): 1190–95.

37. E. Lopez-Garcia et al., The relationship of coffee consumption with mortality, *Annals of Internal Medicine* 148 (2008): 904–14.

38. J. V. Higdon and B. Frei, Coffee and health: A review of recent human research, *Critical Reviews in Food Science and Nutrition* 46, no. 2 (2006): 101–23.

39. F. Afaq et al., Health benefits of tea consumption, in T. Wilson and N. J. Temple, eds., *Beverages in Nutrition and Health* (Totowa, NJ: Humana Press, 2004), 143–56.

40. *Ibid.*; M. Purdy and M. Kestin, The green tea and cancer connection, *Nutrition and Complementary Care Newsletter* 9, no. 3 (2007): 43–45.

41. F. Afaq et al., Health benefits of tea consumption, in T. Wilson and N. J. Temple, eds., *Beverages in Nutrition and Health* (Totowa, NJ: Humana Press, 2004), 143–56.

42. *Ibid.*

43. *Ibid.*; M. Purdy and M. Kestin, The green tea and cancer connection, *Nutrition and Complementary Care Newsletter* 9, no. 3 (2007): 43–45; and D. Welland, The latest international research brews up more big benefits for tea, *Environmental Nutrition* 30, no. 12 (2007): 1, 4.

44. F. Afaq et al., Health benefits of tea consumption, in T. Wilson and N. J. Temple, eds., *Beverages in Nutrition and Health* (Totowa, NJ: Humana Press, 2004), 143–56.

45. T. L. Dillinger et al., Food of the gods: Cure for humanity? A cultural history of the medicinal and ritual use of chocolate, *Journal of Nutrition* 130 (2000): 2057S–72S.

46. *Ibid.*

47. H. H. Schmitz, M. A. Kelm, and J. F. Hammerstone, Effect of cocoa and chocolate beverage consumption on human cardiovascular health, in T. Wilson and N. J. Temple, eds., *Beverages in Nutrition and Health* (Totowa, NJ: Humana Press, 2004), 157–69.

48. J. Rossant, The world's first soft drink, *Saudi Aramco World* (September–October 2005), 36–39.

49. K. Skopitz, History of soft drinks, PageWise (2002), available at www.essortment.com/all/historyofsoft_mtg.htm.

50. L. R. Vartanian, M. B. Schwartz, and K. D. Brownell, Effects of soft drink consumption on nutrition and health: A systematic review and meta-analysis, *American Journal of Public Health* 97, no. 4 (2007): 667–75.

51. S. P. Fowler et al., Fueling the obesity epidemic? Artificially sweetened beverage use and long-term weight gain, *Obesity* (June 5, 2008; e-publication ahead of print), available at www.nature.com/oby/journal/vaop/ncurrent/abs/oby2008284a.html.

52. M. F. Jacobson, Liquid candy: How soft drinks harm the health of Americans, in T. Wilson and N. J. Temple, eds., *Beverages in Nutrition and Health* (Totowa, NJ: Humana Press, 2004), 351–67.

53. L. Bonci, 'Energy' drinks: Help, harm, or hype? *Sports Science Exchange* 15 (2002): 1–4, available at www.gssiweb.com/Article_Detail.aspx?articleid=310.

54. American Dietetic Association, What are "energy drinks" all about? *Journal of the American Dietetic Association* 103, no. 1 (2003): 54.

55. K. M. Zeman, What's the buzz about energy drinks? WebMD (www.webmd.com).

56. American Beverage Association, History of sports drinks (www.ameribev.org).

57. Mintel International Group Ltd., Sports drinks (2005), available at www.the-infoshop.com/study/mt27165_sports_drinks.html.

58. Gatorade Sports Science Institute, Why drink sports beverages? (www.gssiweb.com/Article_Detail.aspx?articleid=434&level=6&topic=2).

59. R. J. Maughan, Sports beverages for optimizing physical performance, in T. Wilson and N. J. Temple, eds., *Beverages in Nutrition and Health* (Totowa, NJ: Humana Press, 2004), 289–305.

60. Gatorade Sports Science Institute, Why drink sports beverages? (www.gssiweb.com/Article_Detail.aspx?articleid=434&level=6&topic=2).

61. *Ibid.*

62. B. Liebman and D. Schardt, H₂No: The scoop on enhanced waters, *Nutrition Action Healthletter* (June 2008), 14–15.

63. *Ibid.*

64. C. C. Freudenrich, *How Alcohol Works* (www.howstuffworks.com/alcohol2.htm).

65. W. F. Bosron, T. Ehrig, and T. K. Li, Genetic factors in alcohol metabolism and alcoholism, *Seminars in Liver Disease* 13, no. 2 (1993): 126–35; and H. Wallgren, Absorption, diffusion, distribution and elimination of ethanol: Effect on biological membranes, in *International Encyclopedia of Pharmacology and Therapeutics* vol. 1 (Oxford: Pergamon, 1970), 161–88.

66. C. C. Freudenrich, *How Alcohol Works* (www.howstuffworks.com/alcohol2.htm); and National Institute on Alcohol Abuse and Alcoholism, *Alcohol Alert* No. 35, PH 371 (Bethesda, MD: National Institute on Alcohol Abuse and Alcoholism, 1997), available at www.niaaa.nih.gov/publications/aa35-text.htm.

67. W. F. Bosron, T. Ehrig, and T. K. Li, Genetic factors in alcohol metabolism and alcoholism. *Seminars in Liver Disease* 13, no. 2 (1993): 126–35; and L. Z. Benet, D. L. Kroetz, and L. B. Sheiner, Pharmacokinetics: The dynamics of drug absorption, distribution, and elimination, in P. B. Molinoff and R. W. Ruddon, eds., *Goodman and Gillman's The Pharmacological Basis of Therapeutics*, 9th ed. (New York: McGraw-Hill, 1996), 3–27.

68. C. C. Freudenrich, *How Alcohol Works* (www.howstuffworks.com/alcohol2.htm); and National Institute on Alcohol Abuse and Alcoholism, *Alcohol Alert* No. 35, PH 371 (Bethesda, MD: National Institute on Alcohol Abuse and Alcoholism, 1997), available at www.niaaa.nih.gov/publications/aa35-text.htm.

69. C. C. Freudenrich, *How Alcohol Works* (www.howstuffworks.com/alcohol2.htm).

70. *Ibid.*

71. W. F. Bosron, T. Ehrig, and T. K. Li, Genetic factors in alcohol metabolism and alcoholism, *Seminars in Liver Disease* 13, no. 2 (1993): 126–35; H. Wallgren, Absorption, diffusion, distribution and elimination of ethanol: Effect on biological membranes, in *International Encyclopedia of Pharmacology and Therapeutics* vol. 1 (Oxford: Pergamon, 1970), 161–88; and A. G. Fraser et al., Individual and intra-individual variability of etha-nol concentration-time profiles: Comparison of ethanol ingestion before and after an evening meal, *British Journal of Clinical Pharmacology* 40 (1996): 387–92.

72. A. Urbano-Márquez et al., The greater risk of alcoholic cardiomyopathy and myopathy in women compared with men, *Journal of the American Medical Association* 274 (1995): 149–54.

73. S. J. Nixon, Cognitive deficits in alcoholic women, *Alcohol Health and Research World* 18, no. 3 (1994): 228–32.

74. National Institute on Alcohol Abuse and Alcoholism, *Alcohol Alert: Alcohol and Women*, No. 10, PH 290 (Bethesda, MD: National Institute on Alcohol Abuse and Alcoholism, 1990), available at www.niaaa.nih.gov/publications/aa10.htm.

75. C. C. Freudenrich, *How Alcohol Works* (www.howstuffworks.com/alcohol2.htm); and National Institute on Alcohol Abuse and Alcoholism, *Alcohol Alert* No. 35, PH 371 (Bethesda, MD: National Institute on Alcohol Abuse and Alcoholism, 1997), available at www.niaaa.nih.gov/publications/aa35-text.htm.

76. R. Spendler, *WebMD Health Guide: Blood Alcohol* (www.webmd.com/content/healthwise/143/35577.htm?printing=true).

77. C. S. Lieber, Metabolic consequence of ethanol, *Endocrinologist* 4, no. 2 (1994): 127–39; National Institute on Alcohol Abuse and Alcoholism, *Alcohol Alert: Alcohol Medication Interactions* No. 27, PH 290 (Bethesda, MD: National Institute on Alcohol Abuse and Alcoholism, 1995), available at www.niaaa.nih.gov/publications/aa27.htm; and M. Black, Acetaminophen hepatotoxicity, *Annual Review of Medicine* 35 (1984): 577–93.

78. National Institute on Alcohol Abuse and Alcoholism, *Alcohol Alert: Alcohol Medication Interactions* No. 27, PH 290 (Bethesda, MD: National Institute on Alcohol Abuse and Alcoholism, 1995), available at www.niaaa.nih.gov/publications/aa27.htm; and L. B. Seeff et al., Acetaminophen hepatotoxicity in alcoholics: A therapeutic misadventure, *Annals of Internal Medicine* 104 (1986): 399–404.

79. S. Andersson et al., Effects of alcohol metabolism, *Alcoholism: Clinical and Experimental Research* 10, no. 6 (1986): 55S–63S; H. I. Wright, J. S. Gavaler, and D. Van Thiel, Effects of alcohol on the male reproductive system, *Alcohol Health & Research World* 15, no. 2 (1991): 110–14; T. J. Cicero and R. D. Bell, Effects of ethanol and acetaldehyde on the biosynthesis of testosterone in the rodent testes, *Biochemical and Biophysical Research Communications* 94, no. 3 (1980): 814–19; D. E. Johnston, Inhibition of testosterone synthesis by ethanol and acetaldehyde, *Biochemical Pharmacology* 30, no. 13 (1981): 1827–30; Y. B. Chiao and D. H. Van Thiel, Biochemical mechanisms that contribute to alcohol-induced hypogonadism in the male, *Alcoholism: Clinical and Experimental Research* 7, no. 2 (1983): 131–34; and N. K. Mello, J. H. Mendelson, and S. K. Tech, An overview of the effects of alcohol on neuroendrocine function in women, in S. Zakhari, ed., *Alcohol and the Endocrine System*, National Institute on Alcohol Abuse and Alcoholism Research Monograph No. 23, NIH Publication No. 93–3533 (Bethesda, MD: National Institute on Alcohol Abuse and Alcoholism, 1993), 139–69.

80. H. I. Wright, J. S. Gavaler, and D. Van Thiel, Effects of alcohol on the male reproductive system, *Alcohol Health & Research World* 15, no. 2 (1991): 110–14; and D. H. Van Thiel, J. Gavaler, and R. Leser, Ethanol inhibition of vitamin A metabolism in the testes: Possible mechanism for sterility in alcoholics, *Science* 186 (1974): 941–42.

81. P. Bannister and M. S. Lowosky, Ethanol and hypogonadism, *Alcohol and Alcoholism* 22, no. 3 (1987): 213–17.

82. C. C. Freudenrich, *How Alcohol Works* (www.howstuffworks.com/alcohol2.htm); and National Institute on Alcohol Abuse and Alcoholism, *Alcohol Alert* No. 35, PH 371 (Bethesda, MD: National Institute on Alcohol Abuse and Alcoholism, 1997), available at www.niaaa.nih.gov/publications/aa35-text.htm.

83. C. C. Freudenrich, *How Alcohol Works* (www.howstuffworks.com/alcohol2.htm).

84. D. M. Goldberg, Health effects of moderate alcohol consumption, *Patient Care* 7, no. 5 (1996): 56–73.

85. M. J. Thun et al., Alcohol consumption and mortality among middle-aged and elderly U.S. adults, *New England Journal of Medicine* 337 (1997): 1705–14.

86. *Ibid.*

87. E. B. Rimm et al., Review of moderate alcohol consumption and reduced risk of coronary heart disease: Is the effect due to beer, wine, or spirits? *British Medical Journal* 312 (1996): 731–36.

88. D. M. Goldberg, Health effects of moderate alcohol consumption, *Patient Care* 7, no. 5 (1996): 56–73.

89. T. A. Pearson, American Heart Association Advisory: Alcohol and heart disease, *Circulation* 94 (1996): 3023–25.

90. T. Gordon et al., Alcohol and high-density lipoprotein cholesterol, *Circulation* 64, suppl. III (1981): III-63–67.

91. P. M. Ridker et al., Association of moderate alcohol consumption and plasma concentration of endogenous tissue-type plasminogen activator, *Journal of the American Medical Association* 272 (1994): 929–33; and S. Renaud and M. de Lorgeril, Wine, alcohol, platelets, and the French paradox for coronary heart disease, *Lancet* 339 (1992): 1523–26.

92. National Institute on Alcohol Abuse and Alcoholism, *Alcohol Alert: Alcohol and Women*, No. 45 (Bethesda, MD: National Institute on Alcohol Abuse and Alcoholism, 1999).

93. T. A. Pearson, American Heart Association Advisory: Alcohol and heart disease, *Circulation* 94 (1996): 3023–25.

94. E. Lehner and J. Lehner, *Folklore and Odysseys of Food & Medicinal Plants* (New York: Tudor Publishing, 1962); and M. Toussant-Samat, *A History of Food* (Cambridge, MA: Blackwell, 1994), as cited in American Dietetic Association, *Food Folklore: Tales and Truths about What We Eat* (Minneapolis, MN: Chronimed Publishing, 1999), 16.

95. NA/AA Recovery Zone, Stages of Addiction (http://nickscape.net/recoveryzone/stagesof.htm).

96. National Institute on Alcohol Abuse and Alcoholism, *Alcohol Alert: Fetal Alcohol Exposure and the Brain*, No. 50 (Bethesda, MD: National Institute on Alcohol Abuse and Alcoholism, 2000).

97. *Ibid.*; K. L. Jones and D. W. Smith, Recognition of the fetal alcohol syndrome in early infancy, *Lancet* 2 (1973): 999–1001.

98. National Institutes of Health, Highlights from the 10th Special Report to Congress, *Alcohol Research and Health* 24, no. 1 (2000); and K. Stratton, C. Howe, and F. Battaglia, eds., *Fetal Alcohol Syndrome: Diagnosis, Epidemiology, Prevention, and Treatment* (Washington, DC: National Academy Press, 1996).

99. National Institutes of Health, Highlights from the 10th Special Report to Congress, *Alcohol Research and Health* 24, no. 1 (2000).

100. National Institute on Alcohol Abuse and Alcoholism, *Alcohol Alert: Fetal Alcohol Exposure and the Brain*, No. 50 (Bethesda, MD: National Institute on Alcohol Abuse and Alcoholism, 2000).

CHAPTER 10

1. Position of the American Dietetic Association: Food and nutrition misinformation, *Journal of the American Dietetic Association* 106 (2006): 601–607; K. M. Flegal, M. D. Carroll, C. L. Ogden, and L.R. Curtin, Prevalence and trends in obesity among U.S. adults, 1999–2008, *Journal of the American Medical Association* 303 (2010): 235–41; and C. L. Ogden et al., Prevalence of high body mass index in U.S. children and adolescents, 2007–2008. *Journal of the American Medical Association* 303 (2010): 242–49.

2. U.S. Department of Health and Human Services, *The Surgeon General's Call to Action to Prevent and Decrease Overweight and Obesity* (Washington, DC: U.S. Department of Health and Human Services, Public Health Service, Office of the Surgeon General, 2001); and E. A. Finkelstein, I. C. Fiebelkorn, and G. Wang, State-level estimates of annual medical expenditures attributable to obesity, *Obesity Research* 12 (2004): 18–24.

3. U.S. Department of Health and Human Services, Office of Disease Prevention and Health Promotion, *Healthy People 2020* (Washington, DC: U.S. Department of Health and Human Services, 2010).

4. The discussion of trends is adapted from M. Fierro, The obesity epidemic—How states can trim the "fat," *Issue Brief* (Washington, DC: National Governors Association Center for Best Practices, June 13, 2002).

5. B. J. Rolls, The supersizing of America: Portion size and the obesity epidemic, *Nutrition Today* 38 (2003): 42–53; National Alliance for Nutrition and Activity (NANA), *From Wallet to Waistline: The Hidden Cost of Super Sizing* (Washington, DC: NANA, June 2002); L. R. Young and M. Nestle, The contribution of expanding portion sizes to the U.S. obesity epidemic, *American Journal of Public Health* 92 (2002): 246–48; S. J. Nielsen and B. M. Popkin, Patterns and trends in food portion sizes, 1977–1998, *Journal of the American Medical Association* 289 (2003): 450–53; M. Nestle, Increasing portion sizes in American diets: More calories, more obesity, *Journal of the American Dietetic Association* 103 (2003): 39–40; and G. L. Austin, L. G. Ogden, and J. Hill, Trends in carbohydrate, fat, and protein intake and association with energy intake in normal-weight, overweight, and obese individuals: 1971–2006, *American Journal of Clinical Nutrition* 93 (2011): 836–43.

6. C. L. Ogden et al., Prevalence of high body mass index in U.S. children and adolescents, 2007–2008. *Journal of the American Medical Association* 303 (2010): 242–49; and C. J. Crespo et al., Television watching, energy intake, and obesity in U.S. children: Results from the third National Health and Nutrition Examination Survey, 1988–1994, *Archives of Pediatrics and Adolescent Medicine* 155 (2001): 360–65.

7. D. Festi et al., Gallbladder motility and gallstone formation in obese patients following very low-calorie diets: Use it (fat) to lose it (well), *International Journal of Obesity and Related Metabolic Disorders* 22 (1998): 592–600.

8. G. A. Bray, The underlying basis for obesity: Relationship to cancer, *Journal of Nutrition* 132 (2002): 3451S–55S.

9. A. Must and S. E. Anderson, Effects of obesity on morbidity in children and adolescents, *Nutrition in Clinical Care* 6 (2003): 4–12; R. S. Strauss and H. A. Pollak, Social marginalization of overweight children, *Archives of Pediatrics and Adolescent Medicine* 157 (2003): 746–52; and M. E. Eisenberg, D. Neumark-Sztainer, and M. Story, Associations of weight-based teasing and emotional well-being among adolescents, *Archives of Pediatrics and Adolescent Medicine* 157 (2003): 733–38.

10. G. A. Bray, *Contemporary Diagnosis and Management of Obesity* (Newtown, PA: Handbooks in Healthcare, 1998).

11. Baylor College of Medicine, Undergoing a BodPod test: Densitometry theory (www.bcm.tmc.edu/bodycomplab/bpodprocesspage.htm); see also G. A. Bray, *Contemporary Diagnosis and Management of Obesity* (Newtown, PA: Handbooks in Healthcare, 1998); and C. J. Hoffman and L. A. Hildebrandt, Use of air displacement plethysmography to monitor body composition: A beneficial tool for dietitians, *Journal of the American Dietetic Association* 101 (2001): 986–87.

12. R. Roubenoff et al., Predicting body fatness: The body mass index versus estimation by bioelectrical impedance, *American Journal of Public Health* 85 (1995): 726–28.

13. P. Bjorntorp, Obesity, *Lancet* 350 (1997): 423–26.

14. The National Heart, Lung, and Blood Institute Expert Panel on the Identification, Evaluation, and Treatment of Overweight and Obesity in Adults, Executive summary of the clinical guidelines on the identification, evaluation, and treatment of overweight and obesity in adults, *Journal of the American Dietetic Association* 98 (1998): 1178–81.

15. R. J. Kuczmarski et al., Varying body mass index cutoff points to describe overweight prevalence among U.S. adults: NHANES III (1988–1994), *Obesity Research* 5 (1997): 542–45.

16. The National Heart, Lung, and Blood Institute Expert Panel on the Identification, Evaluation, and Treatment of Overweight and Obesity in Adults, Executive summary of the clinical guidelines on the identification, evaluation, and treatment of overweight and obesity in adults, *Journal of the American Dietetic Association* 98 (1998): 1178–81; D. Canoy et al., Body fat distribution and risk of coronary heart disease in men and women in the European Prospective Investigation into Cancer and Nutrition in Norfolk Cohort: A population-based prospective study, *Circulation*, 116 (2007): 2933–43.

17. S. Kumanyika et al., Population-based prevention of obesity: The need for comprehensive promotion of healthful eating, physical activity, and energy balance: A scientific statement from American Heart Association Council on Epidemiology and Prevention, Interdisciplinary Committee for Prevention, *Circulation*, 118 (2008): 428–64.

18. E. T. Poehlman and E. S. Horton, Energy needs: Assessment and requirements in humans, in M. E. Shils et al., eds., *Modern Nutrition in Health and Disease*, 10th ed. (Philadelphia: Lippincott Williams & Wilkins, 2006).

19. G. A. Bray, Lifestyle and pharmacological approaches to weight loss: Efficacy and safety, *Journal Clinical Endocrinology and Metabolism*, 93 (2008): S81–S88; J. O. Hill et al., Obesity and the environment: Where do we go from here? *Science* 299 (2003): 853; J. T. Travis, The hunger hormone? *Science News* 161 (2002): 107; E. M. Blass, Biological and environmental determinants of childhood obesity, *Nutrition in Clinical Care* 6

(2003): 13–19; and S. Kumanyika et al., Population-based prevention of obesity: The need for comprehensive promotion of healthful eating, physical activity, and energy balance: A scientific statement from American Heart Association Council on Epidemiology and Prevention, Interdisciplinary Committee for Prevention, *Circulation* 118 (2008): 428–64.

20. A. J. Stunkard et al., A twin study of human obesity, *Journal of the American Medical Association* 256 (1986): 51–54; A. J. Stunkard et al., An adoption study of human obesity, *New England Journal of Medicine* 314 (1986): 193–98; A. J. Stunkard et al., The body mass index of twins who have been reared apart, *New England Journal of Medicine* 322 (1990): 1483–87; J. O. Hill, H. R. Wyatt, and E. L. Melanson, Genetic and environmental contributions to obesity, *Medical Clinics of North America* 84 (2000): 333–46; and T. Rankinen et al., The human obesity gene map: The 2001 update, *Obesity Research* 10 (2002): 196–243.

21. C. Bouchard et al., The response to long-term overfeeding in identical twins, *New England Journal of Medicine* 322 (1990): 1477–82.

22. J. M. Friedman, The function of leptin in nutrition, weight, and physiology, *Nutrition Reviews* 60 (2002): S1.

23. A. Tups, Physiological models of leptin resistance, *Journal of Neuroendocrinology* 21 (2009): 961–71; R. V. Considine, Serum immunoreactive-leptin concentrations in normal weight and obese humans, *New England Journal of Medicine* 334 (1996): 292–95; and J. M. Friedman, Leptin, leptin receptors, and the control of body weight, *Nutrition Reviews* 56 (1998): S38–S41.

24. R. C. Whitaker et al., Predicting obesity in young adulthood from childhood and parental obesity, *New England Journal of Medicine* 337 (1997): 869–73; M. C. Bellizzi and W. H. Dietz, Workshop on childhood obesity: Summary of the discussion, *American Journal of Clinical Nutrition* 70 (1999): S173–S175; and R. J. Hancox, B. J. Milne, and R. Poulton, Association between child and adolescent television viewing and adult health: A longitudinal birth cohort study, *Lancet* 364 (2004): 257–62.

25. F. Xavier Pi-Sunyer, Obesity, in M. E. Shils et al., eds., *Modern Nutrition in Health and Disease*, 10th ed. (Philadelphia: Lippincott Williams & Wilkins, 2006).

26. M. A. McCrory et al., Dietary variety within food groups: Association with energy intake and body fatness in men and women, *American Journal of Clinical Nutrition* 69 (1999): 440–47.

27. K. Patrick et al., Diet, physical activity, and sedentary behaviors as risk factors for overweight in adolescence, *Archives in Pediatric and Adolescent Medicine* 158 (2004): 385–90; and Position of the American Dietetic Association: Weight management, *Journal of the American Dietetic Association* 109 (2009): 330–46.

28. K. D. Brownell and T. A. Wadden, Etiology and treatment of obesity: Understanding a serious, prevalent, and refractory disorder, *Journal of Consulting and Clinical Psychology* 60 (1992): 505–17; Position of the American Dietetic Association: Weight management, *Journal of the American Dietetic Association* 109 (2009): 330–46; and S. Kumanyika et al., Population-based prevention of obesity: The need for comprehensive promotion of healthful eating, physical activity, and energy balance: A scientific statement from American Heart Association Council on Epidemiology and Prevention, Interdisciplinary Committee for Prevention, *Circulation*, 118 (2008): 428–64.

29. *Calorie Control Council National Consumer Survey* (2004); available at www.caloriecontrol.org/press-room/commentary.

30. Adapted from Weighing the diet books, *Nutrition Action Newsletter* (January–February 2004): 3–8; M. Freedman et al., Popular diets: A scientific review, *Obesity Research* 9 (2001): 1S–39S; M. L. Dansinger et al., Comparison of the Atkins, Ornish, Weight Watchers, and Zone diets for weight loss and heart disease risk reduction, *Journal of the American Medical Association* 293 (2005): 43–53; and V. S. Retelny, Fad Diet Review (2005); available at www.foodandhealth.com; Wheat Council, Setting the Record Straight (www.wheatfoods.org/).

31. M. L. Dansinger et al., Comparison of the Atkins, Ornish, Weight Watchers, and Zone diets for weight loss and heart disease risk reduction, *Journal of the American Medical Association* 293 (2005): 43–53.

32. Adapted from American Heart Association, Quick Weight Loss or Fad Diets (http://beta.heart.org/HEARTORG/GettingHealthy/NutritionCenter/Quick-Weight-Loss-or-Fad-Diets_UCM_305970_Article.jsp); Weighing the diet books, *Nutrition Action Newsletter* (January/February 2004): 3–8; Weight-Control Information Network on choosing a safe and successful diet (http://win.niddk.nih.gov/publications/choosing.htm).

33. M. Freedman et al., Popular diets: A scientific review, *Obesity Research* 9 (2001): 1S–39S; National Weight Control Registry (www.nwcr.ws).

34. Y. Ma et al., A dietary quality comparison of popular weight-loss plans, *Journal of the American Dietetic Association* 107 (2007): 1786–91.

35. S. Kumanyika et al., Population-based prevention of obesity: The need for comprehensive promotion of healthful eating, physical activity, and energy balance: A scientific statement from American Heart Association Council on Epidemiology and Prevention, Interdisciplinary Committee for Prevention, *Circulation*, 118 (2008): 428–64; National Center for Chronic Disease Prevention and Health Promotion, Centers for Disease Control and Prevention (CDC), *Obesity and Overweight—A Public Health Epidemic* (www.cdc.gov/obesity/index.html); and G. A. Bray, Lifestyle and pharmacological approaches to weight loss: Efficacy and safety, *Journal Clinical Endocrinology and Metabolism*, 93 (2008): S81–S88.

36. S. Schurgin and R. D. Siegel, Pharmacotherapy of obesity: An update, *Nutrition in Clinical Care* 6 (2003): 27–37; and F. M. Berg, Chromium picolinate: Scam of the hour, *Obesity and Health* 7 (1993): 54–55.

37. Federal Trade Commission, Weighing the Evidence in Diet Ads (www.ftc.gov).

38. Food and Drug Administration, Consumer alert: FDA plans regulation prohibiting sale of ephedra-containing dietary supplements and advises consumers to stop using these products (December 2003); available at www.fda.gov; and M. H. Pittler and E. Ernst, Dietary supplements for body weight reduction: A systematic review, *American Journal of Clinical Nutrition* 79 (2004): 529–36.

39. P. Kurtzweil, Dieter's brews make tea time a dangerous affair, *FDA Consumer* (July–August 1997): 6–11.

40. E. H. Livingston, Obesity and its surgical management, *American Journal of Surgery* 184 (2002): 103–13; E. Lin, Gastric bypass, *Archives of Surgery* 139 (2004): 780–84; H. Buchwald et al., Bariatric surgery: A systematic review and meta-analysis, *Journal of the American Medical Association* 292 (2004): 1724–37; and L. Sjostrom et al., Effects of bariatric surgery on mortality in Swedish obese subjects, *New England Journal of Medicine* 357 (2007): 741–52.

41. J. P. Foreyt and G. K. Goodrick, *Living Without Dieting* (New York: Warner Books, 1992), 28.

42. *Ibid.*, 30.

43. K. D. Brownell, *The LEARN Program for Weight Control* (Dallas, TX: American Health Publishing, 1994), 102–3.

44. J. P. Foreyt and G. K. Goodrick, *Living Without Dieting* (New York: Warner Books, 1992), 43–58.

45. A. Groin et al., Promoting long-term weight control: Does dieting consistency matter? *International Journal of Obesity and Related Metabolic Disorders* 28 (2004): 278–81; J. W. Anderson et al., Long-term weight loss maintenance: A meta-analysis of U.S. studies, *American Journal of Clinical Nutrition* 74 (2001): 579–84; and L. Bren, Losing weight: More than counting calories, *FDA Consumer* 36 (2002): 18–25.

46. M. T. McGuire, R. R. Wing, and J. O. Hill, The prevalence of weight loss maintenance among American adults, *International Journal of Obesity* 23 (1999): 1314–19; and 1,800 calories, 4 miles keep the weight off, *Diabetes Today* (November 17, 2004).

47. M. Shah et al., Comparison of a low-fat, ad libitum complex carbohydrate diet with a low energy diet in moderately obese women, *American Journal of Clinical Nutrition* 59 (1994): 980–84; S. M. Shick et al., Persons successful at long-term weight loss and maintenance continue to consume a low-energy, low-fat diet, *Journal of the American Dietetic Association* 98 (1998): 408–13; and J. P. Foreyt and G. K. Goodrick, Dieting and weight loss: The energy perspective, *Nutrition Reviews* 59 (2001): S25–S28.

48. M. T. McGuire et al., Long-term maintenance of weight loss: Do people who lose weight through various weight loss methods use different methods to maintain their weight? *International Journal of Obesity and Related Metabolic Disorders* 22 (1998): 572–77; and R. R. Wing and J. O. Hill, Successful weight loss maintenance, *Annual Review of Nutrition* 21 (2001): 323.

49. K. Brownell, Yo-yo dieting, in C. C. Cook-Fuller with S. Barrett, eds., *Nutrition 91/92* (Guilford, CT: Dushkin Publishing Group, 1991),

50. R. E. Anderson et al., Changes in bone mineral content in obese dieting women, *Metabolism: Clinical and Experimental* 46 (1997): 857–61; and K. M. Rourke et al., Effect of weight change on bone mass in female adolescents, *Journal of the American Dietetic Association* 103 (2003): 369–72.

51. M. B. Zemel and S. L. Miller, Dietary calcium and dairy modulation of adiposity and obesity risk, *Nutrition Reviews* 62 (2004): 125–31.

52. J. Y. Kim et al., The impacts of obesity on psychological well being: A cross-sectional study about depressive mood and quality of life, *Preventive Medicine and Public Health* 40 (2007): 191–95.

53. H. R. Wyatt and J. O. Hill, Let's get serious about promoting physical activity, *American Journal of Clinical Nutrition* 75 (2002): 449; J. M. Rippe and N. J. Wareham, E. M. van Sluijs, and U. Ekelund, Physical activity and obesity prevention: A review of the current evidence, *Proceedings of the Nutrition Society* 64 (2005): 229–47; R. L. Weinsier et al., Free-living activity energy expenditure in women successful and unsuccessful at maintaining a normal body weight, *American Journal of Clinical Nutrition* 75 (2002): 499; and M. Gilliat-Wimberly et al., Effects of habitual physical activity on the resting metabolic rates and body compositions of women aged 35 to 50 years, *Journal of the American Dietetic Association* 101 (2001): 1181–88.

54. J. O. Prochaska, *Changing for Good* (New York: William Morrow, 1994), 38–50.

55. *Ibid.*, 47.

56. I. Chalmers, *The Great Food Almanac* (San Francisco: Collins Publishers, 1994).

57. G. T. Wilson and R. Shafran, Eating disorders guidelines from NICE, *Lancet* 365 (2005): 79–81; and American Psychiatric Association Workgroup on Eating Disorders, Practice guidelines for the treatment of patients with eating disorders, *American Journal of Psychiatry* 157 (2000): 1–39.

58. D. Neumark-Sztainer and P. J. Hannan, Weight-related behaviors among adolescent girls and boys: Results from a national survey, *Archives of Pediatric and Adolescent Medicine* 154 (2000): 569–77; and D. Neumark-Sztainer et al., Obesity, disordered eating, and eating disorders in a longitudinal study of adolescents: How do dieters fare 5 years later? *Journal of the American Dietetic Association* 106 (2006): 559–68.

59. Position of the American Dietetic Association: Nutrition intervention in the treatment of eating disorders, *Journal of the American Dietetic Association* 111 (2011): 1236–41.

60. Task Force on *DSM-IV*, 307.50 Eating Disorders Not Otherwise Specified, *DSM-IV Draft Criteria* (Washington, DC: American Psychiatric Association, 1993), 2.

61. C. M. Bulik, K. A. Brownley, and J. R. Shapiro, Diagnosis and management of binge eating disorder, *World Psychiatry* 6 (2007): 142–48; and J. D. Latner and C. Clyne, The diagnostic validity of the criteria for binge eating disorder, *International Journal of Eating Disorders* 41 (2008): 1–14.

62. N. I. Hahn and M. M. Woolsey, When food becomes a cry for help, *Journal of the American Dietetic Association* 98 (1998): 395–98.

63. Position of the American Dietetic Association: Nutrition intervention in the treatment of eating disorders, *Journal of the American Dietetic Association* 111 (2011): 1236–41.

64. T. Pryor and W. Wiederman, Personality features and expressed concerns of adolescents with eating disorders, *Adolescence* 33 (1998): 291–98.

65. W. Kaye et al., Serotonin neuronal function and selective serotonin reuptake inhibitor treatment in anorexia and bulimia nervosa, *Biological Psychiatry* 44 (1998): 825–38.

66. A. Gila et al., Subjective body-image dimensions in normal and anorexic adolescents, *British Journal of Medical Psychology* 71 (1998): 175–84.

67. *Ibid.*

68. F. Klapper et al., Psychiatric management of eating disorders, *Nutrition in Clinical Care* 2 (1999): 354–60.

69. D. Williamson, *Assessment of Eating Disorders: Obesity, Anorexia, and Bulimia Nervosa* (New York: Pergamon Press, 1990); and Practice guidelines for eating disorders, *American Journal of Psychiatry* 150 (1993): 212–18.

70. Position of the American Dietetic Association: Nutrition intervention in the treatment of eating disorders, *Journal of the American Dietetic Association* 111 (2011): 1236–41.

71. F. Klapper et al., Psychiatric management of eating disorders, *Nutrition in Clinical Care* 2 (1999): 354–60; and B. Abramovitz and L. L. Birch, Five-year-old girls' ideas about dieting are predicted by their mothers' dieting, *Journal of the American Dietetic Association* 100 (2000): 1157–63.

72. The guidelines listed are from M. Herrin, Dartmouth College Nutrition Education Program; for more information, contact Eating Disorders Awareness and Prevention, Inc. (www.edap.org).

CHAPTER 11

1. C. Rosenbloom and M. Bahns, What can we learn about diet and physical activity from the master athletes? *Nutrition Today* 40 (2005): 267–72; K. Woolf et al., Physical activity is associated with risk factors for chronic disease across adult women's life cycle, *Journal of the American Dietetic Association* 108 (2008): 948–59; and D. E. Warburton, C. W. Nicol, and S. S. Bredin, Health benefits of physical activity: The evidence, *Canadian Medical Association Journal* 174 (2006): 801–9.

2. U.S. Department of Health and Human Services. *2008 Physical Activity Guidelines for Americans* (Washington, DC: Office of Disease Prevention and Health Promotion, 2008), Publication No. U0036.

3. American College of Sports Medicine, *ACSM's Guidelines for Exercise Testing and Prescription*, 8th ed. (Philadelphia: Lippincott Williams & Wilkins, 2008).

4. M. A. F. Singh, Exercise for disease prevention in the geriatric population, *Nutrition in Clinical Care* 4 (2001): 296–305.

5. F. W. Booth et al., Waging war on physical inactivity: Using modern molecular ammunition against an ancient enemy, *Journal of Applied Physiology* 93 (2002): 3–30.

6. To work up a sweat, or not? *Tufts University Health and Nutrition Letter* 20, no. 10 (December 2002): 6.

7. New strength-training guidelines for older people: Too cautious? *Tufts University Health and Nutrition Letter* 20, no. 1 (March 2002): 8.

8. Are you doing all you can to fight sarcopenia? *Tufts University Health and Nutrition Letter* 21, no. 1 (March 2003): 1, 4–5; and C. E. Broeder, The effects of either high-intensity resistance or endurance training on resting metabolic rate, *American Journal of Clinical Nutrition* 55 (1992): 802–10.

9. W. L. Haskell et al., Physical activity and public health: Updated recommendation for adults from the American College of Sports Medicine and the American Heart Association, *Medicine and Science in Sports & Exercise* 39 (2007): 1423–34; and National Center for Health Statistics (NCHS), *Health, United States, 2010*, available at www.cdc.gov/nchs/data/hus/hus10.pdf.

10. M. H. Williams, *Nutrition for Health, Fitness, and Sport*, 8th ed. (Boston: McGraw-Hill, 2007).

11. E. Hultman et al., Work and exercise, in M. E. Shils et al., eds., *Modern Nutrition in Health and Disease*, 10th ed. (Philadelphia: Lippincott Williams & Wilkins, 2006).

12. E. Coleman, Carbohydrate and exercise, in M. Dunford, ed., *Sports Nutrition: A Practice Manual for Professionals*, 4th ed. (Chicago: American Dietetic Association, 2006), 14–32.

13. J. A. Romijn et al., Regulation of endogenous fat and carbohydrate metabolism in relation to exercise intensity and duration, *American Journal of Physiology* 265 (1993): E380–E391; G. A. Brooks and J. Mercier, Balance of carbohydrate and lipid utilization during exercise: The "crossover" concept, *Journal of Applied Physiology* 76 (1994): 2253–61; and J. H. Wilmore, Physical energy: Fuel metabolism, *Nutrition Reviews* 59 (2001): S13.

14. R. A. Fielding and J. A. Parkington, What are the dietary protein requirements of physically active individuals? New evidence on the effects of exercise on protein utilization during post-exercise recovery, *Nutrition in Clinical Care* 5 (2002): 191–96.

15. Position of the American Dietetic Association, Dietitians of Canada, and the American College of Sports Medicine: Nutrition and athletic performance, *Journal of the American Dietetic Association* 109 (2009): 509–27.

16. C. M. Cumming et al., Recreational runners' beliefs and practices concerning water intake, *Journal of Nutrition Education* 26 (1994): 195–97.

17. Position of the American Dietetic Association, Dietitians of Canada, and the American College of Sports Medicine: Nutrition and athletic performance, *Journal of the American Dietetic Association* 109 (2009): 509–27.

18. American College of Sports Medicine, Position stand, Exercise and fluid replacement, *Medicine and Science in Sports and Exercise* 39 (2007): 377–90.

19. C. V. Gisolfi, Fluid balance for optimal performance, *Nutrition Reviews* 54 (1996): S159–S168; and L. Bonci, 'Energy' drinks: Help, harm, or hype? *Sports Science Exchange* 15 (2002): 1–4.

20. J. S. Coombes and K. L. Hamilton, The effectiveness of commercially available sports drinks, *Sports Medicine* 29 (2000): 181–209.

21. American College of Sports Medicine, Position stand, Exercise and fluid replacement, *Medicine and Science in Sports and Exercise* 39 (2007): 377–90.

22. M. F. Bergeron, Sodium: The forgotten nutrient, *Sports Science Exchange* 13 (2000): 1–4.

23. S. L. Williams et al., Antioxidant requirements of endurance athletes: Implications for health, *Nutrition Reviews* 64 (2006): 93–108; and A. K. Adams and T. M. Best, The role of antioxidants in exercise and disease prevention, *The Physician and Sportsmedicine* 30 (2002): 37.

24. N. A. Suedekum and R. J. Dimeff, Iron and the athlete, *Current Sports Medicine Report* 4 (2005): 199–202.

25. S. L. Akabas and K. R. Dolins, Micronutrient requirements of physically active women: What can we learn from iron? *American Journal of Clinical Nutrition* 81 (2005): 1246S–51S.

26. Y. I. Shu and J. D. Haas, Iron depletion without anemia and physical performance in young women, *American Journal of Clinical Nutrition* 66 (1997): 334–41; and S. P. Bourque et al., Twelve weeks of endurance exercise training does not affect iron status in women, *Journal of the American Dietetic Association* 97 (1997): 1116–21.

27. K. A. Beals and A. K. Hill, The prevalence of disordered eating, menstrual dysfunction, and low bone mineral density among U.S. collegiate athletes, *International Journal of Sport Nutrition and Exercise Metabolism* 16 (2006): 1–23; and L. E. Thrash and J. B. Anderson, The female athlete triad: Nutrition, menstrual disturbances, and low bone mass, *Nutrition Today* 35 (2000): 168–74.

28. Position of the American Dietetic Association, Dietitians of Canada, and the American College of Sports Medicine: Nutrition and athletic performance, *Journal of the American Dietetic Association* 109 (2009): 509–27; and M. Dunford and M. Smith, Dietary supplements and ergogenic aids, in M. Dunford, ed., *Sports Nutrition: A Practice Manual for Professionals*, 4th ed. (Chicago: American Dietetic Association, 2006), 116–41.

29. T. H. Murray, The ethics of drugs and sports, in R. H. Strauss, ed., *Drugs and Performance in Sports* (Philadelphia: W. B. Saunders, 1987), 11–21.

30. D. M. Ahrendt, Ergogenic aids: Counseling the athlete, *American Family Physician* 63 (2001): 913; A. Sarubin, *The Health Professional's Guide to Popular Dietary Supplements*, 3rd ed. (Chicago: American Dietetic Association, 2006); and M. Dunford and M. Smith, Dietary supplements and ergogenic aids, in M. Dunford, ed., *Sports Nutrition: A Practice Manual for Professionals*, 4th ed. (Chicago: American Dietetic Association, 2006), 116–41.

31. Position of the American Dietetic Association, Dietitians of Canada, and the American College of Sports Medicine: Nutrition and athletic performance, *Journal of the American Dietetic Association* 109 (2009): 509–27.

32. S. Barrett, Don't buy phony ergogenic aids, *Nutrition Forum* (May–June 1997): 19–21, 24.

33. G. Mirkin, Can bee pollen benefit health? *Journal of the American Medical Association* 262 (1989): 1854.

34. L. S. Walker et al., Chromium picolinate and body composition and muscular performance in wrestlers, *Medicine and Science in Sports and Exercise* 30 (1998): 1730–37; and J. B. Vincent, The potential value and toxicity of chromium picolinate as a nutritional supplement, weight loss agent, and muscle development agent, *Sports Medicine* 33 (2003): 213–30.

35. E. A. Applegate and L. E. Grivetti, Search for the competitive edge: A history of dietary fads and supplements, *Journal of Nutrition* 127 (1997): 869S–73S.

36. P. J. Rasch et al., Protein dietary supplementation and physical performance, *Medicine and Science in Sports* 1 (1969): 195–99.

37. A. B. Parkinson and N. A. Evans, Anabolic androgenic steroids: A survey of 500 users, *Medicine and Science in Sports and Exercise* 38 (2006): 644–51.

38. Committee on Sports Medicine and Fitness, Adolescents and anabolic steroids: A subject review, *Pediatrics* 99 (1997): 443–47.

39. D. C. Nieman, *Fitness and Sports Medicine: An Introduction* (Palo Alto, CA: Bull Publishing, 1990), 246–58.

40. L. Bonci, "Energy" drinks: Help, harm, or hype? *Sports Science Exchange* 15 (2002): 1–4.

41. M. Wahlcrist, FAQ on caffeine and energy drinks, *Nutrition Australia* (2001).

42. A. Izzo and E. Ernst, Interactions between herbal medicines and prescribed drugs: A systematic review, *Drugs* 61 (2005): 2163–75.

43. D. Edell, Are energy drinks safe? *The Athletic Advisor* (2009); available at www.athleticadvisor.com.

CHAPTER 12

1. L. Breslow and N. Breslow, Health practices and disability: Some evidence from Alameda County, *Preventive Medicine* 22 (1993): 86–95; and A. J. Vita et al., Aging, health risks, and cumulative disability, *New England Journal of Medicine* 338 (1998): 1035–41.

2. K. M. Godfrey and D. J. Barker, Fetal nutrition and adult disease, *American Journal of Clinical Nutrition* 71 (2000): 1344S–52S; and M. W. Gillman, Developmental origins of health and disease, *New England Journal of Medicine* 353 (2005): 1848–50.

3. Food and Nutrition Board, Institute of Medicine, *Dietary Reference Intakes for Energy, Carbohydrate, Fiber, Fat, Fatty Acids, Cholesterol, Protein, and Amino Acids* (Washington, DC: National Academies Press, 2002).

4. Position of the American Dietetic Association: Nutrition and lifestyle for a healthy pregnancy outcome, *Journal of the American Dietetic Association* 108 (2008): 553–61; and S. Brundage, Preconception healthcare, *American Family Physician* 65 (2002): 2507.

5. G. Kauwell, Folic acid supplements and fortification affect the risk for neural tube defects, vascular disease, and cancer: Evolving science, *Journal of Nutrition* 133 (2003): 1961S–68S.

6. E. R. Cena et al., Folate intake and food-related behaviors in nonpregnant, low-income women of childbearing age, *Journal of the American Dietetic Association* 108 (2008): 1364–68.

7. J. Erikson, Folic acid and prevention of spina bifida and anencephaly, *Morbidity and Mortality Weekly Report* 51 (2002): 1–3; and A. A. Yates, Dietary Reference Intakes: The new basis for recommendations for calcium and related nutrients, B vitamins, and choline, *Journal of the American Dietetic Association* 98 (1998): 699–706.

8. Committee to Review Dietary Reference Intakes for Vitamin D and Calcium, Food and Nutrition Board, A. C. Ross et al., eds., *Dietary Reference Intakes: Calcium and Vitamin D* (Washington, DC: Institute of Medicine of the National Academies, National Academies Press, 2010); A. C. Ross et al., The 2011 Dietary Reference Intakes for calcium and vitamin D: What dietetics practitioners need to know, *Journal of the American Dietetic Association,* 111 (2011): 524–27; and M. F. Picciano, Pregnancy and lactation: Physiological adjustments, nutritional requirements, and the role of dietary supplements, *Journal of Nutrition* 133 (2003): 1997S–2002S.

9. Position of the American Dietetic Association: Nutrition and lifestyle for a healthy pregnancy outcome, *Journal of the American Dietetic Association* 108 (2008): 553–61.

10. V. Azais-Braesco and G. Pascal, Vitamin A in pregnancy: Requirements and safety limits, *American Journal of Clinical Nutrition* 71 (2000): 1325S.

11. Position of the American Dietetic Association: Nutrition and lifestyle for a healthy pregnancy outcome, *Journal of the American Dietetic Association* 108 (2008): 553–61; and Position of the American Dietetic Association: Nutrient supplementation, *Journal of the American Dietetic Association* 109 (2009): 2073–85.

12. Institute of Medicine and National Research Council of the National Academies, *Weight Gain During Pregnancy: Reexamining the Guidelines* (Washington, DC: National Academies Press, 2009); and Position of the American Dietetic Association and American Society for Nutrition: Obesity, reproduction, and pregnancy outcomes, *Journal of the American Dietetic Association* 109 (2009): 918–27.

13. A. J. Rainville, Pica practices of pregnant women are associated with lower maternal hemoglobin level at delivery, *Journal of the American Dietetic Association* 98 (1998): 293–96.

14. U.S. Department of Health and Human Services, Office of Disease Prevention and Health Promotion, *Healthy People 2020* (Washington, DC: U.S. Department of Health and Human Services, 2010); and S. J. Ventura et al., Trends and variations in smoking during pregnancy and low birthweight: Evidence from the birth certificate, 1990–2000, *Pediatrics* 111 (2003): 1176–80.

15. J. R. Di Franza, C. A. Aligne, and M. Weitzman, Prenatal and postnatal environmental tobacco smoke exposure and children's health, *Pediatrics* 113 (2004): 1007–15.

16. L. W. Eustace, D.H. Kang, and D. Coombs, Fetal alcohol syndrome: A growing concern for health care professionals, *Journal of Obstetrics, Gynecologic, and Neonatal Nursing* 32 (2003): 215–21.

17. American Academy of Pediatrics, Fetal alcohol syndrome and alcohol-related neurodevelopmental disorders, *Pediatrics* 106 (2000): 358–60.

18. Position of the American Dietetic Association: Nutrition and lifestyle for a healthy pregnancy outcome, *Journal of the American Dietetic Association* 108 (2008): 553–61; and Center for Evaluation of Risks to Human Reproduction, (2003), available at http://cerhr.niehs.nih.gov.

19. FDA and EPA announce the revised consumer advisory on methylmercury in fish, *News* (2004), available at www.cfsan.fda.gov.

20. N. I. Hahn and M. Erick, Battling morning (noon and night) sickness: New approaches for treating an age-old problem, *Journal of the American Dietetic Association* 94 (1994): 147–48.

21. G. Dekker and B. Sibai, Primary, secondary, and tertiary prevention of preeclampsia, *Lancet* 357 (2001): 209–15; F. B. Pipkin, Risk factors for preeclampsia, *New England Journal of Medicine* 344 (2001): 926; and B. M. Sibai, Diagnosis and management of gestational hypertension and preeclampsia, *Obstetrics and Gynecology* 102 (2003): 181–92.

22. American Diabetes Association, Position statement: Gestational diabetes mellitus, *Diabetes Care* 26 (2003): S103–105; Report of the Expert Committee on the Diagnosis and Classification of Diabetes Mellitus, *Diabetes Care* 26 (2003): S5–20.

23. M. Story and J. Stang, eds., *Nutrition and the Pregnant Adolescent: A Practical Reference Guide* (Washington, DC: Maternal and Child Health Bureau, Health Resources and Services Administration, 2000).

24. R. J. Trissler, The child within: A guide to nutrition counseling for pregnant teens, *Journal of the American Dietetic Association* 99 (1999): 916–18; see also V. A. Long, T. Martin, and C. Janson-Sand, The Great Beginnings Program: Impact of a nutrition curriculum on nutrition knowledge, diet quality, and birth outcomes in pregnant and parenting teens, *Journal of the American Dietetic Association* 102 (2002): S86–89.

25. Position of the American Dietetic Association: Nutrition and lifestyle for a healthy pregnancy outcome, *Journal of the American Dietetic Association* 108 (2008): 553–61.

26. Position of the American Dietetic Association: Promoting and supporting breastfeeding, *Journal of the American Dietetic Association* 109 (2009): 1926–42; and U.S. Department of Health and Human Services, *The Surgeon General's Call to Action to Support Breastfeeding* (Washington, DC: U.S. Department of Health and Human Services, 2010).

27. R. E. Kleinman, ed., Committee on Nutrition, American Academy of Pediatrics, *Pediatric Nutrition Handbook*, 6th ed. (Elk Grove, IL: American Academy of Pediatrics, 2009).

28. Position of the American Dietetic Association: Promoting and supporting breastfeeding, *Journal of the American Dietetic Association* 109 (2009): 1926–42; and R. Li, V. J. Rock, and L. Grummer-Strawn, Changes in public attitudes toward breastfeeding in the United States, 1999–2003, *Journal of the American Dietetic Association* 107 (2007): 122–27.

29. Office of Women's Health, *Breastfeeding: HHS Blueprint for Action on Breastfeeding* (Washington, DC: Department of Health and Human Services, 2000), 1–21; and U.S. Department of Health and Human Services, Office of Disease Prevention and Health Promotion, *Healthy People 2020* (Washington, DC: U.S. Department of Health and Human Services, 2010).

30. World Health Organization, *Consensus Statement: WHO HIV and Infant Feeding Technical Consultation Held on Behalf of the Inter-agency Task Team (IATT) on Prevention of HIV Infections in Pregnant Women, Mothers and Their Infants* (Geneva: WHO, 2006).

31. R. E. Kleinman, ed., Committee on Nutrition, American Academy of Pediatrics, *Pediatric Nutrition Handbook,* 6th ed. (Elk Grove, IL: American Academy of Pediatrics, 2009).

32. *Ibid.*

33. American Academy of Pediatrics Committee on Nutrition, The use and misuse of fruit juice in pediatrics, *Pediatrics* 107 (2001): 1210–13.

34. W. C. Heird, Nutritional requirements during infancy, in B. A. Bowman and R. M. Russell, *Present Knowledge in Nutrition*, 9th ed. (Washington,

DC: International Life Sciences Institute, 2006); and J. Stang, Improving the eating patterns of infants and toddlers, *Journal of the American Dietetic Association* 106 (2006): 87–89.

35. L. A. Kazal, Prevention of iron deficiency in infants and toddlers, *American Family Physician* 66 (2002): 1217–21.

36. Position of the American Dietetic Association: Dietary guidance for healthy children aged 2 to 11 years, *Journal of the American Dietetic Association* 108 (2008): 1038–47.

37. U.S. Department of Agriculture, Menu planning in the national school lunch program, *Food Program Facts* (Washington, DC: U.S. Department of Agriculture, 2007).

38. Comprehensive School Nutrition Services, Joint Position of ADA, Society for Nutrition Education and School Nutrition Association, *Journal of the American Dietetic Association* 110 (2010): 1738–49; M. Lino et al., The quality of young children's diets, *Family Economics and Nutrition Review* 14 (2002): 52–60; and Center for Nutrition Policy and Promotion, Report card on the diet quality of children ages 2 to 9, *Nutrition Insights* (September 2001).

39. U.S. Department of Health and Human Services, Office of Disease Prevention and Health Promotion, *Healthy People 2020* (Washington, DC: U.S. Department of Health and Human Services, 2010); and E. C. Wilkinson, S. J. Mickle, and J. D. Goldman, Trends in food and nutrient intakes by children in the United States, *Family Economics and Nutrition Review* 14, no. 9 (2002): 56–62.

40. Position of the American Dietetic Association: Dietary guidance for healthy children ages 2 to 11 years, *Journal of the American Dietetic Association* 108 (2008): 1038–47; and Comprehensive School Nutrition Services, Joint Position of ADA, Society for Nutrition Education and School Nutrition Association, *Journal of the American Dietetic Association* 110 (2010): 1738–49.

41. *Ibid.*; L. M. Fiorito, Dairy and dairy-related nutrient intake during middle childhood, *Journal of the American Dietetic Association* 106 (2006): 534–42; P. B. Crawford et al., How discretionary can we be with sweetened beverages for children? *Journal of the American Dietetic Association* 108 (2008): 1440–44; and M. M. Fernandes, The effect of soft drink availability in elementary schools on consumption, *Journal of the American Dietetic Association* 108 (2008): 1445–52.

42. Comprehensive School Nutrition Services, Joint Position of ADA, Society for Nutrition Education and School Nutrition Association, *Journal of the American Dietetic Association* 110 (2010): 1738–49.

43. C. L. Ogden et al., Prevalence of high body mass index in U.S. children and adolescents, 2007–2008, *Journal of the American Medical Association* 303 (2010): 242–49.

44. Centers for Disease Control and Prevention, Iron deficiency—United States, 1999–2000, *Morbidity and Mortality Weekly Report* 51 (2002): 897–99.

45. American Academy of Pediatrics, AAP issues new guidelines on cholesterol screening (July 2008), available at www.aap.org/new/july08lipidscreening.htm; and U.S. Preventive Services Task Force, Screening for lipid disorders in children: US Preventive Services Task Force recommendation statement, *Pediatrics* 120 (2007): 215–19.

46. American Heart Association Scientific Statement: Cardiovascular health in childhood, *Circulation* 106 (2002): 143–60; and American Heart Association, Guidelines for cardiovascular risk reduction for high-risk children, *Circulation* 107 (2003): 1562–66.

47. American Heart Association, Guidelines for cardiovascular risk reduction for high-risk children, *Circulation* 107 (2003): 1562–66.

48. B. A. Spear, Adolescent growth and development, *Journal of the American Dietetic Association* 102 (2002): S23–S29.

49. M. Mascarenhas et al., Adolescence, in B. A. Bowman and R. M. Russell, eds., *Present Knowledge in Nutrition*, 9th ed. (Washington, DC: International Life Sciences Institute, 2006).

50. D. Neumark-Sztainer and P. J. Hannan, Weight-related behaviors among adolescent girls and boys: Results from a national survey, *Archives of Pediatric and Adolescent Medicine* 154 (2000): 569–77; and D. Neumark-Sztainer et al., Obesity, disordered eating, and eating disorders in a longitudinal study of adolescents: How do dieters fare 5 years later? *Journal of the American Dietetic Association* 106 (2006): 559–68.

51. L. A. Lytle, Nutritional issues for adolescents, *Journal of the American Dietetic Association* 102 (2002): S8–S12; F. R. Greer and N. F. Krebs,

Optimizing bone health and calcium intakes of infants, children, and adolescents, *Pediatrics* 117 (2006): 578–85; and K. Glanz and A. Steffen, Development and reliability testing for measures of psychosocial constructs associated with adolescent girls' calcium intake, *Journal of the American Dietetic Association* 108 (2008): 857–61.

52. C. McMahan, S. Gidding, and H. McGill, Coronary heart disease risk factors and atherosclerosis in young people, *Journal of Clinical Lipidology* 2 (2008): 118–26.

53. Centers for Disease Control and Prevention, *Recommendations for Using Fluoride to Prevent and Control Dental Caries in the United States* (August 2001); Centers for Disease Control, Fact Sheet: Preventing Dental Caries (April 2010), available at www.cdc.gov/oralhealth/publications/factsheets/dental_caries.htm.

54. *Ibid.*

55. Position of the American Dietetic Association: Nutrition intervention in the treatment of eating disorders, *Journal of the American Dietetic Association* 111 (2011): 1236–41; and J. Cairns and R. Milne, Eating disorder nutrition counseling: Strategies and education needs of English-speaking dietitians in Canada, *Journal of the American Dietetic Association* 106 (2006): 1087–94.

56. Administration on Aging, *Profile of Older Americans, 2010* (Hyattsville, MD: U.S. Department of Health and Human Services, 2010).

57. D. Christensen, Making sense of centenarians, *Science News* 159 (2001): 156; and N. Wellman, Prevention, prevention, prevention: Nutrition for successful aging, *Journal of the American Dietetic Association* 107 (2007): 741–43.

58. Position Paper of the American Dietetic Association: Nutrition across the spectrum of aging, *Journal of the American Dietetic Association* 105 (2005): 616–33.

59. Administration on Aging, *Profile of Older Americans, 2010* (Hyattsville, MD: U.S. Department of Health and Human Services, 2010).

60. Position of the American Dietetic Association, the American Society for Nutrition, and the Society for Nutrition Education: Food and nutrition programs for community-residing older adults, *Journal of the American Dietetic Association* 110 (2010): 463–72.

61. AMA Conference Proceedings, Unified dietary recommendations, summary of a scientific conference on preventive nutrition: Pediatrics to geriatrics, *Circulation* 100 (1999): 450–55.

62. Position paper of the American Dietetic Association: Nutrition across the spectrum of aging, *Journal of the American Dietetic Association* 105 (2005): 616–33.

63. Administration on Aging, *Profile of Older Americans, 2010* (Hyattsville, MD: U.S. Department of Health and Human Services, 2010).

64. Food and Nutrition Board, Institute of Medicine, *Dietary Reference Intakes for Energy, Carbohydrate, Fiber, Fat, Fatty Acids, Cholesterol, Protein, and Amino Acids* (Washington, DC: National Academies Press, 2002).

65. A. H. Lichtenstein et al., Modified MyPyramid for older adults, *Journal of Nutrition* 138 (2008): 78–82; and T. Peregrin, Getting to know the modified MyPyramid for older adults, *Journal of the American Dietetic Association* 108 (2008): 937–38.

66. Position paper of the American Dietetic Association: Nutrition across the spectrum of aging, *Journal of the American Dietetic Association* 105 (2005): 616–33; and A. H. Lichtenstein et al., Modified MyPyramid for older adults, *Journal of Nutrition* 138 (2008): 78–82.

67. K. M. Fairfield and R. H. Fletcher, Vitamins for chronic disease prevention in adults, *Journal of the American Medical Association* 287 (2002): 3116–26; R. M. Russell and H. W. Baik, Clinical implications of vitamin B$_{12}$ deficiency in the elderly, *Nutrition in Clinical Care* 4 (2001): 214–20; and R. Semba et al., Vitamin D deficiency among older women with and without disability, *American Journal of Clinical Nutrition* 72 (2000): 1529.

68. L. H. Rosenberg, Challenges and opportunities in the translation of the science of vitamins, *American Journal of Clinical Nutrition* 85 (2007): 325S–27S; Position of the American Dietetic Association: Nutrient supplementation, *Journal of the American Dietetic Association* 109 (2009): 2073–85; D. R. Jacobs, M. D. Gross, and L. C. Tapsell, Food synergy: An operational concept for understanding nutrition, *American Journal of Clinical Nutrition* 89 (2009): 1543S–48S; B. B. Timbo et al., Dietary supplements in a national survey: Prevalence of use and reports of adverse events, *Journal of the American Dietetic Association* 106 (2006): 1966–74; and R. S. Sebastian et al., Older adults who use vitamin/mineral supple-

ments differ from nonusers in nutrient intake adequacy and dietary attitudes, *Journal of the American Dietetic Association* 107 (2007): 1322–32.

69. Administration on Aging, *Profile of Older Americans, 2010* (Hyattsville, MD: U.S. Department of Health and Human Services, 2010).

70. *Ibid.*

71. Position paper of the American Dietetic Association: Nutrition across the spectrum of aging, *Journal of the American Dietetic Association* 105 (2005): 616–33.

72. J. Weinberg, Psychologic implications of the nutritional needs of the elderly, *Journal of the American Dietetic Association* 60 (1972): 293–96.

73. P. G. Kittler and K. Sucher, *Food and Culture in America: A Nutrition Handbook,* 2nd ed. (Belmont, CA: Wadsworth, 1998).

74. American Institute for Cancer Research, *Cooking Solo: Cooking Ideas for One or Two That Keep Cancer Prevention in Mind* (Washington, DC: American Institute for Cancer Research, 2001), 9.

75. The following three bulleted tips were adapted from *Healthy Meals: Cooking for One or Two* (Mayo Foundation for Medical Education and Research, September 2007), available at www.mayoclinic.com.

76. N. Wellman, L. Y. Rosenzweig, and J. L. Lloyd, Thirty years of the Older American Nutrition Program, *Journal of the American Dietetic Association* 102 (2002): 348–50; Mathematica Policy Research, Inc., *Serving Elders at Risk, The Older Americans Act Nutrition Programs: National Evaluation of the Elderly Nutrition Program 1993–1995, Volume 1: Title III, Evaluation Findings* (Washington, DC: U.S. Department of Health and Human Services, 1996); and B. E. Millen et al., The Elderly Nutrition Program: An effective national framework for preventive nutrition interventions, *Journal of the American Dietetic Association* 102 (2002): 234–40.

77. The list of advantages is adapted from D. A. Roe, *Geriatric Nutrition,* 3rd ed. (Englewood Cliffs, NJ: Prentice-Hall, 1992), 1–9.

78. Institute of Medicine, *Preventing Childhood Obesity: Health in the Balance* (Washington, DC: National Academies Press, 2005); Childhood obesity in the United States: Facts and Figures (September 2004), available at www.iom.edu.

79. K. M. Flegal et al., Prevalence and trends in obesity among U.S. adults, 1999–2008, *Journal of the American Medical Association* 303 (2010): 235–41; and C. L. Ogden et al., Prevalence of high body mass index in U.S. children and adolescents, 2007–2008, *Journal of the American Medical Association* 303 (2010): 242–49.

80. American Diabetes Association, Diabetes by the Numbers 2011; www.diabetes.org/diabetes-basics/diabetes-statistics.

81. Institute of Medicine, *Preventing Childhood Obesity: Health in the Balance* (Washington, DC: National Academies Press, 2005).

82. Parts of the following discussion are from D. M. Hoelscher and C. McCullum-Gomez, Addressing the obesity epidemic, in M. Boyle and D. Holben, *Community Nutrition in Action: An Entrepreneurial Approach,* 4th ed. (Belmont, CA: Thomson Wadsworth, 2006).

83. *Ibid.*; M. P. Galvez, T. R. Frieden, and P. J. Landrigan, Obesity in the 21st century, *Environmental Health Perspective* 111 (2003): 684–85; and C. B. Ebbeling, D. B. Pawlak, and D. S. Ludwig, Childhood obesity: Public-health crisis, common sense cure, *Lancet* 360 (2002): 473–82.

84. J. T. Dwyer et al., Predictors of overweight and overfatness in a multiethnic pediatric population, *American Journal of Clinical Nutrition* 67 (1998): 602–10; and A. Batada et al., Nine out of 10 food advertisements shown during Saturday morning children's television programming are for foods high in fat, sodium, or added sugars, or low in nutrients *Journal of the American Dietetic Association* 108 (2008): 673–78; C. J. Crespo et al., Television watching, energy intake, and obesity in U.S. children: Results from the third National Health and Nutrition Examination Survey, 1988–1994, *Archives of Pediatrics and Adolescent Medicine* 155 (2001): 360–65.

85. Comprehensive School Nutrition Services, Joint Position of ADA, Society for Nutrition Education and School Nutrition Association, *Journal of the American Dietetic Association* 110 (2010): 1738–49; C. B. Ebbeling, D. B. Pawlak, and D. S. Ludwig, Childhood obesity: Public-health crisis, common sense cure, *Lancet* 360 (2002): 473–82; and G. Wang and W. H. Dietz, Economic burden of obesity in youths aged 6 to 17 years: 1979–1999, *Pediatrics* 109 (2002): E81.

86. P. M. Guenther et al., Evaluation of the Healthy Eating Index—2005, *Journal of the American Dietetic Association* 108 (2008): 1854–64; Center for Nutrition Policy and Promotion, Report card on the diet quality of

children ages 2 to 9, *Nutrition Insights* (September 2001), available at www.cnpp.usda.gov.

87. Position of the American Dietetic Association: Dietary guidance for healthy children ages 2 to 11 years, *Journal of the American Dietetic Association* 108 (2008): 1038–47.

88. S. A. Bowman et al., Effects of fast-food consumption on energy intake and diet quality among children in a national household survey, *Pediatrics* 113 (2004): 112–18, as cited in D. M. Hoelscher and C. McCullum-Gomez, Addressing the obesity epidemic, in M. Boyle and D. Holben, *Community Nutrition in Action: An Entrepreneurial Approach*, 4th ed. (Belmont, CA: Thomson Wadsworth, 2006).

89. Institute of Medicine, *Preventing Childhood Obesity: Health in the Balance* (Washington, DC: National Academies Press, 2005); K. A. Coon et al., Watching television at meals is related to food consumption patterns in children, *Pediatrics* 107 (2001): E7; and A. Batada et al., Nine out of 10 food advertisements shown during Saturday morning children's television programming are for foods high in fat, sodium, or added sugars, or low in nutrients, *Journal of the American Dietetic Association* 108 (2008): 673–78.

90. M. Golan, Parents as the exclusive agents of change in the treatment of childhood obesity, *American Journal of Clinical Nutrition* 67 (1998): 1130–35; S. St. Jeor et al., Family-based interventions for the treatment of childhood obesity, *Journal of the American Dietetic Association* 102 (2002): 640–44.

91. This discussion is adapted in part from M. Boyle and D. Holben, *Community Nutrition in Action: An Entrepreneurial Approach*, 4th ed. (Belmont, CA: Thomson Wadsworth, 2006), 367–69.

92. Comprehensive School Nutrition Services, Joint Position of ADA, Society for Nutrition Education and School Nutrition Association, *Journal of the American Dietetic Association* 110 (2010): 1738–49.

93. K. Anderson et al., *Eat Smart: North Carolina's Recommended Standards for All Foods Available in School* (North Carolina Department of Health and Human Services, North Carolina Division of Public Health, 2004); and Texas Department of Agriculture, Texas public school nutrition policy (www.squaremeals.org/fn/render/parent/channel/0,1253,2348_2350_0_0,00.html).

94. U.S. Department of Health and Human Services, *The Surgeon General's Call to Action to Prevent and Decrease Overweight and Obesity* (Washington, DC: U.S. Department of Health and Human Services, Public Health Service, Office of the Surgeon General, 2001).

95. Comprehensive School Nutrition Services, Joint Position of ADA, Society for Nutrition Education and School Nutrition Association, *Journal of the American Dietetic Association* 110 (2010): 1738–49.

96. *Ibid.*

97. T. Baranowski et al., School-based obesity prevention: A blueprint for taming the epidemic, *American Journal of Health Behavior* 26 (2002): 486–93.

98. D. M. Hoelscher et al., Designing effective interventions for adolescents, *Journal of the American Dietetic Association* 102 (2002): 552–63.

99. Institute of Medicine, *Preventing Childhood Obesity: Health in the Balance* (Washington, DC: National Academies Press, 2005).

CHAPTER 13

1. B. Bruemmer, Food biosecurity, *Journal of the American Dietetic Association* 103 (2003): 687–91.

2. Centers for Disease Control and Prevention, estimates of foodborne illness in the United States (December 2010) available at www.cdc.gov/foodborneburden/index.html.

3. J. Sneed and C. H. Strohbehn, Trends impacting food safety in retail foodservice: Implications for dietetics practice, *Journal of the American Dietetic Association* 108 (2008): 1170–77; and Position of the American Dietetic Association: Food and water safety, *Journal of the American Dietetic Association* 103 (2003): 1203–18.

4. Centers for Disease Control and Prevention, estimates of foodborne illness in the United States, 2010.

5. B. P. Bell et al., A multi-state outbreak of *Escherichia coli* O157:H7-associated bloody diarrhea and hemolytic uremic syndrome from hamburgers: The Washington experience, *Journal of the American Medical Association* 272 (1994): 1349–53; and U.S. Department of Agriculture, Food Safety and Inspection Service, E. coli O157:H7 at a glance, *Food News for Consumers* 10 (1993): 5.

6. R. E. Besser et al., An outbreak of diarrhea and hemolytic uremic syndrome from *Escherichia coli* O157:H7 in fresh-pressed apple cider, *Journal of the American Medical Association* 269 (1993): 2217–19; and R. V. Tauxe and M. Neill, Foodborne infections and food safety, in B. A. Bowman and R. M. Russell, eds., *Present Knowledge in Nutrition*, 9th ed. (Washington, DC: International Life Sciences Institute, 2006), 941–50.

7. S. C. Witt, *Biotechnology, Microbes, and the Environment* (San Francisco: Center for Science Information, 1990), 182–83.

8. L. Kopp-Hoolihan, Prophylactic and therapeutic uses of probiotics: A review, *Journal of the American Dietetic Association* 101 (2001): 229–38, 241; and G. Reid, Probiotics for urogenital health, *Nutrition in Clinical Care* 5 (2002): 3–8.

9. R. S. Horowitz et al., Jin Bu Huan toxicity in children—Colorado 1993, *Morbidity and Mortality Weekly Report* 42 (1993): 633–35.

10. U.S. Department of Health and Human Services, Public Health Service, FDA warns consumers against Nature's Nutrition Formula One (statement issued February 28, 1995); W. K. Jones, Safety of dietary supplements containing ephedrine alkaloids, *FDA Public Meeting Summary* (August 2000), available at www.cfsan.fda.gov/list.html; Food and Drug Administration, *Evidence on the Safety and Effectiveness of Ephedra: Implications for Regulation* (February 28, 2003); and Food and Drug Administration, HHS acts to reduce potential risks of dietary supplements containing ephedra, *FDA News* (February 28, 2003).

11. R. S. Koff, Herbal hepatotoxicity: Revisiting a dangerous alternative, *Journal of the American Medical Association* 273 (1995): 502; and Food and Drug Administration, Center for Food Safety and Applied Nutrition, Kava-containing dietary supplements may be associated with severe liver injury, *Consumer Advisory* (March 25, 2002).

12. Position of the American Dietetic Association: Nutrition and lifestyle for a healthy pregnancy outcome, *Journal of the American Dietetic Association* 108 (2008): 553–61.

13. N. D. Vietmeyer, The preposterous puffer, *National Geographic* 166 (1984): 260–70.

14. L. Bren, Trying to keep mad cow disease out of U.S. herds, *FDA Consumer* (March–April 2001): 12; R. McCarty, Managing bovine spongiform encephalopathy risk in the United States, *Nutrition Today* 37 (2002): 17–18; D. W. K. Acheson, Bovine spongiform encephalopathy (mad cow disease), *Nutrition Today* 37 (2002): 19–25; and Food and Drug Administration, Bovine spongiform encephalopathy (BSE) (May 2003 update), available at www.cfsan.fda.gov/list.html.

15. P. Kurtzweil, Critical steps toward safer seafood, *FDA Consumer*, Publication No. (FDA) 99-2317 (November–December 1997, revised February 1999).

16. Food Safety and Inspection Service, *Food Safety Facts: Barbecue Food Safety* (Washington, DC: U.S. Department of Agriculture, May, 2011).

17. J. Trager, *The Food Chronology* (New York: Henry Holt, 1995), as cited in *Food Folklore: Tales and Truths about What We Eat* (Minneapolis, MN: Chronimed Publishing, 1999).

18. J. F. Rosen and P. Mushak, Primary prevention of childhood lead poisoning—the only solution, *New England Journal of Medicine* 344 (2001): 1470–72.

19. International Food Information Council, *Pesticides and Food Safety* (Washington, DC: International Food Information Council Foundation, 1995), 2.

20. *Ibid.*, 2–4; C. F. Chaisson et al., *Pesticides in Food: A Guide for Professionals* (Chicago: American Dietetic Association, 1991), 4–8.

21. National Academy of Sciences, National Research Council, *Pesticides in the Diets of Infants and Children* (Washington, DC: National Academies Press, 1993); U.S. Environmental Protection Agency, Why children may be especially sensitive to pesticides (January 2003); and B. Weiss, S. Amler, and R. W. Amler, Pesticides, *Pediatrics* 113 (2004): 1030–36.

22. U.S. Department of Agriculture, *The National Organic Program: Background Information* (www.ams.usda.gov/nop).

23. *Ibid.*

24. CNN, In-Depth-Food—Organic Foods. The organic explainer (www.cnn.com/HEALTH/indepth.food/organic/explainer.html).

25. World Health Organization, *Safety and Nutritional Adequacy of Irradiated Food* (Geneva: World Health Organization, 1994), 4–5; and WHO

Technical Report Series 890, *High-Dose Irradiation: Wholesomeness of Food Irradiated with Doses above 10kGy* (Geneva: World Health Organization, 1999).

26. Position of the American Dietetic Association: Food irradiation, *Journal of the American Dietetic Association* 100 (2000): 246–53.

27. *Ibid.*; Community Nutrition Institute, Irradiation of meat begins amidst failure of safe food safety policy, *Nutrition Week* 8 (2000): 1–7.

28. J. Henkel, Genetic engineering: Fast forwarding to the future, *FDA Consumer* (April 1995): 6–11.

29. H. I. Miller, Foods of the future: The new biotechnology and FDA regulation, *Journal of the American Medical Association* 269 (1993): 910–12.

30. P. Abelson, A third technological revolution, *Science* 279 (1998): 5359; D. D. Stadler, A. F. Reeder, and C. H. Strohbehn, Application of genetic engineering to foods, *Topics in Clinical Nutrition* 14 (1999): 39–50; C. McCullum, Food biotechnology in the new millennium: Promises, realities, and challenges, *Journal of the American Dietetic Association* 100 (2000): 1311–15; B. C. Babcock and C. A. Francis, Solving nutrition challenges requires more than new biotechnologies, *Journal of the American Dietetic Association* 100 (2000): 1308–11; and Position of the American Dietetic Association: Biotechnology and the future of food, *Journal of the American Dietetic Association* 106 (2006): 285–93.

31. U.S. Department of Agriculture, Agricultural Research Service, *USDA and Biotechnology* (www.usda.gov).

32. T. Weaver, *USDA Releases New Tomatoes with Increased Beta-Carotene* (Washington, DC: USDA Agricultural Research Service, 1998).

33. A. Moffat, Toting up the harvest of transgenic plants, *Science* 282 (1998): 5397; and U.S. Department of Energy, Office of Science, Human Genome Program, *Genetically Modified Foods and Organisms* (updated March 12, 2003).

34. A. Moffat, Toting up the harvest of transgenic plants, *Science* 282 (1998): 5397; M. A. Mackey and C. R. Santerre, Biotechnology and our food supply, *Nutrition Today* 35 (2000): 120–28; M. Mackey, The application of biotechnology to nutrition: An overview, *Journal of the American College of Nutrition* 21 (2002): 157S; and M. Falk et al., Food biotechnology: Benefits and concerns, *Journal of Nutrition* 132 (2002): 1384–86.

35. J. A. Nordlee et al., Identification of a Brazil-nut allergen in transgenetic soybeans, *New England Journal of Medicine* 334 (1996): 688–92.

36. L. Thompson, Are bioengineered foods safe?, *FDA Consumer* 34 (2000): 18–23.

37. Community Nutrition Institute, Consumers Union calls for GE testing, labeling, *Nutrition Week* 34 (1999): 1–2; C. Silva and R. Leonard, U.S. prepares to OK labels on GE foods, *Nutrition Week* 45 (1999): 1–2; and Community Nutrition Institute, Biotech debate heats up in Boston: Strong grassroots movement forming, *Nutrition Week* 30 (2000): 1–3.

38. Position of the American Dietetic Association: Food insecurity in the United States, *Journal of the American Dietetic Association* 110 (2010): 1368–77; and M. Nord et al., Household food security in the United States, 2009, *Economic Research Report No. ERR-108* (Washington, DC: U.S. Department of Agriculture, Economic Research Service, 2010).

39. Food and Agriculture Organization, *The State of Food Security in the World 2010* (Rome: FAO, 2011).

40. U. Ramakrishnan, Prevalence of micronutrient malnutrition worldwide, *Nutrition Reviews* 60 (2002): S46; and U. Kapil and A. Bhavna, Adverse effects of poor micronutrient status during childhood and adolescence, *Nutrition Reviews* 60 (2002): S84.

41. Community Nutrition Institute, Malnutrition accounts for over half of child deaths worldwide, says UNICEF, *Nutrition Week* 28 (1998): 1–2; and G. H. Brundtland, Nutrition and infection: Malnutrition and mortality in public health, *Nutrition Reviews* 58, suppl. (2000): S1–S6.

42. United Nations Children's Fund, *The State of the World's Children 2001* (New York: UNICEF, 2001).

43. *Ibid.*

44. United Nations Children's Fund, *We the Children: Meeting the Promises of the World Summit for Children* (New York: UNICEF, 2001), available at www.unicef.org.

45. World Health Organization, *The Optimal Duration of Exclusive Breastfeeding: A Systematic Review* (Geneva: WHO, 2002).

46. Position of the American Dietetic Association: Food insecurity in the United States, *Journal of the American Dietetic Association* 110 (2010): 1368–77; and M. Nord et al., Household food security in the United States, 2009, *Economic Research Report No. ERR-108* (Washington, DC: U.S. Department of Agriculture, Economic Research Service, 2010).

47. United States Conference of Mayors, Hunger and Homelessness Survey: *A Status Report on Hunger and Homelessness in America's Cities* (December 2009), available at http://usmayors.org.

48. Position of the American Dietetic Association: Food insecurity in the United States, *Journal of the American Dietetic Association* 110 (2010): 1368–77.

49. Feeding America, *Hunger in America, 2010* (November 2010), available at www.feedingamerica.org.

50. P. Uvin, The state of world hunger, *Nutrition Reviews* 52 (1994): 151–61.

51. P. Foster and H. D. Leathers, *The World Food Problem: Tackling the Causes of Undernutrition in the Third World*, 2nd ed. (Boulder, CO: Lynne Rienner, 1999).

52. Bread for the World Institute, *Hunger 1998: Hunger in a Global Economy* (Silver Spring, MD: Bread for the World Institute, 1997).

53. L. R. Brown, *Who Will Feed China? Wake-Up Call for a Small Planet* (New York: W. W. Norton, 1995); G. Gardner, *Shrinking Fields: Cropland Loss in a World of Eight Billion* (Washington, DC: Worldwatch Institute, 1996); and D. D. Marino, Water and food safety in the developing world: Global implications for health and nutrition of infants and young children, *Journal of the American Dietetic Association* 107 (2007): 1930–34.

54. Oxfam America, *Facts for Action: Women Creating a New World*, No. 3 (Boston: Oxfam America, 1991), 2–3; and A. R. Quisumbing et al., *Women: The Key to Food Security*, Food Policy Report (Washington, DC: International Food Policy Research Institute, 1995).

55. Position of the American Dietetic Association: Addressing world hunger, malnutrition, and food insecurity, *Journal of the American Dietetic Association* 103 (2003).

56. United Nations Children's Fund, *We the Children: Meeting the Promises of the World Summit for Children* (New York: UNICEF, 2001), available at www.unicef.org; and United Nations Children's Fund, *The State of the World's Children 2009* (New York: Oxford University Press, 2010).

57. C. D. Williams, N. Baumslag, and D. B. Jelliffe, *Mother and Child Health: Delivering the Services* (New York: Oxford University Press, 1994).

58. United Nations Children's Fund, *We the Children: Meeting the Promises of the World Summit for Children* (New York: UNICEF, 2001).

59. S. Lewis, Food security, environment, poverty, and the world's children, *Journal of Nutrition Education* 24 (1992): 35–55.

60. United Nations Children's Fund, *We the Children: Meeting the Promises of the World Summit for Children* (New York: UNICEF, 2001); and Millennium Development Goals, 2002, in *Bread for the World Institute, Hunger 2002* (Washington, DC: Bread for the World Institute, 2002).

61. D. Elgin, *Voluntary Simplicity: Toward a Way of Life That Is Outwardly Simple, Inwardly Rich* (New York: William Morrow, 1981), 25.

62. G. McGovern, *The Third Freedom: Ending Hunger in Our Time* (New York: Simon & Schuster, 2001), 11, 14–15.

ANSWERS TO STUDY CARD QUESTIONS

CHAPTER 1
1. Carbohydrate, fat, protein, vitamins, minerals, and water.
2. D
3. D
4. A
5. D
6. A
7. D
8. D
9. A
10. A
11. B
12. D
13. A
14. A
15. B
16. B
17. B
18. A
19. A
20. C

CHAPTER 2
1. **Adequacy**, which means the diet provides the nutrients, fiber, and energy required for healthy living; **Balance**, meaning that no one food or food type is overemphasized at the expense of another; **Calorie control**, which means watching your energy intake and balancing it with the amount of energy you use; **Moderation**, which means having meals that do not contain excessive amounts of any one nutrient; **Variety**, which increases the chances of consuming a good balance of all important nutrients.
2. A
3. D
4. D
5. A
6. B
7. Individual response.
8. D
9. **Lifestyle diseases** are diseases that may result from modern lifestyles that include too little exercise, poor diets, and excessive drinking and smoking. Examples include some forms of cancer, heart disease, diabetes, osteoporosis, high blood pressure, stroke, and others.
10. A
11. A **serving** is a standard amount of food used as a reference to give advice regarding how much food to eat. A **portion** is the amount of food you choose to eat.
12. D
13. A
14. A
15. C
16. B
17. B

CHAPTER 3
1. C
2. A
3. A
4. D
5. D
6. A
7. B
8. D
9. A
10. C
11. A
12. B
13. A
14. D

CHAPTER 4
1. A
2. B
3. A
4. A
5. A
6. A
7. E
8. B
9. A
10. A
11. D
12. A
13. B
14. B
15. B

CHAPTER 5
1. Fats provide a concentrated source of energy, serve as an energy reserve, form the major components of cell membranes, nourish skin and hair, insulate the body from extremes of temperature, and cushion the vital organs to protect them from shock.
2. Fats in foods provide calories, provide satiety, carry fat-soluble vitamins and essential fatty acids, and contribute aroma and flavor.
3. C
4. A
5. E
6. C
7. A
8. C
9. B
10. High LDL-cholesterol; low HDL-cholesterol; high blood pressure; cigarette smoking; diabetes; physical inactivity; obesity; "atherogenic" diet (high in saturated fats); stress; age; gender; genetics.
11. A
12. B

13. A
14. E
15. A
16. Visible fats include fats and oils like butter, the oil in salad dressing, and the fat you trim from a steak. Hidden fats include the fat that marbles a steak, or fats that are hidden in foods like nuts, cheese, biscuits, crackers, doughnuts, cookies, muffins, avocados, olives, fried foods, and chocolate.
17. E
18. E

CHAPTER 6
1. B
2. C
3. A
4. C
5. C
6. A
7. A
8. C
9. C
10. E
11. C
12. D
13. B
14. B
15. C
16. D
17. E
18. D
19. D

CHAPTER 7
1. Essay.
2. D
3. A
4. C
5. C
6. B
7. D
8. A
9. A
10. A
11. E
12. B
13. A
14. B
15. A
16. B
17. B

CHAPTER 8
1. D
2. E
3. B
4. B
5. A
6. B
7. A
8. C
9. D
10. D
11. Sulfur, iodine, and chromium.

12. C
13. E
14. A
15. A
16. E
17. A

CHAPTER 9
1. B
2. B
3. A
4. E
5. C
6. B
7. B
8. A
9. D
10. C
11. B
12. A
13. A
14. A
15. A
16. A
17. B

CHAPTER 10
1. Food portion sizes have increased; fast food and highly processed foods are more prevalent; vending machines sell high-fat, high-sugar foods; adults spend more time in sedentary activities; children spend more time watching TV and playing video games; schools offer fewer physical education classes; communities are more car-dependent.
2. Overweight increases the risk of health issues like high blood pressure, heart disease, high blood cholesterol, diabetes, some types of cancer, liver and gall bladder disease.
3. E
4. Underwater weighing.
5. A
6. B
7. C
8. C
9. B
10. A
11. C
12. D
13. C
14. B
15. A
16. Add high-calorie, nutrient-dense snacks to the diet like peanut butter on whole wheat crackers with skim milk; eat calorie-dense foods at meals; eat more frequently; spend more time eating each meal.
17. A

CHAPTER 11
1. Increased self-confidence; easier weight control; more energy; less stress and anxiety; improved sleep; enhanced immunity; lowered risk of heart disease; lowered risk of certain cancers; stronger bones; lowered risk of diabetes; lowered risk of high blood pressure; increased quality of life; increased independence in life's later years.
2. Sedentary lifestyle; age; family history; cigarette smoking; high blood pressure; high blood cholesterol.

3. A
4. E
5. B
6. C
7. D
8. D
9. A
10. A
11. C
12. B
13. D
14. B
15. C

CHAPTER 12
1. D
2. C
3. D
4. A
5. A
6. D
7. A
8. B
9. D

10. A
11. Undernutrition; iron-deficiency anemia; low calcium intakes; high blood cholesterol; dental caries; eating disorders; obesity.
12. C
13. B
14. B
15. B

CHAPTER 13
1. C
2. A
3. B
4. E
5. E
6. A
7. C
8. E
9. C
10. B
11. B
12. B
13. B
14. D
15. D

TABLE OF FOOD COMPOSITION

This edition of the table of food composition has been updated to reflect current nutrient data for foods, to remove outdated foods, and to add foods that are new to the marketplace.* The nutrient database for this appendix is compiled from a variety of sources, including the current USDA Standard Reference database and manufacturers' data. The USDA database provides data for a wider variety of foods and nutrients than other sources. Because laboratory analysis for each nutrient can be quite costly, manufacturers tend to provide data only for those nutrients mandated on food labels. Consequently, data for their foods are often incomplete; any missing information on this table is designated as a dash. Keep in mind that a dash means only that the information is unknown and should not be interpreted as a zero. A zero means that the nutrient is not present in the food.

Whenever using nutrient data, remember that many factors influence the nutrient contents of foods. These factors include the mineral content of the soil, the diet fed to the animal or the fertilizer used on the plant, the season of harvest, the method of processing, the length and method of storage, the method of cooking, the method of analysis, and the moisture content of the sample analyzed. With so many influencing factors, users should view nutrient data as a close approximation of the actual amount.

For updates, corrections, and a list of 8,000 foods and codes found in the diet analysis software that accompanies this text, visit **www.cengage.com/nutrition** and click on Diet Analysis+.

- **Fats**   Total fats, as well as the breakdown of total fats to saturated, monounsaturated, and polyunsaturated fats, are listed in the table. The fatty acids seldom add up to the total in part due to rounding but also because values are derived from a variety of laboratories and other fatty acid components such as glycerol are not included in these basic categories.

- **Trans *Fats***   *Trans* fat data are listed in the table. Because food manufacturers have been required to report *trans* fats on food labels only since January 1, 2006, much of the data is incomplete. Missing *trans* fat data is designated with a dash.

- **Vitamin A and Vitamin E**   In keeping with the RDA for vitamin A, this appendix presents data for vitamin A in micrograms (μg) RAE. Similarly, because the RDA for vitamin E is based only on the alpha-tocopherol form of vitamin E, this appendix reports vitamin E data in milligrams (mg) alpha-tocopherol, listed on the table as Vit E (mg α).

- **Bioavailability**   Keep in mind that the availability of nutrients from foods depends not only on the quantity provided by a food, but also on the amount absorbed and used by the body—the bioavailability. The bioavailability of folate from fortified foods, for example, is greater than from naturally occurring sources. Similarly, the body can make niacin from the amino acid tryptophan, but niacin values in this table (and most databases) report preformed niacin only.

- **Using the Table**   The foods and beverages in this table have been organized into several categories, which are listed at the head of each right-hand page. Page numbers have been provided, and each group has been color-coded to make it easier to find individual foods.

---

* This food composition table has been prepared by Wadsworth Publishing Company. The nutritional data are supplied by Axxya Systems.

Food Composition

(Computer code is for Cengage Diet Analysis program)   (For purposes of calculations, use "0" for t, <1, <.1, <.01, etc.)

DA+ Code	Food Description	Quantity	Measure	Wt (g)	H₂O (g)	Ener (kcal)	Prot (g)	Carb (g)	Fiber (g)	Fat (g)	Fat Breakdown (g)			
											Sat	Mono	Poly	Trans
	BREADS, BAKED GOODS, CAKES, COOKIES, CRACKERS, CHIPS, PIES													
	Bagels													
8534	Cinnamon and raisin	1	item(s)	71	22.7	194	7.0	39.2	1.6	1.2	0.2	0.1	0.5	—
14395	Multi-grain	1	item(s)	61	—	170	6.0	35.0	1.0	1.5	0.5	0.1	0.4	—
8538	Oat bran	1	item(s)	71	23.4	181	7.6	37.8	2.6	0.9	0.1	0.2	0.3	—
4910	Plain, enriched	1	item(s)	71	25.8	182	7.1	35.9	1.6	1.2	0.3	0.4	0.5	0
4911	Plain, enriched, toasted	1	item(s)	66	18.7	190	7.4	37.7	1.7	1.1	0.2	0.3	0.6	0
	Biscuits													
25008	Biscuits	1	item(s)	41	15.8	121	2.6	16.4	0.5	4.9	1.4	1.4	1.8	—
16729	Scone	1	item(s)	42	11.5	148	3.8	19.1	0.6	6.2	2.0	2.5	1.3	—
25166	Wheat biscuits	1	item(s)	55	21.0	162	3.6	21.9	1.4	6.7	1.9	1.9	2.5	—
	Bread													
325	Boston brown, canned	1	slice(s)	45	21.2	88	2.3	19.5	2.1	0.7	0.1	0.1	0.3	—
8716	Bread sticks, plain	4	item(s)	24	1.5	99	2.9	16.4	0.7	2.3	0.3	0.9	0.9	—
25176	Cornbread	1	piece(s)	55	25.9	141	4.7	18.3	0.9	5.4	2.1	1.4	1.5	0
327	Cracked wheat	1	slice(s)	25	9.0	65	2.2	12.4	1.4	1.0	0.2	0.5	0.2	—
9079	Croutons, plain	¼	cup(s)	8	0.4	31	0.9	5.5	0.4	0.5	0.1	0.2	0.1	—
8582	Egg	1	slice(s)	40	13.9	113	3.8	19.1	0.9	2.4	0.6	0.9	0.4	—
8585	Egg, toasted	1	slice(s)	37	10.5	117	3.9	19.5	0.9	2.4	0.6	1.1	0.4	—
329	French	1	slice(s)	32	8.9	92	3.8	18.1	0.8	0.6	0.2	0.1	0.3	—
8591	French, toasted	1	slice(s)	23	4.7	73	3.0	14.2	0.7	0.5	0.1	0.1	0.2	—
42096	Indian fry, made with lard (Navajo)	3	ounce(s)	85	26.9	281	5.7	41.0	—	10.4	3.9	3.8	0.9	—
332	Italian	1	slice(s)	30	10.7	81	2.6	15.0	0.8	1.1	0.3	0.2	0.4	—
1393	Mixed grain	1	slice(s)	26	9.6	69	3.5	11.3	1.9	1.1	0.2	0.2	0.5	0
8604	Mixed grain, toasted	1	slice(s)	24	7.6	69	3.5	11.3	1.9	1.1	0.2	0.2	0.5	0
8605	Oat bran	1	slice(s)	30	13.2	71	3.1	11.9	1.4	1.3	0.2	0.5	0.5	—
8608	Oat bran, toasted	1	slice(s)	27	10.4	70	3.1	11.8	1.3	1.3	0.2	0.5	0.5	—
8609	Oatmeal	1	slice(s)	27	9.9	73	2.3	13.1	1.1	1.2	0.2	0.4	0.5	—
8613	Oatmeal, toasted	1	slice(s)	25	7.8	73	2.3	13.2	1.1	1.2	0.2	0.4	0.5	—
1409	Pita	1	item(s)	60	19.3	165	5.5	33.4	1.3	0.7	0.1	0.1	0.3	—
7905	Pita, whole wheat	1	item(s)	64	19.6	170	6.3	35.2	4.7	1.7	0.3	0.2	0.7	—
338	Pumpernickel	1	slice(s)	32	12.1	80	2.8	15.2	2.1	1.0	0.1	0.3	0.4	—
334	Raisin, enriched	1	slice(s)	26	8.7	71	2.1	13.6	1.1	1.1	0.3	0.6	0.2	—
8625	Raisin, toasted	1	slice(s)	24	6.7	71	2.1	13.7	1.1	1.2	0.3	0.6	0.2	—
10168	Rice, white, gluten free, wheat free	1	slice(s)	38	—	130	1.0	18.0	0.5	6.0	0	—	—	0
8653	Rye	1	slice(s)	32	11.9	83	2.7	15.5	1.9	1.1	0.2	0.4	0.3	—
8654	Rye, toasted	1	slice(s)	29	9.0	82	2.7	15.4	1.9	1.0	0.2	0.4	0.3	—
336	Rye, light	1	slice(s)	25	9.3	65	2.0	12.0	1.6	1.0	0.2	0.3	0.3	—
8588	Sourdough	1	slice(s)	25	7.0	72	2.9	14.1	0.6	0.5	0.1	0.1	0.2	—
8592	Sourdough, toasted	1	slice(s)	23	4.7	73	3.0	14.2	0.7	0.5	0.1	0.1	0.2	—
491	Submarine or hoagie roll	1	item(s)	135	40.6	400	11.0	72.0	3.8	8.0	1.8	3.0	2.2	—
8596	Vienna, toasted	1	slice(s)	23	4.7	73	3.0	14.2	0.7	0.5	0.1	0.1	0.2	—
8670	Wheat	1	slice(s)	25	8.9	67	2.7	11.9	0.9	0.9	0.2	0.2	0.4	—
8671	Wheat, toasted	1	slice(s)	23	5.6	72	3.0	12.8	1.1	1.0	0.2	0.2	0.4	—
340	White	1	slice(s)	25	9.1	67	1.9	12.7	0.6	0.8	0.2	0.2	0.3	—
1395	Whole wheat	1	slice(s)	46	15.0	128	3.9	23.6	2.8	2.5	0.4	0.5	1.4	—
	Cakes													
386	Angel food, prepared from mix	1	piece(s)	50	16.5	129	3.1	29.4	0.1	0.2	0	0	0.1	—
8772	Butter pound, ready to eat, commercially prepared	1	slice(s)	75	18.5	291	4.1	36.6	0.4	14.9	8.7	4.4	0.8	—
28517	Carrot	1	slice(s)	131	56.6	339	4.8	56.5	1.9	11.1	1.0	5.7	3.8	—
4931	Chocolate with chocolate icing, commercially prepared	1	slice(s)	64	14.7	235	2.6	34.9	1.8	10.5	3.1	5.6	1.2	—
8756	Chocolate, prepared from mix	1	slice(s)	95	23.2	352	5.0	50.7	1.5	14.3	5.2	5.7	2.6	—
393	Devil's food cupcake with chocolate frosting	1	item(s)	35	8.4	120	2.0	20.0	0.7	4.0	1.8	1.6	0.6	—
8757	Fruitcake, ready to eat, commercially prepared	1	piece(s)	43	10.9	139	1.2	26.5	1.6	3.9	0.5	1.8	1.4	—
1397	Pineapple upside down, prepared from mix	1	slice(s)	115	37.1	367	4.0	58.1	0.9	13.9	3.4	6.0	3.8	—
411	Sponge, prepared from mix	1	slice(s)	63	18.5	187	4.6	36.4	0.3	2.7	0.8	1.0	0.4	—
8817	White with coconut frosting, prepared from mix	1	slice(s)	112	23.2	399	4.9	70.8	1.1	11.5	4.4	4.1	2.4	—
8819	Yellow with chocolate frosting, ready to eat, commercially prepared	1	slice(s)	64	14.0	243	2.4	35.5	1.2	11.1	3.0	6.1	1.4	—
8822	Yellow with vanilla frosting, ready to eat, commercially prepared	1	slice(s)	64	14.1	239	2.2	37.6	0.2	9.3	1.5	3.9	3.3	—
	Snack cakes													
8791	Chocolate snack cake, creme filled, with frosting	1	item(s)	50	9.3	200	1.8	30.2	1.6	8.0	2.4	4.3	0.9	—
25010	Cinnamon coffee cake	1	piece(s)	72	22.6	231	3.6	35.8	0.7	8.3	2.2	2.6	3.0	—
16777	Funnel cake	1	item(s)	90	37.6	276	7.3	29.1	0.9	14.4	2.7	4.7	6.1	—
8794	Sponge snack cake, creme filled	1	item(s)	43	8.6	155	1.3	27.2	0.2	4.8	1.1	1.7	1.4	—
	Snacks, chips, pretzels													
29428	Bagel chips, plain	3	item(s)	29	—	130	3.0	19.0	1.0	4.5	0.5	—	—	—
29429	Bagel chips, toasted onion	3	item(s)	29	—	130	4.0	20.0	1.0	4.5	0.5	—	—	—

Chol (mg)	Calc (mg)	Iron (mg)	Magn (mg)	Pota (mg)	Sodi (mg)	Zinc (mg)	Vit A (µg)	Thia (mg)	Vit E (mg α)	Ribo (mg)	Niac (mg)	Vit B6 (mg)	Fola (µg)	Vit C (mg)	Vit B12 (µg)	Sele (µg)
0	13	2.69	19.9	105.1	228.6	0.80	14.9	0.27	0.22	0.19	2.18	0.04	78.8	0.5	0	22.0
0	60	1.08	—	—	310.0	—	0	—	—	—	—	—	—	0	—	—
0	9	2.18	22.0	81.7	360.0	0.63	0.7	0.23	0.23	0.24	2.10	0.03	69.6	0.1	0	24.3
0	63	4.29	15.6	53.3	318.1	1.34	0	0.42	0.07	0.18	2.82	0.04	103.0	0.7	0	16.2
0	65	2.97	15.8	56.1	316.8	0.85	0	0.39	0.07	0.17	2.88	0.04	86.5	0	0	16.6
0	38	0.94	6.0	47.4	206.0	0.20	—	0.16	0.01	0.12	1.20	0.01	31.7	0.1	0.1	7.1
49	79	1.35	7.1	48.7	277.2	0.29	64.7	0.14	0.42	0.15	1.19	0.02	32.3	0	0.1	10.9
0	57	1.21	16.1	81.0	321.1	0.42	—	0.19	0.01	0.14	1.65	0.03	35.3	0.1	0.1	0
0	32	0.94	28.4	143.1	284.0	0.22	11.3	0.01	0.14	0.05	0.50	0.03	5.0	0	0	9.9
0	5	1.02	7.7	29.8	157.7	0.21	0	0.14	0.24	0.13	1.26	0.01	38.9	0	0	9.0
21	94	0.91	10.5	71.5	209.8	0.48	—	0.14	0.32	0.15	1.03	0.04	34.6	1.7	0.2	6.2
0	11	0.70	13.0	44.3	134.5	0.31	0	0.09	—	0.06	0.92	0.08	15.3	0	0	6.3
0	6	0.30	2.3	9.3	52.4	0.06	0	0.04	—	0.02	0.40	0.00	9.9	0	0	2.8
20	37	1.21	7.6	46.0	196.8	0.31	25.2	0.17	0.10	0.17	1.93	0.02	42.0	0	0	12.0
21	38	1.23	7.8	46.6	199.8	0.31	25.5	0.14	0.10	0.16	1.77	0.02	36.3	0	0	12.2
0	14	1.16	9.0	41.0	208.0	0.29	0	0.13	0.05	0.09	1.52	0.03	47.4	0.1	0	8.7
0	11	0.89	7.1	32.2	165.6	0.24	0	0.10	0.04	0.09	1.24	0.02	32.2	0	0	6.8
6	48	3.43	15.3	65.5	279.8	0.29	0	0.36	0.00	0.18	3.91	0.03	103.8	—	0	15.8
0	23	0.88	8.1	33.0	175.2	0.25	0	0.14	0.08	0.08	1.31	0.01	57.3	0	0	8.2
0	27	0.65	20.3	59.8	109.2	0.44	0	0.07	0.09	0.03	1.05	0.06	19.5	0	0	8.6
0	27	0.65	20.4	60.0	109.7	0.44	0	0.06	0.10	0.03	1.05	0.07	16.8	0	0	8.6
0	20	0.93	10.5	44.1	122.1	0.26	0.6	0.15	0.13	0.10	1.44	0.02	24.3	0	0	9.0
0	19	0.92	9.2	33.2	121.0	0.28	0.5	0.12	0.13	0.09	1.29	0.01	18.6	0	0	8.9
0	18	0.72	10.0	38.3	161.7	0.27	1.4	0.10	0.13	0.06	0.84	0.01	16.7	0	0	6.6
0	18	0.74	10.3	38.5	162.8	0.28	1.3	0.09	0.13	0.06	0.77	0.02	13.3	0.1	0	6.7
0	52	1.57	15.6	72.0	321.6	0.50	0	0.35	0.18	0.19	2.77	0.02	64.2	0	0	16.3
0	10	1.95	44.2	108.8	340.5	0.97	0	0.21	0.39	0.05	1.81	0.17	22.4	0	0	28.2
0	22	0.91	17.3	66.6	214.7	0.47	0	0.10	0.13	0.09	0.98	0.04	29.8	0	0	7.8
0	17	0.75	6.8	59.0	101.4	0.18	0	0.08	0.07	0.10	0.90	0.01	27.6	0	0	5.2
0	17	0.76	6.7	59.0	101.8	0.19	0	0.07	0.07	0.09	0.81	0.02	23.5	0.1	0	5.2
0	100	1.08	—	—	140	—	—	0.15	—	0.10	1.20	—	32.0	0	—	—
0	23	0.90	12.8	53.1	211.2	0.36	0	0.13	0.10	0.10	1.21	0.02	35.2	0.1	0	9.9
0	23	0.89	12.5	53.1	210.3	0.36	0	0.11	0.10	0.09	1.09	0.02	29.9	0.1	0	9.9
0	20	0.70	3.9	51.0	175.0	0.18	0	0.10	—	0.08	0.80	0.01	5.3	0	0	8.0
0	11	0.91	7.0	32.0	162.5	0.23	0	0.11	0.05	0.07	1.19	0.03	37.0	0.1	0	6.8
0	11	0.89	7.1	32.2	165.6	0.24	0	0.10	0.04	0.09	1.24	0.02	32.2	0	0	6.8
0	100	3.80	—	128.0	683.0	—	0	0.54	—	0:33	4.50	0.04	—	0	—	42.0
0	11	0.89	7.1	32.2	165.6	0.24	0	0.10	0.04	0.09	1.24	0.02	32.2	0	0	6.8
0	36	0.87	12.0	46.0	130.3	0.30	0	0.09	0.05	0.08	1.30	0.03	21.3	0	0	7.2
0	38	0.94	13.6	51.3	140.5	0.34	0	0.10	0.06	0.09	1.44	0.04	19.8	0	0	7.7
0	38	0.94	5.8	25.0	170.3	0.19	0	0.11	0.06	0.08	1.10	0.02	27.8	0	0	4.3
0	15	1.42	37.3	144.4	159.2	0.69	0	0.13	0.35	0.10	1.83	0.09	29.9	0	0	17.8
0	42	0.11	4.0	67.5	254.5	0.06	0	0.04	0.01	0.10	0.08	0.00	9.5	0	0	7.7
166	26	1.03	8.3	89.3	298.5	0.34	111.8	0.10	—	0.17	0.98	0.03	30.8	0	0.2	6.6
0	65	2.18	23.0	279.6	367.7	0.44	—	0.25	0.01	0.19	1.73	0.10	43.4	4.6	0	14.7
27	28	1.40	21.8	128.0	213.8	0.44	16.6	0.01	0.62	0.08	0.36	0.02	10.9	0.1	0.1	2.1
55	57	1.53	30.4	133.0	299.3	0.65	38.0	0.13	—	0.20	1.08	0.03	25.7	0.2	0.2	11.3
19	21	0.70	—	46.0	92.0	—	—	0.04	—	0.05	0.30	—	2.1	0	—	2.0
2	14	0.89	6.9	65.8	116.1	0.11	3.0	0.02	0.38	0.04	0.34	0.02	8.6	0.2	0	0.9
25	138	1.70	15.0	128.8	366.9	0.35	71.3	0.17	—	0.17	1.36	0.03	29.9	1.4	0.1	10.8
107	26	0.99	5.7	88.8	143.6	0.37	48.5	0.10	—	0.19	0.75	0.03	24.6	0	0.1	11.7
1	101	1.29	13.4	110.9	318.1	0.37	13.4	0.14	0.13	0.21	1.19	0.03	34.7	0.1	0.1	12.0
35	24	1.33	19.2	113.9	215.7	0.39	21.1	0.07	—	0.10	0.79	0.02	14.1	0	0.1	2.2
35	40	0.68	3.8	33.9	220.2	0.16	12.2	0.06	—	0.04	0.32	0.01	17.3	0	0.1	3.5
0	58	1.80	18.0	88.0	194.5	0.52	0.5	0.01	0.54	0.03	0.46	0.07	13.0	1.0	0	1.7
26	55	1.36	9.9	91.9	277.6	0.30	—	0.17	0.23	0.16	1.29	0.02	36.1	0.3	0.1	9.6
62	126	1.90	16.2	152.1	269.1	0.65	49.5	0.23	1.54	0.32	1.86	0.04	50.4	0	0.3	17.7
7	19	0.54	3.4	37.0	155.1	0.12	2.1	0.06	0.50	0.05	0.52	0.01	17.0	0	0	1.3
0	0	0.72	—	45.0	70.0	—	0	—	—	—	—	—	—	0	0	—
0	0	0.72	—	50.0	300.0	—	0	—	—	—	—	—	—	0	0	—

DA+ Code	Food Description	Quantity	Measure	Wt (g)	H₂O (g)	Ener (kcal)	Prot (g)	Carb (g)	Fiber (g)	Fat (g)	Fat Breakdown (g)			
											Sat	Mono	Poly	Trans
BREADS, BAKED GOODS, CAKES, COOKIES, CRACKERS, CHIPS, PIES —Continued														
38192	Chex traditional snack mix	1	cup(s)	45	—	197	3.0	33.3	1.5	6.1	0.8	—	—	—
654	Potato chips, salted	1	ounce(s)	28	0.6	155	1.9	14.1	1.2	10.6	3.1	2.8	3.5	—
8816	Potato chips, unsalted	1	ounce(s)	28	0.5	152	2.0	15.0	1.4	9.8	3.1	2.8	3.5	—
5096	Pretzels, plain, hard, twists	5	item(s)	30	1.0	114	2.7	23.8	1.0	1.1	0.2	0.4	0.4	—
4632	Pretzels, whole wheat	1	ounce(s)	28	1.1	103	3.1	23.0	2.2	0.7	0.2	0.3	0.2	—
4641	Tortilla chips, plain	6	item(s)	11	0.2	53	0.8	7.1	0.6	2.5	0.3	0.8	0.5	0.3
	Cookies													
8859	Animal crackers	12	item(s)	30	1.2	134	2.1	22.2	0.3	4.1	1.0	2.3	0.6	—
8876	Brownie, prepared from mix	1	item(s)	24	3.0	112	1.5	12.0	0.5	7.0	1.8	2.6	2.3	—
25207	Chocolate chip cookies	1	item(s)	30	3.7	140	2.0	16.2	0.6	7.9	2.1	3.3	2.1	—
8915	Chocolate sandwich cookie with extra creme filling	1	item(s)	13	0.2	65	0.6	8.9	0.4	3.2	0.7	2.1	0.3	1.1
14145	Fig Newtons cookies	1	item(s)	16	—	55	0.5	11.0	0.5	1.3	0	—	—	0
8920	Fortune cookie	1	item(s)	8	0.6	30	0.3	6.7	0.1	0.2	0.1	0.1	0	—
25208	Oatmeal cookies	1	item(s)	69	12.3	234	5.7	45.1	3.1	4.2	0.7	1.3	1.8	—
25213	Peanut butter cookies	1	item(s)	35	4.1	163	4.2	16.9	0.9	9.2	1.7	4.7	2.3	—
33095	Sugar cookies	1	item(s)	16	4.1	61	1.1	7.4	0.1	3.0	0.6	1.3	0.9	—
9002	Vanilla sandwich cookie with creme filling	1	item(s)	10	0.2	48	0.5	7.2	0.2	2.0	0.3	0.8	0.8	—
	Crackers													
9012	Cheese cracker sandwich with peanut butter	4	item(s)	28	0.9	139	3.5	15.9	1.0	7.0	1.2	3.6	1.4	—
9008	Cheese crackers (mini)	30	item(s)	30	0.9	151	3.0	17.5	0.7	7.6	2.8	3.6	0.7	—
33362	Cheese crackers, low sodium	1	serving(s)	30	0.9	151	3.0	17.5	0.7	7.6	2.9	3.6	0.7	—
8928	Honey graham crackers	4	item(s)	28	1.2	118	1.9	21.5	0.8	2.8	0.4	1.1	1.1	—
9016	Matzo crackers, plain	1	item(s)	28	1.2	112	2.8	23.8	0.9	0.4	0.1	0	0.2	—
9024	Melba toast	3	item(s)	15	0.8	59	1.8	11.5	0.9	0.5	0.1	0.1	0.2	—
9028	Melba toast, rye	3	item(s)	15	0.7	58	1.7	11.6	1.2	0.5	0.1	0.1	0.2	—
14189	Ritz crackers	5	item(s)	16	0.5	80	1.0	10.0	0	4.0	1.0	—	—	0
9014	Rye crispbread crackers	1	item(s)	10	0.6	37	0.8	8.2	1.7	0.1	0	0	0.1	—
9040	Rye wafer	1	item(s)	11	0.6	37	1.1	8.8	2.5	0.1	0	0	0	—
432	Saltine crackers	5	item(s)	15	0.8	64	1.4	10.6	0.5	1.7	0.2	1.1	0.2	0.5
9046	Saltine crackers, low salt	5	item(s)	15	0.6	65	1.4	10.7	0.5	1.8	0.4	1.0	0.3	—
9052	Snack cracker sandwich with cheese filling	4	item(s)	28	1.1	134	2.6	17.3	0.5	5.9	1.7	3.2	0.7	—
9054	Snack cracker sandwich with peanut butter filling	4	item(s)	28	0.8	138	3.2	16.3	0.6	6.9	1.4	3.9	1.3	—
9048	Snack crackers, round	10	item(s)	30	1.1	151	2.2	18.3	0.5	7.6	1.1	3.2	2.9	—
9050	Snack crackers, round, low salt	10	item(s)	30	1.1	151	2.2	18.3	0.5	7.6	1.1	3.2	2.9	—
9044	Soda crackers	5	tem(s)	15	0.8	64	1.4	10.6	0.5	1.7	0.2	1.1	0.2	0.5
9059	Wheat cracker sandwich with cheese filling	4	item(s)	28	0.9	139	2.7	16.3	0.9	7.0	1.2	2.9	2.6	—
9061	Wheat cracker sandwich with peanut butter filling	4	item(s)	28	1.0	139	3.8	15.1	1.2	7.5	1.3	3.3	2.5	—
9055	Wheat crackers	10	item(s)	30	0.9	142	2.6	19.5	1.4	6.2	1.6	3.4	0.8	—
9057	Wheat crackers, low salt	10	item(s)	30	0.9	142	2.6	19.5	1.4	6.2	1.6	3.4	0.8	—
9022	Whole wheat crackers	7	item(s)	28	0.8	124	2.5	19.2	2.9	4.8	1.0	1.6	1.8	—
	Pastry													
16754	Apple fritter	1	item(s)	17	6.4	61	1.0	5.5	0.2	3.9	0.9	1.7	1.1	—
41565	Cinnamon rolls with icing, refrigerated dough	1	serving(s)	44	12.3	145	2.0	23.0	0.5	5.0	1.5	—	—	2.0
4945	Croissant, butter	1	item(s)	57	13.2	231	4.7	26.1	1.5	12.0	6.6	3.1	0.6	—
9096	Danish, nut	1	item(s)	65	13.3	280	4.6	29.7	1.3	16.4	3.8	8.9	2.8	—
9115	Doughnut with creme filling	1	item(s)	85	32.5	307	5.4	25.5	0.7	20.8	4.6	10.3	2.6	—
9117	Doughnut with jelly filling	1	item(s)	85	30.3	289	5.0	33.2	0.8	15.9	4.1	8.7	2.0	—
4947	Doughnut, cake	1	item(s)	47	9.8	198	2.4	23.4	0.7	10.8	1.7	4.4	3.7	—
9105	Doughnut, cake, chocolate glazed	1	item(s)	42	6.8	175	1.9	24.1	0.9	8.4	2.2	4.7	1.0	—
437	Doughnut, glazed	1	item(s)	60	15.2	242	3.8	26.6	0.7	13.7	3.5	7.7	1.7	—
10617	Toaster pastry, brown sugar cinnamon	1	item(s)	50	5.3	210	3.0	35.0	1.0	6.0	1.0	4.0	1.0	—
30928	Toaster pastry, cream cheese	1	item(s)	54	—	200	3.0	23.0	0	11.0	4.5	—	—	1.5
	Muffins													
25015	Blueberry	1	item(s)	63	29.7	160	3.4	23.0	0.8	6.0	0.9	1.5	3.3	—
9189	Corn, ready to eat	1	item(s)	57	18.6	174	3.4	29.0	1.9	4.8	0.8	1.2	1.8	—
9121	English muffin, plain, enriched	1	item(s)	57	24.0	134	4.4	26.2	1.5	1.0	0.1	0.2	0.5	—
29582	English muffin, toasted	1	item(s)	50	18.6	128	4.2	25.0	1.5	1.0	0.1	0.2	0.5	—
9145	English muffin, wheat	1	item(s)	57	24.1	127	5.0	25.5	2.6	1.1	0.2	0.2	0.5	—
8894	Oat bran	1	item(s)	57	20.0	154	4.0	27.5	2.6	4.2	0.6	1.0	2.4	—
	Granola bars													
38161	Kudos milk chocolate granola bars w/fruit & nuts	1	item(s)	28	—	90	2.0	15.0	1.0	3.0	1.0	—	—	—
38196	Nature Valley banana nut crunchy granola bars	2	item(s)	42	—	190	4.0	28.0	2.0	7.0	1.0	—	—	—
38187	Nature Valley fruit 'n' nut trail mix bar	1	item(s)	35	—	140	3.0	25.0	2.0	4.0	0.5	—	—	—
1383	Plain, hard	1	item(s)	25	1.0	115	2.5	15.8	1.3	4.9	0.6	1.1	3.0	—
4606	Plain, soft	1	item(s)	28	1.8	126	2.1	19.1	1.3	4.9	2.1	1.1	1.5	—
	Pies													
454	Apple pie, prepared from home recipe	1	slice(s)	155	73.3	411	3.7	57.5	2.3	19.4	4.7	8.4	5.2	—
470	Pecan pie, prepared from home recipe	1	slice(s)	122	23.8	503	6.0	63.7	—	27.1	4.9	13.6	7.0	—

Chol (mg)	Calc (mg)	Iron (mg)	Magn (mg)	Pota (mg)	Sodi (mg)	Zinc (mg)	Vit A (µg)	Thia (mg)	Vit E (mg α)	Ribo (mg)	Niac (mg)	Vit B$_6$ (mg)	Fola (µg)	Vit C (mg)	Vit B$_{12}$ (µg)	Sele (µg)
0	0	0.55	—	75.8	621.2	—	0	0.09	—	0.05	1.21	—	12.1	0	—	—
0	7	0.45	19.8	465.5	148.8	0.67	0	0.01	1.91	0.06	1.18	0.20	21.3	5.3	0	2.3
0	7	0.46	19.0	361.5	2.3	0.30	0	0.04	2.58	0.05	1.08	0.18	12.8	8.8	0	2.3
0	11	1.29	10.5	43.8	514.5	0.25	0	0.13	0.10	0.18	1.57	0.03	51.3	0	0	1.7
0	8	0.76	8.5	121.9	57.6	0.17	0	0.12	—	0.08	1.85	0.07	15.3	0.3	0	—
0	19	0.25	15.8	23.2	45.5	0.26	0	0.00	0.46	0.01	0.13	0.02	2.2	0	0	0.7
0	13	0.82	5.4	30.0	117.9	0.19	0	0.10	0.03	0.09	1.04	0.01	30.9	0	0	2.1
18	14	0.44	12.7	42.2	82.3	0.23	42.2	0.03	—	0.05	0.24	0.02	7.0	0.1	0	2.8
13	11	0.69	12.4	62.1	108.8	0.24	—	0.08	0.54	0.06	0.87	0.01	17.8	0	0	4.1
0	2	1.01	4.7	17.8	45.6	0.10	0	0.02	0.25	0.02	0.25	0.00	6.0	0	0	1.1
0	10	0.36	—	—	57.5	—	0	—	0.00	0.01	—	—	—	0	—	—
0	1	0.12	0.6	3.3	21.9	0.01	0.1	0.01	0.00	0.01	0.15	0.00	5.3	0	0	0.2
0	26	1.93	48.8	176.7	311.1	1.42	—	0.26	0.23	0.13	1.35	0.09	34.9	0.3	0	17.4
13	27	0.65	21.1	112.8	154.1	0.46	—	0.08	0.73	0.09	1.85	0.05	23.5	0.1	0.1	4.8
18	5	0.30	1.7	12.2	49.4	0.08	—	0.04	0.28	0.05	0.31	0.01	9.5	0	0	3.1
0	3	0.22	1.4	9.1	34.9	0.04	0	0.02	0.16	0.02	0.27	0.00	5.0	0	0	0.3
0	14	0.76	15.7	61.0	198.8	0.29	0.3	0.15	0.66	0.08	1.63	0.04	26.3	0	0.1	2.3
4	45	1.43	10.8	43.5	298.5	0.33	8.7	0.17	0.01	0.12	1.40	0.16	45.6	0	0.1	2.6
4	45	1.43	10.8	31.8	137.4	0.33	5.1	0.17	0.09	0.12	1.40	0.16	26.7	0	0.1	2.6
0	7	1.04	8.4	37.8	169.4	0.22	0	0.06	0.09	0.08	1.15	0.01	12.9	0	0	2.9
0	4	0.89	7.1	31.8	0.6	0.19	0	0.11	0.01	0.08	1.10	0.03	4.8	0	0	10.5
0	14	0.55	8.9	30.3	124.4	0.30	0	0.06	0.06	0.04	0.61	0.01	18.6	0	0	5.2
0	12	0.55	5.9	29.0	134.9	0.20	0	0.07	—	0.04	0.70	0.01	12.8	0	0	5.8
0	20	0.72	—	10.0	135.0	—	—	—	—	—	—	—	—	0	—	—
0	3	0.24	7.8	31.9	26.4	0.23	0	0.02	0.08	0.01	0.10	0.02	4.7	0	0	3.7
0	4	0.65	13.3	54.5	87.3	0.30	0	0.04	0.08	0.03	0.17	0.03	5.0	0	0	2.6
0	10	0.84	3.3	23.1	160.8	0.12	0	0.01	0.14	0.06	0.78	0.01	20.9	0	0	1.5
0	18	0.81	4.1	108.6	95.4	0.11	0	0.08	0.01	0.06	0.78	0.01	18.6	0	0	2.9
1	72	0.66	10.1	120.1	392.3	0.17	4.8	0.12	0.06	0.19	1.05	0.01	28.0	0	0	6.0
0	23	0.77	15.4	60.2	201.0	0.31	0.3	0.13	0.57	0.07	1.71	0.04	24.1	0	0	3.0
0	36	1.08	8.1	39.9	254.1	0.20	0	0.12	0.60	0.10	1.21	0.01	27.0	0	0	2.0
0	36	1.08	8.1	106.5	111.9	0.20	0	0.12	0.60	0.10	1.21	0.01	27.0	0	0	2.0
0	10	0.84	3.3	23.1	160.8	0.12	0	0.01	0.14	0.06	0.78	0.01	20.9	0	0	1.5
2	57	0.73	15.1	85.7	255.6	0.24	4.8	0.10	—	0.12	0.89	0.07	17.9	0.4	0	6.8
0	48	0.74	10.6	83.2	226.0	0.23	0	0.10	—	0.08	1.64	0.03	19.6	0	0	6.1
0	15	1.32	18.6	54.9	238.5	0.48	0	0.15	0.15	0.09	1.48	0.04	35.1	0	0	1.9
0	15	1.32	18.6	60.9	84.9	0.48	0	0.15	0.15	0.09	1.48	0.04	15.0	0	0	10.1
0	14	0.86	27.7	83.2	184.5	0.60	0	0.05	0.24	0.02	1.26	0.05	7.8	0	0	4.1
14	9	0.26	2.2	22.4	6.8	0.09	7.1	0.03	0.07	0.04	0.23	0.01	6.3	0.2	0.1	2.6
0	—	0.72	—	—	340.1	—	0	—	—	—	—	—	—	0	—	—
38	21	1.15	9.1	67.3	424.1	0.42	117.4	0.22	0.47	0.13	1.24	0.03	50.2	0.1	0.1	12.9
30	61	1.17	20.8	61.8	236.0	0.56	5.9	0.14	0.53	0.15	1.49	0.06	54.0	1.1	0.1	9.2
20	21	1.55	17.0	68.0	262.7	0.68	9.4	0.28	0.24	0.12	1.90	0.05	59.5	0	0.1	9.2
22	21	1.49	17.0	67.2	249.1	0.63	14.5	0.26	0.36	0.12	1.81	0.08	57.8	0	0.2	10.6
17	21	0.91	9.4	59.7	256.6	0.25	17.9	0.10	0.90	0.11	0.87	0.02	24.4	0.1	0.1	4.4
24	89	0.95	14.3	44.5	142.8	0.23	5.0	0.01	0.08	0.02	0.19	0.01	18.9	0	0	1.7
4	26	0.36	13.2	64.8	205.2	0.46	2.4	0.53	—	0.04	0.39	0.03	13.2	0.1	0.1	5.0
0	0	1.80	—	70.0	190.0	—	0	0.15	—	0.17	2.00	0.20	40.0	0	0	—
10	100	1.80	—	—	220.0	—	—	0.15	—	0.17	2.00	—	40.0	0	0.6	—
20	56	1.02	7.8	70.2	289.4	0.28	—	0.17	0.75	0.15	1.25	0.02	34.0	0.4	0.1	8.8
15	42	1.60	18.2	39.3	297.0	0.30	29.6	0.15	0.45	0.18	1.16	0.04	45.6	0	0.1	8.7
0	30	1.42	12.0	74.7	264.5	0.39	0	0.25	—	0.16	2.21	0.02	42.2	0	0	—
0	95	1.36	11.0	71.5	252.0	0.38	0	0.19	0.16	0.14	1.90	0.02	43.5	0.1	0	13.5
0	101	1.63	21.1	106.0	217.7	0.61	0	0.24	0.25	0.16	1.91	0.05	36.5	0	0	16.6
0	36	2.39	89.5	289.0	224.0	1.04	0	0.14	0.37	0.05	0.23	0.09	50.7	0	0	6.3
0	200	0.36	—	—	60.0	—	0	—	—	—	—	—	—	0	—	—
0	20	1.08	—	120.0	160.0	—	—	—	—	—	—	—	—	0	—	—
0	0	0.00	—	—	95.0	—	0	—	—	—	—	—	—	0	—	—
0	15	0.72	23.8	82.3	72.0	0.50	0	0.06	—	0.03	0.39	0.02	5.6	0.2	0	4.0
0	30	0.72	21.0	92.3	79.0	0.42	0	0.08	—	0.04	0.14	0.02	6.8	0	0.1	4.6
0	11	1.73	10.9	122.5	327.1	0.29	17.1	0.22	—	0.16	1.90	0.05	37.2	2.6	0	12.1
106	39	1.80	31.7	162.3	319.6	1.24	100.0	0.22	—	0.22	1.03	0.07	31.7	0.2	0.2	14.6

Food Composition

(Computer code is for Cengage Diet Analysis program)   (For purposes of calculations, use "0" for t, <1, <.1, <.01, etc.)

DA+ Code	Food Description	Quantity	Measure	Wt (g)	H₂O (g)	Ener (kcal)	Prot (g)	Carb (g)	Fiber (g)	Fat (g)	Fat Breakdown (g) Sat	Mono	Poly	Trans
	BREADS, BAKED GOODS, CAKES, COOKIES, CRACKERS, CHIPS, PIES —Continued													
33356	Pie crust mix, prepared, baked	1	slice(s)	20	2.1	100	1.3	10.1	0.4	6.1	1.5	3.5	0.8	—
9007	Pie crust, ready to bake, frozen, enriched, baked	1	slice(s)	16	1.8	82	0.7	7.9	0.2	5.2	1.7	2.5	0.6	—
472	Pumpkin pie, prepared from home recipe	1	slice(s)	155	90.7	316	7.0	40.9	—	14.4	4.9	5.7	2.8	—
	Rolls													
8555	Crescent dinner roll	1	item(s)	28	9.7	78	2.7	13.8	0.6	1.2	0.3	0.3	0.6	—
489	Hamburger roll or bun, plain	1	item(s)	43	14.9	120	4.1	21.3	0.9	1.9	0.5	0.5	0.8	—
490	Hard roll	1	item(s)	57	17.7	167	5.6	30.0	1.3	2.5	0.3	0.6	1.0	—
5127	Kaiser roll	1	item(s)	57	17.7	167	5.6	30.0	1.3	2.5	0.3	0.6	1.0	—
5130	Whole wheat roll or bun	1	item(s)	28	9.4	75	2.5	14.5	2.1	1.3	0.2	0.3	0.6	—
	Sport bars													
37026	Balance original chocolate bar	1	item(s)	50	—	200	14.0	22.0	0.5	6.0	3.5	—	—	
37024	Balance original peanut butter bar	1	item(s)	50	—	200	14.0	22.0	1.0	6.0	2.5	—	—	
36580	Clif Bar chocolate brownie energy bar	1	item(s)	68	—	240	10.0	45.0	5.0	4.5	1.5	—	—	0
36583	Clif Bar crunchy peanut butter energy bar	1	item(s)	68	—	250	12.0	40.0	5.0	6.0	1.5	—	—	0
36589	Clif Luna Nutz over Chocolate energy bar	1	item(s)	48	—	180	10.0	25.0	3.0	4.5	2.5	—	—	0
12005	PowerBar apple cinnamon	1	item(s)	65	—	230	9.0	45.0	3.0	2.5	0.5	1.5	0.5	0
16078	PowerBar banana	1	item(s)	65	—	230	9.0	45.0	3.0	2.5	0.5	1.0	0.5	0
16080	PowerBar chocolate	1	item(s)	65	6.4	230	10.0	45.0	3.0	2.0	0.5	0.5	1.0	0
29092	PowerBar peanut butter	1	item(s)	65	—	240	10.0	45.0	3.0	3.5	0.5	—	—	0
	Tortillas													
1391	Corn tortillas, soft	1	item(s)	26	11.9	57	1.5	11.6	1.6	0.7	0.1	0.2	0.4	—
1669	Flour tortilla	1	item(s)	32	9.7	100	2.7	16.4	1.0	2.5	0.6	1.2	0.5	—
	Pancakes, waffles													
8926	Pancakes, blueberry, prepared from recipe	3	item(s)	114	60.6	253	7.0	33.1	0.8	10.5	2.3	2.6	4.7	—
5037	Pancakes, prepared from mix with egg and milk	3	item(s)	114	60.3	249	8.9	32.9	2.1	8.8	2.3	2.4	3.3	—
1390	Taco shells, hard	1	item(s)	13	1.0	62	0.9	8.3	0.6	2.8	0.6	1.6	0.5	0.6
30311	Waffle, 100% whole grain	1	item(s)	75	32.3	200	6.9	25.0	1.9	8.4	2.3	3.3	2.1	—
9219	Waffle, plain, frozen, toasted	2	item(s)	66	20.2	206	4.7	32.5	1.6	6.3	1.1	3.2	1.5	—
500	Waffle, plain, prepared from recipe	1	item(s)	75	31.5	218	5.9	24.7	1.7	10.6	2.1	2.6	5.1	—
	CEREAL, FLOUR, GRAIN, PASTA, NOODLES, POPCORN													
	Grain													
2861	Amaranth, dry	½	cup(s)	98	9.6	365	14.1	64.5	9.1	6.3	1.6	1.4	2.8	—
1953	Barley, pearled, cooked	½	cup(s)	79	54.0	97	1.8	22.2	3.0	0.3	0.1	0	0.2	—
1956	Buckwheat groats, cooked, roasted	½	cup(s)	84	63.5	77	2.8	16.8	2.3	0.5	0.1	0.2	0.2	—
1957	Bulgur, cooked	½	cup(s)	91	70.8	76	2.8	16.9	4.1	0.2	0	0	0.1	—
1963	Couscous, cooked	½	cup(s)	79	57.0	88	3.0	18.2	1.1	0.1	0	0	0.1	—
1967	Millet, cooked	½	cup(s)	120	85.7	143	4.2	28.4	1.6	1.2	0.2	0.2	0.6	—
1969	Oat bran, dry	½	cup(s)	47	3.1	116	8.1	31.1	7.2	3.3	0.6	1.1	1.3	—
1972	Quinoa, dry	½	cup(s)	85	11.3	313	12.0	54.5	5.9	5.2	0.6	1.4	2.8	—
	Rice													
129	Brown, long grain, cooked	½	cup(s)	98	71.3	108	2.5	22.4	1.8	0.9	0.2	0.3	0.3	—
2863	Brown, medium grain, cooked	½	cup(s)	98	71.1	109	2.3	22.9	1.8	0.8	0.2	0.3	0.3	—
37488	Jasmine, saffroned, cooked	½	cup(s)	280	—	340	8.0	78.0	0	0	0	0	0	0
30280	Pilaf, cooked	½	cup(s)	103	74.0	129	2.1	22.2	0.6	3.3	0.6	1.5	1.0	—
28066	Spanish, cooked	½	cup(s)	244	184.2	241	5.7	50.2	3.3	1.9	0.4	0.6	0.7	0
2867	White glutinous, cooked	½	cup(s)	87	66.7	84	1.8	18.3	0.9	0.2	0	0.1	0.1	—
484	White, long grain, boiled	½	cup(s)	79	54.1	103	2.1	22.3	0.3	0.2	0.1	0.1	0.1	—
482	White, long grain, enriched, instant, boiled	½	cup(s)	83	59.4	97	1.8	20.7	0.5	0.4	0	0.1	0	—
486	White, long grain, enriched, parboiled, cooked	½	cup(s)	79	55.6	97	2.3	20.6	0.7	0.3	0.1	0.1	0.1	—
1194	Wild brown, cooked	½	cup(s)	82	60.6	83	3.3	17.5	1.5	0.3	0	0	0.2	—
	Flour & grain fractions													
505	All purpose flour, self rising, enriched	½	cup(s)	63	6.6	221	6.2	46.4	1.7	0.6	0.1	0	0.2	—
503	All purpose flour, white, bleached, enriched	½	cup(s)	63	7.4	228	6.4	47.7	1.7	0.6	0.1	0	0.2	—
1643	Barley flour	½	cup(s)	56	5.5	198	4.2	44.7	2.1	0.8	0.2	0.1	0.4	—
383	Buckwheat flour, whole groat	½	cup(s)	60	6.7	201	7.6	42.3	6.0	1.9	0.4	0.6	0.6	—
504	Cake wheat flour, enriched	½	cup(s)	69	8.6	248	5.6	53.5	1.2	0.6	0.1	0.1	0.3	—
426	Cornmeal, degermed, enriched	½	cup(s)	69	7.8	255	5.0	54.6	2.8	1.2	0.1	0.2	0.5	0
424	Cornmeal, yellow whole grain	½	cup(s)	61	6.2	221	4.9	46.9	4.4	2.2	0.3	0.6	1.0	—
1978	Dark rye flour	½	cup(s)	64	7.1	207	9.0	44.0	14.5	1.7	0.2	0.2	0.8	—
1644	Masa corn flour, enriched	½	cup(s)	57	5.1	208	5.3	43.5	5.5	2.1	0.3	0.6	1.0	—
1976	Rice flour, brown	½	cup(s)	79	9.4	287	5.7	60.4	3.6	2.2	0.4	0.8	0.8	—
1645	Rice flour, white	½	cup(s)	79	9.4	289	4.7	63.3	1.9	1.1	0.3	0.3	0.3	—
1980	Semolina, enriched	½	cup(s)	84	10.6	301	10.6	60.8	3.2	0.9	0.1	0.1	0.4	—
2827	Soy flour, raw	½	cup(s)	42	2.2	185	14.7	14.9	4.1	8.8	1.3	1.9	4.9	—
1990	Wheat germ, crude	2	tablespoon(s)	14	1.6	52	3.3	7.4	1.9	1.4	0.2	0.2	0.9	—
506	Whole wheat flour	½	cup(s)	60	6.2	203	8.2	43.5	7.3	1.1	0.2	0.1	0.5	—
	Breakfast bars													
39230	Atkins Morning Start apple crisp breakfast bar	1	item(s)	37	—	170	11.0	12.0	6.0	9.0	4.0	—	—	—
10571	Nutri-Grain apple cinnamon cereal bar	1	item(s)	37	—	140	2.0	27.0	1.0	3.0	0.5	2.0	0.5	—

Chol (mg)	Calc (mg)	Iron (mg)	Magn (mg)	Pota (mg)	Sodi (mg)	Zinc (mg)	Vit A (µg)	Thia (mg)	Vit E (mg α)	Ribo (mg)	Niac (mg)	Vit B_6 (mg)	Fola (µg)	Vit C (mg)	Vit B_{12} (µg)	Sele (µg)
0	12	0.43	3.0	12.4	145.8	0.07	0	0.06	—	0.03	0.47	0.01	14.0	0	0	4.4
0	3	0.36	2.9	17.6	103.5	0.05	0	0.04	0.42	0.06	0.39	0.01	8.8	0	0	0.5
65	146	1.96	29.5	288.3	348.8	0.71	660.3	0.14	—	0.31	1.21	0.07	32.6	2.6	0.1	11.0
0	39	0.93	5.9	26.3	134.1	0.18	0	0.11	0.02	0.09	1.16	0.02	31.1	0	0.1	5.5
0	59	1.42	9.0	40.4	206.0	0.28	0	0.17	0.03	0.13	1.78	0.03	47.7	0	0.1	8.4
0	54	1.87	15.4	61.6	310.1	0.53	0	0.27	0.23	0.19	2.41	0.02	54.2	0	0	22.3
0	54	1.86	15.4	61.6	310.1	0.53	0	0.27	0.23	0.19	2.41	0.01	54.2	0	0	22.3
0	30	0.69	24.1	77.1	135.5	0.57	0	0.07	0.26	0.04	1.04	0.06	8.5	0	0	14.0
3	100	4.50	40.0	160.0	180.0	3.75	—	0.37	—	0.42	5.00	0.50	100.0	60.0	1.5	17.5
3	100	4.50	40.0	130.0	230.0	3.75	—	0.37	—	0.42	5.00	0.50	100.0	60.0	1.5	17.5
0	250	4.50	100.0	370.0	150.0	3.00	—	0.37	—	0.25	3.00	0.40	80.0	60.0	0.9	14.0
0	250	4.50	100.0	230.0	250.0	3.00	—	0.37	—	0.25	3.00	0.40	80.0	60.0	0.9	14.0
0	350	5.40	80.0	190.0	190.0	5.25	—	1.20	—	1.36	16.00	2.00	400.0	60.0	6.0	24.5
0	300	6.30	140.0	125.0	100.0	5.25	—	1.50	—	1.70	20.00	2.00	400.0	60.0	6.0	—
0	300	6.30	140.0	190.0	100.0	5.25	0	1.50	—	1.70	20.00	2.00	400.0	60.0	6.0	—
0	300	6.30	140.0	200.0	95.0	5.25	0	1.50	—	1.70	20.00	2.00	400.0	60.0	6.0	5.1
0	300	6.30	140.0	130.0	120.0	5.25	0	1.50	—	1.70	20.00	2.00	400.0	60.0	6.0	—
0	21	0.32	18.7	48.4	11.7	0.34	0	0.02	0.07	0.02	0.39	0.06	1.3	0	0	1.6
0	41	1.06	7.0	49.6	203.5	0.17	0	0.17	0.06	0.08	1.14	0.01	33.3	0	0	7.1
64	235	1.96	18.2	157.3	469.7	0.61	57.0	0.22	—	0.31	1.73	0.05	41.0	2.5	0.2	16.0
81	245	1.48	25.1	226.9	575.7	0.85	82.1	0.22	—	0.35	1.40	0.12	104.6	0.7	0.4	—
0	13	0.25	11.3	29.7	51.7	0.21	0.1	0.03	0.09	0.01	0.25	0.03	9.2	0	0	0.6
71	194	1.60	28.5	171.0	371.3	0.87	48.8	0.15	0.32	0.25	1.47	0.08	28.5	0	0.4	20.0
10	203	4.56	15.8	95.0	481.8	0.35	262.7	0.34	0.64	0.46	5.86	0.68	49.5	0	1.9	8.3
52	191	1.73	14.3	119.3	383.3	0.51	48.8	0.19	—	0.26	1.55	0.04	34.5	0.3	0.2	34.7
0	149	7.40	259.3	356.8	20.5	3.10	0	0.06	—	0.20	1.24	0.20	47.8	4.1	0	—
0	9	1.04	17.3	73.0	2.4	0.64	0	0.06	0.01	0.04	1.61	0.09	12.6	0	0	6.8
0	6	0.67	42.8	73.9	3.4	0.51	0	0.03	0.07	0.03	0.79	0.06	11.8	0	0	1.8
0	9	0.87	29.1	61.9	4.6	0.51	0	0.05	0.01	0.02	0.91	0.07	16.4	0	0	0.5
0	6	0.30	6.3	45.5	3.9	0.20	0	0.05	0.10	0.02	0.77	0.04	11.8	0	0	21.6
0	4	0.75	52.8	74.4	2.4	1.09	0	0.12	0.02	0.09	1.59	0.13	22.8	0	0	1.1
0	27	2.54	110.5	266.0	1.9	1.46	0	0.55	0.47	0.10	0.43	0.07	24.4	0	0	21.2
0	40	3.88	167.4	478.5	4.2	2.62	0.8	0.30	2.06	0.26	1.28	0.40	156.4	0	0	7.2
0	10	0.41	41.9	41.9	4.9	0.61	0	0.09	0.02	0.02	1.49	0.14	3.9	0	0	9.6
0	10	0.51	42.9	77.0	1.0	0.60	0	0.09	—	0.01	1.29	0.14	3.9	0	0	38.0
0	—	2.16	—	—	780.0	—	—	—	—	—	—	—	—	—	—	—
0	11	1.16	9.3	54.6	390.4	0.37	33.0	0.13	0.28	0.02	1.23	0.06	44.3	0.4	0	4.3
0	37	1.52	95.4	330.5	97.1	1.40	—	0.27	0.12	0.05	3.24	0.38	19.0	22.6	0	14.3
0	2	0.12	4.4	8.7	4.4	0.35	0	0.01	0.03	0.01	0.25	0.02	0.9	0	0	4.9
0	8	0.94	9.5	27.7	0.8	0.38	0	0.12	0.03	0.01	1.16	0.07	45.8	0	0	5.9
0	7	1.46	4.1	7.4	3.3	0.40	0	0.06	0.01	0.01	1.43	0.04	57.8	0	0	4.0
0	15	1.43	7.1	44.2	1.6	0.29	0	0.16	0.01	0.01	1.82	0.12	64.0	0	0	7.3
0	2	0.49	26.2	82.8	2.5	1.09	0	0.04	0.19	0.07	1.05	0.11	21.3	0	0	0.7
0	211	2.90	11.9	77.5	793.7	0.38	0	0.42	0.02	0.24	3.64	0.02	122.5	0	0	21.5
0	9	2.90	13.7	66.9	1.2	0.42	0	0.48	0.02	0.30	3.68	0.02	114.4	0	0	21.2
0	16	0.70	45.4	185.9	4.5	1.04	0	0.06	—	0.02	2.56	0.14	12.9	0	0	2.0
0	25	2.42	150.6	346.2	6.6	1.86	0	0.24	0.18	0.10	3.68	0.34	32.4	0	0	3.4
0	10	5.01	11.0	71.9	1.4	0.42	0	0.61	0.01	0.29	4.65	0.02	127.4	0	0	3.4
0	2	2.98	24.1	104.9	4.8	0.48	7.6	0.42	0.10	0.28	3.66	0.12	148.3	0	0	8.0
0	4	2.10	77.5	175.1	21.3	1.10	6.7	0.22	0.24	0.12	2.20	0.18	15.2	0	0	9.4
0	36	4.12	158.7	467.2	0.6	3.58	0.6	0.20	0.90	0.16	2.72	0.28	38.4	0	0	22.8
0	80	4.10	62.7	169.9	2.8	1.00	0	0.80	0.08	0.42	5.60	0.20	132.8	0	0	8.5
0	9	1.56	88.5	228.3	6.3	1.92	0	0.34	0.94	0.06	5.00	0.58	12.6	0	0	—
0	8	0.26	27.6	60.0	0	0.62	0	0.10	0.08	0.02	2.04	0.34	3.2	0	0	11.9
0	14	3.64	39.2	155.3	0.8	0.86	0	0.66	0.20	0.46	5.00	0.08	152.8	0	0	74.6
0	87	2.70	182.0	1067.0	5.5	1.65	2.5	0.24	0.82	0.48	1.83	0.18	146.4	0	0	3.2
0	6	0.90	34.4	128.2	1.7	1.76	0	0.27	—	0.07	0.97	0.18	40.4	0	0	11.4
0	20	2.32	82.8	243.0	3.0	1.74	0	0.26	0.48	0.12	3.82	0.20	26.4	0	0	42.4
0	200	—	—	90.0	70.0	—	—	0.22	—	0.25	3.00	—	—	9.0	—	—
0	200	1.80	8.0	75.0	110.0	1.50	—	0.37	—	0.42	5.00	0.50	40.0	0	—	—

DA+ Code	Food Description	Quantity	Measure	Wt (g)	H₂O (g)	Ener (kcal)	Prot (g)	Carb (g)	Fiber (g)	Fat (g)	Fat Breakdown (g) Sat	Mono	Poly	Trans
	CEREAL, FLOUR, GRAIN, PASTA, NOODLES, POPCORN—Continued													
10647	Nutri-Grain blueberry cereal bar	1	item(s)	37	5.4	140	2.0	27.0	1.0	3.0	0.5	2.0	0.5	—
10648	Nutri-Grain raspberry cereal bar	1	item(s)	37	5.4	140	2.0	27.0	1.0	3.0	0.5	2.0	0.5	—
10649	Nutri-Grain strawberry cereal bar	1	item(s)	37	5.4	140	2.0	27.0	1.0	3.0	0.5	2.0	0.5	—
	Breakfast cereals, hot													
1260	Cream of Wheat, instant, prepared	½	cup(s)	121	—	388	12.9	73.3	4.3	0	0	0	0	0
365	Farina, enriched, cooked with water and salt	½	cup(s)	117	102.4	56	1.7	12.2	0.3	0.1	0	0	0	—
363	Grits, white corn, regular and quick, enriched, cooked with water and salt	½	cup(s)	121	103.3	71	1.7	15.6	0.4	0.2	0	0.1	0.1	—
8636	Grits, yellow corn, regular and quick, enriched, cooked with salt	½	cup(s)	121	103.3	71	1.7	15.6	0.4	0.2	0	0.1	0.1	—
8657	Oatmeal, cooked with water	½	cup(s)	117	97.8	83	3.0	14.0	2.0	1.8	0.4	0.5	0.7	0
5500	Oatmeal, maple and brown sugar, instant, prepared	1	item(s)	198	150.2	200	4.8	40.4	2.4	2.2	0.4	0.7	0.8	—
5510	Oatmeal, ready to serve, packet, prepared	1	item(s)	186	158.7	112	4.1	19.8	2.7	2.0	0.4	0.7	0.8	—
	Breakfast cereals, ready to eat													
1197	All-Bran	1	cup(s)	62	1.3	160	8.1	46.0	18.2	2.0	0.4	0.4	1.3	0
1200	All-Bran Buds	1	cup(s)	91	2.7	212	6.4	72.7	39.1	1.9	0.4	0.5	1.2	0
1199	Apple Jacks	1	cup(s)	33	0.9	130	1.0	30.0	0.5	0.5	0	—	—	0
1204	Cap'n Crunch	1	cup(s)	36	0.9	147	1.3	30.7	1.3	2.0	0.5	0.4	0.3	—
1205	Cap'n Crunch Crunchberries	1	cup(s)	35	0.9	133	1.3	29.3	1.3	2.0	0.5	0.4	0.3	—
1206	Cheerios	1	cup(s)	30	1.0	110	3.0	22.0	3.0	2.0	0	0.5	0.5	0
3415	Cocoa Puffs	1	cup(s)	30	0.6	120	1.0	26.0	0.2	1.0	—	—	—	—
1207	Cocoa Rice Krispies	1	cup(s)	41	1.0	160	1.3	36.0	1.3	1.3	0.7	0	0	—
5522	Complete wheat bran flakes	1	cup(s)	39	1.4	120	4.0	30.7	6.7	0.7	—	—	—	0
1211	Corn Flakes	1	cup(s)	28	0.9	100	2.0	24.0	1.0	0	0	0	0	0
1247	Corn Pops	1	cup(s)	31	0.9	120	1.0	28.0	0.3	0	0	0	0	0
1937	Cracklin' Oat Bran	1	cup(s)	65	2.3	267	5.3	46.7	8.0	9.3	4.0	4.7	1.3	0
1220	Froot Loops	1	cup(s)	32	0.8	120	1.0	28.0	1.0	1.0	0.5	0	0	—
38214	Frosted Cheerios	1	cup(s)	37	—	149	2.5	31.1	1.2	1.2	—	—	—	—
372	Frosted Flakes	1	cup(s)	41	1.1	160	1.3	37.3	1.3	0	0	0	0	0
38215	Frosted Mini Chex	1	cup(s)	40	—	147	1.3	36.0	0	0	0	0	0	0
10268	Frosted Mini-Wheats	1	cup(s)	59	3.1	208	5.8	47.4	5.8	1.2	0	0	0.6	0
38216	Frosted Wheaties	1	cup(s)	40	—	147	1.3	36.0	0.3	0	0	0	0	0
1223	Granola, prepared	½	cup(s)	61	3.3	298	9.1	32.5	5.5	14.7	2.5	5.8	5.6	0
2415	Honey Bunches of Oats honey roasted	1	cup(s)	40	0.9	160	2.7	33.3	1.3	2.0	0.7	1.2	0.1	—
1227	Honey Nut Cheerios	1	cup(s)	37	0.9	149	3.7	29.9	2.5	1.9	0	0.6	0.6	0
2424	Honeycomb	1	cup(s)	22	0.3	83	1.5	19.5	0.8	0.4	0	—	—	0
10286	Kashi whole grain puffs	1	cup(s)	19	—	70	2.0	15.0	1.0	0.5	0	—	—	0
41142	Kellogg's Mueslix	1	cup(s)	83	7.2	298	7.6	60.8	6.1	4.6	0.7	2.4	1.5	0
1231	Kix	1	cup(s)	24	0.5	96	1.6	20.8	0.8	0.4	—	—	—	—
30569	Life	1	cup(s)	43	1.7	160	4.0	33.3	2.7	2.0	0.3	0.6	0.6	0
1233	Lucky Charms	1	cup(s)	24	0.6	96	1.6	20.0	0.8	0.8	—	—	—	—
38220	Multi Grain Cheerios	1	cup(s)	30	—	110	3.0	24.0	3.0	1.0	—	—	—	—
1201	Multi-Bran Chex	1	cup(s)	63	1.3	216	4.3	52.9	8.6	1.6	0	0	0.5	0
13633	Post Bran Flakes	1	cup(s)	40	1.5	133	4.0	32.0	6.7	0.7	0	—	—	—
1241	Product 19	1	cup(s)	30	1.0	100	2.0	25.0	1.0	0	0	0	0	0
32432	Puffed rice, fortified	1	cup(s)	14	0.4	56	0.9	12.6	0.2	0.1	0	—	—	—
32433	Puffed wheat, fortified	1	cup(s)	12	0.4	44	1.8	9.6	0.5	0.1	0	—	—	—
13334	Quaker 100% natural granola oats & honey	½	cup(s)	48	—	220	5.0	31.0	3.0	9.0	3.8	4.1	1.2	—
13335	Quaker 100% natural granola oats, honey & raisins	½	cup(s)	51	—	230	5.0	34.0	3.0	9.0	3.6	3.8	1.1	—
2420	Raisin Bran	1	cup(s)	59	5.0	190	4.0	46.0	8.0	1.0	0	0.1	0.4	—
1244	Rice Chex	1	cup(s)	31	0.8	120	2.0	27.0	0.3	0	0	0	0	0
1245	Rice Krispies	1	cup(s)	26	0.8	96	1.6	23.2	0	0	0	0	0	0
5593	Shredded Wheat	1	cup(s)	49	0.4	177	5.8	40.9	6.9	1.1	0.1	0	0.2	0
1248	Smacks	1	cup(s)	36	1.1	133	2.7	32.0	1.3	0.7	—	—	—	—
1246	Special K	1	cup(s)	31	0.9	110	7.0	22.0	0.5	0	0	0	0	0
3428	Total corn flakes	1	cup(s)	23	0.6	83	1.5	18.0	0.6	0	0	0	0	0
1253	Total whole grain	1	cup(s)	40	1.1	147	2.7	30.7	4.0	1.3	—	—	—	—
1254	Trix	1	cup(s)	30	0.6	120	1.0	27.0	1.0	1.0	—	—	—	—
382	Wheat germ, toasted	2	tablespoon(s)	14	0.8	54	4.1	7.0	2.1	1.5	0.3	0.2	0.9	—
1257	Wheaties	1	cup(s)	36	1.2	132	3.6	28.8	3.6	1.2	—	—	—	—
	Pasta, noodles													
449	Chinese chow mein noodles, cooked	½	cup(s)	23	0.2	119	1.9	12.9	0.9	6.9	1.0	1.7	3.9	—
1995	Corn pasta, cooked	½	cup(s)	70	47.8	88	1.8	19.5	3.4	0.5	0.1	0.1	0.2	—
448	Egg noodles, enriched, cooked	½	cup(s)	80	54.2	110	3.6	20.1	1.0	1.7	0.3	0.5	0.4	0
1563	Egg noodles, spinach, enriched, cooked	½	cup(s)	80	54.8	106	4.0	19.4	1.8	1.3	0.3	0.4	0.3	—
440	Macaroni, enriched, cooked	½	cup(s)	70	43.5	111	4.1	21.6	1.3	0.7	0.1	0.1	0.2	0
2000	Macaroni, tricolor vegetable, enriched, cooked	½	cup(s)	67	45.8	86	3.0	17.8	2.9	0.1	0	0	0	—
1996	Plain pasta, fresh-refrigerated, cooked	½	cup(s)	64	43.9	84	3.3	16.0	—	0.7	0.1	0.1	0.3	—

Chol (mg)	Calc (mg)	Iron (mg)	Magn (mg)	Pota (mg)	Sodi (mg)	Zinc (mg)	Vit A (µg)	Thia (mg)	Vit E (mg α)	Ribo (mg)	Niac (mg)	Vit B6 (mg)	Fola (µg)	Vit C (mg)	Vit B12 (µg)	Sele (µg)
0	200	1.80	8.0	75.0	110.0	1.50	—	0.37	—	0.42	5.00	0.50	40.0	0	0	—
0	200	1.80	8.0	70.0	110.0	1.50	—	0.37	—	0.42	5.00	0.50	40.0	0	0	—
0	200	1.80	8.0	55.0	110.0	1.50	—	0.37	—	0.42	5.00	0.50	40.0	0	0	—
0	862	34.91	21.4	150.8	732.6	0.86	—	1.59	—	1.47	21.55	2.15	431.0	0	0	—
0	5	0.58	2.3	15.1	383.3	0.09	0	0.07	0.01	0.05	0.57	0.01	39.6	0	0	10.6
0	4	0.73	6.1	25.4	269.8	0.08		0.10	0.02	0.07	0.87	0.03	39.9			3.8
0	4	0.73	6.1	25.4	269.8	0.08	2.4	0.10	0.02	0.07	0.87	0.03	39.9	0	0	3.3
0	11	1.05	31.6	81.9	4.7	1.17	0	0.09	0.09	0.02	0.26	0.01	7.0	0	0	6.3
0	26	6.83	49.9	126.4	403.5	1.03	0	1.02	—	0.05	1.56	0.30	42.2	0	0	11.1
0	21	3.96	44.7	112.4	240.9	0.92	0	0.60	—	0.04	0.77	0.18	18.7	0	0	3.8
0	241	10.90	224.4	632.4	150.0	3.00	300.1	1.40	—	1.68	9.16	7.44	800.0	12.4	12.0	5.8
0	57	13.64	186.4	909.1	614.5	4.55	464.5	1.09	1.42	1.27	15.45	6.09	1222.7	18.2	18.2	26.3
0	0	4.50	8.0	30.0	130.0	1.50	150.2	0.37	—	0.42	5.00	0.50	100.0	15.0	1.5	2.4
0	5	6.80	20.0	73.3	266.7	5.00	2.5	0.51	—	0.57	6.68	0.67	133.5	0	0	6.7
0	7	6.53	18.7	73.3	240.0	5.13	2.4	0.51	—	0.57	6.68	0.67	133.7	0	0	6.7
0	100	8.10	40.0	95.0	280.0	3.75	150.3	0.37	—	0.42	5.00	0.50	200.0	6.0	1.5	11.3
0	100	4.50	8.0	50.0	170.0	3.75	0	0.37	—	0.42	5.00	0.50	100.0	6.0	1.5	2.0
0	53	6.00	10.7	66.7	253.3	2.00	200.1	0.49	—	0.56	6.67	0.67	133.3	20.0	2.0	5.8
0	0	24.00	53.3	226.7	280.0	20.00	300.1	2.00	—	2.27	26.67	2.67	533.3	80.0	8.0	4.1
0	0	8.10	3.4	25.0	200.0	0.16	149.8	0.37	—	0.42	5.00	0.50	100.0	6.0	1.5	1.4
0	0	1.80	2.5	25.0	120.0	1.50	150.0	0.37	—	0.42	5.00	0.50	100.0	6.0	1.5	2.0
0	27	2.40	80.0	293.3	200.0	2.00	299.9	0.49	—	0.56	6.67	0.67	133.3	20.0	2.0	14.4
0	0	4.50	8.0	35.0	150.0	1.50	150.1	0.37	—	0.42	5.00	0.50	100.0	15.0	1.5	2.3
0	124	5.60	19.9	68.4	261.3	4.67	—	0.46	—	0.52	6.22	0.62	124.4	7.5	1.9	—
0	0	6.00	3.7	26.7	200.0	0.20	200.1	0.49	—	0.56	6.67	0.67	133.3	8.0	2.0	1.8
0	133	12.00	—	33.3	266.7	4.00	—	0.49	—	0.56	6.67	0.67	266.7	8.0	2.0	—
0	0	16.66	69.4	196.7	5.8	1.74	0	0.43	—	0.49	5.78	0.58	115.7	0	1.7	2.4
0	133	10.80	0	46.7	266.7	10.00	—	1.00	—	1.13	13.33	1.33	533.3	8.0	4.0	—
0	48	2.58	106.8	329.4	15.3	2.45	0.6	0.44	6.77	0.17	1.30	0.17	50.0	0.7	0	17.0
0	0	10.80	21.3	0	253.3	0.40	—	0.49	—	0.56	6.67	0.67	133.0	0	2.0	—
0	124	5.60	39.8	112.0	336.0	4.67	—	0.46	—	0.52	6.22	0.62	248.9	7.5	1.9	8.8
0	0	2.03	6.0	26.3	165.4	1.13	—	0.28	—	0.32	3.74	0.37	75.0	0	1.1	—
0	0	0.36	—	60.0	0	—	0	0.03	—	0.03	0.80	0.00	—	0	0	—
0	48	6.83	74.2	363.3	257.5	5.67	136.7	0.67	6.00	0.67	8.33	3.08	615.0	0.3	9.2	14.4
0	120	6.48	6.4	28.0	216.0	3.00	120.2	0.30	—	0.34	4.00	0.40	160.0	4.8	1.2	4.8
0	149	11.87	41.3	120.0	213.3	5.33	0.9	0.53	—	0.60	7.12	0.71	142.4	0	0	10.7
0	80	3.60	12.8	48.0	168.0	3.00	—	0.30	—	0.34	4.00	0.40	160.0	4.8	1.2	4.8
0	100	18.00	24.0	85.0	200.0	15.00	—	1.50	—	1.70	20.00	2.00	400.0	15.0	6.0	—
0	108	17.50	64.8	237.7	410.6	4.05	171.1	0.40	—	0.45	5.40	0.54	432.2	6.5	1.6	4.9
0	0	10.80	80.0	266.7	280.0	2.00	—	0.49	—	0.56	6.67	0.67	133.3	0	2.0	—
0	0	18.00	16.0	50.0	210.0	15.00	225.3	1.50	—	1.70	20.00	2.00	400.0	60.0	6.0	3.6
0	1	4.43	3.5	15.8	0.4	0.14	0	0.36	—	0.25	4.94	0.01	2.7	0	0	1.5
0	3	3.80	17.4	41.8	0.5	0.28	0	0.31	—	0.21	4.23	0.02	3.8	0	0	14.8
0	61	1.20	51.0	220.0	20.0	1.05	0.5	0.13	—	0.12	0.82	0.07	15.0	0.2	0.1	8.3
0	59	1.20	49.0	250.0	20.0	0.99	0.5	0.13	—	0.12	0.80	0.08	14.1	0.4	0.1	8.8
0	20	10.80	80.0	360.0	360.0	2.25	—	0.37	—	0.42	5.00	0.50	100.0	0	2.1	—
0	100	9.00	9.3	35.0	290.0	3.75	—	0.37	—	0.42	5.00	0.50	200.0	6.0	1.5	1.2
0	0	1.44	12.8	32.0	256.0	0.48	120.1	0.30	—	0.34	4.80	0.40	80.0	4.8	1.2	4.1
0	18	2.90	60.3	179.3	1.1	1.37	0	0.14	—	0.12	3.47	0.18	21.1	0	0	2.0
0	0	0.48	10.7	53.3	66.7	0.40	200.2	0.49	—	0.56	6.67	0.67	133.3	8.0	2.0	17.5
0	0	8.10	16.0	60.0	220.0	0.90	225.1	0.52	—	0.59	7.00	2.00	400.0	21.0	6.0	7.0
0	752	13.53	0	22.6	157.9	11.28	112.8	1.13	22.56	1.28	15.04	1.50	300.8	45.1	4.5	1.2
0	1333	24.00	32.0	120.0	253.3	20.00	200.4	2.00	31.32	2.27	26.67	2.67	533.3	80.0	8.0	1.9
0	100	4.50	0	15.0	190.0	3.75	150.3	0.37	—	0.42	5.00	0.50	100.0	6.0	1.5	6.0
0	6	1.28	45.2	133.8	0.6	2.35	0.7	0.23	2.25	0.11	0.79	0.13	49.7	0.8	0	9.2
0	24	9.72	38.4	126.0	264.0	9.00	180.4	0.90	—	1.02	12.00	1.20	240.0	7.2	3.6	1.7
0	5	1.06	11.7	27.0	98.8	0.31	0	0.13	0.78	0.09	1.33	0.02	20.3	0	0	9.7
0	1	0.18	25.2	21.7	0	0.44	2.1	0.04	—	0.02	0.39	0.04	4.2	0	0	2.0
23	10	1.17	16.8	30.4	4.0	0.52	4.8	0.23	0.13	0.11	1.66	0.03	67.2	0	0.1	19.1
26	15	0.87	19.2	29.6	9.6	0.50	8.0	0.19	0.46	0.09	1.17	0.09	51.2	0	0.1	17.4
0	5	0.90	12.6	30.8	0.7	0.36	0	0.19	0.04	0.10	1.18	0.03	51.1	0	0	18.5
0	7	0.33	12.7	20.8	4.0	0.30	3.4	0.08	0.14	0.04	0.72	0.02	43.6	0	0	13.3
21	4	0.73	11.5	15.4	3.8	0.36	3.8	0.13	—	0.10	0.64	0.02	41.0	0	0.1	—

DA+ Code	Food Description	Quantity	Measure	Wt (g)	H₂O (g)	Ener (kcal)	Prot (g)	Carb (g)	Fiber (g)	Fat (g)	Fat Breakdown (g) Sat	Mono	Poly	Trans
	CEREAL, FLOUR, GRAIN, PASTA, NOODLES, POPCORN—Continued													
1725	Ramen noodles, cooked	½	cup(s)	114	94.5	104	3.0	15.4	1.0	4.3	0.2	0.2	0.2	—
2878	Soba noodles, cooked	½	cup(s)	95	69.4	94	4.8	20.4	—	0.1	0	0	0	—
2879	Somen noodles, cooked	½	cup(s)	88	59.8	115	3.5	24.2	—	0.2	0	0	0.1	—
493	Spaghetti, al dente, cooked	½	cup(s)	65	41.6	95	3.5	19.5	1.0	0.5	0.1	0.1	0.2	—
2884	Spaghetti, whole wheat, cooked	½	cup(s)	70	47.0	87	3.7	18.6	3.2	0.4	0.1	0.1	0.1	—
	Popcorn													
476	Air popped	1	cup(s)	8	0.3	31	1.0	6.2	1.2	0.4	0	0.1	0.2	—
4619	Caramel	1	cup(s)	35	1.0	152	1.3	27.8	1.8	4.5	1.3	1.0	1.6	—
4620	Cheese flavored	1	cup(s)	36	0.9	188	3.3	18.4	3.5	11.8	2.3	3.5	5.5	—
477	Popped in oil	1	cup(s)	11	0.1	64	0.8	5.0	0.9	4.8	0.8	1.1	2.6	—
	FRUIT AND FRUIT JUICES													
	Apples													
952	Juice, prepared from frozen concentrate	½	cup(s)	120	105.0	56	0.2	13.8	0.1	0.1	0	0	0	—
225	Juice, unsweetened, canned	½	cup(s)	124	109.0	58	0.1	14.5	0.1	0.1	0	0	0	—
224	Slices	½	cup(s)	55	47.1	29	0.1	7.6	1.3	0.1	0	0	0	—
946	Slices without skin, boiled	½	cup(s)	86	73.1	45	0.2	11.7	2.1	0.3	0	0	0.1	—
223	Raw medium, with peel	1	item(s)	138	118.1	72	0.4	19.1	3.3	0.2	0	0	0.1	—
948	Dried, sulfured	¼	cup(s)	22	6.8	52	0.2	14.2	1.9	0.1	0	0	0	—
226	Applesauce, sweetened, canned	½	cup(s)	128	101.5	97	0.2	25.4	1.5	0.2	0	0	0.1	—
227	Applesauce, unsweetened, canned	½	cup(s)	122	107.8	52	0.2	13.8	1.5	0.1	0	0	0	—
38492	Crabapples	1	item(s)	35	27.6	27	0.1	7.0	0.9	0.1	0	0	0	—
	Apricot													
228	Fresh without pits	4	item(s)	140	120.9	67	2.0	15.6	2.8	0.5	0	0.2	0.1	—
229	Halves with skin, canned in heavy syrup	½	cup(s)	129	100.1	107	0.7	27.7	2.1	0.1	0	0	0	—
230	Halves, dried, sulfured	¼	cup(s)	33	10.1	79	1.1	20.6	2.4	0.2	0	0	0	—
	Avocado													
233	California, whole, without skin or pit	½	cup(s)	115	83.2	192	2.2	9.9	7.8	17.7	2.4	11.3	2.1	—
234	Florida, whole, without skin or pit	½	cup(s)	115	90.6	138	2.5	9.0	6.4	11.5	2.2	6.3	1.9	—
2998	Pureed	⅛	cup(s)	28	20.2	44	0.5	2.4	1.8	4.0	0.6	2.7	0.5	—
	Banana													
4580	Dried chips	¼	cup(s)	55	2.4	285	1.3	32.1	4.2	18.5	15.9	1.1	0.3	—
235	Fresh whole, without peel	1	item(s)	118	88.4	105	1.3	27.0	3.1	0.4	0.1	0	0.1	—
	Blackberries													
237	Raw	½	cup(s)	72	63.5	31	1.0	6.9	3.8	0.4	0	0	0.2	—
958	Unsweetened, frozen	½	cup(s)	76	62.1	48	0.9	11.8	3.8	0.3	0	0	0.2	—
	Blueberries													
959	Canned in Heavy Syrup	½	cup(s)	128	98.3	113	0.8	28.2	2.0	0.4	0	0.1	0.2	—
238	Raw	½	cup(s)	73	61.1	41	0.5	10.5	1.7	0.2	0	0	0.1	—
960	Unsweetened, frozen	½	cup(s)	78	67.1	40	0.3	9.4	2.1	0.5	0	0.1	0.2	—
	Boysenberries													
961	Canned in heavy syrup	½	cup(s)	128	97.6	113	1.3	28.6	3.3	0.2	0	0	0.1	—
962	Unsweetened, frozen	½	cup(s)	66	56.7	33	0.7	8.0	3.5	0.2	0	0	0.1	—
35576	Breadfruit	1	item(s)	384	271.3	396	4.1	104.1	18.8	0.9	0.2	0.1	0.3	—
	Cherries													
967	Sour red, canned in water	½	cup(s)	122	109.7	44	0.9	10.9	1.3	0.1	0	0	0	—
3000	Sour red, raw	½	cup(s)	78	66.8	39	0.8	9.4	1.2	0.2	0.1	0.1	0.1	—
3004	Sweet, canned in heavy syrup	½	cup(s)	127	98.2	105	0.8	26.9	1.9	0.2	0	0.1	0.1	—
969	Sweet, canned in water	½	cup(s)	124	107.9	57	1.0	14.6	1.9	0.2	0	0	0	—
240	Sweet, raw	½	cup(s)	73	59.6	46	0.8	11.6	1.5	0.1	0	0	0	—
	Cranberries													
3007	Chopped, raw	½	cup(s)	55	47.9	25	0.2	6.7	2.5	0.1	0	0	0	—
1717	Cranberry apple juice drink	½	cup(s)	123	102.6	77	0	19.4	0	0.1	0	0	0.1	—
1638	Cranberry juice cocktail	½	cup(s)	127	109.0	68	0	17.1	0	0.1	0	0	0.1	—
241	Cranberry juice cocktail, low calorie, with saccharin	½	cup(s)	119	112.8	23	0	5.5	0	0	0	0	0	—
242	Cranberry sauce, sweetened, canned	¼	cup(s)	69	42.0	105	0.1	26.9	0.7	0.1	0	0	0	—
	Dates													
244	Domestic, chopped	¼	cup(s)	45	9.1	125	1.1	33.4	3.6	0.2	0	0	0	—
243	Domestic, whole	¼	cup(s)	45	9.1	125	1.1	33.4	3.6	0.2	0	0	0	—
	Figs													
975	Canned in heavy syrup	½	cup(s)	130	98.8	114	0.5	29.7	2.8	0.1	0	0	0.1	—
974	Canned in water	½	cup(s)	124	105.7	66	0.5	17.3	2.7	0.1	0	0	0.1	—
973	Raw , medium	2	item(s)	100	79.1	74	0.7	19.2	2.9	0.3	0.1	0.1	0.1	—
	Fruit cocktail & salad													
245	Fruit cocktail, canned in heavy syrup	½	cup(s)	124	99.7	91	0.5	23.4	1.2	0.1	0	0	0	—
978	Fruit cocktail, canned in juice	½	cup(s)	119	103.6	55	0.5	14.1	1.2	0	0	0	0	—
977	Fruit cocktail, canned in water	½	cup(s)	119	107.6	38	0.5	10.1	1.2	0.1	0	0	0	—
979	Fruit salad, canned in water	½	cup(s)	123	112.1	37	0.4	9.6	1.2	0.1	0	0	0	—
	Gooseberries													
982	Canned in light syrup	½	cup(s)	126	100.9	92	0.8	23.6	3.0	0.3	0	0	0.1	—

Chol (mg)	Calc (mg)	Iron (mg)	Magn (mg)	Pota (mg)	Sodi (mg)	Zinc (mg)	Vit A (µg)	Thia (mg)	Vit E (mg α)	Ribo (mg)	Niac (mg)	Vit B$_6$ (mg)	Fola (µg)	Vit C (mg)	Vit B$_{12}$ (µg)	Sele (µg)
18	9	0.89	8.5	34.5	414.5	0.30	—	0.08	—	0.04	0.71	0.03	4.0	0.1	0	—
0	4	0.45	8.5	33.2	57.0	0.11	0	0.09	—	0.02	0.48	0.03	6.6	0	0	—
0	7	0.45	1.8	25.5	141.7	0.19	0	0.01	—	0.03	0.08	0.01	1.8	0	0	—
0	7	1.00	12.4	51.5	0.5	0.35	0	0.12	0.04	0.07	0.90	0.04	7.8	0	0	40.0
0	11	0.74	21.0	30.8	2.1	0.57	0	0.08	0.21	0.03	0.50	0.06	3.5	0	0	18.1
0	1	0.25	11.5	26.3	0.6	0.25	0.8	0.01	0.02	0.01	0.18	0.01	2.5	0	0	0
2	15	0.61	12.3	38.4	72.5	0.20	0.7	0.02	0.42	0.02	0.77	0.01	1.8	0	0	1.3
4	40	0.79	32.5	93.2	317.4	0.71	13.6	0.04	—	0.08	0.52	0.08	3.9	0.2	0.2	4.3
0	0	0.22	8.7	20.0	116.4	0.34	0.9	0.01	0.27	0.00	0.13	0.01	2.8	0	0	0.2
0	7	0.31	6.0	150.6	8.4	0.05	0	0.00	0.01	0.02	0.05	0.04	0	0.7	0	0.1
0	9	0.46	3.7	147.6	3.7	0.04	0	0.03	0.01	0.02	0.12	0.04	0	1.1	0	0.1
0	3	0.06	2.7	58.8	0.5	0.02	1.6	0.01	0.10	0.01	0.05	0.02	1.6	2.5	0	0
0	4	0.16	2.6	75.2	0.9	0.03	1.7	0.01	0.04	0.01	0.08	0.04	0.9	0.2	0	0.3
0	8	0.16	6.9	147.7	1.4	0.05	4.1	0.02	0.24	0.03	0.12	0.05	4.1	6.3	0	0
0	3	0.30	3.4	96.8	18.7	0.04	0	0.00	0.11	0.03	0.20	0.03	0	0.8	0	0.3
0	5	0.44	3.8	77.8	3.8	0.05	1.3	0.01	0.26	0.03	0.24	0.03	1.3	2.2	0	0.4
0	4	0.14	3.7	91.5	2.4	0.03	1.2	0.01	0.25	0.03	0.22	0.03	1.2	1.5	0	0.4
0	6	0.12	2.5	67.9	0.4	—	0.7	0.01	0.20	0.01	0.03	—	2.0	2.8	0	—
0	18	0.54	14.0	362.6	1.4	0.28	134.4	0.04	1.24	0.05	0.84	0.07	12.6	14.0	0	0.1
0	12	0.38	9.0	180.6	5.2	0.14	80.0	0.02	0.77	0.02	0.48	0.07	2.6	4.0	0	0.1
0	18	0.87	10.5	381.5	3.3	0.12	59.1	0.00	1.42	0.02	0.85	0.05	3.3	0.3	0	0.7
0	15	0.66	33.3	583.0	9.2	0.78	8.0	0.08	2.23	0.16	2.19	0.31	102.3	10.1	0	0.4
0	12	0.19	27.6	403.6	2.3	0.45	8.0	0.02	3.03	0.04	0.76	0.08	40.3	20.0	0	—
0	3	0.14	8.0	134.1	1.9	0.17	1.9	0.01	0.57	0.03	0.48	0.07	22.4	2.8	0	0.1
0	10	0.69	41.8	294.8	3.3	0.40	2.2	0.04	0.13	0.01	0.39	0.14	7.7	3.5	0	0.8
0	6	0.30	31.9	422.4	1.2	0.17	3.5	0.03	0.11	0.08	0.78	0.43	23.6	10.3	0	1.2
0	21	0.45	14.4	116.6	0.7	0.38	7.9	0.01	0.84	0.02	0.47	0.02	18.0	15.1	0	0.3
0	22	0.60	16.6	105.7	0.8	0.19	4.5	0.02	0.88	0.03	0.91	0.05	25.7	2.3	0	0.3
0	6	0.42	5.1	51.2	3.8	0.09	2.6	0.04	0.49	0.07	0.14	0.05	2.6	1.4	0	0.1
0	4	0.20	4.4	55.8	0.7	0.12	2.2	0.03	0.41	0.03	0.30	0.04	4.4	7.0	0	0.1
0	6	0.14	3.9	41.9	0.8	0.05	1.6	0.03	0.37	0.03	0.40	0.05	5.4	1.9	0	0.1
0	23	0.55	14.1	115.2	3.8	0.24	2.6	0.03	—	0.04	0.29	0.05	43.5	7.9	0	0.5
0	18	0.56	10.6	91.7	0.7	0.15	2.0	0.04	0.57	0.02	0.51	0.04	41.6	2.0	0	0.1
0	65	2.07	96.0	1881.6	7.7	0.46	0	0.42	0.38	0.11	3.45	0.38	53.8	111.4	0	2.3
0	13	1.67	7.3	119.6	8.5	0.09	46.4	0.02	0.28	0.05	0.22	0.05	9.8	2.6	0	0
0	12	0.25	7.0	134.1	2.3	0.08	49.6	0.02	0.05	0.03	0.31	0.03	6.2	7.8	0	0
0	11	0.44	11.4	183.4	3.8	0.12	10.1	0.02	0.29	0.05	0.50	0.03	5.1	4.6	0	0
0	14	0.45	11.2	162.4	1.2	0.10	9.9	0.03	0.29	0.05	0.51	0.04	5.0	2.7	0	0
0	9	0.26	8.0	161.0	0	0.05	2.2	0.02	0.05	0.02	0.11	0.04	2.9	5.1	0	0
0	4	0.13	3.3	46.8	1.1	0.05	1.7	0.01	0.66	0.01	0.05	0.03	0.6	7.3	0	0.1
0	4	0.09	1.2	20.8	2.5	0.02	0	0.00	0.15	0.00	0.00	0.00	0	48.4	0	0
0	4	0.13	1.3	17.7	2.5	0.04	0	0.00	0.28	0.00	0.05	0.00	0	53.5	0	0.3
0	11	0.05	2.4	29.6	3.6	0.02	0	0.00	0.06	0.00	0.00	0.00	0	38.2	0	0
0	3	0.15	2.1	18.0	20.1	0.03	1.4	0.01	0.57	0.01	0.06	0.01	0.7	1.4	0	0.2
0	17	0.45	19.1	291.9	0.9	0.12	0	0.02	0.02	0.02	0.56	0.07	8.5	0.2	0	1.3
0	17	0.45	19.1	291.9	0.9	0.12	0	0.02	0.02	0.02	0.56	0.07	8.5	0.2	0	1.3
0	35	0.36	13.0	128.2	1.3	0.14	2.6	0.03	0.16	0.05	0.55	0.09	2.6	1.3	0	0.3
0	35	0.36	12.4	127.7	1.2	0.15	2.5	0.03	0.10	0.05	0.55	0.09	2.5	1.2	0	0.1
0	35	0.36	17.0	232.0	1.0	0.14	7.0	0.06	0.10	0.04	0.40	0.10	6.0	2.0	0	0.2
0	7	0.36	6.2	109.1	7.4	0.09	12.4	0.02	0.49	0.02	0.46	0.06	3.7	2.4	0	0.6
0	9	0.25	8.3	112.6	4.7	0.11	17.8	0.01	0.47	0.02	0.48	0.06	3.6	3.2	0	0.6
0	6	0.30	8.3	111.4	4.7	0.11	15.4	0.02	0.47	0.01	0.43	0.06	3.6	2.5	0	0.6
0	9	0.37	6.1	95.6	3.7	0.10	27.0	0.02	—	0.03	0.46	0.04	3.7	2.3	0	1.0
0	20	0.42	7.6	97.0	2.5	0.14	8.8	0.03	—	0.07	0.19	0.02	3.8	12.6	0	0.5

DA+ Code	Food Description	Quantity	Measure	Wt (g)	H₂O (g)	Ener (kcal)	Prot (g)	Carb (g)	Fiber (g)	Fat (g)	Fat Breakdown (g)			
											Sat	Mono	Poly	Trans
	FRUIT AND FRUIT JUICES —Continued													
981	Raw	½	cup(s)	75	65.9	33	0.7	7.6	3.2	0.4	0	0	0.2	—
	Grapefruit													
251	Juice, pink, sweetened, canned	½	cup(s)	125	109.1	57	0.7	13.9	0.1	0.1	0	0	0	—
249	Juice, white	½	cup(s)	124	111.2	48	0.6	11.4	0.1	0.1	0	0	0	—
3022	Pink or red, raw	½	cup(s)	114	100.8	48	0.9	12.2	1.8	0.2	0	0	0	—
248	Sections, canned in light syrup	½	cup(s)	127	106.2	76	0.7	19.6	0.5	0.1	0	0	0	—
983	Sections, canned in water	½	cup(s)	122	109.6	44	0.7	11.2	0.5	0.1	0	0	0	—
247	White, raw	½	cup(s)	115	104.0	38	0.8	9.7	1.3	0.1	0	0	0	—
	Grapes													
255	American, slip skin	½	cup(s)	46	37.4	31	0.3	7.9	0.4	0.2	0.1	0	0	—
256	European, red or green, adherent skin	½	cup(s)	76	60.8	52	0.5	13.7	0.7	0.1	0	0	0	—
3159	Grape Juice drink, canned	½	cup(s)	125	106.6	71	0	18.2	0.1	0	0	0	0	0
259	Grape Juice, sweetened, with added vitamin C, prepared from frozen concentrate	½	cup(s)	125	108.6	64	0.2	15.9	0.1	0.1	0	0	0	—
3060	Raisins, seeded, packed	¼	cup(s)	41	6.8	122	1.0	32.4	2.8	0.2	0.1	0	0.1	—
987	Guava, raw	1	item(s)	55	44.4	37	1.4	7.9	3.0	0.5	0.2	0	0.2	—
35593	Guavas, strawberry	1	item(s)	6	4.8	4	0	1.0	0.3	0	0	0	0	—
3027	Jackfruit	½	cup(s)	83	60.4	78	1.2	19.8	1.3	0.2	0	0	0.1	—
990	Kiwi fruit or Chinese gooseberries	1	item(s)	76	63.1	46	0.9	11.1	2.3	0.4	0	0	0.2	—
	Lemon													
262	Juice	1	tablespoon(s)	15	13.8	4	0.1	1.3	0.1	0	0	0	0	0
993	Peel	1	teaspoon(s)	2	1.6	1	0	0.3	0.2	0	0	0	0	—
992	Raw	1	item(s)	108	94.4	22	1.3	11.6	5.1	0.3	0	0	0.1	—
	Lime													
269	Juice	1	tablespoon(s)	15	14.0	4	0.1	1.3	0.1	0	0	0	0	—
994	Raw	1	item(s)	67	59.1	20	0.5	7.1	1.9	0.1	0	0	0	—
995	Loganberries, frozen	½	cup(s)	74	62.2	40	1.1	9.6	3.9	0.2	0	0	0.1	—
	Mandarin orange													
1038	Canned in juice	½	cup(s)	125	111.4	46	0.8	11.9	0.9	0	0	0	0	—
1039	Canned in light syrup	½	cup(s)	126	104.7	77	0.6	20.4	0.9	0.1	0	0	0	—
999	Mango	½	cup(s)	83	67.4	54	0.4	14.0	1.5	0.2	0.1	0.1	0	—
1005	Nectarine, raw, sliced	½	cup(s)	69	60.4	30	0.7	7.3	1.2	0.2	0	0.1	0.1	—
	Melons													
271	Cantaloupe	½	cup(s)	80	72.1	27	0.7	6.5	0.7	0.1	0	0	0.1	—
1000	Casaba melon	½	cup(s)	85	78.1	24	0.9	5.6	0.8	0.1	0	0	0	—
272	Honeydew	½	cup(s)	89	79.5	32	0.5	8.0	0.7	0.1	0	0	0	—
318	Watermelon	½	cup(s)	76	69.5	23	0.5	5.7	0.3	0.1	0	0	0	—
	Orange													
14412	Juice with calcium and vitamin D	½	cup(s)	120	—	55	1.0	13.0	0	0	0	0	0	0
29630	Juice, fresh squeezed	½	cup(s)	124	109.5	56	0.9	12.9	0.2	0.2	0	0	0	—
14411	Juice, not from concentrate	½	cup(s)	120	—	55	1.0	13.0	0	0	0	0	0	0
278	Juice, unsweetened, prepared from frozen concentrate	½	cup(s)	125	109.7	56	0.8	13.4	0.2	0.1	0	0	0	—
3040	Peel	1	teaspoon(s)	2	1.5	2	0	0.5	0.2	0	0	0	0	—
273	Raw	1	item(s)	131	113.6	62	1.2	15.4	3.1	0.2	0	0	0	—
274	Sections	½	cup(s)	90	78.1	42	0.8	10.6	2.2	0.1	0	0	0	—
	Papaya, Raw													
16830	Dried, strips	2	item(s)	46	12.0	119	1.9	29.9	5.5	0.4	0.1	0.1	0.1	—
282	Papaya, Raw	½	cup(s)	70	62.2	27	0.4	6.9	1.3	0.1	0	0	0	—
35640	Passion fruit, purple	1	item(s)	18	13.1	17	0.4	4.2	1.9	0.1	0	0	0.1	—
	Peach													
285	Halves, canned in heavy syrup	½	cup(s)	131	103.9	97	0.6	26.1	1.7	0.1	0	0	0.1	—
286	Halves, canned in water	½	cup(s)	122	113.6	29	0.5	7.5	1.6	0.1	0	0	0	—
290	Slices, sweetened, frozen	½	cup(s)	125	93.4	118	0.8	30.0	2.3	0.2	0	0.1	0.1	—
283	Raw, medium	1	item(s)	150	133.3	59	1.4	14.3	2.3	0.4	0	0.1	0.1	—
	Pear													
8672	Asian	1	item(s)	122	107.7	51	0.6	13.0	4.4	0.3	0	0.1	0.1	—
293	D'Anjou	1	item(s)	200	168.0	120	1.0	30.0	5.2	1.0	0	0.2	0.2	—
294	Halves, canned in heavy syrup	½	cup(s)	133	106.9	98	0.3	25.5	2.1	0.2	0	0	0	—
1012	Halves, canned in juice	½	cup(s)	124	107.2	62	0.4	16.0	2.0	0.1	0	0	0	—
291	Raw	1	item(s)	166	139.0	96	0.6	25.7	5.1	0.2	0	0	0	—
1017	Persimmon	1	item(s)	25	16.1	32	0.2	8.4	—	0.1	0	0	0	—
	Pineapple													
3053	Canned in extra heavy syrup	½	cup(s)	130	101.0	108	0.4	28.0	1.0	0.1	0	0	0	—
1019	Canned in juice	½	cup(s)	125	104.0	75	0.5	19.5	1.0	0.1	0	0	0	—
296	Canned in light syrup	½	cup(s)	126	108.0	66	0.5	16.9	1.0	0.2	0	0	0.1	—
1018	Canned in water	½	cup(s)	123	111.7	39	0.5	10.2	1.0	0.1	0	0	0	—
299	Juice, unsweetened, canned	½	cup(s)	125	108.0	66	0.5	16.1	0.3	0.2	0	0	0.1	—
295	Raw, diced	½	cup(s)	78	66.7	39	0.4	10.2	1.1	0.1	0	0	0	—

Chol (mg)	Calc (mg)	Iron (mg)	Magn (mg)	Pota (mg)	Sodi (mg)	Zinc (mg)	Vit A (µg)	Thia (mg)	Vit E (mg α)	Ribo (mg)	Niac (mg)	Vit B_6 (mg)	Fola (µg)	Vit C (mg)	Vit B_{12} (µg)	Sele (µg)
0	19	0.23	7.5	148.5	0.8	0.09	11.3	0.03	0.28	0.02	0.23	0.06	4.5	20.8	0	0.5
0	10	0.45	12.5	202.2	2.5	0.08	0	0.05	0.05	0.03	0.40	0.03	12.5	33.6	0	0.1
0	11	0.25	14.8	200.1	1.2	0.06	1.2	0.05	0.27	0.02	0.25	0.05	12.4	46.9	0	0.1
0	25	0.09	10.3	154.5	0	0.07	66.4	0.04	0.14	0.03	0.23	0.06	14.9	35.7	0	0.1
0	18	0.50	12.7	163.8	2.5	0.10	0	0.04	0.11	0.02	0.30	0.02	11.4	27.1	0	1.1
0	18	0.50	12.2	161.0	2.4	0.11	2.3	0.05	0.11	0.03	0.30	0.02	11.0	26.6	0	1.1
0	14	0.07	10.4	170.2	0	0.08	2.3	0.04	0.15	0.02	0.30	0.05	11.5	38.3	0	1.6
0	6	0.13	2.3	87.9	0.9	0.02	2.3	0.04	0.09	0.02	0.14	0.05	1.8	1.8	0	0
0	8	0.27	5.3	144.2	1.5	0.05	2.3	0.05	0.14	0.05	0.14	0.07	1.5	8.2	0	0.1
0	9	0.16	7.5	41.3	11.3	0.04	0	0.28	0.00	0.44	0.18	0.04	1.3	33.1	0	0.1
0	5	0.13	5.0	26.3	2.5	0.05	0	0.02	0.00	0.03	0.16	0.05	1.3	29.9	0	0.1
0	12	1.06	12.4	340.3	11.6	0.07	0	0.04	—	0.07	0.46	0.07	1.2	2.2	0	0.2
0	10	0.14	12.1	229.4	1.1	0.12	17.1	0.03	0.40	0.02	0.59	0.06	27.0	125.6	0	0.3
0	1	0.01	1.0	17.5	2.2	—	0.3	0.00	—	0.00	0.03	0.00	—	2.2	0	
0	28	0.49	30.5	250.0	2.5	0.35	12.4	0.02	—	0.09	0.33	0.09	11.5	5.5	0	0.5
0	26	0.23	12.9	237.1	2.3	0.10	3.0	0.02	1.11	0.01	0.25	0.04	19.0	70.5	0	0.2
0	1	0.00	0.9	18.9	0.2	0.01	0.2	0.00	0.02	0.00	0.02	0.01	2.0	7.0	0	0
0	3	0.01	0.3	3.2	0.1	0.01	0.1	0.00	0.01	0.00	0.01	0.00	0.3	2.6	0	0
0	66	0.75	13.0	156.6	3.2	0.10	2.2	0.05	—	0.04	0.21	0.11	—	83.2	0	1.0
0	2	0.02	1.2	18.0	0.3	0.01	0.3	0.00	0.03	0.00	0.02	0.01	1.5	4.6	0	0
0	22	0.40	4.0	68.3	1.3	0.07	1.3	0.02	0.14	0.01	0.13	0.02	5.4	19.5	0	0.3
0	19	0.47	15.4	106.6	0.7	0.25	1.5	0.04	0.64	0.03	0.62	0.05	19.1	11.2	0	0.1
0	14	0.34	13.7	165.6	6.2	0.64	53.5	0.10	0.12	0.04	0.55	0.05	6.2	42.6	0	0.5
0	9	0.47	10.1	98.3	7.6	0.30	52.9	0.07	0.13	0.06	0.56	0.05	6.3	24.9	0	0.5
0	8	0.10	7.4	128.7	1.7	0.03	31.4	0.05	0.92	0.04	0.48	0.11	11.6	22.8	0	0.5
0	4	0.19	6.2	138.7	0	0.12	11.7	0.02	0.53	0.02	0.78	0.02	3.5	3.7	0	0
0	7	0.17	9.6	213.6	12.8	0.14	135.2	0.03	0.04	0.01	0.59	0.05	16.8	29.4	0	0.3
0	9	0.29	9.4	154.7	7.7	0.06	0	0.01	0.04	0.03	0.20	0.14	6.8	18.5	0	0.3
0	5	0.15	8.8	201.8	15.9	0.07	2.7	0.03	0.01	0.01	0.37	0.07	16.8	15.9	0	0.6
0	5	0.18	7.6	85.1	0.8	0.07	21.3	0.02	0.04	0.01	0.13	0.03	2.3	6.2	0	0.3
0	175	0.00	12.0	225.0	0	—	0	0.08	—	0.03	0.40	0.06	30.0	36.0	0	—
0	14	0.25	13.6	248.0	1.2	0.06	12.4	0.11	0.05	0.04	0.50	0.05	37.2	62.0	0	0.1
0	10	0.00	12.5	225.0	0	0.06	0	0.08	—	0.03	0.40	0.06	30.0	36.0	0	0.1
0	11	0.12	12.5	236.6	1.2	0.06	6.2	0.10	0.25	0.02	0.25	0.06	54.8	48.4	0	0.1
0	3	0.01	0.4	4.2	0.1	0.01	0.4	0.00	0.01	0.00	0.01	0.00	0.6	2.7	0	0
0	52	0.13	13.1	237.1	0	0.09	14.4	0.11	0.23	0.05	0.36	0.07	39.3	69.7	0	0.7
0	36	0.09	9.0	162.9	0	0.06	9.9	0.07	0.16	0.03	0.25	0.05	27.0	47.9	0	0.4
0	73	0.30	30.4	782.9	9.2	0.21	83.7	0.06	2.22	0.08	0.93	0.05	58.0	37.7	0	1.8
0	17	0.07	7.0	179.9	2.1	0.05	38.5	0.02	0.51	0.02	0.24	0.01	26.6	43.3	0	0.4
0	2	0.28	5.2	62.6	5.0	0.01	11.5	0.00	0.00	0.02	0.27	0.01	2.5	5.4	0	0.1
0	4	0.35	6.6	120.5	7.9	0.11	22.3	0.01	0.64	0.03	0.80	0.02	3.9	3.7	0	0.4
0	2	0.39	6.1	120.8	3.7	0.11	32.9	0.01	0.59	0.02	0.63	0.01	3.7	3.5	0	0.4
0	4	0.46	6.3	162.5	7.5	0.06	17.5	0.01	0.77	0.04	0.81	0.03	3.8	117.8	0	0.5
0	9	0.37	13.5	285.0	0	0.25	24.0	0.03	1.09	0.04	1.20	0.03	6.0	9.9	0	
0	5	0.00	9.8	147.6	0	0.02	0	0.01	0.14	0.01	0.26	0.02	9.8	4.6	0	0.1
0	22	0.50	12.0	250.0	0	0.24	—	0.04	1.00	0.08	0.20	0.03	14.6	8.0	0	1.0
0	7	0.29	5.3	86.5	6.7	0.10	0	0.01	0.10	0.02	0.32	0.01	1.3	1.5	0	0
0	11	0.36	8.7	119.0	5.0	0.11	0	0.01	0.10	0.01	0.25	0.02	1.2	2.0	0	0
0	15	0.28	11.6	197.5	1.7	0.16	1.7	0.02	0.19	0.04	0.26	0.04	11.6	7.0	0	0.2
0	7	0.62	—	77.5	0.3	—	0	—	—	—	—	—	—	16.5	0	—
0	18	0.49	19.5	132.6	1.3	0.14	1.3	0.11	—	0.03	0.36	0.09	6.5	9.5	0	—
0	17	0.35	17.4	151.9	1.2	0.12	2.5	0.12	0.01	0.02	0.35	0.09	6.2	11.8	0	0.5
0	18	0.49	20.2	132.3	1.3	0.15	2.5	0.11	0.01	0.03	0.36	0.09	6.3	9.5	0	0.5
0	18	0.49	22.1	156.2	1.2	0.15	2.5	0.11	0.01	0.03	0.37	0.09	6.2	9.5	0	0.5
0	16	0.39	15.0	162.5	2.5	0.14	0	0.07	0.03	0.03	0.25	0.13	22.5	12.5	0	0.1
0	10	0.22	9.3	84.5	0.8	0.09	2.3	0.06	0.02	0.03	0.39	0.09	14.0	37.0	0	0.1

Food Composition   (Computer code is for Cengage Diet Analysis program)   (For purposes of calculations, use "0" for t, <1, <.1, <.01, etc.)

DA+ Code	Food Description	Quantity	Measure	Wt (g)	H₂O (g)	Ener (kcal)	Prot (g)	Carb (g)	Fiber (g)	Fat (g)	Fat Breakdown (g)			
											Sat	Mono	Poly	Trans
	FRUIT AND FRUIT JUICES —Continued													
1024	Plantain, cooked	½	cup(s)	77	51.8	89	0.6	24.0	1.8	0.1	0.1	0	0	—
300	Plum, raw, large	1	item(s)	66	57.6	30	0.5	7.5	0.9	0.2	0	0.1	0	—
1027	Pomegranate	1	item(s)	154	124.7	105	1.5	26.4	0.9	0.5	0.1	0.1	0.1	—
	Prunes, Dried													
5644	Dried	2	item(s)	17	5.2	40	0.4	10.7	1.2	0.1	0	0	0	—
305	Dried, stewed	½	cup(s)	124	86.5	133	1.2	34.8	3.8	0.2	0	0.1	0	—
306	Juice, canned	1	cup(s)	256	208.0	182	1.6	44.7	2.6	0.1	0	0.1	0	—
	Raspberries													
309	Raw	½	cup(s)	62	52.7	32	0.7	7.3	4.0	0.4	0	0	0.2	—
310	Red, sweetened, frozen	½	cup(s)	125	90.9	129	0.9	32.7	5.5	0.2	0	0	0.1	—
311	Rhubarb, cooked with sugar	½	cup(s)	120	81.5	140	0.5	37.5	2.7	0.1	0	0	0.1	—
	Strawberries													
313	Raw	½	cup(s)	72	65.5	23	0.5	5.5	1.4	0.2	0	0	0.1	—
315	Sweetened, frozen, thawed	½	cup(s)	128	99.5	99	0.7	26.8	2.4	0.2	0	0	0.1	—
16828	Tangelo	1	item(s)	95	82.4	45	0.9	11.2	2.3	0.1	0	0	0	—
	Tangerine													
1040	Juice	½	cup(s)	124	109.8	53	0.6	12.5	0.2	0.2	0	0	0	—
316	Raw	1	item(s)	88	74.9	47	0.7	11.7	1.6	0.3	0	0.1	0.1	—
	VEGETABLES, LEGUMES													
	Amaranth													
1043	Leaves, boiled, drained	½	cup(s)	66	60.4	14	1.4	2.7	—	0.1	0	0	0.1	—
1042	Leaves, raw	1	cup(s)	28	25.7	6	0.7	1.1	—	0.1	0	0	0	—
8683	Arugula leaves, raw	1	cup(s)	20	18.3	5	0.5	0.7	0.3	0.1	0	0	0.1	—
	Artichoke													
1044	Boiled, drained	1	item(s)	120	100.9	64	3.5	14.3	10.3	0.4	0.1	0	0.2	—
2885	Hearts, boiled, drained	½	cup(s)	84	70.6	45	2.4	10.0	7.2	0.3	0.1	0	0.1	—
	Asparagus													
566	Boiled, drained	½	cup(s)	90	83.4	20	2.2	3.7	1.8	0.2	0	0	0.1	—
568	Canned, drained	½	cup(s)	121	113.7	23	2.6	3.0	1.9	0.8	0.2	0	0.3	—
565	Tips, frozen, boiled, drained	½	cup(s)	90	84.7	16	2.7	1.7	1.4	0.4	0.1	0	0.2	—
	Bamboo shoots													
1048	Boiled, drained	½	cup(s)	60	57.6	7	0.9	1.2	0.6	0.1	0	0	0.1	—
1049	Canned, drained	½	cup(s)	66	61.8	12	1.1	2.1	0.9	0.3	0.1	0	0.1	—
	Beans													
1801	Adzuki beans, boiled	½	cup(s)	115	76.2	147	8.6	28.5	8.4	0.1	0	—	—	—
511	Baked beans with franks, canned	½	cup(s)	130	89.8	184	8.7	19.9	8.9	8.5	3.0	3.7	1.1	—
513	Baked beans with pork in sweet sauce, canned	½	cup(s)	127	89.3	142	6.7	26.7	5.3	1.8	0.6	0.6	0.5	0
512	Baked beans with pork in tomato sauce, canned	½	cup(s)	127	93.0	119	6.5	23.6	5.1	1.2	0.5	0.7	0.3	—
1805	Black beans, boiled	½	cup(s)	86	56.5	114	7.6	20.4	7.5	0.5	0.1	0	0.2	—
14597	Chickpeas, garbanzo beans or bengal gram, boiled	½	cup(s)	82	49.4	134	7.3	22.5	6.2	2.1	0.2	0.5	0.9	—
569	Fordhook lima beans, frozen, boiled, drained	½	cup(s)	85	62.0	88	5.2	16.4	4.9	0.3	0.1	0	0.1	—
1806	French beans, boiled	½	cup(s)	89	58.9	114	6.2	21.3	8.3	0.7	0.1	0	0.2	—
2773	Great northern beans, boiled	½	cup(s)	89	61.1	104	7.4	18.7	6.2	0.4	0.1	0	0.2	—
2736	Hyacinth beans, boiled, drained	½	cup(s)	44	37.8	22	1.3	4.0	—	0.1	0.1	0.1	0	—
570	Lima beans, baby, frozen, boiled, drained	½	cup(s)	90	65.1	95	6.0	17.5	5.4	0.3	0.1	0	0.1	—
515	Lima beans, boiled, drained	½	cup(s)	85	57.1	105	5.8	20.1	4.5	0.3	0.1	0	0.1	—
579	Mung beans, sprouted, boiled, drained	½	cup(s)	62	57.9	13	1.3	2.6	0.5	0.1	0	0	0	—
510	Navy beans, boiled	½	cup(s)	91	58.1	127	7.5	23.7	9.6	0.6	0.1	0.1	0.4	0
32816	Pinto beans, boiled, drained, no salt added	½	cup(s)	63	58.8	14	1.2	2.6	—	0.2	0	0	0.1	—
1052	Pinto beans, frozen, boiled, drained	½	cup(s)	47	27.3	76	4.4	14.5	4.0	0.2	0	0	0.1	—
514	Red kidney beans, canned	½	cup(s)	128	99.0	108	6.7	19.9	6.9	0.5	0.1	0.2	0.2	—
1810	Refried beans, canned	½	cup(s)	127	96.1	119	6.9	19.6	6.7	1.6	0.6	0.7	0.2	—
1053	Shell beans, canned	½	cup(s)	123	111.1	37	2.2	7.6	4.2	0.2	0	0	0.1	—
1670	Soybeans, boiled	½	cup(s)	86	53.8	149	14.3	8.5	5.2	7.7	1.1	1.7	4.4	—
1108	Soybeans, green, boiled, drained	½	cup(s)	90	61.7	127	11.1	9.9	3.8	5.8	0.7	1.1	2.7	—
1807	White beans, small, boiled	½	cup(s)	90	56.6	127	8.0	23.1	9.3	0.6	0.1	0.1	0.2	—
575	Yellow snap, string or wax beans, boiled, drained	½	cup(s)	63	55.8	22	1.2	4.9	2.1	0.2	0	0	0.1	—
576	Yellow snap, string or wax beans, frozen, boiled, drained	½	cup(s)	68	61.7	19	1.0	4.4	2.0	0.1	0	0	0.1	—
	Beets													
584	Beet greens, boiled, drained	½	cup(s)	72	64.2	19	1.9	3.9	2.1	0.1	0	0	0.1	—
2730	Pickled, canned with liquid	½	cup(s)	114	92.9	74	0.9	18.5	3.0	0.1	0	0	0	—
581	Sliced, boiled, drained	½	cup(s)	85	74.0	37	1.4	8.5	1.7	0.2	0	0	0.1	—
583	Sliced, canned, drained	½	cup(s)	85	77.3	26	0.8	6.1	1.5	0.1	0	0	0	—
580	Whole, boiled, drained	2	item(s)	100	87.1	44	1.7	10.0	2.0	0.2	0	0	0.1	—
585	Cowpeas or black-eyed peas, boiled, drained	½	cup(s)	83	62.3	80	2.6	16.8	4.1	0.3	0.1	0	0.1	—
	Broccoli													
588	Chopped, boiled, drained	½	cup(s)	78	69.6	27	1.9	5.6	2.6	0.3	0.1	0	0.1	—

APPENDIX E

Chol (mg)	Calc (mg)	Iron (mg)	Magn (mg)	Pota (mg)	Sodi (mg)	Zinc (mg)	Vit A (µg)	Thia (mg)	Vit E (mg α)	Ribo (mg)	Niac (mg)	Vit B_6 (mg)	Fola (µg)	Vit C (mg)	Vit B_{12} (µg)	Sele (µg)
0	2	0.45	24.6	358.1	3.9	0.10	34.7	0.04	0.10	0.04	0.58	0.19	20.0	8.4	0	1.1
0	4	0.11	4.6	103.6	0	0.06	11.2	0.02	0.17	0.02	0.27	0.02	3.3	6.3	0	0
0	5	0.46	4.6	398.9	4.6	0.18	7.7	0.04	0.92	0.04	0.46	0.16	9.2	9.4	0	0.9
0	7	0.16	6.9	123.0	0.3	0.07	6.6	0.01	0.07	0.03	0.32	0.03	0.7	0.1	0	0
0	24	0.51	22.3	398.0	1.2	0.24	21.1	0.03	0.24	0.12	0.90	0.27	0	3.6	0	0.1
0	31	3.02	35.8	706.6	10.2	0.53	0	0.04	0.30	0.17	2.01	0.55	0	10.5	0	1.5
0	15	0.42	13.5	92.9	0.6	0.26	1.2	0.02	0.54	0.02	0.37	0.03	12.9	16.1	0	0.1
0	19	0.81	16.3	142.5	1.3	0.22	3.8	0.02	0.90	0.05	0.28	0.04	32.5	20.6	0	0.4
0	174	0.25	16.2	115.0	1.0	—	—	0.02	—	0.03	0.25	—	—	4.0	0	—
0	12	0.30	9.4	110.2	0.7	0.10	0.7	0.02	0.21	0.02	0.28	0.03	17.3	42.3	0	0.3
0	14	0.59	7.7	125.0	1.3	0.06	1.3	0.01	0.30	0.09	0.37	0.03	5.1	50.4	0	0.9
0	38	0.09	9.5	172.0	0	0.06	10.5	0.08	0.17	0.03	0.26	0.05	28.5	50.5	0	0.5
0	22	0.25	9.9	219.8	1.2	0.04	16.1	0.07	0.16	0.02	0.12	0.05	6.2	38.3	0	0.1
0	33	0.13	10.6	146.1	1.8	0.06	29.9	0.05	0.18	0.03	0.33	0.07	14.1	23.5	0	0.1
0	138	1.49	36.3	423.1	13.9	0.58	91.7	0.01	—	0.09	0.37	0.12	37.6	27.1	0	0.6
0	60	0.65	15.4	171.1	5.6	0.25	40.9	0.01	—	0.04	0.18	0.05	23.8	12.1	0	0.3
0	32	0.29	9.4	73.8	5.4	0.09	23.8	0.01	0.09	0.02	0.06	0.01	19.4	3.0	0	0.1
0	25	0.73	50.4	343.2	72.0	0.48	1.2	0.06	0.22	0.10	1.33	0.09	106.8	8.9	0	0.2
0	18	0.51	35.3	240.2	50.4	0.33	0.8	0.04	0.16	0.07	0.93	0.06	74.8	6.2	0	0.2
0	21	0.81	12.6	201.6	12.6	0.54	45.0	0.14	1.35	0.12	0.97	0.07	134.1	6.9	0	5.5
0	19	2.21	12.1	208.1	347.3	0.48	49.6	0.07	1.47	0.12	1.15	0.13	116.2	22.3	0	2.1
0	16	0.50	9.0	154.8	2.7	0.36	36.0	0.05	1.08	0.09	0.93	0.01	121.5	22.0	0	3.5
0	7	0.14	1.8	319.8	2.4	0.28	0	0.01	—	0.03	0.18	0.06	1.2	0	0	0.2
0	5	0.21	2.6	52.4	4.6	0.43	0.7	0.02	0.41	0.02	0.09	0.09	2.0	0.7	0	0.3
0	32	2.30	59.8	611.8	9.2	2.03	0	0.13	—	0.07	0.82	0.11	139.2	0	0	1.4
8	62	2.24	36.3	304.3	556.9	2.42	5.2	0.08	0.21	0.07	1.17	0.06	38.9	3.0	0.4	8.4
9	75	2.08	41.7	326.4	422.5	1.73	0	0.05	0.03	0.07	0.44	0.07	10.1	3.5	0	6.3
9	71	4.09	43.0	373.2	552.8	6.93	5.1	0.06	0.12	0.05	0.62	0.08	19.0	3.8	0	5.9
0	23	1.80	60.2	305.3	0.9	0.96	0	0.21	—	0.05	0.43	0.05	128.1	0	0	1.0
0	40	2.36	39.4	238.6	5.7	1.25	0.8	0.09	0.28	0.05	0.43	0.11	141.0	1.1	0	3.0
0	26	1.54	35.7	258.4	58.7	0.62	8.5	0.06	0.24	0.05	0.90	0.10	17.9	10.9	0	0.5
0	56	0.95	49.6	327.5	5.3	0.56	0	0.11	—	0.05	0.48	0.09	66.4	1.1	0	1.1
0	60	1.88	44.3	346.0	1.8	0.77	0	0.14	—	0.05	0.60	0.10	90.3	1.2	0	3.6
0	18	0.33	18.3	114.0	0.9	0.16	3.0	0.02	—	0.03	0.20	0.01	20.4	2.2	0	0.7
0	25	1.76	50.4	369.9	26.1	0.49	7.2	0.06	0.57	0.04	0.69	0.10	14.4	5.2	0	1.5
0	27	2.08	62.9	484.5	14.5	0.67	12.8	0.11	0.11	0.08	0.88	0.16	22.1	8.6	0	1.7
0	7	0.40	8.7	62.6	6.2	0.29	0.6	0.03	0.04	0.06	0.51	0.03	18.0	7.1	0	0.4
—	63	2.14	48.2	354.0	0	0.93	0	0.21	0.01	0.06	0.59	0.12	127.4	0.8	0	2.6
0	9	0.41	11.3	61.7	32.1	0.10	0	0.04	—	0.03	0.45	0.03	18.3	3.8	0	0.4
0	24	1.27	25.4	303.6	39.0	0.32	0	0.12	—	0.05	0.29	0.09	16.0	3.3	0	0.7
0	32	1.62	35.8	327.7	330.2	2.09	0	0.13	0.02	0.11	0.57	0.10	25.6	1.4	0	0.6
10	44	2.10	41.7	337.8	378.2	1.48	0	0.03	0.00	0.02	0.39	0.18	13.9	7.6	0	1.6
0	36	1.21	18.4	133.5	409.2	0.33	13.5	0.04	0.04	0.07	0.25	0.06	22.1	3.8	0	2.6
0	88	4.42	74.0	442.9	0.9	0.98	0	0.13	0.30	0.24	0.34	0.20	46.4	1.5	0	6.3
0	131	2.25	54.0	485.1	12.6	0.82	7.2	0.23	—	0.14	1.13	0.05	99.9	15.3	0	1.3
0	65	2.54	60.9	414.4	1.8	0.97	0	0.21	—	0.05	0.24	0.11	122.6	0	0	1.2
0	29	0.80	15.6	186.9	1.9	0.23	2.5	0.05	0.28	0.06	0.38	0.04	20.6	6.1	0	0.3
0	33	0.59	16.2	85.1	6.1	0.32	4.1	0.02	0.03	0.06	0.26	0.04	15.5	2.8	0	0.3
0	82	1.36	49.0	654.5	173.5	0.36	275.8	0.08	1.30	0.20	0.35	0.09	10.1	17.9	0	0.6
0	12	0.46	17.0	168.0	299.6	0.29	1.1	0.01	—	0.05	0.28	0.05	30.6	2.6	0	1.1
0	14	0.67	19.6	259.3	65.5	0.30	1.7	0.02	0.03	0.03	0.28	0.05	68.0	3.1	0	0.6
0	13	1.54	14.5	125.8	164.9	0.17	0.9	0.01	0.02	0.03	0.13	0.04	25.5	3.5	0	0.4
0	16	0.79	23.0	305.0	77.0	0.35	2.0	0.02	0.04	0.04	0.33	0.06	80.0	3.6	0	0.7
0	106	0.92	42.9	344.9	3.3	0.85	33.0	0.08	0.18	0.12	1.15	0.05	104.8	1.8	0	2.1
0	31	0.52	16.4	228.5	32.0	0.35	60.1	0.04	1.13	0.09	0.43	0.15	84.2	50.6	0	1.2

DA+ Code	Food Description	Quantity	Measure	Wt (g)	H₂O (g)	Ener (kcal)	Prot (g)	Carb (g)	Fiber (g)	Fat (g)	Fat Breakdown (g) Sat	Mono	Poly	Trans
	VEGETABLES, LEGUMES—Continued													
590	Frozen, chopped, boiled, drained	½	cup(s)	92	83.5	26	2.9	4.9	2.8	0.1	0	0	0.1	—
587	Raw, chopped	½	cup(s)	46	40.6	15	1.3	3.0	1.2	0.2	0	0	0	—
16848	**Broccoflower, raw, chopped**	½	cup(s)	32	28.7	10	0.9	1.9	1.0	0.1	0	0	0	—
	Brussels sprouts													
591	Boiled, drained	½	cup(s)	78	69.3	28	2.0	5.5	2.0	0.4	0.1	0	0.2	—
592	Frozen, boiled, drained	½	cup(s)	78	67.2	33	2.8	6.4	3.2	0.3	0.1	0	0.2	—
	Cabbage													
595	Boiled, drained, no salt added	1	cup(s)	150	138.8	35	1.9	8.3	2.8	0.1	0	0	0	—
35611	Chinese (pak choi or bok choy), boiled with salt, drained	1	cup(s)	170	162.4	20	2.6	3.0	1.7	0.3	0	0	0.1	—
16869	Kim chee	1	cup(s)	150	137.5	32	2.5	6.1	1.8	0.3	0	0	0.2	—
594	Raw, shredded	1	cup(s)	70	64.5	17	0.9	4.1	1.7	0.1	0	0	0	—
596	Red, shredded, raw	1	cup(s)	70	63.3	22	1.0	5.2	1.5	0.1	0	0	0.1	—
597	Savoy, shredded, raw	1	cup(s)	70	63.7	19	1.4	4.3	2.2	0.1	0	0	0	—
35417	**Capers**	1	teaspoon(s)	4	—	2	0	0	0	0	0	0	0	0
	Carrots													
8691	Baby, raw	8	item(s)	80	72.3	28	0.5	6.6	2.3	0.1	0	0	0.1	—
601	Grated	½	cup(s)	55	48.6	23	0.5	5.3	1.5	0.1	0	0	0.1	0
1055	Juice, canned	½	cup(s)	118	104.9	47	1.1	11.0	0.9	0.2	0	0	0.1	—
600	Raw	½	cup(s)	61	53.9	25	0.6	5.8	1.7	0.1	0	0	0.1	0
602	Sliced, boiled, drained	½	cup(s)	78	70.3	27	0.6	6.4	2.3	0.1	0	0	0.1	—
32725	**Cassava or manioc**	½	cup(s)	103	61.5	165	1.4	39.2	1.9	0.3	0.1	0.1	0	—
	Cauliflower													
606	Boiled, drained	½	cup(s)	62	57.7	14	1.1	2.5	1.4	0.3	0	0	0.1	—
607	Frozen, boiled, drained	½	cup(s)	90	84.6	17	1.4	3.4	2.4	0.2	0	0	0.1	—
605	Raw, chopped	½	cup(s)	50	46.0	13	1.0	2.6	1.2	0	0	0	0	—
	Celery													
609	Diced	½	cup(s)	51	48.2	8	0.3	1.5	0.8	0.1	0	0	0	—
608	Stalk	2	item(s)	80	76.3	13	0.6	2.4	1.3	0.1	0	0	0.1	—
	Chard													
1057	Swiss chard, boiled, drained	½	cup(s)	88	81.1	18	1.6	3.6	1.8	0.1	0	0	0	—
1056	Swiss chard, raw	1	cup(s)	36	33.4	7	0.6	1.3	0.6	0.1	0	0	0	—
	Collard greens													
610	Boiled, drained	½	cup(s)	95	87.3	25	2.0	4.7	2.7	0.3	0	0	0.2	—
611	Frozen, chopped, boiled, drained	½	cup(s)	85	75.2	31	2.5	6.0	2.4	0.3	0.1	0	0.2	—
	Corn													
29614	Yellow corn, fresh, cooked	1	item(s)	100	69.2	107	3.3	25.0	2.8	1.3	0.2	0.4	0.6	—
615	Yellow creamed sweet corn, canned	½	cup(s)	128	100.8	92	2.2	23.2	1.5	0.5	0.1	0.2	0.3	—
612	Yellow sweet corn, boiled, drained	½	cup(s)	82	57.0	89	2.7	20.6	2.3	1.1	0.2	0.3	0.5	—
614	Yellow sweet corn, frozen, boiled, drained	½	cup(s)	82	63.2	66	2.1	15.8	2.0	0.5	0.1	0.2	0.3	—
618	**Cucumber**	¼	item(s)	75	71.7	11	0.5	2.7	0.4	0.1	0	0	0	—
16870	**Cucumber, kim chee**	½	cup(s)	75	68.1	16	0.8	3.6	1.1	0.1	0	0	0	—
	Dandelion greens													
620	Chopped, boiled, drained	½	cup(s)	53	47.1	17	1.1	3.4	1.5	0.3	0.1	0	0.1	—
2734	Raw	1	cup(s)	55	47.1	25	1.5	5.1	1.9	0.4	0.1	0	0.2	—
1066	**Eggplant, boiled, drained**	½	cup(s)	50	44.4	17	0.4	4.3	1.2	0.1	0	0	0	—
621	**Endive or escarole, chopped, raw**	1	cup(s)	50	46.9	8	0.6	1.7	1.5	0.1	0	0	0	—
8784	**Jicama or yambean**	½	cup(s)	65	116.5	49	0.9	11.4	6.3	0.1	0	0	0.1	—
	Kale													
623	Frozen, chopped, boiled, drained	½	cup(s)	65	58.8	20	1.8	3.4	1.3	0.3	0	0	0.2	—
29313	Raw	1	cup(s)	67	56.6	33	2.2	6.7	1.3	0.5	0.1	0	0.2	—
	Kohlrabi													
1072	Boiled, drained	½	cup(s)	83	74.5	24	1.5	5.5	0.9	0.1	0	0	0	—
1071	Raw	1	cup(s)	135	122.9	36	2.3	8.4	4.9	0.1	0	0	0.1	—
	Leeks													
1074	Boiled, drained	½	cup(s)	52	47.2	16	0.4	4.0	0.5	0.1	0	0	0	—
1073	Raw	1	cup(s)	89	73.9	54	1.3	12.6	1.6	0.3	0	0	0.1	—
	Lentils													
522	Boiled	¼	cup(s)	50	34.5	57	4.5	10.0	3.9	0.2	0	0	0.1	—
1075	Sprouted	1	cup(s)	77	51.9	82	6.9	17.0	—	0.4	0	0.1	0.2	—
	Lettuce													
625	Butterhead leaves	11	piece(s)	83	78.9	11	1.1	1.8	0.9	0.2	0	0	0.1	—
624	Butterhead, Boston or Bibb	1	cup(s)	55	52.6	7	0.7	1.2	0.6	0.1	0	0	0.1	—
626	Iceberg	1	cup(s)	55	52.6	8	0.5	1.6	0.7	0.1	0	0	0	—
628	Iceberg, chopped	1	cup(s)	55	52.6	8	0.5	1.6	0.7	0.1	0	0	0.1	—
629	Looseleaf	1	cup(s)	36	34.2	5	0.5	1.0	0.5	0.1	0	0	0	—
1665	Romaine, shredded	1	cup(s)	56	53.0	10	0.7	1.8	1.2	0.2	0	0	0.1	—
	Mushrooms													
15585	Crimini (about 6)	3	ounce(s)	85	—	28	3.7	2.8	1.9	0	0	0	0	0

Chol (mg)	Calc (mg)	Iron (mg)	Magn (mg)	Pota (mg)	Sodi (mg)	Zinc (mg)	Vit A (µg)	Thia (mg)	Vit E (mg α)	Ribo (mg)	Niac (mg)	Vit B6 (mg)	Fola (µg)	Vit C (mg)	Vit B12 (µg)	Sele (µg)
0	30	0.56	12.0	130.6	10.1	0.25	46.9	0.05	1.21	0.07	0.42	0.12	51.5	36.9	0	0.6
0	21	0.33	9.6	143.8	15.0	0.19	14.1	0.03	0.36	0.05	0.29	0.08	28.7	40.6	0	1.1
0	11	0.23	6.4	96.0	7.4	0.20	2.6	0.02	0.01	0.03	0.23	0.07	18.2	28.2	0	0.2
0	28	0.93	15.6	247.3	16.4	0.25	30.4	0.08	0.33	0.06	0.47	0.13	46.8	48.4	0	1.2
0	20	0.37	14.0	224.8	11.6	0.18	35.7	0.08	0.39	0.08	0.41	0.22	78.3	35.4	0	0.5
0	72	0.24	22.5	294.0	12.0	0.30	6.0	0.08	0.20	0.04	0.36	0.16	45.0	56.2	0	0.9
0	158	1.76	18.7	630.7	459.0	0.28	360.4	0.04	0.14	0.10	0.72	0.28	69.7	44.2	0	0.7
0	144	1.26	27.0	379.5	996.0	0.36	288.0	0.06	0.36	0.10	0.80	0.32	88.5	79.6	0	1.5
0	28	0.33	8.4	119.0	12.6	0.12	3.5	0.04	0.10	0.02	0.16	0.08	30.1	25.6	0	0.2
0	31	0.56	11.2	170.1	18.9	0.15	39.2	0.04	0.07	0.05	0.29	0.14	12.6	39.9	0	0.4
0	24	0.28	19.6	161.0	19.6	0.18	35.0	0.05	0.11	0.02	0.21	0.13	56.0	21.7	0	0.6
0	0	0.00	—		140	—	0	—	—	—	—	—	—	0	—	
0	26	0.71	8.0	189.6	62.4	0.13	552.0	0.02	—	0.02	0.44	0.08	21.6	2.1	0	0.7
0	18	0.16	6.6	176.0	37.9	0.13	459.2	0.03	0.36	0.03	0.54	0.07	10.4	3.2	0	0.1
0	28	0.54	16.5	344.6	34.2	0.21	1128.1	0.11	1.37	0.07	0.46	0.26	4.7	10.0	0	0.7
0	20	0.18	7.3	195.2	42.1	0.15	509.4	0.04	0.40	0.04	0.60	0.08	11.6	3.6	0	0.1
0	23	0.26	7.8	183.3	45.2	0.15	664.6	0.05	0.80	0.03	0.50	0.11	10.9	2.8	0	0.5
0	16	0.27	21.6	279.1	14.4	0.35	1.0	0.08	0.19	0.04	0.87	0.09	27.8	21.2	0	0.7
0	10	0.19	5.6	88.0	9.3	0.10	0.6	0.02	0.04	0.03	0.25	0.10	27.3	27.5	0	0.4
0	15	0.36	8.1	125.1	16.2	0.11	0	0.03	0.05	0.04	0.27	0.07	36.9	28.2	0	0.5
0	11	0.22	7.5	151.5	15.0	0.14	0.5	0.03	0.04	0.03	0.26	0.11	28.5	23.2	0	0.3
0	20	0.10	5.6	131.3	40.4	0.07	11.1	0.01	0.14	0.03	0.16	0.04	18.2	1.6	0	0.2
0	32	0.16	8.8	208.0	64.0	0.10	17.6	0.01	0.21	0.04	0.25	0.05	28.8	2.5	0	0.3
0	51	1.98	75.3	480.4	156.6	0.29	267.8	0.03	1.65	0.08	0.32	0.07	7.9	15.8	0	0.8
0	18	0.64	29.2	136.4	76.7	0.13	110.2	0.01	0.68	0.03	0.14	0.03	5.0	10.8	0	0.3
0	133	1.10	19.0	110.2	15.2	0.21	385.7	0.03	0.83	0.10	0.54	0.12	88.4	17.3	0	0.5
0	179	0.95	25.5	213.4	42.5	0.22	488.8	0.04	1.06	0.09	0.54	0.09	64.6	22.4	0	1.3
0	2	0.61	32.0	248.0	242.0	0.48	13.0	0.20	0.09	0.07	1.60	0.06	46.0	6.2	0	0.2
0	4	0.48	21.8	171.5	364.8	0.67	5.1	0.03	0.09	0.06	1.22	0.08	57.6	5.9	0	0.5
0	2	0.36	21.3	173.8	0	0.50	10.7	0.17	0.07	0.05	1.32	0.04	37.7	5.1	0	0.2
0	2	0.38	23.0	191.1	0.8	0.51	8.2	0.02	0.05	0.05	1.07	0.08	28.7	2.9	0	0.6
0	12	0.20	9.8	110.6	1.5	0.14	3.8	0.01	0.01	0.01	0.07	0.03	5.3	2.1	0	0.2
0	7	3.61	6.0	87.8	765.8	0.38	—	0.02	—	0.02	0.34	0.08	17.3	2.6	0	—
0	74	0.95	12.6	121.8	23.1	0.15	179.6	0.07	1.28	0.09	0.27	0.08	6.8	9.5	0	0.2
0	103	1.70	19.8	218.3	41.8	0.22	279.4	0.10	1.89	0.14	0.44	0.13	14.8	19.2	0	0.3
0	3	0.12	5.4	60.9	0.5	0.06	1.0	0.04	0.20	0.01	0.30	0.04	6.9	0.6	0	0
0	26	0.41	7.5	157.0	11.0	0.39	54.0	0.04	0.22	0.03	0.20	0.01	71.0	3.2	0	0.1
0	16	0.78	15.5	194.0	5.2	0.20	1.3	0.02	0.59	0.04	0.25	0.05	15.5	26.1	0	0.9
0	90	0.61	11.7	208.7	9.8	0.11	477.8	0.02	0.59	0.07	0.43	0.05	9.1	16.4	0	0.6
0	90	1.14	22.8	299.5	28.8	0.29	515.2	0.07	—	0.08	0.66	0.18	19.4	80.4	0	0.6
0	21	0.33	15.7	280.5	17.3	0.26	1.7	0.03	0.43	0.02	0.32	0.13	9.9	44.6	0	0.7
0	32	0.54	25.7	472.5	27.0	0.04	2.7	0.06	0.64	0.02	0.54	0.20	21.6	83.7	0	0.9
0	16	0.56	7.3	45.2	5.2	0.02	1.0	0.01	—	0.01	0.10	0.04	12.5	2.2	0	0.3
0	53	1.86	24.9	160.2	17.8	0.10	73.9	0.05	0.81	0.02	0.35	0.20	57.0	10.7	0	0.9
0	9	1.65	17.8	182.7	1.0	0.63	0	0.08	0.05	0.04	0.52	0.09	89.6	0.7	0	1.4
0	19	2.47	28.5	247.9	8.5	1.16	1.5	0.17	—	0.09	0.86	0.14	77.0	12.7	0	0.5
0	29	1.02	10.7	196.4	4.1	0.16	137.0	0.04	0.14	0.05	0.29	0.06	60.2	3.1	0	0.5
0	19	0.68	7.1	130.9	2.7	0.11	91.3	0.03	0.09	0.03	0.19	0.04	40.1	2.0	0	0.3
0	10	0.22	3.8	77.5	5.5	0.08	13.7	0.02	0.09	0.01	0.07	0.02	15.9	1.5	0	0.1
0	10	0.22	3.8	77.5	5.5	0.08	13.7	0.02	0.09	0.01	0.07	0.02	15.9	1.5	0	0.1
0	13	0.31	4.7	69.8	10.1	0.06	133.2	0.02	0.10	0.02	0.13	0.03	13.7	6.5	0	0.2
0	18	0.54	7.8	138.3	4.5	0.13	162.4	0.04	0.07	0.03	0.17	0.04	76.2	13.4	0	0.2
0	0	0.67	—	—	32.6	—	0	—	—	—	—	—	—	0	0	—

DA+ Code	Food Description	Quantity	Measure	Wt (g)	H₂O (g)	Ener (kcal)	Prot (g)	Carb (g)	Fiber (g)	Fat (g)	Sat	Mono	Poly	Trans
	VEGETABLES, LEGUMES—Continued													
8700	Enoki	30	item(s)	90	79.7	40	2.3	6.9	2.4	0.3	0	0	0.1	—
1079	Mushrooms, boiled, drained	½	cup(s)	78	71.0	22	1.7	4.1	1.7	0.4	0	0	0.1	—
1080	Mushrooms, canned, drained	½	cup(s)	78	71.0	20	1.5	4.0	1.9	0.2	0	0	0.1	—
630	Mushrooms, raw	½	cup(s)	48	44.4	11	1.5	1.6	0.5	0.2	0	0	0.1	—
15587	Portabella, raw	1	item(s)	84	—	30	3.0	3.9	3.0	0	0	0	0	0
2743	Shiitake, cooked	½	cup(s)	73	60.5	41	1.1	10.4	1.5	0.2	0	0.1	0	—
	Mustard greens													
2744	Frozen, boiled, drained	½	cup(s)	75	70.4	14	1.7	2.3	2.1	0.2	0	0.1	0	—
29319	Raw	1	cup(s)	56	50.8	15	1.5	2.7	1.8	0.1	0	0	0	—
	Okra													
16866	Batter coated, fried	11	piece(s)	83	55.6	156	2.1	12.7	2.0	11.2	1.5	3.7	5.5	—
32742	Frozen, boiled, drained, no salt added	½	cup(s)	92	83.8	26	1.9	5.3	2.6	0.3	0.1	0	0.1	—
632	Sliced, boiled, drained	½	cup(s)	80	74.1	18	1.5	3.6	2.0	0.2	0	0	0	—
	Onions													
635	Chopped, boiled, drained	½	cup(s)	105	92.2	46	1.4	10.7	1.5	0.2	0	0	0.1	—
2748	Frozen, boiled, drained	½	cup(s)	106	97.8	30	0.8	7.0	1.9	0.1	0	0	0	—
1081	Onion rings, breaded and pan fried, frozen, heated	10	piece(s)	71	20.2	289	3.8	27.1	0.9	19.0	6.1	7.7	3.6	—
633	Raw, chopped	½	cup(s)	80	71.3	32	0.9	7.5	1.4	0.1	0	0	0	—
16850	Red onions, sliced, raw	½	cup(s)	57	50.7	24	0.5	5.8	0.8	0	0	0	0	—
636	Scallions, green or spring onions	2	item(s)	30	26.9	10	0.5	2.2	0.8	0.1	0	0	0	—
16860	Palm hearts, cooked	½	cup(s)	73	50.7	84	2.0	18.7	1.1	0.1	0	0	0.1	—
637	Parsley, chopped	1	tablespoon(s)	4	3.3	1	0.1	0.2	0.1	0	0	0	0	—
638	Parsnips, sliced, boiled, drained	½	cup(s)	78	62.6	55	1.0	13.3	2.8	0.2	0	0.1	0	—
	Peas													
639	Green peas, canned, drained	½	cup(s)	85	69.4	59	3.8	10.7	3.5	0.3	0.1	0	0.1	—
641	Green peas, frozen, boiled, drained	½	cup(s)	80	63.6	62	4.1	11.4	4.4	0.2	0	0	0.1	—
35694	Pea pods, boiled with salt, drained	½	cup(s)	80	71.1	32	2.6	5.2	2.2	0.2	0	0	0.1	—
1082	Peas and carrots, canned with liquid	½	cup(s)	128	112.4	48	2.8	10.8	2.6	0.3	0.1	0	0.2	—
1083	Peas and carrots, frozen, boiled, drained	½	cup(s)	80	68.6	38	2.5	8.1	2.5	0.3	0.1	0	0.2	—
2750	Snow or sugar peas, frozen, boiled, drained	½	cup(s)	80	69.3	42	2.8	7.2	2.5	0.3	0.1	0	0.1	—
640	Snow or sugar peas, raw	½	cup(s)	32	28.0	13	0.9	2.4	0.8	0.1	0	0	0	—
29324	Split peas, sprouted	½	cup(s)	60	37.4	77	5.3	16.9	—	0.4	0.1	0	0.2	—
	Peppers													
644	Green bell or sweet, boiled, drained	½	cup(s)	68	62.5	19	0.6	4.6	0.8	0.1	0	0	0.1	—
643	Green bell or sweet, raw	½	cup(s)	75	69.9	15	0.6	3.5	1.3	0.1	0	0	0	—
1664	Green hot chili	1	item(s)	45	39.5	18	0.9	4.3	0.7	0.1	0	0	0	—
1663	Green hot chili, canned with liquid	½	cup(s)	68	62.9	14	0.6	3.5	0.9	0.1	0	0	0	—
1086	Jalapeno, canned with liquid	½	cup(s)	68	60.4	18	0.6	3.2	1.8	0.6	0.1	0	0.3	—
8703	Yellow bell or sweet	1	item(s)	186	171.2	50	1.9	11.8	1.7	0.4	0.1	0	0.2	—
1087	Poi	½	cup(s)	120	86.0	134	0.5	32.7	0.5	0.2	0	0	0.1	—
	Potatoes													
1090	Au gratin mix, prepared with water, whole milk and butter	½	cup(s)	124	97.7	115	2.8	15.9	1.1	5.1	3.2	1.5	0.2	—
1089	Au gratin, prepared with butter	½	cup(s)	123	90.7	162	6.2	13.8	2.2	9.3	5.8	2.6	0.3	—
5791	Baked, flesh and skin	1	item(s)	202	151.3	188	5.1	42.7	4.4	0.3	0.1	0	0.1	—
645	Baked, flesh only	½	cup(s)	61	46.0	57	1.2	13.1	0.9	0.1	0	0	0	—
1088	Baked, skin only	1	item(s)	58	27.4	115	2.5	26.7	4.6	0.1	0	0	0	—
5795	Boiled in skin, flesh only, drained	1	item(s)	136	104.7	118	2.5	27.4	2.1	0.1	0	0	0.1	—
5794	Boiled, drained, skin and flesh	1	item(s)	150	115.9	129	2.9	29.8	2.5	0.2	0	0	0.1	—
647	Boiled, flesh only	½	cup(s)	78	60.4	67	1.3	15.6	1.4	0.1	0	0	0	—
648	French fried, deep fried, prepared from raw	14	item(s)	70	32.8	187	2.7	23.5	2.9	9.5	1.9	4.2	3.0	—
649	French fried, frozen, heated	14	item(s)	70	43.7	94	1.9	19.4	2.0	3.7	0.7	2.3	0.2	—
1091	Hashed brown	½	cup(s)	78	36.9	207	2.3	27.4	2.5	9.8	1.5	4.1	3.7	—
652	Mashed with margarine and whole milk	½	cup(s)	105	79.0	119	2.1	17.7	1.6	4.4	1.0	2.0	1.2	0.7
653	Mashed, prepared from dehydrated granules with milk, water, and margarine	½	cup(s)	105	79.8	122	2.3	16.9	1.4	5.0	1.3	2.1	1.4	—
2759	Microwaved	1	item(s)	202	145.5	212	4.9	49.0	4.6	0.2	0.1	0	0.1	—
2760	Microwaved in skin, flesh only	½	cup(s)	78	57.1	78	1.6	18.1	1.2	0.1	0	0	0	—
5804	Microwaved, skin only	1	item(s)	58	36.8	77	2.5	17.2	4.2	0.1	0	0	0	—
1097	Potato puffs, frozen, heated	½	cup(s)	64	38.2	122	1.3	17.8	1.6	5.5	1.2	3.9	0.3	—
1094	Scalloped mix, prepared with water, whole milk and butter	½	cup(s)	124	98.4	116	2.6	15.9	1.4	5.3	3.3	1.5	0.2	—
1093	Scalloped, prepared with butter	½	cup(s)	123	99.2	108	3.5	13.2	2.3	4.5	2.8	1.3	0.2	—
	Pumpkin													
1773	Boiled, drained	½	cup(s)	123	114.8	25	0.9	6.0	1.3	0.1	0	0	0	—
656	Canned	½	cup(s)	123	110.2	42	1.3	9.9	3.6	0.3	0.2	0	0	—
	Radicchio													
8731	Leaves, raw	1	cup(s)	40	37.3	9	0.6	1.8	0.4	0.1	0	0	0	—
2498	Raw	1	cup(s)	40	37.3	9	0.6	1.8	0.4	0.1	0	0	0	—

Chol (mg)	Calc (mg)	Iron (mg)	Magn (mg)	Pota (mg)	Sodi (mg)	Zinc (mg)	Vit A (µg)	Thia (mg)	Vit E (mg α)	Ribo (mg)	Niac (mg)	Vit B$_6$ (mg)	Fola (µg)	Vit C (mg)	Vit B$_{12}$ (µg)	Sele (µg)
0	1	0.98	14.4	331.2	2.7	0.54	0	0.16	0.01	0.14	5.31	0.07	46.8	0	0	2.0
0	5	1.35	9.4	277.7	1.6	0.67	0	0.05	0.01	0.23	3.47	0.07	14.0	3.1	0	9.3
0	9	0.61	11.7	100.6	331.5	0.56	0	0.06	0.01	0.01	1.24	0.04	9.4	0	0	3.2
0	1	0.24	4.3	152.6	2.4	0.25	0	0.04	0.01	0.19	1.73	0.05	7.7	1.0	0	4.5
0	39	0.35	—	—	9.9	—	0	—	—	—	—	—	—	0	0	—
0	2	0.31	10.2	84.8	2.9	0.96	0	0.02	0.02	0.12	1.08	0.11	15.2	0.2	0	18.0
0	76	0.84	9.8	104.3	18.8	0.15	265.5	0.03	1.01	0.04	0.19	0.08	52.5	10.4	0	0.5
0	58	0.81	17.9	198.2	14.0	0.11	294.0	0.04	1.12	0.06	0.45	0.10	104.7	39.2	0	0.5
2	54	1.13	32.2	170.8	109.7	0.44	14.0	0.16	1.50	0.12	1.29	0.11	39.6	9.2	0	3.6
0	88	0.61	46.9	215.3	2.8	0.57	15.6	0.09	0.29	0.11	0.72	0.04	134.3	11.2	0	0.6
0	62	0.22	28.8	108.0	4.8	0.34	11.2	0.10	0.21	0.04	0.69	0.15	36.8	13.0	0	0.3
0	23	0.24	11.5	174.3	3.1	0.21	0	0.03	0.02	0.02	0.17	0.12	15.7	5.5	0	0.6
0	17	0.32	6.4	114.5	12.7	0.06	0	0.02	0.01	0.02	0.14	0.06	13.8	2.8	0	0.4
0	22	1.20	13.5	91.6	266.3	0.29	7.8	0.19	—	0.09	2.56	0.05	46.9	1.0	0	2.5
0	18	0.16	8.0	116.8	3.2	0.13	0	0.03	0.01	0.02	0.09	0.09	15.2	5.9	0	0.4
0	13	0.10	5.7	82.4	1.7	0.09	0	0.02	0.01	0.01	0.04	0.08	10.9	3.7	0	0.3
0	22	0.44	6.0	82.8	4.8	0.11	15.0	0.01	0.16	0.02	0.15	0.01	19.2	5.6	0	0.2
0	13	1.23	7.3	1318.4	10.2	2.72	2.2	0.03	0.36	0.12	0.62	0.53	14.6	5.0	0	0.5
0	5	0.23	1.9	21.1	2.1	0.04	16.0	0.00	0.02	0.00	0.05	0.00	5.8	5.1	0	0
0	29	0.45	22.6	286.3	7.8	0.20	0	0.06	0.78	0.04	0.56	0.07	45.2	10.1	0	1.3
0	17	0.80	14.5	147.1	214.2	0.60	23.0	0.10	0.02	0.06	0.62	0.05	37.4	8.2	0	1.4
0	19	1.21	17.6	88.0	57.6	0.53	84.0	0.22	0.02	0.08	1.18	0.09	47.2	7.9	0	0.8
0	34	1.57	20.8	192.0	192.0	0.29	41.6	0.10	0.31	0.06	0.43	0.11	23.2	38.3	0	0.6
0	29	0.96	17.9	127.5	331.5	0.74	368.5	0.09	—	0.07	0.74	0.11	23.0	8.4	0	1.1
0	18	0.75	12.8	126.4	54.4	0.36	380.8	0.18	0.41	0.05	0.92	0.07	20.8	6.5	0	0.9
0	47	1.92	22.4	173.6	4.0	0.39	52.8	0.05	0.37	0.09	0.45	0.13	28.0	17.6	0	0.6
0	14	0.65	7.6	63.0	1.3	0.08	17.0	0.04	0.12	0.02	0.19	0.05	13.2	18.9	0	0.2
0	22	1.34	33.6	228.6	12.0	0.62	4.8	0.12	—	0.08	1.84	0.14	86.4	6.2	0	0.4
0	6	0.31	6.8	112.9	1.4	0.08	15.6	0.04	0.34	0.02	0.32	0.15	10.9	50.6	0	0.2
0	7	0.25	7.5	130.4	2.2	0.09	13.4	0.04	0.27	0.02	0.35	0.16	7.5	59.9	0	0
0	8	0.54	11.3	153.0	3.2	0.13	26.6	0.04	0.31	0.04	0.42	0.12	10.4	109.1	0	0.2
0	5	0.34	9.5	127.2	797.6	0.10	24.5	0.01	0.46	0.02	0.54	0.10	6.8	46.2	0	0.2
0	16	1.28	10.2	131.2	1136.3	0.23	57.8	0.03	0.47	0.03	0.27	0.13	9.5	6.8	0	0.3
0	20	0.85	22.3	394.3	3.7	0.31	18.6	0.05	—	0.04	1.65	0.31	48.4	341.3	0	0.6
0	19	1.06	28.8	219.6	14.4	0.26	3.6	0.16	2.76	0.05	1.32	0.33	25.2	4.8	0	0.8
19	103	0.39	18.6	271.0	543.3	0.29	64.4	0.02	—	0.10	1.16	0.05	8.7	3.8	0	3.3
28	146	0.78	24.5	485.1	530.4	0.85	78.4	0.08	—	0.14	1.22	0.21	13.5	12.1	0	3.3
0	30	2.18	56.6	1080.7	20.2	0.72	2.0	0.12	0.08	0.09	2.84	0.62	56.6	19.4	0	0.8
0	3	0.21	15.3	238.5	3.1	0.18	0	0.06	0.02	0.01	0.85	0.18	5.5	7.8	0	0.2
0	20	4.08	24.9	332.3	12.2	0.28	0.6	0.07	0.02	0.06	1.77	0.35	12.8	7.8	0	0.4
0	7	0.42	29.9	515.4	5.4	0.40	0	0.14	0.01	0.02	1.95	0.40	13.6	17.7	0	0.4
0	13	1.27	34.1	572.0	7.4	0.46	0	0.14	0.01	0.03	2.13	0.44	15.0	18.4	0	—
0	6	0.24	15.6	255.8	3.9	0.21	0	0.07	0.01	0.01	1.02	0.21	7.0	5.8	0	0.2
0	16	1.05	30.8	567.0	8.4	0.39	0	0.08	0.09	0.03	1.34	0.37	16.1	21.2	0	0.4
0	8	0.51	18.2	315.7	271.6	0.26	0	0.09	0.07	0.02	1.55	0.12	19.6	9.3	0	0.1
0	11	0.43	27.3	449.3	266.8	0.37	0	0.13	0.01	0.03	1.80	0.37	12.5	10.1	0	0.4
1	23	0.27	19.9	344.4	349.6	0.31	43.0	0.09	0.44	0.04	1.23	0.25	9.4	11.0	0.1	0.8
2	36	0.21	21.0	164.8	179.5	0.26	49.3	0.09	0.53	0.09	0.90	0.16	8.4	6.8	0.1	5.9
0	22	2.50	54.5	902.9	16.2	0.72	0	0.24	—	0.06	3.46	0.69	24.2	30.5	0	0.8
0	4	0.31	19.4	319.0	5.4	0.25	0	0.10	—	0.01	1.26	0.25	9.3	11.7	0	0.3
0	27	3.44	21.5	377.0	9.3	0.29	0	0.04	0.01	0.04	1.28	0.28	9.9	8.9	0	0.3
0	9	0.41	10.9	199.7	307.2	0.21	0	0.08	0.15	0.02	0.97	0.08	9.0	4.0	0	0.4
14	45	0.47	17.4	252.2	423.7	0.31	43.5	0.02	—	0.06	1.28	0.05	12.4	4.1	0	2.0
15	70	0.70	23.3	463.1	410.4	0.49	0	0.08	—	0.11	1.29	0.22	13.5	13.0	0	2.0
0	18	0.69	11.0	281.8	1.2	0.28	306.3	0.03	0.98	0.09	0.50	0.05	11.0	5.8	0	0.2
0	32	1.70	28.2	252.4	6.1	0.20	953.1	0.02	1.29	0.06	0.45	0.06	14.7	5.1	0	0.5
0	8	0.23	5.2	120.8	8.8	0.25	0.4	0.01	0.90	0.01	0.10	0.02	24.0	3.2	0	0.4
0	8	0.23	5.2	120.8	8.8	0.25	0.4	0.01	0.90	0.01	0.10	0.02	24.0	3.2	0	0.4

DA+ Code	Food Description	Quantity	Measure	Wt (g)	H₂O (g)	Ener (kcal)	Prot (g)	Carb (g)	Fiber (g)	Fat (g)	Sat	Mono	Poly	Trans
	VEGETABLES, LEGUMES—Continued													
657	Radishes	6	item(s)	27	25.7	4	0.2	0.9	0.4	0	0	0	0	—
1099	Rutabaga, boiled, drained	½	cup(s)	85	75.5	33	1.1	7.4	1.5	0.2	0	0	0.1	—
658	Sauerkraut, canned	½	cup(s)	118	109.2	22	1.1	5.1	3.4	0.2	0	0	0.1	—
	Seaweed													
1102	Kelp	½	cup(s)	40	32.6	17	0.6	3.8	0.5	0.2	0.1	0	0	—
1104	Spirulina, dried	½	cup(s)	8	0.4	22	4.3	1.8	0.3	0.6	0.2	0.1	0.2	—
1106	Shallots	3	tablespoon(s)	30	23.9	22	0.8	5.0	—	0	0	0	0	—
	Soybeans													
1670	Boiled	½	cup(s)	86	53.8	149	14.3	8.5	5.2	7.7	1.1	1.7	4.4	—
2825	Dry roasted	½	cup(s)	86	0.7	388	34.0	28.1	7.0	18.6	2.7	4.1	10.5	—
2824	Roasted, salted	½	cup(s)	86	1.7	405	30.3	28.9	15.2	21.8	3.2	4.8	12.3	—
8739	Sprouted, stir fried	½	cup(s)	63	42.3	79	8.2	5.9	0.5	4.5	0.6	1.0	2.5	0
	Soy products													
1813	Soy milk	1	cup(s)	240	211.3	130	7.8	15.1	1.4	4.2	0.5	1.0	2.3	0
2838	Tofu, dried, frozen (koyadofu)	3	ounce(s)	85	4.9	408	40.8	12.4	6.1	25.8	3.7	5.7	14.6	—
13844	Tofu, extra firm	3	ounce(s)	85	—	86	8.6	2.2	1.1	4.3	0.5	0.9	2.8	—
13843	Tofu, firm	3	ounce(s)	85	—	75	7.5	2.2	0.5	3.2	0	0.9	2.3	—
1816	Tofu, firm, with calcium sulfate and magnesium chloride (nigari)	3	ounce(s)	85	72.2	60	7.0	1.4	0.8	3.5	0.7	1.0	1.5	—
1817	Tofu, fried	3	ounce(s)	85	43.0	230	14.6	8.9	3.3	17.2	2.5	3.8	9.7	—
13841	Tofu, silken	3	ounce(s)	85	—	42	3.7	1.9	0	2.3	0.5	—	—	—
13842	Tofu, soft	3	ounce(s)	85	—	65	6.5	1.1	0.5	3.2	0.5	1.1	2.2	—
1671	Tofu, soft, with calcium sulfate and magnesium chloride (nigari)	3	ounce(s)	85	74.2	52	5.6	1.5	0.2	3.1	0.5	0.7	1.8	—
	Spinach													
663	Canned, drained	½	cup(s)	107	98.2	25	3.0	3.6	2.6	0.5	0.1	0	0.2	—
660	Chopped, boiled, drained	½	cup(s)	90	82.1	21	2.7	3.4	2.2	0.2	0	0	0.1	—
661	Chopped, frozen, boiled, drained	½	cup(s)	95	84.5	32	3.8	4.6	3.5	0.8	0.1	0	0.4	—
662	Leaf, frozen, boiled, drained	½	cup(s)	95	84.5	32	3.8	4.6	3.5	0.8	0.1	0	0.4	—
659	Raw, chopped	1	cup(s)	30	27.4	7	0.9	1.1	0.7	0.1	0	0	0	—
8470	Trimmed leaves	1	cup(s)	32	27.5	3	0.9	0	2.8	0.1	—	—	—	—
	Squash													
1662	Acorn winter, baked	½	cup(s)	103	85.0	57	1.1	14.9	4.5	0.1	0	0	0.1	—
29702	Acorn winter, boiled, mashed	½	cup(s)	123	109.9	42	0.8	10.8	3.2	0.1	0	0	0	—
29451	Butternut, frozen, boiled	½	cup(s)	122	106.9	47	1.5	12.2	1.8	0.1	0	0	0	—
1661	Butternut winter, baked	½	cup(s)	102	89.5	41	0.9	10.7	3.4	0.1	0	0	0	—
32773	Butternut winter, frozen, boiled, mashed, no salt added	½	cup(s)	121	106.4	47	1.5	12.2	—	0.1	0	0	0	—
29700	Crookneck and straightneck summer, boiled, drained	½	cup(s)	65	60.9	12	0.6	2.6	1.2	0.1	0	0	0.1	—
29703	Hubbard winter, baked	½	cup(s)	102	86.8	51	2.5	11.0	—	0.6	0.1	0	0.3	—
1660	Hubbard winter, boiled, mashed	½	cup(s)	118	107.5	35	1.7	7.6	3.4	0.4	0.1	0	0.2	—
29704	Spaghetti winter, boiled, drained, or baked	½	cup(s)	78	71.5	21	0.5	5.0	1.1	0.2	0	0	0.1	—
664	Summer, all varieties, sliced, boiled, drained	½	cup(s)	90	84.3	18	0.8	3.9	1.3	0.3	0.1	0	0.1	—
665	Winter, all varieties, baked, mashed	½	cup(s)	103	91.4	38	0.9	9.1	2.9	0.4	0.1	0	0.2	—
1112	Zucchini summer, boiled, drained	½	cup(s)	90	85.3	14	0.6	3.5	1.3	0	0	0	0	—
1113	Zucchini summer, frozen, boiled, drained	½	cup(s)	112	105.6	19	1.3	4.0	1.4	0.1	0	0	0.1	—
	Sweet potatoes													
666	Baked, peeled	½	cup(s)	100	75.8	90	2.0	20.7	3.3	0.2	0	0	0.1	—
667	Boiled, mashed	½	cup(s)	164	131.4	125	2.2	29.1	4.1	0.2	0.1	0	0.1	—
668	Candied, home recipe	½	cup(s)	91	61.1	132	0.8	25.4	2.2	3.0	1.2	0.6	0.1	—
670	Canned, vacuum pack	½	cup(s)	100	76.0	91	1.7	21.1	1.8	0.2	0	0	0.1	—
2765	Frozen, baked	½	cup(s)	88	64.5	88	1.5	20.5	1.6	0.1	0	0	0	—
1136	Yams, baked or boiled, drained	½	cup(s)	68	47.7	79	1.0	18.7	2.7	0.1	0	0	0	—
32785	Taro shoots, cooked, no salt added	½	cup(s)	70	66.7	10	0.5	2.2	—	0.1	0	0	0	—
	Tomatillo													
8774	Raw	2	item(s)	68	62.3	22	0.7	4.0	1.3	0.7	0.1	0.1	0.3	—
8777	Raw, chopped	½	cup(s)	66	60.5	21	0.6	3.9	1.3	0.7	0.1	0.1	0.3	—
	Tomato													
16846	Cherry, fresh	5	item(s)	85	80.3	15	0.7	3.3	1.0	0.2	0	0	0.1	—
671	Fresh, ripe, red	1	item(s)	123	116.2	22	1.1	4.8	1.5	0.2	0	0	0.1	—
675	Juice, canned	½	cup(s)	122	114.1	21	0.9	5.2	0.5	0.1	0	0	0	—
75	Juice, no salt added	½	cup(s)	122	114.1	21	0.9	5.2	0.5	0.1	0	0	0	—
1699	Paste, canned	2	tablespoon(s)	33	24.1	27	1.4	6.2	1.3	0.2	0	0	0.1	—
1700	Puree, canned	¼	cup(s)	63	54.9	24	1.0	5.6	1.2	0.1	0	0	0.1	—
1118	Red, boiled	½	cup(s)	120	113.2	22	1.1	4.8	0.8	0.1	0	0	0.1	—
3952	Red, diced	½	cup(s)	90	85.1	16	0.8	3.5	1.1	0.2	0	0	0.1	—
1120	Red, stewed, canned	½	cup(s)	128	116.7	33	1.2	7.9	1.3	0.2	0	0	0.1	—
1125	Sauce, canned	¼	cup(s)	61	55.6	15	0.8	3.3	0.9	0.1	0	0	0	—

Chol (mg)	Calc (mg)	Iron (mg)	Magn (mg)	Pota (mg)	Sodi (mg)	Zinc (mg)	Vit A (µg)	Thia (mg)	Vit E (mg α)	Ribo (mg)	Niac (mg)	Vit B6 (mg)	Fola (µg)	Vit C (mg)	Vit B12 (µg)	Sele (µg)
0	7	0.09	2.7	62.9	10.5	0.07	0	0.00	0.00	0.01	0.06	0.01	6.8	4.0	0	0.2
0	41	0.45	19.6	277.1	17.0	0.30	0	0.07	0.27	0.04	0.61	0.09	12.8	16.0	0	0.6
0	35	1.73	15.3	200.6	780.0	0.22	1.2	0.03	0.17	0.03	0.17	0.15	28.3	17.3	0	0.7
0	67	1.12	48.4	35.6	93.2	0.48	2.4	0.02	0.32	0.04	0.16	0.00	72.0	1.2	0	0.3
0	9	2.14	14.6	102.2	78.6	0.15	2.2	0.18	0.38	0.28	0.96	0.03	7.1	0.8	0	0.5
0	11	0.36	6.3	100.2	3.6	0.12	18.0	0.02	—	0.01	0.06	0.09	10.2	2.4	—	0.4
0	88	4.42	74.0	442.9	0.9	0.98	0	0.13	0.30	0.24	0.34	0.20	46.4	1.5	0	6.3
0	120	3.39	196.1	1173.0	1.7	4.10	0	0.36	—	0.64	0.90	0.19	176.3	4.0	0	16.6
0	119	3.35	124.7	1264.2	140.2	2.70	8.6	0.08	0.78	0.12	1.21	0.17	181.5	1.9	0	16.4
0	52	0.25	60.4	356.6	8.8	1.32	0.6	0.26	—	0.12	0.69	0.10	79.9	7.5	0	0.4
0	60	1.53	60.0	283.2	122.4	0.28	0	0.14	0.26	0.16	1.23	0.18	43.2	0	0	11.5
0	310	8.27	50.2	17.0	5.1	4.16	22.1	0.42	—	0.27	1.01	0.24	78.2	0.6	0	46.2
0	65	1.16	84.1	—	0	—	0	—	—	—	—	—	—	0	0	—
0	108	1.16	56.1	—	0	—	0	—	—	—	—	—	—	0	0	—
0	171	1.36	31.5	125.9	10.2	0.70	0	0.05	0.01	0.05	0.08	0.06	16.2	0.2	0	8.4
0	316	4.14	51.0	124.2	13.6	1.69	0.9	0.14	0.03	0.04	0.08	0.08	23.0	0	0	24.2
0	56	0.34	33.1	—	4.7	—	0	—	—	—	—	—	—	0	1.7	—
0	108	1.16	35.5	—	0	—	0	—	—	—	—	—	—	0	1.9	—
0	94	0.94	23.0	102.1	6.8	0.54	0	0.04	0.01	0.03	0.45	0.04	37.4	0.2	0	7.6
0	136	2.45	81.3	370.2	28.9	0.48	524.3	0.02	2.08	0.14	0.41	0.11	104.8	15.3	0	1.5
0	122	3.21	78.3	419.4	63.0	0.68	471.6	0.08	1.87	0.21	0.44	0.21	131.4	8.8	0	1.4
0	145	1.86	77.9	286.9	92.2	0.46	572.9	0.07	3.36	0.16	0.41	0.12	115.0	2.1	0	5.2
0	145	1.86	77.9	286.9	92.2	0.46	572.9	0.07	3.36	0.16	0.41	0.12	115.0	2.1	0	5.2
0	30	0.81	23.7	167.4	23.7	0.16	140.7	0.02	0.61	0.06	0.22	0.06	58.2	8.4	0	0.3
0	25	2.13	25.5	134.1	38.0	0.18	—	0.03	—	0.05	0.18	0.07	0	7.5	0	—
0	45	0.95	44.1	447.9	4.1	0.17	21.5	0.17	—	0.01	0.90	0.19	19.5	11.1	0	0.7
0	32	0.68	31.9	322.2	3.7	0.13	50.2	0.12	—	0.01	0.65	0.14	13.5	8.0	0	0.5
0	23	0.70	10.9	161.9	2.4	0.14	203.3	0.06	0.14	0.05	0.56	0.08	19.5	4.3	0	0.6
0	42	0.61	29.6	289.6	4.1	0.13	569.1	0.07	1.31	0.01	0.99	0.12	19.4	15.4	0	0.6
0	23	0.70	10.9	161.2	2.4	0.14	202.4	0.06	—	0.05	0.56	0.08	19.4	4.2	0	0.6
0	14	0.31	13.6	137.1	1.3	0.19	5.2	0.03	—	0.02	0.29	0.07	14.9	5.4	0	0.1
0	17	0.48	22.4	365.1	8.2	0.15	308.0	0.07	—	0.04	0.57	0.17	16.3	9.7	0	0.6
0	12	0.33	15.3	252.5	5.9	0.11	236.0	0.05	0.14	0.03	0.39	0.12	11.8	7.7	0	0.4
0	16	0.26	8.5	90.7	14.0	0.15	4.7	0.02	0.09	0.01	0.62	0.07	6.2	2.7	0	0.2
0	24	0.32	21.6	172.8	0.9	0.35	9.9	0.04	0.12	0.03	0.46	0.05	18.0	5.0	0	0.2
0	23	0.45	13.3	247.0	1.0	0.23	267.5	0.02	0.12	0.07	0.51	0.17	20.5	9.8	0	0.4
0	12	0.32	19.8	227.7	2.7	0.16	50.4	0.04	0.11	0.04	0.39	0.07	15.3	4.1	0	0.2
0	19	0.54	14.5	216.3	2.2	0.22	10	0.05	0.13	0.04	0.43	0.05	8.9	4.1	0	0.2
0	38	0.69	27.0	475.0	36.0	0.32	961.0	0.10	0.71	0.10	1.48	0.28	6.0	19.6	0	0.2
0	44	1.18	29.5	377.2	44.3	0.33	1290.7	0.09	1.54	0.08	0.88	0.27	9.8	21.0	0	0.3
7	24	1.03	10.0	172.6	63.9	0.13	0	0.01	—	0.03	0.36	0.03	10.0	6.1	0	0.7
0	22	0.89	22.0	312.0	53.0	0.18	399.0	0.04	1.00	0.06	0.74	0.19	17.0	26.4	0	0.7
0	31	0.47	18.4	330.1	7.0	0.26	913.3	0.05	0.67	0.04	0.49	0.16	19.3	8.0	0	0.5
0	10	0.35	12.2	455.6	5.4	0.13	4.1	0.06	0.23	0.01	0.37	0.15	10.9	8.2	0	0.5
0	10	0.28	5.6	240.8	1.4	0.37	2.1	0.02	—	0.03	0.56	0.07	2.1	13.2	0	0.7
0	5	0.42	13.6	182.2	0.7	0.15	4.1	0.03	0.25	0.02	1.25	0.03	4.8	8.0	0	0.3
0	5	0.41	13.2	176.9	0.7	0.15	4.0	0.03	0.25	0.02	1.22	0.04	4.6	7.7	0	0.3
0	9	0.22	9.4	201.5	4.3	0.14	35.7	0.03	0.45	0.01	0.50	0.06	12.8	10.8	0	0
0	12	0.33	13.5	291.5	6.2	0.20	51.7	0.04	0.66	0.02	0.73	0.09	18.5	15.6	0	0
0	12	0.52	13.4	278.2	326.8	0.18	27.9	0.06	0.39	0.04	0.82	0.14	24.3	22.2	0	0.4
0	12	0.52	13.4	278.2	12.2	0.18	27.9	0.06	0.39	0.04	0.82	0.14	24.3	22.2	0	0.4
0	12	0.97	13.8	332.6	259.1	0.20	24.9	0.02	1.41	0.05	1.00	0.07	3.9	7.2	0	1.7
0	11	1.11	14.4	274.4	249.4	0.22	16.3	0.01	1.23	0.05	0.91	0.07	6.9	6.6	0	0.4
0	13	0.82	10.8	261.6	13.2	0.17	28.8	0.04	0.67	0.03	0.64	0.10	15.6	27.4	0	0.6
0	9	0.24	9.9	213.3	4.5	0.15	37.8	0.03	0.48	0.01	0.53	0.07	13.5	11.4	0	0
0	43	1.70	15.3	263.9	281.8	0.22	11.5	0.06	1.06	0.04	0.91	0.02	6.4	10.1	0	0.8
0	8	0.62	9.8	201.9	319.6	0.12	10.4	0.01	0.87	0.04	0.59	0.06	6.7	4.3	0	0.1

DA+ Code	Food Description	Quantity	Measure	Wt (g)	H₂O (g)	Ener (kcal)	Prot (g)	Carb (g)	Fiber (g)	Fat (g)	Sat	Mono	Poly	Trans
	VEGETABLES, LEGUMES—Continued													
8778	Sun dried	½	cup(s)	27	3.9	70	3.8	15.1	3.3	0.8	0.1	0.1	0.3	—
8783	Sun dried in oil, drained	¼	cup(s)	28	14.8	59	1.4	6.4	1.6	3.9	0.5	2.4	0.6	—
	Turnips													
678	Turnip greens, chopped, boiled, drained	½	cup(s)	72	67.1	14	0.8	3.1	2.5	0.2	0	0	0.1	—
679	Turnip greens, frozen, chopped, boiled, drained	½	cup(s)	82	74.1	24	2.7	4.1	2.8	0.3	0.1	0	0.1	—
677	Turnips, cubed, boiled, drained	½	cup(s)	78	73.0	17	0.6	3.9	1.6	0.1	0	0	0	—
	Vegetables, mixed													
1132	Canned, drained	½	cup(s)	82	70.9	40	2.1	7.5	2.4	0.2	0	0	0.1	—
680	Frozen, boiled, drained	½	cup(s)	91	75.7	59	2.6	11.9	4.0	0.1	0	0	0.1	—
7489	V8 100% vegetable juice	½	cup(s)	120	—	25	1.0	5.0	1.0	0	0	0	0	0
7490	V8 low sodium vegetable juice	½	cup(s)	120	—	25	0	6.5	1.0	0	0	0	0	0
7491	V8 spicy hot vegetable juice	½	cup(s)	120	—	25	1.0	5.0	0.5	0	0	0	0	0
	Water chestnuts													
31073	Sliced, drained	½	cup(s)	75	70.0	20	0	5.0	1.0	0	0	0	0	0
31087	Whole	½	cup(s)	75	70.0	20	0	5.0	1.0	0	0	0	0	0
1135	**Watercress**	1	cup(s)	34	32.3	4	0.8	0.4	0.2	0	0	0	0	—
	NUTS, SEEDS, AND PRODUCTS													
	Almonds													
32940	Almond butter with salt added	1	tablespoon(s)	16	0.2	101	2.4	3.4	0.6	9.5	0.9	6.1	2.0	—
1137	Almond butter, no salt added	1	tablespoon(s)	16	0.2	101	2.4	3.4	0.6	9.5	0.9	6.1	2.0	—
32886	Blanched	¼	cup(s)	36	1.6	211	8.0	7.2	3.8	18.3	1.4	11.7	4.4	—
32887	Dry roasted, no salt added	¼	cup(s)	35	0.9	206	7.6	6.7	4.1	18.2	1.4	11.6	4.4	—
29724	Dry roasted, salted	¼	cup(s)	35	0.9	206	7.6	6.7	4.1	18.2	1.4	11.6	4.4	—
29725	Oil roasted, salted	¼	cup(s)	39	1.1	238	8.3	6.9	4.1	21.7	1.7	13.7	5.3	—
508	Slivered	¼	cup(s)	27	1.3	155	5.7	5.9	3.3	13.3	1.0	8.3	3.3	0
1138	**Beechnuts, dried**	¼	cup(s)	57	3.8	328	3.5	19.1	5.3	28.5	3.3	12.5	11.4	—
517	**Brazil nuts, dried, unblanched**	¼	cup(s)	35	1.2	230	5.0	4.3	2.6	23.3	5.3	8.6	7.2	—
1166	**Breadfruit seeds, roasted**	¼	cup(s)	57	28.3	118	3.5	22.8	3.4	1.5	0.4	0.2	0.8	—
1139	**Butternuts, dried**	¼	cup(s)	30	1.0	184	7.5	3.6	1.4	17.1	0.4	3.1	12.8	—
	Cashews													
32931	Cashew butter with salt added	1	tablespoon(s)	16	0.5	94	2.8	4.4	0.3	7.9	1.6	4.7	1.3	—
32889	Cashew butter, no salt added	1	tablespoon(s)	16	0.5	94	2.8	4.4	0.3	7.9	1.6	4.7	1.3	—
1140	Dry roasted	¼	cup(s)	34	0.6	197	5.2	11.2	1.0	15.9	3.1	9.4	2.7	—
518	Oil roasted	¼	cup(s)	32	1.1	187	5.4	9.6	1.1	15.4	2.7	8.4	2.8	—
	Coconut, Shredded													
32896	Dried, not sweetened	¼	cup(s)	23	0.7	152	1.6	5.4	3.8	14.9	13.2	0.6	0.2	—
1153	Dried, shredded, sweetened	¼	cup(s)	23	2.9	116	0.7	11.1	1.0	8.3	7.3	0.4	0.1	—
520	Shredded	¼	cup(s)	20	9.4	71	0.7	3.0	1.8	6.7	5.9	0.3	0.1	—
	Chestnuts													
1152	Chinese, roasted	¼	cup(s)	36	14.6	87	1.6	19.0	—	0.4	0.1	0.2	0.1	—
32895	European, boiled and steamed	¼	cup(s)	46	31.3	60	0.9	12.8	—	0.6	0.1	0.2	0.2	—
32911	European, roasted	¼	cup(s)	36	14.5	88	1.1	18.9	1.8	0.8	0.1	0.3	0.3	—
32922	Japanese, boiled and steamed	¼	cup(s)	36	31.0	20	0.3	4.5	—	0.1	0	0	0	—
32923	Japanese, roasted	¼	cup(s)	36	18.1	73	1.1	16.4	—	0.3	0	0.1	0.1	—
4958	**Flax seeds or linseeds**	¼	cup(s)	43	3.3	225	8.4	12.3	11.9	17.7	1.7	3.2	12.6	0
32904	**Ginkgo nuts, dried**	¼	cup(s)	39	4.8	136	4.0	28.3	—	0.8	0.1	0.3	0.3	—
	Hazelnuts or filberts													
32901	Blanched	¼	cup(s)	30	1.7	189	4.1	5.1	3.3	18.3	1.4	14.5	1.7	—
32902	Dry roasted, no salt added	¼	cup(s)	30	0.8	194	4.5	5.3	2.8	18.7	1.3	14.0	2.5	—
1156	**Hickorynuts, dried**	¼	cup(s)	30	0.8	197	3.8	5.5	1.9	19.3	2.1	9.8	6.6	—
	Macadamias													
32905	Dry roasted, no salt added	¼	cup(s)	34	0.5	241	2.6	4.5	2.7	25.5	4.0	19.9	0.5	—
32932	Dry roasted, with salt added	¼	cup(s)	34	0.5	240	2.6	4.3	2.7	25.5	4.0	19.9	0.5	—
1157	Raw	¼	cup(s)	34	0.5	241	2.6	4.6	2.9	25.4	4.0	19.7	0.5	—
	Mixed nuts													
1159	With peanuts, dry roasted	¼	cup(s)	34	0.6	203	5.9	8.7	3.1	17.6	2.4	10.8	3.7	—
32933	With peanuts, dry roasted, with salt added	¼	cup(s)	34	0.6	203	5.9	8.7	3.1	17.6	2.4	10.8	3.7	—
32906	Without peanuts, oil roasted, no salt added	¼	cup(s)	36	1.1	221	5.6	8.0	2.0	20.2	3.3	11.9	4.1	—
	Peanuts													
2807	Dry roasted	¼	cup(s)	37	0.6	214	8.6	7.9	2.9	18.1	2.5	9.0	5.7	—
2806	Dry roasted, salted	¼	cup(s)	37	0.6	214	8.6	7.9	2.9	18.1	2.5	9.0	5.7	—
1763	Oil roasted, salted	¼	cup(s)	36	0.5	216	10.1	5.5	3.4	18.9	3.1	9.4	5.5	—
1884	Peanut butter, chunky	1	tablespoon(s)	16	0.2	94	3.8	3.5	1.3	8.0	1.3	3.9	2.4	—
30303	Peanut butter, low sodium	1	tablespoon(s)	16	0.2	95	4.0	3.1	0.9	8.2	1.8	3.9	2.2	—
30305	Peanut butter, reduced fat	1	tablespoon(s)	18	0.2	94	4.7	6.4	0.9	6.1	1.3	2.9	1.8	—
524	Peanut butter, smooth	1	tablespoon(s)	16	0.3	94	4.0	3.1	1.0	8.1	1.7	3.9	2.3	—
2804	Raw	¼	cup(s)	37	2.4	207	9.4	5.9	3.1	18.0	2.5	8.9	5.7	—
	Pecans													
32907	Dry roasted, no salt added	¼	cup(s)	28	0.3	198	2.6	3.8	2.6	20.7	1.8	12.3	5.7	—

Chol (mg)	Calc (mg)	Iron (mg)	Magn (mg)	Pota (mg)	Sodi (mg)	Zinc (mg)	Vit A (µg)	Thia (mg)	Vit E (mg α)	Ribo (mg)	Niac (mg)	Vit B6 (mg)	Fola (µg)	Vit C (mg)	Vit B12 (µg)	Sele (µg)
0	30	2.45	52.4	925.3	565.7	0.53	11.9	0.14	0.00	0.13	2.44	0.09	18.4	10.6	0	1.5
0	13	0.73	22.3	430.4	73.2	0.21	17.6	0.05	—	0.10	0.99	0.08	6.3	28.0	0	0.8
0	99	0.58	15.8	146.2	20.9	0.10	274.3	0.03	1.35	0.05	0.30	0.13	85.0	19.7	0	0.6
0	125	1.59	21.3	183.7	12.3	0.34	441.2	0.04	2.18	0.06	0.38	0.06	32.0	17.9	0	1.0
0	26	0.14	7.0	138.1	12.5	0.09	0	0.02	0.02	0.02	0.23	0.05	7.0	9.0	0	0.2
0	22	0.86	13.0	237.2	121.4	0.33	475.1	0.04	0.24	0.04	0.47	0.06	19.6	4.1	0	0.2
0	23	0.74	20.0	153.8	31.9	0.44	194.7	0.06	0.34	0.10	0.77	0.06	17.3	2.9	0	0.3
0	20	0.36	12.9	260.0	310.0	0.24	100.0	0.05	—	0.03	0.87	0.17	—	30.0	0	—
0	20	0.36	—	450.0	70.0	—	100.0	0.02	—	0.02	0.75	—	—	30.0	0	—
0	20	0.36	12.9	240.0	360.0	0.24	50.0	0.05	—	0.03	0.88	0.17	—	15.0	0	—
0	7	0.00	—	—	5.0	—	0	—	—	—	—	—	—	2.0	—	—
0	7	0.00	—	—	5.0	—	0	—	—	—	—	—	—	2.0	—	—
0	41	0.06	7.1	112.2	13.9	0.03	54.4	0.03	0.34	0.04	0.06	0.04	3.1	14.6	0	0.3
0	43	0.59	48.5	121.3	72.0	0.49	0	0.02	4.16	0.10	0.46	0.01	10.4	0.1	0	0.8
0	43	0.59	48.5	121.3		0.48	0	0.02	—	0.09	0.46	0.01	10.4	0.1	0	—
0	78	1.34	99.7	249.0	10.2	1.13	0	0.07	8.95	0.20	1.32	0.04	10.9	0	0	1.0
0	92	1.55	98.7	257.4	0.3	1.22	0	0.02	8.97	0.29	1.32	0.04	11.4	0	0	1.0
0	92	1.55	98.7	257.4	117.0	1.22	0	0.02	8.97	0.29	1.32	0.04	11.4	0	0	1.0
0	114	1.44	107.5	274.4	133.1	1.20	0	0.03	10.19	0.30	1.43	0.04	10.6	0	0	1.1
0	71	1.00	72.4	190.4	0.3	0.83	0	0.05	7.07	0.27	0.91	0.03	13.5	0	0	0.7
0	1	1.39	0	579.7	21.7	0.20	0	0.16	—	0.20	0.48	0.38	64.4	8.8	0	4.0
0	56	0.85	131.6	230.7	1.1	1.42	0	0.21	2.00	0.01	0.10	0.03	7.7	0.2	0	671.0
0	49	0.50	35.3	616.7	15.9	0.58	8.5	0.22	—	0.12	4.20	0.22	33.6	4.3	0	8.0
0	16	1.21	71.1	126.3	0.3	0.94	1.8	0.12	—	0.04	0.31	0.17	19.8	1.0	0	5.2
0	7	0.81	41.3	87.4	98.2	0.83	0	0.05	0.15	0.03	0.26	0.04	10.9	0	0	1.8
0	7	0.81	41.3	87.4	2.4	0.83	0	0.05	—	0.03	0.26	0.04	10.9	0	0	1.8
0	15	2.06	89.1	193.5	5.5	1.92	0	0.07	0.32	0.07	0.48	0.09	23.6	0	0	4.0
0	14	1.95	88.0	203.8	4.2	1.73	0	0.12	0.30	0.07	0.56	0.10	8.1	0.1	0	6.5
0	6	0.76	20.7	125.2	8.5	0.46	0	0.01	0.10	0.02	0.13	0.07	2.1	0.3	0	4.3
0	3	0.45	11.6	78.4	60.9	0.42	0	0.01	0.09	0.00	0.11	0.06	1.9	0.2	0	3.9
0	3	0.48	6.4	71.2	4.0	0.21	0	0.01	0.04	0.00	0.11	0.01	5.2	0.7	0	2.0
0	7	0.54	32.6	173.0	1.4	0.33	0	0.05	—	0.03	0.54	0.15	26.1	13.9	0	2.6
0	21	0.80	24.8	328.9	12.4	0.11	0.5	0.06	—	0.03	0.32	0.10	17.5	12.3	0	—
0	10	0.32	11.8	211.6	0.7	0.20	0.4	0.08	0.18	0.05	0.48	0.18	25.0	9.3	0	0.4
0	4	0.19	6.5	42.8	1.8	0.14	0.4	0.04	—	0.01	0.19	0.03	6.1	3.4	0	—
0	13	0.75	23.2	154.8	6.9	0.51	1.4	0.15	—	—	0.24	0.14	21.4	10.1	0	—
0	142	2.13	156.1	354.0	11.9	1.83	0	0.06	0.14	0.06	0.59	0.39	118.4	0.5	0	2.3
0	8	0.62	20.7	390.2	5.1	0.26	21.5	0.17	—	0.07	4.58	0.25	41.4	11.4	0	—
0	45	0.98	48.0	197.4	0	0.66	0.6	0.14	5.25	0.03	0.46	0.17	23.4	0.6	0	1.2
0	37	1.31	51.9	226.5	0	0.74	0.9	0.10	4.58	0.03	0.61	0.18	26.4	1.1	0	1.2
0	18	0.64	51.9	130.8	0.3	1.29	2.1	0.26	—	0.04	0.27	0.06	12.0	0.6	0	2.4
0	23	0.88	39.5	121.6	1.3	0.43	0	0.23	0.19	0.02	0.76	0.12	3.4	0.2	0	3.9
0	23	0.88	39.5	121.6	88.8	0.43	0	0.23	0.19	0.02	0.76	0.12	3.4	0.2	0	3.9
0	28	1.24	43.6	123.3	1.7	0.44	0	0.40	0.18	0.05	0.83	0.09	3.7	0.4	0	1.2
0	24	1.27	77.1	204.5	4.1	1.30	0.3	0.07	—	0.07	1.61	0.10	17.1	0.1	0	1.0
0	24	1.26	77.1	204.5	229.1	1.30	0	0.06	3.74	0.06	1.61	0.10	17.1	0.1	0	2.6
0	38	0.92	90.4	195.8	4.0	1.67	0.4	0.18	—	0.17	0.70	0.06	20.2	0.2	0	—
0	20	0.82	64.2	240.2	2.2	1.20	0	0.16	2.52	0.03	4.93	0.09	52.9	0	0	2.7
0	20	0.82	64.2	240.2	296.7	1.20	0	0.16	2.84	0.03	4.93	0.09	52.9	0	0	2.7
0	22	0.54	63.4	261.4	115.2	1.18	0	0.03	2.49	0.03	4.97	0.16	43.2	0.3	0	1.2
0	7	0.30	25.6	119.2	77.8	0.45	0	0.02	1.01	0.02	2.19	0.07	14.7	0	0	1.3
0	6	0.29	25.4	107.0	2.7	0.47	0	0.01	1.23	0.02	2.14	0.07	11.8	0	0	1.2
0	6	0.34	30.6	120.4	97.2	0.50	0	0.05	1.20	0.01	2.63	0.06	10.8	0	0	1.4
0	7	0.30	24.6	103.8	73.4	0.47	0	0.01	1.44	0.02	2.14	0.09	11.8	0	0	0.9
0	34	1.67	61.3	257.3	6.6	1.19	0	0.23	3.04	0.04	4.40	0.12	87.6	0	0	2.6
0	20	0.78	36.8	118.3	0.3	1.41	1.9	0.12	0.35	0.03	0.32	0.05	4.5	0.2	0	1.1

DA+ Code	Food Description	Quantity	Measure	Wt (g)	H₂O (g)	Ener (kcal)	Prot (g)	Carb (g)	Fiber (g)	Fat (g)	Sat	Mono	Poly	Trans
	NUTS, SEEDS, AND PRODUCTS —Continued													
32936	Dry roasted, with salt added	¼	cup(s)	27	0.3	192	2.6	3.7	2.5	20.0	1.7	11.9	5.6	—
1162	Oil roasted	¼	cup(s)	28	0.3	197	2.5	3.6	2.6	20.7	2.0	11.3	6.5	—
526	Raw	¼	cup(s)	27	1.0	188	2.5	3.8	2.6	19.6	1.7	11.1	5.9	—
12973	Pine nuts or pignolia, dried	1	tablespoon(s)	9	0.2	58	1.2	1.1	0.3	5.9	0.4	1.6	2.9	—
	Pistachios													
1164	Dry roasted	¼	cup(s)	31	0.6	176	6.6	8.5	3.2	14.1	1.7	7.4	4.3	
32938	Dry roasted, with salt added	¼	cup(s)	32	0.6	182	6.8	8.6	3.3	14.7	1.8	7.7	4.4	
1167	Pumpkin or squash seeds, roasted	¼	cup(s)	57	4.0	296	18.7	7.6	2.2	23.9	4.5	7.4	10.9	
	Sesame													
32912	Sesame butter paste	1	tablespoon(s)	16	0.3	94	2.9	3.8	0.9	8.1	1.1	3.1	3.6	—
32941	Tahini or sesame butter	1	tablespoon(s)	15	0.5	89	2.6	3.2	0.7	8.0	1.1	3.0	3.5	—
1169	Whole, roasted, toasted	3	tablespoon(s)	10	0.3	54	1.6	2.4	1.3	4.6	0.6	1.7	2.0	—
	Soy nuts													
34173	Deep sea salted	¼	cup(s)	28	—	119	11.9	8.9	4.9	4.0	1.0	—	—	—
34174	Unsalted	¼	cup(s)	28	—	119	11.9	8.9	4.9	4.0	0	—	—	—
	Sunflower seeds													
528	Kernels, dried	1	tablespoon(s)	9	0.4	53	1.9	1.8	0.8	4.6	0.4	1.7	2.1	—
29721	Kernels, dry roasted, salted	1	tablespoon(s)	8	0.1	47	1.5	1.9	0.7	4.0	0.4	0.8	2.6	—
29723	Kernels, toasted, salted	1	tablespoon(s)	8	0.1	52	1.4	1.7	1.0	4.8	0.5	0.9	3.1	—
32928	Sunflower seed butter with salt added	1	tablespoon(s)	16	0.2	93	3.1	4.4	—	7.6	0.8	1.5	5.0	—
	Trail mix													
4646	Trail mix	¼	cup(s)	38	3.5	173	5.2	16.8	2.0	11.0	2.1	4.7	3.6	—
4647	Trail mix with chocolate chips	¼	cup(s)	38	2.5	182	5.3	16.8	—	12.0	2.3	5.1	4.2	—
4648	Tropical trail mix	¼	cup(s)	35	3.2	142	2.2	23.0	—	6.0	3.0	0.9	1.8	—
	Walnuts													
529	Dried black, chopped	¼	cup(s)	31	1.4	193	7.5	3.1	2.1	18.4	1.1	4.7	11.0	—
531	English or Persian	¼	cup(s)	29	1.2	191	4.5	4.0	2.0	19.1	1.8	2.6	13.8	—
	VEGETARIAN FOODS													
	Prepared													
34222	Brown rice & tofu stir-fry (vegan)	8	ounce(s)	227	244.4	302	16.5	18.0	3.2	21.0	1.7	4.7	13.4	0
34368	Cheese enchilada casserole (lacto)	8	ounce(s)	227	80.3	385	16.6	38.4	4.1	17.8	9.5	6.1	1.1	—
34247	Five bean casserole (vegan)	8	ounce(s)	227	175.8	178	5.9	26.6	6.0	5.8	1.1	2.5	1.9	0
34261	Lentil stew (vegan)	8	ounce(s)	227	227.9	188	11.5	35.9	11.0	0.7	0.1	0.1	0.3	0
34397	Macaroni and cheese (lacto)	8	ounce(s)	227	352.1	391	18.1	37.1	1.0	18.7	9.8	6.0	1.8	0
34238	Steamed rice and vegetables (vegan)	8	ounce(s)	227	222.9	587	11.2	87.7	5.8	23.1	4.1	8.7	9.1	0
34308	Tofu rice burgers (ovo-lacto)	1	piece(s)	218	77.6	435	22.4	68.6	5.6	8.4	1.7	2.4	3.5	—
34276	Vegan spinach enchiladas (vegan)	1	piece(s)	82	59.2	93	4.9	14.5	1.8	2.4	0.3	0.6	1.3	—
34243	Vegetable chow mein (vegan)	8	ounce(s)	227	163.3	166	6.5	22.1	2.0	6.4	0.7	2.7	2.5	0
34454	Vegetable lasagna (lacto)	8	ounce(s)	227	178.9	208	13.7	29.9	2.6	4.1	2.3	1.1	0.3	—
34339	Vegetable marinara (vegan)	8	ounce(s)	252	200.7	104	3.0	16.7	1.4	3.1	0.4	1.4	1.0	0
34356	Vegetable rice casserole (lacto)	8	ounce(s)	227	178.9	238	9.7	24.4	4.0	12.5	4.9	3.5	3.1	—
34311	Vegetable strudel (ovo-lacto)	8	ounce(s)	227	63.1	478	12.0	32.4	2.5	33.8	11.5	16.7	3.9	—
34371	Vegetable taco (lacto)	1	item(s)	85	46.5	117	4.2	13.6	2.9	5.6	2.1	1.9	1.3	—
34282	Vegetarian chili (vegan)	8	ounce(s)	227	191.4	115	5.6	21.4	7.1	1.5	0.2	0.3	0.7	0
34367	Vegetarian vegetable soup (vegan)	8	ounce(s)	227	257.9	111	3.2	16.0	3.2	5.0	1.0	2.1	1.6	0
	Boca burger													
32067	All American flamed grilled patty	1	item(s)	71	—	90	14.0	4.0	3.0	3.0	1.0	—	—	0
32074	Boca Chik n nuggets	4	item(s)	87	—	180	14.0	17.0	3.0	7.0	1.0	—	—	0
32075	Boca meatless ground burger	½	cup(s)	57	—	60	13.0	6.0	3.0	0.5	0	—	—	0
32072	Breakfast links	2	item(s)	45	—	70	8.0	5.0	2.0	3.0	0.5	—	—	0
32071	Breakfast patties	1	item(s)	38	—	60	7.0	5.0	2.0	2.5	0	—	—	0
35780	Cheeseburger meatless burger patty	1	item(s)	71	—	100	12.0	5.0	3.0	5.0	1.5	—	—	0
33958	Original meatless Chik 'n patties	1	item(s)	71	—	160	11.0	15.0	2.0	6.0	1.0	—	—	0
32066	Original patty	1	item(s)	71	—	70	13.0	6.0	4.0	0.5	0	—	—	0
32068	Roasted garlic patty	1	item(s)	71	—	70	12.0	6.0	4.0	1.5	0	—	—	0
37814	Roasted onion meatless burger patty	1	item(s)	71	—	70	11.0	7.0	4.0	1.0	0	—	—	0
	Gardenburger													
37810	BBQ chik'n with sauce	1	item(s)	142	—	250	14.0	30.0	5.0	8.0	1.0	—	—	0
39661	Black bean burger	1	item(s)	71	—	80	8.0	11.0	4.0	2.0	0	—	—	0
39666	Buffalo chick'n wing	3	item(s)	95	—	180	9.0	8.0	5.0	12.0	1.5	—	—	0
39665	Country fried chicken with creamy pepper gravy	1	item(s)	142	—	190	9.0	16.0	2.0	9.0	1.0	—	—	0
37808	Flamed grilled Chik'n	1	item(s)	71	—	100	13.0	5.0	3.0	2.5	0	—	—	0
37803	Garden vegan	1	item(s)	71	—	100	10.0	12.0	2.0	1.0	—	—	—	0
39663	Homestyle classic burger	1	item(s)	71	—	110	12.0	6.0	4.0	5.0	0.5	—	—	0
37807	Meatless breakfast sausage	1	item(s)	43	—	50	5.0	2.0	2.0	3.5	0	—	—	0
37809	Meatless meatballs	6	item(s)	85	—	110	12.0	8.0	4.0	4.5	1.0	—	—	0
37806	Meatless riblets with sauce	1	item(s)	142	—	160	17.0	11.0	4.0	5.0	1.0	—	—	0
29913	Original	1	item(s)	71	—	90	10.0	8.0	3.0	2.0	0.5	—	—	0
39662	Sun-dried tomato basil burger	1	item(s)	71	—	80	10.0	11.0	3.0	1.5	0.5	—	—	0

Chol (mg)	Calc (mg)	Iron (mg)	Magn (mg)	Pota (mg)	Sodi (mg)	Zinc (mg)	Vit A (µg)	Thia (mg)	Vit E (mg α)	Ribo (mg)	Niac (mg)	Vit B$_6$ (mg)	Fola (µg)	Vit C (mg)	Vit B$_{12}$ (µg)	Sele (µg)
0	19	0.75	35.6	114.5	103.4	1.36	1.9	0.11	0.34	0.03	0.31	0.05	4.3	0.2	0	1.1
0	18	0.68	33.3	107.8	0.3	1.23	1.4	0.13	0.70	0.03	0.33	0.05	4.1	0.2	0	1.7
0	19	0.69	33.0	111.7	0	1.23	0.8	0.18	0.38	0.04	0.32	0.06	6.0	0.3	0	1.0
0	1	0.47	21.6	51.3	0.2	0.55	0.1	0.03	0.80	0.02	0.37	0.01	2.9	0.1	0	0.1
0	34	1.29	36.9	320.4	3.1	0.71	4.0	0.26	0.59	0.05	0.44	0.39	15.4	0.7	0	2.9
0	35	1.34	38.4	333.4	129.6	0.73	4.2	0.26	0.61	0.05	0.45	0.40	16.0	0.7	0	3.0
0	24	8.48	303.0	457.4	10.2	4.22	10.8	0.12	0.00	0.18	0.99	0.05	32.3	1.0	0	3.2
0	154	3.07	57.9	93.1	1.9	1.17	0.5	0.04	—	0.03	1.07	0.13	16.0	0	0	0.9
0	21	0.66	14.3	68.9	5.3	0.69	0.5	0.24	—	0.02	0.85	0.02	14.7	0.6	0	0.3
0	94	1.40	33.8	45.1	1.0	0.68	0	0.07	—	0.02	0.43	0.07	9.3	0	0	0.5
0	59	1.07	—	—	148.1	—	0	—	—	—	—	—	—	0	—	—
0	59	1.07	—	—	9.9	—	0	—	—	—	—	—	—	0	—	—
0	7	0.47	29.3	58.1	0.8	0.45	0.3	0.13	2.99	0.03	0.75	0.12	20.4	0.1	0	4.8
0	6	0.30	10.3	68.0	32.8	0.42	0	0.01	2.09	0.02	0.56	0.06	19.0	0.1	0	6.3
0	5	0.57	10.8	41.1	51.3	0.44	0	0.03	—	0.02	0.35	0.07	19.9	0.1	0	5.2
0	20	0.76	59.0	11.5	83.2	0.85	0.5	0.05	—	0.05	0.85	0.13	37.9	0.4	0	—
0	29	1.14	59.3	256.9	85.9	1.20	0.4	0.17	—	0.07	1.76	0.11	26.6	0.5	0	—
2	41	1.27	60.4	243.0	45.4	1.17	0.8	0.15	—	0.08	1.65	0.09	24.4	0.5	0	—
0	20	0.92	33.6	248.2	3.5	0.41	0.7	0.15	—	0.04	0.51	0.11	14.7	2.7	0	—
0	19	0.97	62.8	163.4	0.6	1.05	0.6	0.01	0.56	0.04	0.14	0.18	9.7	0.5	0	5.3
0	29	0.85	46.2	129.0	0.6	0.90	0.3	0.10	0.20	0.04	0.32	0.15	28.7	0.4	0	1.4
0	353	6.34	118.3	501.4	142.2	2.03	—	0.23	0.07	0.14	1.49	0.36	51.8	24.8	0	14.8
39	441	2.44	34.6	191.2	1139.7	1.84	—	0.31	0.05	0.35	2.23	0.11	87.1	20.4	0.4	20.0
0	48	1.78	40.8	364.1	613.6	0.61	—	0.10	0.52	0.07	0.93	0.11	39.4	8.3	0	3.3
0	34	3.23	50.0	548.8	436.5	1.42	—	0.24	0.14	0.16	2.31	0.29	91.4	26.4	0	12.1
43	415	1.71	45.4	267.8	1641.0	2.32	—	0.32	0.27	0.48	2.18	0.13	74.2	0.9	0.8	33.3
0	91	3.31	153.1	810.1	3117.8	2.04	—	0.37	3.03	0.21	6.16	0.64	61.7	35.2	0	18.8
52	467	9.01	89.7	455.6	2449.5	2.06	—	0.27	0.12	0.26	3.43	0.29	106.6	2.0	0.1	43.0
0	117	1.13	40.4	170.5	134.2	0.68	—	0.07	—	0.07	0.53	0.10	52.1	1.8	0	5.1
0	189	3.70	28.0	310.3	372.7	0.76	—	0.13	0.05	0.11	1.43	0.14	45.6	8.0	0	6.5
10	176	1.86	41.9	470.0	759.4	1.14	—	0.26	0.05	0.25	2.49	0.22	64.7	19.0	0.4	21.8
0	17	0.94	19.1	189.9	439.6	0.42	—	0.15	0.55	0.08	1.36	0.12	40.6	23.5	0	10.8
17	190	1.28	29.3	414.2	626.0	1.24	—	0.16	0.35	0.29	2.00	0.19	72.6	56.0	0.2	5.8
29	200	2.15	24.5	181.0	512.1	1.24	—	0.28	0.20	0.31	2.88	0.11	88.3	17.4	0.2	19.7
7	77	0.88	26.3	174.1	280.7	0.59	—	0.08	0.04	0.06	0.49	0.08	38.7	4.6	0	3.0
0	65	1.98	41.0	543.1	390.7	0.74	—	0.14	0.15	0.10	1.31	0.18	59.3	20.3	0	4.4
0	46	1.87	34.9	550.3	729.5	0.56	—	0.13	0.55	0.09	1.99	0.27	47.8	29.9	0	1.4
5	150	1.80	—	—	280.0	—	0	—	—	—	—	—	—	0	—	—
0	40	1.44	—	—	500.0	—	—	—	—	—	—	—	—	0	—	—
0	60	1.80	—	—	270.0	—	0	—	—	—	—	—	—	0	—	—
0	20	1.44	—	—	330.0	—	0	—	—	—	—	—	—	0	—	—
0	20	1.08	—	—	280.0	—	0	—	—	—	—	—	—	0	—	—
5	80	1.80	—	—	360.0	—	—	—	—	—	—	—	—	0	—	—
0	40	1.80	—	—	430.0	—	—	—	—	—	—	—	—	0	—	—
0	60	1.80	—	—	280.0	—	0	—	—	—	—	—	—	0	—	—
0	60	1.80	—	—	370.0	—	0	—	—	—	—	—	—	0	—	—
0	100	2.70	—	—	300.0	—	—	—	—	—	—	—	—	0	—	—
0	150	1.08	—	—	890.0	—	—	—	—	—	—	—	—	0	—	—
0	40	1.44	—	—	330.0	—	—	—	—	—	—	—	—	0	—	—
0	40	0.72	—	—	1000.0	—	—	—	—	—	—	—	—	0	—	—
5	40	1.44	—	—	550.0	—	—	—	—	—	—	—	—	0	—	—
0	60	3.60	—	—	360.0	—	—	—	—	—	—	—	—	0	—	—
0	40	4.50	—	—	230.0	—	—	—	—	—	—	—	—	0	—	—
0	80	1.44	—	—	380.0	—	—	—	—	—	—	—	—	0	—	—
0	20	0.72	—	—	120.0	—	—	—	—	—	—	—	—	0	—	—
0	60	1.80	—	—	400.0	—	—	—	—	—	—	—	—	0	—	—
0	60	1.80	—	—	720.0	—	—	—	—	—	—	—	—	3.6	—	—
0	80	1.08	30.4	193.4	490.0	0.89	—	0.10	—	0.15	1.08	0.08	10.1	1.2	0.1	7.0
5	60	1.44	—	—	260.0	—	—	—	—	—	—	—	—	3.6	—	—

DA+ Code	Food Description	Quantity	Measure	Wt (g)	H₂O (g)	Ener (kcal)	Prot (g)	Carb (g)	Fiber (g)	Fat (g)	Fat Breakdown (g) Sat	Mono	Poly	Trans
	VEGETARIAN FOODS —Continued													
29915	Veggie medley	1	item(s)	71	—	90	9.0	11.0	4.0	2.0	0	—	—	0
	Loma Linda													
9311	Big franks, canned	1	item(s)	51	—	110	11.0	3.0	2.0	6.0	1.0	1.5	3.5	0
9323	Fried Chik'n with gravy	2	piece(s)	80	45.9	150	12.0	5.0	2.0	10	1.5	2.5	5.0	0
9326	Linketts, canned	1	item(s)	35	21.0	70	7.0	1.0	1.0	4.0	0.5	1.0	2.5	0
9336	Redi-Burger patties, canned	1	slice(s)	85	50.5	120	18.0	7.0	4.0	2.5	0.5	0.5	1.5	0
9350	Swiss Stake pattie with gravy, frozen	1	piece(s)	92	65.7	130	9.0	9.0	3.0	6.0	1.0	1.5	3.5	0
9354	Tender Rounds meatball substitute, canned in gravy	6	piece(s)	80	53.9	120	13.0	6.0	1.0	4.5	0.5	1.5	2.5	0
	Morningstar Farms													
33707	America's Original Veggie Dog links	1	item(s)	57	—	80	11.0	6.0	1.0	0.5	0	—	—	0
9362	Better 'n Eggs egg substitute	¼	cup(s)	57	50.3	20	5.0	0	0	0	0	0	0	0
9371	Breakfast bacon strips	2	item(s)	16	6.8	60	2.0	2.0	0.5	4.5	0.5	1.0	3.0	0
9368	Breakfast sausage links	2	item(s)	45	26.8	80	9.0	3.0	2.0	3.0	0.5	1.5	1.0	0
33705	Chik 'n nuggets	4	piece(s)	86	—	190	12.0	18.0	2.0	7.0	1.0	2.0	4.0	0
11587	Chik patties	1	item(s)	71	36.3	150	9.0	16.0	2.0	6.0	1.0	1.5	2.5	0
2531	Garden veggie patties	1	item(s)	67	40.1	100	10.0	9.0	4.0	2.5	0.5	0.5	1.5	0
33702	Spicy black bean veggie burger	1	item(s)	78	—	140	12.0	15.0	3.0	4.0	0.5	1.0	2.5	0
9412	Vegetarian chili, canned	1	cup(s)	230	172.6	180	16.0	25.0	10.0	1.5	0.5	0.5	0.5	0
	Worthington													
9424	Chili, canned	1	cup(s)	230	167.0	280	24.0	25.0	8.0	10.0	1.5	1.5	7.0	0
9436	Diced Chik, canned	¼	cup(s)	55	42.7	50	9.0	2.0	1.0	0	0	0	0	0
9440	Dinner roast, frozen	1	slice(s)	85	53.2	180	14.0	6.0	3.0	11.0	1.5	4.5	5.0	0
9420	Meatless chicken slices, frozen	3	slice(s)	57	38.9	90	9.0	2.0	0.5	4.5	1.0	1.0	2.5	0
36702	Meatless chicken style roll, frozen	1	slice(s)	55	—	90	9.0	2.0	1.0	4.5	1.0	1.0	2.5	0
9428	Meatless corned beef, sliced, frozen	3	slice(s)	57	31.2	140	10.0	5.0	0	9.0	1.0	2.0	5.0	0
9470	Meatless salami, sliced, frozen	3	slice(s)	57	32.4	120	12.0	3.0	2.0	7.0	1.0	1.0	5.0	0
9480	Meatless smoked turkey, sliced	3	slice(s)	57	—	140	10.0	4.0	0	9.0	1.5	2.0	5.0	0
9462	Prosage links	2	item(s)	45	26.8	80	9.0	3.0	2.0	3.0	0.5	0.5	2.0	0
9484	Stakelets patty beef steak substitute, frozen	1	piece(s)	71	41.5	150	14.0	7.0	2.0	7.0	1.0	2.5	3.5	0
9486	Stripples bacon substitute	2	item(s)	16	6.8	60	2.0	2.0	0.5	4.5	0.5	1.0	3.0	0
9496	Vegetable Skallops meat substitute, canned	½	cup(s)	85	—	90	17.0	4.0	3.0	1.0	0	0	0.5	0
	DAIRY													
	Cheese													
1433	Blue, crumbled	1	ounce(s)	28	12.0	100	6.1	0.7	0	8.1	5.3	2.2	0.2	—
884	Brick	1	ounce(s)	28	11.7	105	6.6	0.8	0	8.4	5.3	2.4	0.2	—
885	Brie	1	ounce(s)	28	13.7	95	5.9	0.1	0	7.8	4.9	2.3	0.2	—
34821	Camembert	1	ounce(s)	28	14.7	85	5.6	0.1	0	6.9	4.3	2.0	0.2	—
5	Cheddar, shredded	¼	cup(s)	28	10.4	114	7.0	0.4	0	9.4	6.0	2.7	0.3	—
888	Cheddar or colby	1	ounce(s)	28	10.8	112	6.7	0.7	0	9.1	5.7	2.6	0.3	—
32096	Cheddar or colby, low fat	1	ounce(s)	28	17.9	49	6.9	0.5	0	2.0	1.2	0.6	0.1	—
889	Edam	1	ounce(s)	28	11.8	101	7.1	0.4	0	7.9	5.0	2.3	0.2	—
890	Feta	1	ounce(s)	28	15.7	75	4.0	1.2	0	6.0	4.2	1.3	0.2	—
891	Fontina	1	ounce(s)	28	10.8	110	7.3	0.4	0	8.8	5.4	2.5	0.5	—
8527	Goat cheese, soft	1	ounce(s)	28	17.2	76	5.3	0.3	0	6.0	4.1	1.4	0.1	—
893	Gouda	1	ounce(s)	28	11.8	101	7.1	0.6	0	7.8	5.0	2.2	0.2	—
894	Gruyere	1	ounce(s)	28	9.4	117	8.5	0.1	0	9.2	5.4	2.8	0.5	—
895	Limburger	1	ounce(s)	28	13.7	93	5.7	0.1	0	7.7	4.7	2.4	0.1	—
896	Monterey jack	1	ounce(s)	28	11.6	106	6.9	0.2	0	8.6	5.4	2.5	0.3	—
13	Mozzarella, part skim milk	1	ounce(s)	28	15.2	72	6.9	0.8	0	4.5	2.9	1.3	0.1	—
12	Mozzarella, whole milk	1	ounce(s)	28	14.2	85	6.3	0.6	0	6.3	3.7	1.9	0.2	—
897	Muenster	1	ounce(s)	28	11.8	104	6.6	0.3	0	8.5	5.4	2.5	0.2	—
898	Neufchatel	1	ounce(s)	28	17.6	74	2.8	0.8	0	6.6	4.2	1.9	0.2	—
14	Parmesan, grated	1	tablespoon(s)	5	1.0	22	1.9	0.2	0	1.4	0.9	0.4	0.1	—
17	Provolone	1	ounce(s)	28	11.6	100	7.3	0.6	0	7.5	4.8	2.1	0.2	—
19	Ricotta, part skim milk	¼	cup(s)	62	45.8	85	7.0	3.2	0	4.9	3.0	1.4	0.2	—
18	Ricotta, whole milk	¼	cup(s)	62	44.1	107	6.9	1.9	0	8.0	5.1	2.2	0.2	—
20	Romano	1	tablespoon(s)	5	1.5	19	1.6	0.2	0	1.3	0.9	0.4	0	—
900	Roquefort	1	ounce(s)	28	11.2	105	6.1	0.6	0	8.7	5.5	2.4	0.4	—
21	Swiss	1	ounce(s)	28	10.5	108	7.6	1.5	0	7.9	5.0	2.1	0.3	—
	Imitation cheese													
42245	Imitation American cheddar cheese	1	ounce(s)	28	15.1	68	4.7	3.3	0	4.0	2.5	1.2	0.1	—
53914	Imitation cheddar	1	ounce(s)	28	15.1	68	4.7	3.3	0	4.0	2.5	1.2	0.1	—
	Cottage cheese													
9	Low fat, 1% fat	½	cup(s)	113	93.2	81	14.0	3.1	0	1.2	0.7	0.3	0	—
8	Low fat, 2% fat	½	cup(s)	113	89.6	102	15.5	4.1	0	2.2	1.4	0.6	0.1	—
	Cream cheese													
11	Cream cheese	2	tablespoon(s)	29	15.6	101	2.2	0.8	0	10.1	6.4	2.9	0.4	—
17366	Fat free cream cheese	2	tablespoon(s)	30	22.7	29	4.3	1.7	0	0.4	0.3	0.1	0	—
10438	Tofutti Better than Cream Cheese	2	tablespoon(s)	30	—	80	1.0	1.0	0	8.0	2.0	—	6.0	—

Chol (mg)	Calc (mg)	Iron (mg)	Magn (mg)	Pota (mg)	Sodi (mg)	Zinc (mg)	Vit A (µg)	Thia (mg)	Vit E (mg α)	Ribo (mg)	Niac (mg)	Vit B$_6$ (mg)	Fola (µg)	Vit C (mg)	Vit B$_{12}$ (µg)	Sele (µg)
0	40	1.44	27.0	182.0	290.0	0.46	—	0.07	—	0.08	0.90	0.09	10.6	9.0	0	4.0
0	0	0.77	—	50.0	220.0	—	0	0.22	—	0.10	2.00	0.70	—	0	2.4	—
0	20	1.80	—	70.0	430.0	0.33	0	1.05	—	0.34	4.00	0.30	—	0	2.4	—
0	0	0.36	—	20.0	160.0	0.46	0	0.12	—	0.20	0.80	0.16	—	0	0.9	—
0	0	1.06	—	140.0	450.0	—	0	0.15	—	0.25	4.00	0.40	—	0	1.2	—
0	0	0.72	—	200.0	430.0	—	0	0.45	—	0.25	10.00	1.00	—	0	5.4	—
0	20	1.08	—	80.0	340.0	0.66	0	0.75	—	0.17	2.00	0.16	—	0	1.2	—
0	0	0.72	—	60.0	580.0	—	0							0		
0	20	0.72	—	75.0	90.0	0.60	37.5	0.03	—	0.34	0.00	0.08	24.0	—	0.6	—
0	0	0.36	—	15.0	220.0	0.05	0	0.75	—	0.04	0.40	0.07	—	0	0.2	—
0	0	1.80	—	50.0	300.0	—	0	0.37	—	0.17	7.00	0.50	—	0	3.0	—
0	20	2.70	—	320.0	490.0	—	0	0.52	—	0.25	5.00	0.30	—	0	1.5	—
0	0	1.80	—	210.0	540.0	—	0	1.80	—	0.17	2.00	0.20	—	0	1.2	—
0	40	0.72	—	180.0	350.0	—	—	—	—	—	—	—	—	0	—	—
0	40	1.80	—	320.0	470.0	—	0	—	—	—	0.00	—	—	0	—	—
0	40	3.60	—	660.0	900.0	—	—	—	—	—	—	—	—	0	—	—
0	40	3.60	—	330.0	1130.0	—	0	0.30	—	0.13	2.00	0.70	—	0	1.5	—
0	0	1.08	—	100.0	220.0	0.24	0	0.06	—	0.10	4.00	0.08	—	0	0.2	—
0	20	1.80	—	120.0	580.0	0.64	0	1.80	—	0.25	6.00	0.60	—	0	1.5	—
0	250	1.80	—	250.0	250.0	0.26	0	0.37	—	0.13	4.00	0.30	—	0	1.8	—
0	100	1.08	—	240.0	240.0	—	0	0.37	—	0.13	4.00	0.30	—	0	1.8	—
0	0	1.80	—	130.0	460.0	0.26	0	0.45	—	0.17	5.00	0.30	—	0	1.8	—
0	0	1.08	—	95.0	800.0	0.30	0	0.75	—	0.17	4.00	0.20	—	0	0.6	—
0	60	2.70	—	60.0	450.0	0.23	0	1.80	—	0.17	6.00	0.40	—	0	3.0	—
0	0	1.44	—	50.0	320.0	0.36	0	1.80	—	0.17	2.00	0.30	—	0	3.0	—
0	40	1.08	—	130.0	480.0	0.50	0	1.20	—	0.13	3.00	0.30	—	0	1.5	—
0	0	0.36	—	15.0	220.0	0.05	0	0.75	—	0.03	0.40	0.08	—	0	0.2	—
0	0	0.36	—	10.0	390.0	0.67	0	0.03	—	0.03	0.00	0.01	—	0	0	—
21	150	0.08	6.5	72.6	395.5	0.75	56.1	0.01	0.07	0.10	0.28	0.04	10.2	0	0.3	4.1
27	191	0.12	6.8	38.6	158.8	0.73	82.8	0.00	0.07	0.10	0.03	0.01	5.7	0	0.4	4.1
28	52	0.14	5.7	43.1	178.3	0.67	49.3	0.02	0.06	0.14	0.10	0.06	18.4	0	0.5	4.1
20	110	0.09	5.7	53.0	238.7	0.67	68.3	0.01	0.06	0.14	0.18	0.06	17.6	0	0.4	4.1
30	204	0.19	7.9	27.7	175.4	0.87	74.9	0.01	0.08	0.10	0.02	0.02	5.1	0	0.2	3.9
27	194	0.21	7.4	36.0	171.2	0.87	74.8	0.00	0.07	0.10	0.02	0.02	5.1	0	0.2	4.1
6	118	0.11	4.5	18.7	173.5	0.51	17.0	0.00	0.01	0.06	0.01	0.01	3.1	0	0.1	4.1
25	207	0.12	8.5	53.3	273.6	1.06	68.9	0.01	0.06	0.11	0.02	0.02	4.5	0	0.4	4.1
25	140	0.18	5.4	17.6	316.4	0.81	35.4	0.04	0.05	0.23	0.28	0.12	9.1	0	0.5	4.3
33	156	0.06	4.0	18.1	226.8	0.99	74.0	0.01	0.07	0.05	0.04	0.02	1.7	0	0.5	4.1
13	40	0.53	4.5	7.4	104.3	0.26	81.6	0.02	0.05	0.10	0.12	0.07	3.4	0	0.1	0.8
32	198	0.06	8.2	34.3	232.2	1.10	46.8	0.01	0.06	0.09	0.01	0.02	6.0	0	0.4	4.1
31	287	0.04	10.2	23.0	95.3	1.10	76.8	0.01	0.07	0.07	0.03	0.02	2.8	0	0.5	4.1
26	141	0.03	6.0	36.3	226.8	0.59	96.4	0.02	0.06	0.14	0.04	0.02	16.4	0	0.3	4.1
25	211	0.20	7.7	23.0	152.0	0.85	56.1	0.00	0.07	0.11	0.02	0.02	5.1	0	0.2	4.1
18	222	0.06	6.5	23.8	175.5	0.78	36.0	0.01	0.04	0.08	0.03	0.02	2.6	0	0.2	4.1
22	143	0.12	5.7	21.5	177.8	0.82	50.7	0.01	0.05	0.08	0.02	0.01	2.0	0	0.6	4.8
27	203	0.11	7.7	38.0	178.0	0.79	84.5	0.00	0.07	0.09	0.02	0.01	3.4	0	0.4	4.1
22	21	0.07	2.3	32.3	113.1	0.14	84.5	0.00	—	0.05	0.03	0.01	3.1	0	0.1	0.9
4	55	0.04	1.9	6.3	76.5	0.19	6.0	0.00	0.01	0.02	0.01	0.00	0.5	0	0.1	0.4
20	214	0.14	7.9	39.1	248.3	0.91	66.9	0.01	0.06	0.09	0.04	0.02	2.8	0	0.4	4.1
19	167	0.27	9.2	76.9	76.9	0.82	65.8	0.01	0.04	0.11	0.04	0.01	8.0	0	0.2	10.3
31	127	0.23	6.8	64.6	51.7	0.71	73.8	0.01	0.06	0.12	0.06	0.02	7.4	0	0.2	8.9
5	53	0.03	2.1	4.3	60	0.12	4.8	0.00	0.01	0.01	0.00	0.00	0.4	0	0.1	0.7
26	188	0.15	8.5	25.8	512.9	0.59	83.3	0.01	—	0.16	0.20	0.03	13.9	0	0.2	4.1
26	224	0.05	10.8	21.8	54.4	1.23	62.4	0.01	0.10	0.08	0.02	0.02	1.7	0	0.9	5.2
10	159	0.08	8.2	68.6	381.3	0.73	32.3	0.01	0.07	0.12	0.03	0.03	2.0	0	0.1	4.3
10	159	0.09	8.2	68.6	381.3	0.73	32.3	0.01	0.07	0.12	0.04	0.03	2.0	0	0.1	4.3
5	69	0.15	5.7	97.2	458.8	0.42	12.4	0.02	0.01	0.18	0.14	0.07	13.6	0	0.7	10.2
9	78	0.18	6.8	108.5	458.8	0.47	23.7	0.02	0.02	0.20	0.16	0.08	14.7	0	0.8	11.5
32	23	0.34	1.7	34.5	85.8	0.15	106.1	0.01	0.08	0.05	0.02	0.01	3.8	0	0.1	0.7
2	56	0.05	4.2	48.9	163.5	0.26	83.7	0.01	0.00	0.05	0.04	0.01	11.1	0	0.2	1.5
0	0	0.00	—		135.0	—	0	—	—	—	—	—	—	0	—	—

Food Composition (Computer code is for Cengage Diet Analysis program) (For purposes of calculations, use "0" for t, <1, <.1, <.01, etc.)

DA+ Code	Food Description	Quantity	Measure	Wt (g)	H₂O (g)	Ener (kcal)	Prot (g)	Carb (g)	Fiber (g)	Fat (g)	Fat Breakdown (g) Sat	Mono	Poly	Trans
	DAIRY —Continued													
	Processed cheese													
24	American cheese food, processed	1	ounce(s)	28	12.3	94	5.2	2.2	0	7.1	4.2	2.0	0.3	—
25	American cheese spread, processed	1	ounce(s)	28	13.5	82	4.7	2.5	0	6.0	3.8	1.8	0.2	—
22	American cheese, processed	1	ounce(s)	28	11.1	106	6.3	0.5	0	8.9	5.6	2.5	0.3	—
9110	Kraft deluxe singles pasteurized process American cheese	1	ounce(s)	28	—	108	5.4	0	0	9.5	5.4	—	—	—
23	Swiss cheese, processed	1	ounce(s)	28	12.0	95	7.0	0.6	0	7.1	4.5	2.0	0.2	—
	Soy cheese													
10437	Galaxy Foods vegan grated parmesan cheese alternative	1	tablespoon(s)	8	—	23	3.0	1.5	0	0	0	0	0	0
10430	Nu Tofu cheddar flavored cheese alternative	1	ounce(s)	28	—	70	6.0	1.0	0	4.0	0.5	2.5	1.0	—
	Cream													
26	Half and half cream	1	tablespoon(s)	15	12.1	20	0.4	0.6	0	1.7	1.1	0.5	0.1	—
32	Heavy whipping cream, liquid	1	tablespoon(s)	15	8.7	52	0.3	0.4	0	5.6	3.5	1.6	0.2	—
28	Light coffee or table cream, liquid	1	tablespoon(s)	15	11.1	29	0.4	0.5	0	2.9	1.8	0.8	0.1	—
30	Light whipping cream, liquid	1	tablespoon(s)	15	9.5	44	0.3	0.4	0	4.6	2.9	1.4	0.1	—
34	Whipped cream topping, pressurized	1	tablespoon(s)	3	1.8	8	0.1	0.4	0	0.7	0.4	0.2	0	—
	Sour cream													
30556	Fat free sour cream	2	tablespoon(s)	32	25.8	24	1.0	5.0	0	0	0	0	0	0
36	Sour cream	2	tablespoon(s)	24	17.0	51	0.8	1.0	0	5.0	3.1	1.5	0.2	—
	Imitation cream													
3659	Coffeemate non dairy creamer, liquid	1	tablespoon(s)	15	—	20	0	2.0	0	1.0	0	0.5	0	—
40	Cream substitute, powder	1	teaspoon(s)	2	0	11	0.1	1.1	0	0.7	0.7	0	0	—
904	Imitation sour cream	2	tablespoon(s)	29	20.5	60	0.7	1.9	0	5.6	5.1	0.2	0	—
35972	Non dairy coffee whitener, liquid, frozen	1	tablespoon(s)	15	11.7	21	0.2	1.7	0	1.5	0.3	1.1	0	—
35976	Non dairy dessert topping, frozen	1	tablespoon(s)	5	2.4	15	0.1	1.1	0	1.2	1.0	0.1	0	—
35975	Non dairy dessert topping, pressurized	1	tablespoon(s)	4	2.7	12	0	0.7	0	1.0	0.8	0.1	0	—
	Fluid milk													
60	Buttermilk, low fat	1	cup(s)	245	220.8	98	8.1	11.7	0	2.2	1.3	0.6	0.1	—
54	Lowfat, 1%	1	cup(s)	244	219.4	102	8.2	12.2	0	2.4	1.5	0.7	0.1	—
55	Lowfat, 1%, with nonfat milk solids	1	cup(s)	245	220.0	105	8.5	12.2	0	2.4	1.5	0.7	0.1	—
57	Nonfat, skim or fat free	1	cup(s)	245	222.6	83	8.3	12.2	0	0.2	0.1	0.1	0	—
58	Nonfat, skim or fat free with nonfat milk solids	1	cup(s)	245	221.4	91	8.7	12.3	0	0.6	0.4	0.2	0	—
51	Reduced fat, 2%	1	cup(s)	244	218.0	122	8.1	11.4	0	4.8	3.1	1.4	0.2	—
52	Reduced fat, 2%, with nonfat milk solids	1	cup(s)	245	217.7	125	8.5	12.2	0	4.7	2.9	1.4	0.2	—
50	Whole, 3.3%	1	cup(s)	244	215.5	146	7.9	11.0	0	7.9	4.6	2.0	0.5	—
	Canned milk													
62	Nonfat or skim evaporated	2	tablespoon(s)	32	25.3	25	2.4	3.6	0	0.1	0	0	0	—
63	Sweetened condensed	2	tablespoon(s)	38	10.4	123	3.0	20.8	0	3.3	2.1	0.9	0.1	—
61	Whole evaporated	2	tablespoon(s)	32	23.3	42	2.1	3.2	0	2.4	1.4	0.7	0.1	—
	Dried milk													
64	Buttermilk	¼	cup(s)	30	0.9	117	10.4	14.9	0	1.8	1.1	0.5	0.1	—
65	Instant nonfat with added vitamin A	¼	cup(s)	17	0.7	61	6.0	8.9	0	0.1	0.1	0	0	—
5234	Skim milk powder	¼	cup(s)	17	0.7	62	6.1	9.1	0	0.1	0.1	0	0	—
907	Whole dry milk	¼	cup(s)	32	0.8	159	8.4	12.3	0	8.5	5.4	2.5	0.2	—
909	Goat milk	1	cup(s)	244	212.4	168	8.7	10.9	0	10.1	6.5	2.7	0.4	—
	Chocolate milk													
33155	Chocolate syrup, prepared with milk	1	cup(s)	282	227.0	254	8.7	36.0	0.8	8.3	4.7	2.1	0.5	—
33184	Cocoa mix with aspartame, added sodium and vitamin A, no added calcium or phosphorus, prepared with water	1	cup(s)	192	177.4	56	2.3	10.8	1.2	0.4	0.3	0.1	0	—
908	Hot cocoa, prepared with milk	1	cup(s)	250	206.4	193	8.8	26.6	2.5	5.8	3.6	1.7	0.1	0.2
69	Low fat	1	cup(s)	250	211.3	158	8.1	26.1	1.3	2.5	1.5	0.8	0.1	—
68	Reduced fat	1	cup(s)	250	205.4	190	7.5	30.3	1.8	4.8	2.9	1.1	0.2	—
67	Whole	1	cup(s)	250	205.8	208	7.9	25.9	2.0	8.5	5.3	2.5	0.3	—
70	Eggnog	1	cup(s)	254	188.9	343	9.7	34.4	0	19.0	11.3	5.7	0.9	—
	Breakfast drinks													
10093	Carnation Instant Breakfast classic chocolate malt, prepared with skim milk, no sugar added	1	cup(s)	243	—	142	11.1	21.3	0.7	1.3	0.7	—	—	—
10092	Carnation Instant Breakfast classic French vanilla, prepared with skim milk, no sugar added	1	cup(s)	273	—	150	12.9	24.0	0	0.4	0.4	—	—	—
10094	Carnation Instant Breakfast stawberry sensation, prepared with skim milk, no sugar added	1	cup(s)	243	—	142	11.1	21.3	0	0.4	0.4	—	—	—
10091	Carnation Instant Breakfast strawberry sensation, prepared with skim milk	1	cup(s)	273	—	220	12.5	38.8	0	0.4	0.4	—	—	—
1417	Ovaltine rich chocolate flavor, prepared with skim milk	1	cup(s)	258	—	170	8.5	31.0	0	0	0	0	0	0
8539	**Malted milk, chocolate mix, fortified, prepared with milk**	1	cup(s)	265	215.8	223	8.9	28.9	1.1	8.6	5.0	2.2	0.5	—

Chol (mg)	Calc (mg)	Iron (mg)	Magn (mg)	Pota (mg)	Sodi (mg)	Zinc (mg)	Vit A (µg)	Thia (mg)	Vit E (mg α)	Ribo (mg)	Niac (mg)	Vit B6 (mg)	Fola (µg)	Vit C (mg)	Vit B12 (µg)	Sele (µg)
23	162	0.16	8.8	82.5	358.6	0.90	57.0	0.01	0.06	0.14	0.04	0.02	2.0	0	0.4	4.6
16	159	0.09	8.2	68.6	381.3	0.73	49.0	0.01	0.05	0.12	0.03	0.03	2.0	0	0.1	3.2
27	156	0.05	7.7	47.9	422.1	0.80	72.0	0.01	0.07	0.10	0.02	0.02	2.3	0	0.2	4.1
27	338	0.00	0	33.8	459.0	1.22	114.0	—	—	0.14	—	—	—	0	0.2	—
24	219	0.17	8.2	61.2	388.4	1.02	56.1	0.00	0.09	0.07	0.01	0.01	1.7	0	0.3	4.5
0	60	0.00	—	75.0	97.5	—	—	—	—	—	—	—	—	—	—	—
0	200	0.36	—	—	190.0	—	—	—	—	—	—	—	—	0	—	—
6	16	0.01	1.5	19.5	6.2	0.08	14.6	0.01	0.05	0.02	0.01	0.01	0.5	0.1	0	0.3
21	10	0.00	1.1	11.3	5.7	0.03	61.7	0.00	0.15	0.01	0.01	0.00	0.6	0.1	0	0.1
10	14	0.01	1.4	18.3	6.0	0.04	27.2	0.01	0.08	0.02	0.01	0.01	0.3	0.1	0	0.1
17	10	0.00	1.1	14.6	5.1	0.03	41.9	0.00	0.13	0.01	0.01	0.00	0.6	0.1	0	0.1
2	3	0.00	0.3	4.4	3.9	0.01	5.6	0.00	0.01	0.00	0.00	0.00	0.1	0	0	0
3	40	0.00	3.2	41.3	45.1	0.16	23.4	0.01	0.00	0.04	0.02	0.01	3.5	0	0.1	1.7
11	28	0.01	2.6	34.6	12.7	0.06	42.5	0.01	0.14	0.03	0.01	0.00	2.6	0.2	0.1	0.5
0	0	0.00	—	30.0	0	—	0	0.01	—	0.01	0.20	—	—	0	—	—
0	0	0.02	0.1	16.2	3.6	0.01	0	0.00	0.01	0.00	0.00	0.00	0	0	0	0
0	1	0.11	1.7	46.3	29.3	0.34	0	0.00	0.21	0.00	0.00	0.00	0	0	0	0.7
0	1	0.00	0	28.9	12.0	0.00	0.2	0.00	0.12	0.00	0.00	0.00	0	0	0	0.2
0	0	0.00	0.1	0.9	1.2	0.00	0.3	0.00	0.05	0.00	0.00	0.00	0	0	0	0.1
0	0	0.00	0	0.8	2.8	0.00	0.2	0.00	0.04	0.00	0.00	0.00	0	0	0	0.1
10	284	0.12	27.0	370.0	257.3	1.02	17.2	0.08	0.12	0.37	0.14	0.08	12.3	2.5	0.5	4.9
12	290	0.07	26.8	366.0	107.4	1.02	141.5	0.04	0.02	0.45	0.22	0.09	12.2	0	1.1	8.1
10	314	0.12	34.3	396.9	127.4	0.98	144.6	0.09	—	0.42	0.22	0.11	12.3	2.5	0.9	5.6
5	306	0.07	27.0	382.2	102.9	1.02	149.5	0.11	0.02	0.44	0.23	0.09	12.3	0	1.3	7.6
5	316	0.12	36.8	419.0	129.9	1.00	149.5	0.10	0.00	0.42	0.22	0.11	12.3	2.5	1.0	5.4
20	285	0.07	26.8	366.0	100.0	1.04	134.2	0.09	0.07	0.45	0.22	0.09	12.2	0.5	1.1	6.1
20	314	0.12	34.3	396.9	127.4	0.98	137.2	0.09	—	0.42	0.22	0.11	12.3	2.5	0.9	5.6
24	276	0.07	24.4	348.9	97.6	0.97	68.3	0.10	0.14	0.44	0.26	0.08	12.2	0	1.1	9.0
1	93	0.09	8.6	105.9	36.7	0.28	37.6	0.01	0.00	0.09	0.05	0.01	2.9	0.4	0.1	0.8
13	109	0.07	9.9	141.9	48.6	0.36	28.3	0.03	0.06	0.16	0.08	0.02	4.2	1.0	0.2	5.7
9	82	0.06	7.6	95.4	33.4	0.24	20.5	0.01	0.04	0.10	0.06	0.01	2.5	0.6	0.1	0.7
21	359	0.09	33.3	482.5	156.7	1.21	14.9	0.11	0.03	0.48	0.27	0.10	14.2	1.7	1.2	6.2
3	209	0.05	19.9	289.9	93.3	0.75	120.5	0.07	0.00	0.30	0.15	0.06	8.5	1.0	0.7	4.6
3	214	0.05	20.3	296.0	95.3	0.76	123.1	0.07	0.00	0.30	0.15	0.06	8.7	1.0	0.7	4.7
31	292	0.15	27.2	425.6	118.7	1.06	82.2	0.09	0.15	0.38	0.20	0.09	11.8	2.8	0.2	5.2
27	327	0.12	34.2	497.8	122.0	0.73	139.1	0.11	0.17	0.33	0.67	0.11	2.4	3.2	0.2	3.4
25	251	0.90	50.8	408.9	132.5	1.21	70.5	0.11	0.14	0.46	0.38	0.09	14.1	0	1.1	9.6
0	92	0.74	32.6	405.1	138.2	0.51	0	0.04	0.00	0.20	0.16	0.04	1.9	0	0.2	2.5
20	263	1.20	57.5	492.5	110.0	1.57	127.5	0.09	0.07	0.45	0.33	0.10	12.5	0.5	1.1	6.8
8	288	0.60	32.5	425.0	152.5	1.02	145.0	0.09	0.05	0.41	0.31	0.10	12.5	2.3	0.9	4.8
20	273	0.60	35.0	422.5	165.0	0.97	160.0	0.11	0.10	0.45	0.31	0.06	5.0	0	0.8	8.5
30	280	0.60	32.5	417.5	150.0	1.02	65.0	0.09	0.15	0.40	0.31	0.10	12.5	2.3	0.8	4.8
150	330	0.50	48.3	419.1	137.2	1.16	116.8	0.08	0.50	0.48	0.26	0.12	2.5	3.8	1.1	10.7
9	444	4.00	88.9	631.1	195.6	3.38	—	0.33	—	0.45	4.44	0.44	4.0	26.7	1.3	8.0
9	500	4.50	100.0	665.0	192.0	3.75	—	0.37	—	0.51	5.00	0.49	100.0	30.0	1.5	9.0
9	444	4.00	88.9	568.9	186.7	3.38	—	0.33	—	0.45	4.44	0.44	88.9	26.7	1.3	8.0
9	500	4.47	100.0	665.0	288.0	3.75	—	0.37	—	0.51	5.07	0.50	100.0	30.0	1.5	8.8
5	350	3.60	100.0	—	270.0	3.75	—	0.37	—	—	4.00	0.40	—	12.0	1.2	—
27	339	3.76	45.1	577.7	230.6	1.16	903.7	0.75	0.15	1.31	11.08	1.01	18.6	31.8	1.1	12.5

DA+ Code	Food Description	Quantity	Measure	Wt (g)	H₂O (g)	Ener (kcal)	Prot (g)	Carb (g)	Fiber (g)	Fat (g)	Fat Breakdown (g) Sat	Mono	Poly	Trans
	DAIRY —Continued													
	Milkshakes													
73	Chocolate	1	cup(s)	227	164.0	270	6.9	48.1	0.7	6.1	3.8	1.8	0.2	—
3163	Strawberry	1	cup(s)	226	167.8	256	7.7	42.8	0.9	6.3	3.9	—	—	—
74	Vanilla	1	cup(s)	227	169.2	254	8.8	40.3	0	6.9	4.3	2.0	0.3	—
	Ice cream													
4776	Chocolate	½	cup(s)	66	36.8	143	2.5	18.6	0.8	7.3	4.5	2.1	0.3	—
12137	Chocolate fudge, no sugar added	½	cup(s)	71	—	100	3.0	16.0	2.0	3.0	1.5	—	—	0
16514	Chocolate, soft serve	½	cup(s)	87	49.9	177	3.2	24.1	0.7	8.4	5.2	2.4	0.3	—
16523	Sherbet, all flavors	½	cup(s)	97	63.8	139	1.1	29.3	3.2	1.9	1.1	0.5	0.1	—
4778	Strawberry	½	cup(s)	66	39.6	127	2.1	18.2	0.6	5.5	3.4	—	—	—
76	Vanilla	½	cup(s)	72	43.9	145	2.5	17.0	0.5	7.9	4.9	2.1	0.3	—
12146	Vanilla chocolate swirl, fat free, no sugar added	½	cup(s)	71	—	100	3.0	14.0	2.0	3.0	2.0	—	—	0
82	Vanilla, light	½	cup(s)	76	48.3	125	3.6	19.6	0.2	3.7	2.2	1.0	0.2	—
78	Vanilla, light, soft serve	½	cup(s)	88	61.2	111	4.3	19.2	0	2.3	1.4	0.7	0.1	—
	Soy desserts													
10694	Tofutti low fat vanilla fudge non dairy frozen dessert	½	cup(s)	70	—	140	2.0	24.0	0	4.0	1.0			
15721	Tofutti premium chocolate supreme non dairy frozen dessert	½	cup(s)	70	—	180	3.0	18.0	0	11.0	2.0			
15720	Tofutti premium vanilla non dairy frozen dessert	½	cup(s)	70	—	190	2.0	20.0	0	11.0	2.0			
	Ice milk													
16517	Chocolate	½	cup(s)	66	42.9	94	2.8	16.9	0.3	2.1	1.3	0.6	0.1	—
16516	Flavored, not chocolate	½	cup(s)	66	41.4	108	3.5	17.5	0.2	2.6	1.7	0.6	0.1	—
	Pudding													
25032	Chocolate	½	cup(s)	144	109.7	155	5.1	22.7	0.7	5.4	3.1	1.7	0.2	0
1923	Chocolate, sugar free, prepared with 2% milk	½	cup(s)	133	—	100	5.0	14.0	0.3	3.0	1.5	—	—	—
1722	Rice	½	cup(s)	113	75.6	151	4.1	29.9	0.5	1.9	1.1	0.5	0.1	—
4747	Tapioca, ready to eat	1	item(s)	142	102.0	185	2.8	30.8	0	5.5	1.4	3.6	0.1	—
25031	Vanilla	½	cup(s)	136	109.7	116	4.7	17.6	0	2.8	1.6	0.9	0.2	0
1924	Vanilla, sugar free, prepared with 2% milk	½	cup(s)	133	—	90	4.0	12.0	0.2	2.0	1.5	—	—	—
	Frozen yogurt													
4785	Chocolate, soft serve	½	cup(s)	72	45.9	115	2.9	17.9	1.6	4.3	2.6	1.3	0.2	—
1747	Fruit varieties	½	cup(s)	113	80.5	144	3.4	24.4	0	4.1	2.6	1.1	0.1	—
4786	Vanilla, soft serve	½	cup(s)	72	47.0	117	2.9	17.4	0	4.0	2.5	1.1	0.2	—
	Milk substitutes													
	Lactose free													
16081	Fat free, calcium fortified [milk]	1	cup(s)	240	—	80	8.0	13.0	0	0	0	0	0	0
36486	Low fat milk	1	cup(s)	240	—	110	8.0	13.0	0	2.5	1.5	—	—	—
36487	Reduced fat milk	1	cup(s)	240	—	130	8.0	12.0	0	5.0	3.0	—	—	—
36488	Whole milk	1	cup(s)	240	—	150	8.0	12.0	0	8.0	5.0	—	—	—
	Rice													
10083	Rice Dream carob rice beverage	1	cup(s)	240	—	150	1.0	32.0	0	2.5	0	—	—	—
17089	Rice Dream original rice beverage, enriched	1	cup(s)	240	—	120	1.0	25.0	0	2.0	0	—	—	—
10087	Rice Dream vanilla enriched rice beverage	1	cup(s)	240	—	130	1.0	28.0	0	2.0	0	—	—	—
	Soy													
34750	Soy Dream chocolate enriched soy beverage	1	cup(s)	240	—	210	7.0	37.0	1.0	3.5	0.5	—	—	—
34749	Soy Dream vanilla enriched soy beverage	1	cup(s)	240	—	150	7.0	22.0	0	4.0	0.5	—	—	—
13840	Vitasoy light chocolate soymilk	1	cup(s)	240	—	100	4.0	17.0	0	2.0	0.5	0.5	1.0	—
13839	Vitasoy light vanilla soymilk	1	cup(s)	240	—	70	4.0	10.0	0	2.0	0.5	0.5	1.0	—
13836	Vitasoy rich chocolate soymilk	1	cup(s)	240	—	160	7.0	24.0	1.0	4.0	0.5	1.0	2.5	—
13835	Vitasoy vanilla delite soymilk	1	cup(s)	240	—	120	7.0	13.0	1.0	4.0	0.5	1.0	2.5	—
	Yogurt													
3615	Custard style, fruit flavors	6	ounce(s)	170	127.1	190	7.0	32.0	0	3.5	2.0	—	—	—
3617	Custard style, vanilla	6	ounce(s)	170	134.1	190	7.0	32.0	0	3.5	2.0	0.9	0.1	—
32101	Fruit, low fat	1	cup(s)	245	184.5	243	9.8	45.7	0	2.8	1.8	0.8	0.1	—
29638	Fruit, non fat, sweetened with low calorie sweetener	1	cup(s)	241	208.3	123	10.6	19.4	1.2	0.4	0.2	0.1	0	—
93	Plain, low fat	1	cup(s)	245	208.4	154	12.9	17.2	0	3.8	2.5	1.0	0.1	—
94	Plain, nonfat	1	cup(s)	245	208.8	137	14.0	18.8	0	0.4	0.3	0.1	0	—
32100	Vanilla, low fat	1	cup(s)	245	193.6	208	12.1	33.8	0	3.1	2.0	0.8	0.1	—
5242	Yogurt beverage	1	cup(s)	245	199.8	172	6.2	32.8	0	2.2	1.4	0.6	0.1	—
38202	Yogurt smoothie, nonfat, all flavors	1	item(s)	325	—	290	10.0	60.0	6.0	0	0	0	0	0
	Soy yogurt													
34617	Stonyfield Farm O'Soy strawberry-peach pack organic cultured soy yogurt	1	item(s)	113	—	100	5.0	16.0	3.0	2.0	0	—	—	0
34616	Stonyfield Farm O'Soy vanilla organic cultured soy yogurt	1	item(s)	170	—	150	7.0	26.0	4.0	2.0	0	—	—	0
10453	White Wave plain silk cultured soy yogurt	8	ounce(s)	227	—	140	5.0	22.0	1.0	3.0	0.5	—	—	0
	EGGS													
	Eggs													
99	Fried	1	item(s)	46	31.8	90	6.3	0.4	0	7.0	2.0	2.9	1.2	—

Chol (mg)	Calc (mg)	Iron (mg)	Magn (mg)	Pota (mg)	Sodi (mg)	Zinc (mg)	Vit A (µg)	Thia (mg)	Vit E (mg α)	Ribo (mg)	Niac (mg)	Vit B$_6$ (mg)	Fola (µg)	Vit C (mg)	Vit B$_{12}$ (µg)	Sele (µg)
25	300	0.70	36.4	508.9	252.2	1.09	40.9	0.10	0.11	0.50	0.28	0.05	11.4	0	0.7	4.3
25	256	0.24	29.4	412.0	187.9	0.81	58.9	0.10	—	0.44	0.39	0.10	6.8	1.8	0.7	4.8
27	332	0.22	27.3	415.8	215.8	0.88	56.8	0.06	0.11	0.44	0.33	0.09	15.9	0	1.2	5.2
22	72	0.61	19.1	164.3	50.2	0.38	77.9	0.02	0.19	0.12	0.14	0.03	10.6	0.5	0.2	1.7
10	100	0.36	—	—	65.0	—	—	—	—	—	—	—	—	0	—	—
22	103	0.32	19.0	192.0	43.3	0.45	66.6	0.03	0.22	0.13	0.11	0.03	4.3	0.5	0.3	2.5
0	52	0.13	7.7	92.6	44.4	0.46	9.7	0.02	0.02	0.08	0.07	0.02	6.8	5.6	0.1	1.3
19	79	0.13	9.2	124.1	39.6	0.22	63.4	0.03	—	0.16	0.11	0.03	7.9	5.1	0.2	1.3
32	92	0.06	10.1	143.3	57.6	0.49	85.0	0.03	0.21	0.17	0.08	0.03	3.6	0.4	0.3	1.3
10	100	0.00	—	—	65.0	—	—	—	—	—	—	—	—	0	—	—
21	122	0.14	10.6	158.1	56.2	0.55	97.3	0.04	0.09	0.19	0.10	0.03	4.6	0.9	0.4	1.5
11	138	0.05	12.3	194.5	61.6	0.46	25.5	0.04	0.05	0.17	0.10	0.04	4.4	0.8	0.4	3.2
0	0	0.00	—	8.0	90.0	—	0	—	—	—	—	—	—	0	—	—
0	0	0.00	—	7.0	180.0	—	0	—	—	—	—	—	—	0	—	—
0	0	0.00	—	2.0	210.0	—	0	—	—	—	—	—	—	0	—	—
6	94	0.15	13.1	155.2	40.6	0.36	15.7	0.03	0.05	0.11	0.08	0.02	3.9	0.5	0.3	2.2
16	76	0.05	9.2	136.2	48.5	0.47	90.4	0.02	0.05	0.11	0.06	0.01	3.3	0.1	0.2	1.3
35	149	0.46	31.3	226.7	137.0	0.71	—	0.05	0.00	0.22	0.15	0.06	8.3	1.2	0.5	4.9
10	150	0.72	—	330.0	310.0	—	—	0.06	—	0.26	—	—	—	0	—	—
7	113	0.28	15.8	201.4	66.4	0.52	41.6	0.03	0.05	0.17	0.34	0.06	4.5	0.2	0.2	4.8
1	101	0.15	8.5	130.6	205.9	0.31	0	0.03	0.21	0.13	0.09	0.03	4.3	0.4	0.3	0
35	146	0.17	17.2	188.9	136.4	0.52	—	0.04	0.00	0.22	0.10	0.05	8.0	1.2	0.5	4.6
10	150	0.00	—	190.0	380.0	—	—	0.03	—	0.17	—	—	—	0	—	—
4	106	0.90	19.4	187.9	70.6	0.35	31.7	0.02	—	0.15	0.22	0.05	7.9	0.2	0.2	1.7
15	113	0.52	11.3	176.3	71.2	0.31	55.4	0.04	0.10	0.20	0.07	0.04	4.5	0.8	0.1	2.1
1	103	0.21	10.1	151.9	62.6	0.30	42.5	0.02	0.07	0.16	0.20	0.05	4.3	0.6	0.2	2.4
3	500	0.00	—	—	125.0	—	100.0	—	—	—	—	—	—	0	0	—
10	300	0.00	—	—	125.0	—	100.0	—	—	—	—	—	—	0	—	—
20	300	0.00	—	—	125.0	—	98.2	—	—	—	—	—	—	0	—	—
35	300	0.00	—	—	125.0	—	58.1	—	—	—	—	—	—	0	—	—
0	20	0.72	—	82.5	100.0	—	—	—	—	—	—	—	—	1.2	—	—
0	300	0.00	13.3	60.0	90.0	0.24	—	0.06	—	0.00	0.84	0.07	—	0	1.5	—
0	300	0.00	—	53.0	90.0	—	—	—	—	—	—	—	—	0	1.5	—
0	300	1.80	60.0	350.0	160.0	0.60	33.3	0.15	—	0.06	0.80	0.12	60.0	0	3.0	—
0	300	1.80	40.0	260.0	140.0	0.60	33.3	0.15	—	0.06	0.80	0.12	60.0	0	3.0	—
0	300	0.72	24.0	200.0	140.0	0.90	—	0.09	—	0.34	—	—	24.0	0	0.9	—
0	300	0.72	24.0	200.0	120.0	0.90	—	0.09	—	0.34	—	—	24.0	0	0.9	—
0	300	1.08	40.0	320.0	150.0	0.90	—	0.15	—	0.34	—	—	60.0	0	0.9	—
0	40	0.72	—	320.0	115.0	—	0	—	—	—	—	—	—	0	—	—
15	300	0.00	16.0	310.0	100.0	—	—	—	—	0.25	—	—	—	0	—	—
15	300	0.00	16.0	310.0	100.0	—	—	—	—	0.25	—	—	—	0	—	—
12	338	0.14	31.9	433.7	129.9	1.64	27.0	0.08	0.04	0.39	0.21	0.09	22.1	1.5	1.1	6.9
5	369	0.62	41.0	549.5	139.8	1.83	4.8	0.10	0.16	0.44	0.49	0.10	31.3	26.5	1.1	7.0
15	448	0.19	41.7	573.3	171.5	2.18	34.3	0.10	0.07	0.52	0.27	0.12	27.0	2.0	1.4	8.1
5	488	0.22	46.6	624.8	188.7	2.37	4.9	0.11	0.00	0.57	0.30	0.13	29.4	2.2	1.5	8.8
12	419	0.17	39.2	536.6	161.7	2.03	29.4	0.10	0.04	0.49	0.26	0.11	27.0	2.0	1.3	12.0
13	260	0.22	39.2	399.4	98.0	1.10	14.7	0.11	0.00	0.51	0.30	0.14	29.4	2.1	1.5	—
5	300	2.70	100.0	580.0	290.0	2.25	—	0.37	—	0.42	5.00	0.50	100.0	15.0	1.5	—
0	100	1.08	24.0	5.0	20.0	—	0	0.22	—	0.10	—	0.04	—	0	0	—
0	150	1.44	40.0	15.0	40.0	—	—	0.30	—	0.13	—	0.08	—	0	0	—
0	400	1.44	—	0	30.0	—	0	—	—	—	—	—	—	0	—	—
210	27	0.91	6.0	67.6	93.8	0.55	91.1	0.03	0.56	0.23	0.03	0.07	23.5	0	0.6	15.7

DA+ Code	Food Description	Quantity	Measure	Wt (g)	H₂O (g)	Ener (kcal)	Prot (g)	Carb (g)	Fiber (g)	Fat (g)	Fat Breakdown (g) Sat	Mono	Poly	Trans
	EGGS —Continued													
100	Hard boiled	1	item(s)	50	37.3	78	6.3	0.6	0	5.3	1.6	2.0	0.7	—
101	Poached	1	item(s)	50	37.8	71	6.3	0.4	0	5.0	1.5	1.9	0.7	—
97	Raw, white	1	item(s)	33	28.9	16	3.6	0.2	0	0.1	0	0	0	—
96	Raw, whole	1	item(s)	50	37.9	72	6.3	0.4	0	5.0	1.5	1.9	0.7	—
98	Raw, yolk	1	item(s)	17	8.9	54	2.7	0.6	0	4.5	1.6	2.0	0.7	—
102	Scrambled, prepared with milk and butter	2	item(s)	122	89.2	204	13.5	2.7	0	14.9	4.5	5.8	2.6	0.7
	Egg substitute													
4028	Egg Beaters	¼	cup(s)	61	—	30	6.0	1.0	0	0	0	0	0	0
920	Frozen	¼	cup(s)	60	43.9	96	6.8	1.9	0	6.7	1.2	1.5	3.7	—
918	Liquid	¼	cup(s)	63	51.9	53	7.5	0.4	0	2.1	0.4	0.6	1.0	—
	SEAFOOD													
	Cod													
6040	Atlantic cod or scrod, baked or broiled	3	ounce(s)	85	64.6	89	19.4	0	0	0.7	0.1	0.1	0.2	—
1573	Atlantic cod, cooked, dry heat	3	ounce(s)	85	64.6	89	19.4	0	0	0.7	0.1	0.1	0.2	—
2905	**Eel, raw**	3	ounce(s)	85	58.0	156	15.7	0	0	9.9	2.0	6.1	0.8	—
	Fish fillets													
25079	Baked	3	ounce(s)	84	79.9	99	21.7	0	0	0.7	0.1	0.1	0.3	—
8615	Batter coated or breaded, fried	3	ounce(s)	85	45.6	197	12.5	14.4	0.4	10.5	2.4	2.2	5.3	—
25082	Broiled fish steaks	3	ounce(s)	85	68.1	128	24.2	0	0	2.6	0.4	0.9	0.8	—
25083	Poached fish steaks	3	ounce(s)	85	67.1	111	21.1	0	0	2.3	0.3	0.8	0.7	—
25084	Steamed	3	ounce(s)	85	72.2	79	17.2	0	0	0.6	0.1	0.1	0.2	—
25089	**Flounder, baked**	3	ounce(s)	85	64.4	113	14.8	0.4	0.1	5.5	1.1	2.2	1.4	—
1825	**Grouper, cooked, dry heat**	3	ounce(s)	85	62.4	100	21.1	0	0	1.1	0.3	0.2	0.3	—
	Haddock													
6049	Baked or broiled	3	ounce(s)	85	63.2	95	20.6	0	0	0.8	0.1	0.1	0.3	—
1578	Cooked, dry heat	3	ounce(s)	85	63.1	95	20.6	0	0	0.8	0.1	0.1	0.3	—
1886	**Halibut, Atlantic and Pacific, cooked, dry heat**	3	ounce(s)	85	61.0	119	22.7	0	0	2.5	0.4	0.8	0.8	—
1582	**Herring, Atlantic, pickled**	4	piece(s)	60	33.1	157	8.5	5.8	0	10.8	1.4	7.2	1.0	—
1587	**Jack mackerel, solids, canned, drained**	2	ounce(s)	57	39.2	88	13.1	0	0	3.6	1.1	1.3	0.9	—
8580	**Octopus, common, cooked, moist heat**	3	ounce(s)	85	51.5	139	25.4	3.7	0	1.8	0.4	0.3	0.4	—
1831	**Perch, mixed species, cooked, dry heat**	3	ounce(s)	85	62.3	100	21.1	0	0	1.0	0.2	0.2	0.4	—
1592	**Pacific rockfish, cooked, dry heat**	3	ounce(s)	85	62.4	103	20.4	0	0	1.7	0.4	0.4	0.5	—
	Salmon													
2938	Coho, farmed, raw	3	ounce(s)	85	59.9	136	18.1	0	0	6.5	1.5	2.8	1.6	—
1594	Broiled or baked with butter	3	ounce(s)	85	53.9	155	23.0	0	0	6.3	1.2	2.3	2.3	—
29727	Smoked chinook (lox)	2	ounce(s)	57	40.8	66	10.4	0	0	2.4	0.5	1.1	0.6	—
154	**Sardine, Atlantic with bones, canned in oil**	3	ounce(s)	85	50.7	177	20.9	0	0	9.7	1.3	3.3	4.4	—
	Scallops													
155	Mixed species, breaded, fried	3	item(s)	47	27.2	100	8.4	4.7	—	5.1	1.2	2.1	1.3	—
1599	Steamed	3	ounce(s)	85	64.8	90	13.8	2.0	0	2.6	0.4	1.0	0.8	—
1839	**Snapper, mixed species, cooked, dry heat**	3	ounce(s)	85	59.8	109	22.4	0	0	1.5	0.3	0.3	0.5	—
	Squid													
1868	Mixed species, fried	3	ounce(s)	85	54.9	149	15.3	6.6	0	6.4	1.6	2.3	1.8	—
16617	Steamed or boiled	3	ounce(s)	85	63.3	89	15.2	3.0	0	1.3	0.4	0.1	0.5	—
1570	**Striped bass, cooked, dry heat**	3	ounce(s)	85	62.0	105	19.3	0	0	2.5	0.6	0.7	0.9	—
1601	**Sturgeon, steamed**	3	ounce(s)	85	59.4	111	17.0	0	0	4.3	1.0	2.0	0.7	—
1840	**Surimi, formed**	3	ounce(s)	85	64.9	84	12.9	5.8	0	0.8	0.2	0.1	0.4	—
1842	**Swordfish, cooked, dry heat**	3	ounce(s)	85	58.5	132	21.6	0	0	4.4	1.2	1.7	1.0	—
1846	**Tuna, yellowfin or ahi, raw**	3	ounce(s)	85	60.4	92	19.9	0	0	0.8	0.2	0.1	0.2	—
	Tuna, canned													
159	Light, canned in oil, drained	2	ounce(s)	57	33.9	112	16.5	0	0	4.6	0.9	1.7	1.6	—
355	Light, canned in water, drained	2	ounce(s)	57	42.2	66	14.5	0	0	0.5	0.1	0.1	0.2	—
33211	Light, no salt, canned in oil, drained	2	ounce(s)	57	33.9	112	16.5	0	0	4.7	0.9	1.7	1.6	—
33212	Light, no salt, canned in water, drained	2	ounce(s)	57	42.6	66	14.5	0	0	0.5	0.1	0.1	0.2	—
2961	White, canned in oil, drained	2	ounce(s)	57	36.3	105	15.0	0	0	4.6	0.7	1.8	1.7	—
351	White, canned in water, drained	2	ounce(s)	57	41.5	73	13.4	0	0	1.7	0.4	0.4	0.6	—
33213	White, no salt, canned in oil, drained	2	ounce(s)	57	36.3	105	15.0	0	0	4.6	0.9	1.4	1.9	—
33214	White, no salt, canned in water, drained	2	ounce(s)	57	42.0	73	13.4	0	0	1.7	0.4	0.4	0.6	—
	Yellowtail													
8548	Mixed species, cooked, dry heat	3	ounce(s)	85	57.3	159	25.2	0	0	5.7	1.4	2.2	1.5	—
2970	Mixed species, raw	2	ounce(s)	57	42.2	83	13.1	0	0	3.0	0.7	1.1	0.8	—
	Shellfish, meat only													
1857	Abalone, mixed species, fried	3	ounce(s)	85	51.1	161	16.7	9.4	0	5.8	1.4	2.3	1.4	—
16618	Abalone, steamed or poached	3	ounce(s)	85	40.7	177	28.8	10.1	0	1.3	0.3	0.2	0.2	—
	Crab													
1851	Blue crab, canned	2	ounce(s)	57	43.2	56	11.6	0	0	0.7	0.1	0.1	0.2	—
1852	Blue crab, cooked, moist heat	3	ounce(s)	85	65.9	87	17.2	0	0	1.5	0.2	0.2	0.6	—
8562	Dungeness crab, cooked, moist heat	3	ounce(s)	85	62.3	94	19.0	0.8	0	1.1	0.1	0.2	0.3	—
1860	**Clams, cooked, moist heat**	3	ounce(s)	85	54.1	126	21.7	4.4	0	1.7	0.2	0.1	0.5	—

Chol (mg)	Calc (mg)	Iron (mg)	Magn (mg)	Pota (mg)	Sodi (mg)	Zinc (mg)	Vit A (µg)	Thia (mg)	Vit E (mg α)	Ribo (mg)	Niac (mg)	Vit B$_6$ (mg)	Fola (µg)	Vit C (mg)	Vit B$_{12}$ (µg)	Sele (µg)
212	25	0.59	5.0	63.0	62.0	0.52	84.5	0.03	0.51	0.25	0.03	0.06	22.0	0	0.6	15.4
211	27	0.91	6.0	66.5	147.0	0.55	69.5	0.02	0.48	0.20	0.03	0.06	17.5	0	0.6	15.8
0	2	0.02	3.6	53.8	54.8	0.01	0	0.00	0.00	0.14	0.03	0.00	1.3	0	0	6.6
212	27	0.91	6.0	67.0	70.0	0.55	70.0	0.03	0.48	0.23	0.03	0.07	23.5	0	0.6	15.9
210	22	0.46	0.9	18.5	8.2	0.39	64.8	0.03	0.43	0.09	0.00	0.06	24.8	0	0.3	9.5
429	87	1.46	14.6	168.4	341.6	1.22	174.5	0.06	1.33	0.53	0.09	0.14	36.6	0.2	0.9	27.5
0	20	1.08	4.0	85.0	115.0	0.60	112.5	0.15	—	0.85	0.20	0.08	60.0	0	1.2	—
1	44	1.18	9.0	127.8	119.4	0.58	6.6	0.07	0.95	0.23	0.08	0.08	9.6	0.3	0.2	24.8
1	33	1.32	5.6	207.1	111.1	0.82	11.3	0.07	0.17	0.19	0.07	0.00	9.4	0	0.2	15.6
47	12	0.41	35.7	207.5	66.3	0.49	11.9	0.07	0.68	0.06	2.13	0.24	6.8	0.8	0.9	32.0
47	12	0.41	35.7	207.5	66.3	0.49	11.9	0.07	0.68	0.06	2.13	0.24	6.8	0.9	0.9	32.0
107	17	0.42	17.0	231.3	43.4	1.37	887.0	0.13	3.40	0.03	2.97	0.05	12.8	1.5	2.6	5.5
44	8	0.31	29.1	489.0	86.1	0.48	—	0.02	—	0.05	2.47	0.46	8.1	3.0	1.0	44.3
29	15	1.79	20.4	272.2	452.5	0.37	9.4	0.09	—	0.09	1.78	0.08	14.5	0	0.9	7.7
37	55	0.97	96.7	524.3	62.9	0.49	—	0.05	—	0.08	6.47	0.36	12.6	0	1.2	42.5
32	48	0.85	84.0	455.6	54.7	0.42	—	0.05	—	0.07	5.92	0.33	11.5	0	1.1	37.0
41	12	0.29	24.7	319.3	41.7	0.34	—	0.06	—	0.06	1.89	0.21	6.1	0.8	0.8	32.0
44	19	0.34	47.3	224.7	280.2	0.20	—	0.06	0.40	0.07	2.02	0.18	7.4	2.8	1.6	33.5
40	18	0.96	31.5	404.0	45.1	0.43	42.5	0.06	—	0.01	0.32	0.29	8.5	0	0.6	39.8
63	36	1.15	42.5	339.4	74.0	0.40	16.2	0.03	0.42	0.03	3.94	0.29	6.8	0	1.2	34.4
63	36	1.14	42.5	339.3	74.0	0.40	16.2	0.03	—	0.03	3.93	0.29	11.1	0	1.2	34.4
35	51	0.91	91.0	489.9	58.7	0.45	45.9	0.05	—	0.07	6.05	0.33	11.9	0	1.2	39.8
8	46	0.73	4.8	41.4	522.0	0.31	154.8	0.02	1.02	0.08	1.98	0.10	1.2	0	2.6	35.1
45	137	1.15	21.0	110.0	214.9	0.57	73.7	0.02	0.58	0.12	3.50	0.11	2.8	0.5	3.9	21.4
82	90	8.11	51.0	535.8	391.2	2.85	76.5	0.04	1.02	0.06	3.21	0.55	20.4	6.8	30.6	76.2
98	87	0.98	32.3	292.6	67.2	1.21	8.5	0.06	—	0.10	1.61	0.11	5.1	1.4	1.9	13.7
37	10	0.45	28.9	442.3	65.5	0.45	60.4	0.03	1.32	0.07	3.33	0.22	8.5	0	1.0	39.8
43	10	0.29	26.4	382.7	40.0	0.36	47.6	0.08	—	0.09	5.79	0.56	11.1	0.9	2.3	10.7
40	15	1.02	26.9	376.6	98.6	0.56	—	0.13	1.14	0.05	8.33	0.18	4.2	1.8	2.3	41.0
13	6	0.48	10.2	99.2	1134.0	0.17	14.7	0.01	—	0.05	2.67	0.15	1.1	0	1.8	21.6
121	325	2.48	33.2	337.6	429.5	1.10	27.2	0.04	1.70	0.18	4.43	0.14	10.2	0	7.6	44.8
28	20	0.38	27.4	154.8	215.8	0.49	10.7	0.02	—	0.05	0.70	0.06	17.2	1.1	0.6	12.5
27	20	0.22	45.9	238.0	358.7	0.78	32.3	0.01	0.16	0.05	0.84	0.11	10.2	2.0	1.1	18.2
40	34	0.20	31.5	444.0	48.5	0.37	29.8	0.04	—	0.00	0.29	0.39	5.1	1.4	3.0	41.7
221	33	0.85	32.3	237.3	260.3	1.48	9.4	0.04	—	0.39	2.21	0.04	11.9	3.6	1.0	44.1
227	31	0.62	28.9	192.1	356.2	1.49	8.5	0.01	1.17	0.32	1.69	0.04	3.4	3.2	1.0	43.7
88	16	0.91	43.4	279.0	74.8	0.43	26.4	0.09	—	0.03	2.17	0.29	8.5	0	3.8	39.8
63	11	0.59	29.8	239.7	388.5	0.35	198.9	0.06	0.52	0.07	8.30	0.19	14.5	0	2.2	13.3
26	8	0.22	36.6	95.3	121.6	0.28	17.0	0.01	0.53	0.01	0.18	0.02	1.7	0	1.4	23.9
43	5	0.88	28.9	313.8	97.8	1.25	34.9	0.03	—	0.09	10.02	0.32	1.7	0.9	1.7	52.5
38	14	0.62	42.5	377.6	31.5	0.44	15.3	0.37	0.42	0.04	8.33	0.77	1.7	0.8	0.4	31.0
10	7	0.79	17.6	117.3	200.6	0.51	13.0	0.02	0.49	0.07	7.03	0.06	2.8	0	1.2	43.1
17	6	0.87	15.3	134.3	191.5	0.43	9.6	0.01	0.19	0.04	7.52	0.19	2.3	0	1.7	45.6
10	7	0.78	17.6	117.4	28.3	0.51	0	0.02	—	0.06	7.03	0.06	2.8	0	1.2	43.1
17	6	0.86	15.3	134.4	28.3	0.43	0	0.01	—	0.04	7.52	0.19	2.3	0	1.7	45.6
18	2	0.36	19.3	188.8	224.5	0.26	2.8	0.01	1.30	0.04	6.63	0.24	2.8	0	1.2	34.1
24	8	0.55	18.7	134.3	213.6	0.27	3.4	0.00	0.48	0.02	3.28	0.12	1.1	0	0.7	37.2
18	2	0.36	19.3	188.8	28.3	0.26	0	0.01	—	0.04	6.63	0.24	2.8	0	1.2	34.1
24	8	0.54	18.7	134.4	28.3	0.27	3.4	0.00	—	0.02	3.28	0.12	1.1	0	0.7	37.3
60	25	0.53	32.3	457.6	42.5	0.56	26.4	0.14	—	0.04	7.41	0.15	3.4	2.5	1.1	39.8
31	13	0.28	17.0	238.1	22.1	0.29	16.4	0.08	—	0.02	3.86	0.09	2.3	1.6	0.7	20.7
80	31	3.23	47.6	241.5	502.6	0.80	1.7	0.18	—	0.11	1.61	0.12	11.9	1.5	0.6	44.1
144	50	4.84	68.9	295.0	980.1	1.38	3.4	0.28	6.74	0.12	1.89	0.21	6.0	2.6	0.7	75.6
50	57	0.47	22.1	212.1	188.8	2.27	1.1	0.04	1.04	0.04	0.77	0.08	24.4	1.5	0.3	18.0
85	88	0.77	28.1	275.6	237.3	3.58	1.7	0.08	1.56	0.04	2.80	0.15	43.4	2.8	6.2	34.2
65	50	0.36	49.3	347.0	321.5	4.65	26.4	0.04	—	0.17	3.08	0.14	35.7	3.1	8.8	40.5
57	78	23.78	15.3	534.1	95.3	2.32	145.4	0.12	—	0.36	2.85	0.09	24.7	18.8	84.1	54.4

DA+ Code	Food Description	Quantity	Measure	Wt (g)	H₂O (g)	Ener (kcal)	Prot (g)	Carb (g)	Fiber (g)	Fat (g)	Sat	Mono	Poly	Trans
	SEAFOOD —Continued													
1853	Crayfish, farmed, cooked, moist heat	3	ounce(s)	85	68.7	74	14.9	0	0	1.1	0.2	0.2	0.4	—
	Oysters													
8720	Baked or broiled	3	ounce(s)	85	68.6	89	5.6	3.2	0	5.8	1.3	2.1	1.9	—
152	Eastern, farmed, raw	3	ounce(s)	85	73.3	50	4.4	4.7	0	1.3	0.4	0.1	0.5	—
8715	Eastern, wild, cooked, moist heat	3	ounce(s)	85	59.8	117	12.0	6.7	0	4.2	1.3	0.5	1.6	—
8584	Pacific, cooked, moist heat	3	ounce(s)	85	54.5	139	16.1	8.4	0	3.9	0.9	0.7	1.5	—
1865	Pacific, raw	3	ounce(s)	85	69.8	69	8.0	4.2	0	2.0	0.4	0.3	0.8	—
1854	Lobster, northern, cooked, moist heat	3	ounce(s)	85	64.7	83	17.4	1.1	0	0.5	0.1	0.1	0.1	—
1862	Mussel, blue, cooked, moist heat	3	ounce(s)	85	52.0	146	20.2	6.3	0	3.8	0.7	0.9	1.0	—
	Shrimp													
158	Mixed species, breaded, fried	3	ounce(s)	85	44.9	206	18.2	9.8	0.3	10.4	1.8	3.2	4.3	—
1855	Mixed species, cooked, moist heat	3	ounce(s)	85	65.7	84	17.8	0	0	0.9	0.2	0.2	0.4	—
	BEEF, LAMB, PORK													
	Beef													
4450	Breakfast strips, cooked	2	slice(s)	23	5.9	101	7.1	0.3	0	7.8	3.2	3.8	0.4	—
174	Corned beef, canned	3	ounce(s)	85	49.1	213	23.0	0	0	12.7	5.3	5.1	0.5	—
33147	Cured, thin siced	2	ounce(s)	57	32.9	100	15.9	3.2	0	2.2	0.9	1.0	0.1	—
4581	Jerky	1	ounce(s)	28	6.6	116	9.4	3.1	0.5	7.3	3.1	3.2	0.3	—
	Ground beef													
5898	Lean, broiled, medium	3	ounce(s)	85	50.4	202	21.6	0	0	12.2	4.8	5.3	0.4	—
5899	Lean, broiled, well done	3	ounce(s)	85	48.4	214	23.8	0	0	12.5	5.0	5.7	0.3	0.4
5914	Regular, broiled, medium	3	ounce(s)	85	46.1	246	20.5	0	0	17.6	6.9	7.7	0.6	—
5915	Regular, broiled, well done	3	ounce(s)	85	43.8	259	21.6	0	0	18.4	7.5	8.5	0.5	0.6
	Beef rib													
4241	Rib, small end, separable lean, 0" fat, broiled	3	ounce(s)	85	53.2	164	25.0	0	0	6.4	2.4	2.6	0.2	—
4183	Rib, whole, lean and fat, ¼" fat, roasted	3	ounce(s)	85	39.0	320	18.9	0	0	26.6	10.7	11.4	0.9	—
	Beef roast													
16981	Bottom round, choice, separable lean and fat, ⅛" fat, braised	3	ounce(s)	85	46.2	216	27.9	0	0	10.7	4.1	4.6	0.4	—
16979	Bottom round, separable lean and fat, ⅛" fat, roasted	3	ounce(s)	85	52.4	185	22.5	0	0	9.9	3.8	4.2	0.4	—
16924	Chuck, arm pot roast, separable lean and fat, ⅛" fat, braised	3	ounce(s)	85	42.9	257	25.6	0	0	16.3	6.5	7.0	0.6	—
16930	Chuck, blade roast, separable lean and fat, ⅛" fat, braised	3	ounce(s)	85	40.5	290	22.8	0	0	21.4	8.5	9.2	0.8	—
5853	Chuck, blade roast, separable lean, 0" trim, pot roasted	3	ounce(s)	85	47.4	202	26.4	0	0	9.9	3.9	4.3	0.3	—
4296	Eye of round, choice, separable lean, 0" fat, roasted	3	ounce(s)	85	56.5	138	24.4	0	0	3.7	1.3	1.5	0.1	—
16989	Eye of round, separable lean and fat, ⅛" fat, roasted	3	ounce(s)	85	52.2	180	24.2	0	0	8.5	3.2	3.6	0.3	—
	Beef steak													
4348	Short loin, t-bone steak, lean and fat, ¼" fat, broiled	3	ounce(s)	85	43.2	274	19.4	0	0	21.2	8.3	9.6	0.8	—
4349	Short loin, t-bone steak, lean, ¼" fat, broiled	3	ounce(s)	85	52.3	174	22.8	0	0	8.5	3.1	4.2	0.3	—
4360	Top loin, prime, lean and fat, ¼" fat, broiled	3	ounce(s)	85	42.7	275	21.6	0	0	20.3	8.2	8.6	0.7	—
	Beef variety													
188	Liver, pan fried	3	ounce(s)	85	52.7	149	22.6	4.4	0	4.0	1.3	0.5	0.5	0.2
4447	Tongue, simmered	3	ounce(s)	85	49.2	242	16.4	0	0	19.0	6.9	8.6	0.6	0.7
	Lamb chop													
3275	Loin, domestic, lean and fat, ¼" fat, broiled	3	ounce(s)	85	43.9	269	21.4	0	0	19.6	8.4	8.3	1.4	—
	Lamb leg													
3264	Domestic, lean and fat, ¼" fat, cooked	3	ounce(s)	85	45.7	250	20.9	0	0	17.8	7.5	7.5	1.3	—
	Lamb rib													
182	Domestic, lean and fat, ¼" fat, broiled	3	ounce(s)	85	40.0	307	18.8	0	0	25.2	10.8	10.3	2.0	—
183	Domestic, lean, ¼" fat, broiled	3	ounce(s)	85	50.0	200	23.6	0	0	11.0	4.0	4.4	1.0	—
	Lamb shoulder													
186	Shoulder, arm and blade, domestic, choice, lean and fat,¼" fat, roasted	3	ounce(s)	85	47.8	235	19.1	0	0	17.0	7.2	6.9	1.4	—
187	Shoulder, arm and blade, domestic, choice, lean, ¼" fat, roasted	3	ounce(s)	85	53.8	173	21.2	0	0	9.2	3.5	3.7	0.8	—
3287	Shoulder, arm, domestic, lean and fat, ¼" fat, braised	3	ounce(s)	85	37.6	294	25.8	0	0	20.4	8.4	8.7	1.5	—
3290	Shoulder, arm, domestic, lean, ¼" fat, braised	3	ounce(s)	85	41.9	237	30.2	0	0	12.0	4.3	5.2	0.8	—
	Lamb variety													
3375	Brain, pan fried	3	ounce(s)	85	51.6	232	14.4	0	0	18.9	4.8	3.4	1.9	—
3406	Tongue, braised	3	ounce(s)	85	49.2	234	18.3	0	0	17.2	6.7	8.5	1.1	—
	Pork, cured													
29229	Bacon, Canadian style, cured	2	ounce(s)	57	37.9	89	11.7	1.0	0	4.0	1.3	1.8	0.4	—
161	Bacon, cured, broiled, pan fried or roasted	2	slice(s)	16	2.0	87	5.9	0.2	0	6.7	2.2	3.0	0.7	0

Chol (mg)	Calc (mg)	Iron (mg)	Magn (mg)	Pota (mg)	Sodi (mg)	Zinc (mg)	Vit A (µg)	Thia (mg)	Vit E (mg α)	Ribo (mg)	Niac (mg)	Vit B_6 (mg)	Fola (µg)	Vit C (mg)	Vit B_{12} (µg)	Sele (µg)
117	43	0.94	28.1	202.4	82.5	1.25	12.8	0.03	—	0.06	1.41	0.11	9.4	0.4	2.6	29.1
43	36	5.30	37.4	125.0	403.8	72.22	60.4	0.07	0.98	0.06	1.04	0.04	7.7	2.8	14.7	50.7
21	37	4.91	28.1	105.4	151.3	32.23	6.8	0.08	—	0.05	1.07	0.05	15.3	4.0	13.8	54.1
89	77	10.19	80.8	239.0	358.9	154.45	45.9	0.16	—	0.15	2.11	0.10	11.9	5.1	29.8	60.9
85	14	7.82	37.4	256.8	180.3	28.27	124.2	0.10	0.72	0.37	3.07	0.07	12.8	10.9	24.5	131.0
43	7	4.34	18.7	142.9	90.1	14.13	68.9	0.05	—	0.20	1.70	0.04	8.5	6.8	13.6	65.5
61	52	0.33	29.8	299.4	323.2	2.48	22.1	0.01	0.85	0.05	0.91	0.06	9.4	0	2.6	36.3
48	28	5.71	31.5	227.9	313.8	2.27	77.4	0.25	—	0.35	2.55	0.08	64.6	11.6	20.4	76.2
150	57	1.07	34.0	191.3	292.4	1.17	0	0.11	—	0.11	2.60	0.08	15.3	1.3	1.6	35.4
166	33	2.62	28.9	154.8	190.5	1.32	57.8	0.02	1.17	0.02	2.20	0.10	3.4	1.9	1.3	33.7
27	2	0.71	6.1	93.1	509.2	1.44	0	0.02	0.06	0.05	1.46	0.07	1.8	0	0.8	6.1
73	10	1.76	11.9	115.7	855.6	3.03	0	0.01	0.12	0.12	2.06	0.11	7.7	0	1.4	36.5
23	6	1.53	10.8	243.2	815.9	2.25	0	0.04	0.00	0.10	2.98	0.19	6.2	0	1.5	16.0
14	6	1.53	14.5	169.2	627.4	2.29	0	0.04	0.13	0.04	0.49	0.05	38.0	0	0.3	3.0
58	6	2.00	17.9	266.2	59.5	4.63	0	0.05	—	0.23	4.21	0.23	7.6	0	1.8	16.0
69	12	2.21	18.4	250.0	62.4	5.86	0	0.08	—	0.23	5.10	0.16	9.4	0	1.7	19.0
62	9	2.07	17.0	248.3	70.6	4.40	0	0.02	—	0.16	4.90	0.23	7.6	0	2.5	16.2
71	12	2.30	18.5	242.4	72.4	5.18	0	0.08	—	0.23	4.93	0.17	8.5	0	1.6	18.0
65	16	1.59	21.3	319.8	51.9	4.64	0	0.06	0.34	0.12	7.15	0.53	8.5	0	1.4	29.2
72	9	1.96	16.2	251.7	53.6	4.45	0	0.06	—	0.14	2.85	0.19	6.0	0	2.1	18.7
68	6	2.29	17.9	223.7	35.7	4.59	0	0.05	0.41	0.15	5.05	0.36	8.5	0	1.7	29.3
64	5	1.83	14.5	182.0	29.8	3.76	0	0.05	0.34	0.12	3.92	0.29	6.8	0	1.3	23.0
67	14	2.15	17.0	205.8	42.5	5.93	0	0.05	0.45	0.15	3.63	0.25	7.7	0	1.9	24.1
88	11	2.66	16.2	198.2	55.3	7.15	0	0.06	0.17	0.20	2.06	0.22	4.3	0	1.9	20.9
73	11	3.12	19.6	223.7	60.4	8.73	0	0.06	—	0.23	2.27	0.24	5.1	0	2.1	22.7
49	5	2.16	16.2	200.7	32.3	4.28	0	0.05	0.30	0.15	4.69	0.34	8.5	0	1.4	28.0
54	5	1.98	15.3	193.1	31.5	3.95	0	0.05	0.34	0.13	4.37	0.31	7.7	0	1.5	25.2
58	7	2.56	17.9	233.9	57.8	3.56	0	0.07	0.18	0.17	3.29	0.27	6.0	0	1.8	10.0
50	5	3.11	22.1	278.1	65.5	4.34	0	0.09	0.11	0.21	3.93	0.33	6.8	0	1.9	8.5
67	8	1.88	19.6	294.3	53.6	3.85	0	0.06	—	0.15	3.96	0.31	6.0	0	1.6	19.5
324	5	5.24	18.7	298.5	65.5	4.44	6586.3	0.15	0.39	2.91	14.86	0.87	221.1	0.6	70.7	27.9
112	4	2.22	12.8	156.5	55.3	3.47	0	0.01	0.25	0.25	2.96	0.13	6.0	1.1	2.7	11.2
85	17	1.53	20.4	278.1	65.5	2.96	0	0.08	0.11	0.21	6.03	0.11	15.3	0	2.1	23.3
82	14	1.59	19.6	263.7	61.2	3.79	0	0.08	0.11	0.21	5.66	0.11	15.3	0	2.2	22.5
84	16	1.59	19.6	229.5	64.6	3.40	0	0.07	0.10	0.18	5.95	0.09	11.9	0	2.2	20.3
77	14	1.87	24.7	266.1	72.3	4.47	0	0.08	0.15	0.21	5.56	0.12	17.9	0	2.2	26.4
78	17	1.67	19.6	213.4	56.1	4.44	0	0.07	0.11	0.20	5.22	0.11	17.9	0	2.2	22.3
74	16	1.81	21.3	225.3	57.8	5.13	0	0.07	0.15	0.22	4.89	0.12	21.3	0	2.3	24.2
102	21	2.03	22.1	260.3	61.2	5.17	0	0.06	0.12	0.21	5.66	0.09	15.3	0	2.2	31.6
103	22	2.29	24.7	287.5	64.6	6.20	0	0.06	0.15	0.23	5.38	0.11	18.7	0	2.3	32.1
2130	18	1.73	18.7	304.5	133.5	1.70	0	0.14	—	0.31	3.87	0.19	6.0	19.6	20.5	10.2
161	9	2.23	13.6	134.4	57.0	2.54	0	0.06	—	0.35	3.13	0.14	2.6	6.0	5.4	23.8
28	5	0.38	9.6	195.0	798.9	0.78	0	0.42	0.11	0.09	3.53	0.22	2.3	0	0.4	14.2
18	2	0.22	5.3	90.4	369.6	0.56	1.8	0.06	0.04	0.04	1.76	0.04	0.3	0	0.2	9.9

Food Composition   (Computer code is for Cengage Diet Analysis program)   (For purposes of calculations, use "0" for t, <1, <.1, <.01, etc.)

DA+ Code	Food Description	Quantity	Measure	Wt (g)	H₂O (g)	Ener (kcal)	Prot (g)	Carb (g)	Fiber (g)	Fat (g)	Fat Breakdown (g)			
											Sat	Mono	Poly	Trans
	BEEF, LAMB, PORK—Continued													
35422	Breakfast strips, cured, cooked	3	slice(s)	34	9.2	156	9.8	0.4	0	12.5	4.3	5.6	1.9	—
189	Ham, cured, boneless, 11% fat, roasted	3	ounce(s)	85	54.9	151	19.2	0	0	7.7	2.7	3.8	1.2	—
29215	Ham, cured, extra lean, 4% fat, canned	2	2 ounce(s)	57	41.7	68	10.5	0	0	2.6	0.9	1.3	0.2	—
1316	Ham, cured, extra lean, 5% fat, roasted	3	ounce(s)	85	57.6	123	17.8	1.3	0	4.7	1.5	2.2	0.5	—
16561	Ham, smoked or cured, lean, cooked	1	slice(s)	42	27.6	66	10.5	0	0	2.3	0.8	1.1	0.3	—
	Pork chop													
32671	Loin, blade, chops, lean and fat, pan fried	3	ounce(s)	85	42.5	291	18.3	0	0	23.6	8.6	10	2.6	—
32672	Loin, center cut, chops, lean and fat, pan fried	3	ounce(s)	85	45.1	236	25.4	0	0	14.1	5.1	6.0	1.6	—
32682	Loin, center rib, chops, boneless, lean and fat, braised	3	ounce(s)	85	49.5	217	22.4	0	0	13.4	5.2	6.1	1.1	—
32603	Loin, center rib, chops, lean, broiled	3	ounce(s)	85	55.4	158	21.9	0	0	7.1	2.4	3.0	0.8	0.1
32478	Loin, whole, lean and fat, braised	3	ounce(s)	85	49.6	203	23.2	0	0	11.6	4.3	5.2	1.0	—
32481	Loin, whole, lean, braised	3	ounce(s)	85	52.2	174	24.3	0	0	7.8	2.9	3.5	0.6	—
	Pork leg or ham													
32471	Pork leg or ham, Rump portion, lean and fat, roasted	3	ounce(s)	85	48.3	214	24.6	0	0	12.1	4.5	5.4	1.2	—
32468	Pork leg or ham, Whole, lean and fat, roasted	3	ounce(s)	85	46.8	232	22.8	0	0	15.0	5.5	6.7	1.4	—
	Pork ribs													
32693	Loin, country style, lean and fat, roasted	3	ounce(s)	85	43.3	279	19.9	0	0	21.6	7.8	9.4	1.7	—
32696	Loin, country style, lean, roasted	3	ounce(s)	85	49.5	210	22.6	0	0	12.6	4.5	5.5	0.9	—
	Pork shoulder													
32626	Shoulder, arm picnic, lean and fat, roasted	3	ounce(s)	85	44.3	270	20.0	0	0	20.4	7.5	9.1	2.0	—
32629	Shoulder, arm picnic, lean, roasted	3	ounce(s)	85	51.3	194	22.7	0	0	10.7	3.7	5.1	1.0	—
	Rabbit													
3366	Domesticated, roasted	3	ounce(s)	85	51.5	168	24.7	0	0	6.8	2.0	1.8	1.3	—
3367	Domesticated, stewed	3	ounce(s)	85	50.0	175	25.8	0	0	7.2	2.1	1.9	1.4	—
	Veal													
3391	Liver, braised	3	ounce(s)	85	50.9	163	24.2	3.2	0	5.3	1.7	1.0	0.9	0.3
3319	Rib, lean only, roasted	3	ounce(s)	85	55.0	151	21.9	0	0	6.3	1.8	2.3	0.6	—
1732	Deer or venison, roasted	3	ounce(s)	85	55.5	134	25.7	0	0	2.7	1.1	0.7	0.5	—
	POULTRY													
	Chicken													
29562	Flaked, canned	2	ounce(s)	57	39.3	97	10.3	0.1	0	5.8	1.6	2.3	1.3	—
	Chicken, fried													
29632	Breast, meat only, breaded, baked or fried	3	ounce(s)	85	44.3	193	25.3	6.9	0.2	6.6	1.6	2.7	1.7	—
35327	Broiler breast, meat only, fried	3	ounce(s)	85	51.2	159	28.4	0.4	0	4.0	1.1	1.5	0.9	—
36413	Broiler breast, meat and skin, flour coated, fried	3	ounce(s)	85	48.1	189	27.1	1.4	0.1	7.5	2.1	3.0	1.7	—
36414	Broiler drumstick, meat and skin, flour coated, fried	3	ounce(s)	85	48.2	208	22.9	1.4	0.1	11.7	3.1	4.6	2.7	—
35389	Broiler drumstick, meat only, fried	3	ounce(s)	85	52.9	166	24.3	0	0	6.9	1.8	2.5	1.7	—
35406	Broiler leg, meat only, fried	3	ounce(s)	85	51.5	177	24.1	0.6	0	7.9	2.1	2.9	1.9	—
35484	Broiler wing, meat only, fried	3	ounce(s)	85	50.9	179	25.6	0	0	7.8	2.1	2.6	1.8	—
29580	Patty, fillet or tenders, breaded, cooked	3	ounce(s)	85	40.2	256	14.5	12.2	0	16.5	3.7	8.4	3.7	—
	Chicken, roasted, meat only													
35409	Broiler leg, meat only, roasted	3	ounce(s)	85	55.0	162	23.0	0	0	7.2	1.9	2.6	1.7	—
35486	Broiler wing, meat only, roasted	3	ounce(s)	85	53.4	173	25.9	0	0	6.9	1.9	2.2	1.5	—
35138	Roasting chicken, dark meat, meat only, roasted	3	ounce(s)	85	57.0	151	19.8	0	0	7.4	2.1	2.8	1.7	—
35136	Roasting chicken, light meat, meat only, roasted	3	ounce(s)	85	57.7	130	23.1	0	0	3.5	0.9	1.3	0.8	—
35132	Roasting chicken, meat only, roasted	3	ounce(s)	85	57.3	142	21.3	0	0	5.6	1.5	2.1	1.3	—
	Chicken, stewed													
1268	Gizzard, simmered	3	ounce(s)	85	57.8	124	25.8	0	0	2.3	0.6	0.4	0.3	0.1
1270	Liver, simmered	3	ounce(s)	85	56.8	142	20.8	0.7	0	5.5	1.8	1.2	1.7	0.1
3174	Meat only, stewed	3	ounce(s)	85	56.8	151	23.2	0	0	5.7	1.6	2.0	1.3	—
	Duck													
1286	Domesticated, meat and skin, roasted	3	ounce(s)	85	44.1	287	16.2	0	0	24.1	8.2	11.0	3.1	—
1287	Domesticated, meat only, roasted	3	ounce(s)	85	54.6	171	20	0	0	9.5	3.5	3.1	1.2	—
	Goose													
35507	Domesticated, meat and skin, roasted	3	ounce(s)	85	44.2	259	21.4	0	0	18.6	5.8	8.7	2.1	—
35524	Domesticated, meat only, roasted	3	ounce(s)	85	48.7	202	24.6	0	0	10.8	3.9	3.7	1.3	—
1297	Liver pate, smoked, canned	4	tablespoon(s)	52	19.3	240	5.9	2.4	0	22.8	7.5	13.3	0.4	—
	Turkey													
3256	Ground turkey, cooked	3	ounce(s)	85	50.5	200	23.3	0	0	11.2	2.9	4.2	2.7	—
3263	Patty, batter coated, breaded, fried	1	item(s)	94	46.7	266	13.2	14.8	0.5	16.9	4.4	7.0	4.4	—
219	Roasted, dark meat, meat only	3	ounce(s)	85	53.7	159	24.3	0	0	6.1	2.1	1.4	1.8	—
222	Roasted, fryer roaster breast, meat only	3	ounce(s)	85	58.2	115	25.6	0	0	0.6	0.2	0.1	0.2	—
220	Roasted, light meat, meat only	3	ounce(s)	85	56.4	134	25.4	0	0	2.7	0.9	0.5	0.7	—
1303	Turkey roll, light and dark meat	2	slice(s)	57	39.8	84	10.3	1.2	0	4.0	1.2	1.3	1.0	—
1302	Turkey roll, light meat	2	slice(s)	57	42.5	56	8.4	2.9	0	0.9	0.2	0.2	0.1	0
	PROCESSED MEATS													
	Beef													
1331	Corned beef loaf, jellied, sliced	2	slice(s)	57	39.2	87	13.0	0	0	3.5	1.5	1.5	0.2	—

Chol (mg)	Calc (mg)	Iron (mg)	Magn (mg)	Pota (mg)	Sodi (mg)	Zinc (mg)	Vit A (µg)	Thia (mg)	Vit E (mg α)	Ribo (mg)	Niac (mg)	Vit B$_6$ (mg)	Fola (µg)	Vit C (mg)	Vit B$_{12}$ (µg)	Sele (µg)
36	5	0.67	8.8	158.4	713.7	1.25	0	0.25	0.08	0.12	2.58	0.11	1.4	0	0.6	8.4
50	7	1.13	18.7	347.7	1275.0	2.09	0	0.62	0.26	0.28	5.22	0.26	2.6	0	0.6	16.8
22	3	0.53	9.6	206.4	711.6	1.09	0	0.47	0.09	0.13	3.00	0.25	3.4	0	0.5	8.2
45	7	1.25	11.9	244.1	1023.1	2.44	0	0.64	0.21	0.17	3.42	0.34	2.6	0	0.6	16.6
23	3	0.39	9.2	132.7	557.3	1.07	0	0.28	0.10	0.10	2.10	0.19	1.7	0	0.3	10.7
72	26	0.74	17.9	282.4	57.0	2.71	1.7	0.52	0.17	0.25	3.35	0.28	3.4	0.5	0.7	29.7
78	23	0.77	24.7	361.5	68.0	1.96	1.7	0.96	0.21	0.25	4.76	0.39	5.1	0.9	0.6	33.2
62	4	0.78	14.5	329.1	34.0	1.76	1.7	0.44	—	0.20	3.66	0.26	3.4	0.3	0.4	28.4
56	22	0.57	21.3	291.7	48.5	1.91	0	0.48	0.08	0.18	6.68	0.57	0	0	0.4	38.6
68	18	0.91	16.2	318.1	40.8	2.02	1.7	0.53	0.20	0.21	3.75	0.31	2.6	0.5	0.5	38.5
67	15	0.96	17.0	329.1	42.5	2.10	1.7	0.56	0.17	0.22	3.90	0.32	3.4	0.5	0.5	41.0
82	10	0.89	23.0	318.1	52.7	2.39	2.6	0.63	0.18	0.28	3.95	0.26	2.6	0.2	0.6	39.8
80	12	0.85	18.7	299.4	51.0	2.51	2.6	0.54	0.18	0.26	3.89	0.34	8.5	0.3	0.6	38.5
78	21	0.90	19.6	292.6	44.2	2.00	2.6	0.75	—	0.29	3.67	0.37	4.3	0.3	0.7	31.6
79	25	1.09	20.4	296.8	24.7	3.24	1.7	0.48	—	0.29	3.96	0.37	4.3	0.3	0.7	36.0
80	16	1.00	14.5	276.4	59.5	2.93	1.7	0.44	—	0.25	3.33	0.29	3.4	0.2	0.6	28.6
81	8	1.20	17.0	298.5	68.0	3.46	1.7	0.49	—	0.30	3.66	0.34	4.3	0.3	0.7	32.7
70	16	1.93	17.9	325.7	40.0	1.93	0	0.07	—	0.17	7.17	0.40	9.4	0	7.1	32.7
73	17	2.01	17.0	255.1	31.5	2.01	0	0.05	0.37	0.14	6.09	0.28	7.7	0	5.5	32.7
435	5	4.34	17.0	279.8	66.3	9.55	17983.6	0.15	0.57	2.43	11.18	0.78	281.5	0.9	72.0	16.4
98	10	0.81	20.4	264.5	82.5	3.81	0	0.05	0.30	0.24	6.37	0.23	11.9	0	1.3	9.4
95	6	3.80	20.4	284.9	45.9	2.33	0	0.15	—	0.51	5.70	—	—	0	—	11.0
35	8	0.89	6.8	147.4	408.2	0.79	19.3	0.01	—	0.07	3.58	0.19	2.3	0	0.2	—
67	19	1.05	24.7	222.6	450.2	0.84	—	0.08	—	0.09	10.97	0.46	4.3	0	0.3	—
77	14	0.96	26.4	234.7	67.2	0.91	6.0	0.06	0.35	0.10	12.57	0.54	3.4	0	0.3	22.3
76	14	1.01	25.5	220.3	64.6	0.93	12.8	0.06	0.39	0.11	11.68	0.49	5.1	0	0.3	20.3
77	10	1.13	19.6	194.8	75.7	2.45	21.3	0.06	0.65	0.19	5.13	0.29	8.5	0	0.3	15.6
80	10	1.12	20.4	211.8	81.6	2.73	15.3	0.06	—	0.20	5.22	0.33	7.7	0	0.3	16.7
84	11	1.19	21.3	216.0	81.6	2.53	17.0	0.07	0.38	0.21	5.68	0.33	7.7	0	0.3	16.0
71	13	0.96	17.9	176.9	77.4	1.80	15.3	0.03	0.40	0.10	6.15	0.50	3.4	0	0.3	21.6
49	11	0.75	19.6	244.8	411.4	0.79	4.3	0.09	1.04	0.12	5.99	0.24	24.7	0	0.2	13.9
80	10	1.11	20.4	205.8	77.4	2.43	16.2	0.06	0.22	0.19	5.37	0.31	6.8	0	0.3	18.8
72	14	0.98	17.9	178.6	78.2	1.82	15.3	0.03	0.22	0.10	6.21	0.50	3.4	0	0.3	21.0
64	9	1.13	17.0	190.5	80.8	1.81	13.6	0.05	—	0.16	4.87	0.26	6.0	0	0.2	16.7
64	11	0.91	19.6	200.7	43.4	0.66	6.8	0.05	0.22	0.07	8.90	0.45	2.6	0	0.3	21.9
64	10	1.02	17.9	194.8	63.8	1.29	10.2	0.05	—	0.12	6.70	0.34	4.3	0	0.2	20.9
315	14	2.71	2.6	152.2	47.6	3.75	0	0.02	0.17	0.17	2.65	0.06	4.3	0	0.9	35.0
479	9	9.89	21.3	223.7	64.6	3.38	3385.8	0.24	0.69	1.69	9.39	0.64	491.6	23.7	14.3	70.1
71	12	0.99	17.9	153.1	59.5	1.69	12.8	0.04	0.22	0.13	5.20	0.22	5.1	0	0.2	17.8
71	9	2.29	13.6	173.5	50.2	1.58	53.6	0.14	0.59	0.22	4.10	0.15	5.1	0	0.3	17.0
76	10	2.29	17.0	214.3	55.3	2.21	19.6	0.22	0.59	0.39	4.33	0.21	8.5	0	0.3	19.1
77	11	2.40	18.7	279.8	59.5	2.22	17.9	0.06	1.47	0.27	3.54	0.31	1.7	0	0.3	18.5
82	12	2.44	21.3	330.0	64.6	2.69	10.2	0.07	—	0.33	3.47	0.39	10.2	0	0.4	21.7
78	36	2.86	6.8	71.8	362.4	0.47	520.5	0.04	—	0.15	1.30	0.03	31.2	0	4.9	22.9
87	21	1.64	20.4	229.6	91.0	2.43	0	0.04	0.28	0.14	4.09	0.33	6.0	0	0.3	31.6
71	13	2.06	14.1	258.5	752.0	1.35	9.4	0.09	0.87	0.17	2.16	0.18	38.5	0	0.2	20.8
72	27	1.98	20.4	246.6	67.2	3.79	0	0.05	0.54	0.21	3.10	0.30	7.7	0	0.3	34.8
71	10	1.30	24.7	248.3	44.2	1.48	0	0.03	0.07	0.11	6.37	0.47	5.1	0	0.3	27.3
59	16	1.14	23.8	259.4	54.4	1.73	0	0.05	0.07	0.11	5.81	0.45	5.1	0	0.3	27.3
31	18	0.76	10.2	153.1	332.3	1.13	0	0.05	0.19	0.16	2.72	0.15	2.8	0	0.1	16.6
19	4	0.21	10.8	242.1	590.8	0.50	0	0.01	0.07	0.08	4.05	0.23	2.3	0	0.2	7.4
27	6	1.15	6.2	57.3	540.4	2.31	0	0.00	—	0.06	0.99	0.06	4.5	0	0.7	9.8

DA+ Code	Food Description	Quantity	Measure	Wt (g)	H₂O (g)	Ener (kcal)	Prot (g)	Carb (g)	Fiber (g)	Fat (g)	Fat Breakdown (g)			
											Sat	Mono	Poly	Trans
	PROCESSED MEATS —Continued													
	Bologna													
13459	Beef	1	slice(s)	28	15.1	90	3.0	1.0	0	8.0	3.5	4.3	0.3	—
13461	Light, made with pork and chicken	1	slice(s)	28	18.2	60	3.0	2.0	0	4.0	1.0	2.0	0.4	—
13458	Made with chicken and pork	1	slice(s)	28	15.0	90	3.0	1.0	0	8.0	3.0	4.1	1.1	—
13565	Turkey bologna	1	slice(s)	28	19.0	50	3.0	1.0	0	4.0	1.0	1.1	1.0	0
	Chicken													
7125	Breast, smoked	1	slice(s)	10	—	10	1.8	0.3	0	0.2	0	—	—	—
	Ham													
7127	Deli-sliced, honey	1	slice(s)	10	—	10	1.7	0.3	0	0.3	0.1	—	—	—
7126	Deli-sliced, smoked	1	slice(s)	10	—	10	1.7	0.2	0	0.3	0.1	—	—	—
8614	Beef & pork mortadella, sliced	2	slice(s)	46	24.1	143	7.5	1.4	0	11.7	4.4	5.2	1.4	—
1323	Pork olive loaf	2	slice(s)	57	33.1	133	6.7	5.2	0	9.4	3.3	4.5	1.1	—
1324	Pork pickle & pimento loaf	2	slice(s)	57	34.2	128	6.4	4.8	0.9	9.1	3.0	4.0	1.6	—
	Sausages & frankfurters													
37296	Beerwurst beef, beer salami (bierwurst)	1	slice(s)	29	16.6	74	4.1	1.2	0	5.7	2.5	2.7	0.2	—
37257	Beerwurst pork beer salami	1	slice(s)	21	12.9	50	3.0	0.4	0	4.0	1.3	1.9	0.5	—
35338	Berliner, pork & beef	1	ounce(s)	28	17.3	65	4.3	0.7	0	4.9	1.7	2.3	0.4	—
37298	Bratwurst pork, cooked	1	piece(s)	74	42.3	181	10.4	1.9	0	14.3	5.1	6.7	1.5	—
37299	Braunschweiger pork liver sausage	1	slice(s)	15	8.2	51	2.0	0.3	0	4.5	1.5	2.1	0.5	—
1329	Cheesefurter or cheese smokie, beef & pork	1	item(s)	43	22.6	141	6.1	0.6	0	12.5	4.5	5.9	1.3	—
1330	Chorizo, beef & pork	2	ounce(s)	57	18.1	258	13.7	1.1	0	21.7	8.2	10.4	2.0	—
8600	Frankfurter, beef	1	item(s)	45	23.4	149	5.1	1.8	0	13.3	5.3	6.4	0.5	—
202	Frankfurter, beef & pork	1	item(s)	45	25.2	137	5.2	0.8	0	12.4	4.8	6.2	1.2	—
1293	Frankfurter, chicken	1	item(s)	45	28.1	100	7.0	1.2	0.2	7.3	1.7	2.7	1.7	0.1
3261	Frankfurter, turkey	1	item(s)	45	28.3	100	5.5	1.7	0	7.8	1.8	2.6	1.8	0.4
37275	Italian sausage, pork, cooked	1	item(s)	68	32.0	234	13.0	2.9	0.1	18.6	6.5	8.1	2.2	—
37307	Kielbasa, kolbassa, pork & beef	1	slice(s)	30	18.5	67	5.0	1.0	0	4.7	1.7	2.2	0.5	—
1333	Knockwurst or knackwurst, beef & pork	2	ounce(s)	57	31.4	174	6.3	1.8	0	15.7	5.8	7.3	1.7	—
37285	Pepperoni, beef & pork	1	slice(s)	11	3.4	51	2.2	0.4	0.2	4.4	1.8	2.1	0.3	—
37313	Polish sausage, pork	1	slice(s)	21	11.4	60	2.8	0.7	0	5.0	1.8	2.3	0.5	—
206	Salami, beef, cooked, sliced	2	slice(s)	52	31.2	136	6.5	1.0	0	11.5	5.1	5.5	0.5	—
37272	Salami, pork, dry or hard	1	slice(s)	13	4.6	52	2.9	0.2	0	4.3	1.5	2.0	0.5	—
40987	Sausage, turkey, cooked	2	ounce(s)	57	36.9	111	13.5	0	0	5.9	1.3	1.7	1.5	0.2
8620	Smoked sausage, beef & pork	2	ounce(s)	57	30.6	181	6.8	1.4	0	16.3	5.5	6.9	2.2	0
8619	Smoked sausage, pork	2	ounce(s)	57	32.0	178	6.8	1.2	0	16.0	5.3	6.4	2.1	0.1
37273	Smoked sausage, pork link	1	piece(s)	76	29.8	295	16.8	1.6	0	24.0	8.6	11.1	2.8	—
1336	Summer sausage, thuringer, or cervelat, beef & pork	2	ounce(s)	57	25.6	205	9.9	1.9	0	17.3	6.5	7.4	0.7	—
37294	Vienna sausage, cocktail, beef & pork, canned	1	piece(s)	16	10.4	37	1.7	0.4	0	3.1	1.1	1.5	0.2	—
	Spreads													
1318	Ham salad spread	¼	cup(s)	60	37.6	130	5.2	6.4	0	9.3	3.0	4.3	1.6	—
32419	Pork and beef sandwich spread	4	tablespoon(s)	60	36.2	141	4.6	7.2	0.1	10.4	3.6	4.6	1.5	—
	Turkey													
13604	Breast, fat free, oven roasted	1	slice(s)	28	—	25	4.0	1.0	0	0	0	0	0	0
13606	Breast, hickory smoked fat free	1	slice(s)	28	—	25	4.0	1.0	0	0	0	0	0	0
16049	Breast, hickory smoked slices	1	slice(s)	56	—	50	11.0	1.0	0	0	0	0	0	0
16047	Breast, honey roasted slices	1	slice(s)	56	—	60	11.0	3.0	0	0	0	0	0	0
16048	Breast, oven roasted slices	1	slice(s)	56	—	50	11.0	1.0	0	0	0	0	0	0
7124	Breast, oven-roasted	1	slice(s)	10	—	10	1.8	0.3	0	0.1	0	0	0	—
13567	Turkey ham, 10% water added	2	slice(s)	56	40.9	70	10.0	2.0	0	3.0	0	0.4	0.6	0
37270	Turkey pastrami	1	slice(s)	28	20.3	35	4.6	1.0	0	1.2	0.3	0.4	0.3	—
3262	Turkey salami	2	slice(s)	57	39.1	98	10.9	0.9	0.1	5.2	1.6	1.8	1.4	0
37318	Turkey salami, cooked	1	slice(s)	28	20.4	43	4.3	0.1	0	2.7	0.8	0.9	0.7	—
	BEVERAGES													
	Beer													
866	Ale, mild	12	fluid ounce(s)	360	332.3	148	1.1	13.3	0.4	0	0	0	0	0
686	Beer	12	fluid ounce(s)	356	327.7	153	1.6	12.7	0	0	0	0	0	0
16886	Beer, non alcoholic	12	fluid ounce(s)	360	328.1	133	0.8	29.0	0	0.4	0.1	0	0.2	0
31609	Bud Light beer	12	fluid ounce(s)	355	335.5	110	0.9	6.6	0	0	0	0	0	0
31608	Budweiser beer	12	fluid ounce(s)	355	327.7	145	1.3	10.6	0	0	0	0	0	0
869	Light beer	12	fluid ounce(s)	354	335.9	103	0.9	5.8	0	0	0	0	0	0
31613	Michelob Beer	12	fluid ounce(s)	355	323.4	155	1.3	13.3	0	0	0	0	0	0
31614	Michelob Light beer	12	fluid ounce(s)	355	329.8	134	1.1	11.7	0	0	0	0	0	0
	Gin, rum, vodka, whiskey													
857	Distilled alcohol, 100 proof	1	fluid ounce(s)	28	16.0	82	0	0	0	0	0	0	0	0
687	Distilled alcohol, 80 proof	1	fluid ounce(s)	28	18.5	64	0	0	0	0	0	0	0	0
688	Distilled alcohol, 86 proof	1	fluid ounce(s)	28	17.8	70	0	0	0	0	0	0	0	0
689	Distilled alcohol, 90 proof	1	fluid ounce(s)	28	17.3	73	0	0	0	0	0	0	0	0
856	Distilled alcohol, 94 proof	1	fluid ounce(s)	28	16.8	76	0	0	0	0	0	0	0	0

Chol (mg)	Calc (mg)	Iron (mg)	Magn (mg)	Pota (mg)	Sodi (mg)	Zinc (mg)	Vit A (µg)	Thia (mg)	Vit E (mg α)	Ribo (mg)	Niac (mg)	Vit B$_6$ (mg)	Fola (µg)	Vit C (mg)	Vit B$_{12}$ (µg)	Sele (µg)
20	0	0.36	3.9	47.0	310.0	0.56	0	0.01	—	0.03	0.67	0.04	3.6	0	0.4	—
20	40	0.36	5.6	45.6	300.0	0.45	0	—	—	—	—	—	—	0	—	—
30	20	0.36	5.9	43.1	300.0	0.39	0	—	—	—	—	—	—	0	—	—
20	40	0.36	6.2	42.6	270.0	0.51	0	—	—	—	—	—	—	0	—	—
4	0	0.00	—	—	100.0	—	0	—	—	—	—	—	—	0	—	—
4	0	0.12	—	—	100.0	—	0	—	—	—	—	—	—	0.6	—	—
4	0	0.12	—	—	103.3	—	0	—	—	—	—	—	—	0.6	—	—
26	8	0.64	5.1	75.0	573.2	0.96	0	0.05	0.10	0.07	1.23	0.06	1.4	0	0.7	10.4
22	62	0.30	10.8	168.7	842.9	0.78	34.1	0.16	0.14	0.14	1.04	0.13	1.1	0	0.7	9.3
33	62	0.75	19.3	210.7	740.7	0.95	44.3	0.22	0.22	0.06	1.41	0.23	21.0	4.4	0.3	4.5
18	3	0.44	3.5	66.5	264.9	0.71	0	0.02	0.05	0.03	0.98	0.04	0.9	0	0.6	4.7
12	2	0.15	2.7	53.3	261.0	0.36	0	0.11	0.03	0.04	0.68	0.07	0.6	0	0.2	4.4
13	3	0.32	4.3	80.2	367.7	0.70	0	0.10	—	0.06	0.88	0.05	1.4	0	0.8	4.0
44	33	0.95	11.1	156.9	412.2	1.70	0	0.37	0.01	0.13	2.36	0.15	1.5	0.7	0.7	15.7
24	1	1.42	1.7	27.5	131.5	0.42	641.0	0.03	0.05	0.23	1.27	0.05	6.7	0	3.1	8.8
29	25	0.46	5.6	88.6	465.3	0.96	20.2	0.10	0.10	0.06	1.24	0.05	1.3	0	0.7	6.8
50	5	0.90	10.2	225.7	700.2	1.93	0	0.35	0.12	0.17	2.90	0.30	1.1	0	1.1	12.0
24	6	0.67	6.3	70.2	513.0	1.10	0	0.01	0.09	0.06	1.06	0.04	2.3	0	0.8	3.7
23	5	0.51	4.5	75.2	504.0	0.82	8.1	0.09	0.11	0.05	1.18	0.05	1.8	0	0.6	6.2
43	33	0.52	9.0	90.9	379.8	0.50	0	0.02	0.09	0.11	2.10	0.14	3.2	0	0.2	10.4
35	67	0.66	6.3	176.4	485.1	0.82	0	0.01	0.27	0.08	1.65	0.06	4.1	0	0.4	6.8
39	14	0.97	12.2	206.7	820.8	1.62	6.8	0.42	0.17	0.15	2.83	0.22	3.4	0.1	0.9	15.0
20	13	0.44	4.9	84.4	283.0	0.61	0	0.06	0.06	0.06	0.87	0.05	1.5	0	0.5	5.4
34	6	0.37	6.2	112.8	527.3	0.94	0	0.19	0.32	0.07	1.55	0.09	1.1	0	0.7	7.7
13	2	0.15	2.0	34.7	196.7	0.30	0	0.05	0.00	0.02	0.59	0.04	0.7	0.1	0.2	2.4
15	2	0.29	2.9	37.3	199.3	0.40	0	0.10	0.04	0.03	0.71	0.03	0.4	0.2	0.2	3.7
37	3	1.14	6.8	97.8	592.8	0.92	0	0.04	0.08	0.08	1.68	0.08	1.0	0	1.6	7.6
10	2	0.16	2.8	48.4	289.3	0.53	0	0.11	0.02	0.04	0.71	0.07	0.3	0	0.4	3.3
52	12	0.84	11.9	169.0	377.1	2.19	7.4	0.04	0.10	0.14	3.24	0.18	3.4	0.4	0.7	0
33	7	0.42	7.4	101.5	516.5	0.71	7.4	0.10	0.07	0.06	1.66	0.09	1.1	0	0.3	0
35	6	0.33	6.2	273.9	468.9	0.74	0	0.12	0.14	0.10	1.59	0.10	0.6	0	0.4	10.4
52	23	0.87	14.4	254.6	1136.6	2.13	0	0.53	0.18	0.19	3.43	0.26	3.8	1.5	1.2	16.4
42	5	1.15	7.9	147.4	737.1	1.45	0	0.08	0.12	0.18	2.44	0.14	1.1	9.4	3.1	11.5
14	2	0.14	1.1	16.2	155.0	0.25	0	0.01	0.03	0.01	0.25	0.01	0.6	0	0.2	2.7
22	5	0.35	6.0	90.0	547.2	0.66	0	0.26	1.04	0.07	1.25	0.09	0.6	0	0.5	10.7
23	7	0.47	4.8	66.0	607.8	0.61	15.6	0.10	1.04	0.08	1.03	0.07	1.2	0	0.7	5.8
10	0	0.00	—	—	340.0	—	0	—	—	—	—	—	—	0	—	—
10	0	0.00	—	—	300.0	—	0	—	—	—	—	—	—	0	—	—
25	0	0.72	—	—	720.0	—	0	—	—	—	—	—	—	0	—	—
20	0	0.72	—	—	660.0	—	0	—	—	—	—	—	—	0	—	—
20	0	0.72	—	—	660.0	—	0	—	—	—	—	—	—	0	—	—
4	0	0.06	—	—	103.3	—	0	—	—	—	—	—	—	0	—	—
40	0	0.72	12.3	162.4	700.0	1.44	0	—	—	—	—	—	—	0	—	—
19	3	1.19	4.0	97.8	278.1	0.61	1.1	0.01	0.06	0.07	1.00	0.07	1.4	4.6	0.1	4.6
43	23	0.70	12.5	122.5	569.3	1.31	1.1	0.24	0.13	0.17	2.25	0.24	5.7	0	0.6	15.0
22	11	0.35	6.2	61.2	284.6	0.65	0.6	0.12	0.06	0.08	1.12	0.12	2.8	0	0.3	7.5
0	18	0.07	21.6	90.0	14.4	0.03	0	0.03	0.00	0.10	1.62	0.18	21.6	0	0.1	2.5
0	14	0.07	21.4	96.2	14.3	0.03	0	0.01	0.00	0.08	1.82	0.16	21.4	0	0.1	2.1
0	25	0.21	25.2	28.8	46.8	0.07	—	0.07	0.00	0.18	3.99	0.10	50.4	1.8	0.1	4.3
0	18	0.14	17.8	63.9	9.0	0.10	0	0.03	—	0.10	1.39	0.12	14.6	0	0	4.0
0	18	0.10	21.3	88.8	9.0	0.07	0	0.02	—	0.09	1.60	0.17	21.3	0	0.1	4.0
0	14	0.10	17.7	74.3	14.2	0.03	0	0.01	0.00	0.05	1.38	0.12	21.2	0	0.1	1.4
0	18	0.10	21.3	88.8	9.0	0.07	0	0.02	—	0.09	1.60	0.17	21.3	0	0.1	4.0
0	18	0.14	17.8	63.9	9.0	0.10	0	0.03	—	0.10	1.39	0.12	14.6	0	0	4.0
0	0	0.01	0	0.6	0.3	0.01	0	0.00	—	0.00	0.00	0.00	0	0	0	0
0	0	0.01	0	0.6	0.3	0.01	0	0.00	0.00	0.00	0.00	0.00	0	0	0	0
0	0	0.01	0	0.6	0.3	0.01	0	0.00	0.00	0.00	0.00	0.00	0	0	0	0
0	0	0.01	0	0.6	0.3	0.01	0	0.00	0.00	0.00	0.00	0.00	0	0	0	0
0	0	0.01	0	0.6	0.3	0.01	0	0.00	—	0.00	0.00	0.00	0	0	0	0

DA+ Code	Food Description	Quantity	Measure	Wt (g)	H₂O (g)	Ener (kcal)	Prot (g)	Carb (g)	Fiber (g)	Fat (g)	Sat	Mono	Poly	Trans
	BEVERAGES —Continued													
	Liqueurs													
33187	Coffee liqueur, 53 proof	1	fluid ounce(s)	35	10.8	113	0	16.3	0	0.1	0	0	0	—
3142	Coffee liqueur, 63 proof	1	fluid ounce(s)	35	14.4	107	0	11.2	0	0.1	0	0	0	—
736	Cordials, 54 proof	1	fluid ounce(s)	30	8.9	106	0	13.3	0	0.1	0	0	0	—
	Wine													
861	California red wine	5	fluid ounce(s)	150	133.4	125	0.3	3.7	0	0	0	0	0	0
858	Domestic champagne	5	fluid ounce(s)	150	—	105	0.3	3.8	0	0	0	0	0	0
690	Sweet dessert wine	5	fluid ounce(s)	147	103.7	235	0.3	20.1	0	0	0	0	0	0
1481	White wine	5	fluid ounce(s)	148	128.1	121	0.1	3.8	0	0	0	0	0	0
1811	Wine cooler	10	fluid ounce(s)	300	267.4	159	0.3	20.2	0	0.1	0	0	0	—
	Carbonated													
31898	7 Up	12	fluid ounce(s)	360	321.0	140	0	39.0	0	0	0	0	0	0
692	Club soda	12	fluid ounce(s)	355	354.8	0	0	0	0	0	0	0	0	0
12010	Coca-Cola Classic cola soda	12	fluid ounce(s)	360	319.4	146	0	40.5	0	0	0	0	0	0
693	Cola	12	fluid ounce(s)	368	332.7	136	0.3	35.2	0	0.1	0	0	0	—
2391	Cola or pepper-type soda, low calorie with saccharin	12	fluid ounce(s)	355	354.5	0	0	0.3	0	0	0	0	0	0
9522	Cola soda, decaffeinated	12	fluid ounce(s)	372	333.4	153	0	39.3	0	0	0	0	0	0
9524	Cola, decaffeinated, low calorie with aspartame	12	fluid ounce(s)	355	354.3	4	0.4	0.5	0	0	0	0	0	0
1415	Cola, low calorie with aspartame	12	fluid ounce(s)	355	353.6	7	0.4	1.0	0	0.1	0	0	0	—
1412	Cream soda	12	fluid ounce(s)	371	321.5	189	0	49.3	0	0	0	0	0	0
31899	Diet 7 Up	12	fluid ounce(s)	360	—	0	0	0	0	0	0	0	0	0
12031	Diet Coke cola soda	12	fluid ounce(s)	360	—	2	0	0.2	0	0	0	0	0	0
29392	Diet Mountain Dew soda	12	fluid ounce(s)	360	—	0	0	0	0	0	0	0	0	0
29389	Diet Pepsi cola soda	12	fluid ounce(s)	360	—	0	0	0	0	0	0	0	0	0
12034	Diet Sprite soda	12	fluid ounce(s)	360	—	4	0	0	0	0	0	0	0	0
695	Ginger ale	12	fluid ounce(s)	366	333.9	124	0	32.1	0	0	0	0	0	0
694	Grape soda	12	fluid ounce(s)	372	330.3	160	0	41.7	0	0	0	0	0	0
1876	Lemon lime soda	12	fluid ounce(s)	368	330.8	147	0.2	37.4	0	0.1	0	0	0	—
29391	Mountain Dew soda	12	fluid ounce(s)	360	314.0	170	0	46.0	0	0	0	0	0	0
3145	Orange soda	12	fluid ounce(s)	372	325.9	179	0	45.8	0	0	0	0	0	0
1414	Pepper-type soda	12	fluid ounce(s)	368	329.3	151	0	38.3	0	0.4	0.3	0	0	—
29388	Pepsi regular cola soda	12	fluid ounce(s)	360	318.9	150	0	41.0	0	0	0	0	0	0
696	Root beer	12	fluid ounce(s)	370	330.0	152	0	39.2	0	0	0	0	0	0
12044	Sprite soda	12	fluid ounce(s)	360	321.0	144	0	39.0	0	0	0	0	0	0
	Coffee													
731	Brewed	8	fluid ounce(s)	237	235.6	2	0.3	0	0	0	0	0	0	0
9520	Brewed, decaffeinated	8	fluid ounce(s)	237	234.3	5	0.3	1.0	0	0	0	0	0	0
16882	Cappuccino	8	fluid ounce(s)	240	224.8	79	4.1	5.8	0.2	4.9	2.3	1.0	0.2	—
16883	Cappuccino, decaffeinated	8	fluid ounce(s)	240	224.8	79	4.1	5.8	0.2	4.9	2.3	1.0	0.2	—
16880	Espresso	8	fluid ounce(s)	237	231.8	21	0	3.6	0	0.4	0.2	0	0.2	0
16881	Espresso, decaffeinated	8	fluid ounce(s)	237	231.8	21	0	3.6	0	0.4	0.2	0	0.2	0
732	Instant, prepared	8	fluid ounce(s)	239	236.5	5	0.2	0.8	0	0	0	0	0	0
	Fruit drinks													
29357	Crystal Light sugar free lemonade drink	8	fluid ounce(s)	240	—	5	0	0	0	0	0	0	0	0
6012	Fruit punch drink with added vitamin C, canned	8	fluid ounce(s)	248	218.2	117	0	29.7	0.5	0	0	0	0	0
31143	Gatorade Thirst Quencher, all flavors	8	fluid ounce(s)	240	—	50	0	14.0	0	0	0	0	0	0
260	Grape drink, canned	8	fluid ounce(s)	250	210.5	153	0	39.4	0	0	0	0	0	0
17372	Kool-Aid (lemonade/punch/fruit drink)	8	fluid ounce(s)	248	220.0	108	0.1	27.8	0.2	0	0	0	0	0
17225	Kool-Aid sugar free, low calorie tropical punch drink mix, prepared	8	fluid ounce(s)	240	—	5	0	0	0	0	0	0	0	0
266	Lemonade, prepared from frozen concentrate	8	fluid ounce(s)	248	221.6	99	0.2	25.8	0	0.1	0	0	0	0
268	Limeade, prepared from frozen concentrate	8	fluid ounce(s)	247	212.6	128	0	34.1	0	0	0	0	0	0
14266	Odwalla strawberry 'C' monster smoothie blend	8	fluid ounce(s)	240	—	160	2.0	38.0	0	0	0	0	0	0
10080	Odwalla strawberry lemonade quencher	8	fluid ounce(s)	240	—	110	0	28.0	0	0	0	0	0	0
10099	Snapple fruit punch fruit drink	8	fluid ounce(s)	240	—	110	0	29.0	0	0	0	0	0	0
10096	Snapple kiwi strawberry fruit drink	8	fluid ounce(s)	240	211.2	110	0	28.0	0	0	0	0	0	0
	Slim Fast ready-to-drink shake													
16054	French vanilla ready to drink shake	11	fluid ounce(s)	325	—	220	10.0	40.0	5.0	2.5	0.5	1.5	0.5	—
40447	Optima rich chocolate royal ready-to-drink shake	11	fluid ounce(s)	330	—	180	10.0	24.0	5.0	5.0	1.0	3.5	0.5	0
16055	Strawberries n cream ready-to-drink shake	11	fluid ounce(s)	325	—	220	10.0	40.0	5.0	2.5	0.5	1.5	0.5	—
	Tea													
33179	Decaffeinated, prepared	8	fluid ounce(s)	237	236.3	2	0	0.7	0	0	0	0	0	0
1877	Herbal, prepared	8	fluid ounce(s)	237	236.1	2	0	0.5	0	0	0	0	0	0
735	Instant tea mix, lemon flavored with sugar, prepared	8	fluid ounce(s)	259	236.2	91	0	22.3	0.3	0.2	0	0	0	—
734	Instant tea mix, unsweetened, prepared	8	fluid ounce(s)	237	236.1	2	0.1	0.4	0	0	0	0	0	0
733	Tea, prepared	8	fluid ounce(s)	237	236.3	2	0	0.7	0	0	0	0	0	0
	Water													
1413	Mineral water, carbonated	8	fluid ounce(s)	237	236.8	0	0	0	0	0	0	0	0	0

Chol (mg)	Calc (mg)	Iron (mg)	Magn (mg)	Pota (mg)	Sodi (mg)	Zinc (mg)	Vit A (µg)	Thia (mg)	Vit E (mg α)	Ribo (mg)	Niac (mg)	Vit B₆ (mg)	Fola (µg)	Vit C (mg)	Vit B₁₂ (µg)	Sele (µg)
0	0	0.02	1.0	10.4	2.8	0.01	0	0.00	0.00	0.00	0.05	0.00	0	0	0	0.1
0	0	0.02	1.0	10.4	2.8	0.01	0	0.00	—	0.00	0.05	0.00	0	0	0	0.1
0	0	0.02	0.6	4.5	2.1	0.01	0	0.00	0.00	0.00	0.02	0.00	0	0	0	0.1
0	12	1.43	16.2	170.6	15.0	0.14	0	0.01	0.00	0.04	0.11	0.05	1.5	0	0	—
0	—	—	—	—	—	—	—	—	—	—	—	—	—	—	0	—
0	12	0.34	13.2	135.4	13.2	0.10	0	0.01	0.00	0.01	0.30	0.00	0	0	0	0.7
0	13	0.39	14.8	104.7	7.4	0.18	0	0.01	0.00	0.01	0.15	0.06	1.5	0	0	0.1
0	18	0.75	15.0	129.0	24.0	0.18	—	0.01	0.03	0.03	0.13	0.03	3.0	5.4	0	0.6
0	—	—	—	0.6	75.0	—	—	—	—	—	—	—	—	—	—	—
0	18	0.03	3.5	7.1	74.6	0.35	0	0.00	0.00	0.00	0.00	0.00	0	0	0	0
0	—	—	—	0	49.5	—	0	—	—	—	—	—	—	0	—	—
0	7	0.41	0	7.4	14.7	0.06	0	0.00	0.00	0.00	0.00	0.00	0	0	0	0.4
0	14	0.06	3.5	14.2	56.8	0.11	0	0.00	0.00	0.00	0.00	0.00	0	0	0	0.3
0	7	0.08	0	11.2	14.9	0.03	0	0.00	0.00	0.00	0.00	0.00	0	0	0	0.4
0	11	0.06	0	24.9	14.2	0.03	0	0.02	0.00	0.08	0.00	0.00	0	0	0	0.3
0	11	0.39	3.5	28.4	28.4	0.03	0	0.02	0.00	0.08	0.00	0.00	0	0	0	0
0	19	0.18	3.7	3.7	44.5	0.26	0	0.00	0.00	0.00	0.00	0.00	0	0	0	0
0	—	—	—	77.0	45.0	—	—	—	—	—	—	—	—	—	—	—
0	—	—	—	18.0	42.0	—	0	—	—	—	—	—	—	0	—	—
0	—	—	—	70.0	35.0	—	—	—	—	—	—	—	—	—	—	—
0	—	—	—	30.0	35.0	—	—	—	—	—	—	—	—	—	—	—
0	—	—	—	109.5	36.0	—	0	—	—	—	—	—	—	0	—	—
0	11	0.65	3.7	3.7	25.6	0.18	0	0.00	0.00	0.00	0.00	0.00	0	0	0	0.4
0	11	0.29	3.7	3.7	55.8	0.26	0	0.00	—	0.00	0.00	0.00	0	0	0	0
0	7	0.41	3.7	3.7	33.2	0.14	0	0.00	0.00	0.00	0.05	0.00	0	0	0	0
0	—	—	—	0	70.0	—	—	—	—	—	—	—	—	—	—	—
0	19	0.21	3.7	7.4	44.6	0.36	0	0.00	—	0.00	0.00	0.00	0	0	0	0
0	11	0.14	0	3.7	36.8	0.14	0	0.00	0.00	0.00	0.00	0.00	0	0	0	0.4
0	—	—	—	0	35.0	—	—	—	—	—	—	—	—	—	—	—
0	18	0.18	3.7	3.7	48.0	0.26	0	0.00	0.00	0.00	0.00	0.00	0	0	0	0.4
0	—	—	—	0	70.5	—	0	—	—	—	—	—	—	0	—	—
0	5	0.02	7.1	116.1	4.7	0.04	0	0.03	0.02	0.18	0.45	0.00	4.7	0	0	0
0	7	0.14	11.8	108.9	4.7	0.00	0	0.00	0.00	0.03	0.66	0.00	0	0	0	0.5
12	144	0.19	14.4	232.8	50.4	0.50	33.6	0.04	0.09	0.27	0.13	0.04	7.2	0	0.4	4.6
12	144	0.19	14.4	232.8	50.4	0.50	33.6	0.04	0.09	0.27	0.13	0.04	7.2	0	0.4	4.6
0	5	0.30	189.6	272.6	33.2	0.11	0	0.00	0.04	0.42	12.34	0.00	2.4	0.5	0	0
0	5	0.30	189.6	272.6	33.2	0.11	0	0.00	0.04	0.42	12.34	0.00	2.4	0.5	0	0
0	10	0.09	9.5	71.6	9.5	0.01	0	0.00	0.00	0.00	0.56	0.00	0	0	0	0.2
0	0	0.00	—	160.0	40.0	—	0	—	—	—	—	—	—	0	0	—
0	20	0.22	7.4	62.0	94.2	0.02	5.0	0.05	0.04	0.05	0.05	0.02	9.9	89.3	0	0.5
0	0	0.00	—	30.0	110.0	—	0	—	—	—	—	—	—	0	0	—
0	130	0.17	2.5	30.0	40.0	0.30	0	0.00	0.00	0.01	0.02	0.01	0	78.5	0	0.3
0	14	0.45	5.0	49.6	31.0	0.19	—	0.03	—	0.05	0.04	0.01	4.3	41.6	0	1.0
0	0	0.00	—	10.1	10.1	—	0	—	—	—	—	—	—	6.0	0	—
0	10	0.39	5.0	37.2	9.9	0.05	0	0.01	0.02	0.05	0.04	0.01	2.5	9.7	0	0.2
0	5	0.00	4.9	24.7	7.4	0.02	0	0.01	0.00	0.01	0.02	0.01	2.5	7.7	0	0.2
0	20	0.72	—	0	20.0	—	0	—	—	—	—	—	—	600.0	0	—
0	0	0.00	—	70.0	10.0	—	0	—	—	—	—	—	—	54.0	0	—
0	0	0.00	—	20.0	10.0	—	0	—	—	—	—	—	—	0	0	—
0	0	0.00	—	40.0	10.0	—	0	—	—	—	—	—	—	0	0	—
5	400	2.70	140.0	600.0	220.0	2.25	—	0.52	—	0.59	7.00	0.70	120.0	60.0	2.1	17.5
5	1000	2.70	140.0	600.0	220.0	2.25	—	0.52	—	0.59	7.00	0.70	120.0	30.0	2.1	17.5
5	400	2.70	140.0	600.0	220.0	2.25	—	0.52	—	0.59	7.00	0.70	120.0	60.0	2.1	17.5
0	0	0.04	7.1	87.7	7.1	0.04	0	0.00	0.00	0.03	0.00	0.00	11.9	0	0	0
0	5	0.18	2.4	21.3	2.4	0.09	0	0.02	0.00	0.01	0.00	0.00	2.4	0	0	0
0	5	0.05	2.6	38.9	5.2	0.02	0	0.00	0.00	0.00	0.02	0.00	0	0	0	0.3
0	7	0.02	4.7	42.7	9.5	0.02	0	0.00	0.00	0.01	0.07	0.00	0	0	0	0
0	0	0.04	7.1	87.7	7.1	0.04	0	0.00	0.00	0.03	0.00	0.00	11.9	0	0	0
0	33	0.00	0	0	2.4	0.00	0	0.00	—	0.00	0.00	0.00	0	0	0	0

Food Composition   (Computer code is for Cengage Diet Analysis program)   (For purposes of calculations, use "0" for t, <1, <.1, <.01, etc.)

DA+ Code	Food Description	Quantity	Measure	Wt (g)	H₂O (g)	Ener (kcal)	Prot (g)	Carb (g)	Fiber (g)	Fat (g)	Fat Breakdown (g) Sat	Mono	Poly	Trans
	BEVERAGES—Continued													
33183	Poland spring water, bottled	8	fluid ounce(s)	237	237.0	0	0	0	0	0	0	0	0	0
1821	Tap water	8	fluid ounce(s)	237	236.8	0	0	0	0	0	0	0	0	0
1879	Tonic water	8	fluid ounce(s)	244	222.3	83	0	21.5	0	0	0	0	0	0
	FATS AND OILS													
	Butter													
104	Butter	1	tablespoon(s)	14	2.3	102	0.1	0	0	11.5	7.3	3.0	0.4	—
2522	Butter Buds, dry butter substitute	1	teaspoon(s)	2	—	5	0	2.0	0	0	0	0	0	0
921	Unsalted	1	tablespoon(s)	14	2.5	102	0.1	0	0	11.5	7.3	3.0	0.4	—
107	Whipped	1	tablespoon(s)	9	1.5	67	0.1	0	0	7.6	4.7	2.2	0.3	—
944	Whipped, unsalted	1	tablespoon(s)	11	2.0	82	0.1	0	0	9.2	5.9	2.4	0.3	—
	Fats, cooking													
2671	Beef tallow, semisolid	1	tablespoon(s)	13	0	115	0	0	0	12.8	6.4	5.4	0.5	—
922	Chicken fat	1	tablespoon(s)	13	0	115	0	0	0	12.8	3.8	5.7	2.7	—
5454	Household shortening with vegetable oil	1	tablespoon(s)	13	0	115	0	0	0	13.0	3.4	5.5	2.7	2.2
111	Lard	1	tablespoon(s)	13	0	115	0	0	0	12.8	5.0	5.8	1.4	—
	Margarine													
114	Margarine	1	tablespoon(s)	14	2.3	101	0	0.1	0	11.4	2.1	5.5	3.4	2.1
5439	Soft	1	tablespoon(s)	14	2.3	103	0.1	0.1	0	11.6	1.7	4.4	2.1	3.0
32329	Soft, unsalted, with hydrogenated soybean and cottonseed oils	1	tablespoon(s)	14	2.5	101	0.1	0.1	0	11.3	2.0	5.4	3.5	—
928	Unsalted	1	tablespoon(s)	14	2.6	101	0.1	0.1	0	11.3	2.1	5.2	3.5	—
119	Whipped	1	tablespoon(s)	9	1.5	64	0.1	0.1	0	7.2	1.2	3.2	2.5	—
	Spreads													
54657	I Can't Believe It's Not Butter!, tub, soya oil (non-hydrogenated)	1	tablespoon(s)	14	2.3	103	0.1	0.1	0	11.6	2.8	2.0	5.1	0.1
2708	Mayonnaise with soybean and safflower oils	1	tablespoon(s)	14	2.1	99	0.2	0.4	0	11.0	1.2	1.8	7.6	—
16157	Promise vegetable oil spread, stick	1	tablespoon(s)	14	4.2	90	0	0	0	10.0	2.5	2.0	4.0	—
	Oils													
2681	Canola	1	tablespoon(s)	14	0	120	0	0	0	13.6	1.0	8.6	3.8	0.1
120	Corn	1	tablespoon(s)	14	0	120	0	0	0	13.6	1.8	3.8	7.4	0
122	Olive	1	tablespoon(s)	14	0	119	0	0	0	13.5	1.9	9.9	1.4	—
124	Peanut	1	tablespoon(s)	14	0	119	0	0	0	13.5	2.3	6.2	4.3	—
2693	Safflower	1	tablespoon(s)	14	0	120	0	0	0	13.6	0.8	10.2	2.0	—
923	Sesame	1	tablespoon(s)	14	0	120	0	0	0	13.6	1.9	5.4	5.7	—
128	Soybean, hydrogenated	1	tablespoon(s)	14	0	120	0	0	0	13.6	2.0	5.8	5.1	—
130	Soybean, with soybean and cottonseed oil	1	tablespoon(s)	14	0	120	0	0	0	13.6	2.4	4.0	6.5	—
2700	Sunflower	1	tablespoon(s)	14	0	120	0	0	0	13.6	1.8	6.3	5.0	—
357	**Pam original no stick cooking spray**	1	serving(s)	0	0.2	0	0	0	0	0	0	0	0	
	Salad dressing													
132	Blue cheese	2	tablespoon(s)	30	9.7	151	1.4	2.2	0	15.7	3.0	3.7	8.3	
133	Blue cheese, low calorie	2	tablespoon(s)	32	25.4	32	1.6	0.9	0	2.3	0.8	0.6	0.8	
1764	Caesar	2	tablespoon(s)	30	10.3	158	0.4	0.9	0	17.3	2.6	4.1	9.9	
29654	Creamy, reduced calorie, fat free, cholesterol free, sour cream and/or buttermilk and oil	2	tablespoon(s)	32	23.9	34	0.4	6.4	0	0.9	0.2	0.2	0.5	
29617	Creamy, reduced calorie, sour cream and/or buttermilk and oil	2	tablespoon(s)	30	22.2	48	0.5	2.1	0	4.2	0.6	1.0	2.4	
134	French	2	tablespoon(s)	32	11.7	146	0.2	5.0	0	14.3	1.8	2.7	6.7	
135	French, low fat	2	tablespoon(s)	32	17.4	74	0.2	9.4	0.4	4.3	0.4	1.9	1.6	
136	Italian	2	tablespoon(s)	29	16.6	86	0.1	3.1	0	8.3	1.3	1.9	3.8	
137	Italian, diet	2	tablespoon(s)	30	25.4	23	0.1	1.4	0	1.9	0.1	0.7	0.5	
139	Mayonnaise-type	2	tablespoon(s)	29	11.7	115	0.3	7.0	0	9.8	1.4	2.6	5.3	
942	Oil and vinegar	2	tablespoon(s)	32	15.2	144	0	0.8	0	16.0	2.9	4.7	7.7	
1765	Ranch	2	tablespoon(s)	30	11.6	146	0.1	1.6	0	15.8	2.3	5.2	7.6	
3666	Ranch, reduced calorie	2	tablespoon(s)	30	20.5	62	0.1	2.2	0	6.1	1.1	1.8	2.9	
940	Russian	2	tablespoon(s)	30	11.6	107	0.5	9.3	0.7	7.8	1.2	1.8	4.4	
939	Russian, low calorie	2	tablespoon(s)	32	20.8	45	0.2	8.8	0.1	1.3	0.2	0.3	0.7	
941	Sesame seed	2	tablespoon(s)	30	11.8	133	0.9	2.6	0.3	13.6	1.9	3.6	7.5	
142	Thousand Island	2	tablespoon(s)	32	14.9	118	0.3	4.7	0.3	11.2	1.6	2.5	5.8	
143	Thousand Island, low calorie	2	tablespoon(s)	30	18.2	61	0.3	6.7	0.4	3.9	0.2	1.9	0.8	
	Sandwich spreads													
138	Mayonnaise with soybean oil	1	tablespoon(s)	14	2.1	99	0.1	0.4	0	11.0	1.6	2.7	5.8	0
140	Mayonnaise, low calorie	1	tablespoon(s)	16	10.0	37	0	2.6	0	3.1	0.5	0.7	1.7	
141	Tartar sauce	2	tablespoon(s)	28	8.7	144	0.3	4.1	0.1	14.4	2.2	3.8	7.7	
	SWEETS													
4799	**Butterscotch or caramel topping**	2	tablespoon(s)	41	13.1	103	0.6	27.0	0.4	0	0	0	0	—
	Candy													
1786	Almond Joy candy bar	1	item(s)	45	4.3	220	2.0	27.0	2.0	12.0	8.0	3.3	0.7	0
1785	Bit-O-Honey candy	6	item(s)	40	—	190	1.0	39.0	0	3.5	2.5	—	—	—
33375	Butterscotch candy	2	piece(s)	12	0.6	47	0	10.8	0	0.4	0.2	0.1	0	—

Chol (mg)	Calc (mg)	Iron (mg)	Magn (mg)	Pota (mg)	Sodi (mg)	Zinc (mg)	Vit A (µg)	Thia (mg)	Vit E (mg α)	Ribo (mg)	Niac (mg)	Vit B6 (mg)	Fola (µg)	Vit C (mg)	Vit B12 (µg)	Sele (µg)
0	2	0.02	2.4	0	2.4	0.00	0	0.00	—	0.00	0.00	0.00	0	0	0	0
0	7	0.00	2.4	2.4	7.1	0.00	0	0.00	0.00	0.00	0.00	0.00	0	0	0	0
0	2	0.02	0	0	29.3	0.24	0	0.00	0.00	0.00	0.00	0.00	0	0	0	0
31	3	0.00	0.3	3.4	81.8	0.01	97.1	0.00	0.32	0.01	0.01	0.00	0.4	0	0	0.1
0	0	0.00	0	1.6	120.0	0.00	0	0.00	0.00	0.00	0.00	0.00	0	0	0	—
31	3	0.00	0.3	3.4	1.6	0.01	97.1	0.00	0.32	0.01	0.01	0.00	0.4	0	0	0.1
21	2	0.01	0.2	2.4	77.7	0.01	64.3	0.00	0.21	0.00	0.00	0.00	0.3	0	0	0.1
25	3	0.00	0.2	2.7	1.3	0.01	78.0	0.00	0.26	0.00	0.00	0.00	0.3	0	0	0.1
14	0	0.00	0	0	0	0.00	0	0.00	0.34	0.00	0.00	0.00	0	0	0	0
11	0	0.00	0	0	0	0.00	0	0.00	0.34	0.00	0.00	0.00	0	0	0	0
0	0	0.00	0	0	0	0.00	0	0.00	—	0.00	0.00	0.00	0	0	0	—
12	0	0.00	0	0	0	0.01	0	0.00	0.07	0.00	0.00	0.00	0	0	0	0
0	4	0.01	0.4	5.9	133.0	0.00	115.5	0.00	1.26	0.01	0.00	0.00	0.1	0	0	0
0	4	0.00	0.3	5.5	155.4	0.00	142.7	0.00	1.00	0.00	0.00	0.00	0.1	0	0	0
0	4	0.00	0.3	5.4	3.9	0.00	103.1	0.00	0.98	0.00	0.00	0.00	0.1	0	0	0
0	2	0.00	0.3	3.5	0.3	0.00	115.5	0.00	1.80	0.00	0.00	0.00	0.1	0	0	0
0	2	0.00	0.2	3.4	97.1	0.00	73.7	0.00	0.45	0.00	0.00	0.00	0.1	0	0	0
0	4	0.00	0.3	5.5	155.3	0.00	142.6	0.00	0.72	0.00	0.00	0.00	0.1	0	0	0
8	2	0.06	0.1	4.7	78.4	0.01	11.6	0.00	3.03	0.00	0.00	0.08	1.1	0	0	0.2
0	10	0.18	—	8.7	90.0	—	—	0.00	—	0.00	0.00	—	—	0.6	—	—
0	0	0.00	0	0	0	0.00	0	0.00	2.37	0.00	0.00	0.00	0	0	0	0
0	0	0.00	0	0	0	0.00	0	0.00	1.94	0.00	0.00	0.00	0	0	0	0
0	0	0.07	0	0.1	0.3	0.00	0	0.00	1.93	0.00	0.00	0.00	0	0	0	0
0	0	0.00	0	0	0	0.00	0	0.00	2.11	0.00	0.00	0.00	0	0	0	0
0	0	0.00	0	0	0	0.00	0	0.00	4.63	0.00	0.00	0.00	0	0	0	0
0	0	0.00	0	0	0	0.00	0	0.00	0.19	0.00	0.00	0.00	0	0	0	0
0	0	0.00	0	0	0	0.00	0	0.00	1.10	0.00	0.00	0.00	0	0	0	0
0	0	0.00	0	0	0	0.00	0	0.00	1.64	0.00	0.00	0.00	0	0	0	0
0	0	0.00	0	0	0	0.00	0	0.00	5.58	0.00	0.00	0.00	0	0	0	0
0	0	0.00	0	0.3	1.5	0.01	0.1	0.00	0.00	0.00	0.00	0.00	0	0	0	0
5	24	0.06	0	11.1	328.2	0.08	20.1	0.00	1.80	0.03	0.03	0.01	7.8	0.6	0.1	0.3
0	28	0.16	2.2	1.6	384.0	0.08	—	0.01	0.08	0.03	0.01	0.01	1.0	0.1	0.1	0.5
1	7	0.05	0.6	8.7	323.4	0.03	0.6	0.00	1.56	0.00	0.01	0.00	0.9	0	0	0.5
0	12	0.08	1.6	42.6	320.0	0.05	0.3	0.00	0.21	0.01	0.01	0.01	1.9	0	0	0.5
0	2	0.03	0.6	10.8	306.9	0.01	—	0.00	0.71	0.00	0.01	0.01	0	0.1	0	0.5
0	8	0.25	1.6	21.4	267.5	0.09	7.4	0.01	1.60	0.01	0.06	0.00	0	0	0	0
0	4	0.27	2.6	34.2	257.3	0.06	8.6	0.01	0.09	0.01	0.14	0.01	0.6	0	0	0.5
0	2	0.18	0.9	14.1	486.3	0.03	0.6	0.00	1.47	0.01	0.00	0.01	0	0	0	0.6
2	3	0.19	1.2	25.5	409.8	0.05	0.3	0.00	0.06	0.00	0.00	0.02	0	0	0	2.4
8	4	0.05	0.6	2.6	209.0	0.05	6.2	0.00	0.60	0.01	0.00	0.01	1.8	0	0.1	0.5
0	0	0.00	0	2.6	0.3	0.00	0	0.00	1.46	0.00	0.00	0.00	0	0	0	0.5
1	4	0.03	1.2	8.4	354.0	0.01	5.4	0.00	1.84	0.01	0.00	0.00	0.3	0.1	0	0.1
0	5	0.01	1.5	8.4	413.7	0.01	0.9	0.00	0.72	0.01	0.00	0.00	0.3	0.1	0	0.1
0	6	0.20	3.0	51.9	282.3	0.06	13.2	0.01	0.98	0.01	0.16	0.02	1.5	1.4	0	0.5
2	6	0.18	0	50.2	277.8	0.02	0.6	0.00	0.12	0.00	0.00	0.00	1.0	1.9	0	0.5
0	6	0.18	0	47.1	300.0	0.02	0.6	0.00	1.50	0.00	0.00	0.00	0	0	0	0.5
8	5	0.37	2.6	34.2	276.2	0.08	4.5	0.46	1.28	0.01	0.13	0.00	0	0	0	0.5
0	5	0.27	2.1	60.6	249.3	0.05	4.8	0.01	0.30	0.01	0.13	0.00	0	0	0	0
5	1	0.03	0.1	1.7	78.4	0.02	11.2	0.01	0.72	0.01	0.00	0.08	0.7	0	0	0.2
4	0	0.00	0	1.6	79.5	0.01	0	0.00	0.32	0.00	0.00	0.00	0	0	0	0.3
8	6	0.20	0.8	10.1	191.5	0.05	20.2	0.00	0.97	0.00	0.01	0.07	2.0	0.1	0.1	0.5
0	22	0.08	2.9	34.4	143.1	0.07	11.1	0.01	—	0.03	0.01	0.01	0.8	0.1	0	0
0	18	0.33	30.3	126.5	65.0	0.36	0	0.01	—	0.06	0.21	—	—	0	—	—
0	20	0.00	—	—	150.0	—	0	—	—	—	—	—	—	0	—	—
1	0	0.00	0	0.4	46.9	0.01	3.4	0.00	0.01	0.00	0.00	0.00	0	0	0	0.1

DA+ Code	Food Description	Quantity	Measure	Wt (g)	H$_2$O (g)	Ener (kcal)	Prot (g)	Carb (g)	Fiber (g)	Fat (g)	Sat	Mono	Poly	Trans
	SWEETS —Continued													
1701	Chewing gum, stick	1	item(s)	3	0.1	7	0	2.0	0.1	0	0	0	0	—
33378	Chocolate fudge with nuts, prepared	2	piece(s)	38	2.9	175	1.7	25.8	1.0	7.2	2.5	1.5	2.9	0.1
1787	Jelly beans	15	item(s)	43	2.7	159	0	39.8	0.1	0	0	0	0	—
1784	Kit Kat wafer bar	1	item(s)	42	0.8	210	3.0	27.0	0.5	11.0	7.0	3.5	0.3	0
4674	Krackel candy bar	1	item(s)	41	0.6	210	2.0	28.0	0.5	10.0	6.0	3.9	0.4	0
4934	Licorice	4	piece(s)	44	7.3	154	1.1	35.1	0	1.0	0	0.1	0	—
1780	Life Savers candy	1	item(s)	2	—	8	0	2.0	0	0	0	0	0	—
1790	Lollipop	1	item(s)	28	—	108	0	28.0	0	0	0	0	0	0
4679	M & Ms peanut chocolate candy, small bag	1	item(s)	49	0.9	250	5.0	30.0	2.0	13.0	5.0	5.4	2.1	—
1781	M & Ms plain chocolate candy, small bag	1	item(s)	48	0.8	240	2.0	34.0	1.0	10.0	6.0	3.3	0.3	—
4673	Milk chocolate bar, Symphony	1	item(s)	91	0.9	483	7.7	52.8	1.5	27.8	16.7	7.2	0.6	—
1783	Milky Way bar	1	item(s)	58	3.7	270	2.0	41.0	1.0	10.0	5.0	3.5	0.3	—
1788	Peanut brittle	1 ½	ounce(s)	43	0.3	207	3.2	30.3	1.1	8.1	1.8	3.4	1.9	—
1789	Reese's peanut butter cups	2	piece(s)	51	0.8	280	6.0	19.0	2.0	15.5	6.0	7.2	2.7	0
4689	Reese's pieces candy, small bag	1	item(s)	43	1.1	220	5.0	26.0	1.0	11.0	7.0	0.9	0.4	0
33399	Semisweet chocolate candy, made with butter	½	ounce(s)	14	0.1	68	0.6	9.0	0.8	4.2	2.5	1.4	0.1	—
1782	Snickers bar	1	item(s)	59	3.2	280	4.0	35.0	1.0	14.0	5.0	6.1	2.9	—
4694	Special Dark chocolate bar	1	item(s)	41	0.4	220	2.0	25.0	3.0	12.0	8.0	4.6	0.4	0
4695	Starburst fruit chews, original fruits	1	package(s)	59	3.9	240	0	48.0	0	5.0	1.0	2.1	1.8	—
4698	Taffy	3	piece(s)	45	2.2	179	0	41.2	0	1.5	0.9	0.4	0.1	0.1
4699	Three Musketeers bar	1	item(s)	60	3.5	260	2.0	46.0	1.0	8.0	4.5	2.6	0.3	—
4702	Twix caramel cookie bars	2	item(s)	58	2.4	280	3.0	37.0	1.0	14.0	5.0	7.7	0.5	—
4705	York peppermint pattie	1	item(s)	39	3.9	160	0.5	32.0	0.5	3.0	1.5	1.2	0.1	0
	Frosting, icing													
4760	Chocolate frosting, ready to eat	2	tablespoon(s)	31	5.2	122	0.3	19.4	0.3	5.4	1.7	2.8	0.6	—
4771	Creamy vanilla frosting, ready to eat	2	tablespoon(s)	28	4.2	117	0	19.0	0	4.5	0.8	1.4	2.2	—
17291	Dec-A-Cake variety pack candy decoration	1	teaspoon(s)	4	—	15	0	3.0	0	0.5	0	—	—	—
536	White icing	2	tablespoon(s)	40	3.6	162	0.1	31.8	0	4.2	0.8	2.0	1.2	—
	Gelatin													
13697	Gelatin snack, all flavors	1	item(s)	99	96.8	70	1.0	17.0	0	0	0	0	0	0
2616	Sugar free, low calorie mixed fruit gelatin mix, prepared	½	cup(s)	121	—	10	1.0	0	0	0	0	0	0	0
548	**Honey**	1	tablespoon(s)	21	3.6	64	0.1	17.3	0	0	0	0	0	0
	Jams, jellies													
550	Jam or preserves	1	tablespoon(s)	20	6.1	56	0.1	13.8	0.2	0	0	0	0	—
42199	Jams, preserves, dietetic, all flavors, w/ sodium saccarin	1	tablespoon(s)	14	6.4	18	0	7.5	0.4	0	0	0	0	—
552	Jelly	1	tablespoon(s)	21	6.3	56	0	14.7	0.2	0	0	0	0	—
545	**Marshmallows**	4	item(s)	29	4.7	92	0.5	23.4	0	0.1	0	0	0	0
4800	**Marshmallow cream topping**	2	tablespoon(s)	40	7.9	129	0.3	31.6	0	0.1	0	0	0	0
555	**Molasses**	1	tablespoon(s)	20	4.4	58	0	14.9	0	0	0	0	0	0
4780	**Popsicle or ice pop**	1	item(s)	59	47.5	47	0	11.3	0	0.1	0	0	0	0
	Sugar													
559	Brown sugar, packed	1	teaspoon(s)	5	0.1	17	0	4.5	0	0	0	0	0	0
563	Powdered sugar, sifted	⅓	cup(s)	33	0.1	130	0	33.2	0	0	0	0	0	0
561	White granulated sugar	1	teaspoon(s)	4	0	16	0	4.2	0	0	0	0	0	0
	Sugar substitute													
1760	Equal sweetener, packet size	1	item(s)	1	—	0	0	0.9	0	0	0	0	0	0
13029	Splenda granular no calorie sweetener	1	teaspoon(s)	1	—	0	0	0.5	0	0	0	0	0	0
1759	Sweet N Low sugar substitute, packet	1	item(s)	1	0.1	4	0	0.5	0	0	0	0	0	0
	Syrup													
3148	Chocolate syrup	2	tablespoon(s)	38	11.6	105	0.8	24.4	1.0	0.4	0.2	0.1	0	—
29676	Maple syrup	¼	cup(s)	80	25.7	209	0	53.7	0	0.2	0	0.1	0.1	—
4795	Pancake syrup	¼	cup(s)	80	30.4	187	0	49.2	0	0	0	0	0	0
	SPICES, CONDIMENTS, SAUCES													
	Spices													
807	Allspice, ground	1	teaspoon(s)	2	0.2	5	0.1	1.4	0.4	0.2	0	0	0	—
1171	Anise seeds	1	teaspoon(s)	2	0.2	7	0.4	1.1	0.3	0.3	0	0.2	0.1	—
729	Bakers' yeast, active	1	teaspoon(s)	4	0.3	12	1.5	1.5	0.8	0.2	0	0.1	0	—
683	Baking powder, double acting with phosphate	1	teaspoon(s)	5	0.2	2	0	1.1	0	0	0	0	0	—
1611	Baking soda	1	teaspoon(s)	5	0	0	0	0	0	0	0	0	0	0
8552	Basil	1	teaspoon(s)	1	0.8	0	0	0	0	0	0	0	0	—
34959	Basil, fresh	1	piece(s)	1	0.5	0	0	0	0	0	0	0	0	—
808	Basil, ground	1	teaspoon(s)	1	0.1	4	0.2	0.9	0.6	0.1	0	0	0	—
809	Bay leaf	1	teaspoon(s)	1	0	2	0	0.5	0.2	0.1	0	0	0	—
11720	Betel leaves	1	ounce(s)	28	—	17	1.8	2.4	0	0	—	—	—	—
730	Brewers' yeast	1	teaspoon(s)	3	0.1	8	1.0	1.0	0.8	0	0	0	0	0
11710	Capers	1	teaspoon(s)	5	—	0	0	0	0	0	0	0	0	—
1172	Caraway seeds	1	teaspoon(s)	2	0.2	7	0.4	1.0	0.8	0.3	0	0.2	0.1	—

Chol (mg)	Calc (mg)	Iron (mg)	Magn (mg)	Pota (mg)	Sodi (mg)	Zinc (mg)	Vit A (µg)	Thia (mg)	Vit E (mg α)	Ribo (mg)	Niac (mg)	Vit B₆ (mg)	Fola (µg)	Vit C (mg)	Vit B₁₂ (µg)	Sele (µg)
0	0	0.00	0	0.1	0	0.00	0	0.00	0.00	0.00	0.00	0.00	0	0	0	0
5	22	0.74	20.9	69.5	14.8	0.54	14.4	0.02	0.09	0.03	0.12	0.03	6.1	0.1	0	1.1
0	1	0.05	0.9	15.7	21.3	0.02	0	0.00	0.00	0.01	0.00	0.00	0	0	0	0.5
3	60	0.36	16.4	126.0	30.0	0.51	0	0.07	—	0.22	1.07	0.05	59.6	0	0.1	2.0
3	40	0.36	—	168.8	50.0	—	0	—	—	—	—	—	—	0	—	—
0	0	0.22	2.6	28.2	126.3	0.07	0	0.01	0.07	0.01	0.04	0.00	0	0	0	0
0	0	0.00	—	0	0	—	0	0.00	—	0.00	0.00	—	—	0	—	0
0	0	0.00	—	—	10.8	—	0	0.00	—	0.00	0.00	—	—	0	—	1.0
5	40	0.36	36.5	170.6	25.0	1.13	14.8	0.03	—	0.06	1.60	0.04	17.3	0.6	0.1	1.9
5	40	0.36	19.6	127.4	30.0	0.46	14.8	0.02	—	0.06	0.10	0.01	2.9	0.6	0.1	1.4
22	228	0.82	61.0	398.6	91.9	1.00	0	0.06	—	0.25	0.14	0.10	10.9	2.0	0.4	—
5	60	0.18	19.8	140.1	95.0	0.41	15.1	0.02	—	0.06	0.20	0.02	5.8	0.6	0.2	3.3
5	11	0.51	17.9	71.4	189.2	0.37	16.6	0.05	1.08	0.01	1.12	0.03	19.6	0	0	1.1
3	40	0.72	45.4	217.4	180.0	0.93	0	0.12	—	0.08	2.35	0.07	28.1	0	0.1	2.3
0	20	0.00	18.9	169.9	80.0	0.32	0	0.04	—	0.06	1.22	0.03	12.0	0	0.1	0.8
3	5	0.44	16.3	51.7	1.6	0.23	0.4	0.01	—	0.01	0.06	0.01	0.4	0	0	0.5
5	40	0.36	42.3	—	140.0	1.37	15.3	0.03	—	0.06	1.60	0.05	23.5	0.6	0.1	2.7
0	0	1.80	45.5	136.0	50.0	0.59	0	0.01	—	0.02	0.16	0.01	0.8	0	0	1.2
0	10	0.18	0.6	1.2	0	0.00	0	0.00	—	0.00	0.00	0.00	0	30	0	0.5
4	4	0.00	0	1.4	23.4	0.09	12.2	0.01	0.04	0.01	0.00	0.00	0	0	0	0.3
5	20	0.36	17.5	80.3	110.0	0.33	14.5	0.01	—	0.03	0.20	0.01	0	0.6	0.1	1.5
5	40	0.36	18.5	116.8	115.0	0.45	15.0	0.09	—	0.13	0.69	0.01	13.9	0.6	0.1	1.2
0	0	0.33	23.4	66.1	10.0	0.28	0	0.01	—	0.03	0.31	0.01	1.5	0	0	—
0	2	0.44	6.4	60.0	56.1	0.09	0	0.00	0.48	0.00	0.03	0.00	0.3	0	0	0.2
0	1	0.04	0.3	9.5	51.5	0.01	0	0.00	0.43	0.08	0.06	0.00	2.2	0	0	0
0	0	0.00	—	—	15.0	—	0	—	—	—	—	—	—	0	—	—
0	4	0.01	0.4	5.6	76.4	0.01	44.4	0.00	0.32	0.01	0.00	0.00	0	0	0	0.3
0	0	0.00	—	0	40.0	—	0	—	—	—	—	—	—	0	—	—
0	0	0.00	0	0	50.0	0.00	0	0.00	0.00	0.00	0.00	0.00	0	0	0	—
0	1	0.08	0.4	10.9	0.8	0.04	0	0.00	0.00	0.01	0.02	0.01	0.4	0.1	0	0.2
0	4	0.10	0.8	15.4	6.4	0.01	0	0.00	0.02	0.02	0.01	0.00	2.2	1.8	0	0.4
0	1	0.56	0.7	9.7	0	0.01	0	0.00	0.01	0.00	0.00	0.00	1.3	0	0	0.2
0	1	0.04	1.3	11.3	6.3	0.01	0	0.00	0.00	0.01	0.01	0.00	0.4	0.2	0	0.1
0	1	0.06	0.6	1.4	23.0	0.01	0	0.00	0.00	0.00	0.02	0.00	0.3	0	0	0.5
0	1	0.08	0.8	2.0	32.0	0.01	0	0.00	0.00	0.00	0.03	0.00	0.4	0	0	0.7
0	41	0.94	48.4	292.8	7.4	0.05	0	0.01	0.00	0.00	0.18	0.13	0	0	0	3.6
0	0	0.31	0.6	8.9	4.1	0.08	0	0.00	0.00	0.00	0.00	0.00	0	0.4	0	0.1
0	4	0.03	0.4	6.1	1.3	0.00	0	0.00	0.00	0.00	0.01	0.00	0	0	0	0.1
0	0	0.01	0	0.7	0.3	0.00	0	0.00	0.00	0.00	0.00	0.00	0	0	0	0.2
0	0	0.00	0	0.1	0	0.00	0	0.00	0.00	0.00	0.00	0.00	0	0	0	0
0	0	0.00	0	0	0	0.00	0	0.00	0.00	0.00	0.00	0.00	0	0	0	0
0	0	0.00	—	—	0	—	—	0.00	—	0.00	0.00	—	—	0	0	—
0	0	0.00	—	—	0	—	0	—	0.00	—	—	—	—	0	—	—
0	5	0.79	24.4	84.0	27.0	0.27	0	0.00	0.01	0.01	0.12	0.00	0.8	0.1	0	0.5
0	54	0.96	11.2	163.2	7.2	3.32	0	0.01	0.00	0.01	0.02	0.00	0	0	0	0.5
0	2	0.02	1.6	12.0	65.6	0.06	0	0.00	0.00	0.00	0.00	0.00	0	0	0	0
0	13	0.13	2.6	19.8	1.5	0.01	0.5	0.00	—	0.00	0.05	0.00	0.7	0.7	0	0.1
0	14	0.77	3.6	30.3	0.3	0.11	0.3	0.01	—	0.01	0.06	0.01	0.2	0.4	0	0.1
0	3	0.66	3.9	80.0	2.0	0.25	0	0.09	0.00	0.21	1.59	0.06	93.6	0	0	1.0
0	339	0.51	1.8	0.2	363.1	0.00	0	0.00	0.00	0.00	0.00	0.00	0	0	0	0
0	0	0.00	0	0	1258.6	0.00	0	0.00	0.00	0.00	0.00	0.00	0	0	0	0
0	2	0.02	0.6	2.6	0	0.01	2.3	0.00	0.01	0.00	0.01	0.00	0.6	0.2	0	0
0	1	0.01	0.4	2.3	0	0.00	1.3	0.00	—	0.00	0.00	0.00	0.3	0.1	0	0
0	30	0.58	5.9	48.1	0.5	0.08	6.6	0.00	0.10	0.00	0.09	0.03	3.8	0.9	0	0
0	5	0.25	0.7	3.2	0.1	0.02	1.9	0.00	—	0.00	0.01	0.01	1.1	0.3	0	0
0	110	2.29	—	155.9	2.0	—	—	0.04	—	0.07	0.19	—	—	0.9	0	—
0	6	0.46	6.1	50.7	3.3	0.21	0	0.41	—	0.11	1.00	0.06	104.3	0	0	—
0	—	—	—	—	105.0	—	—	—	—	—	—	—	—	—	0	—
0	14	0.34	5.4	28.4	0.4	0.11	0.4	0.01	0.05	0.01	0.07	0.01	0.2	0.4	0	0.3

TABLE E-1   **Food Composition**

(Computer code is for Cengage Diet Analysis program)   (For purposes of calculations, use "0" for t, <1, <.1, <.01, etc.)

DA+ Code	Food Description	Quantity	Measure	Wt (g)	H₂O (g)	Ener (kcal)	Prot (g)	Carb (g)	Fiber (g)	Fat (g)	Fat Breakdown (g)			
											Sat	Mono	Poly	Trans
	SPICES, CONDIMENTS, SAUCES —Continued													
1173	Celery seeds	1	teaspoon(s)	2	0.1	8	0.4	0.8	0.2	0.5	0	0.3	0.1	—
1174	Chervil, dried	1	teaspoon(s)	1	0	1	0.1	0.3	0.1	0	0	0	0	—
810	Chili powder	1	teaspoon(s)	3	0.2	8	0.3	1.4	0.9	0.4	0.1	0.1	0.2	—
8553	Chives, chopped	1	teaspoon(s)	1	0.9	0	0	0	0	0	0	0	0	—
51420	Cilantro (coriander)	1	teaspoon(s)	0	0.3	0	0	0	0	0	0	0	0	—
811	Cinnamon, ground	1	teaspoon(s)	2	0.2	6	0.1	1.9	1.2	0	0	0	0	0
812	Cloves, ground	1	teaspoon(s)	2	0.1	7	0.1	1.3	0.7	0.4	0.1	0	0.1	—
1175	Coriander leaf, dried	1	teaspoon(s)	1	0	2	0.1	0.3	0.1	0	0	0	0	—
1176	Coriander seeds	1	teaspoon(s)	2	0.2	5	0.2	1.0	0.8	0.3	0	0.2	0	—
1706	Cornstarch	1	tablespoon(s)	8	0.7	30	0	7.3	0.1	0	0	0	0	—
1177	Cumin seeds	1	teaspoon(s)	2	0.2	8	0.4	0.9	0.2	0.5	0	0.3	0.1	—
11729	Cumin, ground	1	teaspoon(s)	5	—	11	0.4	0.8	0.8	0.4	—	—	—	—
1178	Curry powder	1	teaspoon(s)	2	0.2	7	0.3	1.2	0.7	0.3	0	0.1	0.1	—
1179	Dill seeds	1	teaspoon(s)	2	0.2	6	0.3	1.2	0.4	0.3	0	0.2	0	—
1180	Dill weed, dried	1	teaspoon(s)	1	0.1	3	0.2	0.6	0.1	0	0	0	0	—
34949	Dill weed, fresh	5	piece(s)	1	0.9	0	0	0.1	0	0	0	0	0	—
4949	Fennel leaves, fresh	1	teaspoon(s)	1	0.9	0	0	0.1	0	0	—	—	—	—
1181	Fennel seeds	1	teaspoon(s)	2	0.2	7	0.3	1.0	0.8	0.3	0	0.2	0	—
1182	Fenugreek seeds	1	teaspoon(s)	4	0.3	12	0.9	2.2	0.9	0.2	0.1	—	—	—
11733	Garam masala, powder	1	ounce(s)	28	—	107	4.4	12.8	0	4.3	—	—	—	—
1067	Garlic clove	1	item(s)	3	1.8	4	0.2	1.0	0.1	0	0	0	0	—
813	Garlic powder	1	teaspoon(s)	3	0.2	9	0.5	2.0	0.3	0	0	0	0	—
1068	Ginger root	2	teaspoon(s)	4	3.1	3	0.1	0.7	0.1	0	0	0	0	—
1183	Ginger, ground	1	teaspoon(s)	2	0.2	6	0.2	1.3	0.2	0.1	0	0	0	—
35497	Leeks, bulb and lower-leaf, freeze-dried	¼	cup(s)	1	0	3	0.1	0.6	0.1	0	0	0	0	—
1184	Mace, ground	1	teaspoon(s)	2	0.1	8	0.1	0.9	0.3	0.6	0.2	0.2	0.1	—
1185	Marjoram, dried	1	teaspoon(s)	1	0	2	0.1	0.4	0.2	0	0	0	0	—
1186	Mustard seeds, yellow	1	teaspoon(s)	3	0.2	15	0.8	1.2	0.5	0.9	0	0.7	0.2	—
814	Nutmeg, ground	1	teaspoon(s)	2	0.1	12	0.1	1.1	0.5	0.8	0.6	0.1	0	—
2747	Onion flakes, dehydrated	1	teaspoon(s)	2	0.1	6	0.1	1.4	0.2	0	0	0	0	—
1187	Onion powder	1	teaspoon(s)	2	0.1	7	0.2	1.7	0.1	0	0	0	0	—
815	Oregano, ground	1	teaspoon(s)	2	0.1	5	0.2	1.0	0.6	0.2	0	0	0.1	—
816	Paprika	1	teaspoon(s)	2	0.2	6	0.3	1.2	0.8	0.3	0	0	0.2	—
817	Parsley, dried	1	teaspoon(s)	0	0	1	0.1	0.2	0.1	0	0	0	0	—
818	Pepper, black	1	teaspoon(s)	2	0.2	5	0.2	1.4	0.6	0.1	0	0	0	—
819	Pepper, cayenne	1	teaspoon(s)	2	0.1	6	0.2	1.0	0.5	0.3	0.1	0	0.2	—
1188	Pepper, white	1	teaspoon(s)	2	0.3	7	0.3	1.6	0.6	0.1	0	0	0	—
1189	Poppy seeds	1	teaspoon(s)	3	0.2	15	0.5	0.7	0.3	1.3	0.1	0.2	0.9	—
1190	Poultry seasoning	1	teaspoon(s)	2	0.1	5	0.1	1.0	0.2	0.1	0	0	0	—
1191	Pumpkin pie spice, powder	1	teaspoon(s)	2	0.1	6	0.1	1.2	0.3	0.2	0.1	0	0	—
1192	Rosemary, dried	1	teaspoon(s)	1	0.1	4	0.1	0.8	0.5	0.2	0.1	0	0	—
11723	Rosemary, fresh	1	teaspoon(s)	1	0.5	1	0	0.1	0.1	0	0	0	0	—
2722	Saffron powder	1	teaspoon(s)	1	0.1	2	0.1	0.5	0	0	0	0	0	—
11724	Sage	1	teaspoon(s)	1	—	1	0	0.1	0	0	—	—	—	—
1193	Sage, ground	1	teaspoon(s)	1	0.1	2	0.1	0.4	0.3	0.1	0	0	0	—
30189	Salt substitute	¼	teaspoon(s)	1	—	0	0	0	0	0	0	0	0	0
30190	Salt substitute, seasoned	¼	teaspoon(s)	1	—	1	0	0.1	0	0	—	—	—	—
822	Salt, table	¼	teaspoon(s)	2	0	0	0	0	0	0	0	0	0	0
1194	Savory, ground	1	teaspoon(s)	1	0.1	4	0.1	1.0	0.6	0.1	0	—	—	—
820	Sesame seed kernels, toasted	1	teaspoon(s)	3	0.1	15	0.5	0.7	0.5	1.3	0.2	0.5	0.6	—
11725	Sorrel	1	teaspoon(s)	3	—	1	0.1	0.1	0	0	0	—	—	—
11721	Spearmint	1	teaspoon(s)	2	1.6	1	0.1	0.2	0.1	0	0	0	0	—
35498	Sweet green peppers, freeze-dried	¼	cup(s)	2	0	5	0.3	1.1	0.3	0	0	0	0	—
11726	Tamarind leaves	1	ounce(s)	28	—	33	1.6	5.2	0	0.6	—	—	—	—
11727	Tarragon	1	ounce(s)	28	—	14	1.0	1.8	0	0.3	—	—	—	—
1195	Tarragon, ground	1	teaspoon(s)	2	0.1	5	0.4	0.8	0.1	0.1	0	0	0.1	—
11728	Thyme, fresh	1	teaspoon(s)	1	0.5	1	0	0.2	0.1	0	0	0	0	—
821	Thyme, ground	1	teaspoon(s)	1	0.1	4	0.1	0.9	0.5	0.1	0	0	0	—
1196	Turmeric, ground	1	teaspoon(s)	2	0.3	8	0.2	1.4	0.5	0.2	0.1	0	0	—
11995	Wasabi	1	tablespoon(s)	14	10.7	10	0.7	2.3	0.2	0	—	—	—	—
	Condiments													
674	Catsup or ketchup	1	tablespoon(s)	15	10.4	15	0.3	3.8	0	0	0	0	0	—
703	Dill pickle	1	ounce(s)	28	26.7	3	0.2	0.7	0.3	0	0	0	0	—
138	Mayonnaise with soybean oil	1	tablespoon(s)	14	2.1	99	0.1	0.4	0	11.0	1.6	2.7	5.8	0
140	Mayonnaise, low calorie	1	tablespoon(s)	16	10.0	37	0	2.6	0	3.1	0.5	0.7	1.7	—
1682	Mustard, brown	1	teaspoon(s)	5	4.1	5	0.3	0.3	0	0.3	—	—	—	—
700	Mustard, yellow	1	teaspoon(s)	5	4.1	3	0.2	0.3	0.2	0.2	0	0.1	0	0
706	Sweet pickle relish	1	tablespoon(s)	15	9.3	20	0.1	5.3	0.2	0.1	0	0	0	—
141	Tartar sauce	2	tablespoon(s)	28	8.7	144	0.3	4.1	0.1	14.4	2.2	3.8	7.7	—

Chol (mg)	Calc (mg)	Iron (mg)	Magn (mg)	Pota (mg)	Sodi (mg)	Zinc (mg)	Vit A (µg)	Thia (mg)	Vit E (mg α)	Ribo (mg)	Niac (mg)	Vit B₆ (mg)	Fola (µg)	Vit C (mg)	Vit B₁₂ (µg)	Sele (µg)
0	35	0.89	8.8	28.0	3.2	0.13	0.1	0.01	0.02	0.01	0.06	0.01	0.2	0.3	0	0.2
0	8	0.19	0.8	28.4	0.5	0.05	1.8	0.00	—	0.00	0.03	0.01	1.6	0.3	0	0.2
0	7	0.37	4.4	49.8	26.3	0.07	38.6	0.01	0.75	0.02	0.20	0.09	2.6	1.7	0	0.2
0	1	0.01	0.4	3.0	0	0.01	2.2	0.00	0.00	0.00	0.01	0.00	1.1	0.6	0	0
0	0	0.01	0.1	1.7	0.2	0.00	1.1	0.00	0.01	0.00	0.00	0.00	0.2	0.1	0	0
0	23	0.19	1.4	9.9	0.2	0.04	0.3	0.00	0.05	0.00	0.03	0.00	0.1	0.1	0	0.1
0	14	0.18	5.5	23.1	5.1	0.02	0.6	0.00	0.17	0.01	0.03	0.01	2.0	1.7	0	0.1
0	7	0.25	4.2	26.8	1.3	0.02	1.8	0.01	0.01	0.01	0.06	0.00	1.6	3.4	0	0.2
0	13	0.29	5.9	22.8	0.6	0.08	0	0.00	—	0.01	0.03	—	0	0.4	0	0.5
0	0	0.03	0.2	0.2	0.7	0.01	0	0.00	0.00	0.00	0.00	0.00	0	0	0	0.2
0	20	1.39	7.7	37.5	3.5	0.10	1.3	0.01	0.07	0.01	0.09	0.01	0.2	0.2	0	0.1
0	20	—	—	43.6	4.8	—	—	—	—	—	—	—	—	—	—	—
0	10	0.59	5.1	30.9	1.0	0.08	1.0	0.01	0.44	0.01	0.06	0.02	3.1	0.2	0	0.3
0	32	0.34	5.4	24.9	0.4	0.10	0.1	0.01	—	0.01	0.05	0.01	0.2	0.4	0	0.3
0	18	0.48	4.5	33.1	2.1	0.03	2.9	0.00	—	0.00	0.02	0.01	1.5	0.5	0	0
0	2	0.06	0.6	7.4	0.6	0.01	3.9	0.01	0.01	0.00	0.01	0.00	1.5	0.9	0	0
0	1	0.02	—	4.0	0.1	—	—	0.00	—	0.00	0.01	0.00	—	0.3	0	—
0	24	0.37	7.7	33.9	1.8	0.07	0.1	0.01	—	0.01	0.12	0.01	—	0.4	0	—
0	7	1.24	7.1	28.5	2.5	0.09	0.1	0.01	—	0.01	0.06	0.02	2.1	0.1	0	0.2
0	215	9.24	93.6	411.1	27.5	1.07	—	0.09	—	0.09	0.70	—	0	0	0	—
0	5	0.05	0.8	12.0	0.5	0.03	0	0.01	0.00	0.00	0.02	0.03	0.1	0.9	0	0.4
0	2	0.07	1.6	30.8	0.7	0.07	0	0.01	0.01	0.00	0.01	0.08	0.1	0.5	0	1.1
0	1	0.02	1.7	16.6	0.5	0.01	0	0.00	0.01	0.00	0.02	0.01	0.4	0.2	0	0
0	2	0.20	3.3	24.2	0.6	0.08	0.1	0.00	0.32	0.00	0.09	0.01	0.7	0.1	0	0.7
0	3	0.06	1.3	19.2	0.3	0.01	0.1	0.01	—	0.00	0.02	0.01	2.9	0.9	0	0
0	4	0.23	2.8	7.9	1.4	0.03	0.7	0.01	—	0.01	0.02	0.00	1.3	0.4	0	0
0	12	0.49	2.1	9.1	0.5	0.02	2.4	0.00	0.01	0.00	0.01	0.01	1.6	0.3	0	0
0	17	0.32	9.8	22.5	0.2	0.18	0.1	0.01	0.09	0.01	0.26	0.01	2.5	0.1	0	4.4
0	4	0.06	4.0	7.7	0.4	0.04	0.1	0.01	0.00	0.00	0.02	0.00	1.7	0.1	0	0
0	4	0.02	1.5	27.1	0.4	0.03	0	0.01	0.00	0.01	0.02	0.02	2.8	1.3	0	0.1
0	8	0.05	2.6	19.8	1.1	0.04	0	0.01	0.01	0.00	0.01	0.02	3.5	0.3	0	0
0	24	0.66	4.1	25.0	0.2	0.06	5.2	0.01	0.28	0.01	0.09	0.01	4.1	0.8	0	0.1
0	4	0.49	3.9	49.2	0.7	0.08	55.4	0.01	0.62	0.03	0.32	0.08	2.2	1.5	0	0.1
0	4	0.29	0.7	11.4	1.4	0.01	1.5	0.00	0.02	0.00	0.02	0.00	0.5	0.4	0	0.1
0	9	0.60	4.1	26.4	0.9	0.03	0.3	0.00	0.01	0.01	0.02	0.01	0.2	0.4	0	0.1
0	3	0.14	2.7	36.3	0.5	0.04	37.5	0.01	0.53	0.01	0.15	0.04	1.9	1.4	0	0.2
0	6	0.34	2.2	1.8	0.1	0.02	0	0.00	—	0.00	0.01	0.00	0.2	0.5	0	0.1
0	41	0.26	9.3	19.6	0.6	0.28	0	0.02	0.03	0.01	0.02	0.01	1.6	0.1	0	0
0	15	0.53	3.4	10.3	0.4	0.04	2.0	0.00	0.02	0.00	0.04	0.02	2.1	0.2	0	0.1
0	12	0.33	2.3	11.3	0.9	0.04	0.2	0.00	0.01	0.00	0.03	0.01	0.9	0.4	0	0.2
0	15	0.35	2.6	11.5	0.6	0.03	1.9	0.01	—	0.01	0.01	0.02	3.7	0.7	0	0.1
0	2	0.04	0.6	4.7	0.2	0.01	1.0	0.00	—	0.00	0.01	0.00	0.8	0.2	0	—
0	1	0.07	1.8	12.1	1.0	0.01	0.2	0.00	—	0.00	0.01	0.01	0.7	0.6	0	0
0	4	—	1.1	2.7	0	0.01	—	0.00	—	—	—	—	—	—	0	—
0	12	0.19	3.0	7.5	0.1	0.03	2.1	0.01	0.05	0.00	0.04	0.01	1.9	0.2	0	0
0	7	0.00	0	603.6	0.1	—	0	—	—	—	0.05	0.02	—	0.7	0	0.1
0	0	0.00	—	476.3	0.1	—	0	—	—	—	—	—	—	0	—	—
0	0	0.01	0	0.1	581.4	0.00	0	0.00	0.00	0.00	0.00	0.00	0	0	0	0
0	30	0.53	5.3	14.7	0.3	0.06	3.6	0.01	—	—	0.05	0.02	—	0.7	0	0.1
0	3	0.21	9.2	10.8	1.0	0.27	0.1	0.03	0.01	0.01	0.15	0.00	2.6	0	0	0
0	—	—	—	—	0.1	—	—	—	—	—	—	—	—	—	—	—
0	4	0.22	1.2	8.7	0.6	0.02	3.9	0.00	—	0.00	0.01	0.00	2.0	0.3	0	—
0	2	0.16	3.0	50.7	3.1	0.03	4.5	0.01	0.06	0.01	0.11	0.03	3.7	30.4	0	0.1
0	85	1.48	20.2	—	—	—	—	0.06	—	0.02	1.16	—	—	0.9	0	—
0	48	—	14.5	128.1	2.6	0.17	—	0.04	—	—	—	—	—	0.6	0	—
0	18	0.51	5.6	48.3	1.0	0.06	3.4	0.00	—	0.02	0.14	0.03	4.4	0.8	0	0.1
0	3	0.14	1.3	4.9	0.1	0.01	1.9	0.00	—	0.00	0.01	0.00	0.4	1.3	0	0
0	26	1.73	3.1	11.4	0.8	0.08	2.7	0.01	0.10	0.01	0.06	0.01	3.8	0.7	0	0.1
0	4	0.91	4.2	55.6	0.8	0.09	0	0.00	0.06	0.01	0.11	0.04	0.9	0.6	0	0.1
0	13	0.11	—	—	—	—	—	0.02	—	0.01	0.07	—	—	11.2	0	—
0	3	0.07	2.9	57.3	167.1	0.03	7.1	0.00	0.21	0.02	0.21	0.02	1.5	2.3	0	0
0	12	0.10	2.0	26.1	248.1	0.03	2.6	0.01	0.02	0.01	0.03	0.01	0.3	0.2	0	0
5	1	0.03	0.1	1.7	78.4	0.02	11.2	0.01	0.72	0.01	0.00	0.08	0.7	0	0	0.2
4	0	0.00	0	1.6	79.5	0.01	0	0.00	0.32	0.00	0.00	0.00	0	0	0	0.3
0	6	0.09	1.0	6.8	68.1	0.01	0	0.00	0.09	0.00	0.01	0.00	0.2	0.1	0	—
0	3	0.07	2.5	6.9	56.8	0.03	0.2	0.01	0.01	0.00	0.02	0.00	0.4	0.1	0	1.6
0	0	0.13	0.8	3.8	121.7	0.02	9.2	0.00	0.08	0.01	0.03	0.00	0.2	0.2	0	0
8	6	0.20	0.8	10.1	191.5	0.05	20.2	0.00	0.97	0.00	0.01	0.07	2.0	0.1	0.1	0.5

DA+ Code	Food Description	Quantity	Measure	Wt (g)	H₂O (g)	Ener (kcal)	Prot (g)	Carb (g)	Fiber (g)	Fat (g)	Sat	Mono	Poly	Trans
	SPICES, CONDIMENTS, SAUCES —Continued													
	Sauces													
685	Barbecue sauce	2	tablespoon(s)	31	18.9	47	0	11.3	0.2	0.1	0	0	0.1	0
834	Cheese sauce	¼	cup(s)	63	44.4	110	4.2	4.3	0.3	8.4	3.8	2.4	1.6	—
32123	Chili enchilada sauce, green	2	tablespoon(s)	57	53.0	15	0.6	3.1	0.7	0.3	0	0	0.1	0
32122	Chili enchilada sauce, red	2	tablespoon(s)	32	24.5	27	1.1	5.0	2.1	0.8	0.1	0	0.4	0
29688	Hoisin sauce	1	tablespoon(s)	16	7.1	35	0.5	7.1	0.4	0.5	0.1	0.2	0.3	—
1641	Horseradish sauce, prepared	1	teaspoon(s)	5	3.3	10	0.1	0.2	0	1.0	0.6	0.3	0	—
16670	Mole poblano sauce	½	cup(s)	133	102.7	156	5.3	11.4	2.7	11.3	2.6	5.1	3.0	—
29689	Oyster sauce	1	tablespoon(s)	16	12.8	8	0.2	1.7	0	0	0	0	0	—
1655	Pepper sauce or Tabasco	1	teaspoon(s)	5	4.8	1	0.1	0	0	0	0	0	0	—
347	Salsa	2	tablespoon(s)	32	28.8	9	0.5	2.0	0.5	0.1	0	0	0	—
52206	Soy sauce, tamari	1	tablespoon(s)	18	12.0	11	1.9	1.0	0.1	0	0	0	0	—
839	Sweet and sour sauce	2	tablespoon(s)	39	29.8	37	0.1	9.1	0.1	0	0	0	0	—
1613	Teriyaki sauce	1	tablespoon(s)	18	12.2	16	1.1	2.8	0	0	0	0	0	0
25294	Tomato sauce	½	cup(s)	150	132.8	63	2.2	11.9	2.6	1.8	0.2	0.4	0.9	0
728	White sauce, medium	¼	cup(s)	63	46.8	92	2.4	5.7	0.1	6.7	1.8	2.8	1.8	—
1654	Worcestershire sauce	1	teaspoon(s)	6	4.5	4	0	1.1	0	0	0	0	0	0
	Vinegar													
30853	Balsamic	1	tablespoon(s)	15	—	10	0	2.0	0	0	0	0	0	0
727	Cider	1	tablespoon(s)	15	14.0	3	0	0.1	0	0	0	0	0	0
1673	Distilled	1	tablespoon(s)	15	14.3	2	0	0.8	0	0	0	0	0	0
12948	Tarragon	1	tablespoon(s)	15	13.8	2	0	0.1	0	0	0	0	0	0
	MIXED FOODS, SOUPS, SANDWICHES													
	Mixed dishes													
16652	Almond chicken	1	cup(s)	242	186.8	281	21.8	15.8	3.4	14.7	1.8	6.3	5.6	—
25224	Barbecued chicken	1	serving(s)	177	99.3	327	27.1	15.7	0.5	17.1	4.8	6.8	3.8	0
25227	Bean burrito	1	item(s)	149	81.8	326	16.1	33.0	5.6	14.8	8.3	4.7	0.9	—
9516	Beef and vegetable fajita	1	item(s)	223	143.9	397	22.4	35.3	3.1	18.0	5.9	8.0	2.5	—
16796	Beef or pork egg roll	2	item(s)	128	85.2	225	9.9	18.4	1.4	12.4	2.9	6.0	2.6	—
177	Beef stew with vegetables, prepared	1	cup(s)	245	201.0	220	16.0	15.0	3.2	11.0	4.4	4.5	0.5	—
30233	Beef stroganoff with noodles	1	cup(s)	256	190.1	343	19.7	22.8	1.5	19.1	7.4	5.7	4.4	—
16651	Cashew chicken	1	cup(s)	242	186.8	281	21.8	15.8	3.4	14.7	1.8	6.3	5.6	—
30274	Cheese pizza with vegetables, thin crust	2	slice(s)	140	76.6	298	12.7	35.4	2.5	12.0	4.9	4.7	1.6	—
30330	Cheese quesadilla	1	item(s)	54	18.3	190	7.7	15.3	1.0	10.8	5.2	3.6	1.3	—
215	Chicken and noodles, prepared	1	cup(s)	240	170.0	365	22.0	26.0	1.3	18.0	5.1	7.1	3.9	—
30239	Chicken and vegetables with broccoli, onion, bamboo shoots in soy based sauce	1	cup(s)	162	125.5	180	15.8	9.3	1.8	8.6	1.7	3.0	3.1	—
25093	Chicken cacciatore	1	cup(s)	244	175.7	284	29.9	5.7	1.3	15.3	4.3	6.2	3.3	0
28020	Chicken fried turkey steak	3	ounce(s)	492	276.2	706	77.1	68.7	3.6	12.0	3.4	2.9	3.9	—
218	Chicken pot pie	1	cup(s)	252	154.6	542	22.6	41.4	3.5	31.3	9.8	12.5	7.1	—
30240	Chicken teriyaki	1	cup(s)	244	158.3	364	51.0	15.2	0.7	7.0	1.8	2.0	1.7	—
25119	Chicken waldorf salad	½	cup(s)	100	67.2	179	14.0	6.8	1.0	10.8	1.8	3.1	5.2	—
25099	Chili con carne	¾	cup(s)	215	174.4	198	13.7	21.4	7.5	6.9	2.5	2.8	0.5	0
1062	Coleslaw	¾	cup(s)	90	73.4	70	1.2	11.2	1.4	2.3	0.3	0.6	1.2	—
1574	Crab cakes, from blue crab	1	item(s)	60	42.6	93	12.1	0.3	0	4.5	0.9	1.7	1.4	—
32144	Enchiladas with green chili sauce (enchiladas verdes)	1	item(s)	144	103.8	207	9.3	17.6	2.6	11.7	6.4	3.6	1.0	0
2793	Falafel patty	3	item(s)	51	17.7	170	6.8	16.2	—	9.1	1.2	5.2	2.1	—
28546	Fettuccine alfredo	1	cup(s)	244	88.7	279	13.1	46.1	1.4	4.2	2.2	1.0	0.4	0
32146	Flautas	3	item(s)	162	78.0	438	24.9	36.3	4.1	21.6	8.2	8.8	2.3	—
29629	Fried rice with meat or poultry	1	cup(s)	198	128.5	333	12.3	41.8	1.4	12.3	2.2	3.5	5.7	—
16649	General Tso chicken	1	cup(s)	146	91.0	296	18.7	16.4	0.9	17.0	4.0	6.3	5.3	—
1826	Green salad	¾	cup(s)	104	98.9	17	1.3	3.3	2.2	0.1	0	0	0	—
1814	Hummus	½	cup(s)	123	79.8	218	6.0	24.7	4.9	10.6	1.4	6.0	2.6	—
16650	Kung pao chicken	1	cup(s)	162	87.2	434	28.8	11.7	2.3	30.6	5.2	13.9	9.7	—
16622	Lamb curry	1	cup(s)	236	187.9	257	28.2	3.7	0.9	13.8	3.9	4.9	3.3	—
25253	Lasagna with ground beef	1	cup(s)	237	158.4	284	16.9	22.3	2.4	14.5	7.5	4.9	0.8	—
442	Macaroni and cheese, prepared	1	cup(s)	200	122.3	390	14.9	40.6	1.6	18.6	7.9	6.4	2.9	—
29637	Meat filled ravioli with tomato or meat sauce, canned	1	cup(s)	251	198.7	208	7.8	36.5	1.3	3.7	1.5	1.4	0.3	—
25105	Meat loaf	1	slice(s)	115	84.5	245	17.0	6.6	0.4	16.0	6.1	6.9	0.9	0
16646	Moo shi pork	1	cup(s)	151	76.8	512	18.9	5.3	0.6	46.4	6.9	15.8	21.2	—
16788	Nachos with beef, beans, cheese, tomatoes and onions	1	serving(s)	551	253.5	1576	59.1	137.5	20.4	90.8	32.6	41.9	9.4	—
6116	Pepperoni pizza	2	slice(s)	142	66.1	362	20.2	39.7	2.9	13.9	4.5	6.3	2.3	—
29601	Pizza with meat and vegetables, thin crust	2	slice(s)	158	81.4	386	16.5	36.8	2.7	19.1	7.7	8.1	2.2	—
655	Potato salad	½	cup(s)	125	95.0	179	3.4	14.0	1.6	10.3	1.8	3.1	4.7	—
25109	Salisbury steaks with mushroom sauce	1	serving(s)	135	101.8	251	17.1	9.3	0.5	15.5	6.0	6.7	0.8	0
16637	Shrimp creole with rice	1	cup(s)	243	176.6	309	27.0	27.7	1.2	9.2	1.7	3.6	2.9	—

Chol (mg)	Calc (mg)	Iron (mg)	Magn (mg)	Pota (mg)	Sodi (mg)	Zinc (mg)	Vit A (µg)	Thia (mg)	Vit E (mg α)	Ribo (mg)	Niac (mg)	Vit B$_6$ (mg)	Fola (µg)	Vit C (mg)	Vit B$_{12}$ (µg)	Sele (µg)
0	4	0.06	3.8	65.0	349.7	0.04	3.8	0.00	0.20	0.01	0.15	0.01	0.6	0.2	0	0.4
18	116	0.13	5.7	18.9	521.6	0.61	50.4	0.00	—	0.07	0.01	0.01	2.5	0.3	0.1	2.0
0	5	0.36	9.5	125.7	61.9	0.11	—	0.02	0.00	0.02	0.63	0.06	5.7	43.9	0	0
0	7	1.05	11.1	231.3	113.8	0.14	—	0.01	0.00	0.21	0.61	0.34	6.6	0.3	0	0.3
0	5	0.16	3.8	19.0	258.4	0.05	0	0.00	0.04	0.03	0.18	0.01	3.7	0.1	0	0.3
2	5	0.00	0.5	6.7	14.6	0.01	8.0	0.00	0.02	0.01	0.00	0.00	0.5	0.1	0	0.1
1	38	1.81	58.3	280.9	304.8	1.15	13.3	0.06	1.72	0.08	1.84	0.09	15.9	3.4	0.1	1.1
0	5	0.02	0.6	8.6	437.3	0.01	0	0.00	0.00	0.02	0.23	0.00	2.4	0	0.1	0.7
0	1	0.05	0.6	6.4	31.7	0.01	4.1	0.00	0.00	0.00	0.01	0.01	0.1	0.2	0	0
0	9	0.14	4.8	95.0	192.0	0.11	4.8	0.01	0.37	0.01	0.02	0.05	1.3	0.6	0	0.3
0	4	0.43	7.3	38.7	1018.9	0.08	0	0.01	0.00	0.03	0.72	0.04	3.3	0	0	0.1
0	5	0.20	1.2	8.2	97.5	0.01	0	0.00	—	0.01	0.11	0.03	0.2	0.1	0	—
0	5	0.30	11.0	40.5	689.9	0.01	0	0.01	0.00	0.01	0.22	0.01	1.4	0	0	0.2
0	23	1.24	28.9	536.8	268.6	0.36	—	0.08	0.52	0.08	1.64	0.20	23.2	32.0	0	1.0
4	74	0.20	8.8	97.5	221.3	0.25	—	0.04	—	0.11	0.25	0.02	3.1	0.5	0.2	—
0	6	0.30	0.7	45.4	55.6	0.01	0.3	0.00	0.00	0.01	0.03	0.00	0.5	0.7	0	0
0	0	0.00	—	—	0	—	0	—	—	—	—	—	—	0	—	—
0	1	0.03	0.7	10.9	0.7	0.01	0	0.00	0.00	0.00	0.00	0.00	0	0	0	0
0	1	0.09	0	2.3	0.1	0.00	0	0.00	0.00	0.00	0.00	0.00	0	0	0	5.0
0	0	0.07	—	2.3	0.7	—	0	0.07	—	0.07	0.07	—	0.3	0	—	—
41	68	1.86	58.1	539.7	510.6	1.50	31.5	0.07	4.11	0.22	9.57	0.43	26.6	5.1	0.3	13.6
120	26	1.70	32.4	419.7	500.9	2.67	—	0.09	0.01	0.24	6.87	0.40	15.0	7.9	0.3	19.5
38	333	3.01	52.5	447.6	510.6	1.98	—	0.28	0.01	0.30	1.92	0.19	122.9	8.2	0.3	15.9
45	85	3.65	37.9	475.0	756.0	3.52	17.8	0.38	0.80	0.29	5.33	0.39	69.1	23.4	2.1	28.3
74	31	1.68	20.5	248.3	547.8	0.89	25.6	0.32	1.28	0.24	2.55	0.18	38.4	4.0	0.3	17.5
71	29	2.90	—	613.0	292.0	—	—	0.15	0.51	0.17	4.70	—	—	17.0	0	15.0
74	69	3.25	35.8	391.7	816.6	3.63	69.1	0.21	1.25	0.30	3.80	0.21	48.6	1.3	1.8	27.9
41	68	1.86	58.1	539.7	510.6	1.50	31.5	0.07	4.11	0.22	9.57	0.43	26.6	5.1	0.3	13.6
17	249	2.78	28.0	294.0	739.2	1.42	47.6	0.29	1.05	0.33	2.84	0.14	61.6	15.3	0.4	18.6
23	190	1.04	13.5	75.6	469.3	0.86	58.3	0.11	0.43	0.15	0.89	0.02	21.6	2.4	0.1	9.2
103	26	2.20	—	149.0	600.0	—	—	0.05	—	0.17	4.30	—	—	0	—	29.0
42	28	1.19	22.7	299.7	620.5	1.32	81.0	0.07	1.11	0.14	5.28	0.36	16.2	22.5	0.2	12.0
109	47	1.97	40.0	489.3	492.1	2.13	—	0.11	0.00	0.20	9.81	0.57	16.1	14.0	0.3	22.6
156	423	8.79	110.3	1182.9	880.4	6.18	—	0.72	0.00	1.05	20.16	1.20	113.9	2.7	1.3	97.6
68	66	3.32	37.8	390.6	652.7	1.94	259.6	0.39	1.05	0.39	7.25	0.23	80.6	10.3	0.2	27.0
156	51	3.26	68.3	588.0	3208.6	3.75	31.7	0.15	0.58	0.36	16.68	0.88	24.4	2.0	0.5	36.1
42	20	0.82	23.9	202.5	246.5	1.13	—	0.05	0.62	0.09	4.06	0.25	15.8	2.5	0.2	10.7
27	42	2.83	50.6	636.8	864.8	2.36	—	0.15	0.01	0.22	3.18	0.19	58.1	10.3	0.6	7.3
7	41	0.53	9.0	162.9	20.7	0.18	47.7	0.06	—	0.05	0.24	0.11	24.3	29.4	0	0.6
90	63	0.64	19.8	194.4	198.0	2.45	34.2	0.05	—	0.04	1.74	0.10	31.8	1.7	3.6	24.4
27	266	1.07	38.5	251.4	276.3	1.26	—	0.07	0.02	0.16	1.27	0.17	44.6	59.3	0.2	6.0
0	28	1.74	41.8	298.4	149.9	0.76	0.5	0.07	—	0.08	0.53	0.06	47.4	0.8	0	0.5
9	218	1.83	38.4	163.9	472.9	1.24	—	0.41	0.00	0.35	2.85	0.09	96.2	1.6	0.4	38.4
73	146	2.66	61.3	222.9	885.7	3.43	0	0.10	0.10	0.16	3.00	0.26	95.7	0	1.2	36.7
103	38	2.77	33.7	196.0	833.6	1.34	41.6	0.33	1.60	0.18	4.17	0.27	97.0	3.4	0.3	22.0
66	26	1.46	23.4	248.2	849.7	1.40	29.2	0.10	1.62	0.18	6.28	0.28	23.4	12.0	0.2	19.9
0	13	0.65	11.4	178.0	26.9	0.21	59.0	0.03	—	0.05	0.56	0.08	38.3	24.0	0	0.4
0	60	1.91	35.7	212.8	297.7	1.34	0	0.10	0.92	0.06	0.49	0.49	72.6	9.7	0	3.0
65	50	1.96	63.2	427.7	907.2	1.50	38.9	0.15	4.32	0.14	13.22	0.58	42.1	7.5	0.3	23.0
90	38	2.95	40.1	493.2	495.6	6.60	—	0.08	1.29	0.28	8.03	0.21	28.3	1.4	2.9	30.4
68	233	2.22	40.1	420.1	433.6	2.70	—	0.21	0.21	0.29	3.06	0.22	50.4	15.0	0.8	21.4
34	310	2.06	40.0	258.0	784.0	2.06	180.0	0.27	0.72	0.43	2.18	0.08	64.0	0	0.5	30.6
15	35	2.10	20.1	283.6	1352.9	1.28	27.6	0.19	0.70	0.16	2.77	0.14	42.7	21.6	0.4	13.3
85	59	1.87	21.8	300.8	411.7	3.40	—	0.08	0.00	0.27	3.72	0.13	18.7	0.9	1.6	17.9
172	32	1.57	25.7	333.7	1052.5	1.82	49.8	0.49	5.39	0.36	2.88	0.31	21.1	8.0	0.8	30.0
154	948	7.32	242.4	1201.2	1862.4	10.68	259.0	0.29	7.71	0.81	6.39	1.09	148.8	16.0	2.6	44.1
28	129	1.87	17.0	305.3	533.9	1.03	105.1	0.26	—	0.46	6.09	0.11	73.8	3.3	0.4	26.1
36	258	3.14	31.6	352.3	971.7	2.02	49.0	0.37	1.13	0.37	3.77	0.19	64.8	15.6	0.6	22.8
85	24	0.81	18.8	317.5	661.3	0.38	40.0	0.09	—	0.07	1.11	0.17	8.8	12.5	0	5.1
60	74	1.94	23.8	314.5	360.5	3.45	—	0.10	0.00	0.27	3.95	0.13	20.8	0.7	1.6	17.4
180	102	4.68	63.2	413.1	330.5	1.72	94.8	0.29	2.06	0.11	4.75	0.21	75.3	12.9	1.2	49.3

DA+ Code	Food Description	Quantity	Measure	Wt (g)	H₂O (g)	Ener (kcal)	Prot (g)	Carb (g)	Fiber (g)	Fat (g)	Sat	Mono	Poly	Trans
	MIXED FOODS, SOUPS, SANDWICHES—Continued													
	Spaghetti and meat balls with tomato sauce,													
497	prepared	1	cup(s)	248	174.0	330	19.0	39.0	2.7	12.0	3.9	4.4	2.2	—
28585	Spicy thai noodles (pad thai)	8	ounce(s)	227	73.3	221	8.9	35.7	3.0	6.4	0.8	3.3	1.8	—
33073	Stir fried pork and vegetables with rice	1	cup(s)	235	173.6	348	15.4	33.5	1.9	16.3	5.6	6.9	2.6	0
28588	Stuffed shells	2 ½	item(s)	249	157.5	243	15.0	28.0	2.5	8.1	3.1	3.0	1.3	—
16821	Sushi with egg in seaweed	6	piece(s)	156	116.5	190	8.9	20.5	0.3	7.9	2.2	3.2	1.5	—
16819	Sushi with vegetables and fish	6	piece(s)	156	101.6	218	8.4	43.7	1.7	0.6	0.2	0.1	0.2	—
16820	Sushi with vegetables in seaweed	6	piece(s)	156	110.3	183	3.4	40.6	0.8	0.4	0.1	0.1	0.1	—
25266	Sweet and sour pork	¾	cup(s)	249	205.9	265	29.2	17.1	1.0	8.1	2.6	3.5	1.5	0
16824	Tabouli, tabbouleh or tabuli	1	cup(s)	160	123.7	198	2.6	15.9	3.7	14.9	2.0	10.9	1.6	—
25276	Three bean salad	½	cup(s)	99	82.2	95	1.9	9.7	2.6	5.9	0.8	1.4	3.5	0
160	Tuna salad	½	cup(s)	103	64.7	192	16.4	9.6	0	9.5	1.6	3.0	4.2	—
25241	Turkey and noodles	1	cup(s)	319	228.5	270	24.0	21.2	1.0	9.2	2.4	3.5	2.3	—
16794	Vegetable egg roll	2	item(s)	128	89.8	201	5.1	19.5	1.7	11.6	2.5	5.7	2.6	—
16818	Vegetable sushi, no fish	6	piece(s)	156	99.0	226	4.8	49.9	0.2	0.4	0.1	0.1	0.1	—
	Sandwiches													
1744	Bacon, lettuce and tomato with mayonnaise	1	item(s)	164	97.2	341	11.6	34.2	2.3	17.6	3.8	5.5	6.7	—
30287	Bologna and cheese with margarine	1	item(s)	111	45.6	345	13.4	29.3	1.2	19.3	8.1	7.0	2.4	—
30286	Bologna with margarine	1	item(s)	83	33.6	251	8.1	27.3	1.2	12.1	3.7	5.0	2.1	—
16546	Cheese	1	item(s)	83	31.0	261	9.1	27.6	1.2	12.7	5.4	4.2	2.1	—
8789	Cheeseburger, large, plain	1	item(s)	185	78.9	564	32.0	38.5	2.6	31.5	12.5	10.2	1.0	1.8
	Cheeseburger, large, with bacon, vegetables and													
8624	condiments	1	item(s)	195	91.4	550	30.8	36.8	2.5	30.9	11.9	10.6	1.3	1.5
	Club with bacon, chicken, tomato, lettuce and													
1745	mayonnaise	1	item(s)	246	137.5	546	31.0	48.9	3.0	24.5	5.3	7.5	9.4	—
1908	Cold cut submarine with cheese and vegetables	1	item(s)	228	131.8	456	21.8	51.0	2.0	18.6	6.8	8.2	2.3	—
30247	Corned beef	1	item(s)	130	74.9	265	18.2	25.3	1.6	9.6	3.6	3.5	1.0	—
25283	Egg salad	1	item(s)	126	72.1	278	10.7	28.0	1.4	13.5	2.9	4.2	5.0	—
16686	Fried egg	1	item(s)	96	49.7	226	10.0	26.2	1.2	8.6	2.3	3.2	1.9	—
16547	Grilled cheese	1	item(s)	83	27.5	291	9.2	27.9	1.2	15.8	6.0	5.7	3.0	—
16659	Gyro with onion and tomato	1	item(s)	105	68.3	163	12.0	20.0	1.1	3.5	1.3	1.3	0.5	—
1906	Ham and cheese	1	item(s)	146	74.2	352	20.7	33.3	2.0	15.5	6.4	6.7	1.4	—
31890	Ham with mayonnaise	1	item(s)	112	56.3	271	13.0	27.9	1.9	11.6	2.8	4.0	4.0	—
	Hamburger, double patty, large, with condiments													
756	and vegetables	1	item(s)	226	121.5	540	34.3	40.3	—	26.6	10.5	10.3	2.8	—
8793	Hamburger, large, plain	1	item(s)	137	57.7	426	22.6	31.7	1.5	22.9	8.4	9.9	2.1	—
8795	Hamburger, large, with vegetables and condiments	1	item(s)	218	121.4	512	25.8	40.0	3.1	27.4	10.4	11.4	2.2	—
25134	Hot chicken salad	1	item(s)	98	48.4	242	15.2	23.8	1.3	9.2	2.9	2.5	3.0	—
25133	Hot turkey salad	1	item(s)	98	50.1	224	15.6	23.8	1.3	6.9	2.3	1.6	2.5	—
1411	Hotdog with bun, plain	1	item(s)	98	52.9	242	10.4	18.0	1.6	14.5	5.1	6.9	1.7	—
30249	Pastrami	1	item(s)	134	71.2	328	13.4	27.8	1.6	17.7	6.1	8.3	1.2	—
16701	Peanut butter	1	item(s)	93	23.6	345	12.2	37.6	3.3	17.4	3.4	7.7	5.2	—
30306	Peanut butter and jelly	1	item(s)	93	24.2	330	10.3	41.9	2.9	14.7	2.9	6.5	4.4	—
	Roast beef submarine with mayonnaise and													
1909	vegetables	1	item(s)	216	127.4	410	28.6	44.3	—	13.0	7.1	1.8	2.6	—
1910	Roast beef, plain	1	item(s)	139	67.6	346	21.5	33.4	1.2	13.8	3.6	6.8	1.7	—
1907	Steak with mayonnaise and vegetables	1	item(s)	204	104.2	459	30.3	52.0	2.3	14.1	3.8	5.3	3.3	—
25288	Tuna salad	1	item(s)	179	102.2	415	24.5	28.4	1.6	22.4	3.5	6.2	11.4	—
	Turkey submarine with cheese, lettuce, tomato and													
30283	mayonnaise	1	item(s)	277	168.0	529	30.4	49.4	3.0	22.8	6.8	6.0	8.6	—
31891	Turkey with mayonnaise	1	item(s)	143	74.5	329	28.7	26.4	1.3	11.2	2.6	2.6	4.8	—
	Soups													
25296	Bean	1	cup(s)	301	253.1	191	13.8	29.0	6.5	2.3	0.7	0.8	0.5	0
711	Bean with pork, condensed, prepared with water	1	cup(s)	253	215.9	159	7.3	21.0	7.3	5.5	1.4	2.0	1.7	—
713	Beef noodle, condensed, prepared with water	1	cup(s)	244	224.9	83	4.7	8.7	0.7	3.0	1.1	1.2	0.5	—
825	Cheese, condensed, prepared with milk	1	cup(s)	251	206.9	231	9.5	16.2	1.0	14.6	9.1	4.1	0.5	—
826	Chicken broth, condensed, prepared with water	1	cup(s)	244	234.1	39	4.9	0.9	0	1.4	0.4	0.6	0.3	—
25297	Chicken noodle soup	1	cup(s)	286	258.4	117	10.8	10.9	0.9	2.9	0.8	1.1	0.7	—
827	Chicken noodle, condensed, prepared with water	1	cup(s)	241	226.1	60	3.1	7.1	0.5	2.3	0.6	1.0	0.6	0
724	Chicken noodle, dehydrated, prepared with water	1	cup(s)	252	237.3	58	2.1	9.2	0.3	1.4	0.3	0.5	0.4	—
823	Cream of asparagus, condensed, prepared with milk	1	cup(s)	248	213.3	161	6.3	16.4	0.7	8.2	3.3	2.1	2.2	—
824	Cream of celery, condensed, prepared with milk	1	cup(s)	248	214.4	164	5.7	14.5	0.7	9.7	3.9	2.5	2.7	—
708	Cream of chicken, condensed, prepared with milk	1	cup(s)	248	210.4	191	7.5	15.0	0.2	11.5	4.6	4.5	1.6	—
715	Cream of chicken, condensed, prepared with water	1	cup(s)	244	221.1	117	3.4	9.3	0.2	7.4	2.1	3.3	1.5	—
709	Cream of mushroom, condensed, prepared with milk	1	cup(s)	248	215.0	166	6.2	14.0	0	9.6	3.3	2.0	1.8	0.1
	Cream of mushroom, condensed, prepared with													
716	water	1	cup(s)	244	224.6	102	1.9	8.0	0	7.0	1.6	1.3	1.7	—
25298	Cream of vegetable	1	cup(s)	285	250.7	165	7.2	15.2	1.9	8.6	1.6	4.6	1.9	—
16689	Egg drop	1	cup(s)	244	228.9	73	7.5	1.1	0	3.8	1.1	1.5	0.6	—

Chol (mg)	Calc (mg)	Iron (mg)	Magn (mg)	Pota (mg)	Sodi (mg)	Zinc (mg)	Vit A (µg)	Thia (mg)	Vit E (mg α)	Ribo (mg)	Niac (mg)	Vit B$_6$ (mg)	Fola (µg)	Vit C (mg)	Vit B$_{12}$ (µg)	Sele (µg)
89	124	3.70	—	665.0	1009.0	—	81.5	0.25	—	0.30	4.00	—	—	22.0	—	22.0
37	31	1.56	49.4	181.3	591.6	1.05	—	0.18	0.35	0.13	1.82	0.17	44.6	22.6	0.1	3.2
46	38	2.71	33.0	396.9	569.5	2.08	—	0.51	0.38	0.20	5.07	0.30	103.0	18.8	0.4	22.8
30	188	2.26	49.4	403.0	471.5	1.41	—	0.26	0.00	0.26	3.83	0.24	80.0	17.9	0.2	28.9
214	45	1.84	18.7	135.7	463.3	0.98	106.1	0.13	0.67	0.28	1.35	0.13	62.4	1.9	0.7	20.3
11	23	2.15	25.0	202.8	340.1	0.78	45.2	0.26	0.24	0.07	2.76	0.14	76.4	3.6	0.3	13.9
0	20	1.54	18.7	96.7	152.9	0.68	25.0	0.19	0.12	0.03	1.86	0.13	73.3	2.3	0	9.8
74	40	1.76	35.6	619.9	621.8	2.53	—	0.81	0.20	0.37	6.69	0.66	14.7	11.9	0.7	49.6
0	30	1.21	35.2	249.6	796.8	0.48	54.4	0.07	2.43	0.04	1.11	0.11	30.4	26.1	0	0.5
0	26	0.96	15.5	144.8	224.2	0.30	—	0.02	0.88	0.04	0.26	0.04	32.2	10.0	0	2.7
13	17	1.02	19.5	182.5	412.1	0.57	24.6	0.03	—	0.07	6.86	0.08	8.2	2.3	1.2	42.2
77	69	2.56	33.2	400.8	577.1	2.51	—	0.23	0.28	0.30	6.41	0.29	61.4	1.4	1.1	33.4
60	29	1.65	17.9	193.3	549.1	0.48	25.6	0.15	1.28	0.20	1.59	0.09	46.1	5.5	0.2	11.3
0	23	2.38	21.8	157.6	369.7	0.82	48.4	0.28	0.15	0.05	2.44	0.12	85.8	3.7	0	8.1
21	79	2.36	27.9	351.0	944.6	1.08	44.3	0.32	1.16	0.24	4.36	0.21	73.8	9.7	0.2	27.1
40	258	2.38	24.4	215.3	941.3	1.88	102.1	0.31	0.55	0.35	2.97	0.14	59.9	0.2	0.8	20.4
17	100	2.24	16.6	138.6	579.3	1.02	44.8	0.29	0.49	0.21	2.92	0.12	58.1	0.2	0.5	16.0
22	233	2.04	19.9	127.0	733.7	1.24	97.1	0.25	0.47	0.30	2.25	0.05	58.1	0	0.3	13.1
104	309	4.47	44.4	401.5	986.1	5.75	0	0.35	—	0.77	8.26	0.49	109.2	0	2.8	38.9
98	267	4.03	44.9	464.1	1314.3	5.20	0	0.33	—	0.67	8.25	0.47	95.6	1.4	2.4	6.6
71	157	4.57	46.7	464.9	1087.3	1.82	41.8	0.54	1.52	0.40	12.82	0.61	118.1	6.4	0.4	42.3
36	189	2.50	68.4	394.4	1650.7	2.57	70.7	1.00	—	0.79	5.49	0.13	86.6	12.3	1.1	30.8
46	81	3.04	19.5	127.4	1206.4	2.26	2.6	0.23	0.20	0.24	3.42	0.11	59.8	0.3	0.9	31.2
219	85	2.25	18.8	159.0	423.1	0.87	—	0.27	0.12	0.43	2.06	0.15	73.6	0.9	0.6	29.9
206	104	2.79	17.3	117.1	438.7	0.92	89.3	0.26	0.66	0.40	2.26	0.10	79.7	0	0.6	24.2
22	235	2.05	19.9	128.7	763.6	1.26	129.5	0.19	0.72	0.28	2.05	0.05	38.2	0	0.2	13.2
28	47	1.77	22.1	218.4	235.2	2.33	9.5	0.23	0.26	0.20	3.12	0.13	44.1	3.2	0.9	18.3
58	130	3.24	16.1	290.5	770.9	1.37	96.4	0.30	0.29	0.48	2.68	0.20	75.9	2.8	0.5	23.1
34	91	2.47	23.5	210.6	1097.6	1.13	5.6	0.57	0.50	0.26	3.80	0.25	60.5	2.2	0.2	20.2
122	102	5.85	49.7	569.5	791.0	5.67	0	0.36	—	0.38	7.57	0.54	76.8	1.1	4.1	25.5
71	74	3.57	27.4	267.2	474.0	4.11	0	0.28	—	0.28	6.24	0.23	60.3	0	2.1	27.1
87	96	4.92	43.6	479.6	824.0	4.88	0	0.41	—	0.37	7.28	0.32	82.8	2.6	2.4	33.6
39	115	1.88	20.0	172.4	505.1	1.19	—	0.23	0.28	0.22	4.84	0.19	46.4	0.5	0.2	19.9
37	114	1.99	21.3	189.0	494.5	1.07	—	0.22	0.28	0.20	4.26	0.22	46.4	0.5	0.2	23.4
44	24	2.31	12.7	143.1	670.3	1.98	0	0.23	—	0.27	3.64	0.04	48.0	0.1	0.5	26.0
51	80	3.02	22.8	182.2	1364.1	2.70	2.7	0.28	0.26	0.26	4.97	0.14	60.3	0.3	1.0	14.3
0	110	2.92	66.0	226.0	580.3	1.32	0	0.31	2.39	0.23	6.72	0.18	92.1	0	0	13.2
0	94	2.50	56.7	198.1	492.9	1.12	—	0.26	2.01	0.20	5.66	0.15	78.1	0	0	11.2
73	41	2.80	67.0	330.5	844.6	4.38	30.2	0.41	—	0.41	5.96	0.32	71.3	5.6	1.8	25.7
51	54	4.22	30.6	315.5	792.3	3.39	11.1	0.37	—	0.30	5.86	0.26	57.0	2.1	1.2	29.2
73	92	5.16	49.0	524.3	797.6	4.52	0	0.40	—	0.36	7.30	0.36	89.8	5.5	1.6	42.0
59	78	2.97	35.9	316.3	724.7	1.02	—	0.27	0.34	0.26	12.07	0.46	60.8	1.8	2.4	76.9
64	307	4.59	49.9	534.6	1759.0	2.74	74.8	0.49	1.19	0.65	3.91	0.31	121.9	10.5	0.6	42.1
67	100	3.46	34.3	304.6	564.9	3.00	5.7	0.28	0.74	0.32	6.84	0.46	64.4	0	0.3	40
5	79	3.05	61.8	588.8	689.0	1.41	—	0.27	0.02	0.15	3.63	0.23	140.1	3.6	0.2	7.9
3	78	1.89	43.0	371.9	883.0	0.96	43.0	0.08	1.08	0.03	0.52	0.03	30.4	1.5	0	7.8
5	20	1.07	7.3	97.6	929.6	1.51	12.2	0.06	1.22	0.05	1.03	0.03	19.5	0.5	0.2	7.3
48	289	0.80	20.1	341.4	1019.1	0.67	358.9	0.06	—	0.33	0.50	0.07	10.0	1.3	0.4	7.0
0	10	0.51	2.4	209.8	775.9	0.24	0	0.01	0.04	0.07	3.34	0.02	4.9	0	0.2	0
24	25	1.38	16.4	340.1	774.5	0.77	—	0.15	0.02	0.16	5.57	0.13	37.2	1.8	0.3	10.2
12	14	1.59	9.6	53.0	638.7	0.38	26.5	0.13	0.07	0.10	1.30	0.04	19.3	0	0	11.6
10	5	0.50	7.6	32.8	577.1	0.20	2.5	0.20	0.12	0.07	1.08	0.02	17.6	0	0.1	9.6
22	174	0.86	19.8	359.6	1041.6	0.91	62.0	0.10	—	0.27	0.88	0.06	29.8	4.0	0.5	8.0
32	186	0.69	22.3	310.0	1009.4	0.19	114.1	0.07	—	0.24	0.43	0.06	7.4	1.5	0.5	4.7
27	181	0.67	17.4	272.8	1046.6	0.67	178.6	0.07	—	0.25	0.92	0.06	7.4	1.2	0.5	8.0
10	34	0.61	2.4	87.8	985.8	0.63	163.5	0.02	—	0.06	0.82	0.01	2.4	0.2	0.1	7.0
10	164	1.36	19.8	267.8	823.4	0.79	81.8	0.10	1.01	0.29	0.62	0.05	7.4	0.2	0.6	6.0
0	17	1.31	4.9	73.2	775.9	0.24	9.8	0.05	0.97	0.05	0.50	0.00	2.4	0	0	2.9
1	80	1.20	17.5	340.7	787.9	0.56	—	0.12	1.05	0.18	3.32	0.12	39.5	10.7	0.3	4.3
102	22	0.75	4.9	219.6	729.6	0.48	41.5	0.02	0.29	0.19	3.02	0.05	14.6	0	0.5	7.6

DA+ Code	Food Description	Quantity	Measure	Wt (g)	H₂O (g)	Ener (kcal)	Prot (g)	Carb (g)	Fiber (g)	Fat (g)	Sat	Mono	Poly	Trans
											Fat Breakdown (g)			

DA+ Code	Food Description	Quantity	Measure	Wt (g)	H₂O (g)	Ener (kcal)	Prot (g)	Carb (g)	Fiber (g)	Fat (g)	Sat	Mono	Poly	Trans
	MIXED FOODS, SOUPS, SANDWICHES —Continued													
25138	Golden squash	1	cup(s)	258	223.9	145	7.6	20.4	0.4	4.1	0.8	2.2	0.9	—
16663	Hot and sour	1	cup(s)	244	209.7	161	15.0	5.4	0.5	7.9	2.7	3.4	1.1	—
28054	Lentil chowder	1	cup(s)	244	202.8	153	11.4	27.7	12.6	0.5	0.1	0.1	0.2	0
28560	Macaroni and bean	1	cup(s)	246	138.8	146	5.8	22.9	5.1	3.7	0.5	2.2	0.6	0
714	Manhattan clam chowder, condensed, prepared with water	1	cup(s)	244	225.1	73	2.1	11.6	1.5	2.1	0.4	0.4	1.2	—
28561	Minestrone	1	cup(s)	241	185.4	103	4.5	16.8	4.8	2.3	0.3	1.4	0.4	0
717	Minestrone, condensed, prepared with water	1	cup(s)	241	220.1	82	4.3	11.2	1.0	2.5	0.6	0.7	1.1	—
28038	Mushroom & wild rice	1	cup(s)	244	199.7	86	4.7	13.2	1.7	0.3	0	0	0.2	0
828	New England clam chowder, condensed, prepared with milk	1	cup(s)	248	212.2	151	8.0	18.4	0.7	5.0	2.1	0.7	0.6	0
28036	New England style clam chowder	1	cup(s)	244	227.5	61	3.8	8.8	1.8	0.2	0.1	0	0	0
28566	Old country pasta	1	cup(s)	252	183.3	146	6.5	18.3	3.6	4.5	2.0	2.4	0.9	0
725	Onion, dehydrated, prepared with water	1	cup(s)	246	235.7	30	0.8	6.8	0.7	0	0	0	0	—
16667	Shrimp gumbo	1	cup(s)	244	207.2	166	9.5	18.2	2.4	6.7	1.3	2.9	2.0	—
28037	Southwestern corn chowder	1	cup(s)	244	217.8	98	4.9	17.0	2.4	0.5	0.1	0.1	0.2	0
30282	Soybean (miso)	1	cup(s)	240	218.6	84	6.0	8.0	1.9	3.4	0.6	1.1	1.4	—
25140	Split pea	1	cup(s)	165	119.7	72	4.5	16.0	1.6	0.3	0.1	0	0.2	0
718	Split pea with ham, condensed, prepared with water	1	cup(s)	253	206.9	190	10.3	28.0	2.3	4.4	1.8	1.8	0.6	—
726	Tomato vegetable, dehydrated, prepared with water	1	cup(s)	253	238.4	56	2.0	10.2	0.8	0.9	0.4	0.3	0.1	0
710	Tomato, condensed, prepared with milk	1	cup(s)	248	213.4	136	6.2	22.0	1.5	3.2	1.8	0.9	0.3	—
719	Tomato, condensed, prepared with water	1	cup(s)	244	223.0	73	1.9	16.0	1.5	0.7	0.2	0.2	0.2	—
28595	Turkey noodle	1	cup(s)	244	216.9	114	8.1	15.1	1.9	2.4	0.3	1.1	0.7	—
28051	Turkey vegetable	1	cup(s)	244	220.8	96	12.2	6.6	2.0	1.1	0.3	0.2	0.3	0
25141	Vegetable	1	cup(s)	252	228.1	82	5.2	16.5	4.5	0.3	0	0	0.1	0
720	Vegetable beef, condensed, prepared with water	1	cup(s)	244	224.0	76	5.4	9.9	2.0	1.9	0.8	0.8	0.1	—
28598	Vegetable gumbo	1	cup(s)	252	184.5	170	4.4	28.9	3.6	4.7	0.7	3.2	0.5	0
721	Vegetarian vegetable, condensed, prepared with water	1	cup(s)	241	222.7	67	2.1	11.8	0.7	1.9	0.3	0.8	0.7	—
	FAST FOOD													
	Arby's													
36094	Au jus sauce	1	serving(s)	85	—	43	1.0	7.0	0	1.3	0.4	—	—	0.4
751	Beef 'n cheddar sandwich	1	item(s)	195	—	445	22.0	44.0	2.0	21.0	6.0	—	—	1.0
9279	Cheddar curly fries	1	serving(s)	198	—	631	8.0	73.0	7.0	37.4	6.8	—	—	5.7
34770	Chicken breast fillet sandwich, grilled	1	item(s)	233	—	414	32.0	36.0	3.0	17.0	3.0	—	—	0
36131	Chocolate shake, regular	1	serving(s)	397	—	507	13.0	83.0	0	13.0	8.0	—	—	0
36045	Curly fries, large size	1	serving(s)	198	—	631	8.0	73.0	7.0	37.0	7.0	—	—	6.0
36044	Curly fries, medium size	1	serving(s)	128	—	406	5.0	47.0	5.0	24.0	4.0	—	—	4.0
752	Ham 'n cheese sandwich	1	item(s)	167	—	304	23.0	35.0	1.0	7.0	2.0	—	—	0
36048	Homestyle fries, large size	1	serving(s)	213	—	566	6.0	82.0	6.0	37.0	7.0	—	—	5.0
36047	Homestyle fries, medium size	1	serving(s)	142	—	377	4.0	55.0	4.0	25.0	4.0	—	—	4.0
33465	Homestyle fries, small size	1	serving(s)	113	—	302	3.0	44.0	3.0	20.0	4.0	—	—	3.0
9249	Junior roast beef sandwich	1	item(s)	125	—	272	16.0	34.0	2.0	10.0	4.0	—	—	0
9251	Large roast beef sandwich	1	item(s)	281	—	547	42.0	41.0	3.0	28.0	12.0	—	—	2.0
39640	Market Fresh chicken salad with pecans sandwich	1	item(s)	322	—	769	30.0	79.0	9.0	39.0	10.0	—	—	0
39641	Market Fresh Martha's Vineyard salad, without dressing	1	serving(s)	330	—	277	26.0	24.0	5.0	8.0	4.0	—	—	0
34769	Market Fresh roast turkey & Swiss sandwich	1	serving(s)	359	—	725	45.0	75.0	5.0	30.0	8.0	—	—	1.0
9267	Market Fresh roast turkey ranch & bacon sandwich	1	serving(s)	382	—	834	49.0	75.0	5.0	38.0	11.0	—	—	1.0
39642	Market Fresh Santa Fe salad, without dressing	1	serving(s)	372	—	499	30.0	42.0	7.0	23.0	8.0	—	—	2.0
39650	Market Fresh Southwest chicken wrap	1	serving(s)	251	—	567	36.0	42.0	4.0	29.0	9.0	—	—	1.0
37021	Market Fresh Ultimate BLT sandwich	1	item(s)	294	—	779	23.0	75.0	6.0	45.0	11.0	—	—	1.0
750	Roast beef sandwich, regular	1	item(s)	154	—	320	21.0	34.0	2.0	14.0	5.0	—	—	1.0
36132	Strawberry shake, regular	1	serving(s)	397	—	498	13.0	81.0	0	13.0	8.0	—	—	0
2009	Super roast beef sandwich	1	item(s)	198	—	398	21.0	40.0	2.0	19.0	6.0	—	—	1.0
36130	Vanilla shake, regular	1	serving(s)	369	—	437	13.0	66.0	0	13.0	8.0	—	—	0
	Auntie Anne's													
35371	Cheese dipping sauce	1	serving(s)	35	—	100	3.0	4.0	0	8.0	4.0	—	—	0
35353	Cinnamon sugar soft pretzel	1	item(s)	120	—	350	9.0	74.0	2.0	2.0	0	—	—	0
35354	Cinnamon sugar soft pretzel with butter	1	item(s)	120	—	450	8.0	83.0	3.0	9.0	5.0	—	—	0
35372	Marinara dipping sauce	1	serving(s)	35	—	10	0	4.0	0	0	0	0	0	0
35357	Original soft pretzel	1	serving(s)	120	—	340	10.0	72.0	3.0	1.0	0	—	—	0
35358	Original soft pretzel with butter	1	item(s)	120	—	370	10.0	72.0	3.0	4.0	2.0	—	—	0
35359	Parmesan herb soft pretzel	1	item(s)	120	—	390	11.0	74.0	4.0	5.0	2.5	—	—	0
35360	Parmesan herb soft pretzel with butter	1	item(s)	120	—	440	10.0	72.0	9.0	13.0	7.0	—	—	0
35361	Sesame soft pretzel	1	item(s)	120	—	350	11.0	63.0	3.0	6.0	1.0	—	—	0
35362	Sesame soft pretzel with butter	1	item(s)	120	—	410	12.0	64.0	7.0	12.0	4.0	—	—	0

Chol (mg)	Calc (mg)	Iron (mg)	Magn (mg)	Pota (mg)	Sodi (mg)	Zinc (mg)	Vit A (µg)	Thia (mg)	Vit E (mg α)	Ribo (mg)	Niac (mg)	Vit B$_6$ (mg)	Fola (µg)	Vit C (mg)	Vit B$_{12}$ (µg)	Sele (µg)
4	262	0.78	42.4	542.2	515.6	0.88	—	0.16	0.52	0.30	1.14	0.16	32.7	12.5	0.7	6.0
34	29	1.24	19.5	373.3	1561.6	1.43	—	0.26	0.12	0.24	4.96	0.20	14.6	0.5	0.4	19.3
0	48	4.38	59.3	626.2	26.7	1.57	—	0.24	0.06	0.12	1.87	0.32	176.3	16.1	0	3.5
0	59	1.90	32.4	275.9	531.0	0.51	—	0.16	0.37	0.12	1.44	0.10	58.2	9.1	0	8.8
2	27	1.56	9.8	180.6	551.4	0.87	48.8	0.02	1.22	0.03	0.77	0.09	9.8	3.9	3.9	9.0
0	62	1.70	29.9	287.5	442.7	0.42	—	0.09	0.23	0.09	0.70	0.07	47.3	13.3	0	3.5
2	34	0.91	7.2	313.3	911.0	0.74	118.1	0.05	—	0.04	0.94	0.09	36.2	1.2	0	8.0
0	30	1.38	27.3	376.9	283.8	1.00	—	0.06	0.07	0.23	3.21	0.14	18.1	3.7	0.1	4.7
17	169	3.00	29.8	456.3	887.8	0.99	91.8	0.20	0.54	0.43	1.96	0.17	22.3	5.2	11.9	10.9
3	89	1.30	29.2	503.7	256.8	0.52	—	0.06	0.02	0.10	1.30	0.15	24.2	11.8	3.0	3.8
5	57	2.43	51.9	500.4	355.7	0.78	—	0.21	0.01	0.14	2.64	0.20	80.0	22.0	0.1	10.3
0	22	0.12	9.8	76.3	851.2	0.12	0	.03	0.02	0.03	0.15	0.06	0	0.2	0	0.5
51	105	2.85	48.8	461.2	441.6	0.90	80.5	0.18	1.90	0.12	2.52	0.19	85.4	17.6	0.3	14.9
1	83	1.03	26.3	434.4	217.4	0.57	—	0.08	0.09	0.13	1.81	0.21	32.1	39.6	0.2	1.7
0	65	1.87	36.0	362.4	988.8	0.86	232.8	0.06	0.96	0.16	2.61	0.15	57.6	4.6	0.2	1.0
0	28	1.26	29.8	328.1	602.4	0.52	—	0.10	0.00	0.07	1.50	0.16	49.5	8.1	0	0.7
8	23	2.27	48.1	399.7	1006.9	1.31	22.8	0.14	—	0.07	1.47	0.06	2.5	1.5	0.3	8.0
0	20	0.60	10.1	169.5	334.0	0.20	10.1	0.06	0.43	0.09	1.26	0.06	12.7	3.0	0.1	2.0
10	166	1.36	29.8	466.2	711.8	0.84	94.2	0.09	0.44	0.31	1.35	0.15	5.0	15.6	0.6	9.2
0	20	1.31	17.1	273.3	663.7	0.29	24.4	0.04	0.41	0.07	1.23	0.10	0	15.4	0	6.1
26	28	1.40	24.1	223.8	395.9	0.75	—	0.21	0.01	0.12	2.85	0.15	45.0	7.1	0.1	13.7
21	38	1.48	25.0	423.4	348.7	0.99	—	0.09	0.01	0.09	3.64	0.27	23.9	10.6	0.2	10.3
0	40	2.08	39.1	681.5	670.3	0.67	—	0.16	0.00	0.09	2.70	0.26	36.3	22.4	0	2.2
5	20	1.09	7.3	168.4	773.5	1.51	190.3	0.03	0.58	0.04	1.00	0.07	9.8	2.4	0.3	2.7
0	56	1.85	38.9	360.2	518.4	0.64	—	0.18	0.64	0.08	1.77	0.17	57.6	21.9	0	4.1
0	24	1.06	7.2	207.3	814.6	0.45	171.1	0.05	1.39	0.04	0.90	0.05	9.6	1.4	0	4.3
0	0	—	—	—	1510.0	—	—	—	—	—	—	—	—	—	—	—
51	80	3.96	—	—	1274.0	—	—	—	—	—	—	—	—	1.8	—	—
0	80	3.24	—	—	1476.0	—	—	—	—	—	—	—	—	9.6	—	—
9	90	3.06	—	—	913.0	—	—	—	—	—	—	—	—	10.8	—	—
34	510	0.54	—	—	357.0	—	—	—	—	—	—	—	—	5.4	—	—
0	80	3.24	—	—	1476.0	—	—	—	—	—	—	—	—	9.6	—	—
0	50	1.98	—	—	949.0	—	—	—	—	—	—	—	—	6.0	—	—
35	160	2.70	—	—	1420.0	—	—	—	—	—	—	—	—	1.2	—	—
0	50	1.62	—	—	1029.0	—	—	—	—	—	—	—	—	12.6	—	—
0	30	1.08	—	—	686.0	—	—	—	—	—	—	—	—	8.4	—	—
0	30	0.90	—	—	549.0	—	—	—	—	—	—	—	—	6.6	—	—
29	60	3.06	—	—	740.0	—	0	—	—	—	—	—	—	0	—	—
102	70	6.30	—	—	1869.0	—	0	—	—	—	—	—	—	0.6	—	—
74	180	4.32	—	—	1240.0	—	—	—	—	—	—	—	—	30.0	—	—
72	200	1.62	—	—	454.0	—	—	—	—	—	—	—	—	33.6	—	—
91	360	5.22	—	—	1788.0	—	—	—	—	—	—	—	—	10.2	—	—
109	330	5.40	—	—	2258.0	—	—	—	—	—	—	—	—	11.4	—	—
59	420	3.60	—	—	1231.0	—	—	—	—	—	—	—	—	36.6	—	—
88	240	4.50	—	—	1451.0	—	—	—	—	—	—	—	—	7.8	—	—
51	170	4.68	—	—	1571.0	—	—	—	—	—	—	—	—	16.8	—	—
44	60	3.60	—	—	953.0	—	0	—	—	—	—	—	—	0	—	—
34	510	0.72	—	—	363.0	—	—	—	—	—	—	—	—	6.6	—	—
44	70	3.78	—	—	1060.0	—	—	—	—	—	—	—	—	6.0	—	—
34	510	0.36	—	—	350.0	—	—	—	—	—	—	—	—	5.4	—	—
10	100	0.00	—	—	510.0	—	—	—	—	—	—	—	—	0	—	—
0	20	1.98	—	—	410.0	—	0	—	—	—	—	—	—	0	—	—
25	30	2.34	—	—	430.0	—	—	—	—	—	—	—	—	0	—	—
0	0	0.00	—	—	180.0	—	0	—	—	—	—	—	—	0	—	—
0	30	2.34	—	—	900.0	—	0	—	—	—	—	—	—	0	—	—
10	30	2.16	—	—	930.0	—	—	—	—	—	—	—	—	0	—	—
10	80	1.80	—	—	780.0	—	—	—	—	—	—	—	—	1.2	—	—
30	60	1.80	—	—	660.0	—	—	—	—	—	—	—	—	1.2	—	—
0	20	2.88	—	—	840.0	—	0	—	—	—	—	—	—	0	—	—
15	20	2.70	—	—	860.0	—	—	—	—	—	—	—	—	0	—	—

DA+ Code	Food Description	Quantity	Measure	Wt (g)	H₂O (g)	Ener (kcal)	Prot (g)	Carb (g)	Fiber (g)	Fat (g)	Sat	Mono	Poly	Trans
	FAST FOOD—Continued													
35364	Sour cream & onion soft pretzel	1	item(s)	120	—	310	9.0	66.0	2.0	1.0	0	—	—	0
35366	Sour cream & onion soft pretzel with butter	1	item(s)	120	—	340	9.0	66.0	2.0	5.0	3.0	—	—	0
35373	Sweet mustard dipping sauce	1	serving(s)	35	—	60	0.5	8.0	0	1.5	1.0	—	—	0
35367	Whole wheat soft pretzel	1	item(s)	120	—	350	11.0	72.0	7.0	1.5	0	—	—	0
35368	Whole wheat soft pretzel with butter	1	item(s)	120	—	370	11.0	72.0	7.0	4.5	1.5	—	—	0
	Boston Market													
34978	Butternut squash	¾	cup(s)	143	—	140	2.0	25.0	2.0	4.5	3.0	—	—	0
35006	Caesar side salad	1	serving(s)	71	—	40	3.0	3.0	1.0	20.0	2.0	—	—	1.5
35013	Chicken Carver sandwich with cheese and sauce	1	item(s)	321	—	700	44.0	68.0	3.0	29.0	7.0	—	—	0
34979	Chicken gravy	4	ounce(s)	113	—	15	1.0	4.0	0	0.5	0	—	—	0
35053	Chicken noodle soup	¾	cup(s)	283	—	180	13.0	16.0	1.0	7.0	2.0	—	—	0
34973	Chicken pot pie	1	item(s)	425	—	800	29.0	59.0	4.0	49.0	18.0	—	—	7.0
35054	Chicken tortilla soup with toppings	¾	cup(s)	227	—	340	12.0	24.0	1.0	22.0	7.0	—	—	0
35007	Cole slaw	¾	cup(s)	125	—	170	2.0	21.0	2.0	9.0	2.0	—	—	0
35057	Cornbread	1	item(s)	45	—	130	1.0	21.0	0	3.5	1.0	—	—	1.0
34980	Creamed spinach	¾	cup(s)	191	—	280	9.0	12.0	4.0	23.0	15.0	—	—	0
34998	Fresh vegetable stuffing	1	cup(s)	136	—	190	3.0	25.0	2.0	8.0	1.0	—	—	0
34991	Garlic dill new potatoes	¾	cup(s)	156	—	140	3.0	24.0	3.0	3.0	1.0	—	—	0
34983	Green bean casserole	¾	cup(s)	170	—	60	2.0	9.0	2.0	2.0	1.0	—	—	0
34982	Green beans	¾	cup(s)	91	—	60	2.0	7.0	3.0	3.5	1.5	—	—	0
34984	Homestyle mashed potatoes	¾	cup(s)	221	—	210	4.0	29.0	3.0	9.0	6.0	—	—	0
34985	Homestyle mashed potatoes and gravy	1	cup(s)	334	—	225	5.0	33.0	3.0	9.5	6.0	—	—	0
34988	Hot cinnamon apples	¾	cup(s)	145	—	210	0	47.0	3.0	3.0	0	—	—	0
34989	Macaroni and cheese	¾	cup(s)	221	—	330	14.0	39.0	1.0	12.0	7.0	—	—	0.5
51193	Market chopped salad with dressing	1	item(s)	563	—	580	11.0	31.0	9.0	48.0	9.0	—	—	1.0
34970	Meatloaf	1	serving(s)	218	—	480	29.0	23.0	2.0	33.0	13.0	—	—	0
39383	Nestle Toll House chocolate chip cookie	1	item(s)	78	—	370	4.0	49.0	2.0	19.0	9.0	—	—	0
34965	Quarter chicken, dark meat, no skin	1	item(s)	134	—	260	30.0	2.0	0	13.0	4.0	—	—	0
34966	Quarter chicken, dark meat, with skin	1	item(s)	149	—	280	31.0	3.0	0	15.0	4.5	—	—	0
34963	Quarter chicken, white meat, no skin or wing	1	item(s)	173	—	250	41.0	4.0	0	8.0	2.5	—	—	0
34964	Quarter chicken, white meat, with skin and wing	1	item(s)	110	—	330	50.0	3.0	0	12.0	4.0	—	—	0
34968	Roasted turkey breast	5	ounce(s)	142	—	180	38.0	0	0	3.0	1.0	—	—	0
35011	Seasonal fresh fruit salad	1	serving(s)	142	—	60	1.0	15.0	1.0	0	0	0	0	0
51192	Spinach with garlic butter sauce	1	serving(s)	170	—	130	5.0	9.0	5.0	9.0	6.0	—	—	0
34969	Spiral sliced holiday ham	8	ounce(s)	227	—	450	40.0	13.0	0	26.0	10.0	—	—	0
35003	Steamed vegetables	1	cup(s)	136	—	50	2.0	8.0	3.0	2.0	0	—	—	0
35005	Sweet corn	¾	cup(s)	176	—	170	6.0	37.0	2.0	4.0	1.0	—	—	0
35004	Sweet potato casserole	¾	cup(s)	198	—	460	4.0	77.0	3.0	17.0	6.0	—	—	0
	Burger King													
29731	Biscuit with sausage, egg & cheese	1	item(s)	191	—	610	20.0	33.0	1.0	45.0	15.0	—	—	1.0
14249	Cheeseburger	1	item(s)	133	—	330	17.0	31.0	1.0	16.0	7.0	—	—	0.5
14251	Chicken sandwich	1	item(s)	219	—	660	24.0	52.0	4.0	40.0	8.0	—	—	2.5
3808	Chicken Tenders, 8 pieces	1	serving(s)	123	—	340	19.0	21.0	0.5	20.0	5.0	—	—	3.0
14259	Chocolate shake, small	1	item(s)	315	—	470	8.0	75.0	1.0	14.0	9.0	—	—	0
29732	Croissanwich with sausage & cheese	1	item(s)	106	37.2	370	14.0	23.0	0.5	25.0	9.0	12.7	3.3	2.0
14261	Croissanwich with sausage, egg & cheese	1	item(s)	159	71.4	470	19.0	26.0	0.5	32.0	11.0	15.8	6.1	2.5
3809	Double cheeseburger	1	item(s)	189	—	500	30.0	31.0	1.0	29.0	14.0	—	—	1.5
14244	Double Whopper sandwich	1	item(s)	373	—	900	47.0	51.0	3.0	57.0	19.0	—	—	2.0
14245	Double Whopper with cheese sandwich	1	item(s)	398	—	990	52.0	52.0	3.0	64.0	24.0	—	—	2.5
14250	Fish Filet sandwich	1	item(s)	250	—	630	24.0	67.0	4.0	30.0	6.0	—	—	2.5
14255	French fries, medium, salted	1	serving(s)	116	—	360	4.0	41.0	4.0	20.0	4.5	—	—	4.5
14262	French toast sticks (5)	1	serving(s)	112	37.6	390	6.0	46.0	2.0	20.0	4.5	10.6	2.9	4.5
14248	Hamburger	1	item(s)	121	—	290	15.0	30.0	1.0	12.0	4.5	—	—	0
14263	Hash brown rounds, small	1	serving(s)	75	27.1	230	2.0	23.0	2.0	15.0	4.0	—	—	5.0
14256	Onion rings, medium	1	serving(s)	91	—	320	4.0	40.0	3.0	16.0	4.0	—	—	3.5
39000	Tendercrisp chicken sandwich	1	item(s)	286	—	780	25.0	73.0	4.0	43.0	8.0	—	—	4.0
37514	TenderGrill chicken sandwich	1	item(s)	258	—	450	37.0	53.0	4.0	10.0	2.0	—	—	0
14258	Vanilla shake, small	1	item(s)	296	—	400	8.0	57.0	0	15.0	9.0	—	—	0
1736	Whopper sandwich	1	item(s)	290	—	670	28.0	51.0	3.0	39.0	11.0	—	—	1.5
14243	Whopper with cheese sandwich	1	item(s)	315	—	760	33.0	52.0	3.0	47.0	16.0	—	—	1.5
	Carl's Jr													
33962	Carl's bacon Swiss crispy chicken sandwich	1	item(s)	268	—	750	31.0	91.0	—	28.0	28.0	—	—	—
10801	Carl's Catch Fish sandwich	1	item(s)	215	—	560	19.0	58.0	2.0	27.0	7.0	—	1.9	—
10862	Carl's Famous Star hamburger	1	item(s)	254	—	590	24.0	50.0	3.0	32.0	9.0	—	—	—
10785	Charbroiled chicken club sandwich	1	item(s)	270	—	550	42.0	43.0	4.0	23.0	7.0	—	2.9	—
10866	Charbroiled chicken salad	1	item(s)	437	—	330	34.0	17.0	5.0	7.0	4.0	—	1.0	—
10855	Charbroiled Santa Fe Chicken sandwich	1	item(s)	266	—	610	38.0	43.0	4.0	32.0	8.0	—	—	—
10790	Chicken stars (6 pieces)	6	item(s)	85	—	260	13.0	14.0	1.0	16.0	4.0	—	1.6	—
34864	Chocolate shake, small	1	serving(s)	595	—	540	15.0	98.0	0	11.0	7.0	—	—	—

APPENDIX E

Chol (mg)	Calc (mg)	Iron (mg)	Magn (mg)	Pota (mg)	Sodi (mg)	Zinc (mg)	Vit A (µg)	Thia (mg)	Vit E (mg α)	Ribo (mg)	Niac (mg)	Vit B6 (mg)	Fola (µg)	Vit C (mg)	Vit B12 (µg)	Sele (µg)
0	30	1.98	—	—	920.0	—	—	—	—	—	—	—	—	0	—	—
10	40	2.16	—	—	930.0	—	—	—	—	—	—	—	—	0	—	—
40	0	0.00	—	—	120.0	—	0	—	—	—	—	—	—	0	—	—
0	30	1.98	—	—	1100.0	—	0	—	—	—	—	—	—	0	—	—
10	30	2.34	—	—	1120.0	—	—	—	—	—	—	—	—	0	—	—
10	59	0.80	—	—	35.0	—	—	—	—	—	—	—	—	22.2	—	—
0	60	0.43	—	—	75.0	—	—	—	—	—	—	—	—	5.4	—	—
90	211	2.85	—	—	1560.0	—	—	—	—	—	—	—	—	15.8	—	—
0	0	0.00	—	—	570.0	—	0	—	—	—	—	—	—	0	—	—
55	0	1.07	—	—	220.0	—	—	—	—	—	—	—	—	1.8	—	—
115	40	4.50	—	—	800.0	—	—	—	—	—	—	—	—	1.2	—	—
45	123	1.32	—	—	1310.0	—	—	—	—	—	—	—	—	18.4	—	—
10	41	0.48	—	—	270.0	—	—	—	—	—	—	—	—	24.5	—	—
5	0	0.71	—	—	220.0	—	0	—	—	—	—	—	—	0	—	—
70	264	2.84	—	—	580.0	—	—	—	—	—	—	—	—	9.5	—	—
0	41	1.48	—	—	580.0	—	—	—	—	—	—	—	—	2.5	—	—
0	0	0.85	—	—	120.0	—	0	—	—	—	—	—	—	14.3	—	—
5	20	0.72	—	—	620.0	—	—	—	—	—	—	—	—	2.4	—	—
0	43	0.38	—	—	180.0	—	—	—	—	—	—	—	—	5.1	—	—
25	51	0.46	—	—	660.0	—	—	—	—	—	—	—	—	19.2	—	—
25	100	0.59	—	—	1230.0	—	—	—	—	—	—	—	—	24.9	—	—
0	16	0.28	—	—	15.0	—	—	—	—	—	—	—	—	0	—	—
30	345	1.65	—	—	1290.0	—	—	—	—	—	—	—	—	0	—	—
10	—	—	—	—	2010.0	—	—	—	—	—	—	—	—	—	—	—
125	140	3.77	—	—	970.0	—	—	—	—	—	—	—	—	1.8	—	—
20	0	1.32	—	—	340.0	—	—	—	—	—	—	—	—	0	—	—
155	0	1.52	—	—	260.0	—	0	—	—	—	—	—	—	0	—	—
155	0	2.14	—	—	660.0	—	0	—	—	—	—	—	—	0	—	—
125	0	0.89	—	—	480.0	—	0	—	—	—	—	—	—	0	—	—
165	0	0.78	—	—	960.0	—	0	—	—	—	—	—	—	0	—	—
70	20	1.80	—	—	620.0	—	0	—	—	—	—	—	—	0	—	—
0	16	0.29	—	—	20.0	—	—	—	—	—	—	—	—	29.5	—	—
20	—	—	—	—	200.0	—	—	—	—	—	—	—	—	—	—	—
140	0	1.73	—	—	2230.0	—	0	—	—	—	—	—	—	0	—	—
0	53	0.46	—	—	45.0	—	—	—	—	—	—	—	—	24.0	—	—
0	0	0.43	—	—	95.0	—	—	—	—	—	—	—	—	5.8	—	—
20	44	1.18	—	—	210.0	—	—	—	—	—	—	—	—	9.8	—	—
210	250	2.70	—	—	1620.0	89.9	—	—	—	—	—	—	—	0	—	—
55	150	2.70	—	—	780.0	—	—	0.24	—	0.31	4.17	—	—	1.2	—	—
70	64	2.89	—	—	1440.0	—	—	0.50	—	0.32	10.29	—	—	0	—	—
55	20	0.72	—	—	960.0	—	—	0.14	—	0.11	10.93	—	—	0	—	—
55	333	0.79	—	—	350.0	—	—	0.11	—	0.61	0.26	—	—	2.7	0	—
50	99	1.78	20.1	217.3	810.0	1.51	—	0.34	1.03	0.33	4.33	—	—	0	0.6	22.2
180	146	2.63	28.6	313.2	1060.0	2.08	—	0.38	1.66	0.51	4.72	0.28	—	0	1.1	38.0
105	250	4.50	—	—	1030.0	—	—	0.26	—	0.44	6.37	—	—	1.2	—	—
175	150	8.07	—	—	1090.0	—	—	0.39	—	0.59	11.05	—	—	9.0	—	—
195	299	8.08	—	—	1520.0	—	—	0.39	—	0.66	11.03	—	—	9.0	—	—
60	101	3.62	—	—	1380.0	—	—	—	—	—	—	—	—	3.6	—	—
0	20	0.71	—	—	590.0	—	0	0.15	—	0.48	2.30	—	—	8.9	—	—
0	60	1.80	21.3	124.3	440.0	0.57	—	0.31	0.98	0.19	2.88	0.05	—	0	0	13.7
40	80	2.70	—	—	560.0	—	—	0.25	—	0.28	4.25	—	—	1.2	—	—
0	0	0.36	—	—	450.0	—	0	0.11	0.83	0.06	1.35	0.17	—	1.2	—	—
0	100	0.00	—	—	460.0	—	0	0.14	—	0.09	2.32	—	—	0	—	—
75	79	4.43	—	—	1730.0	—	—	—	—	—	—	—	—	8.9	—	—
75	57	6.82	—	—	1210.0	—	—	—	—	—	—	—	—	5.7	—	—
60	348	0.00	—	—	240.0	—	—	0.11	—	0.63	0.21	—	—	2.4	0	—
51	100	5.38	—	—	1020.0	—	—	0.38	—	0.43	7.30	—	—	9.0	—	—
115	249	5.38	—	—	1450.0	—	—	0.38	—	0.51	7.28	—	—	9.0	—	—
80	200	5.40	—	—	1900.0	—	—	—	—	—	—	—	—	2.4	—	—
80	150	2.70	—	—	990.0	—	60.0	—	—	—	—	—	—	2.4	—	—
70	100	4.50	—	—	910.0	—	—	—	—	—	—	—	—	6.0	—	—
95	200	3.60	—	—	1330.0	—	—	—	—	—	—	—	—	9.0	—	—
75	200	1.80	—	—	880.0	—	—	—	—	—	—	—	—	30.0	—	—
100	200	3.60	—	—	1440.0	—	—	—	—	—	—	—	—	9.0	—	—
35	19	1.02	—	—	470.0	—	0	—	—	—	—	—	—	0	—	—
45	600	1.08	—	—	360.0	—	0	—	—	—	—	—	—	0	—	—

DA+ Code	Food Description	Quantity	Measure	Wt (g)	H₂O (g)	Ener (kcal)	Prot (g)	Carb (g)	Fiber (g)	Fat (g)	Fat Breakdown (g)			
											Sat	Mono	Poly	Trans
	FAST FOOD—Continued													
10797	Crisscut fries	1	serving(s)	139	—	410	5.0	43.0	4.0	24.0	5.0	—	—	—
10799	Double Western Bacon cheeseburger	1	item(s)	308	—	920	51.0	65.0	2.0	50	21.0	—	6.6	—
14238	French fries, small	1	serving(s)	92	—	290	5.0	37.0	3.0	14.0	3.0	—	—	—
10798	French toast dips without syrup, 5 pieces	1	serving(s)	155	—	370	8.0	49.0	0	17.0	5.0	—	1.4	—
10802	Onion rings	1	serving(s)	128	—	440	7.0	53.0	3.0	22.0	5.0	—	0.8	—
34858	Spicy chicken sandwich	1	item(s)	198	—	480	14.0	48.0	2.0	26.0	5.0	—	—	—
34867	Strawberry shake, small	1	serving(s)	595	—	520	14.0	93.0	0	11.0	7.0	—	—	—
10865	Super Star hamburger	1	item(s)	348	—	790	41.0	52.0	3.0	47.0	14.0	—	—	—
38925	The Six Dollar burger	1	item(s)	429	—	1010	40.0	60.0	3.0	66.0	26.0	—	—	—
10818	Vanilla shake, small	1	item(s)	398	—	314	10.0	51.5	0	7.4	4.7	—	—	—
10770	Western Bacon cheeseburger	1	item(s)	225	—	660	32.0	64.0	2.0	30.0	12.0	—	4.8	—
	Chick Fil-A													
38746	Biscuit with bacon, egg & cheese	1	item(s)	163	—	470	18.0	39.0	1.0	26.0	9.0			3.0
38747	Biscuit with egg	1	item(s)	135	—	350	11.0	38.0	1.0	16.0	4.5			3.0
38748	Biscuit with egg & cheese	1	item(s)	149	—	400	14.0	38.0	1.0	21.0	7.0			3.0
38753	Biscuit with gravy	1	item(s)	192	—	330	5.0	43.0	1.0	15.0	4.0			4.0
38752	Biscuit with sausage, egg & cheese	1	item(s)	212	—	620	22.0	39.0	2.0	42.0	14.0			3.0
38771	Carrot & raisin salad	1	item(s)	113	—	170	1.0	28.0	2.0	6.0	1.0			0
38761	Chargrilled chicken Cool Wrap	1	item(s)	245	—	390	29.0	54.0	3.0	7.0	3.0			0
38766	Chargrilled chicken garden salad	1	item(s)	275	—	180	22.0	9.0	3.0	6.0	3.0			0
38758	Chargrilled chicken sandwich	1	item(s)	193	—	270	28.0	33.0	3.0	3.5	1.0			0
38742	Chicken biscuit	1	item(s)	145	—	420	18.0	44.0	2.0	19.0	4.5			3.0
38743	Chicken biscuit with cheese	1	item(s)	159	—	470	21.0	45.0	2.0	23.0	8.0			3.0
38762	Chicken Caesar Cool Wrap	1	item(s)	227	—	460	36.0	52.0	3.0	10.0	6.0			0
38757	Chicken deluxe sandwich	1	item(s)	208	—	420	28.0	39.0	2.0	16.0	3.5			0
38764	Chicken salad sandwich on wheat bun	1	item(s)	153	—	350	20.0	32.0	5.0	15.0	3.0			0
38756	Chicken sandwich	1	item(s)	170	—	410	28.0	38.0	1.0	16.0	3.5			0
38768	Chick-n-Strip salad	1	item(s)	327	—	400	34.0	21.0	4.0	20.0	6.0			0
38763	Chick-n-Strips	4	item(s)	127	—	300	28.0	14.0	1.0	15.0	2.5			0
38770	Cole slaw	1	item(s)	128	—	260	2.0	17.0	2.0	21.0	3.5			0
38776	Diet lemonade, small	1	cup(s)	255	—	25	0	5.0	0	0	0	0	0	0
38755	Hashbrowns	1	serving(s)	84	—	260	2.0	25.0	3.0	17.0	3.5			1.0
38765	Hearty breast of chicken soup	1	cup(s)	241	—	140	8.0	18.0	1.0	3.5	1.0			0
38741	Hot buttered biscuit	1	item(s)	79	—	270	4.0	38.0	1.0	12.0	3.0			3.0
38778	IceDream, small cone	1	item(s)	135	—	160	4.0	28.0	0	4.0	2.0			0
38774	IceDream, small cup	1	serving(s)	227	—	240	6.0	41.0	0	6.0	3.5			0
38775	Lemonade, small	1	cup(s)	255	—	170	0	41.0	0	0.5	0			0
38777	Nuggets	8	item(s)	113	—	260	26.0	12.0	0.5	12.0	2.5			0
38769	Side salad	1	item(s)	108	—	60	3.0	4.0	2.0	3.0	1.5			0
38767	Southwest chargrilled salad	1	item(s)	303	—	240	25.0	17.0	5.0	8.0	3.5			0
40481	Spicy chicken cool wrap	1	serving(s)	230	—	380	30.0	52.0	3.0	6.0	3.0			0
38772	Waffle potato fries, small, salted	1	serving(s)	85	—	270	3.0	34.0	4.0	13.0	3.0			1.5
	Cinnabon													
39572	Caramellata Chill with whipped cream	16	fluid ounce(s)	480	—	406	10.0	61.0	0	14.0	8.0		—	0
39571	Cinnabon Bites	1	serving(s)	149	—	510	8.0	77.0	2.0	19.0	5.0		—	5.0
39570	Cinnabon Stix	5	item(s)	85	—	379	6.0	41.0	1.0	21.0	6.0		—	4.0
39567	Classic roll	1	item(s)	221	—	813	15.0	117.0	4.0	32.0	8.0		—	5.0
39568	Minibon	1	item(s)	92	—	339	6.0	49.0	2.0	13.0	3.0		—	2.0
39573	Mochalatta Chill with whipped cream	16	fluid ounce(s)	480	—	362	9.0	55.0	0	13.0	8.0		—	0
39569	Pecanbon	1	item(s)	272	—	1100	16.0	141.0	8.0	56.0	10.0		—	5.0
	Dairy Queen													
1466	Banana split	1	item(s)	369	—	510	8.0	96.0	3.0	12.0	8.0		—	0
38552	Brownie Earthquake®	1	serving(s)	304	—	740	10.0	112.0	0	27.0	16.0		—	0.5
38561	Chocolate chip cookie dough blizzard®, small	1	item(s)	319	—	720	12.0	105.0	0	28.0	14.0		—	2.5
1464	Chocolate malt, small	1	item(s)	418	—	640	15.0	111.0	1.0	16.0	11.0		—	0.5
38541	Chocolate shake, small	1	item(s)	397	—	560	13.0	93.0	1.0	15.0	10.0		—	0.5
17257	Chocolate soft serve	½	cup(s)	94	—	150	4.0	22.0	0	5.0	3.5		—	0
1463	Chocolate sundae, small	1	item(s)	163	—	280	5.0	49.0	0	7.0	4.5		—	0
1462	Dipped cone, small	1	item(s)	156	—	340	6.0	42.0	1.0	17.0	9.0	4.0	3.0	1.0
38555	Oreo cookies blizzard, small	1	item(s)	283	—	570	11.0	83.0	0.5	21.0	10.0		—	2.5
38547	Royal Treats Peanut Buster® Parfait	1	item(s)	305	—	730	16.0	99.0	2.0	31.0	17.0		—	0
17256	Vanilla soft serve	½	cup(s)	94	—	140	3.0	22.0	0	4.5	3.0		—	0
	Domino's													
31606	Barbeque buffalo wings	1	item(s)	25	—	50	6.0	2.0	0	2.5	0.5		—	—
31604	Breadsticks	1	item(s)	30	—	115	2.0	12.0	0	6.3	1.1		—	—
37551	Buffalo Chicken Kickers	1	item(s)	24	—	47	4.0	3.0	0	2.0	0.5		—	—
37548	CinnaStix	1	item(s)	30	—	123	2.0	15.0	1.0	6.1	1.1		—	—
37549	Dot, cinnamon	1	item(s)	28	7.6	99	1.9	14.9	0.7	3.7	0.7		—	—
31605	Double cheesy bread	1	item(s)	35	—	123	4.0	13.0	0	6.5	1.9		—	—

Chol (mg)	Calc (mg)	Iron (mg)	Magn (mg)	Pota (mg)	Sodi (mg)	Zinc (mg)	Vit A (µg)	Thia (mg)	Vit E (mg α)	Ribo (mg)	Niac (mg)	Vit B$_6$ (mg)	Fola (µg)	Vit C (mg)	Vit B$_{12}$ (µg)	Sele (µg)
0	20	1.80	—	—	950.0	—	0	—	—	—	—	—	—	12.0	—	—
155	300	7.20	—	—	1730.0	—	—	—	—	—	—	—	—	1.2	—	—
0	0	1.08	—	—	170.0	—	0	—	—	—	—	—	—	21.0	—	—
3	0	0.00	—	—	470.0	—	0	0.25	—	0.23	2.00	—	—	0	—	—
0	20	0.72	—	—	700.0	—	0	—	—	—	—	—	—	3.6	—	—
40	100	3.60	—	—	1220.0	—	—	—	—	—	—	—	—	6.0	—	—
45	600	0.00	—	—	340.0	—	0	—	—	—	—	—	—	0	—	—
130	100	7.20	—	—	980.0	—	—	—	—	—	—	—	—	9.0	—	—
145	279	4.29	—	—	1960.0	—	—	—	—	—	—	—	—	16.7	—	—
30	401	0.00	—	—	234.0	—	0	—	—	—	—	—	—	0	—	—
85	200	5.40	—	—	1410.0	—	60.0	—	—	—	—	—	—	1.2	—	—
270	150	2.70	—	—	1190.0	—	—	—	—	—	—	—	—	0	—	—
240	80	2.70	—	—	740.0	—	—	—	—	—	—	—	—	0	—	—
255	150	2.70	—	—	970.0	—	—	—	—	—	—	—	—	0	—	—
5	60	1.80	—	—	930.0	—	0	—	—	—	—	—	—	0	—	—
300	200	3.60	—	—	1360.0	—	—	—	—	—	—	—	—	0	—	—
10	40	0.36	—	—	110.0	—	—	—	—	—	—	—	—	4.8	—	—
65	200	3.60	—	—	1020.0	—	—	—	—	—	—	—	—	6.0	—	—
65	150	0.72	—	—	620.0	—	—	—	—	—	—	—	—	30	—	—
65	80	2.70	—	—	940.0	—	—	—	—	—	—	—	—	6.0	—	—
35	60	2.70	—	—	1270.0	—	0	—	—	—	—	—	—	0	—	—
50	150	2.70	—	—	1500.0	—	—	—	—	—	—	—	—	0	—	—
80	500	3.60	—	—	1350.0	—	—	—	—	—	—	—	—	1.2	—	—
60	100	2.70	—	—	1300.0	—	—	—	—	—	—	—	—	2.4	—	—
65	150	1.80	—	—	880.0	—	—	—	—	—	—	—	—	0	—	—
60	100	2.70	—	—	1300.0	—	—	—	—	—	—	—	—	0	—	—
80	150	1.44	—	—	1070.0	—	—	—	—	—	—	—	—	6.0	—	—
65	40	1.44	—	—	940.0	—	—	—	—	—	—	—	—	0	—	—
25	60	0.36	—	—	220.0	—	—	—	—	—	—	—	—	36.0	—	—
0	0	0.36	—	—	5.0	—	0	—	—	—	—	—	—	15.0	—	—
5	20	0.72	—	—	380.0	—	—	—	—	—	—	—	—	0	—	—
25	40	1.08	—	—	900.0	—	—	—	—	—	—	—	—	0	—	—
0	60	1.80	—	—	660.0	—	0	—	—	—	—	—	—	0	—	—
15	100	0.36	—	—	80.0	—	—	—	—	—	—	—	—	0	—	—
25	200	0.36	—	—	105.0	—	—	—	—	—	—	—	—	0	—	—
0	0	0.36	—	—	10.0	—	0	—	—	—	—	—	—	15.0	—	—
70	40	1.08	—	—	1090.0	—	0	—	—	—	—	—	—	0	—	—
10	100	0.00	—	—	75.0	—	—	—	—	—	—	—	—	15.0	—	—
60	200	1.08	—	—	770.0	—	—	—	—	—	—	—	—	24.0	—	—
60	200	3.60	—	—	1090.0	—	—	—	—	—	—	—	—	3.6	—	—
0	20	1.08	—	—	115.0	—	0	—	—	—	—	—	—	1.2	—	—
46	—	—	—	—	187.0	—	—	—	—	—	—	—	—	—	—	—
35	—	—	—	—	530.0	—	—	—	—	—	—	—	—	—	—	—
16	—	—	—	—	413.0	—	—	—	—	—	—	—	—	—	—	—
67	—	—	—	—	801.0	—	—	—	—	—	—	—	—	—	—	—
27	—	—	—	—	337.0	—	—	—	—	—	—	—	—	—	—	—
46	—	—	—	—	252.0	—	—	—	—	—	—	—	—	—	—	—
63	—	—	—	—	600.0	—	—	—	—	—	—	—	—	—	—	—
30	250	1.80	—	—	180.0	—	—	—	—	—	—	—	—	15.0	—	—
50	250	1.80	—	—	350.0	—	—	—	—	—	—	—	—	0	—	—
50	350	2.70	—	—	370.0	—	—	—	—	—	—	—	—	1.2	—	—
55	450	1.80	—	—	340.0	—	—	—	—	—	—	—	—	2.4	—	—
50	450	1.44	—	—	280.0	—	—	—	—	—	—	—	—	2.4	—	—
15	100	0.72	—	—	75.0	—	—	—	—	—	—	—	—	0	—	—
20	200	1.08	—	—	140.0	—	—	—	—	—	—	—	—	0	—	—
20	200	1.08	—	—	130.0	—	—	—	—	—	—	—	—	1.2	—	—
40	350	2.70	—	—	430.0	—	—	—	—	—	—	—	—	1.2	—	—
35	300	1.80	—	—	400.0	—	—	—	—	—	—	—	—	1.2	—	—
15	150	0.72	—	—	70.0	—	—	—	—	—	—	—	—	0	—	—
26	10	0.36	—	—	175.5	—	—	—	—	—	—	—	—	0	—	—
0	0	0.72	—	—	122.1	—	—	—	—	—	—	—	—	0	—	—
9	0	0.00	—	—	162.5	—	—	—	—	—	—	—	—	0	—	—
0	0	0.72	—	—	111.4	—	—	—	—	—	—	—	—	0	—	—
0	6	0.59	—	—	85.7	—	—	—	—	—	—	—	—	0	—	—
6	40	0.72	—	—	162.3	—	—	—	—	—	—	—	—	0	—	—

DA+ Code	Food Description	Quantity	Measure	Wt (g)	H₂O (g)	Ener (kcal)	Prot (g)	Carb (g)	Fiber (g)	Fat (g)	Fat Breakdown (g) Sat	Mono	Poly	Trans
	FAST FOOD —Continued													
31607	Hot buffalo wings	1	item(s)	25	—	45	5.0	1.0	0	2.5	0.5	—	—	—
	Domino's Classic hand tossed pizza													
31573	America's favorite feast, 12"	1	slice(s)	102	—	257	10.0	29.0	2.0	11.5	4.5	—	—	—
31574	America's favorite feast, 14"	1	slice(s)	141	—	353	14.0	39.0	2.0	16.0	6.0	—	—	—
37543	Bacon cheeseburger feast, 12"	1	slice(s)	99	—	273	12.0	28.0	2.0	13.0	5.5	—	—	—
37545	Bacon cheeseburger feast, 14"	1	slice(s)	137	—	379	17.0	38.0	2.0	18.0	8.0	—	—	—
37546	Barbeque feast, 12"	1	slice(s)	96	—	252	11.0	31.0	1.0	10.0	4.5	—	—	—
37547	Barbeque feast, 14"	1	slice(s)	131	—	344	14.0	43.0	2.0	13.5	6.0	—	—	—
31569	Cheese, 12"	1	slice(s)	55	—	160	6.0	28.0	1.0	3.0	1.0	—	—	0
31570	Cheese, 14"	1	slice(s)	75	—	220	8.0	38.0	2.0	4.0	1.0	—	—	0
37538	Deluxe feast, 12"	1	slice(s)	201	101.8	465	19.5	57.4	3.5	18.2	7.7	—	—	—
37540	Deluxe feast, 14"	1	slice(s)	273	138.4	627	26.4	78.3	4.7	24.1	10.2	—	—	—
31685	Deluxe, 12"	1	slice(s)	100	—	234	9.0	29.0	2.0	9.5	3.5	—	—	—
31694	Deluxe, 14"	1	slice(s)	136	—	316	13.0	39.0	2.0	12.5	5.0	—	—	—
31686	Extravaganzza, 12"	1	slice(s)	122	—	289	13.0	30.0	2.0	14.0	5.5	—	—	—
31695	Extravaganzza, 14"	1	slice(s)	165	—	388	17.0	40.0	3.0	18.5	7.5	—	—	—
31575	Hawaiian feast, 12"	1	slice(s)	102	—	223	10.0	30.0	2.0	8.0	3.5	—	—	—
31576	Hawaiian feast, 14"	1	slice(s)	141	—	309	14.0	41.0	2.0	11.0	4.5	—	—	—
31687	Meatzza, 12"	1	slice(s)	108	—	281	13.0	29.0	2.0	13.5	5.5	—	—	—
31696	Meatzza, 14"	1	slice(s)	146	—	378	17.0	39.0	2.0	18.0	7.5	—	—	—
31571	Pepperoni feast, extra pepperoni & cheese, 12"	1	slice(s)	98	—	265	11.0	28.0	2.0	12.5	5.0	—	—	—
31572	Pepperoni feast, extra pepperoni & cheese, 14"	1	slice(s)	135	—	363	16.0	39.0	2.0	17.0	7.0	—	—	—
31577	Vegi feast, 12"	1	slice(s)	102	—	218	9.0	29.0	2.0	8.0	3.5	—	—	—
31578	Vegi feast, 14"	1	slice(s)	139	—	300	13.0	40.0	3.0	11.0	4.5	—	—	—
	Domino's thin crust pizza													
31583	America's favorite, 12"	1	slice(s)	72	—	208	8.0	15.0	1.0	13.5	5.0	—	—	—
31584	America's favorite, 14"	1	slice(s)	100	—	285	11.0	20.0	2.0	18.5	7.0	—	—	—
31579	Cheese, 12"	1	slice(s)	49	—	137	5.0	14.0	1.00	7.0	2.5	—	—	—
31580	Cheese, 14"	1	slice(s)	68	27.0	214	8.8	19.0	1.4	11.4	4.6	2.9	2.5	—
31688	Deluxe, 12"	1	slice(s)	70	—	185	7.0	15.0	1.0	11.5	4.0	—	—	—
31697	Deluxe, 14"	1	slice(s)	94	—	248	10.0	20.0	2.0	15.0	5.5	—	—	—
31689	Extravaganzza, 12"	1	slice(s)	92	—	240	11.0	16.0	1.0	15.5	6.0	—	—	—
31698	Extravaganzza, 14"	1	slice(s)	123	—	320	14.0	21.0	2.0	20.5	8.0	—	—	—
31585	Hawaiian, 12"	1	slice(s)	71	—	174	8.0	16.0	1.0	9.5	3.5	—	—	—
31586	Hawaiian, 14"	1	slice(s)	100	—	240	11.0	21.0	2.0	13.0	5.0	—	—	—
31690	Meatzza, 12"	1	slice(s)	78	—	232	11.0	15.0	1.0	15.0	6.0	—	—	—
31699	Meatzza, 14"	1	slice(s)	104	—	310	14.0	20.0	2.0	20	8.0	—	—	—
31581	Pepperoni, extra pepperoni & cheese, 12"	1	slice(s)	68	—	216	9.0	14.0	1.0	14.0	5.5	—	—	—
31582	Pepperoni, extra pepperoni & cheese, 14"	1	slice(s)	93	—	295	13.0	20.0	1.0	19.0	7.5	—	—	—
31587	Vegi, 12"	1	slice(s)	71	—	168	7.0	15.0	1.0	9.5	3.5	—	—	—
31588	Vegi, 14"	1	slice(s)	97	—	231	10.0	21.0	2.0	13.5	5.0	—	—	—
	Domino's Ultimate deep dish pizza													
31596	America's favorite, 12"	1	slice(s)	115	—	309	12.0	29.0	2.0	17.0	6.0	—	—	—
31702	America's favorite, 14"	1	slice(s)	162	—	433	17.0	42.0	3.0	23.5	8.0	—	—	—
31590	Cheese, 12"	1	slice(s)	90	—	238	9.0	28.0	2.0	11.0	3.5	—	—	—
31591	Cheese, 14"	1	slice(s)	128	53.9	351	14.5	41.0	2.9	13.2	5.2	3.8	2.5	—
31589	Cheese, 6"	1	item(s)	215	—	598	22.9	68.4	3.9	27.6	9.9	—	—	—
31691	Deluxe, 12"	1	slice(s)	122	—	287	11.0	29.0	2.0	15.0	5.0	—	—	—
31700	Deluxe, 14"	1	slice(s)	156	—	396	15.0	42.0	3.0	20.0	7.0	—	—	—
31692	Extravaganzza, 12"	1	slice(s)	136	—	341	14.0	30.0	2.0	19.0	7.0	—	—	—
31701	Extravaganzza, 14"	1	slice(s)	186	—	468	20.0	43.0	3.0	25.5	9.5	—	—	—
31599	Hawaiian, 12"	1	slice(s)	114	—	275	12.0	30.0	2.0	13.0	5.0	—	—	—
31600	Hawaiian, 14"	1	slice(s)	162	—	389	17.0	43.0	3.0	18.0	6.5	—	—	—
31693	Meatzza, 12"	1	slice(s)	121	—	333	14.0	29.0	2.0	19.0	7.0	—	—	—
31703	Meatzza, 14"	1	slice(s)	167	—	458	19.0	42.0	3.0	25.0	9.5	—	—	—
31593	Pepperoni, extra pepperoni & cheese, 12"	1	slice(s)	110	—	317	13.0	29.0	2.0	17.5	6.5	—	—	—
31594	Pepperoni, extra pepperoni & cheese, 14"	1	slice(s)	155	—	443	18.0	42.0	3.0	24.0	9.0	—	—	—
31602	Vegi, 12"	1	slice(s)	114	—	270	11.0	30.0	2.0	13.5	5.0	—	—	—
31603	Vegi, 14"	1	slice(s)	159	—	380	15.0	43.0	3.0	18.0	6.5	—	—	—
31598	With ham and pineapple tidbits, 6"	1	item(s)	430	—	619	25.2	69.9	4.0	28.3	10.2	—	—	—
31595	With Italian sausage, 6"	1	item(s)	430	—	642	24.8	69.6	4.2	31.1	11.3	—	—	—
31592	With pepperoni, 6"	1	item(s)	430	—	647	25.1	68.5	3.9	32.0	11.7	—	—	—
31601	With vegetables, 6"	1	item(s)	430	—	619	23.4	70.8	4.6	28.7	10.1	—	—	—
	In-n-Out Burger													
34391	Cheeseburger with mustard & ketchup	1	serving(s)	268	—	400	22.0	41.0	3.0	18.0	9.0	—	—	0.5
34374	Cheeseburger	1	serving(s)	268	—	480	22.0	39.0	3.0	27.0	10.0	—	—	0.5
34390	Cheeseburger, lettuce leaves instead of buns	1	serving(s)	300	—	330	18.0	11.0	3.0	25.0	9.0	—	—	0
34377	Chocolate shake	1	serving(s)	425	—	690	9.0	83.0	0	36.0	24.0	—	—	1.0
34375	Double-Double cheeseburger	1	serving(s)	330	—	670	37.0	39.0	3.0	41.0	18.0	—	—	1.0

Chol (mg)	Calc (mg)	Iron (mg)	Magn (mg)	Pota (mg)	Sodi (mg)	Zinc (mg)	Vit A (µg)	Thia (mg)	Vit E (mg α)	Ribo (mg)	Niac (mg)	Vit B$_6$ (mg)	Fola (µg)	Vit C (mg)	Vit B$_{12}$ (µg)	Sele (µg)
26	10	0.36	—	—	254.5	—	—	—	—	—	—	—	—	1.2	—	—
																—
22	100	1.80	—	—	625.5	—	—	—	—	—	—	—	—	0.6	—	—
31	140	2.52	—	—	865.5	—	—	—	—	—	—	—	—	0.6	—	—
27	140	1.80	—	—	634.0	—	—	—	—	—	—	—	—	0	—	—
38	190	2.52	—	—	900.0	—	—	—	—	—	—	—	—	0	—	—
20	140	1.62	—	—	600.0	—	—	—	—	—	—	—	—	0.6	—	—
27	190	2.16	—	—	831.5	—	—	—	—	—	—	—	—	0.6	—	—
0	0	1.80	—	—	110.0	—	0	—	—	—	—	—	—	0	—	—
0	0	2.70	—	—	150.0	—	0	—	—	—	—	—	—	0	—	—
40	199	3.56	—	—	1063.1	—	—	—	—	—	—	—	—	1.4	—	—
53	276	4.84	—	—	1432.2	—	—	—	—	—	—	—	—	1.8	—	—
17	100	1.80	—	—	541.5	—	—	—	—	—	—	—	—	0.6	—	—
23	130	2.34	—	—	728.5	—	—	—	—	—	—	—	—	1.2	—	—
28	140	1.98	—	—	764.0	—	—	—	—	—	—	—	—	0.6	—	—
37	190	2.70	—	—	1014.0	—	—	—	—	—	—	—	—	1.2	—	—
16	130	1.62	—	—	546.5	—	—	—	—	—	—	—	—	1.2	—	—
23	180	2.34	—	—	765.0	—	—	—	—	—	—	—	—	1.2	—	—
28	130	1.80	—	—	739.5	—	—	—	—	—	—	—	—	0	—	—
37	190	2.52	—	—	983.5	—	—	—	—	—	—	—	—	0	—	—
24	130	1.62	—	—	670.0	—	70.9	—	—	—	—	—	—	0	—	—
33	180	2.34	—	—	920.0	—	104.7	—	—	—	—	—	—	0	—	—
13	130	1.62	—	—	489.0	—	—	—	—	—	—	—	—	0.6	—	—
18	180	2.34	—	—	678.0	—	—	—	—	—	—	—	—	0.6	—	—
23	100	0.90	—	—	533.0	—	—	—	—	—	—	—	—	2.4	—	—
32	140	1.26	—	—	736.5	—	—	—	—	—	—	—	—	3.0	—	—
10	90	0.54	—	—	292.5	—	60.0	—	—	—	—	—	—	1.8	—	—
14	151	0.48	17.7	125.1	338.0	0.02	64.6	0.05	1.01	0.07	0.69	—	—	2.4	0.5	24.1
19	100	0.90	—	—	449.0	—	—	—	—	—	—	—	—	2.4	—	—
24	130	1.08	—	—	601.0	—	—	—	—	—	—	—	—	3.6	—	—
29	140	1.08	—	—	671.5	—	—	—	—	—	—	—	—	2.4	—	—
38	190	1.44	—	—	886.5	—	—	—	—	—	—	—	—	3.6	—	—
17	130	0.72	—	—	454.0	—	—	—	—	—	—	—	—	3.0	—	—
24	180	0.90	—	—	637.5	—	—	—	—	—	—	—	—	3.6	—	—
29	140	0.90	—	—	647.0	—	—	—	—	—	—	—	—	1.8	—	—
38	190	1.26	—	—	865.5	—	—	—	—	—	—	—	—	2.4	—	—
26	130	0.72	—	—	577.0	—	80.0	—	—	—	—	—	—	1.8	—	—
35	80	1.08	—	—	792.5	—	105.8	—	—	—	—	—	—	2.4	—	—
14	130	0.72	—	—	396.5	—	—	—	—	—	—	—	—	2.4	—	—
19	180	1.08	—	—	550.5	—	—	—	—	—	—	—	—	3.0	—	—
25	120	2.34	—	—	796.5	—	—	—	—	—	—	—	—	0.6	—	—
34	170	3.24	—	—	1110.0	—	—	—	—	—	—	—	—	0.6	—	—
11	110	1.98	—	—	555.5	—	70.0	—	—	—	—	—	—	0	—	—
18	189	3.78	32.0	209.9	718.1	1.75	99.8	0.29	1.13	0.31	5.44	—	—	0	0.6	45.6
36	295	4.67	—	—	1341.4	—	174.0	—	—	—	—	—	—	0.5	—	—
20	120	2.16	—	—	712.0	—	—	—	—	—	—	—	—	1.2	—	—
26	170	3.06	—	—	974.5	—	—	—	—	—	—	—	—	1.2	—	—
31	160	2.52	—	—	934.5	—	—	—	—	—	—	—	—	1.2	—	—
40	220	3.42	—	—	1260.0	—	—	—	—	—	—	—	—	1.2	—	—
19	150	1.98	—	—	717.0	—	—	—	—	—	—	—	—	1.2	—	—
26	210	2.88	—	—	1011.0	—	—	—	—	—	—	—	—	1.8	—	—
31	160	2.34	—	—	910.5	—	—	—	—	—	—	—	—	0	—	—
40	220	3.24	—	—	1230.0	—	—	—	—	—	—	—	—	0.6	—	—
27	150	2.16	—	—	840.5	—	86.5	—	—	—	—	—	—	0	—	—
37	220	3.06	—	—	1166.0	—	115.4	—	—	—	—	—	—	0.6	—	—
15	150	2.16	—	—	659.5	—	—	—	—	—	—	—	—	0.6	—	—
21	220	3.06	—	—	924.0	—	—	—	—	—	—	—	—	1.2	—	—
43	298	4.84	—	—	1497.8	—	—	—	—	—	—	—	—	1.5	—	—
45	302	4.89	—	—	1478.1	—	—	—	—	—	—	—	—	0.6	—	—
47	299	4.81	—	—	1523.7	—	167.9	—	—	—	—	—	—	0.6	—	—
36	307	5.10	—	—	1472.5	—	—	—	—	—	—	—	—	4.7	—	—
60	200	3.60	—	—	1080.0	—	—	—	—	—	—	—	—	12.0	—	—
60	200	3.60	—	—	1000.0	—	—	—	—	—	—	—	—	9.0	—	—
60	200	2.70	—	—	720.0	—	—	—	—	—	—	—	—	12.0	—	—
95	300	0.72	—	—	350.0	—	—	—	—	—	—	—	—	0	—	—
120	350	5.40	—	—	1440.0	—	—	—	—	—	—	—	—	9.0	—	—

DA+ Code	Food Description	Quantity	Measure	Wt (g)	H₂O (g)	Ener (kcal)	Prot (g)	Carb (g)	Fiber (g)	Fat (g)	Sat	Mono	Poly	Trans
	FAST FOOD —Continued													
34393	Double-Double cheeseburger with mustard & ketchup	1	serving(s)	330	—	590	37.0	41.0	3.0	32.0	17.0	—	—	1.0
34392	Double-Double cheeseburger, lettuce leaves instead of buns	1	serving(s)	362	—	520	33.0	11.0	3.0	39.0	17.0	—	—	1.0
34376	French fries	1	serving(s)	125	—	400	7.0	54.0	2.0	18.0	5.0	—	—	0
34373	Hamburger	1	item(s)	243	—	390	16.0	39.0	3.0	19.0	5.0	—	—	0
34389	Hamburger with mustard & ketchup	1	serving(s)	243	—	310	16.0	41.0	3.0	10.0	4.0	—	—	0
34388	Hamburger, lettuce leaves instead of buns	1	serving(s)	275	—	240	13.0	11.0	3.0	17.0	4.0	—	—	0
34379	Strawberry shake	1	serving(s)	425	—	690	9.0	91.0	0	33.0	22.0	—	—	0.5
34378	Vanilla shake	1	serving(s)	425	—	680	9.0	78.0	0	37.0	25.0	—	—	1.0
	Jack in the Box													
30392	Bacon ultimate cheeseburger	1	item(s)	338	—	1090	46.0	53.0	2.0	77.0	30.0	—	—	3.0
1740	Breakfast Jack	1	item(s)	125	—	290	17.0	29.0	1.0	12.0	4.5	—	—	0
14074	Cheeseburger	1	item(s)	131	—	350	18.0	31.0	1.0	17.0	8.0	—	—	1.0
14106	Chicken breast strips, 4 piece	4	piece(s)	201	—	500	35.0	36.0	3.0	25.0	6.0	—	—	6.0
37241	Chicken club salad, plain, without salad dressing	1	serving(s)	431	—	300	27.0	13.0	4.0	15.0	6.0	—	—	0
14064	Chicken sandwich	1	item(s)	145	—	400	15.0	38.0	2.0	21.0	4.5	—	—	2.5
14111	Chocolate ice cream shake, small	1	serving(s)	414	—	880	14.0	107.0	1.0	45.0	31.0	—	—	2.0
14073	Hamburger	1	item(s)	118	—	310	16.0	30.0	1.0	14.0	6.0	—	—	1.0
14090	Hash browns	1	serving(s)	57	—	150	1.0	13.0	2.0	10.0	2.5	—	—	3.0
14072	Jack's Spicy Chicken sandwich	1	item(s)	270	—	620	25.0	61.0	4.0	31.0	6.0	—	—	3.0
1468	Jumbo Jack hamburger	1	item(s)	261	—	600	21.0	51.0	3.0	35.0	12.0	—	—	1.5
1469	Jumbo Jack hamburger with cheese	1	item(s)	286	—	690	25.0	54.0	3.0	42.0	16.0	—	—	1.5
14099	Natural cut french fries, large	1	serving(s)	196	—	530	8.0	69.0	5.0	25.0	6.0	—	—	7.0
14098	Natural cut french fries, medium	1	serving(s)	133	—	360	5.0	47.0	4.0	17.0	4.0	—	—	5.0
1470	Onion rings	1	serving(s)	119	—	500	6.0	51.0	3.0	30.0	6.0	—	—	10
33141	Sausage, egg & cheese biscuit	1	item(s)	234	—	740	27.0	35.0	2.0	55.0	17.0	—	—	6.0
14095	Seasoned curly fries, medium	1	serving(s)	125	—	400	6.0	45.0	5.0	23.0	5.0	—	—	7.0
14077	Sourdough Jack	1	item(s)	245	—	710	27.0	36.0	3.0	51.0	18.0	—	—	3.0
37249	Southwest chicken salad, plain, without salad dressing	1	serving(s)	488	—	300	24.0	29.0	7.0	11.0	5.0	—	—	0
14112	Strawberry ice cream shake, small	1	serving(s)	417	—	880	13.0	105.0	0	44.0	31.0	—	—	2.0
14078	Ultimate cheeseburger	1	item(s)	323	—	1010	40.0	53.0	2.0	71.0	28.0	—	—	3.0
14110	Vanilla ice cream shake, small	1	serving(s)	379	—	790	13.0	83.0	0	44.0	31.0	—	—	2.0
	Jamba Juice													
31645	Aloha Pineapple smoothie	24	fluid ounce(s)	730	—	500	8.0	117.0	4.0	1.5	1.0	—	—	—
31646	Banana Berry smoothie	24	fluid ounce(s)	719	—	480	5.0	112.0	4.0	1.0	0	—	—	—
31656	Berry Lime Sublime smoothie	24	fluid ounce(s)	728	—	460	3.0	106.0	5.0	2.0	1.0	—	—	—
31647	Carribean Passion smoothie	24	fluid ounce(s)	730	—	440	4.0	102.0	4.0	2.0	1.0	—	—	—
38422	Carrot juice	16	fluid ounce(s)	472	—	100	3.0	23.0	0	0.5	0	—	—	—
31648	Chocolate Moo'd smoothie	24	fluid ounce(s)	634	—	720	17.0	148.0	3.0	8.0	5.0	—	—	—
31649	Citrus Squeeze smoothie	24	fluid ounce(s)	727	—	470	5.0	110.0	4.0	2.0	1.0	—	—	—
31651	Coldbuster smoothie	24	fluid ounce(s)	724	—	430	5.0	100.0	5.0	2.5	1.0	—	—	—
31652	Cranberry Craze smoothie	24	fluid ounce(s)	793	—	460	6.0	104.0	4.0	0.5	0	—	—	—
31654	Jamba Powerboost smoothie	24	fluid ounce(s)	738	—	440	6.0	105.0	6.0	1.0	0	—	—	—
38423	Lemonade	16	fluid ounce(s)	483	—	300	1.0	75.0	0	0	0	0	0	0
31657	Mango-a-go-go smoothie	24	fluid ounce(s)	690	—	440	3.0	104.0	4.0	1.5	0.5	—	—	—
38424	Orange juice, freshly squeezed	16	fluid ounce(s)	496	—	220	3.0	52.0	0.5	1.0	0	—	—	—
38426	Orange/carrot juice	16	fluid ounce(s)	484	—	160	3.0	37.0	0	1.0	0	—	—	—
31660	Orange-a-peel smoothie	24	fluid ounce(s)	726	—	440	8.0	102.0	5.0	1.5	0	—	—	—
31662	Peach Pleasure smoothie	24	fluid ounce(s)	720	—	460	4.0	108.0	4.0	2.0	1.0	—	—	—
31665	Protein Berry Pizzaz smoothie	24	fluid ounce(s)	710	—	440	20.0	92.0	5.0	1.5	0	—	—	—
31668	Razzmatazz smoothie	24	fluid ounce(s)	730	—	480	3.0	112.0	4.0	2.0	1.0	—	—	—
31669	Strawberries Wild smoothie	24	fluid ounce(s)	725	—	450	6.0	105.0	4.0	0.5	0	—	—	—
38421	Strawberry Tsunami smoothie	24	fluid ounce(s)	740	—	530	4.0	128.0	4.0	2.0	1.0	—	—	—
38427	Vibrant C juice	16	fluid ounce(s)	448	—	210	2.0	50.0	1.0	0	0	0	0	0
38428	Wheatgrass juice, freshly squeezed	1	ounce(s)	28	—	5	0.5	1.0	0	0	0	0	0	0
	Kentucky Fried Chicken (KFC)													
31850	BBQ baked beans	1	serving(s)	136	—	220	8.0	45.0	7.0	1.0	0	—	—	0
31853	Biscuit	1	item(s)	57	—	220	4.0	24.0	1.0	11.0	2.5	—	—	3.5
51223	Boneless Fiery Buffalo Wings	6	item(s)	211	—	530	30.0	44.0	3.0	26.0	5.0	—	—	2.5
39386	Boneless Honey BBQ Wings	6	item(s)	213	—	570	30.0	54.0	5.0	26.0	5.0	—	—	2.5
51224	Boneless Sweet & Spicy Wings	6	item(s)	203	—	550	30.0	50.0	3.0	26.0	5.0	—	—	2.5
31851	Cole slaw	1	serving(s)	130	—	180	1.0	22.0	3.0	10.0	1.5	—	—	0
31842	Colonel's Crispy Strips	3	item(s)	151	—	370	28.0	17.0	1.0	20.0	4.0	—	—	2.5
31849	Corn on the cob	1	item(s)	162	—	150	5.0	26.0	7.0	3.0	1.0	—	—	0
51221	Double Crunch sandwich	1	item(s)	213	—	520	27.0	39.0	3.0	29.0	5.0	—	—	1.5
3761	Extra Crispy chicken, breast	1	item(s)	162	—	370	33.0	10.0	2.0	22.0	5.0	—	—	1.5
3762	Extra Crispy chicken, drumstick	1	item(s)	60	—	150	12.0	4.0	0	10.0	2.5	—	—	1.0

Chol (mg)	Calc (mg)	Iron (mg)	Magn (mg)	Pota (mg)	Sodi (mg)	Zinc (mg)	Vit A (µg)	Thia (mg)	Vit E (mg α)	Ribo (mg)	Niac (mg)	Vit B$_6$ (mg)	Fola (µg)	Vit C (mg)	Vit B$_{12}$ (µg)	Sele (µg)
115	350	5.40	—	—	1520.0	—	—	—	—	—	—	—	—	12.0	—	—
120	350	4.50	—	—	1160.0	—	—	—	—	—	—	—	—	12.0	—	—
0	20	1.80	—	—	245.0	—	0	—	—	—	—	—	—	0	—	—
40	40	3.60	—	—	650.0	—	—	—	—	—	—	—	—	9.0	—	—
35	40	3.60	—	—	730.0	—	—	—	—	—	—	—	—	12.0	—	—
40	40	2.70	—	—	370.0	—	—	—	—	—	—	—	—	12.0	—	—
85	300	0.00	—	—	280.0	—	—	—	—	—	—	—	—	0	—	—
90	300	0.00	—	—	390.0	—	—	—	—	—	—	—	—	0	—	—
140	308	7.38	—	540.0	2040.0	—	—	—	—	—	—	—	—	0.6	—	—
220	145	3.48	—	210.0	760.0	—	—	—	—	—	—	—	—	3.5	—	—
50	151	3.61	—	270.0	790.0	—	40.2	—	—	—	—	—	—	0	—	—
80	18	1.60	—	530.0	1260.0	—	—	—	—	—	—	—	—	1.1	—	—
65	280	3.35	—	560.0	880.0	—	—	—	—	—	—	—	—	50.4	—	—
35	100	2.70	—	240.0	730.0	—	—	—	—	—	—	—	—	4.8	—	—
135	460	0.47	—	840.0	330.0	—	—	—	—	—	—	—	—	0	—	—
40	100	3.60	—	250.0	600.0	—	0	—	—	—	—	—	—	0	—	—
0	10	0.18	—	190.0	230.0	—	0	—	—	—	—	—	—	0	—	—
50	150	1.80	—	450.0	1100.0	—	—	—	—	—	—	—	—	9.0	—	—
45	164	4.92	—	380.0	940.0	—	—	—	—	—	—	—	—	9.8	—	—
70	234	4.20	—	410.0	1310.0	—	—	—	—	—	—	—	—	8.4	—	—
0	20	1.42	—	1240.0	870.0	—	0	—	—	—	—	—	—	8.9	—	—
0	19	1.01	—	840.0	590.0	—	0	—	—	—	—	—	—	5.6	—	—
0	40	2.70	—	140.0	420.0	—	40.0	—	—	—	—	—	—	18.0	—	—
280	88	2.36	—	310.0	1430.0	—	—	—	—	—	—	—	—	0	—	—
0	40	1.80	—	580.0	890.0	—	—	—	—	—	—	—	—	0	—	—
75	200	4.50	—	430.0	1230.0	—	—	—	—	—	—	—	—	9.0	—	—
55	274	4.10	—	670.0	860.0	—	—	—	—	—	—	—	—	43.8	—	—
135	466	0.00	—	750.0	290.0	—	—	—	—	—	—	—	—	0	—	—
125	308	7.39	—	480.0	1580.0	—	—	—	—	—	—	—	—	0.6	—	—
135	532	0.00	—	750.0	280.0	—	—	—	—	—	—	—	—	0	—	—
5	200	1.80	60.0	1000.0	30.0	0.30	—	0.37	—	0.34	2.00	0.60	60.0	102.0	0	1.4
0	200	1.44	40.0	1010.0	115.0	0.60	—	0.09	—	0.25	0.80	0.70	24.0	15.0	0.2	1.4
5	200	1.80	16.0	510.0	35.0	0.30	—	0.06	—	0.25	6.00	0.70	140.0	54.0	0	1.4
5	100	1.80	24.0	810.0	60.0	0.30	—	0.09	—	0.25	5.00	0.50	100.0	78.0	0	1.4
0	150	2.70	80.0	1030.0	250.0	0.90	—	0.52	—	0.25	5.00	0.70	80.0	18.0	0	5.6
30	500	1.08	60.0	810.0	380.0	1.50	—	0.22	—	0.76	0.40	0.16	16.0	6.0	1.5	4.2
5	100	1.80	80.0	1170.0	35.0	0.30	—	0.37	—	0.34	1.90	0.60	100.0	180.0	0	1.4
5	100	1.08	60.0	1260.0	35.0	16.50	—	0.37	—	0.34	3.00	0.40	121.5	1302.0	0	1.4
0	250	1.44	16.0	500.0	50.0	0.30	—	0.03	—	0.25	5.00	0.60	120.0	54.0	0	1.4
0	1200	1.80	480.0	1070.0	45.0	16.50	—	5.55	—	6.12	68.00	7.40	640.0	288.0	10.8	77.0
0	20	0.00	8.0	200.0	10.0	0.00	—	0.03	—	0.17	14.00	1.80	320.0	36.0	0	0
5	100	1.08	24.0	780.0	50.0	0.30	—	0.15	—	0.25	5.00	0.70	120.0	72.0	0	1.4
0	60	1.08	60.0	990.0	0	0.30	—	0.45	—	0.13	2.00	0.20	160.0	246.0	0	0
0	100	1.80	60.0	1010.0	125.0	0.60	—	0.45	—	0.25	3.00	0.50	120.0	132.0	0	2.8
0	250	1.80	80.0	1380.0	160.0	0.90	—	0.45	—	0.42	2.00	0.50	140.0	240.0	0.6	1.4
5	100	0.72	32.0	740.0	60.0	0.30	—	0.06	—	0.25	4.00	0.60	80.0	18.0	0	1.4
0	1100	2.62	60.0	650.0	240.0	0.58	—	0.08	—	0.17	1.20	0.70	58.3	60.0	0	5.6
5	150	1.80	32.0	810.0	70.0	0.30	—	0.09	—	0.34	6.00	1.00	160.0	60.0	0	1.4
5	250	1.80	40.0	1050.0	180.0	0.90	—	0.12	—	0.34	0.80	0.40	40.0	60.0	0.6	1.4
5	100	1.08	24.0	480.0	10.0	0.30	—	0.06	—	0.34	14.00	1.80	320.0	90.0	0	1.4
0	20	1.08	40.0	720.0	0	0.30	—	0.30	—	0.10	1.60	0.40	80.0	678.0	0	0
0	0	1.80	8.0	80.0	0	0.00	0	0.03	—	0.03	0.40	0.04	16.0	3.6	0	2.8
0	100	2.70	—	—	730.0	—	—	—	—	—	—	—	—	1.2	—	—
0	40	1.80	—	—	640.0	—	—	—	—	—	—	—	—	0	—	—
65	40	1.80	—	—	2670.0	—	—	—	—	—	—	—	—	1.2	—	—
65	40	1.80	—	—	2210.0	—	—	—	—	—	—	—	—	1.2	—	—
65	60	1.80	—	—	2000.0	—	—	—	—	—	—	—	—	1.2	—	—
5	40	0.72	—	—	270.0	—	—	—	—	—	—	—	—	12.0	—	—
65	40	1.44	—	—	1220.0	—	0	—	—	—	—	—	—	1.2	—	—
0	60	1.08	—	—	10.0	—	—	—	—	—	—	—	—	6.0	—	—
55	100	2.70	—	—	1220.0	—	—	—	—	—	—	—	—	6.0	—	—
85	20	2.70	—	—	1020.0	—	—	—	—	—	—	—	—	1.2	—	—
55	0	1.44	—	—	300.0	—	0	—	—	—	—	—	—	0	—	—

DA+ Code	Food Description	Quantity	Measure	Wt (g)	H₂O (g)	Ener (kcal)	Prot (g)	Carb (g)	Fiber (g)	Fat (g)	Fat Breakdown (g)			
											Sat	Mono	Poly	Trans
	FAST FOOD —Continued													
3763	Extra Crispy chicken, thigh	1	item(s)	114	—	290	17.0	16.0	1.0	18.0	4.0	—	—	1.5
3764	Extra Crispy chicken, whole wing	1	item(s)	52	—	150	11.0	11.0	1.0	7.0	1.5	—	—	0
51218	Famous Bowls mashed potatoes with gravy	1	serving(s)	531	—	720	26.0	79.0	6.0	34.0	9.0	—	—	3.5
51219	Famous Bowls rice with gravy	1	serving(s)	384	—	610	25.0	67.0	5.0	27.0	8.0	—	—	2.5
31841	Honey BBQ chicken sandwich	1	item(s)	147	—	290	23.0	40.0	2.0	4.0	1.0	—	—	0
31833	Honey BBQ wing pieces	6	item(s)	157	—	460	27.0	26.0	3.0	27.0	6.0	—	—	2.0
10859	Hot wings pieces	6	piece(s)	134	—	450	26.0	19.0	2.0	30.0	7.0	—	—	2.0
42382	KFC Snacker sandwich	1	serving(s)	119	—	320	14.0	29.0	2.0	17.0	3.0	—	—	1.0
31848	Macaroni & cheese	1	serving(s)	136	—	180	8.0	18.0	0	8.0	3.5	—	—	1.0
31847	Mashed potatoes with gravy	1	serving(s)	151	—	140	2.0	20.0	1.0	5.0	1.0	—	—	0.5
10825	Original Recipe chicken, breast	1	item(s)	161	—	340	38.0	9.0	2.0	17.0	4.0	—	—	1.0
10826	Original Recipe chicken, drumstick	1	item(s)	59	—	140	13.0	3.0	0	8.0	2.0	—	—	0.5
10827	Original Recipe chicken, thigh	1	item(s)	126	—	350	19.0	7.0	1.0	27.0	7.0	—	—	1.0
10828	Original Recipe chicken, whole wing	1	item(s)	47	—	140	10.0	4.0	0	9.0	2.0	—	—	0.5
51222	Oven roasted Twister chicken wrap	1	item(s)	269	—	520	30.0	46.0	4.0	23.0	3.5	—	—	0
31844	Popcorn chicken, small or individual	1	item(s)	114	—	370	19.0	21.0	2.0	24.0	4.5	—	—	2.5
31852	Potato salad	1	serving(s)	128	—	180	2.0	22.0	2.0	9.0	1.5	—	—	0
10845	Potato wedges, small	1	serving(s)	102	—	250	4.0	32.0	3.0	12.0	2.0	—	—	1.5
31839	Tender Roast chicken sandwich with sauce	1	item(s)	236	—	430	37.0	29.0	2.0	18.0	3.5	—	—	0
	Long John Silver													
39392	Baked cod	1	serving(s)	101	—	120	22.0	1.0	0	4.5	1.0	—	—	0
3777	Batter dipped fish sandwich	1	item(s)	177	—	470	18.0	48.0	3.0	23.0	5.0	—	—	4.5
37568	Battered fish	1	item(s)	92	—	260	12.0	17.0	0.5	16.0	4.0	—	—	4.5
37569	Breaded clams	1	serving(s)	85	—	240	8.0	22.0	1.0	13.0	2.0	—	—	2.5
37566	Chicken plank	1	item(s)	52	—	140	8.0	9.0	0.5	8.0	2.0	—	—	2.5
39404	Clam chowder	1	item(s)	227	—	220	9.0	23.0	0	10.0	4.0	—	—	1.0
39398	Cocktail sauce	1	ounce(s)	28	—	25	0	6.0	0	0	0	0	0	0
3770	Coleslaw	1	serving(s)	113	—	200	1.0	15.0	3.0	15.0	2.5	1.8	4.1	0
39400	French fries, large	1	item(s)	142	—	390	4.0	56.0	5.0	17.0	4.0	—	—	5.0
3774	Fries, regular	1	serving(s)	85	—	230	3.0	34.0	3.0	10.0	2.5	—	—	3.0
3779	Hushpuppy	1	piece(s)	23	—	60	1.0	9.0	1.0	2.5	0.5	—	—	1.0
3781	Shrimp, batter-dipped, 1 piece	1	piece(s)	14	—	45	2.0	3.0	0	3.0	1.0	—	—	1.0
39399	Tartar sauce	1	ounce(s)	28	—	100	0	4.0	0	9.0	1.5	—	—	—
39395	Ultimate Fish sandwich	1	item(s)	199	—	530	21.0	49.0	3.0	28.0	8.0	—	—	5.0
	McDonald's													
50828	Asian salad with grilled chicken	1	item(s)	362	—	290	31.0	23.0	6.0	10.0	1.0	—	—	0
2247	Barbecue sauce	1	item(s)	28	—	45	0	11.0	0	0	0	0	0	0
737	Big Mac hamburger	1	item(s)	219	—	560	25.0	47.0	3.0	30.0	10.0	—	—	1.5
29777	Caesar salad dressing	1	package(s)	44	—	150	1.0	5.0	0	13.0	2.5	—	—	—
38391	Caesar salad with grilled chicken, no dressing	1	serving(s)	278	230.6	181	26.4	10.5	3.1	6.0	2.9	1.7	0.8	0.2
38393	Caesar salad without chicken, no dressing	1	serving(s)	190	170.4	84	6.0	8.1	3.0	3.9	2.2	0.9	0.3	0.1
738	Cheeseburger	1	item(s)	119	—	310	15.0	35.0	1.0	12.0	6.0	—	—	1.0
29775	Chicken McGrill sandwich	1	item(s)	213	—	400	27.0	38.0	3.0	16.0	3.0	—	—	0
1873	Chicken McNuggets, 6 piece	6	item(s)	96	—	250	15.0	15.0	0	15.0	3.0	—	—	1.5
3792	Chicken McNuggets, 4 piece	4	item(s)	64	—	170	10.0	10.0	0	10.0	2.0	—	—	1.0
29774	Crispy chicken sandwich	1	item(s)	232	121.8	500	27.0	63.0	3.0	16.0	3.0	5.7	7.4	1.5
743	Egg McMuffin	1	item(s)	139	76.8	300	17.0	30.0	2.0	12.0	4.5	3.8	2.5	0
742	Filet-O-Fish sandwich	1	item(s)	141	—	400	14.0	42.0	1.0	18.0	4.0	—	—	1.0
2257	French fries, large	1	serving(s)	170	—	570	6.0	70.0	7.0	30.0	6.0	—	—	8.0
1872	French fries, small	1	serving(s)	74	—	250	2.0	30.0	3.0	13.0	2.5	—	—	3.5
33822	Fruit 'n Yogurt Parfait	1	item(s)	149	111.2	160	4.0	31.0	1.0	2.0	1.0	0.2	0.1	0
739	Hamburger	1	item(s)	105	—	260	13.0	33.0	1.0	9.0	3.5	—	—	0.5
2003	Hash browns	1	item(s)	53	—	140	1.0	15.0	2.0	8.0	1.5	—	—	2.0
2249	Honey sauce	1	item(s)	14	—	50	0	12.0	0	0	0	0	0	0
38397	Newman's Own creamy caesar salad dressing	1	item(s)	59	32.5	190	2.0	4.0	0	18.0	3.5	4.6	9.6	0
38398	Newman's Own low fat balsamic vinaigrette salad dressing	1	item(s)	44	29.1	40	0	4.0	0	3.0	0	1.0	1.2	0
38399	Newman's Own ranch salad dressing	1	item(s)	59	30.1	170	1.0	9.0	0	15.0	2.5	9.0	3.7	0
1874	Plain Hotcakes with syrup and margarine	3	item(s)	221	—	600	9.0	102.0	2.0	17.0	4.0	—	—	4.0
740	Quarter Pounder hamburger	1	item(s)	171	—	420	24.0	40.0	3.0	18.0	7.0	—	—	1.0
741	Quarter Pounder hamburger with cheese	1	item(s)	199	—	510	29.0	43.0	3.0	25.0	12.0	—	—	1.5
2005	Sausage McMuffin with egg	1	item(s)	165	82.4	450	20.0	31.0	2.0	27.0	10.0	10.9	4.6	0.5
50831	Side salad	1	item(s)	87	—	20	1.0	4.0	1.0	0	0	0	0	0
	Pizza Hut													
39009	Hot chicken wings	2	item(s)	57	—	110	11.0	1.0	0	6.0	2.0	—	—	0.3
14025	Meat Lovers hand tossed pizza	1	slice(s)	118	—	300	15.0	29.0	2.0	13.0	6.0	—	—	0.5
14026	Meat Lovers pan pizza	1	slice(s)	123	—	340	15.0	29.0	2.0	19.0	7.0	—	—	0.5
31009	Meat Lovers stuffed crust pizza	1	slice(s)	169	—	450	21.0	43.0	3.0	21.0	10.0	—	—	1.0
14024	Meat Lovers thin 'n crispy pizza	1	slice(s)	98	—	270	13.0	21.0	2.0	14.0	6.0	—	—	0.5

Chol (mg)	Calc (mg)	Iron (mg)	Magn (mg)	Pota (mg)	Sodi (mg)	Zinc (mg)	Vit A (µg)	Thia (mg)	Vit E (mg α)	Ribo (mg)	Niac (mg)	Vit B$_6$ (mg)	Fola (µg)	Vit C (mg)	Vit B$_{12}$ (µg)	Sele (µg)
95	20	2.70	—	—	700.0	—	—	—	—	—	—	—	—	—	—	—
45	20	1.08	—	—	340.0	—	—	—	—	—	—	—	—	0	—	—
35	200	5.40	—	—	2330.0	—	—	—	—	—	—	—	—	6.0	—	—
35	200	4.50	—	—	2130.0	—	—	—	—	—	—	—	—	6.0	—	—
60	80	2.70	—	—	710.0	—	—	—	—	—	—	—	—	2.4	—	—
140	40	1.80	—	—	970.0	—	—	—	—	—	—	—	—	21.0	—	—
115	40	1.44	—	—	990.0	—	—	—	—	—	—	—	—	1.2	—	—
25	60	2.70	—	—	690.0	—	—	—	—	—	—	—	—	2.4	—	—
15	150	0.72	—	—	800.0	—	—	—	—	—	—	—	—	1.2	—	—
0	40	1.44	—	—	560.0	—	—	—	—	—	—	—	—	1.2	—	—
135	20	2.70	—	—	960.0	—	—	—	—	—	—	—	—	6.0	—	—
70	20	1.08	—	—	340.0	—	—	—	—	—	—	—	—	0	—	—
110	20	2.70	—	—	870.0	—	—	—	—	—	—	—	—	1.2	—	—
50	20	1.44	—	—	350.0	—	0	—	—	—	—	—	—	1.2	—	—
60	40	6.30	—	—	1380.0	—	—	—	—	—	—	—	—	15.0	—	—
25	40	1.80	—	—	1110.0	—	0	—	—	—	—	—	—	—	—	—
5	0	0.36	—	—	470.0	—	—	—	—	—	—	—	—	6.0	—	—
0	20	1.08	—	—	700.0	—	0	—	—	—	—	—	—	0	—	—
80	80	2.70	—	—	1180.0	—	—	—	—	—	—	—	—	9.0	—	—
90	20	0.72	—	—	240.0	—	—	—	—	—	—	—	—	0	—	—
45	60	2.70	—	—	1210.0	—	—	—	—	—	—	—	—	2.4	—	—
35	20	0.72	—	—	790.0	—	—	—	—	—	—	—	—	4.8	—	—
10	20	1.08	—	—	1110.0	—	0	—	—	—	—	—	—	0	—	—
20	0	0.72	—	—	480.0	—	0	—	—	—	—	—	—	2.4	—	—
25	150	0.72	—	—	810.0	—	—	—	—	—	—	—	—	0	—	—
0	0	0.00	—	—	250.0	—	—	—	—	—	—	—	—	0	—	—
20	40	0.36	—	222.7	340.0	0.70	—	0.07	—	0.08	2.34	—	—	18.0	—	—
0	0	0.00	—	—	580.0	—	0	—	—	—	—	—	—	24.0	—	—
0	0	0.00	—	370.0	350.0	0.30	0	0.09	—	0.01	1.60	—	—	15.0	—	—
0	20	0.36	—	—	200.0	—	0	—	—	—	—	—	—	0	—	—
15	0	0.00	—	—	160.0	—	0	—	—	—	—	—	—	1.2	—	—
15	0	0.00	—	—	250.0	—	0	—	—	—	—	—	—	0	—	—
60	150	2.70	—	—	1400.0	—	—	—	—	—	—	—	—	4.8	—	—
65	150	3.60	—	—	890.0	—	—	—	—	—	—	—	—	54.0	—	—
0	0	0.00	—	55.0	260.0	—	—	—	—	—	—	—	—	0	—	—
80	250	4.50	—	400.0	1010.0	—	—	—	—	—	—	—	—	1.2	—	—
10	40	0.18	—	30.0	400.0	—	—	—	—	—	—	—	—	0.6	—	—
67	178	1.77	—	708.9	767.3	—	—	0.15	—	0.19	10.62	—	127.9	29.2	0.2	—
10	163	1.15	17.1	410.4	157.7	—	—	0.08	—	0.07	0.40	—	102.6	26.8	0	0.4
40	200	2.70	—	240.0	740.0	—	60.0	—	—	—	—	—	—	1.2	—	—
70	150	2.70	—	510.0	1010.0	—	—	—	—	—	—	—	—	6.0	—	—
35	20	0.72	—	240.0	670.0	—	—	—	—	—	—	—	—	1.2	—	—
25	0	0.36	—	160.0	450.0	—	—	—	—	—	—	—	—	1.2	—	—
60	80	3.60	62.6	526.6	1380.0	1.53	41.8	0.46	2.27	0.39	12.85	—	104.4	6.0	0.4	—
230	300	2.70	26.4	218.2	860.0	1.59	—	0.36	0.82	0.51	4.31	0.20	109.8	1.2	0.9	—
40	150	1.80	—	250.0	640.0	—	36.2	—	—	—	—	—	—	0	—	—
0	20	1.80	—	—	330.0	—	0	—	—	—	—	—	—	9.0	—	—
0	20	0.72	—	—	140.0	—	0	—	—	—	—	—	—	3.6	—	—
5	150	0.67	20.9	248.8	85.0	0.53	0	0.06	—	0.17	0.35	—	19.4	9.0	0.3	—
30	150	2.70	—	210.0	530.0	—	5.0	—	—	—	—	—	—	1.2	—	—
0	0	0.36	—	210.0	290.0	—	0	—	—	—	—	—	—	1.2	—	—
0	0	0.00	—	0	0	—	0	—	—	—	—	—	—	0	—	—
20	61	0.00	3.0	16.0	500.0	0.20	—	0.01	15.43	0.02	0.01	0.64	2.4	0	0.1	0.1
0	4	0.00	1.3	8.8	730.0	0.01	—	0.00	0.00	0.00	0.00	0.00	0	2.4	0	0
0	40	0.00	1.8	70.4	530.0	0.03	0	0.01	—	0.08	0.01	0.02	0.6	0	0	0.2
20	150	2.70	—	280.0	620.0	—	—	—	—	—	—	—	—	0	—	—
70	150	4.50	—	390.0	730.0	—	10.0	—	—	—	—	—	—	1.2	—	—
95	300	4.50	—	440.0	1150.0	—	100.0	—	—	—	—	—	—	1.2	—	—
255	300	3.60	29.7	282.2	950.0	2.01	—	0.43	0.82	0.56	4.83	0.24	—	0	1.2	—
0	20	0.72	—	—	10.0	—	—	—	—	—	—	—	—	15.0	—	—
70	0	0.36	—	—	450.0	—	—	—	—	—	—	—	—	0	—	—
35	150	1.80	—	—	760.0	—	—	—	—	—	—	—	—	6.0	—	—
35	150	2.70	—	—	750.0	—	—	—	—	—	—	—	—	6.0	—	—
55	250	2.70	—	—	1250.0	—	—	—	—	—	—	—	—	9.0	—	—
35	150	1.44	—	—	740.0	—	—	—	—	—	—	—	—	6.0	—	—

DA+ Code	Food Description	Quantity	Measure	Wt (g)	H₂O (g)	Ener (kcal)	Prot (g)	Carb (g)	Fiber (g)	Fat (g)	Sat	Mono	Poly	Trans
	FAST FOOD—Continued													
14031	Pepperoni Lovers hand tossed pizza	1	slice(s)	113	—	300	15.0	30.0	2.0	13.0	7.0	—	—	0.5
14032	Pepperoni Lovers pan pizza	1	slice(s)	118	—	340	15.0	29.0	2.0	19.0	7.0	—	—	0.5
31011	Pepperoni Lovers stuffed crust pizza	1	slice(s)	163	—	420	21.0	43.0	3.0	19.0	10.0	—	—	1.0
14030	Pepperoni Lovers thin 'n crispy pizza	1	slice(s)	92	—	260	13.0	21.0	2.0	14.0	7.0	—	—	0.5
10834	Personal Pan pepperoni pizza	1	slice(s)	61	—	170	7.0	18.0	0.5	8.0	3.0	—	—	1.0
10842	Personal Pan supreme pizza	1	slice(s)	77	—	190	8.0	19.0	1.0	9.0	3.5	—	—	1.0
39013	Personal Pan Veggie Lovers pizza	1	slice(s)	69	—	150	6.0	19.0	1.0	6.0	2.0	—	—	0.5
14028	Veggie Lovers hand tossed pizza	1	slice(s)	118	—	220	10.0	31.0	2.0	6.0	3.0	—	—	0.3
14029	Veggie Lovers pan pizza	1	slice(s)	119	—	260	10.0	30.0	2.0	12.0	4.0	—	—	0.3
31010	Veggie Lovers stuffed crust pizza	1	slice(s)	172	—	360	16.0	45.0	3.0	14.0	7.0	—	—	0.5
14027	Veggie Lovers thin 'n crispy pizza	1	slice(s)	101	—	180	8.0	23.0	2.0	7.0	3.0	—	—	0.5
39012	Wing blue cheese dipping sauce	1	item(s)	43	—	230	2.0	2.0	0	24.0	5.0	—	—	1.0
39011	Wing ranch dipping sauce	1	item(s)	43	—	210	0.5	4.0	0	22.0	3.5	—	—	0.5
	Starbucks													
38052	Cappuccino, tall	12	fluid ounce(s)	360	—	120	7.0	10.0	0	6.0	4.0	—	—	—
38053	Cappuccino, tall nonfat	12	fluid ounce(s)	360	—	80	7.0	11.0	0	0	0	0	0	0
38054	Cappuccino, tall soymilk	12	fluid ounce(s)	360	—	100	5.0	13.0	0.5	2.5	0	—	—	—
38059	Cinnamon spice mocha, tall nonfat w/o whipped cream	12	fluid ounce(s)	360	—	170	11.0	32.0	0	0.5	—	—	—	—
38057	Cinnamon spice mocha, tall w/ whipped cream	12	fluid ounce(s)	360	—	320	10.0	31.0	0	17.0	11.0	—	—	—
38051	Espresso, single shot	1	fluid ounce(s)	30	—	5	0	1.0	0	0	0	0	0	0
38088	Flavored syrup, 1 pump	1	serving(s)	10	—	20	0	5.0	0	0	0	0	0	0
32562	Frappuccino bottled coffee drink, mocha	9 ½	fluid ounce(s)	298	—	190	6.0	39.0	3.0	3.0	2.0	—	—	—
32561	Frappuccino coffee drink, all bottled flavors	9 ½	fluid ounce(s)	281	—	190	7.0	35.0	0	3.5	2.5	—	—	—
38073	Frappuccino, mocha	12	fluid ounce(s)	360	—	220	5.0	44.0	0	3.0	1.5	—	—	—
38067	Frappuccino, tall caramel w/o whipped cream	12	fluid ounce(s)	360	—	210	4.0	43.0	0	2.5	1.5	—	—	—
38070	Frappuccino, tall coffee	12	fluid ounce(s)	360	—	190	4.0	38.0	0	2.5	1.5	—	—	—
39894	Frappuccino, tall coffee, light blend	12	fluid ounce(s)	360	—	110	5.0	22.0	2.0	1.0	0	—	—	—
38071	Frappuccino, tall espresso	12	fluid ounce(s)	360	—	160	4.0	33.0	0	2.0	1.5	—	—	—
39897	Frappuccino, tall mocha, light blend	12	fluid ounce(s)	360	—	140	5.0	28.0	3.0	.1.5	0	—	—	—
39887	Frappuccino, tall Strawberries & Creme, w/o whipped cream	12	fluid ounce(s)	360	—	330	10.0	65.0	0	3.5	1.0	—	—	—
38063	Frappuccino, tall Tazo chai creme w/o whipped cream	12	fluid ounce(s)	360	—	280	10.0	52.0	0	3.5	1.0	—	—	—
38066	Frappuccino, tall Tazoberry	12	fluid ounce(s)	360	—	140	0.5	36.0	0.5	0	0	0	0	0
38065	Frappuccino, tall Tazoberry Crème	12	fluid ounce(s)	360	—	240	4.0	54.0	0.5	1.0	0	—	—	—
38080	Frappuccino, tall vanilla w/o whipped cream	12	fluid ounce(s)	360	—	270	10.0	51.0	0	3.5	1.0	—	—	—
39898	Frappuccino, tall white chocolate mocha, light blend	12	fluid ounce(s)	360	—	160	6.0	32.0	2.0	2.0	1.0	—	—	—
38074	Frappuccino, tall white chocolate w/o whipped cream	12	fluid ounce(s)	360	—	240	5.0	48.0	0	3.5	2.5	—	—	—
39883	Java Chip Frappuccino, tall w/o whipped cream	12	fluid ounce(s)	360	—	270	5.0	51.0	1.0	7.0	4.5	—	—	—
33111	Latte, tall w/ nonfat milk	12	fluid ounce(s)	360	335.3	120	12.0	18.0	0	0	0	0	0	0
33112	Latte, tall w/ whole milk	12	fluid ounce(s)	360	—	200	11.0	16.0	0	11.0	7.0	—	—	—
33109	Macchiato, tall caramel w/ nonfat milk	12	fluid ounce(s)	360	—	170	11.0	30.0	0	1.0	0	—	—	—
33110	Macchiato, tall caramel w/ whole milk	12	fluid ounce(s)	360	—	240	10.0	28.0	0	10.0	6.0	—	—	—
33107	Mocha coffee drink, tall nonfat, w/o whip cream	12	fluid ounce(s)	360	—	170	11.0	33.0	1.0	1.5	0	—	—	—
38089	Mocha syrup	1	serving(s)	17	—	25	1.0	6.0	0	0.5	0	—	—	—
33108	Mocha, tall mocha w/ whole milk	12	fluid ounce(s)	360	—	310	10.0	32.0	1.0	17.0	10.0	—	—	—
38042	Steamed apple cider, tall	12	fluid ounce(s)	360	—	180	0	45.0	0	0	0	—	—	—
38087	Tazo chai black tea, soymilk, tall	12	fluid ounce(s)	360	—	190	4.0	39.0	0.5	2.0	0	—	—	—
38084	Tazo chai black tea, tall	12	fluid ounce(s)	360	—	210	6.0	36.0	0	5.0	3.5	—	—	—
38083	Tazo chai black tea, tall nonfat	12	fluid ounce(s)	360	—	170	6.0	37.0	0	0	0	0	0	0
38076	Tazo iced tea, tall	12	fluid ounce(s)	360	—	60	0	16.0	0	0	0	0	0	0
38077	Tazo tea, grande lemonade	16	fluid ounce(s)	480	—	120	0	31.0	0	0	0	0	0	0
38045	Vanilla creme steamed nonfat milk, tall w/whipped cream	12	fluid ounce(s)	360	—	260	11.0	33.0	0	8.0	5.0	—	—	—
38046	Vanilla crème steamed soymilk, tall w/whipped cream	12	fluid ounce(s)	360	—	300	8.0	37.0	1.0	12.0	6.0	—	—	—
38044	Vanilla creme steamed whole milk, tall w/whipped cream	12	fluid ounce(s)	360	—	330	10.0	31.0	0	18.0	11.0	—	—	—
38090	Whipped cream	1	serving(s)	27	—	100	0	2.0	0	9.0	6.0	—	—	—
38062	White chocolate mocha, tall nonfat w/o whipped cream	12	fluid ounce(s)	360	—	260	12.0	45.0	0	4.0	3.0	—	—	—
38061	White chocolate mocha, tall w/ whipped cream	12	fluid ounce(s)	360	—	410	11.0	44.0	0	20.0	13.0	—	—	—
38048	White hot chocolate, tall nonfat w/o whipped cream	12	fluid ounce(s)	360	—	300	15.0	51.0	0	4.5	3.5	—	—	—
38050	White hot chocolate, tall soymilk w/whipped cream	12	fluid ounce(s)	360	—	420	11.0	56.0	1.0	16.0	9.0	—	—	—
38047	White hot chocolate, tall w/whipped cream	12	fluid ounce(s)	360	—	460	13.0	50.0	0	22.0	15.0	—	—	—

Chol (mg)	Calc (mg)	Iron (mg)	Magn (mg)	Pota (mg)	Sodi (mg)	Zinc (mg)	Vit A (µg)	Thia (mg)	Vit E (mg α)	Ribo (mg)	Niac (mg)	Vit B$_6$ (mg)	Fola (µg)	Vit C (mg)	Vit B$_{12}$ (µg)	Sele (µg)
40	200	1.80	—	—	710.0	—	57.7	—	—	—	—	—	—	2.4	—	—
40	200	2.70	—	—	700.0	—	57.7	—	—	—	—	—	—	2.4	—	—
55	300	2.70	—	—	1120.0	—	—	—	—	—	—	—	—	3.6	—	—
40	200	1.44	—	—	690.0	—	58.0	—	—	—	—	—	—	2.4	—	—
15	80	1.44	—	—	340.0	—	38.5	—	—	—	—	—	—	1.4	—	—
20	80	1.86	—	—	420.0	—	—	—	—	—	—	—	—	3.6	—	—
10	80	1.80	—	—	280.0	—	—	—	—	—	—	—	—	3.6	—	—
15	150	1.80	—	—	490.0	—	—	—	—	—	—	—	—	9.0	—	—
15	150	2.70	—	—	470.0	—	—	—	—	—	—	—	—	9.0	—	—
35	250	2.70	—	—	980.0	—	—	—	—	—	—	—	—	9.0	—	—
15	150	1.44	—	—	480.0	—	—	—	—	—	—	—	—	9.0	—	—
25	20	0.00	—	—	550.0	—	0	—	—	—	—	—	—	0	—	—
10	0	0.00	—	—	340.0	—	0	—	—	—	—	—	—	0	—	—
25	250	0.00	—	—	95.0	—	—	—	—	—	—	—	—	1.2	0	—
3	200	0.00	—	—	100.0	—	—	—	—	—	—	—	—	0	0	—
0	250	0.72	—	—	75.0	—	—	—	—	—	—	—	—	0	0	—
5	300	0.72	—	—	150.0	—	—	—	—	—	—	—	—	0	0	—
70	350	1.08	—	—	140.0	—	—	—	—	—	—	—	—	2.4	0	—
0	0	0.00	—	—	0	—	0	—	—	—	—	—	—	0	0	—
0	0	0.00	—	—	0	—	0	—	—	—	—	—	—	0	0	—
12	219	1.08	—	530.0	110.0	—	—	—	—	—	—	—	—	0	—	—
15	250	0.36	—	510.0	105.0	—	—	—	—	—	—	—	—	0	—	—
10	150	0.72	—	—	180.0	—	—	—	—	—	—	—	—	0	0	—
10	150	0.00	—	—	180.0	—	—	—	—	—	—	—	—	0	0	—
10	150	0.00	—	—	180.0	—	—	—	—	—	—	—	—	0	—	—
0	150	0.00	—	—	220.0	—	—	—	—	—	—	—	—	0	—	—
10	100	0.00	—	—	160.0	—	—	—	—	—	—	—	—	0	—	—
0	150	0.72	—	—	220.0	—	—	—	—	—	—	—	—	0	—	—
3	350	0.00	—	—	270.0	—	—	—	—	—	—	—	—	21.0	—	—
3	350	0.00	—	—	270.0	—	—	—	—	—	—	—	—	3.6	0	—
0	0	0.00	—	—	30.0	—	0	—	—	—	—	—	—	0	0	—
0	150	0.00	—	—	125.0	—	0	—	—	—	—	—	—	1.2	0	—
3	350	0.00	—	—	370.0	—	—	—	—	—	—	—	—	3.6	0	—
3	150	0.00	—	—	250.0	—	—	—	—	—	—	—	—	0	—	—
10	150	0.00	—	—	210.0	—	—	—	—	—	—	—	—	0	0	—
10	150	1.44	—	—	220.0	—	—	—	—	—	—	—	—	0	—	—
5	350	0.00	39.8	—	170.0	1.35	—	0.12	—	0.47	0.36	0.13	17.5	0	1.3	—
45	400	0.00	46.6	—	160.0	1.28	—	0.12	—	0.54	0.34	0.14	16.8	2.4	1.2	—
5	300	0.00	—	—	160.0	—	—	—	—	—	—	—	—	1.2	—	—
30	300	0.00	—	—	135.0	—	—	—	—	—	—	—	—	2.4	—	—
5	300	2.70	—	—	135.0	—	—	—	—	—	—	—	—	0	—	—
0	0	0.72	—	—	0	—	0	—	—	—	—	—	—	0	0	—
55	300	2.70	—	—	115.0	—	—	—	—	—	—	—	—	0	—	—
0	0	1.08	—	—	15.0	—	0	—	—	—	—	—	—	0	0	—
0	200	0.72	—	—	70.0	—	—	—	—	—	—	—	—	0	0	—
20	200	0.36	—	—	85.0	—	—	—	—	—	—	—	—	1.2	0	—
5	200	0.36	—	—	95.0	—	—	—	—	—	—	—	—	0	0	—
0	0	0.00	—	—	0	—	0	—	—	—	—	—	—	0	0	—
0	0	0.00	—	—	15.0	—	0	—	—	—	—	—	—	4.8	0	—
35	350	0.00	—	—	170.0	—	—	—	—	—	—	—	—	0	—	—
30	400	1.44	—	—	130.0	—	—	—	—	—	—	—	—	0	0	—
65	350	0.00	—	—	140.0	—	—	—	—	—	—	—	—	0	0	—
40	0	0.00	—	—	10.0	—	—	—	—	—	—	—	—	0	0	—
5	400	0.00	—	—	210.0	—	—	—	—	—	—	—	—	0	0	—
70	400	0.00	—	—	210.0	—	—	—	—	—	—	—	—	2.4	0	—
10	450	0.00	—	—	250.0	—	—	—	—	—	—	—	—	0	0	—
35	500	1.44	—	—	210.0	—	—	—	—	—	—	—	—	0	0	—
75	500	0.00	—	—	250.0	—	—	—	—	—	—	—	—	3.6	0	—

DA+ Code	Food Description	Quantity	Measure	Wt (g)	H₂O (g)	Ener (kcal)	Prot (g)	Carb (g)	Fiber (g)	Fat (g)	Sat	Mono	Poly	Trans
											Fat Breakdown (g)			

FAST FOOD —Continued

Subway

DA+ Code	Food Description	Quantity	Measure	Wt (g)	H₂O (g)	Ener (kcal)	Prot (g)	Carb (g)	Fiber (g)	Fat (g)	Sat	Mono	Poly	Trans
15842	Cheese steak sandwich, 6", wheat bread	1	item(s)	250	—	360	24.0	47.0	5.0	10.0	4.5	—	—	0
40478	Chicken & bacon ranch sandwich, 6", white or wheat bread	1	serving(s)	297	—	540	36.0	47.0	5.0	25.0	10.0	—	—	0.5
38622	Chicken & bacon ranch wrap with cheese	1	item(s)	257	—	440	41.0	18.0	9.0	27.0	10.0	—	—	0.5
32045	Chocolate chip cookie	1	item(s)	45	—	210	2.0	30.0	1.0	10.0	6.0	—	—	0
32048	Chocolate chip M&M cookie	1	item(s)	45	—	210	2.0	32.0	0.5	10.0	5.0	—	—	0
32049	Chocolate chunk cookie	1	item(s)	45	—	220	2.0	30.0	0.5	10.0	5.0	—	—	0
4024	Classic Italian B.M.T. sandwich, 6", white bread	1	item(s)	236	—	440	22.0	45.0	2.0	21.0	8.5	—	—	0
15838	Classic tuna sandwich, 6", wheat bread	1	item(s)	250	—	530	22.0	45.0	4.0	31.0	7.0	—	—	0.5
15837	Classic tuna sandwich, 6", white bread	1	item(s)	243	—	520	21.0	43.0	2.0	31.0	7.5	—	—	0.5
16397	Club salad, no dressing and croutons	1	item(s)	412	—	160	18.0	15.0	4.0	4.0	1.5	—	—	0
3422	Club sandwich, 6", white bread	1	item(s)	250	—	310	23.0	45.0	2.0	6.0	2.5	—	—	0
4030	Cold cut combo sandwich, 6", white bread	1	item(s)	242	—	400	20.0	45.0	2.0	17.0	7.5	—	—	0.5
34030	Ham & egg breakfast sandwich	1	item(s)	142	—	310	16.0	35.0	3.0	13.0	3.5	—	—	0
3885	Ham sandwich, 6", white bread	1	item(s)	238	—	310	17.0	52.0	2.0	5.0	2.0	—	—	0
3888	Meatball marinara sandwich, 6", wheat bread	1	item(s)	377	—	560	24.0	63.0	7.0	24.0	11.0	—	—	1.0
4651	Meatball sandwich, 6", white bread	1	item(s)	370	—	550	23.0	61.0	5.0	24.0	11.5	—	—	1.0
15839	Melt sandwich, 6", white bread	1	item(s)	260	—	410	25.0	47.0	4.0	15.0	5.0	—	—	—
32046	Oatmeal raisin cookie	1	item(s)	45	—	200	3.0	30.0	1.0	8.0	4.0	—	—	0
16379	Oven-roasted chicken breast sandwich, 6", wheat bread	1	item(s)	238	—	330	24.0	48.0	5.0	5.0	1.5	—	—	0
32047	Peanut butter cookie	1	item(s)	45	—	220	4.0	26.0	1.0	12.0	5.0	—	—	0
4655	Roast beef sandwich, 6", wheat bread	1	item(s)	224	—	290	19.0	45.0	4.0	5.0	2.0	—	—	0
3957	Roast beef sandwich, 6", white bread	1	item(s)	217	—	280	18.0	43.0	2.0	5.0	2.5	—	—	0
16378	Roasted chicken breast, 6", white bread	1	item(s)	231	—	320	23.0	46.0	3.0	5.0	2.0	—	—	0
34028	Southwest steak & cheese sandwich, 6", Italian bread	1	item(s)	271	—	450	24.0	48.0	6.0	20.0	6.0	—	—	0
4032	Spicy Italian sandwich, 6", white bread	1	item(s)	220	—	470	20.0	43.0	2.0	25.0	9.5	—	—	0
4031	Steak & cheese sandwich, 6", white bread	1	item(s)	243	—	350	23.0	45.0	3.0	10.0	5.0	—	—	0
32050	Sugar cookie	1	item(s)	45	—	220	2.0	28.0	0.5	12.0	6.0	—	—	0
40477	Sweet onion chicken teriyaki sandwich, 6", white or wheat bread	1	serving(s)	281	—	370	26.0	59.0	4.0	5.0	1.5	—	—	0
38623	Turkey breast & bacon melt wrap with chipotle sauce	1	item(s)	228	—	380	31.0	20.0	9.0	24.0	7.0	—	—	0
15834	Turkey breast & ham sandwich, 6", white bread	1	item(s)	227	—	280	19.0	45.0	2.0	5.0	2.0	—	—	0
16376	Turkey breast sandwich, 6", white bread	1	item(s)	217	—	270	17.0	44.0	2.0	4.5	2.0	—	—	0
15841	Veggie Delite sandwich, 6", wheat bread	1	item(s)	167	—	230	9.0	44.0	4.0	3.0	1.0	—	—	0
16375	Veggie Delite, 6", white bread	1	item(s)	160	—	220	8.0	42.0	2.0	3.0	1.5	—	—	0
32051	White chip macadamia nut cookie	1	item(s)	45	—	220	2.0	29.0	0.5	11.0	5.0	—	—	0

Taco Bell

DA+ Code	Food Description	Quantity	Measure	Wt (g)	H₂O (g)	Ener (kcal)	Prot (g)	Carb (g)	Fiber (g)	Fat (g)	Sat	Mono	Poly	Trans
29906	7-Layer burrito	1	item(s)	283	—	490	17.0	65.0	9.0	18.0	7.0	—	—	1.0
744	Bean burrito	1	item(s)	198	—	340	13.0	54.0	8.0	9.0	3.5	—	—	0.5
749	Beef burrito supreme	1	item(s)	248	—	410	17.0	51.0	7.0	17.0	8.0	—	—	1.0
33417	Beef Chalupa Supreme	1	item(s)	153	—	380	14.0	30.0	3.0	23.0	7.0	—	—	0.5
34474	Beef Gordita Baja	1	item(s)	153	—	340	13.0	29.0	4.0	19.0	5.0	—	—	0
29910	Beef Gordita Supreme	1	item(s)	153	—	310	14.0	29.0	3.0	16.0	6.0	—	—	0.5
2014	Beef soft taco	1	item(s)	99	—	200	10.0	21.0	3.0	9.0	4.0	—	—	0
10860	Beef soft taco supreme	1	item(s)	135	—	250	11.0	23.0	3.0	13.0	6.0	—	—	0.5
34472	Chicken burrito supreme	1	item(s)	248	—	390	20.0	49.0	6.0	13.0	6.0	—	—	0.5
33418	Chicken Chalupa Supreme	1	item(s)	153	—	360	17.0	29.0	2.0	20.0	5.0	—	—	0
34475	Chicken Gordita Baja	1	item(s)	153	—	320	17.0	28.0	3.0	16.0	3.5	—	—	0
29909	Chicken quesadilla	1	item(s)	184	—	520	28.0	40.0	3.0	28.0	12.0	—	—	0.5
29907	Chili cheese burrito	1	item(s)	156	—	390	16.0	40.0	3.0	18.0	9.0	—	—	1.5
10794	Cinnamon twists	1	serving(s)	35	—	170	1.0	26.0	1.0	7.0	0	—	—	0
29911	Grilled chicken Gordita Supreme	1	item(s)	153	—	290	17.0	28.0	2.0	12.0	5.0	—	—	0
14463	Grilled chicken soft taco	1	item(s)	99	—	190	14.0	19.0	1.0	6.0	2.5	—	—	—
29912	Grilled Steak Gordita Supreme	1	item(s)	153	—	290	15.0	28.0	2.0	13.0	5.0	—	—	0
29904	Grilled steak soft taco	1	item(s)	128	—	270	12.0	20.0	2.0	16.0	4.5	—	—	0
29905	Grilled steak soft taco supreme	1	item(s)	135	—	235	13.0	21.0	1.0	11.0	6.0	—	—	—
2021	Mexican pizza	1	serving(s)	216	—	530	20.0	42.0	7.0	30.0	8.0	—	—	1.0
29894	Mexican rice	1	serving(s)	131	—	170	6.0	23.0	1.0	11.0	3.0	—	—	0
10772	Meximelt	1	serving(s)	128	—	280	15.0	22.0	3.0	14.0	7.0	—	—	0.5
2011	Nachos	1	serving(s)	99	—	330	4.0	32.0	2.0	21.0	3.5	—	—	2.0
2012	Nachos Bellgrande	1	serving(s)	308	—	770	19.0	77.0	12.0	44.0	9.0	—	—	3.0
2023	Pintos 'n cheese	1	serving(s)	128	—	150	9.0	19.0	7.0	6.0	3.0	—	—	0.5
34473	Steak burrito supreme	1	item(s)	248	—	380	18.0	49.0	6.0	14.0	7.0	—	—	0.5
33419	Steak Chalupa Supreme	1	item(s)	153	—	360	15.0	28.0	2.0	21.0	6.0	—	—	0
747	Taco	1	item(s)	78	—	170	8.0	13.0	3.0	10.0	3.5	—	—	0
2015	Taco salad with salsa, with shell	1	serving(s)	548	—	840	30.0	80.0	15.0	45.0	11.0	—	—	1.5

Chol (mg)	Calc (mg)	Iron (mg)	Magn (mg)	Pota (mg)	Sodi (mg)	Zinc (mg)	Vit A (µg)	Thia (mg)	Vit E (mg α)	Ribo (mg)	Niac (mg)	Vit B_6 (mg)	Fola (µg)	Vit C (mg)	Vit B_{12} (µg)	Sele (µg)
35	150	8.10	—	—	1090.0	—	—	—	—	—	—	—	—	18.0	—	—
90	250	4.50	—	—	1400.0	—	—	—	—	—	—	—	—	21.0	—	—
90	300	2.70	—	—	1680.0	—	—	—	—	—	—	—	—	9.0	—	—
15	0	1.08	—	—	150.0	—	—	—	—	—	—	—	—	0	—	—
10	20	1.00	—	—	100.0	—	—	—	—	—	—	—	—	0	—	—
10	0	1.00	—	—	100.0	—	—	—	—	—	—	—	—	0	—	—
55	150	2.70	—	—	1770.0	—	—	—	—	—	—	—	—	16.8	—	—
45	100	5.40	—	—	1030.0	—	—	—	—	—	—	—	—	21.0	—	—
45	100	3.60	—	—	1010.0	—	—	—	—	—	—	—	—	16.8	—	—
35	60	3.60	—	—	880.0	—	—	—	—	—	—	—	—	30.0	—	—
35	60	3.60	—	—	1290.0	—	—	—	—	—	—	—	—	13.8	—	—
60	150	3.60	—	—	1530.0	—	—	—	—	—	—	—	—	16.8	—	—
190	80	4.50	—	—	720.0	—	66.7	—	—	—	—	—	—	3.6	—	—
25	60	2.70	—	—	1375.0	—	—	—	—	—	—	—	—	13.8	—	—
45	200	7.20	—	—	1610.0	—	—	—	—	—	—	—	—	36.0	—	—
45	200	5.40	—	—	1590.0	—	—	—	—	—	—	—	—	31.8	—	—
45	150	5.40	—	—	1720.0	—	—	—	—	—	—	—	—	24.0	—	—
15	20	1.08	—	—	170.0	—	—	—	—	—	—	—	—	0	—	—
45	60	4.50	—	—	1020.0	—	—	—	—	—	—	—	—	18.0	—	—
15	20	0.72	—	—	200.0	—	—	—	—	—	—	—	—	0	—	—
20	60	6.30	—	—	920.0	—	—	—	—	—	—	—	—	18.0	—	—
20	60	4.50	—	—	900.0	—	—	—	—	—	—	—	—	13.8	—	—
45	60	2.70	—	—	1000.0	—	—	—	—	—	—	—	—	13.8	—	—
45	150	8.10	—	—	1310.0	—	—	—	—	—	—	—	—	21.0	—	—
55	60	2.70	—	—	1650.0	—	—	—	—	—	—	—	—	16.8	—	—
35	150	6.30	—	—	1070.0	—	—	—	—	—	—	—	—	13.8	—	—
15	0	0.72	—	—	140.0	—	—	—	—	—	—	—	—	0	—	—
50	80	4.50	—	—	1220.0	—	—	—	—	—	—	—	—	24.0	—	—
50	200	2.70	—	—	1780.0	—	—	—	—	—	—	—	—	6.0	—	—
25	60	2.70	—	—	1210.0	—	—	—	—	—	—	—	—	13.8	—	—
20	60	2.70	—	—	1000.0	—	—	—	—	—	—	—	—	13.8	—	—
0	60	4.50	—	—	520.0	—	—	—	—	—	—	—	—	18.0	—	—
0	60	2.70	—	—	500.0	—	—	—	—	—	—	—	—	13.8	—	—
15	20	0.72	—	—	160.0	—	—	—	—	—	—	—	—	0	—	—
25	250	5.40	—	—	1350.0	—	—	—	—	—	—	—	—	15.0	—	—
5	200	4.50	—	—	1190.0	—	5.9	—	—	—	—	—	—	4.8	—	—
40	200	4.50	—	—	1340.0	—	9.9	—	—	—	—	—	—	6.0	—	—
40	150	2.70	—	—	620.0	—	—	—	—	—	—	—	—	3.6	—	—
35	100	2.70	—	—	780.0	—	—	—	—	—	—	—	—	2.4	—	—
40	150	2.70	—	—	620.0	—	—	—	—	—	—	—	—	3.6	—	—
25	100	1.80	—	—	630.0	—	—	—	—	—	—	—	—	1.2	—	—
40	150	2.70	—	—	650.0	—	—	—	—	—	—	—	—	3.6	—	—
45	200	4.50	—	—	1360.0	—	—	—	—	—	—	—	—	9.0	—	—
45	100	2.70	—	—	650.0	—	—	—	—	—	—	—	—	4.8	—	—
40	100	1.80	—	—	800.0	—	—	—	—	—	—	—	—	3.6	—	—
75	450	3.60	—	—	1420.0	—	—	—	—	—	—	—	—	1.2	—	—
40	300	1.80	—	—	1080.0	—	—	—	—	—	—	—	—	0	—	—
0	0	0.37	—	—	200.0	0	—	—	—	—	—	—	—	0	—	—
45	150	1.80	—	—	650.0	—	—	—	—	—	—	—	—	4.8	—	—
30	100	1.08	—	—	550.0	—	14.6	—	—	—	—	—	—	1.2	—	—
40	100	2.70	—	—	530.0	—	—	—	—	—	—	—	—	3.6	—	—
35	100	2.70	—	—	660.0	—	—	—	—	—	—	—	—	3.6	—	—
35	120	1.44	—	—	565.0	—	29.2	—	—	—	—	—	—	3.6	—	—
40	350	3.60	—	—	1000.0	—	—	—	—	—	—	—	—	4.8	—	—
15	100	1.44	—	—	790.0	—	—	—	—	—	—	—	—	3.6	—	—
40	250	2.70	—	—	880.0	—	—	—	—	—	—	—	—	2.4	—	—
3	80	0.71	—	—	530.0	—	0	—	—	—	—	—	—	0	—	—
35	200	3.60	—	—	1280.0	—	—	—	—	—	—	—	—	4.8	—	—
15	150	1.44	—	—	670.0	—	—	—	—	—	—	—	—	3.6	—	—
35	200	4.50	—	—	1250.0	—	9.9	—	—	—	—	—	—	9.0	—	—
40	100	2.70	—	—	530.0	—	—	—	—	—	—	—	—	3.6	—	—
25	80	1.08	—	—	350.0	—	—	—	—	—	—	—	—	1.2	—	—
65	450	7.20	—	—	1780.0	—	—	—	—	—	—	—	—	12.0	—	—

DA+ Code	Food Description	Quantity	Measure	Wt (g)	H₂O (g)	Ener (kcal)	Prot (g)	Carb (g)	Fiber (g)	Fat (g)	Fat Breakdown (g)			
											Sat	Mono	Poly	Trans
	FAST FOOD —Continued													
14459	Taco supreme	1	item(s)	113	—	210	9.0	15.0	3.0	13.0	6.0	—	—	0
748	Tostada	1	item(s)	170	—	230	11.0	27.0	7.0	10.0	3.5	—	—	0.5
	CONVENIENCE MEALS													
	Banquet													
29961	Barbeque chicken meal	1	item(s)	281	—	330	16.0	37.0	2.0	13.0	3.0	—	—	—
14788	Boneless white fried chicken meal	1	item(s)	286	—	310	10.0	21.0	4.0	20.0	5.0	—	—	—
29960	Fish sticks meal	1	item(s)	207	—	470	13.0	58.0	1.0	20.0	3.5	—	—	—
29957	Lasagna with meat sauce meal	1	item(s)	312	—	320	15.0	46.0	7.0	9.0	4.0	—	—	—
14777	Macaroni and cheese meal	1	item(s)	340	—	420	15.0	57.0	5.0	14.0	8.0	—	—	—
1741	Meatloaf meal	1	item(s)	269	—	240	14.0	20.0	4.0	11.0	4.0	—	—	—
39418	Pepperoni pizza meal	1	item(s)	191	—	480	11.0	56.0	5.0	23.0	8.0	—	—	—
33759	Roasted white turkey meal	1	item(s)	255	—	230	14.0	30.0	5.0	6.0	2.0	—	—	—
1743	Salisbury steak meal	1	item(s)	269	196.9	380	12.0	28.0	3.0	24.0	12.0	—	—	—
	Budget Gourmet													
1914	Cheese manicotti with meat sauce entree	1	item(s)	284	194.0	420	18.0	38.0	4.0	22.0	11.0	6.0	1.3	—
1915	Chicken with fettucini entree	1	item(s)	284	—	380	20.0	33.0	3.0	19.0	10.0	—	—	—
3986	Light beef stroganoff entrée	1	item(s)	248	177.0	290	20.0	32.0	3.0	7.0	4.0	—	—	—
3996	Light sirloin of beef in herb sauce entrée	1	item(s)	269	214.0	260	19.0	30.0	5.0	7.0	4.0	2.3	0.3	—
3987	Light vegetable lasagna entrée	1	item(s)	298	227.0	290	15.0	36.0	4.8	9.0	1.8	0.9	0.6	—
	Healthy Choice													
9425	Cheese French bread pizza	1	item(s)	170	—	340	22.0	51.0	5.0	5.0	1.5	—	—	—
9306	Chicken enchilada suprema meal	1	item(s)	320	251.5	360	13.0	59.0	8.0	7.0	3.0	2.0	2.0	—
3821	Familiar Favorites lasagna bake with meat sauce entrée	1	item(s)	255	—	270	13.0	38.0	4.0	7.0	2.5	—	—	—
13744	Familiar Favorites sesame chicken with vegetables and rice entrée	1	item(s)	255	—	260	17.0	34.0	4.0	6.0	2.0	2.0	2.0	—
9316	Lemon pepper fish meal	1	item(s)	303	—	280	11.0	49.0	5.0	5.0	2.0	1.0	2.0	—
9322	Traditional salisbury steak meal	1	item(s)	354	250.3	360	23.0	45.0	5.0	9.0	3.5	4.0	1.0	—
9359	Traditional turkey breasts meal	1	item(s)	298	—	330	21.0	50.0	4.0	5.0	2.0	1.5	1.5	—
	Stouffers													
2313	Cheese French bread pizza	1	serving(s)	294	—	380	15.0	43.0	3.0	16.0	6.0	—	—	—
11138	Cheese manicotti with tomato sauce entrée	1	item(s)	255	—	360	18.0	41.0	2.0	14.0	6.0	—	—	—
2366	Chicken pot pie entrée	1	item(s)	284	—	740	23.0	56.0	4.0	47.0	18.0	12.4	10.5	—
11116	Homestyle baked chicken breast with mashed potatoes and gravy entrée	1	item(s)	252	—	270	21.0	21.0	2.0	11.0	3.5	—	—	—
11146	Homestyle beef pot roast and potatoes entrée	1	item(s)	252	—	260	16.0	24.0	3.0	11.0	4.0	—	—	—
11152	Homestyle roast turkey breast with stuffing and mashed potatoes entrée	1	item(s)	273	—	290	16.0	30.0	2.0	12.0	3.5	—	—	—
11043	Lean Cuisine Comfort Classics baked chicken and whipped potatoes and stuffing entrée	1	item(s)	245	—	240	15.0	34.0	3.0	4.5	1.0	2.0	1.0	0
11046	Lean Cuisine Comfort Classics honey mustard chicken with rice pilaf entrée	1	item(s)	227	—	250	17.0	37.0	1.0	4.0	1.0	1.0	1.0	0
9479	Lean Cuisine Deluxe French bread pizza	1	item(s)	174	—	310	16.0	44.0	3.0	9.0	3.5	0.5	0.5	0
360	Lean Cuisine One Dish Favorites chicken chow mein with rice	1	item(s)	255	—	190	13.0	29.0	2.0	2.5	0.5	1.0	0.5	0
11054	Lean Cuisine One Dish Favorites chicken enchilada Suiza with Mexican-style rice	1	serving(s)	255	—	270	10.0	47.0	3.0	4.5	2.0	1.5	1.0	0
9467	Lean Cuisine One Dish Favorites fettucini alfredo entree	1	item(s)	262	—	270	13.0	39.0	2.0	7.0	3.5	2.0	1.0	0
11055	Lean Cuisine One Dish Favorites lasagna with meat sauce entrée	1	item(s)	298	—	320	19.0	44.0	4.0	7.0	3.0	2.0	0.5	0
	Weight Watchers													
11164	Smart Ones chicken enchiladas suiza entrée	1	item(s)	255	—	340	12.0	38.0	3.0	10.0	4.5	—	—	—
39763	Smart Ones chicken oriental entrée	1	item(s)	255	—	230	15.0	34.0	3.0	4.5	1.0	—	—	—
11187	Smart Ones pepperoni pizza	1	item(s)	198	—	400	22.0	58.0	4.0	9.0	3.0	—	—	—
39765	Smart Ones spaghetti bolognese entrée	1	item(s)	326	—	280	17.0	43.0	5.0	5.0	2.0	—	—	—
31512	Smart Ones spicy szechuan style vegetables & chicken	1	item(s)	255	—	220	11.0	34.0	4.0	5.0	1.0	—	—	—
	BABY FOODS													
787	Apple juice	4	fluid ounce(s)	127	111.6	60	0	14.8	0.1	0.1	0	0	0	—
778	Applesauce, strained	4	tablespoon(s)	64	56.7	26	0.1	6.9	1.1	0.1	0	0	0	—
779	Bananas with tapioca, strained	4	tablespoon(s)	60	50.4	34	0.2	9.2	1.0	0	0	0	0	—
604	Carrots, strained	4	tablespoon(s)	56	51.7	15	0.4	3.4	1.0	0.1	0	0	0	—
770	Chicken noodle dinner, strained	4	tablespoon(s)	64	54.8	42	1.7	5.8	1.3	1.3	0.4	0.5	0.3	—
801	Green beans, strained	4	tablespoon(s)	60	55.1	16	0.7	3.8	1.3	0.1	0	0	0	—
910	Human milk, mature	2	fluid ounce(s)	62	53.9	43	0.6	4.2	0	2.7	1.2	1.0	0.3	—
760	Mixed cereal, prepared with whole milk	4	ounce(s)	113	84.6	128	5.4	18.0	1.5	4.0	2.2	1.2	0.4	—
772	Mixed vegetable dinner, strained	2	ounce(s)	57	50.3	23	0.7	5.4	0.8	0	—	—	0	—
762	Rice cereal, prepared with whole milk	4	ounce(s)	113	84.6	130	4.4	18.9	0.1	4.1	2.6	1.0	0.2	—
758	Teething biscuits	1	item(s)	11	0.7	44	1.0	8.6	0.2	0.6	0.2	0.2	0.1	—

Chol (mg)	Calc (mg)	Iron (mg)	Magn (mg)	Pota (mg)	Sodi (mg)	Zinc (mg)	Vit A (µg)	Thia (mg)	Vit E (mg α)	Ribo (mg)	Niac (mg)	Vit B6 (mg)	Fola (µg)	Vit C (mg)	Vit B12 (µg)	Sele (µg)
40	100	1.08	—	—	370.0	—	—	—	—	—	—	—	—	3.6	—	—
15	200	1.80	—	—	730.0	—	—	—	—	—	—	—	—	4.8	—	—
																—
50	40	1.08	—	—	1210.0	—	0	—	—	—	—	—	—	4.8	—	—
45	80	1.44	—	—	1200.0	—	—	—	—	—	—	—	—	18.0	—	—
55	20	1.44	—	—	710.0	—	—	—	—	—	—	—	—	0	—	—
20	100	2.70	—	—	1170.0	—	—	—	—	—	—	—	—	0	—	—
20	150	1.44	—	—	1330.0	—	0	—	—	—	—	—	—	0	—	—
30	0	1.80	—	—	1040.0	—	0	—	—	—	—	—	—	0	—	—
35	150	1.80	—	—	870.0	—	0	—	—	—	—	—	—	0	—	—
25	60	1.80	—	—	1070.0	—	—	—	—	—	—	—	—	3.6	—	—
60	40	1.44	—	—	1140.0	—	0	—	—	—	—	—	—	0	—	—
85	300	2.70	45.4	484.0	810.0	2.29	—	0.45	—	0.51	4.00	0.22	30.7	0	0.7	—
85	100	2.70	—	—	810.0	—	—	0.15	—	0.42	6.00	—	—	0	—	—
35	40	1.80	38.9	280.0	580.0	4.71	—	0.17	—	0.36	4.28	0.27	18.9	2.4	2.5	—
30	40	1.80	57.7	540.0	850.0	4.81	—	0.15	—	0.29	5.53	0.37	38.4	6.0	1.6	—
15	283	3.03	78.5	420.0	780.0	1.39	—	0.22	—	0.45	3.13	0.32	74.8	59.1	0.2	—
10	350	3.60	—	—	600.0	—	—	—	—	—	—	—	—	0	—	—
30	40	1.44	—	—	580.0	—	—	—	—	—	—	—	—	3.6	—	—
20	100	1.80	—	—	600.0	—	—	—	—	—	—	—	—	0	—	—
35	18	0.72	—	—	580.0	—	—	—	—	—	—	—	—	12.0	—	—
35	20	0.36	—	—	580.0	—	—	—	—	—	—	—	—	30.0	—	—
45	80	2.70	—	—	580.0	—	—	—	—	—	—	—	—	21.0	—	—
35	40	1.80	—	—	600.0	—	—	—	—	—	—	—	—	0	—	—
30	200	1.80	—	230.0	660.0	—	—	—	—	—	—	—	—	2.4	—	—
70	250	1.44	—	550.0	920.0	—	—	—	—	—	—	—	—	6.0	—	—
65	150	2.70	—	—	1170.0	—	—	—	—	—	—	—	—	2.4	—	—
55	20	0.72	—	490.0	770.0	—	0	—	—	—	—	—	—	0	—	—
35	20	1.80	—	800.0	960.0	—	—	—	—	—	—	—	—	6.0	—	—
45	40	1.08	—	490.0	970.0	—	—	—	—	—	—	—	—	3.6	—	—
25	40	1.16	—	500.0	650.0	—	—	—	—	—	—	—	—	3.6	—	—
30	64	0.38	—	370.0	650.0	—	—	—	—	—	—	—	—	0	—	—
20	150	2.70	—	300.0	700.0	—	—	—	—	—	—	—	—	15.0	—	—
25	40	0.72	—	380.0	650.0	—	—	—	—	—	—	—	—	2.4	—	—
20	150	0.72	—	350.0	510.0	—	—	—	—	—	—	—	—	2.4	—	—
15	200	0.72	—	290.0	690.0	—	0	—	—	—	—	—	—	0	—	—
30	250	1.47	—	610.0	690.0	—	—	—	—	—	—	—	—	2.4	—	—
40	200	0.72	—	—	800.0	—	—	—	—	—	—	—	—	2.4	—	—
35	40	0.72	—	—	790.0	—	—	—	—	—	—	—	—	6.0	—	—
15	200	1.08	—	401.0	700.0	—	69.1	—	—	—	—	—	—	4.8	—	—
15	150	3.60	—	—	670.0	—	—	—	—	—	—	—	—	9.0	—	—
10	40	1.44	—	—	890.0	—	—	—	—	—	—	—	—	0	—	—
0	5	0.72	3.8	115.4	3.8	0.03	1.3	0.01	0.76	0.02	0.10	0.03	0	73.4	0	0.1
0	3	0.12	1.9	45.4	1.3	0.01	0.6	0.01	0.36	0.02	0.04	0.02	1.3	24.5	0	0.2
0	3	0.12	6.0	52.8	5.4	0.04	1.2	0.01	0.36	0.02	0.08	0.04	3.6	10.0	0	0.4
0	12	0.20	5.0	109.8	20.7	0.08	320.9	0.01	0.29	0.02	0.25	0.04	8.4	3.2	0	0.1
10	17	0.40	9.0	89.0	14.7	0.32	70.4	0.03	0.12	0.04	0.44	0.04	7.0	0.2	0	2.4
0	23	0.40	12.0	87.6	3.0	0.12	10.8	0.02	0.04	0.04	0.20	0.02	14.4	0.2	0	0
9	20	0.02	1.8	31.4	10.5	0.10	37.6	0.01	0.04	0.02	0.10	0.01	3.1	3.1	0	1.1
12	249	11.82	30.6	225.7	53.3	0.80	28.4	0.49	—	0.65	6.54	0.07	12.5	1.4	0.3	—
—	12	0.18	6.2	68.6	4.5	0.08	77.1	0.01	—	0.02	0.28	0.04	4.5	1.6	0	0.4
12	271	13.82	51.0	215.5	52.2	0.72	24.9	0.52	—	0.56	5.90	0.12	9.1	1.4	0.3	4.0
0	11	0.39	3.9	35.5	28.4	0.10	3.1	0.02	0.02	0.05	0.47	0.01	5.4	1.0	0	2.6

Glossary

absorption the passage of nutrients or substances into cells or tissues; nutrients pass into intestinal cells after digestion and then into the circulatory system (for example, into the bloodstream).

Acceptable Macronutrient Distribution Range (AMDR) a range of intakes for a particular energy source (carbohydrates, fat, protein) that is associated with a reduced risk of chronic disease while providing adequate intakes of essential nutrients.

accreditation approval; in the case of hospitals or university departments, approval by a professional organization of the educational program offered. There are phony accrediting agencies; the genuine ones are listed in a directory called *Accredited Institutions of Postsecondary Education.*

acesulfame-K (AY-see-sul-fame) a derivative of acetoacetic acid approved for use in the United States in 1988. Because it is not metabolized by the body, acesulfame-K does not contribute calories and is excreted from the body unchanged. It is currently found in more than 100 international products, including chewing gum, gelatins, nondairy creamers, powdered drink mixes, and puddings.

acetaldehyde (ass-et-AL-duh-hide) a substance into which drinking alcohol (ethanol) is metabolized.

acid–base balance equilibrium between acid and base concentrations in the body fluids.

acidosis (a-sih-DOSE-is) blood acidity above normal, indicating excess acid.

acids compounds that release hydrogens in a watery solution; acids have a low pH.

activities of daily living (ADL) include bathing, dressing, grooming, transferring from bed to chair, going to the bathroom, and feeding oneself.

added sugars sugars and other caloric sweeteners that are added to foods during processing or preparation. Added sugars do not include naturally occurring sugars such as those found in milk and fruits.

adequacy characterizes a diet that provides all of the essential nutrients, fiber, and energy (calories) in amounts sufficient to maintain health.

Adequate Intake (AI) the average amount of a nutrient that appears to be adequate for individuals when there is not sufficient scientific research to calculate an RDA. The AI exceeds the EAR and possibly the RDA.

ADL *see* activities of daily living.

adverse reaction an unusual response to food, including food allergies and food intolerances.

aerobic requiring oxygen.

aflatoxin a poisonous toxin produced by molds.

agave (uh-GAH-vee) a plant with spiny-margined leaves and flower.

age-related macular degeneration (AMD) oxidative damage to the central portion of the eye (the *macula*) that allows you to focus and see details clearly (peripheral vision remains unimpaired). The carotenoids lutein and zeaxanthin (found in broccoli, Brussels sprouts, kale, spinach, corn, lettuce, and peas) may protect normal macular function.

AI *see* Adequate Intake.

AIDS (acquired immune deficiency syndrome) an immune system disorder caused by the human immunodeficiency virus (HIV).

alcohol clear, colorless, volatile liquid; the most commonly ingested form is ethyl alcohol or ethanol.

alcohol abuse continued use of alcohol despite negative psychological, social, family, employment, or school problems caused by the alcohol use.

alcohol dehydrogenase a liver enzyme that mediates the metabolism of alcohol.

alcohol dependency (alcoholism) a condition marked by compulsive, uncontrollable drinking that has negative effects on the affected person's physical health, family relationships, and social health.

alcoholic hepatitis inflammation and injury to the liver due to excess alcohol consumption.

alkalosis (al-kuh-LOW-sis) blood alkalinity above normal.

alternative sweeteners nutritive (calorie-containing) sweeteners such as fructose, sorbitol, mannitol, and xylitol.

amaranth a golden-colored grain.

AMDR *see* Acceptable Macronutrient Distribution Range.

amenorrhea cessation of menstruation associated with strenuous athletic training.

amine (a-MEEN) group the nitrogen-containing portion of an amino acid.

amino (uh-MEEN-oh) acids building blocks of protein; each is a compound with an amine group at one end, an acid group at the other, and a distinctive side chain.

anabolic steroids synthetic male hormones (related to testosterone) that stimulate growth of body tissues, with many adverse effects.

anaerobic not requiring oxygen.

anaphylaxis (an-uh-fuh-LACKS-is) a potentially fatal reaction to a food allergen that causes reduced oxygen supply to the heart and other body tissues. Symptoms include difficulty breathing; low blood pressure; pale skin; a weak, rapid pulse; and loss of consciousness.

anemia any condition in which the blood is unable to deliver oxygen to the cells of the body. Examples include a shortage or abnormality of the red blood cells. Many nutrient deficiencies and diseases can cause anemia.

anorexia nervosa literally, "nervous lack of appetite," a disorder (usually seen in teenage girls) involving self-starvation to the extreme (*an* = without; *orexis* = appetite).

antibodies large proteins of the blood and body fluids that are produced by one type of immune cell in response to invasion of the body by unfamiliar molecules (mostly foreign proteins). Antibodies inactivate the foreign substances and so protect the body. The foreign substances are called *antigens.*

antioxidant a compound that protects other compounds from oxygen by itself reacting with oxygen; that is, it helps prevent damage done to the body as a result of chemical reactions involving oxygen.

antioxidant nutrients vitamins and minerals that protect other compounds from damaging reactions involving oxygen by themselves reacting with oxygen. The antioxidant nutrients are vitamin C, vitamin E, and

beta-carotene. The mineral selenium also has a role in antioxidant reactions in the body.

appendicitis inflammation and/or infection of the appendix, a sac protruding from the large intestine.

appetite the psychological desire to find and eat food, experienced as a pleasant sensation, often in the absence of hunger.

appropriate technology a technology that utilizes locally abundant resources in preference to locally scarce resources. For developing countries, which usually have a large labor force and little capital, the appropriate technology would therefore be labor intensive.

arousal as used in this context, heightened activity of certain brain centers associated with excitement and anxiety.

artesian water or artesian well water water drawn from a well that taps a confined water-bearing rock or rock formation.

artificial sweeteners nonnutritive sugar replacements such as acesulfame-K, aspartame, neotame, saccharin, and sucralose.

aspartame a dipeptide containing the amino acids aspartic acid and phenylalanine and used in the United States and Canada since 1981. Although it is digested as protein and supplies calories, it is so sweet that only small amounts, which contribute negligible calories, are needed to sweeten foods. Thus, it is classified as a nonnutritive sweetener. Aspartame is often sold under the trade name NutraSweet, and it is also blended with lactose and an anticaking agent and sold commercially as Equal.

atherosclerosis (ATH-er-oh-skler-OH-sis) a type of cardiovascular disease characterized by the formation of fatty deposits, or plaques, in the inner walls of the arteries.

atrophic gastritis an age-related condition characterized by the stomach's inability to produce enough acid, which in turn leads to vitamin B_{12} deficiencies.

atrophy a decrease in size in response to disuse.

balance a feature of a diet that provides a number of types of foods in harmony with one another, such that foods rich in one nutrient do not crowd out of the diet foods that are rich in another nutrient.

basal metabolic rate (BMR) the rate at which the body spends energy to support its basal metabolism. The BMR accounts for the largest component of a person's daily energy (calorie) needs.

basal metabolism the sum total of all the chemical activities of the cells necessary to sustain life, exclusive of voluntary activities; that is, the ongoing activities of the cells when the body is at rest.

bases compounds that accept hydrogens from a watery solution; bases have a high pH.

bee pollen mixture of bee saliva, plant nectar, and pollen touted falsely to enhance athletic performance. May cause allergic reactions in people with sensitivity to bee stings and honey allergies.

behavior modification a process developed by psychologists for helping people make lasting behavior changes.

beriberi the thiamin deficiency disease, characterized by irregular heartbeat, paralysis, and extreme wasting of muscle tissue.

beta-carotene an orange pigment found in plants that is converted into vitamin A inside the body. Beta-carotene is also an antioxidant.

bialy a flat breakfast roll that is softer than a bagel.

bicarbonate a chemical that neutralizes acid; a secretion of the pancreas.

bifidus factor (BIFF-id-us) a factor in colostrum and breast milk that favors growth in the infant's intestinal tract of the "friendly" bacteria *Lactobacillus bifidus*; colonization with friendly bacteria means that other, less desirable intestinal inhabitants will not flourish.

bile a mixture of compounds, including cholesterol, made by the liver, stored in the gallbladder, and secreted into the small intestine. Bile emulsifies lipids to ready them for enzymatic digestion and helps transport them into the intestinal wall cells.

binders in foods, chemical compounds that can combine with nutrients (especially minerals) to form complexes the body cannot absorb. For example, *phytates* (FYE-tates) in whole grains and *oxalates* (in vegetables) may bind iron and calcium and interfere with their absorption to a minor extent.

bing thin pancakes.

binge-eating disorder an eating disorder characterized by uncontrolled chronic episodes of overeating (compulsive overeating) without other symptoms of eating disorders. Typically, the episodes of binge eating occur at least twice a week for a period of six months or more.

bioelectrical impedance estimation of body fat content made by measuring how quickly electrical current is conducted through the body.

biological value (BV) a measure of protein quality, assessed by determining how well a given food or food mixture supports nitrogen retention.

biotechnology the use of biological systems or living organisms to make or modify products. Includes traditional methods used in making products such as wine, beer, yogurt, and cheese; cross-breeding to enhance crop production; and modification of living plants, animals, and fish through the manipulation of genes (genetic engineering).

bisphosphonates drugs that decrease the risk of fractures by acting on the bone-dismantling cells (osteoclasts) and inhibiting their resorption of bone tissue; examples are Fosamax, Actonel, and Boniva.

black-eyed peas small, oval-shaped, creamy-white peas with a black spot. They have a vegetable flavor with mealy texture. Use in salads with rice and greens.

black, Cuban, or turtle beans medium-size, black-skinned oval beans that have a rich, sweet taste. They are best served in Mexican and Latin American dishes or thick soups and stews.

BMI *see* body mass index.

BMR *see* basal metabolic rate.

body mass index (BMI) an index of a person's weight in relation to height that correlates with total body fat content.

bok choy a vegetable with broad, white or greenish-white stalks and dark-green leaves; also called *Chinese chard*.

bolillo (buh-LEE-oh) a roll-like bread often used instead of tortillas or to make sandwiches.

bone density a measure of bone strength that reflects the degree of bone mineralization. The DEXA (dual-energy X-ray absorptiometry) bone-density test compares your bone density to that of a healthy young adult.

buffers compounds that help keep the acidity (amount of acid) or alkalinity (amount of base) of a solution constant.

bulimia nervosa, bulimarexia (byoo-LEE-me-uh, byoo-lee-ma-REX-ee-uh) binge eating (literally, "eating like an ox"), combined with an intense fear of becoming fat and usually followed by self-induced vomiting or taking laxatives (*buli* = ox).

burritos warm flour tortillas stuffed with a mixture of egg, meat, beans, and/or avocado

BV *see* biological value.

caffeine a type of compound, called a methylxanthine, found in coffee beans, cola nuts, cocoa beans, and tea leaves. A central nervous system stimulant, caffeine's effects include increasing the heart rate, boosting urine production, and raising the metabolic rate.

caffeine dependence syndrome dependence on caffeine characterized by at least three of the four following criteria: withdrawal symptoms such as headache and fatigue; caffeine consumption despite knowledge that it may be causing harm; repeated, unsuccessful attempts to cut back on caffeine; and tolerance to caffeine.

calcitonin a hormone used as a drug to decrease the rate of bone loss in osteoporosis. Administered as a nasal spray (Miacalcin) or by injection, calcitonin works by inhibiting the bone resorption activity of osteoclasts.

calorie the unit used to measure energy.

calorie control control of consumption of energy (calories); a feature of a sound diet plan.

capillaries minute, weblike blood vessels that connect arteries to veins and permit transfer of materials between blood and tissues.

carbohydrates compounds made of single sugars or multiple sugars and composed of carbon, hydrogen, and oxygen atoms (*carbo* = carbon (C); *hydrate* = water (H_2O)).

cardiovascular conditioning or **training effect** the effect of regular exercise on the cardiovascular system; includes improvements in heart, lung, and muscle function and increased blood volume; also called *cardiorespiratory endurance* or *cardiorespiratory training*.

carnitine (KAR-nih-teen) a compound synthesized in the body from two amino acids (lysine and methionine) and required in fat metabolism. Falsely claimed to increase the use of fatty acids and spare glycogen during exercise, delay fatigue, and decrease body fat.

carotenoids (kah-ROW-teh-noyds) a group of pigments (yellow, orange, and red) found in foods.

cassava a starchy root that is never eaten raw because it must be cooked to eliminate its bitter smell.

cellophane noodles thin, translucent noodles made from mung beans.

cells the smallest units in which independent life can exist. All living things are single cells or organisms made of cells.

central obesity excess fat on the abdomen and around the trunk. *Peripheral obesity* is excess fat on the arms, thighs, hips, and buttocks.

certified organic foods crops or livestock grown and processed according to USDA regulations concerning use of pesticides, herbicides, fungicides, fertilizers, preservatives, other synthetic chemicals, growth hormones, antibiotics, or other drugs.

challah (KAH-luh) an egg-containing yeast bread, often braided, typically served on the Sabbath and holidays.

chapatti a flat, unleavened bread made with finely milled whole-meal flour.

cherimoya a fruit with a rough green outer skin and sherbet-like flesh.

chickpeas *see* garbanzo beans.

chilaquiles (chee-lah-KEE-lehs) tortilla casserole often made with eggs or meat.

chiles relleños (CHEE-lehs Ray-EN-yose) roasted mild green chili peppers stuffed with cheese, dipped in egg batter, and fried.

Chinese broccoli a green leafy vegetable often stir-fried; also called *Chinese kale.*

chitterlings (chitlins) pig intestine.

chlorophyll the green pigment of plants that traps energy from sunlight and uses this energy in photosynthesis (the synthesis of carbohydrate by green plants).

cholesterol (koh-LESS-ter-all) one of the sterols, manufactured in the body for a variety of purposes.

choline a nonessential nutrient used by the body to synthesize various compounds, including the phospholipid lecithin; the body can make choline from the amino acid methionine.

chorizo (chuh-REE-zoh) spicy beef or pork sausage.

choy sum a bright green vegetable commonly stir fried; also called *field mustard* or *Chinese flowering cabbage.*

chromium picolinate Chromium is an essential component of the glucose tolerance factor, which facilitates the action of insulin in the body. Picolinate is a natural derivative of the amino acid tryptophan. Falsely promoted to increase muscle mass, decrease body fat, enhance energy, and promote weight loss.

chylomicron (KIGH-loh-MY-cron) a type of lipoprotein that transports newly digested fat—mostly triglyceride—from the intestine through lymph and blood.

cirrhosis a chronic, degenerative disease of the liver in which the liver cells become infiltrated with fibrous tissues; blood flow through the liver is obstructed, causing back pressure and

eventually leading to coma and death unless the cause of the disease is removed; the most common cause of cirrhosis is chronic alcohol abuse.

club soda artificially carbonated water containing added salts and minerals.

coenzyme Q10 a lipid made by the body and used by cells in energy metabolism; falsely touted to increase exercise performance and stamina in athletes; potential antioxidant role. May increase oxygen use and stamina in patients with heart disease, but no significant effect is seen in healthy athletes.

coenzymes enzyme helpers; small molecules that interact with enzymes and enable them to do their work. Many coenzymes are made from water-soluble vitamins.

cofactor a mineral element that, like a coenzyme, works with an enzyme to facilitate a chemical reaction.

collagen (KAHL-uh-jen) the characteristic protein of connective tissue (*kolla* = glue; *gennan* = to produce).

colon cancer cancer of the large intestine (colon), the terminal portion of the digestive tract.

colostrum (kuh-LAHS-trum) a milk-like secretion from the breast, rich in protective factors, present during the first day or so after delivery and before milk appears.

complementary proteins two or more food proteins whose amino acid assortments complement each other in such a way that the essential amino acids limited in or missing from each are supplied by the others.

complete proteins proteins containing all the essential amino acids in the right proportion relative to need. The *quality* of a food protein is judged by the proportions of essential amino acids that it contains relative to our needs. Animal and soy proteins are the highest in quality.

complex carbohydrates long chains of sugars (glucose) arranged as starch or fiber; also called *polysaccharides* (*poly* = many; *saccharide* = sugar unit).

constipation hardness and dryness of bowel movements associated with discomfort in passing them.

contaminants potentially dangerous substances, such as lead, that can accidentally get into foods.

contamination iron iron found in foods as the result of contamination by inorganic iron salts from iron cookware, iron-containing soils, and the like.

control group a group of individuals with characteristics that match those of the group being treated in an intervention study but who receive a sham treatment or no treatment at all.

correlation a simultaneous change in two factors, such as a decrease in blood pressure with regular aerobic activity (a direct or positive correlation) or the decrease in incidence of bone fractures with increasing calcium intakes (an inverse or negative correlation).

correspondence school a school from which courses can be taken and degrees granted by mail (or online). Schools that are accredited offer respectable courses and degrees.

cortex an outer covering; in the brain, that part in which conscious thought takes place.

cortical bone the dense outer ivory-like layer of bone that provides an exterior shell over trabecular bone.

creatine A nitrogen-containing substance made by the body that combines with phosphate to form the high-energy compound creatine phosphate (CP). CP is stored and used by muscle for ATP production. Creatine is used as a dietary supplement to increase lean body mass and strength. Some (not all) studies show that creatine may increase CP content in muscles and improve short-term (< 30 seconds) strenuous exercise performance (for instance, sprinting, weight lifting). Long-term effects are unknown.

cretinism (CREE-tin-izm) severe mental and physical retardation of an infant caused by iodine deficiency during pregnancy.

cross-contamination the inadvertent transfer of bacteria from one food to another that occurs, for instance, by chopping vegetables on the same cutting board that was used to skin poultry.

cross-reaction the reaction of one antigen with antibodies developed against another antigen.

culture knowledge, beliefs, customs, laws, morals, art, and literature acquired by members of a society and passed along to succeeding generations.

daily limit for empty calories the balance of calories remaining in a person's energy allowance, after accounting for the number of calories needed to meet recommended nutrient intakes through consumption of nutrient-dense foods in low-fat or no-added-sugar forms.

Daily Value the amount of fat, sodium, fiber, and other nutrients health experts say should make up a healthful diet. The "% Daily Values" that appear on food labels tell you the

percentage of a nutrient that a serving of the food contributes to a healthful diet.

degenerative disease chronic disease characterized by deterioration of body organs as a result of misuse and neglect. Poor eating habits, smoking, lack of exercise, and other lifestyle habits often contribute to degenerative diseases including heart disease, cancer, osteoporosis, and diabetes.

dehydroepiandrosterone *see* DHEA.

denaturation a change in the shape of a protein brought about by heat, alcohol, acids, bases, salts of heavy metals, or other agents.

dental caries decay of the teeth; tooth cavities (*caries* = rottenness).

dental plaque a colorless film, consisting of bacteria and their byproducts, that is constantly forming on the teeth.

designer estrogens (selective estrogen receptor modulators—SERMs) drugs that act on *estrogen receptors* in osteoblasts to promote an increase in bone mass; an example is raloxifene (Evista).

DHEA (dehydroepiandrosterone) a precursor of the hormones testosterone and estrogen; falsely promoted to increase production of testosterone, build muscle, burn fat, and delay the effects of aging. Long-term effects unknown; self-supplementation not recommended.

diabetes (dye-uh-BEET-eez) a disorder (technically termed *diabetes mellitus*) characterized by insufficiency or relative ineffectiveness of insulin, which renders a person unable to regulate the blood glucose level normally.

Dietary Reference Intakes (DRI) a set of reference values for energy and nutrients that can be used for planning and assessing diets for healthy people.

dietetic technician, registered (DTR) nutrition professionals who work as part of health care and food service management teams. DTRs have completed at least a two-year associate's degree at a U.S. regionally accredited college or university and an accredited dietetic technician program, and passed a national examination.

digestion the process by which foods are broken down into smaller, absorbable products.

digestive system the body system composed of organs and glands associated with the ingestion and processing of food for absorption of nutrients into the body.

dim sum dumplings stuffed with pork, shrimp, beef, sweet paste, or preserves and steamed or fried.

dipeptides (dye-PEP-tides) protein fragments two amino acids long. A peptide is a strand of amino acids.

diploma mill a correspondence school that grinds out degrees (sometimes worth no more than the paper they are printed on) the way a grain mill grinds out flour.

disordered eating eating food as an outlet for emotional stress rather than in response to internal physiological cues.

diuretics (dye-you-RET-iks) medications and other substances causing increased water excretion (*dia* = through; *ouron* = urine).

diverticulosis (dye-ver-tik-you-LOH-siss) outpocketings of weakened areas of the intestinal wall that can rupture, like blowouts in a tire, causing dangerous infections.

DRI *see* Dietary Reference Intakes.

drugs substances that can modify one or more of the body's functions.

dysentery (DISS-en-terry) an infection of the digestive tract that causes diarrhea.

EAR *see* Estimated Average Requirement.

eating disorder general term for several conditions (anorexia nervosa, bulimia nervosa, binge-eating disorder) that exhibit an excessive preoccupation with body weight, a fear of body fatness, and a distorted body image.

eclampsia a severe extension of preeclampsia characterized by convulsions; may lead to coma.

edema (eh-DEEM-uh) swelling of body tissue caused by leakage of fluid from the blood vessels, seen in (among other conditions) protein deficiency.

EER *see* Estimated Energy Requirement.

electrolytes compounds that partially dissociate in water to form ions; examples are sodium, potassium, and chloride.

empty-calorie foods a phrase used to indicate that a food supplies calories but negligible nutrients.

emulsifier a substance that mixes with both fat and water and can break fat globules into small droplets, thereby suspending fat in water.

endocrine (EN-doh-krin) a term to describe a gland secreting or a hormone being secreted into the blood.

endurance the ability to sustain an effort for a long time. One type, **muscle endurance**, is the ability of a muscle to contract repeatedly within a given time without becoming exhausted. Another type, **cardiovascular**

endurance, is the ability of the cardiovascular system to sustain effort over a period of time.

energy the capacity to do work, such as moving or heating something.

energy metabolism all reactions by which the body obtains and expends the energy from food or body stores.

enriched refers to the process by which the B vitamins thiamin, riboflavin, niacin, folic acid, and the mineral iron are added to refined grains and grain products at levels specified by law.

enterotoxin a toxic compound, produced by microorganisms, that harms mucous membranes, as in the gastrointestinal tract (*entero* = intestine).

enzyme a protein catalyst. A *catalyst* is a compound that facilitates (speeds up the rate of) a chemical reaction without itself being altered in the process.

epidemiological study a study of a population that searches for possible correlations between nutrition factors and health patterns over time.

epithelial (ep-ih-THEE-lee-ul) tissue the cells that form the outer surface of the body and line the body cavities and the principal passageways leading to the exterior. Examples include the cornea, digestive tract lining, respiratory tract lining, and skin. The epithelial cells produce mucus to protect these tissues from bacteria and other potentially harmful substances. Without this mucus, infections become more likely.

ergogenic aids anything that helps increase the capacity to work or exercise (*ergo* = work; *genic* = give rise to).

essential amino acids amino acids that cannot be synthesized by the body or that cannot be synthesized in amounts sufficient to meet physiological need.

essential fatty acid a fatty acid that cannot be synthesized in the body in amounts sufficient to meet physiological need.

essential nutrients compounds that must be obtained from food because they cannot be synthesized by the body in amounts sufficient to meet physiological needs.

Estimated Average Requirement (EAR) the amount of a nutrient that is estimated to meet the requirement for the nutrient in half of the people of a specific age and gender. The EAR is used in setting the RDA.

Estimated Energy Requirement (EER) the average calorie intake that is predicted to maintain energy balance in a healthy adult of a defined age, gender, weight, height, and level of physical activity, consistent with good health.

estrogen a major female hormone; important in connection with nutrition because it maintains calcium balance and because its secretion abruptly declines at menopause.

estrogen replacement therapy (ERT) administration of estrogen to replace the natural hormone that declines with menopause. Because ERT may increase the risk of uterine and breast cancer, *hormone replacement therapy* (the administration of a combination of estrogen with the hormone progesterone) is often used. However, the FDA recommends that HRT and ERT be considered only for women with significant risk of osteoporosis that outweighs the risk of the drugs (increased risk for breast cancer, stroke, and heart disease).

ethnic cuisine the traditional foods eaten by the people of a particular culture.

exchange lists lists of foods with portion sizes specified. The foods on a single list are similar with respect to nutrient and calorie content and thus can be mixed and matched in the diet.

exocrine (EX-oh-krin) glands that secrete chemical messengers through a duct into a body cavity or onto the surface of the skin; *exo* means "out."

experimental group the participants in a study who receive the real treatment or intervention under investigation.

external cue theory the theory that some people eat in response to such external factors as the presence of food or the time of day rather than to such internal factors as hunger.

famine widespread lack of access to food caused by natural disasters, political factors, or war; characterized by a large number of deaths due to starvation and malnutrition.

fat cell theory states that during the growing years, fat cells respond to overfeeding by producing additional fat cells; the number of fat cells eventually becomes relatively fixed, and overfeeding from this point on causes the body to enlarge existing fat cells.

fats lipids that are solid at normal room temperature.

fatty acids basic units of fat composed of chains of carbon atoms with an acid group at one end and hydrogen atoms attached all along their length.

Feeding America a national food-banking network to which most food banks belong.

fetal alcohol syndrome (FAS) the cluster of symptoms seen in an infant or child whose mother consumed excess alcohol during pregnancy; includes retarded growth, impaired development of the central nervous system, and facial malformations. A lesser condition—called *fetal alcohol effect (FAE)*—causes learning impairment and other more subtle abnormalities in infants exposed to alcohol during pregnancy.

fibers the indigestible residues of food, composed mostly of polysaccharides. The best known of the fibers are cellulose, hemicellulose, pectin, and gums.

First Amendment the amendment to the U.S. Constitution that guarantees freedom of the press.

fitness the body's ability to meet physical demands; composed of four components: flexibility, strength, muscle endurance, and cardiovascular endurance.

flexibility the ability to bend or extend without injury; depends on the elasticity of the muscles, tendons, and ligaments and on the condition of the joints.

fluid balance distribution of fluid among body compartments.

fluorosis (floor-OH-sis) discoloration of the teeth from ingestion of too much fluoride during tooth development.

foam cells cells from the immune system containing scavenged oxidized LDL-cholesterol; thought to initiate arterial plaque formation.

food additive any substance added to food, including substances used in production, processing, treatment, packaging, transportation, or storage.

food allergen a substance in food (usually a protein) that is seen by the body as harmful and causes the immune system to mount an allergic reaction.

food allergy an adverse reaction to an otherwise harmless substance that involves the body's immune system.

food aversion a strong desire to avoid a particular food.

food banks nonprofit community organizations that collect surplus commodities from the government as well as edible but often unmarketable foods from private industry for use by nonprofit charities, institutions, and feeding programs at nominal cost.

food group plan a diet-planning tool, such as MyPlate, that groups foods according to similar origin and nutrient content and then specifies the amount of food a person should eat from each group.

food insecurity the inability to acquire or consume an adequate quality or sufficient quantity of food in socially acceptable ways, or the uncertainty that one will be able to do so.

food intolerance a general term for any adverse reaction to a food or food component that does not involve the body's immune system.

food intoxication illness caused by eating food that contains a harmful toxin or chemical.

food recovery such activities as salvaging perishable produce from grocery stores and wholesale food markets; rescuing surplus prepared food from restaurants, corporate cafeterias, and caterers; and collecting non-perishable, canned, or boxed processed food from manufacturers, supermarkets, or people's homes. The items recovered are donated to hungry people.

food security access by all people at all times to enough food for an active and healthy life. Food security has two aspects: ensuring that adequate food supplies are available and ensuring that households whose members suffer from undernutrition have the ability to acquire food, either by producing it themselves or by being able to purchase it.

foodborne illness illness occurring as a result of ingesting food or water contaminated with a poisonous substance, such as a toxin or chemical *(food intoxication)* or an infectious agent, such as bacteria, viruses, or parasites *(foodborne infection)*; commonly called *food poisoning.*

foodborne infection illness caused by eating a food containing bacteria or other microorganisms capable of growing and thriving in a person's tissues.

fortified term for foods to which nutrients have been added. Typically, commonly eaten foods are chosen for fortification with added nutrients to help prevent a deficiency (iodized salt, milk with vitamin D) or to reduce the risk of chronic disease (juices with added calcium).

fortified foods foods to which nutrients have been added, either because they were not already present or because they were present in insignificant amounts; foods to which manufacturers have added 10 percent or more of the Daily Value for a particular nutrient. Examples: margarine with added vitamin A, milk with added vitamin D, certain brands of orange juice with added calcium, and breakfast cereals with added nutrients and nonnutrients.

free radicals highly toxic compounds created in the body as a result of chemical reactions that involve oxygen. Environmental pollutants such as cigarette smoke and ozone also prompt the formation of free radicals.

free range a method of producing livestock where animals are allowed to roam freely instead of being contained in any manner; this term is not regulated and can be used to imply the animal product has been produced more humanely than it actually has been. USDA certifies "free range" chickens as chickens raised for their meat that have access to the outdoors; there is no USDA certification for "free range" beef, pork, or other non-poultry products; free-range eggs also have no legal definition.

fructose (FROOK-toce) fruit sugar; the sweetest of the single sugars.

functional tolerance actual change in sensitivity to a drug; often results in hallucinations and convulsions when alcohol is withheld.

fusion cuisine a term used to describe food that combines the elements of two or more cuisines—say, European and Oriental—to create a new one.

galactose (guh-LACK-toce) a single sugar that occurs bonded to glucose in the sugar of milk.

garbanzo beans or chickpeas large, round, tan-colored legumes that have a nutty flavor and crunchy texture. Use in soups and stews and puréed for dips.

gefilte fish (guh-FIL-tuh) a chopped fish mixture often made with pike and whitefish as well as matzoh crumbs, eggs, and seasonings.

gene a unit of a cell's inheritance; made of a chemical, DNA, that is copied faithfully so that every time the cell divides, both its offspring get identical copies. Genes direct the cells' machinery to make the proteins that form each cell's structures and enable the cell to do its work.

genetic engineering the use of biotechnology to alter the genes of a plant in an effort to create a new plant with different traits; also called *genetic modification.* In some cases, a plant's genes may be deleted or altered, or genes may be introduced from different organisms or species. The foods or crops produced are called genetically engineered (GE) or genetically modified (GM) foods or crops.

genetically modified organisms (GMOs)/ genetically engineered organisms (GEOs) organisms from bacteria, plants, or animals that have been genetically changed in a laboratory through DNA technology. Genetically engineered plants have been developed

to increase nutritional value, be resistant to pests, herbicides, or harsh environmental conditions like drought, and for other purposes.

gestational diabetes the appearance of abnormal glucose tolerance during pregnancy.

ghee a clarified butter without any milk solids or water.

GI *see* glycemic index.

ginkgo biloba (GING-coe bye-LOW-buh) an herb used to treat any number of conditions, including fatigue and failing memory.

ginseng (gin-SING) a collective term used to describe several species of plants, belonging to the genus Panax, that contain bioactive compounds in their roots. Falsely touted to enhance exercise endurance and boost energy.

gleaning the harvesting of excess food from farms, orchards, and packing houses to feed the hungry.

glucagon (GLOO-kuh-gahn) a hormone released by the pancreas that signals the liver to release glucose into the bloodstream.

glucose (GLOO-koce) the building block of carbohydrate; a single sugar used in both plant and animal tissues as quick energy. A single sugar is known as a *monosaccharide (mono = one).*

glutamic acid (gloo-TAM-ik) a nonessential amino acid found in plant and animal tissue; used as a flavor-intensifying seasoning.

glutinous rice short-grained, opaque white rice that turns sticky when cooked.

glycemic effect (also called *glycemic response*) the effect of food on a person's blood glucose and insulin response; how fast and how high the blood glucose rises and how quickly the body responds by bringing it back to normal.

glycemic index (GI) a ranking of foods based on their potential to raise blood glucose levels.

glycerol (GLISS-er-all) an organic compound that serves as the backbone for triglycerides.

glycogen (GLEYE-ko-jen) a polysaccharide composed of chains of glucose, manufactured in the body and stored in liver and muscle. As a storage form of glucose, liver glycogen can be broken down by the liver to maintain a constant blood glucose level when carbohydrate intake is inadequate.

GOBI an acronym formed from the elements of UNICEF's Child Survival campaign: **g**rowth charts, **o**ral rehydration therapy, **b**reast milk, and **i**mmunization.

goiter (GOY-tur) enlargement of the thyroid gland caused by iodine deficiency.

GRAS (Generally Recognized As Safe) list a list of ingredients, established by the FDA, that had long been in use and were believed safe. The list is subject to revision as new facts become known.

grass fed refers to livestock allowed to forage on pasture grass; not legally defined.

grazing eating small amounts of food at intervals throughout the day rather than or in addition to eating regular meals.

Great Northern beans a medium-white, kidney-shaped bean. Enjoy the delicate flavor and firm texture in salads, soups, and main dishes.

green soybeans (edamame) these large soybeans are harvested when the beans are still green and sweet and can be served as a snack or a main vegetable dish, after boiling in water for 15 to 20 minutes. They are high in protein and fiber and contain no cholesterol.

grits coarsely ground cornmeal.

ground water water that comes from an underground body of water that does not come into contact with any surface water.

guarana (gwuh-RAH-nuh) an herb chemically equivalent to caffeine.

guava a sweet, juicy fruit with green or yellow skin and red or yellow flesh.

hard water water that has a high mineral (calcium and magnesium) content.

hazard state of danger; used to refer to any circumstance in which harm is possible.

HDL (high-density lipoprotein) carries cholesterol in the blood back to the liver for recycling or disposal.

health claim a statement on the food label linking the nutritional profile of a food to a reduced risk of a particular disease, such as osteoporosis or cancer. Manufacturers must adhere to strict government guidelines when making such claims.

health fraud conscious deceit practiced for profit, such as the promotion of a false or an unproven product or therapy.

health promotion helping people achieve their maximum potential for good health.

heat stroke an acute and dangerous reaction to heat buildup in the body, requiring emergency medical attention; also called *sun stroke*.

heavy metals any of a number of mineral ions, such as mercury and lead, so named because of their relatively high atomic weight. Many heavy metals are poisonous.

heme (HEEM) iron the iron-holding part of the hemoglobin protein, found in meat, fish, and poultry. About 40 percent of the iron in meat, fish, and poultry is bound into heme. Meat, fish, and poultry also contain a factor (MFP factor) other than heme that promotes the absorption of iron, even of the iron from other foods eaten at the same time as the meat.

hemoglobin (HEEM-uh-glow-bin) the oxygen-carrying protein of the blood; found in the red blood cells.

hemorrhoids (HEM-or-oyds) swollen, hardened (varicose) veins in the rectum, usually caused by the pressure resulting from constipation.

high-density lipoprotein *see* HDL.

histamine a substance released by cells of the immune system during an allergic reaction to an antigen; causes inflammation, itching, hives, dilation of blood vessels, and a drop in blood pressure.

HMB (beta-hydroxy-beta-methylbutyrate) a metabolite of the branched-chain amino acid leucine that is promoted to increase muscle mass and strength by preventing muscle damage or speeding up muscle repair during resistance training. More research on long-term safety and effectiveness is needed.

hominy hulled, dried corn kernels.

homocysteine (hoe-moe-SIS-teen) a chemical that appears to be toxic to the blood vessels of the heart. High blood levels of homocysteine have been associated with low blood levels of vitamin B_{12}, vitamin B_6, and folate.

hormone a chemical messenger, secreted by one organ (a gland) in response to a condition in the body, that acts on another organ or organs to change that condition.

hormone free generally refers to products free from recombinant bovine somatotropin (rBST), a hormone synthetically produced from recombinant DNA administered to dairy cattle to increase milk production.

hunger the physiological drive to find and eat food, experienced as an unpleasant sensation.

hydrogenation (high-drah-jen-AY-shun) the process of adding hydrogen to unsaturated fat to make it more solid and more resistant to chemical change.

hydrolyzed vegetable protein (HVP) a protein obtained from any vegetable, including soybeans. The protein is broken down into amino acids by a chemical process called *acid hydrolysis*. HVP is a flavor enhancer that can be used in soups, broths, sauces, gravies, flavoring and spice blends, canned and frozen vegetables, and meats and poultry.

hyperglycemia an abnormally high blood glucose concentration, often a symptom of diabetes.

hypertension sustained high blood pressure (*hyper* = too much; *tension* = pressure).

hypertrophy an increase in size in response to use.

hypoglycemia (HIGH-po-gligh-SEEM-ee-uh) an abnormally low blood glucose concentration, below 60 to 70 mg/100 mL.

hyponatremia (HIGH-poe-na-TREE-mee-ah) a decreased concentration of sodium in the blood.

hypothalamus (high-poh-THALL-ah-mus) a part of the brain that senses a variety of conditions in the blood, such as temperature, salt content, and glucose content, and then signals other parts of the brain or body to change those conditions when necessary.

immunity specific disease resistance derived from the immune system's memory of prior exposure to specific disease agents and its ability to mount a swift response against them.

incidental food additives (indirect additives) substances that accidentally get into food as a result of contact with it during growing, processing, packaging, storing, or some other stage before the food is consumed.

incomplete protein a protein lacking or low in one or more of the essential amino acids.

ingredients list a listing of the ingredients in a food, with items listed in descending order of predominance by weight. All food labels are required to bear an ingredients list.

inorganic being or composed of matter other than plant or animal.

inositol (ih-NAH-sih-tall) once identified in association with the B vitamins; a type of sugar (related to glucose) found synthesized in the body and many foods.

insoluble fiber includes the fiber types called cellulose, hemicellulose, and lignin. Insoluble fibers do not dissolve in water.

insulin a hormone secreted by the pancreas in response to high blood glucose levels; it assists cells in drawing glucose from the blood.

integrated pest management the use of biological controls, crop rotation, genetic engineering, and other tactics to reduce chemical use in the growing and production of crops.

intentional food additives substances intentionally added to food; examples include nutrients, colors, spices, and herbs.

intervention study a population study examining the effects of a treatment on experimental subjects compared to a control group.

intestinal flora the normal bacterial inhabitants of the digestive tract (*flora* = plant inhabitants).

intrinsic factor a compound made in the stomach that is necessary for the body's absorption of vitamin B_{12}.

ions (EYE-onz) electrically charged particles, such as sodium (positively charged) and chloride (negatively charged).

iron overload a condition in which the body contains more iron than it needs or can handle; excess iron is toxic and can damage the liver. The most common cause of iron overload is the genetic disorder hemochromatosis.

iron-deficiency anemia a reduction of the number and size of red blood cells and a loss of their color because of iron deficiency.

irradiation the process of exposing a substance to low doses of radiation, using gamma rays, X-rays, or electricity (electron beams) to kill insects, bacteria, and other potentially harmful microorganisms.

jicama (HE-cah-mah) a crisp root vegetable of the bean family that is tan outside and white inside and is always eaten raw; as popular in Mexico as the potato is in the United States.

jujube (JOO-joob) Chinese date.

kasha cracked buckwheat, barley, millet, or wheat that is served as a cooked cereal or potato substitute.

kefir a fermented milk made by adding *Lactobacillus acidophilus* and other bacteria that break down lactose.

ketosis (key-TOE-siss) an adaptation of the body to prolonged (several days) fasting or carbohydrate restriction; body fat is converted to ketones, which can be used as fuel for some brain cells. Caused by abnormal amounts of ketone bodies in the blood and urine; ketone bodies are produced from the incomplete breakdown of fat when glucose is unavailable for the brain and nerve cells.

kidney beans these familiar beans are large, red, and kidney shaped (the white variety is called cannellini). They have a bland taste and soft texture but tough skins. Use in chili, bean stews, and Mexican dishes (for red) or Italian dishes (for white).

knish a potato pastry filled with ground meat, potato, or kasha.

kola nut an herb chemically equivalent to caffeine.

kosher fit, proper, or in accordance with religious law.

kwashiorkor (kwash-ee-OR-core) a deficiency disease caused by inadequate protein in the presence of adequate food energy.

lacto-ovovegetarian Milk and milk products and eggs are included in this diet, but meat, poultry, fish, and seafood are excluded.

lacto-vegetarian Milk and milk products are included in this diet, but meat, poultry, fish, seafood, and eggs are excluded.

lactoferrin (lak-toe-FERR-in) a factor in breast milk that binds and helps absorb iron and keeps it from supporting the growth of the infant's intestinal bacteria.

lactose a double sugar composed of glucose and galactose; commonly known as *milk sugar*.

lactose intolerance inability to digest lactose as a result of a lack of the necessary enzyme lactase.

lassi a milkshake-like drink made from yogurt.

LBW *see* low birthweight.

LDL (low-density lipoprotein) carries cholesterol (much of it synthesized in the liver) to body cells. A high blood cholesterol level usually reflects high LDL.

lecithin (LESS-ih-thin) a phospholipid, a major constituent of cell membranes; manufactured by the liver and also found in many foods.

legumes (LEG-yoomz or leh-GYOOMZ) plants of the bean and pea family having roots with nodules containing bacteria that can trap nitrogen from the air in the soil and make it into compounds that become part of the seed. The seeds are rich in high-quality protein compared with those of most other plant foods.

lentils small, flat, and round legumes. Usually brown, lentils can also be green, pink, or red. They have a mild taste with firm texture. Best used when combined with grains or vegetables in salads, soups, or stews.

leptin an appetite-suppressing hormone produced in the fat cells that communicates information about the body's fat content to the brain (*leptos* means slender).

lifestyle diseases conditions that may be aggravated by modern lifestyles that include too little exercise, poor diets, and excessive drinking and smoking. Lifestyle diseases are also referred to as *diseases of affluence*.

lima or butter beans limas are soft and mealy in texture. They are flat, oval shaped, and white tinged with green. The smaller variety has a milder taste. Use in soups and stews.

limiting amino acid a term given to the essential amino acid in shortest supply (relative to the body's need) in a food protein; it therefore *limits* the body's ability to make its own proteins.

linoleic (lin-oh-LAY-ik) acid, linolenic (lin-oh-LEN-ik) acid polyunsaturated fatty acids, essential for human beings.

lipids a family of compounds that includes triglycerides (fats and oils), phospholipids (lecithin), and sterols (cholesterol).

lipoprotein lipase (LPL) an enzyme located on the surfaces of fat cells that enables the cell to convert blood triglycerides into fatty acids and glycerol to be pulled into the cell for reassembly and storage as body fat.

lipoproteins (LYE-poh-PRO-teenz) clusters of lipids associated with protein that serve as transport vehicles for lipids in blood and lymph. The four main types of lipoproteins are chylomicrons, VLDL, LDL, and HDL.

liposuction a type of surgery (also called lipectomy) that vacuums out fat cells that have accumulated, typically in the buttocks and thighs.

litchi (LEE-chee) small, round fruits with orange-red skin and opaque white flesh; also called *litchee* or *lychee*.

liver the large, lobed organ that lies under the ribs and filters the blood, in the process removing, processing, and readying for redistribution many of its materials.

longan a small round fruit with smooth brown skin and clear pulp.

low birthweight (LBW) a birthweight of 5½ lb (2,500 g) or less; used as a predictor of poor health in the newborn and as a probable indicator of poor nutrition status of the mother during and/or before pregnancy. Normal birthweight for a full-term baby is 6½ to 8¾ lb (about 3,000 to 4,000 g).

low-density lipoprotein *see* LDL.

lox smoked salmon.

LPL *see* lipoprotein lipase.

lymph (LIMF) the fluid outside the circulatory system that bathes the cells; derived from the blood by being pressed through the capillary walls; similar to the blood in composition but without red blood cells; this body fluid transports the products of fat digestion toward the heart and eventually drains back into the bloodstream.

macrobiotic diet extremely restrictive diet based on metaphysical beliefs and consisting mostly of legumes, whole grains, and certain vegetables.

mad cow disease also called *bovine spongiform encephalopathy* (BSE); a rare and fatal degenerative disease first diagnosed in 1986 in cattle in the United Kingdom. The bovine disease may be passed to humans who eat the meat of infected animals and may lead to death due to brain and nerve damage.

major mineral an essential mineral nutrient found in the human body in amounts greater than 5 grams.

malnutrition any condition caused by an excess, deficiency, or imbalance of calories or nutrients; the impairment of health resulting from a relative deficiency or excess of food energy and specific nutrients necessary for health.

maltose a double sugar composed of two glucose units.

mantou steamed bread.

marasmus (muh-RAZ-mus) an energy deficiency disease; starvation.

margin of safety from a food-safety standpoint, the *margin* is a zone between the maximum amount of a substance that appears to be safe and the amount allowed in the food supply.

matzoh a cracker-like bread eaten most often at Passover.

meat alternatives meat alternatives made from soybeans contain soy protein or tofu and other ingredients mixed together to simulate various kinds of meat. These meat alternatives are sold as frozen, canned, or dried foods.

meat replacements textured vegetable protein products formulated to look and taste like meat, fish, or poultry; also called meat analogs.

menopause the time of life at which a woman's menstrual cycle ceases, usually at about 45 to 50 years of age.

metabolic syndrome a cluster of interrelated symptoms, including obesity, high blood pressure, abnormal blood lipids, and insulin resistance; highly associated with development of type 2 diabetes and heart disease.

metabolic tolerance increased efficiency of removing high levels of alcohol from the blood due to long-term exposure; may lead to more drinking and possible addiction.

metabolism (meh-TAB-o-lism) collective term for all the chemical and physical reactions occurring in living cells, including the reactions by which the body obtains and uses energy from foods.

microvilli (MY-croh-VILL-ee, MY-croh-VILL-eye) tiny hairlike projections on each cell of the intestinal tract lining that can trap nutrient particles and translocate them into the cells (singular: *microvillus*).

mineral water water that is drawn from an underground source and contains at least 250 parts per million of dissolved solids. If the water contains between 250 and 500 parts per million total dissolved solids, the statement "low mineral content" must appear. If it contains more than 1,500 parts per million, the statement "high mineral content" must appear. If a cup of the water contains at least 20 milligrams of calcium, 0.36 milligram of iron, or 5 milligrams of sodium, the product must carry nutrition labeling.

minerals naturally occurring, inorganic, chemical elements; serve as structural components of the body and are used in many vital body processes.

miso a rich, salty condiment that characterizes the essence of Japanese cooking. A smooth paste, miso is made from soybeans and a grain such as rice, plus salt and a mold culture, and then aged in cedar vats for 1 to 3 years. Miso should be refrigerated. Use miso to flavor soups, sauces, dressings, marinades, and pâtés.

moderation the attribute of a diet that provides no unwanted constituent in excess.

monoglyceride (mahn-oh-GLISS-er-ide) a glycerol molecule with one fatty acid attached to it. A *diglyceride* is a glycerol molecule with two fatty acids attached to it.

monounsaturated fatty acid a fatty acid containing one point of unsaturation; found mostly in vegetable oils such as olive, canola, and peanut.

mucus (MYOO-kuss) a thick, slippery coating of the intestinal tract lining (and other body linings) that protects the cells from exposure to digestive juices. The adjective form is *mucous*, and the coating is often called the *mucous membrane*.

multinational corporations international companies with direct investments and/or operative facilities in more than one country. U.S. oil and food companies are examples.

natural most foods labeled natural, except for meat and poultry, are not subject to government regulations; this term applies broadly to foods minimally processed and free of synthetic preservatives, artificial sweeteners, colors, flavors, and other artificial additives, hydrogenated oils, stabilizers, and emulsifiers.

neotame a zero-calorie, heat-stable sweetener approved in 2002 that is a derivative of the dipeptide composed of aspartic acid and phenylalanine. It is 7,000 to 13,000 times as sweet as sugar, depending on how it is used. Unlike aspartame, it is not metabolized to phenylalanine, and thus no special labeling for people with PKU is required.

neural tube defects any of a number of birth defects in the orderly formation of the neural tube during early gestation. Both the brain and the spinal cord develop from the neural tube, and defects cause various central nervous system disorders. The two main types are spina bifida (incomplete closure of the bony casing around the spinal cord) and anencephaly (a partially or completely missing brain). Because the neural tube closes before the sixth week of pregnancy, women are advised to consume adequate folate from as early as 3 months before conception.

neurotoxin a poisonous compound that disrupts the nervous system (*neuro* = nerve).

night blindness slow recovery of vision following flashes of bright light at night; an early symptom of vitamin A deficiency.

non-GMO term used to imply a food contains no ingredients made from GMO/GEO products; not legally defined and therefore cannot be enforced.

nondairy soy frozen dessert nondairy frozen desserts are made from soy milk or soy yogurt. Soy ice cream is one of the most popular desserts made from soybeans and can be found in many grocery stores.

nonheme iron the iron found in plant foods.

nursing bottle syndrome (also called *baby bottle tooth decay*) decay of all the upper and sometimes the back lower teeth that occurs in infants given carbohydrate-containing liquids when they sleep. The syndrome can also develop in toddlers given bottles of liquid to carry around and sip all day.

nutrient content claims claims such as "low-fat" and "low-calorie" used on food labels to help consumers who don't want to scrutinize the Nutrition Facts panel get an idea of a food's nutritional profile. These claims must adhere to specific definitions set forth by the Food and Drug Administration.

nutrient dense refers to a food that supplies large amounts of nutrients relative to the number of calories it contains. The higher the level of nutrients and the fewer the number of calories, the more nutrient dense the food is.

nutrients substances obtained from food and used in the body to promote growth, maintenance, and repair. The nutrients include carbohydrate, fat, protein, vitamins, minerals, and water.

nutrition the study of foods, their nutrients and other chemical components, their actions and interactions in the body, and their influence on health and disease.

Nutrition Facts panel a detailed breakdown of the nutritional content of a serving of a food that must appear on virtually all packaged foods sold in the United States.

nutritional yeast a fortified food supplement containing B vitamins, iron, and protein; can be used to improve the quality of a vegetarian diet.

nutritionist a person who claims to be capable of advising people about their diets. Some nutritionists are registered dietitians, whereas others are self-described experts whose training is questionable.

o-LDL *see* oxidized LDL-cholesterol.

obesity body weight that is high enough above normal weight to constitute a health hazard; conventionally defined as weight 20 percent or more above the desirable weight for height, or a BMI of 30 or greater.

oils lipids that are liquid at normal room temperature.

olestra an artificial fat derived from sucrose and fatty acids combined in such a way that the body cannot break them down. Sold under the brand name Olean, olestra does not contribute calories to food.

omega-3 fatty acids polyunsaturated fatty acids that have their endmost double bonds after the third carbon in the chain.

omega-6 fatty acids polyunsaturated fatty acids that have their endmost double bonds after the sixth carbon in the chain.

oral rehydration therapy (ORT) the treatment of dehydration (usually due to diarrhea caused by infectious disease) with an oral solution; ORT as developed by UNICEF is intended to enable a mother to mix a simple solution for her child from substances that she has at home.

organic legal term that applies to a food and how the food was produced; crops must be grown without use of synthetic pesticides, bioengineered genes, petroleum-based fertilizers, and sewage sludge-based fertilizers; organic livestock must have access to outdoors and be given no antibiotics or growth hormones; organic foods may not be irradiated.

organic halogens compounds that contain one or more of a class of atoms called *halogens*, including fluorine, chlorine, iodine, or bromine.

Oriental radish a large, cylindrically shaped vegetable with smooth skin; also called *daikon*.

osteoblast a bone-building cell; responsible for formation of bone.

osteoclast a bone-destroying cell; responsible for resorption and removal of bone.

osteomalacia (os-tee-oh-muh-LAY-shuh) the disease resulting from vitamin D deficiency in adults and characterized by softening of the bones. (Its counterpart in children is called rickets.) Symptoms include bowed legs and a curved spine.

osteoporosis (OSS-tee-oh-pore-OH-sis) also known as *adult bone loss;* a disease in which the bones become porous and fragile (*osteo* = bones; *poros* = porous).

overload an extra physical demand placed on the body. A principle of training is that for a body system to improve, its workload must be increased by increments over time.

overnutrition calorie or nutrient overconsumption severe enough to cause disease or increased risk of disease; a form of malnutrition.

overweight conventionally defined as weight between 10 and 20 percent above the desirable weight for height, or a body mass index (BMI) of 25.0 through 29.9.

ovovegetarian Eggs are included in this diet, but milk and milk products, meat, poultry, fish, and seafood are excluded.

oxidized LDL-cholesterol (o-LDL) the cholesterol in LDLs that is attacked by reactive oxygen molecules inside the walls of the arteries; o-LDL is taken up by scavenger cells and deposited in plaque.

pancreas a gland that secretes the endocrine hormone insulin and also produces the exocrine secretions that aid digestion in the small intestine. (See also *endocrine; exocrine*.)

parathyroid hormone a hormone used as a drug (teriparatide) to stimulate new bone formation; administered by injection once a day in the thigh or abdomen. Daily injections of teriparatide (brand name Forteo) lead to increased bone mineral density.

pasteurization the process of sterilizing food via heat treatment.

peak bone mass the highest bone density achieved for an individual; accumulated over the first three decades of life; typically occurs by 30 years of age. After age 30, bone resorption slowly begins to exceed bone formation.

pellagra (pell-AY-gruh) niacin deficiency characterized by diarrhea, inflammation of the skin, and, in severe cases, mental disorders and death.

PEM *see* protein-energy malnutrition.

peptide bond a bond that connects one amino acid with another.

periodontal disease inflammation or degeneration of the tissues that surround and support the teeth.

peristalsis (pair-ih-STALL-sis) the wavelike squeezing motions of the stomach and intestines that push their contents along the digestive tract.

pesticides chemicals intentionally applied to plants, including foods, to prevent or eliminate pest damage. *Pests* include all living organisms that destroy or spoil foods: bacteria, molds and fungi, insects, and rats and other rodents, to name a few.

pH the concentration of hydrogen ions. The lower the pH, the stronger the acid: pH 2 is a strong acid, pH 7 is neutral, and a pH above 7 is alkaline (base).

phenylketonuria (FEN-il-KEY-toe-NYOO-ree-ah) or PKU an inborn error of metabolism, detectable at birth, in which the body lacks the enzyme needed to convert the amino acid phenylalanine to the amino acid tyrosine. If not detected and treated, derivatives of phenylalanine accumulate in the blood and tissues, where they can cause severe damage, including mental retardation.

phosphate salt a salt with claims for improved endurance. Found to increase a substance in red blood cells (diphosphoglycerate) and enhance the cell's ability to deliver oxygen to muscle cells and reduce levels of disabling lactic acid in elite athletes. However, more research on safety and efficacy of phosphate loading is needed. Excess can cause loss of bone calcium.

phospholipid (FOSS-foe-LIP-id) a lipid similar to a triglyceride but containing phosphorus; one of the three main classes of lipids.

phytochemicals (FIGHT-oh-KEM-ih-kulz) physiologically active compounds found in plants that are not essential nutrients but appear to help promote health and reduce risk for cancer, heart disease, and other conditions. Also called *phytonutrients (phyto* = plant).

pica the craving of nonfood items such as clay, ice, and laundry starch; does not appear to be limited to any particular geographic area, race, sex, culture, or social status.

pigment a molecule capable of absorbing certain wavelengths of light. Pigments in the eye permit us to perceive different colors.

PIH *see* pregnancy-induced hypertension.

pinto beans medium-size, oval beans that are mottled beige and brown and have an earthy flavor. They are most often used in Mexican dishes, such as refried beans, stews, or dips.

PKU *see* phenylketonuria.

placebo (plah-SEE-bo) a sham treatment given to a control group; an inert, harmless "treatment" that the group's members cannot recognize as different from the real thing. Using a placebo minimizes the chance that an effect of the treatment will appear to have occurred due to the healing effect of the belief in the treatment (known as the placebo effect), rather than the treatment itself.

placenta (pluh-SEN-tuh) the organ inside the uterus in which the mother's and fetus's circulatory systems intertwine and in which exchange of materials between maternal and fetal blood takes place. The fetus receives nutrients and oxygen across the placenta; the mother's blood picks up carbon dioxide and other waste materials to be excreted via her lungs and kidneys.

plantain a greenish, starchy banana; because it is starchy even when ripe, it is never eaten raw and is usually pan-fried.

polysaccharide a long chain of 10 or more glucose molecules linked together in straight or branched chains; another term for *complex carbohydrates.*

polyunsaturated fatty acid (sometimes abbreviated PUFA) a fatty acid in which two or more points of unsaturation occur; found in nuts and vegetable oils such as safflower, sunflower, and soybean as well as in fatty fish.

postnatal after birth.

poverty the state of having too little money to meet minimum needs for food, clothing, and shelter. The U.S. Department of Agriculture defined the poverty level in the United States as an annual income of $22,350 for a family of four in 2011.

precursor a compound that can be converted into another compound. For example, beta-carotene is a precursor of vitamin A (*pre* = before; *cursor* = runner, forerunner).

preeclampsia a condition characterized by hypertension, fluid retention, and protein in the urine.

preformed vitamin A vitamin A in its active form.

pregnancy-induced hypertension (PIH) high blood pressure that develops during the second half of pregnancy.

prenatal prior to birth.

protein quality a measure of the essential amino acid content of a protein relative to the essential amino acid needs of the body.

protein synthesis the process by which cells assemble amino acids into proteins. All individuals are unique because of minute differences in the ways their body proteins are made. The instructions for making all the proteins in our bodies are transmitted in the genetic information we receive at conception.

protein-energy malnutrition (PEM) also called protein-calorie malnutrition (PCM); the world's most widespread malnutrition problem; includes both kwashiorkor and marasmus as well as the states in which they overlap.

protein-sparing a description of the effect of carbohydrate and fat, which, by being available to yield energy, allow amino acids to be used to build body proteins.

proteins compounds composed of atoms of carbon, hydrogen, oxygen, and nitrogen and arranged as strands of amino acids. Some amino acids also contain atoms of sulfur.

psyllium seed husk, an ingredient in certain cereals and bulk-forming laxatives; contains both soluble and insoluble fibers.

purified water also known as *demineralized water, distilled water, deionized water, or reverse osmosis water;* water from which all minerals have been removed, thereby eliminating the possibility that the minerals might corrode pipes or appliances (such as a steam iron).

purslane a leafy vegetable that can be used in salads or cooked like spinach.

pylorus (pie-LORE-us) muscle that regulates the opening of the bottom of the stomach.

pyruvate (pie-ROO-vate) a three-carbon compound derived from the breakdown of glucose for energy in the body. Falsely promoted to increase fat burning and endurance. Side effects include intestinal gas and diarrhea, which could interfere with performance.

quackery A quack is a person who practices health fraud (quack = to boast loudly).

queso blanco, fresco, or Mexicano soft white cheese made of part-skim milk.

RAE *see* retinol activity equivalents.

rBST free refers to milk (dairy products) produced from cattle that have not been treated with recombinant BST (bovine somatotropin hormone); all milk contains some BST as it is produced in pituitary glands of cattle.

RD *see* registered dietitian.

Recommended Dietary Allowance (RDA) the average daily amount of a nutrient that is sufficient to meet the nutrient needs of nearly all (97–98 percent) healthy individuals of a specific age and gender.

red beans this versatile bean is a medium-size, dark-red oval. The taste and texture are similar to kidney beans. Use in soups and stews, and serve with rice.

reference dose the estimated amount of a chemical that could be consumed daily without causing harmful effects.

reference protein egg-white protein, the standard to which other proteins are compared to determine protein quality.

refined grains grains that have been milled to remove the chaff, bran, and germ from the grain—leaving only the endosperm—in order to yield a finer texture and improved shelf life. The process also removes dietary fiber, iron and other minerals, and many B vitamins. The recommended amount of refined grains is no more than 3 ounce-equivalents per day.

registered dietitian (RD) a professional who has graduated from a program of dietetics accredited by the Commission on Accreditation for Dietetics Education (CADE) of the American Dietetic Association (ADA), has completed an internship program or the equivalent to gain practical skills, has passed a registration examination, and maintains competencies through continuing education.

regulation a legal mandate that must be obeyed. Failure to follow a regulation brings about legal consequences.

requirement the minimum amount of a nutrient that will prevent the development of deficiency symptoms. Requirements differ from the RDA and AI, which include a substantial margin of safety to cover the requirements of different individuals.

retina (RET-in-uh) the paper-thin layer of light-sensitive cells lining the back of the inside of the eye.

retinal (RET-in-al) one of the active forms of vitamin A that functions in the pigments of the eye. Other active forms of vitamin A include retinol and retinoic acid.

retinol one of the active forms of vitamin A.

retinol activity equivalents (RAE) a measure of the amount of retinol the body will derive from a food containing preformed vitamin A or beta-carotene and other vitamin A precursors. Note that some tables list vitamin A in terms of *International Units (IU).* (See Appendix A for methods of converting from one measure to another.)

rice sticks flat, opaque, wide noodles made from rice flour.

rice vermicelli (vur-mi-SEL-ee) thin white noodles made from rice flour.

rickets a disease that occurs in children as a result of vitamin D deficiency; characterized by abnormal growth of bone, which in turn leads to bowed legs and an outward-bowed chest.

risk for pesticide residues, the harm a substance may cause. Scientists estimate risk by assessing the amount of a chemical that each person in a population might consume over time (also called *exposure*) and by considering how toxic the substance might be (*toxicity*).

roti an unleavened flatbread, resembling a tortilla, usually made from whole-wheat flour.

royal jelly substance secreted from the saliva glands of honeybees; provides food for all young larvae and is the only food for larvae that will develop into queen bees. It has been incorrectly claimed to impart extra energy and to have restorative properties; can be potentially hazardous for individuals who may suffer allergic reactions.

saccharin a zero-calorie sweetener discovered in 1879 and used in the United States since the turn of the century. A possible link to bladder cancer led to saccharin being banned as a food additive in Canada, although it is available there as a tabletop sweetener. Saccharin is the sweetening agent in Sweet'N Low.

salt a pair of charged mineral particles, such as sodium (Na^+) and chloride (Cl^-), that associate together. In water, they dissociate and help carry electric current; that is, they become *electrolytes*.

satiety (sah-TIE-eh-tee) the feeling of fullness or satisfaction experienced following a meal.

saturated fatty acid a fatty acid carrying the maximum possible number of hydrogen atoms (having no points of unsaturation). Saturated fats are found in animal foods such as meat, poultry, and full-fat dairy products as well as in tropical oils such as palm and coconut.

schmaltz chicken fat.

scurvy the vitamin C-deficiency disease characterized by bleeding gums, tooth loss, and, in severe cases, death.

selective estrogen receptor modulators (SERMs) *see* designer estrogens.

seltzer tap water injected with carbon dioxide and containing no added salts.

semivegetarian Some but not all groups of animal-derived products, such as meat, poultry, fish, seafood, eggs, milk, and milk products, are included in this diet.

serving the standard amount of food used as a reference to give advice regarding how much to eat (such as a 1-cup serving of milk).

set-point theory theory that the body tends to maintain a certain weight by adjusting hunger, appetite, and food energy intake on the one hand and metabolism (energy output) on the other so that a person's conscious efforts to alter weight may be foiled.

simple carbohydrates (sugars) the single sugars (*monosaccharides*) and the pairs of sugars (*disaccharides*) linked together.

Simplesse the trade name for a protein-based, low-calorie fat replacer, approved by the FDA for use in foods such as frozen desserts; cannot be used for frying or baking.

skinfold test a method for measuring body fat in which the thickness of a fold of skin on the back of the arm (the triceps), below the shoulder blade (subscapular), or in other areas is measured with an instrument called a caliper.

small intestine a long, tubular organ of digestion that is the site of nutrient absorption.

social group a group of people, such as a family, who depend on one another and share a set of norms, beliefs, values, and behaviors.

sodium bicarbonate baking soda is touted to buffer lactic acid in the body and thereby reduce pain and improve maximal-level anaerobic performance. High sodium load may cause diarrhea in users. Effects of repeated ingestion are unknown.

soft water water containing a high sodium concentration and little or no calcium or magnesium.

solid fats fats that are usually solid at room temperature. Most solid fats contain saturated fats, cholesterol, and/or *trans* fats. Examples include butter, beef fat, chicken fat, pork fat (lard), stick margarine, and shortening. These fats may be visible or may be a constituent of foods such as milk, cheese, meats, or baked products.

soluble fiber includes the fiber types called pectin, gums, mucilages, some hemicelluloses, and algal substances (for example, carrageenan). Soluble fibers either dissolve or swell when placed in water.

sopa rice or pasta that is fried and cooked in consommé.

soy cheese cheese made from soy milk. Its creamy texture makes it an easy substitute for sour cream or cream cheese. It can be found in a variety of flavors in many food stores.

soy flour flour made from roasted soybeans ground into a fine powder. Soy flour gives a protein boost to recipes. Soy flour is gluten-free, so yeast-raised breads made with soy flour are denser in texture.

soy milk, soy beverages soybeans that are soaked, ground fine, and strained produce a fluid called *soybean milk*. Soy milk is an excellent source of high-quality protein and B vitamins.

soy nuts whole soybeans that have been soaked in water and then baked until browned. High in protein and isoflavones, soy nuts are similar in texture and flavor to peanuts.

soy protein, texturized *texturized soy protein* usually refers to products made from texturized soy flour. Texturized soy flour is made by running defatted soy flour through an extrusion cooker, which allows the creation of many different forms and sizes. When hydrated, it has a chewy texture. It is widely used as a meat extender.

soy yogurt yogurt made from soy milk. Its creamy texture makes it an easy substitute for sour cream or cream cheese. Soy yogurt can be found in a variety of flavors in many food stores.

soy-nut butter made from roasted, whole soy nuts, which are then crushed and blended with soy oil and other ingredients. Soy-nut butter has a slightly nutty taste, contains significantly less fat than peanut butter, and provides many other nutritional benefits.

soybeans as soybeans mature in the pod, they ripen into a hard, dry bean. Most soybeans are yellow. However, brown and black varieties are also available. Whole soybeans can be cooked and used in sauces, stews, and soups.

sparkling bottled water water in which carbon dioxide (the ingredient that makes soda pop bubbly) is naturally present; that is, carbonation is not added from an outside source.

split peas green or yellow, these small, halved peas supply an earthy flavor with a mealy texture. They are best used in soups and with rice or grains.

sports anemia a temporary condition of low blood hemoglobin level, associated with the early stage of athletic training.

spring water water derived from an underground formation from which water flows naturally to the surface of the earth and to which minerals have not been added or taken away. It may be collected either at the spring itself or through a hole tapping the underground formation feeding the spring.

staple grain a grain used frequently in the diet; for example, corn (in Mexico) or rice (in Asia).

starch a plant polysaccharide composed of hundreds of glucose molecules, digestible by human beings.

sterols (STEER-alls) lipids with a structure similar to that of cholesterol; one of the three main classes of lipids.

stevia (STEE-vee-uh) a small, shrubby perennial with small white flowers; native to Paraguay and Brazil. The plant's sweetness appears to be concentrated in the leaves.

strength the ability of muscles to work against resistance.

stress fracture bone damage or breakage caused by stress on bone surfaces during exercise.

strict vegetarian/vegan All animal-derived foods, including meat, poultry, fish, seafood, eggs, milk, and milk products are excluded from this diet.

sucralose a nonnutritive sweetener approved in 1998 as a tabletop sweetener and for use in a variety of desserts, confections, and nonalcoholic beverages. Sucralose is a noncaloric, heat-stable sweetener derived from a chlorinated form of sugar. Although sucralose is made from sugar, the body does not recognize it as sugar, and the sucralose molecule is excreted in the urine essentially unchanged.

sucrose (SOO-krose) a double sugar composed of glucose and fructose. A double sugar is known as a *disaccharide* (*di* = two).

sugar alcohols include mannitol, sorbitol, isomalt, xylitol; can be derived from fruits or commercially produced from dextrose and are absorbed more slowly and metabolized differently than other sugars in the human body. The sugar alcohols are not readily used by bacteria ordinarily found in the mouth and therefore are associated with less cavity formation. Although the sugar alcohols are used as sugar substitutes, they do add calories (about 1.5 to 3 calories per gram) to a food product.

target heart rate the heartbeat rate that will achieve a cardiovascular conditioning effect for a given person: fast enough to push the heart, but not so fast as to strain it.

taro a starchy vegetable with brown, hairy skin and a pink-purple interior.

tempeh a chunky, tender soybean cake; a traditional Indonesian food. Whole soybeans, sometimes mixed with another grain such as rice or millet, are fermented into a rich cake of soybeans with a smoky, nutty flavor. Tempeh can be marinated and grilled and added to soups, casseroles, or chili.

tofu and tofu products *tofu*, also known as *soybean curd*, is a soft, cheese-like food made by curdling fresh, hot soy milk with a coagulant. Tofu is a bland product that easily absorbs the flavors of other ingredients with which it is cooked. Tofu is rich in high-quality protein and B vitamins and is low in sodium.

Tolerable Upper Intake Level (UL) the maximum amount of a nutrient that is unlikely to pose any risk of adverse health effects to most healthy people. The UL is not intended to be a recommended level of intake.

tolerance the maximum amount of a particular substance allowed in food.

tonic water artificially carbonated water with added sugar and/or high-fructose corn syrup, sodium, and quinine.

toxicants poisons; that is, agents that cause physical harm or death when present in large amounts.

toxicity the ability of a substance to harm living organisms. All substances are toxic if present in high enough concentrations.

trabecular (tra-BECK-you-lar) **bone** the lacy inner network of calcium-containing crystals; spongelike in appearance, it supports the bone's structure.

trace mineral an essential mineral nutrient found in the human body in amounts less than 5 grams.

trans fat a type of fatty acid created when an unsaturated fat is hydrogenated. Found primarily in margarines, shortenings, commercial frying fats, and baked goods, trans fatty acids have been implicated by research as culprits in heart disease.

transgenetics the process of transferring genes from one species to another unrelated species.

transport proteins proteins that carry nutrients and other molecules in body fluids. Some transport proteins reside in cell membranes and act as "pumps" by picking up compounds on one side of the membrane and releasing them on the other side as needed.

triglycerides (try-GLISS-er-ides) the major class of dietary lipids, including fats and oils. A triglyceride is made up of three units known as *fatty acids* and one unit called *glycerol.*

trimester one-third of the normal duration of pregnancy; the first trimester is 0 to 13 weeks, the second is 13 to 26 weeks, and the third trimester is 26 to 40 weeks.

tripeptides (try-PEP-tides) protein fragments three amino acids long.

UL *see* Tolerable Upper Intake Level.

under-5-mortality rate (U5MR) the number of children who die before the age of 5 for every 1,000 live births.

undernutrition severe underconsumption of calories or nutrients, leading to disease or increased susceptibility to disease; a form of malnutrition.

underwater weighing a measure of density and volume; the less a person weighs underwater compared to the person's out-of-water weight, the greater the proportion of body fat (fat is less dense or more buoyant than lean tissue).

underweight weight 10 percent or more below the desirable weight for height, or a BMI of less than 18.5.

UNICEF the United Nations International Children's Emergency Fund, now referred to as the United Nations Children's Fund.

unique radiolytic products substances unique to irradiated food and apparently created during the process of irradiation.

unsaturated fatty acid a fatty acid with one or more points of unsaturation. Unsaturated fats are found in foods from both plant and animal sources. Unsaturated fatty acids are further divided into monounsaturated fatty acids and polyunsaturated fatty acids.

unspecified eating disorders condition in which a person exhibits some but not all of the criteria for specific eating disorders.

urea (yoo-REE-uh) the principal nitrogen excretion product of metabolism, generated mostly by the removal of amine groups from unneeded amino acids or from those amino acids being sacrificed to a need for energy.

variety a feature of a diet in which different foods are used for the same purposes on different occasions; the opposite of *monotony.*

veins blood vessels that carry used blood from the tissues back to the heart.

very-low-density lipoprotein *see* VLDL.

villi (VILL-ee, VILL-eye) fingerlike projections of the sheet of cells that line the GI tract; the villi make the surface area much greater than it would otherwise be (singular: *villus*).

vitamin a potent, indispensable compound that performs various bodily functions that promote growth and reproduction and maintain health. Vitamins are *essential* nutrients required in minute amounts and *organic*, meaning that they contain or are related to carbon compounds. Contrary to popular belief, vitamins do not supply calories. (*vita* = life; *amine* = containing nitrogen)

VLDL (very-low-density lipoprotein) carries fats packaged or made by the liver to various tissues in the body.

waist circumference a measure used to assess abdominal (visceral) fat; excess fat in the abdomen increases a person's risk of health problems.

water provides the medium for life processes.

well water ground water derived from a rock formation by way of a hole bored, drilled, or otherwise constructed in the ground.

white navy beans small, white, oval beans that are best used in soups and stews and as baked beans.

whole food a food that is altered as little as possible from the plant or animal tissue from which it was taken; examples are milk, oats, potatoes, and apples.

whole grain refers to a grain that is milled in its entirety (all but the husk), not refined. Whole grains include wheat, corn, rice, rye, oats, amaranth, barley, buckwheat, sorghum, and millet. Two others, bulgur and couscous, are processed from wheat grains.

yard-long beans thin, tender string beans that grow to as long as 18 inches.

yerba maté an herb chemically equivalent to caffeine.

yohimbe (yoh-HIM-bee) touted as an aphrodisiac, a cure-all for sexual dysfunction; also claimed to be able to alter body mass, muscle mass, or exercise performance when combined with resistance training. None of these claims have been proven by research.

zapote (sah-POH-tay) an apple-sized fruit with green skin and black flesh.

Index

The Nutrients in Foods

Learning Objective 1: *List the six classes of nutrients.*

- Nutrition is the study of foods, their nutrients and other chemical components, their actions and interactions in the body, and their influences on health and disease.
- Nutrients are substances obtained from food and used in the body to promote growth, maintenance, and repair.
- There are six classes of nutrients: carbohydrates, fats, protein, vitamins, minerals, and water.
- Essential nutrients must be obtained from food because the body cannot make them for itself.
- There are three energy-yielding nutrients: carbohydrates, fats, and proteins.
- Vitamins are organic, essential nutrients vital for life but needed only in relatively minute amounts.
- Minerals are inorganic compounds, some of which are essential nutrients.
- Water provides the medium for life processes.

Nutrition and Health Promotion

- Malnutrition is any condition caused by an excess, deficiency, or imbalance of calories or nutrients.
- Overnutrition is a form of malnutrition where calorie or nutrient overconsumption is severe enough to cause disease or increased risk of disease.
- Degenerative diseases are chronic diseases characterized by the deterioration of body organs as a result of misuse and neglect. Poor eating habits, smoking, lack of exercise, and other lifestyle habits often contribute to degenerative diseases including heart disease, cancer, osteoporosis, and diabetes.

A National Agenda for Improving Nutrition and Health

- Lifestyle behaviors reflect personal choices, habits, and customs that can influence health.
- Health promotion helps people achieve their maximum potential for good health.
- *Healthy People 2020* is a plan of action for national health promotion and disease prevention.

Understanding Our Food Choices

Learning Objective 2: *Differentiate between appetite and hunger.*

Learning Objective 3: *Identify different factors that affect food choices.*

- The act of eating is complex: we derive many benefits, both physical and emotional, when we eat foods.
- Factors that influence food choices include:
  - Hunger, appetite, and food habits
  - Nutrition knowledge, health beliefs or concerns, and practices
  - Availability, convenience, and economy of certain foods
  - Advertising and the media
  - Early experiences, social interactions, and cultural traditions
  - Personal preferences, taste, and psychological needs
  - Values, such as political views, environmental concerns, and religious beliefs
- Hunger is the physiological need for food.
- Appetite is the psychological desire to eat, which is often but not always accompanied by hunger.
- Undernutrition is a form of malnutrition caused by severe underconsumption of calories or nutrients, leading to disease or increased susceptibility to disease.

FIGURE 1-1

Caloric Values of Carbohydrate, Protein, Fat, and Alcohol

	Calories per gram
Carbohydrate	4
Protein	4
Fat	9
Alcohol	7

The Nutrients in Foods (pp. 5–7)

1. **Name the six classes of nutrients.**

2. **Which of the following is a definition of "nutrients"?**
   a. Calories
   b. MyPlate
   c. Substances in food used by the body for digestion
   d. Chemicals in food used by the body for growth and health
   e. All of the above

3. **The term *essential* describes a nutrient that:**
   a. The body cannot make for itself
   b. The body must obtain from food
   c. Cannot be found in food
   d. A and B
   e. A and C

4. **Nonessential nutrients are:**
   a. Needed for normal growth or health
   b. Excreted from the body
   c. Must be obtained from the foods we eat
   d. All of the above

5. **A food contains 15 grams of carbohydrate, 3 grams of protein, and 1 gram of fat. How many calories does this food contain?**
   a. 19 calories
   b. 76 calories
   c. 79 calories
   d. 81 calories

6. **How many calories are supplied by vitamins?**
   a. 0
   b. 4
   c. 7
   d. 9

7. **Which nutrient is the medium for all bodily processes?**
   a. Protein
   b. Fat
   c. Minerals
   d. Water

8. **The energy-yielding nutrients are:**
   a. Protein, fats, and minerals
   b. Fats, carbohydrates, and water
   c. Vitamins, minerals, and water
   d. Protein, carbohydrates, and fats
   e. Fats, carbohydrates, and vitamin

Nutrition and Health Promotion (pp. 7–10)

9. **Overnutrition and undernutrition are forms of malnutrition.**
   a. True
   b. False

10. **Malnutrition can result from which of the following?**
    a. Poor food choices
    b. Selecting only nutrient-dense foods
    c. A balanced diet
    d. All of the above

11. **Nutritional deficiency diseases are health problems in the U.S.**
    a. True
    b. False

12. **Which of the following is a cause of death in which diet can play a part?**
    a. Heart disease
    b. Cancer
    c. High blood pressure
    d. All of the above

A National Agenda for Improving Nutrition and Health (pp. 10–14)

13. **Lifestyle behaviors can change the probable length and quality of life.**
    a. True
    b. False

14. **Since the 1980s, prevalence of overweight has:**
    a. Increased
    b. Decreased
    c. Stayed the same

15. **Half of U.S. adults engage in 30 minutes or more of leisure-time activity each day.**
    a. True
    b. False

Understanding Our Food Choices (pp. 15–19)

16. **Hunger and appetite are the same thing.**
    a. True
    b. False

17. **The desire to eat after smelling fresh-baked bread right out of the oven is an example of:**
    a. Hunger
    b. Appetite
    c. Satiety
    d. Personal preference

18. **A protective mechanism to keep the body alive is:**
    a. Hunger
    b. Appetite
    c. Satiety
    d. Personal preference

19. **An abundant availability of food can be just as detrimental to health as scarceness of food.**
    a. True
    b. False

20. **"Healthful foods don't taste good" is an example of:**
    a. *Healthy People 2010*
    b. Advertising
    c. Perceived barriers to healthful eating
    d. Cultural factor

ABCs of Eating for Health

Learning Objective 1: *List and describe characteristics of planning a healthy diet.*

- A healthy diet has the following characteristics: adequacy, balance, calorie control, moderation, and variety. It should also contain foods that fit your personality, family and cultural traditions, lifestyle, and budget.

Nutrient Recommendations

Learning Objective 2: *Briefly describe the various guidelines that can be used to plan a healthy diet: Dietary Reference Intake (DRI), Dietary Guidelines for Americans, MyPlate, and Food Labels.*

- **Dietary Reference Intakes (DRI)** are reference values for energy and nutrients that can be used for planning and assessing diets for healthy people.
- **Estimated Average Requirement (EAR)** is the amount of a nutrient estimated to meet the requirement for the nutrient in half of the people of a specific age and gender. The EAR is used for setting the RDA.
- **Recommended Dietary Allowance (RDA)** is the average daily amount of a nutrient that is sufficient to meet the nutrient needs of nearly all (97–98 percent) healthy individuals of a specific age and gender.
- **Adequate Intake (AI)** is the average amount of a nutrient that appears to be adequate for individuals when there is not sufficient scientific research to calculate an RDA. The AI exceeds the EAR and possibly the RDA.
- **Tolerable Upper Intake Level (UL)** is the maximum amount of a nutrient that is unlikely to pose any risk of adverse health effects to most healthy people. The UL is not intended to be a recommended level of intake.
- **Estimated Energy Requirement (EER)** is the average calorie intake that is predicted to maintain energy balance in a healthy adult of a defined age, gender, weight, height, and level of physical activity consistent with good health.

- **Acceptable Macronutrient Distribution Range (AMDR)** is the range of intakes for carbohydrates, fat, and protein that is associated with reduced risk of chronic disease while providing adequate intakes of essential nutrients.

The Challenge of Dietary Guidelines

- The *Dietary Guidelines for Americans* provide scientific-based advice to promote health and reduce risk for chronic diseases through diet and physical activity.
- They focus on variety, calorie control, moderation, and nutrient density while emphasizing physical activity to maintain a healthy weight.

Introducing the MyPlate Diet-Planning Tool

- MyPlate illustrates the five food groups using a familiar mealtime visual: a place setting. It is designed to remind Americans to eat healthfully; it is not intended to change consumer behavior.

How Well Do We Eat?

- In general, Americans eat too many calories and too much solid fat, added sugars, refined grains, and sodium.
- Too few vegetables, fruits, whole grains, and fat-free or low-fat milk and milk products are consumed.

More Tools for Diet Planning

- **Food labels**, by law, must contain five components: the name of the food (statement of identity); the name and address of manufacturer, packer, or distributor; the net contents of the package; the ingredients list; and the Nutrition Facts panel.
- **Nutrient content claims** such as "low-fat" and "low-calorie" must adhere to specific definitions set forth by the Food and Drug Administration.
- **Health claims** are statements on food labels linking the nutritional profile of a food to a reduced risk of a particular disease. Manufacturers must adhere to strict government guidelines when making such claims.

FIGURE 2-1

Nutrient Density of Selected Beverages

Understanding the concept of nutrient density will help you locate foods that are high in the amount of nutrients provided per calorie. Compare the nutrient contributions made by the three beverages shown here. The figures show that 1% low-fat milk and whole milk are good sources of several essential nutrients, whereas the cola beverage simply provides calories (from sugar). Note that the low-fat milk is more nutrient dense per 100 calories than the whole milk.

[a] The Daily Values, used on food labels, are based on 2,000 calories a day for adults and children over 4 years old and are useful for making comparisons among foods in terms of their nutrient composition.
[b] A good source of this nutrient (more than 10 percent of the recommended intake).

ABCs of Eating for Health (pp. 30–32)

1. **Name the diet planning characteristics and briefly describe why each characteristic is important in diet planning.**

2. **A healthful diet can be described by which of the following?**
   a. Balance, moderation, and variety
   b. Variety, moderation, and physical activity
   c. Eat good foods, stay away from bad foods
   d. "Think outside the bun"

3. **Nutrient density refers to foods that:**
   a. Are rich in protein
   b. Contain a mixture of carbohydrate, protein, and fat
   c. Carry the USDA nutrition labeling
   d. Are rich in nutrients but relatively low in calories

Nutrient Recommendations (pp. 32–35)

4. **The Dietary Reference Intakes (DRI) for nutrients are generally:**
   a. The minimum amounts that average people need
   b. More than twice as high as anyone needs
   c. Designed to prevent deficiency disease in half the population
   d. Designed to be adequate for almost all healthy people
   e. The same for all ages for all nutrients

5. **Foods that have nutrients added to enhance their nutrient content are:**
   a. Fortified
   b. Enriched
   c. Nutrient dense
   d. Calorie dense

6. **When it comes to nutrient intake:**
   a. More is better.
   b. Too much can be dangerous.
   c. It does not matter as long as a person gets enough protein.

The Challenge of Dietary Guidelines (pp. 35–37)

7. **Review the *Dietary Guidelines for Americans*. What changes, if any, would you need to make to achieve these recommendations?**

8. **Which of the following describes an action that applies to the *Dietary Guidelines for Americans*?**
   a. Enjoy your food but eat less.
   b. Avoid oversized/supersized portions.
   c. Switch to fat-free or low-fat (1%) milk.
   d. All of these would apply to the *Dietary Guidelines for Americans*.

Introducing the MyPlate Diet-Planning Tool (pp. 37–48)

9. **Define lifestyle diseases.**

10. **Foods in a given food group in MyPlate are equivalent in their content of:**
    a. Vitamins and minerals
    b. Carbohydrate and fat content
    c. Protein and energy content
    d. Fiber content

How Well Do We Eat? (pp. 49–50)

11. **What is the difference between a serving and a portion?**

12. **Typical Americans eat:**
    a. Too few calories
    b. Too few fruits/vegetables and whole grains
    c. Too many fruits/vegetables and fiber
    d. Too many calories and solid fat

More Tools for Diet Planning (pp. 51–58)

13. **Which of the following is a convenient way to determine appropriate portion sizes and avoid consuming extra calories?**
    a. Your hand
    b. Your foot
    c. Both feet and hands
    d. Whatever you are served is the correct portion size.

14. **Food labels list ingredients in:**
    a. Descending order of predominance (by weight)
    b. Ascending order of predominance (by weight)
    c. Alphabetical order
    d. Manufacturers are not required to list ingredients

15. **Nonnutritive substances in plants that possess health-protective benefits are examples of:**
    a. Calories
    b. Alcohol
    c. Phytochemicals
    d. Antioxidant nutrients

16. **A person can eat all of the low-fat and fat-free foods he/she wants because they are calorie-free.**
    a. True
    b. False

17. **A statement linking the nutritional profile of a food to a reduced risk of a particular disease is known as a:**
    a. Nutrient content claim
    b. Health claim
    c. Disease reducer
    d. Composition claim

The Digestive System

Learning Objective 1: *Define essential nutrients.*

Learning Objective 2: *Follow the flow of food through the digestive system.*

Learning Objective 3: *Identify the role of each organ involved in digestion, including accessory organs.*

- Essential nutrients are compounds that must be obtained from food because they cannot be synthesized by the body in amounts sufficient to meet physiological needs.
- The body's cells need nutrients 24 hours a day.
- Hunger signals originate in the hypothalamus.
- The gastrointestinal tract provides the body with a constant supply of water and nutrients.
- Digestion begins in the mouth.
- The esophagus is strictly a passageway for food from the mouth to the stomach: No digestion takes place.
- The stomach carries out three functions during digestion:
  - Storing food until it can be processed;
  - Mixing food and gastric secretions to form chyme; and
  - Controlling movement of chyme into the small intestine at a rate suitable for digestion.
- The majority of digestion and absorption takes place in the small intestine.
- The primary functions of the large intestine are to absorb dissolved minerals and water and to eliminate waste products.
- Organs outside the digestive tract also participate in the digestive process.
  - The liver determines the metabolic fate of every nutrient digested and absorbed in addition to producing bile.
  - The gallbladder is attached to the liver and stores and concentrates bile.
  - The pancreas is an endocrine and exocrine organ that functions to regulate blood glucose levels and aid in digestion.

Metabolism

- All nutrients leave the digestive system via the circulatory system and are transported to the liver, where they are processed.
- The liver processes these nutrients.
  - Carbohydrates are converted to glucose or glycogen, and excess is converted to fat.
  - Amino acids are converted to make proteins, and excess is used to make glucose and/or fat.
  - Fatty acids and glycerol are reassembled into larger fats and are packaged with protein for transportation to other parts of the body.
- These new products of liver metabolism are released into the bloodstream to be circulated to all other cells of the body.

Other Systems

- Most of the body's work is done automatically and is finely regulated to achieve a state of well-being.
- Each system in the body contributes to the welfare of the whole by doing specialized work.

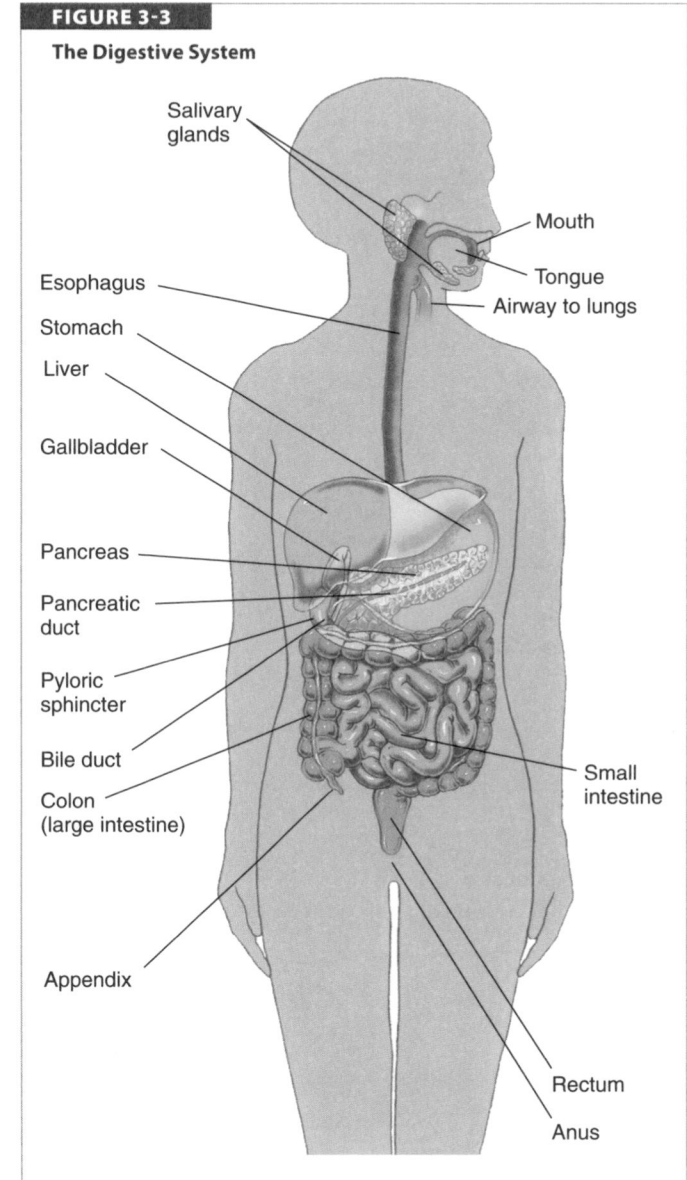

FIGURE 3-3

The Digestive System

Salivary glands
Mouth
Esophagus
Tongue
Airway to lungs
Stomach
Liver
Gallbladder
Pancreas
Pancreatic duct
Pyloric sphincter
Bile duct
Colon (large intestine)
Small intestine
Appendix
Rectum
Anus

The Digestive System (pp. 70–84)

1. It has been awhile since your last meal and you are feeling hungry. Which part of the brain is sending these signals?
   a. Cortex
   b. Spinal cord
   c. Hypothalamus
   d. Pancreas

2. Xavier Breth has just consumed a bagel. Trace the order in which the bagel will move through the organs of the GI tract.
   a. Mouth → stomach → small intestine → large intestine
   b. Mouth → stomach → liver → small intestine
   c. Stomach → pancreas → small intestine → esophagus
   d. Mouth → stomach → large intestine → small intestine

3. Food is pushed through the esophagus by:
   a. Peristalsis
   b. Bolus
   c. Pylorus
   d. Sphincters

4. Once the stomach has completed its part of the digestive process, it passes its contents along to the small intestine. Which of the following regulates this step?
   a. Peristalsis
   b. Bolus
   c. Pylorus
   d. Sphincters

5. Which of the following protects the stomach from the damaging effects of the hydrochloric acid?
   a. Chyme
   b. Hydrochloric acid
   c. Peristalsis
   d. Mucus

6. Which of the following is the correct order in which food empties from the stomach?
   a. Liquids, carbohydrates, proteins, fats
   b. Fats, carbohydrates, proteins, liquids
   c. Protein, liquids, carbohydrates, fats
   d. Carbohydrates, fats, liquids, protein

7. Bile functions as:
   a. A digestive enzyme
   b. An emulsifier
   c. An enzyme
   d. Lymph

8. Bicarbonate:
   a. Acts as a protein catalyst
   b. Is made from cholesterol
   c. Is stored in the gallbladder
   d. Neutralizes acid from the stomach

9. Villi and microvilli are fingerlike and hairlike projections that line the small intestine. Their purpose is to:
   a. Increase surface area of the small intestine
   b. Neutralize hydrochloric acid from the stomach
   c. Increase peristalsis
   d. Speeds up the chemical reactions of enzymes

10. Nutrients are absorbed in the:
    a. Esophagus
    b. Stomach
    c. Small intestine
    d. Large intestine

11. Colonic bacteria synthesize absorbable vitamin K.
    a. True
    b. False

12. Intestinal gas is produced by:
    a. Bile
    b. Bacterial fermentation
    c. Chyme
    d. Vitamin K

13. Metabolism takes place in the:
    a. Liver
    b. Gallbladder
    c. Large intestine
    d. Kidneys

14. Virtually infinite storage for energy is found in:
    a. Glycogen
    b. Glucose
    c. Catabolism
    d. Fat cells

The Carbohydrates: Sugar, Starch, and Fiber

The Body's Need for Carbohydrates

Learning Objective 1: *Describe how carbohydrates are used in the body.*

- The primary function of carbohydrates is to provide the body with energy; they are the preferred source of energy for certain body systems (e.g., brain and nervous system).

Carbohydrate Basics

- Carbohydrates are almost exclusively found in plant foods.
- Milk is the only animal-derived food that contains a significant amount of carbohydrate.
- Carbohydrates are divided into two categories: complex and simple.

The Simple Carbohydrates: Monosaccharides and Disaccharides

Learning Objective 2: *Differentiate between simple and complex carbohydrates.*

- There are two categories of simple carbohydrates (or simple sugars): monosaccharides and disaccharides.
- Monosaccharides can be found in three variations: glucose, fructose, and galactose.
- Disaccharides can also be found in three variations: sucrose, maltose, and lactose.

The Complex Carbohydrate Starch

Learning Objective 2: *Differentiate between simple and complex carbohydrates.*

Learning Objective 3: *Define the terms* refined, enriched, *and* fortified.

- Starch and fiber are the two categories of complex carbohydrates.
- All starchy foods are plant foods, such as grains and legumes (dried beans and peas).
- Grain products can be refined, enriched, fortified, or whole grain.
- Refining grains removes most of the grain's dietary fiber, iron, other minerals, and many B vitamins.
- By law, enriched grains must have the B vitamins thiamin, riboflavin, niacin, and folic acid, as well as the mineral iron, added at specified levels.
- Whole grains are processed in their entirety (except the husk), and are not refined.

The Complex Carbohydrate Fiber

- Fiber comes from parts of plants: leaves, stems, and seeds.
- Most fibers cannot be digested by the human digestive system like starches can.
- Insoluble fiber stimulates the muscles of the digestive tract, thus preventing constipation, hemorrhoids, and diverticulosis.
- Soluble fibers reduce risk of heart disease and improve diabetes control.

Guidelines for Choosing Carbohydrates

- Nutrition goals and guidelines emphasize consumption of complex carbohydrates such as fruits, vegetables, and whole grains.

How the Body Handles Carbohydrates

- Glucose is the basic carbohydrate unit each cell uses for energy.
- Enzymes break down disaccharides and complex carbohydrates until only glucose molecules remain.

Hypoglycemia and Diabetes

- Hypoglycemia is the term used for abnormally low blood glucose levels.
- Hypoglycemia can be the result of going too long between meals, but some individuals experience hypoglycemia as a result of their bodies secreting abnormal amounts of insulin.
- Diabetes is an absence, insufficiency, or ineffectiveness of insulin, resulting in the inability to regulate blood glucose levels normally.
- Elevated glucose levels (hyperglycemia) over a period of time can result in loss of vision, cardiovascular disease, stroke, amputation, kidney failure, and damage to nerve cells in the hands and feet.

Sugar and Health

- The only health problem proven to be directly related to sugar is tooth decay.
- Artificial sweeteners sweeten foods without adding calories.
- To be sold or used in the U.S., artificial sweeteners must be approved by the Food and Drug Administration.

TABLE 4-9	Distinctions between the Two Major Types of Diabetes	
	Type 1 Diabetes	**Type 2 Diabetes**
Incidence	5%–10% of cases	90%–95% of cases
Age of Onset	< 30	> 40*
Insulin Deficiency	Yes; pancreas unable to make insulin to meet needs.	In some cases there may be insufficient insulin, or cells may be unresponsive to insulin.
Risk Factors	Genetic predisposition plus environmental factor (for example, viral infection)	Genetic predisposition plus obesity, family history, aging
Treatment	Insulin injections, diet, and physical activity	Weight loss, diet, and physical activity; oral drugs or insulin sometimes needed.

*The incidence of type 2 diabetes is increasing among children. In the majority of cases, the children are overweight or obese and have a family history of type 2 diabetes.

The Body's Need for Carbohydrates (p. 98)

1. **The primary role of carbohydrates in the human body is to:**
   a. Provide the body with energy
   b. Fatten the body
   c. Make foods taste good
   d. None of the above

Carbohydrate Basics (pp. 98–99)

2. **How many grams of carbohydrate would it take to equal 32 calories?**
   a. 7
   b. 8
   c. 23
   d. 87

The Simple Carbohydrates: Monosaccharides and Disaccharides (pp. 99–100)

3. **Sucrose is an example of a:**
   a. Disaccharide
   b. Polysaccharide
   c. Triglyceride
   d. Protein

4. **When fructose and glucose are bonded together they form:**
   a. Sucrose
   b. Galactose
   c. Lactose
   d. Maltose

5. **Which of the following is a carbohydrate composed of 2 monosaccharide units (sucrose, lactose, or maltose) that have been chemically linked?**
   a. Disaccharide
   b. Polysaccharide
   c. Starch
   d. Fiber

The Complex Carbohydrate Starch (pp. 100–102)

6. **Starches are a classification of:**
   a. Carbohydrates
   b. Vitamins
   c. Proteins
   d. Minerals

The Complex Carbohydrate Fiber (pp. 102–104)

7. **You have probably heard that dietary fiber is good for you. Which of these foods is a good source?**
   a. Baby carrots
   b. Oatmeal
   c. Fresh fruit
   d. Dried beans
   e. All of the above

8. **Which of the following is *not* true about fiber?**
   a. It cannot be digested by humans.
   b. It provides four calories per gram.
   c. It is found only in plants.
   d. It can be soluble or insoluble.

9. **Which of the following usable parts of a wheat kernel is rich in fiber?**
   a. Bran
   b. Germ
   c. Endosperm
   d. Husk

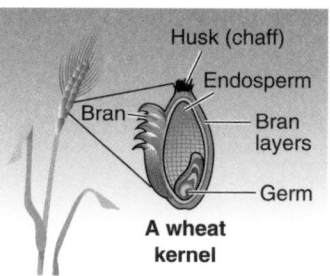

A wheat kernel

10. **What happens to carbohydrates that are not digested (fiber) in the small intestine?**
    a. They pass into the large intestine and are eliminated.
    b. They pass into the large intestine and are absorbed there.
    c. They are absorbed and converted to fat.
    d. They are absorbed and converted to glycogen.

Guidelines for Choosing Carbohydrates (pp. 104–108)

11. **What are the main sources of complex carbohydrates in a person's diet?**
    a. Leafy green vegetables
    b. Dairy foods
    c. Sugary sweets
    d. Seeds, beans, and grains

How the Body Handles Carbohydrates (pp. 108–112)

12. **The body's form of stored glucose is called glycogen.**
    a. True
    b. False

13. **If a person uses up the body's reserve supply of glycogen and still does not eat, glucose (sugar) comes from the:**
    a. Heart
    b. Liver
    c. Bloodstream
    d. Bones

14. **A hormone that lowers high levels of glucose in the blood is:**
    a. Glycogen
    b. Insulin
    c. Diabetes
    d. Adrenaline

Hypoglycemia and Diabetes (pp. 112–116)

Sugar and Health (pp. 117–120)

15. **Which of the following statements is/are true about sugar?**
    a. Sugar causes hyperactivity in children.
    b. Sugar contributes to tooth decay.
    c. Sugar causes diabetes.
    d. All of the above are true about sugar.

The Lipids: Fats and Oils

A Primer on Fats

- Fat is the body's chief form of energy storage, and there is virtually no limit to the amount of fat we can store.
- Fats in food provide satiety by slowing the emptying rate of the stomach.
- Fats in foods carry fat-soluble vitamins (A, D, E, K) along with essential fatty acids.

A Closer View of Fats

- Triglycerides are the major class of dietary fats (fats and oils) and make up approximately 95 percent of the lipids in foods and the human body.
- Saturated fatty acids are found in all animal foods (meat, poultry, full-fat dairy products) as well as tropical oils (palm and coconut).

Characteristics of Fat in Foods

- Animal fats tend to have a higher proportion of saturated fats, and plant foods tend to have higher proportions of monounsaturated and/or polyunsaturated fatty acids.
- Fats with a high percentage of saturated or trans fatty acids are solid at room temperature.
- Fats that are more unsaturated are usually liquid at room temperature.
- The process of hydrogenation causes unsaturated fats to become solid fats, which affects their health benefits.

The Other Members of the Lipid Family: Phospholipids and Sterols

- Lecithin and other phospholipids are important components of cell membranes.
- Phospholipids help fats travel back and forth across the lipid-containing membrane.
- Lecithin acts as an emulsifier and is used as a food additive.
- Cholesterol (a sterol) is found only in animal foods and is also made in the human body. Cholesterol also enables our bodies to convert sunlight on our skin into vitamin D.

How the Body Handles Fat

- The body makes chylomicrons from triglycerides and other lipids. This allows fat to travel in the blood.
- Two types of lipoproteins (low-density lipoproteins and high-density lipoproteins) have implications for the heart and blood vessels.
- Low-density lipoproteins (LDL) carry cholesterol to body cells.
- High-density lipoproteins (HDL) carry cholesterol in the blood back to the liver for recycling or disposal.

"Good" versus "Bad" Cholesterol

- The underlying cause of heart disease is atherosclerosis, a narrowing of the arteries caused by buildup of cholesterol-containing plaque in arterial walls.
- Elevated LDL ("bad" cholesterol) concentrations in the blood are a risk factor for heart disease.
- Elevated HDL ("good" cholesterol) concentrations protect us from heart disease.

Fat in the Diet

- The type of fat consumed is more important in influencing risk for heart disease than total amount of fat in the diet.
- Total fat in the diet should not exceed 20–35 percent of a day's total caloric intake. Most of that fat should come from polyunsaturated and monounsaturated fat sources: fish, nuts, and vegetable oils.
- Saturated fat in the diet should contribute no more than 10 percent of daily calories.
- *Trans* fats should be kept as low as possible.

Fat Can Be Healthy

- Fats from plant sources tend to be healthier than fats from animal sources.
- Olive oil keeps the heart healthy by raising the amount of HDL-cholesterol in the blood and lowering LDL-cholesterol.

The *Trans* Fatty Acid Controversy—Is Butter Better?

- *Trans* fats are produced when unsaturated fats in foods are hydrogenated. Shortenings, baked goods, commercial frying fat used to make French fries, and other fried menu items have been found to raise levels of LDL-cholesterol and lower HDL-cholesterol in the blood.
- Food labels are required to indicate the amount of *trans* fat a product contains.

Understanding Fat Replacers

- Reduced-fat foods *still* contain calories and are *not* a replacement for a healthful diet rich in whole grains, fruits, and vegetables.

Animal products
Candy: 1%
Breakfast cereal: 1%
Salad dressing: 3%
Household shortening: 4%
Potato chips, corn chips, popcorn: 5%
Fried potatoes: 8%
Margarine
21%
40%
17%
Cakes, cookies, crackers, pies, bread, etc.

FIGURE 5-10

Major Sources of *Trans* Fatty Acids in the U.S. Diet

Trans fatty acids occur naturally in meat, poultry, and dairy products. They show up in packaged and fast foods to which hydrogenated vegetable oils have been added. Note that high levels of *trans* fatty acids go hand in hand with foods containing solid fats. Thus, if you choose foods prepared with little or no solid fats, you will consume minimal amounts of *trans* fatty acids.

SOURCE: FDA Consumer, 2003.

A Primer on Fats (pp. 128–129)

1. List the functions of fat in the body.
2. List the functions of fat in foods.
3. How many calories are provided by fat in your roommate's most recent meal?

   Three pieces of fried chicken (40 g fat)
   Mashed potatoes with gravy (6 g fat)
   Coleslaw (5 g fat)
   Roll with butter (5 g fat)
   Soft drink

   a. 56 calories
   b. 224 calories
   c. 504 calories
   d. There is not enough information to answer this question.
4. Which of the following is a function of fat?
   a. Most concentrated source of energy for the body
   b. Provides essential amino acids for growth and repair
   c. Carries water-soluble vitamins
   d. All of the above are functions of fat.
   e. You can't fool me . . . none of the above are functions of fat.

A Closer View of Fats (pp. 129–131)

5. If the source of a fat is from an animal, then it likely contains:
   a. Saturated fat
   b. Monounsaturated fat
   c. Polyunsaturated fat
   d. Cholesterol
   e. Both A and D
6. Your family knows you are taking this nutrition course and wants you to give them a list of good fats to eat and a list of bad fats to avoid. What do you tell them?
   a. We are only learning about fats that are bad. I will email a list soon.
   b. That's easy . . . we've learned that all fats that taste good are bad for us, and all fats that taste terrible are good for us.
   c. There are no good or bad fats.
   d. I will bring the lists home next weekend.

Characteristics of Fat in Foods (pp. 131–132)

7. Which fats are liquid at room temperature?
   a. Unsaturated fats
   b. Saturated fats
   c. Both A and B
   d. None of the above

The Other Members of the Lipid Family: Phospholipids and Sterols (pp. 132–133)

8. Which of the following groups of foods DO NOT contain cholesterol?
   a. Cheese, meat, and eggs
   b. Canola oil, peanut oil, and lard
   c. Avocados, bananas, and beans
   d. All of these foods contain cholesterol.
   e. None of the food contain cholesterol.
9. The most effective way to lower blood cholesterol is to:
   a. Reduce monounsaturated fat intake
   b. Reduce saturated fat intake
   c. Increase saturated fat intake
   d. Decrease exercise

How the Body Handles Fat (pp. 133–136)

10. List the leading risk factors for heart disease.

"Good" versus "Bad" Cholesterol (pp. 136–137)

11. Which of the following is "good" cholesterol?
    a. HDL-cholesterol
    b. LDL-cholesterol
    c. VLDL-cholesterol
    d. There is no such thing as "good" cholesterol.

Fat in the Diet (pp. 138–140)

12. It seems everyone is talking about (dietary) fat. The best way to control fat in your diet is to:
    a. Cut out all meat and dairy products.
    b. Use "balance, variety, and moderation."
    c. Go on the Ginny Craig diet.
    d. None of the above
13. Excessive fat in the diet is associated with the development of:
    a. Heart disease
    b. Tooth decay
    c. Osteoporosis
    d. All of the above
14. Atherosclerosis can occur:
    a. In the coronary arteries, causing poor circulation to parts of the heart
    b. In the arteries of the legs, causing pain and muscle cramps
    c. In the arteries in the neck, decreasing blood flow to the brain
    d. In the arteries in the brain, leading to stroke
    e. All of the above
15. Cholesterol is produced in the liver.
    a. True
    b. False

Fat Can Be Healthy (pp. 140–145)

16. Describe the difference between visible fats/oils and hidden fats/oils.
17. A deficiency of fat in the diet may result in which of the following?
    a. Flaky dermatitis
    b. Decreased growth
    c. Deficiency of fat-soluble vitamins
    d. A and D only
    e. A, B, and C

Understanding Fat Replacers (pp. 146–147)

18. Which of the following is true about Olean (olestra)?
    a. It is approved for use by the Food and Drug Administration (FDA).
    b. It does provide calories but not as many as fat provides.
    c. It cannot be digested, therefore it does not add calories to foods.
    d. It cannot be used for cooking.
    e. A and C only

The Proteins and Amino Acids

What Proteins Are Made Of
- Nitrogen is the component that makes protein different from carbohydrates and fats.
- Amino acids make up the chemical backbone of proteins.
- Proteins are formed by strands of amino acids that differ from each other.
- The body can make about half of the necessary amino acids (known as nonessential amino acids) for itself. There are 11 nonessential amino acids.
- Essential amino acids must be obtained from food. There are nine essential amino acids. If they are not supplied in the diet, the body will not be able to make the necessary proteins.

The Functions of Body Proteins
- No new living tissue can be built without protein, because protein is part of every cell in the body.
- The proteins in our bodies function as enzymes, antibodies, hormones, transport vehicles, oxygen carriers, tendons and ligaments, scars, cores of bones and teeth, filaments of hair, fingernails, and more.

How the Body Handles Protein
- Amino acids in foods are broken apart during digestion. The body then reassembles these individual amino acids into proteins that are used by the body.
- If an essential amino acid is missing, the building of the protein will stop.
- The body cannot store amino acids for future use.
- If there are amino acids that are not used to build body proteins, the liver will remove the amine group to be excreted by the kidneys.

Protein Quality of Foods
- Complete proteins contain all essential amino acids in the correct proportion relative to the body's needs.
- Quality of food protein is judged by the proportion of essential amino acids contained relative to the body's needs. Animal and soy proteins are the highest in quality.
- Incomplete proteins are lacking or low in one or more of the essential amino acids.
- The limiting amino acid is the essential amino acid in shortest supply (relative to the body's need) in a food protein, therefore "limiting" the body's ability to make its own proteins.
- Complementary proteins are two or more food proteins whose amino acid assortments complement each other in such a way as to provide all essential amino acids.

Recommended Protein Intakes
- Recommended protein intakes can be stated two ways:
  - As a percentage of total calories (10–35 percent of total caloric intake).
  - As an absolute number (grams per day). The recommended daily protein allowance for a healthy adult is 0.8 grams/kg of desirable body weight.

Protein and Health
- Protein-energy malnutrition (PEM) includes both kwashiorkor and marasmus.
- Kwashiorkor is caused by inadequate protein consumption in the presence of adequate food energy.
- Marasmus is a deficiency of protein and calories brought about by starvation (inadequate food energy).
- High protein intake in humans results in increased calcium excretion.
- The higher a person's intake of animal-protein sources (meats), the more likely the diet will be low in fruits, vegetables, and whole grains resulting in deficiencies of other nutrients.
- Protein consumed in excess of the body's needs results in the extra protein being converted to energy and stored as body fat.
- There are no known benefits from consuming excess protein.
- A diet high in animal protein is often high in saturated fats.

The Vegetarian Diet
- Vegetarian diets can be healthful during all stages of life with the correct planning.
- Semivegetarians eat plant foods and some but not all groups of animal-derived products, such as meat, poultry, fish, seafood, eggs, milk, and milk products.
- Lacto-vegetarians eat plant foods, as well as milk and milk products, but avoid poultry, fish, and seafood.
- Lacto-ovovegetarians eat plant foods, milk, milk products, and eggs, but meat, poultry, fish, and seafood are excluded.
- Ovovegetarians eat plant foods and eggs but avoid milk and milk products, meat, poultry, fish, and seafood.
- Strict vegetarians (or vegans) eat plant foods only, including honey.
- A macrobiotic diet is a way of life as well as a way to eat. It incorporates natural, organic, plant-based foods with tenets of Zen Buddhism.
- Adequate amounts of amino acids can be obtained from a plant-based diet when a variety of foods is routinely consumed on a daily basis.
- Plant-based protein foods are higher in fiber, richer in certain vitamins and minerals, and lower in fat than animal-based proteins.
- Research has found that people who follow vegetarian diets have lower rates of heart disease, cancer, diabetes, and obesity.

What Proteins Are Made Of (pp. 162–163)

1. _____ is a chemical compound made up of chains of amino acids.
   a. Fat
   b. Protein
   c. Energy
   d. Fiber

2. The chemical element that makes proteins different from carbohydrates and fats is:
   a. Oxygen
   b. Carbon
   c. Nitrogen
   d. Hydrogen

3. A nonessential amino acid:
   a. Is not necessary in the diet
   b. Must be supplied by the diet
   c. Can be made from carbohydrates in the body
   d. Can be made from fats in the body
   e. Both C and D

4. If the diet is lacking an essential amino acid:
   a. The body cells will synthesize it.
   b. Proteins will be made but they will lack that particular amino acid.
   c. Protein synthesis will be limited.
   d. Health will not be affected as long as other nutrients are adequate.

The Functions of Body Proteins (pp. 163–166)

5. The basic building block of a protein is called a(n):
   a. Fatty acid
   b. Monosaccharide
   c. Amino acid
   d. Keto acid

6. The primary function of protein is to provide:
   a. Building material for body tissues
   b. Building material for fat
   c. Storage for fat
   d. Energy

How the Body Handles Protein (pp. 166–167)

7. Which of the following statements is *not* true about protein?
   a. The primary role of protein is to provide calories.
   b. Most Americans eat well *above* the RDA for protein each day.
   c. High (animal) protein intakes increase the body's excretion of calcium.
   d. The primary role of protein is to provide the building material for body tissues.

Protein Quality of Foods (pp. 167–169)

8. Combining two plant sources of protein:
   a. Cannot be done
   b. Does not taste good
   c. May provide complete proteins
   d. Makes the protein more expensive

9. What is the only plant food that qualifies as a complete protein for children and adults?
   a. Corn
   b. Broccoli
   c. Soybeans
   d. Wheat
   e. Baked beans

10. Which of the following foods contain complete proteins?
    a. Beef
    b. Milk
    c. Eggs
    d. Kidney beans
    e. A, B, and C

11. A protein is said to be "complete" when:
    a. It supplies 100 percent of the RDA.
    b. It is found in unprocessed foods.
    c. All essential amino acids are present and in proper proportion.
    d. It satisfies your appetite.

Recommended Protein Intakes (p. 169)

12. Which of the following statements is true?
    a. Most people in the U.S. suffer from a protein deficiency.
    b. It is recommended that we consume 48 percent of our calories from protein.
    c. Most of the protein in the U.S. diet comes from plant sources.
    d. Vegetarian diets can be healthy diets.

Protein and Health (pp. 169–172)

13. A diet high in protein generally corresponds to a diet that is also too high in:
    a. Carbohydrates
    b. Fat
    c. Water
    d. Chocolate
    e. None of the above

14. Eating more protein than the body needs has absolutely no adverse effects.
    a. True
    b. False

The Vegetarian Diet (pp. 172–179)

15. The best advice for someone who is a vegetarian is to:
    a. Find out how much of each essential amino acid is required daily.
    b. Determine how much of each amino acid is contained in favorite foods.
    c. Use a wide variety of whole foods in generous servings.
    d. A and B only

16. Which of the following is a proven benefit of vegetarianism?
    a. It can cure cancer.
    b. It will give a person inner peace.
    c. It can cure AIDS.
    d. It can lower the risk of heart disease.

17. Which of the following foods would a lacto-ovovegetarian eat?
    a. Milk
    b. Fish
    c. Lamb chops
    d. Eggs
    e. A and D

18. Which nutrient(s) is/are most likely missing from a *vegan* diet?
    a. Vitamin B_{12}
    b. Vitamin D
    c. Iron
    d. A, B, and C
    e. There is never a problem getting all essential vitamins and minerals on a vegan diet.

19. Which of the following diseases/conditions may be a possible risk of consuming a vegetarian diet?
    a. Edema
    b. Type 2 diabetes
    c. High blood pressure
    d. Iron deficiency anemia
    e. All of the above

Turning Back the Clock

- Vitamins are potent compounds that perform many tasks in the body to promote growth and reproduction and to maintain health and life. They are essential in helping the body make use of the calories consumed in foods.

The Two Classifications of Vitamins

- Scientists have identified 13 vitamins, each with its own special role in the body.
- Vitamins are classified as either water-soluble (they dissolve in water) or fat-soluble (they dissolve in fat).
- Water-soluble vitamins:
  - The eight B vitamins are thiamin, riboflavin, niacin, B_6, folate, B_{12}, pantothenic acid, and biotin.
  - Vitamin C is also called ascorbic acid.
  - Water-soluble vitamins act as coenzymes in the body.
  - Water-soluble vitamins can be easily destroyed during food storage, processing, and preparation.
- Fat-soluble vitamins:
  - The fat-soluble vitamins are A, E, D, and K; they are generally found in fats and oils of food.
  - Fat-soluble vitamins are stored in the liver and in body fat, making it possible for large amounts of fat-soluble vitamins to build up to toxic levels and cause undesirable side effects.
  - Fat-soluble vitamins are absorbed with the aid of fats in food and bile produced by the liver.

The Antioxidant Vitamins

- Antioxidants help prevent damage done to the body as a result of chemical reactions involving oxygen, which lead to the creation of highly toxic compounds called free radicals. Free radicals can cause severe cell injury and may contribute to development of chronic diseases such as heart disease and cancer.
- Antioxidant nutrients (vitamins C and E and beta-carotene) are used by the body as a defense mechanism to protect against potential damage from free radicals.
- Beta-carotene is a carotenoid and the precursor of vitamin A. Beta-carotene, which humans convert into vitamin A, is found in orange and dark-green leafy vegetables.
- Beta-carotene is found in plant foods, while preformed vitamin A is found in animal foods.

Vitamins for Healthy Blood

- Folate and vitamin B_{12} assist with formation of new blood cells.
- B_6 helps make hemoglobin, the compound that carries oxygen in the blood.
- Folate is important in the synthesis of DNA and the formation of red blood cells.
- Intrinsic factor (produced in the stomach) is necessary for the absorption and utilization of vitamin B_{12}.
- Vitamin K is necessary for the ability of blood to clot, which can mean the difference between life and death. Vitamin K is obtained both in the diet and via intestinal bacteria.

Vitamins for Healthy Bones

- Vitamins D, K, and C are essential for healthy bones.
- Vitamin D assists in the absorption of dietary calcium and makes calcium and phosphorus available in the blood.
- Vitamin D can be synthesized in the body when the skin is exposed to sunlight. Sunlight on skin transforms a cholesterol compound in skin into a vitamin D precursor. The liver alters this compound, and the kidneys alter it further to produce the active form of vitamin D.
- Vitamin K works in conjunction with vitamin D to synthesize bone protein that helps regulate calcium levels in the blood.

Vitamins and Energy Metabolism

- Many vitamins (and minerals) are essential for the body to metabolize carbohydrates, proteins, and fats into energy.
- Thiamin acts as a coenzyme in reactions that release energy from carbohydrates.
- Riboflavin acts as a coenzyme in energy-releasing reactions; it also helps prepare fatty acids and amino acids for breakdown.
- Niacin is part of a coenzyme vital to producing energy from foods.
- Pantothenic acid and biotin are necessary to synthesize coenzymes active in energy metabolism and in many body systems.

Vitamin-Like Compounds

- Nonvitamins are substances important to cell membranes and cellular activities, but they are not essential in human diets because our bodies can make these compounds as needed.

Phytonutrients in Foods: The Phytochemical Superstars

- Phytochemicals are nonnutritive substances in plants that possess health-protective benefits. They are the compounds that give plants their colors and distinctive aromas.
- Phytochemicals are not nutrients, but they may perform important functions as antioxidants, decreasing blood pressure and cholesterol, preventing cataracts, reducing menopause symptoms, reducing osteoporosis risk, and slowing/reducing certain cancers.
- Pure extracts of phytochemicals (supplements) are less effective than phytochemicals in whole foods.

Turning Back the Clock (pp. 190–191)

1. **Describe how vitamin deficiency diseases were discovered.**

2. **Scurvy is the result of a deficiency of which vitamin?**
   a. Vitamin D
   b. Vitamin E
   c. Riboflavin
   d. Vitamin C

The Two Classifications of Vitamins (pp. 191–194)

3. **Why are fat-soluble vitamins toxic if consumed in large amounts?**
   a. They can be stored in the body and reach toxic levels.
   b. The kidneys can absorb fat-soluble vitamins from the diet.
   c. Fat-soluble vitamins are changed into toxins when they enter the body.
   d. The body can make its own fat-soluble vitamins so excess in the diet is toxic.

4. **The water-soluble vitamins include:**
   a. A and C
   b. B and E
   c. B and C
   d. B and K
   e. All vitamins are fat-soluble

5. **When a fat-soluble vitamin is consumed in excess of body needs, the excess is usually:**
   a. Not absorbed
   b. Excreted in the urine
   c. Stored in the liver and adipose tissue
   d. Excreted in the breath

The Antioxidant Vitamins (pp. 194–199)

6. **With exposure to sunlight, vitamin C can be made by humans.**
   a. True
   b. False

7. **Beta-carotene:**
   a. Has no established RDA
   b. Is found in fruits and vegetables
   c. Can be converted to Vitamin A in the body
   d. All of these are true.

Vitamins for Healthy Blood (pp. 199–205)

8. **What vitamin may prevent neural tube defects in newborns when consumed in adequate amounts before and during pregnancy?**
   a. Folic acid
   b. Thiamin
   c. Vitamin C
   d. Beta-carotene

9. **Which nutrient(s) is (are) most likely missing from a *vegan* diet?**
   a. Vitamin B_{12}
   b. Vitamin C
   c. Phytochemicals
   d. A, B, and C

10. **When you cut yourself, which vitamin comes to your rescue?**
    a. Vitamin K
    b. Vitamin B_{12}
    c. Vitamin E
    d. Vitamin A

Vitamins for Healthy Bones (pp. 205–208)

11. **Vitamin D helps your body use calcium more effectively. You can get vitamin D from:**
    a. Being outdoors in the sunlight
    b. Fortified milk
    c. Broccoli
    d. Whole wheat bread
    e. Both A and B

12. **Vitamin D aids in absorption of:**
    a. Iron
    b. Calcium
    c. Magnesium
    d. Phosphorous

Vitamins and Energy Metabolism (pp. 208–210)

13. **Vitamins provide _____ calories per gram.**
    a. 0
    b. 9
    c. 4
    d. Variable—it depends on the vitamin

Vitamin-Like Compounds (pp. 211)

14. **Nonvitamins are essential nutrients.**
    a. True
    b. False

Phytonutrients in Foods: The Phytochemical Superstars (pp. 211–213)

15. **Phytochemicals are chemical substances in plants.**
    a. True
    b. False

16. **Supplements of phytonutrients are as beneficial to health as eating the actual fruit or vegetable.**
    a. True
    b. False

17. **What would you tell someone who thinks that taking supplements to prevent cancer provides the same level of protection as eating 5 or more servings of fruits and vegetables each day?**
    a. It would be worthwhile to consume larger amounts of vitamins and minerals than he/she already takes, just to be sure he/she is getting enough.
    b. Scientists suspect there are other chemicals in foods besides vitamins that protect us from cancer.
    c. Vitamins are an easy way to get needed nutrients and therefore well worth the expense.

The Two Classifications of Minerals

- Minerals are inorganic compounds that are classified as either major minerals or trace minerals.
- Major minerals occur in large quantities in the body and are needed in the daily diet in relatively large amounts.
- Trace minerals occur in the body in minute quantities and are needed in smaller amounts in the daily diet.

Minerals for Healthy Bones

- The minerals needed for healthy bones are calcium, phosphorus, magnesium, and fluoride.
- Vitamins D, C, and K—along with hormones, collagen, calcium, phosphorus, magnesium, and fluoride—are necessary for the growth and maintenance of healthy bones.
- Calcium deficits during growing years and adulthood contribute to gradual bone loss, or osteoporosis, which can cripple a person later in life.
- The body has limited ability to absorb calcium, so taking more frequent, smaller amounts increases calcium absorption.
- Phosphorus in the body is combined with calcium in the bones and teeth.
- Magnesium promotes resistance to tooth decay by helping hold calcium in tooth enamel.
- Only a trace amount of fluoride occurs in the body; it protects children's teeth from decay and makes the bones of older people resistant to adult bone loss.
- Drinking water is the usual source of fluoride in the diet. Many brands of bottled water do not contain fluoride.

Minerals for Healthy Blood

- Iron deficiency is the most common and widespread nutritional disorder in the world.
- Every living cell contains iron bound into hemoglobin in red blood cells to help transport oxygen from lungs to tissues throughout the body.
- Women require more iron in their diets than men because some of their iron stores (blood) are depleted each month through menstruation.
- There are two types of iron in foods: heme iron found in animal foods, and nonheme iron found in plant and animal foods. Heme iron is much more reliably absorbed than nonheme iron.
- Vitamin C promotes iron absorption and can triple the amount of nonheme iron absorbed from foods eaten at the same meal.
- Zinc is also found in every cell of the body and plays a major role in the regulation of cell multiplication and growth; normal metabolism of protein, carbohydrate, fat, and alcohol; and disposal of damaging free radicals.
- Copper plays an important role in a variety of metabolic and physiological processes like making red blood cells, manufacturing collagen, healing wounds, and maintaining the sheaths around nerve fibers.

Minerals and Energy Metabolism

- Chromium is involved in carbohydrate and lipid metabolism.
- Sulfur is a component of the hormone insulin, which regulates blood glucose levels. It is also part of the B vitamins thiamin and biotin.
- Iodine's principal role in human nutrition is as an integral part of the thyroid hormones that regulate body temperature, metabolic rate, reproduction, and growth.
- A deficiency of iodine results in an enlarged thyroid gland, a condition called goiter.
- Selenium is necessary for the synthesis of the active form of the thyroid hormone that regulates the body's metabolic processes. Selenium is also an antioxidant nutrient.

Water, Minerals, and Fluid Balance

- Water is indispensable to life itself and is essential for health. It is the nutrient most needed by the body.
- Water accounts for about 60 percent of body weight.
- Water helps transport nutrients necessary to nourish cells and carries away waste products. Water also acts as a "shock absorber" in joints and around the spinal cord; lubricates the digestive tract, as well as all tissues moistened with mucus; surrounds and cushions unborn children; and plays a key role in maintaining body temperature.
- Sodium, potassium, phosphorus, and chloride are examples of electrolytes—dissolved substances in blood and body fluids that carry electric charges. Electrolytes influence distribution of fluids among various body compartments.

Sodium, Potassium, and Chloride

- Sodium is part of sodium chloride (table salt). The use of highly salted foods can contribute to high blood pressure in those genetically susceptible.
- The more processed a food is, the more sodium it contains, though it might not always taste salty.
- Potassium is critical in maintaining the heartbeat, water balance, and cell integrity.
- Sodium and potassium are lost along with water in dehydration, which can be dangerous.
- Chloride helps maintain acid-base balance and is part of the hydrochloric acid produced in the stomach.

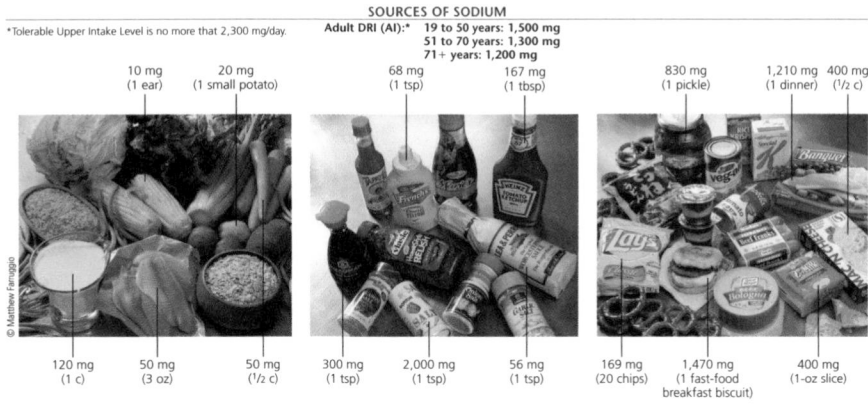

SOURCES OF SODIUM

*Tolerable Upper Intake Level is no more that 2,300 mg/day.

Adult DRI (AI):* 19 to 50 years: 1,500 mg
51 to 70 years: 1,300 mg
71+ years: 1,200 mg

10 mg (1 ear) 20 mg (1 small potato) 68 mg (1 tsp) 167 mg (1 tbsp) 830 mg (1 pickle) 1,210 mg (1 dinner) 400 mg (½ c)

120 mg (1 c) 50 mg (3 oz) 50 mg (½ c) 300 mg (1 tsp) 2,000 mg (1 tsp) 56 mg (1 tsp) 169 mg (20 chips) 1,470 mg (1 fast-food breakfast biscuit) 400 mg (1-oz slice)

© Matthew Farruggio

The Two Classifications of Minerals (pp. 230–231)

1. **Minerals are divided into two categories. Name the categories:**
   a. Essential and non-essential
   b. Primary and secondary
   c. Fat-soluble and water-soluble
   d. Trace and major
   e. None of the above

2. **Which statement(s) describes the role of major minerals in the body?**
   a. They give teeth and bone their rigidity and strength.
   b. They regulate body processes.
   c. They provide energy.
   d. They help maintain acid-base balance.
   e. A, B, and D only

Minerals for Healthy Bones (pp. 232–238)

3. **Calcium absorption is aided by:**
   a. Oxalate and phytate
   b. Vitamin D
   c. Phosphorus and fiber
   d. All of the above

4. **Calcium is the only mineral involved in bone health.**
   a. True
   b. False

5. **What happens when a person's calcium intake is low?**
   a. Calcium is released from bones into the bloodstream to be used by the rest of the body.
   b. The body stops growing.
   c. Calcium is absorbed by bones.
   d. Osteoporosis will develop within the week.

6. **Calcium from animal and plant sources is absorbed with the same efficiency.**
   a. True
   b. False

Minerals for Healthy Blood (pp. 238–242)

7. **Iron from animal sources is absorbed better than iron from plant sources.**
   a. True
   b. False

8. **Vitamin C will increase the absorption of _____ when included in the same meal.**
   a. Vitamin B$_{12}$
   b. Niacin
   c. Iron
   d. Vitamin K
   e. Zinc

9. **Which food provides iron in the most absorbable form?**
   a. Spinach
   b. Rice
   c. Iron-fortified baby cereal
   d. Steak
   e. Peas

10. **Who is at *least* risk for iron-deficiency anemia?**
   a. Tina (female), who is 17 and a vegetarian.
   b. Tom (male), who loves meat and potatoes.
   c. Mary (female), who is 62 and past menopause.
   d. Tom and Mary
   e. Tina and Mary

Minerals and Energy Metabolism (pp. 242–243)

11. **List the minerals involved in energy metabolism.**

12. **What determines the amount of iodine in foods?**
   a. How long a food is cooked
   b. The order in which the foods are eaten: meat first, fruits last
   c. Amount of iodine in the soil where plants are grown or animals grazed
   d. How large their goiter is

Water, Minerals, and Fluid Balance (pp. 243–245)

13. **Functions of water in the body include:**
   a. Speeding up weight loss
   b. Providing an energy source
   c. Helping regulate body temperature
   d. Providing a medium for chemical reactions
   e. C and D only

14. **Strictly speaking, water can be classified as a nutrient.**
   a. True
   b. False

15. **Dehydration is a deficiency of:**
   a. Water
   b. Protein
   c. Carbohydrates
   d. Vitamins and minerals

Sodium, Potassium, and Chloride (pp. 245–248)

16. **From which sources might people unintentionally consume excessive sodium?**
   a. Meats
   b. Fresh vegetables
   c. Processed foods
   d. Fast food restaurants
   e. C and D

17. **A lifelong intake of foods low in sodium and high in potassium protects against high blood pressure.**
   a. True
   b. False

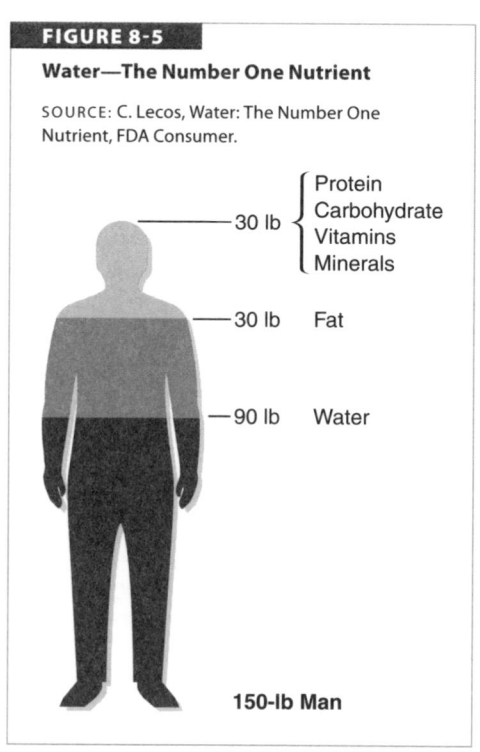

FIGURE 8-5

Water—The Number One Nutrient

SOURCE: C. Lecos, Water: The Number One Nutrient, FDA Consumer.

30 lb — Protein, Carbohydrate, Vitamins, Minerals

30 lb — Fat

90 lb — Water

150-lb Man

Water

- Adults are advised to consume 1.0 to 1.5 milliliters of water from all sources for each calorie *expended*.
- The FDA does not require bottled water to meet higher standards for quality than water from the tap.
- Bottled water is economically and environmentally expensive. Demand for bottled water increases the use of nonrenewable resources to make plastic bottles, which then creates an energy demand for recycling used bottles.

Juices

- The best nutritional choice for juices is 100 percent fruit/vegetable juice. "Juice drinks," "fruit-flavored drinks," or "juice blends" contain little real juice and provide quite a bit of added sugar.
- The darker or more colorful a juice is, the more nutritious it is.

Milk

- Milk supplies significant amounts of protein, calcium, phosphorus, riboflavin, vitamins A and D, potassium, magnesium, vitamins B$_6$ and B$_{12}$, and zinc.
- The term *milk* is also applied to non-animal beverages such as soy milk.

Caffeinated Beverages: Coffee, Tea, Cocoa, and Soft Drinks

- Caffeine is found naturally in coffee beans, cola nuts, cocoa beans, and tea leaves. It is a central nervous system stimulant, and can increase heart rate, urine production, and metabolic rate. Caffeine is often added to energy drinks.
- Consuming caffeine on a daily basis can be habit forming and result in caffeine dependence syndrome.
- Pregnant women should limit caffeine intake because caffeine crosses the placenta and enters the fetus.
- Coffee consumption has been shown to reduce risk of cardiovascular disease, stroke, alcohol-related cirrhosis, and liver cancers.
- Tea leaves are rich in phytochemicals, antioxidants, and caffeine.
- The health benefits of tea can be found on page 270. These health benefits appear to be dose dependent (drinking more provides greater benefits).
- Cocoa contains minerals, fats, proteins, antioxidants, phytochemicals, and caffeine, and may positively influence cardiovascular health. Dark chocolate is more beneficial than milk chocolate.
- Soft drinks are beverages that do not contain alcohol and are typically carbonated, sweetened (sugar, high fructose corn syrup, or noncaloric), and served cold.
- Health issues connected with (sweetened) soft drink consumption are listed on pages 271–272.

Functional Beverages

- Energy drinks are nonalcoholic beverages containing purportedly energy-enhancing ingredients. Their primary ingredients are sugar and caffeine.
- Energy drinks are not regulated by the Food and Drug Administration; therefore, manufacturers of energy drinks do not have to prove the drink is safe or effective.
- The level of caffeine found in most energy drinks is unsuitable for children.
- Sports drinks are formulated to replace fluids and electrolytes (minerals) lost through sweat; they also provide energy for muscles.
- "Enhanced waters" contain added vitamins, minerals, amino acids, antioxidants, and/or fiber.
- Water-soluble vitamins added to water will flush out of the body along with the water. Fat-soluble vitamins are almost impossible to absorb without the presence of fat.

Alcohol

- Alcohol is a clear, colorless liquid that burns easily and is most commonly ingested in the form of ethyl alcohol or ethanol (EtOH).
- Alcohol is a sedative, and the extent to which central nervous system function is impaired is directly related to amount of alcohol in the blood.
- Health benefits from alcohol consumption depend on the amount of alcohol consumed.
- Alcohol is metabolized by the enzyme alcohol dehydrogenase. Until all alcohol consumed has been metabolized, it circulates in the blood and affects the brain and other tissues.
- Alcohol is metabolized more slowly than it is absorbed. A healthy average person can eliminate ½ ounce of alcohol per hour. Rate of alcohol metabolism depends, in part, on the amount of alcohol dehydrogenase formed in the liver. The amount of alcohol dehydrogenase produced in the liver varies among individuals.
- The risks associated with low to moderate alcohol consumption outweigh health benefits.
- Alcoholism is a dependency on alcohol characterized by craving, loss of control, physical dependence and withdrawal symptoms, and tolerance.
- Alcoholics Anonymous (AA) is perhaps the best-known method for dealing with alcohol addiction.

FIGURE 9-5

Alcohol's Effects on the Brain

Alcohol is rightly termed an anesthetic because it puts brain centers to sleep in this order: first the cortex, then the emotion-governing centers, then the centers that govern muscular control, and finally the deep centers that control respiration and heartbeat.

Next most sensitive: voluntary muscle control and emotion-governing centers (sensory area)

Most sensitive: judgment and reasoning (motor and speech area)

Last to be affected: respiration and heart action

Water (pp. 262–265)

1. **Bottled water is less expensive than tap water.**
   a. True
   b. False

2. **Drinking bottled water is always safer and healthier than drinking water from the tap.**
   a. True
   b. False

Juices (pp. 265–266)

3. **Which of the following contains the highest amount of fiber?**
   a. Grapes
   b. Grape juice
   c. Grape-flavored drink
   d. Wine from grapes

4. **100 percent fruit/vegetable juices contain:**
   a. Antioxidants
   b. Phytochemicals
   c. Vitamins
   d. Minerals
   e. All of the above

Milk (pp. 266–267)

5. **Which of the following statements about milk is false?**
   a. Nutritional composition of milk varies widely between mammal species.
   b. Milk is a nutrient-dense food.
   c. More milk is consumed than any other beverage.
   d. All of these are false.
   e. None of these are false.

Caffeinated Beverages: Coffee, Tea, Cocoa, and Soft Drinks (pp. 267–272)

6. **Caffeine is a central nervous system depressant.**
   a. True
   b. False

7. **Consuming caffeine on a daily basis:**
   a. Is recommended for pregnant women
   b. Is habit forming
   c. Can lead to health problems

8. **Antioxidants and phytochemicals can be found in:**
   a. Coffee, teas, and cocoa
   b. Soft drinks and teas
   c. Coffee and high-fructose corn syrup

Functional Beverages (pp. 272–275)

9. **Which of the following agencies regulates energy drinks?**
   a. EPA
   b. USDA
   c. FDA
   d. None of the above

10. **Sports drinks are formulated to:**
    a. Provide vitamins and minerals
    b. Provide caffeine for alertness
    c. Replace fluids and electrolytes lost through sweat
    d. Provide fiber

11. **It is very beneficial to use water enhanced with vitamins, minerals, amino acids, herbs, or fiber.**
    a. True
    b. False

Alcohol (pp. 275–285)

12. **Nearly all alcohol ingested is *metabolized* in the:**
    a. Liver
    b. Stomach
    c. Gall bladder
    d. Large intestine
    e. Small intestine

13. **Drinking alcohol may be associated with an *increased* risk of breast cancer.**
    a. True
    b. False

14. **Your roommate is pregnant and wants to celebrate her soon-to-be status as a mom. She knows you are taking this nutrition class and just studied alcohol and nutrition. She wants to know how many beers she can safely drink.**
    a. No alcohol is the safest.
    b. She can drink beer, but no "hard" liquor.
    c. A pregnant woman can safely consume 1 alcoholic beverage a day.
    d. A pregnant woman can safely consume 2 alcoholic beverages a day.

15. **The liver can metabolize _____ drink of alcohol per hour.**
    a. 1
    b. 3
    c. A 12-pack
    d. Heck, the liver can metabolize a whole case of beer in an hour!
    e. The liver's ability to metabolize alcohol is limitless.

16. **All of the following can influence the absorption and metabolism of alcohol except:**
    a. Type of alcohol
    b. Gender
    c. Ethnicity
    d. The presence of food in the stomach
    e. B, C, and D

17. **Which of the following is correct concerning alcohol's metabolism in a man and woman of the same height and weight?**
    a. They have equal alcohol metabolizing abilities.
    b. The woman has a lower capacity for metabolizing alcohol than the man.
    c. The man has a lower capacity for metabolizing alcohol than the woman.
    d. It is not yet known.

Study Card 10
Weight Management

A Closer Look at Obesity
- Obesity has been described as "an epidemic."
- The leading causes of obesity and overweight are excess energy intake and physical inactivity.
- Food portion sizes and obesity rates have grown at the same rate.

Problems Associated with Weight
- Too little weight can also cause health problems. An underweight person may experience prolonged periods of physiological stress or injury, menstrual irregularities, infertility, or osteoporosis.
- The physical risks from overweight/obesity are greater for some people than others, depending on inherited susceptibilities to chronic diseases.

What Is a Healthy Weight?
- People who are obese have too much body fat. Desirable body fat levels are 12–20 percent body fat for most men and 20–30 percent body fat for most women.
- Lean tissue is denser than fat (adipose) tissue.
- The ways in which body fat is distributed, such as central (apple shaped) distribution or lower body (pear shaped) distribution, have health implications.
- Health risk is associated with three factors: body weight, amount and location of body fat, and current health status.
- BMI (body mass index) is an index of weight in relation to height, but it does not account for location of fat in the body; a muscular person with a low percentage of body fat may have a high BMI.

Energy Balance
- Fat is stored when more food energy is consumed in a day than is utilized (by metabolism and other activities). Stored fat is lost only if less food is consumed in a day than is utilized.
- Not as much energy is necessary for muscular activity as is required for basal metabolism. On average, it accounts for only about 30 percent of total energy expenditure.
- A small amount of energy (about 5–10 percent) is required to digest, absorb, and metabolize food.
- Total energy needs = basal metabolism + physical activity + thermic effect of foods.

Causes of Obesity
- Some theories that try to explain reasons for obesity are:
  - Set-point theory
  - Leptin
  - Fat cell theory
- Environmental factors linked with obesity suggest people overeat or underexercise as a result of factors in their surroundings, such as the availability of food and lack of opportunity for vigorous physical activity.

- The hunger drive is programmed by heredity, but appetite is influenced by learned responses (e.g., eating ice cream when sad or hurt).
- Hunger is physiological, whereas appetite is psychological.

Weight Gain and Loss
- When more calories are consumed than needed, excess weight is gained.
- When fewer calories are consumed than needed, the body draws on stored supplies of nutrients.
- Glycogen stores are exhausted in less than 24 hours without food. Muscles and other organs can use stored fat as fuel, but the nervous system cannot.
- Fasting or starving people remain alive only until their body fat is gone or until half their lean tissue is gone, whichever comes first.
- Two common types of surgery performed for weight loss are gastric bypass and gastric banding.

Successful Weight-Loss Strategies
- The "secret" to successful weight loss (losing weight and keeping it off) is healthful eating habits, exercise, and behavior change.
- The most effective weight loss is gradual weight loss.
- Increasing physical activity is an important part of a successful weight loss program.

Weight-Gain Strategies
- A person who wants to gain weight is faced with some of the same challenges as someone who wants to lose weight.
- One should engage in consistent physical exercise while eating nutritious foods (but not at the same time) containing enough calories to support weight gain.

Breaking Old Habits
- Breaking old patterns of behavior and developing new ones involves a number of stages before reaching the point at which you're able to change for good:
  - Accept the belief he/she *can* change.
  - A person must *want* to change.
  - Take action ("just do it").
  - Maintain behavior change by planning for inevitable lapses that will occur over time.
- Behavior modification is a process that helps people make lasting behavior changes.
- Eating records can help with weight loss.

A Closer Look at Obesity (pp. 292–293)

1. Describe societal trends that have increased opportunities for consumption of excess calories and decreased opportunities for physical activity.

Problems Associated with Weight (pp. 293–294)

2. List health problems associated with weight.

3. Being overweight/obese increases the risk for various health conditions. Which of the following conditions is affected by overweight/obesity?
   a. Orthopedic conditions
   b. Respiratory disorders
   c. Type 2 diabetes
   d. Pregnancy
   e. All of the above

What Is a Healthy Weight? (pp. 294–298)

4. Densitometry is the most reliable way to determine percent body fat. Another name for densitometry is _____.

5. Bioelectrical impedance is based on the fact that:
   a. Fat is a poor conductor of electricity.
   b. Fat weighs more than water.
   c. Infrared light is deflected by muscle and not by fat.
   d. Muscle impedes electric current flow more than fat.

6. Distribution of body fat is often defined using analogies to fruit. Which of the following body shapes is a risk factor for chronic diseases?
   a. Pear shape
   b. Apple shape
   c. Banana shape
   d. Kumquat shape

7. BMI assesses body weight using a measure of a person's weight in relation to height that correlates with total body fat content. BMI stands for:
   a. Balance, moderation, improvement
   b. Body maintaining inactivity
   c. Body mass index
   d. Bodies moving ideally

Energy Balance (pp. 298–300)

8. The basic theory behind what causes overweight/obesity is:
   a. Emotional eating
   b. Hormone dysfunction
   c. Energy intake greater than energy expenditure
   d. Lack of effective medical treatment

9. Vital organs (lungs, heart, kidneys, etc.) require energy to function. Which of the following describes this type of energy?
   a. Thermic effect of food
   b. Basal energy metabolism
   c. Total energy expenditure
   d. Physical activity related to energy expenditure

10. It takes energy to digest the foods we eat. This increased energy expenditure over and above basal metabolism that can be measured for several hours following a meal is:
    a. Thermic effect of food
    b. Basal energy metabolism
    c. Total energy expenditure
    d. Physical activity related to energy expenditure

Causes of Obesity (pp. 301–303)

11. On the whole, obesity is caused by which of the following factors?
    a. Genetic
    b. Cultural
    c. Multifactorial
    d. Environmental

12. The body's physiological need for food is experienced as:
    a. Thirst
    b. Satiety
    c. Appetite
    d. Hunger

13. According to set-point theory, the body reacts when weight or fatness falls below the level perceived to be appropriate. How does the body react?
    a. Experience a greater desire for high fat foods
    b. Develop a new, lower set point for body weight
    c. Experience an overall increase in hunger drive
    d. Decrease physical activity level until weight is regained

Weight Gain and Loss (pp. 303–313)

14. If you weigh yourself every day, the weight gain/loss you see on the scale always reflects gain/loss of body fat.
    a. True
    b. False

Successful Weight-Loss Strategies (pp. 313–319)

15. Regular exercise can help people increase lean body mass and reduce fat tissue.
    a. True
    b. False

Weight-Gain Strategies (pp. 319–320)

16. Describe the strategies for gaining weight.

Breaking Old Habits (pp. 320–324)

17. To maintain a healthy weight, the most important goals are related to changes in:
    a. Behaviors
    b. Food intake
    c. Physical activity
    d. Supplement use

FIGURE 10-4

How the Body Expends Energy

The body expends most of its energy on *basal metabolism*—maintaining basic physiological processes such as breathing, heartbeat, and other involuntary activities. The second largest amount of energy is expended for voluntary *physical activities*—an amount that will vary by activity level. A minor amount of energy is also used for the *thermic effect of food*—the energy needed to digest, absorb, and process the food you eat.

Thermic effect of food 5%–10%

Basal metabolism 60%–65%

Physical activity 25%–35%

Nutrition, Physical Activity, and Fitness

Getting Started on Lifetime Fitness

- Health benefits occur with at least 2 hours and 30 minutes a week of moderate-intensity physical activity or 1 hour and 15 minutes of vigorous-intensity aerobic activity. Additional benefits occur with more physical activity.
- The total amount of activity is more important than what the activity is.
- Total fitness incorporates aerobic activity, strength training, and stretching.

TABLE 11-1	Benefits of Physical Activity

- Increased self-confidence
- Easier weight control
- More energy
- Less stress and anxiety
- Improved sleep
- Enhanced immunity
- Lowered risk of heart disease
- Lowered risk of certain cancers
- Stronger bones
- Lowered risk of diabetes
- Lowered risk of high blood pressure
- Increased quality of life
- Increased independence in life's later years

The Components of Fitness

- Physical conditioning is a planned program of physical activity directed toward improving the function of a particular body system.
- Muscle responds to extra physical demand (overload).
- Strength is the ability of muscles to work against resistance. Resistance training prevents injury by increasing muscle strength.
- Flexibility (range of motion) is critical for fitness.
- Endurance is defined as the ability to sustain an effort for a long time. There are two types of endurance: muscle and cardiovascular.

Energy for Physical Activity

- In addition to oxygen, muscles need fuel, especially during activity.
- When ample oxygen is present, muscles use carbohydrate and fat as energy by means of aerobic (requiring oxygen) metabolism.
- When exertion becomes great enough that energy demands are greater than the oxygen supply, muscles must rely on anaerobic (without oxygen) metabolism for energy.
- The anaerobic metabolic pathway can only use carbohydrate for energy, drawing heavily on the body's limited stores of carbohydrates. This process allows lactic acid to build up in muscles.
- Aerobic exercise burns fat and extracts carbohydrates more efficiently, thereby conserving carbohydrate stores in the body.
- Aerobic exercise improves endurance of the lungs and muscles.
- The effect of regular exercise on the cardiovascular system includes improvements in heart, lung, and muscle function.
- Heart rate must be elevated to the target heart rate to develop cardiac conditioning.

Fuels for Physical Activity

- The energy-producing pathways require oxygen, glucose, and fatty acids.
- During physical activity, glucose is supplied to muscles from stores of glycogen in the liver and the muscles themselves. Longer or more intense activity uses more glucose.
- Trained muscles are able to use more fat and conserve their glucose.
- Fat used during exercise comes from fatty deposits from the whole body, especially those areas with the greatest amounts to spare.

Protein Needs for Fitness

- Even though muscle protein breaks down during heavy exercise, muscle growth escalates after exercise.
- Muscles use *available* amino acids to repair and build.

TABLE 11-3	Protein Recommendations for Athletes	
		Recommendations (g/kg/day)
RDA for adults		0.8
Endurance athletes		1.2–1.4
Resistance training (bodybuilders, strength athletes)		1.2–1.7

SOURCE: Position of the American Dietetic Association, Dietitians of Canada, and the American College of Sports Medicine: Nutrition and athletic performance, *Journal of the American Dietetic Association* 109 (2009): 509–27.

Fluids and Physical Activity

- Water is critical during exercise.
- Water helps reduce the body heat generated by working muscles by sweating. The more intense an activity, the more heat is generated, the more sweat is produced, and the more water is required by the body.
- Rehydration must take place before, during, and after exercising.
- Weighing before and after an activity indicates how much water is lost during exercise. To replace the lost fluid, one pound equals roughly 2 cups of fluid.

Vitamins and Minerals for Physical Activity

- How well the body utilizes energy from foods depends upon the supply of vitamins and minerals in the diet.
- B vitamins govern the energy-producing reactions of metabolism. Needs for B vitamins increase proportionally with energy expenditure.
- Iron transports oxygen to cells in the body. Inadequate iron stores in the body can compromise the ability to carry oxygen through the body.
- Weight-bearing activities (running, walking, dancing, rope skipping, strength training) encourage bone development.
- Inadequate calcium intake can lead to stress fractures.

Getting Started on Lifetime Fitness (pp. 334–336)

1. List the benefits of physical activity.
2. List the major risk factors for heart disease.

The Components of Fitness (pp. 336–340)

3. Your cousin's mother's sister-in-law knows you are taking this class and wants you to tell her how to become physically fit. Which of the following will she need?
   a. Flexibility
   b. Roller blades
   c. A spectacular body
   d. Membership in a gym
   e. All of the above

4. Your roommate knows there are three components of fitness, but can't remember which of the following is not a component of fitness. Can you help (which is *not* a component of fitness)?
   a. Flexibility
   b. Muscle strength
   c. Muscle endurance
   d. Cardiovascular endurance
   e. Lung volume

5. Overall improvement of the cardiovascular system is known as:
   a. Stroke volume
   b. Cardiovascular conditioning
   c. O$_2$ consumption
   d. Vital capacity
   e. Circulatory improvement

6. You've gone home for the weekend and your younger brother wants to know how to achieve cardiovascular fitness. Since you are taking this class, you tell him a person must exercise (work) at a specific heart rate which is called the:
   a. Training load
   b. Pulse rate
   c. Target heart rate
   d. Exercise rate
   e. Maximum heart rate

Energy for Physical Activity (pp. 340–342)

7. Glucose is the primary fuel for intense activity. It comes primarily from:
   a. Oxygen
   b. Fat stores
   c. Protein stores
   d. Glycogen stores
   e. A and D

8. Fitness enables the heart muscle, blood vessels, and lungs to work at maximum efficiency. What is this called?
   a. Cardiovascular flexibility
   b. Cardiovascular strength
   c. Muscular endurance
   d. Cardiovascular endurance

9. Which of the following fuel sources provide energy for exercising muscles?
   a. Glucose and fatty acids
   b. Protein and fatty acids
   c. Protein and glucose
   d. Vitamin supplements

Fuels for Physical Activity (pp. 342–348)

10. You are getting ready to run a half-marathon and you want your diet to provide an edge. Which of the following nutrients will do this?
    a. Carbohydrates
    b. Protein
    c. Sugar
    d. Fat

11. It takes about _____ minutes of aerobic activity for our bodies to switch to fat as a fuel.
    a. 5
    b. 10
    c. 20
    d. 40

Protein Needs for Fitness (p. 348)

12. Sol Ataire is an athlete who trains 6 days a week. He has read that athletes need more protein. Although he eats animal protein sources at most meals and drinks skim milk regularly, he has started using a protein shake twice a day. Is this shake necessary? Why or why not?
    a. No, the shake is not needed. Athletes do not need any more protein than sedentary individuals.
    b. No, the shake is not needed. Most athletes such as Sol consume sufficient amounts of protein to meet their extra needs of 1.0 to 1.7g/kg body weight.
    c. Yes, the shake is needed. The protein shake provides better quality protein than protein contained in foods of animal origin.
    d. Yes, the shake is needed. Athletes, such as Sol, need substantially more protein than sedentary individuals to meet additional protein needs of twice the RDA of 56 grams per day.

Fluids and Physical Activity (pp. 348–350)

13. Your sister is training for a marathon and is exercising in hot, humid weather. She weighted herself before and after a training run, and found she had lost 5 pounds. Approximately how much water loss does this represent?
    a. Half a cup
    b. 4 cups
    c. 6 cups
    d. 10 cups

14. A sports drink needs to leave the digestive tract rapidly and enter circulation where it is needed. For this to happen, the drink should contain between _____ percent and _____ percent carbohydrate by volume.
    a. 1, 3
    b. 4, 8
    c. 10, 15
    d. 20, 30

Vitamins and Minerals for Physical Activity (pp. 350–352)

15. Which of the following minerals is a component of the body's oxygen transporting system?
    a. Zinc
    b. Phosphorus
    c. Iron
    d. Magnesium

The Life Cycle: Conception through the Later Years

Pregnancy: Nutrition for the Future

- Nutrients reach the fetus through the placenta.
- A woman with poor nutritional habits during pregnancy can impair the health of her future grandchild.
- Infants born to malnourished mothers are more likely to develop illness, birth defects, and retarded mental and physical development.
- Nutrient needs are higher during pregnancy and lactation than at any other time in adult life.
- Nutrients deserving special attention during pregnancy include protein, folate, iron, zinc, calcium, and vitamins.
- Weight gain and adequate nutrition are required to support the health of the mother and development of the fetus. Weight gain recommendations are based on a mother's pre-pregnancy weight for height or body mass index (BMI).
- Low birthweight (LBW) infants are either born premature or have suffered growth failure in the uterus. Low birthweight is 5½ pounds (2500 g) or less.
- Excessive weight gain during pregnancy increases the risk of complications during labor and delivery as well as of postpartum obesity.
- Harmful eating practices during pregnancy include dieting (to lose weight) during pregnancy, smoking, consuming alcohol, drug use (other than alcohol), and eating nonfood items like dirt or clay (pica) that have no nutritional value.
- Common nutrition-related problems experienced during pregnancy are morning sickness and constipation.
- Adolescents who become pregnant are nutritionally at risk and require early intervention and special care during pregnancy.
- Successful lactation requires adequate nutrition on the mother's part. A nursing mother produces 30 ounces of milk per day (on average), requiring approximately 400 additional calories each day.

TABLE 12-1	Recommended Weight Gain for Pregnant Women
Weight Category*	**Recommended Gain (lb)**
Underweight	28–40
Normal weight	25–35
Overweight	15–25
Obesity	11–20

*Underweight is defined as BMI < 18.5; normal weight as BMI 18.5 to 24.9; overweight as BMI 25.0 to 29.9; and obesity as BMI ≥ 30.0.

SOURCE: Adapted from Institute of Medicine and National Research Council of the National Academies, *Weight Gain During Pregnancy: Reexamining the Guidelines,* (Washington, D.C.: The National Academies Press, 2009).

Healthy Infants

- The growth of children is the major indicator of their health and nutritional status.
- Breast milk perfectly meets the nutrient needs of the infant.
- The American Academy of Pediatrics recommends that infants receive exclusive breastfeeding for the first 6 months of life, and breastfeeding with solid foods for at least 12 months.

- Infant formulas are manufactured to approximate nutrient composition of breast milk, but cannot provide the same immunologic protection as breast milk.
- Water is critical. The younger the child, the greater the percentage of body weight is water.
- Solid foods can be added to an infant's diet at about 6 months of age if the child can sit upright (with support) and control head movements.
- Solid foods introduced too early (before 6 months of age) may result in the development of food allergies.
- Feeding cow's milk to a child earlier than is recommended can cause iron deficiency.
- New foods should be introduced one at a time to facilitate the detection of allergies.

Early and Middle Childhood

- Growth rate slows after a child's first birthday, but the body continues to change. Appetite decreases noticeably around the age of 1, in line with the reduction in growth rate.
- Diet quality tends to decline from childhood to adolescence, especially with decreased consumption of vegetables, fruits, and milk, and increased intake of soft drinks.

The Importance of Teen Nutrition

- Marked changes take place in physical, intellectual, and emotional growth and development between the ages of 10–18 years in girls and 12–20 years in boys.
- Nutrition-related problems among adolescents include overweight and obesity, undernutrition, iron-deficiency anemia (especially in girls), low dietary calcium intakes, high blood cholesterol levels, dental caries, and eating disorders.

Nutrition in Later Life

- An adult's health profile is substantially determined by exercise and a healthful diet.
- Most nutritional needs of older adults are the same as for younger adults, except that calorie needs decrease (reduced basal metabolism), and the needs for certain nutrients (i.e., calcium, vitamins D, C, B_{12}, and B_6) actually increase with aging.
- Poverty and isolation impair the nutritional status of many older adults.

Looking Ahead and Aging Wisely

- The ability of an older adult to function well varies from person to person and depends upon the following factors:
  - Genetic potential for extended longevity.
  - Continued desire for new knowledge and new experiences.
  - Socialization, intimacy, and family integrity.
  - Eating a healthy diet.
  - Avoiding substance abuse.
  - Acceptable living arrangements.
  - Financial independence.
  - Access to health care.

Pregnancy: Nutrition for the Future (pp. 362–370)

1. The placenta is:
   a. The developing infant from the eighth week after conception until birth
   b. Another name for the mother's uterus
   c. The infant from birth until the first birthday
   d. The organ from which the infant receives nourishment

2. Your sister, Irene, went to the doctor for her first prenatal visit today. Her physician told her to be sure to consume folate in adequate amounts before and early in pregnancy, but she can't remember why. She knows you are taking this class and asks you what folate helps prevent:
   a. Morning sickness
   b. Diarrhea
   c. Neural tube defects in newborns
   d. Cold feet

3. Which of the following is the most influential indicator of an infant's future health status?
   a. Mother's weight before pregnancy
   b. Amount of weight gained during pregnancy
   c. Mother's nutrition status prior to pregnancy
   d. Birthweight of the infant

4. Your roommate is pregnant and wants to celebrate her soon-to-be status as a mom. She knows you are taking this nutrition class and just studied alcohol and lifecycle nutrition. She wants to know how many beers she can safely drink.
   a. No alcohol is the safest.
   b. She can drink beer, but no "hard" liquor.
   c. A pregnant woman can safely consume 1 alcoholic beverage a day.
   d. A pregnant woman can safely consume 2 alcoholic beverages a day.

5. Sue is visiting her pregnant cousin Anna and notices that she is eating laundry starch and clay when watching TV at night. What is going on?
   a. Anna is experiencing pica.
   b. Anna is showing symptoms of high blood pressure.
   c. Anna is constipated and this will alleviate the problem.
   d. Sue should mind her own business.

Healthy Infants (pp. 370–375)

6. Which of the following is *not* a benefit of breastfeeding?
   a. Lower cost
   b. Ideal serving size
   c. Fewer allergies
   d. Easy to accommodate to lifestyle/work schedules

7. What advice can you give parents to help their infant enjoy food?
   a. The parents should respect the infant's natural signals of hunger and satiety.
   b. The parents should ignore the infant during mealtimes so he/she will focus only on food.
   c. The parent should argue at the dinner table to distract the infant.
   d. They should force the infant to eat.

8. Alcohol consumed by a mother does not pass through breast milk.
   a. True
   b. False

9. When your sister Irene has her baby (your niece), the nurse told her foods should be introduced to an infant one at a time but didn't explain why. You just finished reading this chapter and can tell her why!
   a. It prevents overfeeding.
   b. The swallowing reflex is not under voluntary control.
   c. Immunological protection hasn't been developed.
   d. Any allergic reactions can be detected.

Early and Middle Childhood (pp. 376–380)

10. Your niece, Hilda, drinks a lot of milk and does not consume much else. You tell your sister, Irene, that Hilda is likely to show signs of:
    a. Anemia
    b. Ariboflavinosis
    c. Hyperkeratosis
    d. Rickets
    e. Pellagra

The Importance of Teen Nutrition (pp. 380–382)

11. Describe the nutrition-related problems of adolescents.

12. Your cousin Elvis is a teenager. What is the *best* way to be sure he is growing normally?
    a. Compare his heights and weights to his friends.
    b. Use growth charts from the doctor's office.
    c. Compare his heights and weights with previous heights and weights.
    d. Weigh him and measure his height weekly.

Nutrition in Later Life (pp. 382–390)

13. Your grandparents play bridge weekly at their condo and the discussion around the table always seems to be about food. They know you are taking this nutrition course and they want to know which of the following statements made at the bridge table is true regarding nutrition and aging.
    a. Older adults need more calories, but fewer vitamins and minerals.
    b. Older adults need fewer calories, but more vitamins and minerals.
    c. Nothing changes nutritionally as we age.

Looking Ahead and Aging Wisely (pp. 348–350)

14. An environmental factor that influences nutritional status later in life includes all of the following *except*:
    a. Lack of cooking facilities
    b. Lack of appetite
    c. Lack of transportation
    d. Lack of access to health services

15. When you visit your grandparents, you notice several differences from the last time you saw them. Which of the following statements about biological changes connected to aging is true?
    a. These changes are occurring at the same rate in both grandparents.
    b. Changes have occurred in both, but the rate of change has been different (probably because your grandmother ran marathons when she was younger).
    c. The changes could have been prevented by megadoses of vitamin E.
    d. The changes have occurred faster in your grandfather than in your grandmother.

Foodborne Illnesses and the Agents That Cause Them

- Foodborne illness is one of the greatest public health and food industry concerns.
- Foodborne illness is often mistaken as "the flu" and is often never reported to local health agencies.
- Symptoms of foodborne illness are diarrhea, nausea, abdominal pain, or vomiting without fever or upper respiratory distress.
- Disease-producing microorganisms thrive in our environment.
- Pasteurization and canning (both heat treatments) kill or inactivate bacteria or their spores. Freezing, dehydrating, and refrigeration stops or slows bacterial growth.
- Most foodborne illnesses are caused by the mishandling of food or the faulty cooking or storage of food, either in food service establishments or at home.
- The two types of foodborne diseases are intoxications and infections. Food intoxications occur when a chemical or toxin transmitted by food causes illness. Foodborne infections are caused by eating food contaminated by bacteria, viruses, or parasites capable of growing in the body.
- Mold is another potential microbial food contaminant. Certain molds produce poisonous compounds call aflatoxins that are very toxic to the liver.
- Some plants contain natural toxicants, especially herbs.

Safe Food Storage and Preparation

- Most cases of food poisoning that occur in the home or at a restaurant are caused by improper food storage or handling.

Pesticides and Other Chemical Contaminants

- Some risk of contamination from undesirable substances is unavoidable.
- Toxicity is the ability of a substance to harm living organisms. All substances are toxic if presented in high enough concentrations.
- A hazard is a state of danger and is used to refer to any circumstance in which harm is possible.
- Lead contamination builds gradually. Lead paint in older housing, ceramic hollowware, lead crystal, and foil capsules on wine bottles are potential sources.
- Pesticides are used to prevent, destroy, or repel harmful pests, but can inadvertently harm wildlife, people, and other species. Their place in farming can boost crop yields, which keeps food prices down. Pesticides are regulated by the Environmental Protection Agency (EPA).

Food Additives

- Intentional food additives are used to enhance the color, flavor, texture, or stability of foods, or even to improve their nutritional value.
- Incidental food additives accidentally get into food as a result of contact with them during growing, processing, packaging, storing, or some other stage before the food is consumed.
- Food additives are regulated by the Food and Drug Administration (FDA).

- The GRAS (generally recognized as safe) list includes ingredients, established by the FDA, that had long been in use and were believed safe.

New Technologies on the Horizon

- Irradiation involves exposing food to low doses of radiation to destroy insects and several types of bacteria, including *Salmonella*.
- Genetic engineering uses biotechnology to alter the genes of a plant to create a new plant with different traits (genetic modifications). Plants' genes may be deleted or altered, or genes from different organisms or species are introduced. Food or crops produced by genetic engineering are called genetically engineered (GE) or genetically modified (GM) foods or crops.
- Biotechnology uses the biological systems of living organisms to make, modify, or enhance living plants, animals, and fish through the manipulation of genes (genetic engineering).
- Transgenetics is the process of transferring genes from one species to another unrelated species. This process raises many ethical questions, especially regarding the insertion of animal genes into a plant species. Some people are concerned about the potential impact of organisms that would have not evolved under normal conditions.
- The long-term environmental consequences of genetically engineered plants are still unknown.

TABLE 13-2	Microbial Food Agents: Organisms That Can Bug You

Disease and Organism That Causes It

Bacteria:

Botulism (*Botulinum* toxin produced by *Clostridium botulinum* bacteria)

Campylobacteriosis (*Campylobacter jejuni*)

Listeriosis (*Listeria monocytogenes*)

Perfringens food poisoning (*Clostridium perfringens*)

Salmonellosis (*Salmonella* bacteria)

E. coli infection (shiga toxin-producing *E. coli* [STEC] released by *Escherichia coli* O157:H7 and others)

Shigellosis (*Shigella* bacteria)

Staphylococcal food poisoning (staphylococcal toxin produced by *Staphylococcus aureus* bacteria)

Vibrio infection (*Vibrio vulnificus*)

Parasite:

Cryptosporidiosis (*Cryptosporidium parvum*)

Cyclosporiasis (*Cyclospora cayetanensis*)

Trichinosis (*Trichinella spiralis*)

Giardiasis or "Traveler's Diarrhea" (*Giardia duodenalis*, formerly *G. lamblia*)

Virus:

Hepatitis (hepatitis A virus)

"Stomach flu"* Noroviruses (Norwalk-type viruses)

*A misnomer; unrelated to influenza.

SOURCES: Adapted from A. Hecht, *The Unwelcome Dinner Guest: Preventing Foodborne Illness* (Washington, DC: U.S. Government Printing Office, DHHS Publication No. (FDA) 91-2244); and Economic Research Service, *Economics of Foodborne Disease: Food and Pathogens*, available at www.ers.usda.gov.

Foodborne Illnesses and the Agents That Cause Them (pp. 400–406)

1. **The greatest hazard in the U.S. food supply is:**
   a. Food additives
   b. Pesticides
   c. Microorganisms that cause food poisoning
   d. The Food TV channel

2. **Which of the following microbial pathogens is one of the leading causes of foodborne illness in the United States?**
   a. *Campylobacter jejuni*
   b. *Clostridium botulinum*
   c. *Cryptosporidium parvum*
   d. *Vibrio vulnificus*
   e. *Listeria monocytogenes*

3. **Most *E. coli* outbreaks have been linked to:**
   a. Unpasteurized milk
   b. Undercooked hamburger
   c. Eggs
   d. Apples
   e. Rice

4. **Common symptoms of foodborne illness are:**
   a. Diarrhea
   b. Nausea and vomiting
   c. Abdominal pain
   d. Taking this class
   e. A, B, and C only

5. **Which of the following are signs that indicate canned foods may be contaminated?**
   a. Bulging
   b. Leaking
   c. Spurting when opened
   d. You saw your instructor put the can back on the shelf in the grocery store.
   e. A, B, and C only

Safe Food Storage and Preparation (pp. 406–410)

6. **The best place to defrost frozen meat, seafood, or poultry is the:**
   a. Refrigerator
   b. Sink
   c. Counter top
   d. Dashboard of your car on a sunny day

7. **Leftovers can be kept at room temperature for:**
   a. no longer than 30 minutes
   b. no longer than 60 minutes
   c. no longer than 2 hours
   d. Indefinitely

8. **Don't be a dope, wash with soap. When would it be important to wash your hands?**
   a. After cleaning the cat litter
   b. After going to the bathroom or changing diapers
   c. Between handling uncooked foods and cooked foods
   d. When eating food that won't be cooked (like carrot sticks)
   e. All of the above

9. **You and your roommates are getting ready to make chicken and vegetable stir-fry using the same cutting board for both. You suggest using two cutting boards, one for the chicken and one for the ready-to-eat foods. Your roommates say they don't want to have to clean two cutting boards, so they will just use one. What should you do?**
   a. Don't eat the stir-fry and laugh at your roommates when they get sick.
   b. Wipe the board with a paper towel in between uses.
   c. Offer to wash the board yourself because you know the importance of preventing cross contamination.
   d. You should never argue with your roommates. . . . do whatever they want.

10. **The best way to tell if a food is spoiled is to taste it.**
    a. True
    b. False

Pesticides and Other Chemical Contaminants (pp. 410–416)

11. **Why are pesticides used?**
    a. To act as preservatives in our food supply
    b. To kill insects, rodents, or other living things
    c. To enhance the flavor of soups and salad dressings
    d. To stimulate plant growth beyond the normal rate

Food Additives (pp. 416–420)

12. **Substances widely used for many years without apparent ill effects are on the _____ list.**
    a. FDA
    b. GRAS
    c. Delaney
    d. Additive Safety
    e. Preferred Additive

New Technologies on the Horizon (pp. 382–390)

13. **The process of altering foods through genetic manipulation is called:**
    a. Irradiation.
    b. Biotechnology.
    c. Transmutation.
    d. Organic food production.

14. **Food that has been irradiated will have this symbol on the label:**

    a.

    b.

    c.

    d.

15. **Which of the following is *not* an example of genetically modified food?**
    a. Seedless grapes
    b. Nectarines
    c. Milk
    d. The latest energy drink to hit the market

Dietary Reference Intakes (DRI)

The Dietary Reference Intakes (DRI) include two sets of values that serve as goals for nutrient intake—Recommended Dietary Allowances (RDA) and Adequate Intakes (AI). The RDA reflect the average daily amount of a nutrient considered adequate to meet the needs of most healthy people. If there is insufficient evidence to determine an RDA, an AI is set.

In addition to the values that serve as goals for nutrient intakes (presented in the tables on these two pages), the DRI include a set of values called Tolerable Upper Intake Levels (UL). The UL represent the maximum amount of a nutrient that appears safe for most healthy people to consume on a regular basis. Turn the page for a listing of the UL for selected vitamins and minerals.

Estimated Energy Requirements (EER), Recommended Dietary Allowances (RDA), and Adequate Intakes (AI) for Water, Energy, and the Energy Nutrients

Age (yr)	Reference BMI (kg/m²)	Reference height, cm (in)	Reference weight, kg (lb)	Water[a] AI (L/day)	Energy EER[b] (cal/day)	Carbohydrate RDA (g/day)	Total fiber AI (g/day)	Total fat AI (g/day)	Linoleic acid AI (g/day)	Linolenic acid[c] AI (g/day)	Protein RDA (g/day)[d]	Protein RDA (g/kg/day)
Males												
0–0.5	—	62 (24)	6 (13)	0.7[e]	570	60	—	31	4.4	0.5	9.1	1.52
0.5–1	—	71 (28)	9 (20)	0.8[f]	743	95	—	30	4.6	0.5	13.5	1.5
1–3[g]	—	86 (34)	12 (27)	1.3	1046	130	19	—	7	0.7	13	1.1
4–8[g]	15.3	115 (45)	20 (44)	1.7	1742	130	25	—	10	0.9	19	0.95
9–13	17.2	144 (57)	36 (79)	2.4	2279	130	31	—	12	1.2	34	0.95
14–18	20.5	174 (68)	61 (134)	3.3	3152	130	38	—	16	1.6	52	0.85
19–30	22.5	177 (70)	70 (154)	3.7	3067[h]	130	38	—	17	1.6	56	0.8
31–50				3.7	3067[h]	130	38	—	17	1.6	56	0.8
>50				3.7	3067[h]	130	30	—	14	1.6	56	0.8
Females												
0–0.5	—	62 (24)	6 (13)	0.7[e]	520	60	—	31	4.4	0.5	9.1	1.52
0.5–1	—	71 (28)	9 (20)	0.8[f]	676	95	—	30	4.6	0.5	13.5	1.5
1–3[g]	—	86 (34)	12 (27)	1.3	992	130	19	—	7	0.7	13	1.1
4–8[g]	15.3	115 (45)	20 (44)	1.7	1642	130	25	—	10	0.9	19	0.95
9–13	17.4	144 (57)	37 (81)	2.1	2071	130	26	—	10	1.0	34	0.95
14–18	20.4	163 (64)	54 (119)	2.3	2368	130	26	—	11	1.1	46	0.85
19–30	21.5	163 (64)	57 (126)	2.7	2403[i]	130	25	—	12	1.1	46	0.8
31–50				2.7	2403[i]	130	25	—	12	1.1	46	0.8
>50				2.7	2403[i]	130	21	—	11	1.1	46	0.8
Pregnancy												
1st trimester				3.0	+0	175	28	—	13	1.4	+25	1.1
2nd trimester				3.0	+340	175	28	—	13	1.4	+25	1.1
3rd trimester				3.0	+452	175	28	—	13	1.4	+25	1.1
Lactation												
1st 6 months				3.8	+330	210	29	—	13	1.3	+25	1.1
2nd 6 months				3.8	+400	210	29	—	13	1.3	+25	1.1

NOTE: For all nutrients, values for infants are AI. Dashes indicate that values have not been determined.
[a] The water AI includes drinking water, water in beverages, and water in foods; in general, drinking water and other beverages contribute about 70 to 80 percent, and foods, the remainder. Conversion factors: 1 L = 33.8 fluid oz; 1 L = 1.06 qt; 1 cup = 8 fluid oz.
[b] The Estimated Energy Requirement (EER) represents the average dietary energy intake that will maintain energy balance in a healthy person of a given gender, age, weight, height, and physical activity level. The values listed are based on an "active" person at the reference height and weight and at the midpoint ages for each group until age 19. Chapter 2 (page 43), Chapter 10 (page 300), and www.ChooseMyPlate.gov provide tools to determine Estimated Energy Requirements.
[c] The linolenic acid referred to in this table and text is the omega-3 fatty acid known as alpha-linolenic acid.
[d] The values listed are based on reference body weights.
[e] Assumed to be from human milk.
[f] Assumed to be from human milk and complementary foods and beverages. This includes approximately 0.6 L (~3 cups) as total fluid including formula, juices, and drinking water.
[g] For energy, the age groups for young children are 1–2 years and 3–8 years.
[h] For males, subtract 10 calories per day for each year of age above 19.
[i] For females, subtract 7 calories per day for each year of age above 19.

SOURCE: Adapted from the *Dietary Reference Intakes* series, National Academies Press. Copyright 1997, 1998, 2000, 2001, 2002, 2004, 2005, by the National Academy of Sciences.

Recommended Dietary Allowances (RDA) and Adequate Intakes (AI) for Vitamins

Age (yr)	Thiamin RDA (mg/day)	Riboflavin RDA (mg/day)	Niacin RDA (mg/day)[a]	Biotin AI (μg/day)	Pantothenic acid AI (mg/day)	Vitamin B6 RDA (mg/day)	Folate RDA (μg/day)[b]	Vitamin B12 RDA (μg/day)	Choline AI (mg/day)	Vitamin C RDA (mg/day)	Vitamin A RDA (μg/day)[c]	Vitamin D RDA (IU/day)[d]	Vitamin E RDA (mg/day)[e]	Vitamin K AI (μg/day)
Infants														
0–0.5	0.2	0.3	2	5	1.7	0.1	65	0.4	125	40	400	400 (10 μg)	4	2.0
0.5–1	0.3	0.4	4	6	1.8	0.3	80	0.5	150	50	500	400 (10 μg)	5	2.5
Children														
1–3	0.5	0.5	6	8	2	0.5	150	0.9	200	15	300	600 (15 μg)	6	30
4–8	0.6	0.6	8	12	3	0.6	200	1.2	250	25	400	600 (15 μg)	7	55
Males														
9–13	0.9	0.9	12	20	4	1.0	300	1.8	375	45	600	600 (15 μg)	11	60
14–18	1.2	1.3	16	25	5	1.3	400	2.4	550	75	900	600 (15 μg)	15	75
19–30	1.2	1.3	16	30	5	1.3	400	2.4	550	90	900	600 (15 μg)	15	120
31–50	1.2	1.3	16	30	5	1.3	400	2.4	550	90	900	600 (15 μg)	15	120
51–70	1.2	1.3	16	30	5	1.7	400	2.4	550	90	900	600 (15 μg)	15	120
>70	1.2	1.3	16	30	5	1.7	400	2.4	550	90	900	800 (20 μg)	15	120
Females														
9–13	0.9	0.9	12	20	4	1.0	300	1.8	375	45	600	600 (15 μg)	11	60
14–18	1.0	1.0	14	25	5	1.2	400	2.4	400	65	700	600 (15 μg)	15	75
19–30	1.1	1.1	14	30	5	1.3	400	2.4	425	75	700	600 (15 μg)	15	90
31–50	1.1	1.1	14	30	5	1.3	400	2.4	425	75	700	600 (15 μg)	15	90
51–70	1.1	1.1	14	30	5	1.5	400	2.4	425	75	700	600 (15 μg)	15	90
>70	1.1	1.1	14	30	5	1.5	400	2.4	425	75	700	800 (20 μg)	15	90
Pregnancy														
≤18	1.4	1.4	18	30	6	1.9	600	2.6	450	80	750	600 (15 μg)	15	75
19–30	1.4	1.4	18	30	6	1.9	600	2.6	450	85	770	600 (15 μg)	15	90
31–50	1.4	1.4	18	30	6	1.9	600	2.6	450	85	770	600 (15 μg)	15	90
Lactation														
≤18	1.4	1.6	17	35	7	2.0	500	2.8	550	115	1200	600 (15 μg)	19	75
19–30	1.4	1.6	17	35	7	2.0	500	2.8	550	120	1300	600 (15 μg)	19	90
31–50	1.4	1.6	17	35	7	2.0	500	2.8	550	120	1300	600 (15 μg)	19	90

NOTE: For all nutrients, values for infants are AI.
[a]Niacin recommendations are expressed as niacin equivalents (NE), except for recommendations for infants younger than 6 months, which are expressed as preformed niacin.
[b]Folate recommendations are expressed as dietary folate equivalents (DFE).
[c]Vitamin A recommendations are expressed as retinol activity equivalents (RAE).
[d]Vitamin D recommendations are expressed as cholecalciferol and assumes minimal sun exposure.
[e]Vitamin E recommendations are expressed as α-tocopherol.

Recommended Dietary Allowances (RDA) and Adequate Intakes (AI) for Minerals

Age (yr)	Sodium AI (mg/day)	Chloride AI (mg/day)	Potassium AI (mg/day)	Calcium RDA (mg/day)	Phosphorus RDA (mg/day)	Magnesium RDA (mg/day)	Iron RDA (mg/day)	Zinc RDA (mg/day)	Iodine RDA (μg/day)	Selenium RDA (μg/day)	Copper RDA (μg/day)	Manganese AI (mg/day)	Fluoride AI (mg/day)	Chromium AI (μg/day)	Molybdenum RDA (μg/day)
Infants															
0–0.5	120	180	400	200	100	30	0.27	2	110	15	200	0.003	0.01	0.2	2
0.5–1	370	570	700	260	275	75	11	3	130	20	220	0.6	0.5	5.5	3
Children															
1–3	1000	1500	3000	700	460	80	7	3	90	20	340	1.2	0.7	11	17
4–8	1200	1900	3800	1000	500	130	10	5	90	30	440	1.5	1.0	15	22
Males															
9–13	1500	2300	4500	1300	1250	240	8	8	120	40	700	1.9	2	25	34
14–18	1500	2300	4700	1300	1250	410	11	11	150	55	890	2.2	3	35	43
19–30	1500	2300	4700	1000	700	400	8	11	150	55	900	2.3	4	35	45
31–50	1500	2300	4700	1000	700	420	8	11	150	55	900	2.3	4	35	45
51–70	1300	2000	4700	1000	700	420	8	11	150	55	900	2.3	4	30	45
>70	1200	1800	4700	1200	700	420	8	11	150	55	900	2.3	4	30	45
Females															
9–13	1500	2300	4500	1300	1250	240	8	8	120	40	700	1.6	2	21	34
14–18	1500	2300	4700	1300	1250	360	15	9	150	55	890	1.6	3	24	43
19–30	1500	2300	4700	1000	700	310	18	8	150	55	900	1.8	3	25	45
31–50	1500	2300	4700	1000	700	320	18	8	150	55	900	1.8	3	25	45
51–70	1300	2000	4700	1200	700	320	8	8	150	55	900	1.8	3	20	45
>70	1200	1800	4700	1200	700	320	8	8	150	55	900	1.8	3	20	45
Pregnancy															
≤18	1500	2300	4700	1300	1250	400	27	12	220	60	1000	2.0	3	29	50
19–30	1500	2300	4700	1000	700	350	27	11	220	60	1000	2.0	3	30	50
31–50	1500	2300	4700	1000	700	360	27	11	220	60	1000	2.0	3	30	50
Lactation															
≤18	1500	2300	5100	1300	1250	360	10	13	290	70	1300	2.6	3	44	50
19–30	1500	2300	5100	1000	700	310	9	12	290	70	1300	2.6	3	45	50
31–50	1500	2300	5100	1000	700	320	9	12	290	70	1300	2.6	3	45	50

Tolerable Upper Intake Levels (UL) for Vitamins

Age (yr)	Niacin (mg/day)[a]	Vitamin B$_6$ (mg/day)	Folate (μg/day)[a]	Choline (mg/day)	Vitamin C (mg/day)	Vitamin A (μg/day)[b]	Vitamin D (IU/day)	Vitamin E (mg/day)[c]
Infants								
0–0.5	—	—	—	—	—	600	1000 (25 μg)	—
0.5–1	—	—	—	—	—	600	1500 (38 μg)	—
Children								
1–3	10	30	300	1000	400	600	2500 (63 μg)	200
4–8	15	40	400	1000	650	900	3000 (75 μg)	300
Adolescents								
9–13	20	60	600	2000	1200	1700	4000 (100 μg)	600
14–18	30	80	800	3000	1800	2800	4000 (100 μg)	800
Adults								
19–70	35	100	1000	3500	2000	3000	4000 (100 μg)	1000
>70	35	100	1000	3500	2000	3000	4000 (100 μg)	1000
Pregnancy								
≤18	30	80	800	3000	1800	2800	4000 (100 μg)	800
19–50	35	100	1000	3500	2000	3000	4000 (100 μg)	1000
Lactation								
≤18	30	80	800	3000	1800	2800	4000 (100 μg)	800
19–50	35	100	1000	3500	2000	3000	4000 (100 μg)	1000

[a]The UL for niacin and folate apply to synthetic forms obtained from supplements, fortified foods, or a combination of the two.
[b]The UL for vitamin A applies to the preformed vitamin only.

[c]The UL for vitamin E applies to any form of supplemental α-tocopherol, fortified foods, or a combination of the two.

Tolerable Upper Intake Levels (UL) for Minerals

Age (yr)	Sodium (mg/day)	Chloride (mg/day)	Calcium (mg/day)	Phosphorus (mg/day)	Magnesium (mg/day)[d]	Iron (mg/day)	Zinc (mg/day)	Iodine (μg/day)	Selenium (μg/day)	Copper (μg/day)	Manganese (mg/day)	Fluoride (mg/day)	Molybdenum (μg/day)	Boron (mg/day)	Nickel (mg/day)
Infants															
0–0.5	—[e]	—[e]	1000	—	—	40	4	—	45	—	—	0.7	—	—	—
0.5–1	—[e]	—[e]	1500	—	—	40	5	—	60	—	—	0.9	—	—	—
Children															
1–3	1500	2300	2500	3000	65	40	7	200	90	1000	2	1.3	300	3	0.2
4–8	1900	2900	2500	3000	110	40	12	300	150	3000	3	2.2	600	6	0.3
Adolescents															
9–13	2200	3400	3000	4000	350	40	23	600	280	5000	6	10	1100	11	0.6
14–18	2300	3600	3000	4000	350	45	34	900	400	8000	9	10	1700	17	1.0
Adults															
19–50	2300	3600	2500	4000	350	45	40	1100	400	10,000	11	10	2000	20	1.0
51–70	2300	3600	2000	4000	350	45	40	1100	400	10,000	11	10	2000	20	1.0
>70	2300	3600	2000	3000	350	45	40	1100	400	10,000	11	10	2000	20	1.0
Pregnancy															
≤18	2300	3600	3000	3500	350	45	34	900	400	8000	9	10	1700	17	1.0
19–50	2300	3600	2500	3500	350	45	40	1100	400	10,000	11	10	2000	20	1.0
Lactation															
≤18	2300	3600	3000	4000	350	45	34	900	400	8000	9	10	1700	17	1.0
19–50	2300	3600	2500	4000	350	45	40	1100	400	10,000	11	10	2000	20	1.0

[d]The UL for magnesium applies to synthetic forms obtained from supplements or drugs only.
[e]Source of intake should be from human milk (or formula) and food only.

NOTE: An Upper Limit was not established for vitamins and minerals not listed and for those age groups listed with a dash (—) because of a lack of data, not because these nutrients are safe to consume at any level of intake. All nutrients can have adverse effects when intakes are excessive.

SOURCE: Adapted with permission from the *Dietary Reference Intakes* series, National Academies Press. Copyright 1997, 1998, 2000, 2001, 2002, 2005 by the National Academy of Sciences. Courtesy of the National Academies Press, Washington, D.C.

How to Read a Food Label

You can use the food label to help you make informed food choices for healthy eating practices. The food label allows you to compare similar products, determine the nutritional value of the foods you choose, and can increase your awareness of the links between good nutrition and reduced risk of chronic diet-related diseases.

The package must always state the product name and the name and address of the manufacturer.

Claims such as "Good Source" or "High" can appear on labels but must follow legal definitions.

Approved health claims may be made but only in terms of total diet.

The net contents are shown in weight, measure, or count.

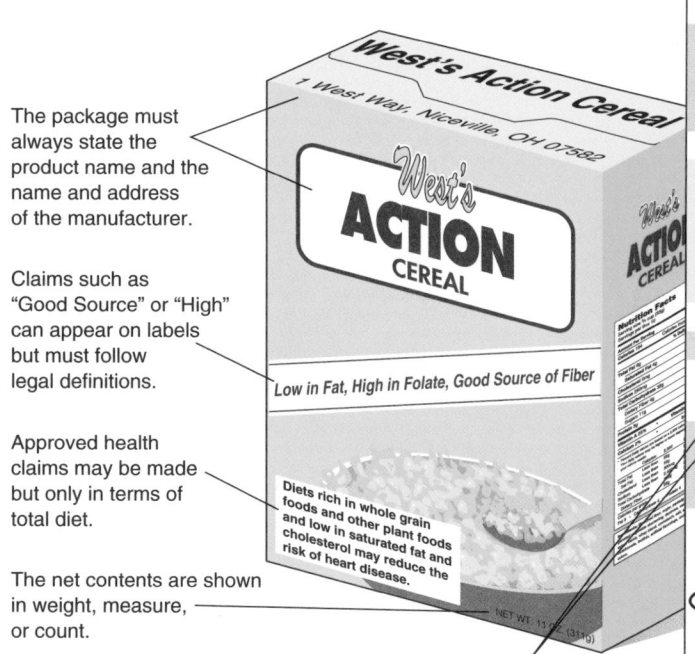

The only vitamins required to appear on the Nutrition Facts panel are vitamins A and C. If a manufacturer makes a nutrition claim about another vitamin, however, the amount of that nutrient in a serving of the product must also be stated on the panel. For instance, the cereal shown is touted as "High in Folate," so the percentage of the recommended intake for folate in a serving of the cereal (25%) is stated on the label. The manufacturer has the option of listing any other nutrients as well.

Nutrition Facts

Serving size ¾ cup (55g)
Servings per Box 5

Amount Per Serving		
Calories 167	Calories from Fat 27	
		% Daily Value*
Total Fat 3g		**5%**
Saturated Fat 1g		**5%**
Trans Fat 0g		
Cholesterol 0mg		**0%**
Sodium 250mg		**10%**
Total Carbohydrate 32g		**11%**
Dietary Fiber 4g		**16%**
Sugars 11g		
Protein 3g		

Vitamin A	0%
Vitamin C	15%
Calcium	10%
Iron	25%
Vitamin D	0%
Thiamin	25%
Riboflavin	25%
Niacin	25%
Folate	25%
Phosphorus	15%
Magnesium	10%
Zinc	6%
Copper	8%

*Percent Daily values are based on a 2,000 calorie diet. Your daily values may be higher or lower depending on your calorie needs:

		Calories:	2,000	2,500
Total Fat	Less than		65g	80g
Sat Fat	Less than		20g	25g
Cholesterol	Less than		300mg	300mg
Sodium	Less than		2,400mg	2,400mg
Total Carbohydrate			300g	375g
Dietary Fiber			25g	30g

Calories per gram
Fat 9 · Carbohydrate 4 · Protein 4

Calorie/gram reminder

Ingredients in descending order of predominance by weight

Ingredients: Whole oats, milled corn, enriched wheat flour (contains niacin, reduced iron, thiamin mononitrate, riboflavin, folic acid), dextrose, maltose, high fructose corn syrup, brown sugar, partially hydrogenated cottonseed oil, coconut oil, walnuts, vitamin C (sodium ascorbate), vitamin A (palmitate), iron.

Start here

Serving size, number of servings per container, and calorie information

Limit these nutrients.

Information on sodium is required on food labels.

Get enough of these nutrients.

Guide to the % Daily Value: Quantities of nutrients per serving and percentage of Daily Value for nutrients based on a 2,000 calorie energy intake.
   5% or less is low.
   10% or more is good.
   20% or more is high.

Calcium and iron are required on the label. Look for foods that provide 10% or more of these minerals.

Reference values

This allows comparison of some values for nutrients in a serving of the food with the needs of a person requiring 2,000 or 2,500 calories per day.

Daily Values (DV) Used on Food Labels[a]

Daily Values[b]

Food Component	Amount
protein[c]	50 g
fat	65 g[d]
saturated fat	20 g
cholesterol	300 mg[e]
total carbohydrate	300 g
fiber	25 g
sodium	2,400 mg
potassium	3,500 mg

Daily Values

Nutrient	Amount	Nutrient	Amount
Thiamin	1.5 mg	Calcium	1,000 mg
Riboflavin	1.7 mg	Iron	18 mg
Niacin	20 mg	Zinc	15 mg
Biotin	300 μg	Iodine	150 μg
Pantothenic Acid	10 mg	Copper	2 mg
Vitamin B$_6$	2 mg	Chromium	120 μg
Folate	400 μg[f]	Selenium	70 μg
Vitamin B$_{12}$	6 μg	Molybdenum	75 μg
Vitamin C	60 mg	Manganese	2 mg
Vitamin A	5,000 IU[g]	Chloride	3,400 mg
Vitamin D	400 IU[g]	Magnesium	400 mg
Vitamin E	30 IU[g]	Phosphorus	1,000 mg
Vitamin K	80 μg		

[a]Based on 2,000 calories a day for adults and children over 4 years old.
[b]Formerly the U.S. RDA, based on National Academy of Sciences' 1968 Recommended Dietary Allowances.
[c]The Daily Values for protein vary for different groups of people: children 1 to 4 years: 16 g; infants under 1 year: 14 g; pregnant women: 60 g; nursing mothers: 65 g.

[d](g) grams
[e](mg) milligrams
[f](μg) micrograms
[g]Equivalent values for the nutrients expressed as IU are: vitamin A, 1,500 RAE (assumes a mixture of 40% retinol and 60% beta-carotene); vitamin D, 10 μg; vitamin E, 20 mg.

Body Mass Index (BMI)

Height	18	19	20	21	22	23	24	25	26	27	28	29	30	31	32	33	34	35	36	37	38	39	40
												Body Weight (pounds)											
4'10"	86	91	96	100	105	110	115	119	124	129	134	138	143	148	153	158	162	167	172	177	181	186	191
4'11"	89	94	99	104	109	114	119	124	128	133	138	143	148	153	158	163	168	173	178	183	188	193	198
5'0"	92	97	102	107	112	118	123	128	133	138	143	148	153	158	163	168	174	179	184	189	194	199	204
5'1"	95	100	106	111	116	122	127	132	137	143	148	153	158	164	169	174	180	185	190	195	201	206	211
5'2"	98	104	109	115	120	126	131	136	142	147	153	158	164	169	175	180	186	191	196	202	207	213	218
5'3"	102	107	113	118	124	130	135	141	146	152	158	163	169	175	180	186	191	197	203	208	214	220	225
5'4"	105	110	116	122	128	134	140	145	151	157	163	169	174	180	186	192	197	204	209	215	221	227	232
5'5"	108	114	120	126	132	138	144	150	156	162	168	174	180	186	192	198	204	210	216	222	228	234	240
5'6"	112	118	124	130	136	142	148	155	161	167	173	179	186	192	198	204	210	216	223	229	235	241	247
5'7"	115	121	127	134	140	146	153	159	166	172	178	185	191	198	204	211	217	223	230	236	242	249	255
5'8"	118	125	131	138	144	151	158	164	171	177	184	190	197	203	210	216	223	230	236	243	249	256	262
5'9"	122	128	135	142	149	155	162	169	176	182	189	196	203	209	216	223	230	236	243	250	257	263	270
5'10"	126	132	139	146	153	160	167	174	181	188	195	202	209	216	222	229	236	243	250	257	264	271	278
5'11"	129	136	143	150	157	165	172	179	186	193	200	208	215	222	229	236	243	250	257	265	272	279	286
6'0"	132	140	147	154	162	169	177	184	191	199	206	213	221	228	235	242	250	258	265	272	279	287	294
6'1"	136	144	151	159	166	174	182	189	197	204	212	219	227	235	242	250	257	265	272	280	288	295	302
6'2"	141	148	155	163	171	179	186	194	202	210	218	225	233	241	249	256	264	272	280	287	295	303	311
6'3"	144	152	160	168	176	184	192	200	208	216	224	232	240	248	256	264	272	279	287	295	303	311	319
6'4"	148	156	164	172	180	189	197	205	213	221	230	238	246	254	263	271	279	287	295	304	312	320	328
6'5"	151	160	168	176	185	193	202	210	218	227	235	244	252	261	269	277	286	294	303	311	319	328	336
6'6"	155	164	172	181	190	198	207	216	224	233	241	250	259	267	276	284	293	302	310	319	328	336	345

Under-weight (<18.5)	Healthy Weight (18.5–24.9)	Overweight (25–29.9)	Obese (≥30)

Find your height along the left-hand column and look across the row until you find the number that is closest to your weight. The number at the top of that column identifies your BMI. The area shaded in green represents healthy weight ranges.

Key Physical Activity Guidelines for Adults

Type of Physical Activity	Examples	Duration/Frequency	Benefits
Moderate–intensity aerobic activity	■ Walking briskly (3–4 mph) ■ Water aerobics ■ Bicycling (≤ 10 mph) ■ Tennis (doubles) ■ Ballroom or line dancing ■ Active play (volleyball, basketball, softball, Ping-Pong) ■ Active recreation (canoeing, hiking, rollerblading)	■ Accumulate at least 150 minutes a week of moderate-intensity, or 75 minutes a week of vigorous-intensity aerobic activity, or an equivalent combination of moderate- and vigorous-intensity activity.	Improves overall health; helps manage body weight and prevent gradual unhealthy weight gain in adulthood
Vigorous–intensity aerobic activity	■ Racewalking, jogging, or running ■ Swimming laps, rowing ■ Tennis (singles) ■ Aerobic dancing, stair climbing ■ Bicycling (≥10 mph) ■ Jumping rope, hiking uphill ■ Sports (soccer, ice or field hockey, basketball) ■ Cross-country skiing	■ For more extensive health benefits, adults should increase their aerobic activity to 300 minutes (5 hours) a week of moderate-intensity, or 150 minutes a week of vigorous-intensity aerobic activity, or the equivalent.	Helps lower blood pressure and cholesterol levels and reduce risk for heart disease, diabetes, obesity, and certain types of cancer
Muscle and bone strengthening activities	■ Push-ups, pull-ups ■ Sit-ups ■ Resistance exercises with exercise bands, weight machines, hand-held weights	20–45 minutes per session, 2–3 nonconsecutive days per week (8 or more exercises, 1–3 sets, 8–12 reps)	Maintains muscle mass; promotes strong bones; reduces symptoms of arthritis
Stretching	■ Standing or seated toe touch ■ Overhead reach ■ Yoga	Hold each positioned stretch for 20–30 seconds; do on most, preferably all, days of the week	Reduces risk for injuries and falls; increases muscle and joint flexibility

SOURCE: Adapted from Department of Health and Human Services, *2008 Physical Activity Guidelines for Americans,* ODPHP Publication No. U0036 (Washington, DC: Department of Health and Human Services), October 2008. You can find more information about the advice on physical activity at www.health.gov/paguidelines

MyPlate Icon and Summary of the 2010 MyPlate Food Guidance System

Make half your plate fruits and vegetables

» Choose fresh, frozen, canned, or dried fruits and vegetables.

» Eat red, orange, and dark-green vegetables, such as tomatoes, sweet potatoes, and broccoli, in main and side dishes.

» Use fruit as snacks, salads, or desserts.

» Keep raw, cut-up vegetables handy for quick snacks.

» Choose whole or cut-up fruits more often than fruit juice.

Switch to skim or 1% milk

» They have the same amount of calcium and other essential nutrients as whole milk, but less fat and calories.

Make at least half your grains whole

» Choose 100% whole-grain cereals, breads, crackers, rice, and pasta.

» Check the ingredients list on food packages to find whole-grain foods.

Vary your protein food choices

» Choose a variety of foods including seafood, beans and peas, nuts, lean meats, poultry, and eggs.

» Keep meat and poultry portions small and lean.

» Try grilling, broiling, poaching, or roasting. These methods do not add extra fat.

Cut back on foods high in solid fats, added sugars, and salt

• Choose foods and drinks with little or no added sugars.

• Look out for salt (sodium) in foods you buy.

• Eat fewer foods that are high in solid fats.

Eat the right amount of calories for you

• Enjoy your food, but eat less.

• Cook more often at home, where you are in control of what's in your food.

• When eating out, choose lower calorie menu options.

Get your personal daily calorie limit at www.ChooseMyPlate.gov and keep that number in mind when deciding what to eat.

Be physically active your way

Pick activities that you like and start by doing what you can, at least 10 minutes at a time. Every bit adds up, and the health benefits increase as you spend more time being active.

Reprinted with permission from Learning ZoneXpress, Inc www.learningzonexpress.com

MyPyramid: Steps to a Healthier You

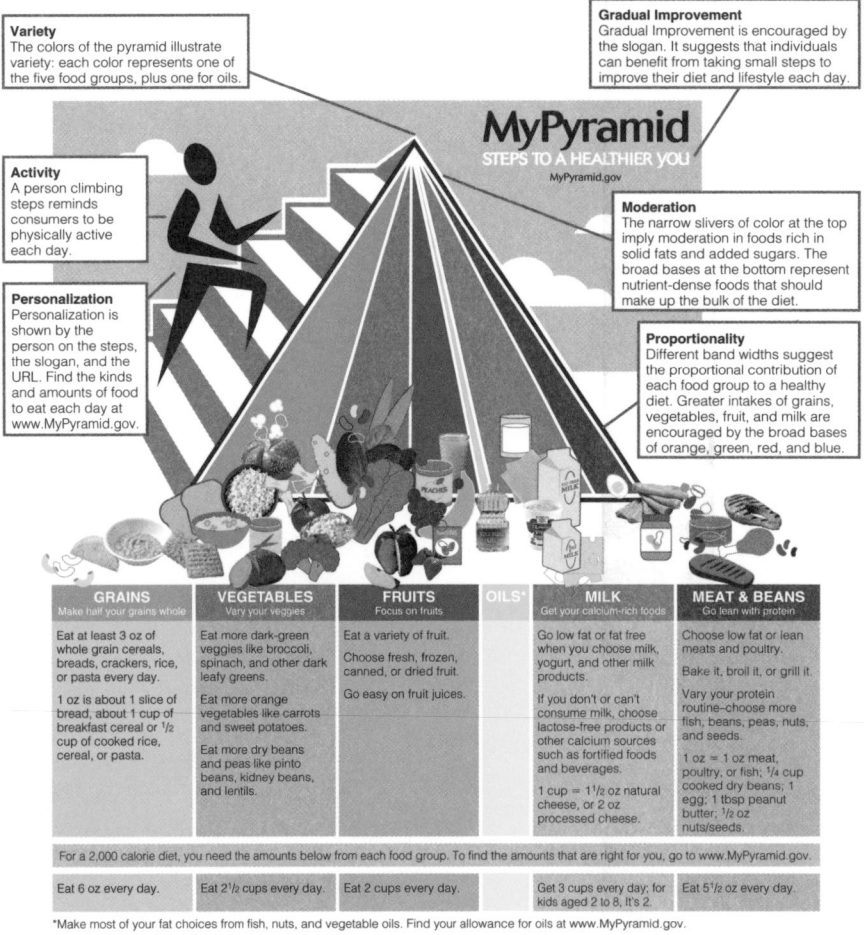

Variety
The colors of the pyramid illustrate variety: each color represents one of the five food groups, plus one for oils.

Gradual Improvement
Gradual Improvement is encouraged by the slogan. It suggests that individuals can benefit from taking small steps to improve their diet and lifestyle each day.

Activity
A person climbing steps reminds consumers to be physically active each day.

Moderation
The narrow slivers of color at the top imply moderation in foods rich in solid fats and added sugars. The broad bases at the bottom represent nutrient-dense foods that should make up the bulk of the diet.

Personalization
Personalization is shown by the person on the steps, the slogan, and the URL. Find the kinds and amounts of food to eat each day at www.MyPyramid.gov.

Proportionality
Different band widths suggest the proportional contribution of each food group to a healthy diet. Greater intakes of grains, vegetables, fruit, and milk are encouraged by the broad bases of orange, green, red, and blue.

GRAINS Make half your grains whole	VEGETABLES Vary your veggies	FRUITS Focus on fruits	OILS*	MILK Get your calcium-rich foods	MEAT & BEANS Go lean with protein
Eat at least 3 oz of whole grain cereals, breads, crackers, rice, or pasta every day.	Eat more dark-green veggies like broccoli, spinach, and other dark leafy greens.	Eat a variety of fruit. Choose fresh, frozen, canned, or dried fruit.		Go low fat or fat free when you choose milk, yogurt, and other milk products.	Choose low fat or lean meats and poultry. Bake it, broil it, or grill it.
1 oz is about 1 slice of bread, about 1 cup of breakfast cereal or ½ cup of cooked rice, cereal, or pasta.	Eat more orange vegetables like carrots and sweet potatoes. Eat more dry beans and peas like pinto beans, kidney beans, and lentils.	Go easy on fruit juices.		If you don't or can't consume milk, choose lactose-free products or other calcium sources such as fortified foods and beverages. 1 cup = 1½ oz natural cheese, or 2 oz processed cheese.	Vary your protein routine—choose more fish, beans, peas, nuts, and seeds. 1 oz = 1 oz meat, poultry, or fish; ¼ cup cooked dry beans; 1 egg; 1 tbsp peanut butter; ½ oz nuts/seeds.

For a 2,000 calorie diet, you need the amounts below from each food group. To find the amounts that are right for you, go to www.MyPyramid.gov.

Eat 6 oz every day.	Eat 2½ cups every day.	Eat 2 cups every day.		Get 3 cups every day; for kids aged 2 to 8, it's 2.	Eat 5½ oz every day.

*Make most of your fat choices from fish, nuts, and vegetable oils. Find your allowance for oils at www.MyPyramid.gov.

SOURCE: U.S. Department of Agriculture, 2005.